AF752886

Biodegradable and Biocompatible Polymer Nanocomposites

Processing, Characterization, and Applications

Biodegradable and Biocompatible Polymer Nanocomposites

Processing, Characterization, and Applications

Edited by

Kalim Deshmukh

Chemical Processes and Biomaterials, New Technologies—Research Centre, University of West Bohemia, Plzeň, Czech Republic

Mayank Pandey

Department of Electronics, Kristu Jayanti College (Autonomous), Bangalore, India

ELSEVIER

Elsevier
Radarweg 29, PO Box 211, 1000 AE Amsterdam, Netherlands
The Boulevard, Langford Lane, Kidlington, Oxford OX5 1GB, United Kingdom
50 Hampshire Street, 5th Floor, Cambridge, MA 02139, United States

Copyright © 2023 Elsevier Inc. All rights reserved.

No part of this publication may be reproduced or transmitted in any form or by any means, electronic or mechanical, including photocopying, recording, or any information storage and retrieval system, without permission in writing from the publisher. Details on how to seek permission, further information about the Publisher's permissions policies and our arrangements with organizations such as the Copyright Clearance Center and the Copyright Licensing Agency, can be found at our website: www.elsevier.com/permissions.

This book and the individual contributions contained in it are protected under copyright by the Publisher (other than as may be noted herein).

Notices
Knowledge and best practice in this field are constantly changing. As new research and experience broaden our understanding, changes in research methods, professional practices, or medical treatment may become necessary.

Practitioners and researchers must always rely on their own experience and knowledge in evaluating and using any information, methods, compounds, or experiments described herein. In using such information or methods they should be mindful of their own safety and the safety of others, including parties for whom they have a professional responsibility.

To the fullest extent of the law, neither the Publisher nor the authors, contributors, or editors, assume any liability for any injury and/or damage to persons or property as a matter of products liability, negligence or otherwise, or from any use or operation of any methods, products, instructions, or ideas contained in the material herein.

ISBN: 978-0-323-91696-7

For Information on all Elsevier publications
visit our website at https://www.elsevier.com/books-and-journals

Publisher: Matthew Deans
Acquisitions Editor: Ana Claudia A. Garcia
Editorial Project Manager: Aera F. Gariguez
Production Project Manager: Prem Kumar Kaliamoorthi
Cover Designer: Mark Rogers

Typeset by MPS Limited, Chennai, India

Working together
to grow libraries in
developing countries
www.elsevier.com • www.bookaid.org

Contents

List of contributors

Kafeel Ahmad
Civil Engineering Department, Jamia Millia Islamia, New Delhi, India

Perihan Kubra Akman
Food Engineering Department, Chemical and Metallurgical Engineering Faculty, Yildiz Technical University, Esenler, Istanbul, Turkey

Alice Alex
Department of Chemistry, CMS College, Kottayam, Kerala, India

John Amalraj
Laboratory of Materials Science, Instituto de Química de Recursos Naturales, Universidad de Talca, Talca, Maule, Chile

Md. Khairul Amin
Chemistry Discipline, Khulna University, Khulna, Bangladesh; School of Chemistry, University of Edinburgh, Edinburgh, United Kingdom

Asok Aparna
Department of Chemistry, Amrita VishwaVidyapeetham, Amritapuri, Clappana, Kerala, India

S.L. Aravind
Department of Mechanical Engineering, The National Institute of Engineering, Mysuru, Karnataka, India

A.K. Bajpai
Bose Memorial Research Laboratory, Department of Chemistry, Govt. Autonomous Science College, Jabalpur, Madhya Pradesh, India

Jaydip Bhaliya
Department of Applied Chemistry, School of Science, ITM SLS Baroda University, Vadodara, Gujarat, India

Khodidas Bhanderi
Industrial Chemistry Department, Institute of Science & Technology for Advanced Studies and Research, Vallabh Vidyanagar, Anand, Gujarat, India

Rachid Bouhfid
Mohammed VI Polytechnic University, Hay Moulay Rachid, Ben Guerir, Morocco

Nilave Chakraborty
Department of Chemistry & Biochemistry, Auburn University, AL, United States

Kumar Chandrasekaran
Centre for Ocean Research (DST—FIST Sponsored Centre), MoES—Earth Science & Technology Cell, Sathyabama Institute of Science and Technology, Chennai, Tamil Nadu, India

Saravanan Chandrasekaran
Department of Chemistry, School of Engineering, Presidency University, Bangalore, Karnataka, India

Maria-Beatrice Coltelli
Department of Civil and Industrial Engineering, University of Pisa, Pisa, Italy; Interuniversity Consortium of Materials Science and Technology (INSTM), Florence, Italy; Academy of History of Health Care Art, Rome, Italy

S.M. Darshan
Department of Mechanical Engineering, The National Institute of Engineering, Mysuru, Karnataka, India; Department of Mechanical Engineering, CHRIST (Deemed to be University), Bengaluru, Karnataka, India

Narayan Ch. Das
Rubber Technology Centre, Indian Institute of Technology Kharagpur, Kharagpur, West Bengal, India

Laila El Foujji
Moroccan Foundation for Advanced Science, Innovation and Research (MAScIR), Composites et Nanocomposites Center, Rabat Design Center, Madinat El Irfane, Rabat, Morocco

Mehrez E. El-Naggar
Textile Research and Technology Institute, National Research Centre, Dokki, Cairo, Egypt

Gigi George
Department of Chemistry, CMS College, Kottayam, Kerala, India

B. Harshavardhan
Department of Mechanical Engineering, The National Institute of Engineering, Mysuru, Karnataka, India

Hemalata Jena
School of Mechanical Engineering, KIIT Deemed to be University, Bhubaneswar, Odisha, India

Jeimin Joshi
Industrial Chemistry Department, Institute of Science & Technology for Advanced Studies and Research, Vallabh Vidyanagar, Anand, Gujarat, India

Swarnalatha Kalayar
Department of Chemistry, Manonmaniam Sundaranar University, Tirunelveli, Tamil Nadu, India

Hafsa Kanwal
Department of Pharmacy, Faculty of Biological Sciences, Quaid-i-Azam University, Islamabad, Pakistan

Niranjan Karak
Department of Chemical Sciences, Tezpur University, Napaam, Tezpur, Assam, India

Mostafa Khodakarami
Department of Chemical and Materials Engineering, University of Alberta, Edmonton, AB, Canada

Ushna Laraib
Department of Pharmacy, College of Pharmacy, University of Sargodha, Sargodha, Pakistan

Andrea Lazzeri
Department of Civil and Industrial Engineering, University of Pisa, Pisa, Italy; Interuniversity Consortium of Materials Science and Technology (INSTM), Florence, Italy

Sreekanth M.S.
Department of Manufacturing Engineering, School of Mechanical Engineering, Vellore Institute of Technology, Vellore, Tamil Nadu, India

Arvnabh Mishra
Department of Industrial Chemistry, Institute of Science & Technology for Advanced Studies & Research (ISTAR), CVM University, V.V. Nagar, Gujarat, India

Momina
Civil Engineering Department, Jamia Millia Islamia, New Delhi, India

Pierfrancesco Morganti
Academy of History of Health Care Art, Rome, Italy

Krishnendu Nath
Rubber Technology Centre, Indian Institute of Technology Kharagpur, Kharagpur, West Bengal, India

Poonam Negi
School of Pharmaceutical Sciences, Shoolini University of Biotechnology and Management Sciences, Solan, Himachal Pradesh, India

Gautam M. Patel
Department of Applied Chemistry, School of Science, ITM SLS Baroda University, Vadodara, Gujarat, India

Jigar Patel
Industrial Chemistry Department, Institute of Science & Technology for Advanced Studies and Research, Vallabh Vidyanagar, Anand, Gujarat, India

Parimal R. Patel
Department of Chemistry, Government Science College, Pardi, Valsad, Gujarat, India

Aleesha K. Paulose
Department of Chemistry, CMS College, Kottayam, Kerala, India

Meera Popaliya
Department of Industrial Chemistry, Institute of Science & Technology for Advanced Studies & Research (ISTAR), CVM University, V.V. Nagar, Gujarat, India

Laly A. Pothan
Department of Chemistry, CMS College, Kottayam, Kerala, India

Priyabrata Pradhan
Mechanical Department, Synergy Institute of Engineering and Technology, Dhenkanal, Odisha, India

Abhilash Purohit
Mechanical Department, Synergy Institute of Engineering and Technology, Dhenkanal, Odisha, India

Radha D. Pyarasani
Centro de Investigación de Estudios Avanzados del Maule (CIEAM), Vicerrectoria de Investigacion y Postgrado, Universidad Católica del Maule, Talca, Maule, Chile

Abou el kacem Qaiss
Mohammed VI Polytechnic University, Hay Moulay Rachid, Ben Guerir, Morocco

Mohd Rafatullah
School of Industrial Technology, Universiti Sains Malaysia, Penang, Malaysia

Ariful Rahaman
Department of Manufacturing Engineering, School of Mechanical Engineering, Vellore Institute of Technology, Vellore, Tamil Nadu, India

Mohammad Mizanur Rahman Khan
Department of Chemistry, Shahjalal University of Science and Technology, Sylhet, Bangladesh

Sithara Raj
Department of Chemistry, CMS College, Kottayam, Kerala, India

Huda R.M. Rashdan
Chemistry of Natural and Microbial Products Department, Pharmaceutical and Drug Industries Research Institute, National Research Centre, Dokki, Cairo, Egypt

Fakhara Sabir
Institute of Pharmaceutical Technology and Regularity Affairs, University of Szeged, Szeged, Hungary

Subramanian Saigeetha
Department of Biotechnology, School of Biosciences and Technology, Vellore Institute of Technology, Vellore, Tamil Nadu, India

Aditi Saikia
Department of Chemistry, University of Science and Technology, Baridua, Meghalaya, India

Antony V. Samrot
School of Bioscience, Faculty of Medicine, Bioscience and Nursing, MAHSA University, Selangor, Jenjarom, Malaysia

Appukuttan Saritha
Department of Chemistry, Amrita VishwaVidyapeetham, Amritapuri, Clappana, Kerala, India

A.S. Sethulekshmi
Department of Chemistry, Amrita VishwaVidyapeetham, Amritapuri, Clappana, Kerala, India

Vraj Shah
Department of Applied Chemistry, School of Science, ITM SLS Baroda University, Vadodara, Gujarat, India

Shifana Shareef
Department of Chemistry, CMS College, Kottayam, Kerala, India

Nagarajan Shobana
Department of Biotechnology, School of Bio and Chemical Engineering, Sathyabama Institute of Science and Technology, Chennai, Tamil Nadu, India

A.K. Shrivastava
School of Studies in Physics, Jiwaji University, Gwalior, Madhya Pradesh, India

E.R. Simge
Biochemistry Department, Institute of Science, Ege University, Izmir, Turkey

Rashmi Singh
Department of Physics, Starex University, Gurugram, Haryana, India

Sunish K. Sugunan
Department of Chemistry, CMS College, Kottayam, Kerala, India

B. Suresha
Department of Mechanical Engineering, The National Institute of Engineering, Mysuru, Karnataka, India

Vaishali Suthar
Applied Chemistry Department, Faculty of Technology and Engineering, The M. S. University of Baroda, Kalabhavan, Vadodara, Gujarat, India

Sumangala T.P.
Department of Physics, School of Advanced Sciences, Vellore Institute of Technology, Vellore, Tamil Nadu, India

Fatih Tornuk
Food Engineering Department, Chemical and Metallurgical Engineering Faculty, Yildiz Technical University, Esenler, Istanbul, Turkey

Gopika Venu
Department of Chemistry, Amrita VishwaVidyapeetham, Amritapuri, Clappana, Kerala, India

Hongbo Zeng
Department of Chemical and Materials Engineering, University of Alberta, Edmonton, AB, Canada

CHAPTER 1

Introduction to biodegradable and biocompatible polymer nanocomposites: synthesis, structure, fundamental properties, biocompatibility, and biodegradability

Aditi Saikia[1] and Niranjan Karak[2]

[1]Department of Chemistry, University of Science and Technology, Baridua, Meghalaya, India [2]Department of Chemical Sciences, Tezpur University, Napaam, Tezpur, Assam, India

Chapter Outline

1.1 Introduction

Over the last few decades, tremendous advances in science and technology have resulted in a steady increase in the global yearly production of synthetic polymers [1]. About 400 million tons of polymers are manufactured each year due to their extensive commercial uses. However, because these synthetic polymers are resistant to physical and chemical degradation, they contribute to land filling issues and

Biodegradable and Biocompatible Polymer Nanocomposites. DOI: https://doi.org/10.1016/B978-0-323-91696-7.00007-6
Copyright © 2023 Elsevier Inc. All rights reserved.

environmental problems [2]. According to recent statistics, it is estimated that these synthetic polymers generate approximately 50% waste, with approximately 20% of this waste being recycled. Recycling helped to reduce polymer waste from 100% in the 1980s to 50% in 2017 [3,4]. Because only 20% of waste is recyclable, an alternative method was researched to help further reduce the waste disposal problem. A new concept of biodegradability was introduced in the late 1970s. Biodegradability is the only solution to land filling problems, and it was discovered that polymers with hydrolysable linkages have suitable biodegradability [3–7]. Some commercially important biodegradable polymers are poly (lactic acid) (PLA), poly(glycolic acid) (PGA), poly(butyl succinate) (PBS), poly(hydroxyalkanoate) (PHA) and their copolymers [8–11]. The synthetic polymers, on the other hand, are derived from estimated fossil reserves that are nearing depletion. They also have a negative impact on the environment and human health. Concerning these issues, a plethora of renewable resources are being thoroughly investigated to design environmentally benign biodegradable polymers, either through polymerization of natural moieties or by incorporating bio-based raw materials into the conventional synthetic polymeric chain [5]. The first research effort to develop biodegradable polymers began in the 1980s. Following that, several modified and advanced methods were published in the literature. The combined approach of biogasification and composting of solid biodegradable waste has sparked tremendous commercial interest [6]. These biodegradable polymers are primarily derived from renewable resources such as starch, cellulose, bacterial polyesters, and even certain synthetic polymers. Thus, various polymers such as polyurethane, polyester, epoxy, poly(ethylene terephthalate) (PET), and PLA have been developed using natural resources such as vegetable oils, polysaccharides, terpenes, lignin, and so on, which help to mitigate the environmental problems associated with fossil fuels [6,12]. These bio-based biodegradable polymers, however, require extensive optimizations in their properties such as physical, mechanical, biodegradation, chemical resistance, etc. during fabrication. Conventional bio-based polymers are not sufficiently resistant to thermal and harsh chemical environments. Furthermore, their mechanical and barrier properties are insufficient for high-end applications [6–8]. To address such issues, composites with bio-based fillers such as wood powder, natural fibers, etc., were developed [7,8]. These biodegradable composites have processing limitations, as well as the degradation of some of the pristine polymers' integral properties. In this context, the use of nanotechnology in the development of polymer nanocomposites aids in overcoming the limitations of conventional polymer composite systems. The fabricated nanocomposites were reported to have some striking improvements in properties while retaining the inherent properties of the pristine polymers. The use of nanomaterials is increasing dramatically in the field of current research due to their appealing combination of thermal, mechanical, physical, and chemical properties that offer potential applications in a variety of fields [7–9]. The wide range of properties provided by these nanomaterials can be tuned depending on their size, shape, and structure [10]. In this venture, the wide arena of different nanomaterials like carbon dots (CD), carbon nanotubes (CNT), graphene oxide (GO), reduced GO (rGO), montmorillonite (MMT), organically modified MMT (OMMT), fullerenes, nanofibers, and quantum dots (QD) are widely employed in the fabrication of the polymer nanocomposites [8–13]. *Ex-situ* or *in-situ* incorporation of these nanomaterials in pristine biodegradable polymers resulted in biodegradable polymeric nanocomposites with enhanced biodegradability and physicomechanical properties [8]. In this vein, Zhuang et al. reported PLA/titanium dioxide (TiO_2) nanocomposites made by *in-situ* polymerization with varying TiO_2 contents. The nanocomposites exhibited enthralling photocatalytic activity, hydrophilicity, and appealing antibacterial properties. Ultraviolet (UV) irradiation studies confirmed that the reported composites degraded faster as the amount of TiO_2 increased [14]. Similarly, Shimpi et al.

reported that *Pseudomonas aeruginosa* bacterial strain biodegraded polystyrene/PLA and polystyrene/PLA/OMMT nanocomposites of various compositions [15].

Biocompatible polymers are one of the most notable categories of bio-based polymers. Biocompatible polymers are bio-based polymers that are compatible with the biological systems of living animals, including humans. These biocompatible polymers are said to have a wide range of applications, including tissue engineering and controlled drug delivery systems, as well as medical implants such as dialysis, tubing, etc. [13]. The requirement of the polymeric surface to interact with a biological system is the most important criterion for biocompatible polymers. Several studies have been conducted in this area, and significant efforts have been made to develop such advanced biocompatible polymers. Continuous efforts are made to improve methods, materials, and devices in biomedical applications. However, because most developed biomaterials have not been properly optimized, they lack the necessary properties for the desired interface with biological systems. As a result, novel biocompatible materials must be developed to address these issues in the field of biomedical applications. In this venture, the combination of biocompatible and biodegradable polymers with appropriate nanomaterials created a new class of biomaterials with great potential for biomedical devices, scaffolds, health-care sensors, and drug delivery systems [8]. Lin et al. used developed nanofibers containing hydroxyapatite (HA) from lactic acid for bone mineralization in a study [16]. The nanofibers can be used as a scaffold for osteoblast growth and differentiation. Duarah and Karak reported a rapid self-expandable stent made of a bio-based polyurethane nanocomposite and a CD/silver nanohybrid in another study [17]. This stent demonstrated hematocompatibility and antimicrobial properties in vitro, making it suitable for use in endoscopic surgeries. As a result of the growing importance of biodegradable and biocompatible polymers, this chapter delves into the synthesis, structures, fundamental properties, biocompatibility, and biodegradability of such bio-based polymer nanocomposites.

1.2 Synthesis, fabrication, and structure

1.2.1 Synthesis of biodegradable and biocompatible synthetic polymers

Condensation polymerization, ring opening polymerization (ROP), and other traditional methods are used to create biodegradable polymers. The selection of appropriate reactant moieties with the required functionalities is required for the synthesis of a suitable biocompatible polymer. Table 1.1 summarizes the preparative routes of a few such biodegradable polymers.

These polymers can also be modified by combining two or more biodegradable polymers using various chemical methods such as nitration, alkylation, graft polymerization, and so on. Some of these synthetic methods are briefly illustrated in the following section.

1.2.1.1 Synthesis of polymers via ROP

The ROP is the most commonly used synthetic method for the synthesis of polymers, particularly synthetic biodegradable polymers. This method can combine cationic, anionic, and free radical polymerization. ROP is used in the commercial synthesis of synthetic polymers such as PCL, PLA, and others. It is a chain-growth polymerization technique in which an active polymer terminus

Table 1.1 Structure, preparative technique and catalyst used for a few industrially important synthetic biodegradable and biocompatible polymers.

Polymer	Structure	Catalyst/initiator	Preparative technique	Refs.
Poly(lactic acid) (PLA)	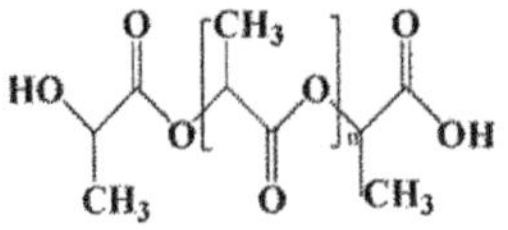	Sn $(Oct)_2$	ROP	[18–20]
Poly(glycolic acid) (PGA)	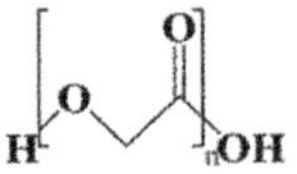	Sb_2O_3, zinc lactate, etc.	Polycondensation and ROP	
Polycaprolactone (PCL)	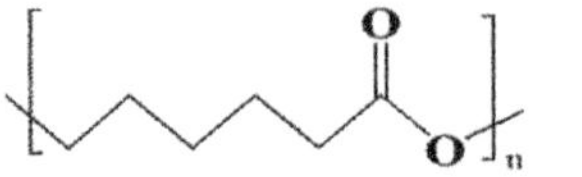	SbO_2	ROP	
Poly(butyl succinate) (PBS)	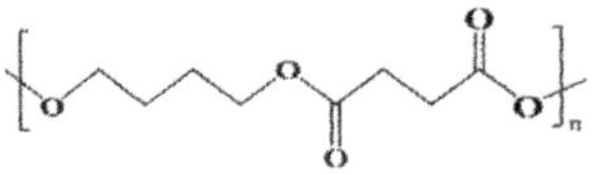	1,4-Butanediol	Polycondensation	[20–22]
Poly(hydroxy-alkanoate)s (PHA)	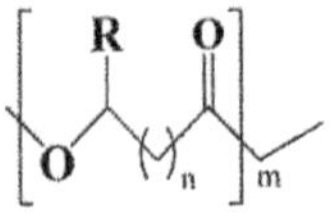 = $CH_3.C_2H_5$ etc.	Tetrabutylaluminoxane, cobalt salens	Polycondensation by *Cupriavidus necator*, *B. Subtilis*	

attacks highly stained aliphatic cyclic monomers, resulting in the formation of polymer. Free radical, anionic, or cationic active polymer termini are all possible. Because this method involves the release of strain from a cyclic monomer, the overall enthalpy change is negative. It is the most commonly used industrial route for PCL synthesis. In the synthesis of PCL, different metal catalysts can be used for polymerization however, stannous octoate is one of the most commonly used metal catalysts for this method as shown in Table 1.1 [23]. Zhang et al. reported ROP of ε-caprolactone using benzyl alcohol initiator and squaric acid (a commercial organic acid used as the organocatalyst) as the catalyst to find environmentally friendly alternatives. As shown in Fig. 1.1 [24], the reaction was carried out under controlled conditions at a temperature of 80°C. However, literature has also reported the use of various enzymes for this polymerization method due to their eco-friendliness and ability to work in mild conditions. Antartica Candida Lipase B is a dominant enzyme in the lipase family that has been identified as the best polymerization catalyst [25].

1.2.1.2 Step-growth polymerization of monomers

The polymerization method that results in polymer by step by step of the chains is the polymerization of monomers via step-growth process. The addition of monomers occurs in a sequential, step-by-step manner. Step-growth can take place in solution, melt, or under interfacial conditions. When compared to the other two types, melt polymerization produces polymers with a higher percentage of purity. Polyurethane, which is synthesized using the step-growth method, is one of the most versatile polymers [23]. One of the most widely used conventional step-growth methods for the synthesis of polyurethane is diisocynate polymerization with diol. Polyurethanes were produced by a

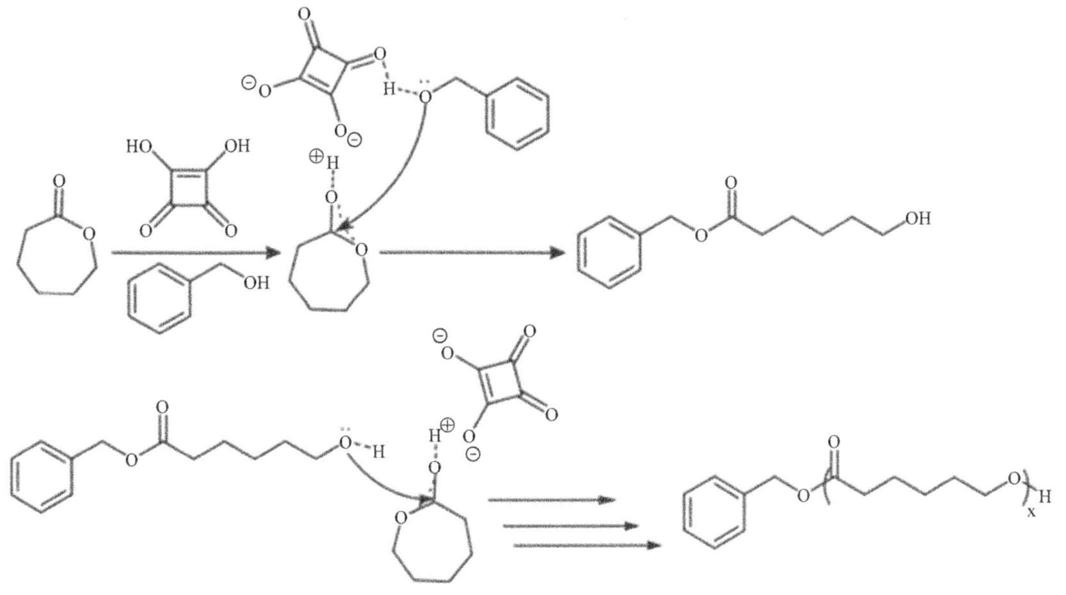

FIGURE 1.1

ROP of ε-caprolactone using squaric acid as the catalyst.

Reproduced with permission from J. Zhang, L. Xu, W. Xiao, Y. Chen, Z. Dong, J. Xu, C. Lei, Ring-opening polymerization of ε-caprolactone with recyclable and reusable squaric acid organocatalyst. Eur. Polym. J, 2021:157:110643. https://doi.org/10.1016/j.eurpolymj.2021.110643. Copyright 2021, Elsevier.

step-growth polymerization reaction between D-mannitol-1,2:5,6-dicarbonate and diamine, according to Prömpers et al. [26]. Gogoi et al. also synthesized biodegradable and bio-based waterborne polyurethane via step-growth polymerization reactions using tannic acid along with isophorone diisocyanate, polyethylene glycol and triethylamine as shown in Fig. 1.2 [27].

1.2.1.3 Modification of monomers

Natural or synthetic polymers can be modified using various chemical processes to improve or decrease their hydrophobicity. Hydrophilic polymers with hydrolysable linkages biodegrade quickly. Bio-based hyperbranched epoxy resins are one such example. Because of their high hydrophobicity, conventional epoxy resins are nonbiodegradable. Several efforts are being made to improve these resins' biodegradability. Saikia and Karak reported hyperbranched epoxy resins with high biodegradability in a study [28]. The resins were created by modifying the conventional reactants by incorporating bio-based reactants such as castor oil monoglyceride and sorbitol as the branching moiety, in addition to epichlorohydrin and bisphenol A, as shown in Fig. 1.3 [28].

Natural polymers such as chitin, starch, cellulose, and others are also modified, according to literature. Natural cellulose fibers are highly hydrophilic and polar in nature, which makes processing difficult. As a result, surface modification of cellulose is required to increase its versatility in various applications. In this regard, Baiardo et al. investigated and reported chemically modified cellulose fibers with

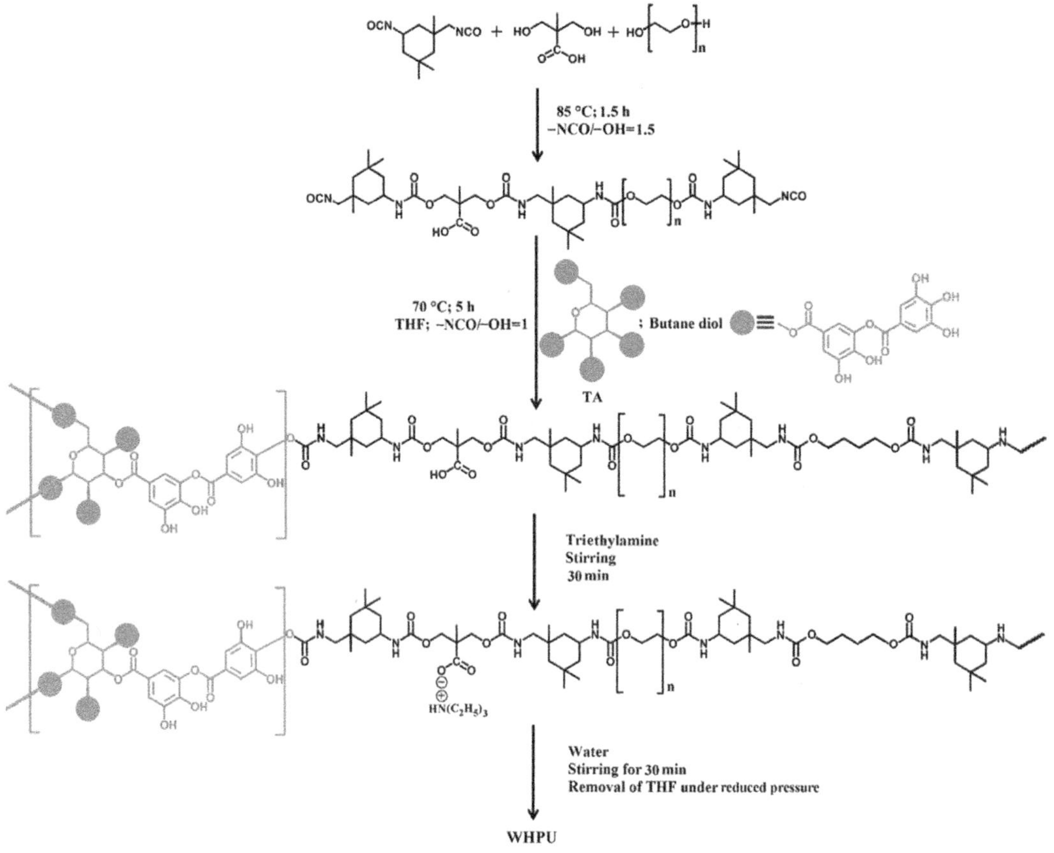

FIGURE 1.2

Synthesis of waterborne polyurethane (WHPU) using bio-based moieties.

Reproduced with permission from S. Gogoi, N. Karak, Bio-based biodegradable waterborne hyperbranched polyurethane as an ecofriendly sustainable material. ACS Sustain. Chem. Eng. 2014: 2(12):2730–8. https://doi.org/10.1021/sc5006022. Copyright 2014, ACS Publications.

acceptable hydrophilicity and polarity via esterification and etherification approaches. The fibers were esterified with pyridine, acyl chloride, and carbon tetrachloride, while the etherification was accomplished with alcohol and ethyl iodide in the presence of a base [29]. Another significant modification through graft polymerization of acrylate on chitosan was reported by Molatlhegi et al. [30]. This graft polymerization was initiated by using ceric ammonium nitrate as initiator Fig. 1.4.

1.2.2 Fabrication of polymer nanocomposites

The fabrication of nanocomposites addresses the limitations of conventional composite systems. As a result, the demand for highly efficient polymeric systems is constantly increasing. To meet these

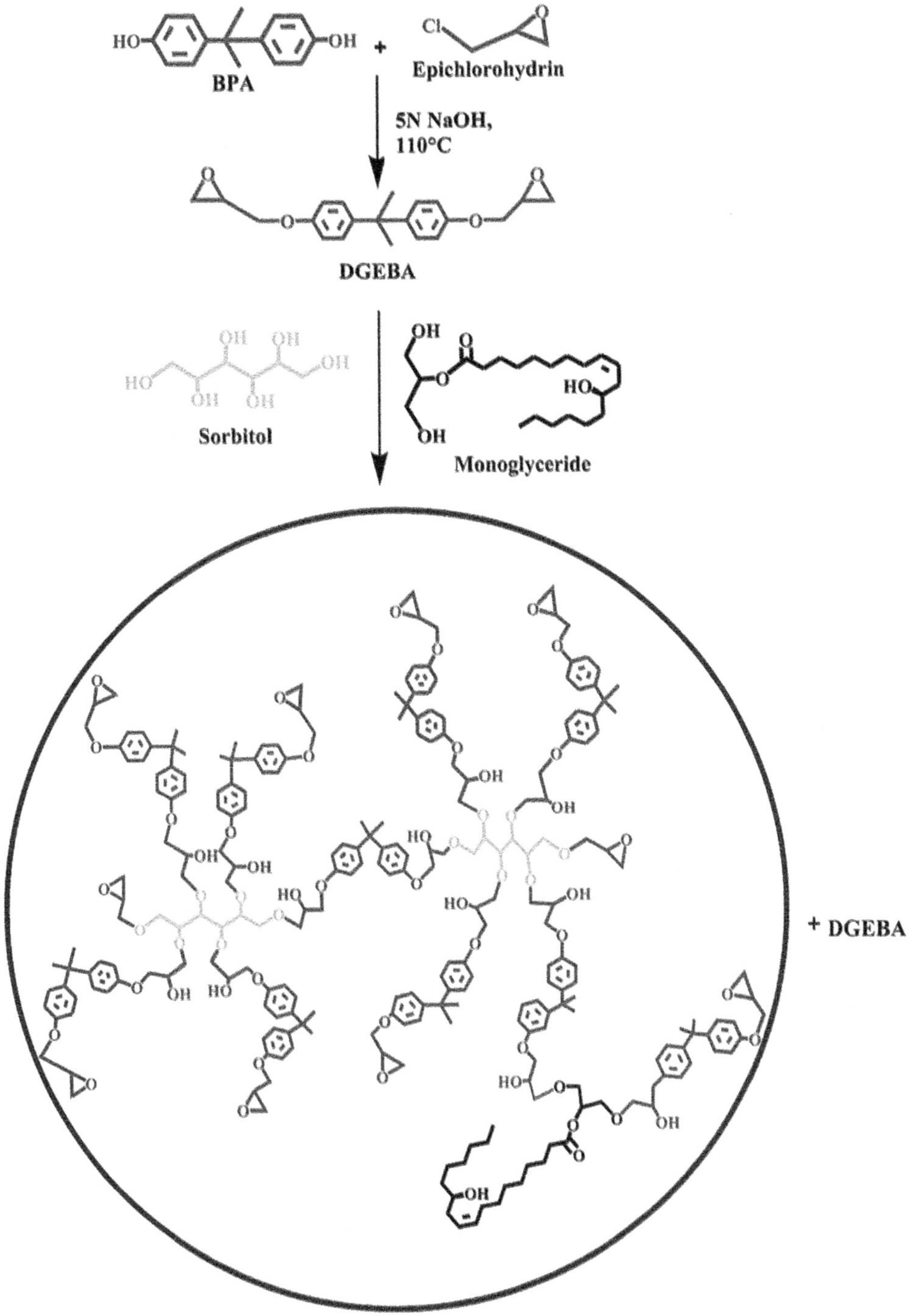

FIGURE 1.3

Synthesis of hyperbranched bio-based epoxy resin using bio-based moieties.

Reproduced with permission from A. Saikia, N. Karak, Renewable resource based thermostable tough hyperbranched epoxy thermosets as sustainable materials. Polym. Degrad. Stab. 2017;135:8–17. https://doi.org/10.1016/j.polymdegradstab.2016.11.014. Copyright 2017, Elsevier.

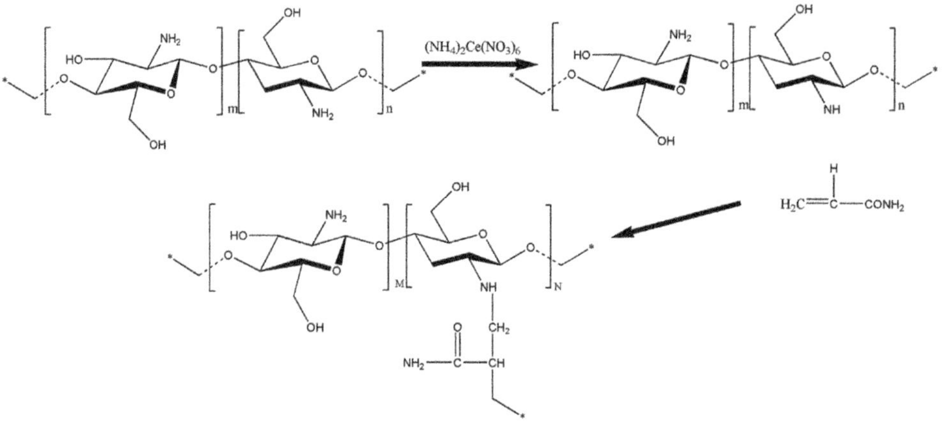

FIGURE 1.4

Surface modification of chitosan via graft polymerization.

Reproduced with permission from O. Molatlhegi, L. Alagha, Adsorption characteristics of chitosan grafted copolymer on kaolin. Appl. Clay Sci. 2017;150:342–53. https://doi.org/10.1016/j.clay.2017.09.032. Copyright 2017, Elsevier.

requirements, nanocomposites are manufactured in two ways: in-situ and ex-situ. The polymerization of the monomer/prepolymer occurs in the presence of the nanomaterials in the in-situ fabrication technique. Under suitable reaction conditions, these monomers/prepolymers have very low viscosity at first, allowing for high diffusibility and smooth mixing of the nanomaterials. This allows for proper dispersion and morphology control of the resulting nanocomposites. Lin et al. reported nanosilica epoxy-polyurethane nanocomposites using this method in 2014 [31]. In this method, epoxy and acrylate are copolymerized in situ in the presence of silica sol. This resinous mixture is then treated with isocynate and anhydride cured to produce polyurethane nanocomposites, as shown in Fig. 1.5.

The second method is the ex-situ method, which involves allowing nanomaterials to swell for a period of time before dispersing them using ultrasonic forces and or mechanical mixing. These swollen and dispersed nanomaterials are added to the polymer solution externally. The ability of the solvent or miscible solvent mixture to disperse the nanomaterials as well as solubilize the polymer influences the choice of suitable solvent or miscible solvent mixture. To obtain the thermoplastic or thermosetting nanocomposite, the solvent is later removed or evaporated using precipitation or heat. Because the presence of the solvent allows deagglomeration and easy intercalation of the polymer chains in between the agglomerates, this method is also known as solvent intercalation. One such example of this technique is the fabrication of poly(vinyl alcohol) (PVA)/chitosan nanocomposite films with GO/nickel oxide (NiO) nanohybrid as shown in Fig. 1.6. In this study, a chitosan and PVA blend was first prepared in distilled water, and then externally added well dispersed swollen GO/NiO nanomaterials, followed by casting and solvent evaporation [32].

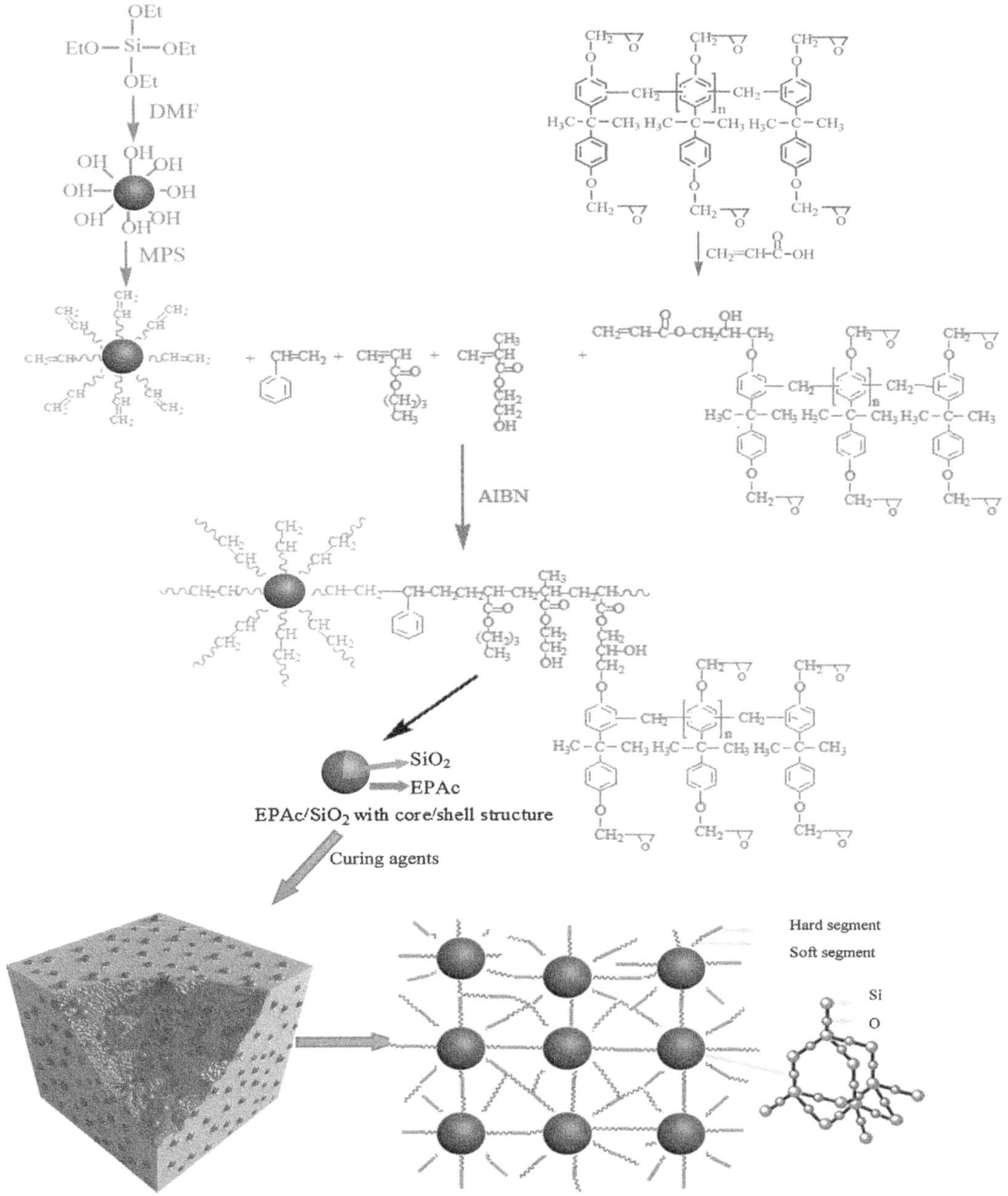

FIGURE 1.5

Reaction mechanism showing in-situ polymerization of epoxy with acrylate in presence of silica sol.

Reproduced with permission from J. Lin, W. Xu, C. Zheng, P. Zhang, B. Huang, N. Guo, L. Jin, Synthesis and properties of epoxy-polyurethane/silica nanocomposites by a novel sol method and in-situ solution polymerization route. Appl Surf Sci 2014;303:67 – 30. https://doi.org/10.1007/s10904-020-01869-z. Copyright 2014. Elsevier.

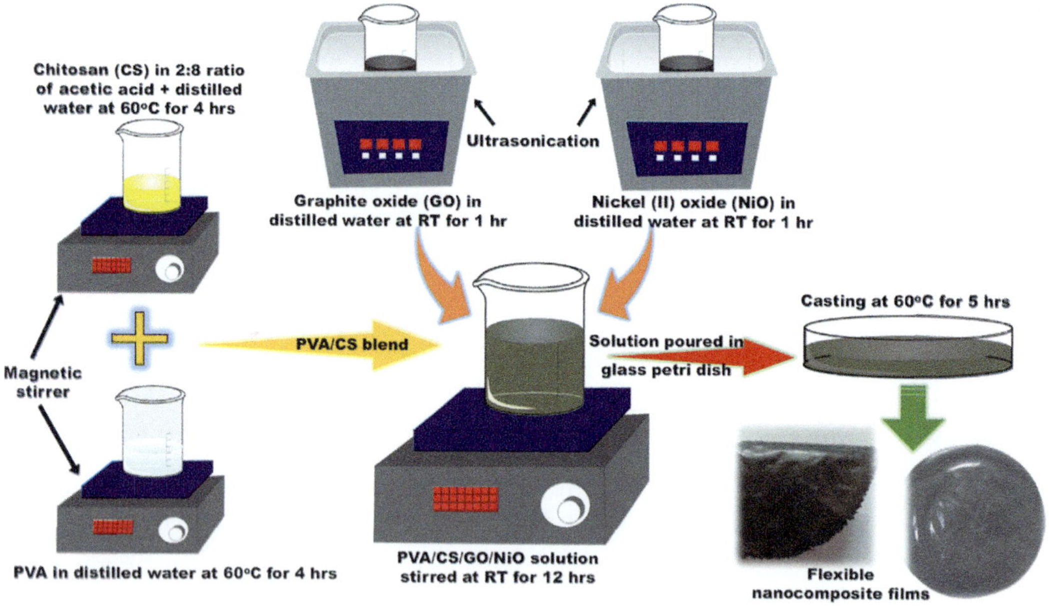

FIGURE 1.6

Ex-situ preparation technique of poly(vinyl alcohol)/chitosan/GO/NiO nanocomposite films.

Reproduced with permission P. Rani, M.B. Ahamed, K. Deshmukh, Structural, dielectric and EMI shielding properties of polyvinyl alcohol/chitosan blend nanocomposites integrated with graphite oxide and nickel oxide nanofillers. J Mater Sci: Mater Electron 2021;32:764–79. https://doi.org/10.1007/s10854-020-04855-w. Copyright 2021, Springer.

However, the melt-mixing technique is the most commonly used in industry, where mechanical mixing of the molten polymer is performed using an extruder, injection molding, or kneader. The nanomaterials are intercalated or exfoliated under shear force during melt conditions. Several reports in the literature stated the use of this technique in the fabrication of clay/polymer nanocomposites. Liu et al. created Eudragit RS nanocomposites with MMT or OMMT using a melt extrusion process and investigated their interactions and drug release [33]. Eudragit RS is a commercially available copolymer composed of acrylic-methacrylic acid esters and quaternary ammonium groups. Although this melt processing technique produces only partially exfoliated nanocomposites, it is the most preferred fabrication technique for polymer nanocomposites by industries because it is cost effective, environmentally friendly (solvent free), and faster.

1.3 Types and features of biodegradable and biocompatible polymers and their nanocomposites

Polymeric degradable biomaterials are widely used in medical devices such as scaffolds, interim prostheses, surgical stents, and drug delivery vehicles. As a result, these biodegradable polymers can be synthetically or naturally derived, and they can be degradable enzymatically or nonenzymatically. The sections that follow provide a brief description of various biodegradable and biocompatible polymeric nanocomposites, both synthetic and natural.

1.3.1 Synthetic biodegradable polymers

PLA, PGA, PBS, PCL, PHA, and their copolymers are examples of synthetic biodegradable polymers that are synthesized using various polymerization methods and have good processing flexibility.

1.3.1.1 Poly(lactic acid) and poly(glycolic acid)

PLA is an aliphatic thermoplastic with a biodegradable backbone. It is a bioderived synthetic polymer with high biodegradability, biocompatibility, and strength. It was discovered in the mid-1950s through the ROP of lactide, a lactic acid derivative [18]. Lactic acid is a chiral molecule with an asymmetric carbon atom that is obtained through the traditional fermentation method. It is easily converted into various compounds such as cyclic esters and lactides composed of lactic acid dimers. Because lactic acid is chiral in nature, it can exist as one of three optically isomeric lactides: L-lactide, D-lactide, or meso-lactide. Lactide racemic mixture can also be obtained. As a result, it should be noted that these lactides have two asymmetric carbon centers. Each of these lactides polymerizes separately to produce PLA in its D, L, or D-L form. The polymerization process is subjected to optimized temperature, pressure, or time conditions. Furthermore, many factors influence the polymerization process, including the presence of acid-water, as well as the concentration and type of initiator used. Metal-based initiators, such as stannous octanoate, or alcohol-based initiators are commonly used. The initiator is chosen primarily for its application. Only by continuously removing water from the batch can high molecular weight PLA be obtained, as the presence of the latter may result in hydrolysis of the polymer chains [34]. However, because of its biocompatibility, PLA has been widely used in biomedical applications. Chang reported semipermeable PLA microcapsules encapsulating enzymes, vaccines, hormones, and other substances in 1956. The microcapsules were discovered to be nontoxic in the body. The incorporation of various nanomaterials and their combinations with the polymer matrix produced significant results in thermal and mechanical properties, broadening its applicability. In this regard, Zhuang et al. created nanocomposites of PLA with varying TiO_2 proportions using an in-situ polymerization method [14]. The uniformity in dispersion of TiO_2 nanoparticles was observed in the fabricated nanocomposites consisting of TiO_2 less than 3 wt.%. This consistency was reflected in their improved mechanical and thermal properties. Furthermore, the UV and solution degradation experiments revealed that the nanocomposites' biostatic activity and biodegradability improved significantly. PLA nanofibers have also demonstrated extremely high biocompatibility. In a study PLA nanofibers containing chitosan and HA were synthesized by Lin et al. [16]. The synthesized nanocomposite was able to replicate natural bone functions in both structure and composition and was thus found to be compatible with bone regeneration. Skaria et al. reported microneedle arrays made of PLA nanocomposites with carboxyl functionalized multiwalled carbon nanotubes (MWCNT) as a dermal biosensor in 2019. Aside from biomedical applications, the PLA nanocomposite was also used as a food packaging material [35]. Mahmoodi et al. also reported a PLA nanocomposite with a dye (organic)–clay (inorganic) hybrid nanopigment with improved properties [36].

Another extensively researched biodegradable aliphatic polyester that is biocompatible and widely used in medical applications is PGA. The ROP of glycolic acid resulted in the formation of PGA [18,19]. According to reports, PGA is more biodegradable than existing polyesters, and it is critical that biodegradable PGA polymers have a high molecular weight and good mechanical strength for use in packaging [37]. The potential use of PGA as a biodegradable packaging material was first explored in 2005 by Peter M. Bonutti [38]. Bonutti created a packaging film with two layers of PGA and a reactive chemical layer in between. These films were used to protect medical supplies and perishable foods. The packaging film contains a self-indicating color that develops due to air exposure when one layer degrades, indicating the expiry of the package's shelf-life. Samadi et al. investigated the biodegradability, mechanical, and rheological properties of PLA-*co*-PGA copolymer blends in 2019 [39]. Blends prepared using the melt processing technique produced more methane under anaerobic conditions after degreasing the molecular weight and increasing the PLGA content. Furthermore, the blends had lower ductility, modulus, and tensile strength when compared to pure PLA. Several studies have reported the use of PGA and its composites in various biomedical applications. The most important use of PGA in biomedical applications is as a scaffold for tissue culture or nerve regeneration. Patrascu et al. synthesized hyaluronic acid (hyaluronan) with PGA scaffolds and tested their in-vitro chondrogenic potential with mesenchymal stem cells (MSCs) in a rabbit articular cartilage defect model. The loaded scaffolds produced positive results, indicating that they can be used successfully for stem cell-mediated cartilage renewal [40]. In 2013, Kobayashi et al. fabricated biocompatible nanocomposites of PGA by combining with collagen [41]. These nanocomposites were used as scaffolds in animal models, and the results showed that within 5 days of implantation, the scaffolds were completely occupied and vascularized. Ashammakhi et al. reported a self-reinforced polyglycolide membrane made of braided and knitted PGA fibers glued on one side by PGA [42]. When implanted subcutaneously in rats over a 4-week period, the synthesized membrane demonstrated high strength retention and was completely biocompatible. With the incorporation of different metal-based nanoparticles, the copolymers of PGA with PLA, also known as PLGA polymer, demonstrated excellent antimicrobial, activity, cytotoxicity, or biocompatibility. Wu and colleagues created nanocomposites of TiO_2 and PLGA with optimized disinfection efficiency by varying the proportions of TiO_2 in PLGA [43]. The nanocomposites demonstrated excellent antibacterial and wound healing properties. Thus, the biodegradability of PGA can be successfully tuned to its biomedical properties by decorating it with appropriate copolymers, nanomaterials, or a combination of the two.

1.3.1.2 Polycaprolactone

PCL is a biocompatible and environmentally friendly synthetic polyester with a variety of properties such as high flexibility, low glass transition temperature (Tg), high permeability, and high flexibility. This polymer is synthesized by ROP of ε-caprolactone monomer in the presence of an appropriate catalyst. Because the polymer's properties are unsuitable for advanced applications, different nanocomposites were created by incorporating the appropriate nanomaterial based on their applications [18,19]. Lepoittevin et al. prepared PCL/clay nanocomposites in 2002 using dibutyltin dimethoxide as an initiator or catalyst in in-situ ROP of ε-caprolactone. The polymerization was co-initiated by alcohol-containing OMMT, which controlled the molecular weight. Clay dispersion and exfoliation were entirely dependent on the hydrophobicity of the alkyl ammonium; hydroxylation of alkyl ammonium resulted in clay exfoliation [44]. Maiti and his colleagues reported the solution route fabrication of nanocomposites of PCL with layered silicate in another paper published in 2009. They created two different organically modified nanoclays and investigated their properties. The biodegradability of nanohybrids improved significantly and is solely dependent on the type of media used. This enhancement is solely due to the crystallization behavior and organic modification of nanoclays [45]. The self-healing characteristics of PCL nanocomposites with polydopamine-capped rGO was reported by Wang et al. [46]. When exposed to near-infrared (NIR) light, the nanocomposites rapidly reduce the melting temperature of the polymer; thus, NIR light induces the nanocomposites' self-healing properties [46]. PCL nanocomposites are also widely used in tissue engineering as nanocomposites. Nezakati et al. reported conductive POSS-PCL/graphene nanocomposites with ultra-low percolation thresholds for use as a neural tissue engineering scaffold for neurosurgery [47]. PCL/zein/gum electrospun nanocomposites were also reported by Rad et al. [48]. The fabricated nanocomposites demonstrated hydrophilicity, porosity greater than 77%, increased cell viability, and antibacterial activity, indicating that they can be used as a scaffold for skin tissue engineering (Fig. 1.7). As a result, PCL and its nanocomposites may be useful as biodegradable and biocompatible polymeric materials.

1.3.1.3 Poly(butyl succinate)

PBS has received a lot of attention because of its unique properties, which include easy melt processability, good biodegradability, and thermo-chemical resistance. It is created through the polycondensation reaction of 1,4-butanediol and succinic acid. PBS's biocompatibility and osteoconductive properties have also been extensively studied, as it can promote both in vitro and

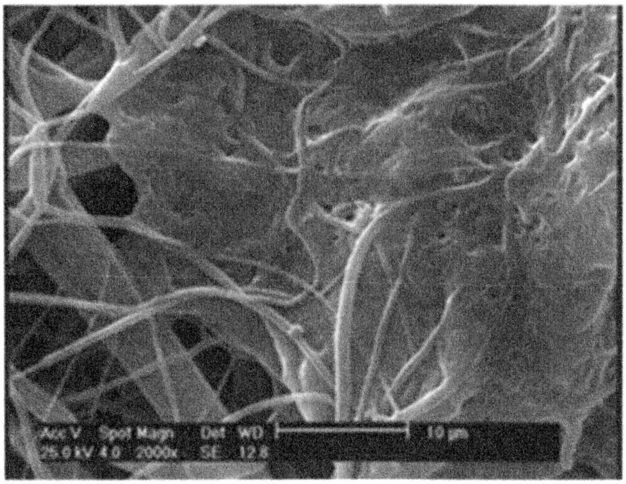

FIGURE 1.7

SEM morphological images showing L929 fibroblasts cell migration of cells inside PCL/Zein/GA (P9) scaffold after a period of 24 h culture.

Reproduced with permission from Z.P. Rad, J. Mokhtari, M. Abbasi, Fabrication and characterization of PCL/zein/gum arabic electrospun nanocomposite scaffold for skin tissue engineering. Mater. Sci. Eng. C, 2018;93:356–366. https://www.sciencedirect.com/science/article/pii/S0928493118300845. Copyright 2018, Elsevier.

in vivo proliferations as well as tissue cellular adhesion [49]. Phua et al. investigated the biodegradability of PBS/OMMT nanocomposites using the soil burial testing method. Maleic anhydride-grafted PBS (PBS-g-MA) was used as a compatibilizer to create these nanocomposites from PBS and OMMT [21]. The nanocomposites were tested in organic humus compost under controlled conditions for 180 days. According to the findings, the weight loss of neat PBS was greater than that of nanocomposites due to the presence of OMMT, which induces enhanced barrier properties to the polymeric substrate. The mechanical properties of PBS nanocomposites were found to be significantly reduced as a result of biodegradation. El-Rahman et al. recently modified high-density polyethylene (HDPE) by incorporating PBS and calcium carbonate nanoparticles [50]. The study reported improved biodegradability, which was confirmed by SEM morphological analysis. The nonbiodegradable HDPE reported a weight loss% of 0.265 wt.% after 8 weeks upon the incorporation of PBS and calcium carbonate nanoparticles. Huang et al. investigated the biocompatibility of PBS in a study where bimodal-open PBS/cellulose nanocrystal (CNC) nanocomposites were successfully used as scaffolds in tissue engineering [51]. In vitro degradation of the nanocomposites was higher (13.74%) than that of the neat PBS (4.5%). Furthermore, 3T3 fibroblast cell culture studies confirmed fast proliferation, indicating the nanocomposites' good biocompatibility in scaffold applications. Shirali et al. created biocompatible nanocomposites from PBS-*co*-ethylene terephthalate/HA using in-situ polymerization in 2017. Because the nanocomposites showed significant proliferation and osteogenic differentiation of human MSCs, they could be used as scaffolds for bone tissue engineering [52]. Thus, the biodegradable and biocompatible PBS has great potential as a novel biomaterial [52].

1.3.1.4 Poly(hydroxyalkanoate)

Another biodegradable and water-insoluble aliphatic polyester is PHA. In 1925, a microbiologist named Lemoigne isolated a polyester of 3-hydroxybutyric acid from *Bacillus megaterium* (*B. megaterium*) culture. In general, it is beta-hydroxyalkanoic acids, as opposed to PLA, which is an alpha-hydroxyalkanoic acid. It has recently received a lot of attention in the food packaging industry. Poly(3-hydroxybutyrate) (PHB) and copolymer PHB valerate are the most commonly used forms of PHA in food packaging due to their excellent perm-selectivity, mechanical, and thermal properties. Furthermore, PHA has some limitations, including low barrier effect, high brittleness, and low melt stability, as well as processing difficulty, which can be overcome by using nanomaterials [18]. Pascual created various nanocomposites of PHB and PHB valerate with varying zinc oxide (ZnO) nanoparticle compositions. The addition of ZnO improved the barrier, mechanical, thermal, and antibacterial properties. Both Gram-positive and Gram-negative bacterial strains were resistant to the nanocomposites' antibacterial activity [22]. In another approach reported by Fabra et al. where they fabricated bio-nanocomposite of PHB bacterial cellulose nanowhiskers (BCNW) showed improved barrier properties against water and oxygen with a record of 46% and 95%, respectively, with a loading of 15 wt.% BCNW [53]. These studies demonstrated the potential of PHA nanocomposites in food packaging due to their inherent antibacterial properties. PHA has played an important role in biomedical applications, as current research focuses on developing bio-nanocomposites with enhanced antibacterial activity against multidrug-resistant bacteria. Mukheem et al. created antibacterial nanocomposites by encapsulating molybdenum disulfide (MoS_2) nanoparticles in PHA and chitosan [54]. The nanocomposites' in-vitro antibacterial activity and host cytotoxicity reports revealed effective antibacterial activity against both Gram-positive and Gram-negative bacteria, indicating improved biocompatibility. As a result, such PHA nanocomposites demonstrated their potential and suitability for antibacterial applications as well as future biomedical studies.

1.3.2 Natural biodegradable polymers

Natural biodegradable polymers have also been extensively researched due to their inherent biocompatibility and biodegradability. These natural polymers, however, have very poor mechanical, thermal, and chemical resistance properties, which can be improved by the addition of specific nanomaterials. The following are some of the most extensively researched natural polymers and nanocomposites.

1.3.2.1 Starch

In biodegradable nanocomposites, starch is a common biopolymer matrix. It comes in two varieties: amylose and amylopectin. Amylose has a linear structure and exists in a tightly bound helical structure, making it insoluble in cold water. Amylopectin, on the other hand, is highly branched, loosely packed, and thus easily digestible. During the plasticization process, the amylopectin's strongly bound structure opens up the helical amylose to produce a suitable material for processing and molding. The processed amylose molecular weight increases from 5000 to 340,000 g/mol, whereas in case of amylopectin it is about 80,000,000 g/mol; as stated by BeMiller and Wistler in 2009 [18]. The properties of the resulting high viscosity starch are solely determined by the proportions of amylose and amylopectin. Nasseri and Mohammadi presented a comparative study on the effect of cellulose nanowhiskers (CNW) and starch on the plasticized starch reinforcement effect in their paper on starch-based nanocomposites. When compared to CNW-based nanocomposites, the fabricated nanocomposites containing quasi-spherical nanoparticles of starch demonstrated a higher degree of reinforcement [55]. This effect is attributed to the starch nanoparticles' efficient hydrogen bonding interactions with the starch matrix. Mirab et al. created a gradient-structured highly porous starch-based nanocomposite scaffold suitable for bone tissue engineering. Using cryogenic freeze-casting and freeze-drying methods, a highly porous starch/PVA-based nanocomposite was created by incorporating different nanomaterials such as citric acid, cellulose nanofibers (CNFs), and HA. The in-vitro study found that HA nanoparticles effectively induce apatite mineralization. Furthermore, the nanocomposites' cell line assay demonstrated controlled biodegradability and good osteoblast cell cytocompatibility [56]. Another study created starch-based nanocomposites by immobilizing silver nanoparticles on cellulose nanofibrils, which demonstrated high mechanical strength, antimicrobial activity against *Escherichia coli* and *Staphylococcus aureus* (*S. aureus*), and biocompatibility with 87.71% cell viability [57]. Thus, modified starch films containing various nanomaterials can be a suitable candidate for various biodegradable packaging materials and biocompatible materials.

1.3.2.2 Cellulose

Cellulose is one of the most abundant polysaccharides found in plants, leaves, fruits, bacteria, and other organisms. It is widely used in a variety of applications such as food packaging, paper, filtration, and biomaterials. Cellulosic nanocomposites are less expensive and biodegradable, and cellulosic fibers have a high aspect ratio. When used as reinforcing material, orientated fibers provide high stress dissipation properties to the nanocomposites [58]. However, the presence of numerous hydroxyl groups frequently resulted in aggregation at high concentrations when used as nanofillers, so the concentration of CNFs in composites is an important factor in nanocomposites. The two main sources of nanocellulosic materials used to create cellulosic polymer nanocomposites are microfibrillated cellulose (MFC) and bacterial cellulose [59]. Jonoobi et al. reported in 2015 on various approaches used in the literature for isolating nanocellulose from nature. They also discussed the important properties associated with various types of cellulosic nanomaterials derived from various natural sources [60]. Cellulosic nanomaterials are divided into four categories: CNFs, CNW, cellulose nanofibrils, and CNCs, which differ in aspect ratio and are produced using different methods. CNFs, as the name implies, are long nanofibers with a high flexibility and aspect ratio. CNC, on the other hand, are needle-shaped short-length crystals with nanoscale diameter and lengths of up to 500 nm. Cellulosic biopolymers are widely used in a variety of applications such as foams, films, and hydrogels [61–63]. This adaptability is due to the various chemical and physical modifications used in conjunction with other nanomaterials or polymers. The most important properties of cellulose are its biodegradability and biocompatibility, both of which can be improved through these modifications [64]. In this regard, Aguilar et al. used CNFs that had been chemically oxidized by 2,2,6,6-tetramethylpiperidine-1-oxyl (TEMPO) and physically dual-crosslinked by epichlorohydrin (ECH), followed by carboxylate ion formation using Ca^{2+} salts. The fabricated films demonstrated good transparency (90% transmittance), excellent tensile strength, and high thermal stability, making them suitable for the production of environmentally friendly bioplastics [65]. Saikia et al. created transparent nanocomposites with CNFs incorporated inside the matrix of bio-based hyperbranched epoxy resins in another study. The nanocomposites also demonstrated good biodegradability, as well as high mechanical and thermostability properties [66]. The biocompatibility of CNFs was also investigated in another study by the same group, in which a nanohybrid of CNFs with polyaniline nanofibers (PAni) and CD was prepared and demonstrated excellent glucose sensing ability [67]. Because of the presence of CNFs in the matrix, the nanocomposites of the nanohybrid with bio-based hyperbranched epoxy resin demonstrated good hemocompatibility and biodegradability. In another study, Barua et al. prepared CNFs coated with copper-copper oxide nanoparticles that demonstrated excellent antimicrobial activity and compatibility with both peripheral blood mononuclear cells and mammalian red blood cells (RBCs) [68]. Thus, these studies demonstrated the biocompatibility and biodegradability of CNFs modified with various nanomaterials and polymer matrixes.

1.3.2.3 Chitin and chitosan

Chitin, like starch and cellulose, is a naturally abundant polymer. It is a water-insoluble linear aminopolysaccharide. Chitin is made up of 2-acetamido-2-deoxy-β-D-glucose, also known as *N*-acetylglucosamine, and is thus an amino analog of cellulose [69]. Purified chitin is deacetylated in the presence of 5% NaOH to produce chitosan. The difference in their chemical structures results in differences in their physical and chemical properties. Chitosan, for example, dissolves in acetic acid, whereas chitin dissolves in trichloroacetic acid. However, at a specific pH, chitosan is found to be water soluble, as opposed to chitin, which is inherently water insoluble [70]. Various literatures on modified chitosan films have recently been reported [70]. The synergistic effects of cellulose and chitosan increased the strength of these films. Aside from excellent biocompatibility and biodegradability, chitin and chitosan also have anticarcinogenic, hemostatic, bacteriostatic, and other properties [70,71]. As a result, they can be used effectively as drug delivery systems.

Miyazaki et al. prepared tablets of chitosan and alginate polymer encapsulating diltiazem drug and studied its release behavior when administered orally and sublingually to rabbits via injection [72]. The rate of release can be controlled by adjusting the chitosan-alginate mixture ratio. The results showed that when administered sublingually, the drug bioavailability was approximately 69.6%, compared to 30.4% when administered orally. Bhattarai et al. reported the use of chitosan in scaffolds, where they created nanofibrous scaffolds of chitosan and polyethylene (ethylene oxide). These scaffolds have shown enough promise for controlled drug release as well as tissue engineering [73]. Thus, all these results demonstrated the use of chitin and chitosan in different applications.

1.3.2.4 Collagen and gelatin

Natural polymers play an important role in biomedical applications. As a result, they are also obtained from animal sources. Collagen is a natural polymer derived from animals. It is a protein composed of three polypeptide chains found in animal extracellular matrix [74]. It is extracted from the bones, tendons, cartilage, or skin of animals and is widely accepted due to its biocompatibility, biocompatibility, low stimulation, and high cell adhesion [75,76]. It is primarily employed as a bioadhesive and in drug delivery systems. Collagen has numerous biomedical applications, including delivery systems, burn treatment, and scaffolds [77]. Luque et al. reported new hybrid collagen bioadhesives, in which they synthesized acrylic-collagen latexes with film forming capability via in-situ emulsion polymerization of acrylic acid and butyl acrylate in the presence of hydrolyzed collagen species [78]. The neutralization degree and moisture content of these nanocomposites can be adjusted to control their adhesive properties. Mebert et al. reported another significant approach in which they created a multifunctional nanocomposite of collagen hydrogels and incorporated core-shell silica particles combined with sodium rifamycin and gentamicin sulfate. When tested in vivo for cutaneous infection, the nanocomposites demonstrated controlled drug delivery and antibacterial properties, as well as no skin inflammation [79].

Gelatin is another naturally derived polymer which is obtained from the collagen hydrolysis [80]. When dissolved in water, it can form both a colloid and a hydrogel if the conditions are controlled properly. Many gelatin applications have been reported, including biodegradable packaging films, scaffolds, drug-delivery systems, and so on [81]. Azizi-Lalabadi et al. reported gelatin and PVA nanocomposite films with functional nanoparticles 4A-zeolite loaded TiO_2 and ZnO nanoparticles [82]. The fabricated films improved significantly in tensile strength and antibacterial activity and can thus be used successfully as food packaging films. Several studies on the use of gelatin nanocomposites as scaffolds in tissue engineering have been published. In this regard, Purohit et al. reported in 2019 that nanocomposites of GO incorporated in the matrix of gelatin and alignate effectively enhance bone regeneration [83]. The scaffolds demonstrated increased compressive strength, swelling, and slow biodegradation rates. As a result, these gelatin nanocomposite scaffolds demonstrated potential as food packaging materials as well as effective biomaterials.

1.3.2.5 Bio-based polymer nanocomposites

Bio-based nanocomposites are another type of polymeric material that can be used in a variety of applications. These bio-based polymers are created by modifying synthetic polymers with natural substituents such as starch, glycerol, vegetable oils, sorbitol, and so on. In addition to conventional reactants, bio-based moieties are frequently used as reactants or branching units, which induce hydrophilic functionalities that play an important role in inhibiting biodegradability and biocompatibility to the resulting polymer matrix. Recent advances in the use of bio-based moieties as branching units have resulted in the formation of highly branched polymers known as "hyperbranched polymers," which are globular shaped polymers with a large number of surface functional moieties [66,67]. Barua et al. reported a nanocomposite scaffold for bone regeneration by incorporating OMMT into the matrix of glycerol-based hyperbranched epoxy resin in a study [84]. The crosslinked films demonstrated excellent antimicrobial activity as well as the required biocompatibility (Fig. 1.8) by promoting dermatocyte proliferation without toxicity. In 2015, De et al. reported a transparent nanocomposite with antimicrobial properties created by combining biocide-immobilized OMMT with CD and Cu_2O nanohybrids [85]. In vitro adhesion and antimicrobial activity of bio-based poly(ester amide) (HBPEA) nanocomposites with functionalized MWCNT were excellent (Fig. 1.9). As a result, these nanocomposites were discovered to be suitable as burn wound dressing materials [86]. Significant enzyme activity was demonstrated by cynamide immobilized poly(ethylene glycol) (PEG) assisted with magnetic iron oxide nanoparticles, indicating their suitability as a biocompatible material [87]. Hazarika et al. reported nanocomposites of functionalized GO-based waterborne citric acid modified polyester with superior biodegradability and catalytic activity against the *P. aeruginosa* bacterial strain [88]. As a result, these bio-based nanocomposites with various nanomaterials can be effectively used as high-performing sustainable materials for a wide range of applications.

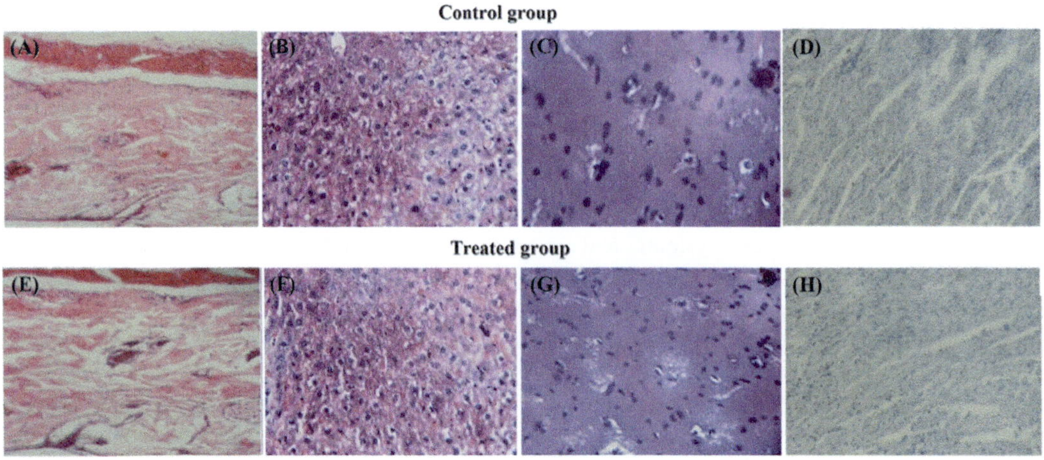

FIGURE 1.8

Histological results for (A) skin, (B) liver, (C) brain, and (D) heart sections of the control group and (E), (F), (G) and (H) represents the treated groups of the same organs, respectively.

Reproduced with permission from S. Barua, N. Dutta, S. Karmakar, P. Chattopadhyay, L. Aidew, A.K. Buragohain, N. Karak, Biocompatible high performance hyperbranched epoxy/clay nanocomposite as an implantable material. Biomed. Mater. 2014;9 (2):025006. https://iopscience.iop.org/article/10.1088/1748-6041/9/2/025006. Copyright 2014, IOP.

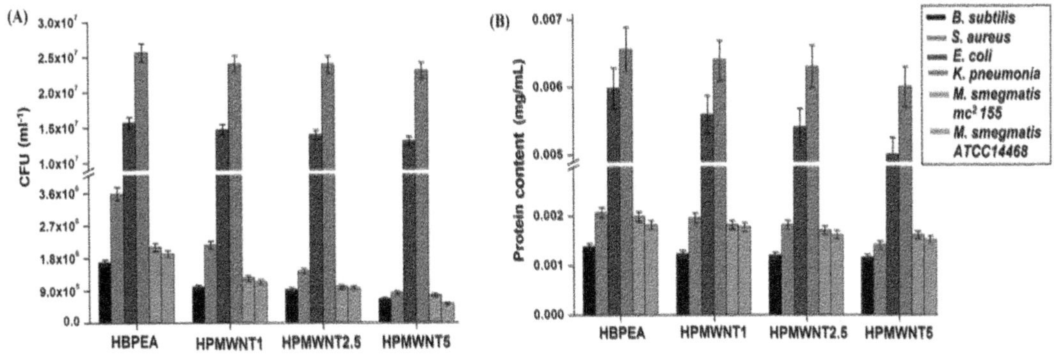

FIGURE 1.9

Comparative antibacterial study in terms of (A) colony-forming units (CFU) enumerated from, and (B) protein adsorbed on pristine bio-based polymer (HBPEA), and its nanocomposites (HBPEAM1, HBPEAM2.5 and HBPEAM5; digits indicate weight% of MWCNT in the matrix) toward *B. subtilis* (*red*), *S. aureus* (*green*), *E. coli* (maroon), *K. pneumonia* (red), *M. smegmatis* mc2 155 (green) and *M. smegmatis* ATCC14468 (blue).

Reproduced with permission from S. Pramanik, R. Konwarh, N. Barua, A.K. Buragohain, N. Karak, Bio-based hyperbranched poly (ester amide)–MWCNT nanocomposites: multimodalities at the biointerface. Biomater. Sci. 2014;2(2):192–202. https://pubs.rsc.org/EN/content/articlelanding/2014/bm/c3bm60170f. Copyright 2014, The Royal Society of Chemistry.

1.4 Fundamental properties

The properties of biodegradable polymer nanocomposites are determined by the type of polymer, functional moieties, fabrication conditions, and nanomaterial selection. This section discusses some of the fundamental properties of these polymer nanocomposites.

1.4.1 Mechanical properties

Mechanical properties of the polymers substantially improved after incorporation of suitable nanomaterials (≤5 wt.%) because of their large surface area and the surface functional groups that provide sufficient interaction with the polymer matrix. Tensile strength, elongation, toughness, scratch resistance, and impact strength are among the mechanical properties. These mechanical properties are investigated using a Universal Testing Machine (UTM), which pulls rectangular strips of the desired dimension of the cured thermosets at a specific speed while applying a load of 0.5–10 kN [66,89]. The other mechanical properties like impact and scratch resistances of such polymeric nanocomposites are determined by impact tester using standard ASTM D5628–16 method and scratch tester using ASTM D7027–13. De and Karak reported a significant increase in the tensile strength (up to 57 MPa) by increasing the loading of bentonite nanoclay from 1 to 3 wt.%, as shown in Fig. 1.10. However, excessive loading of nanomaterials may result into some agglomeration, as observed from the report when the amount was increased to 5 wt.% [90]. Han et al. reported iron chloride functionalized nanocellulose and PVA nanocomposites that are conductive, biocompatible, and self-healing. Due to the efficient combination of weak H-bonds and coordination bonds, the nanocomposites improved mechanical strength by up to 409% and toughness by up to 407.1% when compared to pristine PVA matrix [91]. Dehnavi et al. reported PGA/collagen

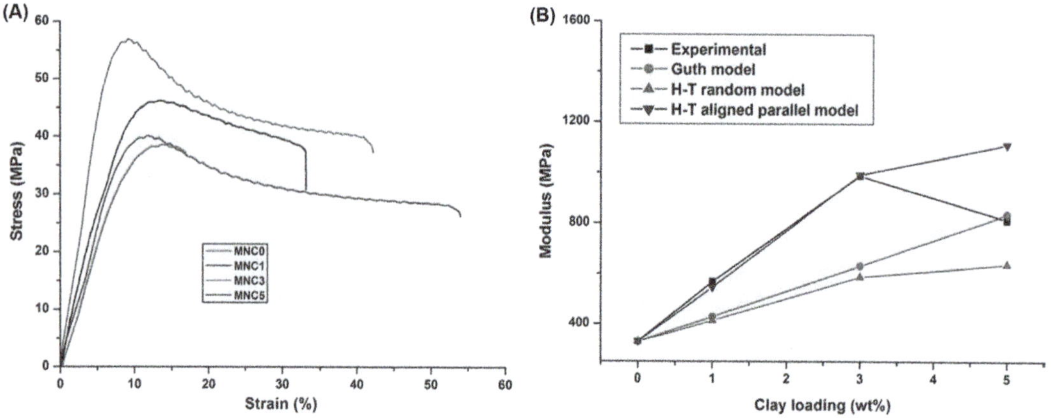

FIGURE 1.10

(A) Stress–strain profiles of pristine epoxy thermoset and nanocomposites, (B) plots of tensile modulus of predicted and experimental values of nanocomposites.

Reproduced with permission from B. De, N. Karak, Tough hyperbranched epoxy/poly (amido-amine) modified bentonite thermosetting nanocomposites. J Appl Polym Sci 2014;131:40327 (1–8). https://doi.org/10.1002/app.40327. Copyright, 2014, Wiley.

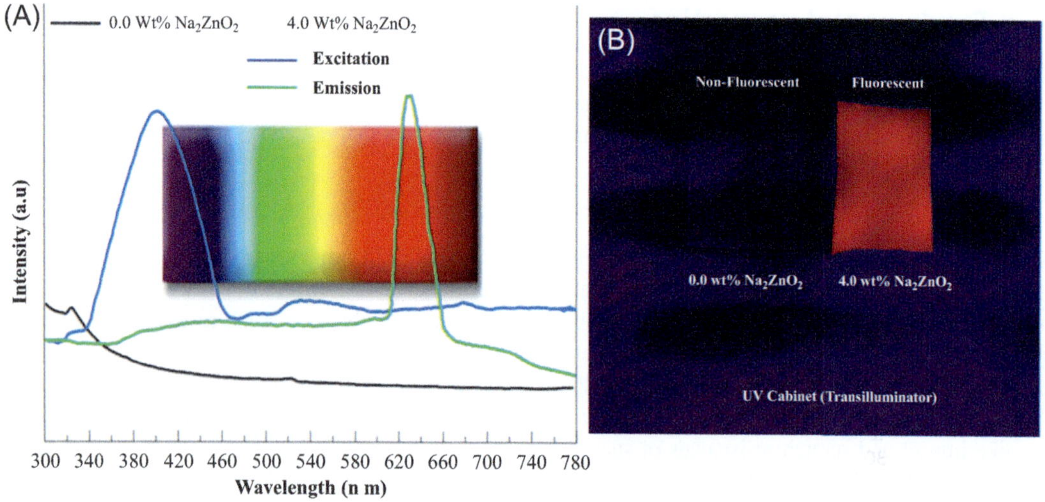

FIGURE 1.11

PVA/Na_2ZnO_2 nanocomposite showing (a) synchronous excitation and emission behaviours and; (b) visible orange-red emission images obtained from UV-transilluminator.

https://doi.org/10.1021/am402415g. Reproduced with permission from Shivanna S, Subramani NK, Swamy K, Nagaraj SK, Muthuraj JRB, Siddaramaiah H. Orange-red fluorescent polymer nanocomposite films with large stokes shift: An opto-electronic exercise. J Lumin 2019:208: 488-494. https://doi.org/10.1016/j.jlumin.2018.12.041.

nanocomposites with nano bio-glass in another study. Mechanical properties testing revealed a significant improvement in elongation of up to 28% compared to 2.84% for pure PGA [92].

1.4.2 Optical properties

The optical properties of polymeric nanocomposites include transparency and fluorescence. Transparency is typically measured using a UV-vis spectrophotometer as the percentage of light transmittance through the sample in the range of 200–800 nm. Saikia et al. in 2019 reported optically transparency for CNFs and hyperbranched bio-based epoxy thermosetting nanocomposites [66]. The incorporation of suitable nanomaterials such as CD, sodium zincate (Na_2ZnO_2), TiO_2 or oxides of CD can induce significant luminescence to the polymer matrix. Polymer nanocomposites with orange red fluorescence were reported by Shivanna et al.] by incorporating Na_2ZnO_2 in PVA via aqueous solution casting approach as shown in Fig 1.11 [93]. Besides fluorescence the developed polymer nanocomposites exhibited high UVA screening abilities, self-cleaning attributes and high thermal stability.

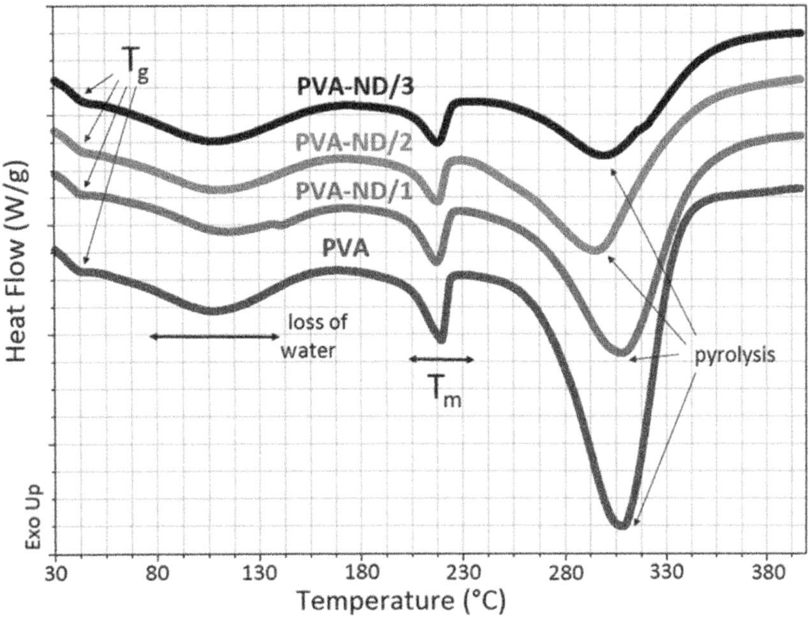

FIGURE 1.12

DSC thermogram of PVA and its nanocomposites at different weight percentages of nanodiamond (ND).

Reproduced with permission from T. Remiš, P. Bělský, T. Kovářík, J. Kadlec, M. Ghafouri Azar, R. Medlín, V. Vavruňková, K. Deshmukh, K.K. Sadasivuni, Study on structure, thermal behavior and viscoelastic properties of nanodiamond-reinforced poly (vinyl alcohol) nanocomposites. Polymers. 2021;13:1426. https://doi.org/10.3390/polym13091426. Copyright 2021, MDPI. This is an open access article distributed under the Creative Commons Attribution License (https://creativecommons.org/licenses/by/4.0/).

1.4.3 Thermal properties

A thermogravimetric analyzer is used to assess the thermal stability of polymer nanocomposites (TGA). Thermal properties of the nanocomposites were reported, including degradation patterns, onset, end-set, and peaks of degradation temperatures under controlled atmospheric conditions. The composition, cross-linking density, dispersion, and interactions of the nanomaterials with the polymer matrix influence the thermostability of the nanocomposite. According to Saikia and Karak, bio-based epoxy thermosetting nanocomposites were fabricated by incorporating CNFs/PAni/CD nanohybrid, and they demonstrated a significant increase in thermal stability after the incorporation of the nanohybrid [66]. The peak degradation temperature was found to be 343°C, compared to 313°C for the neat epoxy thermoset. This is due to the presence of nanofibers with a high aspect ratio, as well as CD with a wide range of surface functionalities, which provide additional strength and stability to the polymer matrix by forming physico-chemical bonds with it. Aside from thermographs of TGA, where thermostability, weight losses at different temperatures, degradation pattern, weight residue at a specific temperature, and so on can be obtained, there is another well-known thermal study, carried out by Differential Scanning Calorimetry, that also provides sufficient insight into the thermal characteristics of a polymer and its nanocomposites (DSC). The later study yielded information such as melting peak temperature (melting point, Tm) for crystalline polymeric materials, polymer Tg, enthalpy changes with temperature, and so on. Furthermore, these thermal studies provided data on the thermal changes in the polymer matrix caused by the incorporation of nanomaterials and their loading. Remis et al. created PVA nanocomposites with nanodiamond, and their thermal studies revealed that the incorporation of nanodiamond into

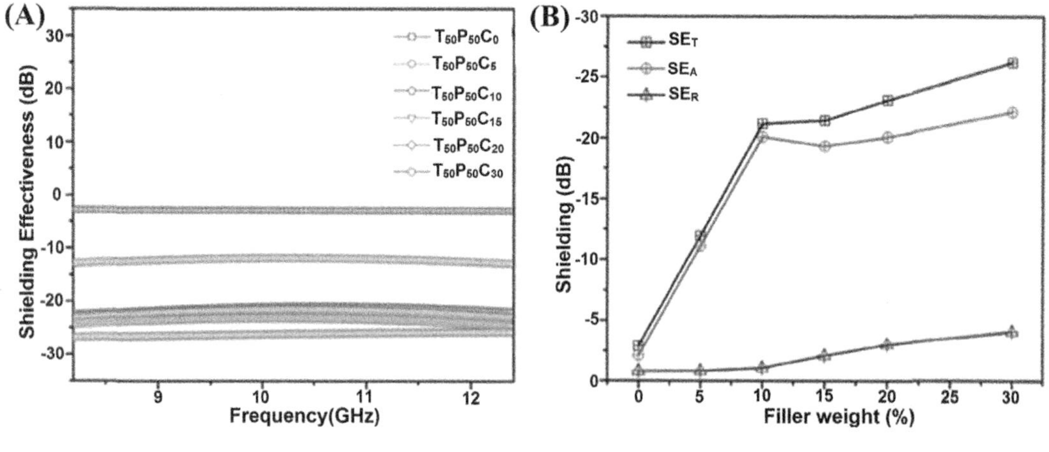

FIGURE 1.13

(A) Electromagnetic interference shielding (EMI) of TPU/PLA/VCB nanocomposites: total shielding effectiveness (SET) versus frequency (GHz) in ranging from 8.2 to 12.4 GHz that is X-band region, and (B) graphical representation of shielding effectiveness due to reflection (SER), absorption (SEA) along with total shielding effectiveness (SET) of PLA/TPU/VCB nanocomposites at different filler loading.

Reproduced with permission from K. Nath, S. Ghosh, S.K. Ghosh, P. Das, N.C. Das, Facile preparation of light-weight biodegradable and electrically conductive polymer based nanocomposites for superior electromagnetic interference shielding effectiveness. J Appl Polym Sci 2021:138: e5051. https://doi.org/10.1002/app.50514.

the PVA matrix resulted in sufficient enthalpy changes [7]. The PVA matrix has a T_g of 42°C which enhanced to 43.8°C upon increasing the nanodiamond content of 1.9 wt.% in PVA matrix. This is due to partially crosslinked structure of PVA with nanodiamond as shown in Fig. 1.12.

1.4.4 Electromagnetic interference shielding property

The rapid expansion of the electronic industries over the last three decades has resulted in a greater reliance on electrical equipment. However, the operation of these electrical devices necessitates higher frequencies, but the presence of electromagnetic interference (EMI) signals interferes with their proper operation. As a result, various polymer nanocomposites containing conductive nanofillers such as nanoclay, carbon black, graphene nanoplatelets, GO, or metal oxides such as CdO, NiO, WO_3, TiO_2, and others have been reported to provide effective EMI shielding [7,94–96].

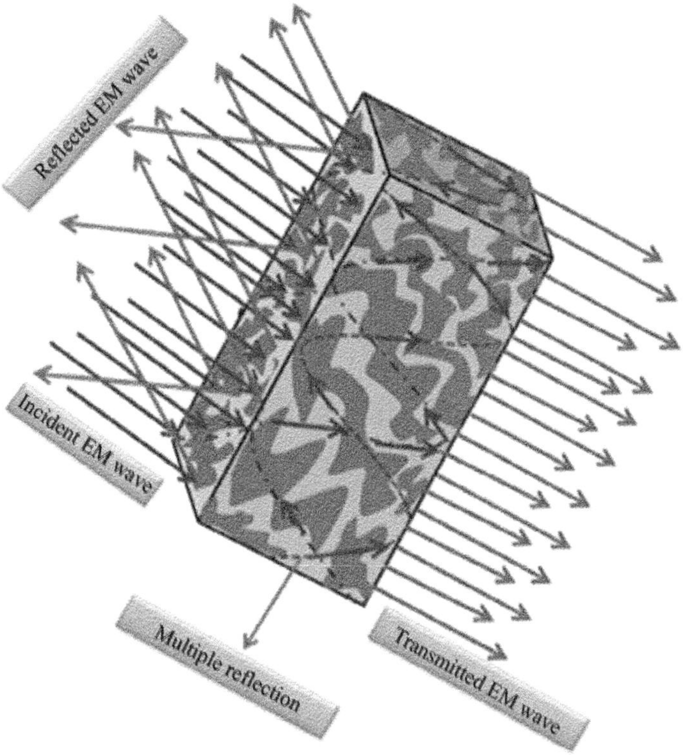

FIGURE 1.14

Typical EMI shielding phenomena in VCB-based binary blend polymer nanocomposite: electromagnetic interference.

Reproduced with permission from K. Nath, S. Ghosh, S.K. Ghosh, P. Das, N.C. Das, Facile preparation of light-weight biodegradable and electrically conductive polymer based nanocomposites for superior electromagnetic interference shielding effectiveness. J Appl Polym Sci 2021:138: e5051. https://doi.org/10.1002/app.50514.

Conductive fillers are required because they induce both reflection and absorption to the polymer matrix, resulting in effective EMI shielding. Das and his coworkers investigated the shielding properties of nanocomposites of poly(lactic acid) (PLA)/ester linkage based thermoplastic polyurethane (TPU) with conductive carbonblack Vulcan XC-72 (VCB) at different filler loading [97]. According to the findings, the presence of conducting carbon black filler in the PLA/TPU matrix improves its EMI shielding capacity in the X band when compared to the pristine matrix, as shown in Fig. 1.13. Again, Fig. 1.14 shows that the used biodegradable EMI shielding material effectively absorbs or reflects the maximum portion of electromagnetic waves and transmits a very little.

1.4.5 Biodegradability

The incorporation of bio-based materials into the fabrication of polymer nanocomposites aids in imbuing different biological properties into hydrophobic polymeric materials. As previously discussed, the need for biodegradation of polymeric products is critical for addressing environmental issues. The biodegradability of polymeric films is commonly tested in laboratories using the McFarland turbidity method, which involves measuring the optical density (OD) of the microorganism growth medium in the presence of degraded films with a UV-vis spectrophotometer [65].

FIGURE 1.15

SEM images showing biodegraded surfaces of pristine hyperbranched epoxy thermoset: (A) control, (B) degraded by *P. aeruginosa* and (C) degraded by *B. subtilis*; and nanocomposite with GO thermosets: (D) control, (E) degraded by *P. aeruginosa* and (F) degraded by *B. subtilis*.

Reproduced with permission from P. Baruah, N. Karak, Bio-based tough hyperbranched epoxy/graphene oxide nanocomposite with enhanced biodegradability attribute. Polym Degrad Stab 2016;129:26–33. https://www.sciencedirect.com/science/article/pii/S0141391016300738. Copyright 2016, Elsevier.

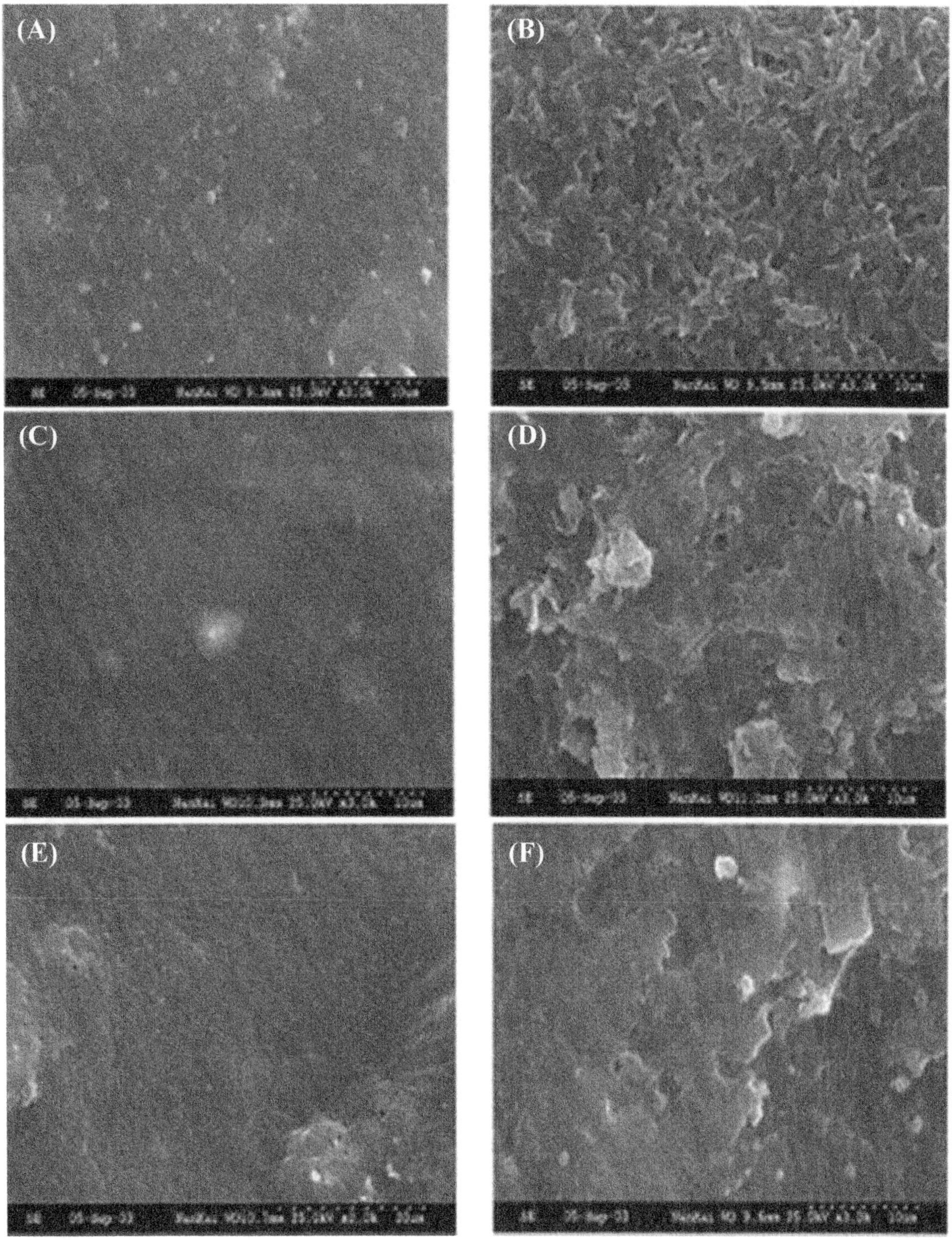

FIGURE 1.16

Scanning electron microscopy images of pristine polymer of composition (100/0) (A) before and (B) after biodegradation; images of nanocomposites with composition of 100/3 (C) before and (D) after biodegradation; and composition of 100/5 (E) before and (F) after biodegradation for 350 h each.

Reproduced with permission from S. Wang, C. Song, G. Chen, T. Guo, J. Liu, B. Zhang, S. Takeuchi, Characteristics and biodegradation properties of poly (3-hydroxybutyrate-co-3-hydroxyvalerate)/organophilic montmorillonite (PHBV/OMMT) nanocomposite. Polym. Degrad. Stab. 2005;87:69–76. https://doi.org/10.1016/j.polymdegradstab.2004.07.008. Copyright 2005, Elseiver.

The degradation process is divided into two stages. The polymeric chains are broken down into smaller subatomic masses in the first stage, which is followed by bio-osmosis by microorganisms and mineralization. Hydrophilic polymers and their nanocomposites biodegrade at a faster rate than hydrophobic polymeric materials. However, the extent of biodegradability has been reported to significantly improve after the incorporation of appropriate nanomaterials [97,98]. This is supported by a study published in 2016 by Baruah and Karak, who discovered that the biodegradability of glycerol-based epoxy/GO nanocomposites was significantly higher than that of pristine epoxy [99]. This study contradicts previous publications on the antibacterial activity of GO. Thus, by selecting a suitable combination of nanomaterial and polymer matrix, the rate of biodegradability can be increased, as shown in Fig. 1.15. Wang et al. reported biodegradable PHBV/OMMT nanocomposites, which was investigated in soil suspension [100]. Up to a certain level of OMMT concentration, an increase in biodegradability was observed (Fig. 1.16). The biodegradability tends to decrease after this concentration of OMMT, which could be due to clay's antimicrobial activity. Thus, biodegradability is entirely dependent on the careful selection and concentration of both the nanomaterial and the polymer's reactant moieties.

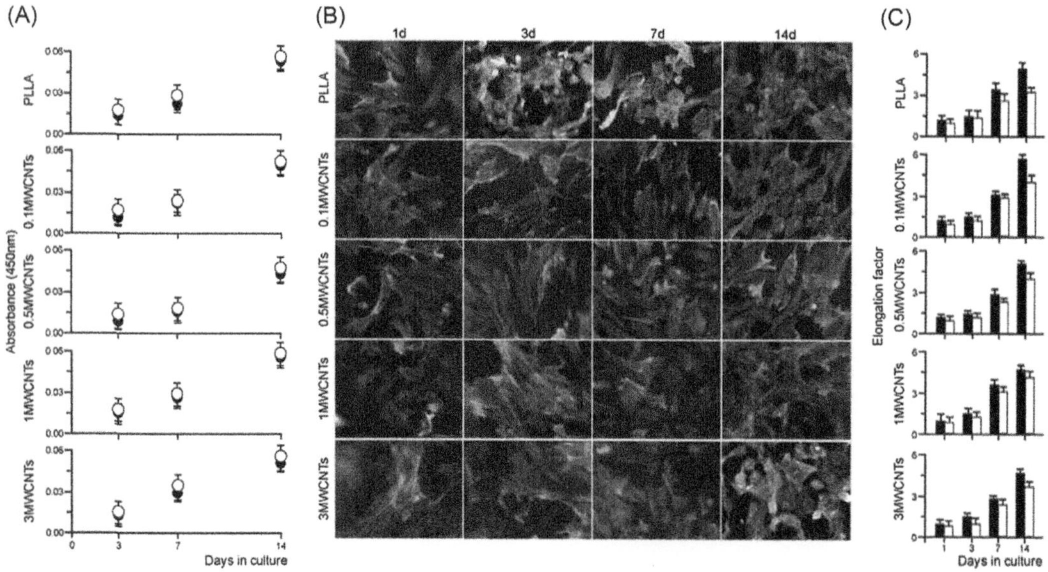

FIGURE 1.17

Stem cell interaction and culture shown by the PLA/MWCNT nanocomposites at different MWCNT loadings demonstrating (A) XTT viability assay; (B) representative images of stem cell seeded on substrates with different composition and (C) elongation factor measured for cell cultures on substrate at different time intervals.

Reproduced with permission from E. Lizundia, J.R., Sarasua, F. D'Angelo, A. Orlacchio, S. Martino, J.M. Kenny, I. Armentano, Biocompatible poly (L-lactide)/MWCNT nanocomposites: morphological characterization, electrical properties, and stem cell interaction. Macromol Biosci 2012;12:870–881. https://doi.org/10.1002/mabi.201200008. Copyright 2012, Wiley.

1.4.6 Biocompatibility

Biocompatibility strictly refers to the harmonious interaction between mammalian cells and polymers and their nanocomposites when used as stents, scaffolds, or drug delivery agents in the human or animal body. Biocompatible polymeric materials used in medical devices are classified into three types: (1) microphase-separated surfaces, (2) hydrophilic surfaces, and (3) bio-membrane-mimicking surfaces. Surface formation is influenced by several factors such as surface functionalities, surface charge, topography, and wettability. In today's medical applications, hemocompatibility refers to cell separation from blood as well as the growth of endothelial cells on artificial blood vessels, implants, and stents. As a result, several techniques are used to control the intermediate water content, resulting in effective blood contact by regulating protein adsorption followed by cell attachment. As a result, well-defined biocompatible polymers must be synthesized by precisely controlling the primary architecture of polymers such as molecular weight, monomer distribution, the functionality of molecular side chains, chain structure, branching, and so on. This can be accomplished by modifying the side chains by introducing more hydrophilic groups, changing the backbone of the polymer chain, or introducing some spaces in the side chains. These changes will result in a more defined hydrophilic chemical structure with improved biocompatibility. The presence of zwitterionic species in the polymer structure has undoubtedly improved the biocompatibility of the polymer chains, according to the reported literature. Poly(serine methacrylate) was recently identified as a new zwitterionic polymer family composed of an amino acid, L-serine, as the side-chain moiety [101]. The incorporation of suitable nanomaterials in the matrix can also result in the structural reformation of the polymers for improved

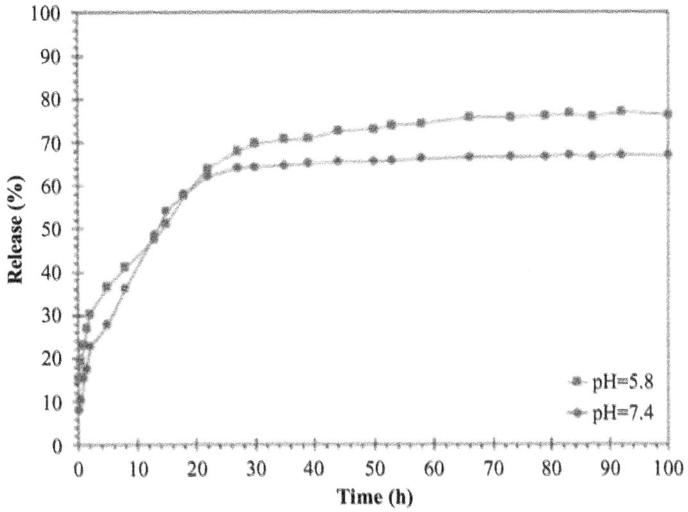

FIGURE 1.18

Drug release profiles of 5-Fluorouracil in PVA/magnetic FeO/GO nanocomposite at pH = 5.8 and 4.

Reproduced with permission from M. Aliabadi, H. Shagholani, Synthesis of a novel biocompatible nanocomposite of graphene oxide and magnetic nanoparticles for drug delivery. Int. J. Biol. Macromol. 2017;98:287–91. https://www.sciencedirect.com/science/article/pii/S0141813016327313. Copyright 2017, Elsevier.

biocompatibility. Lizundiae et al. reported biocompatible nanocomposites of PLA/MWCNT. These nanocomposites are found to be suitable substrates for primary stem cell culture, as supported by the results of Fig. 1.17 [102]. By incorporating GO and magnetic nanoparticles, Aliabadi et al. created novel nanocomposites of biocompatible polymers such as PVA and chitosan [103]. In acidic medium, the nanocomposites demonstrated high drug entrapment efficiency and a faster drug delivery rate of 5-fluorouracil (Fig. 1.18). As a result of these findings, it is clear that the polymer with improved biocompatibility can be used for advanced biomedical applications.

1.5 Conclusions and future perspectives

Biodegradability and biocompatibility are important requirements for polymers and nanocomposites in biomedical applications, among other things. Regardless of the nature or origin of the virgin polymers, various types of polymer nanocomposites have been successfully used in biomedical applications after improving their biocompatibility and biodegradability features. Thus, this chapter provided a fundamental understanding of various types of synthetic and natural polymers, as well as their respective nanocomposites. The nanostructure, in particular, provides advantageous bio-nano interfaces with biological systems, which ultimately aids in the biomedical applications of polymeric nanocomposites. Furthermore, the biodegradability of polymeric products is critical for environmentally friendly disposal after their useful service life. The biocompatibility of these polymer nanocomposites is entirely dependent on surface properties, which have a significant impact on cell interaction features. These surfaces' effective interactions result in active in-vivo biomedical applications of these polymer nanocomposites. As a result, this chapter is an embodiment of various types and properties of biodegradable and biocompatible polymers for biomedical applications.

This chapter describes in detail the various biodegradable and biocompatible polymer nanocomposites. Despite significant effort on biodegradable synthetic polymers, bio-based or natural polymers have not been widely commercialized due to processing difficulties. As a result, if appropriate low-cost and simple processing alternatives are adopted, there will be ample scope for large-scale natural polymer production in various engineering devices. In the future, such biodegradable and biocompatible polymer nanocomposites could be used to fabricate quick access health monitoring devices in the form of strips attached to the body for monitoring virus attacks, blood-glucose levels, or blood pressure. As a result, many unexplored or undervalued corners of natural and synthetic biodegradable polymers can be explored in the future by designing appropriate polymeric nanocomposites.

References

[1] M. Shimao, Biodegradation of plastics, Curr. Opin. Biotechnol. 12 (2001) 242–247.
[2] M.T. Taghizadeh, N. Nalbandi, A. Bahadori, Stabilizing effect of epoxidized sunflower oil as a secondary stabilizer for Ca/Hg stabilized PVC, eXPRESS Polym. Lett. 2 (2020) 65–76.
[3] J.S.C. Viera, M.R.C. Marques, M.C. Nazareth, P.C. Jimenez, L.B. Castro, On replacing single-use plastic with so-called biodegradable ones: the case with straws, Env. Sci. Policy 106 (2020) 177–181.

[4] I. Vroman, L. Tighzert, Biodegradable polymers, Materials. 2 (2009) 307–344.
[5] S. Mishra, N.R. Talele, Filler effect of potato starch and urea on degradation of linear low density polyethylene composites, Polym. Plast. Technol. Eng. 41 (2002) 361–381.
[6] J. Kyrikou, D. Briassoulis, Biodegradation of agricultural plastic films: a critical review, J. Polym. Env. 15 (2007) 125–150.
[7] T. Remiš, P. Bělský, T. Kovářík, J. Kadlec, M. Ghafouri Azar, R. Medlín, et al., Study on structure, thermal behavior and viscoelastic properties of nanodiamond-reinforced poly (vinyl alcohol) nanocomposites, Polymers 13 (9) (2021) 1426.
[8] D. Elvers, C.H. Song, A. Steinbüchel, J. Leker, Technology trends in biodegradable polymers: evidence from patent analysis, Polym. Rev. 56 (2016) 584–606.
[9] E. Fortunati, I. Armentano, Q. Zhou, A. Iannonia, E. Saino, L. Visai, et al., Multifunctional bionanocomposite films of poly(lactic acid), cellulose nanocrystals and silver nanoparticles, Carbohydr. Polym. 87 (2012) 1596–1605.
[10] V. Karagkiozaki, S. Logothetidis, Horizons in clinical, Nanomedicine (2014) 39.
[11] A.E. Nell, L. Mädler, D. Velegol, T. Xia, E.M.V. Hoek, P. Somasundaran, et al., Understanding biophysicochemical interactions at the nano-bio interface, Nat. Mater. 8 (2009) 543–557.
[12] J. Sahari, S.M. Sapuan, Natural fibre reinforced biodegradable polymer composites, Rev. Adv. Mater. Sci. 30 (2011) 166–174.
[13] J. Wang, L. Wang, Z. Zhou, H. Lai, P. Xu, L. Liao, et al., Biodegradable polymer membranes applied in guided bone/tissue regeneration: a review, Polymers 8 (4) (2016) 115.
[14] W. Zhuang, J. Liu, J.H. Zhang, H.B. Xing, J. Shen, Preparation, characterization and properties of TiO_2/PLA nanocomposites by in situ polymerization, Polym. Compos. 3 (2009) 1074–1080.
[15] N. Shimpi, M. Borane, S. Mishra, M. Kadam, Biodegradation of polystyrene (PS)-poly(lactic acid) (PLA) nanocomposites using *Pseudomonas aeruginosa*, Macromol. Res. 20 (2) (2012) 181–187.
[16] C.C. Lin, S.J. Fu, Y.C. Lin, I.K. Yang, Y. Gu, Chitosan-coated electrospun PLA fibers for rapid mineralization of calcium phosphate, Int. J. Biol. Macromol. 68 (2014) 39–47.
[17] R. Duarah, Y.P. Singh, P. Gupta, B.B. Mandal, N. Karak, High performance bio-based hyperbranched polyurethane/carbon dot-silver nanocomposite: a rapid self-expandable stent, Biofabrication 8 (4) (2016) 045013.
[18] J.C. Middleton, A.J. Tipton, Synthetic biodegradable polymers as orthopedic devices, Biomaterials 21 (2000) 2335–2346.
[19] Y. Ikada, H. Tsuji, Biodegradable polyesters for medical and ecological applications, Macromol. Rapid Commun. 21 (2000) 117–132.
[20] A.R. McLauchlin, N.L. Thomas, Biodegradable polymer nanocomposites, Adv. Polym. Nanocomposites (2012) 398–430.
[21] Y.J. Phua, N.S. Lau, K. Sudesh, W.S. Chow, Z.M. Ishak, Biodegradability studies of poly (butylene succinate)/organo-montmorillonite nanocomposites under controlled compost soil conditions: effects of clay loading and compatibiliser, Polym. Degrad. Stab. 97 (2012) 1345–1354.
[22] A.M. Díez-Pascual, Biodegradable food packaging nanocomposites based on ZnO-reinforced polyhydroxyalkanoates, Food Packaging, 2017, pp. 185–221.
[23] A.B. Balaji, H. Pakalapati, M. Khalid, R. Walvekar, H. Siddiqui, Natural and synthetic biocompatible and biodegradable polymers, in: N.G. Shimpi (Ed.), Biodegradable and Biocompatible Polymer Composites, 2018, pp. 3–2.
[24] J. Zhang, L. Xu, W. Xiao, Y. Chen, Z. Dong, J. Xu, et al., Ring-opening polymerization of ε-caprolactone with recyclable and reusable squaric acid organocatalyst, Eur. Polym. J. 157 (2021) 110643.
[25] M. Hunsen, A. Azim, H. Mang, S.R. Wallner, A. Ronkvist, W. Xie, et al., A cutinase with polyester synthesis activity, Macromolecules 40 (2) (2017) 148–150.

[26] G. Prömpers, H. Keul, H. Höcker, Polyurethanes with pendant hydroxy groups: polycondensation of D-mannitol-1, 2: 5, 6-dicarbonate with diamines, Des. Monomers Polym. 8 (2005) 547–569.
[27] S. Gogoi, N. Karak, Biobased biodegradable waterborne hyperbranched polyurethane as an ecofriendly sustainable material, ACS Sustain. Chem. Eng. 2 (2014) 2730–2738.
[28] A. Saikia, N. Karak, Renewable resource based thermostable tough hyperbranched epoxy thermosets as sustainable materials, Polym. Degrad. Stab. 135 (2017) 8–17.
[29] M. Baiardo, G. Frisoni, M. Scandola, A. Licciardello, Surface chemical modification of natural cellulose fibers, J. Appl. Polym. Sci. 83 (2002) 38–45.
[30] O. Molatlhegi, L. Alagha, Adsorption characteristics of chitosan grafted copolymer on kaolin, Appl. Clay Sci. 150 (2017) 342–353.
[31] J. Lin, W. Xu, C. Zheng, P. Zhang, B. Huang, N. Guo, et al., Synthesis and properties of epoxy-polyurethane/silica nanocomposites by a novel sol method and in-situ solution polymerization route, Appl. Surf. Sci. 303 (2014) 67–75.
[32] P. Rani, M.B. Ahamed, K. Deshmukh, Structural, dielectric and EMI shielding properties of polyvinyl alcohol/chitosan blend nanocomposites integrated with graphite oxide and nickel oxide nanofillers, J. Mater. Sci. Mater Electron. 32 (2021) 764–779.
[33] X. Liu, X. Lu, Y. Su, E. Kun, F. Zhang, Clay-polymer nanocomposites prepared by reactive melt extrusion for sustained drug release, Pharmaceutics 12 (2020) 51.
[34] M. Vert, G. Schwarch, J. Coudane, Present and future of PLA polymers, in: A. Albertsson, S. Huang (Eds.), Degradable Polymers, Recycling, and Plastics Waste Management, first ed., Marcel Dekker Inc., New York, 1995.
[35] E. Skaria, B.A. Patel, M.S. Flint, K.W. Ng, Poly (lactic acid)/carbon nanotube composite microneedle arrays for dermal biosensing, Anal. Chem. 91 (2019) 4436–4443.
[36] A. Mahmoodi, S. Ghodrati, M. Khorasani, High-strength, low-permeable, and light-protective nanocomposite films based on a hybrid nanopigment and biodegradable PLA for food packaging applications, ACS Omega 4 (2019) 14947–14954.
[37] K.J. Jem, B. Tan, The development and challenges of poly (lactic acid) and poly (glycolic acid), Adv. Ind. Eng. Polym. Res. 3 (2) (2020) 60–70.
[38] United States (12) Patent Application Publication (10) Pub. No.: US 2012/0259295 A1 US 20120259295A1 Bonutti United States Pat., 8,399,077; 2013.
[39] K. Samadi, M. Francisco, S. Hegde, C.A. Diaz, T.A. Trabold, E.M. Dell, et al., Mechanical, rheological and anaerobic biodegradation behavior of a Poly (lactic acid) blend containing a Poly (lactic acid)-co-poly (glycolic acid) copolymer, Polym. Degrad. Stab. 170 (2019) 109018.
[40] J.M. Patrascu, J.P. Kruger, H.G. Boss, A.K. Ketzmar, U. Freymann, M. Sittinger, Polyglycolic acid-hyaluronan scaffolds loaded with bone marrowderived mesenchymal stem cells show chondrogenic differentiation in vitro and cartilage repair in the rabbit model, J. Biomed. Mater. Res. B Appl. Biomater. 101 (2013) 1310–1320.
[41] H. Kobayashi, D. Terada, Y. Yokoyama, D.W. Moon, Y. Yasuda, H. Koyama, et al., Vascular-inducing poly (glycolic acid)-collagen nanocomposite-fiber scaffold, J. Biomed. Nanotechnol. 9 (2013) 1318–1326.
[42] N. Ashammakhi, E.A. Makela, K. Vihtonen, P. Rokkanen, H. Kuisma, P. Tormala, Biomaterials 16 (1995) 135–138.
[43] J.Y. Wu, C.W. Li, C.H. Tsai, C.W. Chou, D.R. Chen, G.J. Wang, Synthesis of antibacterial TiO_2/PLGA composite biofilms, Nanomed. Nanomed. Nanotechnol. 10 (2014) e1097–e1107.
[44] B. Lepoittevin, N. Pantoustier, M. Devalckenaere, M. Alexandre, D. Kubies, C. Calberg, et al., Poly (ε-caprolactone)/clay nanocomposites by in-situ intercalative polymerization catalyzed by dibutyltin dimethoxide, Macromolecules 35 (2002) 8385–8390.

[45] N.K. Singh, B.D. Purkayastha, J.K. Roy, R.M. Banik, M. Yashpal, G. Singh, et al., Nanoparticle-induced controlled biodegradation and its mechanism in poly (ε-caprolactone), ACS Appl. Mater. Interfaces 2 (2010) 69–81.
[46] Y. Wang, T. Li, P. Ma, S. Zhang, M. Du, W. Dong, et al., Graphene-assisted fabrication of poly (ε-caprolactone)-based nanocomposites with high mechanical properties and self-healing functionality, N. J. Chem. 42 (2018) 10348–10356.
[47] T. Nezakati, A. Tan, J. Lim, R.D. Cormia, S.H. Teoh, A.M. Seifalian, Ultra-low percolation threshold POSS-PCL/graphene electrically conductive polymer: neural tissue engineering nanocomposites for neurosurgery, Mater. Sci. Eng. C. 104 (2019) 109915.
[48] Z.P. Rad, J. Mokhtari, M. Abbasi, Fabrication and characterization of PCL/zein/gum arabic electrospun nanocomposite scaffold for skin tissue engineering, Mater. Sci. Eng. C. 93 (2018) 356–366.
[49] G.X. Chen, E.S. Kim, J.S. Yoon, Poly (butylene succinate)/twice functionalized organoclay nanocomposites: preparation, characterization and properties, J. Appl. Polym. Sci. 98 (2005) 1727–1732.
[50] A. El-Rahman, K.M. Ali, S.F.A. Khalil, A.I. Khalil, S. Kandil, Influence of poly (butylene succinate) and calcium carbonate nanoparticles on the biodegradability of high density-polyethylene nanocomposites, J. Polym. Res. 27 (2020) 1–21.
[51] A. Huang, X. Peng, L. Geng, L. Zhang, K. Huang, B. Chen, et al., Electrospun poly (butylene succinate)/cellulose nanocrystals bio-nanocomposite scaffolds for tissue engineering: preparation, characterization and in vitro evaluation, Polym. Test. 71 (2018) 101–109.
[52] H. Shirali, M. Rafizadeh, F. Afshar Taromi, E. Jabbari, Fabrication of in situ polymerized poly (butylene succinate-co-ethylene terephthalate)/hydroxyapatite nanocomposite to fibrous scaffolds for enhancement of osteogenesis, J. Biomed. Mater. Res. A 105 (9) (2017) 2622–2631.
[53] M.J. Fabra, A. López-Rubio, J. Ambrosio-Martín, J.M. Lagaron, Improving the barrier properties of thermoplastic corn starch-based films containing bacterial cellulose nanowhiskers by means of PHA electrospun coatings of interest in food packaging, Food Hydrocoll. 61 (2017) 261–268.
[54] A. Mukheem, S. Shahabuddin, N. Akbar, A. Anwar, N.M. Sarih, K. Sudesh, et al., Fabrication of biopolymer polyhydroxyalkanoate/chitosan and 2D molybdenum disulfide-doped scaffolds for antibacterial and biomedical applications, Appl. Microbiol. Biotechnol. 104 (2020) 3121–3131.
[55] R. Nasseri, N. Mohammadi, Starch-based nanocomposites: a comparative performance study of cellulose whiskers and starch nanoparticles, Carbohydr. Polym. 106 (2014) 432–439.
[56] F. Mirab, M. Eslamian, R. Bagheri, Fabrication and characterization of a starch-based nanocomposite scaffold with highly porous and gradient structure for bone tissue engineering, Biomed. Phys. Eng. Express 4 (2018) 055021.
[57] T. Yuan, J. Zeng, B. Wang, Z. Cheng, W. Gao, J. Xu, et al., Silver nanoparticles immobilized on cellulose nanofibrils for starch-based nanocomposites with high antibacterial, biocompatible, and mechanical properties, Cellulose 28 (2021) 855–869.
[58] M.D. Sanchez-Garcia, E. Gimenez, J.M. Lagaron, Morphology and barrier properties of solvent cast composites of thermoplastic biopolymers and purified cellulose fibers, Carbohydr. Polym. 71 (2008) 235–244.
[59] A. Dufresne, Nanocellulose: a new ageless bionanomaterial, Mater. Today 16 (2013) 220–227.
[60] M. Jonoobi, J. Harun, A.P. Mathew, M.Z.B. Hussein, K. Oksman, Preparation of cellulose nanofibers with hydrophobic surface characteristics, Cellulose 17 (2010) 299–307.
[61] D. Trache, M. Hazwan Hussin, M.K. Haafiz, V.K. Thakur, Recent progress in cellulose nanocrystals: sources and production, Nanoscale 9 (2017) 1763–1786.
[62] R. Li, J. Du, Y. Zheng, Y. Wen, X. Zhang, W. Yang, et al., Ultralightweight cellulose foam material: preparation and properties, Cellulose 24 (3) (2017) 1417–1426.

[63] J. Pang, X. Liu, X. Zhang, Y. Wu, R. Sun, Fabrication of cellulose film with enhanced mechanical properties in ionic liquid 1-allyl-3-methylimidazolium chloride (AmimCl), Materials 6 (2013) 1270–1284.
[64] A. Kaiser, W.J. Stark, R.N. Grass, Rapid production of a porous cellulose acetate membrane for water filtration using readily available chemicals, J. Chem. Educ. 94 (2017) 483–487.
[65] N.M. Aguilar, F. Arteaga-Cardona, M.E. de Anda Reyes, J.J. Gervacio-Arciniega, U. Salazar-Kuri, Magnetic bioplastics based on isolated cellulose from cotton and sugarcane bagasse, Mater. Chem. Phys. 238 (2019) 12191.
[66] A. Saikia, N. Debbarma, N. Karak, Renewable resource based hyperbranched epoxy thermosetting nanocomposite with cellulose nanofiber as a sustainable material, Cellulose 26 (2019) 4743–4755.
[67] A. Saikia, N. Karak, Cellulose nanofiber-polyaniline nanofiber-carbon dot nanohybrid and its nanocomposite with sorbitol based hyperbranched epoxy: physical, thermal, biological and sensing properties, Colloids Surf. Physicochem. Eng. Asp. 584 (2020) 124049.
[68] S. Barua, G. Das, L. Aidew, A.K. Buragohain, N. Karak, Copper–copper oxide coated nanofibrillar cellulose: a promising biomaterial, RSC Adv. 3 (35) (2013) 14997–15004.
[69] C. Saikia, P. Gogoi, T.K. Maji, Chitosan: a promising biopolymer in drug delivery applications, J. Mol. Genet. Med. S 4 (006) (2015) 899–910.
[70] M. Rinaudo, Chitin and chitosan: properties and applications, Prog. Polym. Sci. 31 (2006) 603–632.
[71] G. Crini, P.M. Badot, Application of chitosan, a natural amino polysaccharide, for dye removal from aqueous solutions by adsorption processes using batch studies: a review of recent literature, Prog. Polym. Sci. 33 (2008) 399–447.
[72] S. Miyazaki, N. Kawasaki, T. Nakamura, M. Iwatsu, T. Hayashi, W.M. Hou, et al., Oral mucosal bioadhesive tablets of pectin and HPMC: in vitro and in vivo evaluation, Int. J. Pharm. 204 (2000) 127–132.
[73] N. Bhattarai, D. Edmondson, O. Veiseh, F.A. Matsen, M. Zhang, Electrospun chitosan-based nanofibers and their cellular compatibility, Biomaterials 26 (2005) 6176–6184.
[74] J. Myllyharju, Prolyl 4-hydroxylases, the key enzymes of collagen biosynthesis, Matrix Biol. 22 (2003) 15–24.
[75] A. Lynn, I. Yannas, W. Bonfield, Antigenicity and immunogenicity of collagen, J. Biomed. Mater. Res. B Appl. Biomater. 71 (2004) 343–354.
[76] P.B. Malafaya, G.A. Silva, R.L. Reis, Natural-origin polymers as carriers and scaffolds for biomolecules and cell delivery in tissue engineering applications. origin as and for and delivery tissue applications, Adv. Drug. Deliv. Rev. 59 (2007) 207–233.
[77] S. Lo, M.B. Fauzi, Current update of collagen nanomaterials-fabrication, characterisation and its applications: a review, Pharmaceutics 13 (3) (2021) 316–325.
[78] G.C. Luque, R. Stürtz, M.C. Passeggi Jr, L.M. Gugliotta, V.D. Gonzalez, R.J. Minari, New hybrid acrylic/collagen nanocomposites and their potential use as bio-adhesives, Int. J. Adhes. Adhes 100 (2020) 102624.
[79] A.M. Mebert, G.S. Alvarez, R. Peroni, C. Illoul, C. Hélary, T. Coradin, et al., Collagen-silica nanocomposites as dermal dressings preventing infection in vivo, Mater. Sci. Eng. C. 93 (2018) 170–177.
[80] S. Shanmugam, R. Manavalan, D. Venkappayya, K. Sundramoorthy, V.M.S. Mounnissamy, T. Ayyappan, Natural polymers and their applications, Natl Prod. Radiance 4 (2005) 478–481.
[81] A. Marino, C. Tonda-Turo, D. De Pasquale, F. Ruini, G. Genchi, S. Nitti, et al., Gelatin/nanoceria nanocomposite fibers as antioxidant scaffolds for neuronal regeneration, Biochim. Biophys. Acta Gen. Subj. 186 (2) (2017) 386–395.
[82] M. Azizi-Lalabadi, M. Alizadeh-Sani, B. Divband, A. Ehsani, D.J. McClements, Nanocomposite films consisting of functional nanoparticles (TiO_2 and ZnO) embedded in 4A-Zeolite and mixed polymer matrices (gelatin and polyvinyl alcohol), Int. Food Res. J. 137 (2020) 109716.

[83] S.D. Purohit, R. Bhaskar, H. Singh, I. Yadav, M.K. Gupta, N.C. Mishra, Development of a nanocomposite scaffold of gelatin–alginate–graphene oxide for bone tissue engineering, Int. J. Biol. Macromol. 133 (2019) 592–602.

[84] S. Barua, N. Dutta, S. Karmakar, P. Chattopadhyay, L. Aidew, A.K. Buragohain, et al., Biocompatible high performance hyperbranched epoxy/clay nanocomposite as an implantable material, Biomed. Mater. 9 (2) (2014) 025006.

[85] B. De, M. Kumar, B.B. Mandal, N. Karak, An in situ prepared photo-luminescent transparent biocompatible hyperbranched epoxy/carbon dot nanocomposite, RSC Adv. 5 (91) (2015) 74692–74704.

[86] S. Pramanik, R. Konwarh, N. Barua, A.K. Buragohain, N. Karak, Bio-based hyperbranched poly (ester amide)–MWCNT nanocomposites: multimodalities at the biointerface, Biomater. Sci. 2 (2) (2014) 192–202.

[87] R. Konwarh, N. Karak, S.K. Rai, A.K. Mukherjee, Polymer-assisted iron oxide magnetic nanoparticle immobilized keratinase, Nanotechnology 20 (22) (2009) 225107.

[88] D. Hazarika, K. Gupta, M. Mandal, N. Karak, High-performing biodegradable waterborne polyester/functionalized graphene oxide nanocomposites as an eco-friendly material, ACS Omega 3 (2) (2018) 2292–2303.

[89] V.B. Kumar, A.K. Sahu, A.S. Mohsin, X. Li, A. Gedanken, Refractive-index tuning of highly fluorescent carbon dots, ACS Appl. Mater. Interfaces. 30 (2017) 28930–28938.

[90] B. De, N. Karak, Tough hyperbranched epoxy/poly (amido-amine) modified bentonite thermosetting nanocomposites, J. Appl. Polym. Sci. 131 (2014) 11.

[91] L. Han, S. Cui, H.Y. Yu, M. Song, H. Zhang, N. Grishkewich, et al., Self-healable conductive nanocellulose nanocomposites for biocompatible electronic skin sensor systems, ACS Appl. Mater. Interfaces 11 (47) (2019) 44642–44651.

[92] N. Dehnavi, K. Parivar, V. Goodarzi, A. Salimi, M.R. Nourani, Systematically engineered electrospun conduit based on PGA/collagen/bioglass nanocomposites: the evaluation of morphological, mechanical, and bio-properties, Polym. Adv. Technol. 30 (9) (2019) 2192–2206.

[93] S. Shivanna, N.K. Subramani, K. Swamy, S.K. Nagaraj, J.R.B. Muthuraj, H. Siddaramaiah, Orange-red fluorescent polymer nanocomposite films with large stokes shift: An opto-electronic exercise, J Lumin, 208 (2019) 488–494.

[94] A.H. Kumar, M.B. Ahamed, K. Deshmukh, M.S. Sirajuddeen, Morphology, dielectric and emi shielding characteristics of graphene nanoplatelets, montmorillonite nanoclay and titanium dioxide nanoparticles reinforced polyvinylidenefluoride nanocomposites, J. Inorg. Organomet. Polym. Mater. 31 (5) (2021) 2003–2016.

[95] P. Rani, B. Ahamed, K. Deshmukh, Dielectric and electromagnetic interference shielding properties of zeolite 13X and carbon black nanoparticles based PVDF nanocomposites, J. Appl. Polym. Sci. 138 (13) (2021) 50107.

[96] P.L. Reddy, K. Deshmukh, K. Chidambaram, B. Ahamed, K.K. Sadasivuni, D. Ponnamma, et al., Effect of poly ethylene glycol (PEG) on structural, thermal and photoluminescence properties of CdO nanoparticles for optoelectronic applications, Mater. Today: Proc. 9 (2019) 175–183.

[97] K. Nath, S. Ghosh, S.K. Ghosh, P. Das, N.C. Das, Facile preparation of light-weight biodegradable and electrically conductive polymer based nanocomposites for superior electromagnetic interference shielding effectiveness, J. Appl. Polym. Sci. 138 (2021) e5051.

[98] K. Vasanthi, Biodegradable polymers - a review, Polym. Sci. (2017) 3–7.

[99] P. Baruah, N. Karak, Bio-based tough hyperbranched epoxy/graphene oxide nanocomposite with enhanced biodegradability attribute, Polym. Degrad. Stab. 129 (2016) 26–33.

[100] S. Wang, C. Song, G. Chen, T. Guo, J. Liu, B. Zhang, et al., Characteristics and biodegradation properties of poly (3-hydroxybutyrate-co-3-hydroxyvalerate)/organophilic montmorillonite (PHBV/OMMT) nanocomposite, Polym. Degrad. Stab. 87 (1) (2005) 69–76.

[101] M. Tanaka, K. Sato, E. Kitakami, S. Kobayashi, T. Hoshiba, K. Fukushima, Design of biocompatible and biodegradable polymers based on intermediate water concept., Polym. J. 47 (2) (2015) 114–121.

[102] E. Lizundia, J.R. Sarasua, F. D'Angelo, A. Orlacchio, S. Martino, J.M. Kenny, I. Armentano, Biocompatible poly (L-lactide)/MWCNT nanocomposites: Morphological characterization, electrical properties, and stem cell interaction., Macromol. Biosci. 12 (7) (2012) 870–881.

[103] M. Aliabadi, H. Shagholani, Synthesis of a novel biocompatible nanocomposite of graphene oxide and magnetic nanoparticles for drug delivery., Int. J. Biol. Macromol. 98 (2017) 287–291.

CHAPTER

2

Processing methods of polymer nanocomposites: influence of processing parameters, nanofiller nature, size, and shape on their functional properties

Asok Aparna, Gopika Venu, A.S. Sethulekshmi and Appukuttan Saritha

Department of Chemistry, Amrita VishwaVidyapeetham, Amritapuri, Clappana, Kerala, India

Chapter Outline

Biodegradable and Biocompatible Polymer Nanocomposites. DOI: https://doi.org/10.1016/B978-0-323-91696-7.00010-6
Copyright © 2023 Elsevier Inc. All rights reserved.

2.1 Introduction

Nanocomposites are basically heterogeneous materials with at least one-phase dimension in the nanometer range. Because of their larger surface area and aspect ratio, nanocomposites have superior properties to macrocomposites. There are different type of nanocomposites based on the corresponding phases [1] such as metal nanocomposites [2], ceramic nanocomposites [3], polymer nanocomposites etc. [4]. In the case of polymer nanocomposites, it can be either bio-based [5] or biodegradable [6] where the matrix will be the polymer in which the reinforcement nanofillers are evenly distributed. Polymer nanocomposites are gaining popularity due to their exceptional properties such as mechanical, thermal, electrical, barrier, and so on. Because of the depletion of petroleum resources and environmental concerns, biodegradable composites are gaining prominence [1]. Commonly used polymers for polymer composites synthesis are polyethylene (PE) [7], polybutadiene [8], polystyrene (PS) [9] etc. which are nonbiodegradable and reuse or recycling of these materials is found to be challenge to researchers all over the world. Despite their widespread use in everything from everyday household items to industrial and automotive applications, composites made from these polymers contribute to nonbiodegradable chemical waste all over the world. The only way to deal with these chemical wastes is to use biodegradable and biocompatible composites derived from renewable sources. However, scientists have discovered that the most difficult problem they face is the synthesis of 100% efficient biocomposites as an alternative to commercial ones without compromising the properties. As a result, there are no biodegradable or biocompatible polymer composites that meet the end-to-end applications [1].

The introduction of biodegradable nanocomposites has resulted in significant changes not only in chemical waste management but also in the medical field [10]. The main advantage of these materials is its nanosize filler inclusion which gives better mechanical properties [11], strength [12] and biodegradability [1]. This has opened great opportunities for biodegradable polymer composites especially in drug delivery [13], tissue engineering [14], dental applications [15] etc. Polymer nanocomposites are also used extensively in the electronic industry. The use of biodegradable materials in the electronics industry has resulted in a significant improvement in the control of plastic pollution caused by electronic waste, which is a novel approach [6]. Addition of reinforcement fillers like clay [16], cellulose [17], graphene [18] and other nanomaterials also contributes to better thermo-mechanical properties of polymer nanocomposites. Thus, the selection of nanofiller is also very important for the synthesis of polymer composites to get the most out of them [19]. Processing techniques, the type of nanofiller used, and other factors all influence the synthesis of biopolymer composites with improved properties. The addition of nanofillers improves the overall properties of biopolymer nanocomposites, such as better dispersion, size, shape, and even the nature of the selected nanofiller, which can tailor the composites' properties [20]. This chapter focuses on the latest developments in the field of polymer nanocomposites, their properties, and various types of processing techniques, with a particular emphasis on bionanocomposites. This chapter also discusses the impact of nanofillers on the properties of polymer composites, specifically the size, shape, and nature of the fillers.

2.2 Polymer nanocomposites

Polymers have become an unavoidable component of human life. Polymers and polymer composites are widely used in transportation, aviation, adhesives, the painting industry, electronics, and other fields. Polymer properties can be tailored by adding appropriate fillers, making them useful for a wide range of diverse and unique applications. When the polymers chosen are bio-derived, the demand for such composites will be at an all-time high [21]. In the field of polymer composites, the term "nano" is very important in defining the properties of the fabricated composites. Because of the nanoscale dimensions of the filler, the surface area of these composites will be very large, making them unique in many ways. When polymer chains and reinforced nanofillers are combined homogeneously, they can produce materials with superior properties and strength. Numerous studies have been conducted on polymer composites and the effect of nanofillers on composite properties [22–24]. Ohlan et al. [25] synthesized polyaniline-based polymer nanocomposites using barium ferrite and titanium dioxide (TiO_2) as the nanofiller by in-situ polymerization technique. They examined the microwave absorption properties of the fabricated composites and found them to be excellent. The findings also showed that the intrinsic properties of the nanofillers had a significant impact on the adsorption properties of the composites. The incorporation of barium ferrite, in particular, had a significant impact on the composite's permittivity, permeability, and so on. The surfactant used was dodecyl benzene sulfonic acid (DBSO), which can also function as a dopant. Because of the nanosize of the fillers, the addition of these nanofillers was sufficient for imparting interfacial dipolar polarization as well as higher anisotropic energy. Vitale et al. [26] reported the fabrication of gold nanoparticle (NP)-reinforced metal poly-yne poly[platinum-bis(tributhylphosphine) diethynylbiphenyl](Pt-DEBP)-based polymer. Two colloidal synthesis methods were used to functionalize gold NPs with dodecanethiol: one with two phases and one with one phase. The transmission electron microscopy (TEM) images of the one-phase composites revealed no aggregation of the gold NP. Aggregation, however, was discovered in the two-phase synthesized composites. As a result, one-phase synthesized composites have more dominant properties than two-phase synthesized composites. The possibility of synthesizing zinc oxide (ZnO) nanopowders and using it as reinforcement filler was carried out by Matei et al. [27]. They used coprecipitation to create the nanopowders, which were then characterized using scanning electron microscopy (SEM). When dispersed in the polymer matrix, silver (Ag) was added as a dopant to the ZnO nanopowder to induce antibacterial properties. The polymer matrix used was poly vinyl alcohol (PVA), and grafting improved the composite's stability and particle dispersion. Agglomeration of NPs was not observed in the composite, indicating that the filler and matrix were more compatible. Antibacterial tests confirmed that the silver-doped ZnO retained its antibacterial properties after being incorporated into the polymer matrix.

Polymer nanocomposites with embedded magnetic materials have found widespread use in the electronics industry, particularly in electromagnetic shielding devices. Wilson et al. [28] used poly (methyl methacrylate) (PMMA) as the polymer matrix to conduct a study on the electrical properties of polymer nanocomposites containing varying concentrations of iron NPs. The microwave plasma method was used to create the NPs, and the composite was created by melt blending. The addition of different concentrations of iron NPs resulted in consistent ferromagnetism. The excellent compatibility and dispersion of nanofiller in the polymer matrix was the primary cause of the composites' improved electrical properties, making them suitable for electromagnetic shielding applications.

Atom beam cosputtering synthesis of Ag NPs reinforced polyethylene terephthalate (PET) matrix was demonstrated by Mishra et al. [29]. TEM analysis (Fig. 2.1) was used to study the microstructure of the composite which showed the irregular structure of the silver particle. Because

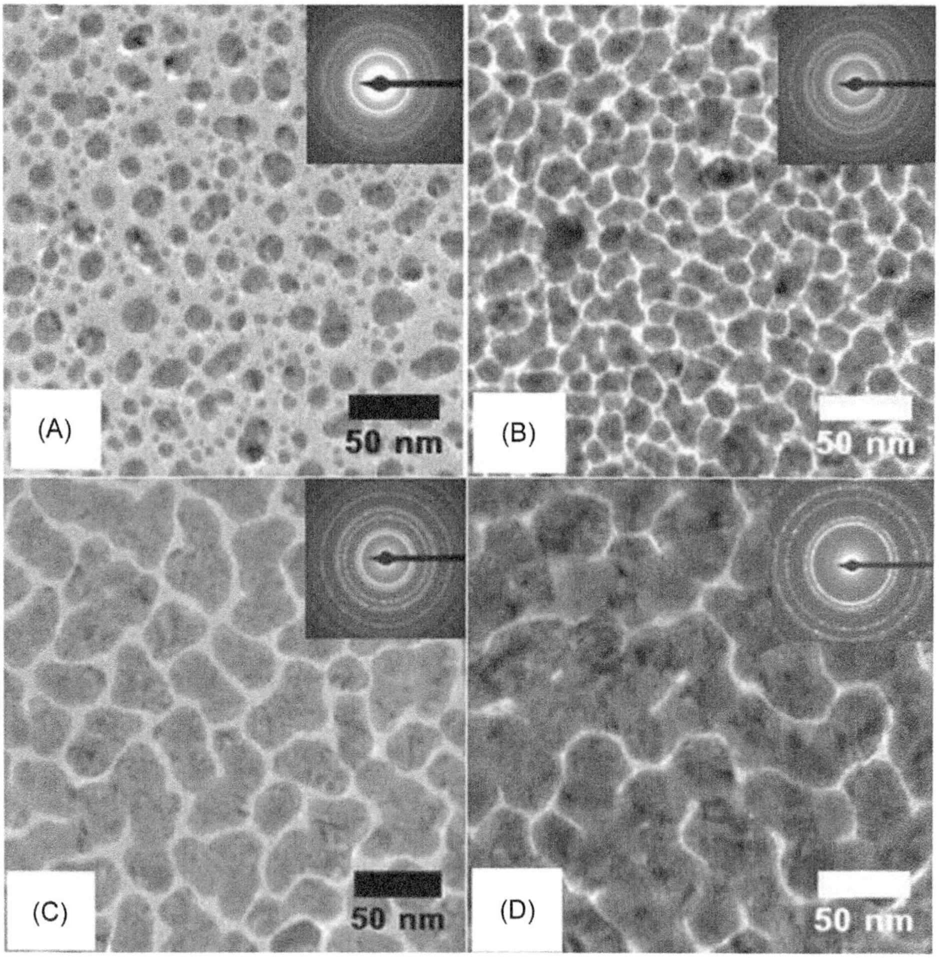

FIGURE 2.1

(A–D)—Transmission electron microscopy images of Ag/PET nanocomposites showing evolution of Ag nanostructures with increase in Ag at.% (from 30% to 75%). The selected area electron diffraction patterns corresponding to each metal fractions are shown as inset figures in 4(A–D) [29].

Reproduced with permission from American Scientific Publishers. Copyright 2010.

of the presence of Ag NPs, these materials could be useful in solar type absorbers because they have a broad surface plasmon resonance that extends up to the infrared region. Electrodes and separating materials benefit from conducting polymer nanocomposites. Such type of chiral polymeric composites, which are soluble in water, was synthesized via template method by McCarthy et al. [30]. He reported that the addition of a chiral agent, such as acid, could be used to synthesize chiral polyaniline-based nanocomposite. At the molecular level, infrared spectroscopy was used to examine the proper mixing of polyaniline and the template. The chirality analysis resulted in a fixed polyaniline derived backbone conformation because the chirality could be varied after polymerization.

Gelves et al. [31] conducted a study based on conductive polymer nanocomposites which were found to be light weight and highly conducting. The composite was created by mixing miscible solvents with polyaniline and copper wire, followed by precipitation (MSMP). This method does not require the use of any surfactants or surface modifiers. A cell-like nanostructure was observed in the polymer phase. Even at lower concentration of the nanowire, the composites showed excellent electrical conductivities of 104 S/m. The copper nanowires' high aspect ratio, as well as their interconnected nature, contribute to electrical percolation, resulting in a critical exponent factor of 2.9. Because these materials performed well, they can be used in electromagnetic shielding applications or devices.

Electrochemical oxidation-based polymerization of thiophene, done in the presence of nickel NPs to form polythiophene/Ni (PT/Ni)-based nanocomposites was reported by Pascariu et al. [32]. Polyvinylpyrollidone was chosen as the protecting agent for the hydrazine-based chemical reduction of nickel chloride to produce nickel NPs. From the morphological analysis of the sample, two main nickel structures was found—small Ni particles that ranges 6–20 nm in diameter and Ni clusters, which are larger and spherical that ranges between 50 and 100 nm. When the Ni content was increased, corresponding enhancement in electrical conductivity was observed up to 1.25×10^{-2} cm^{-1}. Fig. 2.2 represents the SEM images of the PT and PT/Ni nanocomposites. Zare et al. [33] analyzed the conductive properties of polyaniline/dextrin nanocomposites synthesized in situ. The conductivity increased as the aniline concentration increased. Furthermore, these nanocomposites demonstrated improved antioxidant properties and were capable of removing heavy metal ions from aqueous solutions. Based on soil analysis, these composites were found to be biodegradable, with degradation rates ranging from 30.18% to 74.52%. These composites can be used in biomedical devices, packaging, sensors, and other applications. Ponnamma et al. [34] analyzed the piezo-elctrical properties of ternary polyvinylidene-fluoride hexafluoropropylene (PVDF-HFP)-based nanocomposites prepared by simple solution casting. As reinforcement fillers, NP-sized ceramic-based barium titanate (BT) and hexagonal boron nitride nanosized layers were used. From the piezo-electric studies, the output voltage was found to be 2.4 V along with high dielectric constant that is, 45. The dielectric loss was 7.8. The studies revealed that the crystalline structure of the BT contributed to the composites' high piezoelectricity. The combined effect of the diphasic reinforcing filler was also attributed to the excellent thermal and dielectrical properties.

2.3 Methods of processing

For improved mechanical and thermal properties, NPs are used as reinforcement fillers in polymer composites. Carbon nanotubes (CNT), graphene, silicon, and other NPs are commonly used as nanofillers. The processing of polymer nanocomposites varies greatly depending on the polymer matrix. However, the processing method should also take into account factors such as aggregation, dispersion, compatibility, and functionalization, among others, in order to produce polymer nanocomposites with improved properties. Melt mixing, in-situ polymerization, template method, solution method, and other techniques are commonly used for processing polymer nanocomposites [35].

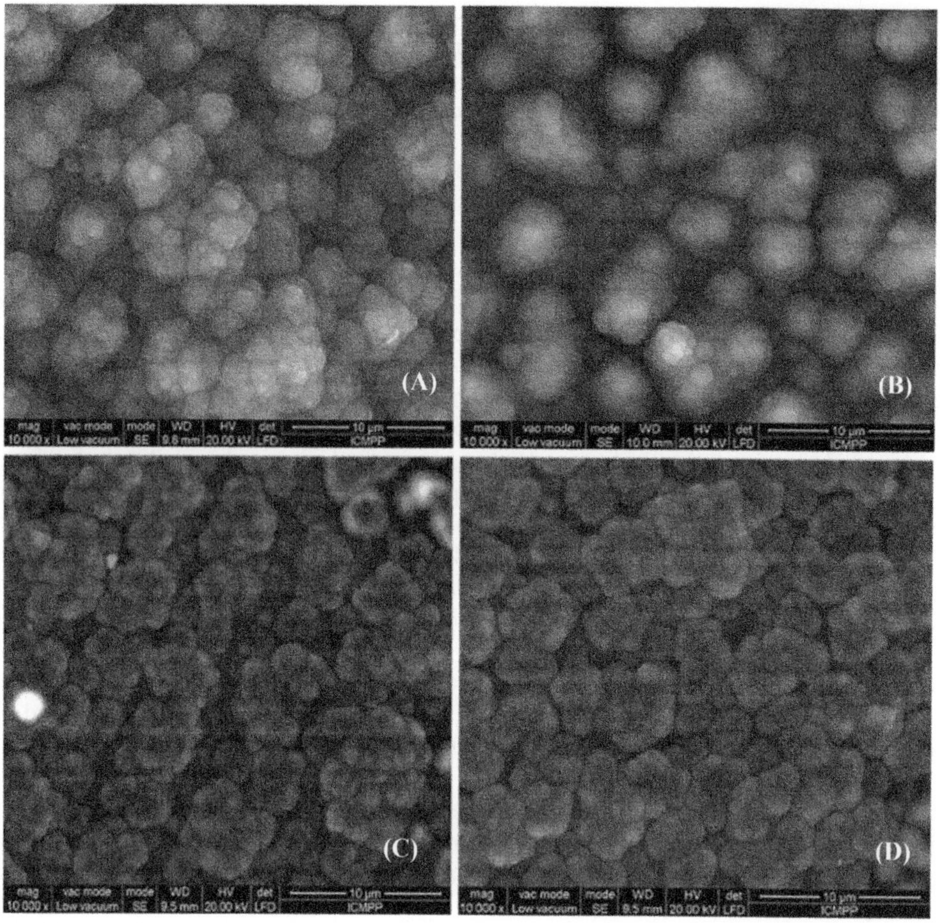

FIGURE 2.2

SEM images of PT (A), PT–Ni (S4) (B), PT–Ni (S5) (C) and PT–Ni (S6) (D) – (the scale bar is 10 μm) [32].

Reproduced with permission from Elsevier. Copyright 2014.

2.3.1 Melt mixing

The polymer and host mixture is annealed statistically or under shear above the softening point in this technique [36]. Melt mixing is useful for insoluble polymers, particularly thermoplastic polymers such as polypropylene, polycarbonate, high-density polyethylene (HDPE), and others. This method is used in a variety of techniques, including extrusion, injection, and blow molding, because it is free of solvents and other contaminants [35]. Sukumaran et al. [37] prepared multiwalled carbon nanotube (MWCNT)/acrylonitrile butadiene-based polymer nanocomposites by both methods that is, melt mixing and solution mixing. Electrical conductivity properties of solution-casted and melt-mixed composites were investigated and characterized. The study demonstrated that simply mixing the polymer melts could improve conductivity. However, the percolation threshold of melt-mixed composites was found to be lower than that of solution-casted composites. Because these composites have higher conductivity even at low filler loadings, they can be used as 3D printing materials. The incorporation of nanocomposite into the melted polymer may result in a decrease in viscosity, which may lead to polymer degradation when high shear is applied. Kamkar et al. [38] synthesized PS/MWCNT-based composites using solution casting and melt mixing. For the first time, the effect of mixing methods on viscoelastic response and EMI shielding was investigated. The MWCNT network was discovered to be interlocked, which improves EMI shielding over the cluster-based MWCNT prepared through solution mixing. Di et al. [39] synthesized biodegradable polycaprolactone (PCL)-based polymer nanocomposites by melt mixing using organoclay as the nanofiller. For the exfoliation of organoclay, the melt mixing process employs the use of an internal mixer. The composite constituents were first blended dry in a predetermined amount before being melt mixed in the mixer.

ElAchaby et al. [40] mentioned the melt mixing of polypropylene based nanocomposites using graphene as the nanofiller. Nanocomposites were created using a twin screw corotating extruder. When the amount of graphene nanosheets (GNS) increased, so did the crystallinity and melting temperature. Bikiaris et al. [41] prepared isotactic polypropylene (iPP) and silicon dioxide-based nanocomposites by melt mixing method using a twin screw corotating extruder. Mechanical properties such as tensile strength at break and Young's modulus were found to improve after silica NPs were added. Fine dispersion of silicon dioxide NPs in the polymer matrix was observed, as well as improved adhesion between the two material interfaces. Effect of two types of graphene nanoplatelets (GNPs) addition on low-density polyethylene (LLDPE) was investigated by Khanam et al. [42] in 2016. Addition of GNPs having grades C and M, improved the thermal conductivity (TC) of the LLDPE, in which more pronounced enhancement was observed in M-GNPs than C-GNPs. Fig. 2.3 shows the GNPs/LLDPE nanocomposites synthesis through melt mixing method.

2.3.2 Solution mixing

One of the most common processing methods for polymer nanocomposites is solution mixing. The first step in this method was to disperse the nanofiller and swell the polymer in a suitable solvent. The polymer solution and nanofiller dispersion were then mixed together using stirring, ultrasonication, or high speed mixing. The final step is to cast the above polymer-nanofiller dispersion in a mold, followed by solvent evaporation. Solvents that are commonly used include chloroform, toluene, acetone, dimethylformamide, and others. This method allows for better dispersion of nanofillers. However, effective solvent removal at the end of this method remains a major issue. Furthermore, because this method uses a large amount of solvent, the high cost of solvents and environmental issues associated with its disposal are disadvantages of this technique [43].

The choice of solvent is critical in determining the final properties of polymer nanocomposites. Saritha et al. [44] prepared chlorobutyl rubber (CIIR)-clay nanocomposites through solution casting method and evaluated the impact of solvent-clay and solvent-rubber parameters on the final properties of the nanocomposites. For the polymer-nanofiller dispersion, they chose toluene, xylene, cyclohexane, chloroform, and tetra hydro furan. The samples were cast in a petri dish and dried at room temperature after mixing both solutions with stirring. After mixing the solution, they used a two-roll mill to add additional compounding ingredients. When cyclohexane was used as the

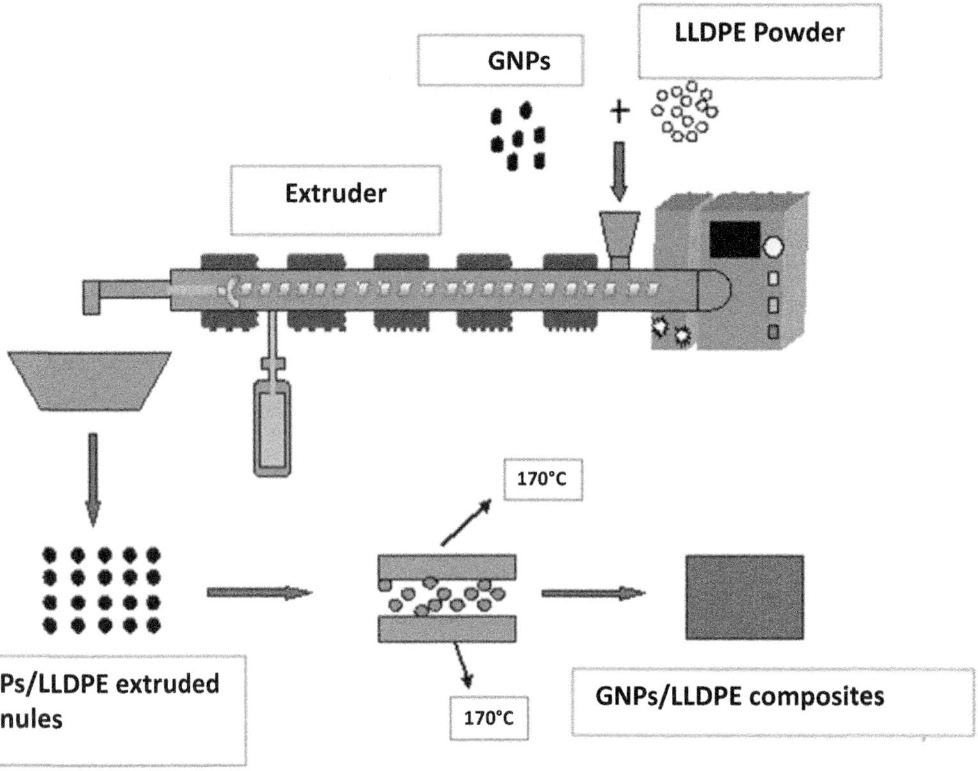

FIGURE 2.3

Schematic diagram of melt mix processing of GNPs/LLDPE nanocomposites [42].

Reproduced with permission from Taylor and Francis. Copyright 2016.

solvent, they obtained better mechanical and gas barrier properties because the difference between CIIR-solvent and clay-solvent interaction parameters was smaller in the case of cyclohexane. Tsai et al. [45] used hexane for MoS_2 exfoliation and then mixed this with CIIR in hexane for nanocomposite preparation. Later, they studied the mechanical, barrier properties of the CIIR-MoS_2 nanocomposites. MoS_2 exfoliation and CIIR-MoS_2 nanocomposites preparation is illustrated in Fig. 2.4.

2.3.3 In-situ polymerization

The NPs, such as CNT or silicates, are first dispersed in the monomer liquid in this type of polymerization. This method is used for polymers that are both insoluble and thermally unstable and cannot be processed via solution or melt polymerization. The polymerization process begins between the intercalated sheets that are layered under the application of appropriate heat radiation in the presence of a catalyst or an initiator. Mitchell et al. [46] evaluated the nonisothermal crystallization behavior of PCL and single-walled CNT by in-situ polymerization, which suggested that

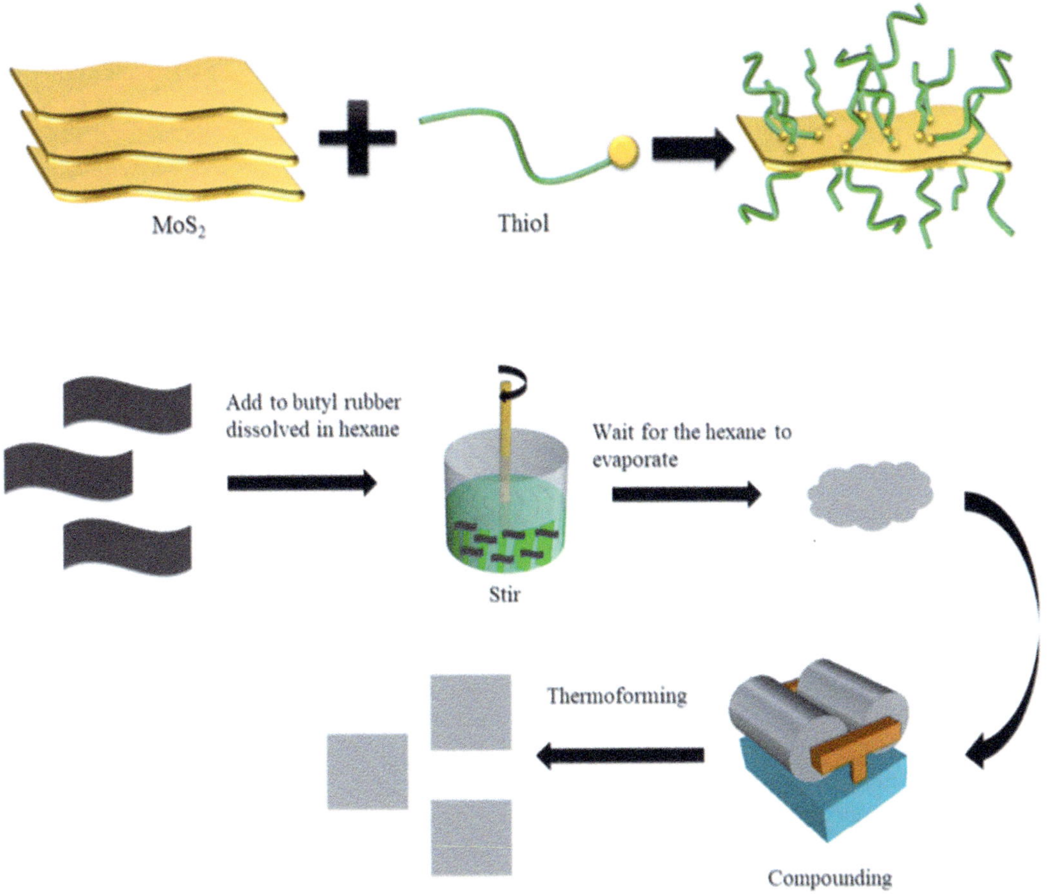

FIGURE 2.4

Schematic illustration for exfoliation modification of MoS_2 and corresponding production of chlorobutyl rubber-based nanocomposites [45].

Reproduced with permission from MDPI. Copyright 2018.

they are better nucleators in PCL matrix. The study found that increasing the amount of nanotubes improved nucleation activity. Kuila et al. [47] reported the in-situ emulsion polymerization of poly methyl methacrylate-based polymer nanocomposites using graphene as the reinforcement filler. PMMA was created through the simple emulsion polymerization of the monomer, methyl methacrylate. The physicochemical properties of the fabricated composites were investigated, which suggests that the increase in glass transition temperature was caused by a strong interaction between the oxygen functionality in graphene and the polar groups in PMMA. Ramesan et al. [48] synthesized polypyrrole-based nanocomposites in situ via chemical oxidation polymerization using silver-doped TiO_2 as the nanofiller. To characterize the composites, various techniques such as fourier-transform infrared spectroscopy (FTIR), differential scanning calorimetry (DSC), thermogravimetric analysis (TGA), SEM, and others were used. The results indicated that increasing the concentration of the filler improved the glass transition temperature (Tg), as well as electrical conductivity, dielectric constant, and other properties. The Ag-doped TiO_2 was created using the sol-gel method. Tripathi et al. [49] described the influence of filler concentration on PMMA-based nanocomposites using reduced graphene oxide as the filler. They chose three techniques for the development of polymer composites: in-situ polymerized methyl methcrylate (MMA) in the presence of filler, bulk polymerization of MMA in the presence of PMMA beads and the filler, and finally in-situ polymerization of methyl methacrylate in the presence of filler via sheet casting. Regardless of the method of preparation, the study found that the properties of the composite were dependent on the filler concentration. Trujillo et al. [50] carried out a work on in-situ polymerization of PE CNT-based composites. The effect of filler concentration and structure on composite properties was investigated, as well as a nucleation study. In comparison to neat high-density polyethylene, CNT-reinforced in-situ polymerized composites demonstrated better nucleation regardless of CNT structure, resulting in faster crystallization of in-situ composites at isothermal conditions. Okamoto et al. [51] prepared clay/PMMA and clay/PS-based composites by means of in situ intercalative free-radical polymerization from lipophilized smectic clays. The edged interaction between the silicate layers resulted in flocculation behavior.

2.3.4 Template method

This method employs a template to create various shaped nanocomposite materials such as layered or hexagonal, among others. The polymer solute will act as a template, facilitating the formation of nanocomposite. This method is commonly used in the case of mesoporous materials [36]. Wong et al. [52] reported the fabrication of emulsion template polymerized macroporous polymer composites using 2-D reduced graphene oxide. According to this study, emulsion templating is a versatile approach for the synthesis of macroporous composites with tuneable porous structures. The study also demonstrated that this method can be used to convert 2D-reduced graphene oxide materials into conducting and stiffened 3D macroporous materials. This method is commonly used in the case of mesoporous materials. Gan et al. [53] evaluated the properties of conductive polymer nanocomposites based on polypyrrole and polyacrylamide/chitosan (PPyPAM/CS). The molecular template for controlled in-situ polymerization was an interpenetrating chitosan network. The results demonstrated that the biocompatibility and electroactivity of the composites were improved.

2.4 Waste-mediated synthesis of polymer nanocomposites

In a research work by Hassan and group [54] mentioned about the preparation of bio-calcium carbonate ($CaCO_3$) NPs from waste chicken eggshells and incorporated it into Bioplast GS 2189 thermoplastic polymer. They created the NPs using a top-down approach that included cleaning and drying waste eggshells, grinding to generate fine powder, reducing the size through mechanical attrition technique, standardization using Retsch high speed shaker, and final size reduction via sonochemical technique. The Bio-$CaCO_3$/Bioplast GS 2189 nanocomposites were created using the melt extrusion method. To completely disperse bio-$CaCO_3$ NPs in the polymer system, they used a solution mixing method. Interestingly in another work [55], cellulose nanocrystals (NC) were generated from red algae waste using alkali treatment, bleaching treatment followed by acid hydrolysis and introduced it into PVA matrix. Alkali treatment was done using 4 wt.% NaOH solution and bleaching was performed using a mixture containing equal amount of acetate buffer and aq. sodium chlorite. It was mechanically stirred with 64 wt.% sulfuric acid for different times and the obtained samples were introduced into PVA matrix through solvent casting method. The preparation of reinforcing filler from waste material is illustrated in Fig. 2.5.

Adegbotolu et al. [56] mentioned about the possibility of incorporating nanoclays, which was produced from oil-based drilling fluid onto polyamide 6 polymer to prepare efficient nanocomposites. An interesting work by Verma et al. [57] mentioned the generation of carbon black powder from waste tires and the preparation of epoxy nanocomposite using the prepared carbon black

FIGURE 2.5

Overall steps of CNC extraction at different hydrolysis times and digital images of the obtained products [55].

Reproduced with permission from Elsevier. Copyright 2017.

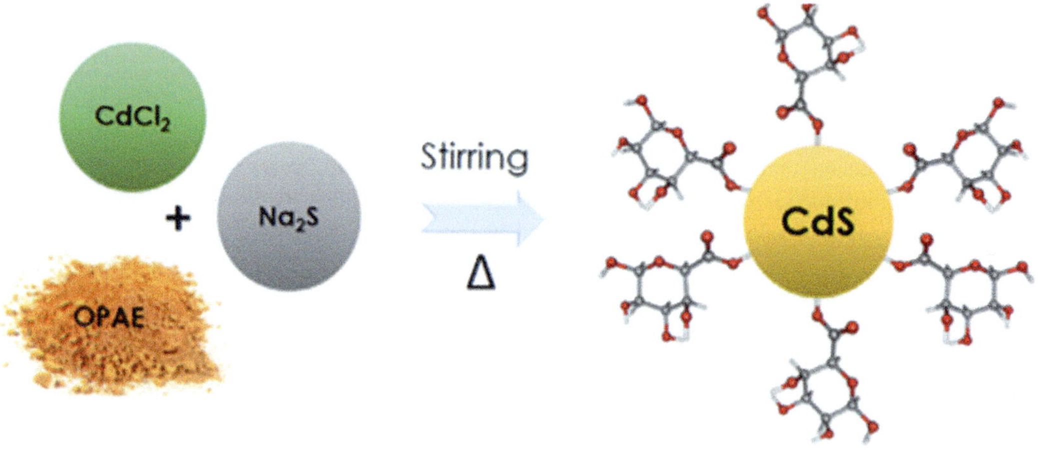

FIGURE 2.6

Mechanism of CdS NPs formation using OPAE as a capping or stabilizing agent [58].

Reproduced with permission from Wiley. Copyright 2020.

filler. Because various environmental issues arise when scrap tires are dumped into nature, the recycling process and conversion of these tires to a useful filler material has gained prominence in recent years. They prepared the black powdered chunks through pyrolysis and reduction of size to nanolevel was performed through ball milling process for 8 hours. The carbon black and epoxy resin were heated and cooled while being stirred with a magnetic stirrer. After that they added the hardener followed by stirring for about 5 minutes, casted and cured. Another waste material having no commercial value is orange peel. Reddy and team [58] used orange peel extract to generate CdS NPs through a green synthesis approach. Orange peels contain pectin biomolecules with numerous hydroxyl groups and hence by adding orange peel aq. extract (OPAE) along with $CdCl_2$ and NaS_2 solutions, it can make interactions with hydroxyl groups of CdS particles. It acts as a capping or stabilizing agent to control the size of CdS NPs. The mechanism is represented in Fig. 2.6. They prepared the PVA nanocomposite through solution casting technique which is shown in Fig. 2.7.

2.5 Properties of polymer nanocomposites

In recent years, there has been a steady increase in researchers' interest in improving the properties of polymeric materials so that they can be used in a variety of fields. Incorporation of inorganic or organic nanostructures into polymer matrix, particularly those that are biodegradable and biocompatible, improves mechanical, thermal, antibacterial, and barrier properties, allowing for applications in biomedical, packaging, coatings, automotive, and other fields.

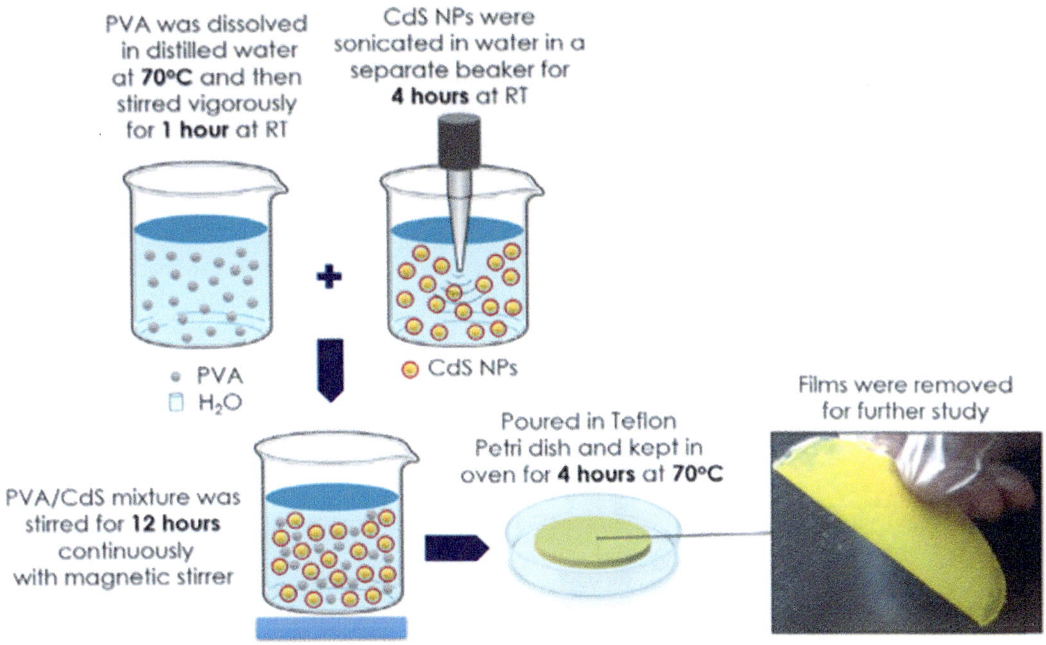

FIGURE 2.7

Protocol followed for the preparation of PVA/CdS nanocomposite films [58].

Reproduced with permission from Wiley. Copyright 2020.

2.5.1 Mechanical properties

Mechanical properties of polymer nanocomposites can be evaluated using tensile tests or dynamic mechanical analysis, such as modulus of elasticity, tensile strength, elongation, and hardness (DMA) [11]. Paramitha and team [59] evaluated the mechanical properties of poly lactic acid (PLA) nanocomposite fabricated by incorporating different kinds of clay such as commercial montmorillonite (MMT) and commercial bentonite. Even at low clay content, the PLA/clay nanocomposite showed an increase in Young's modulus and tensile strength when compared to pure PLA. Tensile strength improved by 23.25% uponmixing 0.5 wt.% bentonite owing to the better dispersion of clay which leads to the enhancement in interfacial area between clay and PLA resultingin an easy transfer of stress. However, the mechanical properties of nanocomposites decrease due to the filler–filler interaction and the increase in clay amount. They confirmed partial clay exfoliation while using the melt extrusion technique. When NPs are incorporated into biodegradable polymer materials, the surface area and surface activity of nanocomposites improve, and montmorillonite becomes a strong candidate due to the increased aspect ratio and increased capacity for expansion [60]. Dong and colleagues described the mechanical properties of a modified MMT clay nanocomposite incorporating poly (3-hydroxybutyrate-*co*-3-hydroxyhexanoate) (PHBH) [61]. The fabricated nanocomposite showed an interesting improvement in young's modulus (220.05%), tensile strength (158.25%), and elongation at break (58.1%) due to improved interfacial interaction between polymer and modified filler, as well as even distribution of

MMT, which resulted in easy stress transfer. Optimal mechanical strength was achieved when the nanofiller was modified with 5 wt.% silane coupling agent (3-aminopropyltriethoxysilane (KH550)) due to the drop in the surface energy of MMT. There is also an improvement in the mechanical properties of a PLA nanocomposite containing molybdenum disulfide modified with mercapto-ethylamine and PLA chains (MoS_2-g-PLA) [62]. They noticed the enhancement of storage modulus in both glass and rubbery state when the content of MoS_2-g-PLA was 0.5 wt.% due to the better interface interaction and compatibility between MoS_2-g-PLA nanosheet and PLA material.

Another group of researchers [63] introduced organoclay to a biodegradable PCL/cellulose acetate butyrate (CAB) blend resultingin the development of nanocomposite with improved mechanical properties. An enhancement of 45% in tensile modulus was identified for nanocomposite with 5 wt.% nanofiller along with an enhancement of 16.5% in yield strength. Furthermore, increasing the nanoclay loading improved break elongation by 6%. They stated that mechanical property improvement in the fabricated nanocomposites is due to better dispersion of nanofiller with an intercalated structure and improved compatibility between PCL and CAB phases. Frone et al. [64] noticed an improvement in the storage modulus of biodegradable polymer blend nanocomposite fabricated using PLA, poly(3-hydroxybutyrate) (PHB) and cellulose NC in both glass as well as rubber state. They discovered that nanocomposite films had a higher storage modulus above Tg due to the formation of cross-linked structures, as well as improved interfacial adhesion of NC with PLA/PHB, which was further enhanced in the presence of dicumyl peroxide (DCP). Similarly the improvement in the modulus of elasticity, tensile stress as well as elongation at break of poly (butylene adipate-*co*-terephthalate) (PBAT) nanocomposite incorporated with octadecylisocyanate functionalized cellulose nanocrystals (MCNC) is shown in Table 2.1 [65]. Modulus of elasticity shows a hike with nanocrystal content and a maximum was obtained with 7 wt.% filler content, which was actually a 30% increase in comparison with pure PBAT polymer owing to the absorbance of stress by MCNC. Nanodiamond (ND) insertion improved the mechanical properties of the PVA matrix [66]. The addition of NDs improved tensile strength and Young's modulus, implying that the nanofiller adhered better to the matrix. This adhesion resulted from the formation of hydrogen bonds between the hydroxyl groups on PVA and the oxygen-containing groups on the ND surface.

2.5.2 Thermal properties

TC is an assessment of a material's ability to conduct heat, whereas thermal diffusivity is an assessment of how quickly a substance can react to temperature changes [67]. TC, Tg, melting and

Table 2.1 Results of tensile tests for neat PBAT and PBAT reinforced with MCNC [65].

Sample	*E* (MPa)	σ (MPa)	ε (%)
Neat PBAT	55 ± 2	22 ± 4	544 ± 113
3 wt.% MCNC/PBAT	56 ± 5	21 ± 3	654 ± 79
5 wt.% MCNC/PBAT	65 ± 4	12 ± 1	577 ± 132
7 wt.% MCNC/PBAT	71 ± 5	11 ± 1	471 ± 68

E, *modulus of elasticity;* σ, *tensile stress;* ε, *elongation at break.*
Reproduced with permission from Springer.

crystallization temperatures, and other properties of polymer nanocomposites can be evaluated using TGA and DSC analysis. A group of researchers [68] described the thermal properties of PLA nanocomposites fabricated with graphene nanoplatelets (GNP) and MWCNT separately. They observed an enhancement in the TC and thermal diffusivity with a raise in filler content and was prominent in case of nanocomposite containing 6 wt.% GNP, which showed a value of 0.577 (W/mK) and 0.485, respectively. TC increased by 181% and thermal diffusivity increased by 214% when compared to the pure matrix, owing to graphene's superior TC and improved adhesive nature between filler and matrix, which can reduce phonon scattering. Because of the presence of free electrons, phonons generally serve as thermal carriers in polymeric materials [69].

Similarly, Bicy and collegues [70] developed PCL nanocomposites by incorporating functionalized CNT (f-CNT) nanofillers using electrospinning method and observed an improvement in thermal properties, confirming the ability of f-CNT to become a heterogeneous nucleating agent. This lowers the melt temperature while increasing the percentage crystallinity of the polymeric material. In the case of PCL/f-CNT composites, DSC analysis revealed a shift in the melting peak to a lower temperature value, while the crystallization temperature increased by 27%. The filler's nucleation effect resulted in the formation of imperfect crystals, which melted first and released energy, assisting the melting of ordered crystals. Decrease in lamellar thickness from 16 to 10 nm after the incorporation nanofillers confirm the importance of filler in the growth of crystals.

Another study found that DSC and TGA analysis revealed an increased crystallization temperature and degradation temperature of Ag NP-incorporated PHB composite [71]. TGA analysis revealed a degradation temperature of 340°C, while DSC results revealed crystallization, melting, and glass transition temperatures of 212.60°C, 398.72°C, and 29.35°C, respectively. They discovered that the presence of several nucleation sites provided by NPs, as well as the reinforcement by Ag NPs, resulted in an increase in melting and crystallization temperature. Zheng and colleagues reported a 0.3°C decrease in Tg and a 10°C increase in crystallization in the case of nanohydroxyapatite (n-HA) filled PLA nanocomposite reinforced with glass fiber [72]. DSC curves of the PLA-nHA nanocomposites with various HA content are shown in Fig. 2.8. Thermal properties of polymer composites were improved by the addition of organo-montmorillonite clay (OMMT), which was confirmed in a study by Luo et al. [73]. The melting temperature was more declined in nanocomposite containing 2.5 wt.% nanofillers due to the strong interaction of organo-montmorillonite and functional moieties of polymer, which implies higher crystallization. They noted an increase in the thermal stability of PLA/OMMT nanocomposite owing to the generation of more exfoliated particles and an increase in nanofiller incorporation.

2.5.3 Electrical properties

Most polymer matrices are insulating by nature, but they can be converted into conductive polymers by incorporating suitable conducting nanofillers such as graphene, CNT, carbon black, and so on [74]. The finished electrically conductive nanocomposites can be used instead of metal. The mechanism underlying this conduction is the formation of percolated structures in polymer matrices. Nakaramontri et al. [75] adopted latex and melt mixing methods for the preparation of CNT-filled NR nanocomposites and evaluated the electrical properties of the resultant nanocomposite. They functionalized the CNT surface with a coupling agent in their study to improve NR–CNT interactions. Nanocomposites prepared using the latex mixing method had better electrical properties than those prepared using the

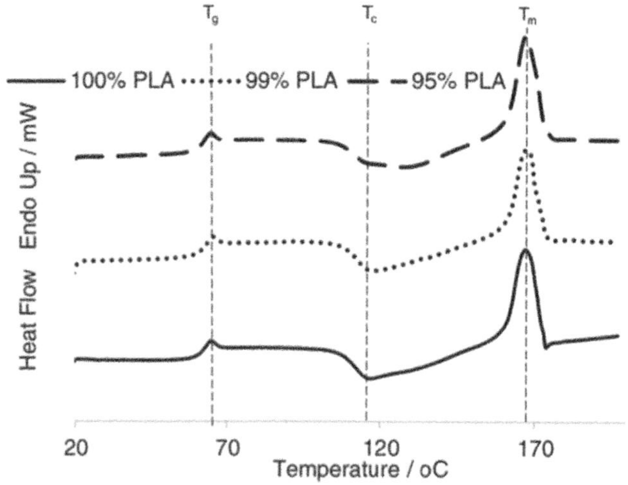

FIGURE 2.8

DSC thermograms of PLA-nHA nanocomposites, containing different wt.% HA content [72].

Reproduced with permission from Wiley. Copyright 2020.

melt mixing method. The lowest percolation threshold concentration of the CNT in latex–CNT composites was around 0.55 phr.

MWCNT decorated with ZnO NPs was used to reinforce the epoxy matrix, yielding an epoxy nanocomposite with improved electrical conductivity [76]. When compared to epoxy with pristine MWCNT, ZnO decorated MWCNT containing epoxy matrix demonstrated higher electrical conductivity (P-MWCNT). This could be because the MWCNT contains semiconductive ZnO quantum dots (QDs). Defects on the CNT surface act as scattering spots, reducing the mean free path of electrons and charge carrier number density, resulting in lower electrical conductivity. When the ZnO QDs are tightly bound to the CNT surface, contact inhibition of CNT junctions decreases and the number density of charge carriers in the polymer matrix inceases by reducing the influence of surface defects on electron scattering, thereby enhancing the electrical conductivity. Fig. 2.9 depicts the electrical conductivities of the P-MWCNT and ZnO/P-MWCNT composite films.

2.5.4 Dielectric properties

Dielectric properties of (Polyvinylidenefuoride) PVDF/GNPs/MMT/TiO_2 nanocomposite films were described by Kumar et al. [77]. They observed a dielectric constant value of 210.92 at 50 Hz for the fabricated PVDF nanocomposite, whereas it was only about 47.11 for the bare PVDF film, indicating a fivefold improvement in the case of PVDF nanocomposite due to better interaction of reinforcing filler and polymer system, as well as uniform filler dispersion. PVDF/GNPs/MMT/TiO_2 nanocomposite's improved conducting nature resulted in a dielectric loss tangent value of 10.4. Similarly Wu et al. [78] fabricated PLA nanocomposite by incorporating TiO_2 decorated MWCNT

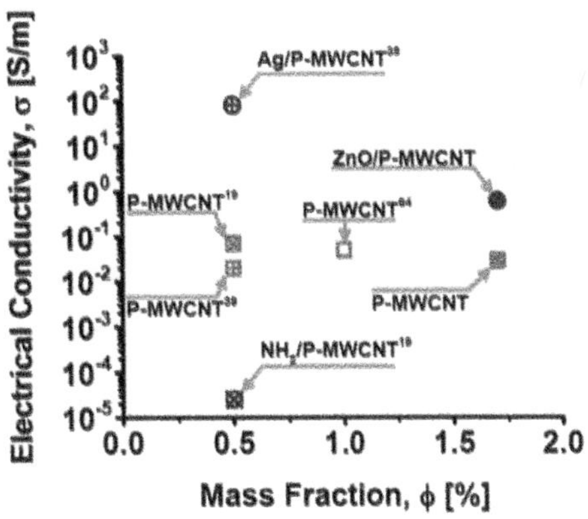

FIGURE 2.9

Electrical conductivities of P-MWCNT and ZnO/P-MWCNT composite films, along with other literature values [76].

Reproduced with permission from Elesevier. Copyright 2016.

by solution mixing method and hot compression process. A dielectric constant of 26.6 as well as a low value of dielectric loss (0.2) was observed for the sample containing 5 wt.% MWCNTs@TiO_2. Dielectric properties are provided in Fig. 2.10.

They stated that the dielectric constant was 8.3 times higher than that of pure PLA. The isolation effect caused by the TiO_2 decoration on the surface of MWCNTs resulted in an increase in distance between MWCNTs in the polymer matrix as well as a decrease in direct contact, which further suppressed dielectric loss. In another work by Rani et al. [79], fabrication of PVDF nanocomposites incorporated with Zeolite13X and carbon black nanoparticles (CBNPs), having excellent dielectric abilities was reported. A value of 76.3 for dielectric constant was obtained for PVDF nanocomposite containing 10 wt.% of CBNPs and a reduction in dielectric properties was observed with an enhancement in frequency.

2.5.5 EMI shielding properties

Electromagnetic interference (EMI) shielding properties of (polyvinylidenefuoride) PVDF/GNPs/MMT/TiO_2 nanocomposite films were pointed out by Kumar et al. [77]. EMI shielding effectiveness of 12.6 dB was noticed in case of developed PVDF nanocomposite incorporated with 2.5% of GNP, 1% of MMT as well as 11.5% of TiO_2 due to the high aspect ratio of reinforcing fillers and conductive interconnected network of fillers with PVDF system. Similarly Rani et al. [79] mentioned about the fabrication of PVDF nanocomposites incorporated with Zeolite13X and carbon black nanoparticles (CBNPs) with excellent EMI shielding abilities. EMI shielding effectiveness of PVDF nanocomposite in X-band frequency region showed an increase in the range of (−2.0 dB) − (−3.7 dB) after adding Zeolite13X and further increased to −11.1 dB after adding CBNPs, which

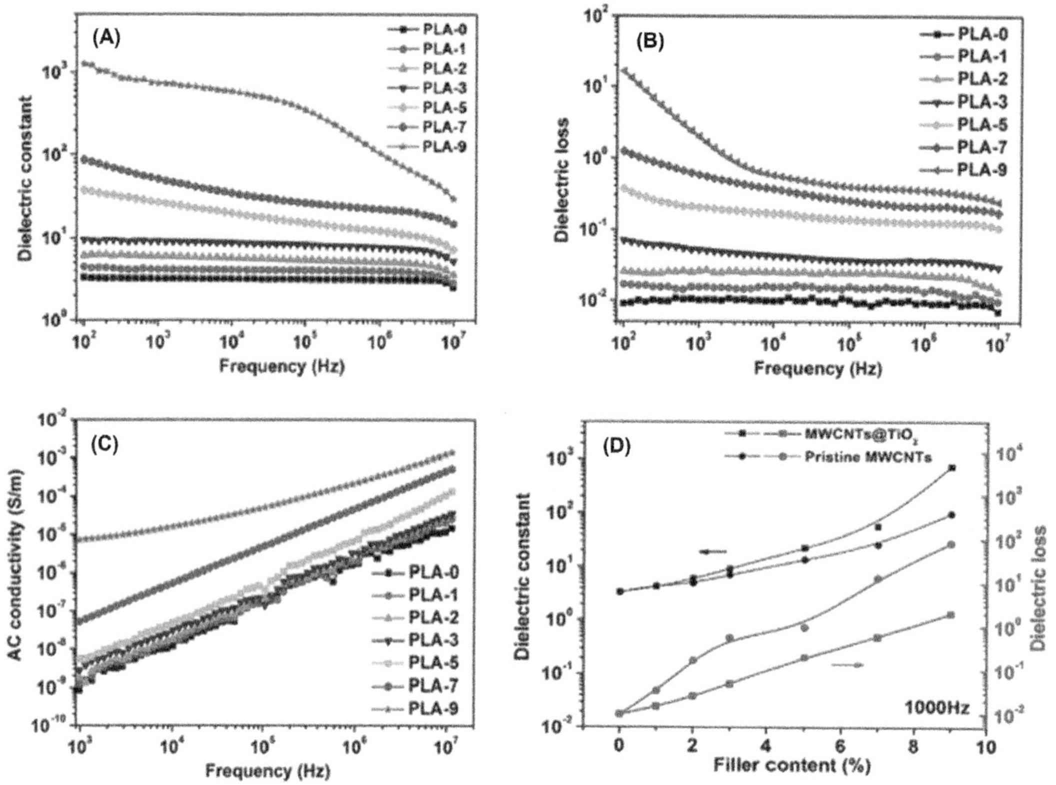

FIGURE 2.10

Dielectric property of PLA/MWCNTs@TiO_2 nanocomposites: (A) dielectric constant, (B) dielectric loss, (C) AC conductivity, and (D) dielectric perfermance of various PLA nanocomposites as a function of filler percentage at 1000 Hz [78].

Reproduced with permission from Elsevier. Copyright 2019.

is illustrated in Fig. 2.11. This is due to the better dispersion and formation of interconnecting network by the fillers that can be understood from the SEM images shown in Fig. 2.12.

In another research work [80], PVA/chitosan/GO/nickel oxide (NiO) nanocomposites were fabricated and observed a dielectric constant value of 1315.5 as well as an EMI shielding effectiveness value of 12 dB in the Ku-band region. Because of the conductive fillers (GO and NiO), an electric network was formed in the polymeric system, resulting in numerous active sites as well as interfaces with multiple internal reflections, leading to an improvement in EMI shielding effectiveness.

2.5.6 Antimicrobial properties

Because polymer nanocomposites have numerous applications in the biomedical and food packaging industries, determining their antimicrobial properties is becoming increasingly important.

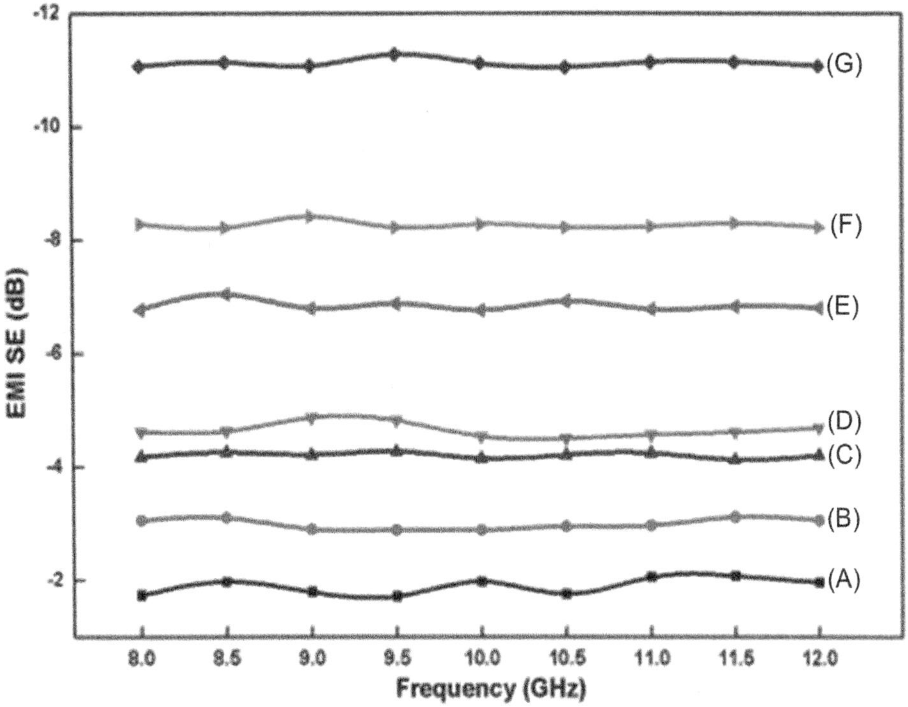

FIGURE 2.11

EMI SE in the X-band frequency region for (A) pure PVDF (B) PVDF/Zeolite13X (90/10) nanocomposite film, (C) PVDF/Zeolite13X/CBNPs nanocomposite films with 2 wt.%, (D) 4 wt.%, (E) 6 wt.%, (F) 8 wt.% CBNPs and (G) PVDF/CBNPs (90/10) nanocomposite [79].

Reproduced with permission from Wiley. Copyright 2020.

Venkatesan and coworkers [81] prepared PBAT nanocomposites containing different wt.% of silver oxide (Ag_2O) and revealed its antibacterial properties against *Klebsiella pneumonia* as well as *Staphylococcus aureus*. During their study, the pure matrix film showed no particular microbial inhibition zone while the polymer film containing Ag nanoparticles exhibited characteristic inhibition zone against *K. pneumonia* and *S. aureus* with a diameter of 11.0–19.7 mm against *K. pneumonia* and 11.0–14.0 mm against *S. aureus* corresponding to 0–10 wt.% NP loading. Another group described the antibacterial property of Ag nanoparticles incorporated chitosan (AgNP/chitosan (CS)) nanocomposites fabricated in situ [82]. They clearly noticed the bacterial inhibition zone created by AgNP/chitosan films against *S. aureus* and *E. coli*, which was not seen in pure or plasma-treated CS films. Because of the thinness of the negatively charged lipopolysaccharide layer in the cell wall, the size of the zones generated against *E. coli* increased with the amount of Ag in the polymer matrix. The reactive oxygen species (ROS) generated in the NP-incorporated CS films migrated to the surface, causing the potential of the bacterial plasma

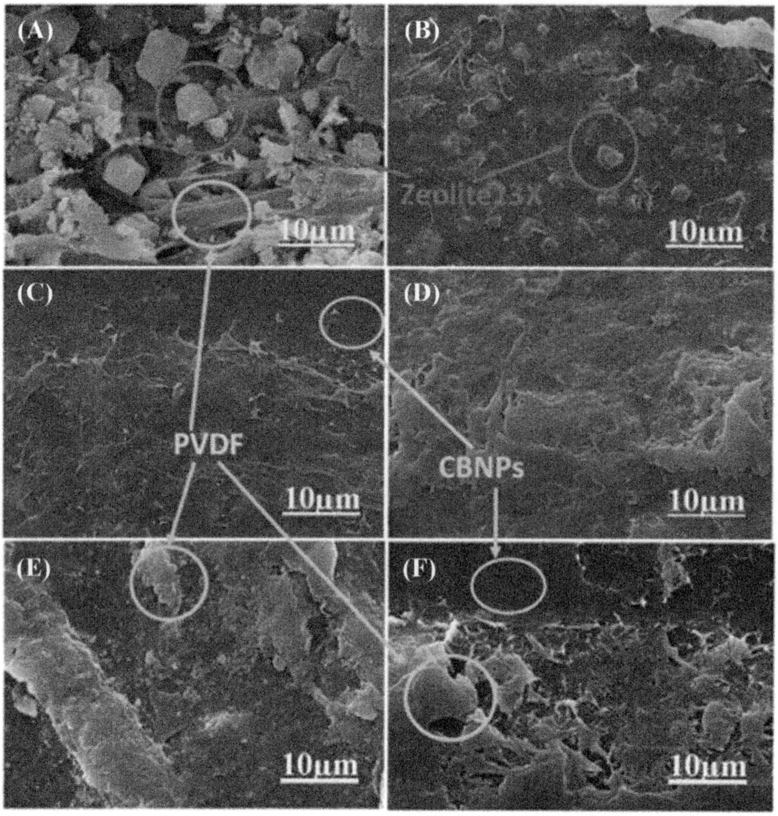

FIGURE 2.12

SEM micrographs of (A) PVDF/Zeolite13X (90/10) nanocomposite film (B) PVDF/Zeolite13X/CBNPs nanocomposite films with 2 wt.% (C) 4 wt.% (D) 6 wt.% (E) 8 wt.% CBNPs and (F) PVDF/CBNPs (90/10) nanocomposite film [79].

Reproduced with permission from Wiley. Copyright 2020.

membrane to destabilize. They discovered a decrease in intracellular adenosine triphosphate, which causes *E. coli* and *S. aureus* cells to die. Shankar and team [83] incorporated ZnO NPs into PLA matrix and developed a polymer nanocomposite film with improved antibacterial activity against *E. coli* and *Listeria monocytogenes*. The amount of *L. monocytogenes* in the nanocomposite was enhanced during incubation around 6 hours due to the slow release of NPs and decreased thereafter whereas in case of *E. coli*, bacterial growth remained unchanged for 3 hours and then decreased. They described that either the straight interaction between ZnO NPs and cell walls, or the interference of Zn^{2+} ions when replicating DNA and synthesizing proteins, or the generation of ROS will be the reason for the destruction of bacteria. Table 2.2 summarizes the details of antibacterial action and application regarding various polymer nanocomposites as described above.

Table 2.2 Details of antibacterial action and application regarding various polymer nanocomposites.

No.	Polymer nanocomposite	Antimicrobial test used	Affected microorganism	Application	Refs.
1	PBAT/Ag_2O	ASTM D-2149 standard method	*K. pneumonia* and *S. aureus*	Food packaging	[81]
2	AgNP/CS	CLSI method	*S. aureus* and *E. coli*	Antibacterial coatings, would dressings and membranes for water sanitation	[82]
3	PLA/ZnO NPs	Viable colony count method	*E. coli* and *L. monocytogenes*	Food packaging, biomedical	[83]
4	Clay and ZnO nanocomposites	Agar diffusion method and death curve	*E. coli* and *S. aureus*	Food packaging	[84]
5	PU/Ag	Agar-well diffusion method	*E. coli* and *S. aureus*	–	[85]

2.6 Factors affecting the properties of nanocomposites

2.6.1 Processing parameters

Various processing methods are used in the fabrication of polymer nanocomposites, and their impact on the properties of the nanocomposites varies. Each processing technique yields different results, indicating that it is tailored to a specific polymeric system [86]. The mixing time of the matrix and filler, as well as the processing methods used, all have an impact on the final mechanical properties of polymer nanocomposites. When compared to extrusion molding, fibers in compression molding were oriented at 0 degree, which resulted in improved mechanical properties [87]. The properties of MWCNT/PS nanocomposite generated using solvent mixing method were found to be better than those generated using melt mixing method [38]. More intriguingly, various processing conditions/parameters such as mixing time, mixing speed, temperature, and so on will influence the properties.

An interesting study by D'Anna and group [88] mentioned about the effect of process parameters on the properties of PLA/PHB blend nanocomposite incorporated with clay in which they used three various screw profiles along with various speed for rotating screw. Storage modulus increased by 20% at two specific screw profiles compared to others due to better filler dispersion, resulting in intercalated structures. In addition, an arrest of the relaxing process of polymer chain and high shear stress was reported with screw profiles 1 and 3 as shown in Fig. 2.13. Various processing conditions such as 240°C and 260°C, 5 and 10 minutes of extrusion time, 50 and 100 rpm of shear rate were implemented to know its influence on the properties of poly (butylene terephthalate)/GNP nanocomposites [89]. Lower electrical and TC values were observed when higher mixing time, temperature, and shear rate were combined, owing to a decrease in GNP aspect ratio, indicating the importance of short mixing time and low shear rate in the enhancement of conductivity.

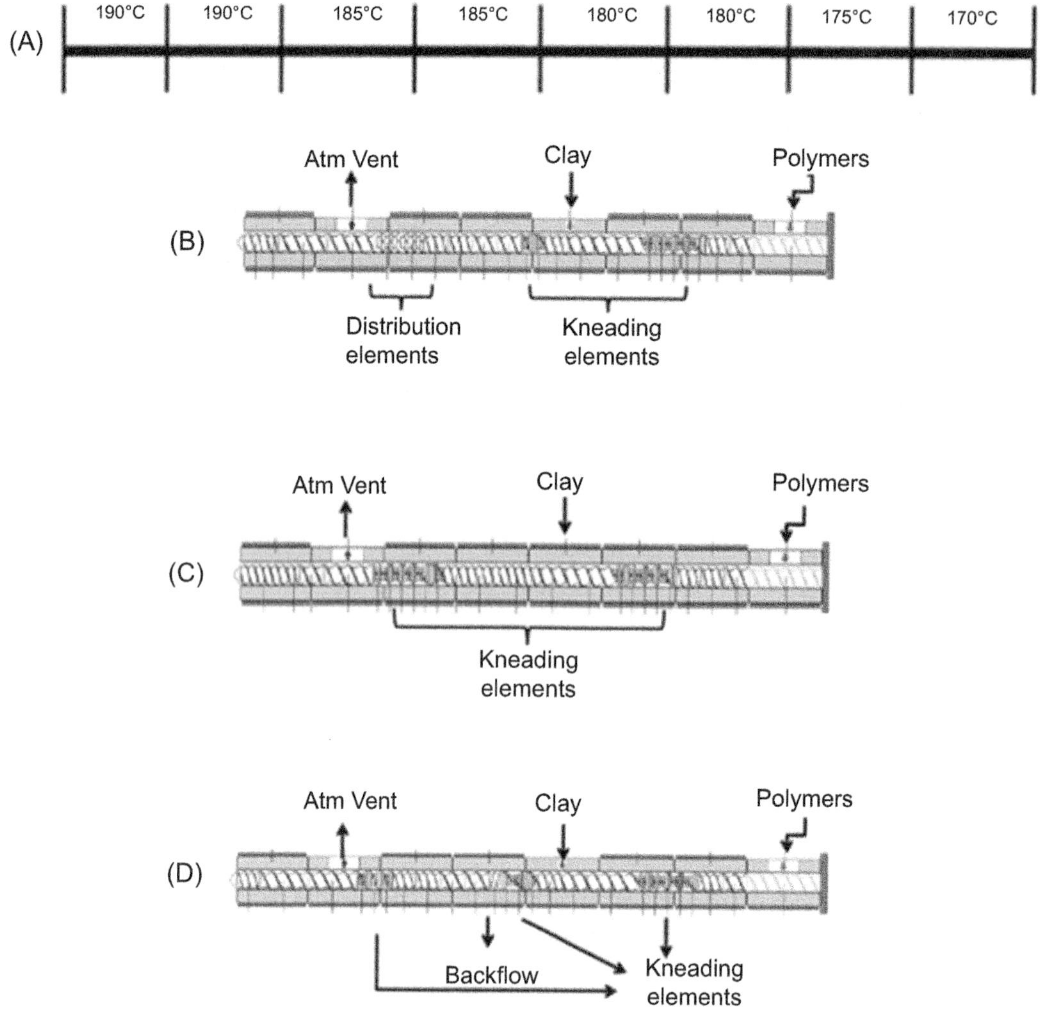

FIGURE 2.13

(A) Temperature profile, (B) Screw profile 1, (C) Screw profile 2, (D) Screw profile 3 [88].

Reproduced with permission from Springer. Copyright 2021.

Coppola and team [90] described the effect of printing temperature in the 3D printing technique for the fabrication of composites with PLA and clay. They performed melt compounding of PLA 4032D and PLA 2003D grades with 4 wt.% Cloisite 30B and followed the formation of feedstock filaments. Three printing temperatures were used for both PLA grades, and DMA nanocomposites showed an improvement in storage modulus at medium and higher temperatures than the glass transition temperature. TGA was used to confirm that the addition of clay improved thermal stability. This study revealed the diverse behavior of nanocomposites prepared at various printing temperatures. Because of differences in the morphology of nanocomposites as well as the macromolecular structure of the polymer, there is an increase in elastic modulus with an increase in printing temperature for PLA 4032D + clay compared to PLA 2003D + clay.

Similarly, Enqvist et al. [91] pointed out the effect of rotational speed and time in the ball milling technique for the fabrication of ultra-high molecular weight polyethylene (UHMWPE)/MWCNT nanocomposites. More importantly, they concluded that rotational speed, rather than rotational time, is critical in reducing cluster size and improving MWCNT dispersion. An increase in rotation speed resulted in a better flattening process of polymer particles. It was also reported that PBT/nanoSb_2O_3 composites prepared using high energy ball milling (HEBM) had better mechanical properties than those prepared using high speed rotation (HSR) [92]. Garofalo and colleagues emphasized the importance of draw ratio in the blown film coextrusion method for the fabrication of polyamide (PA)/PE nanocomposite film incorporated with layered silicates for packaging applications [93]. The effect of barrel temperature and residence time during the injection molding technique on the thermal, mechanical, and structural properties of HDPE/TiO_2 nanocomposite was discussed [94]. With an increase in barrel temperature and time, a decrease in degradation temperature and an increase in crystallization rate were observed. In addition, tensile strength varied from 22.53 to 26.30 MPa and Young's modulus got increased with temperature. They recommended a temperature below 300°C and above 150°C for effortless flow of the system as well as for improved properties.

2.6.2 Nanofiller

Nanofillers are materials with at least one dimension in the nano regime that can be incorporated into polymeric materials to improve their properties significantly. The primary challenge is to achieve complete NP dispersion, and the uniform dispersive nature of fillers can result in a larger area of interface between the nanocomposite components [36]. The nature, size, and shape of the added nanofillers, as well as the concentration of the nanosized filler, all contributed to the final character of polymer nanocomposites [95].

2.6.2.1 Nature of the filler

The nature of nanofillers is a major factor that influences their reinforcing ability. To prepare nanocomposites with polymers, a variety of nanofillers such as graphene, CNT, silicates, boron nitride, nanocellulose, nanolignin, and others can be used. The nanofiller's nature can predict the final property of the nanocomposite. This means that while graphene in natural rubber may improve electrical properties, layered silicates may improve mechanical properties in natural rubber. CNT, graphene, graphene oxide, carbon nano onions, fullerenes, and other members of the carbon family However, the reinforcing ability of these materials varies depending on the polymer. Liang and team [96] created a PVA/graphene nanocomposite using a simple and environmentally friendly method in which water was used as a solvent. The mechanical properties of the PVA/graphene nanocomposite were significantly improved by the addition of very small amounts of filler. With the addition of 0.7 wt.% of graphene oxide, a 62% enhancement in Young's modulus and 76% raise in tensile strength were observed. MoS_2 has a structure similar to graphene but different electrical properties. PVA/MoS_2 nanocomposites were created using a solvent blending method, with MoS2 uniformly dispersed and partially exfoliated in the polymer matrix [97]. Thermal analysis techniques confirmed the improvement in Tg and thermal decomposition temperature of the PVA nanocomposite. Furthermore, these nanocomposites demonstrated an excellent improvement in fire resistance properties. All of these advancements can be attributed to the strong interaction between polymer and filler, the homogeneous dispersion of MoS2 in PVA, and the barrier effects of layered MoS_2.

2.6.2.2 Size of the filler

In a recent study [98], by performing combined experiments and modeling, single-layer graphene (SLG) and GNP in various aspect ratios were incorporated in PS composite material to study the influence of size on the damping properties of polymer nanocomposite. The damping behavior of the GNP-incorporated PS nanocomposite was improved by 70%, whereas the damping behavior of the SLG-incorporated PS nanocomposite was reduced. The micromechanical damping model results revealed a decrease in damping coefficients with decreasing aspect ratio. Gao and team [99] took GNP with large lateral size (GNP-L) as well as small (GNP-S) to evaluate its effect on the properties of PLA composites. They noticed the agglomeration of $\geq$ 7 wt.% GNP-L loading from electron microscopy and X-ray diffraction (XRD), which was not in case of small-sized GNP. The young's modulus was enhanced 10% by GNP-S whereas 24% by GNP-L owing to its greater influence in crystallization. Large-sized GNP-incorporated PLA nanocomposites exhibited embrittlement and low electrical percolation threshold. Conductivity was enhanced while using GNP-S with respect to GNP-L and heat distortion temperature (HDT) was improved up to 13°C in case of large-sized GNP-incorporated nanocomposites. Interestingly, the effect of nanofiller size on the interfacial or interphase properties and thereby the effect of mechanical properties were discussed using different equations in another study [100]. The NPs having a radius of 10 nm and density of 2 g/cm^3 can generate 250 m^2 interfacial area in the polymeric system, and they pointed out that the tensile strength will increase with a decrease on nanofillers size. Additionally, they considered the agglomerated NPs as large-sized NPs and explained its negative effect on interphase properties as well as tensile strength even at higher thickness of interface. Even with the same thickness of interphase, the interphase volume fraction (ϕ_1) attained the higher value for small NPs and goes to zero value when the radius was higher than 30 nm indicating the role of particle size. Bi and coworkers [101] mentioned about the effect of nanofillers size on the dielectric performance and thereby energy storage of polyvinylidenefluoride (PVDF) nanocomposite. BT having 92.3, 17.8, and 5.9 nm sizes are incorporated separately into polymer matrix and noticed the formation of uniform microstructures in case of nanocomposite with small BT NPs. The melting temperature showed a slight increase from 164.02°C to 166.48°C at 20 vol.% when the size decreased from 92.3 to 5.9 nm indicating the ability of ultrafine fillers to encourage normal arrangement of polyvinylidenefluoride. Because of the lower intrinsic dielectric permittivity of ultrafine BT NPs, permittivity values decreased as filler size decreased. Because of the weakened concentration of electric filed present in interface and dielectric uniformity caused by the lower defect size of nanofillers, breakdown strength got increased to 220 kV/mm for nanocomposite loaded with 5.9 nm-sized filler from 80 kV/mm for nanocomposite loaded with 92.3 nm-sized filler. The greater thermal intensification was observed with larger NPs in another group of study [102]. Alumina NPs having 27–43, 30–40, 150, and 200 nm sizes were incorporated in polyester and epoxy resin matrix and an enhanced TC was noticed for nanocomposite with nanofillers of 200 and 150 nm sizes owing to the thermal resistance present in the interface between polymeric material and nanofillers or the generation of bigger agglomerates. For a constant matrix and volume concentration, the density results are nearly identical for all samples. Furthermore, they stated that variations in the geometry of NPs as well as interactions caused a higher thermal intensification for PNC with 27–43 nm NPs compared to those with 30–40 nm NPs.

2.6.2.3 Shape of the filler

A group of researchers described the fabrication and property studies of PLA nanocomposites containing GNP and MWCNT [68]. They investigated the effect of MWCNT and GNP filler shape on the thermal properties of PLA nanocomposites. TC of PLA/GNP composites increased by 181% for 6 wt.% GNP content, whereas thermal conductivity PLA/MWCNT increased by only 48% due to the shape of the MWCNT, which resulted in a poor thermal conduction network. They found that the thermal diffusivity of PLA/GNP nanocomposites increased by 214% when compared to PLA/MWCNT nanocomposites, which they attributed to the sheet structure of the GNP filler. Xia and colleagues [103] investigated the impact of nanoplatelet and nanosphere fillers on the mechanical properties of polymer nanocomposite by incorporating poly (ethylene glycol) (PEG) grafted silica nanosphere on PLA matrix (NS@PLA) and nanoplatelets on PLA matrix (NP@PLA). Uniform distribution of filler in polymeric material was confirmed using SEM and tensile tests were carried out in both NS@PLA and NP@PLA systems. Elongation at break value by NP@PLA was 279% enhanced compared to neat PLA for 2.4 vol.% NP loading whereas it is 185% enhanced in case of NS@PLA indicating the importance of filler shape. NP increased the shear yielding of PLA material by generating several dimples on the fracture surface. Because of the dense microcracks in NP@PLA, which were sparse in NS@PLA, nanoplatelets cause widespread plastic deformation of the polymeric matrix. The effect on storage modulus and creep behavior is similar. Fig. 2.14 represents the damage zone of both nanocomposites.

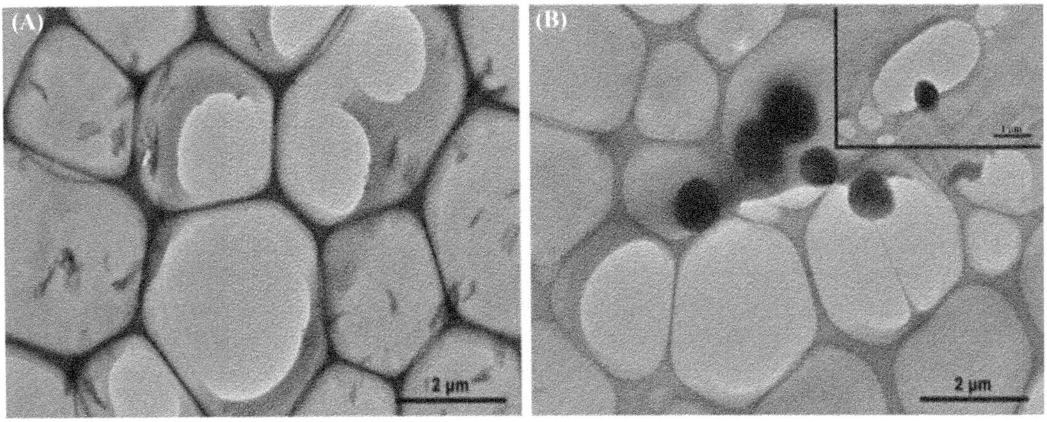

FIGURE 2.14

TEM images of damage zone of nanocomposites: (A) NP@PLA, (B) NS@PLA [103].

Reproduced with permission from Elsevier. Copyright 2019.

Gas transport properties, thermal stability and ultraviolet (UV) shielding was greatly influenced by rod-shaped ZnO NP incorporated in PLA nanocomposite (PLA/ZnO_r) compared to spherical-shaped NP incorporated in PLA matrix (PLA/ZnO_s) [104]. Thermal degradation temperature decreased from 330.8°C to 278.8°C in case of 10 wt.% rod-shaped ZnO NP-loaded PLA composite. ZnO NP has the ability to catalyze degradation reaction and was more pronounced in PLA/ZnO_r for amount up to 0.5 wt.%. TEM analysis revealed homogeneous dispersion of rod-shaped ZnO in the PLA material and aggregation of spherical-shaped ZnO in the matrix, resulting in enhanced UV-Vis shielding by PLA/ZnO_r. Gas transport properties improved even more when ZnO_r was added, leading to the conclusion that better dispersion leads to better properties. Cha and coworkers [105] compared the effects of CNT and GNP on the mechanical properties of epoxy composites. Due to the variation in the geometry of CNT and GNP, they observed an improvement in Young's modulus of 71%, tensile strength of 23%, and fracture toughness of 124% for nanocomposite incorporated with melamine functionalized GNP (M-GNP/epoxy). 2-D GNP has a larger interface than one-dimensional CNT, allowing for easy load transfer and the development of improved crack bridging. GNP with a higher aspect ratio and surface area generates more reinforcing effect than CNT, resulting in improved properties. Similarly, Li et al. [106] used molecular models generated with equal weight percentages of both fillers in the polymeric material to describe the greater improvement in mechanical properties provided by GNS compared to CNT. The pull-out and strain constant processes resulted in an 18% increase in Young's modulus, an 8.7% increase in tensile strength, and a 5% increase in surface crack energy for polymer matrix with GNS compared to those with CNT. Crack propagation is delayed in GNP/PMMA nanocomposite, as confirmed by van der Waals energy calculations, and increased interfacial shear stress indicates improved performance.

2.7 Polymer nanocomposite coatings

Ansari and team [107] created a nanocomposite coating of chitosan and hydroxyapatite (CS/HA) to improve in vitro bioactivity of Ti6Al4V alloy, which is important in hard tissue repair. The effect of HA on morphology and roughness, as well as contact angle and adhesion strength, was studied. They noticed a uniform HA dispersion for the sample containing 10 wt.% of HA along with reduced surface roughness, however increased adhesion strength along with improved surface roughness for sample containing 60 wt.% HA. Because of the generation of extra nucleation sites during simulated body fluid treatment (21 days), more apatite was precipitated with increased HA addition, indicating the bioactive nature of the prepared coating. The authors clearly explained the implanting applications of fabricated coatings. Similarly in another work [108], a PCL/fluoride substituted-hydroxyapatite (FHA) nanocomposite coating was created to improve the anticorrosive nature and in vitro bioactivity of the Ti6Al4V substrate, which can then be used for dental implant applications. They used an in-situ sol-gel process to create coatings with homogeneously dispersed FHA and no cracks. Similarly to the previous work, authors observed an increase in FHA amount to improve roughness, adhesion strength, hydrophilicity, and in vitro bioactivity of the coating. The sample containing 10% FHA had improved anticorrosive properties, and no cytotoxicity was detected using the MTT assay. pH level variations and Ca and P ions concentrations of coated

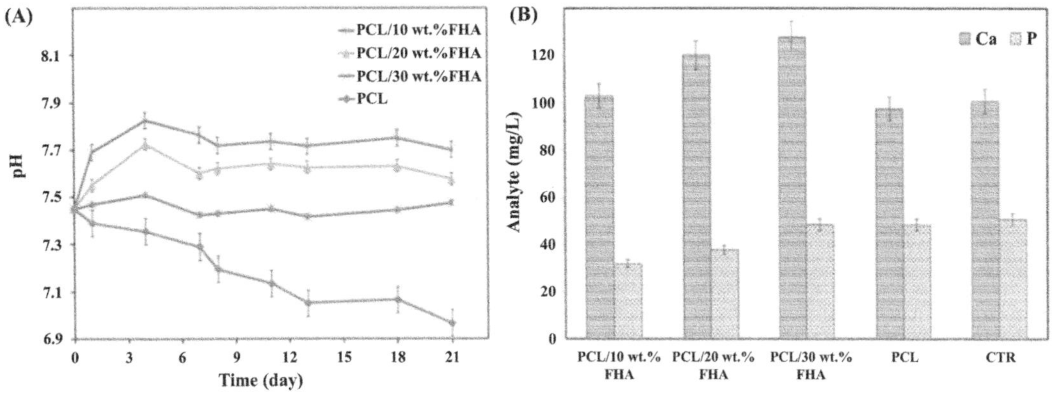

FIGURE 2.15

(A) pH of the SBF solution, after soaking the pure PCL and PCL/FHA coated specimens at different time intervals (B) Ca and P ions concentration of the pure PCL and PCL/FHA coated specimens as well as SBF solution as a control after 14 days [108].

Reproduced with permission from Elsevier. Copyright 2020.

specimens after SBF treatment are provided in Fig. 2.15. Parvizifard and group [109] described their efforts to create PHB-Chitosan/MWCNTs nanocomposite coatings that can be used to improve the properties of bioceramic scaffolds. They coated the prepared product on nanobioglass (nBG)-titania ($nTiO_2$) scaffolds created using the foam replication method, and structural analysis confirmed the presence of higher interconnected porous nature even with 1 wt.% loading of CNT. A 30-fold increase in compressive strength of a 1 wt.% filler containing nanocomposite coated scaffold was observed, as well as an improvement in surface roughness. Scaffolds coated with PHB-Cs/MWCNTs demonstrated reduced contact angle, increased pH, faster degradation, and increased apatite-like formation.

Another group of researchers described the incorporation of cellulose NPs (CNs) on biodegradable PLA coating to improve anticorrosive and mechanical properties [110]. They dip-coated all samples containing varying amounts of CNs on AZ31 magnesium alloy and tested their anticorrosive properties with SBF solution. PLA-CNs nanocomposite coating incorporated with 5 wt.% of CNs showed improved anticorrosion performance owing to increased corrosion potential as well as hydrophobic nature. The tendency of CNs to agglomerate and the decrease in the adhesive nature of the coating decreased the anticorrosive nature of the coating as CNs addition increased. The coating with CNs, on the other hand, had better hardness and mechanical behavior. Rajitha et al. investigated the corrosion application of PCL nanocomposite coating incorporated with graphene-based NPs that are GO and functionalized GO (FGO) [111]. They used 8-bromo- [1,2,4] triazolo [1,5-a]pyridin-2-amine (BTP) functionalization and both samples were coated on mild steel. Electrochemical impedance spectroscopy (EIS) and potentiodynamic polarization measurements confirmed that the FGO-PCL nanocomposite coating has a better anticorrosive nature, as well as improved barrier properties due to its better dispersion in the polymer system. They also noticed that the FGO nanosheet improved the adhesion strength of the PCL system.

2.8 Conclusion and future perspectives

In an era when polymer nanocomposites with competing properties and applications are being developed, researchers are being forced to delve deeply into the various processing methods and processing conditions to improve the properties of polymer nanocomposites. Various properties such as thermal, mechanical, and antibacterial properties, particularly in the case of biodegradable and biocompatible polymer nanocomposites, were attempted, and the factors influencing these properties are discussed in this chapter. The majority of the properties are improved by carefully implementing different processing conditions such as mixing/rotation time, temperature, and speed. The incorporation of nanofillers of varying nature, size, and shape can significantly improve the properties. Many studies are now focusing on the effect of graphitic nanomaterial size and shape on the properties of nanocomposites. However, there has been little effort to understand the influence of processing parameters and nanofiller structure on property variation in biodegradable and biocompatible polymer nanocomposites, which could be an interesting research avenue in the future. As a result, there is much more to be discovered in the field of polymer nanocomposites to scale them up for industrial applications. Theoretical modeling and extensive research will be useful in determining the best parameters, allowing nanocomposites to be tailored for specific applications in the future.

Acknowledgement

The authors would like to thank Amrita VishwaVidyapeetham, Amritapuri for the funding.

References

[1] S.S. Bari, A. Chatterjee, S. Mishra, Biodegradable polymer nanocomposites: an overview, Polym. Rev. 56 (2016) 287–328.
[2] S.K. Pillalamarri, F.D. Blum, A.T. Tokuhiro, M.F. Bertino, One-pot synthesis of polyaniline – metal nanocomposites, Chem. Mater. 17 (2005) 5941–5944.
[3] M. Sternitzke, Structural ceramic nanocomposites, J. Eur. Ceram. Soc. 17 (1997) 1061–1082.
[4] H. Kim, A.A. Abdala, C.W. Macosko, Graphene/polymer nanocomposites, Macromolecules 43 (2010) 6515–6530.
[5] M. Haq, R. Burgueño, A.K. Mohanty, M. Misra, Bio-based polymer nanocomposites from UPE/EML blends and nanoclay: development, experimental characterization and limits to synergistic performance, Compos. A Appl. Sci. Manuf. 42 (2011) 41–49.
[6] Y. Guo, X. Zuo, Y. Xue, et al., Engineering thermally and electrically conductive biodegradable polymer nanocomposites, Compos. B Eng. 189 (2020) 107905.
[7] K.H. Wang, M.H. Choi, C.M. Koo, et al., Synthesis and characterization of maleated polyethylene/clay nanocomposites, Polym. (Guildf.) 42 (2001) 9819–9826.
[8] S. Su, D.D. Jiang, C.A. Wilkie, Polybutadiene-modified clay and its nanocomposites, Polym. Degrad. Stab. 84 (2004) 279–288.
[9] X. Fu, S. Qutubuddin, Synthesis of polystyrene–clay nanocomposites, Mater. Lett. 42 (2000) 12–15.

[10] R.A. Hule, D.J. Pochan, Polymer nanocomposites for biomedical applications, MRS Bull. 32 (2007) 354–358.

[11] S.C. Tjong, Structural and mechanical properties of polymer nanocomposites, Mater. Sci. Eng. R. Rep. 53 (2006) 73–197.

[12] W. Peng, S. Rhim, Y. Zare, K.Y. Rhee, Effect of "Z" factor for strength of interphase layers on the tensile strength of polymer nanocomposites, Polym. Compos. 40 (2019) 1117–1122.

[13] V.H. DeLeon, T.D. Nguyen, M. Nar, et al., Polymer nanocomposites for improved drug delivery efficiency, Mater. Chem. Phys. 132 (2012) 409–415.

[14] M. Silva, F.N. Ferreira, N.M. Alves, M.C. Paiva, Biodegradable polymer nanocomposites for ligament/tendon tissue engineering, J. Nanobiotechnol. 18 (2020) 1–33.

[15] A.J. Monjarás-Ávila, G. Sanchez-Olivares, F. Calderas, et al., Sodium montmorillonite concentration effect on Bis-GMA/TEGDMA resin to prepare clay polymer nanocomposites for dental applications, Appl. Clay Sci. 196 (2020) 105755.

[16] N. Tripathi, A.K. Thakur, A. Shukla, D.T. Marx, Dielectric, transport and thermal properties of clay based polymer-nanocomposites, Polym. Eng. Sci. 58 (2018) 220–227.

[17] A. Dufresne, Cellulose-based composites and nanocomposites, Monomers, Polymers and Composites from Renewable Resources, Elsevier, 2008, pp. 401–418.

[18] J.R. Potts, D.R. Dreyer, C.W. Bielawski, R.S. Ruoff, Graphene-based polymer nanocomposites, Polym. (Guildf.) 52 (2011) 5–25.

[19] E.I. Akpan, X. Shen, B. Wetzel, K. Friedrich, Design and synthesis of polymer nanocomposites, Polymer Composites with Functionalized Nanoparticles, Elsevier, 2019, pp. 47–83.

[20] B. Sharma, P. Malik, P. Jain, Biopolymer reinforced nanocomposites: a comprehensive review, Mater. Today Commun. 16 (2018) 353–363.

[21] K.I. Winey, R.A. Vaia, Polymer nanocomposites, MRS Bull. 32 (2007) 314–322.

[22] A.J. Crosby, J. Lee, Polymer nanocomposites: the "nano" effect on mechanical properties, Polym. Rev. 47 (2007) 217–229.

[23] A.S. Sethulekshmi, J.S. Jayan, S. Appukuttan, K. Joseph, Insights into the Reinforcibility and multifarious role of WS2 in polymer matrix, J. Alloy. Compd. 160107 (2021).

[24] J.S. Jayan, S. Appukuttan, K. Joseph, MoS_2: advanced nanofiller for reinforcing polymer matrix, Phys. E Low-Dimensional Syst. Nanostruct. 114716 (2021).

[25] A. Ohlan, K. Singh, A. Chandra, et al., Conjugated polymer nanocomposites: synthesis, dielectric, and microwave absorption studies, J. Appl. Phys. 106 (2009) 44305.

[26] F. Vitale, L. Mirenghi, E. Piscopiello, et al., Gold nanoclusters–organometallic polymer nanocomposites: synthesis and characterization, Mater. Sci. Eng. C. 27 (2007) 1300–1304.

[27] A. Matei, I. Cernica, O. Cadar, et al., Synthesis and characterization of ZnO–polymer nanocomposites, Int. J. Mater. Form. 1 (2008) 767–770.

[28] J.L. Wilson, P. Poddar, N.A. Frey, et al., Synthesis and magnetic properties of polymer nanocomposites with embedded iron nanoparticles, J. Appl. Phys. 95 (2004) 1439–1443.

[29] Y.K. Mishra, S. Mohapatra, V.S.K. Chakravadhanula, et al., Synthesis and characterization of Ag-polymer nanocomposites, J. Nanosci. Nanotechnol. 10 (2010) 2833–2837.

[30] P.A. McCarthy, J. Huang, S.-C. Yang, H.-L. Wang, Synthesis and characterization of water-soluble chiral conducting polymer nanocomposites, Langmuir 18 (2002) 259–263.

[31] G.A. Gelves, M.H. Al-Saleh, U. Sundararaj, Highly electrically conductive and high performance EMI shielding nanowire/polymer nanocomposites by miscible mixing and precipitation, J. Mater. Chem. 21 (2011) 829–836.

[32] P. Pascariu, A. Airinei, M. Grigoras, et al., Metal–polymer nanocomposites based on Ni nanoparticles and polythiophene obtained by electrochemical method, Appl. Surf. Sci. 352 (2015) 95–102.

[33] E.N. Zare, M.M. Lakouraj, Biodegradable polyaniline/dextrin conductive nanocomposites: synthesis, characterization, and study of antioxidant activity and sorption of heavy metal ions, Iran. Polym. J. 23 (2014) 257–266.

[34] D. Ponnamma, M.A.A. Al-Maadeed, Influence of $BaTiO_3$/white graphene filler synergy on the energy harvesting performance of a piezoelectric polymer nanocomposite, Sustain. Energy Fuels 3 (2019) 774–785.

[35] J. Gou, J. Zhuge, F. Liang, Processing of polymer nanocomposites, Manufacturing Techniques for Polymer Matrix Composites (PMCs), Elsevier, 2012, pp. 95–119.

[36] K. Müller, E. Bugnicourt, M. Latorre, et al., Review on the processing and properties of polymer nanocomposites and nanocoatings and their applications in the packaging, automotive and solar energy fields, Nanomaterials 7 (2017) 74.

[37] S.K. Sukumaran, T. Kobayashi, S. Takeda, et al., Electrical conductivity and linear rheology of multiwalled carbon nanotube/acrylonitrile butadiene styrene polymer nanocomposites prepared by melt mixing and solution casting, J. Electrochem. Soc. 166 (2019) B3091.

[38] M. Kamkar, S.M.N. Sultana, S.P. Pawar, et al., The key role of processing in tuning nonlinear viscoelastic properties and microwave absorption in CNT-based polymer nanocomposites, Mater. Today Commun. 24 (2020). 101010.

[39] Y. Di, S. Iannace, E. Di Maio, L. Nicolais, Nanocomposites by melt intercalation based on polycaprolactone and organoclay, J. Polym. Sci. B Polym. Phys. 41 (2003) 670–678.

[40] M. El Achaby, F. Arrakhiz, S. Vaudreuil, et al., Mechanical, thermal, and rheological properties of graphene-based polypropylene nanocomposites prepared by melt mixing, Polym. Compos. 33 (2012) 733–744.

[41] D.N. Bikiaris, A. Vassiliou, E. Pavlidou, G.P. Karayannidis, Compatibilisation effect of PP-g-MA copolymer on iPP/SiO_2 nanocomposites prepared by melt mixing, Eur. Polym. J. 41 (2005) 1965–1978.

[42] P. Noorunnisa Khanam, M.A. AlMaadeed, M. Ouederni, et al., Effect of two types of graphene nanoplatelets on the physico–mechanical properties of linear low-density polyethylene composites, Adv. Manuf. Polym. Compos. Sci. 2 (2016) 67–73.

[43] M.T. Byrne, Y.K. Gun'ko, Recent advances in research on carbon nanotube–polymer composites, Adv. Mater. 22 (2010) 1672–1688.

[44] A. Saritha, K. Joseph, Role of solvent interaction parameters in tailoring the properties of chlorobutyl rubber nanocomposites, Polym. Compos. 37 (2016) 353–359.

[45] C.-Y. Tsai, S.-Y. Lin, H.-C. Tsai, Butyl rubber nanocomposites with monolayer MoS_2 additives: structural characteristics, enhanced mechanical, and gas barrier properties, Polymers (Basel), 10, 2018, p. 238.

[46] C.A. Mitchell, R. Krishnamoorti, Non-isothermal crystallization of in situ polymerized poly (ε-caprolactone) functionalized-SWNT nanocomposites, Polym. (Guildf.) 46 (2005) 8796–8804.

[47] T. Kuila, S. Bose, P. Khanra, et al., Characterization and properties of in situ emulsion polymerized poly (methyl methacrylate)/graphene nanocomposites, Compos. A Appl. Sci. Manuf. 42 (2011) 1856–1861.

[48] M.T. Ramesan, V. Santhi, B.K. Bahuleyan, M.A. Al-Maghrabi, Structural characterization, material properties and sensor application study of in situ polymerized polypyrrole/silver doped titanium dioxide nanocomposites, Mater. Chem. Phys. 211 (2018) 343–354.

[49] S.N. Tripathi, P. Saini, D. Gupta, V. Choudhary, Electrical and mechanical properties of PMMA/reduced graphene oxide nanocomposites prepared via in situ polymerization, J. Mater. Sci. 48 (2013) 6223–6232.

[50] M. Trujillo, M.L. Arnal, A.J. Müller, et al., Thermal and morphological characterization of nanocomposites prepared by in-situ polymerization of high-density polyethylene on carbon nanotubes, Macromolecules 40 (2007) 6268–6276.

[51] M. Okamoto, S. Morita, H. Taguchi, et al., Synthesis and structure of smectic clay/poly (methyl methacrylate) and clay/polystyrene nanocomposites via in situ intercalative polymerization, Polym. (Guildf.) 41 (2000) 3887–3890.

[52] L.L.C. Wong, S. Barg, A. Menner, et al., Macroporous polymer nanocomposites synthesised from high internal phase emulsion templates stabilised by reduced graphene oxide, Polym. (Guildf.) 55 (2014) 395–402.

[53] D. Gan, L. Han, M. Wang, et al., Conductive and tough hydrogels based on biopolymer molecular templates for controlling in situ formation of polypyrrole nanorods, ACS Appl. Mater. Interfaces 10 (2018) 36218–36228.

[54] T.A. Hassan, V.K. Rangari, S. Jeelani, Value-added biopolymer nanocomposites from waste eggshell-based $CaCO_3$ nanoparticles as fillers, ACS Sustain Chem. Eng. 2 (2014) 706–717.

[55] M. El Achaby, Z. Kassab, A. Aboulkas, et al., Reuse of red algae waste for the production of cellulose nanocrystals and its application in polymer nanocomposites, Int. J. Biol. Macromol. 106 (2018) 681–691.

[56] U.V. Adegbotolu, J. Njuguna, P. Pollard, et al., Waste to want: polymer nanocomposites using nanoclays extracted from oil based drilling mud waste, in: IOP Conference Series: Materials Science and Engineering. IOP Publishing, 2014, p. 12023.

[57] A. Verma, K. Baurai, M.R. Sanjay, S. Siengchin, Mechanical, microstructural, and thermal characterization insights of pyrolyzed carbon black from waste tires reinforced epoxy nanocomposites for coating application, Polym. Compos. 41 (2020) 338–349.

[58] P.L. Reddy, K. Deshmukh, T. Kovářík, et al., Green chemistry mediated synthesis of cadmium sulphide/polyvinyl alcohol nanocomposites: assessment of microstructural, thermal, and dielectric properties, Polym. Compos. 41 (2020) 2054–2067.

[59] T. Paramitha, V. Wonoputri, D.S.D. Sitompul, et al., Properties of clays reinforced PLA nanocomposites by melt extrusion technique, Malaysian J. Fundam. Appl. Sci. 16 (2020) 453–457.

[60] B. Zhu, Y. Wang, H. Liu, et al., Effects of interface interaction and microphase dispersion on the mechanical properties of PCL/PLA/MMT nanocomposites visualized by nanomechanical mapping, Compos. Sci. Technol. 190 (2020) 108048.

[61] J. Dong, W. Zhou, Y. Su, X. Ma, Enhanced mechanical, thermal, and barrier properties of poly (3-hydroxybutyrate-co-3-hydroxyhexanoate)/montmorillonite nanocomposites using silane coupling agent, Polym. Compos. 41 (2020) 4538–4549.

[62] P. Chen, X. Liang, Y. Xu, et al., Enhanced thermal and mechanical properties of PLA/MoS_2 nanocomposites synthesized via the in-situ ring-opening polymerization, Appl. Surf. Sci. 440 (2018) 1143–1149.

[63] B. Zhu, X. Wang, Q. Zeng, et al., Enhanced mechanical properties of biodegradable poly (ε-caprolactone)/cellulose acetate butyrate nanocomposites filled with organoclay, Compos. Commun. 13 (2019) 70–74.

[64] A.N. Frone, D. Batalu, I. Chiulan, et al., Morpho-Structural, thermal and mechanical properties of PLA/PHB/cellulose biodegradable nanocomposites obtained by compression molding, extrusion, and 3D printing, Nanomaterials 10 (2020) 51.

[65] I.F. Pinheiro, F.V. Ferreira, G.F. Alves, et al., Biodegradable PBAT-based nanocomposites reinforced with functionalized cellulose nanocrystals from Pseudobombax munguba: rheological, thermal, mechanical and biodegradability properties, J. Polym. Env. 27 (2019) 757–766.

[66] T. Remiš, P. Bělský, T. Kovářík, et al., Study on structure, thermal behavior and viscoelastic properties of nanodiamond-reinforced poly (vinyl alcohol) nanocomposites, Polymers (Basel), 13, 2021, p. 1426.

[67] X. Xu, J. Chen, J. Zhou, B. Li, Thermal conductivity of polymers and their nanocomposites, Adv. Mater. 30 (2018) 1705544.

[68] E. Ivanov, R. Kotsilkova, H. Xia, et al., PLA/graphene/MWCNT composites with improved electrical and thermal properties suitable for FDM 3D printing applications, Appl. Sci. 9 (2019) 1209.

[69] J. Lee, C.Y. Koh, J.P. Singer, et al., 25th anniversary article: ordered polymer structures for the engineering of photons and phonons, Adv. Mater. 26 (2014) 532–569.
[70] K. Bicy, V.G. Geethamma, N. Kalarikkal, et al., Poly(ε-caprolactone)/functionalized-carbon nanotube electrospun nanocomposites: crystallization and thermal properties, Macromolecular Symposia, Wiley Online Library, 2018, p. 1800140.
[71] A. Jayakumar, K. Prabhu, L. Shah, P. Radha, Biologically and environmentally benign approach for PHB-silver nanocomposite synthesis and its characterization, Polym. Test. 81 (2020) 106197.
[72] Y. Zheng, G. Fu, B. Wang, et al., Physico-chemical, thermal, and mechanical properties of PLA-nHA nanocomposites: effect of glass fiber reinforcement, J. Appl. Polym. Sci. 137 (2020) 49286.
[73] Y. Luo, H.Y. Liu, G.Z. Zhang, et al., Morphological and thermal properties of PLA/OMMT nanocomposites prepared via vane extruder, in: IOP Conference Series: Materials Science and Engineering. IOP Publishing, 2017, p. 12018.
[74] Y.S. Wang, M.A. O'gurkis, J.T. Lindt, Electrical properties of exfoliated-graphite filled polyethylene composites, Polym. Compos. 7 (1986) 349–354.
[75] Y. Nakaramontri, C. Nakason, C. Kummerlöwe, N. Vennemann, Enhancement of electrical conductivity and filler dispersion of carbon nanotube filled natural rubber composites by latex mixing and in situ silanization, Rubber Chem. Technol. 89 (2016) 272–291.
[76] S.A. Hawkins, H. Yao, H. Wang, H.-J. Sue, Tensile properties and electrical conductivity of epoxy composite thin films containing zinc oxide quantum dots and multi-walled carbon nanotubes, Carbon N. Y. 115 (2017) 18–27.
[77] A.H. Kumar, M.B. Ahamed, K. Deshmukh, M.S. Sirajuddeen, Morphology, dielectric and emi shielding characteristics of graphene nanoplatelets, montmorillonite nanoclay and titanium dioxide nanoparticles reinforced polyvinylidenefluoride nanocomposites, J. Inorg. Organomet. Polym. Mater. 31 (2021) 2003–2016.
[78] W. Wu, T. Liu, D. Zhang, et al., Significantly improved dielectric properties of polylactide nanocomposites via TiO_2 decorated carbon nanotubes, Compos. A Appl. Sci. Manuf. 127 (2019) 105650.
[79] P. Rani, B. Ahamed, K. Deshmukh, Dielectric and electromagnetic interference shielding properties of zeolite 13X and carbon black nanoparticles based PVDF nanocomposites, J. Appl. Polym. Sci. 138 (2021) 50107.
[80] P. Rani, M.B. Ahamed, K. Deshmukh, Structural, dielectric and EMI shielding properties of polyvinyl alcohol/chitosan blend nanocomposites integrated with graphite oxide and nickel oxide nanofillers, J. Mater. Sci. Mater. Electron. 32 (2021) 764–779.
[81] R. Venkatesan, N. Rajeswari, A. Tamilselvi, Antimicrobial, mechanical, barrier, and thermal properties of bio-based poly (butylene adipate-co-terephthalate)(PBAT)/Ag_2O nanocomposite films for packaging application, Polym. Adv. Technol. 29 (2018) 61–68.
[82] D. Sun, J. Turner, N. Jiang, et al., Atmospheric pressure microplasma for antibacterial silver nanoparticle/chitosan nanocomposites with tailored properties, Compos. Sci. Technol. 186 (2020) 107911.
[83] S. Shankar, L.-F. Wang, J.-W. Rhim, Incorporation of zinc oxide nanoparticles improved the mechanical, water vapor barrier, UV-light barrier, and antibacterial properties of PLA-based nanocomposite films, Mater. Sci. Eng. C. 93 (2018) 289–298.
[84] C. Rodrigues, D. Francieli, M.D.L. Pier, et al., Mechanical, thermal and antimicrobial properties of chitosan-based-nanocomposite with potential applications for food packaging, J. Polym. Env. 28 (2020) 1216–1236.
[85] H. Deka, N. Karak, R.D. Kalita, A.K. Buragohain, Bio-based thermostable, biodegradable and biocompatible hyperbranched polyurethane/Ag nanocomposites with antimicrobial activity, Polym. Degrad. Stab. 95 (2010) 1509–1517.
[86] S. Fu, Z. Sun, P. Huang, et al., Some basic aspects of polymer nanocomposites: a critical review, Nano Mater. Sci. 1 (2019) 2–30.

[87] R. Verma, M.J. Rathod, R.K. Goyal, A review on the effect of processing techniques and functionalization of filler on mechanical properties of polymer nanocomposites, in: IOP Conference Series: Materials Science and Engineering, 2020, IOP Publishing, p. 12031.

[88] A. D'Anna, R. Arrigo, A. Frache, Rheology, morphology and thermal properties of a PLA/PHB/clay blend nanocomposite: the influence of process parameters, J. Polym. Env. (2021) 1–12.

[89] S. Colonna, M.M. Bernal, G. Gavoci, et al., Effect of processing conditions on the thermal and electrical conductivity of poly (butylene terephthalate) nanocomposites prepared via ring-opening polymerization, Mater. Des. 119 (2017) 124–132.

[90] B. Coppola, N. Cappetti, L. Di Maio, et al., 3D printing of PLA/clay nanocomposites: influence of printing temperature on printed samples properties, Materials (Basel), 11, 2018, p. 1947.

[91] E. Enqvist, D. Ramanenka, P.A.A.P. Marques, et al., The effect of ball milling time and rotational speed on ultra high molecular weight polyethylene reinforced with multiwalled carbon nanotubes, Polym. Compos. 37 (2016) 1128–1136.

[92] W. Yang, J. Xu, L. Niu, et al., Effects of high energy ball milling on mechanical and interfacial properties of PBT/nano-Sb_2O_3 composites, J. Adhes. Sci. Technol. 32 (2018) 291–301.

[93] E. Garofalo, P. Scarfato, L. Di Maio, L. Incarnato, Tuning of co-extrusion processing conditions and film layout to optimize the performances of PA/PE multilayer nanocomposite films for food packaging, Polym. Compos. 39 (2018) 3157–3167.

[94] A.-H.I. Mourad, M.S. Mozumder, A. Mairpady, et al., On the injection molding processing parameters of HDPE-TiO_2 nanocomposites, Mater. (Basel) 10 (2017) 85.

[95] O.M. Sanusi, A. Benelfellah, D.N. Bikiaris, N. Ait Hocine, Effect of rigid nanoparticles and preparation techniques on the performances of poly (lactic acid) nanocomposites: a review, Polym. Adv. Technol. 32 (2021) 444–460.

[96] J. Liang, Y. Huang, L. Zhang, et al., Molecular-level dispersion of graphene into poly (vinyl alcohol) and effective reinforcement of their nanocomposites, Adv. Funct. Mater. 19 (2009) 2297–2302.

[97] K. Zhou, S. Jiang, C. Bao, et al., Preparation of poly (vinyl alcohol) nanocomposites with molybdenum disulfide (MoS_2): structural characteristics and markedly enhanced properties, Rsc Adv. 2 (2012) 11695–11703.

[98] S. Sarikaya, T.C. Henry, M. Naraghi, Graphene size and morphology: peculiar effects on damping properties of polymer nanocomposites, Exp. Mech. 60 (2020) 753–762.

[99] Y. Gao, O.T. Picot, E. Bilotti, T. Peijs, Influence of filler size on the properties of poly (lactic acid) (PLA)/graphene nanoplatelet (GNP) nanocomposites, Eur. Polym. J. 86 (2017) 117–131.

[100] M.A. Ashraf, W. Peng, Y. Zare, K.Y. Rhee, Effects of size and aggregation/agglomeration of nanoparticles on the interfacial/interphase properties and tensile strength of polymer nanocomposites, Nanoscale Res. Lett. 13 (2018) 1–7.

[101] M. Bi, Y. Hao, J. Zhang, et al., Particle size effect of $BaTiO_3$ nanofillers on the energy storage performance of polymer nanocomposites, Nanoscale 9 (2017) 16386–16395.

[102] D.C. Moreira, N.R. Braga Jr, L.A. Sphaier, L.C.S. Nunes, Size effect on the thermal intensification of alumina-filled nanocomposites, J. Compos. Mater. 50 (2016) 3699–3707.

[103] S. Xia, X. Liu, J. Wang, et al., Role of poly (ethylene glycol) grafted silica nanoparticle shape in toughened PLA-matrix nanocomposites, Compos. B Eng. 168 (2019) 398–405.

[104] E. Lizundia, M.C. Penayo, A. Guinault, et al., Impact of ZnO nanoparticle morphology on relaxation and transport properties of PLA nanocomposites, Polym. Test. 75 (2019) 175–184.

[105] J. Cha, J. Kim, S. Ryu, S.H. Hong, Comparison to mechanical properties of epoxy nanocomposites reinforced by functionalized carbon nanotubes and graphene nanoplatelets, Compos. B Eng. 162 (2019) 283–288.

[106] Y. Li, S. Wang, Q. Wang, M. Xing, A comparison study on mechanical properties of polymer composites reinforced by carbon nanotubes and graphene sheet, Compos. B Eng. 133 (2018) 35–41.

[107] Z. Ansari, M. Kalantar, A. Soriente, et al., In-situ synthesis and characterization of chitosan/hydroxyapatite nanocomposite coatings to improve the bioactive properties of Ti_6Al_4V substrates, Mater. (Basel) 13 (2020) 3772.

[108] Z. Ansari, M. Kalantar, M. Kharaziha, et al., Polycaprolactone/fluoride substituted-hydroxyapatite (PCL/FHA) nanocomposite coatings prepared by in-situ sol-gel process for dental implant applications, Prog. Org. Coat. 147 (2020) 105873.

[109] M. Parvizifard, S. Karbasi, Physical, mechanical and biological performance of PHB-Chitosan/MWCNTs nanocomposite coating deposited on bioglass based scaffold: potential application in bone tissue engineering, Int. J. Biol. Macromol. 152 (2020) 645–662.

[110] N. Rahimi Roshan, H. Hassannejad, A. Nouri, Corrosion and mechanical behaviour of biodegradable PLA-cellulose nanocomposite coating on AZ31 magnesium alloy, Surf. Eng. 37 (2021) 236–245.

[111] K. Rajitha, K.N. Mohana, Application of modified graphene oxide–polycaprolactone nanocomposite coating for corrosion control of mild steel in saline medium, Mater. Chem. Phys. 241 (2020) 122050.

CHAPTER

3

Spectroscopic and microscopic analysis of biodegradable and biocompatible polymer nanocomposites

Sithara Raj, Alice Alex, Aleesha K. Paulose, Shifana Shareef, Sunish K. Sugunan, Gigi George and Laly A. Pothan

Department of Chemistry, CMS College, Kottayam, Kerala, India

Chapter Outline

Biodegradable and Biocompatible Polymer Nanocomposites. DOI: https://doi.org/10.1016/B978-0-323-91696-7.00013-1
Copyright © 2023 Elsevier Inc. All rights reserved.

3.1 Introduction

Polymer nanocomposites have emerged as a vibrant research theme in the field of materials science and engineering over the last two decades. Polymer nanocomposites are materials created by fusing polymers and nanomaterials. In general, the incorporation of nanomaterials in polymer matrices has a synergistic effect that leads to an improvement in the mechanical, thermal, and electrical properties of the individual constituents of the composite, which frequently present new sets of properties. In terms of engineering properties, polymer nanocomposites typically outperform traditional filled and composite polymeric systems [1]. Polymer nanocomposites containing nanofillers perform better, owing to a reduction in filler size and an increase in interfacial area.

Among the various categories of polymer nanocomposites, biodegradable and biocompatible nanocomposites (BPNCs) based on biopolymers, which are derived from green resources, have of late garnered much attention owing to their biocompatibility, environmental friendliness, and exotic physical and mechanical properties. As such, numerous naturally occurring materials have been used to develop BPNCs. BPNCs based on cellulose are good examples, given their low cost of production, abundance, and biodegradability [2,3]. In general, BPNCs are created by incorporating nanofillers into biopolymers while maintaining effective biodegradability and biocompatibility [4]. To name a few, BPNCs play important roles in biomedical applications, as nanocarriers, in microbial treatments and tissue engineering, and in electronics. Because of their exceptional effectiveness, BPNCs have been widely used in drug delivery applications [5]. Spectroscopic and microscopic techniques enable the characterization and in-depth analysis of polymer nanocomposites at the molecular level. Spectroscopic characterization at the molecular level aids in understanding composite properties down to quantum dimensions, and thus aids in material development [6]. Microscopic techniques, on the other hand, allow for a thorough examination and comprehension of material properties over a wide magnification range. Microscopic methods can provide certain details, mostly related to surface and morphological properties, by obtaining an image of the substance at the required resolution and examining the sample's micro or nanoscale structure [7]. The goal of this book chapter is to show how spectroscopic and microscopic methods can be used to analyze the molecular and bulk level properties of BPNCs. Several leading articles are cited in this review, which will not only help to understand how recent developments and the use of various spectroscopic and microscopic methods define the direction of research in BPNCs, but will also help to understand how information gained from these studies is fundamental and practical in the development of BPNCs with appealing properties.

3.2 Spectroscopic analysis

Spectroscopy is a versatile technique that makes use of light's interaction with matter. According to the energy conservation principle, each quantum of photons consumed or radiated from a sample matrix must correspond to a finite physical or chemical change in the sample. These changes are typically centered on parameters that describe sample characteristics ranging from the quantum regime to the ensemble level. As a result, accurate observations of physical or chemical changes in a sample exposed to electromagnetic radiation can provide insights into the details of its structure-property relationship. According to the findings reported in a number of research articles, combinations of various spectroscopic techniques, when used either alone or in conjunction with other analysis techniques, are potential tools for probing the relationship between material properties and their structure, composition, and morphology.

Given the growing interest in BPNCs, it is critical to fully comprehend their properties. Among the many techniques available, spectroscopy stands out as a sensitive and adaptable method for characterizing the properties of polymers, fillers, and nanocomposites. Spectroscopy is useful for revealing both steady-state and dynamic properties at the mesoscopic and molecular levels, and it aids in the development of polymer nanocomposites with the desired structure, morphology, and properties [6]. Among the various spectroscopic methods available for detailed characterization of BPNCs are [8]. This chapter will go over techniques like Atomic Absorption Spectroscopy (AAS), Circular Dichroism (CD) Spectroscopy, Electron Paramagnetic Resonance (EPR) Spectroscopy, Atomic Emission Spectroscopy (AES), Energy-Dispersive Spectroscopy (EDS), Fluorescence Spectroscopy, Fourier Transform Infrared (FTIR) Spectroscopy, Gamma-Ray Spectroscopy.

3.2.1 Atomic absorption spectroscopy

Atomic absorption spectroscopy (AAS) is a popular quantitative technique for analyzing metallic and nonmetallic components. Through two steps, the techniques employ the principle of measuring the amount of an absorbing species formed at an analytical wavelength. This entails atomization of the analyte molecule and subsequent radiation absorption. Atomization is accomplished primarily through the use of flame or electrochemical methods. Because of its selectivity, sensitivity, and longer pathlength, graphite furnaces are primarily used for atomization [9]. An AAS system, represented in Fig. 3.1, usually

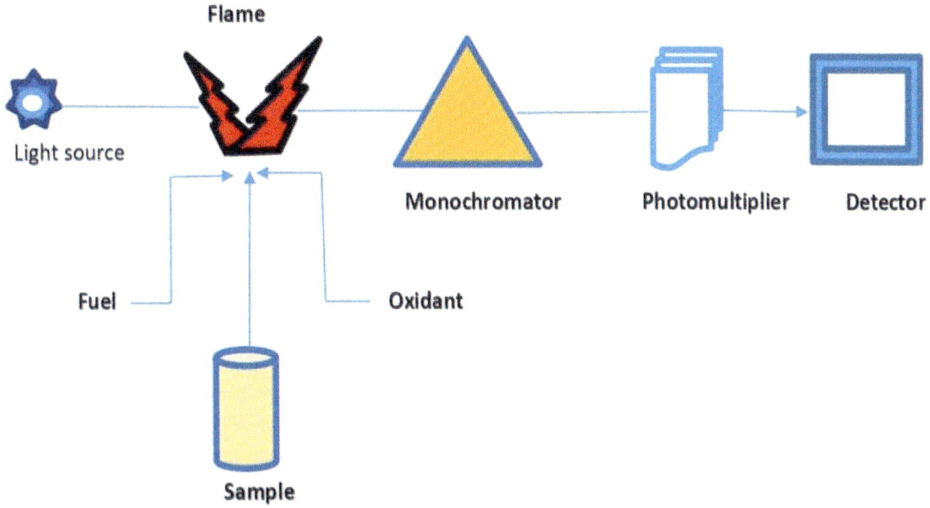

FIGURE 3.1

Working principle of atomic absorption spectroscopy.

comprises an empty cathode light, a pneumatic nebulizer with low flame which separates the sample into its constituent atoms, an ordinary grating monochromator, and a photomultiplier tube detector [10].

Following electrothermal atomization, some of the atoms absorb energy from the characteristic radiation of a suitable frequency that passes through the atoms' flame and are excited to higher electronic energy levels, while the majority of the atoms remain in the ground state. AAS is fundamentally based on Beer-Law, Lambert's which relates the extent of atom light absorption to atom concentration in the sample. Here are some recent studies that have used AAS as a diagnostic tool in analyzing the properties of BPNCs.

Faycal et al. created a biodegradable nanocomposite with antibacterial activity for use in food packaging [11]. The antibacterial film was created using a solvent casting method with poly(caprolactone) (PCL) and silver loaded kaolinite (PCL/Ag-KT). In a slightly acidic aqueous solution, the as-made PCL/Ag-KT film demonstrated excellent antibacterial activity. According to the research, silver ions are responsible for the antibacterial activity. AAS was used to monitor the controlled release of silver ions from the films into the aqueous medium, which was discovered to be diffusion-controlled in solution. Shameli et al. created a silver/poly(lactic acid) nanocomposite by reacting the nanoparticles in PLA with sodium borohydride as the reducing agent [12]. The Ag/PLA-NC film was found to have potent antibacterial properties against both gram-positive and gram-negative bacteria. The studies revealed that Ag/PLA-NC films have a wide range of medical applications. The AAS analysis of this system revealed that it causes the stable and continuous liberation of silver ions, which inhibits bacterial growth.

3.2.2 Circular dichroism spectroscopy

Circular dichroism (CD) spectroscopy is a powerful tool for examining the similarity of molecules, chiral discrimination, protein and nucleic acid conformation analysis, structural analysis of biomolecules, and other applications [10,13,14]. CD spectroscopy takes advantage of a sample's unequal absorption of right and left plane polarized light, resulting in circularly polarized light. Seema et al. created a nanocomposite out of montmorillonite (Mt) and polylactic-*co*-glycolic acid with the intention of using it as an insulin carrier [15]. The authors discovered that Mt-PLGA exhibited remarkable low pH stability and muco-adhesivity, which aided in better insulin oral drug delivery. The CD spectra of native insulin, insulin-PLGA, and Mt-insulin-PLGA released from PLGA and Mt-PLGA resembled the spectrum of native insulin, with its characteristic peaks at 210 and 225 nm due to the α-helical structure, which confirmed that the secondary structure of insulin was preserved during the stages of synthesis of the nanocomposite.

Vibrational circular dichroism (VCD) is the differential absorption of left circularly polarized light and right circularly polarized light in the infrared spectral region (4000–800 cm^{-1}) [16]. This method has been used successfully to determine the absolute configuration of a wide range of medium-sized chiral molecules. The VCD spectrometers available today are mainly based on FTIR methods using a Michelson interferometer, which are capable of producing high quality spectral data in the mid-IR region (2000–800 cm^{-1}) with excellent spectral resolution (~4 cm^{-1}). Fig. 3.2 depicts the schematic diagram of a VCD spectrometer. Only IR light in a specific range is transmitted by the optical fillers, and it is then polarized as it passes through a wire grid polarizer. The radiation is then passed through a photoelastic modulator (PEM), which generates elliptically polarized radiation by mixing left and right circularly polarized infrared light. The differentially absorbed left

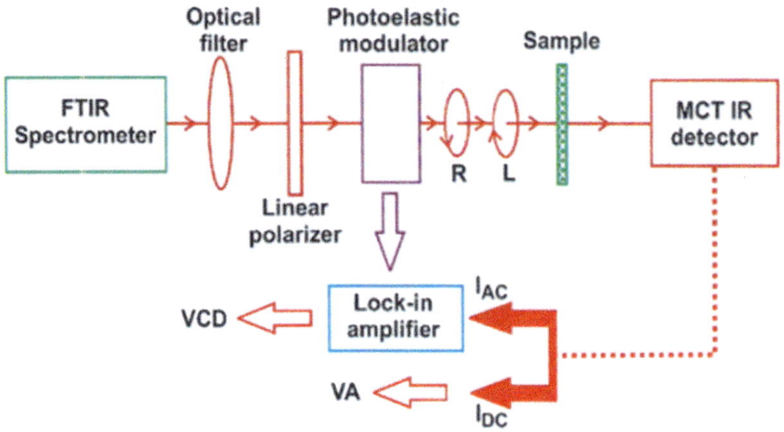

FIGURE 3.2

Representative outline of a VCD spectrometer [16]. https://s100.copyright.com/order/266dba01-3d5b-4ba7-891b-a19de6f9fb2a.

Reproduced with permission from Springer. Copyright 2010.

and right circularly polarized light is then detected by a high sensitivity detector, such as a liquid nitrogen cooled diode photoconductor.

Santhosh kumar et al. created biodegradable chitosan/graphene oxide bio-nanocomposites for use in biosensors, biotechnology, and biomedicine [17]. The CD spectrum of pure chitosan and chitosan/graphene oxide in aqueous solution was obtained at three different pH values: 6.31, 4.29, and 4.23. The spectra showed a strong negative band at 210 nm corresponding to an $n-\pi^*$ electronic transition, which is independent of pH and exhibits the highest intensity for pure chitosan. On the other hand, it was discovered that as the pH of the chitosan/graphene oxide solution varied, the strength of this negative band decreased. Kan To immobilize proteins and enzymes in the development of amperometric biosensors, Zhan et al. created a matrix using poly(methacrylic acid-*co*-acrylamide) (P(MAA-*co*-AAM)) noncovalently linked to MWCNT (multiwalled carbon nanotube) [18]. When Mb(myoglobin) was used as a model protein, the CD spectra of native Mb and Mb-P (MAA-*co*-AAM) exhibited similar spectral patterns, indicating that the protein's native structure was preserved even after immobilization in the matrix.

Yi Fang et al. created a glucose biosensor by immobilizing glucose oxidase (GOD) on top of a chitosan-polypyrrole (CS-PPy) nanocomposite, which was then coated on top of a glassy carbon electrode (GCE) [19]. According to research, CS-PPy is a specially modified nanocomposite with biocompatibility, biodegradability, and nontoxicity. The CD spectrum of glucose oxidase (GOD) in phosphate buffer saline (PBS) showed one positive band at 203 nm, indicating $\pi-\pi^*$ transition of the amide group in the glucose oxidase peptide chain, giving two bands at 209 and 219 nm, representing $\pi-\pi^*$ and $\mathrm{n}-\pi^*$ transition, respectively, of the amide group in GOD. Differentially absorbed left and right circularly polarized light are depicted in the CD spectra as shown in Fig. 3.3.

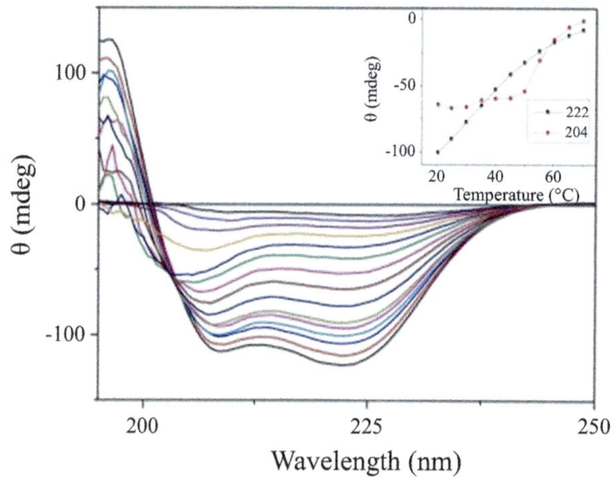

FIGURE 3.3

The CD spectra of poly(glutamic acid) [20].

Reproduced with permission from MDPI. Copyright 2012. Open access from MDPI.

When CS-PPy is present, the strength of the band at 203 nm was found to decrease, illustrating the interaction of amino acid residue of GOD with the CS-PPy nanocomposites. Furthermore, there was a slight increase in the strength of the negative bands at 209 and 219 with added CS-PPy, indicating that CS-PPy had little effect on GOD's secondary structure.

3.2.3 Electron paramagnetic resonance spectroscopy

The interaction of unpaired electron spin density in a sample with microwave radiation is studied using electron paramagnetic resonance spectroscopy (EPRS). A magnetic moment is the property of an electron spinning around its own axis. When an unpaired electron in a molecular system is exposed to a strong enough magnetic field, the electron-Zeeman interaction divides the electronic energy levels into two with different energies [21]. Absorption or emission of microwave radiation at appropriate frequencies can cause transitions between these energy states. Complex EPR spectral patterns result from hyperfine interactions between unpaired electrons and adjacent nuclei, as well as zero-field interactions among unpaired electrons. Nonetheless, they contain a wealth of information about the molecular system being studied. When compared to other spectroscopic techniques, the EPRS technique is extremely sensitive. The primary application of EPR spectroscopy is to investigate metal complexes, organic radicals, and organic molecules in triplet electronic states [22]. A typical EPR spectrometer, represented in Fig. 3.4, contains a light source which produces microwave radiation of suitable frequencies. Continuous or pulsed radiation is directed towards a sample placed in a resonance cavity at the center of an electromagnet. A phase-sensitive detector is then used to detect the microwave signal left by the sample.

Pawlicka et al. successfully developed a chitosan-based biomembrane doped with copper perchlorate and containing acetic acid and/or glycerol [24]. The EPR analysis provided the detailed

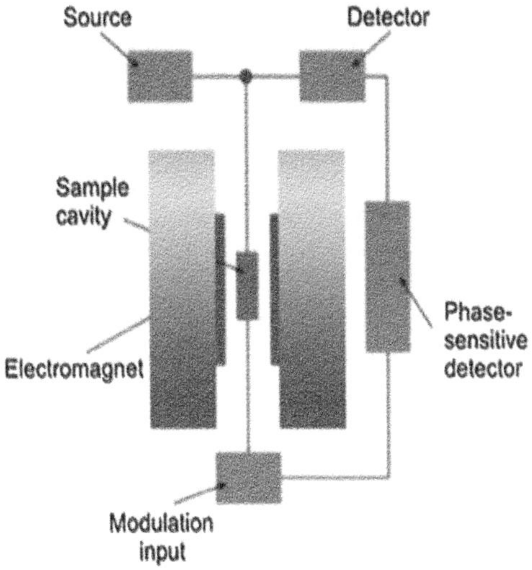

FIGURE 3.4

Diagram showing the working principle of an EPR spectrometer [23]. https://s100.copyright.com/order/e68fca5f-2fa2-44fc-9783-04e814eb42bd.

Reproduced with permission from Elsevier. Copyright 2017.

characterization of the biomembrane and is shown in Fig. 3.5. The Cu^{2+} ions are concentrated at axially strained sites in the membrane, according to the EPR analysis. The coordination of the Cu^{2+} ions with nitrogen and oxygen atoms was revealed by the Hamilton spin parameters. The covalent nature of the chitosan biomembrane was revealed by the thorough analysis of Cu^{2+} binding parameters.

Sidra Iftekhar et al. successfully developed a novel nanocomposite, GA-g-PAM/SiO_2, from gum Arabic (GA) grafted polyacrylamide-based silica nanocomposite using the in-situ radical graft copolymerization method to extract rare earth element (REE) ions from an aqueous solution [25]. The mechanism of formation of various free radicals generated during the formation of GA-g-PAM was investigated using EPR spectroscopy. The trapping agent used in the study was 5,5-Dimethyl-1-pyrroline *N*-oxide (DMPO), and the free radical source was potassium persulfate (KPS). The EPR analysis revealed that no free radicals were formed when DMPO was added to GA. The addition of KPS results in the formation of SO_4^- radicals, which react with the GA backbone to form oxygen-centered radicals. These radicals then react with AA chains to form GA-g-PAM, and the intensity of EPR lines due to oxygen-centered radicals was found to decrease with time as the reaction progressed. The study also revealed that the reaction ended with the complete conversion of SO_4^- radicals into OH radicals.

3.2.4 Atomic emission spectroscopy

Among the various spectroscopic techniques, atomic emission spectroscopy (AES) is an important tool. When an atom or molecule is excited to a higher energy level by means such as absorption of

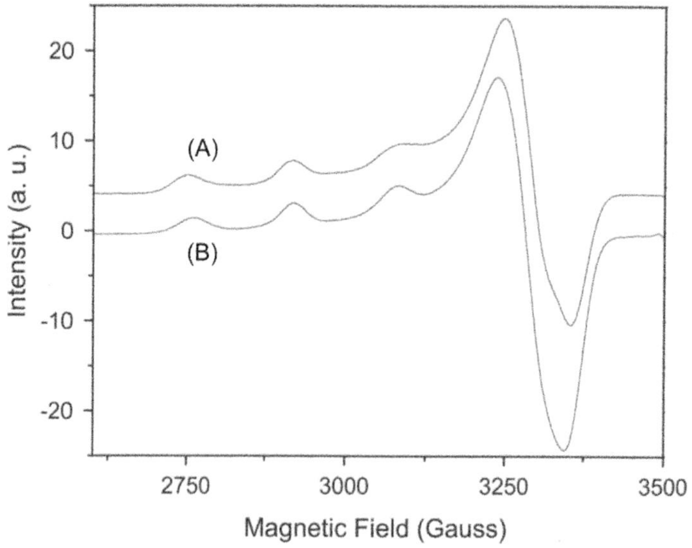

FIGURE 3.5

EPR spectrum of chitosan doped with copper perchlorate, containing 5% acetic acid with (B) and without (A) glycerol [24]. https://s100.copyright.com/order/63c68ea2-06fa-410d-ae0c-1ea3d7875af5.

Reproduced with permission from Elsevier. Copyright 2013.

suitable energy radiation, the system's finite lifetime causes it to decay back to the lower energy level by emitting radiation. AES can be used as a quantitative and qualitative technique because the emitted radiation has characteristic wavelengths and analyte concentrations can be estimated by evaluating the light absorptive and emissive properties of the sample (Fig. 3.6).

Using excitation sources such as flame, arc, spark, and plasma, the AES can convert samples into free atoms. The photons emitted by the atoms are then dispersed by a grating and directed through a monochromator onto a photodiode.

Na YIN et al. created a nanocomposite of hydroxyapatite (Hap)/bacterial cellulose (BC) with polyvinylpyrrolidone (PVP), and the nanocomposite was primarily characterized using AES spectroscopy [27]. AES analysis revealed that the ratio of calcium to phosphorus abundance (Ca/P) in the composite is 1.67, a value that qualifies the composite as a bone substitute material. Furthermore, AES measurements revealed that the incorporation of PVP increased the rate of nucleation of the apatite—BC system during composite formation.

ICP-AES (inductively-coupled plasma) is the most commonly used AES technique in practice for elemental analysis [10]. A significant and perhaps obvious advantage of AES is that it lends itself well to multivariate analysis. AES is used in a variety of fields, including biopsy procedures, mineral analysis, crude oil analysis, metal determination in food, metal determination in serum, and so on [28,29]. Ming-Feng Chiang et al. successfully synthesized a biodegradable nanocomposite based on PLLA (poly (L-lactide)) and Mg/Al layered double hydroxide (Mg/Al-LDH) by coprecipitation [30]. The thermal stability and degradation activation energy of PLLA and PLLA-LDH were

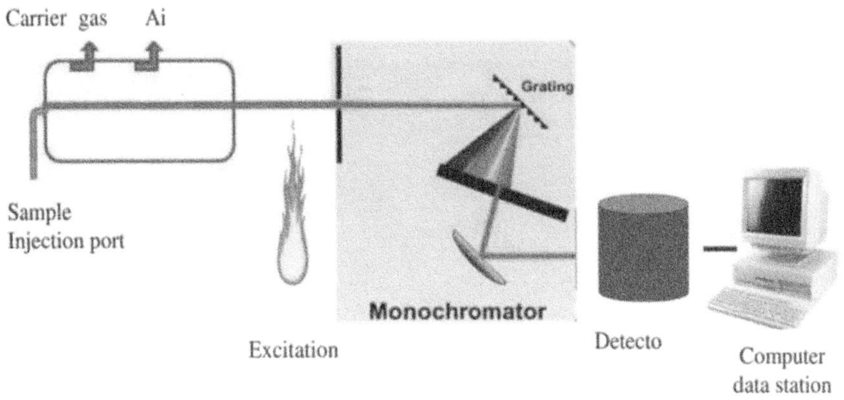

FIGURE 3.6

Represented atomic emission spectrometer [26]. https://s100.copyright.com/order/e19e214e-f236-4c77-b9c8-f8747e4f2ecc.

Reproduced with permission from Elsevier. Copyright 2019.

assessed using AES by analyzing the concentration of Mg/Al metal in the PLLA/P-LDH nanocomposite. These studies revealed that the metal content of the prepared nanocomposite influences its thermal and degradation stability.

3.2.5 Energy dispersive X-ray spectroscopy

The elemental analysis of samples is performed using energy dispersive X-ray spectroscopy (EDXS). A primary beam of X-rays or electrons is used in EDXS to knock out an inner shell electron in the atom, leaving a hole. This hole can be occupied by an electron from the atom's outer shell, but energy conservation requires this electron to emit the excess energy. The fundamental principle of EDS is that each element has a distinct atomic structure, resulting in distinct sets of peaks corresponding to the relaxation of its outer shell electrons. EDX spectroscopy is commonly used for chemical characterization of a sample as well as estimating the relative abundance of elements (with an atomic number greater than boron) in the sample [31]. EDXS is routinely used for material evaluation, identification, contamination, quality control screening, etc., as represented in Fig. 3.7.

Through a chemical in-situ polymerization process, Nafees et al. successfully synthesized a novel bio-nanocomposite (C/PPy-GO) containing biodegradable polymer (PPy) and graphene oxide (GO) [33]. This bio-nanocomposite was shown to have excellent biodegradability against food-borne pathogens as well as photocatalytic activity against industrial effluent dyes. SEM coupled EDX spectroscopy was used to determine the elemental composition of the developed C/PPy-GO, and the spectra of C/PPy and C/PPy-GO revealed that the C/PPy-GO contains more carbon than C.

Tippabattinijayaramudu et al. used a green method to create a novel biodegradable gold nanocomposite hydrogel from acrylamide (AM) and wheat protein isolate (WPI) because the gold nanoparticles were created from a plant extract [34]. According to the findings, the prepared composite,

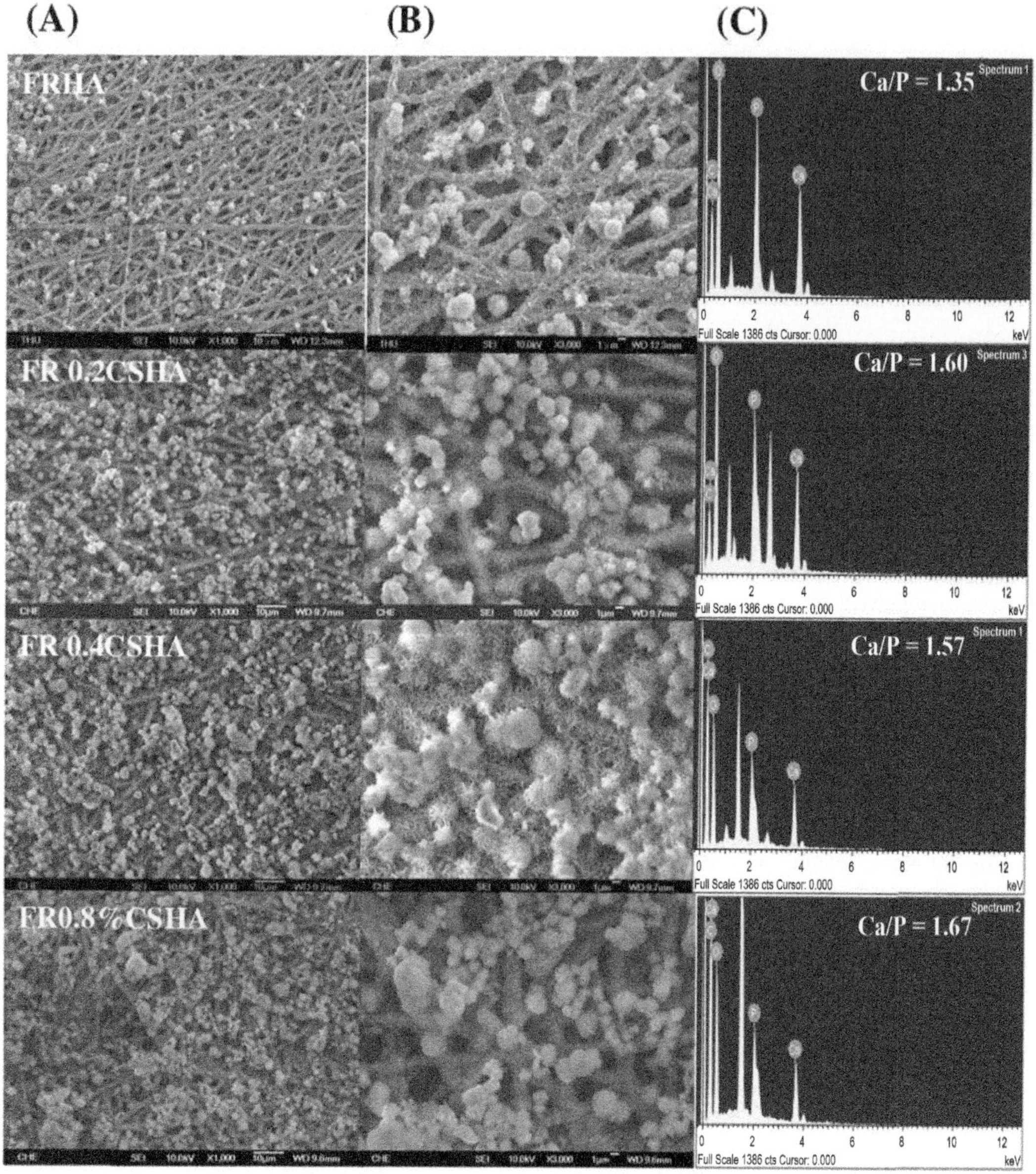

FIGURE 3.7

SEM images of different structures (A and B) of calcium phosphate on the nanofiber surface; EDX spectra (C) which analyses the calcium:phosphate ratio [32]. https://s100.copyright.com/order/cdb77b8d-40cd-401b-9574-fc0b2ca26856.

Reproduced with permission from Pubmed. Copyright 2014.

P(WPI-AM), possessed significant antibacterial properties. The presence of gold particles in unloaded and gold-loaded hydrogels was confirmed by the samples' EDXS spectra. The spectrum analysis shows that P(WPI-AM) has a strong gold peak whose intensity is proportional to the concentration of gold. Khanadanlou et al. developed a rice straw/polymer nanocomposite, RS/Fe_3O_4/PCL-NC, based on polycaprolactone and Fe_3O_4 through a casting method [35]. According to the findings, the RS/Fe_3O_4/PCL-NC film has potent antibacterial properties. Data from EDX analysis of RS and RS/Fe_3O_4 revealed that when RS is coated with Fe_3O_4 nanoparticles, an additional Fe peak can be seen in the EDX spectrum in addition to the carbon and oxygen peaks. According to the authors, the EDXS provided direct evidence for Fe_3O_4 adsorption on the surface of RS.

3.2.6 Fluorescence spectroscopy

Fluorescence spectroscopy is an emission spectroscopic technique for analyzing a sample's fluorescence properties. This tool can provide information at a nanoscopic level with exceptional sensitivity in the case of polymer nanocomposites. The fluorescence can be attributed to a polymer, a nanofiller, or both. This section will go over the fluorescence spectra and corresponding information of some biodegradable PNCs.

The schematic outline of a spectrofluorometer is shown in Fig. 3.8. A light source's broad frequency radiation in the UV-vis spectral region is passed through an excitation monochromator to select a single excitation wavelength in this technique. The sample excited by this single wavelength emits photons across a wide wavelength range, which are collected by an emission monochromator and then detected by a photomultiplier tube (PMT).

Santosh et al. synthesized bio-based chitosan/poly (vinyl alcohol) /ZnO (CS/PVA/ZnO) nanocomposite through a green approach; all the materials used for the synthesis were environment-friendly [36]. They fabricated a series of CS/PVA/ZnO containing varying amounts of ZnO and studied photoluminescent spectra of CS/PVA/ZnO films as well as CS/PVA films. The

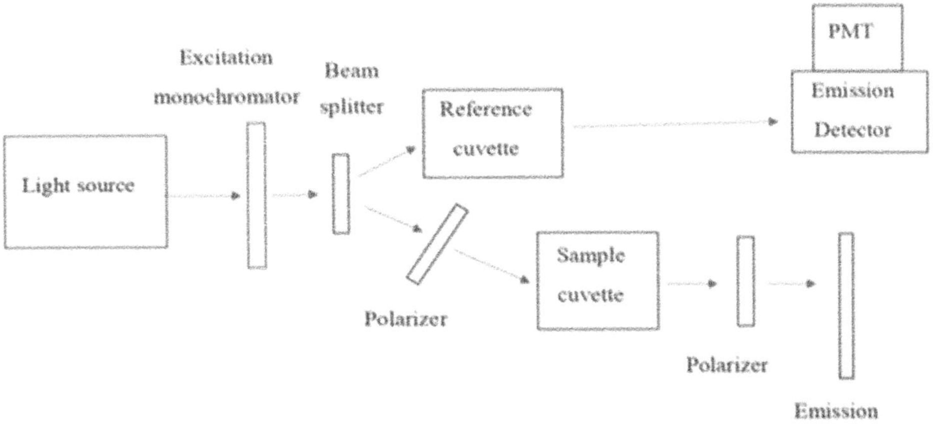

FIGURE 3.8

Schematic illustration of the working principle of spectrofluorometer.

photoluminescence emission spectra of PVA, CS/PVA, CS/PVA/ZnO-10, CS/PVA/ZnO-50 and CS/PVA/ZnO-70 were obtained. In the spectra of CS/PVA, a band was observed at 394 nm, whereas for CS-centered NCs, an emission band in the range of 400–440 nm was seen. For various amounts of ZnO NPs, the spectra revealed a series of wavelengths. PVA has no photoluminescent properties, whereas CS-centered nanocomposite has an increase in wavelength. Similarly, CS embedded PVA demonstrated an increase in wavelength, indicating successful PNC formation.

Nafees et al. created a new bio-nanocomposite composed of cellulose/polypyrrole-graphene oxide (C/PPy-GO) NCs. C/PPy-GO nanocomposites are biodegradable and biocompatible, which bodes well for a more advanced era of PNCs [33]. The PPy and C/PPy-GO fluorescence spectra are obtained. The rate of recombination was found to increase the intensity of emission [37]. The intensity peak of biodegradable PPy was found to be lower than that of as-prepared PNCs based on the spectra. This indicates that the recombination rate in C/PPy/GO PNCs is faster. The introduction of GO into the PPy is the previous reason for the higher recombination rate. As a result, the band gap characterized both recombination and sample intensity. The greater the band gap, the greater the possibility of recombination rate.

3.2.7 Fourier transform-infrared spectroscopy

Infrared Fourier transform (FTIR) Spectroscopy is a universal analytical technique based on a Michelson interferometer (the heart of FTIR spectrometer). All of the infrared frequencies characteristic of the allowed vibrational modes of the material under consideration are encoded in the resulting signal. The Fourier transform infrared (FTIR) method is a workhorse for interpreting biodegradable and biocompatible polymer nanocomposites. FTIR can detect conformational changes associated with polymer chain backbone bending and twisting by decoding the nature and dynamics of the material's characteristic vibrational bands [38].

The fact that FTIR spectroscopy is also known as an interferometric method demonstrates its superiority. The interferometer consists of mainly three active components; a beam splitter, a moving mirror and a fixed mirror as shown in the Fig. 3.9. The beam splitter divides the beam that is emitted from the source and reflected off the mirrors. For one beam, the fixed mirror produces a fixed path length, whereas the moving mirror produces a constantly changing path length. Thus, the interferometer signal is the result of two beams interfering with each other. The sample's interferogram is used for further analysis. The relationship between the signal in the time and frequency domain is defined by performing the Fourier transform on the interferogram.

In this technique, a light source's broad frequency radiation in the UV-Visible spectral region is passed through an excitation monochromator to select a single excitation wavelength.

Rui et al. used FTIR to investigate a poly (ε-caprolactone) (PCL)/TiO_{22} PNCs to confirm the successful incorporation of TiO_2 into PCL. They discovered that the crystallinity of PCL acts as a barrier to the polymer's rapid degradation. They discovered that the crystallinity of PCL acts as a barrier to the polymer's rapid degradation [40]. They proposed an in-situ sol-gel process for the synthesis of (PCL)/TiO_2 PNCs that could quench crystallization from PCL chains. The following information was obtained from the spectra of pristine (PCL) and PCL/TiO_2 PNCs. A series of bands observed between 3100 and 3700 cm^{-1} were attributed to the stretching vibration of hydroxyl groups (Ti-OH) of the PNCs. The imperfect poly condensation resulting from the TBT was the reason for the existence of hydroxyl groups. The characteristic bands at 2870 and

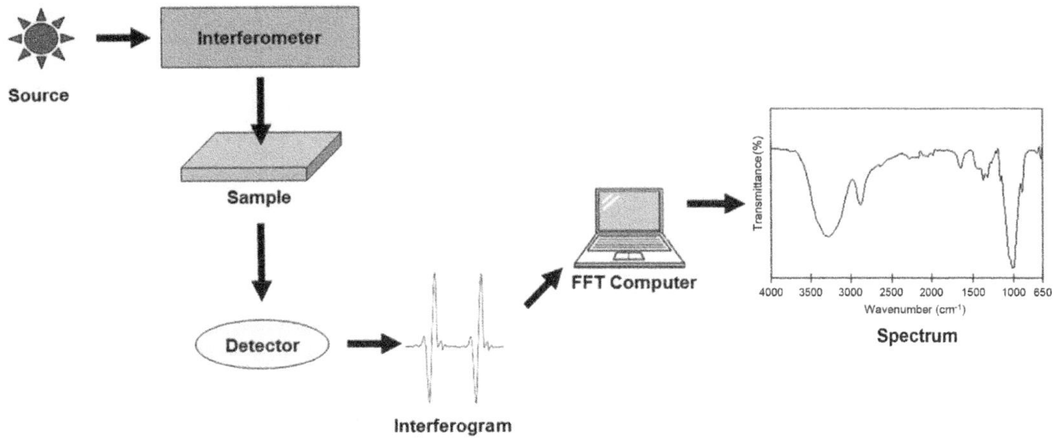

FIGURE 3.9

Working principle of Fourier transform infrared spectrometer [39]. https://s100.copyright.com/order/f2f9f68b-4730-4d0d-91e3-79ddb71b5f76.

Reproduced with permission from Elsevier. Copyright 2017.

2954 cm^{-1} were related to the stretching vibration of C-H bond contributed by the hydrocarbon part of the pristine polymer (PCL). The vibration from TiO_2 was confirmed from a number of broad bands lying in the range of 450–915 cm^{-1}. Thus, these characteristic bands revealed that both PCL and inorganic networks of TiO_2 are interlocked.

Wei et al. created collagen-infused poly(L-lactic acid)-*co*-poly(ε-caprolactone) nanofibers [P (LLA-CL)]. They also looked into the biocompatibility of these nanofibers [41]. They chose human coronary artery endothelial cells to test their biocompatibility (HCAECs). Because it has morphological and chemical similarities with the extracellular matrix, it has opened the door to new possibilities (ECM). According to their findings, the excellent biocompatibility provided by the collagen/P(LLA-CL) nanofibers aided the biological functions of HCAECs. The collagen-P (LLA-CL) mixture solution was electrospun with a suitable solvent, namely 1, 1, 1, 3, 3, 3-hexafluoro-2 propanol, during the fabrication process (HFP). The presence of collagen molecules in the PNC was discovered using attenuated total reflectance-Fourier transform infrared (FTIR-ATR) spectroscopy. The following peaks are obtained from the IR spectra of collagen containing P (LLA-CL). The characteristic peaks in the spectra of blended nanofibers were attributed to the type-1 collagen at wave numbers 1652 and 1537 cm^{-1}, respectively. These peaks represented the amide I and amide II band of the collagen molecules. Also, the P (LLA-CL) peak at 1734 cm^{-1} and another peak at 3310 cm^{-1} corresponding to the N-H stretch, showcased the incorporation of collagen into the polymer, and hence a successful synthesis of collagen-blended polymer nanocomposite was confirmed.

Rezai et al. used a sol-gel process to create hydroxyapatite/polycaprolactone NCs. Because of its chemical similarity to natural bone, hydroxyapatite (HA) is required in this PNC for human bone regeneration [42]. Nonetheless, due to its poor mechanical properties, HA is not an ideal

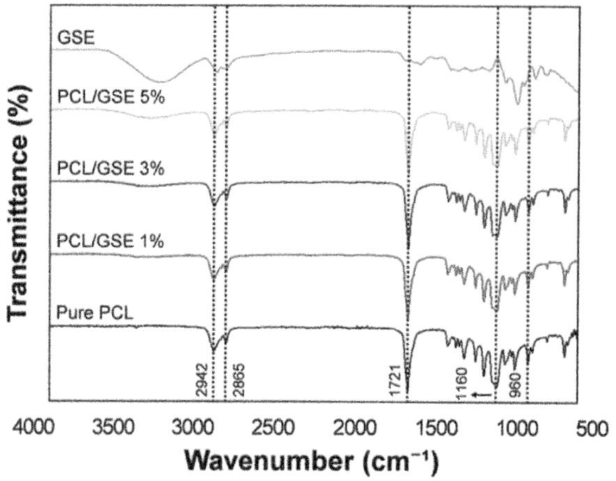

FIGURE 3.10

FTIR spectrum of biodegradable polycaprolactone film [43].

Reproduced with permission from Nature. Open access from Nature. Copyright 2019.

candidate for use as a bone substitute. To overcome this limitation, PCL was introduced as a solid support. A salt leaching method was used to prepare a HA/PCL scaffold. FTIR spectrum of biodegradable polycaprolactone film is depicted in Fig. 3.10.

Yazhang et al. developed a biomimetic poly (citric acid-ocatnediol-polyethylene glycol) (PCE)-graphene (PCEG) nanocomposite [44]. PCEG nanocomposites are essential in promoting myogenic functions. PCE was chosen as the biomaterial because it exhibits excellent biomimetic elastomeric behavior. Polycitrate (PC) in particular can mimic the elastomeric behavior of extracellular matrix (ECM). RGO was mixed with PCE along with a cross-linking agent, 1, 6-hexamethylene diisocyanate (HDI). The following conclusions are drawn from the FTIR-ATR spectroscopic study of PCEG nanocomposite. An intense peak formed at 1742 cm^{-1} in the PCEG spectrum revealed an ester bond formed in the PCE. Peaks from RGO's C = C vibration were observed at 1626 cm^{-1}. A series of peaks ranging from 996 to 1196 cm^{-1} reflected the epoxy groups from RGO. The intensities of these peaks were discovered to increase as RGO weight percentage increased. This indicated that the PCEG nanocomposite had been successfully manufactured.

Jingjing et al. investigated the biodegradation of chitosan-graphene oxide (CS-GO) nanocomposite using wastewater bacterial community [45]. Chitosan was chosen as the biopolymer because it is both biodegradable and inexpensive. Chitosan was added to the GO suspension during the synthesis of the CS-GS nanocomposite. FTIR spectroscopy was used to characterize CS-GO nanocomposites. After 3 days, a sharp peak at 1647 cm^{-1} was obtained and linked to the CS. This peak was attributed to vibrational changes in amide 1 groups. This amide functional group was produced by microbial growth on chitosan films. The rate of biodegradation was discovered to be proportional to the secretion of an enzyme called chitosanases. This resulted in a higher peak at 1647 cm^{-1} [46].The decrease in the intensity of the peak at 1150 cm^{-1} indicated the breaking of glycosidic bridges of the glycosidic units. The spectra of pure chitosan and CS-GS-0.25% after 3 and 6 days of microbial interaction were similar. However, after a few days of microbial interaction, the spectra of 0.6% GO incorporated CS showed discernible changes. After 6 days, a distinctive peak at 1735 cm^{-1} was obtained. The presence of GO on the film's surface was revealed by C = 0 stretching the vibration of GO. The C = O stretching from the carboxyl in GO was attributed to this peak. The appearance of GO revealed chitosan biodegradation. The study found that chitosan degrades faster than graphene oxide.

Ehsan and colleagues created a conductive polypyrrole/dextrin (PPy/Dex) nanocomposite. When conductive polypyrrole was added to biodegradable Dex, PNCs of PPy and Dex with improved properties were formed [47]. The FTIR spectra of dextrin, PPy, and PPy/Dex nanocomposites were obtained. The spectrum of Dex consists of a broad peak at 3410 cm^{-1}. The peak observed at 2916 cm^{-1} reflected the C-H stretching mode of the pyrenoid ring of dextrin [48]. The vibrational changes happening in the C-O-C bridge of dextrin formed a peak at 1151 cm^{-1} [47,48]. The stretching vibration of C-O was revealed by a peak at 1602 cm^{-1}. The IR band in the spectrum of polypyrrole observed at 1533 cm^{-1} substantiated the stretching vibration of C = C bond, and the stretching vibration of the N-H bond was illustrated by the emergence of a band at 3400 cm^{-1} [49]. FTIR spectra of biodegradable carboxymethyl cellulose-based material is given below (Fig. 3.11).

3.2.8 Mass spectroscopy

The analyte's molecular mass and fragmentation pattern are predicted by mass spectroscopy. Compounds are ionized and then fragmented to produce different daughter ions in this method. Various mass spectroscopic methods have been used to analyze polymer nanocomposite over the last two decades. The mass spectra of several PNCs, as well as the information extracted from them, will be discussed in this section.

Marin et al. synthesized poly (vinyl alcohol)/cellulose nanocrystals (PVA/CNCs) composite. PVA has high tensile strength and flexibility [51]. A solution casting technique was used to create the PVA/CNCs composite. A series of PNC films with varying proportions of cellulose nanocrystals were created and named: PVA/CNCs (4), (8), (12), and (15). Mass spectroscopic techniques were combined with a thermal technique to confirm the formation of PVA/CNCs. Thermal degradation products are analyzed using this method. Most of the degradation processes produced water as an end product. The degradation of cellulose particles produced CO_2 and SO_2. Acetaldehyde and CO_2 were obtained from the degradation of PVA. The corresponding spectra of the produced compounds are presented in Fig. 3.12. Thus, the presence of both polymer (PVA) and reinforcing element (CNCs) are verified in the product (PVA/CNCs).

Niloofar et al. determined the antimicrobial activity of chitosan-blended poly(vinyl alcohol)/gelatin/Zataria Multiflora (CS/PVA/Gel/ZM) [52]. Because of its inherent microbial activity, naturally derived ZM essential oil is a significant component of the PNC. A series of nanofiber films with varying ZM compositions were created. To determine the amount of ZM and characterize the as-prepared CS/PVA/Gel/ZM PNCs, gas chromatography-mass spectroscopy was used. A set of 29 compounds were identified from the GC-Ms spectral analysis, and these 29 compounds constitute the majority of the ZM oil.

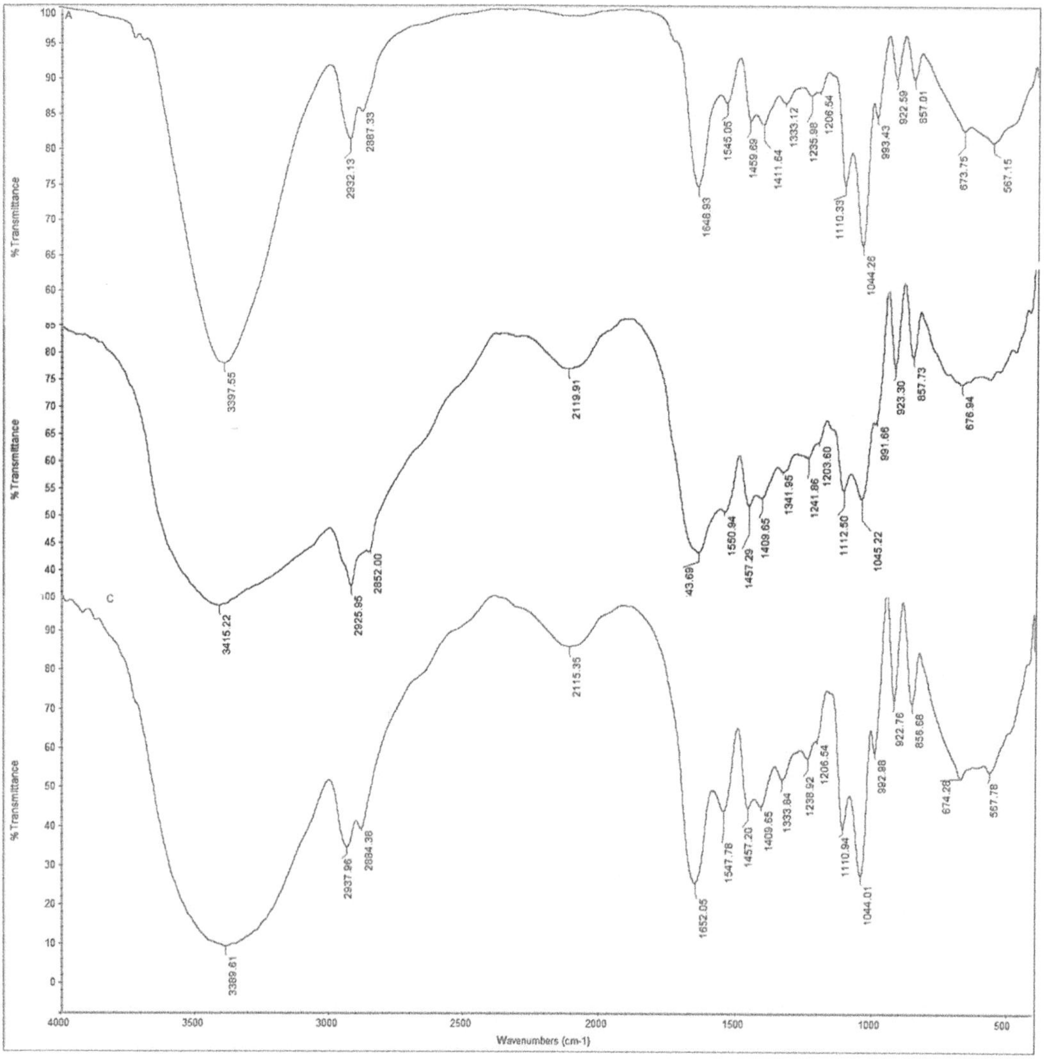

FIGURE 3.11

FTIR spectra of biodegradable carboxymethyl cellulose-based material [50].

Reproduced with permission from Nature. Open access from Nature. Copyright 2020.

3.2.9 Mossbauer spectroscopy

Mossbauer spectroscopy is an interdisciplinary spectroscopic technique that involves gamma-ray photon absorption and subsequent transition. It is especially useful for analyzing the changes caused by metal-based compounds found in polymer nanocomposites (PNCs). This section discusses Mossbauer spectroscopy and metal fillers in biodegradable and biocompatible PNCs.

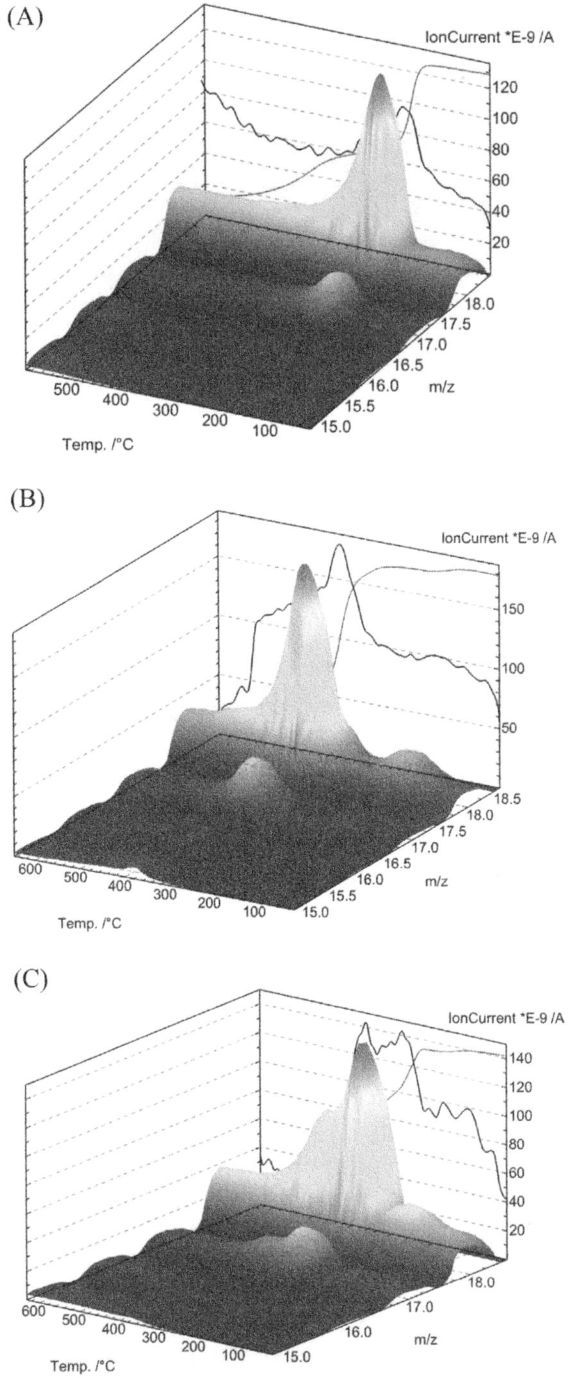

FIGURE 3.12

Mass spectra of the thermal decomposition products of CNCs (A), PVA/CNCs-8 (B), and 8 PVA/CNCs-25 (C) in the range of m/z values from 15 to 19 amu [51]. https://s100.copyright.com/order/412475b0-2a6f-4e90-bccf-32c5876381d0.

Reproduced with permission from Elsevier. Copyright 2016.

Navokoya et al. scrutinized the dependence of iron oxide on the magnetic behavior of PVA/Fe_3O_4NCs. They used nanosize magnetite (Fe_3O_4) and PVA as the reinforcement element and polymer matrix, respectively [53]. Iron oxide compositions varying in volume concentration were incorporated onto the surface of PVA. The Mossbauer spectrum with a doublet was produced by the lowest volume concentration of iron oxide in PVA. This doublet's parameters included an isomer shift, δ, and quadrupole splitting. These doublet parameters were attributed to the supermagnetic behavior of iron oxide nanoparticles [54,55].

Fang et al. created iron/polytetrafluoroethylene (Fe/PTFE) nanocomposites and investigated their Mossbauer properties. The nanocomposite was prepared using ball milling and a mixture of α-Fe and PTFE powder [56]. During the ball milling process, three different components were determined at several stages. They were FeF_2, FeF and Fe_3C. Mossbauer spectroscopy was used to investigate these components. The Mossbauer spectra of α-Fe/PTFE for 0–400 hours and spectra of Fe/PTFE milled for about 400 hours are obtained. To prove the formation of FeF_2 from Fe^{2+}, the spectra of Fe/PTFE (140 hours) were taken at room temperature and low temperature. The doublet corresponding to Fe^{2+} appeared as a sextet in this spectrum, in accordance with the investigation by Ramsamy [57]. This sextet was attributed to the antiferromagnetic behavior of FeF_2. Two sextets and one doublet were expected for samples milled for 200 and 250 hours; however, the spectra showed only one singlet. The characteristic isomer shifts associated with this singlet were 0.58 and 0.49 mm/s, respectively. As described by Grenech, the singlet could be associated with the formation of Fe [58]. These characteristics indicated the appearance of a new phase. This new component is associated with an isomer shift and an average magnetic hyperfine field. As the milling time increased, the C-C bonds in the PTFE were broken, and it served as the source of C in Fe_3C.57 Fe Mossbauer spectra of graphite oxide based polyaniline (GO-PANI)composiotes are given in Fig. 3.13.

3.2.10 Nuclear magnetic resonance (NMR) spectroscopy

Atom nuclei are subjected to absorption by radiofrequency radiation in NMR. For the semiquantitative characterization of PNCs, the NMR technique is extremely useful.

Wenshou et al. developed an efficient method for producing poly(ester urethane) (PEU)/polyhedral oligomeric silesquioxane (POSS) nanocomposites [60]. A macro level structure was created using the solution polymerization method. The improved material properties suggest that PEU/POSS nanocomposites are a promising class of biomedical candidates. Using proton NMR, hydrogen from water produced a peak at 2.5 ppm. The proton from POSS was reflected in the NMR sharp peak at 0.9 ppm. MDI protons contributed a peak at 7.3 ppm. The urethane component of PEU contains a number of hydrogens, resulting in the appearance of a sharp peak at 9.5 ppm. A peak at 2.5 ppm revealed the protons from the dimethyl sulfoxide reagent.

Hsiang et al. synthesized nanocomposites containing maleic acid (MA)-grafted poly [(butylene adipate)-*co*-terephthalate](g-PBAT) [61]. The reinforcing element was layered zinc phenylphosphonate (m-PPZn). Because of its promising biodegradability and aromatic units that provide enhanced mechanical properties, PBAT is an ideal polymer. The presence of a terephthalate group was revealed in the NMR spectrum of PBAT by a peak at 8.11 ppm. Another peak at 2.29 ppm may be associated with a proton from the adipate part of the PBAT matrix. The incorporation of MA into PBAT was confirmed from the ^{13}C NMR spectrum as it exhibited a peak that is typical of $C=O$ in MA. Thus, the complete grafting of MA to PBAT was demonstrated using NMR spectroscopy.

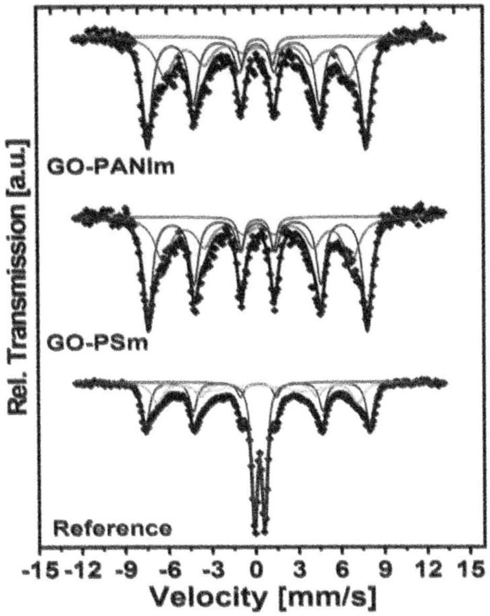

FIGURE 3.13

Mossbauer spectra of iron oxide nanoparticles [59].

Reproduced with permission from MDPI. Open access from MDPI. Copyright 2015.

First-generation poly(pro-E2) biomaterial scaffolds were studied using 1H NMR as given in Fig. 3.14 [62].

Yazhang et al. developed poly(citric acid-ocatnediol-polyethylene glycol)(PCE)-graphene (PCEG) nanocomposites [44]. PCE was chosen as the biomaterial because it exhibits excellent biomimetic elastomeric behavior. Polycitrate, in particular, was able to mimic the elastomeric behavior of an extracellular matrix. A cross-linking agent, namely 1, 6-hexamethylene diisocyanate (HDI), was used to prepare PCEG PNCs. The proton NMR spectra of polycitrate contained a series of sharp peaks at 1.26, 1.54, and 3.98 ppm, corresponding to the protons of PCEG PNCs' 1,8-octanediol. A set of peaks between 2.62 and 2.90 ppm confirmed the presence of hydrogens from citric acid. PCE has an additional methylene peak in its spectrum. This peak was observed at 3.50 ppm and was related to the PEG portion of the PCEG, indicating successful PCE synthesis.

3.2.11 X-ray photoelectron spectroscopy

X-ray Photoelectron Spectroscopy (XPS) is a potential characterizing tool in which an X-ray photon beam causes electrons to be ejected from a sample. XPS provides quantitative and qualitative data on the elemental composition of matter, particularly solid surfaces as represented in Fig. 3.15. Additionally, this tool can determine the oxidation state of elements.

Chenlu et al. created a poly(lactic acid) (PLA)/graphene nanocomposite, with graphene created via pressurized oxidation of graphite followed by multiplex reduction of GO [64]. PLA was chosen as the

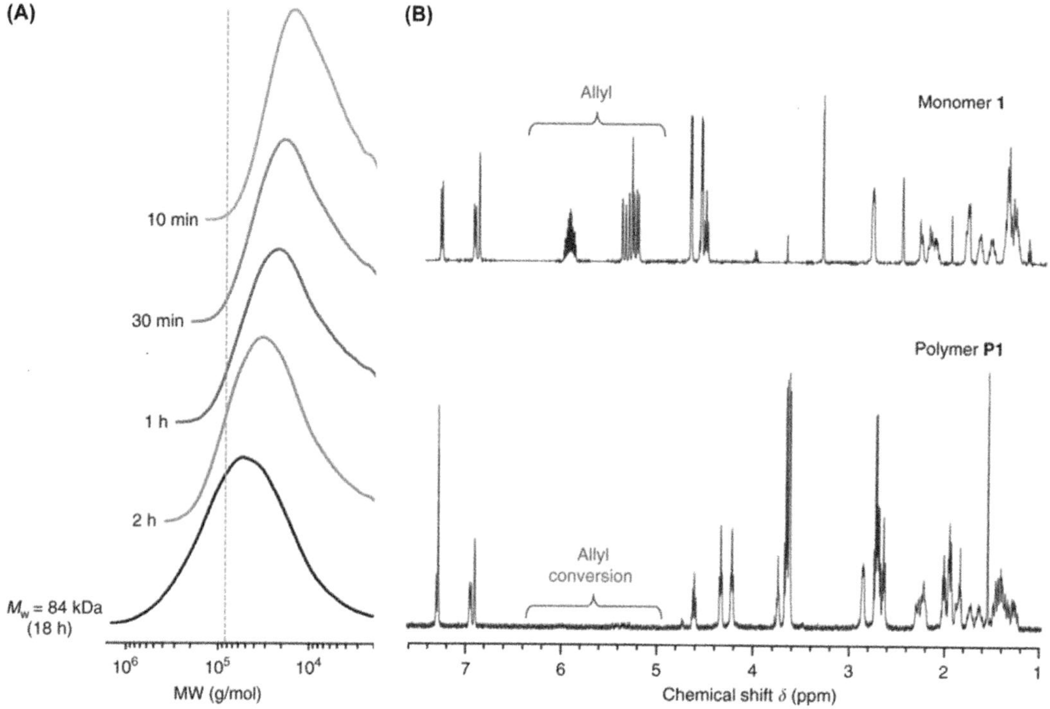

FIGURE 3.14

Validation of a successful synthesis of polymer using (A) GPC as a function of polymerization time and (B) ^1H NMR spectra of the monomer and the polymer [62].

Reproduced with permission from Nature. Open access from Nature. Copyright 2019.

biopolymer because it is a biodegradable and environmentally friendly polymer. The C/O atomic ratio for GO was determined by XPS analysis of graphene and GO to be 68%/32%, confirming the successful oxidation of graphite. For graphene and graphene oxide, the splitting of the C 1s curves substantiated the existence of a number of functional groups that have carbon atoms in common.

Manik and colleagues created bio-based silica/carbon hybrid nanoparticles (SCNPs). The biodegradable polymer ecoflex was used to create ecoflex/SCNPs biocomposites [65]. Rice husk was used to prepare silica because it contains a high concentration of silicon and oxygen. A 3-D printing technique was used to create the ecoflex/ SCNPs. It was possible to obtain the XPS spectrum of rice husk ash SCNPs. The major elements found in rice husk ash were discovered. In decreasing order of composition, they are C, O, and Si.

Mishra et al. developed a sophisticated method for producing Ag-polyethylene terephthalate (Ag/PET) nanocomposites with varying Ag atomic fractions (30%–75%). An atom beam cosputtering method was used for this [66]. To study the chemical composition of nanocomposite films, an XPS study was carried out on the sputtered PET and cosputtered Ag-PET as well as PET bulk foil. In the spectra of cosputtered Ag-PET nanocomposite, the presence of carbon was revealed from the C 1s peak. The

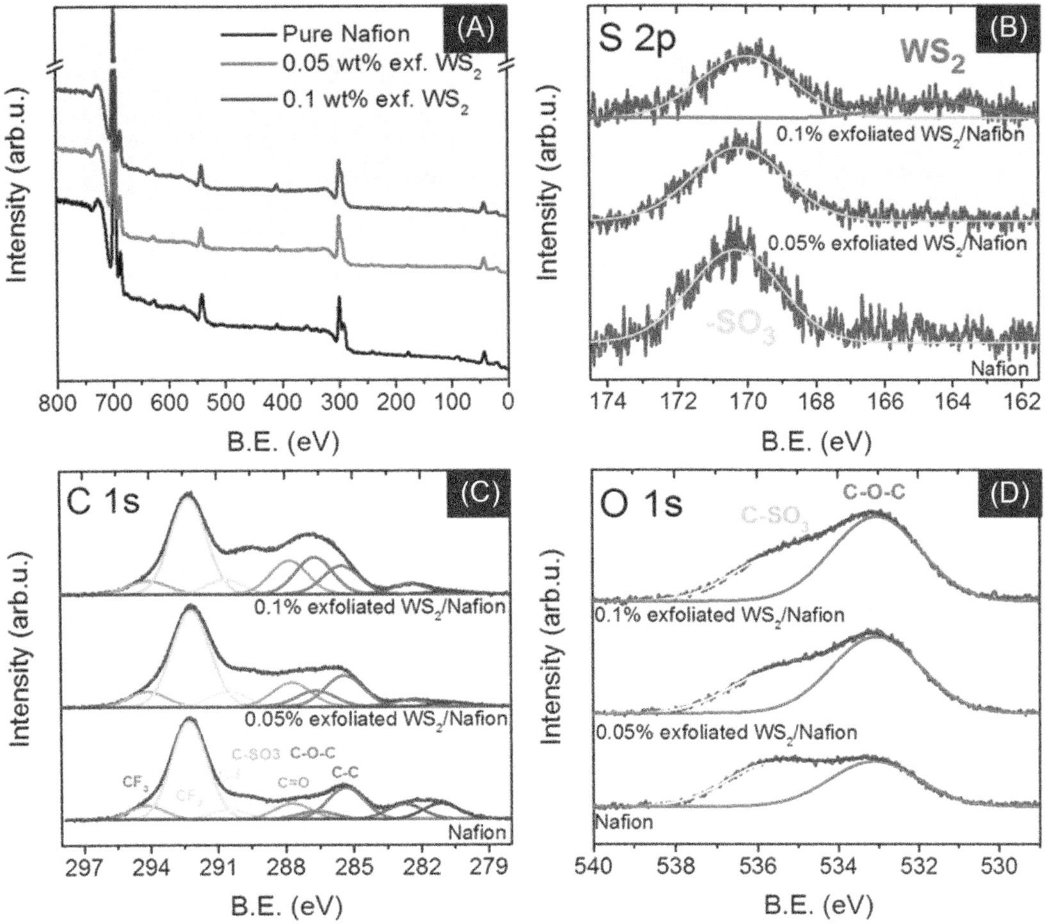

FIGURE 3.15

Nafion-TMD Nanocomposites; (A) spectra of all the samples; (B) spectra showing—SO_3 bonding in Nafion, and weak W-S bonding; (C, D) spectra showing improved C-SO_3, C-O-C and C = O peaks attributing to the successful incorporation of WS_2 with the polymer matrix [63].

Reproduced with permission from Elsevier. Open access from Elsevier. Copyright 2017.

appearance of $3d_{3/2}$ and $3d_{5/2}$ peaks was attributed to the existence of Ag. Thus, XPS results obtained from cosputtered Ag-PET nanocomposite demonstrated the presence of C and Ag in the nanocomposite.

3.2.12 Raman spectroscopy

The method is based on inelastically scattered radiation properties. The scattered radiation determines the vibrational modes. Raman spectroscopy is particularly useful for structural investigation and verification of successful nanoparticle incorporation onto the surface of selected polymers. It is extremely useful for determining the presence of carbon-based nanoparticles in polymers [64].

Min et al. presented an effective electropolymerization process for the development of polypyrrole films rooted with Co nanoparticles (PPy/Co) [67]. At room temperature, PPy/Co PNCs are super magnets with negative magnetoresistance in external magnetic fields. The Raman spectroscopic technique was used to confirm the successful incorporation of Co into PPy. The Raman spectra of the initial polymer, PPy, and Co incorporated PPy are depicted in Fig. 3.16. The C = C stretching of the pyrrole ring was illustrated by a peak at 1570 cm^{-1}. The C-C in-ring stretching was disclosed by a peak at 1322 cm^{-1}. From the C-C interring stretching of PPy, a peak emerged at 1412 cm^{-1}. Its CH in-plane bending was illustrated by a peak around 1044 cm^{-1}. The peaks at 935 and 974 cm^{-1} were assigned to the PPy ring deformation. However, the peak observed at 935 cm^{-1} was very weak in the spectrum of PPy/Co PNCs, and this interpreted the bonding between Co NPs and amine group of polypyrrole [68].

Chenlu et al. developed an excellent method for the fabrication of the poly(lactic acid)(PLA)/graphene nanocomposites [64]. Graphite was first oxidized to graphite oxide (GO) by pressurized oxidation before being reduced by a multiplex reduction to form graphene. A PLA/graphene

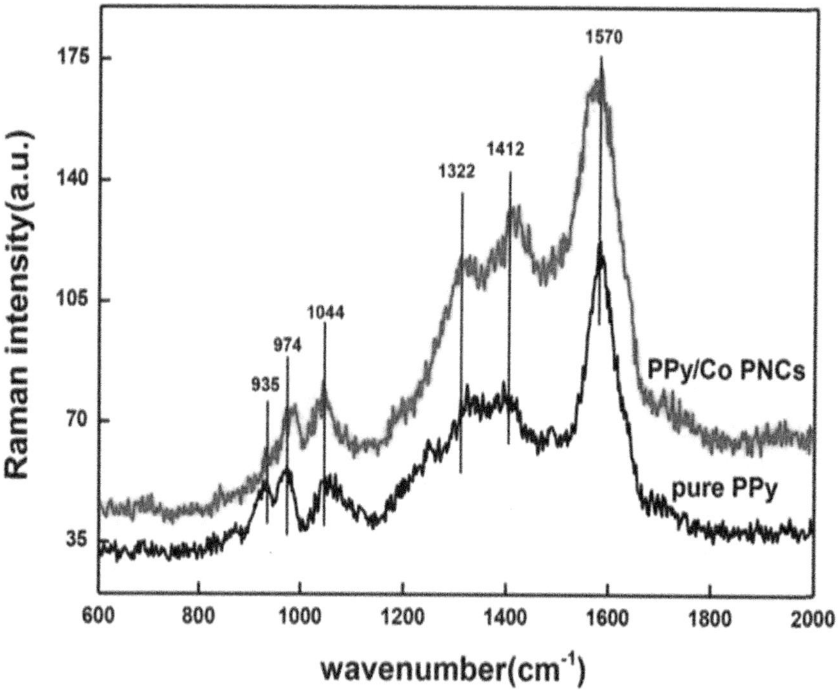

FIGURE 3.16

Raman spectra of the polymer and PNCs [67]. https://s100.copyright.com/order/a5e4727a-84bc-4dc9-b0fd-030ab197a365.

Reproduced with permission from Elsevier. Copyright 2017.

nanocomposite was created using a melt blending technique. Raman spectroscopy was used to investigate the internal changes associated with GO and graphene. From the spectrum of GO, a G band was observed at 1590 cm^{-1}. Similarly, a D band was observed at 1350 cm^{-1}. Nevertheless, for graphene, the peak positions remained unchanged, while the D/G ratio showed an appreciable increase. This difference in intensity ratio could be attributed to the reduction of GO to graphene.

3.2.13 Ultraviolet-visible (UV-Vis) spectroscopy

UV-vis spectroscopy is a technique in which UV-vis radiation interacts with matter, causing electronic transitions. The optical properties of a few nanofillers as well as the optical response of the polymer can be studied using this method. This section discusses the UV-vis spectra and related optical properties of a few biodegradable PNCs.

Biplab et al. prepared a poly(3-hexylthiophene) (P3HT)-montmorillonite clay nanocomposite [69]. P3HTs are a type of conducting polymer that is widely used. The melt-cooled samples were prepared using the solvent cast method. The obtained samples were examined. Various solvent cast PNCs' UV-vis absorption spectra were obtained. The spectra showed a common peak at 272 nm, indicating the $\pi - \pi^*$ transition occurring in the thiophene ring, a peak at 516 nm, and it was related to the π electron conjugation. Nonetheless, from the spectra of PNCs, a 20 nm blue shift from pristine P3HT spectra at 516 nm was observed. In addition, a blue shift of 10 nm was observed on the prominent peak of melt-cooled P3HT in comparison with that of solvent cast P3HT films. The P3HT chains in solvent cast samples were discovered to be relatively straight. The observed blue shift in PNCs was attributed to a decrease in conjugation length caused by two mechanisms: intrachain mechanisms caused by possible conformation changes and side group disorder, and interchain mechanisms caused by crystallization, gelation, induced rigidity, and so on. UV-vis spectra of poly(lactic acid)based biodegradable nanocomposites coatings are illustrated in Fig. 3.17.

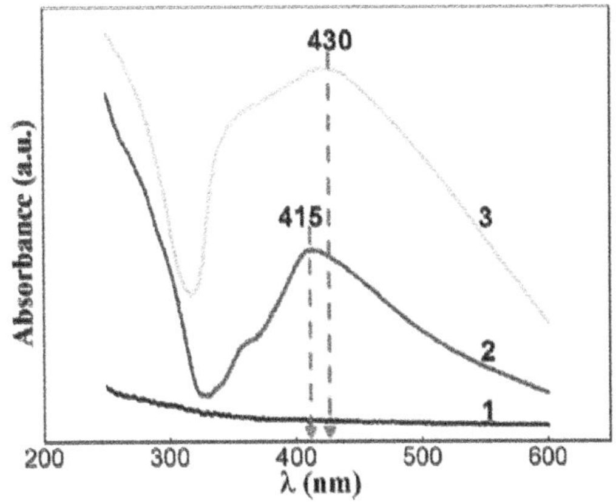

FIGURE 3.17

UV-vis absorption spectra of nanocomposites [70]. https://s100.copyright.com/order/43c95021-d24d-4468-9ba5-00a815a9c1de.

Reproduced with permission from Elsevier. Copyright 2018.

Richard et al. investigated the drug-releasing performance of chitosan-graphene oxide (CGO) nanocomposite [71]. Biodegradable chitosan-GO nanocomposites outperformed pure chitosan in terms of drug delivery capacity. To investigate the effect of CGO on drug delivery capacity, they used fluorescein sodium (FL). The required concentrations of nanofiller (GO) (0.25, 0.5, 1.2, and 5 wt.%) were mixed with chitosan solution. UV-vis spectra were obtained for pristine FL, GO, and FL incorporated GO-FL. A broad peak that was observed at 233 nm could be related to the $\pi - \pi^*$ transition in C = C bonds. In the spectrum of GO, a weak peak was observed at 300 nm, reflecting the $n - \pi^*$ transitions in C = O bonds. By comparing the peaks of FL, FL loaded GO and GO, the following observations were obtained. First, the peaks resulted from the electronic transition in FL were observed at 450 and 480 nm, which are characteristic of FL, but the FL-GO curve showed a shift in wavelength to 445 and 475 nm, indicating interaction between FL and GO. Second, another peak (shoulder peak) in the FL curve was discovered to be slightly displaced in the GO-FL curve. As a result, the complete integration of FL into GO was confirmed.

Prakash et al. developed a method for the preparation of chitosan/Al_2O_3 PNCs. For the preparation of chitosan/Al_2O_3 nanocomposite, an ultrasonication method was employed [72]. In this method, 10 wt.% of Al_2O_3 nanoparticles were introduced into the chitosan solution and it was subjected to different periods of sonication (15, 30, 45, and 60 minutes). They also studied the effect of sonication on the chitosan/Al_2O_3 PNC samples. UV-vis spectra of the sonicated films were obtained. The optical band gap of the membranes was determined to be about 2.2 eV. When the sonication period was increased, the band gap increased. Similarly, the transmission spectra demonstrated that as the sonication period increased, the percentage of transmission increased, demonstrating the successful use of spectroscopy in process optimization.

3.3 Microscopic analysis

Microscopic techniques are the best choice for examining samples that cannot be examined with the naked eye. The evolution of microscopes began with a single magnified lens and has now led to the most advanced electron microscopic techniques. Microscopic techniques undeniably play important roles in the in-depth analysis of materials to nanometer dimensions because they provide a highly magnified picture of the sample's physical, chemical, and structural properties. The available information ranges from a few millimeters to a few nanometers, depending on the techniques used and the resolution provided. Optical microscopy, electron microscopy, and scanning probe microscopy are the most commonly used microscopic techniques [7]. Other emerging techniques include infrared, digital holographic, laser, and photo acoustic microscopy. The effectiveness of the techniques used is determined by the magnification resolution and the contrast of the image obtained. A plethora of recently published data suggests that various types of microscopic techniques are useful in analyzing the microscopic level details of biocompatible polymer nanocomposites. Biocompatible polymer nanocomposites are being characterized using a variety of techniques (BPNs). The nanostructure of BPNs was successfully analyzed using electron microscopy. The aggregate size can be measured to the nanometer scale. Scanning Probe Microscopy, in addition to electron and optical microscopy, is the most reliable method for measuring particle size on the nanometer scale.

3.3.1 Electron microscopy

Electron microscopy is a highly invasive technique. It achieves excellent results in determining the morphology, crystal and defect structure composition of materials, as well as their electronic properties [73]. The resolution of the images is well agreed upon by electron microscopes. As a result, they are widely used in biological material imaging. Scanning electron microscopes (SEM) and transmission electron microscopes (TEM) are the two types of electron microscopes (TEM).

3.3.1.1 Scanning electron microscopy

Scanning electron microscopes (SEM) employ a high-energy electron beam that strikes the sample's surface. The dissipation of the electrons' kinetic energy in the beam is recorded as signals, which are sent to the detector to generate micrographs. The signals generated would effectively convey information about the topography and crystal structure of the sample. The imaging is accomplished through the use of secondary electrons emitted from the surface, which are reflected and detected by at least one detector in the instrument [73]. SEM studies have been widely employed in the field of biopolymers and nanocomposites [74,75]. The dispersion and agglomeration of nanofillers used in nanocomposites were effectively characterized by SEM, this was clearly depicted in Fig. 3.18 [77]. Polyhedral oligomeric silsesquioxane was reinforced with polyhedral oligomeric silsesquioxane (L-lactic acid). These nanocomposites were primarily influenced in the field of tissue engineering, specifically in the replacement of diseased tissue. The mechanical properties of the PLLA polymer were improved by POSS nanoparticles. Biodegradable and biocompatible PLLA/POSS-$(PLLA)_{32}$ nanocomposites were synthesized by means of solution casting. These synthesized biofilms could be categorized by 2, 4, 6, and 8 wt.% of $(PLLA)_{32}$ content in nanocomposites. The electrospun fibers of pure PLLA and PLLA/POSS-$(PLLA)_{32}$ 100%, 6% nanocomposites were observed using SEM. The SEM micrographs as given Fig. 3.19 indicated that fresh PLA and PLLA/POSS-$(PLLA)_{32}$ have a similar diameter of 350 nm. The observation revealed that the outer plane of pure PLLA was clean, but the nanocomposites were distributed with elliptical apertures on the surface. It was porous and hard, and of higher quality than pure PLLA. The porous structure of PLLA/POSS may have resulted from the assorted nucleation of POSS nanoparticles on the PLLA surface. It would undoubtedly hasten crystallization and cause crystal phase separation, resulting in porous fibers [78].

Hybrid nanocomposites of wood and coir fiber, with MMT nanoclay and polypropylene (pp) as the polymer matrix, were biodegradable and biocompatible. Six types of nanocomposites were prepared, such as wood/pp, coir/pp, wood/pp/mmt, coir/pp/mmt, wood/coir/pp, wood/coir/pp/mmt hybrid nanocomposites. SEM studies were performed to get details of the surface morphology. It was observed that wood/pp and coir/pp nanocomposites could be distinguished by their roughened surface. This result was also similar to wood/coir/pp hybrid composites. It would result in a reduction in the quality of mechanical properties. This could be due to the fragile interfacial adhesion between the fibers and the polymer matrix. This weakness, however, could be overcome by the addition of nanoclays. It was also discovered that hybrid nanocomposites strongly adhered to the interfaces and had a very smooth surface. This would lead to better mechanical properties [79].

Hap/BC nanocomposites were created by combining bacterial cellulose (BC) and hydroxyapatite (Hap). Bacterial cellulose combined with hydroxyl apatite nanoparticles improves biodegradability, biocompatibility, and mechanical properties, promoting new bone formation. SEM was used to

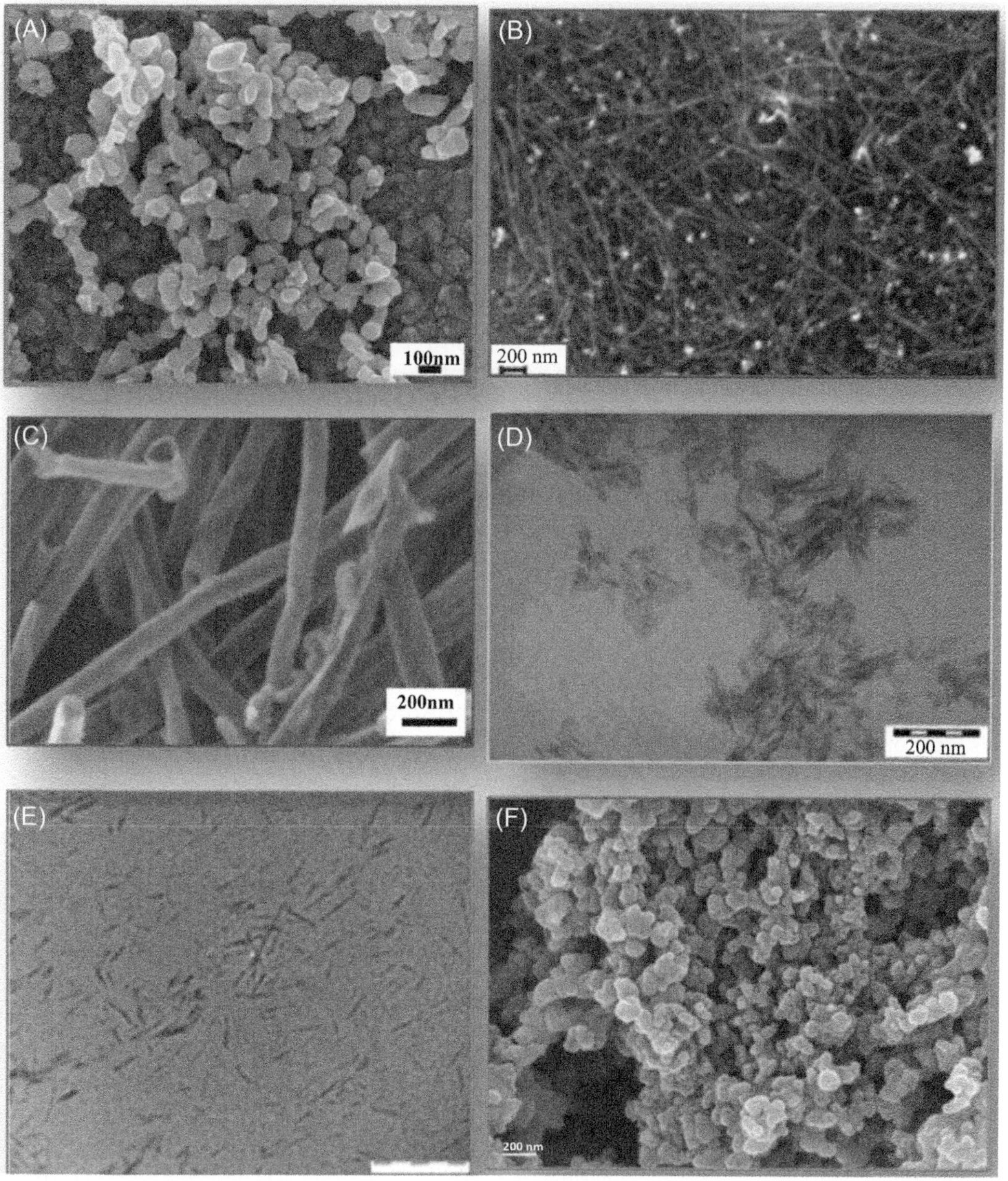

FIGURE 3.18

Electron microscopy images of the main used nanofillers: silver nanoparticles (A), single-walled carbon nanotubes (B), carbon nanofibers (C), nanohydroxyapatite (D), cellulose nanocrystals (E), and lignin nanoparticles (F) [76].

Reproduced with permission from MDPI. Open access from MDPI. Copyright 2018.

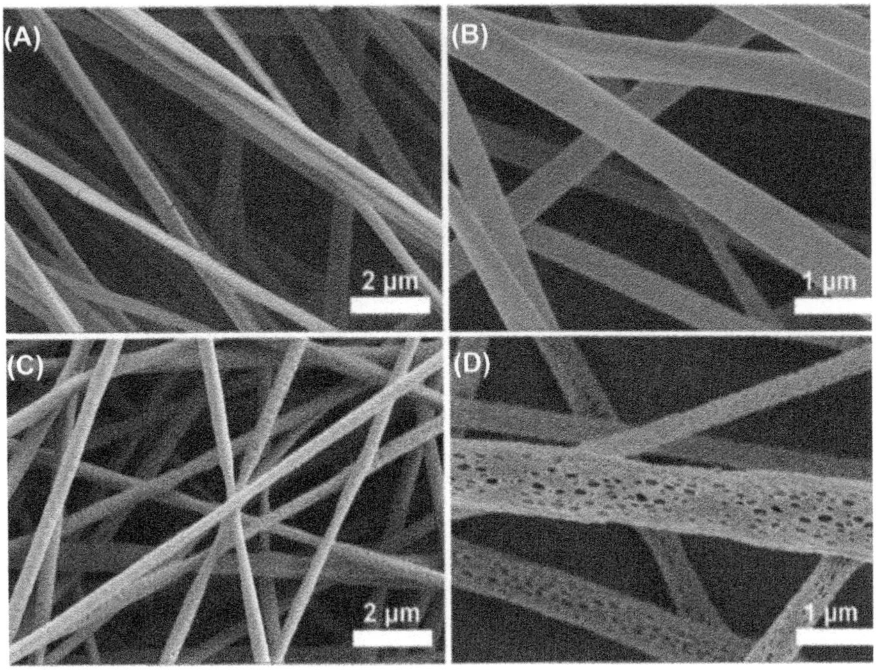

FIGURE 3.19

(A, B) Pure PLLA (C, D) PLLA/ POSS-$(PLLA)_{32}$600, 6% nanocomposite [78]. https://s100.copyright.com/order/1d852f87-6a7b-4786-846a-44a75f7cebba.

Reproduced with permission from Elsevier. Copyright 2019.

investigate Hap/BC nanocomposites. SEM was also used to examine phosphorylated BC. The samples were analyzed after soaking in 1.5% SBF. Unphosphorylated BC fibers were found to be surrounded by spheres of HAp particles, and Hap was found to be distributed unevenly throughout the BC fibers. In the micrographs of phosphorylated BC/HAp nanocomposites, the homogeneous distribution of Hap was clearly visible. The Hap spheres covered the entire sample both internally and externally [80].

Single-walled carbon nanotubes (SWCNT) opened up opportunities in nanotechnology. SWCNT incorporated with PLLA offers applications in the area of tissue engineering. PLLA/SWCNT and PLLA/SWCNT-COOH nanocomposites were desired nanocomposites in this study. The nanocomposites were visualized using SEM. The micrograph functionalized differently from the PLLA/SWCNT. PLLA/SWCNT could be characterized by better dispersal over the polymer matrix. SWCNT-COOH tubes were fragmented by the presence of COOH group. SWCNT, on the other hand, could be visualized as bundles of agglomerated tubes. SEM was also used to examine the cell morphology on PLLA nanocomposites. Cells preferred materials reinforced with PLLA and nanotubes. The cells were widely distributed along the nanocomposites. It was an unequivocal sign of biocompatibility [81]. Polymeric scaffolds are widely used in orthopedic surgery. The application of nanocomposites enhanced its role in this field. The performance of HBPu in conjunction with multiwalled carbon nanotubes was investigated in this study. Under SEM, porous and nonporous nanocomposites were clearly visible. The location of pores within the HBpu matrix was of particular interest. With a porosity of 39.53%, it was possible to see that the pores were distributed evenly. The observed pore size of the nanocomposite was in the range of 200–330 μm. This data was consistent with the biomedical data [82]. Chitin/silk fibroin/TiO_2 nanocomposites have been widely used in wound dressing, as well as other biomedical applications. These nanocomposites were investigated to determine their hemostatic ability in the human body. TiO_2 nanoparticles were used because they are nontoxic and readily available. SEM was used to prepare and analyze the scaffolds. The micrographs were examined to determine pore size and porosity. The layered structure was visible in pure chitin and silk fibroin. A 3D porous surface indicated chitin/silk/TiO_2 nanocomposites. The rough pore wall allowed for easy cell penetration, growth, and adhesion. Porosity decreased as TiO_2 content increased, possibly due to pore homogeneity caused by the interaction of polymer matrix and TiO_2 nanofillers. The observed porosity met the parameters for wound dressing. The size of the pores was 100–150 μm and 150–250 μm, which would agree with the above fact [83].

3.3.1.2 Transmission electron microscopy

Transmission electron microscopy (TEM) is a surface characterization technique that uses high electron beams, similar to SEM. The sample is allowed to pass through high electron beams. The presence of lattice planes within the crystal lattice causes scattering and diffraction of high-energy electrons. This phase identification is followed by elemental analysis and surface characterization [84]. The efficacy of the electrons from the gun was correlated to the nature of the surface of the sample. Some regions would deflect the electrons and some would allow them to pass [73]. In-depth analysis of BPNs was effectively done by TEM, by means of morphological, crystallographic and elemental information [85].

Because of their high biocompatibility and mechanical properties, chitosan-polylactic acid/hydroxyapatite nanocomposites have been widely used. These CS-PLA/HAp composites were widely used in the field of bone tissue engineering. These nanocomposites were created in the field. TEM was used to perform quantitative elemental analysis on the microstructure. The inorganic phases of the CS-PLA/HAp nanocomposites were characterized by TEM as represented in Fig. 3.20. The nanocomposites underwent an oxidation reaction and the organic phase was clearly separated from it. The size of the rod-shaped Hap particles inside the nanocomposites were found to be about 10 nm. The particles were found to be finely crystalline in nature. The adhesive properties of both the inorganic and organic phases would allow nanocomposites to improve mechanical strength. There is a possibility of nonmixing during the mixing of Hap particles with organic polymer matrix, resulting in the noncrystalline zone. As a result, the diffused spots would be visible in the micrograph as well [86].

TiO_2/poly(L-lactide-*co*-ε-caprolactide)(PLCL) nanocomposites are widely used in the biomedical field. TiO_2 nanoparticles have excellent physical and bioactive properties. Solution casting techniques were used to create nanocomposites based on PLCL and g-TiO_2 nanoparticles. TEM was used to characterize the morphology of g-TiO_2/PLCL nanocomposites. PLCL was mixed with both grafted and nongrafted TiO_2 nanoparticles. The distribution of nongrafted TiO_2 nanocomposites in the polymer matrix was easily identified in the TEM image. Aggregations of nanoparticles were more complex and difficult to determine than individual particles. Grafted TiO_2 particles, on the other hand, could be easily identified by a significant dispersion in the matrix. On the surface, the g-TiO_2 particles were stacked with the PCL polymer chains [37].

FIGURE 3.20

TEM micrographs of (A) nanocomposite (B) crystal lattice [86]. https://s100.copyright.com/order/ed8eec79-113d-4d07-92f2-0e50694d7d61.

Reproduced with permission from Elsevier. Copyright 2009.

Gelatin/hydroxyapatite nanocomposites were created as part of an investigation into their potential for appropriate responses of osteoblast cells. Gelatin-Hap nanocomposites were effective in promoting blood circulation and cell growth. The formation of gelatin-Hap nanocomposites was studied using TEM. TEM was used to examine hap nanocomposites and gelatin composites. 400–500 μm was the pore size of the nanocomposites. The alignment of Hap nanocrystals that vanished in the gelatin matrix was investigated. The composites had an elongated structure, indicating that the Hap nanoparticles were oriented in the crystal framework in this manner. The orientation of Hap nanocrystals was thought to be affected by the matrix's long protein network. The crystallization was not well defined, which resulted in the diffused ring pattern seen in the TEM images. The particles in conventional composites were distorted, and an agglomeration of Hap particles was stacked on the scaffolds [87].

Poly(L-lactic acid)/layered silicate nanocomposites met the biocompatibility and mechanical strength parameters in applications in the biomedical industry. Three types of nanoclays (cloisite 30B, cloisite 25A, cloisite 15A) were dispersed in the PLLA matrix. The resulting nanocomposites were studied using TEM. The TEM micrographs showed clay nanofillers throughout the PLLA matrix. The nanocomposites were filled with 10 wt.% each of cloisite 30B, cloisite 25A and cloisite 15A. Intercalated clay tactoids were highly magnified using TEM. The distance between two adjacent lines was measured. PLLA-cloisite 25A was distinguished by exfoliated platelets of clay layers. This was also the case for intercalated tactoids. Exfoliation of the PLLA/30B10 nanosystem was observed. Furthermore, the distribution of nanoparticles revealed uniformity throughout the PLLA matrix. As seen in the TEM, the nanoparticles were not bundled [88].

Poly(ε-caprolactone)/poly(lactic acid)/hydroxyapatite (PCA/PLA/HAp) nanocomposites are bioactive and they can be used in place of ruptured tissues in the human body. The polymer matrix in this study was a blend of two biodegradable and biocompatible polymers. TEM clearly revealed the crystalline nanostructure. The HAp nanocrystals were stacked in the interphase of the PCL-PLA matrix. The concentration of HAp nanofillers was inversely related to the size of the PLA spheres. This observation was made due to the clustering of nanoparticles within the polymer matrix. Because of the HA nanoparticle's affinity for the PLA polymer phase, agglomeration occurred inside the matrix, forming a network-like structure. The micrographs would show the densely packed HA nanoparticles [89].

The food packaging and biomedical industries are not the only ones that require biodegradable and biocompatible polymer nanocomposites. It has also been discovered to reduce the amount of electronic waste generated. The PBAT/PLA blend contained graphene nanoparticles. The topology of BPNs has been observed using TEM. The nanofiller used in this case was GH5. Five micrographs of GH5 10 wt.% and PLA ranges from 100% to 0% were examined. When there was no PLA content, GH5 dispersed more in the PBAT phase. The GH5 was replaced by a GC750, and a TEM was performed. In comparison to GH5, GC750 had a poor arrangement throughout the polymer matrix. GH5 has stacked within the PBAT phase as the PLA content of the matrix increases. An image of 75 wt.% PLA in the polymer matrix showed an inversion in the dispersed phase. Long rod structures were formed by the aggregated GH5. The micrographs clearly distinguished the affinity of GH5 to the PBAT phase rather than the PLA phase [90].

3.3.2 Scanning probe microscopy/atomic force microscopy

Scanning probe microscopy (SPM) is a powerful technique for imaging various surfaces. SPM includes atomic force microscopy (AFM). Surfaces can be imaged in three dimensions using this technique. It could also detect mechanical strength, adhesion strength, and atomic forces. In AFM, the contact and tapping modes were used. In contact mode, the probe's tip always makes contact with the surface. In tapping mode, on the other hand, the tip of the probe intermittently touches the surface. FM does not necessitate any special operating conditions. In normal sample conditions, it performs admirably. FM has a higher resolution than OM, SEM, or TEM [7]. The complex structure formation inside BPNs was clearly observed using AFM (Fig. 3.21).

AFM analysis of PLA fibers infused with TiO_2 nanoparticles effectively explained drug behavior within desired cells. AFM was used to characterize the nanostructure level. AFM micrographs of daunorubicin-treated PLA and PLA/TiO_2 nanocomposites were obtained. The presence of enhanced nanocrystals inside the PLA chain distinguished PLA/TiO_2 nanocomposites containing daunorubicin. The size of the nanocrystals was found to be 32 ± 1.62 nm. A small precipitate can also be found beneath the nanocrystals. On the PLA chain, TiO_2 nanoparticles and drug molecules were aggregated as spheres. This interaction was founded on the potential electrostatic interactions between TiO_2 particles, drug molecules, and PLA fibers [92]. Chitosan/montmorillonite/hydroxyapatite nanoparticles are well-known tools in the manufacture of bone tissue. In nature, it is very biocompatible. AFM was used effectively for nanostructure characterization. The samples under consideration were Chi/MMT, Chi/HAp, and Chi/MMT/HAp.

The polymer region was represented by the darker zones in the micrographs, while the brighter zones were represented by the nanofillers. The presence of nanofillers in the chitosan polymer matrix was discovered. MMT fibers lengthened at a range of 500 nm and widened at a range of 200 nm. On the other hand, 50–100 nm was the range of Hap nanoparticles in Chi/Hap nanocomposites. It was evenly distributed across the chitosan polymer matrix. Separately from the Chi/MMT/Hap

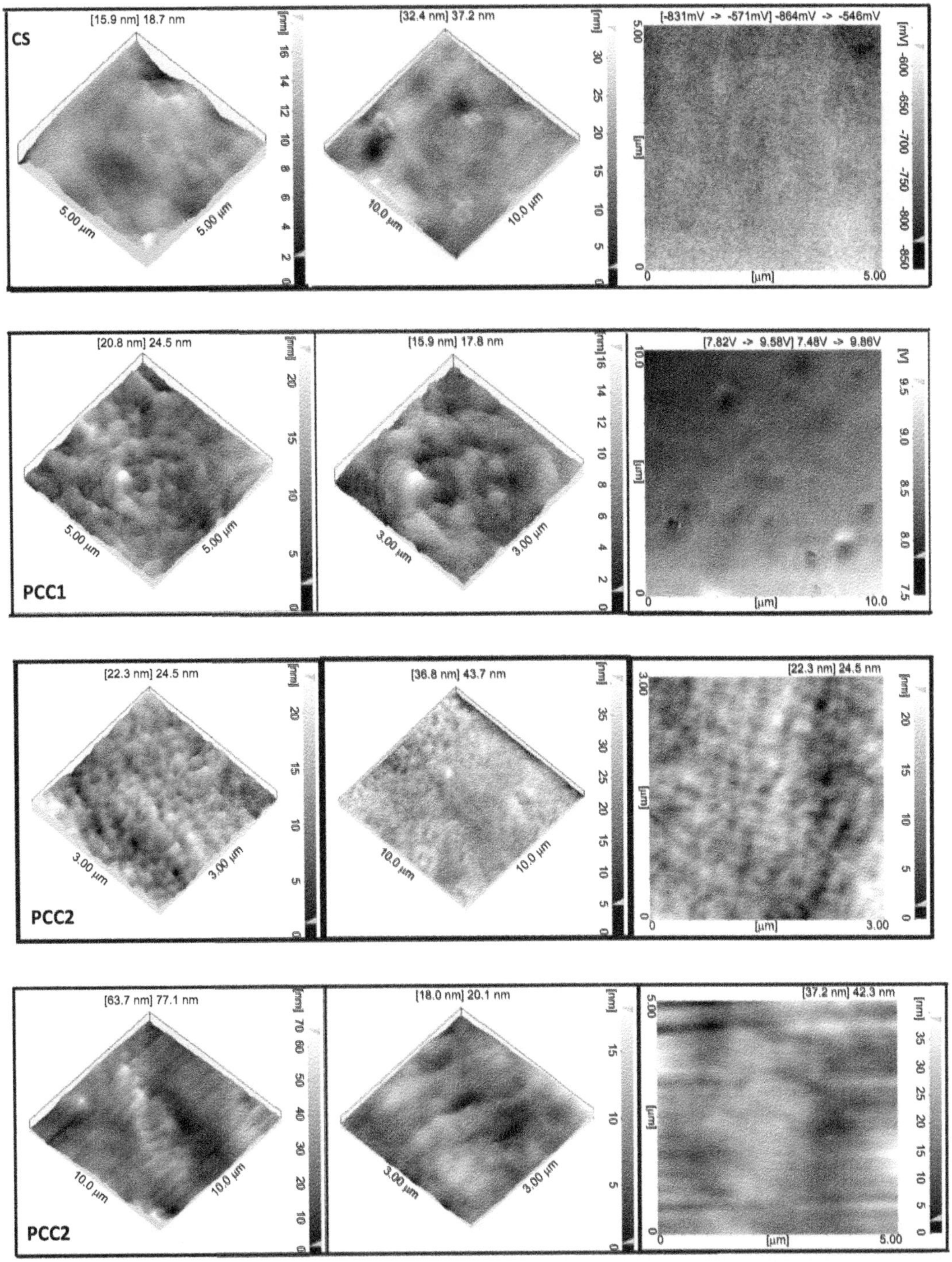

FIGURE 3.21

AFM of antibacterial food packaging [91]. https://s100.copyright.com/order/27a99363-832f-4194-9a95-2d21fd1024ee.

Reproduced with permission from Springer Nature. Copyright 2021.

nanocomposites, MMT and Hap nanoparticles were observed as rod-like and globular. Because of their small particle size, MMT and Hap particles dispersed more widely. The polymer matrix was discovered to strongly interact with the inorganic particles, which may account for the improved dispersion within the polymer matrix [93]. AFM study of poly(ε-caprolactone) based grafted cellulose nano whiskers provided a good contrast image. Polymer matrix was incorporated with 8 wt.% AFM was used to investigate CNWr. The needle shape of the cellulose nano whiskers had to be maintained. The CNWr nanoparticles were not distributed evenly. On the contrary, they appear in the PCL matrix as large aggregates. This could be due to electrostatic attraction between CNW-CNW particles. The morphology of CAPA6500/8 wt.% CNWr −g-PCL nanocomposites was totally at variance with the above observation. The CNWr nanocrystals were not found as aggregates on the surface of the matrix. On the other hand, they were homogeneously dispersed throughout the matrix. The size of the nanospheres was about 40 nm and they appeared perpendicular to the PCL matrix. The PCL grafting on CNWr was effectively worked out to control the surface assembly of nanospheres [94]. The incorporation of Au or Ag nanoparticles into chitosan would significantly alter the study's properties in a positive way. AFM was used to examine the topology of chitosan/Au nanocomposites and chitosan/Ag nanocomposites. Chitosan nanoparticles with 30, 60, 120, and 240 ppm Au and Ag were synthesized. When pure chitosan was mixed, the topography did not change significantly. The brighter spots indicated the presence of nanoparticles. The micrographs clearly showed the uniform distribution of Ag/Au nanoparticles within the matrix. The addition of nanoparticles increased the size of the harder domains. 26 nm was the average size of the aggregates found in the Chit/Au -60 ppm nanocomposites. However, 42 nm was observed in the case of Chit/Ag 120 ppm nanocomposites. Average roughness was in the range of 0.3–0.6 nm [95].

Poly(lactic acid)/poly(-caprolactone) carbon nanotube bio-nanocomposites were widely used in the fabrication of tissue engineering scaffolds. The nanocomposites PLA/PCL/MWCNT were created. Pure PLA and PCL samples were also collected. Various samples were subjected to AFM analysis. The pure PLA represented smooth morphology with 3.7 nm thickness. However, it was 17.7 nm in the case of pure PCL. PLA/PCL blends were also investigated. PCL had a slight affinity for PLA. The voids were clearly visible within the PLA phase. The formation of voids was caused by the deprivation of PCL from PLA caused by cryo-microtome cutting. The addition of MWCNT changed the topography of the polymer blend significantly. The voids vanished. The LA in PCL/PLA/MWCNT had the same flatness as pure PLA. Between the PCL-PLA interphases, MWCNT was stacked. It would significantly improve polymer compatibility. The true width of the brighter spot was found to be 40–60 nm range [96].

3.3.3 Optical microscopy

Because of its simplicity, it is the oldest but most widely used microscopic technique. This technique can be used in both reflected (episcopy) and transmitted (diascopy) modes. The sample is examined using a light beam emitted from a suitable source. An optical microscope's basic components include a light source, condenser lens, objective lens, projected lens, and detector [84]. Basic information such as spherulitic growth and the issuing of nanofillers through the desired polymer matrix could be obtained from episcopic technique and the insightful observations on the microstructure can be obtained using the diascopic technique represented in Fig. 3.22. The observed magnification is up to 1500x in studies and the theoretical resolution is in the range of 200–300 nm

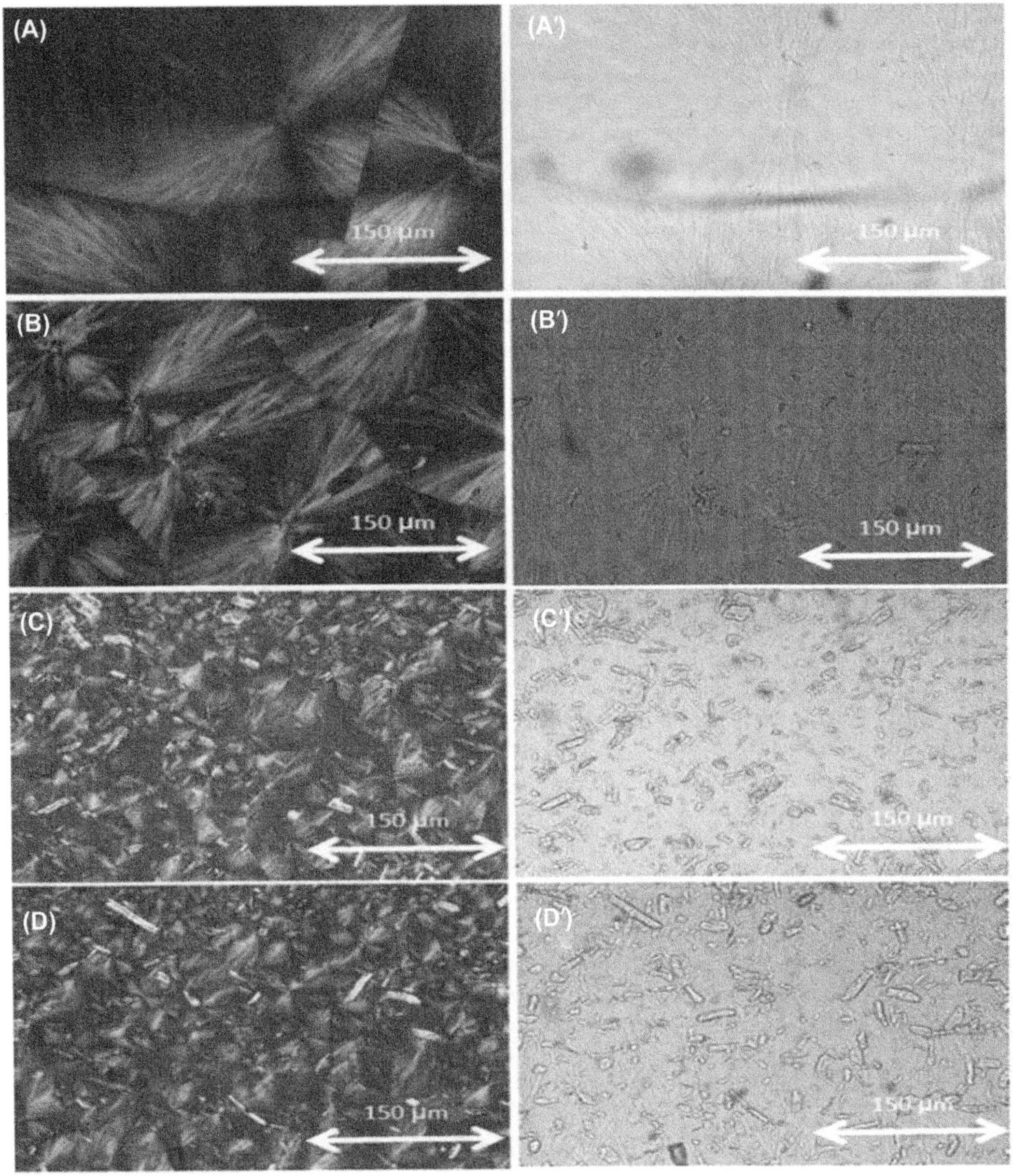

FIGURE 3.22

Polarized optical microscopy of spherulite of pure PLLA and its composite. (A) Neat PLLA, (B) biocomposite 1, (C) biocomposite 2, (D) biocomposite 3; without optical polarizer (A′) Neat PLLA, (B′) biocomposite 1 (C′) biocomposite 2, (D′) biocomposite 3 [97].

Reproduced with permission from Nature. Open access from Nature. Copyright 2017.

(lateral mode) with the axial mode in the 500–700 nm range [7]. Provided here are some insightful data of OM analysis of BPNs.

Polyhydroxybutyrate (PHB) is a well-known biodegradable polymer derived from natural resources like sugar and molasses. To investigate the improvement of PHB's physical and mechanical properties following the addition of two types of nanofillers, montmorillonite cloisite-Na^+ and cloisite C15A were incorporated into the PHB polymer matrix. The OM technique could be used to assess the differences in properties. Polarized optical microscopy (POM) was used to investigate the morphological properties of PHB/MMT nanocomposites. The images were created by continuously measuring the diameter of the spherulite over time. According to the micrographs, CNa^+ and C25A have the same crystallization morphology. The films were melted and micrographs were taken to differentiate the two nanocomposites. It was clear that CNa^+ dispersion had an agglomerated arrangement, which was not seen in C15A nanocomposites. The conclusion was that clay C15A has a large interlayer distance in PHB, which would result in better dispersion in the matrix. Isothermal kinetic analysis was carried out using POM. The spherulitic growth rate (G) was calculated using the spherulitic diameter vs. time curve. The slope of the curve was used to calculate radial growth. From the curve of G vs. T_c it was evident that C15A nanocomposites have improved crystallization in PHB [98].

PHBV polymer associated with inorganic nanotubes of tungsten disulfide (WS_2) was synthesized and it was compressed into thin film. POM was effectively able to characterize the spherulitic structure of the PHBV/INT-WS_2 nanocomposites. The comparison of fresh PHBV and PHBV/INT-WS_2 was carried out under the dynamic crystallization rates of 10°c min and 2°c min. 0.5 and 1.0 wt.% nanocomposites, and clean PHBV samples possessed ring-banded spherulites and it resulted from stress-induced or nonlamellar twisting. After cooling to room temperature, it revealed concentric bands and Maltese cross shapes. Following an increase in the wt.% of INT-WS_2 (0.5% and 1.0%), there was a disruption in the aggregation of spherulites. However, it could be the result of increased nucleation and crystal growth of PHBV [99].

In terms of physical properties and enhanced processability, polylactide/PLA is a widely discussed biodegradable polymer. PLA was filled with clay nanocomposites that had previously been provided organically. Clay nanoparticles include 3 wt.% of Na^+ montmorillonite [MC] and organo-modified montmorillonite [NC]. At room temperature, the crystalline structure of neat PLA and micro and nanocomposites derived from PLA was evaluated using POM. The large spherulite in the pure PLA matrix has a diameter of 330 m, indicating a low nucleation density. On the other hand, the spherulitic size was decreased to 150 μm in the case of PLA/clay nanocomposites. The nanofillers played an unavoidable role in the polymer matrix nucleation. The number of inclusions in the matrix has an inverse relationship with the size of the spherulites. These inclusions rearranged the lamellar regions in general [100].

Poly(ε-caprolactone) (PCL) is another widely discussed polymer that holds a unique place in the food packaging industry due to its ability to provide desirable material properties such as biodegradability and biocompatibility. The mechanical and physical properties of the PCL polymer were triggered by PCL nanocomposites. Natural layered silicates were mixed into the PCL matrix. Unmodified sepiolite and two types of modified sepiolite were used to create PCL nanocomposites. Polymer backbone grafting was used to create the modified sepiolites, which were then subjected to ionic exchange. The optical microscope was used to examine the surface morphology of PCl-SEPx, PCl-SEPSIx, and PCl-SEPHDAx. The micrographs of the two sides of the BPNs were discovered to be dissimilar. The air-contact side was about to describe the morphology of the spherulites. The diameter of the spherulites gradually decreased as the content of the nanofillers incorporated increased the nucleational density. The boundary was undetected due to spherulitic impingement. The diameter of spherulites distinguished both modified and unmodified sepiolites. The diameter of PCL-SEPSI nanocomposites was greater than that of PCL-SEP nanocomposites. This observation could be explained by the nucleation and proper placement of nanofillers throughout the PCL matrix [101].

The promise of starch c-clay nanocomposites in meeting the requirement in environmentally friendly BPNs was intriguing. This work was done primarily to investigate the differences in the desirable mechanical and physical properties of two different starch-clay nanocomposites. Melt compounding was used to incorporate organically modified O-clay and pristine clay (P-clay) into the starch polymer matrix. The presence of POM was discovered in the topological study of starch-clay nanocomposites. POM micrographs were taken during gelatinization at temperatures of 20°C, 100°C, and 110°C. Because of the large spherulite size, the birefringence of the starch granules in the micrographs made the early stages easy to identify. When the temperature reached 100°C, the birefringence began to blur and then vanished completely. However, even after gelatinization, the starch granules retained their shape [102].

MWCNTs (multiwalled carbon nanotubes) have been extensively studied for their mechanical, thermal, and electrical properties. PCL/MWCNT nanocomposites were biodegradable and biocompatible, which would be beneficial in the food packaging industry. The optical micrographs of PCL and PCL/MWCNT nanocomposites were compared at 90°C and then cooled to room temperature. The Maltese cross ring bands can be seen in the pristine PCL matrix. When 1 wt.% PCL/MWCNT was indicated by the increase in the nucleation density, it would result in an increase in the number of spherulites. The PCL /MWCNT exhibited a greater nucleation rate than the pure PCL [103].

3.3.4 Confocal laser scanning microscopy

The confocal laser scanning microscope (CLSM) is a type of optical microscope distinguished from other types by its high resolution. The sample was scanned using light emitted through a pinhole aperture. A series of optical images were collected. Regardless of specimen thickness, CLSM provides a three-dimensional level of information on cell biology. It might be able to remove background noise from the focal plane [7]. The nanometer level information was not practically obtained from CLSM [104,105].

Novel chitosan/agarose/hydroxyapatite nanocomposites could be used to create scaffolds for bone tissue engineering. It was a sturdy biocompatible scaffold that aided bone healing. The biodegradable polymer matrix used here is chitosan/agarose. CLSM was used to investigate surface morphology and cell interaction. The absorption of proteins was visualized using CLSM as given in Fig. 3.23. The chitosan-based new polymeric scaffolds were filled with MC3T3-E1 cells and nFoB 1.19 osteoblasts. The stained cells were scrutinized. Fibrinogen was easily converted to fibrin fibrils at blood plasma sites with low concentrations of nano HA particles. This could be due to the electrostatic attraction of the protein chain's polar groups and the chit/agarose/Hap nanocomposites [106].

Lactose-modified chitosan was combined with silica nanoparticles, which have potent antimicrobial properties. These nanocomposites were used as coatings for methyl acrylic thermosets, which have important applications in dentistry and orthopedics. CLSM was used to characterize the surface characterization and dispersion of nanocomposites over cells. The BisGMA/TEGDMA

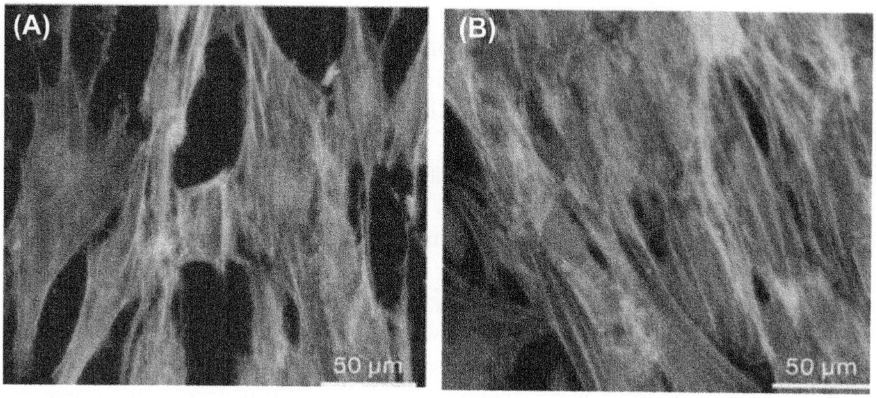

FIGURE 3.23

CLSM micrographs of human primary fibroblasts spread on (A) uncoated and (B) chitlac-nAg coated thermosets [106]. https://s100.copyright.com/order/613ff2bf-e762-4b3d-8825-5ace9fe84e2d.

Reproduced with permission from Elsevier. Copyright 2011.

thermoset was coated with chit/Ag nanocomposites. CLSM was used to stain and analyze osteoblast or fibroblast cells. Chitlac-nAg nanocomposites coated on thermoset, as well as uncoated thermoset, were cultured with MG63 cells for 72 h. The topography of the cells in coated and uncoated thermosets indicated healthy growth. The grooves on the row thermosets were filled in an orderly fashion with actin filaments. CLSM was also used to stain and visualize SaOS-2 (osteoblast cells). The cells were evenly distributed across the polysaccharide matrix. Cell mitosis was observed to result in healthy cell proliferation [107].

Polymeric hydrogels are a recent breakthrough with applications in cartilage tissue engineering. A coating of TiO_2 nanoparticle-based nanocomposites improved the surface activity of the hydrogels. PVA/PVP was the appropriate polymer matrix used in this case. CLSM was used to investigate bioactivity and cell adhesion. For a better understanding of bacterial inhibition against nanocomposites, micrographs of PVP/PVA and PVP/PVA-TiO_2 biofilms were examined. The microbes under consideration were *S. aureus* and *E. coli*. Bacterial inhibition was approximately 75% in PVA/PVP-TiO_{22} nanocomposites. The increased addition of TiO_{22} nanofillers throughout the film improved cytoplasmic membrane binding. This would aid in the prevention of bacterial proliferation [108].

3.3.5 Photoacoustic microscopy

The conversion of light energy into ultrasound energy is the underlying principle of photoacoustic microscopy (PAM). A sample was irradiated with a brief burst of light. This would result in ultrasound waves. Photoacoustic images have deep applications in biomedical imaging represented in Fig. 3.24. PAM enables the molecular and functional imaging of living cells [91]. PAM is classified into two types. The first is acoustic resolution PAM (AR- PAM), and the second is optical resolution PAM (OR-PAM) [110]. The main advantages of PAM inside the living system are deep imaging and compatible functioning. The most prominent fields of interest are neuroscience and cancer biology [111].

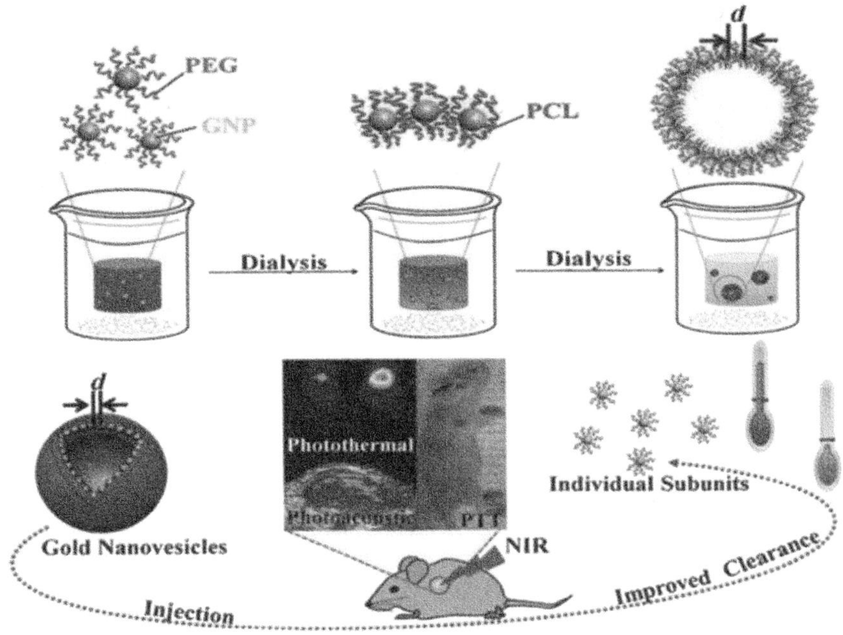

FIGURE 3.24

Schematic illustration of using PAM [109]. https://s100.copyright.com/order/1b7b1fbc-6db7-4296-a7ba-1aaa1d8bd069.

Reproduced with permission from Elsevier. Copyright 2016.

The PLGA matrix was infused with single-walled carbon nanotubes. PLGA was a biodegradable and biocompatible polymer that was used in tissue regeneration. The SWNT/PLGA nanocomposites were designed to be biocompatible within living cells. The multiscale photoacoustic microscope was used to characterize the surface of the nanocomposites. SWNT was responsible for generating signals from PAM. The porosity and pore size of the scaffold were determined using the acquired images. PAM indisputably imaged the porous surface of the wet SWNT/PLGA nanocomposites. For OR-PAM the porosity was found to be 86.5% ± 1.2% and the pore size was 153 ± 15 μm (Fig. 3.19B). However, AR-PAM demonstrated the porosity as 81.8% ± 4.4% and pore size as 167 ± 24 μm. The pore size of the scaffolds was completely responsible for cellular infiltration, exchange of essential elements, and waste management among the scaffolds and living cells. As a result, the larger the pore size, the greater the efficiency of the scaffolds [112].

3.4 Conclusion

Only a few polymeric materials have been studied for biodegradability and biocompatibility. Both microscopic and spectroscopic techniques can be used to drive structural and morphological characterization, optimize synthetic and dynamic processes, and develop technological know-how for developing new bio-nanocomposite materials with tailored properties. As a result, concurrent scientific exploration using microscopic and spectroscopic tools has the potential to strengthen this field of study. It was discovered from both microscopic and spectroscopic information of composites available in the literature that useful strategies for embedding biodegradability and biocompatibility in a given material include polymer backbone modification via cross-linking, copolymerization, and blending. Thus, various microscopic and spectroscopic techniques that aid in the chemical and physical identification of biodegradable and biocompatible nanocomposites have been identified in this study. The prospects for spectroscopic and microscopic analysis of biodegradable polymer nanocomposites are promising in the future. Though each spectroscopic technique is discussed in this chapter separately, the future potential endowed by the interdisciplinary and complementary nature of these techniques in studying the properties of bio-nanocomposites should not be overlooked.

References

[1] N. Karak, in: N. Karak (Ed.), Chapter 1 - Fundamentals of Nanomaterials and Polymer Nanocomposites, Elsevier, 2019, pp. 1–45.

[2] S.S. Bari, A. Chatterjee, S. Mishra, Biodegradable polymer nanocomposites: an overview, Polym. Rev. 56 (2016) 287–328.

[3] S.J. Eichhorn, Cellulose nanowhiskers: promising materials for advanced applications, Soft Matter. 7 (2011) 303–315.

[4] H. Liu, R. Jian, H. Chen, X. Tian, C. Sun, J. Zhu, et al., Application of biodegradable and biocompatible nanocomposites in electronics: current status and future directions, Nanomaterials 9 (2019) 950.

[5] S. Kumar, M. Nehra, N. Dilbaghi, K. Tankeshwar, K.-H. Kim, Recent advances and remaining challenges for polymeric nanocomposites in healthcare applications, Prog. Polym. Sci. 80 (2018) 1–38.

[6] L. Bokobza, Spectroscopic techniques for the characterization of polymer nanocomposites: a review, Polym. (Basel) 10 (2018) 7.

[7] A. Venkateshaiah, V.V.T. Padil, M. Nagalakshmaiah, S. Waclawek, M. Černík, R.S. Varma, Microscopic techniques for the analysis of micro and nanostructures of biopolymers and their derivatives, Polym. (Basel) 12 (2020) 512.

[8] K.M. Dean, P. Sangwan, C. Way, M.A.L. Nikolic, Biodegradability characterization of polymer nanocomposites, Charact. Tech. Polym. Nanocompos. 3 (2012).

[9] D. Ponnamma, D. Rouxel, S. Thomas, Spectroscopy—introducing the advantages and application areas in polymer nanocomposites, Spectroscopy of Polymer Nanocomposites, Elsevier, 2016, pp. 1–14.

[10] E. Rosenberg, U. Panne, Atomic absorption spectrometry (AAS) and atomic emission spectrometry (AES), Handb. Spectrosc. (2003) 421–496.

[11] F. Benhacine, A. Ouargli, A.S. Hadj-Hamou, Preparation and characterization of novel food packaging materials based on biodegradable PCL/Ag-Kaolinite nanocomposites with controlled release properties, Polym. Technol. Mater. 58 (2019) 328–340.

[12] K. Shameli, M.Bin Ahmad, W. Md Zin Wan Yunus, N.A. Ibrahim, R.A. Rahman, M. Jokar, et al., Silver/poly (lactic acid) nanocomposites: preparation, characterization, and antibacterial activity, Int. J. Nanomed. 5 (2010) 573–579.

[13] D.H.A. Corrêa, C.H.I. Ramos, The use of circular dichroism spectroscopy to study protein folding, form and function, Afr. J. Biochem. Res. 3 (2009) 164–173.

[14] L. Whitmore, B.A. Wallace, Protein secondary structure analyses from circular dichroism spectroscopy: methods and reference databases, Biopolym. Orig. Res. Biomol. 89 (2008) 392–400.

[15] S. Lal, A. Perwez, M.A. Rizvi, M. Datta, Design and development of a biocompatible montmorillonite PLGA nanocomposites to evaluate in vitro oral delivery of insulin, Appl. Clay Sci. 147 (2017) 69–79.

[16] G. Yang, Y. Xu, Vibrational circular dichroism spectroscopy of chiral molecules, Electron. Magn. Prop. Chiral Mol. Supramol. Archit. (2010) 189–236.

[17] S. Kumar, J. Koh, Physiochemical and optical properties of chitosan based graphene oxide bionanocomposite, Int. J. Biol. Macromol. 70 (2014) 559–564.

[18] K. Zhan, H. Liu, H. Zhang, Y. Chen, H. Ni, M. Wu, et al., A facile method for the immobilization of myoglobin on multi-walled carbon nanotubes: poly(methacrylic acid-co-acrylamide) nanocomposite and its application for direct bio-detection of H_2O_2, J. Electroanal. Chem. 724 (2014) 80–86.

[19] Y. Fang, Y. Ni, G. Zhang, C. Mao, X. Huang, J. Shen, Biocompatibility of CS–PPy nanocomposites and their application to glucose biosensor, Bioelectrochemistry 88 (2012) 1–7.

[20] L. Mendonça, F. Hache, Nanosecond T-jump experiment in poly(glutamic acid): a circular dichroism study, Int. J. Mol. Sci. 13 (2012).

[21] V.I. Krinichnyi, EPR spectroscopy of polymer: fullerene nanocomposites, Spectroscopy of Polymer Nanocomposites, Elsevier, 2016, pp. 202–275.

[22] D. Hinderberger, EPR spectroscopy in polymer science, EPR Spectrosc. (2011) 67–89.

[23] K.C. Khulbe, A.F. Ismail, T. Matsuura, in: N. Hilal, A.F. Ismail, T. Matsuura, D.B.T.-M.C. Oatley-Radcliffe (Eds.), Chapter 3 - Electron Paramagnetic Resonance (EPR) Spectroscopy, Elsevier, 2017, pp. 47–68.

[24] A. Pawlicka, R.I. Mattos, C.E. Tambelli, I.D.A. Silva, C.J. Magon, J.P. Donoso, Magnetic resonance study of chitosan bio-membranes with proton conductivity properties, J. Memb. Sci. 429 (2013) 190–196.

[25] S. Iftekhar, V. Srivastava, A. Casas, M. Sillanpää, Synthesis of novel GA-g-PAM/SiO_2 nanocomposite for the recovery of rare earth elements (REE) ions from aqueous solution, J. Clean. Prod. 170 (2018) 251–259.

[26] R. Thirumdas, M. Janve, K. Siliveru, A. Kothakota, 10 - Determination of food quality using atomic emission spectroscopy. Wang XBT-ET for FQ in: J. Zhong (Ed.), Woodhead Publ. Ser. Food Sci. Technol. Nutr, Woodhead Publishing, 2019, pp. 175–192.

[27] Y.I.N. Na, S. Chen, Y. Ouyang, T. Lian, J. Yang, H. Wang, Biomimetic mineralization synthesis of hydroxyapatite bacterial cellulose nanocomposites, Prog. Nat. Sci. Mater. Int. 21 (2011) 472–477.

[28] W.E. Braselton, K.J. Stuart, T.P. Mullaney, T.H. Herdt, Biopsy mineral analysis by inductively coupled plasma—atomic emission spectroscopy with ultrasonic nebulization, J. Vet. Diagnostic Investig. 9 (1997) 395–400.

[29] J. Nelson, G. Gilleland, L. Poirier, D. Leong, P. Hajdu, F. Lopez-Linares, Elemental analysis of crude oils using microwave plasma atomic emission spectroscopy, Energy Fuels 29 (2015) 5587–5594.

[30] M.F. Chiang, T.M. Wu, Synthesis and characterization of biodegradable poly(l-lactide)/layered double hydroxide nanocomposites, Compos. Sci. Technol. 70 (2010) 110–115.

[31] M. Abd Mutalib, M.A. Rahman, M.H.D. Othman, A.F. Ismail, J. Jaafar, Scanning electron microscopy (SEM) and energy-dispersive X-ray (EDX) spectroscopy, Membr. Charact, Elsevier, 2017, pp. 161–179.

[32] C.-C. Lin, S.-J. Fu, Y.-C. Lin, I.-K. Yang, Y. Gu, Chitosan-coated electrospun PLA fibers for rapid mineralization of calcium phosphate, Int. J. Biol. Macromol. 68 (2014) 39–47.

[33] N. Ahmad, S. Sultana, A. Azam, S. Sabir, M.Z. Khan, Novel bio-nanocomposite materials for enhanced biodegradability and photocatalytic activity, N. J. Chem. 41 (2017) 10198–10207.

[34] T. Jayaramudu, G.M. Raghavendra, K. Varaprasad, R. Sadiku, K.M. Raju, Development of novel biodegradable Au nanocomposite hydrogels based on wheat: for inactivation of bacteria, Carbohydr. Polym. 92 (2013) 2193–2200.

[35] R. Khandanlou, M.B. Ahmad, K. Shameli, E. Saki, K. Kalantari, Studies on properties of rice straw/polymer nanocomposites based on polycaprolactone and Fe_3O_4 nanoparticles and evaluation of antibacterial activity, Int. J. Mol. Sci. 15 (2014) 18466–18483.

[36] S. Kumar, B. Krishnakumar, Sobral AJFN, Koh J. Bio-based (chitosan/PVA/ZnO) nanocomposites film: thermally stable and photoluminescence material for removal of organic dye, Carbohydr. Polym. 205 (2019) 559–564.

[37] X. Lu, X. Lü, J. Wang, Z. Sun, Y. Tong, Preparation and shape memory properties of TiO_2/PLCL biodegradable polymer nanocomposites, Trans. Nonferrous Met. Soc. China 23 (2013) 120–127.

[38] M.C. Chang, J. Tanaka, FT-IR study for hydroxyapatite/collagen nanocomposite cross-linked by glutaraldehyde, Biomaterials 23 (2002) 4811–4818.

[39] M.A. Mohamed, J. Jaafar, A.F. Ismail, M.H.D. Othman, M.A. Rahman, in: N. Hilal, A.F. Ismail, T. Matsuura, D.B.T.-M.C. Oatley-Radcliffe (Eds.), Chapter 1 - Fourier Transform Infrared (FTIR) Spectroscopy, Elsevier, 2017, pp. 3–29.

[40] R. Li, K. Nie, X. Shen, S. Wang, Biodegradable polyester hybrid nanocomposites containing titanium dioxide network and poly(ε-caprolactone): synthesis and characterization, Mater. Lett. 61 (2007) 1368–1371.

[41] W. He, T. Yong, W.E. Teo, Z. Ma, S. Ramakrishna, Fabrication and endothelialization of collagen-blended biodegradable polymer nanofibers: potential vascular graft for blood vessel tissue engineering, Tissue Eng. 11 (2005) 1574–1588.

[42] A. Rezaei, M.R. Mohammadi, In vitro study of hydroxyapatite/polycaprolactone (HA/PCL) nanocomposite synthesized by an in situ sol–gel process, Mater. Sci. Eng. C. 33 (2013) 390–396.

[43] J.S. Lyu, J.-S. Lee, J. Han, Development of a biodegradable polycaprolactone film incorporated with an antimicrobial agent via an extrusion process, Sci. Rep. 9 (2019) 20236.

[44] Y. Du, J. Ge, Y. Li, P.X. Ma, B. Lei, Biomimetic elastomeric, conductive and biodegradable polycitrate-based nanocomposites for guiding myogenic differentiation and skeletal muscle regeneration, Biomaterials 157 (2018) 40–50.

[45] J. Fan, C.D. Grande, D.F. Rodrigues, Biodegradation of graphene oxide-polymer nanocomposite films in wastewater, Env. Sci. Nano 4 (2017) 1808–1816.

[46] D. Klemenčič, B. Simončič, B. Tomšič, B. Orel, Biodegradation of silver functionalised cellulose fibres, Carbohydr. Polym. 80 (2010) 426–435.

[47] E. Nazarzadeh Zare, M. Mansour Lakouraj, M. Mohseni, Biodegradable polypyrrole/dextrin conductive nanocomposite: synthesis, characterization, antioxidant and antibacterial activity, Synth. Met. 187 (2014) 9–16.

[48] D. Predoi, A study on iron oxide nanoparticles coated with dextrin obtained by coprecipitation, Dig. J. Nanomater. Biostruct. 2 (2007) 169–173.

[49] T.K. Vishnuvardhan, V.R. Kulkarni, C. Basavaraja, S.C. Raghavendra, Synthesis, characterization and a.c. conductivity of polypyrrole/Y_2O_3 composites, Bull. Mater. Sci. 29 (2006) 77–83.

[50] J.S. Yaradoddi, N.R. Banapurmath, S.V. Ganachari, M.E.M. Soudagar, N.M. Mubarak, S. Hallad, et al., Biodegradable carboxymethyl cellulose based material for sustainable packaging application, Sci. Rep. 10 (2020) 21960.

[51] M.I. Voronova, O.V. Surov, S.S. Guseinov, V.P. Barannikov, A.G. Zakharov, Thermal stability of polyvinyl alcohol/nanocrystalline cellulose composites, Carbohydr. Polym. 130 (2015) 440–447.

[52] N.T. Ardekani, M. Khorram, K. Zomorodian, S. Yazdanpanah, H. Veisi, H. Veisi, Evaluation of electrospun poly (vinyl alcohol)-based nanofiber mats incorporated with Zataria multiflora essential oil as potential wound dressing, Int. J. Biol. Macromol. 125 (2019) 743–750.

[53] A.A. Novakova, V.Y. Lanchinskaya, A.V. Volkov, T.S. Gendler, T.Y. Kiseleva, M.A. Moskvina, et al., Magnetic properties of polymer nanocomposites containing iron oxide nanoparticles, J. Magn. Magn Mater. 258–259 (2003) 354–357.

[54] B.S. Clausen, S. Mørup, H. Topsøe, R. Candia, E.J. Jensen, A. Baranski, et al., Mössbauer studies of the activation of ammonia synthesis catalysts, Le. J. Phys. Colloq. 37 (1976) C6–245.

[55] F. Bødker, S. Mørup, S. Linderoth, Surface effects in metallic iron nanoparticles, Phys. Rev. Lett. 72 (1994) 282.

[56] F.-Z. Qian, B. Melhachat, C.-X. Chen, K.-Y. Jiang, Z.-J. Zhao, X.-L. Yang, A Mössbauer study of iron/polytetrafluoroethylene nanocomposites prepared by high-energy ball milling, Nucl. Sci. Tech. 17 (2006) 139–142.

[57] S. Ramasamy, J. Jiang, H. Gleiter, R. Birringer, U. Gonser, Investigation of nanocrystalline FeF_2 by Mössbauer spectroscopy, Solid. State Commun. 74 (1990) 851–855.

[58] H. Guérault, M. Tamine, J.M. Greneche, Mössbauer study of nanostructured iron fluoride powders, J. Phys. Condens. Matter 12 (2000) 9497.

[59] K. Rekos, Z.-C. Kampouraki, C. Sarafidis, V. Samanidou, E. Deliyanni, Graphene oxide based magnetic nanocomposites with polymers as effective bisphenol–a nanoadsorbents, Materials (Basel), 12, 2019, p. 1987.

[60] W. Wang, Y. Guo, J.U. Otaigbe, The synthesis, characterization and biocompatibility of poly(ester urethane)/polyhedral oligomeric silesquioxane nanocomposites, Polym. (Guildf.) 50 (2009) 5749–5757.

[61] H.-T. Wang, J.-M. Wang, T.-M. Wu, Synthesis and characterization of biodegradable aliphatic–aromatic nanocomposites fabricated using maleic acid-grafted poly[(butylene adipate)-co-terephthalate] and organically modified layered zinc phenylphosphonate, Polym. Int. 68 (2019) 1531–1537.

[62] A.R. D'Amato, D.L. Puhl, S.A.T. Ellman, B. Balouch, R.J. Gilbert, E.F. Palermo, Vastly extended drug release from poly(pro-17β-estradiol) materials facilitates in vitro neurotrophism and neuroprotection, Nat. Commun. 10 (2019) 4830.

[63] M.S. Loeian, D.A. Ziolkowska, F. Khosravi, J.B. Jasinski, B. Panchapakesan, Exfoliated WS2-nafion composite based electromechanical actuators, Sci. Rep. 7 (2017) 14599.

[64] C. Bao, L. Song, W. Xing, B. Yuan, C.A. Wilkie, J. Huang, et al., Preparation of graphene by pressurized oxidation and multiplex reduction and its polymer nanocomposites by masterbatch-based melt blending, J. Mater. Chem. 22 (2012) 6088–6096.

[65] M.C. Biswas, S. Jeelani, V. Rangari, Influence of biobased silica/carbon hybrid nanoparticles on thermal and mechanical properties of biodegradable polymer films, Compos. Commun. 4 (2017) 43–53.

[66] Y.K. Mishra, S. Mohapatra, V.S.K. Chakravadhanula, N.P. Lalla, V. Zaporojtchenko, D.K. Avasthi, et al., Synthesis and characterization of Ag-polymer nanocomposites, J. Nanosci. Nanotechnol. 10 (2010) 2833–2837.

[67] M. Gao, L. Lu, X. Xu, Z. Yao, Y. Jiang, Electropolymerization and multifunctional properties of novel polypyrrole films embedded with Co nanoparticles, Electrochim. Acta 228 (2017) 522–527.

[68] X. Yuan, X. Zeng, H.-J. Zhang, Z.-F. Ma, C.-Y. Wang, Improved performance of proton exchange membrane fuel cells with p-toluenesulfonic acid-doped co-PPy/C as cathode electrocatalyst, J. Am. Chem. Soc. 132 (2010) 1754–1755.

[69] B.K. Kuila, A.K. Nandi, Physical, mechanical, and conductivity properties of poly(3-hexylthiophene)–montmorillonite clay nanocomposites produced by the solvent casting method, Macromolecules 37 (2004) 8577–8584.

[70] C. He, Q. Chen, M.A. Yarmolenko, A.A. Rogachev, D.G. Piliptsou, X. Jiang, et al., Structure and antibacterial activity of PLA-based biodegradable nanocomposite coatings by electron beam deposition from active gas phase, Prog. Org. Coat. 123 (2018) 282–291.

[71] R. Justin, B. Chen, Characterisation and drug release performance of biodegradable chitosan–graphene oxide nanocomposites, Carbohydr. Polym. 103 (2014) 70–80.

[72] B. Prakash, M.A. Jothirajan, S. Umapathy, V. Amala, Synthesis and characterization of biodegradable ultrasonicated films made from chitosan/Al_2O_3 polymer nanocomposites, Phys. Procedia 49 (2013) 84–91.

[73] F. Adams, C. Barbante, Electron-based imaging techniques, Compr. Anal. Chem., 69, Elsevier, 2015, pp. 269–313.

[74] S. Mun, H.C. Kim, M. Yadave, J. Kim, Graphene oxide–gellan gum–sodium alginate nanocomposites: synthesis, characterization, and mechanical behavior, Compos. Interfaces 22 (2015) 249–263.

[75] A. Wittmar, D. Vorat, M. Ulbricht, Two step and one step preparation of porous nanocomposite cellulose membranes doped with TiO_2, RSC Adv. 5 (2015) 88070–88078.

[76] I. Armentano, D. Puglia, F. Luzi, C.R. Arciola, F. Morena, S. Martino, et al., Nanocomposites based on biodegradable polymers, Mater 11 (2018).

[77] S. Mafirad, M.R. Mehrnia, P. Zahedi, S. Hosseini, Chitosan-based nanocomposite membranes with improved properties: effect of cellulose acetate blending and TiO_2 nanoparticles incorporation, Polym. Compos. 39 (2018) 4452–4466.

[78] L. Huang, J. Tan, W. Li, L. Zhou, Z. Liu, B. Luo, et al., Functional polyhedral oligomeric silsesquioxane reinforced poly (lactic acid) nanocomposites for biomedical applications, J. Mech. Behav. Biomed. Mater. 90 (2019) 604–614.

[79] M.S. Islam, M.B. Ahmad, M. Hasan, S.A. Aziz, M. Jawaid, M.K.M. Haafiz, et al., Natural fiber-reinforced hybrid polymer nanocomposites: effect of fiber mixing and nanoclay on physical, mechanical, and biodegradable properties, BioResources 10 (2015) 1394–1407.

[80] Y.Z. Wan, Y. Huang, C.D. Yuan, S. Raman, Y. Zhu, H.J. Jiang, et al., Biomimetic synthesis of hydroxyapatite/bacterial cellulose nanocomposites for biomedical applications, Mater. Sci. Eng. C. 27 (2007) 855–864.

[81] I. Armentano, L. Marinucci, M. Dottori, S. Balloni, E. Fortunati, M. Pennacchi, et al., Novel poly (L-lactide) PLLA/SWNTs nanocomposites for biomedical applications: material characterization and biocompatibility evaluation, in: J. Biomater (Ed.), Polymer Science, 22, 2011, pp. 541–556.

[82] B. Das, P. Chattopadhyay, D. Mishra, T.K. Maiti, S. Maji, R. Narayan, et al., Nanocomposites of bio-based hyperbranched polyurethane/funtionalized MWCNT as non-immunogenic, osteoconductive, biodegradable and biocompatible scaffolds in bone tissue engineering, J. Mater. Chem. B 1 (2013) 4115–4126.

[83] M.G. Mehrabani, R. Karimian, R. Rakhshaei, F. Pakdel, H. Eslami, V. Fakhrzadeh, et al., Chitin/silk fibroin/TiO_2 bio-nanocomposite as a biocompatible wound dressing bandage with strong antimicrobial activity, Int. J. Biol. Macromol. 116 (2018) 966–976.

[84] S. Ebnesajjad, Surface and material characterization techniques, Handb. Adhes. Surf. Prep, Elsevier, 2011, pp. 31–48.

[85] B.T. Tomoda, P.H. Yassue-Cordeiro, J.V. Ernesto, P.S. Lopes, L.O. Péres, C.F. da Silva, et al., Chapter 3—Characterization of biopolymer membranes and films: physicochemical, mechanical, barrier, and biological properties, Biopolymer Membranes and Films, Elsevier Inc., Amsterdam, 2020, pp. 67–95.

[86] X. Cai, H. Tong, X. Shen, W. Chen, J. Yan, J. Hu, Preparation and characterization of homogeneous chitosan–polylactic acid/hydroxyapatite nanocomposite for bone tissue engineering and evaluation of its mechanical properties, Acta Biomater. 5 (2009) 2693–2703.

[87] H.-W. Kim, H.-E. Kim, V. Salih, Stimulation of osteoblast responses to biomimetic nanocomposites of gelatin–hydroxyapatite for tissue engineering scaffolds, Biomaterials 26 (2005) 5221–5230.

[88] V. Krikorian, D.J. Pochan, Poly (L-lactic acid)/layered silicate nanocomposite: fabrication, characterization, and properties, Chem. Mater. 15 (2003) 4317–4324.

[89] S. Shojaei, M. Nikuei, V. Goodarzi, M. Hakani, H.A. Khonakdar, M.R. Saeb, Disclosing the role of surface and bulk erosion on the viscoelastic behavior of biodegradable poly (ε-caprolactone)/poly (lactic acid)/hydroxyapatite nanocomposites, J. Appl. Polym. Sci. 136 (2019) 47151.

[90] Y. Guo, X. Zuo, Y. Xue, J. Tang, M. Gouzman, Y. Fang, et al., Engineering thermally and electrically conductive biodegradable polymer nanocomposites, Compos. Part. B Eng. 189 (2020) 107905.

[91] F. Zahiri Oghani, K. Tahvildari, M. Nozari, Novel antibacterial food packaging based on chitosan loaded zno nano particles prepared by green synthesis from nettle leaf extract, J. Inorg. Organomet. Polym. Mater. 31 (2021) 43–54.

[92] C. Chen, G. Lv, C. Pan, M. Song, C. Wu, D. Guo, et al., Poly (lactic acid)(PLA) based nanocomposites—a novel way of drug-releasing, Biomed. Mater. 2 (2007) L1.

[93] K.S. Katti, D.R. Katti, R. Dash, Synthesis and characterization of a novel chitosan/montmorillonite/hydroxyapatite nanocomposite for bone tissue engineering, Biomed. Mater. 3 (2008) 34122.

[94] A.-L. Goffin, J.-M. Raquez, E. Duquesne, G. Siqueira, Y. Habibi, A. Dufresne, et al., Poly (ε-caprolactone) based nanocomposites reinforced by surface-grafted cellulose nanowhiskers via extrusion processing: morphology, rheology, and thermo-mechanical properties, Polym. (Guildf.) 52 (2011) 1532–1538.

[95] S. Hsu, Y.-B. Chang, C.-L. Tsai, K.-Y. Fu, S.-H. Wang, H.-J. Tseng, Characterization and biocompatibility of chitosan nanocomposites, Colloids Surf. B Biointerfaces 85 (2011) 198–206.

[96] S. Zhang, H. Liu, J. Gou, J. Ying, Y. Wang, C. Liu, et al., Quantitative nanomechanical mapping on poly (lactic acid)/poly (ε-caprolactone)/carbon nanotubes bionanocomposites using atomic force microscopy, Polym. Test. 77 (2019) 105904.

[97] A.M. El-hadi, Increase the elongation at break of poly (lactic acid) composites for use in food packaging films, Sci. Rep. 7 (2017) 46767.

[98] D.A. D'Amico, L.B. Manfredi, V.P. Cyras, Relationship between thermal properties, morphology, and crystallinity of nanocomposites based on polyhydroxybutyrate, J. Appl. Polym. Sci. 123 (2012) 200–208.

[99] T. Silverman, M. Naffakh, C. Marco, G. Ellis, Morphology and thermal properties of biodegradable poly(hydroxybutyrate-co-hydroxyvalerate)/tungsten disulphide inorganic nanotube nanocomposites, Mater. Chem. Phys. 170 (2016) 145–153.

[100] M. Pluta, A. Galeski, M. Alexandre, M. Paul, P. Dubois, Polylactide/montmorillonite nanocomposites and microcomposites prepared by melt blending: structure and some physical properties, J. Appl. Polym. Sci. 86 (2002) 1497–1506.

[101] M.S. Nikolic, R. Petrovic, D. Veljovic, V. Cosovic, N. Stankovic, J. Djonlagic, Effect of sepiolite organomodification on the performance of PCL/sepiolite nanocomposites, Eur. Polym. J. 97 (2017) 198–209.

[102] Q.-X. Zhang, Z.-Z. Yu, X.-L. Xie, K. Naito, Y. Kagawa, Preparation and crystalline morphology of biodegradable starch/clay nanocomposites, Polym. (Guildf.) 48 (2007) 7193–7200.

[103] E.-C. Chen, T.-M. Wu, Isothermal crystallization kinetics and thermal behavior of poly(ε-caprolactone)/multi-walled carbon nanotube composites, Polym. Degrad. Stab. 92 (2007) 1009–1015.

[104] N.A. Shazali, N.E. Zaidi, H. Ariffin, L.C. Abdullah, F. Ghaemi, J.M. Abdullah, et al., Characterization and cellular internalization of spherical cellulose nanocrystals (CNC) into normal and cancerous fibroblasts, Mater (2019) 12.

[105] A. Ur-Rehman, N.M. Khan, F. Ali, H. Khan, Z.U. Khan, A.K. Jan, et al., Kinetics study of biopolymers mixture with the help of confocal laser scanning microscopy, J. Food Process. Eng. 39 (2016) 533–541.

[106] P. Kazimierczak, A. Benko, M. Nocun, A. Przekora, Novel chitosan/agarose/hydroxyapatite nanocomposite scaffold for bone tissue engineering applications: comprehensive evaluation of biocompatibility and osteoinductivity with the use of osteoblasts and mesenchymal stem cells, Int. J. Nanomed. 14 (2019) 6615–6630.

[107] A. Travan, E. Marsich, I. Donati, M. Benincasa, M. Giazzon, L. Felisari, et al., Silver–polysaccharide nanocomposite antimicrobial coatings for methacrylic thermosets, Acta Biomater. 7 (2011) 337–346.

[108] L. Cao, X. Wu, Q. Wang, J. Wang, Biocompatible nanocomposite of TiO_2 incorporated bi-polymer for articular cartilage tissue regeneration: a facile material, J. Photochem. Photobiol. B Biol. 178 (2018) 440–446.

[109] Q. Chen, J. Wen, H. Li, Y. Xu, F. Liu, S. Sun, Recent advances in different modal imaging-guided photothermal therapy, Biomaterials 106 (2016) 144–166.

[110] S. Jeon, J. Kim, D. Lee, J.W. Baik, C. Kim, Review on practical photoacoustic microscopy, Photoacoustics 15 (2019) 100141.

[111] A.M. Smith, M.C. Mancini, S. Nie, Second window for in vivo imaging, Nat. Nanotechnol. 4 (2009) 710–711.

[112] X. Cai, B.S. Paratala, S. Hu, B. Sitharaman, L.V. Wang, Multiscale photoacoustic microscopy of single-walled carbon nanotube-incorporated tissue engineering scaffolds, Tissue Eng. Part. C. Methods 18 (2011) 310–317.

CHAPTER

4 Dielectric properties, thermal analysis, and conductivity studies of biodegradable and biocompatible polymer nanocomposites

Hemalata Jena[1], Priyabrata Pradhan[2] and Abhilash Purohit[2]

[1]*School of Mechanical Engineering, KIIT Deemed to be University, Bhubaneswar, Odisha, India* [2]*Mechanical Department, Synergy Institute of Engineering and Technology, Dhenkanal, Odisha, India*

Chapter Outline

4.1 Introduction

4.1.1 Biodegradable polymer

Biodegradable polymers were first introduced in 1980, ensuring product reliability without negatively impacting the environment. In a natural system, biodegradation begins with microorganisms or enzymes [1]. Biodegradable polymers derived from renewable or natural resources are gradually replacing conventional synthetic or petroleum-based polymers due to inherent properties such as biodegradability and biocompatibility. Nonbiodegradable polymers pose an environmental risk and are produced using petroleum resources. We cannot continue to ignore the ecological issues that the world is currently facing; therefore, the concepts of green (safe) chemistry and sustainable development are required to develop biodegradable materials for its application. The development of biodegradable polymers for a variety of applications has piqued the interest of scientists and technologists. Polymer spur gears are made from thermoplastic biodegradable polymers such as polylactide (PLA) and polybutylene succinate (PBS) [2]. In the sectors of food production,

Biodegradable and Biocompatible Polymer Nanocomposites. DOI: https://doi.org/10.1016/B978-0-323-91696-7.00011-8
Copyright © 2023 Elsevier Inc. All rights reserved.

agriculture, cosmetics, medical care, tissue-engineering applications and pharmaceuticals, biodegradable polymers are widely used. In this context, Pachence et al. [3] have explained the qualities of the biodegradable polymers in tissue engineering products and these are: (1) minimal or nonexistent toxicity of biodegradable polymers; (2) drug delivery compatibility; (3) mechanical characteristics that can handle long-term function without conflict as well as short-term function; (4) production viability, including sufficient amounts of the bulk polymer; and (5) the capacity to include polymer into the design of final products.

The term "biodegradable" ensures that products will decompose harmlessly in the environment. The American Society for Testing and Materials (ASTM) defines biodegradable materials as those that degrade by 70% in 30 days under anaerobic conditions (ASTM D5526-18 and ASTM D5511-18). ASTM D6868-19 and ASTM D6400-19, on the other hand, require a ninety percent mineralization breakdown of the substance into carbon dioxide in 180 days under aerobic conditions. Biodegradable polymers are organic polymers that are derived biologically or synthetically or from different sources as shown in Fig. 4.1 [4]. They are composed of monomeric components that are joined together to form larger forms that exhibit their distinct characteristics. Biodegradable polymers are derived from a variety of sources, including agro-resource biomass and bio-derived monomers [5]. These polymers degrade into CO_2, humus, biomass, and water. Biodegradable polymers are derived primarily from five sources: (1) biomass products, (2) animals and insects, (3) microorganism extracts, (4) bio-derived monomers, and (5) petrochemical products [6]. However,

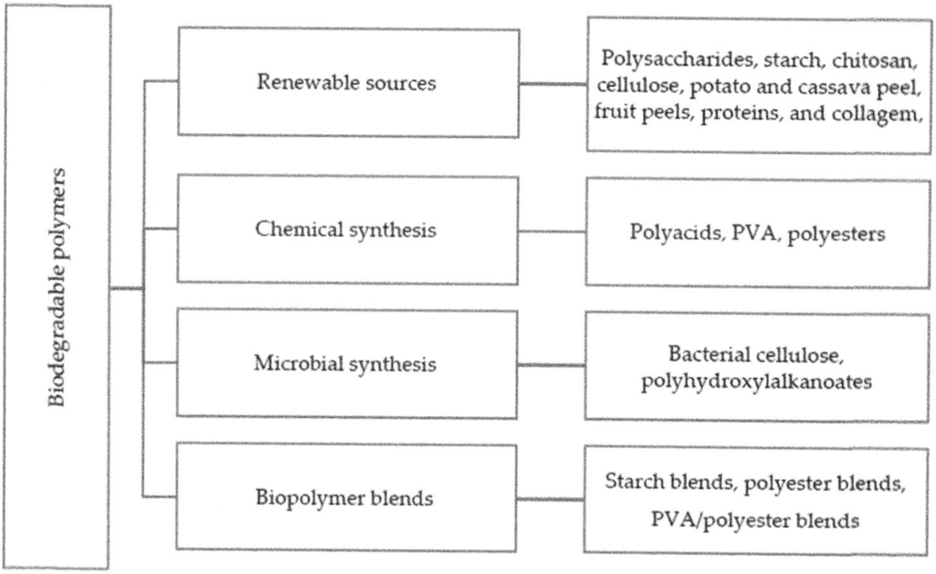

FIGURE 4.1

Biodegradable polymers and their production processes.

Adapted from C. Maraveas, Production of sustainable and biodegradable polymers from agricultural waste, Polymers (Basel), 12, 5, 2020, doi:10.3390/POLYM12051127. Copyright 2020, MDPI.

biodegradable polymers are classified based on their extraction method, such as natural-based polymers and synthetic polymers. Natural polymers are further subdivided into polynucleotides, polyamides, and polysaccharides. Plant-derived polysaccharide polymers include cellulose, starch, and alginate, whereas animal-derived polymers include chitosan, collagen, and silk fibroin, among others [7]. These natural-based polymers have a number of advantages, including improved mechanical properties, lower toxicity, high biodegradability and biocompatibility, and low cost. The most important biopolymer extracted from biomass products such as cotton and trees is cellulose. According to spectroscopic analysis, the chemical structure of cellulose contains a large number of hydroxyl (OH) groups. Cellulose exists as nanofibers and nanocrystals on the nanoscale [8].

4.1.2 Biodegradable polymer nanocomposite

Polymers can be used in place of traditional materials like glass and metals. Polymer composites have several advantages, including a high strength-to-weight ratio, lightweight, low cost, and so on [9–12]. The properties of these biodegradable polymers can be significantly improved by the addition of several nanoparticles. To create nanocomposites, nanoparticles are reinforced into polymers. Polymer nanocomposites are prepared by using several composite fabrication methods such as:

1. Solution casting technique: This technique is used if the dissolution occurs in an appropriate solvent in the presence of nanoparticles.
2. In situ polymerization: In this technique, polymerization of nanoparticle occurs without involving nanoparticles.
3. Melt mixing: Polymers are directly melt-mixed with nanoparticles.

BPNCs are primarily composed of two materials: biodegradable polymers such as polylactic acid (PLA), polycaprolactone (PCL), polyvinyl alcohol (PVA), polyvinyl pyrrolidone (PVP), polybutylene adipate-terephthalate (PBAT), thermoplastic starch (TPS), and polyhydroxyalkanoate (PHA) that are reinforced with other nanoparticles [8]. By incorporating these nanoparticles into biodegradable polymers, bioplastics with improved properties and degradation are created. Nanoparticles enhance the polymer matrix's functionality by providing antimicrobial and smart properties. As a result, different nanoparticles are added to the polymer based on the needs of specific applications. BPNCs are most commonly found in packaging, biomedical, and smart applications [13]. PLA-reinforced multiwall carbon nanotube composites are used to make piezo-resistive vapor sensors, whereas PLA composites containing Fe_3O_4 are used to make shape-memory devices [14]. Carbon black (CB), graphene, and Fe_3O_4 nanoparticles are examples of high electrical conductive fillers that not only improve the mechanical and fire retardant properties of the polymer but also include tailored electrical and thermal properties [15]. However, the addition of nanoparticles can sometimes degrade the properties of the polymer. The addition of calcium oxide and magnesium oxide to PLA composites reduces their thermomechanical properties [16]. Similarly, other nanofillers like organo-modified Beidellite nanoclays in PBS [17], cellulose nanowhiskers (CNW), polypyrrole (PPy), and silver nanoparticles [18] nanocomposites are prepared. Fig. 4.2 shows the SEM image of pure PBS, 1–3 wt.%, of organo-modified Beidellite nanoclays in PBS nanocomposite [17]. It shows that the presence of nanofiller reduces the number of microvoids in PBS polymer. This indicates that the particle has a favorable affinity for the PBS matrix and exhibits better dispersion in the matrix, but agglomerations can be observed due to very small particles [17]. TEM

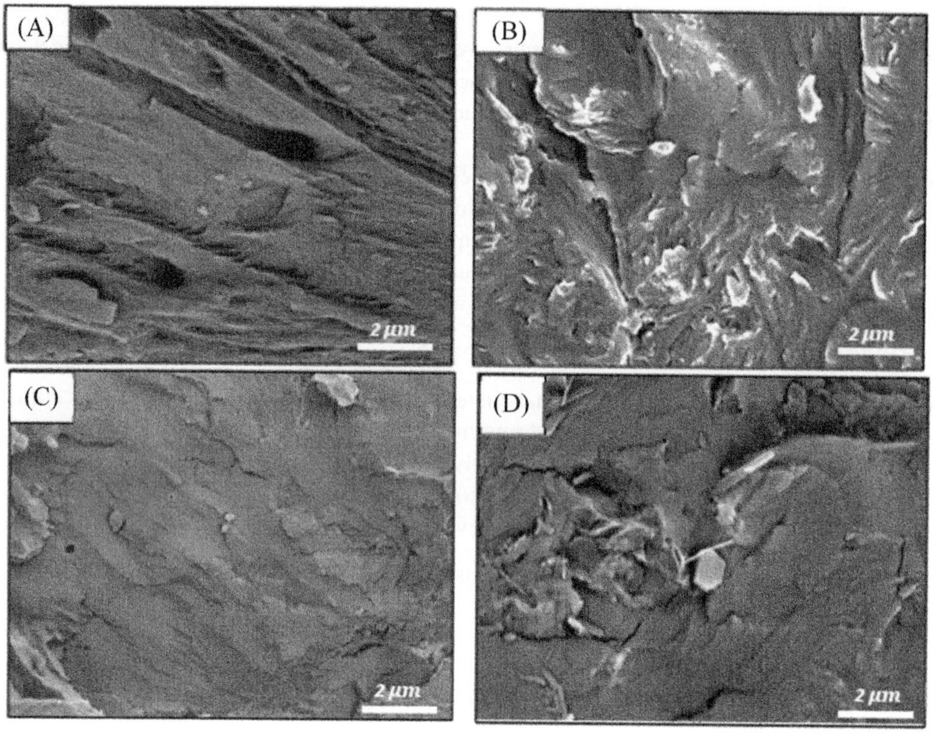

FIGURE 4.2

SEM images of (A) PBS, (B) 1 wt.%, (C) 2 wt.% and (D) 3 wt.% of organo-modified Beidellite nanoclays in PBS nanocomposite [17].

Reproduced with permission from Elsevier. Copyright 2021.

images are given in Fig. 4.3. Fig. 4.3A shows a TEM image of pure CNW. Fig. 4.3B–D shows the creation of PPy nanoparticles of spherical shape having an average diameter of 9–23 nm. Fig. 4.3E and F shows the uniform distribution Ni_2O_3 nanoparticles in CNW/PPy/AgNPs nanocomposites [18]. The shape and distribution of nanoparticles determine the mechanical, thermal, electrical, and dielectric properties of polymer composites.

Surface modification of nanoparticles is also important during the fabrication of the biopolymer nanocomposite. PLA composites reinforced with untreated ZnO particles show degradation at melt-processing temperatures [19]. Later surface modification of ZnO particles with silane improves the PLA matrix's physio-mechanical properties [20]. Several researchers have reported that the degradation process differs from that of unfilled biopolymer decomposition. The degradation of BPNCs is primarily determined by the filler type and the chemical modification process. The addition of spent coffee nanoparticles improved the mechanical properties of PBS composites while shortening the polymer's biodegradation time [21]. Bondeson and Oksman used compression moulding to create PLA nanocomposites filled with cellulose [22]. The tensile strength of the nanocomposites decreases as filler content increases, which could be attributed to a poor interfacial bond between

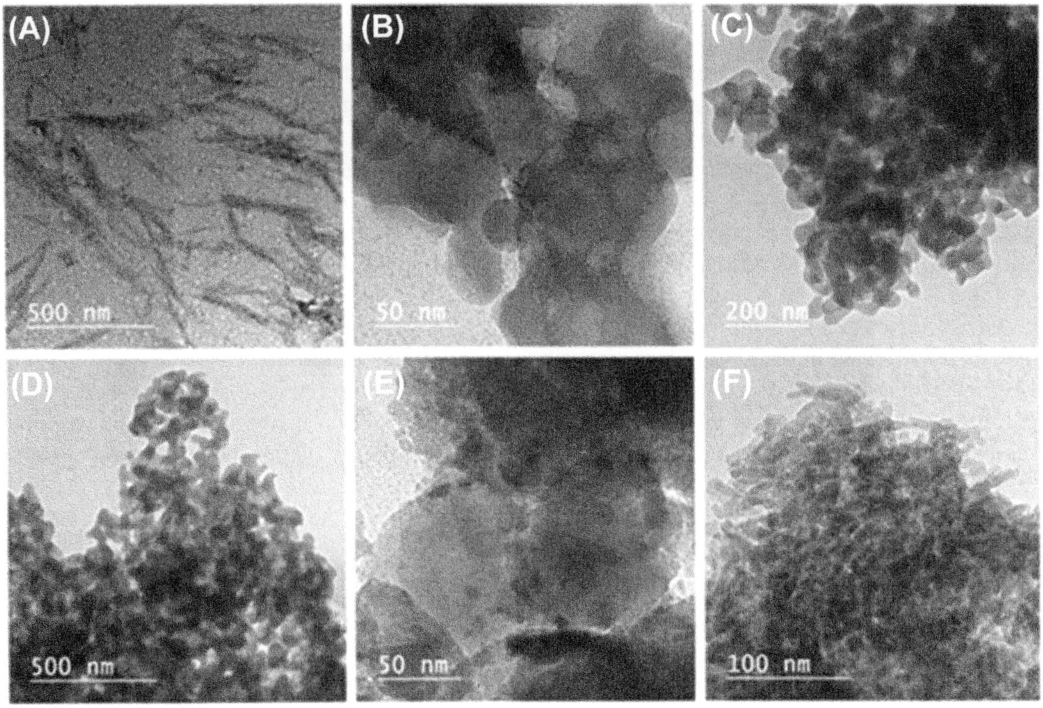

FIGURE 4.3

TEM images of (A) virgin CNW, (B, C, D), CNW/PPy/AgNPs and (E, F) CNW/PPy/AgNPs with Ni_2O_3 [18].

Reproduced with permission from Elsevier. Copyright 2021.

polymer and reinforcement. Lee et al. investigated the mechanical and thermal properties of injection-molded wood flour-reinforced PLA nanocomposites [23]. With the addition of wood flour nanofillers, the tensile modulus of PLA nanocomposites increases from 62.5% to 169.5%. Recently, Qu et al. [24] have developed PLA nanocomposites filled with a compatibilizer and cellulose fibrils. Nowadays, cellulose from agricultural products is popular as a biopolymer that can replace polymers derived from petroleum products. Nanoclay, pure cellulose acetate (80 wt.%), and triethyl citrate plasticizer (20 wt.%) are used to make nanocomposites. It demonstrates that adding 10% filler content improves tensile strength and modulus by approximately 75% and 180%, respectively [10]. Another study created antimicrobial packaging material by reinforcing montmorillonite (MMT) into methylcellulose [25]. Epoxidized soybean oil and DGEBA are also used to make bio-nanocomposites (diglycidyl ether of bisphenol-A). 30% soybean oil and 70% epoxy resin are combined with nanoclay in concentrations of 1%, 3%, 5%, and 7% by weight [26]. The mechanical properties of the nanocomposites (tensile, flexural, and impact strength) increase monotonically with the addition of filler content. Azeredo et al. [27] previously developed nanocomposites by reinforcing cellulose nanofibers into mango puree-based edible films that can be used in packaging. The study shows that the tensile strength of the bio-composites increases from 4.06 to 8.09 MPa due to the inclusion of cellulose nanofibers which may be due to their good interactions with the mango puree. Using the vacuum-assisted resin transfer molding process, halloysite nanotubes (HNT) are reinforced into soya-based polyurethane to create E-glass hybrid nanocomposites. The results of these composites' interlaminar shear strength tests show that shear strength increases with the addition of 2.4 wt.% HNT [28].

Several studies are also being conducted on the combination of two or more polymers in order to achieve biodegradable polymers. Following mechanical and degradability testing, these polymers are recommended for use as matrices in composite manufacturing. These polymers include starch/PLA blends, PCL/PVA blends, and PBS/cellulose acetate blends [29,30]. Another naturally occurring polyester used in BPN produced by numerous bacteria in nature is polyhydroxyl butyrate. Chitosan, a naturally occurring polymer, is extracted from the cell walls of fungi and microorganisms [31,32]. The addition of cellulose nanofiber improves the mechanical behavior of chitosan-based nanocomposites, but chitosan filed with carbon fiber does not improve mechanical properties significantly [33]. According to the findings of the study, the primary challenge in creating biopolymer-nanocomposites is the compatibility of the hydrophobic nature of the polymer matrix and the hydrophilic characteristics of the nanoparticles, which results in poor bonding between matrix and reinforcement and a reduction in mechanical properties. Several compatibilizers, including titanium-derived mixture (TDM), maleated polyethylene (MAPE), and maleic anhydride polypropylene (MAPP), are incorporated into the polymer matrix to improve adhesion. Another issue with biopolymers is that their mechanical properties, brittleness, and water resistance are poor due to large particle sizes, which can be improved by using plasticizers such as glycerol. It has been discovered that the presence of plasticizers causes an increase in sugar content, which repairs the adhesion properties. However, PLA has a higher mechanical strength and is less toxic. However, the application of PLA nanocomposites is limited due to their low ductility and modulus at temperatures above the glass transition. As a result, modifiers are added to the PLA matrix to increase stiffness at high temperatures. Due to the high cost of production for other biopolymers such as chitosan, cellulose, and plant-based oils, PLA is now widely used to prepare nanocomposites among all biopolymers [34]. These biopolymer nanocomposites have a wide range of applications, including door handles and panels, mirror housings, engine covers, trunk liners and headrests, roofs, and so on. These composites are also being researched for potential applications such as power tool housings, impellers and covers for mobile phones, vacuum cleaner blades, and so on. These composites' chemical resistance and surface appearance make them suitable for a variety of packaging applications [35].

4.2 Properties of biodegradable polymer nanocomposites

4.2.1 Dielectric property

When an electric field is applied to a dielectric material, it can become polarized. Because dielectric materials are insulators, no electric charges can flow through them. This material exhibits significant electrical polarizability due to the accumulation of electric charges on dielectric materials. How well a material performs is determined by the dielectric constant (k), which describes the substance's ability to store electrical charge. Materials with high k values can be used as gate dielectrics in supercapacitors, memory cells, and metal-oxide-silicon (MOS) transistors. Low-*k* materials, on the other hand, are critical as electrical insulators [36]. High-performance polymer-based dielectric capacitors have received a lot of attention in optoelectronic and microelectronic devices due to their exceptional mechanical qualities, high power density, and excellent cycle stability. To describe the dielectric constant of a material, energy storage density is used. The energy storage density (Ue) is typically proportional to the dielectric constant (k), vacuum permittivity (ε_0) of 8.85×10^{-12} Fm^{-1} and the applied electric field squared (E^2) according to the following relation [37];

$$\mathrm{Ue} = \frac{1}{2} k \varepsilon_0 E^2$$

The following formula is also used to calculate the nanocomposites' dielectric constant (k) by using parallel plate capacitors [38]

$$C = \varepsilon_0 k \frac{A}{D}$$

where D is the sample's thickness, A its area, C its parallel capacitance, ε_0 is the dielectric constant of vacuum (8.85×10^{-12} Fm^{-1}) and k its dielectric constant. Another way to calculate the dielectric constant from the complex permittivity, which is denoted as ε^*, determines the type of resistance of the produced nanocomposite to the electric field as given below [39]:

$$\varepsilon^* = \varepsilon' - j\varepsilon''$$

where ε' is the dielectric constant known as the real part of complex permittivity and ε'' is dielectric loss known as the imaginary part of complex permittivity. The complex impedance can be found as

$$Z = Z' - jZ''$$

The values of ε' for the samples are calculated from the following equation:

$$\varepsilon' = \frac{Z''}{2\pi f C_0 Z^2}$$

The dielectric loss (the imaginary part of complex permittivity) ε'' is typically expressed in terms of the loss tangent $\tan(\delta)$, which is the proportion of energy stored in a substance to energy lost in that material as given below [40]:

$$\tan(\delta) = \frac{\varepsilon''}{\varepsilon'}$$

A novel method must be developed to increase the breakdown strength, dielectric constant, and discharge energy density of dielectric films. Because increasing the energy storage density necessitates consideration of both the breakdown strength and the dielectric constant. Polymers are typically unsuitable for capacitive devices due to their low dielectric constant. The dielectric properties of polymers can be significantly improved by adding conductive filler to the polymer matrix. The conductivity difference between the fillers and matrix is one of the factors that contributes to the high dielectric constant of polymer-based nanocomposites [41].

Filler plays an important role in enhancing the biodegradable polymer's dielectric property. The dielectric constant of the dielectric substance reflects the material's polarization capacity. As a result, to improve the polarization ability of any composite film, a filler with a high ultrahigh dielectric constant and a good dielectric response should be added to the polymer. Some important fillers that are widely used to improve the dielectric property of biodegradable polymers are copper calcium titanate (CCTO) and boron nitride (BN) [37]. In chitin, BN plays an important role in improving the breakdown strength, while CCTO nanofiber has improved the dielectric constant. The dielectric constant of both BN and CCTO composite films has significantly increased. Because there is little dielectric loss, the discharge energy density is high. Deshmukh et al. [42] have added Al_2O_3 nanoparticles in cellulose acetate (CA) and found improved dielectric constant after adding nanofiller. Kashi et al. [43] have incorporated graphene nanoparticles (GNP) (0, 3, 6, 9, 12, 15 wt.%) in PLA and Polybutylene adipate-*co*-terephthalate (PBT) matrix. and found that the addition of nanoparticles improves the dielectric constant and dielectric loss on both matrix cases at a frequency of 10 GHz as shown in Fig. 4.4. Similarly, Mollik et al. [41] have added reduced graphene oxide (0.5, 1 wt.%) in plasticized starch (PS) and found that the dielectric constant is high for all wt.% of nanofiller at low-frequency range and dielectric constant is found to be maximum at 1 wt.% of nanofiller as shown in Fig. 4.5A [41]. Dielectric constants at 100 Hz are 195, 4000, and 60,000 at- 0%, 0.5%, 1% filler loading. The increase in dielectric constant with increasing wt.% of rGO can be attributed to several factors, including the number of polar functional groups in rGO, the size of the sheets, and the multilayer structure of rGO. Fig. 4.5B shows the dielectric loss (tan δ) is higher at 1.0 wt.% rGO at the frequency range of 10^5 Hz. They blamed it on the presence of lattice defects in their sample. Similarly, lower frequencies exhibit greater dielectric loss than higher frequencies. The Debye relaxation model is also developed to validate the experimental data of the dielectric constant in order to understand it, as shown in Fig. 4.6, and from the figure, it can be found that at high frequencies, fitting is poorly correlated with the experimental data.

Many different types of nanofillers are frequently tested with the biodegradable matrix of the PVP/PEO blend combination. It is only possible because the PVP and PEO macromolecules are highly compatible, resulting in excellent electrical, thermal, and mechanical properties. In terms of low dielectric loss values and nanofiller concentration dependent dielectric permittivity, different nanofillers such as TiO_2, silica, ZnO, and SnO_2 have demonstrated synergism in the PVP/PEO

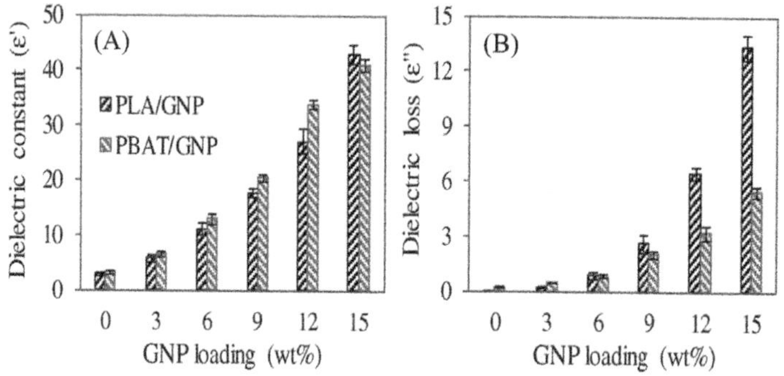

FIGURE 4.4

(A) Dielectric constant (real part ε') and (B) dielectric loss (imaginary parts ε'') of permittivity of PLA-GNP and PBAT-GNP [43].

Reproduced with permission from Elsevier. Copyright 2021.

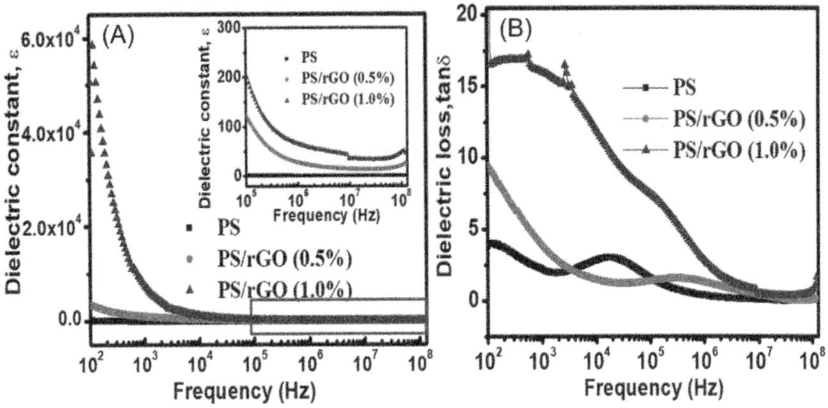

FIGURE 4.5

(A) Dielectric constant (real part ε') and (B) dielectric loss (imaginary parts ε'') of PS/rGO nanocomposites [41].

Reproduced with permission from Elsevier. Copyright 2021.

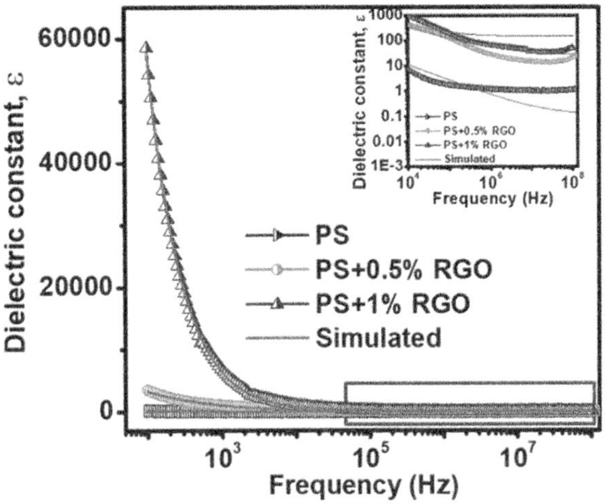

FIGURE 4.6

Dielectric constant versus frequency [41].

Reproduced with permission from Elsevier. Copyright 2021.

blend [44–47]. In another study, organo-modified Beidellite nanoclays (1–3 wt.%) are added to PBS at a frequency range 10^{-1} to 10^6 and temperature −5°C and 90°C and found that the dependency of dielectric property with frequency as well as temperature, and the dielectric measurements shows a better capacitance at 2 wt.% of filler [17]. Table 4.1 gives the summary of the dielectric property of polymer nanocomposite and its temperature and frequency dependency. It also

Table 4.1 Dielectric spectroscopy of biodegradable polymers and their composites.

Sl. no.	Biodegradable polymers	Filler material	Experimental parameter I. Frequency II. Temperature	Dielectric constant (maximum)	Dielectric loss	Refs.
1	Plasticized starch (PS)	Reduced graphene oxide (rGO)-0, 0.5, 1 wt.%	10^2–108 Hz room temperature	60,000 at 1 wt.% of rGO loading, 100 Hz	16 at 1 wt. % of rGO loading, 100 Hz	[41]
2	Regenerated chitin (RCH)	Boron nitride nanosheets (BNNS) and copper calcium titanate nanofiber (CCTO) of (2, 5, 8, 12 wt.%)	10^2–106 Hz	9.64 at 12 wt.% CCTO in RCH/ BNNS, 1 KHz	0.02242 at 12 wt.% CCTO in RCH/ BNNS, 1 KHz	[37]
3	Polyimide	Graphene oxide (0, 0.25, 0.5, 0.75, 1.0 wt.%)	1–10^3 KHz	8.5 at 1.0 wt.% GO, 1000 Hz	0.12 at 1.0 wt.% GO, 1000 Hz	[48]
4	PLA	Graphene nanoplatelet (0, 3, 6, 9, 12, 15 wt.%)	10 GHz	42 at 15 wt.% of GNP	13 at 15 wt.% of GNP	[43]
5	PBT	Graphene nanoplatelet (0, 3, 6, 9, 12, 15 wt.%)	10 GHz	40 at 15 wt.% of GNP	5 at 15 wt. % of GNP	[43]
6	Methylcellulose (MC)	Lead sulfide (LS) 0.05, 0.1, 0.15, 0.2, 0.25, 0.3 M	10^2–10^6 Hz 30 and 80 degrees	2.44 at 0.3 LS, 10^2 Hz, 30 degrees	0.9 at 0.3 LS, 10^2 Hz, 30 degrees	[49]
7	PANI	Reduced graphene oxide (rGO) and sol-gel derived calcium copper titanate (CCTO)	10^2–10^6 Hz room temperature	3.85 10^2 Hz	2.4 10^2 Hz	[39]
8	PVP and PVA	Reduced graphene oxide (rGO) (5, 10, 15, 20, 25 wt.%)	10 Hz to 0.1 MHz room temperature	4.17 at 15 wt.%	12 at 15 wt.%	[50]

demonstrates that the addition of nanofiller increases the dielectric constant as well as dielectric loss, with the maximum value found at higher filler loading.

4.2.2 Thermal properties

Different techniques, such as differential scanning calorimetry (DSC), thermogravimetric analysis (TGA), and differential thermal analysis, can be used to investigate the thermal properties of biopolymer nanocomposites (BPN) (DTA). TGA, in conjunction with FT-IR spectroscopy, is also used to investigate the thermal degradation process of composites. The thermal properties of biobased silica/carbon (SCNPs) hybrid nanocomposites are investigated using the polymer Ecoflex [51]. DSC analysis reveals that the glass transition temperature (Tg) and crystallization temperature of the composites improve with filler content when compared to virgin polymer. This improvement could be attributed to improved cross-linking of inorganic nanoparticles with polymer matrix. The thermogravimetric analysis of these composites determines their thermal stability. The pristine SCNPs filled with SiO_2 nanoparticles exhibit exceptional thermal stability up to 750°C, which could be attributed to the uniform dispersion of silica-carbon nanoparticles, which leads to a decrease in oxygen permeability. Carbon nanotubes (CNTs) are reinforced in PLA/poly(ethylene oxide) blends to create biodegradable nanocomposites [52]. TGA results show that the inclusion of CNT increases the thermal stability of the blend up to 68°C, and the melting temperature of PLA crystals improves due to the nucleation effect of CNT on PLA resin. The PLA composites' crystallization enthalpy is also determined. PLA/PEO blend has higher enthalpy than virgin PLA, indicating that PEO facilitates PLA crystallization. The melting temperature of PLA and PEO decreases due to miscibility, according to DSC analysis. The glass transition and cold crystallization temperatures of neat PLA are 60°C and 95°C, respectively. However, due to the melting of PEO overlapped PLA, the glass transition temperature of the composites shows little variation. Polyaniline (PANI) is a conductive polymer that is gaining popularity among scientists due to its high electrical conductivity and low density. Nepomuceno et al. [53] combine cellulose nanocrystals (CNC) and PANI using in situ polymerization to create a nanostructured material. It is observed from TGA test results that thermal stability and time of polymerization improve with CNC content which signals that cellulose nanostructure acts as a backbone to PANI deposition. At 100°C, the PANI samples experience their first mass loss. It could be caused by the loss of volatile elements in the material. The second mass loss occurs between 150°C and 320°C due to dopant loss. The final mass loss occurs around 550°C due to total polymer bond rupture.

Cadmium sulfide (CdS) quantum dots (QDs) are reinforced into chitosan (CTS) via in situ hydrothermal methods, resulting in CdS @CTS [54]. The incorporation of heavy metal ions improves the thermal stability of the CdS @CTS composite over pure CTS. Other studies have found that the thermal decomposition of CdS @CTS composite increased to 60°C when compared to pure chitosan. Abutalib and Rajeh [55] have used the solution casting technique to prepare polymer films of carboxymethyl cellulose (CMC) doped with ZnO/Co nanoparticles. Three stages are involved in the decomposition of pure CMC and CMC/ZnO/Co polymer electrolyte films. At 100°C, the first degradation occurred due to the freezing of moisture and other volatile materials in the polymer. The second stage of degradation occurs between the temperatures of 100°C and 250°C. It could be related to the decomposition of the polymer's main chain. The third stage is observed at 250°C–350°C, where the maximum mass loss occurred due to polymeric matrix degradation. Mofokeng and Luyt [56] have investigated the thermal stability of melt-mixed PLA/PCL mix nanocomposites reinforced with TiO_2 nanoparticles and observed that neat PCL possess better thermal stability compared to PLA but the activation energy of degradation of PLA is higher than PCL, which shows that thermal degradation rate is subjected to the temperature. Blending reduces the thermal stability of PCL and PLA, which is further improved by the reinforcement of TiO_2 nanoparticles. Another study found that blending PLA and PHBV blends improved the thermal stability of poly(hydroxybutyrate-*co*-valerate) (PHBV), but decreased the thermal stability of neat PLA. The addition of TiO_2 nanoparticles improves the thermal stability of both polymers in the blending process. According to the degradation kinetics results, the inclusion of TiO_2 NPs delayed the volatilization of the degradation products, and this was dependent on the interaction with the polymer. The activation energy is the least amount of energy required to initiate thermal degradation. PLA has a higher activation energy than PCL, indicating that the degradation rates of PLA are strongly temperature dependent.

Ravi et al. [57] have prepared lithium ion conducting polymer electrolyte films using PCL along with lithium thiocyanate (LiSCN) salt. DSC analysis is used to determine the melting temperature (Tm) of the polymer electrolyte. DSC is a technique for investigating what happens to nanocomposites when they are heated. DSC curves were obtained in the temperature range of 30°C–95°C for various compositions of PCL with lithium thiocyanate (LiSCN) salt complexes. One sharp endothermic peak at 64°C for pure PCL indicates its melting temperature, which may be attributed to PCL crystalline melting. With the addition of salt, the melting temperature decreases, which may result in a reduction in the size of spherulites. The melting temperature decreases monotonically as the salt percentage increases, indicating crystallite suppression in the PCL: LiSCN complexes. A comparable observation [58]. The melting temperature (Tm) of PP/clay nanocomposites is around 160°C, which is slightly lower than the temperature of 162.8°C for pure PP. It shows that adding clay layers causes more crystal defects, and clay nanocomposites had a lower degree of crystallization than pure PP. Fig. 4.7 shows the DSC curve of PLA and 3 wt.% CNT nanocomposites, which indicates that annealed PLA/CNT nanocomposites have higher glass transition (Tg) and melting (Tm) temperatures [14]. The DSC demonstrated that the change caused by this thermal treatment of PLA/CNT is responsible for the significant changes in vapor sensing behavior.

Shen et al. [59] have fabricated solvent-free spherical cellulose nanocrystal fluids (CNCfs), which are then embedded into PLA membrane using an electro-spinning process. PLA/CNCfs fibrous membrane shows speedy heat dissipation, the thermal conductivity of 0.27 W/mK along with the good antibacterial activity. PLA/CNCfs fibrous membranes also have improved mechanical properties, allowing them to be used as medical dressings and antibacterial textiles. Sun et al. [60] have fabricated PLA/poly(butylene adipate-*co*-butylene terephthalate) composites reinforced with carbon nanofibers (CNF) using a melt extrusion-stretching method. The thermal conductivity

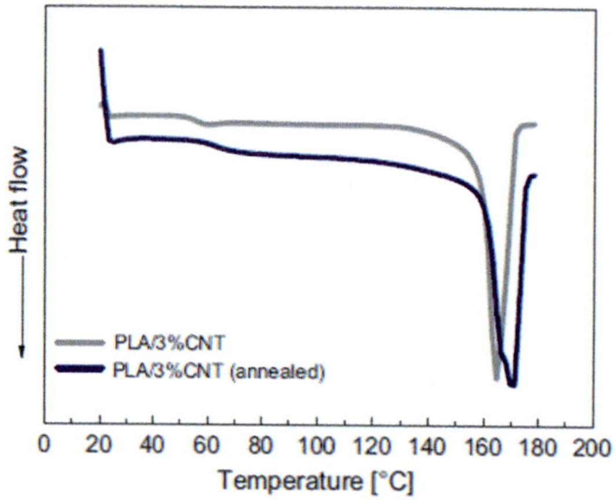

FIGURE 4.7

DSC curve of PLA/3%CNT and annealed PLA/3%CNT [14].

Reproduced with permission from Elsevier. Copyright 2012.

of PLA/PBAT/CNF blend composites reaches to 1.53 W/mK at 10 wt.% of CNF content. The creation of a CNF thermal conductive network in the blend composites is the primary reason for this advancement. Zou et al. [61] have studied the thermal behavior of a novel eco-friendly material which is prepared with chitosan microspheres (CSM) and hydroxyl-functionalized hexagonal boron nitride as shown in Fig. 4.8. The thermal conductivity of OH-h-BN/CS nanocomposites is recorded as 0.94 $\pm$ 0.02 W/mK at 50 wt.% filler loading which is increased by about 502% for OH-h-BN/CSM nanocomposites with honeycomb structure for the same filler content.

Li et al. [62] in another study have fabricated polyoxymethylene (POM)/MWCNT and POM/GNP composites by assisting miscible poly(L-lactide) (PLLA). Both the composites present strong thermal properties, for example, the in-plane thermal conductivity of PMGNP48 increased by about 1298% compared to neat POM. Fig. 4.9 shows the TGA curves of PMCNT and PMGNP samples. The thermal stability of both composites is found to be lower than POM [62]. Similarly, TGA curves for LA/nZnO composites (Fig. 4.10a) and PHBV/nZnO composites (Fig. 4.10b) are shown in the temperature region between RT and 600°C. The figures indicated that whereas PHBV exhibited no relationship on the nZnO concentration, PLA's onset and peak degradation temperatures are pushed towards lower temperatures with the rise of nZnO concentration. It is concluded that nZnO has a considerably greater impact on the PLA matrix than PHBV one. Guo et al. [63] suggested a biodegradable polymer-based nanocomposite in which graphene nanoplatelets (GNPs) are reinforced into the blending of highly ductile biodegradable polymer-poly(butylene adipate-*co*-butylene terephthalate) (PBAT) and PLA. The highest thermal conductivity is achieved for the nanocomposites with 40 wt.% of GNPs loading and better mechanical properties are also achieved at this filler loading which may be due to the strong bonding between GNPs and PBAT. The thermal conductivity of PBAT/PLA/GH5 nanocomposites is found to be higher compared to PBAT/GH5 as shown in Fig. 4.11 [63].

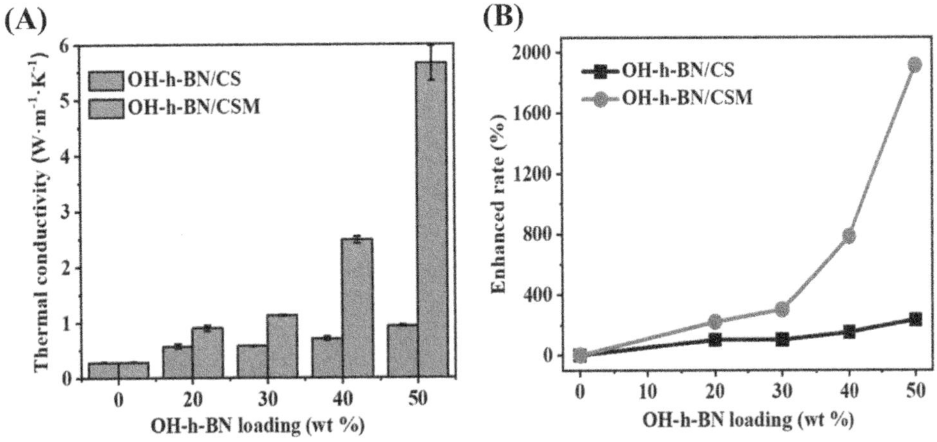

FIGURE 4.8

(A) Thermal conductivity. (B) Enhanced rate of OH-h-BN loading on CS and CSM [61].

Reproduced with permission from Elsevier. Copyright 2021.

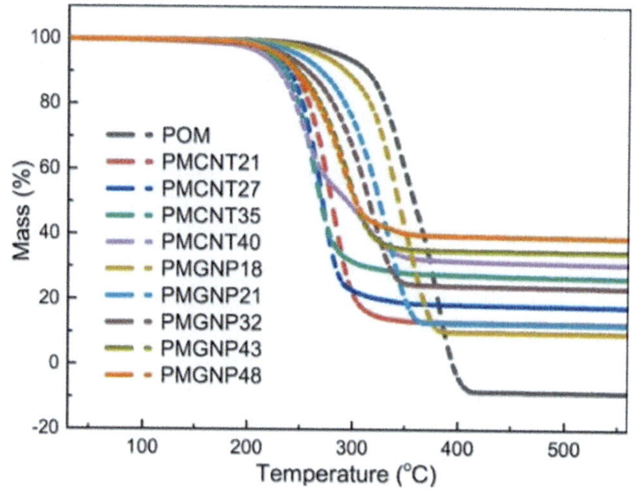

FIGURE 4.9

TGA curves of PMCNT nanocomposites at different concentrations [62].

Reproduced with permission from Elsevier. Copyright 2021.

4.2.3 Electrical conductivity studies

The study of biopolymer systems has gained popularity due to a variety of intriguing applications, including bioplastics, electronics, coating, medicine, and food packaging. Because of its properties such as biodegradability, safety, electrical conductivity, and renewable, biopolymer is essential in the production of environmentally friendly products. Biopolymers are generally poor electrical conductors. Inorganic nanofillers are used to improve the electrical conductivity of the biopolymer matrix. These fillers differ significantly from traditional biopolymers and exhibit unexpected properties [64]. Polymers such as polyimide (PI), polypropylene (PP), and polyethylene terephthalate (PET) are commonly found in electronic devices. These traditional polymers are harmful to the environment because they are not biodegradable. Biodegradable polymers, on the other hand, have high metabolization properties and are environmentally friendly. They are widely used in wearable devices and implantable chips because of this property [10]. Except for conducting polymers, most polymers are poor electrical conductors (CPs). Polymers have low electrical conductivity due to the presence of captive electrons and covalent bonds. CPs outperform traditional conductive materials in terms of flexibility, cost, and weight. CPs are of two types (1) structural CPs (SCPs): e.g., polypyrrole (PPy), polyphenyl acetylene (PPA), and PANI, and (2) CPs composites (CPCs): reinforcing the polymers with conductive fillers, e.g., carbon nanotubes (CNTs), graphene (G) and CB [10]. Fig. 4.12 shows the addition of graphene nanoparticles (GNP) improves the electrical conductivity of the PLA and PBAT polymers [43]. The electrical conductivity of both polymers increases as GNP increases. It could be due to the inclusion of more conductive filler in the polymer matrix, which creates a conductive path between the filler's close contact or electron transmission through nearby particles (noncontact of the fillers). PLA-GNP has greater electrical conductivity than PBAT-GNP. This disparity is due to the different levels of GNP dispersion in the two matrices (GNPs disperse better in PBAT than in PLA).

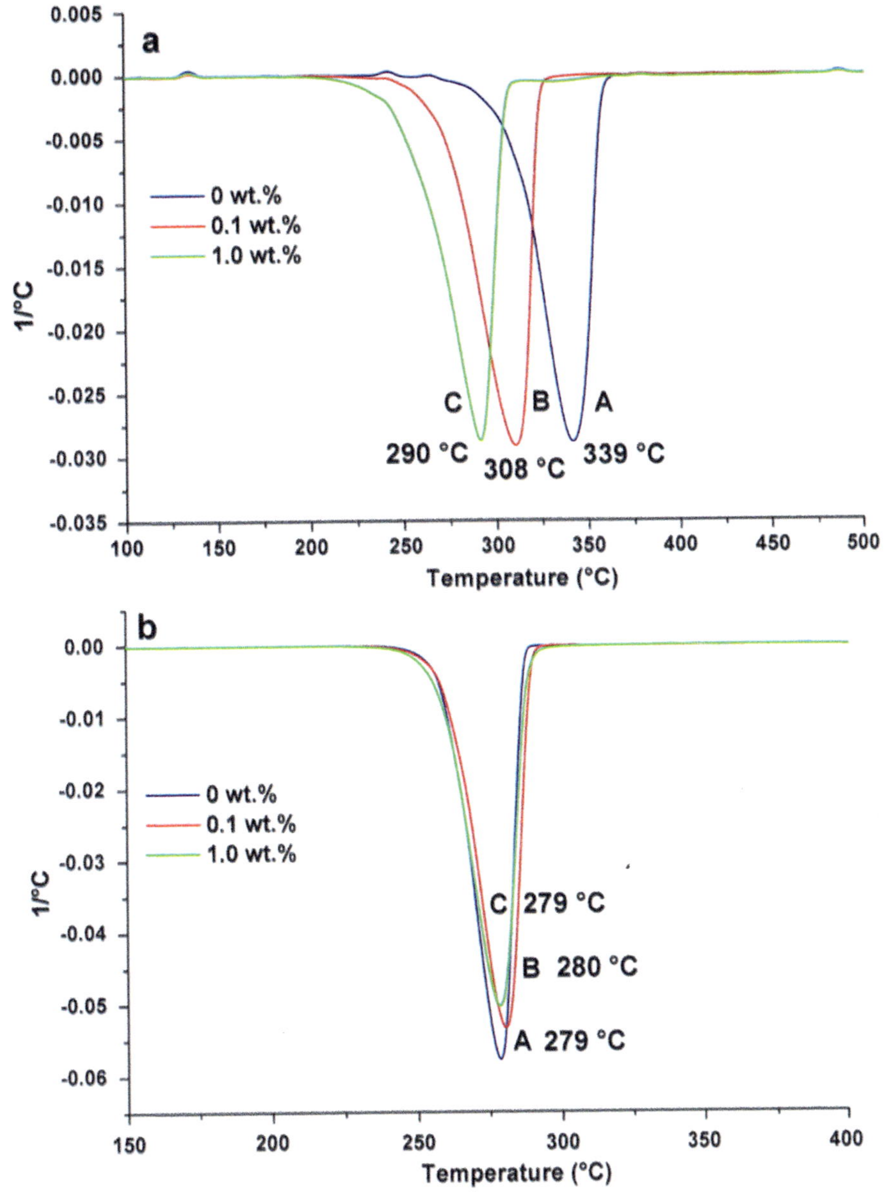

FIGURE 4.10

TGA curves of (a) PLA/ZnO composites and (b) PHBV/ZnO composites.

Adapted from A. Anžlovar, A. Kržan, and E. Žagar, Degradation of PLA/ZnO and PHBV/ZnO composites prepared by melt processing, Arab. J. Chem., 2017, doi:10.1016/j.arabjc.2017.07.001. Reproduced with permission from Elsevier. Copyright 2018.

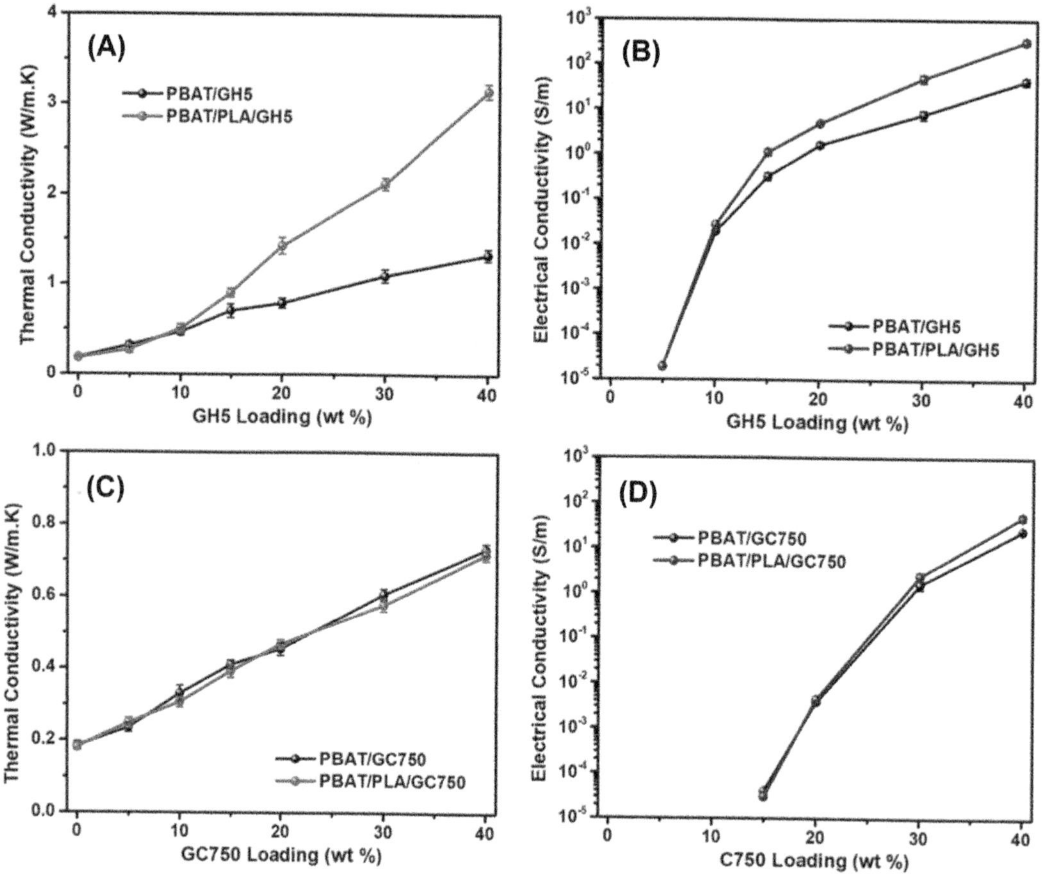

FIGURE 4.11

(A) Thermal conductivity of the PBAT/GH5 and PBAT/PLA/GH5 nanocomposites. (B) Electrical conductivity of the PBAT/GH5 and PBAT/PLA/GH5 nanocomposites. (C) Thermal conductivity of the PBAT/GC750 and PBAT/PLA/GC750 nanocomposites. (D) Electrical conductivity of the PBAT/GC750 and PBAT/PLA/GC750 nanocomposites [63].

Reproduced with permission from Elsevier. Copyright 2020.

Another limitation of biodegradable polymers in a variety of applications is their low electrical conductivity. The addition of small-volume fraction conductive nanofillers to biodegradable polymers can improve their conductivity. The electrical property improves as particle size decreases from micro to nano. Adding conductive nanofillers with a high aspect ratio, such as graphene or carbon nanotubes, can result in a low percolation threshold. It could be because of the creation of a well-connected conductive network. Graphene nanoplatelets (GNPs) (40 wt.%) filled PBAT composites showed electrical conductivity of 3.15 W/mK [63]. It is observed that the incorporation of GNPs increases the electrical conductivity for PCL/polyhedral oligomeric silsesquioxane (PCL-POSS) [65], and chitosan/PLA (25/75 wt.%/wt.%)

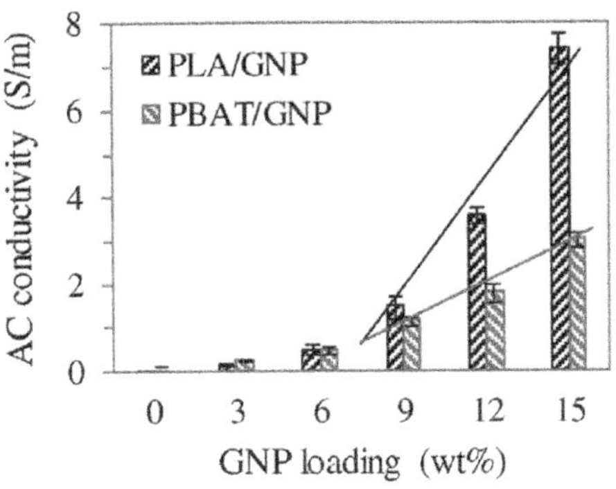

FIGURE 4.12

Electrical conductivity of GNP in PLA and PBAT [43].

Reproduced with permission from Elsevier. Copyright 2016.

[66]. The electrical conductivity of PVA increases with an increase in starch and glycerol content [67]. A similar study showed that the addition of a plasticizer (glycerol) improves the conductivity of the chitosan\$Al_2O_3$\$AgNO_3$ system [68]. Gold nanoparticles (Au NPs) as a reinforcing phase can enhance the conductivity of PLA [69]. PLA filled with indium tin oxide (ITO) and Au NPs electrode is effectively used to know the presence of leukemia cancer cells [69]. The conductivity of PLA can also be increased by the addition of silver nanowires (Ag NPs). These Ag NPs filled PLA is used as electrodes for the organic light emitting diodes (LEDs) [70]. The electrical conductivity of PPy/cellulose nanoparticles can be improved by adding Ni_2O_3−AgNPs [18].

A similar study found that when chitosan is reinforced with Ag NPs, its electrical conductivity improves [71]. It is difficult to achieve proper wetting and good interfacial bonding between conductive nanofiller and polymer matrix when fabricating a conductive nanocomposite. The nanoparticles must be distributed uniformly in the polymer matrix to achieve excellent conductivity. However, nanoparticles are typically oriented randomly, which affects conductivity. The conductivity of numerous polymers, such as PLA [72], PCL [73,74], chitosan-PCL and PP is improved when reinforced with CNTs [75]. The addition of short ramie fiber to the PLA matrix improved the distribution of nanocarbon particles like CNTs and CBs [72]. The conductivity of CNT-filled PCL composites was improved through polymerization and in-situ processing [76]. The melt-spun technique is used to create PLA-based composites filled with 4% CNTs, resulting in a variety of sensing devices used to monitor temperature ranging from 20°C to 80°C [74]. Fig. 4.13 indicates a comparison between the electrical conductivity of MWCNT-filled polyoxymethylene (POM) composites known as PMCNT and GNP-filled POM composites known as PMGNP. The conductivity of MWCNT and GNP is found to be dependent on the filler loading. PMCNT composites had higher conductivity than PMGNP composites [62].

Many attempts have been made to improve degradable polymer electrical conductivity by incorporating conductive metal oxides or metallic nanoparticles. The electrical conductivity of PLA-based nanocomposites is enhanced by incorporating graphene [43] and diamond/graphite particles [77]. Fig. 4.14A shows the orientation of graphene (GH5) in PBAT matrix where a high filler loading is required to create the percolated channel in the matrix. Fig. 4.14B shows the incorporation of PLA concentrates the GH5 within the PBAT which forms a conductive pathway (red arrow marks)

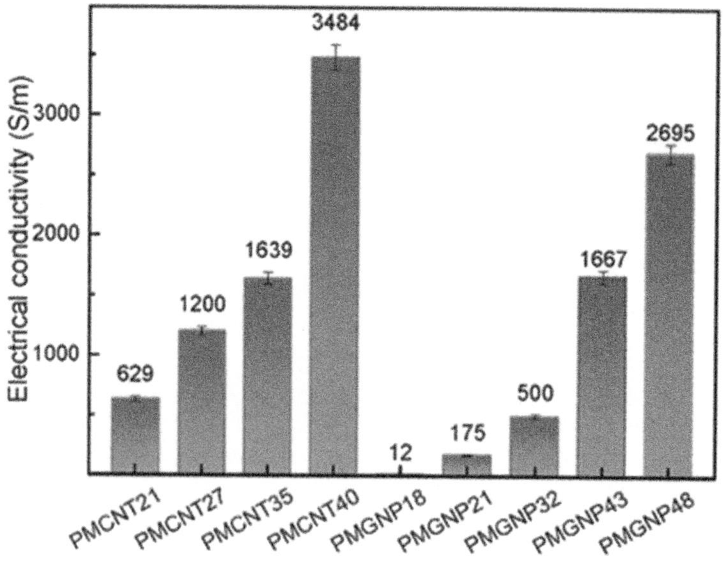

FIGURE 4.13

Electrical conductivity of PMCNT and PMGNP composites [62].

Reproduced with permission from Elsevier. Copyright 2021.

in the matrix. The electrical and thermal conductivity of the nanocomposite both drop when the PLA percentage of the matrix reaches 50%, which is attributed to the loss of PBAT phase continuity as shown result in Fig. 4.15. The cause is that at PBAT:PLA = 25:75 (higher PLA content), GH5 are aggregated and isolated in the dispersed PBAT matrix. The GH5 is prevented from forming a percolated network within the matrix by this dispersion tendency.

The electrical conductivity of clay nanoparticles filled PANI/bacterial cellulose (BC) improves to 0.49 S/cm with the addition of 5 wt.% of clay. This composite shows 16-fold increase in electrical conductivity than the neat polymer (Fig. 4.16) [78]. At highest wt.% of clay shows the highest electrical conductivity of the nanocomposite. Figs. 4.17 and 4.18 shows the frequency dependence on electrical conductivity. Fig. 4.17 shows the increment of electrical conductivity with respect to the amount of rGO nanofillers in the polymer matrix. Over the whole frequency range, the nanocomposite's value of conductivity is greater than the PS's. At 0.5 wt.% of rGO, the conductivity increases with a frequency very slowly, but at 1 wt.% of rGO, it increases exponentially [41]. Fig. 4.18 shows the electrical conductivity of PLA and PBAT nanocomposites over the frequency range, which increases with an increase in frequency. It is because increasing frequency improves particle connections and increases tunneling current. As a result, the composite's conductivity rises [43]. In another study, MWCN-filled PLA biopolymer composites exhibited excellent conductivity and are used as vapor sensors [14] and Graphene in MWCN-filled PLA Composites is used in FDM 3D printing applications due to their excellent in electrical and thermal property [79]. When different weight percentages of CNT are added to PHBV/PLA, the electrical properties improve [80]. Biodegradable polymers reinforced with different conductive nanofillers (such as; graphene,

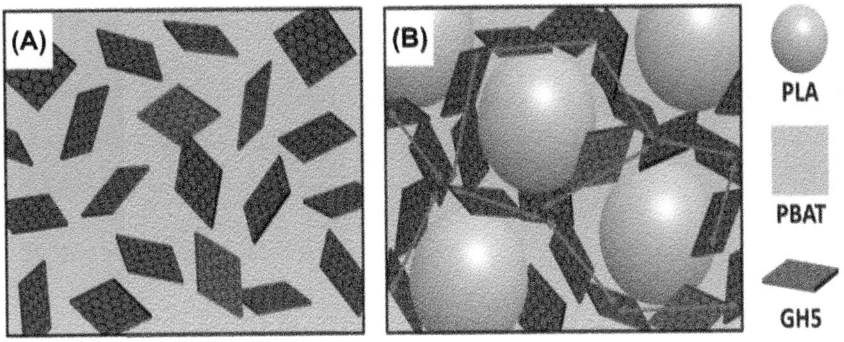

FIGURE 4.14

Graphene (GH5) dispersion in (A) PBAT matrix, and (B) PBAT/PLA blend matrix (PBAT: PLA = 75:25) [63].

Reproduced with permission from Elsevier. Copyright 2020.

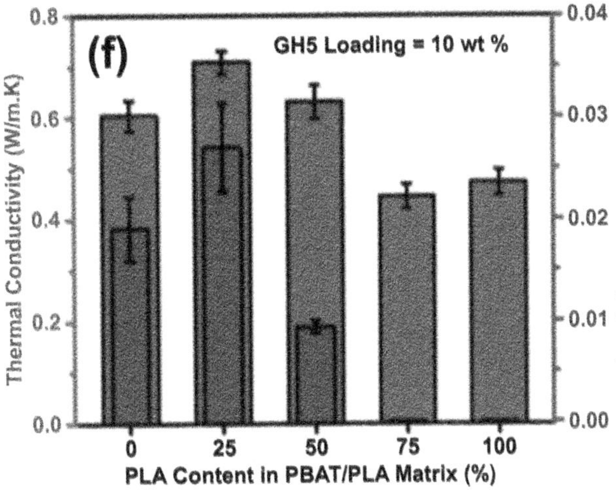

FIGURE 4.15

Thermal and electrical conductivity of PLA content in PBAT matrix at 10 wt.% of Graphene (GH5) loading [63].

Reproduced with permission from Elsevier. Copyright 2020.

CB, nanofibers, Fe_3O_4, and CNTs) improve the thermal-mechanical, fire-retardant, and nucleating properties and also show good electrical conductivity [81–86]. Biopolymer matrices containing various conductive metal nano- or microparticles are of interest to a variety of scientific fields [87–89]. Many researchers have studied the effect of silver nanoparticles (Ag-NP) and carbon nanotubes (CNTs) on the multifunctional conductivity of PCL biopolymer composites. It has been discovered that Ag-NP acts as conductive linkages between nanotube bundles, makes it easier to

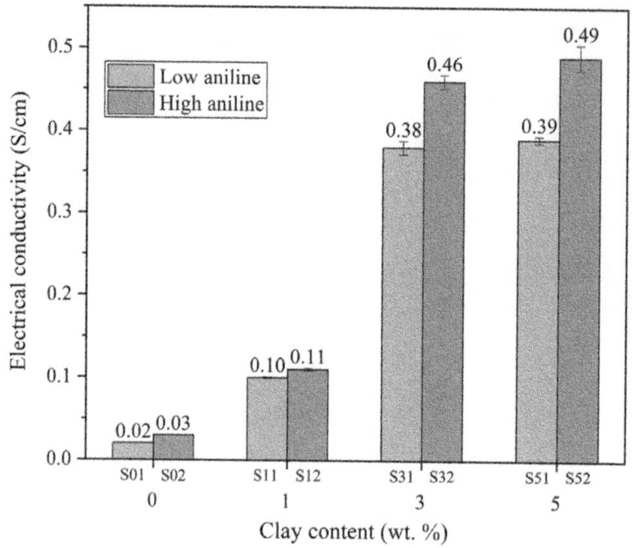

FIGURE 4.16

Electrical conductivity of clay nanoparticles filled PANI/bacterial cellulose (BC) [78].

Reproduced with permission from Elsevier. Copyright 2021.

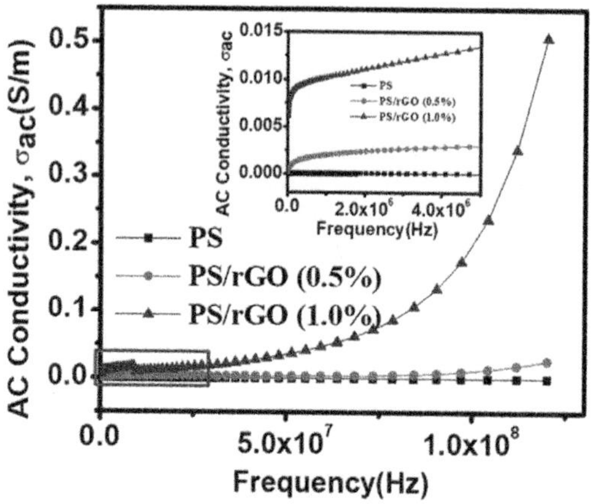

FIGURE 4.17

Electrical conductivity of PS and rGO versus frequency [41].

Reproduced with permission from Elsevier. Copyright 2021.

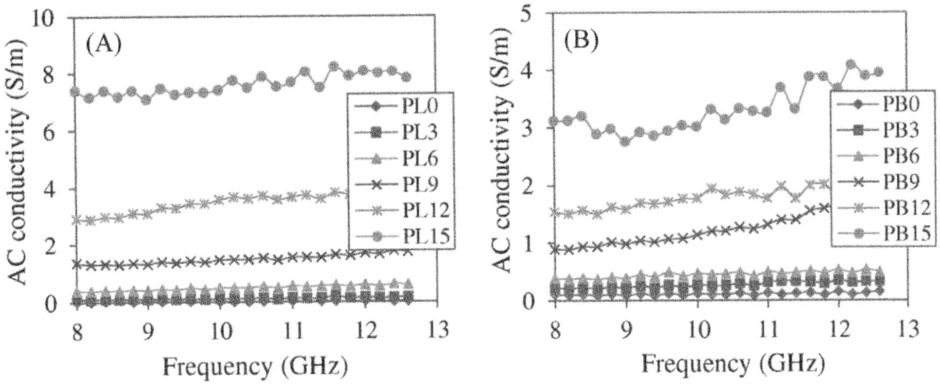

FIGURE 4.18

(A) Electrical conductivity of PLA/GNP and (B) PBAT/GNP nanocomposites versus frequency [43].

create conductive passageways when single-walled carbon nanotubes are present, and facilitates electron movement [90]. Biological tests are conducted on PCL + 15% silver, PCL + 1% CNTs, and PCL + 15% silver + 1% CNTs. The suitability of the composite conductive coatings as it is demonstrated during the traditional period that basic human bone marrow mesenchyme system cells exhibit equal vitality and cell-material interaction.

The oxidative polymerization technique is used to investigate the impact of combining poly(ethylene glycol) (PEG) with multifunctional conductive PPy [91]. It was discovered that the electrical conductivity of pure PPy increased. The number of additives, the electrical conductivity of PPy, and the exciter appeared to be effectively correlated. Similarly, increasing the PANI content increases the conductivity of the nanocomposite. In that figure, lower conductivity is seen at high concentrations of CNC in PANI, along with the conductivity value for each content of the PANI. A method for creating three-component composite materials of polyethylene (PE) substrate modified with grafted polyacrylic acid (PAA) and PPy has been proposed by Ilicheva et al. [92]. Ammonium persulfate or iron chloride (III) are used as oxidizers. The presence of PE film, which is strongly linked to PPy via PAA, causes the composite materials to have improved mechanical properties. The resulting materials have a high mechanical strength as well as electrical and ionic conductivities.

By employing a surface-initiated "grafting-from" technique, Ku et al. [93] have developed water-soluble SWCNT-poly-acetylene nanocomposites. Loaded single-walled carbon nanotubes account for 21% of the SWNT grafted ionic poly-acetylenes. High water solubility (8 mg/mL) is observed in the SWNT-poly-acetylene combination. Their findings indicate that the doped single-walled carbon nanotube poly-acetylene composites had electrical conductivity at room temperature which is between 1023 and 1024 S/cm. Electrical conductivity of doped SWNT poly-acetylene composites is two orders of magnitude higher than previously observed values for poly-acetylene composites. They also discovered that the composites are extremely soluble in water. The core-shell structure is developed by Uygun et al. [94] using a combination of PTh (Polythiophene) polymer and titanium dioxide (TiO_2) nanoparticles. It is made by oxidatively polymerizing thiophene with iron chloride and incorporating three different surfactants: cationic, nonionic, and anionic. The electrical properties of the nanocomposites are investigated in the presence of surfactant materials, with the PTh/TiO_2 anionic regime exhibiting the best semiconductor properties.

The direct current conductivity of PEDOT [poly(3,4-ethylenedioxythiophene)] doped with PSS (polystyrene sulfonate) is determined by Kim et al. [95] using various organic solvents. They reported that changing the solvent conductivity from 0.8 to 80 S/cm improved the DC in the PEDOT/PSS mixture. Once several natural solvents like THF, DMSO, and DMF are utilized, the temperature shows that the PEDOT/PSS system reaches the necessary system. A detailed analysis of the electrical characteristics of conductive composites comprising silver and the conducting polymer-PANI has been provided by Varga et al. [96]. Composites are formed by oxidizing PANI with silver nitrate ($AgNO_3$) and a variety of acids that act as doping agents. A semiconducting behavior is observed under the threshold of approximately 21–27 vol.% silver in composite, and a metallic behavior is observed above. It also demonstrates how conductivity changes with temperature.

4.3 Summary

This chapter discusses the dielectric, thermal, and electrical conductivity properties of biodegradable and biocompatible polymer nanocomposites. Biodegradable polymer composites are widely used due to their numerous environmental advantages over synthetic polymers. The compatibility of the hydrophobic nature of the polymer matrix and the hydrophilic characteristics of nanoparticles is the main challenge in making biopolymer-nanocomposites. The mechanical and other properties are due to the bonding between polymers and nanofillers. There are various methods for increasing the interfacial adhesion between nanofillers and matrix. Nanofillers also have a significant impact on the dielectric property of the biodegradable polymer. As the filler weight percentage increases, so does the dielectric constant in polymer composites, reaching a maximum at higher filler content. The dielectric constant is normally observed to be highest in the low frequency range and lowest in the higher frequency range. The same pattern can be seen in the case of dielectric loss. Biodegradable polymers reinforced with various conductive nanofillers improve not only thermal properties but may also exhibit good electrical conductivity. The current chapter is a thorough but not in-depth examination of the biodegradable nanocomposite. Hence, the present work can be expanded in future:

- The effect of the manufacturing process on composite on dielectric, thermal, and electrical properties should be explored.
- Process parameters of each test should be carefully examined for their influence and the optimum parameter should be discussed.
- Many other properties, such as mechanical and tribological properties should have been explored elaborately.

References

[1] H.P. Felgueiras, T.D. Tavares, M.T.P. Amorim, Biodegradable, spun nanocomposite polymeric fibrous dressings loaded with bioactive biomolecules for an effective wound healing: a review, IOP Conf. Ser. Mater. Sci. Eng. Available from: https://doi.org/10.1088/1757-899X/634/1/012033.

[2] H. Jena, J.K. Katiyar, Tribo-analysis of polymer composite in spur gear, Tribological Applications of Composite Materials, Springer, 2021, pp. 309–329. Available from: http://doi.org/10.1007/978-981-15-9635-3_12.

[3] J.M. Pachence, M.P. Bohrer, J. Kohn, Biodegradable polymers, Princ. Tissue Eng. (2007) 323–339. Available from: https://doi.org/10.1016/B978-012370615-7/50027-5.

[4] C. Maraveas, Production of sustainable and biodegradable polymers from agricultural waste, Polymers (Basel) 12 (5) (2020). Available from: http://doi.org/10.3390/POLYM12051127.

[5] F. Ruggero, R. Gori, C. Lubello, Methodologies to assess biodegradation of bioplastics during aerobic composting and anaerobic digestion: a review, Waste Manag. Res 37 (10) (2019) 959–975. Available from: https://doi.org/10.1177/0734242X19854127.

[6] W. Amass, A. Amass, B. Tighe, A review of biodegradable polymers: uses, current developments in the synthesis and characterization of biodegradable polyesters, blends of biodegradable polymers, and recent advances in biodegradation studies, Polym. Int. 47 (89) (1998) 144.

[7] H. Nasser, A.P. Mathew, Cellulose – metal organic frameworks (CelloMOFs) hybrid materials and their multifaceted applications: a review, Coord. Chem. Rev. 451 (2022). Available from: https://doi.org/10.1016/j.ccr.2021.214263.

[8] S.N.A.M. Jamil, K. Abdan, A.A. Shamsuri, A brief review on the influence of ionic liquids on the mechanical, thermal, and chemical properties of biodegradable polymer composites, Polym. (Basel) 13 (2021) 1–14.

[9] Samson O. Adeosun, G.I. Lawal, S.A. Balogun, E.I. Akpan, Review of green polymer nanocomposites, J. Mine. Mater. Charact. Eng. 11 (4) (2012) 385–416.

[10] K.M. Zia, N. Akram, S. Tabasum, A. Noreen, M.U. Akbar, Processing of bio-based polymers for industrial and medical applications, Processing Technology for Bio-Based Polymers, Elsevier, 2021, pp. 191–238.

[11] Q. Yao, A.B. Gómez, J. Su, V. Pascanu, Y. Yun, H. Zheng, et al., Series of highly stable isoreticular lanthanide metal-organic frameworks with expanding pore size and tunable luminescent properties, Chem. Mater. 27 (15) (2015) 5332–5339.

[12] R.B. Seymour, C.E. Carraher, Mechanical properties of polymers, Structure—Property Relationships in Polymers, Springer, Boston, MA, 1984. Available from: https://doi.org/10.1007/978-1-4684-4748-4_51.

[13] M. Martina, D.W. Hutmacher, Biodegradable polymers applied in tissue engineering research: a review, Polym. Int. 157 (2) (2007) 145–157. Available from: https://doi.org/10.1002/pi.2108.

[14] B. Kumar, M. Castro, J.F. Feller, Poly (lactic acid)-multi-wall carbon nanotube conductive biopolymer nanocomposite vapour sensors, Sens. Actuat. B. Chem 161 (1) (2012) 621–628. Available from: https://doi.org/10.1016/j.snb.2011.10.077.

[15] X. Zheng, S. Zhou, Y. Xiao, X. Yu, X. Li, P. Wu, Shape memory effect of poly(D,L-lactide)/Fe_3O_4 nanocomposites by inductive heating of magnetite particles, Colloids Surf. B. 71 (2009) 67–72. Available from: https://doi.org/10.1016/j.colsurfb.2009.01.009.

[16] Y. Fan, H. Nishida, T. Mori, Y. Shirai, T. Endo, Thermal degradation of poly(L-lactide): effect of alkali earth metal oxides for selective L,L-lactide formation, J. Polym. 45 (2004) 1197–1205. Available from: https://doi.org/10.1016/j.polymer.2003.12.058.

[17] I. Ezzeddine, N. Ghorbel, M. Ilsouk, M. Arous, M. Lahcini, F.Z. Bouharras, et al., Dielectric and thermal characteristics of Beidellite nanoclay-reinforced poly (butylene succinate, Mater. Chem. Phys. 258 (2021) 123855. Available from: https://doi.org/10.1016/j.matchemphys.2020.123855.

[18] A.M. El-Nahrawy, A.B. Abou Hammad, T.A. Khattab, A. Haroun, S. Kamel, Development of electrically conductive nanocomposites from cellulose nanowhiskers, polypyrrole and silver nanoparticles assisted with Nickel(III) oxide nanoparticles, React. Funct. Polym. 149 (2020). Available from: https://doi.org/10.1016/j.reactfunctpolym.2020.104533.

[19] R. Pantani, G. Gorrasi, G. Vigliotta, M. Murariu, P. Dubois, PLA-ZnO nanocomposite films: water vapor barrier properties and specific end-use characteristics, Eur. Polym. J. (2013). Available from: https://doi.org/10.1016/j.eurpolymj.2013.08.005.

[20] A. Anžlovar, A. Kržan, E. Žagar, Degradation of PLA/ZnO and PHBV/ZnO composites prepared by melt processing, Arab. J. Chem. (2017). Available from: https://doi.org/10.1016/j.arabjc.2017.07.001.

[21] G. Gaidukova, O. Platnieks, A. Aunins, A. Barkane, Spent coffee waste as a renewable source for the production of sustainable poly(butylene succinate) biocomposites from a circular economy perspective, RSC Adv 11 (2021) 18580–18589. Available from: https://doi.org/10.1039/d1ra03203h.

[22] D. Bondeson, K. Oksman, Dispersion and characteristics of surfactant modified cellulose whiskers nanocomposites, Compos. Interfaces 14 (2007) 617–630. Available from: https://doi.org/10.1163/156855407782106519.

[23] S.Y. Lee, I.A. Kang, G.H. Doh, H.G. Yoon, B.D. Park, Q. Wu, Thermal and mechanical properties of wood flour/talc-filled polylactic acid composites: effect of filler content and coupling treatment, J. Thermoplast. Compos. Mater. 21 (3) (2008) 209–223. Available from: https://doi.org/10.1177/0892705708089473.

[24] P. Qu, Y. Gao, G. Wu, L. Zhang, Nanocomposites of poly(lactic acid) reinforced with cellulose nanofibrils, Bioresources 5 (3) (2010) 1811–1823.

[25] V.K. Modi, Y. Shrives, C. Sharma, P.K. Sen, Review on green polymer nanocomposite and their applications, Int. J. Innov. Res. Sci. Eng. Tech. 3 (11) (2014) 17651–17656. Available from: https://doi.org/10.15680/IJIRSET.2014.0311079.

[26] Z. Liu, S.Z. Erhan, J. Xu, Preparation, characterization and mechanical properties of epoxidized soybean oil / clay nanocomposites, Polymer 46 (2005) 10119–10127. Available from: https://doi.org/10.1016/j.polymer.2005.08.065.

[27] Henriette M.C. Azeredo, Luiz Henrique C. Mattoso, Delilah Wood, Tina G. Williams, Roberto J. Avena-Bustillos, Tara H. McHugh, Nanocomposite edible films from mango puree reinforced with cellulose nanofibers, J. Food Sci. 74 (5) (2009) 31–35. Available from: https://doi.org/10.1111/j.1750-3841.2009.01186.x.

[28] J.S. Tate, A.T. Akinola, D. Kabakov, Bio-based nanocomposites: an alternative to traditional composites, J. Technol. Stud. 35 (1) (2009) 25–32.

[29] T. Ke, X. Sun, Effects of moisture content and heat treatment on the physical properties of starch and poly(lactic acid) blends, J. Appl. Polym. Sci. 81 (12) (2001) 3069–3082.

[30] T. Uesaka, K. Nakane, S. Maeda, T. Ogihara, N. Ogata, Structure and physical properties of poly(butylene succinate)/cellulose acetate blends, Polymer 41 (2000) 8449–8454.

[31] J.P. Zheng, P. Li, Y.L. Ma, K. De Yao, Gelatin / montmorillonite hybrid nanocomposite. I. Preparation and properties, J. Appl. Polym. Sci. 86 (5) (2002) 1189–1194. Available from: https://doi.org/10.1002/app.11062.

[32] Y. Han, S. Lee, K. Ho, I. Park, Preparation and characterization of chitosan–clay nanocomposites with antimicrobial activity, J. Phys. Chem. Solids 71 (4) (2010) 464–467. Available from: https://doi.org/10.1016/j.jpcs.2009.12.012.

[33] A. Takegawa, M. Murakami, Y. Kaneko, J. Kadokawa, Preparation of chitin / cellulose composite gels and films with ionic liquids, Carbohydr. Polym. 79 (1) (2010) 85–90. Available from: https://doi.org/10.1016/j.carbpol.2009.07.030.

[34] H. Jeevan Rao, S. Singh, P. Janaki Ramulu, Basant K. Agarwal, Characterization techniques and evolution of natural polymer nanofiber characterization (NPNFC), Adv. Eng. Mater. (2021) 173–186. Available from: https://doi.org/10.1007/978-981-33-6029-7.

[35] P. Henrique, C. Camargo, K.G. Satyanarayana, F. Wypych, Nanocomposites: synthesis, structure, properties and new application opportunities, Mat. Res. 12 (1) (2009) 1–39.

[36] Venkata Ramana, L.S. Sundar, Zafar Said, A.C.M. Sousa, 3 - Thermophysical, electrical, magnetic, and dielectric properties of hybrid nanofluids, Micro and Nano Technologies, Hybrid Nanofluids, Elsevier, 2022, pp. 65−92. Available from: https://doi.org/10.1016/B978-0-323-85836-6.00003-X.

[37] X. Li, J. Wang, H. Chen, C. Xiong, Z. Shi, Q. Yang, Flexible dielectric nanocomposite films based on chitin/boron nitride/copper calcium titanate with high energy density, Compos. Part. A Appl. Sci. Manuf. 149 (2021) 106554. Available from: https://doi.org/10.1016/j.compositesa.2021.106554.

[38] T.T. Grove, M.F. Masters, R.E. Miers, T.T. Grove, M.F. Masters, R.E. Miers, Determining dielectric constants using a parallel plate capacitor determining dielectric constants using a parallel plate capacitor, Am. J. Phys 52 (2005) (2014) 1−6. Available from: https://doi.org/10.1119/1.1794757.

[39] J. Mohammed, B.F. Abubakar, K.U. Yerima, H. Hamisu, U.T. Isma, Biodegradable polymer modified rGO / PANI / CCTO nanocomposites: structural and dielectric properties, Mater. Today Proc. 5 (14) (2018) 28462−28469. Available from: https://doi.org/10.1016/j.matpr.2018.10.133.

[40] H.T. Ahmed, O.G. Abdullah, Preparation and composition optimization of PEO: MC polymer blend films to enhance, Polymers (Basel) 11 (5) (2019) 853−864.

[41] S.I. Mollik, R.B. Alam, M.R. Islam, Significantly improved dielectric properties of bio-compatible starch / reduced graphene oxide nanocomposites, Synth. Met. 271 (2021) 116624. Available from: https://doi.org/10.1016/j.synthmet.2020.116624.

[42] K. Deshmukh, M. Basheer Ahamed, R.R. Deshmukh, S.K. Khadheer Pasha, K.K. Sadasivuni, A. Reddy Polu, et al., Newly developed biodegradable polymer nanocomposites of cellulose acetate and Al_2O_3 nanoparticles with enhanced dielectric performance for embedded passive applications, J. Mater. Sci. Mater. Electron. 28 (1) (2017) 973−986. Available from: https://doi.org/10.1007/s10854-016-5616-9.

[43] S. Kashi, R.K. Gupta, T. Baum, N. Kao, S.N. Bhattacharya, Dielectric properties and electromagnetic interference shielding effectiveness of graphene-based biodegradable nanocomposites, Mater. Des. 109 (2016) 68−78. Available from: https://doi.org/10.1016/j.matdes.2016.07.062.

[44] P. Dhatarwal, R.J. Sengwa, S. Choudhary, Multifunctional (PVP/PEO)/SnO_2 nanocomposites of tunable optical and dielectric properties, Optik 221 (2020) 1−10.

[45] R. Jeewan, S. Shobhna, P. Dhatarwal, Nonlinear optical and dielectric properties of TiO_2 nanoparticles incorporated PEO / PVP blend matrix based multifunctional polymer nanocomposites, J. Mater. Sci. Mater. Electron. (2019). Available from: https://doi.org/10.1007/s10854-019-01587-4.

[46] S. Choudhary, R.J. Sengwa, Investigation on structural and dielectric properties of silica nanoparticles incorporated poly (ethylene oxide)/ poly (vinyl pyrrolidone) blend matrix based nanocomposites, J. Inorg. Organomet. Polym. Mater. (2018). Available from: https://doi.org/10.1007/s10904-018-1034-1.

[47] S. Choudhary, Structural, optical, dielectric and electrical properties of (PEO−PVP)−ZnO nanocomposites, J. Phys. Chem. Solids (2018). Available from: https://doi.org/10.1016/j.jpcs.2018.05.017.

[48] P. Liu, Z. Yao, J. Zhou, Mechanical, thermal and dielectric properties of graphene oxide / polyimide resin composite, High. Perform. Polym. 28 (9) (2015). Available from: https://doi.org/10.1177/0954008315613558.

[49] A.H. Mohamad, O. Gh, S.R. Saeed, Effect of very fine nanoparticle and temperature on the electric and dielectric properties of MC-PbS polymer nanocomposite film, Results Phys 16 (2020). Available from: https://doi.org/10.1016/j.rinp.2019.102898.

[50] M. Sadiq, M.M. Hasan Raza, A.K. Singh, S.K. Chaurasia, M. Zulfequar, A. Arya, et al., Dielectric properties and ac conductivity behavior of rGO incorporated PVP-PVA blended polymer nanocomposites films, Mater. Today Proc. 49 (2022) 3164−3169. Available from: https://doi.org/10.1016/j.matpr.2020.11.169.

[51] V.R. Manik, C. Biswas, Shaik Jeelani, Influence of biobased silica/carbon hybrid nanoparticles on thermal and mechanical properties of biodegradable polymer films, Compos. Commun. 4 (2017) 43−53. Available from: https://doi.org/10.1016/j.coco.2017.04.005.

[52] K. Behera, Y. Chang, M. Yadav, F. Chiu, Enhanced thermal stability, toughness, and electrical conductivity of carbon nanotube-reinforced biodegradable poly(lactic acid)/poly(ethylene oxide) blend-based nanocomposites, Polym. (Guildf.) (2019) 122002. Available from: https://doi.org/10.1016/j.polymer.2019.122002.

[53] N.C. Nepomuceno, A.A.A. Seixas, E.S. Medeiros, Evaluation of conductivity of nanostructured polyaniline / cellulose nanocrystals (PANI / CNC) obtained via in situ polymerization, J. Solid. State Chem 302 (2021). Available from: https://doi.org/10.1016/j.jssc.2021.122372.

[54] Y.O. Al-ghamdi, H.N. Abdelhamid, Chitosan-CdS quantum dots biohybrid for highly selective interaction with copper(II) ions, ACS Omega 7 (24) (2022) 21014–21024. Available from: https://doi.org/10.1021/acsomega.2c01793.

[55] M.M. Abutalib, A. Rajeh, Structural, thermal, optical and conductivity studies of Co / ZnO nanoparticles doped CMC polymer for solid state battery applications, Polym. Test. 91 (2020) 106803. Available from: https://doi.org/10.1016/j.polymertesting.2020.106803.

[56] J.P. Mofokeng, A.S. Luyt, Morphology and thermal degradation studies of melt-mixed poly(lactic acid) (PLA)/poly(ε-caprolactone) (PCL) biodegradable polymer blend nanocomposites with TiO_2 as filler, Polym. Test. (2015). Available from: https://doi.org/10.1016/j.polymertesting.2015.05.007.

[57] M. Ravi, S. Song, K. Gu, J. Tang, Z. Zhang, Electrical properties of biodegradable poly(ε-caprolactone): lithium thiocyanate complexed polymer electrolyte films, Mater. Sci. Eng. B (2015) 1–10. Available from: https://doi.org/10.1016/j.mseb.2015.02.003.

[58] L. Wang, A. He, Microstructure and thermal properties of polypropylene/clay nanocomposites with $TiCl_4$/$MgCl_2$/clay compound catalyst, J. Nanomater. 2015 (2015). Available from: https://doi.org/10.1155/2015/591038.

[59] H. Shen, Y. Li, W. Yao, S. Yang, L. Yang, F. Pan, et al., Solvent-free cellulose nanocrystal fluids for simultaneous enhancement of mechanical properties, thermal conductivity, moisture permeability and antibacterial properties of polylactic acid fibrous membrane, Compos. Part. B 222 (2021) 109042. Available from: https://doi.org/10.1016/j.compositesb.2021.109042.

[60] Y. Sun, Z. Zhang, K.S. Moon, C.P. Wong, Glass transition and relaxation behavior of epoxy nanocomposites, J. Polym. Sci. Part. B Polym. Phys. 42 (21) (2004) 3849–3858. Available from: https://doi.org/10.1002/polb.20251.

[61] Q. Zou, S.-wei Xiong, M.-ying Jiang, Li-ye Chen, K. Zheng, Pei-gen Fu, et al., Highly thermally conductive and eco-friendly OH-h-BN/chitosan nanocomposites by constructing a honeycomb thermal network, Carbohydr. Polym. 266 (2021) 118127. Available from: https://doi.org/10.1016/j.carbpol.2021.118127.

[62] J. Li, Y. Wang, T.N. Yue, Y.N. Gao, Y.D. Shi, J.B. Shen, et al., Robust electromagnetic interference shielding, joule heating, thermal conductivity, and anti-dripping performances of polyoxymethylene with uniform distribution and high content of carbon-based nanofillers, Compos. Sci. Technol. 206 (2021) 108681. Available from: https://doi.org/10.1016/j.compscitech.2021.108681.

[63] Y. Guo, X. Zuo, Y. Xue, J. Tang, M. Gouzman, Y. Fang, et al., "Engineering thermally and electrically conductive biodegradable polymer nanocomposites, Compos. Part. B 189 (2020) 107905. Available from: https://doi.org/10.1016/j.compositesb.2020.107905.

[64] H.S. Abdo, A.A. Elzatahry, H.F. Alharbi, K.A. Khalil, Electrical Conductivity Behavior of Biopolymer Composites, Elsevier Inc., 2017.

[65] T. Nezakati, A. Tan, J. Lim, R.D. Cormia, S.H. Teoh, A.M. Seifalian, Ultra-low percolation threshold POSS-PCL/graphene electrically conductive polymer: neural tissue engineering nanocomposites for neurosurgery, Mater. Sci. Eng. C. 104 (2019) 109915. Available from: https://doi.org/10.1016/j.msec.2019.109915.

[66] A. Paydayesh, S.R. Mousavi, S. Estaji, H.A. Khonakdar, M.A. Nozarinya, Functionalized graphene nanoplatelets/poly (lactic acid)/chitosan nanocomposites: mechanical, biodegradability, and electrical conductivity properties, Polym. Compos. 43 (1) (2022) 411–421. Available from: https://doi.org/10.1002/pc.26385.

[67] O.A. Bin-dahman, D. Khastgir, M.A. Al-harthi, Electrical and dielectric properties of poly(vinyl alcohol)/starch/ graphene nanocomposites, Can. J. Chem. Eng. 96 (2018) 903–911. Available from: https://doi.org/10.1002/cjce.22999.

[68] J.M. Hadi, B.S. Aziz, M.M. Nofal, S.A. Hussein, M.H. Hafiz, M.A. Brza, et al., Electrical, dielectric property and electrochemical performances of plasticized silver ion-conducting chitosan-based polymer nanocomposites, Membranes (Basel) 10 (7) (2020) 1–22. Available from: https://doi.org/10.3390/membranes10070151.

[69] X. Wu, H. Jiang, J. Zheng, X. Wang, Z. Gu, C. Chen, Highly sensitive recognition of cancer cells by electrochemical biosensor based on the interface of gold nanoparticles / polylactide nanocomposites, J. Electroanal. Chem. 656 (1–2) (2011) 174–178. Available from: https://doi.org/10.1016/j.jelechem.2010.11.035.

[70] J. Wang, J. Yu, D. Bai, Z. Li, H. Liu, Y. Li, et al., Biodegradable, flexible, and transparent conducting silver nanowires/polylactide film with high performance for optoelectronic devices, Polymers (Basel) 12 (3) (2020). Available from: https://doi.org/10.3390/polym12030604.

[71] G.B. Patel, N.L. Singh, F. Singh, P.K. Kulriya, "Effect of swift heavy ions irradiation on physicochemical and dielectric properties of chitosan and chitosan-Ag nanocomposites, Radiat. Phys. Chem. 181 (2021) 109288. Available from: https://doi.org/10.1016/j.radphyschem.2020.109288.

[72] Y. Li, H. Liu, K. Dai, G. Zheng, C. Liu, J. Chen, et al., Tuning of vapor sensing behaviors of eco-friendly conductive polymer composites utilizing ramie fiber, Sens. Actuators B Chem. 221 (2015) 1279–1289. Available from: https://doi.org/10.1016/j.snb.2015.07.100.

[73] S. Parandeh, M. Kharaziha, F. Karimzadeh, An eco-friendly triboelectric hybrid nanogenerators based on graphene oxide incorporated polycaprolactone fibers and cellulose paper, Nano Energy 59 (2019) 412–421. Available from: https://doi.org/10.1016/j.nanoen.2019.02.058.

[74] I. Krucińska, B. Surma, M. Chrzanowski, E. Skrzetuska, M. Puchalski, "Application of melt-blown technology for the manufacture of temperature-sensitive nonwoven fabrics composed of polymer blends PP / PCL loaded with multiwall carbon nanotubes, J. Appl. Polym. Sci. 127 (2) (2012) 1–10. Available from: https://doi.org/10.1002/app.37834.

[75] V.K. Rana, S. Akhtar, S. Chatterjee, S. Mishra, R.P. Singh, C.S. Ha, Chitosan and chitosan-co-poly (-caprolactone) grafted multiwalled carbon nanotube transducers for vapor sensing, J. Nanosci. Nanotechnol. 14 (3) (2014) 2425–2435. Available from: https://doi.org/10.1166/jnn.2014.8498.

[76] M. Castro, J. Lu, S. Bruzaud, B. Kumar, J.F. Feller, Carbon nanotubes/poly(ε-caprolactone) composite vapour sensors, Carbon N. Y 47 (8) (2009) 1930–1942. Available from: https://doi.org/10.1016/j.carbon.2009.03.037.

[77] J. Fal, K. Bulanda, M. Oleksy, J. Sobczak, J. Shi, M. Liu, High AC and DC electroconductivity of scalable and economic graphite – diamond polylactide nanocomposites, Mater 14 (11) (2021).

[78] M.H. Salehi, H. Golbaten-Mofrad, S.H. Jafari, V. Goodarzi, M. Entezari, M. Hashemi, et al., Electrically conductive biocompatible composite aerogel based on nanofibrillated template of bacterial cellulose/polyaniline/nano-clay, Int. J. Biol. Macromol. 173 (2021) 467–480. Available from: https://doi.org/10.1016/j.ijbiomac.2021.01.121.

[79] I. Evgeni, R. Kotsilkova, H. Xia, Y. Chen, R.K. Donato, K. Donato, et al., PLA/graphene/MWCNT composites with improved electrical and thermal properties suitable for FDM 3D printing applications, Appl. Sci. 9 (6) (2019). Available from: https://doi.org/10.3390/app9061209.

[80] A.P.B. Silva, L.S. Montagna, F.R. Passador, M.C. Rezende, A.P. Lemes, Biodegradable nanocomposites based on PLA / PHBV blend reinforced with carbon nanotubes with potential for electrical and electromagnetic applications, Exp. Polym. Letter. 15 (10) (2021) 987–1003.

[81] I. Armentano, D. Puglia, F. Luzi, C.R. Arciola, F. Morena, S. Martino, et al., Nanocomposites based on biodegradable polymers, Mater. (Basel) 11 (5) (2018) 1–27. Available from: https://doi.org/10.3390/ma11050795.

[82] T. Glaskova-kuzmina, O. Starkova, S. Gaidukovs, O. Platnieks, G. Gaidukova, Durability of biodegradable polymer nanocomposites, Polymer, 13, 3375. https://doi.org/10.3390/polym13193375.

[83] A. Mtibe, M.P. Motloung, J. Bandyopadhyay, S.S. Ray, Synthetic biopolymers and their composites: advantages and limitations—an overview, Macromol. Rapid Commun. 42 (15) (2021) 1–28. Available from: https://doi.org/10.1002/marc.202100130.

[84] E. Chiellini, A. Corti, S. D'Antone, R. Solaro, Biodegradation of poly (vinyl alcohol) based materials, Prog. Polym. Sci. 28 (6) (2003) 963–1014.

[85] K. Van De Velde, P. Kiekens, Biopolymers: overview of several properties and consequences on their applications, Polym. Test. 21 (2002) 433–442.

[86] S.S. Bari, A. Chatterjee, S. Mishra, Biodegradable polymer nanocomposites: an overview, Polym. Rev. 56 (2) (2016) 287–328. Available from: https://doi.org/10.1080/15583724.2015.1118123.

[87] S.K. Battacarya, Metal-filled polymers-properties and applications, J. Polym. Sci. Part. C: Polym. Letter. 25 (6) (1987) 360.

[88] P. Fakultet, B. Luka, Optimization of ellulose arylate and grafted 4-vinylpyridine and 1-vinylimidazole synthesis, Vaso Bojani 64 (6) (2010) 529–535. Available from: https://doi.org/10.2298/HEMIND100401043B.

[89] G. Pinto, A. Maaroufi, R. Benavente, M. Peren, "Electrical Conductivity of urea – formaldehyde – cellulose composites loaded with copper, Polym. Compos. 32 (2) (2011). Available from: https://doi.org/10.1002/pc.

[90] E. Fortunati, F.D. Angelo, S. Martino, A. Orlacchio, J.M. Kenny, I. Armentano, Carbon nanotubes and silver nanoparticles for multifunctional conductive biopolymer composites, Carbon N. Y 49 (7) (2011) 2370–2379. Available from: https://doi.org/10.1016/j.carbon.2011.02.004.

[91] H.C. Kang, K.E. Geckeler, Enhanced electrical conductivity of polypyrrole prepared by chemical oxidative polymerization: effect of the preparation technique and polymer additive, Polymer 41 (2000) 6931–6934.

[92] N.S. Ilicheva, N.K. Kitaeva, V.R. Duflot, V.I. Kabanova, Synthesis and properties of electroconductive polymeric composite material based on polypyrrole,", ISRN Polym. Sci. (2012). Available from: https://doi.org/10.5402/2012/320316.

[93] B.-C. Ku, D.K. Kim, J.S. Lee, A. Blumstein, J. Kumar, L.A. Samuelson, Synthesis and properties of water soluble single-walled carbon nanotube graft ionic polyacetylene nanocomposites, Polym. Compos. 30 (12) (2009) 1817–1824. Available from: https://doi.org/10.1002/pc.20754.

[94] A. Uygun, O. Turkoglu, S. Sen, E. Ersoy, A.G. Yavuz, G.G. Batir, The electrical conductivity properties of polythiophene/TiO_2 nanocomposites prepared in the presence of surfactants, Curr. Appl. Phys. 9 (4) (2009) 866–871. Available from: https://doi.org/10.1016/j.cap.2008.08.005.

[95] J.Y. Kim, J.H. Jung, D.E. Lee, J. Joo, Enhancement of electrical conductivity of poly(3, 4-ethylenedioxythiophene)/poly(4-styrenesulfonate) by a change of solvents, Synth. Met. 126 (2–3) (2002) 311–316. Available from: https://doi.org/10.1016/S0379-6779(01)00576-8.

[96] M. Varga, J. Proke, Electrical conductivity of polyaniline – silver nanocomposites, WDS'12 Proceedings of Contributed Papers, Part III, pp. 52–57, 2012.

CHAPTER

5 Thermomechanical and viscoelastic properties of biodegradable and biocompatible polymer nanocomposites

B. Suresha[1], S.M. Darshan[1,2], S.L. Aravind[1] and B. Harshavardhan[1]

[1]Department of Mechanical Engineering, The National Institute of Engineering, Mysuru, Karnataka, India

[2]Department of Mechanical Engineering, CHRIST (Deemed to be University), Bengaluru, Karnataka, India

Chapter Outline

5.1 Introduction

5.1.1 Biodegradable polymers

Polymers are popular in industrial applications due to a number of advantages. These benefits include low weight, ease of processing, and chemical and environmental resistance. It is difficult to treat polymers after they have been used due to their resistance to the surrounding environment. Polymers can survive in natural environments for extremely long periods of time without degrading. Biopolymer-based polymer matrices are becoming increasingly popular. It's possible that this

Biodegradable and Biocompatible Polymer Nanocomposites. DOI: https://doi.org/10.1016/B978-0-323-91696-7.00004-0
Copyright © 2023 Elsevier Inc. All rights reserved.

is due to people's concern about the depletion of fossil fuels, which are the primary sources of the monomers from which the majority of synthetic polymers have been produced up to this point. Because nonbiodegradable polymers such as polystyrene, polypropylene, polyethylene, and poly (methyl methacrylate) are commonly used in conventional applications, recycling or reusing them is a difficult task [1]. This practice generates massive amounts of nonbiodegradable waste all over the world. The composites made with these polymers are also disposed of in the waste. The majority of biodegradable polymers come from other suitable renewable sources. The use of biodegradable polymers in the production of polymeric composites is fascinating. Nonetheless, replacing biodegradable polymers with synthetic polymers poses a number of difficult challenges. This category's primary characteristics are low thermal stability and mechanical properties. Many scientists are now interested in biodegradable polymers because they have the potential to reduce synthetic waste through biodegradation. After a period of time, these materials degrade into molecules that are safe for the environment. These microorganisms in the environment use available carbon dioxide and water to biodegrade such polymers, which can then be used for a variety of applications [2].

Ester, amide, and ether bonds are the building blocks of biodegradable polymers. According to the structure and chemical processes of production, biodegradable polymers could be divided into two broad categories as shown in Fig. 5.1. One of these categories is agro-polymers, which are made from biomass [3]. The other component is made up of bio-polyesters, which are either produced naturally by microorganisms or manufactured artificially using natural/synthetic monomers. Polysaccharides (starches in wood or potatoes) and proteins are two examples of agro-polymers (plant or animal based). Polysaccharides are composed of glycosidic bonds. These bonds are formed by the elimination of water from the binding of hemiacetal saccharide to alcohol. Amino acids, the building blocks of proteins, each have their own set of functional groups. Condensation reactions recombine such amino acids so that they can form peptide bonds composed of amide-based functional groups [4]. Polylactic acid and poly-hydroxybutyrate are two examples of bio-polyesters.

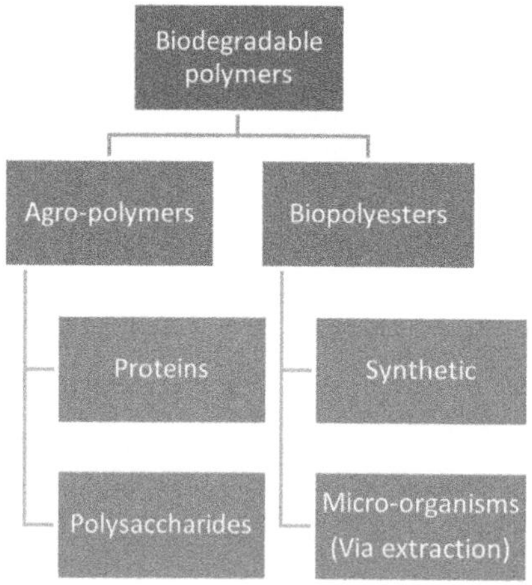

FIGURE 5.1

Classification of biodegradable polymers based on occurrence and its structure.

5.1.2 Biocompatible polymers

The biocompatibility of biomedical polymeric materials whose surfaces interact with biological systems is an important feature. Numerous studies have concentrated on examining phenomena like these interactions among polymeric surface and organism. "The potential of any material to function in a particular application while eliciting an adequate reaction from the host organism" is one definition of biocompatibility that is frequently put into use. However, interactions between materials and biological systems occur in a wide range of applications and molecular processes. A biocompatible system and its exact measurement cannot be precisely defined or quantified. Furthermore, the criterion for evaluating biocompatibility should be whether the material will be accepted by a living body. Tissue scaffolding, tissue culture, implantable constrained drug delivery devices, dialysis and catheter tubing, envelops for cardiac system transplantation, wildlife vaccinations, artificial grafts, and coatings of medical equipment are all potential applications for biocompatible polymeric systems [5].

5.2 Natural/synthetic biocompatible and biodegradable polymers

The term "natural biopolymers" refers to a class of macromolecular compounds found in nature in a wide range of organisms. Living microorganisms are responsible for the synthesis of biopolymers, which are made up of repeating units linked together by covalent bonds. Microorganisms such as algae, fungi, and bacteria contain biopolymers. Animals contain complex polymers such as hydrocarbons, fats, proteins, and nucleic acid. Starches, cellulose, oils, and even polyesters can be found in plants and lower organisms. In addition to their biocompatibility and biodegradability, the ability of these polymeric materials to be synthetically altered is contributing to their growing importance in the industrial sector. Biopolymers are extracted in a variety of ways in order to make them more affordable to the general public [6].

Polylactic acid (PLA) is one of the few polymers derived from bio-renewable resources. As a result, they qualify as biopolymers. They can be produced in large quantities from cassava roots or corn starch via chemical and enzyme polymerization. Various chemical methods can be used to create a diverse range of biodegradable and biologically compatible polymers [7]. Esters, diacids, anhydrides, and amides are a few examples of such chemical compositions found in polymers. The presence of poor hydrolyzable links in the backbones of synthetic bio-based polymers is the primary contributor to biodegradability. They decompose enzymatically or chemically into monomer units, of which only a few are biologically acceptable to human bodies, emphasizing their importance in a variety of biomedical fields.

5.3 Surface modification methods of biodegradable and biocompatible polymers

Polymers are the most diverse class of bio-materials and are the subject of extensive research for a wide range of application fields due to their unique properties such as viscosity, malleability, and thermo-mechanical properties. Polymers can be found naturally or synthesized. Biopolymers such as chitosan and collagen are biodegradable, abundant, and highly biocompatible. Several biodegradable aliphatic polyesters are available, including polybutylene succinate (PBS), poly(3-hydroxybutyrate) (PHB), and poly(-caprolactone) (PCL). These biodegradable polymers undergo a transformation in the presence of soil microorganisms, causing them to degrade into carbon dioxide and water. As a result, these polymers are thought to be environmentally friendly. The surface properties of biopolymers have a significant impact on their thermal, mechanical, biodegradability, and biocompatibility properties. Surface properties include molecule permeation, hydrophobicity and hydrophilicity, molecule adsorption, ionic charge, adhesive strength, chemical reaction kinetics, friction, and so on. As a result, any change in surface properties may have an effect on the bulk properties of the polymer, resulting in a broader range of polymer applications. Nonetheless, the most important characteristic in defining the implementation is the surface's hydrophobicity/hydrophilicity or wettability, which is associated with chemical compositions, adsorption, surface roughness, and polymer adhesion [8].

A variety of methods for changing polymer surfaces have been developed over the years due to the importance of surface modification, and are illustrated in Fig. 5.2. These procedures are classified into three types: mechanical, physicochemical, and biological techniques. The primary goal of physicochemical methodology is to improve surface hydrophilicity by attaching coupling agents to hydrophilic chains. Furthermore, the desorption process removes any impurities that may be present on the polymer surface. This results in a greater number of ions on the polymer's surface, which can then be coupled with coupling agents. This contributes to the improvement of interactions between hydrophobic polymers and hydrophilic fillers, resulting in the production of homogeneous polymeric biocomposites. Mechanical methods increase friction and surface roughness by modifying the surface topography with roughening methods. This modification also improves the biopolymer's adhesion and wettability properties [9]. A biological technique that includes a

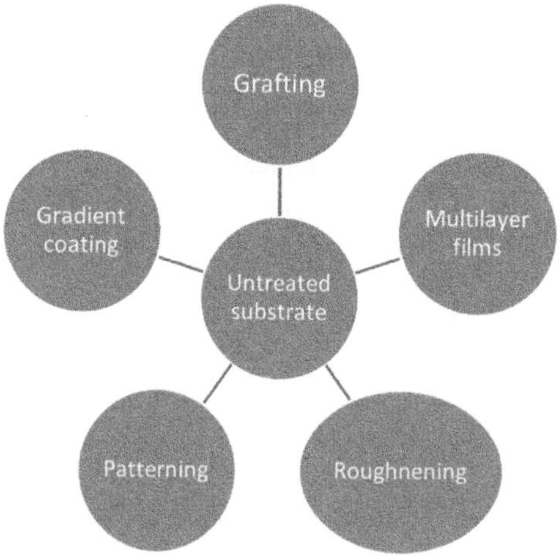

FIGURE 5.2

Various surface modification methodologies.

compound that acts as an interconnector improves bimolecular interactions between polymers and biomolecules. The surface modification has a significant impact on the thermo-mechanical properties of the BBPCs.

5.4 Natural fiber-reinforced hybrid polymer composites' fundamental properties

Natural fiber-reinforced hybrid polymer composites are frequently used as metal material substitutes in the automotive, structural, and aerospace industries. Because they combine two or more reinforcing components with a polymer matrix, hybrid composites can be stiffer and more powerful. The hybrid reinforcement can have a variety of properties depending on its source (natural and natural, natural and synthetic), and scale (micro/nano). In theory, if there is no physical or chemical contact between the reinforcing fibers, the characteristics of hybrid composites must adhere to the law of hybrid mixtures, which is illustrated below [10].

$$P_H = P_1 V_1 + P_2 V_2 \tag{5.1}$$

where P_H stands for the estimated property of the hybrid composite, P_1 and P_2 stand for the properties of composites with distinct fiber/filler reinforcements 1 and 2, respectively. Also, V_1 and V_2 stand for the volume fractions of fiber/fillers respectively in the hybrid composite.

The polymer matrix, size of the reinforcement, type of reinforcement, and binding strength are all factors that differentiate hybrid composites. They are frequently divided into two categories based on matrix materials: (1) thermoset (polyester, epoxy) and (2) thermoplastic (polycarbonate, polypropylene, and acrylonitrile butadiene styrene). Because of its intended biodegradability, the use of environmentally conscious matrices such as starch, cellulose, poly (lactic acid), and biodegradable epoxy has recently been investigated. Furthermore, fiber-reinforced biopolymers may help to reduce carbon footprint. A novel category group of hybrid composites known as nanocomposites has been developed based on the scale of the reinforcing agent (if any reinforcing component used in the matrix is in nano level; 1–100 nm). By incorporating nanoparticles, nanotubes, and lamellar nanostructures into a polymer matrix, nanocomposites with improved thermomechanical properties are created [11].

5.5 Bio-nanocomposites

Bio-nanocomposites are an emerging category of bio-hybrid materials which exhibit dimensions in nanoscale range (1–100 nm) eventually resulting from unique combination of natural polymers like proteins, nucleic acids as well as polysaccharides with nano metric inorganic solids. Theng coined the term "bio-nanocomposites" (also known as biocomposites, green composites, nanobiocomposites, and bio-hybrids) in 1970. They can be used in a variety of applications due to their multidimensional properties such as biodegradability, antimicrobial activity, and biocompatibility. They also have better thermo-mechanical properties than micro or macro composites [12]. These inorganic fractions are composed of finely ground solids derived from either synthetic or natural

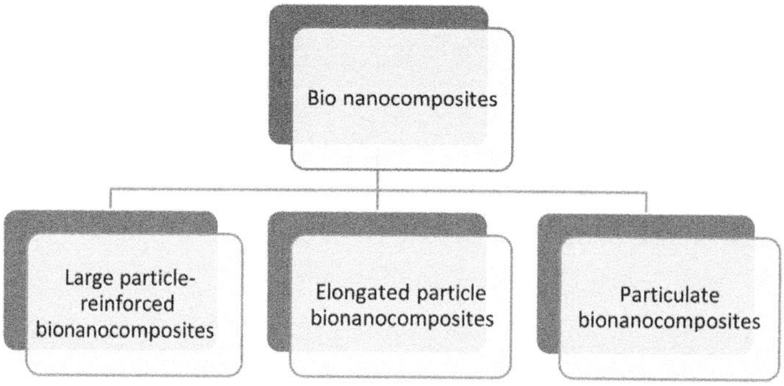

FIGURE 5.3

Types of bio-nanocomposites.

sources. Phosphates, clays, and carbonates are examples of such solids. The major challenge posed by bio-nanocomposites, as with multiphased materials, is to develop materials with superior performance. This can be accomplished by carefully managing the properties of the constituents that make up the biocomposites. Biopolymers' biodegradable and biocompatible properties, combined with the thermo-mechanical properties of their inorganic counterparts, serve as a link between structural and functional materials. In recent years, a significant amount of research has gone into the process of creating nanocomposites from bio-renewable polymeric materials. The majority of bio-nanocomposites research focuses on polycaprolactone (PCL), polysaccharides, proteins, and polylactic acid (PLA) with layered silicates in the smectite group. Bio-nanocomposites of this type are widely used in food packaging, electronics, tissue engineering, drug delivery, and regenerative medicine [13]. Recent research has shown that using micro fibrous clay minerals such as palygorskite and sepiolite in the process of developing clay-polymer bio-nanocomposites can result in quite an intriguing reinforcement in biopolymer [14].

According to the shape, size, origin of the reinforcement, and type of matrix, bio-nanocomposites can be divided into three main categories: particulate, elongated particle (cellulose nanofibrils, carbon nanotubes), as well as layered structure (exfoliated and intercalated nanocomposites), as shown in Fig. 5.3. The properties of organic and inorganic materials are typically quite different from one another. Ceramics and glass are examples of inorganic materials that are hard and have poor impact resistance, whereas organic polymers exhibit resilient behavior. Organic polymers typically have significant limitations, such as a proclivity for heat instability and natural degradation with increasing age. Overall, inorganic species are known to have excellent thermal and mechanical stability, as well as excellent optical properties. Materials that are a hybrid of organic and inorganic components may then have desirable property combinations.

5.6 Dynamic mechanical analysis

It is an analytical technique used to assess the viscoelastic property of a material, polymer, or composite by measuring strain or stress arising from dynamically fluctuating strain or stress applied to a

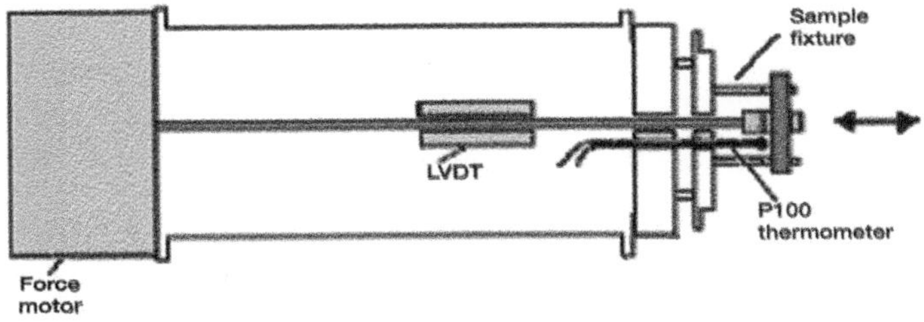

FIGURE 5.4

Schematic of dynamic mechanical analysis.

sample. It describes the properties of a material as they change as a result of stress, time, temperature, frequency, environment, or a combination of these factors. This method involves repeatedly deforming a sample with a small amount of deformation. This allows researchers to investigate how materials respond to stress, temperature, frequency, and other factors. Thermomechanical analysis (TMA) involves applying a constant static force to a material, which is then observed to change as temperature or time changes, as schematically shown in Fig. 5.4. DMA measures variations in damping and stiffness by applying an oscillating force to the sample at a predetermined frequency [15].

While TMA provides the thermal expansion coefficient, DMA data is used to obtain modulus information. Transitions are recognized by both, but DMA is far more sensitive. This technique uses a sinusoidal deformation to deform a sample with a specific shape. A controlled stress or strain can be applied to the sample. Following that, the polymer sample will distort to a predetermined degree under a known stress. The amount of material deformation is determined by the stiffness of the material. The force motor generates sinusoidal waves, which are then transmitted to the sample via a drive shaft. Concerns have long been raised about this drive shaft's compliance and the impact of any stabilizing bearing used to keep it in place [16]. This method produces typical parameters such as the storage modulus (E' or G'), loss modulus (E'' or G'), and damping factor, which are critical for composite materials [17]. The storage modulus of a material specifies both its stiffness and its elastic properties. It is critical to remember that, while this value and Young's modulus may be related, they are not the same. The loss modulus, which is related to the sample's thermal energy dissipation, indicates the viscous reaction of a material. This factor is usually related to the material's "internal friction." It is most affected by movements, phase transition processes, heterogeneities, and molecular configurations [18].

The storage modulus and loss modulus as seen in Fig. 5.5 could be explained by the behavior of a jumping ball that dissipates and stores energy. The damping factor ($\tan\delta$), which is shown in Fig. 5.6 and represents the ratio of loss modulus to storage modulus ($\tan\delta = E''/E'$), is an important quantity generated using the same way. The complex modulus (E^*), which is the sum of E' and E'', measures the material's resistance to deformation [19]. Because materials bend easily when subjected to external force, more flexible materials have lower damping factor levels, whereas nonflexible materials have the opposite effect. As the interfacial bonding between the nanofillers, fiber, and matrix components of fiber-reinforced polymer nanocomposites increases, so will the damping factor. Because interlocking molecular structures move together, internal motion is constrained and less energy is lost, resulting in a higher loss modulus to storage modulus ratio [20].

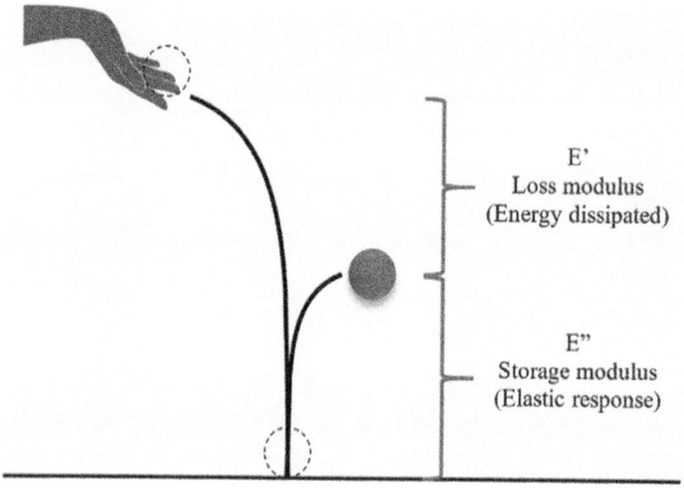

FIGURE 5.5

Depiction of the storage modulus and loss modulus [19].

Reproduced with permission from Elsevier. Copyright 2022.

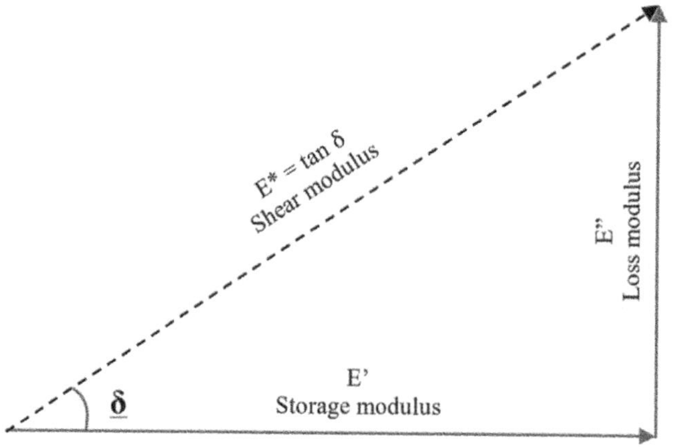

FIGURE 5.6

Relationship among loss modulus, storage modulus and damping factor [19].

Reproduced with permission from Elsevier. Copyright 2022.

5.6.1 Glass transition temperature

DMA is primarily used to determine the glass transition temperature (Tg). Tg denotes the temperature at which an amorphous and thermosetting polymer transitions from a stiffer to a rubber state [21]. Tg does not represent the melting temperature (Tm), but rather the point at which melting

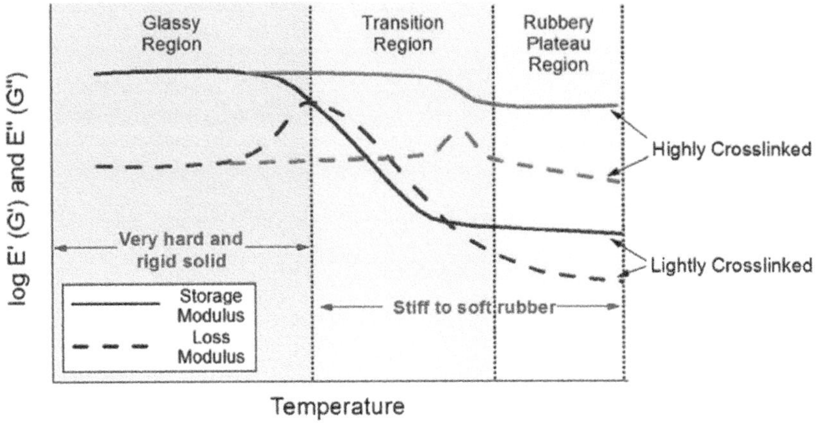

FIGURE 5.7

Storage as well as loss modulus of cross-linked composites [19].

Reproduced with permission from Elsevier. Copyright 2022.

begins, which occurs frequently in polymers with a definite crystalline structure. The number of crosslinks in a polymer has a large influence on its Tg. In thermosets, high crosslink densities limit molecular movements, increasing the energy required to facilitate segmental mobility during the glass transition. The modulus, damping factor, and Tg are all increased as a result of this behavior. Crosslink density had a clear influence on a material's rubbery and transitional states but not on its glassy area. As a result, the modulus levels of moderately and heavily cross-linked polymers are comparable in the DMA curve near the glassy area. Heavy cross-linking polymers, on the other hand, would have significantly higher modulus values at the rubbery and transitional portions of the curve, indicating a stiffness molecular arrangement [22].

Composites with elevated/low crosslink densities are represented in Fig. 5.7. Because amorphous thermoplastics have elastic backbones, their Tg levels are lower, whereas plastics with stiff, hard, and inflexible molecular structures have higher Tg levels. Furthermore, the degree of branching (DB) of polymers influences Tg; polymers with a higher DB have a higher Tg. Tg and Tm both exist in polymers with amorphous and crystalline structures, with Tg always being lower than Tm. Polymers can be used in a variety of applications because they can exist in two different functional states. Several polymers, including polystyrene and polymethyl methacrylate in stiff or brittle forms, could be used below their Tg. Because the molecular links are interlocked and molecular mobility is restricted in this state [23]. Because of the material's toughness, they were useful for packaging, utensil furniture, and laboratory apparatus. Poly-isoprene and poly-isobutylene, on the other hand, are used above their Tg in their rubbery state. The polymer's molecular chains are mobilized in the rubbery phase as opposed to the stiff phase, resulting in high flexibility and softness [24]. Lubricants, sealants, and adhesives are examples of products that use this property. The DMA methodology outperforms other methods for determining Tg, with sensitivity up to 100 times greater than differential scanning calorimetry (DSC) [25]. There are three different methods to calculate Tg using DMA, that is, measuring the temperature at (1) the start of E', (2) the peak of E^*,

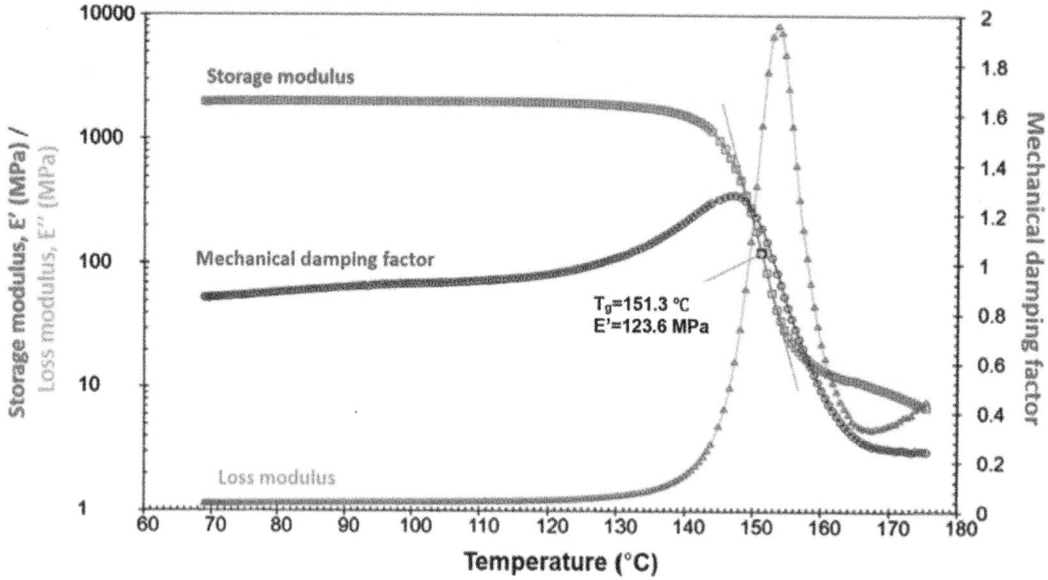

FIGURE 5.8

Storage modulus curve for the determination of Tg [26].

Reproduced with permission from Elsevier. Copyright 2022.

and (3) the highest point of tanδ curve. An illustration of Tg determination using the DMA storage modulus curve is shown in Fig. 5.8 [26].

5.6.2 Applications of dynamic mechanical analysis

Although dynamic mechanical analysis (DMA) was designed to study metal deformation, it is also useful for characterizing polymers and polymer matrix composites. This is because it has the potential to provide extraordinary insights into the molecular structure and stiffness of polymers [27]. Researchers frequently prefer the DMA methodology to temperature-only based techniques due to its exceptional responsiveness and precision in identifying macroscopic processes and molecular relaxation. DMA is used in a variety of experiments in material engineering research, including dynamic temperature, time, and frequency scanning, creep recovery, dynamic and static stress strain, and stress relaxation. Tg, thermal expansivity, loss and storage moduli, damping, load capacity, toughness, ductility, stiffness, and crosslink density were discovered using these methods. DMA-characterized manufactured materials, as stated in the published work, are primarily intended for use in the biomedical, military, packaging, electrical, and chemical industries [28]. The DMA method permits for extremely accurate as well as sensitive identification of the loss modulus and storage modulus, tanδ and Tg of heterogeneous polymer matrix composites. In general, the DMA is an essential tool for understanding the characteristics of BBPCs.

5.7 Thermo-mechanical and viscoelastic behavior of biodegradable and biocompatible polymer nanocomposites

DMA was recently used to thoroughly investigate the thermo-mechanical behavior of several nano-filled BBPCs. Marras et al. [29] investigated the temperature dependence of the storage modulus (E') of nanocomposite films based on PCL in comparison to pure PCL. According to DMA results, nano-filled (organically modified montmorilonite, Cloisite 25A-Cl25A) composites have a storage modulus that is noticeably higher than pure polymers across the entire temperature range, as shown in Fig. 5.9A. This effect can be attributed to the higher aspect ratios of the highly dispersed silicate

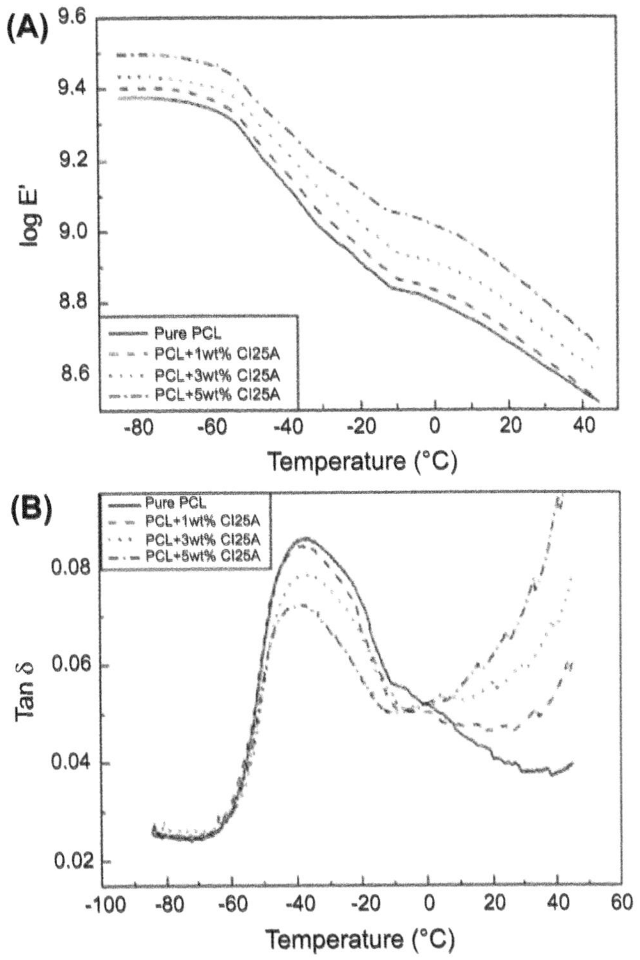

FIGURE 5.9

Influence of modified clay concentration on: (A) storage modulus and (B) tanδ of PCL nanocomposite films [29].

Reproduced with permission from Elsevier. Copyright 2008.

platelets as well as their effective interactions with the polymer. The augmentation becomes stronger as the temperature rises, while the E' rises with the concentration of clay. This property demonstrates the improved thermo-mechanical stability of nanofilled BBPCs at elevated temperatures. Fig. 5.9B illustrates the loss tangent ($\tan\delta$) changes with temperature. Since the segmental movements of the intercalated matrix of PCL were affected by the nano clays, the peaks of $\tan\delta$ for the PCL-based nanocomposites have somewhat shifted to low temperature within the glass transition area. The magnitude of peak falls as the clay content increases. The dynamics of polymers actually deviate significantly from the aggregate when they are constrained in its geometry with a typical size of less than 10 nm [30]. A similar behavior could be observed in polymers contained within pores, between layers of foreign particles or different phases (such as crystalline) within the same materials, or even between independent layers [31]. Several studies on the confinement implications of polymers interpenetrating in montmorillonite layers have revealed that the Tg is extremely difficult to detect or suppress [32]. According to Masenelli-Varlot et al. [33], the causes for the Tg drop is the addition of surfactants that improved the chain mobility. Tg of polymeric nanocomposites can be increased or decreased by modifying the interaction of the polymer and nanofiller. When compared to bulk values, favorable interactions may result in Tg increases rather than declinations [34].

A bio-nanocomposite is made up of biologically generated materials, biopolymers, or at least one type of nanoscale particle [35]. The majority of hybrid bio-nanocomposites studied are reinforced with multiscale particles. A class of composite materials known as "multiscale hybrid composites" contain reinforcing elements with length scales ranging from macro to nano [36]. A thorough investigation into the DMA of polyester composites reinforced with hybridized sugar palm yarn and glass fiber was carried out by Nurazzi et al. [37]. According to the findings, adding alkali-treated sugar palm yarn significantly increased stiffness as evidenced by the rise in storage modulus shown in Fig. 5.10 with various fiber concentrations.

The dynamic mechanical properties of Kevlar/Cocos nucifera sheath (CS) bio-composites with graphene nanoplatelets (GNPs) have been investigated [38]. It is widely accepted that switching from synthetic to natural fiber can have a negative impact on composite properties. GNPs were added to mono composites to improve thermal and viscoelastic performance. While the GNPs' was applied with varied compositions ranging between 0% and 0.75 wt.%, the Kevlar as well as CS fibers were put in a specific order. According to the DMA, the loss and storage moduli of S3G3 composite (CS/in a 1:1 ratio with 0.75 wt.% of GNPs) were nearly identical to those of S1G0 (Kevlar/epoxy). It also demonstrated improved compatibility with lower tan *d* among the multi scaled filler and matrix. As a result, at the optimal composition, Kevlar reinforced polymer composites can be successfully replaced by the hybrid bio-nanocomposite. Understanding the rheological behaviors of polymer nanocomposites is critical for gaining a thorough understanding of the processability as well as structure-property correlation for such materials. Rheological measurements involving shear were conducted at a steady shear rate of 200 s^{-1}. The plot of viscosity versus time of neat silk fibroin as well as SF/modified clay-Benzalkonium chloride (BAC) (1%, 3%, and 5%) mixtures is depicted in Fig. 5.11 [39]. It has been observed that neat silk solution gradually loses some of its viscosity over time. The viscosity of a nanofilled composite containing 1% modified montmorillonite (MMT) clay is lower and nearly constant over time when compared to neat silk fibroin solution. When 3% clay is added, the solution has even lower viscosity and is not affected by time. When 5% clay filled SF solution is compared to neat silk solution, the viscosity increases significantly and then decreases over time. This phenomenon is primarily caused by shear thinning,

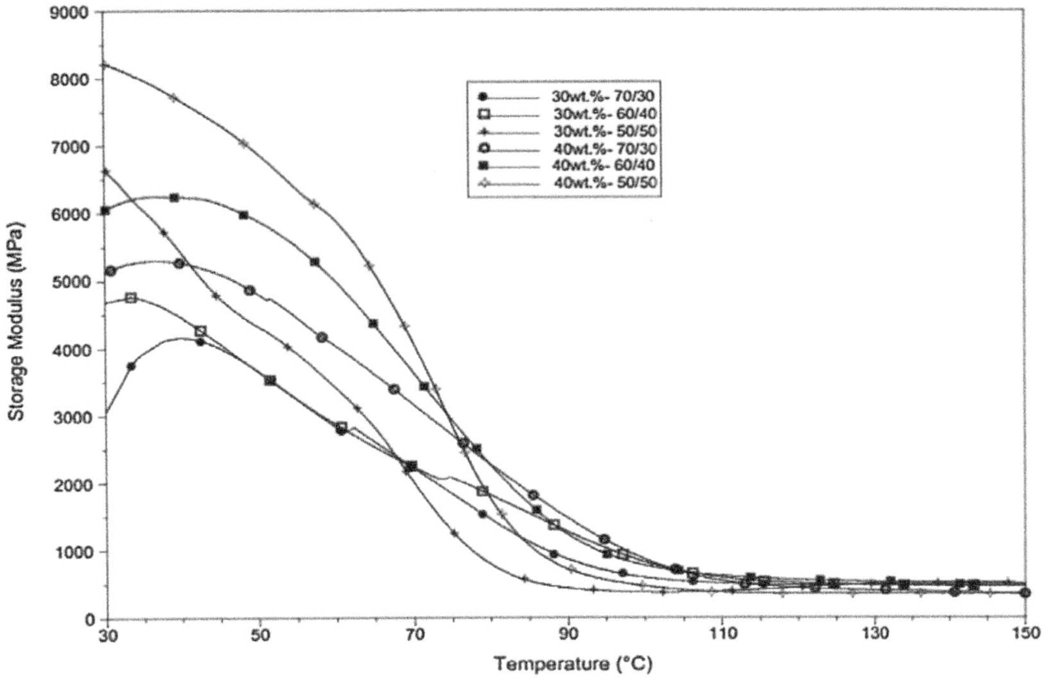

FIGURE 5.10

Storage modulus of treated sugar palm yarn fiber hybrid composites [37].

Reproduced with permission from Elsevier. Copyright 2020.

which is caused by straightening of polymeric chains in solution along the direction of applied shear stresses, allowing for improved flow and viscosity reduction [40]. When nano clay was added, the viscosity of the polymeric matrix increased, possibly due to improved clay gallery delamination and increased interactions between polymer chains and clay particles.

DMA was used to investigate the effect of temperature on PLA G' during nanocomposites production under a variety of experimental conditions [41]. The relationship between tan and temperature for various organically modified synthetic fluorine mica (OMSFM)/PLA nanocomposites as well as pure PLA is shown in Fig. 5.12. All nanocomposites tested within the investigated temperature range had higher G' than plain PLA, demonstrating that OMSFM has a significant impact on the elastic properties of unfilled PLA. Under Tg, G' increased significantly in all nanocomposites. Because of the mechanical strengthening of the silicate layers and the prolonged intercalation at high temperatures, all nanocomposites improved in G' more than the plain PLA matrix. When materials reached Tg, the reinforcing effect of silicate layers became apparent due to the limited mobility of polymer chains, which was augmented by the unprecedented rise in G'.

A very tiny amount of oligo(-caprolactone) (o-PCL) loading was used to create clay-incorporated PLA nanocomposites by Ray et al. [42]. To investigate the compatibilizer's effects on morphological and thermomechanical properties. The names and compositions of the various kinds

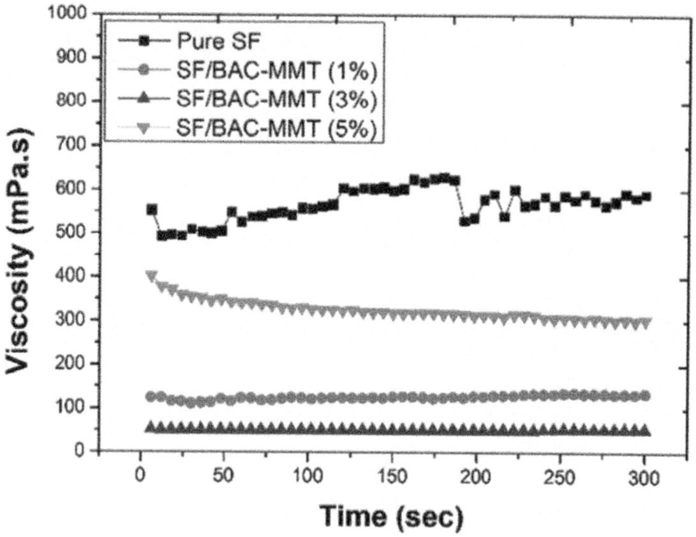

FIGURE 5.11

Analysis of viscosity versus time plots of pure SF as well as SF/BAC-MMT clay nanocomposite films [39].

Reproduced with permission from Elsevier. Copyright 2021.

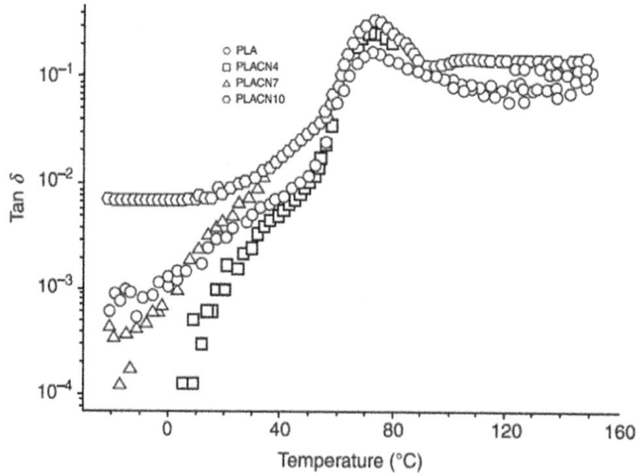

FIGURE 5.12

Temperature versus tanδ plots of neat PLA as well as PLA/OMSFM nanocomposites (PLACNs). The number subsequent to PLACN designates the loading of OMSFM [41].

Reproduced with permission from Elsevier. Copyright 2013.

Table 5.1 Compositions of various PLACNs based on PLA, o-PCL and OMMT composites [41].

Composites	Composition by wt.%		
	PLA		PLA
PLACN-2	97	–	3
PLACN-3	95	–	5
PLACN-4	93	–	7
PLACN-5	94.8	0.2	5
PLACN-6	94.5	0.5	5
PLACN-7	93	2	5
PLA	92	3.1	5
PLA-1	100	–	–
PLA-2	99.8	0.2	–
PLA-3	99.5	0.5	–
PLACN-2	98	2	–

Reproduced with permission from Elsevier. Copyright 2013.

of bio-nanocomposites are summarized in Table 5.1. The increase in G' over the investigated temperature range was seen in both PLA/OMSFM systems and pure PLA, demonstrating that OMMT has a significant impact on the elastic properties of pure PLA as shown in Fig. 5.13. The addition of a small amount of o-PCL did not cause a significant broadening or shift in tan curves. Furthermore, a significant increase in G' over Tg was observed, demonstrating the stiffness improvements caused by the distributed flocculated nanoparticles' significant anisotropy.

To evaluate the compatibility, dispersion state, and degree of reinforcement of organoclay-Cloisite 15A (C15A) and Cloisite 30B (C30B) inside the PCL matrix, the viscoelastic properties of all material systems were evaluated at the working temperatures [43]. For the duration of the experiment, rheological tests are conducted at a frequency of 1 Hz and a deformation of 0.5% to check the melting stability of all samples. Fig. 5.14A and B depict the viscosity and storage modulus vs test time plots for neat PCL and nano loaded composites. Because the complicated viscosity measurements and G' do not change over time, the samples remain stable and no structural or chemical changes occur. The viscosity and storage modulus as a function of test time plots for neat PCL as well as nanofiller filled composites are depicted in Fig. 5.14A and B. Because the G' and complex viscosity measurements do not change over time, the samples remain stable, and no structural or chemical changes occurred during the experimental period [44].

The rheological properties of OMMT-PCL mixtures containing 3 mass percentage of C30B and C15A exhibit a different viscoelastic behavior than neat PCL and PCL-Gr blends. In fact, a secondary peak forms within the terminal area, indicating a transition from pseudo-liquid to pseudo-solid behavior, as well as an improvement in storage modulus. These results are attributed to the development of three-dimensional (3D) networking as well as the excellent dispersion that resulted in exfoliated and intercalated nanocomposites [45]. When ternary OMMT/Gr-PCL compositions are made with cloisite and graphene filler compositions, the rheological behavior of the PCL is significantly altered, resulting in higher storage modulus values.

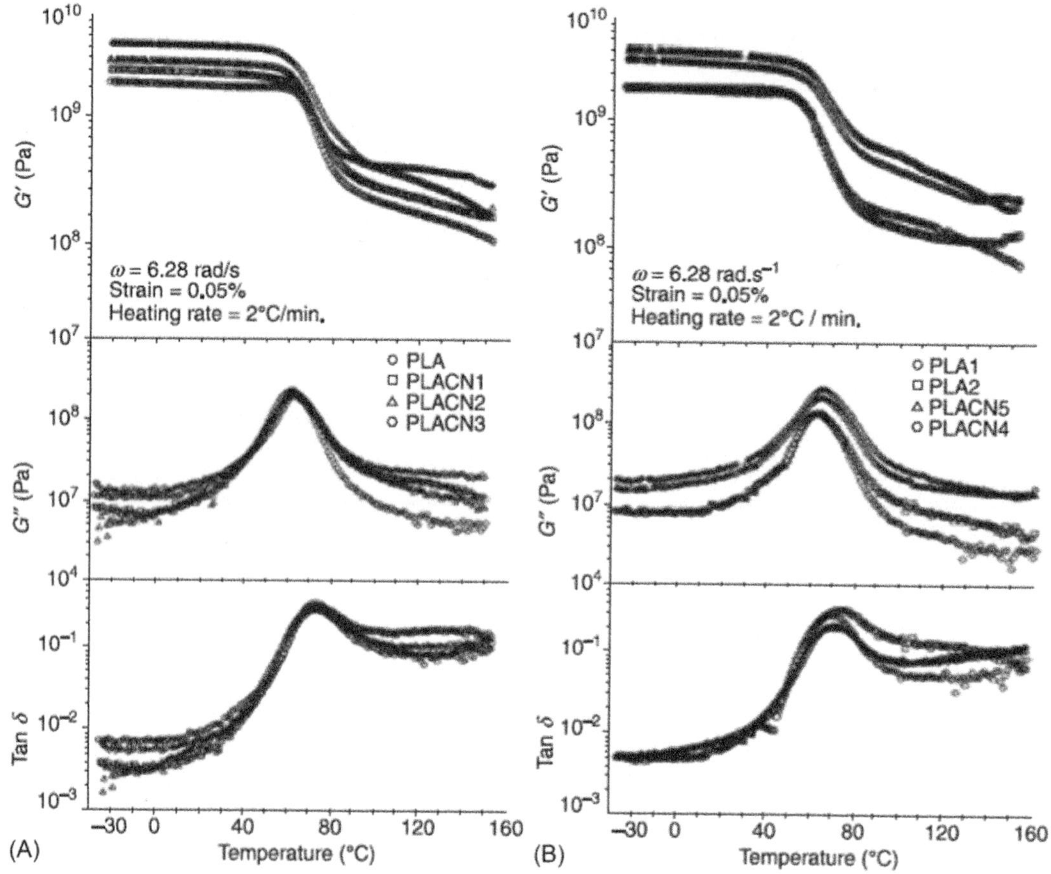

FIGURE 5.13

Storage modulus (G'), loss modulus (G'') and tanδ for PLACNs and the corresponding matrices: (A) without o-PCL; and (B) with o-PCL [41].

Reproduced with permission from Elsevier. Copyright 2013.

A technique called dynamic mechanical thermal analysis (DMTA) was utilized to look into the relaxation of biopolymer films as shown in Fig. 5.15. Two distinct transitions were easily distinguished in the tanδ curves among all chitosan (A) films [46]. Sample designation is represented in Table 5.2. It is believed that the poor transition, which is centered at around 47°C, is a β-relaxation caused by the movement of chitosan's lateral groups or side chains interacting with tiny molecules like water through hydrogen bonding. The chitosan α-transition may be linked to a more noticeable shift with a maximum temperature of around 119°C. In accordance with Quijada-Garrido et al. [47], the α-relaxations were explained by the glass transitions involving torsional oscillations among 2 glucopyranose rings over a glycosidic oxygen as well as the rearranging of hydrogen bonds. The tanδ curves among all CMC/chitosan (B) films resembled to that of chitosan (A) films.

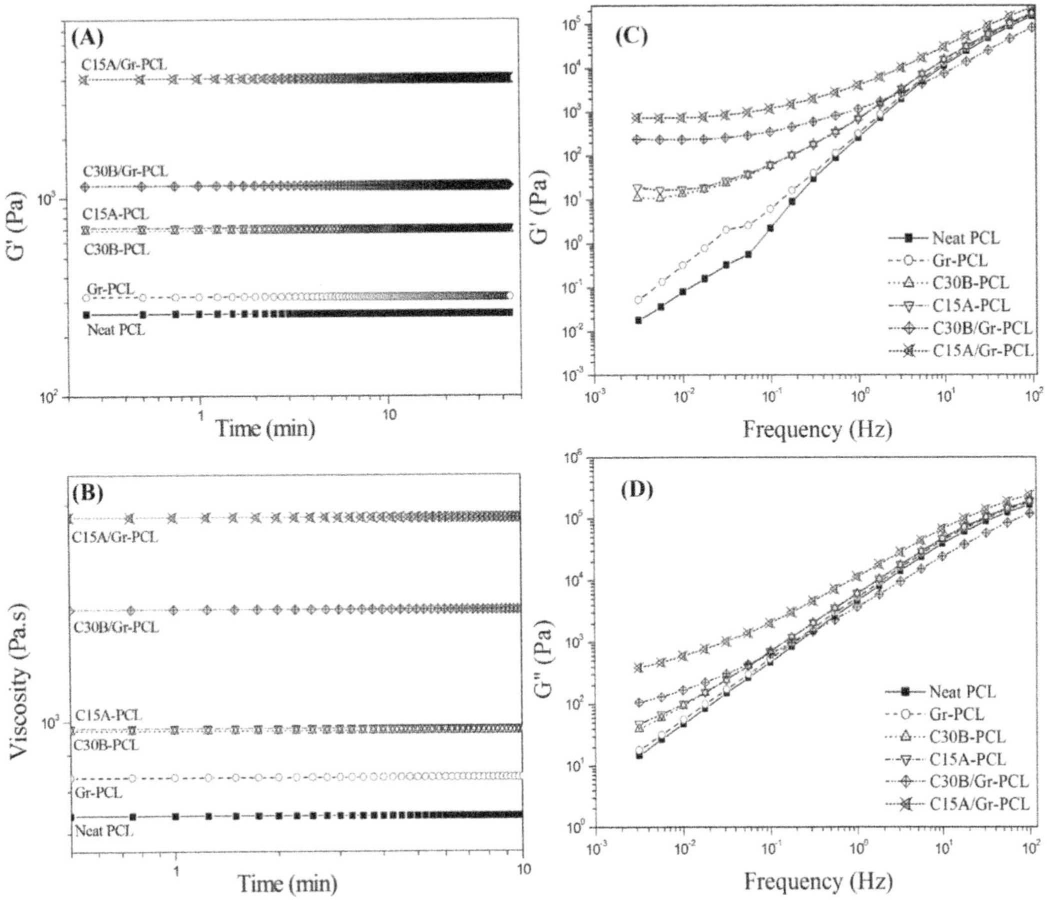

FIGURE 5.14

Storage modulus (A) and complex viscosity (B) as a function of time for neat PCL and hybrid materials; Storage modulus (C) and loss modulus (D) of PCL based nanocomposites with frequency sweep [43].

Reproduced with permission from Elsevier. Copyright 2018.

The β-relaxation shifted to elevated temperatures (at around −37°C) for B/M-F, B/S-F, and B/Ms-F, indicating that the movements of the lateral group or side chain of bio-based polymers were more constrained as a result of the intercalation of the two biopolymers. A lower Tg of ~96°C as well as a significantly larger peak intensity for B-F films relative to chitosan (A) film suggest that CMC had a plasticizing impact upon chitosan and caused the system to behave more viscously (less elastic behavior). For B/S-F, B/M-F, as well as B/Ms-F, the inclusion of nano clay raised the Tg of B-F to approximately 107°C−110°C. The interactions among chitosan as well as CMC are disturbed by the addition of nano clay. The biopolymer films were also subjected to frequency scanning at room temperature in the range of 0.01 Hz until 20 Hz. E' remained rising

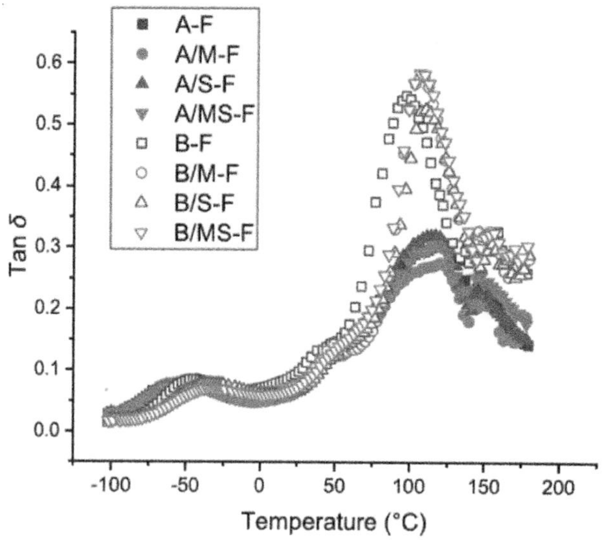

FIGURE 5.15

DMTA curves for different biopolymer films [46].

Reproduced with permission from Elsevier. Copyright 2020.

Table 5.2 Compositions of biopolymer films [46].

Biopolymer films	Chitosan (g)	Carboxymethyl cellulose-CMC (g)	MMT (g)	Sepiolite-SPT (g)	2 M formic acid (mL)
A-F	45	–	–	–	90 + 25
A/M-F	45	–	0.6750	–	90 + 25
A/S-F	45	–	–	0.6750	90 + 25
A/Ms-F	45	–	0.3375	0.3375	90 + 25
B-F	22.5	22.5	–	–	90 + 25
B/M-F	22.5	22.5	0.6750	–	90 + 25
B/S-F	22.5	22.5	–	0.6750	90 + 25
B/Ms-F	22.5	22.5	0.3375	0.3375	90 + 25

Reproduced with permission from Elsevier. Copyright 2020.

with the frequency for all the test samples, correlating to the viscoelastic nature of polymers. A log-log chart revealed similar slopes for the E' versus frequency curve, indicating that the nano clays had no effect on the viscoelasticity of the test samples.

DMA under tension mode is used to examine the thermomechanical characteristics of polyvinyl alcohol (PVA)/cellulose nanocrystal (CNC)/GO nanocomposites as shown in Fig. 5.16A and B. The nanocomposites were represented as PVA/(CNC_x/GO_y) wherein the x and y indicate the ratio

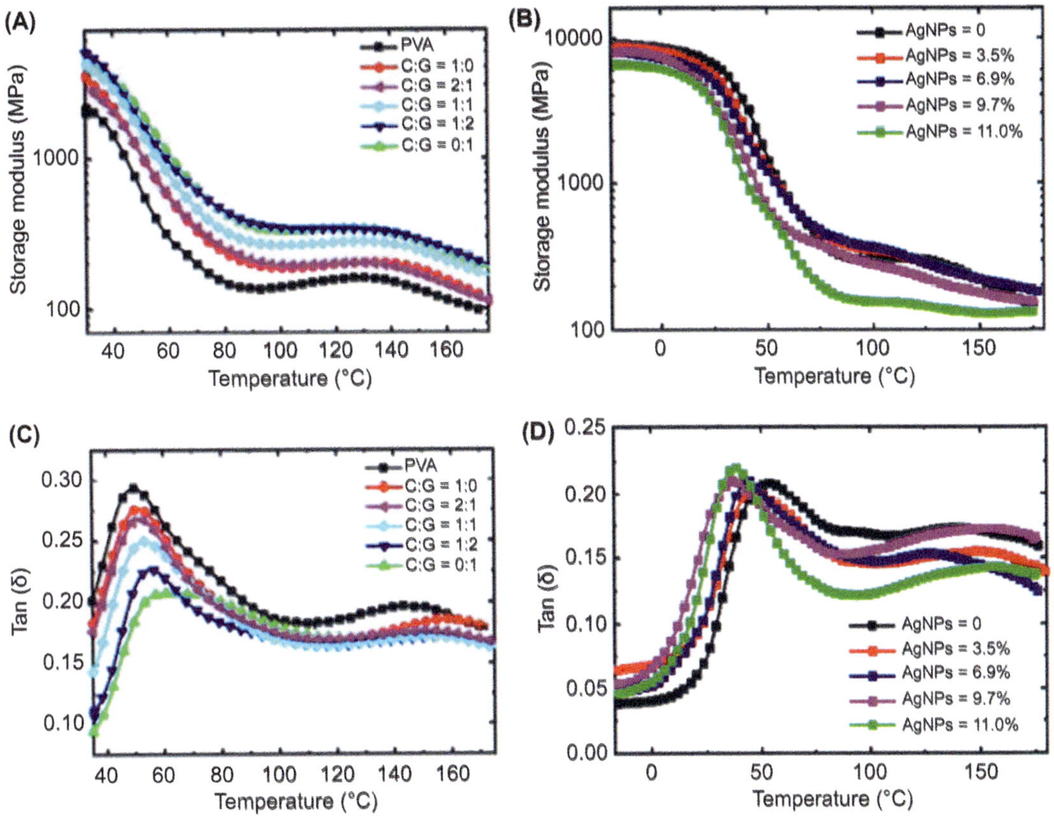

FIGURE 5.16

Thermo-mechanical properties of (A, B) PVA/CNC/GO and (C, D) PVA/CNC_1/GO_2/Ag nanocomposites [48].

Reproduced with permission from Elsevier. Copyright 2020.

of CNC as well as GO in the fraction of 3 wt.%. The storage modulus (G') reveals the material's rigidity or stiffness. Plots of G' of PVA/CNC/GO composites were constructed as response to temperatures between 30°C and 175°C. Owing to improved hydrogen bonding, the PVA/CNC_1/GO_2 composites showed a notable increase in E' between 2115 MPa for neat PVA film toward 4981 MPa at 30°C. Due to enhanced internal movements of PVA as well as the creation of free volume, all the curves exhibited a similar fluctuation trend that shows a reduction in G' as a response to temperature. The viscoelasticity behavior of materials is demonstrated by the loss factor, tanδ. A greater tanδ value denotes a more viscous substance, however it actually exhibits more elastic behavior. In relative to neat PVA composite, the PVA/CNC/GO biocomposite exhibit reduced tanδ values at 30°C, which suggests that 1D CNC as well as 2D GO formed hydrogen bonds with PVA matrix Fig. 5.16C and D. The distinct peaks at 58°C indicate the Tg of PVA/CNC/GO nanocomposites. The change in Tg caused by the addition of CNC/GO demonstrates how CNC/GO and PVA interact through hydrogen bonds.

Table 5.3 Storage moduli of bio-polyurethanes [49].

Biopolymers composites with different weight average molecular weights	Storage modulus at different temperature (GPa)		
	30°C	50°C	120°C
10,000	2.1	1.6	–
16,000	3.0	2.8	2.1
18,000	2.1	2.0	1.5
8000	1.8	1.7	1.4
10,000	3.4	3.2	2.7

The thermo-mechanical characteristics of the PVA/CNC_1/GO_2/silver (Ag) nanofilled composites were investigated. Plots of the G' of PVA/CNC_1/GO_2/Ag composites were obtained as a response to temperature between 30°C and 175°C. Hydrogen bonding's got disrupted by the immersion in a water/ethanol solution causing the maximum storage modulus of the PVA/CNC_1/GO_2/Ag composite to significantly decrease, from 4981 MPa to around 2686 MPa at 30°C. The removal of plasticizer or even a trace quantity of water with in PVA/CNC_1/GO_2/Ag nanocomposite films causes the Tg to change from 55.9°C to around 37.5°C when Ag nanoparticles are present in an amount of 9.7 wt.%. DMA was used to investigate the viscoelastic properties of bio-polyeurethane composites synthesized with different molecular weights [49]. The E' of the same at different temperatures are given in Table 5.3. The DMA results demonstrated that the polyurethanes' structures had a substantial impact on their characteristics. Due to the polymer's intrinsic hard segment formed from isosorbide, storage modulus of biopolymers with molecular weight of 3 wt.% was rapidly reduced at about 120°C. Up to 140°C, the storage modulus of biopolymers with molecular weight of 4 wt.% remained constant before progressively starting to decline. Biopolymers with molecular weight of 5 wt.% retained its storage modulus at a temperature of 170°C. The addition of the bisphenol A and diphenylmethane groups has increased heat resistance by around 20°C–50°C compared to biopolymers with molecular weight of 3 wt.%. In contrast to the 3 wt.% biopolymer, the addition of bisphenol A and diphenylmethane groups increased the stiffness of polymers with weight average molecular weights of 8000 and 10,000.

DMA was performed at various concentrations on nanofilled banana fiber-reinforced biocomposites [50]. The storage modulus of the nanocomposites was found to be slightly higher with the addition of nanoparticles. The transition temperature of neat composite was found to be 65°C, whereas composites with banana nanoparticles had a transition temperature of 70°C–75°C. The E' of a banana nanoparticle-reinforced composite increases as particle loading increases in the glassy and rubbery regions. Furthermore, in comparison to the neat composite sample, the Tg of the banana nanoparticle composites is slightly shifted to a higher temperature with a wider range of the transition region. Recent researches suggest that the thermo-mechanical properties of BBPCs could be increased by the addition of synthetic or natural additives. Table 5.4 provides a summary of the impact of the fiber + filler weight percent in TCE (Thermoplastic copolyester elastomer) composites on the DMA results. The E', E'' as well as tanδ of the fiber and particulate reinforced TCE + PTFE (polytetrafluoroethylene) composite materials were assessed using DMA as shown in Fig. 5.17. These plots demonstrate the material's elastic properties and shed light on the stiffness of

Table 5.4 Dynamic mechanical properties of TCE + PTFE filled composites [51].

TCE + fiber + filler (wt.%)	Storage modulus (MPa)	Loss modulus (MPa)	Tanδ	Glass transition (Tg)
85 + 15	2505.4	93.7	0.075	24.65
68 + (12 + 20)	3151.1	109.17	0.060	24.65
60 + (20 + 20)	4381.6	171.26	0.0.076	23.56

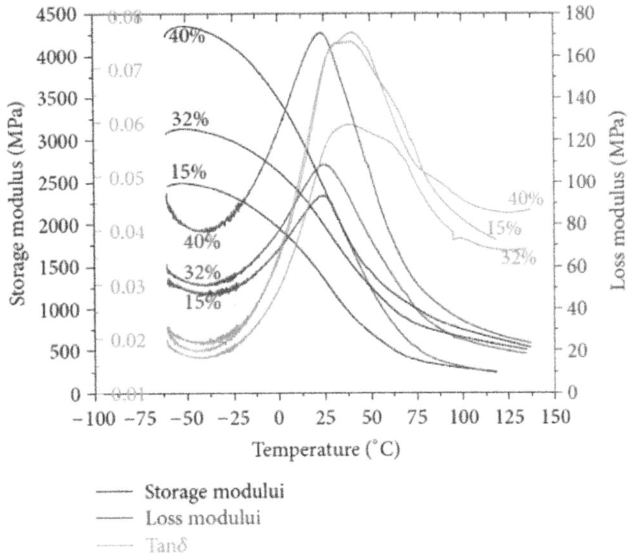

FIGURE 5.17

DMA thermogram of TCE + PTFE-filled composites [51].

Adapted from Hindawi Publishing Corporation. Copyright 2014.

the composites. When the temperature is 60°C, the E' of hybrid TCE + PTFE composites is large and decreases dramatically as the temperature rises. The E' in the glass transition region noticeably decreases due to the components' increased mobility and reduction in their tight packing arrangement. The loss modulus to storage modulus ratio, or tanδ, is a measure of internal damping. The main contributing factors of composite damping are the matrix and filler properties, the interphase, frictional damping caused by delamination at the filler-matrix interface, and energy loss in the vicinity of matrix cracks. The movement of side groups, low molecular-weight units, and molecular chains within rubbers is linked to the damping peak, which typically appears at the glass transition zone. As a result, as the damping peak increases, so does the degree of molecular mobility.

A solution-cast poly(lactide) starch matrix (PLA-ST) was prepared and mixed with polyfunctionalized graphene oxide in a study similar to the one described above. To improve interfacial

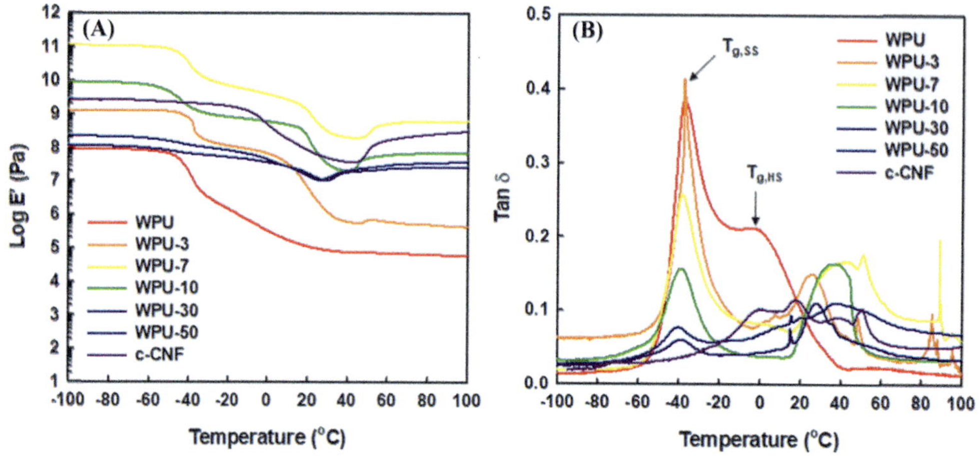

FIGURE 5.18

(A) Storage moduli and (B) tanδ variations of neat WPU, neat c-CNF, and nanocomposite films as a function of temperature [53].

Reproduced with permission from Elsevier. Copyright 2019.

compatibility, graphene oxide (f-GO) was grafted with maleic anhydride and modified with dodecylamine. According to the DMA results, the addition of f-GO to the PLA-ST matrix significantly increased the Tg. These findings suggest that f-GO influences the major relaxations of PLA chains. PLA chain movement was also amplified and suppressed by f-GO amplification. The storage modulus of the samples was another thermomechanical parameter studied. E' of the PLA-ST composite with 0.2 wt.% f-GO content increased as the frequency increased from 1 to 10 Hz. The glass transition temperature also changed to a higher value at 5 Hz because of the delayed relaxation of the PLA chains at relatively high frequencies [52].

Kim et al. investigated the effects of waterborne polyurethane (WPU)-based nanocomposite films on carboxymethylated cellulose nanofibrils on their microstructural, thermal, and mechanical properties (c-CNF). A one-of-a-kind interaction between c-CNF and WPU hard segments was discovered. Increasing the c-CNF content by 7 wt.% dramatically improved the storage modulus of the nanocomposite films and the glass transition temperature of the WPU hard segment [53]. From Fig. 5.18, it is clear that the E' remained constant for nanocomposite films with c-CNF levels between 1 and 50 wt.% between 100°C and 50°C, gradually decreasing at 50°C. This was due to the increased mobility of the chain of the WPU's hard segments and the partial plasticizing impact of water molecules tightly attached to the films, as well as the glassy to rubbery transition of the WPU component at 0°C. When the temperature surpassed 50°C, bound water molecules began to escape. Storage modules have been shown to decrease at higher c-CNF contents of 12–50 wt.%, but significantly increase at 0–10 wt.% c-CNF content in nanocomposite sheets. The higher storage moduli for nanocomposite films are assumed to be 1%–7% by weight. The loadings of c-CNF compared to the pure WPU film were due to the enhancing effect of the c-CNF loading in the WPU matrix, as shown in Fig. 5.19.

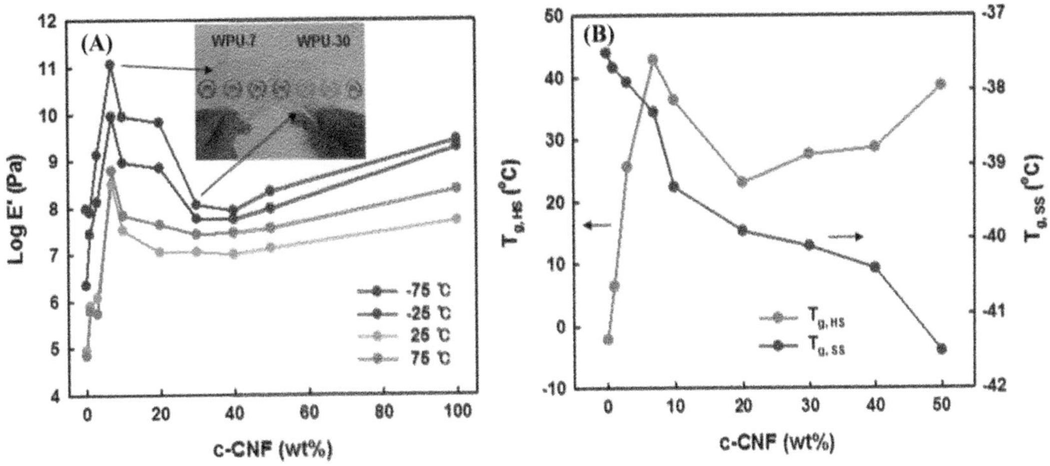

FIGURE 5.19

(A) Storage moduli of WPU/c-CNF nanocomposite films at different temperatures (B) glass transition temperatures of the soft and hard segments of WPU matrix in the nanocomposite films as a function of c-CNF content [53].

Reproduced with permission from Elsevier. Copyright 2019.

Another study reports on the thermally activated shape memory behavior of biodegradable nanocomposites based on a PLA/PCL mixture reinforced with various types of cellulose nanocrystals. At the same transition temperature of 55°C and different percentage strains of 50, 100, and 150, these were compared to the pure matrix. The inclusion of various CNC-based nanofillers had an effect on the compatibility of the PLA/PCL blend, according to the study. Because the nanofiller was localized at the blend interface, the addition of CNC-g-poly L-L-lactic acid (PLLA) to the blend matrix improved interfacial adhesion. To improve the thermomechanical properties of the matrix, additives were added. Polylactic acid (PLA) nanocomposites and perlite nanoparticles have been created using the melt mixing technique [54]. It appears from Fig. 5.20 that the G' and the loss modulus G″ were increased with the addition of 5% nanoperlite but decreased to 7% nanoperlite. Furthermore, the G' and G″ did not intersect, indicating that the nanocomposites exhibited no viscoelastic behavior during processing. In dynamic frequency tests, all nanocomposites exhibit the same drift, while nanoperlite exhibits a sharp 7% drop.

Diego et al. [55] investigated the thermos mechanical characteristics of chitosan-based nanocomposite. CH for chitosan films is labeled as CH, and CH-CNC1h, CH-CNC2h, and CH-CNC3h for chitosan-based nanocomposite films, indicating the concentration of nanocrystal. As illustrated in Fig. 5.21, the neat CH films as well as its nanocomposite films made with various concentration exhibit temperature dependency on the peaks of loss tangent (tanδ) as well as the storage modulus (E' or G'). Based on the dynamic mechanical spectrum, the Tg is represented by the initial transition, which is connected to hydration of macromolecule's side chain (β-relaxation), as well as the relaxation transition [56,57]. Both of these transitions occurred at −6°C as well as 90°C. The dynamic mechanical spectra of nanocomposites showed a considerable reduction in E' as well as E'' modulus within

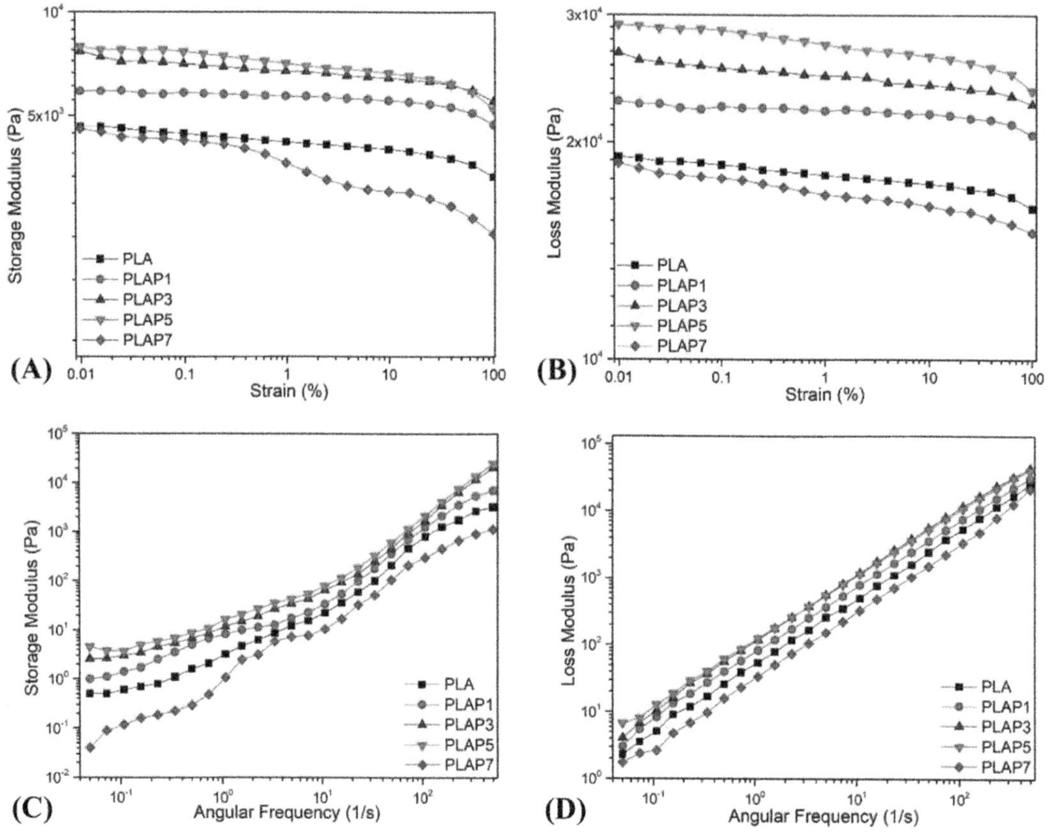

FIGURE 5.20

Rheological properties of PLA/nanoperlite (P) nanocomposites (A) *G′* versus strain (B) G″ versus strain (C) *G′* versus frequency (D) *G″* versus frequency [54].

Reproduced with permission from Elsevier. Copyright 2019.

relaxation region, which was attributed to the Tg as well as the magnitude of the peak variation associated cellulose nanocrystals (CNC) loading. The degree of CNC dispersion within chitosan matrix could be determined using DMA tests. According to Dufresne [58], a rise in E' as well as a shift in tanδ peak to elevated temperature is a sign of a well-distributed CNC with effective matrix interactions. However, if tanδ peaks moves to low temperatures, this might be a sign of a poor interaction among the matrix and CNC. The patterns for the nanofilled composites prepared with CNC after 1 hour of hydrolysis indicated that the tanδ peak had shifted to low temperatures. With increased CNC concentration in nanocomposites, an intensity drops as well as a widening of the major relaxation peaks were seen Fig. 5.21, indicating that segment particulate involvement in the transition. The addition of CNC improved the storage modulus of the nanofilled composite, demonstrating the elastic response of nanocomposites under deformation. The composites containing 0.5 wt.% as well as 0.75 wt.% of

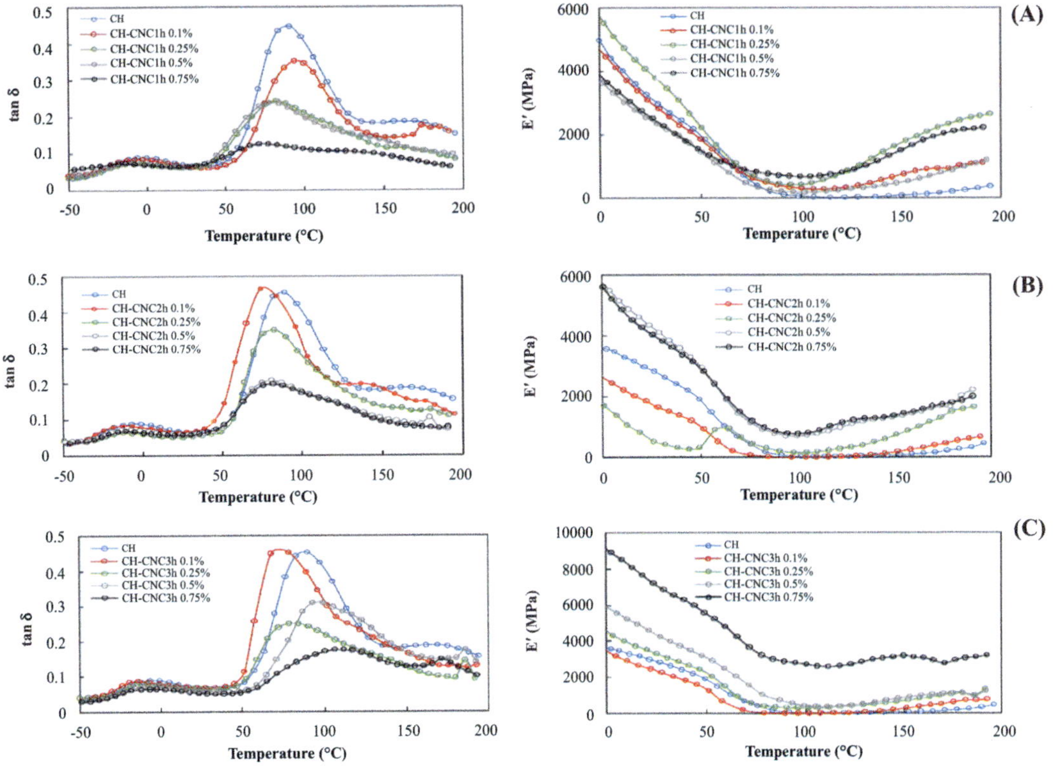

FIGURE 5.21

The temperature V speak of loss tangent delta (tanδ) and the storage modulus (E') of the pristine CH films and their nanocomposite films [55].

Reproduced with permission from Elsevier. Copyright 2020.

nanocelluloses that were produced after two and three hours of hydrolysis had the maximum modulus. According to Abraham et al. [59], the higher E' values account for the reinforcing impact of cellulose nanoparticles at temperatures well below Tg. According to Li et al. [60] as well as Corcione and Frigione [61], the interfacial interactions would increase the E' values of nanofilled composites by limiting the thermo-molecular movements of chitosan that is surrounded by CNC particles. This demonstrates that adding CNC to a chitosan matrix had a significant reinforcing effect. Matrix reinforcement could explain the rise in E' values within the plateaued area. Due to the larger size of the CNC and its uneven distribution within the chitosan, such behavior was not observed in the case of CH-CNC1h [62]. Li et al. [63] highlighted similar justifications. When the CNC particles are evenly distributed and small in size, the surface area and interactions increase, making the change more noticeable. According to the findings, adding natural fibers/fillers to the matrix improves the thermo-mechanical and rheological properties of the samples. The review of the viscoelastic properties derived from DMA revealed that they were superior or at least equivalent to those of artificial or synthetic polymer composites.

5.8 Applications of biodegradable and biocompatible polymer nanocomposites

Polymers are viscoelastic materials with both solid and liquid mechanical properties. As a result, the BBPCs will have the properties of both solids and liquids. Thermal analysts routinely measure the dynamic mechanical properties of polymers for a variety of reasons. Polymer composites with viscoelastic properties are widely used in biomedical applications. PLA, Plyglycolic acid (PGA), and Polylactic-*co*-gycolide (PLGA) are biocompatible polymer composites that are widely used in a variety of thermo-mechanical applications.

The viscoelastic properties of polymeric composites cause the shape memory effect. This effect is also commonly used in biomedical applications. The shape memory effect of chitosan/poly(L-lactide) composites was studied, and it was discovered that PLLA, as a shape memory polymer, cannot withstand large deformation strains [64,65]. Another biomedical application for polymer viscoelasticity is that it can be used to create artificial tendons and ligaments inside the human body [66].

Since the invention of vulcanized rubber, fillers have evolved into highly advanced nanomaterials with precisely tailored interfaces. As a result, advances in BBPC research have resulted in a plethora of distinct bio-composites for a wide range of applications ranging from biosensors to complex medical equipment. Because of their versatility and improved thermo-mechanical properties, bio-composites have carved out a significant niche as bio-materials [67]. Because nanofillers have large specific surface areas, they can form excellent contacts in polymeric matrix. This is possibly the most important property of nanofillers, making them ideal materials for potential biomedical polymeric composites. Furthermore, the effectiveness of bio-materials in medical equipment is heavily influenced by the following underlying factors: reinforcement young's modulus, filler concentration, thermo-mechanical properties, and biodegradation rate [68].

Biosensor applications include chemical sensors and gas sensors. Among the nanofillers used are metal oxide based nanowires, carbon nanotubes, nanoscale palladium, platinum, copper, nickel, silver, and gold particles [69]. Changes in conductivity caused by chemical or gas interactions with polymers/nanofiller, changes in pH, electro-chromic and thermo-mechanical changes, catalytic activity, and biological recognition could all drive the sensing potential of these BBPCs. According to a National Institutes of Health survey on denture material, there are more than a hundred million people who are missing teeth, and there is a high demand for implant dentistry. Teeth are the most durable materials in the human body, primarily composed of hydroxyapatite ($Ca_5(PO_4)_3$) (OH). BBPCs have sparked a lot of interest due to their wide range of characteristics and processing options depending on the conditions encountered in the body. These are widely used as biocomposite restorative materials and implants in dentistry [70].

Bone is a bio-based extracellular matrix composed of organic/inorganic complex minerals (60%), matrix (30%), and 10% water. Because of its high strength and elastic properties, it is ideal for framed structures [71]. Hospitalizations for fracture-related bone injuries, as well as grafting operations, are becoming a worldwide issue. Several materials are now used to replace/repair bone, and they must have appropriate mechanical properties, a regulated rate of degradation, and acceptable biological activity in terms of interactions with human tissues [72]. BBPCs made of hydroxyapatite ($Ca_{10}(PO_4)^6(OH)_2$) have been investigated for bone healing and implantation. The incorporation of nanofillers in BBPCs improves functionality in terms of electrical conductivity, thermo-mechanical properties, and mechanical properties of the biopolymers. For example, silicate-based fillers provide minerals required for the maintenance of a variety of physiological functions [73]. The use of biodegradable bone screws eliminates the need for additional surgery to remove metallic fixing devices. Another advantage is the availability of bioactive fillers, which can be used to either prevent bacterial infection around the implant or to stimulate tissue development and eventual bone reconstruction via implant biodegradation. Biocompatibility and nanosized features are critical for using bio-nanocomposites in drug delivery applications. Furthermore, drug release capabilities have a high potential for wound-dressing treatments due to their high water absorption, noncytotoxicity, and muco-adhesivity [74].

5.9 Conclusions

Synthesis and characterization of BBPCs are critical because disposing of pure polymer materials is a time-consuming process. TMA and DMA thermo-mechanical characterization techniques can be used effectively to characterize BBPCs. The review of DMA analysis of BBPCs revealed that they are thermally stable, and the addition of bio-materials to polymers improves their thermal, viscoelastic, and rheological properties. The addition of nanofillers to BBPCs increased the storage modulus, loss modulus, and damping capabilities. Furthermore, DMA studies on BBPCs have shed light on the viscoelastic properties of these materials, making them suitable for tendon and ligament implants in humans. It was discovered that research data on biocompatible composites was scarce, and it is critical that research on BBPCs be accelerated to the next stage so that these materials can be widely used in tissue/tendon transplantations in humans and other biomedical applications.

Regulating biological responses to composites used to make surgical implants and scaffolds for tissue engineering could open up new strategic avenues for the development of multifunctional bio-based materials with clinical applications. In comparison to the number of trials conducted with metallic implants, the number of documented clinical findings for BBPCs' bone implants is small. Clinical data for biocompatible and biodegradable polymers, on the other hand, are critical for prospective composite design approaches. The advancements of BBPCs open up new avenues for meeting specific demands in developing technologies such as biosensors, scaffolds for tissue engineering, surgical implants, matrix synthesis, tissue regeneration, drug delivery, and thermo-mechanical applications.

References

[1] Haina, S. Gul, M. Awais, S. Jabeen, M. Farooq, Recent trends in preparation and applications of biodegradable polymer composites, J. Renew. Mater. 8 (2020) 1305–1326. Available from: https://doi.org/10.32604/jrm.2020.010037.

[2] S.O. Prakash, P. Sahu, M. Madhan, A. Johnson Santhosh, A review on natural fibre-reinforced biopolymer composites: properties and applications, Int. J. Polym. Sci. (2022). Available from: https://doi.org/10.1155/2022/7820731.

[3] L. Avérous, E. Pollet, Environmental silicate nano-biocomposites, Green. Energy Technol. 50 (2012). Available from: https://doi.org/10.1007/978-1-4471-4108-2.

[4] I. Yalcin-enis, M. Kucukali-ozturk, H. Sezgin, Nanoscience and biotechnology for environmentalapplications, 2019. Available from: http://link.springer.com/10.1007/978-3-319-97922-9.

[5] U. Arif, S. Haider, A. Haider, N. Khan, A.A. Alghyamah, N. Jamila, et al., Biocompatible polymers and their potential biomedical applications: a review, Curr. Pharm. Des. 25 (2019) 3608–3619. Available from: https://doi.org/10.2174/1381612825999191011105148.

[6] A.B. Balaji, H. Pakalapati, M. Khalid, et al., Natural and synthetic biocompatible and biodegradable polymers, Biodegrad. Biocomp. Polym. Compos. Process. Prop. Appl. (2017) 3–32. Available from: https://doi.org/10.1016/B978-0-08-100970-3.00001-8.

[7] G. Rajeshkumar, S. Arvindh Seshadri, G.L. Devnani, M.R. Sanjay, S. Siengchin, J. Prakash Maran, et al., Environment friendly, renewable and sustainable poly lactic acid (PLA) based natural fiber reinforced composites – a comprehensive review, J. Clean. Prod. 310 (2021). Available from: https://doi.org/10.1016/j.jclepro.2021.127483.

[8] F.M. Michael, M. Khalid, R. Walvekar, et al., Surface modification techniques of biodegradable and biocompatible polymers, Biodegrad. Biocomp. Polym. Compos. Process. Prop. Appl. (2017) 33–54. Available from: https://doi.org/10.1016/B978-0-08-100970-3.00002-X.

[9] A. Rudolph, M. Teske, S. Illner, V. Kiefel, K. Sternberg, N. Grabow, et al., Surface modification of biodegradable polymers towards better biocompatibility and lower thrombogenicity, PLoS One 10 (2015). Available from: https://doi.org/10.1371/journal.pone.0142075.

[10] R.A. Ilyas, H.A. Aisyah, A.H. Nordin, N. Ngadi, M.Y.M. Zuhri, M.R.M. Asyraf, et al., Natural-fiber-reinforced chitosan, chitosan blends and their nanocomposites for various advanced applications, Polym. (Basel) 14 (2022). Available from: https://doi.org/10.3390/polym14050874.

[11] M.A.A. Ahmad, M.S. Abdul Majid, M.J.M. Ridzuan, M.N. Mazlee, A.G. Gibson, Dynamic mechanical analysis and effects of moisture on mechanical properties of interwoven hemp/polyethylene terephthalate (PET) hybrid composites, Constr. Build. Mater. 179 (2018) 265–276. Available from: https://doi.org/10.1016/j.conbuildmat.2018.05.227.

[12] M. Ramesh, K. Palanikumar, K.H. Reddy, Mechanical property evaluation of sisal-jute-glass fiber reinforced polyester composites, Compos. B Eng. 48 (2013) 1–9. Available from: https://doi.org/10.1016/j.compositesb.2012.12.004.

[13] T. Prodpran, S. Benjakul, A. Artharn, Properties and microstructure of protein-based film from round scad (*Decapterus maruadsi*) muscle as affected by palm oil and chitosan incorporation, Int. J. Biol. Macromol. 41 (2007) 605–614. Available from: https://doi.org/10.1016/j.ijbiomac.2007.07.020.

[14] J. Puiggalí, R. Katsarava, Bionanocomposites, Clay-Polymer Nanocomposites (2017) 239–272. Available from: https://doi.org/10.1016/B978-0-323-46153-5.00007-0.

[15] G.E. Anderton, The determination of dynamic properties of polymers and composites, J. Mech. Work. Technol. 4 (1980) 91. Available from: https://doi.org/10.1016/0378-3804(80)90009-1.

[16] D. Nedelcu, S.N. Mazurchevici, R.I. Popa, N.M. Lohan, D. Maldonado-Cortés, C. Carausu, Tribological and dynamical mechanical behavior of prototyped PLA-based polymers, Mater. (Basel) 13 (2020). Available from: https://doi.org/10.3390/MA13163615.

[17] A. Shrivastava, Plastic properties and testing, Introd. Plast. Eng., 2018, 49–110. Available from: https://doi.org/10.1016/b978-0-323-39500-7.00003-4.

[18] Fluoropolymer Applications in the Chemical Processing Industries, 2018. Available from: https://doi.org/10.1016/c2015-0-04419-7.

[19] N.I.N. Haris, M.Z. Hassan, R.A. Ilyas, M.A. Suhot, S.M. Sapuan, R. Dolah, et al., Dynamic mechanical properties of natural fiber reinforced hybrid polymer composites: a review, J. Mater. Res. Technol. 19 (2022) 167–182. Available from: https://doi.org/10.1016/j.jmrt.2022.04.155.

[20] W.M. Ji, L.W. Zhang, Diamond nanothread reinforced polymer composites: ultra-high glass transition temperature and low density, Compos. Sci. Technol. 183 (2019). Available from: https://doi.org/10.1016/j.compscitech.2019.107789.

[21] Introduction to Plastics Engineering, 2018. Available from: https://doi.org/10.1016/c2014-0-03688-x.

[22] X. Luo, S. Xie, J. Liu, H. Hu, J. Jiang, W. Huang, et al., The relationship between the degree of branching and glass transition temperature of branched polyethylene: experiment and simulation, Polym. Chem. 5 (2014) 1305–1312. Available from: https://doi.org/10.1039/c3py00896g.

[23] M.J. Suriani, S.M. Sapuan, C.M. Ruzaidi, D.S. Nair, R.A. Ilyas, Flammability, morphological and mechanical properties of sugar palm fiber/polyester yarn-reinforced epoxy hybrid biocomposites with magnesium hydroxide flame retardant filler, Text. Res. J. 91 (2021) 2600–2611. Available from: https://doi.org/10.1177/00405175211008615.

[24] N.M. Nurazzi, M.R.M. Asyraf, S. Fatimah Athiyah, S.S. Shazleen, S.A. Rafiqah, M.M. Harussani, et al., A review on mechanical performance of hybrid natural fiber polymer composites for structural applications, Polym. (Basel) 13 (2021). Available from: https://doi.org/10.3390/polym13132170.

[25] H. Khoramishad, M.V. Mousavi, Hybrid polymer composite materials, AIP Conf. Proc., 2019. Available from: https://doi.org/10.1063/1.5123100.

[26] M.R.M. Asyraf, M. Rafidah, A. Azrina, M.R. Razman, Dynamic mechanical behaviour of kenaf cellulosic fibre biocomposites: a comprehensive review on chemical treatments, Cellulose 28 (2021) 2675–2695. Available from: https://doi.org/10.1007/s10570-021-03710-3.

[27] D. Mohana Krishnudu, D. Sreeramulu, P. Venkateshwar Reddy, A study of filler content influence on dynamic mechanical and thermal characteristics of coir and luffa cylindrica reinforced hybrid composites, Constr. Build. Mater. 251 (2020). Available from: https://doi.org/10.1016/j.conbuildmat.2020.119040.

[28] S.S. Chee, M. Jawaid, M.T.H. Sultan, O.Y. Alothman, L.C. Abdullah, Thermomechanical and dynamic mechanical properties of bamboo/woven kenaf mat reinforced epoxy hybrid composites, Compos. B Eng. 163 (2019) 165–174. Available from: https://doi.org/10.1016/j.compositesb.2018.11.039.

[29] S.I. Marras, K.P. Kladi, I. Tsivintzelis, I. Zuburtikudis, C. Panayiotou, Biodegradable polymer nanocomposites: the role of nanoclays on the thermomechanical characteristics and the electrospun fibrous structure, Acta Biomater. 4 (2008) 756–765. Available from: https://doi.org/10.1016/j.actbio.2007.12.005.

[30] M. Alcoutlabi, G.B. McKenna, Effects of confinement on material behaviour at the nanometre size scale, J. Phys. Condens. Matter. 17 (2005). Available from: https://doi.org/10.1088/0953-8984/17/15/R01.

[31] J.A. Forrest, K. Dalnoki-Veress, The glass transition in thin polymer films, Adv. Colloid Interface Sci. 94 (2001) 167–195. Available from: https://doi.org/10.1016/S0001-8686(01)00060-4.

[32] D.B. Zax, D.K. Yang, R.A. Santos, H. Hegemann, E.P. Giannelis, E. Manias, Dynamical heterogeneity in nanoconfined poly(styrene) chains, J. Chem. Phys. 112 (2000) 2945–2951. Available from: https://doi.org/10.1063/1.480867.

[33] K. Masenelli-Varlot, E. Reynaud, G. Vigier, J. Varlet, Mechanical properties of clay-reinforced polyamide, J. Polym. Sci. B Polym. Phys. 40 (2002) 272–283. Available from: https://doi.org/10.1002/polb.10088.

[34] J.L. Keddie, R.A.L. Jones, R.A. Cory, Interface and surface effects on the glass-transition temperature in thin polymer films, Faraday Discuss 98 (1994) 219–230. Available from: https://doi.org/10.1039/FD9949800219.

[35] N.I.N. Haris, M.Z. Hassan, R.A. Ilyas, M.A. Suhot, S.M. Sapuan, R. Dolah, et al., Dynamic mechanical properties of natural fiber reinforced hybrid polymer composites: a review, J. Mater. Res. Technol. 19 (2022) 167–182. Available from: https://doi.org/10.1016/j.jmrt.2022.04.155.

[36] S. Rana, S. Parveen, R. Fangueiro, Multiscale composites for aerospace engineering, Adv. Compos. Mater. Aerosp. Eng. (2016) 265–293. Available from: https://doi.org/10.1016/b978-0-08-100037-3.00009-2.

[37] N.M. Nurazzi, A. Khalina, S.M. Sapuan, R.A. Ilyas, S.A. Rafiqah, Z.M. Hanafee, Thermal properties of treated sugar palm yarn/glass fiber reinforced unsaturated polyester hybrid composites, J. Mater. Res. Technol. 9 (2020) 1606–1618. Available from: https://doi.org/10.1016/j.jmrt.2019.11.086.

[38] N. Jesuarockiam, M. Jawaid, E.S. Zainudin, M.T. Hameed Sultan, R. Yahaya, Enhanced thermal and dynamic mechanical properties of synthetic/natural hybrid composites with graphene nanoplateletes, Polymers (Basel) 11 (2019). Available from: https://doi.org/10.3390/polym11071085.

[39] R. Yadav, R. Purwar, U. Bose, Studies of rheological, electrical, and optical properties of silk fibroin/Benzalkonium chloride modified MMT clay nanocomposite films, Mater. Today Proc. (2021) 4998–5004. Available from: https://doi.org/10.1016/j.matpr.2021.01.424.

[40] A. Matsumoto, A. Lindsay, B. Abedian, D.L. Kaplan, Silk fibroin solution properties related to assembly and structure, Macromol. Biosci. 8 (2008) 1006–1018. Available from: https://doi.org/10.1002/mabi.200800020.

[41] S. Sinha Ray, Dynamic mechanical properties of environmentally friendly polymer nanocomposites using biodegradable polymer matrices and clay/carbon nanotube (CNT) reinforcements, Environ. Friendly Polym. Nanocomposites (2013) 269–294. Available from: https://doi.org/10.1533/9780857097828.2.269.

[42] S. Sinha Ray, P. Maiti, M. Okamoto, K. Yamada, K. Ueda, New polylactide/layered silicate nanocomposites. 1. Preparation, characterization, properties, Macromolecules 35 (2002) 3104–3110. Available from: https://doi.org/10.1021/ma011613e.

[43] B.S. Bouakaz, A. Habi, Y. Grohens, I. Pillin, Effect of combinations of nanofillers on rheology-structure relations in biodegradable poly(ε-caprolactone) nanocomposites, Appl. Clay Sci. 161 (2018) 35–47. Available from: https://doi.org/10.1016/j.clay.2018.04.006.

[44] K. Rinawa, S.N. Maiti, R. Sonnier, J.M. Lopez Cuesta, Dynamic. rheological studies and applicability of time–temperature superposition principle for PA12/SEBS-g-MA blends, Polym. Bull. 72 (2015) 3305–3324. Available from: https://doi.org/10.1007/s00289-015-1467-4.

[45] K. Issaadi, A. Habi, Y. Grohens, I. Pillin, Effect of the montmorillonite intercalant and anhydride maleic grafting on polylactic acid structure and properties, Appl. Clay Sci. 107 (2015) 62–69. Available from: https://doi.org/10.1016/j.clay.2015.01.004.

[46] P. Chen, F. Xie, F. Tang, T. McNally, Thermomechanical-induced polyelectrolyte complexation between chitosan and carboxymethyl cellulose enabling unexpected hydrolytic stability, Compos. Sci. Technol. 189 (2020). Available from: https://doi.org/10.1016/j.compscitech.2020.108031.

[47] I. Quijada-Garrido, B. Laterza, J.M. Mazón-Arechederra, J.M. Barrales-Rienda, Characteristic features of chitosan/glycerol blends dynamics, Macromol. Chem. Phys. 207 (2006) 1742–1751. Available from: https://doi.org/10.1002/macp.200600230.

[48] H. Bai, Z. Liang, D. Wang, J. Guo, S. Zhang, P. Ma, et al., Biopolymer nanocomposites with customized mechanical property and exceptionally antibacterial performance, Compos. Sci. Technol. 199 (2020). Available from: https://doi.org/10.1016/j.compscitech.2020.108338.

[49] M. Kato, C.H. Lee, Preparation and properties of biopolymers based on isosorbide (1, 4: 3, 6-dianhydro-D-sorbitol), 2012.

[50] P. Surya Nagendra, V.V.S. Prasad, K. Ramji, A study on dynamic mechanical analysis of natural nano banana particle filled polymer matrix composites, Mater. Today Proc. (2017) 9081–9086. Available from: https://doi.org/10.1016/j.matpr.2017.07.262.

[51] H. Rajashekaraiah, S. Mohan, P.K. Pallathadka, S. Bhimappa, Dynamic mechanical analysis and three-body abrasive wear behaviour of thermoplastic copolyester elastomer composites, Adv. Tribol. 2014 (2014). Available from: https://doi.org/10.1155/2014/210187.

[52] P. Wang, S.W. Tang, F. Sheng, J. Cai, P. Fei, A. Nawaz, et al., Crystallization, thermal stability, barrier property, and aging resistance application of multi-functionalized graphene oxide/poly(lactide)/starch nanocomposites, Int. J. Biol. Macromol. 132 (2019) 1208–1220. Available from: https://doi.org/10.1016/j.ijbiomac.2019.03.183.

[53] M.S. Kim, K.M. Ryu, S.H. Lee, Y.C. Choi, Y.G. Jeong, Influences of cellulose nanofibril on microstructures and physical properties of waterborne polyurethane-based nanocomposite films, Carbohydr. Polym. 225 (2019). Available from: https://doi.org/10.1016/j.carbpol.2019.115233.

[54] S. Toghi Aval, S.M. Davachi, R. Sahraeian, Y. Dadmohammadi, B. Shiroud Heidari, J. Seyfi, et al., Nanoperlite effect on thermal, rheological, surface and cellular properties of poly lactic acid/nanoperlite nanocomposites for multipurpose applications, Polym. Test. 91 (2020). Available from: https://doi.org/10.1016/j.polymertesting.2020.106779.

[55] D.A. Marín-Silva, S. Rivero, A. Pinotti, Chitosan-based nanocomposite matrices: development and characterization, Int. J. Biol. Macromol. 123 (2019) 189–200. Available from: https://doi.org/10.1016/j.ijbiomac.2018.11.035.

[56] M. Mucha, A. Pawlak, Thermal analysis of chitosan and its blends, Thermochim. Acta 427 (2005) 69–76. Available from: https://doi.org/10.1016/j.tca.2004.08.014.

[57] S. Rivero, M.A. García, A. Pinotti, Heat treatment to modify the structural and physical properties of chitosan-based films, J. Agric. Food Chem. 60 (2012) 492–499. Available from: https://doi.org/10.1021/jf204077k.

[58] A. Dufresne, Dynamic mechanical analysis of the interphase in bacterial polyester/cellulose whiskers natural composites, Compos. Interfaces 7 (2000) 53–67. Available from: https://doi.org/10.1163/156855400300183588.

[59] E. Abraham, B. Deepa, L.A. Pothan, M. John, S.S. Narine, S. Thomas, et al., Physicomechanical properties of nanocomposites based on cellulose nanofibre and natural rubber latex, Cellulose 20 (2013) 417–427. Available from: https://doi.org/10.1007/s10570-012-9830-1.

[60] Q. Li, J. Zhou, L. Zhang, Structure and properties of the nanocomposite films of chitosan reinforced with cellulose whiskers, J. Polym. Sci. B Polym. Phys. 47 (2009) 1069–1077. Available from: https://doi.org/10.1002/polb.21711.

[61] C.E. Corcione, M. Frigione, Characterization of nanocomposites by thermal analysis, Mater. (Basel) 5 (2012) 2960–2980. Available from: https://doi.org/10.3390/ma5122960.

[62] Y. Wang, H. Tian, L. Zhang, Role of starch nanocrystals and cellulose whiskers in synergistic reinforcement of waterborne polyurethane, Carbohydr. Polym. 80 (2010) 665–671. Available from: https://doi.org/10.1016/j.carbpol.2009.10.043.

[63] W. Li, J. Yue, S. Liu, Preparation of nanocrystalline cellulose via ultrasound and its reinforcement capability for poly(vinyl alcohol) composites, Ultrason. Sonochem. 19 (2012) 479–485. Available from: https://doi.org/10.1016/j.ultsonch.2011.11.007.

[64] T. Peixoto, M.C. Paiva, A.T. Marques, M.A. Lopes, Potential of graphene–polymer composites for ligament and tendon repair: a review, Adv. Eng. Mater. 22 (2020). Available from: https://doi.org/10.1002/adem.202000492.

[65] Q. Meng, J. Hu, K. Ho, F. Ji, S. Chen, The shape memory properties of biodegradable chitosan/poly (l-lactide) composites, J. Polym. Env. 17 (2009) 212–224. Available from: https://doi.org/10.1007/s10924-009-0141-z.

[66] N. Goonoo, A. Bhaw-Luximon, G.L. Bowlin, D. Jhurry, An assessment of biopolymer- and synthetic polymer-based scaffolds for bone and vascular tissue engineering, Polym. Int. 62 (2013) 523–533. Available from: https://doi.org/10.1002/pi.4474.

[67] K. Rezwan, Q.Z. Chen, J.J. Blaker, A.R. Boccaccini, Biodegradable and bioactive porous polymer/inorganic composite scaffolds for bone tissue engineering, Biomaterials 27 (2006) 3413–3431. Available from: https://doi.org/10.1016/j.biomaterials.2006.01.039.

[68] N.G. Shimpi, Biodegradable and biocompatible polymer composites: processing, properties and applications, 2017. Available from: https://doi.org/10.1016/C2015-0-05524-1.

[69] R.K. Saini, A.K. Bajpai, E. Jain, Advances in bionanocomposites for biomedical applications, Biodegrad. Biocompatible Polym. Compos. Process. Prop. Appl. (2017) 379–399. Available from: https://doi.org/10.1016/B978-0-08-100970-3.00013-4.

[70] M. Mozafari, Handbook of biomaterials biocompatibility, 2020. Available from: https://doi.org/10.1016/B978-0-08-102967-1.00033-5.

[71] H. Bundela, A.K. Bajpai, Designing of hydroxyapatite-gelatin based porous matrix as bone substitute: correlation with biocompatibility aspects, Express Polym. Lett. 2 (2008) 201–213. Available from: https://doi.org/10.3144/expresspolymlett.2008.25.

[72] A. Fritsch, C. Hellmich, L. Dormieux, Ductile sliding between mineral crystals followed by rupture of collagen crosslinks: experimentally supported micromechanical explanation of bone strength, J. Theor. Biol. 260 (2009) 230–252. Available from: https://doi.org/10.1016/j.jtbi.2009.05.021.

[73] Y. Okazaki, Y. Abe, K. Yasuda, K. Hiasa, I. Hirata, Osteoclast response to bioactive surface modification of hydroxyapatite, Open. J. Stomatol. 04 (2014) 340–344. Available from: https://doi.org/10.4236/ojst.2014.47047.

[74] T.L. do A. Montanheiro, R.G. Ribas, L.S. Montagna, B.R.C. de Menezes, V.M. Schatkoski, K.F. Rodrigues, et al., A brief review concerning the latest advances in the influence of nanoparticle reinforcement into polymeric-matrix biomaterials, J. Biomater. Sci. Polym. Ed. 31 (2020) 1869–1893. Available from: https://doi.org/10.1080/09205063.2020.1781527.

CHAPTER 6

Eco-friendly polymer nanocomposites based on bio-based fillers: preparation, characterizations and potential applications

Khodidas Bhanderi[1], **Jeimin Joshi**[1], **Vaishali Suthar**[2], **Vraj Shah**[3], **Gautam M. Patel**[3] **and Jigar Patel**[1]

[1]Industrial Chemistry Department, Institute of Science & Technology for Advanced Studies and Research, Vallabh Vidyanagar, Anand, Gujarat, India [2]Applied Chemistry Department, Faculty of Technology and Engineering, The M. S. University of Baroda, Kalabhavan, Vadodara, Gujarat, India [3]Department of Applied Chemistry, School of Science, ITM SLS Baroda University, Vadodara, Gujarat, India

Chapter Outline

Biodegradable and Biocompatible Polymer Nanocomposites. DOI: https://doi.org/10.1016/B978-0-323-91696-7.00015-5
Copyright © 2023 Elsevier Inc. All rights reserved.

6.1 Introduction

Polymer was invented a few decades ago because it has excellent properties and is less expensive and easier to manufacture than metal-based products. In response to demand, additional metal alloys and polymeric products were synthesized, thereby improving human life. However, as a result of the use of these polymeric products, the problem of pollution has also increased, primarily during the manufacturing of these polymers and their applications. Many disadvantages of synthetic polymers include the release of vapors and toxic gases after improper disposal. To address this challenge, governments in several countries are imposing stringent restrictions on the widespread use of polyethylene or synthetic polymers and encouraging the use of sustainable and environmentally friendly alternatives. The aforementioned topic introduces a new field of research: bio-based polymeric materials and biopolymer composite materials. Due to issues such as polymer dependency on fossil fuels and pollution caused by plastic wastes in the environment, a new advanced research area has emerged in the field of environmentally or eco-friendly polymer composites [1]. This chapter focuses on an overview of bio-based composite materials, comparisons with conventional composite materials, various processes for producing bio-based polymer composites, characterization and applications of several bio-based fillers reinforced into eco-friendly polymer composites. Finally, the scope and prospects of bio-based polymer composites were discussed [1]. Fig. 6.1 illustrates the overall outlook of the chapter.

Polymer composites are materials created by combining two or more components. Their sole purpose is to produce a product of higher quality than the originals. One component is known as the matrix phase, while the other is known as the reinforcement phase or dispersion phase. The matrix denotes the fundamental phase, which is inherently continuous. The matrix phase is more ductile and less stiff in general. It is responsible for retaining the scattered phase. In the second phase, the matrix incorporates a discontinuous shape (or phases). The following phase is known as the dispersed phase. Because the scattered phase is generally stronger than the matrix, it is often referred to as the reinforcement phase. A Composite is created by fusing two or more different types of components. They are commonly composed of a polymer matrix and an organic or inorganic support material. Phases are formed when these materials combine. Individual components are combined to create a compound of higher quality than the originals. In other words, the goal is to create a resource that can fill gaps left by pure polymers. However, there is a distinction between polymer blends and composites. Composites are formed when a new substance with properties that differ from the individual properties of the constituents is created. Composite materials are made up of at least two components with distinctly dissimilar physicochemical properties. Composite

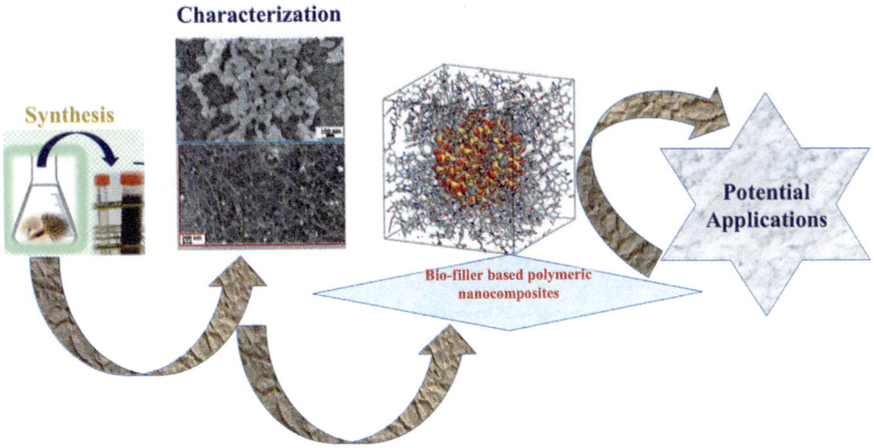

FIGURE 6.1

Strategic overview of bio-filler based polymeric nanocomposites.

materials include lignocellulosic (straw) in sludge, wood (cellulose fibers reinforced in hemicellulose with the binder lignin), bones (soft protein collagen combined with the hard mineral apatite), and pearlite (ferrite combined with cementite) [2–6]. Blends, on the other hand, are physical mixtures of at least two different polymers (miscible or immiscible), the extent of which depends on the thermodynamic interaction between the two materials, despite the fact that they are both mixtures. Both aim to develop materials that are more durable than pure polymers.

The synthesis of novel polymeric compounds focuses on specific obstacles such as environmental concerns and market competition, both of which necessitate new, novel, cost/performance effective, and bio-based/biodegradable materials. The use of fillers becomes critical in this situation. In today's competitive environment, introducing fillers into a polymeric matrix to achieve desired characteristics is a faster and more cost-effective method [7]. In recent years, environmental protection has been the most important global issue, providing much-needed motivation to develop an alternative method of using our natural treasures [8].

The term "bio-based polymers" refers to polymer substances derived from renewable sources that can be modified to create products with plastic-like properties for a specific purpose. Fats and oils, lignin, polynucleotides, polyisoprenoids, polysaccharides like cellulose, chitosan, and starch, proteins like wool, silk, and gelatin, and polymers made from monomer components derived from renewable sources are all examples of bio-based polymers found in the environment. Natural polymers, also known as biopolymers, are created by plants and other living things via complex biochemical pathways that consume carbon dioxide in order to keep resources sustainable. These polymers are then destroyed and recycled. Because of the slow rate of biodegradation of polymers derived from natural sources, such as rubber, lignin, and humus, there is little need to be concerned about their environmental effects because they will eventually be mineralized. If this concept is applicable to synthetic and semisynthetic polymeric materials, as well as their hybrid composites, the development of environmentally friendly plastics may open up new and exciting possibilities.

Biopolymers include "artificial polymers" derived from naturally occurring polymers such as cellulose esters and ethers, as well as polylactate derived from starch-derived lactic acid. Other polymeric materials, such as bacterial polysaccharides and a variety of polyesters synthesized by native and genetically modified microorganisms, have yet to be thoroughly investigated [8]. Currently, there is a unique category of bio-based composites being developed by combining the inherent strength and capability of fibers and nanoparticles with native eco-friendly polymeric materials. Because of its long lifespan, environmental advantages, and complete biodegradability [9].

6.2 Classifications of composite materials

Distinct types of composite materials are used in our daily lives, and they are classified into different sectors depending on their qualities, applications, and synthesis techniques, as illustrated in Fig. 6.2.

6.2.1 Composites with different matrix phase

First, (1) polymer matrix composites (PMCs): PMCs are inexpensive. They are made by mounting and connecting thin polymer and fiber layers. PMCs are made of thermosetting or thermoplastic materials that are reinforced with carbon, glass, Kevlar, and metal fibers. Thermosets are more widely used than thermoplastics due to their greater strength and heat resistance. Thermosets are made by combining resin and hardener. Laminar structures are most commonly used by stacking and connecting thin layers of fiber and polymer up to the required thickness. PMCs are low-cost composites due to their ease of handling and simple fabrication processes [3]. (2) Ceramic matrix composites (CMCs): CMCs are a type of ceramic that was developed to address the brittleness issue in monolithic ceramics. Specialized processing methods such as gas- or liquid-phase approaches are used in the synthesis of CMCs. Fibers

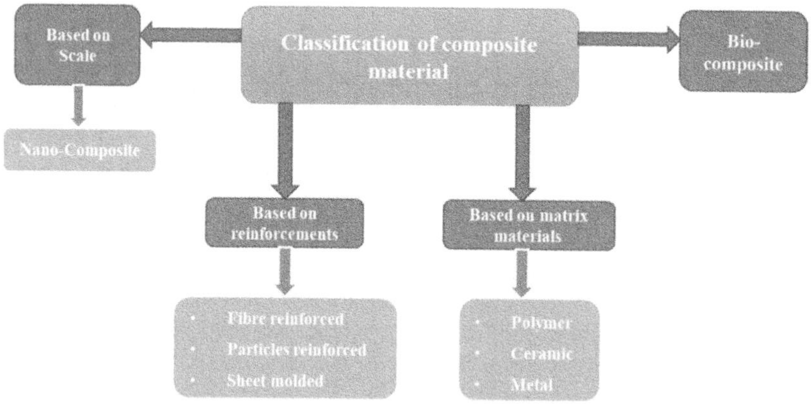

FIGURE 6.2

Classification of composite materials [3].

Reproduced with permission from Elsevier. Copyright 2019.

and matrix are made from gaseous or liquid precursors in inverted composites. The majority of CMCs, a type of ceramic, are made up of carbon, silicon carbide (SiC), aluminum oxide (Al_2O_3), and silicon nitride (SiN) fibers contained in a ceramic matrix structure. They are designed to counteract the brittleness of monolithic ceramics, which is a disadvantage. CMCs are referred to as inverse composites because the matrix's failure strain is less than the failure strain of the fibers. Specialized processing methods such as gas- or liquid-phase approaches are used in the synthesis of CMCs. The internal phase and matrix are formed around the fibers using gaseous or liquid precursors in this method [3]. (3) Metal matrix composites (MMCs): MMCs must have a metallic matrix, which is primarily made of aluminum (Al), magnesium (Mg), copper (Cu), and titanium (Ti), with the reinforcement consisting of scattered ceramics such as oxides and carbides or metallic fibers (i.e., tungsten, molybdenum, lead). Reinforcement contributes a small percentage of the total volume of the composite material, approximately 50%. MMCs with aluminum as the primary reinforcing material are used in the automotive and aerospace industries [3].

6.2.2 Composite with different reinforcements

In this first lets discuss about fiber, particle, and sheet one-by-one. Fibers containing a dispersed phase of synthetic fibers such as carbon, basalt, glass, and Kevlar within a composite structure demonstrated improved material properties such as great stiffness, strength, and resistance to chemical, temperature, and wear in fiber-reinforced composite materials [3]. When particle-based composites are linked to fiber-based composites, they are less effective at strengthening. Road surfaces are one application for particle embedded composites with high wear resistance. The hardness of cement is greatly increased when gravel is used as reinforcing filler [3]. Sheet-based composites, also known as sheet-molding composites (SMC), are compression-molded thermoset molding materials that contain glass reinforcement [10]. Unsaturated resins and long glass fiber are combined to create an extremely strong molding composite [3].

6.2.3 Composite based on scale

The composite based on scale here refers to nanocomposites. Desired material qualities can be achieved by mixing two or more different materials at the nanoscale, resulting in the creation of a brand-new substance known as a nanocomposite. Unintercalated, intercalated, and exfoliated nanocomposites are commonly classified as such. These are created through melt compounding, in situ polymerization, template synthesis, and polymer intercalation [11,12]. Before we get into the bio-based composite, let's take a look at some of the negative effects of conventional polymer composites (PC). These materials have several advantages over metal and virgin polymer products, including higher modulus and glass transition temperatures. It will be possible to modify characteristics such as low thermal expansion, high corrosion resistance, flexibility, freedom of design forms, high fatigue resistance, ease of molding, and bonding to a variety of materials or other composite materials for a variety of applications. Traditional composites have some disadvantages when compared to bio-composite materials, such as being more expensive, requiring specialized manufacturing techniques, and not being easily recyclable or environmentally friendly. Furthermore, composite materials are more expensive and difficult to repair than metals. The basic materials for composite materials are expensive. Because of the limitations of traditional PC mentioned above, it is necessary to develop environmentally friendly, bio-based PC [6].

Furthermore, due to the need for biodegradable, environmentally friendly materials, researchers are forced to look for new opportunities in the manufacturing of bio-composites. The bio-composite made of sugar palm fibers reinforced in a sago starch matrix improves thermal stability, tensile and tearing strength, water absorption propensity, and durability. Bio-nanocomposites made at the nanoscale have shown promise in a variety of applications, including scaffolds for tissue engineering and bio-packaging. Ginger fiber's antibacterial properties have been successfully used to maintain the quality of food packaging [13].

6.3 Bio-composite materials

The term "bio-composites" refers to a material combination made of natural fibers or biopolymers (natural fibers) that can be used as reinforcement. When used as composite reinforcement, natural fibers have several advantages, including high specific strength and modulus, lower density, lower cost, lower pollution emissions, and the ability to decompose naturally [14]. A naturally occurring composite is composed of cellulose (wood fibers) and lignin (a chemical). Lignin is a natural glue matrix that bonds and stabilizes the fibers, giving wood its strength. Plywood is a composite material made of natural and man-made materials. Laminated wood sheets are stronger than raw wood because thin layers of wood veneer are glued together.

Bio-based products contain components derived from biodegradable sources. These novel bio-based materials are gradually taking the place of crude oil-based polymers. Bio-renewable resources include biomass feedstock, biofibers, and biopolymers. Hemp and flax are two popular biofibers used as reinforcements in the production of bio-based composites [15]. Bio-based composites can include a nonregional bio-renewable component. It's possible that the matrix, the filler, or both are to blame. These various composites necessitate a variety of manufacturing techniques. These bio-based composites are made up of various fillers and reinforcements for a variety of applications. Eco-friendly polymer composites (EPC) have several advantages, one of which is that they can convert waste into a truly beneficial product while minimizing environmental impact. This is a free or low-cost sustainable method of converting trash to treasure. Wood flour, rice husk, eggshells, and a variety of industrial and agricultural wastes are examples [16].

Sheldon and colleagues classified bio-composite raw materials into two types: first generation and second generation components. Wood is the first generation raw material, and its increased use can contribute to deforestation and biodiversity loss, whereas lignocellulosic wastes collected from foodstuff, forestry, and crop residue are the second-generation raw materials. Until now, most 2nd generation waste has been used in the production of bio-composites, though commercialization has been limited [17].

6.3.1 Advantages of bio-based composites materials

Bio-composites materials have distinct and valuable properties such as material recyclability, lightweight nature, and cost-effectiveness [18]. Bio-composite elements derived from renewable natural sources, such as bio-based polymers and fillers, can be used to replace nonrenewable oil-based plastics. Bio-composites have a lower environmental impact, making them safer for humans and other life forms. Furthermore, the vast majority of bio-composites are recyclable and reusable. According to Joshi and colleagues, one of the most notable benefits of bio-composites is that they provide a manageable option for disposal after use [19]. Using bio-fillers to reinforce oil-based plastic composites may also improve biodegradability. Bio-fillers improve polymer strength and modulus while also increasing the biodegradability of the overall composite system (when used in conjunction with a coupling agent). Bio-composite materials are less expensive, easier to prepare, and degrade faster than traditional composites. Small density, more efficient energy recovery, excellent strength and rigidity from renewable matrix and reinforcement sources [20]. As previously discussed in this chapter, bio-based composites are made up of two phases: matrix phase and dispersion phase or reinforcement phase. The reinforcement phase is made up of fillers, and in this case, the fillers are bio-fillers. The following section goes over bio-fillers and matrix in more detail.

Fillers are added to polymeric materials to improve or develop specific aspects of the composite material while using less of the more expensive polymer ingredient. Millions of tonnes of fillers are used each year in a variety of applications around the world, including paper, plastics, rubber, paints, varnishes, adhesives, and sealants. As a result of recent environmental concerns, which have reignited interest in natural, renewable resource-based, and compostable materials, bio-fillers are being used more frequently as fillers in PC. Proteins and polysaccharides are common organic fillers, also known as agropolymers [21]. In comparison to traditional inorganic fillers, these materials have some advantages due to their renewable nature, low density, nonabrasive properties, and sufficient strength and stiffness. As a result, they can be used as reinforcement in certain composite applications [21].

Mineral or metallic fillers and organic fillers are the two types of fillers. Calcium carbonate ($CaCO_3$), silver nanoparticles (AgNPs), and mineral or metallic fillers such as clay and nanoclay are among the most commonly used inorganic reinforcing agents in bio-composites. Nanoparticles are used as fillers in composite materials, as are other organic fillers recovered from waste. Whether bio-based or not, the majority of organic fillers used in green composites are cellulose-rich and derived from renewable resources. Natural cellulosic fibers such as sisal, hemp, jute, bamboo, kenaf, and plant tissue fiber are frequently used as fillers in polymer composite materials as alternatives to glass fiber because of their lower density, lower cost, and renewable nature [22]. Green composite materials have been thoroughly researched and used in the transportation and construction industries, with applications ranging from insulation panels and window frames to railroad sleepers and various automotive parts, as well as a number of other less expensive products such as gardening tools, agricultural mulch, and packaging [23]. Because these organic fillers may be derived from agro-industrial or forestry byproducts and leftovers, there is a growing interest in biodegradable fillers and renewable materials, also known as sustainable materials [23].

Furthermore, fibers derived from leaves of the plants and stems that are grown for their fibers, as well as byproducts of agri-food production such as wheat husk [24], straw of rice [25], sugarcane bagasse [26], malt bagasse, and [27], banana leaves [28], have been examined as bio-composite fillers [29]. Furthermore, bagasse, peel, root and tuber starch, microalgae, algae, and byproducts have all been documented [30]. Sabai grass (*Eulaliopsis binata*), Cogon grass (*Imperata cylindrica*), and Napier grass (*Pennisetum purpureum*) have all recently been tested as composite fillers, and the results have been promising [31]. Recently, novel green bio-composites were developed by combining waste materials from tea brewing, starch-rich potato washing slurries, and coffee silverskin

[32,33]. Sesame, rapeseed, peanut, and sunflower oil cakes, for example, have been investigated as fillers for bio-composite films and foams [34,35]. The main sources of organic fillers are shown in Fig. 6.3, along with some samples of each [36]. Keratin and chitin, derived from animal feed waste such as shrimp shells or chicken wings, are used as fillers in composites [37].

6.3.2 Bio-based fillers reinforcement in nonbio-based polymers

Wood-based waste materials are one of the readily available bio-based resources that can be easily used as reinforcement in bio-based composites, with a rapid industrial use trend emerging over the last 10 years in a market dominated by the use of inorganic fillers. Wood flour [39] and wood fiber [40] are plentiful, simple to research with waste items from the wood industry. Furthermore, waste paper sludge (WPS) and ink-eliminated sludge (IES) have been used as filler and additive in composites made of high-density polyethylene (HDPE) and wood flour in an effort to develop more environmentally acceptable materials. The addition of WPS to HDPE/wood flour composites has been shown to improve flexural properties, which can be further improved by the addition of a compatibilizer and reinforced maleated anhydride grafted polyethylene (MAPE) [39]. When WPS and IES were combined, swelling and water absorption were both reduced [39]. With an ideal

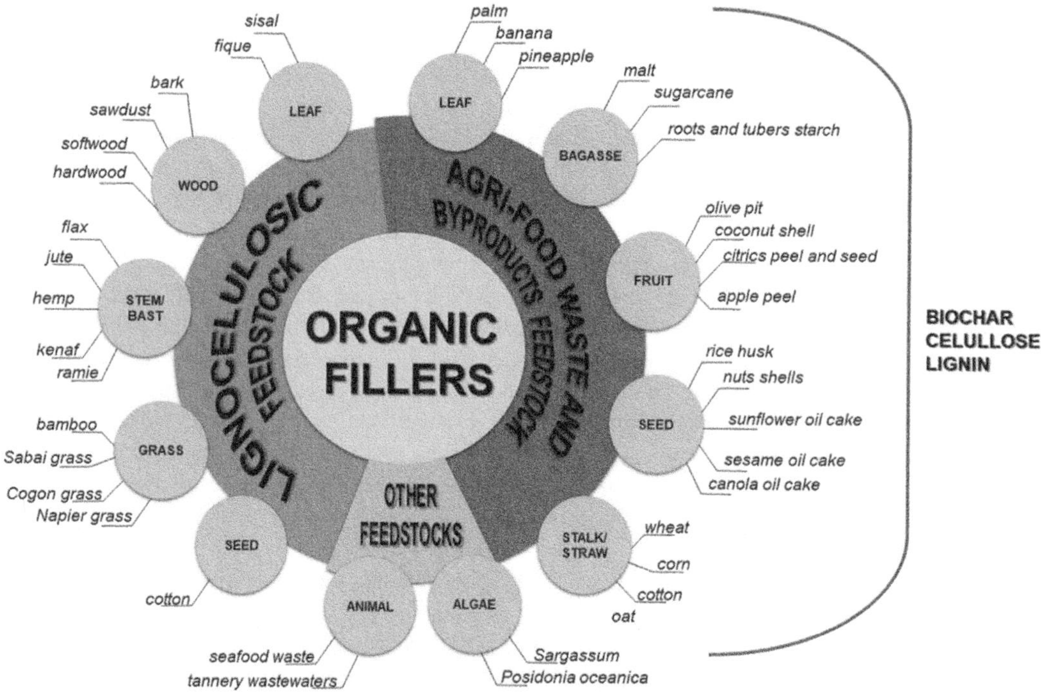

FIGURE 6.3

Main sources of organic fillers, with some examples [38].

Reproduced with permission from Springer Nature. Copyright 2021.

loading of 60%, IES-incorporated composites had better overall physical and mechanical properties than WPS-added composites, which can be used to make bio-based composites [39].

6.3.2.1 Bio-based polymer and bio-based reinforcement

Natural fibers used in bio-composites have several advantages over traditional polymeric materials and are increasingly being researched. Natural polymeric materials derived from renewable sources are less expensive to produce and better for the environment. Composites with excellent flexibility and dampening properties have been created using pure cellulose, hemp, keratin, and flax fibers [15,41].

6.4 Synthesis methods for composited material

Numerous techniques are used in the preparation or synthesis of bio-composites, but three types of composite manufacturing techniques are primarily used. Within these molding categories, there are various processing techniques, each with its own set of advantages. Extrusion, injection molding, compression molding, machine presses, filament winding, pultrusion, sheet molding, and resin transfer molding are the most common [42]. Before delving into each method, consider the benefits and drawbacks of some popular synthesis techniques used to create polymeric nanocomposites (Table 6.1).

Table 6.1 Rewards and drawbacks of synthesis techniques of bio-based polymeric nanocomposites.

Method	Rewards	Downsides
Melt blending	• Very compatible process • Eco-friendly procedures	• High maintenance • High cost
Solution-casting	• Greater film thickness and surface uniformity • Wide range of film thickness. • Excellent flatness and dimensional stability	• The solution should be stable • The polymer must be soluble in a volatile solvent
Electrospinning	• Low cost, simple handling, a basic solution, and material with controlled diameter can all be obtained	• Limited control pore size • Electric current rate should always be maintained
Electrodeposition	• Short synthesis time; no unspecified byproducts	• Nanocomposites with poor active sites may be created; however, the requisite polymeric composite size cannot be reached
In situ polymerization	• Use of low-cost materials • Possibility of automation • Integration with various heating and curing methods	• Expensive equipment • Limited time to complete the polymerization process • Complex process requires real-time monitoring

6.4.1 Injection molding and compression molding

Ujendra Kumar Komal used a different approach, creating bio-composites from banana fiber and polylactic acid using three distinct processing techniques, as shown in Fig. 6.4. During the experimental investigation, bio-composites containing varying amounts of banana fiber were created. illustrates the direct injection moulding approach, in which polylactic acid (PLA) pellets and banana fibers were aggressively mixed and fed straight into the hopper of an injection molding machine, and bio-composite test specimens were manufactured in accordance with the American Society for Testing and Materials (ASTM) standard [43].

6.4.2 Electrospinning technique

Xiang fang Peng created a prepaid bio-composite material by combining cellulose nanocrystals (CNC) as a filler (reinforcement) with poly (butylene succinate) (PBS) polymeric material as a Metrix. A bio-nanocomposite of poly (butylene succinate) (PBS)/cellulose nanocrystals (CNC) scaffold was created using an electrospinning procedure and a solvent mixture of chloroform (CF) and methanol. To get well-dispersed PBS/CNC suspensions, PBS and CNC powder were dissolved in a mixture of CF and methanol with magnetic stirring at room temperature, followed by ultrasonic treatment. Fig. 6.5 depicts the full process of producing a bio-nanocomposite [44].

6.4.3 Solution-casting method

Anand Babu Perumal et al. developed a novel composite material (poly vinyl alcohol (PVA)/chitosan (CS)/CNC) by solvent casting and evaporating polyvinyl alcohol, chitosan, and cellulose nanocrystals. At the boiling point, PVA was dissolved in distilled water while being stirred. CS was also dissolved in water and acetic acid. After cooling and mixing, the PVA and CS solutions were sonicated. As a plasticizer, polyethylene glycol (PEG) was added to the aforementioned mixture and constantly stirred. The nanocomposites were created by adding the required amount to the ultrasonically dispersed PVA/CS mixture in distilled water. The resulting homogeneous PVA/CS/CNC solutions were cast onto a Petri plate [45].

6.4.4 Process hand lay-up technique equipped with vacuum bagging system and vacuum infusion

Rusdi and Nurazzi et al. worked for the fabrication procedures shown in Fig. 6.6 included a vacuum bagging system technique in which bacterial cellulose (BC) was isolated from Nata de Coco and utilized as a green reinforcing material hybrid with Kevlar reinforced epoxy composites. The BC/Kevlar reinforced epoxy composite was created using a hand lay-up process and a vacuum bagging system. The hybrid composite fabrication was created by cutting the BC sheet and Kevlar fabric. To facilitate composite removal, the surface is then coated with a silicone release agent. Both chemicals were combined by stirring. Brushing was used to spread the resin mixture evenly. A roller was also used to remove any voids. A sample was carefully placed in a vacuum bag film and sealed with tape. It was critical to stretch the plastic sheet evenly when using the vacuum pump to avoid creases on the sample's surface. The performance of the hybrid BC/Kevlar reinforced epoxy composites was evaluated using a variety of methodologies [46].

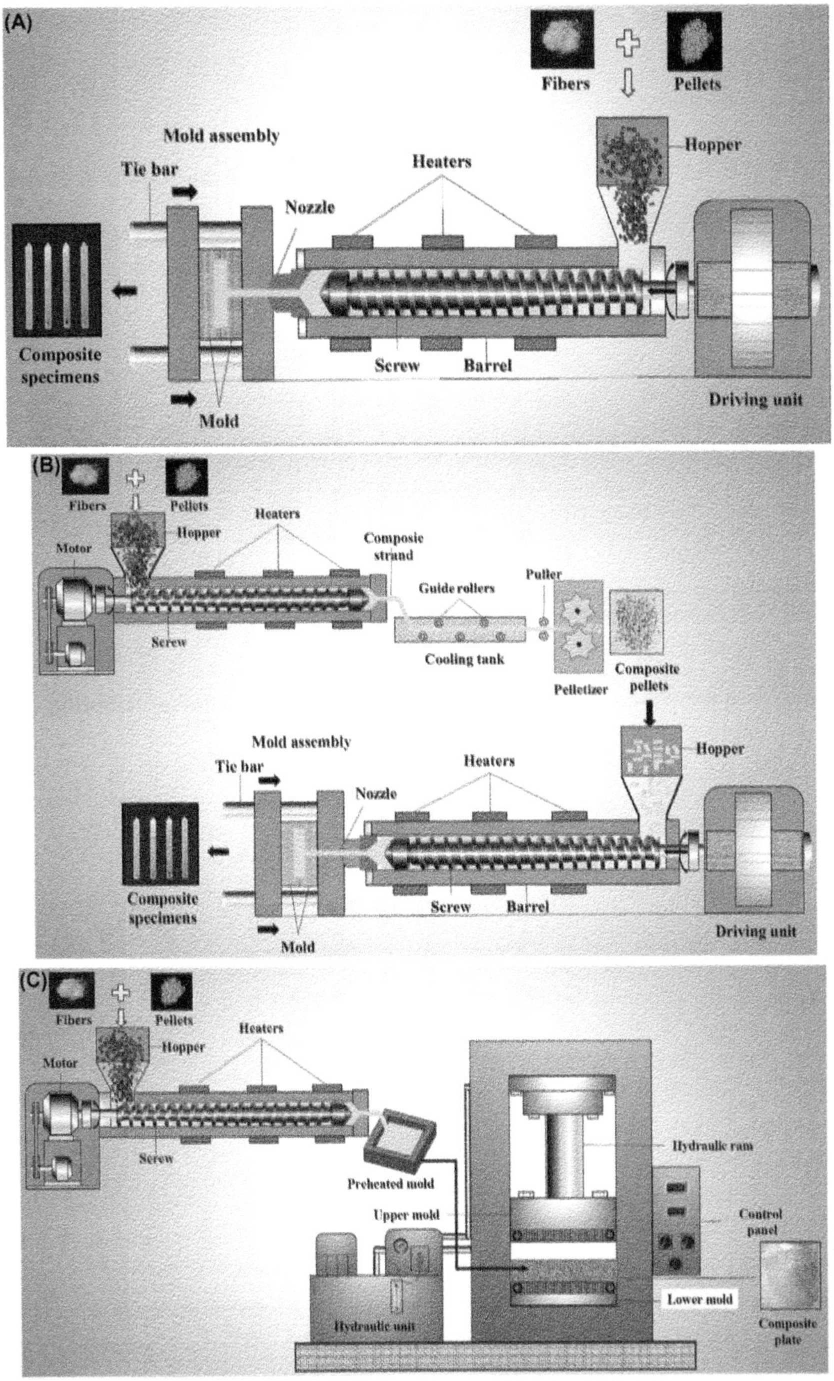

FIGURE 6.4

Schematic diagram of (A) direct injection moulding, (B) EIM, and (C) ECM [43].

Reproduced with permission from Elsevier. Copyright 2020.

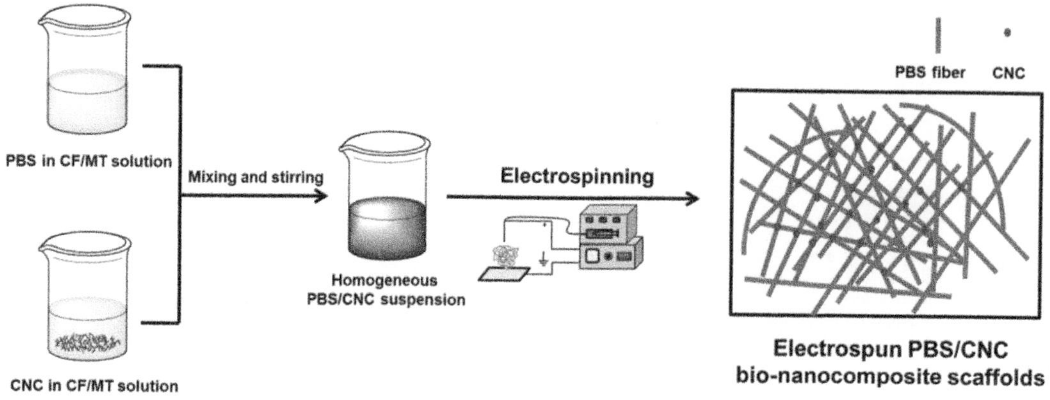

FIGURE 6.5

Schematic diagram of the experimental process of making PBS and PBS/CNC scaffolds by using electrospinning method [44].

Reproduced with permission from Elsevier. Copyright 2018.

Kristiina Oksman created composite material using vacuum infusion. A technique for producing cellulose bio-composites was developed using vacuum infusion. Porous cellulose nanofiber networks were created using ice-templating. To begin, as shown in Fig. 6.7A, cellulose nanofibers (CNF) aerogels were created by ice-templating. The aerogels were vacuum-infused with a bio-epoxy resin to create cellulose nanocomposites (Fig. 6.7B). The aerogel sample was placed on a metal plate and covered with a plate-attached plastic sheet. After degassing a premixed epoxy/hardener mixture until no significant number of bubbles emerged, the preform was packed with a vacuum pump along the freezing direction. The temperature of the mold was then raised and held there for curing. After cooling, the mold was removed from the hot plate, and the nanocomposite sample was demolded [47].

6.4.5 Hot press method

Ji-Won Park used phenyl silane as a surface hydrophobic treatment and fiber carding/melt pressing to create a novel composite material with a higher lyocell content. A PLA matrix was used to improve the composite's environmental qualities, and a carding method was used to place the fibers in the composite. Fig. 6.8 depicts an overall schematic representation of the composite material production process using PLA and lyocell. The fibers were combined using a bale opener in the following weight ratios before being fed into the carding machine: 1:9, 3:7, 5:5, 7:3, and 9:1. The PLA/lyocell bio-composite was created using a carding machine. Before hot pressing, composites were prepared using the carding method [48].

6.5 Characterizations of different bio-filler based polymeric nanocomposites

Composite material characterization is an important step in the product development and manufacturing processes. Quality is maintained through physical and chemical controls. Mechanical, optical, thermal, and

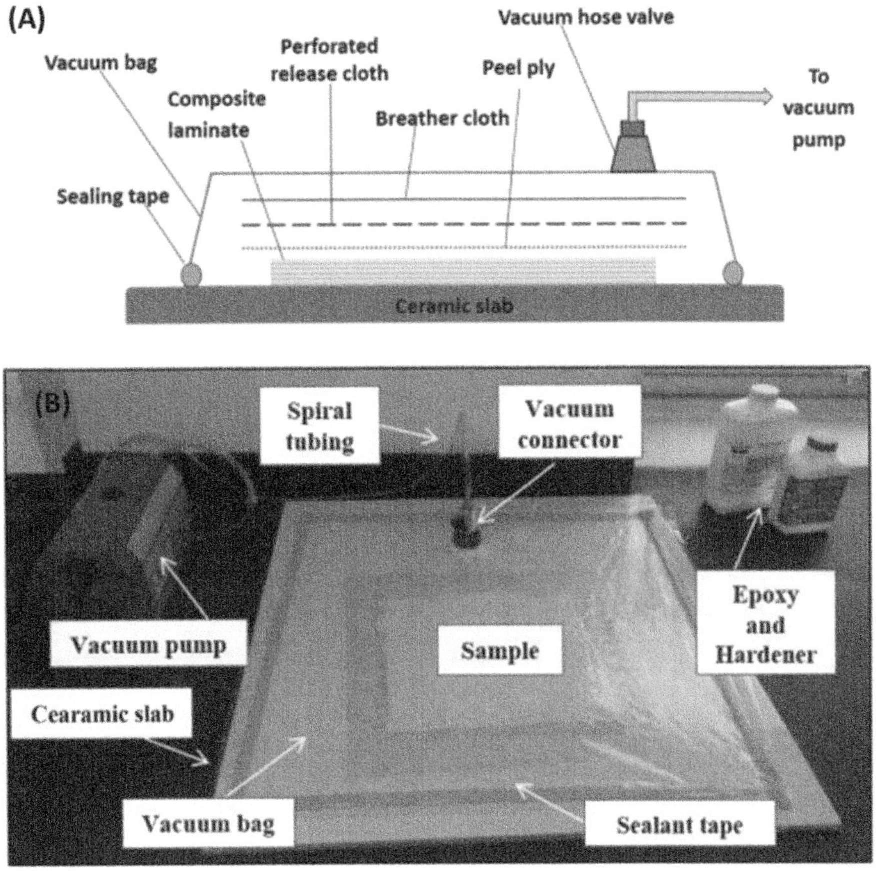

FIGURE 6.6

Schematic diagram of BC/Kevlar reinforced epoxy composite preparation method [46].

Reproduced with permission from Elsevier. Copyright 2022.

electrical properties, as well as exposure, flammability, surface, and barrier properties, are all taken into account. Characterization of bio-composite materials includes scanning electron microscope (SEM), transmission electron microscope (TEM), fiber length distribution, notched impact strength, mechanical characteristics, impact strength, thermo-gravimetric analysis, degradation time, sample particle size distribution, thermo-gravimetric analysis (TGA). Different characterization techniques are illustrated in Fig. 6.9 [49].

6.5.1 Characterization of polyaniline-based bio-composites

The SEM analysis was performed to evaluate the micron-scale textural properties of adsorbents; it is commonly used to determine how particles are arranged on surfaces. Fig. 6.10A–D shows the obtained micrographs of the as-synthesized walnut shells, polyaniline-modified walnut shells,

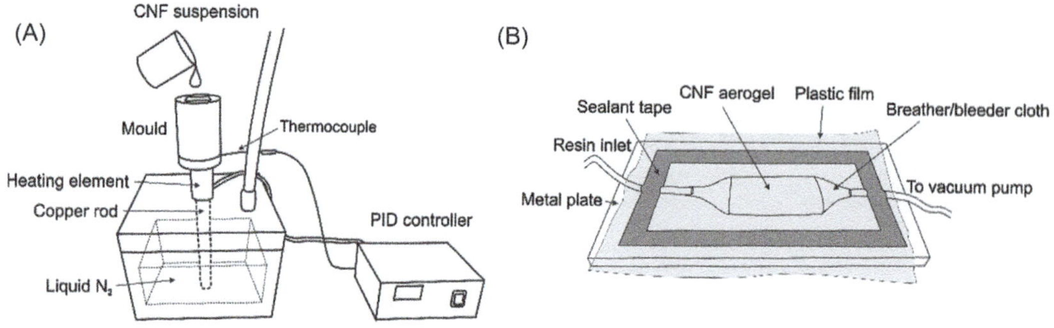

FIGURE 6.7

A schematic representation of the composite production method (A) the ice-templating method and (B) the vacuum infusion [47].

Reproduced with permission from Elsevier. Copyright 2019.

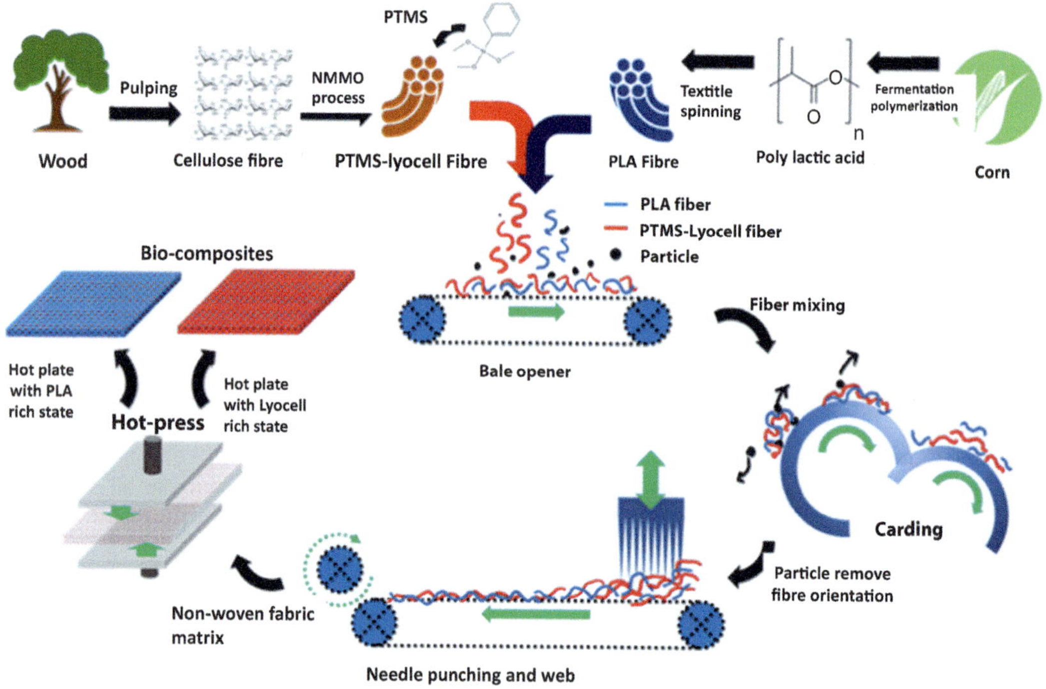

FIGURE 6.8

Entire schematic representation of the composite material production process employing PLA and lyocell [48].

Reproduced with permission from Elsevier. Copyright 2019.

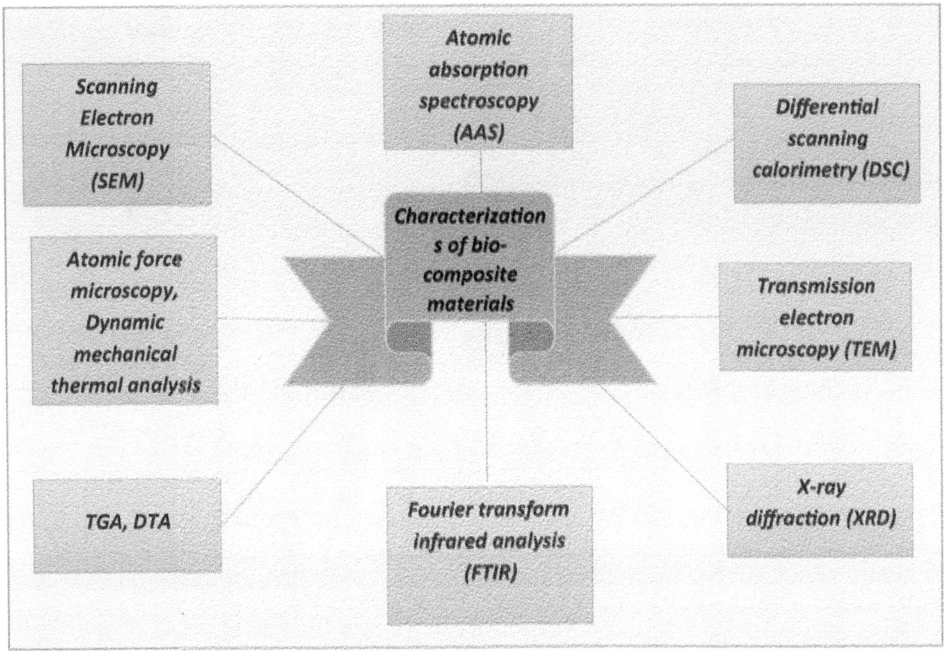

FIGURE 6.9

Different characterization techniques.

almond shells, and polyaniline-modified almond. The raw materials almond shells (Fig. 6.10A) and walnut shells (Fig. 6.10C) particles are shown in the SEM images to have a heterogeneous and porous texture surface, whereas the SEM images of the bio-composites was predicated on the assumption that the Walnut Shells and Almond Shells surface were well shielded by particles of polyaniline, leading to the formation of pores within the textures of the bio-composites. These pores facilitate the penetration and diffusion of Orange-G dye molecules into the internal surfaces of polyaniline-modified walnut shells and polyaniline-modified almonds, respectively. The data from the energy-dispersive X-ray spectroscopy (EDS) examination of the prepared almond shells, walnut shells, polyaniline-modified almond, and polyaniline-modified walnut shells are shown in Fig. 6.10E–H. The relevant tables contain the quantitative elemental compositions. The raw materials EDS analysis spectrum (Fig. 6.10E–G) reveals that carbon and oxygen make up the majority of the elements in these shells. The atoms belonging to the raw adsorbents (C and O) along with nitrogen (N), chloride (Cl), and sulfur (S), which come from PANI, HCl, and $Na_2S_2O_8$, correspondingly, are all present, as shown in Fig. 6.10F–H.

The Fourier transform infrared (FTIR) is an essential tool for identifying the distinctive functional. The FTIR spectrum of prepared adsorbents are shown in (Fig. 6.10J and K), and the strong peak with a frequency of 3590 cm^{-1} was attributed to the hydroxyl functionalities with OH^- group stretching vibration. The peak at 2920 cm^{-1} is due to the CH-stretching vibrations that are presented in methyl and methylene. The aromatic cycle of lignin's C = C stretch is represented by the peak at 1515 cm^{-1}.

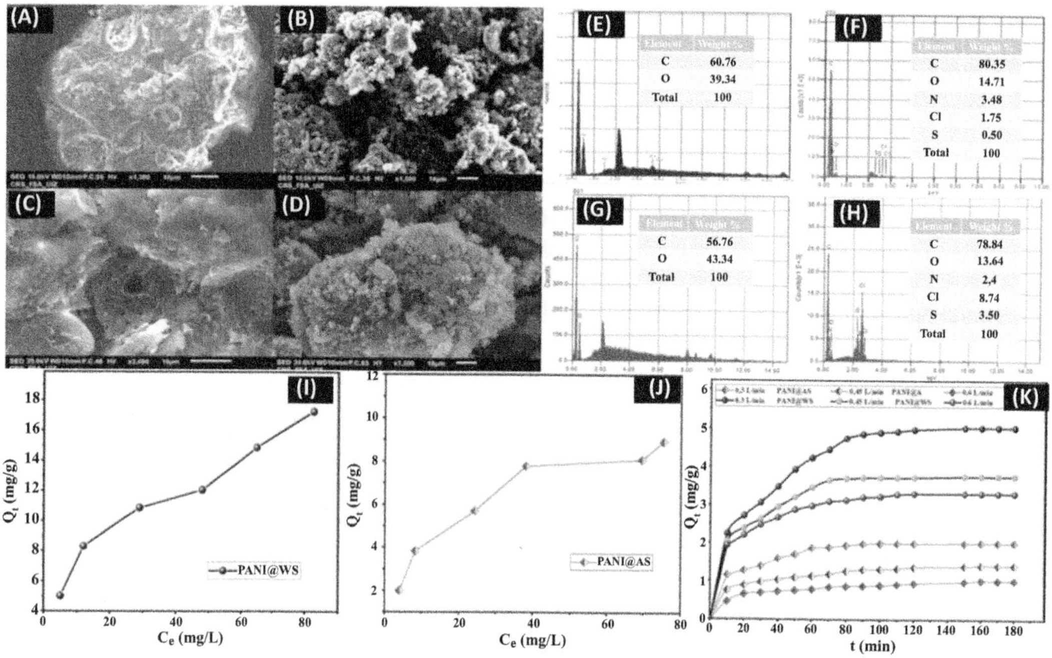

FIGURE 6.10

SEM images of walnut shells (WS) (A), polyaniline-modified walnut shells (B), almond shells (AS) (C), and polyaniline-modified almond shells (D) EDS elemental analysis of walnut shells (E), polyaniline-modified walnut shells (F), almond shells (G), and polyaniline-modified almond shells (H) Fourier transform infrared (FTIR) spectrum of walnut shells, polyaniline-modified walnut shells (I), effect of the initial concentration of OG dye on the removal by the PANI@WS (75 mg) (J), almond shells, polyaniline-modified almond (K) bio-composites before adsorption [50].

Copyright 2021. Reproduced with permission from Elsevier.

The C = C vibration bond stretching may be responsible for the peak located about 1595 cm^{-1}. The peak at 1044 cm^{-1} represents the stretching of C-O in phenols, alcohols, or ether groups [51]. As opposed to this, a band that is centered at 3500 cm^{-1} and is brought on by the hydrogen's extension vibrations in the O-H hydroxyl groups can be seen in the S spectrometer (Fig. 6.10K) [52]. It is confirmed that the walnut shells (WS) and almond shells (AS) surfaces of the polyaniline were well ornamented. Additionally, when the raw materials have undergone polymerization, amine groups can be seen on their surface with a high bandwidth at 3554 cm^{-1} owing to N-H stretching vibration.

Adsorbent doses are a critical chemical factor because they determine the adsorbent's capacity for a given amount of adsorbent under working conditions. In this work, the effects of polyaniline-modified WS and polyaniline-modified almond shell levels having the range from 75 to 225 mg on the adsorption of the Orange-G dye were examined. It should be noted that high mass ratios were chosen due to the poor performance of the polyaniline-modified AS bio-composite at low mass ratios. With an increase in the adsorbent dose ratio, less Orange-G dye is adsorbed (Fig. 6.10I)

[53]. Further, the effect of pH on Orange-G dye removal was investigated in the current work by altering the pH values (6, 2, and 10). At natural pH (pH = 6), the adsorbed quantities by polyaniline-modified WS and polyaniline-modified AS are negligible, as illustrated in Fig. 6.10J. The zero charge points of polyaniline-modified WS and polyaniline-modified AS were found 3.8 and 3.6, correspondingly (Fig. 6.10K) [50].

6.5.2 Polyvinyl alcohol/chitosan bio-nanocomposite films reinforced with cellulose nanocrystals (PVA/CC/CNC composite)

In 2018, Sella Muthu et al. and colleagues described the solution-casting method for the synthesis of PVA/CC/CNC composites. The hydrolysis method was used to create bio-based PC or CNC from rice straw in this process. The isolated CNC had a rod-like shape with particles measuring about 15 nm in diameter. Following that, the morphological features, thermal and mechanical properties of PVA/CS films reinforced with various concentrations (1, 3, and 5 wt.%) of CNC were evaluated. It was discovered that adding RS-CNC to the PVA/CS films enhanced their tensile strength (98.15 MPa) and thermal stability. The bio-nanocomposites had the same level of transparency as the PVA/CS blend film, indicating that the CNC were dispersed at the nanoscale level. The PVA/CS/CNC bio-nanocomposite film was also naturally antifungal and antibacterial. Furthermore, Fig. 6.11A depicts the field emission scanning electron microscopy (FESEM) analysis of a PVA/CS/CNC composite, revealing that the particles are uniformly distributed with optimal diameter and length. Aside from the morphological data,

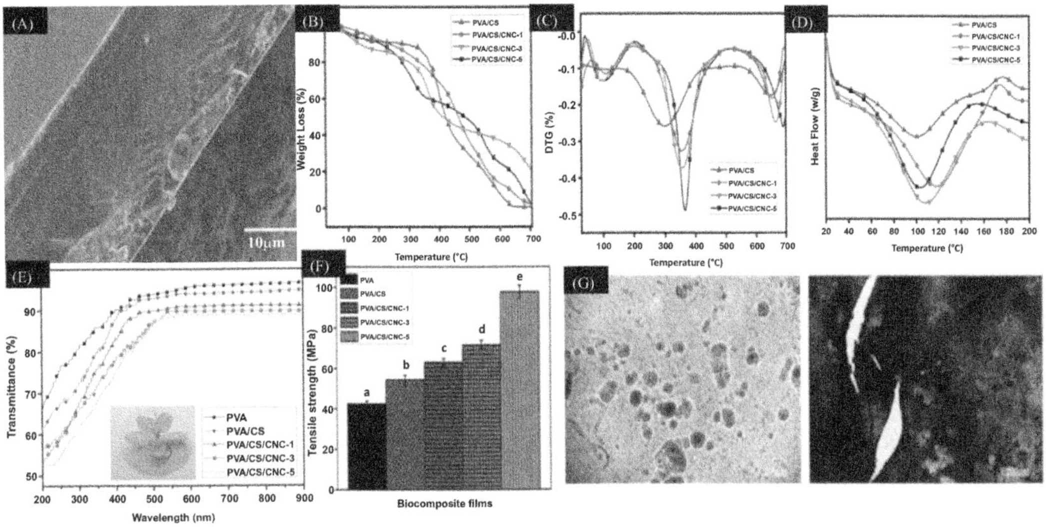

FIGURE 6.11

(A) FESEM analysis of PVA/CC/CNC composite, (B) thermal analysis of bio-composite films, (C) derivative thermogravimetry characterization of bio-composite films, (D) differential scanning calorimetry analysis of bio-composite films, (E) UV-spectroscopy analysis of four bio-composite films, (F) Tensile strength analysis data, (G) Optical microscopy data over the period of 180 days [45].

Fig. 6.11B depicts the thermal analysis of four composites, with the desired composite exhibiting exceptionally good properties in terms of weight loss percentage versus temperature. Fig. 6.11C and D also describes the derivative thermogravimetry and differential scanning calorimetry analyses, respectively. Both produce the best results in terms of the desired data from such analysis. Furthermore, Fig. 6.11E depicts a graphical representation of UV-spectroscopy results, from which it is clear that PVA/CS/CNC-5 composites received the highest transmission with respect to wavelength. Following that, the greater tensile strength was also obtained through PVA/CS/CNC-5 composite that was more than 80 MPa in number with respect to the type of bio-composite films (Fig. 6.11F). Additionally, Fig. 6.11G illustrates the optical microscopy data to check the change in morphological outlook over the period of 180 days from the date of synthesis [45,54–67].

6.5.3 Ti-HA bio-composite foams reinforced with hydroxyapatite (HA)

Shbeh et al. created the bio-composite by reinforcing it with hydroxyapatite (HA). Fig. 6.12A and B represents SEM photographs of Ti-HA composite foams after mechanical testing. Ti foams

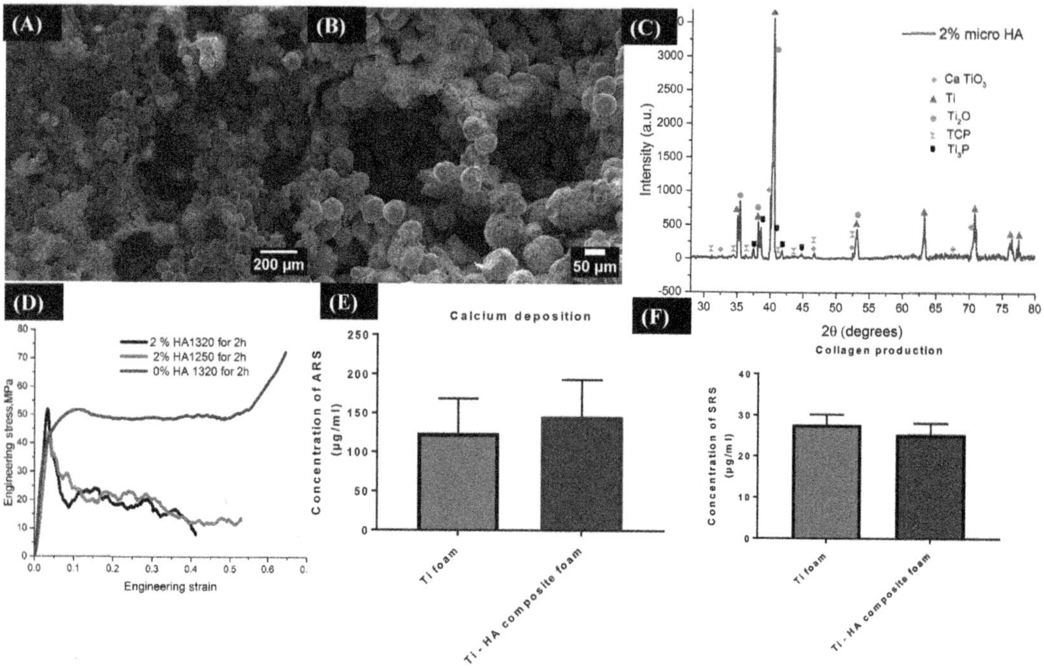

FIGURE 6.12

(A and B) SEM images of Ti-HA composite foams containing 7% micro-HA after sintering (C) XRD image of Ti-HA composite foam containing 2% micro-HA sintered at 1320 degrees for 2 h (D) Compression test results for Ti-HA composite foam with 2% micro-HA Mean SD for (E) MLO-A5 cell viability on Ti foam and Ti-HA composite foams after 28 days (F) calcium staining [71].

Reproduced with permission from Elsevier. Copyright 2019.

showed some plastic deformation [68]. Fig. 6.12C represents brittle behavior of Ti-HA composite by XRD analysis. The crystal phases noticed in the specimen with the 2% micro-HA were Ti, $CaTiO_3$, Ti_2O, TCP, and Ti_3P which are similar to the crystal phases found in Ti-HA dense composites as stated in the literature. Balbinotti et al. [69] stated that phases containing Ca are partially detached during polishing of Ti-HA composites, representing the flaw of these phases to support the load. Calcium titanate is a biocompatible material used to stimulate apatite creation and increases the HA electrical conductivity and equal polarizability like natural bone. Also Ti_2O promotes and induces the nucleation and creation of apatite in Ti-HA dense composites after 2 hours soaking in simulated body fluid [70].

Fig. 6.12D exhibits the same strain rate and the same result as a foam created with 52% KCl without HA. The yield strength of Ti-HA composite foams sintered at 1320°C for 2 hours with 2% micro-HA is more than that of composites sintered with 7% micro-HA and greater than that of foams prepared without HA. The volume percentages of porosity for samples manufactured with 2% micro-HA and without HA were essentially the same, that is, 621.2%, with an average pore size of 348 m and a roundness of approximately 0.73. This conclusion is consistent with the findings in the literature, which show that the progressive addition of HA causes a slight increase in porosity. The specimens' porosity was reduced by up to 29% [72].

Fig. 6.12E depicts MLO-A5 cell viability on Ti foam and Ti-HA composite foams after 28 days. Viability improved significantly from day 4 to day 7, with similar values reported on samples with and without HA, indicating strong biocompatibility with good cell proliferation in both samples. Collagen staining on day 28 of cell culture is depicted in Fig. 6.12F. However, the differences between the samples were not statistically significant. The presence of HA in conjunction with Ti is said to promote bone growth, increased alkaline phosphatase activity, and collagen production [73,74].

6.6 Applications of various bio-composite materials

Bio-composites are materials made up of a matrix (resin) reinforced with natural fibers such as newspaper and magazine paper fibers, as well as various nonwood and wood fibers. These composite materials are superior in terms of mechanical, thermal, and biodegradability. Because of the exceptional properties demonstrated in Fig. 6.13, composite materials have a wide range of applications, including automotive, military, aviation, marine, construction, and packing. Bio-composite materials are used as sensors, gas detection devices, and various industrial safety equipment in the food industry [75–77].

6.6.1 Biomedical applications

Today, biopolymer composites are widely used in the biomedical field, including tissue engineering and wound repair. According to [78], bones, blood vessels, and oral tissues can be created through tissue engineering using nanofibrous materials made of polymers. Wound dressings for a variety of novel bio-composites layers and hydrogels with controllable medication or active chemical release have been described. Varaprasad and colleagues discuss multiple studies on the most recent

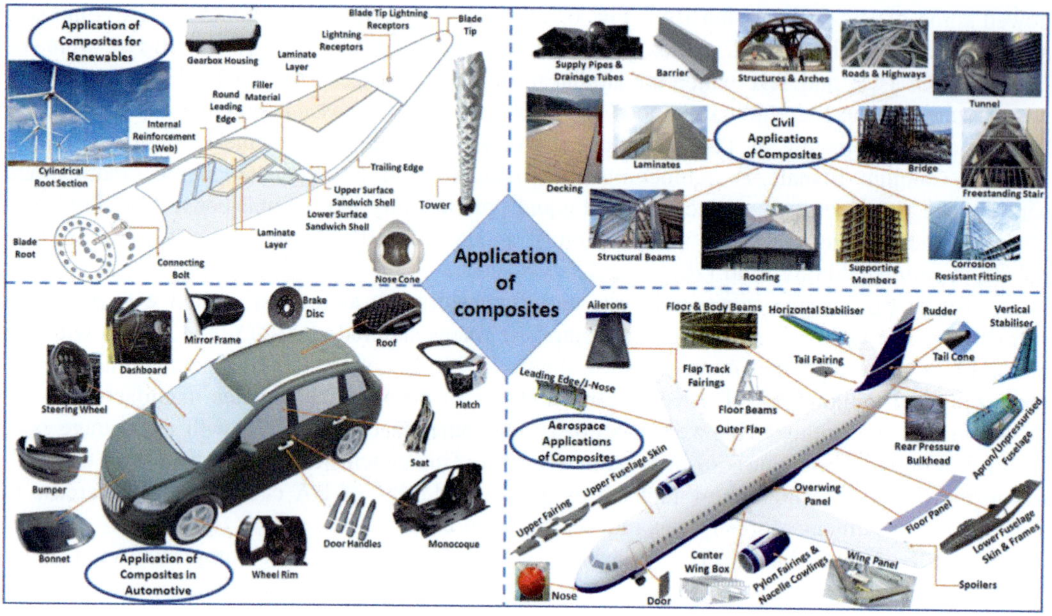

FIGURE 6.13

Application of bio-composite materials in different sectors [75].

Reproduced with permission from MDP. Copyright 2021.

alginate-based wound dressing substances in their review [79]. Alginates and their derivatives have a homeostatic quality, which allows them to be used to treat draining injuries, pressure ulcers, cutaneous, surgical, or dehisced wounds, as well as infectious and postoperative wounds. Hyaluronate-based wound dressings stand out due to their flexibility, biocompatibility, and bacteriostatic properties. Because human cells can cluster and form around the polymer biocompatibility, biocompatible and bio-composite scaffolds accelerate tissue integration.

Originally, the scaffold's sole purpose would have been to provide support for the cell, but scaffolds can now be supplied with biochemicals to promote cell growth. Tissue engineering considers the creation of synthetic tissue that is physiologically functioning, so technological advancements in this field are ongoing. In this regard, Gauvin et al. [80] emphasized that research on cell-seeded scaffolds had yielded novel chemicals and manufacturing procedures, resulting in well-engineered, biocompatible systems. Another technological innovation in tissue engineering is the use of nano-sized biomaterials, such as nanoparticles and nanocomposites, as well as organic–inorganic hybrid polymers, to produce scaffolds for organ regeneration. According to Iqbal et al. [81] and Khan et al. [82], scaffolds have been created using a variety of man-made and natural polymeric materials and their composite materials for guided tissue regeneration (GTR), bone and cartilage regeneration, controlled drug release, nerve regeneration, tooth structure regeneration, and guided tissue engineering. Electrospinning, foaming, and 3D printing have all been used to investigate bio-composites scaffolds, with PLA being the most popular bioplastic for these applications [83].

Regarding microfabrication technologies, according to Iqbal et al. [81], techniques like as lithography, bioprinting, micromolding, and photolithography were mentioned as being more prevalent and successful for generating biomaterials and tissue-engineered constructs by providing cells with a microenvironment that closely mimics their natural structure, that is, micro and nanotechnologies cannot only replicate complex cell-scale processes, but also allow for the formation of three-dimensional (3D) structures [38].

6.6.2 Bone cements

Bone performs vital functions in the body, such as protecting and maintaining vital organs. These are in charge of both mineral storage and the production of red blood cells (RBCs) and white blood cells (WBCs). This vital function emphasizes the importance of bone in humans (Homo sapiens). Many people have bone irregularities caused by trauma, metabolic abnormalities, or tumors, and bone replacement is highly recommended. As a result, there has been a lot of interest in the development of bone repair products. There are currently a variety of bone regeneration therapies available depending on the specific state.

Several products used in bone healing are discussed. Biomaterials were used to incorporate bone design and dynamics into the design of prosthetics, plates, bone transplantation, and bone cements. Bone grafting is a surgical treatment that is associated with the restoration of missing bone and is used for severe bone fractures that pose a serious risk to the patients or do not heal completely. Bone replacement grafts provide structural support for the development, maturation, and remodeling of clots, which results in the formation of new bone in bone defects [84].

6.6.3 Automobile industry

Despite the fact that they are rarely commercialized, they are useful in informing the industry about the potential of specific components and concepts. Because of innovative manufacturing techniques and technologies, the car weighs less than 15% less. The scientists predicted that the bio-fore car would serve as a model for future manufacturing and technological advancements in automobile production. Bio-composite materials was used in various automobile sectors shown in Figs. 6.14 and 6.15 [85].

6.6.4 Construction and textile industry

Bio-composites are used in the construction industry to make windows, doors, window frames, roofs, roof tiles, and floor mats. Use for load-bearing materials such as tanks, pipes, beams, and floor slabs. Bio-composites are also used in the restoration and repair of various structural components. Because of their superior thermal and acoustical properties, natural fiber composites are used as insulation and soundproofing solutions. When compared to other binders, hemp, lime, and concrete have demonstrated superior noise attenuation. Before choosing any bio-composite as a building material, life cycle assessment, endurance traits, and environmental factors must be considered. Low weight and mechanical properties comparable to synthetic composites are critical for building applications. Natural fiber composites have enormous potential in the textile industry for use in ropes, sacks, bags, and clothing. Furthermore, many countries are employing bio-composite

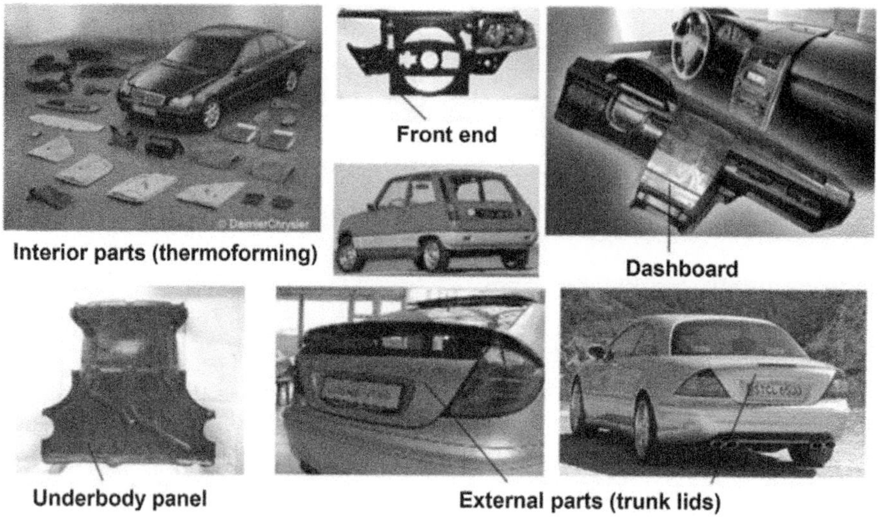

FIGURE 6.14

Parts of cars which are synthesized from bio-composites [86].

Reproduced with permission from Elsevier. Copyright 2021.

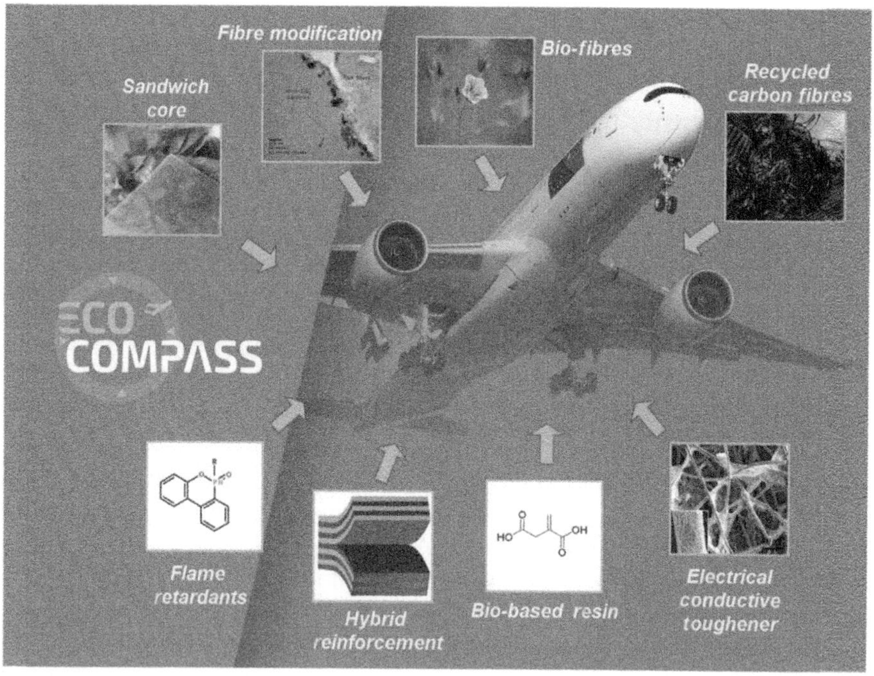

FIGURE 6.15

Bio-composites used in airplane [87].

Reproduced with permission from Elsevier. Copyright 2018.

materials to address environmental issues. Many companies are investing in bio-composites due to anticipated demand [88].

6.6.5 Marine applications

Excessive moisture absorption is an issue for bio-composites, which is controlled by vegetal fibers. It is critical in a marine ecosystem to keep bio-composites' natural fiber and matrix from deteriorating. A protective film of a comparable biopolymer on a bio-composite may improve the ability of the fiber and matrix interface to reduce weight gain. The mechanical and thermal properties of the bio-composites after soaking reveal that the covering layers increase the bio-composites' durability by preventing matrix hydrolysis [89]. DuPont's specially designed marine composite with Kevlar helps to provide the optimum blend of resilience, stiffness, and lightweight qualities for a variety of marine applications. This improves the greater surveillance and service boat acceleration that composites from renewable and sustainable materials can achieve by boosting engine power. These Kevlar-infused composites are not only more resistant to damage, lighter but also harder, and perform better when subjected to hydrodynamic fatigue stress. According to Davies' study of the ecological degradation of composites for marine structures, the use of composite materials in highly stacked marine parts, such as composite propellers or tidal turbine blades, is growing and necessitates a thorough understanding of stress and seawater. In contrast to the conceptual framework, there is a scarcity of experimental data, and the current method for completing a few investigations requires the use of specific test instruments. More research in this area is desperately needed [90].

6.7 Other novel applications for bio-based composites

Several novel solutions for bio-based composites and nanocomposites, based primarily on agricultural byproducts and green nanoparticles, have been documented. In addition, some new studies on the use of bio-based bio-composites and nanocomposites for wastewater treatment and contaminant removal have recently been published [38,91]. Finally, bio-based composites and nanocomposite components show promise in rigid and biodegradable electronic parts and power storage devices [92]. Thiangtham et al. created bio-composite membranes out of microcrystalline cellulose (MCC) isolated from sugarcane agricultural waste and mixed with PLA/PBS matrices [38]. Only a few bio-composites have been commercialized and investigated. A large number of bio-composites are currently being researched and developed. Innovative manufacturing processes and techniques are being developed to produce bio-composites at a lower cost. Bio-composites are most commonly used in nonstructural, nonload-bearing applications. Natural fibers are abundant in developing countries, but a lack of resources prevents them from being used in composite materials or the development of novel processing techniques, whereas developed nations in Europe and Asia are at the forefront of bio-composites development. Given the benefits of these bio-composites, a few issues such as cost reduction, dependable performance, and unsatisfactory mechanical qualities must be addressed before large-scale production can begin. Despite these challenges, bio-composites hold enormous promise for use in a wide range of applications. The goal is to achieve properties similar to those of conventional composites [88].

6.7.1 Sports goods

The use of bamboo composites in sports is well documented [93]. Reinforcing composites like this have been used to make surfboards, baseballs, and polo balls, among other things. Bamboo fibers are ideal for use in sportswear due to their softness, moisture permeability, and thermal properties. These items are lightweight, with distinctive designs and a waterproof surface. In multilayered boards, an epoxy matrix with a specialty over glass boards is used. This specialization allows for greater rigidity and mechanical duration.

6.7.2 Military

Materials such as nanocomposite and bio-composite have emerged as the subject of intense military research in recent years due to their exceptional mechanical properties, low cost, and lightweight. Because of the possibility of contact with harsh climatic conditions such as critical temperatures, exposure to ultraviolet light (UV), humidity, and alkaline environments, the performance of materials used for such applications must be carefully examined. Bulletproof shields for military use were created using Ramie single fiber and reinforced epoxy. Ramie fibers, which are mostly found in China, are made from naturally occurring sources and are both environmentally friendly and almost weightless. Numerous studies have been published on the use of polymers derived from starch to reinforce Ramie fibers in protective headgear. Composites manufactured in this manner have many advantages, one of which is that they are far lighter than composites prepared using conventional techniques, and the manufacturing technique is also less complicated. Marsyahyo [94] created bulletproof sheets by incorporating Ramie reinforcing fibers into an epoxy-based matrix; while materials were determined only to be acceptable for the restricted grade of bulletproof panels, it was significantly less expensive and lighter than synthetic fibers [95].

6.8 Conclusion and future aspects

The primary benefits provided by bio fiber-based polymeric materials are lightness, biodegradability, and environmental friendliness; these are ultimately the primary reasons for their popularity among producers and customers. With growing concern about finite petroleum resources, the environment, and the ecosystem, researchers and engineers are now faced with the challenge of developing environmentally friendly composite materials or bio-composite materials. This chapter delves into the specifics of several bio-composite materials and their manufacturing processes, as well as a variety of bio-composite material characterization methodologies. Bio-composites are already widely used in a broader range of industries, including automobiles, airplanes, and biomedicine, generating increased interest in their production. The majority of bio-composites research have shown that they have the potential to replace synthetic fiber-reinforced PC in a variety of technical applications.

Bio-composite materials are becoming more popular. Another industry that is expected to grow is wood-based fillers. This is because of the low cost and the ability to strengthen. In the automotive and construction industries, wood flour-filled composites are used in structural applications such as fencing, decking, outdoor furniture, window components, roofing items, door panels, and so on. Continuous advances in understanding of the chemistry related to the interaction of the reinforcement and matrix have paved the way for an increase in the application areas of various eco-friendly materials. The next challenges, however, will be in the processing, characterization, and mechanisms that control the behavior of these advanced materials. The most important factor in a composite material's performance is the contact between the reinforcing fiber/particulate and the matrix. The development of environmentally friendly and sustainable polymeric composite materials made from renewable resources, recyclable, and biodegradable bio-fillers has increased global awareness of environmental issues.

References

[1] B. Middleton, Composites: manufacture and application, Design and Manufacture of Plastic Components for Multifunctionality, Elsevier, 2016, pp. 53–101. Available from: http://doi.org/10.1016/B978-0-323-34061-8.00003-X.

[2] Composite Terms and Classifications | MATSE 81: Materials in Today's World. Accessed March 9, 2022. https://www.e-education.psu.edu/matse81/node/2214.

[3] D.K. Rajak, D.D. Pagar, R. Kumar, C.I. Pruncu, Recent progress of reinforcement materials: a comprehensive overview of composite materials, J. Mater. Res. Technol 8 (6) (2019) 6354–6374. Available from: https://doi.org/10.1016/j.jmrt.2019.09.068.

[4] M.M. Dawoud, M.H. Saleh, Introductory chapter: background on composite materials, in: M. El-Din, H. Saleh, M. Koller (Eds.), Characterizations of Some Composite Materials, IntechOpen, 2019. Available from: http://doi.org/10.5772/intechopen.80960.

[5] S.J. Christian, Natural fibre-reinforced noncementitious composites (biocomposites), Nonconventional and Vernacular Construction Materials, Elsevier, 2016, pp. 111–126. Available from: http://doi.org/10.1016/B978-0-08-100038-0.00005-6.

[6] Q. Zhang, H. Cai, W. Yi, et al., Biocomposites from organic solid wastes derived biochars: a review, Materials 13 (18) (2020) 3923. Available from: https://doi.org/10.3390/ma13183923.

[7] V. Fombuena, L. Bernardi, O. Fenollar, T. Boronat, R. Balart, Characterization of green composites from biobased epoxy matrices and bio-fillers derived from seashell wastes, Mater. Des 57 (2014) 168–174. Available from: https://doi.org/10.1016/j.matdes.2013.12.032.

[8] E. Chiellini, P. Cinelli, F. Chiellini, S.H. Imam, Environmentally degradable bio-based polymeric blends and composites, Macromol. Biosci. 4 (3) (2004) 218–231. Available from: https://doi.org/10.1002/mabi.200300126.

[9] M. Avella, A. Buzarovska, M. Errico, G. Gentile, A. Grozdanov, Eco-challenges of bio-based polymer composites, Materials 2 (3) (2009) 911–925. Available from: https://doi.org/10.3390/ma2030911.

[10] K. Fang, Encapsulation process technology, Encapsulation Technologies for Electronic Applications, Elsevier, 2019, pp. 123–181. Available from: http://doi.org/10.1016/B978-0-12-811978-5.00003-1.

[11] V. Shah, J. Bhaliya, G.M. Patel, K. Deshmukh, Advances in polymeric nanocomposites for automotive applications: a review, Polym. Adv. Techs 17 (2022). Available from: https://doi.org/10.1002/pat.5771. pat.5771.

[12] G.M. Patel, V. Shah, M. Vora, Synthesis techniques of shape-memory polymer composites, in: M.R. Maurya, K.K. Sadasivuni, J.J. Cabibihan, S. Ahmad, S. Kazim (Eds.), *Shape Memory* Composites Based on Polymers and Metals for 4D Printing, Springer International Publishing, 2022, pp. 115–153. Available from: http://doi.org/10.1007/978-3-030-94114-7_7.

[13] H. Abral, J. Ariksa, M. Mahardika, et al., Transparent and antimicrobial cellulose film from ginger nanofiber, Food Hydrocoll 98 (2020) 105266. Available from: https://doi.org/10.1016/j.foodhyd.2019.105266.

[14] R. Dewi, N. Sylvia, M. Riza, The effect of rice husk and saw dusk filler on mechanical property of bio composite from sago starch, Int. J. Eng. Sci. Inf. Technol. 1 (3) (2021) 98–103. Available from: https://doi.org/10.52088/ijesty.v1i3.113.

[15] A. Murawski, R. Diaz, S. Inglesby, K. Delabar, R.L. Quirino, Synthesis of bio-based polymer composites: fabrication, fillers, properties, and challenges, in: K.K. Sadasivuni, D. Ponnamma, M. Rajan, B. Ahmed, M.A.S.A. Al-Maadeed (Eds.), *Polymer Nanocomposites in Biomedical Engineering*. Lecture Notes in Bioengineering, Springer International Publishing, 2019, pp. 29–55. Available from: http://doi.org/10.1007/978-3-030-04741-2_2.

[16] V.S. Aigbodion, E.G. Okonkwo, E.T. Akinlabi, Eco-friendly polymer composite: state-of-arts, opportunities and challenge, in: Inamuddin, S. Thomas, R. Kumar Mishra, A.M. Asiri (Eds.), Sustainable Polymer Composites and Nanocomposites, Springer International Publishing, 2019, pp. 1233–1265. Available from: http://doi.org/10.1007/978-3-030-05399-4_42.

[17] S. Ramakrishna, Z.M. Huang, Biocomposites, Reference Module in Materials Science and Materials Engineering, Elsevier, 2016. B9780128035818010000. Available from: https://doi.org/10.1016/B978-0-12-803581-8.00965-6.

[18] A. Kumar, T. Jyske, V. Möttönen, Properties of injection molded biocomposites reinforced with wood particles of short-rotation aspen and willow, Polymers 12 (2) (2020) 257. Available from: https://doi.org/10.3390/polym12020257.

[19] S.V. Joshi, L.T. Drzal, A.K. Mohanty, S. Arora, Are natural fiber composites environmentally superior to glass fiber reinforced composites, Compos. Part. A Appl. Sci. Manuf. 35 (3) (2004) 371–376. Available from: https://doi.org/10.1016/j.compositesa.2003.09.016.

[20] G. Simon, E. Petinakis, L. Yu, K. De, Natural fibre bio-composites incorporating poly(lactic acid), in: M. Masuelli (Ed.), Fiber Reinforced Polymers - The Technology Applied for Concrete Repair, InTech, 2013. Available from: http://doi.org/10.5772/52253.

[21] A. Mousa, G. Heinrich, U. Wagenknecht, Bio-based fillers, in: S. Kobayashi, K. Müllen (Eds.), Encyclopedia of Polymeric Nanomaterials, Springer Berlin Heidelberg, 2014, pp. 1–4. Available from: http://doi.org/10.1007/978-3-642-36199-9_285-1.

[22] F. Versino, O.V. López, M.A. García, Green biocomposites for packaging applications, in: M.T. Hameed Sultan, M.S.A. Majid, M.R.M. Jamir, A.I. Azmi, N. Saba (Eds.), *Biocomposite Materials*. Composites Science and Technology, Springer Singapore, 2021, pp. 1–30. Available from: http://doi.org/10.1007/978-981-33-4091-6_1.

[23] V. Florencia, O.V. López, M.A. García, Exploitation of by-products from cassava and ahipa starch extraction as filler of thermoplastic corn starch, Compos. Part. B Eng 182 (2020) 107653. Available from: https://doi.org/10.1016/j.compositesb.2019.107653.

[24] Š. Hýsek, M. Podlena, H. Bartsch, C. Wenderdel, M. Böhm, Effect of wheat husk surface pre-treatment on the properties of husk-based composite materials, Ind. Crop. Products 125 (2018) 105–113. Available from: https://doi.org/10.1016/j.indcrop.2018.08.035.

[25] M. Bassyouni, S. WaheedUl Hasan, 13 - The use of rice straw and husk fibers as reinforcements in composites, in: O. Faruk, M. Sain (Eds.), Biofiber Reinforcements in Composite Materials, Woodhead Publishing, 2015, pp. 385–422. Available from: http://doi.org/10.1533/9781782421276.4.385.

[26] V. Guna, M. Ilangovan, C. Hu, K. Venkatesh, N. Reddy, Valorization of sugarcane bagasse by developing completely biodegradable composites for industrial applications, Ind. Crop. Products 131 (2019) 25–31. Available from: https://doi.org/10.1016/j.indcrop.2019.01.011.

[27] L.R.P.F. Mello, S. Mali, Use of malt bagasse to produce biodegradable baked foams made from cassava starch, Ind. Crop. Products 55 (2014) 187–193. Available from: https://doi.org/10.1016/j.indcrop.2014.02.015.

[28] A. El Barnossi, F. Moussaid, A. Iraqi Housseini, Tangerine, banana and pomegranate peels valorisation for sustainable environment: a review, Biotechnol. Rep 29 (2021) e00574. Available from: https://doi.org/10.1016/j.btre.2020.e00574.

[29] D.C.M. Ferreira, G. Molina, F.M. Pelissari, Biodegradable trays based on cassava starch blended with agroindustrial residues, Compos. Part. B Eng 183 (2020) 107682. Available from: https://doi.org/10.1016/j.compositesb.2019.107682.

[30] G.R. de Carvalho, G.S. Marques, L.M. de Matos Jorge, R.M.M. Jorge, Cassava bagasse as a reinforcement agent in the polymeric blend of biodegradable films, J. Appl. Polym. Sci 136 (12) (2019) 47224. Available from: https://doi.org/10.1002/app.47224.

[31] R. Jumaidin, M.A.A. Khiruddin, Z. Asyul Sutan Saidi, M.S. Salit, R.A. Ilyas, Effect of cogon grass fibre on the thermal, mechanical and biodegradation properties of thermoplastic cassava starch biocomposite, Int. J. Biol. Macromol. 146 (2020) 746–755. Available from: https://doi.org/10.1016/j.ijbiomac.2019.11.011.

[32] G. Oliveira, I. Gonçalves, A. Barra, C. Nunes, P. Ferreira, M.A. Coimbra, Coffee silverskin and starch-rich potato washing slurries as raw materials for elastic, antioxidant, and UV-protective biobased films, Food Res. Int 138 (2020) 109733. Available from: https://doi.org/10.1016/j.foodres.2020.109733.

[33] M. Liu, M. Arshadi, F. Javi, P. Lawrence, S.M. Davachi, A. Abbaspourrad, Green and facile preparation of hydrophobic bioplastics from tea waste, J. Clean. Prod 276 (2020) 123353. Available from: https://doi.org/10.1016/j.jclepro.2020.123353.

[34] K.R. Sumesh, K. Kanthavel, V. Kavimani, Peanut oil cake-derived cellulose fiber: extraction, application of mechanical and thermal properties in pineapple/flax natural fiber composites, Int. J. Biol. Macromol. 150 (2020) 775–785. Available from: https://doi.org/10.1016/j.ijbiomac.2020.02.118.

[35] F. Versino, O.V. López, M.A. García, Sunflower oil industry by-product as natural filler of biocomposite foams for packaging applications, J. Polym. Env 29 (6) (2021) 1869–1879. Available from: https://doi.org/10.1007/s10924-020-01981-8.

[36] A. Khosravi, A. Fereidoon, M.M. Khorasani, et al., Soft and hard sections from cellulose-reinforced poly(lactic acid)-based food packaging films: a critical review, Food Packaging Shelf Life 23 (2020) 100429. Available from: https://doi.org/10.1016/j.fpsl.2019.100429.

[37] G. Sanchez-Olivares, A. Sanchez-Solis, F. Calderas, J. Alongi, Keratin fibres derived from tannery industry wastes for flame retarded PLA composites, Polym. Degrad. Stab 140 (2017) 42–54. Available from: https://doi.org/10.1016/j.polymdegradstab.2017.04.011.

[38] F. Ortega, F. Versino, O.V. López, M.A. García, Biobased composites from agro-industrial wastes and by-products. emergent mater. Published online November 25, 2021. Available from: https://doi.org/10.1007/s42247-021-00319-x.

[39] B. Mirzaei, Y. Hamzeh, A. Ashori, Effects of waste paper sludge on the physico-mechanical properties of high density polyethylene/wood flour composites, J. Polym. Env 19 (1) (2011) 120–124. Available from: https://doi.org/10.1007/s10924-010-0255-3.

[40] N.M. Stark, Wood fiber derived from scrap pallets used in polypropylene composites, Forest Prod. J. (1999) 8.

[41] M. John, S. Thomas, Biofibres and biocomposites, Carbohydr. Polym 71 (3) (2008) 343–364. Available from: https://doi.org/10.1016/j.carbpol.2007.05.040.

[42] R. Payal, Reliable natural-fibre augmented biodegraded polymer composites, in: Inamuddin, S. Thomas, R. Kumar Mishra, A.M. Asiri (Eds.), Sustainable Polymer Composites and Nanocomposites, Springer International Publishing, 2019, pp. 961–975. Available from: http://doi.org/10.1007/978-3-030-05399-4_33.

[43] U.K. Komal, M.K. Lila, I. Singh, PLA/banana fiber based sustainable biocomposites: a manufacturing perspective, Compos. Part. B Eng 180 (2020) 107535. Available from: https://doi.org/10.1016/j.compositesb.2019.107535.

[44] A. Huang, X. Peng, L. Geng, et al., Electrospun poly (butylene succinate)/cellulose nanocrystals bio-nanocomposite scaffolds for tissue engineering: preparation, characterization and in vitro evaluation, Polym. Test 71 (2018) 101–109. Available from: https://doi.org/10.1016/j.polymertesting.2018.08.027.
[45] A.B. Perumal, P.S. Sellamuthu, R.B. Nambiar, E.R. Sadiku, Development of polyvinyl alcohol/chitosan bio-nanocomposite films reinforced with cellulose nanocrystals isolated from rice straw, Appl. Surf. Sci 449 (2018) 591–602. Available from: https://doi.org/10.1016/j.apsusc.2018.01.022.
[46] R.A.A. Rusdi, N.A. Halim, N.M. Nurazzi, et al., The effect of layering structures on mechanical and thermal properties of hybrid bacterial cellulose/Kevlar reinforced epoxy composites, Heliyon 8 (6) (2022) e09442. Available from: https://doi.org/10.1016/j.heliyon.2022.e09442.
[47] T. Nissilä, M. Hietala, K. Oksman, A method for preparing epoxy-cellulose nanofiber composites with an oriented structure, Compos. Part. A Appl. Sci. Manuf 125 (2019) 105515. Available from: https://doi.org/10.1016/j.compositesa.2019.105515.
[48] J.W. Park, T.H. Lee, J.H. Back, S.W. Jang, H.J. Kim, M. Skrifvars, Phenyl silane treatment and carding process to improve the mechanical, thermal, and water-absorption properties of regenerated cellulose lyocell/polylactic acid bio-composites, Compos. Part. B Eng 167 (2019) 387–395. Available from: https://doi.org/10.1016/j.compositesb.2019.02.064.
[49] P.K. Bajpai, F. Ahmad, V. Chaudhary, Processing and characterization of bio-composites, in: L.M.T. Martínez, O.V. Kharissova, B.I. Kharisov (Eds.), Handbook of Ecomaterials, Springer International Publishing, 2019, pp. 123–139. Available from: http://doi.org/10.1007/978-3-319-68255-6_98.
[50] A. Imgharn, H. ighnih, A. Hsini, et al., Synthesis and characterization of polyaniline-based biocomposites and their application for effective removal of Orange G dye using adsorption in dynamic regime, Chem. Phys. Lett 778 (2021) 138811. Available from: https://doi.org/10.1016/j.cplett.2021.138811.
[51] J.S. Cao, J.X. Lin, F. Fang, M.T. Zhang, Z.R. Hu, A new absorbent by modifying walnut shell for the removal of anionic dye: kinetic and thermodynamic studies, Bioresour. Technol 163 (2014) 199–205. Available from: https://doi.org/10.1016/j.biortech.2014.04.046.
[52] O. Koysuren, Improving ultraviolet light photocatalytic activity of polyaniline/silicon carbide composites by Fe-doping, J. Appl. Polym. Sci 137 (14) (2020) 48524. Available from: https://doi.org/10.1002/app.48524.
[53] A. Stavrinou, C.A. Aggelopoulos, C.D. Tsakiroglou, Exploring the adsorption mechanisms of cationic and anionic dyes onto agricultural waste peels of banana, cucumber and potato: adsorption kinetics and equilibrium isotherms as a tool, J. Environ. Chem. Eng 6 (6) (2018) 6958–6970. Available from: https://doi.org/10.1016/j.jece.2018.10.063.
[54] E. Fortunati, P. Benincasa, G.M. Balestra, et al., Revalorization of barley straw and husk as precursors for cellulose nanocrystals extraction and their effect on PVA_CH nanocomposites, Ind. Crop. Products 92 (2016) 201–217. Available from: https://doi.org/10.1016/j.indcrop.2016.07.047.
[55] E. Fortunati, D. Puglia, F. Luzi, C. Santulli, J.M. Kenny, L. Torre, Binary PVA bio-nanocomposites containing cellulose nanocrystals extracted from different natural sources: part I, Carbohydr. Polym 97 (2) (2013) 825–836. Available from: https://doi.org/10.1016/j.carbpol.2013.03.075.
[56] J. Bonilla, E. Fortunati, L. Atarés, A. Chiralt, J.M. Kenny, Physical, structural and antimicrobial properties of poly vinyl alcohol–chitosan biodegradable films, Food Hydrocoll 35 (2014) 463–470. Available from: https://doi.org/10.1016/j.foodhyd.2013.07.002.
[57] M. Roohani, Y. Habibi, N.M. Belgacem, G. Ebrahim, A.N. Karimi, A. Dufresne, Cellulose whiskers reinforced polyvinyl alcohol copolymers nanocomposites, Eur. Polym. J 44 (8) (2008) 2489–2498. Available from: https://doi.org/10.1016/j.eurpolymj.2008.05.024.
[58] S. Singh, K.K. Gaikwad, Y.S. Lee, Antimicrobial and antioxidant properties of polyvinyl alcohol bio composite films containing seaweed extracted cellulose nano-crystal and basil leaves extract, Int. J. Biol. Macromol. 107 (2018) 1879–1887. Available from: https://doi.org/10.1016/j.ijbiomac.2017.10.057.

[59] S. El-Sayed, K.H. Mahmoud, A.A. Fatah, A. Hassen, DSC, TGA and dielectric properties of carboxymethyl cellulose/polyvinyl alcohol blends, Phys. B Condens. Matter. 406 (21) (2011) 4068–4076. Available from: https://doi.org/10.1016/j.physb.2011.07.050.

[60] M. El Achaby, N. El Miri, A. Aboulkas, et al., Processing and properties of eco-friendly bio-nanocomposite films filled with cellulose nanocrystals from sugarcane bagasse, Int. J. Biol. Macromolecules 96 (2017) 340–352. Available from: https://doi.org/10.1016/j.ijbiomac.2016.12.040.

[61] K. Lewandowska, Miscibility and thermal stability of poly(vinyl alcohol)/chitosan mixtures, Thermochim. Acta 493 (1-2) (2009) 42–48. Available from: https://doi.org/10.1016/j.tca.2009.04.003.

[62] Z. Shen, D.P. Kamdem, Development and characterization of biodegradable chitosan films containing two essential oils, Int. J. Biol. Macromol. 74 (2015) 289–296. Available from: https://doi.org/10.1016/j.ijbiomac.2014.11.046.

[63] K. Xu, C. Liu, K. Kang, et al., Isolation of nanocrystalline cellulose from rice straw and preparation of its biocomposites with chitosan: physicochemical characterization and evaluation of interfacial compatibility, Compos. Sci. Technol 154 (2018) 8–17. Available from: https://doi.org/10.1016/j.compscitech.2017.10.022.

[64] F.S. Kittur, K.V. Harish Prashanth, K. Udaya Sankar, R.N. Tharanathan, Characterization of chitin, chitosan and their carboxymethyl derivatives by differential scanning calorimetry, Carbohydr. Polym 49 (2) (2002) 185–193. Available from: https://doi.org/10.1016/S0144-8617(01)00320-4.

[65] S.B. Bahrami, S. SalahshoorKordestani, H. Mirzadeh, P. Mansoori, Poly(vinyl alcohol) - chitosan blends: preparation, mechanical and physical properties, Iran. Polym. J. (Engl. Ed.) 12 (2003) 139–146.

[66] N. El Miri, K. Abdelouahdi, M. Zahouily, et al., Bio-nanocomposite films based on cellulose nanocrystals filled polyvinyl alcohol/chitosan polymer blend, J. Appl. Polym. Sci 132 (22) (2015). Available from: https://doi.org/10.1002/app.42004.

[67] F. Luzi, E. Fortunati, G. Giovanale, A. Mazzaglia, L. Torre, G.M. Balestra, Cellulose nanocrystals from Actinidia deliciosa pruning residues combined with carvacrol in PVA_CH films with antioxidant/antimicrobial properties for packaging applications, Int. J. BiolMacromol 104 (Pt A) (2017) 43–55. Available from: https://doi.org/10.1016/j.ijbiomac.2017.05.176.

[68] M.M. Shbeh, R. Goodall, Single- and multi-layered porous titanium via metal injection moulding, Adv. Mater. Lett 8 (4) (2017) 500–505. Available from: https://doi.org/10.5185/amlett.2017.7052.

[69] P. Balbinotti, E. Gemelli, G. Buerger, et al., Microstructure development on sintered Ti/HA biocomposites produced by powder metallurgy, Mat. Res 14 (2011) 384–393. Available from: https://doi.org/10.1590/S1516-14392011005000044.

[70] C.Q. Ning, Y. Zhou, In vitro bioactivity of a biocomposite fabricated from HA and Ti powders by powder metallurgy method, Biomaterials 23 (14) (2002) 2909–2915. Available from: https://doi.org/10.1016/S0142-9612(01)00419-7.

[71] M. Shbeh, Z.J. Wally, M. Elbadawi, et al., Incorporation of HA into porous titanium to form Ti-HA biocomposite foams, J. Mech. Behav. Biomed. Mater 96 (2019) 193–203. Available from: https://doi.org/10.1016/j.jmbbm.2019.04.043.

[72] L. Zhang, Z.Y. He, Y.Q. Zhang, Y.H. Jiang, R. Zhou, Enhanced in vitro bioactivity of porous NiTi–HA composites with interconnected pore characteristics prepared by spark plasma sintering, Mater. & Des 101 (2016) 170–180. Available from: https://doi.org/10.1016/j.matdes.2016.03.128.

[73] P.A. Ramires, A. Romito, F. Cosentino, E. Milella, The influence of titania/hydroxyapatite composite coatings on in vitro osteoblasts behaviour, Biomaterials 22 (12) (2001) 1467–1474. Available from: https://doi.org/10.1016/S0142-9612(00)00269-6.

[74] T.L. Arinzeh, T. Tran, J. Mcalary, G. Daculsi, A comparative study of biphasic calcium phosphate ceramics for human mesenchymal stem-cell-induced bone formation, Biomaterials 26 (17) (2005) 3631–3638. Available from: https://doi.org/10.1016/j.biomaterials.2004.09.035.

[75] R. Gupta, D. Mitchell, J. Blanche, et al., A review of sensing technologies for non-destructive evaluation of structural composite materials, J. Compos. Sci 5 (12) (2021) 319. Available from: https://doi.org/10.3390/jcs5120319.

[76] G. Patel, V. Pillai, P. Bhatt, S. Mohammad, Application of nanosensors in the food industry, Nanosensors for Smart Cities, Elsevier, 2020, pp. 355–368. Available from: http://doi.org/10.1016/B978-0-12-819870-4.00020-7.

[77] G. Patel, V. Pillai, M. Vora, Liquid phase exfoliation of two-dimensional materials for sensors and photocatalysis—a review, J. Nanosci. Nanotechnol. 19 (8) (2019) 5054–5073. Available from: https://doi.org/10.1166/jnn.2019.16933.

[78] A. Al-Enizi, M. Zagho, A. Elzatahry, Polymer-based electrospun nanofibers for biomedical applications, Nanomaterials 8 (4) (2018) 259. Available from: https://doi.org/10.3390/nano8040259.

[79] K. Varaprasad, T. Jayaramudu, V. Kanikireddy, C. Toro, E.R. Sadiku, Alginate-based composite materials for wound dressing application: a mini review, Carbohydr. Polym 236 (2020) 116025. Available from: https://doi.org/10.1016/j.carbpol.2020.116025.

[80] R. Gauvin, Y.C. Chen, J.W. Lee, et al., Microfabrication of complex porous tissue engineering scaffolds using 3D projection stereolithography, Biomaterials 33 (15) (2012) 3824–3834. Available from: https://doi.org/10.1016/j.biomaterials.2012.01.048.

[81] N. Iqbal, A.S. Khan, A. Asif, M. Yar, J.W. Haycock, I.U. Rehman, Recent concepts in biodegradable polymers for tissue engineering paradigms: a critical review, Int. Mater. Rev 64 (2) (2019) 91–126. Available from: https://doi.org/10.1080/09506608.2018.1460943.

[82] S. Khan, M. Ul-Islam, M. Ikram, et al., Preparation and structural characterization of surface modified microporous bacterial cellulose scaffolds: a potential material for skin regeneration applications in vitro and in vivo, Int. J. Biol. Macromol. 117 (2018) 1200–1210. Available from: https://doi.org/10.1016/j.ijbiomac.2018.06.044.

[83] L. Dejob, B. Toury, S. Tadier, L. Grémillard, C. Gaillard, V. Salles, Electrospinning of in situ synthesized silica-based and calcium phosphate bioceramics for applications in bone tissue engineering: a review, Acta Biomater. 123 (2021) 123–153. Available from: https://doi.org/10.1016/j.actbio.2020.12.032.

[84] A. Nauth, J. Lane, J.T. Watson, P. Giannoudis, Bone graft substitution and augmentation, J. Orthopaedic Trauma 29 (12) (2015) S34–S38. Available from: https://doi.org/10.1097/BOT.0000000000000464.

[85] O. Akampumuza, M. WambuaPaul, A. Ahmed, W. Li, X.H. Qin, Review of the applications of biocomposites in the automotive industry, Polym. Compos 38 (11) (2017) 2553–2569. Available from: https://doi.org/10.1002/pc.23847.

[86] A. Muhammad, M.d.R. Rahman, R. Baini, M.K. Bin Bakri, Applications of sustainable polymer composites in automobile and aerospace industry, Advances in Sustainable Polymer Composites, Elsevier, 2021, pp. 185–207. Available from: http://doi.org/10.1016/B978-0-12-820338-5.00008-4.

[87] J. Bachmann, X. Yi, H. Gong, et al., Outlook on ecologically improved composites for aviation interior and secondary structures, CEAS Aeronaut. J 9 (3) (2018) 533–543. Available from: https://doi.org/10.1007/s13272-018-0298-z.

[88] M. Zwawi, A review on natural fiber bio-composites, surface modifications and applications, Molecules 26 (2) (2021) E404. Available from: https://doi.org/10.3390/molecules26020404.

[89] A. Le Duigou, J.M. Deux, P. Davies, C. Baley, Protection of flax/PLLA biocomposites from seawater ageing by external layers of PLLA, Int. J. Polym. Sci 2011 (2011) 1–8. Available from: https://doi.org/10.1155/2011/235805.

[90] T.R.K. Reddy, H.J. Kim, J.W. Park, Renewable biocomposite properties and their applications, in: M. Poletto (Ed.), Composites from Renewable and Sustainable Materials, InTech, 2016. Available from: http://doi.org/10.5772/65475.

[91] V. Javanbakht, R. Shafiei, Preparation and performance of alginate/basil seed mucilage biocomposite for removal of eriochrome black T dye from aqueous solution, Int. J. Biol. Macromol. 152 (2020) 990–1001. Available from: https://doi.org/10.1016/j.ijbiomac.2019.10.185.

[92] S. Thiangtham, J. Runt, N. Saito, H. Manuspiya, Fabrication of biocomposite membrane with microcrystalline cellulose (MCC) extracted from sugarcane bagasse by phase inversion method, Cellulose 27 (3) (2020) 1367–1384. Available from: https://doi.org/10.1007/s10570-019-02866-3.

[93] A. Subie, A. Mouritz, O. Troynikov, Sustainable design and environmental impact of materials in sports products, Sports Technol. 2 (3-4) (2009) 67–79. Available from: https://doi.org/10.1080/19346182.2009.9648504.

[94] E. Marsyahyo, Jamasri, H.S.B. Rochardjo, Soekrisno, Preliminary investigation on bulletproof panels made from ramie fiber reinforced composites for NIJ level II, IIA, and IV, J. Ind. Text. 39 (1) (2009) 13–26. Available from: https://doi.org/10.1177/1528083708098913.

[95] M.S. Ali, A.A. Al-Shukri, M.R. Maghami, C. Gomes, Nano and bio-composites and their applications: a review, IOP Conf. Ser. Mater. Sci. Eng. 1067 (1) (2021) 012093. Available from: https://doi.org/10.1088/1757-899X/1067/1/012093.

CHAPTER

7

Functionalized nanoparticles-based polymer nanocomposites: synthesis, characterizations, and biodegradability aspects

Fakhara Sabir[1], Hafsa Kanwal[2], Ushna Laraib[3] and E.R. Simge[4]

[1]Institute of Pharmaceutical Technology and Regularity Affairs, University of Szeged, Szeged, Hungary [2]Department of Pharmacy, Faculty of Biological Sciences, Quaid-i-Azam University, Islamabad, Pakistan [3]Department of Pharmacy, College of Pharmacy, University of Sargodha, Sargodha, Pakistan [4]Biochemistry Department, Institute of Science, Ege University, Izmir, Turkey

Chapter Outline

Biodegradable and Biocompatible Polymer Nanocomposites. DOI: https://doi.org/10.1016/B978-0-323-91696-7.00017-9
Copyright © 2023 Elsevier Inc. All rights reserved.

7.1 Introduction

Polymer nanocomposites (PNCs) are important areas of current nanotechnology research, and the investigation field covers a wide range of topics. Examples include nanoelectronics, polymeric bionanomaterials, reinforced PNCs, nanocomposite-based drug delivery devices, and so on. PNCs use a novel range of fillers to improve polymer function and utility while retaining the manufacturing and processing flexibility that plastics, thermosets, and resins provide. PNCs, in particular, have demonstrated success in overcoming normally incompatible property combinations. Fillers that are nanoscopic provide advantages that result in improved performance for filled polymers and composites. The main reasons for these advantages are the smaller filler size and the resulting increase in area. The additive's size could be reduced by up to three orders of magnitude when compared to typical alternatives [1].

Nanocomposites are generally made up of single-walled carbon nanotubes (SWCNTs), double-walled carbon nanotube (DWCNT), multiwalled carbon nanotube (MWCNT), graphite nanoplatelets (GnPs), carbon nanofibers (CNFs), montmorillonite (MMT), polyhedral oligomeric silesquioxanes and nanoclay and contain at least one phase with a particle size of 1–100 nm [2–4]. However, silicon dioxide (SiO_2), aluminum oxide (Al_2O_3), titanium dioxide (TiO_2) and silica nanoparticles are less frequently used in nanocomposites [5,6]. Improved mechanical properties; the inclusion of nanoparticles with small sizes and low loading properties in typical fiber-reinforced nanocomposites increases fracture, stiffness, toughness, and shear strength [4,7,8]. When compared to stainless steel and chopped carbon fibers, carbon nanotubes (CNTs), CNFs, and GnPs demonstrated good electrical conductivity in nanocomposite components at lower encapsulating levels [8,9]. Good flame retardancy [4]; manufacturing procedures for polymer matrix composites (PMCs) developed by the use of CNFs, CNTs, GnPs, nanoclay, will improve char production, boost char toughness, minimize the polymer matrix's flammability, and promote heat transfer away from the combustion chamber [9,10]. Because of their low loading and small size, nanoparticles have less of an impact on part surface quality. Because nanoparticles have a smaller mean diameter than other particulates, they have less warpage than additive materials and are less susceptible to shear [4,9,11]. Because of the random dispersion, part warpage is less likely. Various synthetic polymer materials, such as plastics, fibers, and synthetic rubber, were developed in the previous century and are now widely used in a variety of fields, including packing, construction, hospital, and agricultural equipment [9,12]. These synthetic polymer materials are extremely important in general. As a result, developing and employing biodegradable polymers is the most comprehensive approach to resolving this quandary. With this backdrop, biodegradable polymer production is quite an issue for far too long. Number of wide range of biodegradable polymers have been developed [9,13,14], and based on their sources, polymers of biodegradable nature are grouped into three classes: (1) synthetic polymers, including aliphatic polyesters like polylactic acid (PLA) [15,16], polycaprolactone (PCL) [14,17], poly-p-dixanone (PPDO) [16], poly-butylene succinate (PBS) [18,19] and poly (ethylene succinate) (PES) [20,21]; (2) polyesters made by microbes, which refer to several forms of poly-hydroxyalkanoates, as poly-3-hydroxybutyrate-*co*-3-hydroxyvalerate (PHBV), polyhydroxybutyrate (PHB), (3) starch, cellulose, chitin, chitosan, lignin, and proteins are examples of polymers derived from natural resources [22–24]. Biodegradable polymers have advanced at a breakneck pace, and current work strategies in this field are well developed. However, they are still a long way from replacing older nonbiodegradable polymers. The main reason for this is that these materials have unfavorable properties such as low mechanical strength, excessive hydrophilicity, and poor storage stability. There are two major methods for producing polymer nanoparticles [25,26]. The most common method for creating polymer/nanoparticle composites is to incorporate nanoscale particles into a polymer matrix. Alternately, polymer materials can be modified at the nanoscale [26,27]. Fig. 7.1 represent the classification of biodegradable polymers applied for the fabrication of PNCs. We discussed biodegradable nanocomposites, their synthesis, functionalization, biodegradability aspects of the functionalized nanoparticles in the properties of the nanocomposites, and their characterization in this chapter.

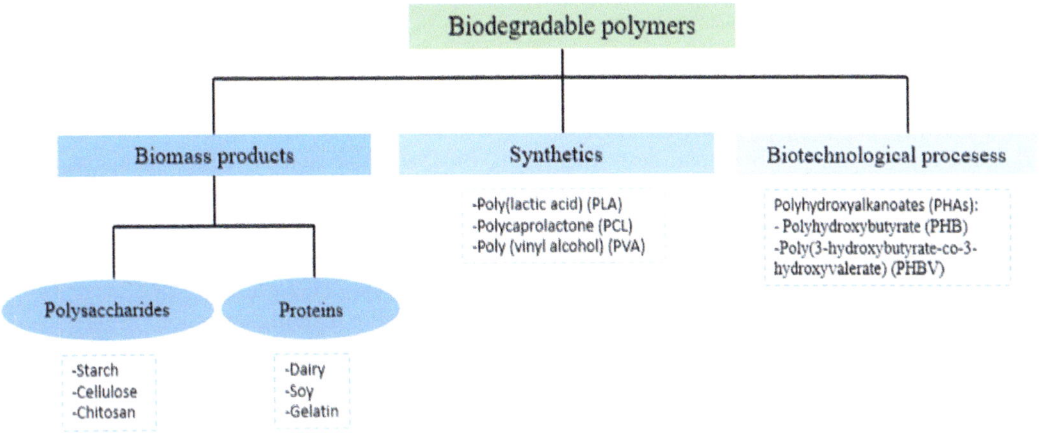

FIGURE 7.1

Categorization of biodegradable polymers [28].

Adapted from MDPI foods. https://www.mdpi.com/2304-8158/10/6/1256/htm. Copyright 2021.

7.2 Nonbiodegradable and biodegradable PNCs

7.2.1 Nonbiodegradable nanocomposites

The vast majority of commercial polymers on the market today are made from nonbiodegradable materials. These polymers were created using nonrenewable resources such as petroleum. When they reach the end of their useful life, they become difficult to use and accumulate as waste, causing severe environmental damage [29,30]. Each year, nearly 15 million and 30 million tons of nonbiodegradable polymeric waste are generated in the United States and Western Europe, respectively. The vast majority of plastics are usually discarded in junkyards, where they remain undisturbed for decades without degrading. Polymers, textiles, and other chemicals comprise approximately 6%–7% of all petroleum products [30,31]. Petroleum products are currently consumed 100,000 times faster than other bio-based products derived from natural sources. At the current rate of consumption, current petroleum reserves are expected to last only 50–60 years. In recent years, these circumstances have prompted the development of natural biodegradable-based polymeric nanocomposites [30]. Polymers that are not biodegradable, such as polypropylene (PP), polystyrene, polyethylene, and polymethylmethacrylate, make recycling and reusing difficult [32]. As a result, products made from nonbiodegradable materials have been manufactured on a large scale. Polymers and their derivatives are also used to make composites, which are then discarded. The use of biodegradable polymers in polymer composite technology is exciting because most biodegradable polymers come from other renewable sources [27]; nevertheless, substituting commercial synthetic polymers that endure some significant challenges [33]. Brittleness, limited and low stability, and poor barrier properties are just a few examples of common issues [32].

7.2.2 Biodegradable polymeric nanocomposites

PNCs with nanoscale filler dispersion have been generated with varying degrees of success using the major commercial polymer matrices. As a result, biodegradable and bio-based nanocomposites with improved properties and the addition of various nanoscale elements have been used and manufactured [3,6]. This system incorporates biopolymers derived from biotechnology resources (poly (hydroxyalkanoates), poly(lactic acid), agricultural origin (proteins and carbohydrates), and petrochemical sources (PCL)). Organic and inorganic nanofillers of various types were used. Nanostructures include carbon nanostructures, nanohydroxyapatite, nanolignin (NL), metallic nanoparticles, and nanocellulose [10]. (1) Silver nanoparticles (AgNPs), (2) SWCNTs (3) CNFs (4) NHA (nanohydroxyapatite) (5) CNCs (cellulose nanocrystals) (6) and lignin nanoparticles are the principle nanofillers employed in the study shown in Fig. 7.2 [34].

In the bacterial composites (BC) network, polyvinyl alcohol (PVA) penetration and pore filling are excellent. When hot pressed, this allows for the creation of consistent composite structures with excellent properties. Previously, BC-polyethylene glycol (PEG) composites with similar surface topography were reported. The formation of H-bonds (hydrogen bonds) between the hydroxyl groups (OH groups) of BC and PEG and PVA can cause structural changes by displacing the water surrounding the poly-glucosan chains [11,35]. Fig. 7.3A and B are scanning electron microscope (SEM) illustration of freeze-dried BC and freeze-dried BC (50%) PVA nanocomposites, respectively [30]. The BC network has excellent PVA penetration and pore filling. When hot pressed, this provides an excellent opportunity to create consistent composite structures with exceptional properties. Previously, BC-PEG composites with similar surface topography were reported [36,37]. Aside from synthetic polymers, natural polymers that are renewable and biodegradable play an important role. Starch, cellulose, chitin, gelatin, chitosan, and protein are common examples [38].

7.2.3 Natural biodegradable nanocomposites

7.2.3.1 Starch nanocomposites

Starch is a low-cost, easily accessible natural polymer. It can be converted into thermoplastic under certain conditions and in the presence of a plasticizer. However, its low water resistance, combined with other mechanical properties, limits its applications. Copolymerization and blending techniques

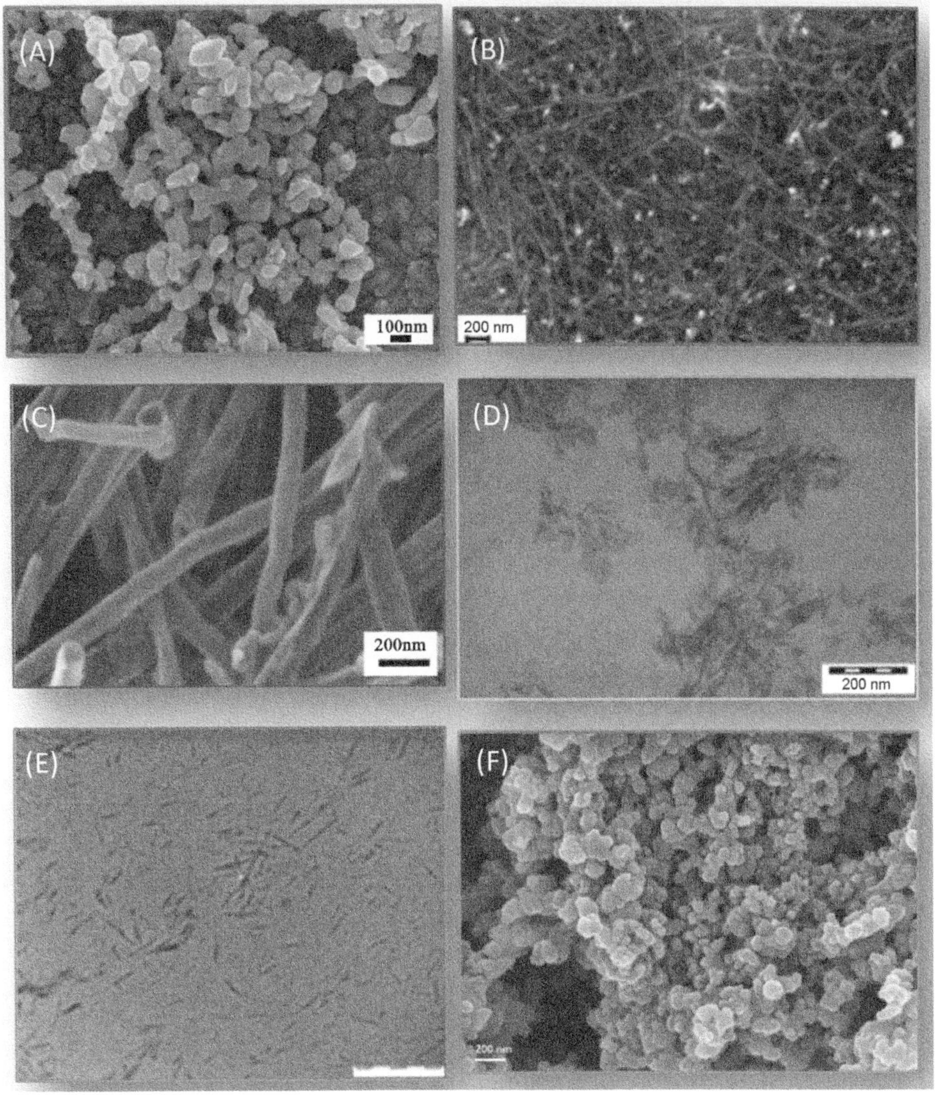

FIGURE 7.2

(A) AgNPs, (B) SWCNTs, (C) CNFs, (D) NHA, (E) CNCs and (F) lignin nanoparticles are the most commonly utilized nanofillers [34].

Adapted from MDPI materials. https://www.mdpi.com/1996-1944/11/5/795. Copyright 2018.

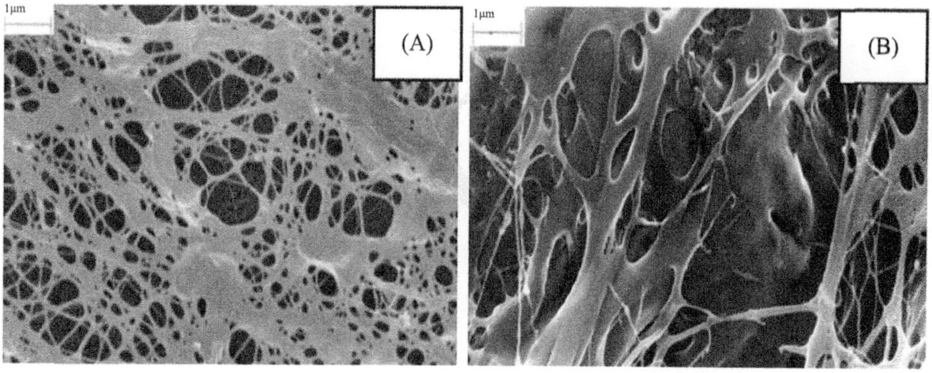

FIGURE 7.3

Illustration (A) SEM micrograph of freeze-dried BC (B) freeze-dried BC-PVA nanocomposites [30]. https://www.tandfonline.com/doi/abs/10.1080/15583724.2014.896018?tab = permissions&scroll = top.

have been used to alter its properties. Recently, the production of starch nanocomposites has been highlighted as a cost-effective way to improve barrier and mechanical properties. Because of its high efficiency and low-cost, layered silicate is a material of choice for incorporation into thermoplastic starch (TPS) [39,40].

Chen and colleagues created a series of TPS/clay composites with varying clays and clay loads. They discovered that hectorite, MMT, and natural smectite clay readily formed TPS nanocomposites. Characterization results showed partial exfoliation in the case of untreated hectorite and conventional composites in the case of TPS-treated hectorite. At high clay loadings, MMT and glycerol plasticized starch also interacted. When clay was present, the elastic modulus of TPS increased in all cases. The elastic moduli of nanocomposites were higher than those of typical nanocomposites. Instead of blowing, a casting technique was used to create starch/clay nanocomposites for food packaging. M. Avella and colleagues demonstrated this in their research work [41].

7.2.3.2 Cellulose nanocomposites

Cellulose has gained popularity as a feedstock for the development of biodegradable materials capable of replacing petroleum-based polymers. It is a natural resource that is plentiful. Gindl et al. created cellulose nanocomposite by combining the advantages of nanofiber reinforcement for optimal properties. Although raw cellulose processing is difficult due to its lack of thermoplasticity and insufficient solubility, esterification of raw cellulose produces derivatives that exhibit CAB (cellulose butyrate), CA (cellulose acetate), CAB (cellulose butyrate), and cellulose propionate as thermoplastic materials (CAP). CA has gained popularity among these derivatives due to its superior optimal clarity and increased toughness. The addition of a plasticizer to CA solves the problem of its melting point range being closer to its decomposition temperature [42,43]. Researchers could successfully construct CA nanocomposites using triethyl citrate (TEC) as a plasticizer in combination with Cloisite 30B (organoclay) and as a compatibilizer maleic anhydride-grafted cellulose acetate butyrate (CAB-g-MA). In contrast to nanocomposites without any compatibilizer, atomic force microscopy of these hybrid nanocomposites revealed that nanocomposites with 5 wt.

% of compatibilizer exhibited good exfoliation. This, combined with clay matrix dispersion, resulted in improved mechanical properties [43,44].

7.2.3.3 Chitosan nanocomposites

Chitin, after cellulose, is one of the most abundant bio-derived renewable resources. Chitosan is extracted after chitin has been deacetylated. It is widely used as a drug carrier and biomedical material due to its nontoxic and biocompatible properties. Chitosan is soluble in aqueous solutions of various acids, but it cannot form micelles in water. The presence of reactive amino and OH groups allows for modification of this material via graft reactions and ionic interactions [45,46]. Chitosan composites containing MMT, hydroxyapatite, and magnetite have been developed for use in medicine. Shen et al., for example, used in situ hybridization to create chitosan nanocomposite rods from hydroxyapatite. This technique also overcomes the problem of nanoscale particle aggregation in the polymer matrix. The mechanical properties of these composite rods improved significantly after characterization. These rods have characteristics that allow them to be 2–3 times stronger than some bone substitutes. As a result, these nanocomposite rods could be used in the internal fixation of bone fractures [47,48].

7.2.3.4 Plant oil nanocomposites

Plant oils and proteins, both of which are natural resources, have been used to create novel green materials. For the synthesis of green polymeric materials from plant oil, new approaches such as nanocomposite synthesis have emerged. For example, in acidic conditions, 3-GPTMS (3-glycidoxypropyltrimethoxysilane) was combined with epoxidized plant oil to produce plant oil/silica nanocomposites based transparent coatings. The incorporation of silica resulted in improved mechanical properties, flexibility, biodegradability, and hardness in nanocomposites [49,50].

7.2.3.5 Protein nanocomposites

Proteins can form amorphous three-dimensional structures. Noncovalent interactions add to the stability of these structures. Proteins are also being used to create new green materials. However, protein properties such as hydrophilicity, structural heterogeneity, and thermal stability have a significant impact on the functional properties of green materials. Proteins such as peanut protein, soy protein, corn zein, and wheat protein are thought to be good choices for environmentally friendly plastic materials [51,52]. Soy isolates are a popular commercial soy protein product among scientists due to their high protein content (90%). Soy protein-based plastics have low flexibility and tensile strength, which are typically compensated for by the use of reinforcement fillers and plasticizers. Fillers such as lignin, nanoclay, and chitin whisk have already been studied and demonstrated improved results [52].

Zhang et al. successfully prepared soy protein isolate nanocomposites using hydroxypropyl alkaline lignin (HPL). As a compatibilizer, glutaraldehyde was used. Following the synthesis of nanocomposites, they are compressed and molded into plastic sheets. Microstructural analysis revealed nanoscale dispersion of HPL, which significantly reinforced the soy protein isolate material [53]. The transition temperature (Tg) of produced sheets enhanced from 62.5°C to 70.4°C when HPL concentration rises from 0 to 6 wt.%. As compared to soy protein isolate sheets, the tensile strength of HPL-based protein soy isolate enhanced from 8.4 to 23.1 MPa when HPL and glutaraldehyde content was 6 and 3.3 wt.%, respectively [54,55].

7.2.4 Synthetic biodegradable nanocomposites of aliphatic polyesters

Because of their importance as biodegradable materials, nanocomposites with aliphatic polyesters have piqued the interest of researchers. Various nanocomposites are being developed using various nanoparticles, such as silver, gold, carbon nanotubes, layered silicates, magnetite nanoparticles, and so on. Significant research is being conducted on polyester/layered silicate nanocomposites, particularly PLA/layered silicates. In the following sections, we will look at nanocomposites made of various polymers and nanoparticles [56]. Among these polyesters are biopolymers derived from bio-derived monomers, such as PLA, and biopolymers derived from derived monomers and a few derivatives of Poly-hydroxyalkanoates (PHAs) derived from microorganisms, such as PLA. Graphene oxide (GO), oligomeric silsesquioxanes (POSS), CNTs, POSS, gold, silver, silica, TiO_2, and other organic and inorganic particles have been used in previous research [57,58].

7.2.4.1 PLA nanocomposites

PLA piqued the interest of most researchers due to its exceptional biodegradability, adaptability, and durability. It is possible to combine the PLA matrix with a variety of nanoparticles to improve the material's properties, particularly its thermal and mechanical functionality, and thus increase its usefulness [59–61]. Zhuang et al. used titanium dioxide (TiO_2) in various ratios to perform in situ polymerization for the development of (PLA)/TiO_2 nanocomposites formulations. TiO_2 nanoparticles with a greater aspect ratio have a variety of desirable features, including excellent photo catalytic activity, good magnetic capabilities, and enticing hydrophilic and antimicrobial activity [62]. It was analyzed that nano-TiO_2 < 3 wt.%, shows ideal properties of nanocomposites with increased dispersion in PLA mixture uniformly and stability is also improved. The solution degradation and UV irradiation and studies indicated that increasing the amount of TiO_2 improves nanocomposites deterioration. The addition of PLA/TiO_2-based nanocomposite having nanoparticles of TiO_2 resulted in significant bacteriostatic activity [63].

Nekhamanurak et al. analyzed the impact of adding calcium carbonate ($CaCO_3$) nanoparticles on PLA fracture behavior and mechanical properties. The nanoparticles can increase the elastic modulus as the SiO_2 contents on nanofiller enhance, the effect of nano-$CaCO_3$ that can cause the notched impact strength, elongation at break, elastic modulus enhancement illustrating the better nanoparticles interactions with coated PLA matrix nanocomposites [64].

Basilissi and his colleagues used the bulk ring opening polymerization technique (ROP) to create PLA-silica nanocomposites after treating nanosilica with two distinct oraganosilane agents. This step is required due to incompatibility between silica and PLA, which may result in material property loss. The loading capacity of silica samples layered with saline is increased, which may increase the crystalline property of nanocomposites. Thermal stability studies help to explain why PLA/silica nanocomposites are so stable [27,65]. Researchers investigated the antibacterial activity of a PLA-cellulose-Ag nanocomposite for culinary and hygienic applications. According to the findings, the substance has high stability and antibacterial efficacy against *Escherichia coli* and *Staphylococcus aureus* [66,67]. Because of their specific surface area and high conduction qualities, functionalized MWCNTs (multiwalled carbon nanotubes) can improve the thermal stability of poly-L-lactide (PLLA) [68]. The PLLA was coupled after in situ ROP with MWCNT-NH_2 and octanoate (act as stimulator) MWCNT-g-PLLA of LA (L-lactide oligomers). Crystallinity and thermal stability of nanocomposite is improved via increasing the amount of MWCNT-NH_2

(MWCNT-g-PLLA). For graphene diffusion, the melt blending synthesis process for PLA/graphene nanocomospite was chosen [69,70].

Pseudomonas aeruginosa was used to conduct biodegradation studies on polystyrene/poly(lactic acid) PS/PLA/organophilic montmorillonite (OMMT) and PS/PLA nanocomposites at various concentrations [71]. The degradation was accelerated by increasing the amount of OMMT. The increased rate of disintegration was attributed to OMMT's improved d-spacing, which made it easier for bacteria to secrete within the polymer chains. Furthermore, the MMT increased hydrophilicity as a result of increased d-spacing, allowing for easy bacterial growth inside the polymer matrix, and increased the rate of disintegration by absorbing water into the polymer chains [72].

7.2.4.2 Polycaprolactone nanocomposites

Polycaprolactone nanocomposites (PCL) is a widely used type of aliphatic polyester. Under ideal conditions, ROP of -caprolactone has generally resulted in its synthesis. This polyester has higher permeability, crystallinity, break elongation, flexibility, and a lower glass transition temperature than other polyesters [73]. There is still room for improvement before using it in various applications. PCL nanocomposites are created for this purpose by diffusing various nanoparticles. Liu and his colleagues investigated the impact of nano-$CaCO_3$ on the mechanical and thermal properties of the PCL matrix [74]. The addition of nano-$CaCO_3$ to the PCL matrix enhance its mechanical properties while also acting as a crystallization nucleating agent. The PCL/$CaCO_3$ nanocomposite was made via a chemical foaming process. With changing $CaCO_3$ loading amounts, there has been modification in the arrangement of cells in foams of nanocomposite [75]. With a rise in $CaCO_3$ concentration, thickness of cell wall increased, and mean size of cell reached its lowest point at a $CaCO_3$ concentration of 5 wt.%. The adjustment of cell structure can controlled the mechanical properties of PCL/$CaCO_3$ foams.

The tensile strength of concrete of PCL/$CaCO_3$ nanocomposites foams of equal density was also increased as the $CaCO_3$ concentration was raised [75]. Wu et al. investigated the degradability and compressive strength of PCL composites containing the phosphate-solubilizing bacteria Bacillus species PG01. The PG01 strain would disrupt PCL composites, and all cells are made up of PCL-loaded capsules [76].

7.2.4.3 Poly-p-dioxanone nanocomposites

Poly-p-dioxanone nanocomposites (PPDO) has better mechanical properties than PLA and PCL, as well as greater biocompatibility and biodegradability. It is one of the few biodegradable aliphatic polyesters with high stiffness and excellent fluidity [77]. Many researchers have proposed using PPDO to make laminate, adhesives, nonwoven materials, molded sheets, and coatings [27]. PPDO has a number of significant flaws, including a low crystallization rate, low melt strength, and hydrophobicity, all of which reduce its market demand significantly [57,78]. Furthermore, the melt strength of the polypropylene montmorillonite (PPDO)/MMT nanocomposites increased dramatically when compared to raw PPDO.

Because of the low crystallization and melting power, preparing a fine layer of nanocomposite using the blowing process is a difficult task. It was possible to create a biodegradable PPDO/MMT thin film with good mechanical properties by adding clay and blowing technique. CNTs are preferred for polymer material reinforcement because of their unique high aspect ratio, atomic structures, and excellent mechanical properties such as flexibility and strength [79].

7.2.4.4 Polybutyl succinate nanocomposites

Polybutyl succinate (PBS) is a biopolymer with unique properties such as high biodegradability, melt processability, and thermo-chemical resistance. Polycondensation of 1,4-butanediol and succinic acid is a common method for producing this aliphatic polyester. Twice functionalized organoclay (TFC) and C25A clay nanoparticles were added twice to PBS systems, resulting in nanocomposites with improved favorable properties, similar to other biodegradable polymers [80]. Ray and colleagues used melt intercalation to create a chain of PBS/OMLS nanocomposites using MMT modified with quaternary hexadecyl tri-*n*-butylphosphonium bromide and octadecyl ammonium chloride [27,81].

Chen and Evans developed a novel method for improving the connections between the PBS and organoclay, Cloisite 25A, by grafting epoxy groups on C25A in the presence of (glycidoxypropyl) trimethoxysilane to produce a twice functionalized clay (C25A) (TFC)> [82]. TFC was found to have higher nucleation activity for PBS crystallization than C25A.

Someya et al. created PBS/layered silicate nanocomposites using the melt-intercalation method. The results of the trials using the five different organo-modified MMTs and unmodified MMT were inconclusive; no significant changes in mechanical properties were identified during the trials [83].

Choi and his colleagues documented a variety of new biodegradable aliphatic polyesters biodegradable (BAP/OMMT) aliphatic polyesters developed from diols (ethylene glycol and 1,4-butanediol) adipic and succinic acid (dicarboxylix acids) as well as O-MMT by combining melt intercalation and solvent-casting processes. BAP/OMMT nanocomposites containing 6% OMMT were successfully prepared using the melt-intercalation process. It was discovered that adding OMMT to both composites increased their mechanical strength. According to the findings, the addition of OMMT has a positive effect on the thermo-physical properties [84,85].

7.2.4.5 PHAs nanocomposites

PHAs are biodegradable polymers that can be extracted from a wide range of microorganisms. PHAs have a number of drawbacks, such as brittleness and melt instability, which have a significant impact on their applicability in various industries. When PHB was substituted with poly 3-hydroxybutyrate-*co*-3-hydroxyvalerate (PHBV), these limits were partially restored. PHBV is well-known for its ability to replace traditional polymers in an environmentally beneficial manner. PHBV currently has a number of disadvantages, including its high cost, slow manufacturing rate, and higher crystallization rate. Researchers devised various approaches to address constraints, such as loading nanoparticles into the matrix to modify the properties of PHBV [86].

Choi et al. developed a melt-intercalation process for PHBV/MMT nanocomposites using Cloisite 30B as an organoclay. The intercalated structural shape was achieved, resulting in increased thermal stability and tensile strength [87]. The findings revealed that the nucleation of the organo-clay (nondispersed) was also responsible for increasing the temperature and rate of crystallization, which resulted in a change in tensile strength and thermal stability [27]. Wang et al. investigated the degradation of PHBV/OMMT nanocomposites [87]. They discovered that the biodegradability of PHBV/OMMT nanocomposites in soil suspension decreased as the amount of OMMT in the nanocomposites increased. They created PHBV/graphene nanocomposites using the solution casting method. Mechanical properties improved significantly in both dynamic and static tensile testing. According to differential thermal analysis (DTA) and soil degradation analysis, the application of equally dispersed graphene nanoparticles raises the temperature of PHBV without affecting biodegradability [88].

7.2.4.6 Miscellaneous nanocomposites

Aside from the aliphatic polyester nanocomposites mentioned above, which include a biodegradable polymer matrix, numerous efforts are being made to improve the biodegradability and other properties of synthetic polymers (from nonbiodegradable origin) [27,89].

7.3 Synthesis of functionalized nanoparticle-based polymeric nanocomposites

PNCs have been encapsulated into the polymer matrix in recent years to fabricate them [4]. While nanocomposites processing procedures are fundamentally diverse, they all must handle the following challenges that have a direct impact on the nanocomposite's properties: dispersion, alignment, and functionalization. PNCs are commonly processed by applying the following methods [90,91]:

Melt-mixing: Melt-mixing has been in use for nanoparticles and thermoplastic polymers [4] done by using extrusion, internal mixing, injection molding, and blow molding are examples of classic purifying procedures. These techniques are used because they are less time consuming, have greater coherence, and are more accessible. Another advantage of these processes is that no solvents are used, and the possibility of adulteration is low [92].

In situ polymerization: This technique is used to enhance the diffusion and dispersion among the phases. The focused benefit of this technique is the provision of strengthening effect even at molecular level [93].

Solution processing: This method of diffusion and distribution is beneficial because of providing the final form with low viscosity. Furthermore, pretreatment is required because the preparation of nanoparticles at the micro level is required for subsequent processing. Pretreatment is performed to remove contaminations, deagglomerate nanoparticles for proper distribution, and make it chemically active to improve interaction between nanoparticles and polymers [90].

7.3.1 Processing of solution

To begin, a polymer solution was used to facilitate nanoparticle diffusion [4]. This solution of nanoparticles and polymers is then mixed using a magnetic stirrer or another automated mixing method such as sonication or shear mixing. In sonication, two types of instruments are used: high power sonication, which uses horns or tips, and mild power sonication, which can be used in a water bath. Nanoparticles are typically not easily diffused into most solvents due to their high expect ratio. A surfactant added to nanoparticles can improve their dispersion properties. Once the nanoparticles are evenly distributed, the polymer solution is poured into the mold [94]. The solvent is then allowed to evaporate in order to form a thin film/layer of composite on the nanoparticles. The solution processing technique is commonly used for the preparation of films or sheets containing nanocomposites. The solvent used in this technique is determined by the properties of the polymers [95,96].

The solution processing technique was used by Jin et al. to disperse a large amount of MWCNTs in a polymer mixture. The results demonstrated MWCNTs' high entrapment capacity in a polymer solution [97]. In that study, MWCNTs were generated using an arc discharge process and diffused in chloroform using a sonication procedure. The polyhydroxyaminoether (PHAE) polymer matrix is sonicated in a previously prepared dispersion mixture, that is, MWCNT-chloroform solution [4]. The well-distributed mixture was placed in teflon and dried under the flame hood. This method is used for nanocomposites in which 50% of the total solution weight quantity of MWCNTs is required [98]. Another study's evidence explains the dispersion properties of nanocomposites that can be synthesized using a solution processing method. With the help of magnetic stirring, expanded graphite (EG) was embedded in a stiffened Bisphenol A diglycidyl ether (DGEBA) epoxy resin, and a stimulator was also added to the mixture by continuous stirring slowly followed by degasing [98,99]. The final product, a nanocomposite, was created by pouring the solution into a teflon mold. Sonication was used to ensure that the EG diffused into the acetone. Sonication was also used to add DGEBA epoxy resin. Following that, acetone was completely removed from the mixture by evaporation on a hot plate. The thickener and stimulator were also added, and the sample was then degassed, cast, and cured [100]. When nanocomposites prepared by these two processes were examined under a light microscope, the quality of nanocomposites synthesized through solution processing was higher than that of nanocomposites developed through direct mixing. For clay-containing nanocomposites, the solution processing method is widely used. Clay is composed primarily of MMT and silica. MMT is composed of tetrahedral silicate and octahedral alumina layers. The silicate layer contains hexagonal meshwork arranged by the bonding of SiO_4 groups and formed an array of subunits of Si_4O_{10}. While in octahedral layer, aluminum atoms are sandwiched between the hydroxyl and oxygen ions. The octahedral layer's oxygen is linked to the silicate in the other layer. Naturally, MMT clay have Na^+ ion. One sandwiched layer have width of about 0.96 nm and have a chemical formula Na1/3(Al5/3Mg/3) Si_4O_{10} $(OH)_2$ [4]. As the clay nature is hydrophilic, they are treated organically before the synthesis of nanocomposites. This is done by replacing the Na^+ cations with the required organic cations [101]. Different organically modified cations have been manufactures like Cloisite 20A and Inc. is also a natural MMT redesigned by addition of ammonium salt (quaternary). The product is designed to improve the reinforcing effect, linear thermal expansion coefficient, heat deflection temperature, and resistance properties of polymers [4]. Because of the weak bonding between these layers, nanoclays can be diffused as a thin layer with the addition of an adequate amount of polymer soluble solvent or any other aiding material. The polymer solution is then dispersed to fill the spaces in the silicate layers. Allowing the solvent to precipitate allows the silicate layers to rejoin, entrapping the polymer to form a multilayered assembly. There are two types of outcome structures: intercalated and exfoliated [102].

The intercalated structure was formed by a self-assembled, excellently arranged multilayered integrity. The spaces of 2–3 nm between the two consecutive silicate layers is filled with long polymer chain. When these spaces become too large, the interaction between the polymer cations and the silicate layer becomes weak, and the firm structure of nanocomposites fails to form. The disorganized, intercalated, or exfoliated summary can be examined using X-ray diffraction (XRD). The XRD results show similar peaks to those seen in natural nanoclays. However, the intercalated assembly peaks appeared earlier than the natural nanoclay [103,104]. Rao and Pochan investigated the physical and chemical properties of solution-processed clay-containing nanocomposites. Centrifuging and mixing were used to distribute 3 wt.% of Cloisite Na + (MMT clay) in water [4]. This method was chosen to improve material purification. Even after several months, the colloidal mixture remained stable. By using emulsion polymerization the copolymer latex (PBSMaSO$_3$) was formulated by mixing the liquid clay. After preparing the highly dispersed solution, it was transferred to the teflon mold and allowed to dry at room temperature [4]. It was scrapped and vacuum dried after the synthesis of nanocomposites in thin layer form. Again the PBSMaSO$_3$ quality clay based nanocomposite was examined through XRD and results interpreted if loading would be done with 4 wt.% clay then the exfoliated structure was formulated [4]. Moreover, increasing from > 4 wt.% clay will be resulted into an intercalated type of nanocomposites [105,106].

7.3.2 In situ polymerization process

In this process, the polymerization is triggered by heating or irradiation, as well as the addition of an appropriate stimulator, surfactant with organic catalytic properties. When polymerization is completed, the polymers arrange themselves around the nanoparticles based on their surface functionality. Thermally sensitive polymers are prepared using this method for nanocomposites. Various chain transfer polymerizations, such as radical, anionic, and ring opening, are used depending on the polymer molecular weight and distribution [4,107,108]. This method has the advantage of allowing for high nanoparticle loading capacity on polymers with good diffusion properties. Toyota Motor Corp used this method for the first time in the 1990s to create nanocomposites (clay/polyamide-6) [109]. Zeng and Lee then used the process to create poly (methyl methacrylate) and polystyrene/clay nanocomposites. It is certain that a less hydrophobic and more polar initiator-monomer combination was preferred to achieve good dispersion quality [4]. Furthermore, adding polymerization groups to the clay's surface would improve dispersion quality. With the clay contents of 5 wt.%, the PS/clay nanocomposites and exfoliated polymethyl methacrylate (PMMA) were generated successfully [4,110]. Kim et al. used the same method to create nanocomposites from MWCNTs and high-density polyethylene (HDPE). Begin the preparation by attaching a metallocene catalytic mix to the walls of the nanotubes. The polymer branches on the walls were then created using surface-initiated polymerization. The established nanotubes were evenly distributed and packed with polyethylene (PE) nanoparticles using a combination of existing analysis techniques: transmission electron microscopy (TEM) and SEM [4,111].

7.3.3 Melt process

When dealing with an insoluble polymer matrix, the melting process is the preferred method. This method is useful for creating nanocomposites from thermoplastic polymers such as HDPE, polyamide-6, polycarbonate, polypropylene (PP), and polystyrene. It could entrap large amounts of nanoparticles in a polymer mixture. This process includes a critical step in which thermoplastic polymers melt when heated and then regain their functional properties after cooling. Taking this into consideration, amorphous and semicrystalline polymers can be processed by melting [112].

In this process, the polymer pallets are softened first to produce a thick solution, and then nanoparticles are mixed into it using high shear mixing or extrusion. Nanocomposites are manufactured using methods such as injection molding, compression molding, or extrusion. The clinical step in this process is the addition of nanoparticles to the molten polymer because of the risk of degradation due to high shearing, so process optimization is required. Because this method is simple and takes less time, it can be used for large-scale industrial manufacturing. Polymers have the ability to integrate the interlayer gaps in clay particles when using this method to prepare clay-containing nanocomposites. The dispersion method capable of scraping away the layers (clay) [4]. Similar to the preceding method, the nature of synthesized nanocomposites (either intercalated or exfoliated) is determined by the affinity of clay and polymer matrix [112,113].

7.4 Nanocomposites functionalization

Agglomeration of dispersed particles occurs as Van der Wals forces increase. As a result, such materials must be introduced between CNTs and polymers to improve dispersion and interfacial interaction, thereby preventing agglomeration. This is known as functionalization. CNT functionalization can be accomplished through defect functionalization, covalent functionalization, or noncovalent functionalization [114].

7.4.1 Defect functionalization

The purification of CNTs is done by oxidative method to remove carbon from raw material [115]. Another study revealed that pure single-walled nanotubes (SWNTs) contain carbon in–COOH group form [116]. The oxidation of SWNTs is done in nanotube with 100–300 nm length [117]. Mawhinney et al. investigated the defect site density with the amount of released CO_2 (g) and CO (g) by heating at 1273K temperature. The findings revealed that almost 5% of carbon is present on the defect site [118]. The fraction of acidic sites in purified SWNTs was estimated to be around 1%–3% using the acid base titration. Unfortunately, the faulty sites formed at the CNT surfaces by this method are extremely rare, and the polymer/CNT composites exhibit poor dispersion. Furthermore, for organic group covalent bonding, the conversion of acid chlorides and reaction with amine can be used to synthesize amides. CNTs' organic phase solubility improved after functionalization. By combining the acidified chloride group with amines at elevated temperatures for four days, the SWNTs will be soluble in organic solvents [119,120].

7.4.2 Noncovalent functionalization

Noncovalent functionalization of CNTs is more feasible because it improves CNT physical properties [120]. Surfactants and other bio-macromolecules or polymers are used to perform this type of functionalization. Surfactants can be used to transport liquid media in order to find nondestructive purification methods [120,121]. In this scenario, the hydrophobic components of matched micelles surround the nanotube [120]. When the hydrophobic half of the amphiphilic contains an aromatic group, the interaction becomes stronger. Anionic, cationic, and nonionic surfactants can effectively disperse CNTs in water [122,123]. To prevent CNT aggregation in water anionic surfactants such as sodium dodecylbenzene sulfonate (NaDDBS) and sodium dodecylsulfate (SDS) [124,125] are widely used. The nature of the interaction between CNTs and surfactant is determined by the surfactant.

Because of the lack of a benzene group, SDS has a weaker bond with the surface of nanotubes than NaDDBS and Triton-X100. Benzene groups stack on the surface of graphite, which significantly improves surfactant molecule surface coverage and binding to graphite [126]. As its long alkyl chain and head group [120], NaDDBS diffuses efficiently than Triton-X100. The adsorption of several surfactants onto the nanotube surfaces is shown in the illustration (Fig. 7.4) [126].

Protein and DNA can be immobilized on CNTs without the use of a surfactant [126]. The hydrophobic portions of the proteins most likely aid in adsorption. A selective controlled method for protein immobilization has been developed to noncovalently modify the SWNTs [120,127]. After noncovalent functionalization, carbon nanotubes are easily soluble in organic or water media [128]. The compatibility of CNTs with polymer can be increased to increase solubility [129]. A wrapping technique is used for this purpose, in which polymers are wrapped around CNTs to form a complex [129,130]. Balu and colleagues created an improved conduction mixture of nanotube-polymer in organic solvent with SWNTs. With the help of noncovalent loading technique to fabricate and design the SWNTs [120] coated with block polystyrene-poly acrylic acid micelles [131]. Wrapping improves the dispersion of SWNTs in different solvents, such as nonpolar or polar solvents [120]. Nonwrapping methods have also been used to investigate CNT dispersion and dissolution in various fluids [132]. Nativ-Roth et al. investigated the fundamental problem of wrapping the nanotube and ensuring that the molecules face the least amount of noncovalent attractive force, resulting in a low load transfer capacity as a filler in a nanocomposites [120,133].

7.4.3 Covalent functionalization

In C atom presence of Sp^2 hybridization, make CNTs more reactive than those having graphene layer, thus enhance the ability of bonding covalently [134]. The translational integrity of CNTs is modified via alteration of carbon atoms (C-C) from sp^2 to sp^3 in the case of covalent

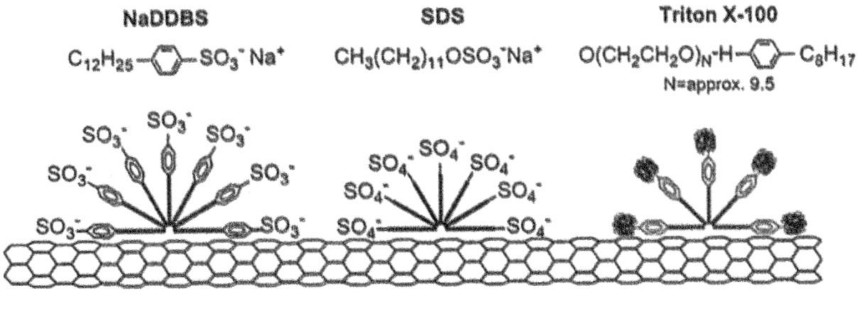

FIGURE 7.4

illustration of surfactant adsorption onto nanotube surface [126].

Reproduced with permission from ACS publications. https://pubs.acs.org/doi/10.1021/nl025924u, https://s100.copyright.com/AppDispatchServlet. Copyright 2004.

functionalization, and CNTs features regarding electrical and transport, can be altered [120,135]. CNT functionalization, on the other hand, can improve polymer solubility and dispersion in respective organic phases [120].

The surface of nanotubes is modified by a carboxylic acid group through covalent functionalization or by adding reagents to the side walls of the nanotubes. COOH and −OH are functional groups formed by oxidizing CNTs in nitric acid, sulfuric acid, oxygen, acid, and hydrogen peroxide solutions [136,137]. The number of −COOH groups on CNTs is proportional to the acidic treatment temperature and duration [138]. The oxidation processes and oxidizing agents used have an effect on the amount of −COOH and −OH functionality produced [139]. The presence of −COOH or −OH groups on the nanotube surface aids in the adhesion of organic molecules or inorganic molecules, which is required for nanotube solubility. Organic liquids dissolve functionalized carbon nanotubes. If the acid chloride group reacts with amine at high temperatures for four days, the SWNTs will be soluble in organic solvents [137]. NTs with chemically modified side walls were used to improve solubilization and dispersion in polymer matrices and solvents [140]. At various temperatures, elemental fluorine was passed through the side walls of SWNTs. The dissolution of fluorinated SWNTs in dimethyl formamide or isopropanol was aided by ultrasonication. Fluorinated SWNTs reaction [120] the chloroform soluble alkyl lithium or Grignard reagent can be transformed to side wall alkylated SWNTs [141−143].

Polymer grafting is used to functionalize CNTs in the processing of polymer/CNT nanocomposites. Two strategies have been established for covalent grafting of polymers to nanotubes: "grafting to" and "grafting from." Chemical processes such as radical coupling, amidation, and esterification [120], and others are used to attach as-prepared or commercially available polymer molecules to the CNT surface in the "grafting to" technique. To make composites using this method, the polymer must have reactive functional groups [144,145].

Fu et al. reported using the "grafting to" method to functionalize the CNTs. CNTs were refluxed to convert thionyl chloride to carboxylic acid groups, which were then converted to acyl-chlorides. The acylchloride moieties on the surface of CNTs were esterified using dendritic PEG polymer hydroxyl groups [146].

7.5 Aspects of functionalized nanoparticles on the properties of the nanocomposites

Polymeric nanocomposites fabrication employed contribution of both nanofibers or nanoparticles [147] and metallic nanocomposites to improve mechanical, electric, and optical properties and to control the electrodeposition. Alumina is less expensive than other nanomaterials, including carbon nanotubes, and it can be functionalized for the synthesis of nanocomposite materials [60,147]. Vinyl ester resin was chosen because it is a thermosetting cured resin with a chemically resistant morphology. The nanocomposites created by incorporating alumina nanoparticles into a vinyl ester resin matrix have applications in the fabrication of materials such as marine vessels and electrodeposition tanks [60,65,147]. The greater tensile strength is the current difficulty in nanocomposite development due to local stress within the nanocomposite. In other words, the nature of the bonds determines how a material responds to stress. Poor coupling between the polymer matrix and filler will result in the formation of artificial breaches. The mechanical properties of the composites suffer as a result of the simple mixing [67,70].

A properly designed interphase, on the other hand, could increase the strength and hardness of the composites while also making the nanocomposites more stable in difficult environments. The interfacial interaction between the nanoparticle and the polymer matrix determines the quality and properties of a nanocomposite. Surface modification of nanoparticles with surfactant is critical not only for nanoparticle stabilization but also for improving polymer compatibility [147,148].

Merlin Theodore et al. investigated the effect of MWCNTs on the thermal, morphological, and mechanical properties of epoxy PNCs. Chemical functionalization of MWCNT by oxidation (MWCNT-COOH) and direct fluorination was established using Fourier transformed spectroscopy (FTIR), Raman spectroscopy, and thermogravimetric analysis (TGA) (MWCNT-F). Epoxy-based nanocomposites were made with a 1 wt.% loading of MWCNT using in situ polymerization. Nanocomposites made with MWCNT-COOH had a 25.5% improvement in ultimate flexural strength and a 54.8% increase in flexural modulus when compared to the clean epoxy system [116].

To increase the performance of the polyoxymethylene (POM) homopolymer, Yang Liu et colleagues developed maleic MAH-g-POE anhydride-grafted polyolefin elastomer and AgNPs AgNPs as penetration additives. A melt-mixing approach was used to synthesize ternary POM/Ag/MAH-g-POE nanocomposites with different AgNPs and MAH-g-POE concentrations. The additives' effects on the ternary nanocomposites' microstructure, thermal stability, crystallization behavior, mechanical properties, and dynamic mechanical thermal properties were investigated. The compatibility between POM and Ag/MAH-g-POE and POM matrix. It was concluded that the MAH-g-POE played a role in binding the AgNPs and POM matrix and enhanced the interfacial adhesion between two components and enhanced their interfacial adhesion. Moreover, the ternary nanocomposites of AgNPs inclusion and MAH-g-POE to the POM/Ag/MAH-gPOE greatly improved their heat stability, crystallization characteristics, and mechanical capabilities [149]. Fig. 7.5 shows SEM images of nanocomposites of POM/Ag/MAH-g-POE. The SEM picture of POM in Fig. 7.5A demonstrates that there were no nanoparticles in the matrix and that the surface was smooth. However, after eliminating the MAHg-POE-dispersed phase using xylene etching, an apparent multiphase structure could be seen, as illustrated in Fig. 7.5B–F, where the holes represent the deleted MAHg-POE component. The MAHg-POE may play role in the AgNPs and POM, that results in good interfacial adhesion in matrix of POM and AgNPs. Furthermore, the number and size of pores in nanocomposites of POM/Ag/MAH-g-POE increased as the Ag/MAH-g-POE contents increased from 1% to 20%, and the pores were well disseminated in the POM matrix, indicating that Ag/MAH-g-POE had high compatibility with POM [150,151].

Fig. 7.6A and B shows the mechanical characteristics of POM/Ag nanocomposites [149]. POM had tensile strength of 50.03 MPa, elongation at break of 48.05%, and Young's modulus of 1293.91 MPa, respectively. However, when the quantity of Ag nanoparticles increased, the tensile strength and Young's modulus of POM/Ag nanocomposites increased at first, then decreased. Tensile strength and Young's modulus recorded higher values of 56.12 and 1450.39 MPa, respectively, when the Ag nanoparticles content was 1 wt.%. On the other hand, the POM/Ag nanocomposite elongation at break decreased at first and then increased as the AgNPs content increased [152,153].

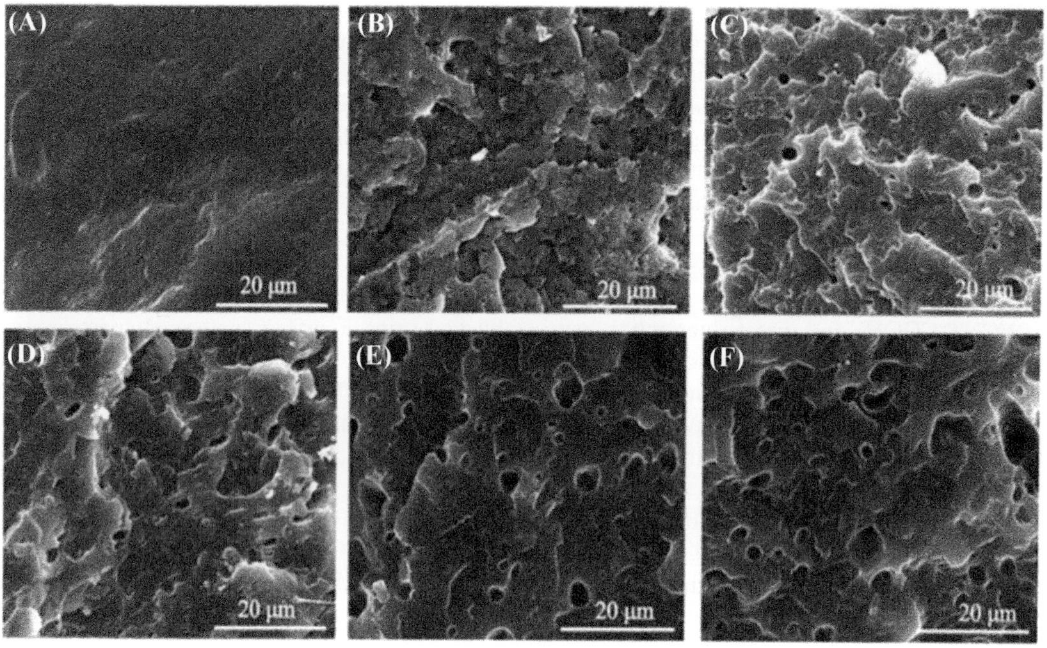

FIGURE 7.5

SEM illustrations of (A) polyoxymethylene (POM), (B) IPAM, (C) 2PAM, (D) 3PAM, (E) 4PAM, (F) 5PAM [149].

Reproduced with permission from MDPI polymers. https://pubs.acs.org/doi/10.1021/cm0110627, https://s100.copyright.com/AppDispatchServlet. Copyright 2021.

Marya Raji et al., performed melt compounding with several grafted clays, such as MMT, halloysite, and sepiolite [154], yielded PP/clay nanocomposites. These clays were employed as nanofillers for PP at varying concentrations (1–5 wt.%) after being organically changed by grafting of two organosilanes, namely 3-aminopropyltriethoxysilane (APTES) and vinyltrimethoxysilane (VTMS) [154,155]. The physicochemical properties of organosilane-modified clay-type were investigated using various structural, thermal, and morphological examination routes. All clays were successfully intercalated utilizing silane molecules, as evidenced by a decrease in the size of clay nanoparticles, a change in their chemical composition, and an increase in their d-spacing [154], as mentioned in findings [156]. The efficacy of the silylation procedure as a good way to improve the matrix-filler interaction was demonstrated by comparing the mechanical parameters of clay nanocomposites before and after grafting with organosilanes [157]. Finally, it was discovered that adding nanoorganoclay to silane grafted nanocomposites improves their properties by allowing for better interfacial adhesion between the organoclay and the PP as well as greater spatial dispersion-distribution. The PP/organosilane-grafted clay nanocomposites could be employed in industrial sector [158], with potential applications in the automotive and construction area [154].

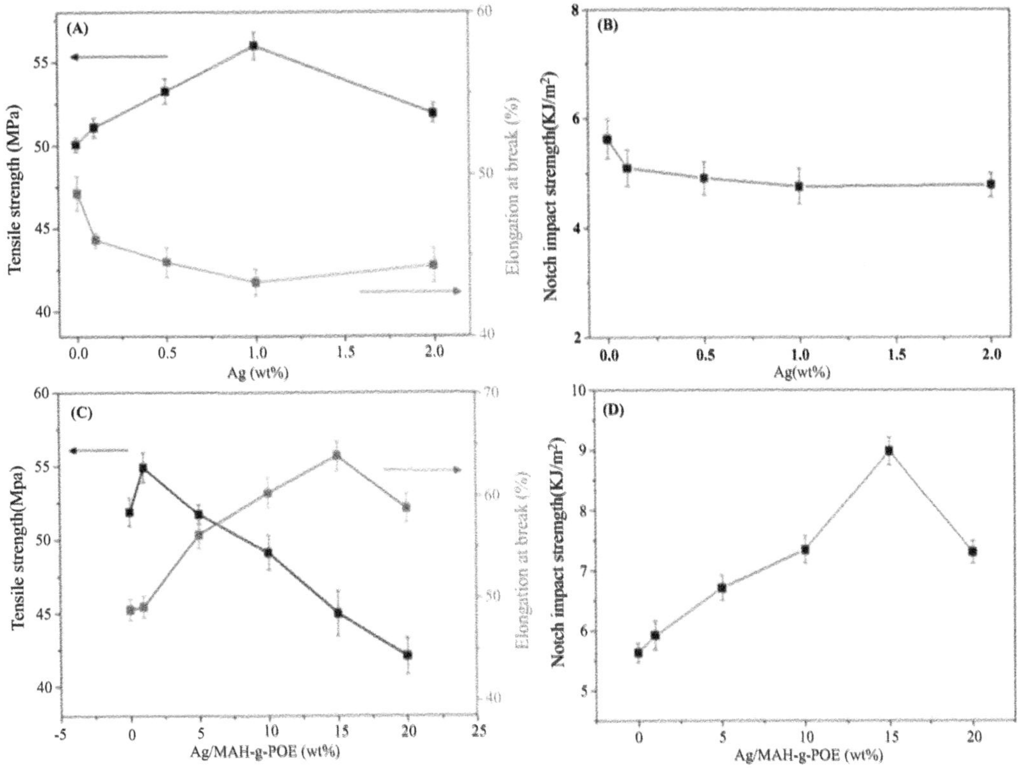

FIGURE 7.6

POM/Ag nanocomposites' tensile strength and elongation at break (A), as well as their impact strength; POM/Ag/MAH-g-POE ternary nanocomposites' (B) tensile strength and elongation at break, (C) POM/Ag/MAH-g-POE ternary nanocomposites' tensile strength and elongation at break (D) [149].

Reproduced with permission from MDPI polymers. https://pubs.acs.org/doi/abs/10.1021/jp202449k, https://s100.copyright.com/AppDispatchServlet. Copyright 2021.

7.6 Biodegradability aspects of functionalized nanoparticle-based PNCs

Biodegradable polymer materials, as well as nanofillers (such as natural fibers), are important because they can be used in environmentally friendly enterprises. Polymers that degrade naturally serve a vital role in addressing polymer-related public health issues and maintaining ecological equilibrium. Composites supplemented with diverse nanofillers can overcome the intrinsic disadvantages of biodegradable polymers, such small processing windows, low electrical and thermal features [80]. These biodegradable polymer composites can be used in a variety of applications due to their large surface area and higher aspect ratio. Because aliphatic polyesters are so important in the field of biodegradable materials, scientists are becoming increasingly interested in their nanocomposites [159]. Layered titanates and silicates, silver, maghemite, fluorine mica, and gold nanoparticles (AuNPs) are increasingly being used to create new nanocomposites. The modification matrices cover nearly all biodegradable aliphatic polyesters. Among them, polyesters/layered silicate nanocomposites, particularly PLA/layered silicate nanocomposites, have received extensive research. Natural biodegradable polymers include starch, chitosan, chitin, gelatin, protein, and cellulose. These components were created using bio-based materials, chemical synthesis, and microorganisms [160,161].

These polymers have been widely used and have significantly aided the development of biomedical areas due to their biocompatibility and biodegradability. As biotechnology and medical technology advance, biomedical materials must now meet higher standards [162]. There is a high demand for novel biodegradable polymers with specialized properties. Biodegradable polymers can be classified as natural or synthetic polymers, depending on the source. Synthetic biodegradable polymers have increasingly adaptable and diverse biomedical applications due to their tailorable designs or alterations [163].

Aleksandra Buzarovska et al., used the solution blending approach to develop the modified TiO_2 and biodegradable poly(lactic acid) (PLA) nanoparticles (with propanoic acid). Differential scanning calorimetry (DSC), SEM, FTIR, and puncture measurements were performed on nanocomposite films containing 0.5%, 2%, 5%, and 10% nanoparticles, respectively. Only with low TiO_2 loadings (less than 2 wt.%) improved and the dispersion become apparent. The TiO_2 filled modified and unmodified composites, there was an increase in crystallinity (from 7.4% to 34%). Puncture testing revealed that the films become more brittle as the nanoparticle content increased [164].

E Liu, M Zhang et al., use the solution blending approach to make PLA (biodegradable) and modified TiO_2 nanoparticles (with propanoic acid). SEM, DSC, FTIR, and puncture measurements were utilized for analysis of nanocomposite films composed of 0.5%, 2%, 5, and 10% nanoparticles. Only with low TiO_2 loadings (less than 2 wt.%) improved dispersion become apparent. Composite filled with unmodified and modified TiO_2, there was an increase in crystallinity (from 7.4% to 34%). Puncture testing revealed that the films become more brittle as the nanoparticle content increased [165].

Antonella Marra et al., used the twin-screw mixer and compression molding to make poly(L-lactic acid)–TiO_2 nanocomposite films. Modification of the TiO_2 surface with fluorocarbons plasma treatment [158] increased the compatibility of the polymer matrix and nanoparticles. The weight percentages of modified functionalized TiO_2 (fTi) employed in this study were 2 and 5. For comparison, unmodified TiO_2 nanoparticles were used at the same percent. The nanocomposites' molecular mass, structure, morphology, thermal stability, mechanical properties, and UV absorbance were all investigated [158,166]. The molecular weight of PLA was shown to be slightly altered by TiO_2 nanoparticles during mixing, with the phenomenon being more pronounced for functionalized nanoparticles (fTi). In the PLA matrix, all of the films had good nanoparticle dispersion [158,167]. However, nanocomposites containing modified nanoparticles demonstrated improved interfacial contact, as evidenced by improved mechanical properties [158]. Furthermore, all of the nanocomposites improved their thermal stability and increased their UV absorption area [167]. Increased TiO_2 content and at a given composition increased both UV absorption and thermal stability for nanocomposites containing modified TiO_2 [167,168]. CNTs have excellent mechanical, electrical, and magnetic properties, as well as a nanometer-scale diameter and a high aspect ratio, which makes them an effective reinforcing agent for high strength polymer composites [169]. CNTs, on the other hand, are extremely difficult to scatter and align in a polymer matrix because Van der Waals interactions cause them to form stable bundles. The most difficult aspects of producing CNT-reinforced composites are ensuring efficient CNT dispersion in a polymer matrix, assessing dispersion, and aligning and controlling CNTs in the matrix [170]. PLA, PCL, PPDO, PBS, and PHAs nanocomposites are the most promising biodegradable materials, not only because of their high biodegradability, compatibility, and strength, but also because they can be manufactured using renewable resources [169].

7.7 Characterization of functionalized nanoparticle-based polymeric nanocomposites

Exfoliation degree and its effects in nanocomposites were studied using various techniques such as XRD, TEM, and SEM, among others. However, it should be noted that high precision in equipment analysis, as well as other limitations, can lead to misunderstandings about the morphology of nanocomposites [171].

7.7.1 XRD

The XRD technique can be used to assess the spacing of clay layers. This change in clay morphology caused by intercalation or exfoliation into polymer galleries is also investigated. XRD shows the normal layered structure of these clays as peak features. These peaks represent d-spacing and/or platelet separation in the clay structure. Peaks shift towards lower diffraction angles as d-spacing increases. This technique reveals the entry of polymer into the gaps between clay platelets. However, it does not provide results for layer stacking. Exfoliation can be indicated by the absence or intensity of peaks. A group of researchers created epoxy/glass/OMMT nanocomposites and discovered that the OMMT characteristic peak completely disappeared in XRD characterization. Their findings indicated the incorporation of exfoliated silica platelets in an epoxy matrix [172,173]. XRD analysis revealed the fluoro-organic mmt and the C18-mmt resulting fluoro-organic mmt (f-mmt) (Fig. 7.7). In the same figure, the XRD patterns of nanocomposites (unfunctionalized-PP/f-mmt) created by extrusion and melt intercalation are shown [173]. A favorable thermodynamics for nanocomposites has been shown via melt-intercalation hybrid (i.e., annealing a f-mmt and polymer physical mixture). Moreover, the application of mechanical shear encourages the f-mmt layers to exfoliate and disperse [173].

7.7.2 TEM and AFM

A lower angle of 2 h is critical in determination of spacing between layers in XRD [171]. As a result, absence of peaks show the [171] intercalation with wide spaces instead of complete exfoliation. TEM is a very useful and highly specific approach for analyzing clay exfoliation due to its ability to image at the nanoscale. It also measures intercalation and exfoliation directly. Wetzel created nanocomposites with various nanoparticles and used TEM to investigate their dispersion within the matrix. The polymeric matrix showed a consistent distribution of nanoparticles as well as minor agglomerates [171,173]. Barreca and his colleagues used a combination of FE-SEM and AFM to investigate the system's nanoorganization and morphology [174]. Pure CuO morphology (Fig. 7.8A) characterized by the stable existence of dispersed nanoflowers (with size of 600 nm) that result in a greater surface area, a feature for the dispersion of AuNPs and TiO_2, matches well

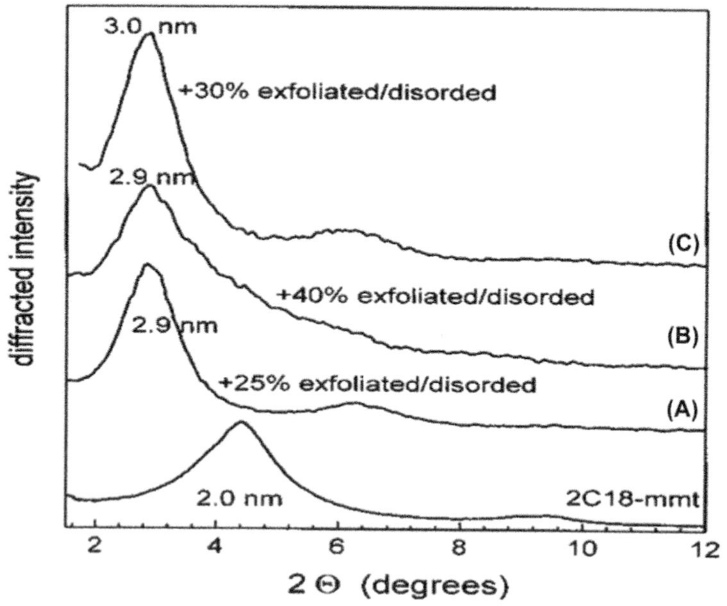

FIGURE 7.7

X-ray diffraction patterns of functionalized-PP/2C18-mmt nanocomposites and a dimethyldioctadecyl ammonium-modified montmorillonite (2C18-MMT). MS (methylstyrene) (1 mol.%) (A), MAH (maleic anhydride) (0.5 mol.%) (B) OH-containing styrene (0.5 mol.%) and (C) hydroxyl-containing styrene 0.5 mol.% are the functional groups utilized [173].

Reproduced with permission from ACS publications. https://pubs.acs.org/doi/10.1021/cm0110627. Copyright 2001.

with earlier findings [174]. In fact, no surface alteration that involve collapse/aggregation could be found after the deposition of Au and TiO_2 (Fig. 7.8B), verifying the capabilities of the employed synthetic approach in CuO nano deposits functionalization without surface abnormalities [174]. The similar results have been searched for Au/CO_3O_4 and ZnO/TiO_2 previously. Surface corrugation could also be seen in AFM images (Fig. 7.8C and D), that showed the porous and globular consistency with minor changes between pure CuO and $CuOTiO_2Au$ specimens. All of the examined samples had an rms roughness of 60 nm, which corresponded to this observation [174].

7.7.3 SEM

SEM is a technique that uses an intense electron beam to scan the sample and generate images. Surface polishing is used to make the sample conductive, which is then used to disperse clay in various enlargements [171]. The structure phenomena and the study of key features of the incorporated polymer play an important role. The mechanical characteristics of polymers are determined by modifications in shape confirmation as well as the movement of individual polymer molecules. Fig. 7.9 shows the SEM micrograph of PMMA composites that were developed via radical- free methyl methacrylate (MMA) polymerization in the presence of Al_2O_3 NPs [175].

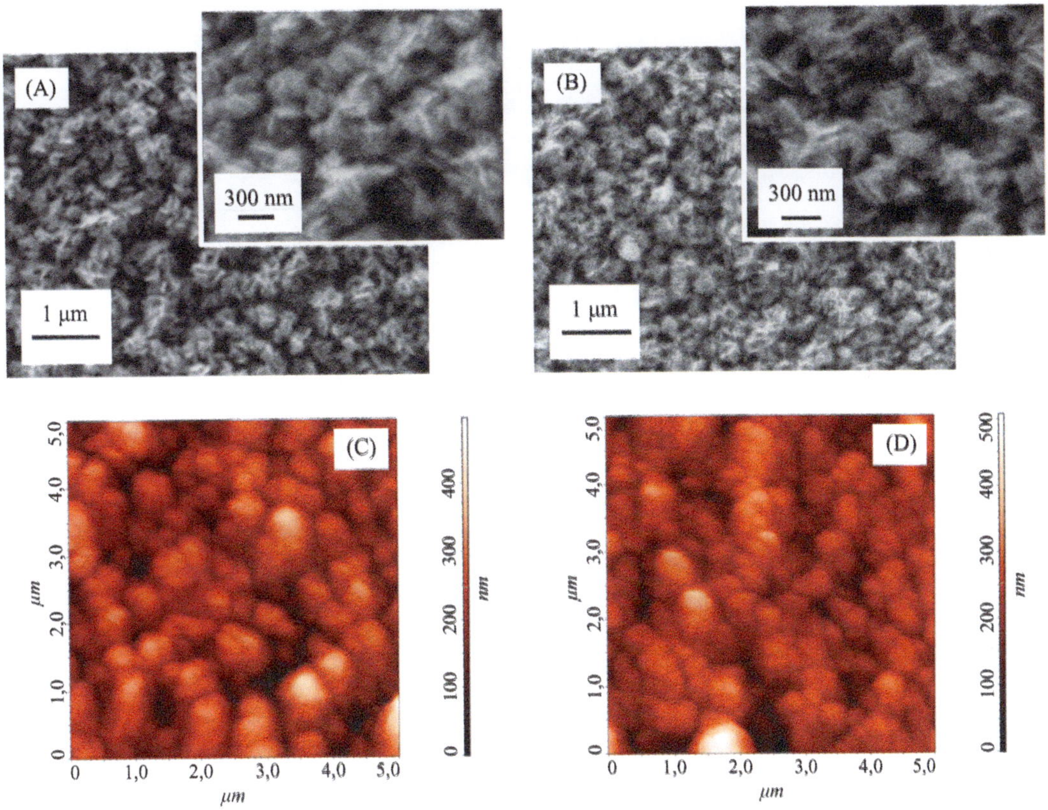

FIGURE 7.8

Copper oxide (CuO) (A and C) and $CuOTiO_2Au$ (B and D) nanocomposites are shown in straight image of Atomic force microscopy (AFM) and FE-SEM micrographs. The insets provide higher magnification FE-SEM images [174].

Reproduced with permission from ACS publications. https://pubs.acs.org/doi/10.1021/jp202449k. Copyright 2011.

7.7.4 Thermogravimetric analysis (TGA) and DSC

TGA was used to evaluate the physical and chemical properties of the materials. TGA and differential thermogravimetry analysis (DTGA): the former records changes in sample weight as a function of time or temperature, while the latter simply measures temperature differences. DSC is a technique that is similar to DTA. DSC is an analytical tool that can aid in the thermal characterization of PNCs. This is also useful for determining the glass transition temperature (Tg) of PNCs and polymers. A higher Tg temperature indicates the presence of inorganic materials in the polymer

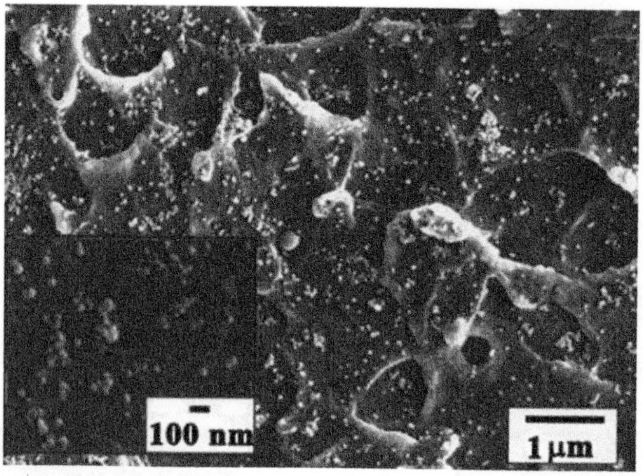

FIGURE 7.9

SEM scans of a 38 nm 2.2 wt.% alumina/PMMA sample's surface coated with glycidoxy propyltrimethoxysilane [175].

Reproduced with permission from John Wiley & Sons. https://onlinelibrary.wiley.com/doi/10.1002/polb.20297. Copyright 2004..

matrix [39,176]. Tomáš Remiš and colleagues investigated the TGA and DTGA curves of a PVA sample with three distinct degradation phases. Step I (25°C–140°C) clearly demonstrated the evaporation of water from the film (a 5% weight loss). The dehydration of PVA, which was accompanied by the formation of polyene structures, was attributed to Step II (200°C–310°C), which resulted in a 65% weight loss. Polyene structures were further converted to low-molecular-weight products in Step III (350°C–530°C), with a weight decrease of 25% [70]. The addition of NDs (nanodiamonds) in the PVA (polyvinylalcohol) structure significantly altered the PVA-ND nanocomposites' degradation profile (Fig. 7.10). The weight loss of the PVA-ND films (Fig. 7.10) was thus slowed down, compared to the pure PVA film (Steps II–III degradation). This could be explained first by the presence of ND that did not decompose in Steps II and III, which reduced the heights of the two PVA decomposition steps; second, the shift of Step II to higher temperatures could have contributed to the stabilization of alcoholic groups of PVA against dehydration by the H-bonding (hydrogen linkage) between the oxygen-containing functional moiety of ND particles and the hydroxyl groups of PVA [177,178]. The DTGA illustrated in PVA-ND/3 nanocomposite, the three-step degradation diagram found for PVA was modified to four stages for the PVA-ND nanocomposite films.

Because the ND nanoparticles were incorporated into the PVA structure, the degradation was milder in the PVA-NDs nanocomposites than in the PVA. The evaporation of water caused the first stage of degradation within 30°C–145°C temperature range. The second stage went from 190°C to 380°C. PVA-ND/1 (5% of the ND) and had the maximum weight reduction of 65%, while PVA-ND/3 had the lowest weight loss of 50% (28% of the ND) [179].

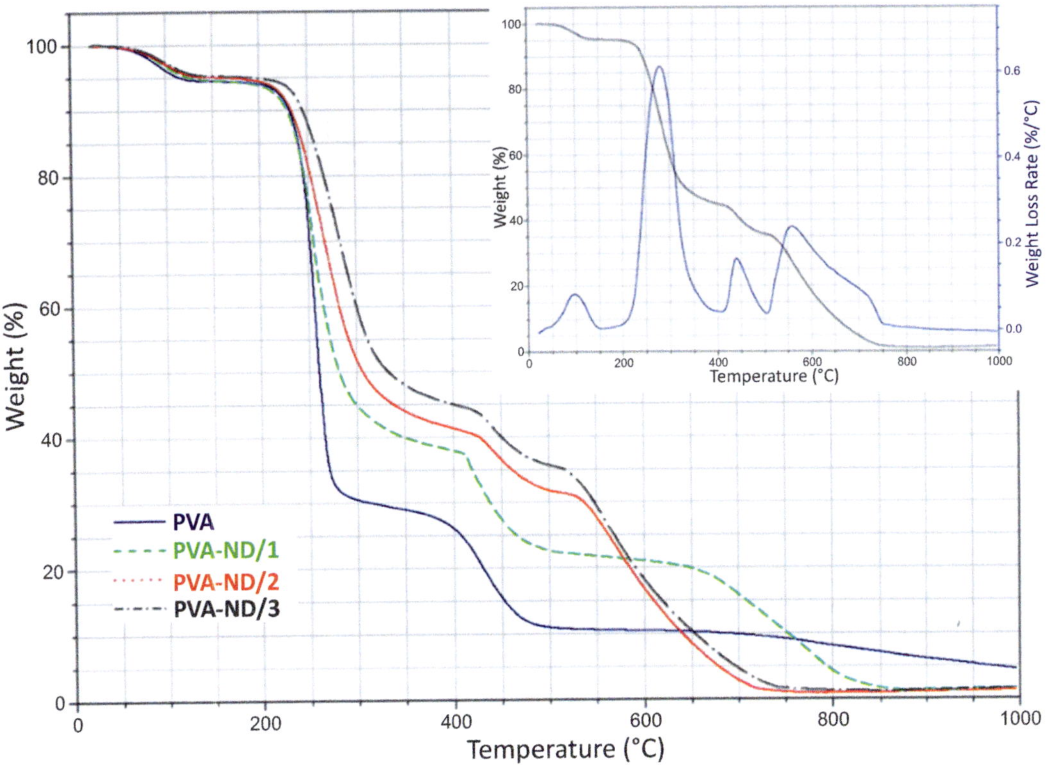

FIGURE 7.10

(A) Thermograms of pure PVA film and nanocomposites of PVA-ND [70].

Adapted from MDPI polymers. Copyright 2021.

The heating curves of DSC for the PVA and PVA-ND samples with different ND encapsulation are illustrated in Fig. 7.11. PVA glass transition (Tg) is detected between 41°C and 45°C. The release of water caused a broad peak in temperature range upto 170°C. Melting temperature of PVA (Tm) was next, as evidenced by a short, reasonably sharp peak with a maximum at 218°C, which faded in intensity as the ND concentration in the samples increased. The thermal pyrolysis of the PVA was represented by the rightmost big endothermic peak in the temperature range of 240°C–350°C, with a heat consumption of 690 J/g for PVA and 495, 347, and 260 J/g for PVA-ND/1, PVA-ND/2, and PVA-ND/3 n, respectively. The Tg increased somewhat as the number of NDs in the PVA matrix increased, from 42°C for PVA to 42.6°C, 43°C, and 43.8°C for PVA-ND/1, PVA-ND/2, and PVA-ND/3, respectively. The partial crosslinking of the PVA by the NDs most likely produced the rise in Tg. The increase in Tg was in line with the dynamic mechanical analysis (DMA) results, which also showed an increase in Tg [70].

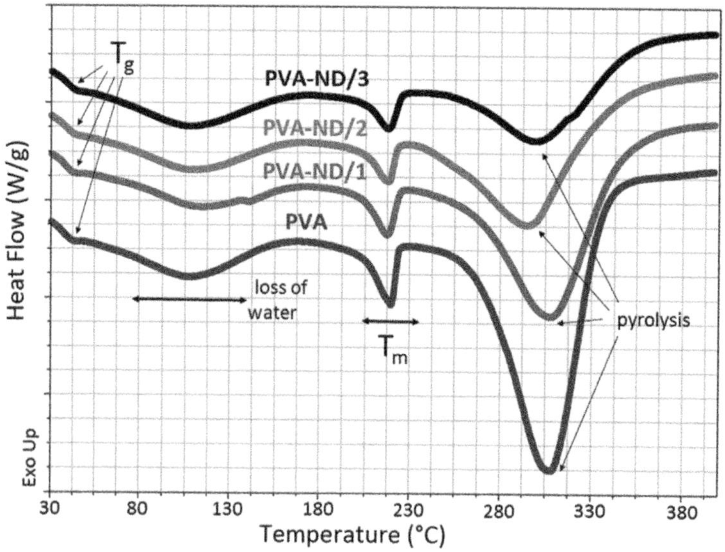

FIGURE 7.11

The DTGA curve of the PVA-ND/3 nanocomposites is shown in the inset graph (together with the original TGA curve) DSC curves of PVA-ND nanocomposites and PVA film [70].

Adapted from MDPI polymers. Copyright 2021.

7.8 Applications of functionalized nanoparticles-based nanocomposites

When the surface area to volume ratio is increased, electro-catalyst efficiency and performance improve significantly. Aggregation and instability, on the other hand, can occur in an attempt to reduce particle size in order to increase surface area. In general, polymer coating or an inorganic matrix can be used to prevent metal oxide catalyst particle aggregation. Because transition metals are thermodynamically unstable in acidic media, noble metal compounds are used as electro-catalysts in water splitting. Single noble metal oxides ruthenium (Ru) oxide and iridium (Ir) oxide have applications in water electrolysis. In comparison to Ir nanoparticles, which demonstrated sufficient stability and high oxygen evolution reaction (OER) activity, Ru nanoparticles are unstable. Both characteristics are greatly influenced by the conditions of synthesis and the chemical/structural existence of iron oxide species [180,181].

Researchers were able to create a polymer-supported graphene-cadmium sulfide composite catalyst. Graphene was created by reducing GO, which was then coated with cadmium sulfide (CdS) using a solvo-thermal method. Under visible light, composite catalysts were used. PSGM/rGO/CdS-3 had the highest hydrogen production rate of any sample prepared, and it was 19.3 times higher than samples mixed physically. The combination of holes and photo-induced electrons was suppressed by loading CdS on graphene. The graphene-based composite catalyst demonstrated good stability and efficacy for photocatalytic hydrogen synthesis, according to the results [171]. The graphene structure supported by polymer prevented aggregation between single layer graphene sheets as well as aggregation of CdS nanoparticles [182,183].

The depletion of fossil fuels has resulted in an intensifying energy crisis, necessitating the development of energy storage systems for renewable resources such as wind and solar energy. Energy storage systems must be able to generate and store energy efficiently, quickly, and in a stable manner [171]. Hydro pumped storage, compressed air energy storage, fly wheel and thermal reserves are important electrochemical energy storage systems (EESS) [171]. With respect to PNC, lithium-ion batteries are popular rechargeable batteries that offer various benefits e.g., long cycle life, low toxicity, high working voltage and huge power. Examples of materials that are used as anode include silicon, metal oxide and graphite whereas examples of cathode materials include $LiCoO_2$, $Li[NiCoAl]O_2$, $LiMn_2O_4$, $LiFePO_4$ and $Li[MnNiCo]O_2$ [184,185].

The disadvantages of dye-sensitized solar cells include brittle glass electrodes, instability, leakage, high weight, and rigidity. Certain DSC constituents can be replaced with PNCs to reduce costs while improving DSC performance. Polymer electrolytes, catalyst materials, and flexible substrates are a few examples. By replacing glass electrodes with polyethylene naphthalate (PEN) or polyethylene terephthalate (PET) electrodes, the impact and adaptability resistance of a DSC can be increased. Polymer electrolytes can be used to solve leakage and volatility problems. Conducting polymers can replace platinum, which is very expensive, in the catalyst layer. This resulted in the development of versatile solar systems that are light in weight and offer cost effectiveness in solar industry applications such as mobile phones, watches, and ID cards [186].

7.9 Conclusion

Nanocomposites have enormous potential for investigating new functions not seen in traditional materials. One of the most promising areas of research is nanocomposites. They have distinct characteristics such as production feasibility, light weight, and ease of manufacture. The functionalization of PNCs can improve the efficiency of nanocomposites. The increased demand for biodegradable materials in the world requires researchers to design and fabricate products with superior properties to current synthetic materials. The incorporation of nanoparticles into biodegradable matrices and their functionalization results in the creation of nanocomposites with enhanced properties. Various polymer materials, in combination with nanoparticles, can improve the barrier, thermal stability, melt strength, and mechanical properties of nanocomposites. In short, functionalized biodegradable nanocomposites outperformed conventional composites in terms of efficiency. Nanocomposites are being promoted as the future material of choice in a wide range of industries, including medicine, energy, engineering, and agriculture.

References

[1] K.I. Winey, R.A. Vaia, Polymer nanocomposites, MRS Bull. 32 (4) (2007) 314–322.

[2] B. Ates, S. Koytepe, A. Ulu, C. Gurses, V.K. Thakur, Chemistry, structures, and advanced applications of nanocomposites from biorenewable resources, Chem. Rev. 120 (17) (2020) 9304–9362.

[3] A.H. Kumar, M.B. Ahamed, K. Deshmukh, M.S. Sirajuddeen, Morphology, dielectric and emi shielding characteristics of graphene nanoplatelets, montmorillonite nanoclay and titanium dioxide nanoparticles reinforced polyvinylidenefluoride nanocomposites, J. Inorg. Organo. Met. Polym. Mater. 31 (5) (2021) 2003–2016.

[4] J. Gou, J. Zhuge, F. Liang, Processing of polymer nanocomposites, Manufacturing Techniques for Polymer Matrix Compos. (PMCs), Elsevier, 2012, pp. 95–119.

[5] X. Sun, C. Huang, L. Wang, L. Liang, Y. Cheng, W. Fei, Y. Li, Recent progress in graphene/polymer nanocomposites, Adv. Mater. 33 (6) (2021) 2001105.

[6] P. Rani, M.B. Ahamed, K. Deshmukh, Structural, dielectric and EMI shielding properties of polyvinyl alcohol/chitosan blend nanocomposites integrated with graphite oxide and nickel oxide nanofillers, J. Mater. Sci. Mater. Electron. 32 (1) (2021) 764–779.

[7] J. Parameswaranpillai, J.A. Gopi, S. Radoor, C.D.M. Dominic, S. Krishnasamy, K. Deshmukh, N. Hameed, N.V. Salim, N. Sienkiewicz, Turning waste plant fibers into advanced plant fiber reinforced polymer composites: a comprehensive review, Comp. Part C: Open Access 10 (2023) 100333.

[8] E. Omanović-Mikličanin, A. Badnjević, A. Kazlagić, M. Hajlovac, Nanocomposites: a brief review, Health Technol. 10 (1) (2020) 51–59.

[9] M. Tabandeh-Khorshid, A. Kumar, E. Omrani, C. Kim, P. Rohatgi, Synthesis, characterization, and properties of graphene reinforced metal-matrix nanocomposites, Compos. Part. B: Eng. 183 (2020) 107664.

[10] P. Rani, B. Ahamed, K. Deshmukh, Dielectric and electromagnetic interference shielding properties of zeolite 13X and carbon black nanoparticles based PVDF nanocomposites, J. Appl. Polym. Sci. 138 (13) (2021) 50107.

[11] A. Muzaffar, M.B. Ahamed, K. Deshmukh, S.K. Pasha, Dielectric properties and electromagnetic interference shielding studies of nickel oxide and tungsten oxide reinforced polyvinylchloride nanocomposites, Polym-Plast Tech. Mat. 59 (15) (2020) 1667–1678.

[12] P.L. Reddy, K. Deshmukh, K. Chidambaram, B. Ahamed, K.K. Sadasivuni, D. Ponnamma, R. Lakshmipathy, D. Dayananda, S.K.K. Pasha, Effect of poly ethylene glycol (PEG) on structural, thermal and photoluminescence properties of CdO nanoparticles for optoelectronic applications, Mater. Today Proc. 9 (2019) 175–183.

[13] M. Schmid, A. Benz, C. Stinga, D. Samain, K.P. Zeyer, Fundamental investigations regarding barrier properties of grafted PVOH layers, Int. J. Polym. Sci. (2012).

[14] M. Schmid, S. Sängerlaub, O. Miesbauer, V. Jost, J. Werthan, C. Stinga, D. Samain, C. Stramm, K. Noller, K. Müller, Water repellence and oxygen and water vapor barrier of PVOH-coated substrates before and after surface esterification, Polymers 6 (11) (2014) 2764–2783.

[15] K. Müller, E. Bugnicourt, M. Latorre, M. Jorda, Y. Echegoyen Sanz, J.M. Lagaron, O. Miesbauer, A. Bianchin, S. Hankin, U. Bölz, G. Pérez, M. Jesdinszki, M. Lindner, Z. Scheuerer, S. Castelló, M. Schmid, Review on the processing and properties of polymer nanocomposites and nanocoatings and their applications in the packaging, automotive and solar energy fields, Nanomaterials 7 (4) (2017) 74.

[16] C.P. McAdam, N. Hudson, J. Liggat, R. Pethrick, Synthesis and characterization of nylon 6/clay nanocomposites prepared by ultrasonication and in situ polymerization, J. Appl. Polym. Sci. 108 (4) (2008) 2242–2251.

[17] M. Pandey, G.M. Joshi, K. Deshmukh, J. Ahmad, Impedance spectroscopy and conductivity studies of $CdCl_2$ doped polymer electrolyte, Adv. Mater. Lett. 6 (2) (2015).

[18] K.M. Holder, B.R. Spears, M.E. Huff, M.A. Priolo, E. Harth, J.C. Grunlan, Stretchable gas barrier achieved with partially hydrogen-bonded multilayer nanocoating, Macromol. Rapid Commun. 35 (10) (2014) 960–964.

[19] M. Joshi, R. Khanna, R. Shekhar, K. Jha, Chitosan nanocoating on cotton textile substrate using layer-by-layer self-assembly technique, J. Appl. Polym. Sci. 119 (5) (2011) 2793–2799.

[20] D.R. Paul, L.M. Robeson, Polymer nanotechnology: nanocomposites, Polymer 49 (15) (2008) 3187–3204.

[21] H.D. Wagner, R.A. Vaia, Nanocomposites: issues at the interface, Mater. Today 7 (11) (2004) 38–42.

[22] C.C. Okpala, Nanocomposites—an overview, Int. J. Eng. 8 (11) (2013) 17–23.

[23] R.A. Vaia, E.P. Giannelis, Polymer nanocomposites: status and opportunities, MRS Bull. 26 (5) (2001) 394–401.

[24] M. Pandey, G.M. Joshi, K. Deshmukh, M. Khutia, N.N. Ghosh, Optimized AC conductivity correlated to structure, morphology and thermal properties of PVDF/PVA/Nafion composites, Ionics 20 (10) (2014) 1427–1433.
[25] A. Hashim, A. Hadi, Novel pressure sensors made from nanocomposites (biodegradable polymers–metal oxide nanoparticles): fabrication and characterization, Ukr. J. Phys. 63 (8) (2018). 754-754.
[26] J.K. Pandey, A.P. Kumar, M. Misra, A.K. Mohanty, L.T. Drzal, R. Palsingh, Recent advances in biodegradable nanocomposites, J. Nanosci. Nanotechnol. 5 (4) (2005) 497–526.
[27] S.S. Bari, A. Chatterjee, S. Mishra, Biodegradable polymer nanocomposites: an overview, Polym. Rev. 56 (2) (2016) 287–328.
[28] E. Hernández-García, M. Vargas, C. González-Martínez, A. Chiralt, Biodegradable antimicrobial films for food packaging: effect of antimicrobials on degradation, Foods 10 (6) (2021) 1256.
[29] T. Ojeda, Polymers and the environment, Polym. Sci. (2013) 23.
[30] K. Qiu, Biobased and Biodegradable Polymer Nanocomposites, Cornell University, 2012.
[31] V. Shah, J. Bhaliya, G.M. Patel, K. Deshmukh, Advances in polymeric nanocomposites for automotive applications: a review, Polym. Adv. Technol. 33 (2022) 3023–3048.
[32] R.C. Thompson, C.J. Moore, F.S. Vom Saal, S.H. Swan, Plastics, the environment and human health: current consensus and future trends, Philos. Trans. R. Soc. London. Ser. B, Biol. Sci. 364 (1526) (2009) 2153–2166.
[33] M.E. Grigore, Methods of recycling, properties and applications of recycled thermoplastic polymers, Recycling 2 (4) (2017) 24.
[34] I. Armentano, D. Puglia, F. Luzi, C.R. Arciola, F. Morena, S. Martino, L. Torre, Nanocomposites based on biodegradable polymers, Materials 11 (5) (2018) 795.
[35] A. Chinnathambi, Synthesis and characterization of spinel FeV_2O_4 coupled ZnO nanoplates boosted white light photocatalysis and antibacterial applications, J. Alloy. Compd. 890 (2022) 161742.
[36] T.N. Ghosh, S.S. Pradhan, S.K. Sarkar, A.K. Bhunia, On the incorporation of the various reduced graphene oxide into poly (vinyl alcohol) nano-compositions: comparative study of the optical, structural properties and magnetodielectric effect, J. Mater. Sci. Mater. Electron. (2021) 1–22.
[37] S.B. Aziz, M.H. Hamsan, W.O. Karim, M.F.Z. Kadir, M.A. Brza, O.G. Abdullah, High proton conducting polymer blend electrolytes based on chitosan: dextran with constant specific capacitance and energy density, Biomolecules 9 (7) (2019) 267.
[38] S.A. McGlashan, P.J. Halley, Preparation and characterisation of biodegradable starch-based nanocomposite materials, Polym. Int. 52 (11) (2003) 1767–1773.
[39] K.-K. Yang, X.-L. Wang, Y.-Z. Wang, Progress in nanocomposite of biodegradable polymer, J. Ind. Eng. Chem. 13 (4) (2007) 485–500.
[40] M. Avella, J.J. De Vlieger, M.E. Errico, S. Fischer, P. Vacca, M.G. Volpec, Biodegradable starch/clay nanocomposite films for food packaging applications, Food Chem. 93 (3) (2005) 467–474.
[41] S. Alfei, B. Marengo, G. Zuccari, Nanotechnology application in food packaging: a plethora of opportunities versus pending risks assessment and public concerns, Food Res. Intern. 137 (2020) 109664.
[42] W. Gindl, J. Keckes, All-cellulose nanocomposite, Polymer 46 (23) (2005) 10221–10225.
[43] H.-M. Park, X. Liang, A.K. Mohanty, M. Misra, L.T. Drzal, Effect of compatibilizer on nanostructure of the biodegradable cellulose acetate/organoclay nanocomposites, Macromolecules 37 (24) (2004) 9076–9082.
[44] F. Hussain, M. Hojjati, M. Okamoto, R.E. Gorga, Polymer-matrix nanocomposites, processing, manufacturing, and application: an overview, J. Compos. Mater. 40 (17) (2006) 1511–1575.
[45] Q. Hu, B. Li, M. Wang, J. Shen, Preparation of orientated magnetite/chitosan nano-composite material via in-situ precipitation, Chem. J. Chin. Univ. Chinese Ed. 26 (10) (2005) 1960.
[46] L. Kong, Y. Gao, W. Cao, Y. Gong, N. Zhao, X. Zhang, Preparation and characterization of nanohydroxyapatite/chitosan composite scaffolds, J. Biomed. Mater. Res. Part A 75A (2) (2005) 275–282.

[47] Q. Hu, B. Li, M. Wang, J. Shen, Preparation and characterization of biodegradable chitosan/hydroxyapatite nanocomposite rods via in situ hybridization: a potential material as internal fixation of bone fracture, Biomaterials 25 (5) (2004) 779–785.

[48] X. Shen, H. Tong, T. Jiang, Z. Zhu, P. Wan, J. Hu, Homogeneous chitosan/carbonate apatite/citric acid nanocomposites prepared through a novel in situ precipitation method, Compos. Sci. Technol. 67 (11–12) (2007) 2238–2245.

[49] H. Uyama, M. Kuwabara, T. Tsujimoto, M. Nakano, A. Usuki, S. Kobayashi, Green nanocomposites from renewable resources: plant oil – clay hybrid materials, Chem. Mater. 15 (13) (2003) 2492–2494.

[50] T. Tsujimoto, H. Uyama, S. Kobayashi, Green nanocomposites from renewable resources: biodegradable plant oil-silica hybrid coatings, Macromol. Rapid Commun. 24 (12) (2003) 711–714.

[51] X. Shen, X. Mo, R. Moore, S.J. Frazier, T. Iwamoto, J.M. Tomich, X.S. Sun, Adhesion and structure properties of protein nanomaterials containing hydrophobic and charged amino acids, J. Nanosci. Nanotechnol. 6 (3) (2006) 837–844.

[52] D. Ponnamma, H. Parangusan, K. Deshmukh, P. Kar, A. Muzaffar, S.K.K. Pasha, M.B. Ahamed, M.A. A. Al-Maadeed, Green synthesized materials for sensor, actuator, energy storage and energy generation: a review, Polym. Plast. Technol. Mater. 59 (2020) 1–59.

[53] J. Zhang, E. Manias, C.A. Wilkie, Polymerically modified layered silicates: an effective route to nanocomposites, J. Nanosci. Nanotechnol. 8 (4) (2008) 1597–1615.

[54] M. Dai, J. Haghpanah, N. Singh, E.W. Roth, A. Liang, R.S. Tu, J.K. Montclare, Artificial protein block polymer libraries bearing two SADs: effects of elastin domain repeats, Biomacromolecules 12 (12) (2011) 4240–4246.

[55] V. Ibrahim, G. Mamo, P.-J. Gustafsson, R. Hatti-Kaul, Production and properties of adhesives formulated from laccase modified Kraft lignin, Ind. Crop. Prod. 45 (2013) 343–348.

[56] S.S. Ray, M. Okamoto, Biodegradable polylactide and its nanocomposites: opening a new dimension for plastics and composites, Macromol. Rapid Commun. 24 (14) (2003) 815–840.

[57] D.N. Bikiaris, Nanocomposites of aliphatic polyesters: an overview of the effect of different nanofillers on enzymatic hydrolysis and biodegradation of polyesters, Polym. Degrad. Stab. 98 (9) (2013) 1908–1928.

[58] S.H. Othman, Bio-nanocomposite materials for food packaging applications: types of biopolymer and nano-sized filler, Agric. Agric. Sci. Procedia 2 (2014) 296–303.

[59] V. Vatanpour, S.S. Madaeni, A.R. Khataee, E. Salehi, S. Zinadini, H.A. Monfared, TiO_2 embedded mixed matrix PES nanocomposite membranes: Influence of different sizes and types of nanoparticles on antifouling and performance, Desalination 292 (2012) 19–29.

[60] M. Khutia, G.M. Joshi, K. Deshmukh, M. Pandey, Preparation of modified polymer blend and electrical performance, Comp. Interf. 22 (3) (2015) 167–178.

[61] E. Aawani, N. Memarian, H.R. Dizaji, Synthesis and characterization of reduced graphene oxide–V_2O_5 nanocomposite for enhanced photocatalytic activity under different types of irradiation, J. Phys. Chem. Solids 125 (2019) 8–15.

[62] H. Zheng, H. Xue, Y. Zhang, Z. Shen, A glucose biosensor based on microporous polyacrylonitrile synthesized by single rare-earth catalyst, Biosens. Bioelectron. 17 (6–7) (2002) 541–545.

[63] J. Zhang, Y. Kan, L. Gu, C. Wang, Y. Zhang, Graphite carbon nitride and its composites for medicine and health applications, Chem. Asian J. (2021).

[64] B. Nekhamanurak, P. Patanathabutr, N. Hongsriphan, Thermal–mechanical property and fracture behaviour of plasticised PLA–$CaCO_3$ nanocomposite, Plast. Rubber Compos. 41 (4–5) (2012) 175–179.

[65] M. Khutia, G.M. Joshi, K. Deshmukh, M. Pandey, Optimization of dielectric constant of polycarbonate/polystyrene modified blend by ceramic metal oxide, Polym. Plast. Technol. Eng. 54 (4) (2015) 383–389.

[66] M. Radetić, V. Ilić, V. Vodnik, S. Dimitrijević, P. Jovančić, Z. Šaponjić, J.M. Nedeljković, Antibacterial effect of silver nanoparticles deposited on corona-treated polyester and polyamide fabrics, Polym. Adv. Technol. 19 (12) (2008) 1816–1821.

[67] M. Pandey, G.M. Joshi, S. Bhattacharya, Influence of $K_2Ti_6O_{13}$ on dielectric and barrier properties of polymer, Ionics 22 (3) (2016) 445–449.

[68] Q. Lan, J. Yu, J. He, F.H. Maurer, J. Zhang, Thermal behavior of poly(L-lactide) having low l-isomer content of 94% after compressed CO_2 treatment, Macromolecules 43 (20) (2010) 8602–8609.

[69] C. Bao, L. Song, W. Xing, B. Yuan, C.A. Wilkie, J. Huang, Y. Guo, Y. Hu, Preparation of graphene by pressurized oxidation and multiplex reduction and its polymer nanocomposites by masterbatch-based melt blending, J. Mater. Chem. 22 (13) (2012) 6088–6096.

[70] T. Remiš, P. Bělský, T. Kovářík, J. Kadlec, M. Ghafouri Azar, R. Medlín, V. Vavruňková, K. Deshmukh, K.K. Sadasivuni, Study on structure, thermal behavior and viscoelastic properties of nanodiamond-reinforced poly (vinyl alcohol) nanocomposites, Polymers 13 (9) (2021) 1426.

[71] N. Shimpi, M. Borane, S. Mishra, M. Kadam, Biodegradation of polystyrene (PS)-poly (lactic acid) (PLA) nanocomposites using *Pseudomonas aeruginosa*, Macromol. Res. 20 (2) (2012) 181–187.

[72] S.M. de Andrade e Silva, M.R.d O. Carrilho, L. Marquezini Junior, F.C.P. Garcia, A.P. Manso, M.C. Alves, R.M. de Carvalho, Effect of an additional hydrophilic versus hydrophobic coat on the quality of dentinal sealing provided by two-step etch-and-rinse adhesives, J. App. Oral. Sci. Rev. FOB 17 (3) (2009) 184–189.

[73] S. Sharma, P. Sudhakara, J. Singh, R.A. Ilyas, M.R.M. Asyraf, M.R. Razman, Critical review of biodegradable and bioactive polymer composites for bone tissue engineering and drug delivery applications, Polymers 13 (16) (2021) 2623.

[74] H.S. Mahadevaswamy, B. Suresha, Role of nano-$CaCO_3$ on mechanical and thermal characteristics of pineapple fibre reinforced epoxy composites, Mater. Today Proc. 22 (2020) 572–579.

[75] H.-C. Dang, W.-C. Nie, X.-L. Wang, W.-T. Wang, F. Song, Y.-Z. Wang, Dandelion-like $CaCO_3$ microspheres: ionic liquid-assisted biomimetic synthesis and in situ fabrication of poly(ε-caprolactone)/$CaCO_3$ composites with high performance, RSC Adv. 4 (96) (2014) 53380–53386.

[76] K.-J. Wu, C.-S. Wu, J.-S. Chang, Biodegradability and mechanical properties of polycaprolactone composites encapsulating phosphate-solubilizing bacterium Bacillus spPG01 Process. Bio. Chem. 42 (4) (2007) 669–675.

[77] K.-K. Yang, X.-L. Wang, Y.-Z. Wang, Poly(p-dioxanone) and its copolymers, J. Macromol. Sci. Part. C. 42 (3) (2002) 373–398.

[78] O. Vincent, R. Suprakas Sinha, Nano-biocomposites based on synthetic aliphatic polyesters and nanoclay, Prog. Mater. Sci. 62 (2014) 1–57.

[79] I. Chiulan, A.N. Frone, C. Brandabur, D.M. Panaitescu, Recent advances in 3D printing of aliphatic polyesters, Bio. Eng. 5 (1) (2018) 2.

[80] J.-M. Raquez, R. Narayan, P. Dubois, Recent advances in reactive extrusion processing of biodegradable polymer-based compositions, Macromol. Mater. Eng. 293 (6) (2008) 447–470.

[81] S.S. Ray, S. Vaudreuil, A. Maazouz, M. Bousmina, Dispersion of multi-walled carbon nanotubes in biodegradable poly(butylene succinate) matrix, J. Nanosci. Nanotech. 6 (7) (2006) 2191–2195.

[82] G.X. Chen, E.S. Kim, J.S. Yoon, Poly(butylene succinate)/twice functionalized organoclay nanocomposites: preparation, characterization, and properties, J. App. Polym. Sci. 98 (4) (2005) 1727–1732.

[83] Y. Someya, T. Nakazato, N. Teramoto, M. Shibata, Thermal and mechanical properties of poly (butylene succinate) nanocomposites with various organo-modified montmorillonites, J. App. Polym. Sci 91 (3) (2004) 1463–1475.

[84] S.T. Lim, C.H. Lee, H.B. Kim, H.J. Choi, M.S. Jhon, Polymer/organoclay nanocomposites with biodegradable aliphatic polyester and its blends: preparation and characterization, e-Polymers 4 (1) (2004).

[85] J. Kim, T.K. Shin, H.J. Choi, M.S. Jhon, Miscibility of biodegradable synthetic aliphatic polyester and poly (epichlorohydrin) blends, Polymer 40 (24) (1999) 6873–6876.

[86] C.Y. Tang, D. Chen, C.P. Tsui, P. Uskokovic, P.H. Yu, M.C. Leung, Nonisothermal melt-crystallization kinetics of hydroxyapatite-filled poly (3-hydroxybutyrate) composites, J. App. Poly. Sci. 102 (6) (2006) 5388–5395.

[87] W.M. Choi, T.W. Kim, O.O. Park, Y.K. Chang, J.W. Lee, Preparation and characterization of poly (hydroxybutyrate-co-hydroxyvalerate)–organoclay nanocomposites, J. App. Polym. Sci. 90 (2) (2003) 525–529.

[88] S. Wang, C. Song, G. Chen, T. Guo, J. Liu, B. Zhang, S. Takeuchi, Characteristics and biodegradation properties of poly (3-hydroxybutyrate-co-3-hydroxyvalerate)/organophilic montmorillonite (PHBV/OMMT) nanocomposite, Polym. Degrad. Stab. 87 (1) (2005) 69–76.

[89] J. Puiggalí, L.J. del Valle, R. Katsarava, Other miscellaneous materials and their nanocomposites, Nanomaterials Polymer Nanocomposites, Elsevier, 2019, pp. 353–398.

[90] R. Jannapu Reddy, Preparation, Characterization and Properties Of Injection Molded Graphene Nanocomposites, Wichita State University, 2010.

[91] H. Xia, Q. Wang, Preparation of conductive polyaniline/nanosilica particle composites through ultrasonic irradiation, J. App. Poly. Sci. 87 (11) (2003) 1811–1817.

[92] B. Sharma, S. Mahajan, R. Chhibber, R. Mehta, Glass fiber reinforced polymer-clay nanocomposites: processing, structure and hygrothermal effects on mechanical properties, Procedia Chem. 4 (2012) 39–46.

[93] M. Alexandre, P. Dubois, Polymer-layered silicate nanocomposites: preparation, properties and uses of a new class of materials, Mater. Sci. Eng. R Rep. 28 (1–2) (2000) 1–63.

[94] M. Tanahashi, Development of fabrication methods of filler/polymer nanocomposites: with focus on simple melt-compounding-based approach without surface modification of nanofillers, Materials 3 (3) (2010) 1593–1619.

[95] H.Y. Ma, L.F. Tong, Z.B. Xu, Z.P. Fang, Functionalizing carbon nanotubes by grafting on intumescent flame retardant: nanocomposite synthesis, morphology, rheology, and flammability, Advan. Funct. Mater. 18 (3) (2008) 414–421.

[96] X.-W. Hu, S. Liu, B.-T. Qu, X.-Z. You, Starfish-shaped $Co_3O_4/ZnFe_2O_4$ hollow nanocomposite: synthesis, supercapacity, and magnetic properties, ACS Appl. Mater. Interfaces 7 (18) (2015) 9972–9981.

[97] H. Jing, E. Sahle-Demessie, G.A. Sorial, Inhibition of biofilm growth on polymer-MWCNTs composites and metal surfaces, Sci. Total. Environ. 633 (2018) 167–178.

[98] A.A. Koval'chuk, V.G. Shevchenko, Effect of carbon nanotube functionalization on the structural and mechanical properties of polypropylene/MWCNT composites, Macromolecules 41 (20) (2008) 7536–7542.

[99] B.P. Frank, D.P. Durkin, E.R. Caudill, L. Zhu, D.H. White, M.L. Curry, J.A. Pederson, D.H. Fairbrother, Impact of silanization on the structure, dispersion properties, and biodegradability of nanocellulose as a nanocomposite filler, ACS Appl. Nano Mater. 1 (12) (2018) 7025–7038.

[100] G. Mago, D.M. Kalyon, F.T. Fisher, Nanocomposites of polyamide-11 and carbon nanostructures: development of microstructure and ultimate properties following solution processing, J. Polym. Sci. Part. B: Polym. Phys. 49 (18) (2011) 1311–1321.

[101] J. Ma, J. Xu, J.-H. Ren, Z.-Z. Yu, Y.-W. Mai, A new approach to polymer/montmorillonite nanocomposites, Polymer 44 (16) (2003) 4619–4624.

[102] Y. Shen, Y. Lin, C.W. Nan, Interfacial effect on dielectric properties of polymer nanocomposites filled with core/shell-structured particles, Adv. Funct. Mater. 17 (14) (2007) 2405–2410.

[103] V. Mittal, Polymer layered silicate nanocomposites: a review, Materials 2 (3) (2009) 992–1057.

[104] N. Ismail, A. Ismail, A. Mustafa, A. Zulhairun, F. Aziz, N. Bolong, A.R. Razali, Polymer clay nanocomposites for gas separation: a review, Environ. Cont. Rev. 2 (1) (2019) 01–05.

[105] Y. Rao, Engineering of interface in nanocomposites based on PU polymers, Polyurethane Polymers, Elsevier, 2017, pp. 73–133.

[106] M. Auad, S. Nutt, V. Pettarin, P.M. Frontini, Synthesis and properties epoxy-phenolic clay nanocomposites, eXPRESS Polymer Letters (2007).

[107] F.d C. Fim, J.M. Guterres, N.R. Basso, G.B. Galland, Polyethylene/graphite nanocomposites obtained by in situ polymerization, J. Polym. Sci. A Polym. Chem. 48 (3) (2010) 692–698.

[108] X.-L. Xie, Q.-X. Liu, R.K.-Y. Li, X.-P. Zhou, Q.-X. Zhang, Z.-Z. Yu, Y.W. Mai, Rheological and mechanical properties of PVC/$CaCO_3$ nanocomposites prepared by in situ polymerization, Polymer 45 (19) (2004) 6665–6673.

[109] M.A. Tasdelen, Van Camp, W. Goethals, E. Dubois, P. Du Prez, F. Yagci, Y. Polytetrahydrofuran/clay nanocomposites by in situ polymerization and "click" chemistry processes, Macromolecules 41 (16) (2008) 6035–6040.

[110] C. Zeng, L.J. Lee, Poly (methyl methacrylate) and polystyrene/clay nanocomposites prepared by in-situ polymerization, Macromolecules 34 (12) (2001) 4098–4103.

[111] Y.H. Kim, D.K. Lee, H.G. Cha, C.W. Kim, Y.C. Kang, Y.S. Kang, Preparation and characterization of the antibacterial Cu nanoparticle formed on the surface of SiO_2 nanoparticles, J. Phys. Chem. B 110 (49) (2006) 24923–24928.

[112] B. AlMangour, D. Grzesiak, J.-M. Yang, Scanning strategies for texture and anisotropy tailoring during selective laser melting of TiC/316L stainless steel nanocomposites, J. All. Comp. 728 (2017) 424–435.

[113] Q. Han, R. Setchi, F. Lacan, D. Gu, S.L. Evans, Selective laser melting of advanced Al-Al_2O_3 nanocomposites: simulation, microstructure and mechanical properties, Mater. Sci. Eng. A 698 (2017) 162–173.

[114] P.-C. Ma, N.A. Siddiqui, G. Marom, J.-K. Kim, Dispersion and functionalization of carbon nanotubes for polymer-based nanocomposites: a review, Compos. A App. Sci. Man. 41 (10) (2010) 1345–1367.

[115] K. Yang, L. Feng, H. Hong, W. Cai, Z. Liu, Preparation and functionalization of graphene nanocomposites for biomedical applications, Nat. Protoc. 8 (12) (2013) 2392–2403.

[116] M. Theodore, M. Hosur, J. Thomas, S. Jeelani, Influence of functionalization on properties of MWCNT–epoxy nanocomposites, Mater. Sci. Eng. A 528 (3) (2011) 1192–1200.

[117] P.C. Ma, J.-K. Kim, B.Z. Tang, Effects of silane functionalization on the properties of carbon nanotube/epoxy nanocomposites, Comp. Sci. Tech. 67 (14) (2007) 2965–2972.

[118] D.B. Mawhinney, V. Naumenko, A. Kuznetsova, J.T. Yates Jr, J. Liu, R. Smalley, *S*urface defect site density on single walled carbon nanotubes by titration, Chem. Phys. Lett. 324 (1–3) (2000) 213–216.

[119] I.-Y. Jeon, D.W. Chang, N.A. Kumar, J.-B. Baek, Functionalization of carbon nanotubes, Carbon Nanotubes Polym. Nanocompos. (2011) 91–110.

[120] N.G. Sahoo, S. Rana, J.W. Cho, L. Li, S.H. Chan, Polymer nanocomposites based on functionalized carbon nanotubes, Prog. Poly. Sci. 35 (7) (2010) 837–867.

[121] R. Rafiee, R. Pourazizi, Influence of CNT functionalization on the interphase region between CNT and polymer, Comp. Mater. Sci. 96 (2015) 573–578.

[122] J. Paredes, M. Burghard, Dispersions of individual single-walled carbon nanotubes of high length, Langmuir 20 (12) (2004) 5149–5152.

[123] V.C. Moore, M.S. Strano, E.H. Haroz, R.H. Hauge, R.E. Smalley, J. Schmidt, Y. Talmon, Individually suspended single-walled carbon nanotubes in various surfactants, Nano Lett. 3 (10) (2003) 1379–1382.

[124] G. Sáfar, H. Ribeiro, L. Malard, F. Plentz, C. Fantini, A. Santos, G. de Freitas-Silva, Y.M. Idemori, Optical study of porphyrin-doped carbon nanotubes, Chem. Phys. Lett 462 (1–3) (2008) 109–111.

[125] Y. Tan, D.E. Resasco, Dispersion of single-walled carbon nanotubes of narrow diameter distribution, J. Phys. Chem. B 109 (30) (2005) 14454–14460.

[126] M. Islam, E. Rojas, D. Bergey, A. Johnson, A. Yodh, High weight fraction surfactant solubilization of single-wall carbon nanotubes in water, Nano Lett. 3 (2) (2003) 269–273.

[127] V. Lordi, N. Yao, Molecular mechanics of binding in carbon-nanotube−polymer composites, J. Mater. Res. 15 (12) (2000) 2770−2779.
[128] K. Yurekli, C.A. Mitchell, R. Krishnamoorti, Small-angle neutron scattering from surfactant-assisted aqueous dispersions of carbon nanotubes, J.A. Chem. Soc. 126 (32) (2004) 9902−9903.
[129] R. Bandyopadhyaya, E. Nativ-Roth, O. Regev, R. Yerushalmi-Rozen, Stabilization of individual carbon nanotubes in aqueous solutions, Nano Lett. 2 (1) (2002) 25−28.
[130] P. Imin, F. Cheng, A. Adronov, Supramolecular complexes of single walled carbon nanotubes with conjugated polymers, Poly. Chem. 2 (2) (2011) 411−416.
[131] Y. Kang, T.A. Taton, Micelle-encapsulated carbon nanotubes: a route to nanotube composites, J.A. Chem. S 125 (19) (2003) 5650−5651.
[132] R. Shvartzman-Cohen, Y. Levi-Kalisman, E. Nativ-Roth, R. Yerushalmi-Rozen, Generic approach for dispersing single-walled carbon nanotubes: the strength of a weak interaction, Langmuir 20 (15) (2004) 6085−6088.
[133] E. Nativ-Roth, R. Shvartzman-Cohen, C. Bounioux, M. Florent, D. Zhang, I. Szleifer, R. Yerushalmi-Rozen, Physical adsorption of block copolymers to SWNT and MWNT: a nonwrapping mechanism, Macromolecules 40 (10) (2007) 3676−3685.
[134] H. Park, J. Zhao, J.P. Lu, Effects of sidewall functionalization on conducting properties of single wall carbon nanotubes, Nano Lett. 6 (5) (2006) 916−919.
[135] X. Zhang, T. Sreekumar, T. Liu, S. Kumar, Properties and structure of nitric acid oxidized single wall carbon nanotube films, J. Phys. Chem. C. B 108 (42) (2004) 16435−16440.
[136] A. Kuznetsova, D.B. Mawhinney, V. Naumenko, J.T. Yates Jr, J. Liu, R. Smalley, Enhancement of adsorption inside of single-walled nanotubes: opening the entry ports, Chem. Phys. Lett. 321 (3−4) (2000) 292−296.
[137] V. Georgakilas, K. Kordatos, M. Prato, D.M. Guldi, M. Holzinger, A. Hirsch, Organic functionalization of carbon nanotubes, J. Am. Chem. Soc. 124 (5) (2002) 760−761.
[138] V. Datsyuk, M. Kalyva, K. Papagelis, J. Parthenios, D. Tasis, A. Siokou, I. Kallitsis, C. Galiotis, Chemical oxidation of multiwalled carbon nanotubes, Carbon 46 (6) (2008) 833−840.
[139] I. Vasiliev, S.A. Curran, Cross linking of thiolated carbon nanotubes: an ab initio study, J. App. Phy 102 (2) (2007) 024317.
[140] S. Niyogi, M. Hamon, H. Hu, B. Zhao, P. Bhowmik, R. Sen, M.E. Itkis, R.C. Haddon, Chemistry of single-walled carbon nanotubes, Account. Chem. Res. 35 (12) (2002) 1105−1113.
[141] G.R. Berdiyorov, H. Hamoudi, Doping-enhanced current rectification in carbon nanotube−metal junctions for rectenna applications, ACS Omega 5 (1) (2019) 189−196.
[142] Y. Liu, Y. Zhang, X. Wang, Z. Wang, W. Lai, X. Zhang, X. Liu, Excellent microwave absorbing property of multiwalled carbon nanotubes with skin−core heterostructure formed by outer dominated fluorination, J. Phys. Chem. C. 122 (11) (2018) 6357−6367.
[143] D. Sakyi-Arthur, S.Y. Mensah, K.W. Adu, K.A. Dompreh, R. Edziah, N.G. Mensah, C. Jebuni-Adanu, Induced hall-like current by acoustic phonons in semiconductor fluorinated carbon nanotube, World J. Condens. Matter. Phys. 10 (02) (2020) 71.
[144] P. Bhagabati, M. Rahaman, D. Khastgir, Morphology and spectroscopy of polymer−carbon composites, Carbon-Containing Polymer Composites, Springer, 2019, pp. 295−338.
[145] D. Atre, S. Pant, Charged carbon nanotubes, Carbon Nanotubes Recent. Prog. (2018) 109.
[146] R. Dubey, D. Dutta, A. Sarkar, P. Chattopadhyay, Functionalized carbon nanotubes: synthesis, properties and applications in water purification, drug delivery, Mater. Biomed. Sci. Nanoscale Adv. (2021).
[147] Z. Guo, T. Pereira, O. Choi, Y. Wang, H.T. Hahn, Surface functionalized alumina nanoparticle filled polymeric nanocomposites with enhanced mechanical properties, J. Mater. Chem. 16 (27) (2006) 2800−2808.

[148] L. Guerrini, R.A. Alvarez-Puebla, N. Pazos-Perez, Surface modifications of nanoparticles for stability in biological fluids, Materials 11 (7) (2018) 1154.
[149] Y. Liu, X. Zhang, Q. Gao, H. Huang, Y. Liu, M. Min, L. Wang, Structure and properties of polyoxymethylene/silver/maleic anhydride-grafted polyolefin elastomer ternary nanocomposites, Polymers 13 (12) (2021) 1954.
[150] H. Zheng, H. Xue, Y. Zhang, Z. Shen, Effect of silver nanoparticles on the microstructure, non-isothermal crystallization behavior and antibacterial activity of polyoxymethylene, Polymers 12 (2) (2020) 424.
[151] Y. Zeng, Y. Liu, L. Wang, H. Huang, X. Zhang, Y. Liu, G. Qi, M. Min, Y. Li, Effect of silver nanoparticles on the melting behavior, isothermal crystallization kinetics and morphology of polyoxymethylene, Crystals 10 (7) (2020) 594.
[152] A. Grigaloviča, R.M. Meri, J. Zicāns, J. Grabis, A. Aniskevich, Aging of nanosized ZnO modified polyoxymethylene blends with ethylene–octene copolymer, Polym. Plast. Technol. Eng. 54 (11) (2015) 1201–1206.
[153] C. Pagano, R. Surace, V. Bellantone, F. Baldi, I. Fassi, Mechanical characterisation and replication quality analysis of micro-injected parts made of carbon nanotube/polyoxymethylene nanocomposites, J. Comp. Mater. 52 (5) (2018) 645–657.
[154] M. Raji, M.E.M. Mekhzoum, D. Rodrigue, R. Bouhfid, Effect of silane functionalization on properties of polypropylene/clay nanocomposites, Compos. B Eng. 146 (2018) 106–115.
[155] C. Cazan, A. Enesca, L. Andronic, Synergic effect of TiO_2 filler on the mechanical properties of polymer nanocomposites, Polymers 13 (12) (2021) 2017.
[156] S. Kalia, A. Dufresne, B.M. Cherian, B.S. Kaith, L. Avérous, J. Njuguna, E. Nassiopoulos, Cellulose-based bio-and nanocomposites: a review, Intern. J. Polym. Sci. (2011) 2011.
[157] R. Bouhfid, M. Jawaid, Graphene and Nanoparticles Hybrid Nanocomposites: From Preparation to Applications, Springer, 2021.
[158] N. Saba, P.M. Tahir, M. Jawaid, A review on potentiality of nano filler/natural fiber filled polymer hybrid composites, Polymers 6 (8) (2014) 2247–2273.
[159] N.B. Singh, S. Agarwal, Nanocomposites: an overview, Emerg. Mater. Res. 5 (1) (2016) 5–43.
[160] A. Dasari, Z.-Z. Yu, Y.-W. Mai, Polymer Nanocomposites: Towards Multi-Functionality, Springer, 2016.
[161] G.E. Luckachan, C. Pillai, Biodegradable polymers-a review on recent trends and emerging perspectives, J. Poly. Environ. 19 (3) (2011) 637–676.
[162] R. Song, M. Murphy, C. Li, K. Ting, C. Soo, Z. Zheng, Current development of biodegradable polymeric materials for biomedical applications, Drug. Des. Dev. 12 (2018) 3117.
[163] A. Buzarovska, PLA nanocomposites with functionalized TiO_2 nanoparticles, Polym. Plast. Technol. Eng. 52 (3) (2013) 280–286.
[164] M. Kaseem, K. Hamad, Z. Ur Rehman, Review of recent advances in polylactic acid/TiO_2 composites, Materials 12 (22) (2019) 3659.
[165] Y. Luo, Z. Lin, G. Guo, Biodegradation assessment of poly (lactic acid) filled with functionalized titania nanoparticles (PLA/TiO(2)) under compost conditions, Nano. Res. lett. 14 (1) (2019). 56–56.
[166] A. Marra, C. Silvestre, A.P. Kujundziski, D. Chamovska, D. Duraccio, Preparation and characterization of nanocomposites based on PLA and TiO_2 nanoparticles functionalized with fluorocarbons, Polym. Bullt. 74 (8) (2017) 3027–3041.
[167] R. Sigwadi, T. Mokrani, P. Msomi, F. Nemavhola, The effect of sulfated zirconia and zirconium phosphate nanocomposite membranes on fuel-cell efficiency, 2021. doi:10.3390/polym14020263.
[168] H.H. So, J.W. Cho, N.G. Sahoo, Effect of carbon nanotubes on mechanical and electrical properties of polyimide/carbon nanotubes nanocomposites, Euro. Polym. J. 43 (9) (2007) 3750–3756.
[169] S. Abazari, A. Shamsipur, H.R. Bakhsheshi-Rad, A.F. Ismail, S. Sharif, M. Razzaghi, S. Ramakrishna, F. Berto, Carbon nanotubes (CNTs)-reinforced magnesium-based matrix composites: a comprehensive review, Materials 13 (19) (2020) 4421.

[170] T.A. Saleh, N.P. Shetti, M.M. Shanbhag, K.R. Reddy, T.M. Aminabhavi, Recent trends in functionalized nanoparticles loaded polymeric composites: an energy application, Mater. Sci. Ener. Technol. 3 (2020) 515–525.
[171] B. Sharma, R. Chhibber, R. Mehta, Seawater ageing of glass fiber reinforced epoxy nanocomposites based on silylated clays, Polym. Degrad. Stab. 147 (2018) 103–114.
[172] A. Tcherbi-Narteh, D. Watson, M. Zaheeruddin, D. Tobias, M. Hosur, S. Jeelani. Comparison of seawater effects on mechanical and viscoelastic properties of epoxy composites with glass, carbon and their hybrid fabric reinforcements.
[173] E. Manias, A. Touny, L. Wu, K. Strawhecker, B. Lu, T. Chung, Polypropylene/montmorillonite nanocomposites. Review of the synthetic routes and materials properties, Chem. Mater. 13 (10) (2001) 3516–3523.
[174] D. Barreca, G. Carraro, E. Comini, A. Gasparotto, C. Maccato, C. Sada, G. Sberveglieri, E. Tondello, Novel synthesis and gas sensing performances of $CuO-TiO_2$ nanocomposites functionalized with Au nanoparticles, J. Phys. Chem. C. 115 (21) (2011) 10510–10517.
[175] B.J. Ash, R.W. Siegel, L.S. Schadler, Glass-transition temperature behavior of alumina/PMMA nanocomposites, J. Polym. Sci. Part B: Polym. Phys. 42 (2004) 4371–4383.
[176] P. Saha, J. Adhikari, K. Deshmukh, M.B. Ahamed, J.J. Cabibihan, Recent advances in mechanical properties of biopolymer composites: a review, Polym. Comp. 41 (2020) 32–59.
[177] L. Li, X. Xu, L. Liu, P. Song, Q. Cao, Z. Xu, Z. Fang, H. Wang, Water governs the mechanical properties of poly (vinyl alcohol), Polymer 213 (2021) 123330.
[178] M. Jonoobi, J. Harun, A.P. Mathew, K. Oksman, Mechanical properties of cellulose nanofiber (CNF) reinforced polylactic acid (PLA) prepared by twin screw extrusion, Comp. Sci. Technol. 70 (12) (2010) 1742–1747.
[179] Y. Lattach, J.F. Rivera, T. Bamine, A. Deronzier, J.-C. Moutet, Iridium oxide–polymer nanocomposite electrode materials for water oxidation, ACS Appl. Mater. Interfaces 6 (15) (2014) 12852–12859.
[180] S. Pan, J. Li, Z. Wen, R. Lu, Q. Zhang, H. Jin, L. Zhang, Y. Chen, S. Wang, Materials for photo (electro) chemical applications: dimensionality, heterojunction, and performance, Advan. Ener. Mater. (2021) 2004002.
[181] S. Alhammadi, V.R. Minnam Reddy, S. Gedi, H. Park, M.S. Sayed, J.-J. Shim, W.K. Kim, Performance of graphene–cds hybrid nanocomposite thin film for applications in Cu (In, Ga) Se2 solar cell and h2 production, Nanomaterials 10 (2) (2020) 245.
[182] E. Roy, S. Patra, A. Tiwari, R. Madhuri, P.K. Sharma, Introduction of selectivity and specificity to graphene using an inimitable combination of molecular imprinting and nanotechnology, Biosens. Bioelectron. 89 (2017) 234–248.
[183] T. Reier, M. Oezaslan, P. Strasser, Electrocatalytic oxygen evolution reaction (OER) on Ru, Ir, and Pt catalysts: a comparative study of nanoparticles and bulk materials, ACS Catal. 2 (8) (2012) 1765–1772.
[184] S. Cherevko, T. Reier, A.R. Zeradjanin, Z. Pawolek, P. Strasser, K.J. Mayrhofer, Stability of nanostructured iridium oxide electrocatalysts during oxygen evolution reaction in acidic environment, Electrochem. Comm. 48 (2014) 81–85.
[185] A. Manthiram, A.V. Murugan, A. Sarkar, T. Muraliganth, Nanostructured electrode materials for electrochemical energy storage and conversion, Ener. Environ. Sci. 1 (6) (2008) 621–638.
[186] T. Brown, F. De Rossi, F. Di Giacomo, G. Mincuzzi, V. Zardetto, A. Reale, A. Di Carlo, Progress in flexible dye solar cell materials, processes and devices, J. Mater. Chem. A 2 (28) (2014) 10788–10817.

CHAPTER 8

Green-synthesized nanoparticle-based polymer nanocomposites: synthesis, characterizations, and applications

Antony V. Samrot[1], Subramanian Saigeetha[2], Nagarajan Shobana[3] and Kumar Chandrasekaran[4]

[1]School of Bioscience, Faculty of Medicine, Bioscience and Nursing, MAHSA University, Selangor, Jenjarom, Malaysia [2]Department of Biotechnology, School of Biosciences and Technology, Vellore Institute of Technology, Vellore, Tamil Nadu, India [3]Department of Biotechnology, School of Bio and Chemical Engineering, Sathyabama Institute of Science and Technology, Chennai, Tamil Nadu, India [4]Centre for Ocean Research (DST—FIST Sponsored Centre), MoES—Earth Science & Technology Cell, Sathyabama Institute of Science and Technology, Chennai, Tamil Nadu, India

Chapter Outline

Biodegradable and Biocompatible Polymer Nanocomposites. DOI: https://doi.org/10.1016/B978-0-323-91696-7.00012-X
Copyright © 2023 Elsevier Inc. All rights reserved.

8.1 Introduction

Nanotechnology is one of the most recently researched fields due to its numerous applications. Nanomaterials are nothing but composition of atoms/molecules of size less than 100 nm. It has high surface area: volume ratio [1]. Nanoparticles are classified into inorganic, organic, metal, or carbon nanoparticles based on their compositions [2]. Several types of nanoparticles have been synthesized from various sources, and each have variety of applications [3]. These nanostructured materials are typically small in size and are released into the environment, where they become toxic to nature. However, not all nanoparticles are toxic. One of the main criteria for determining the toxicity of the produced nanoparticles is their chemical composition. Some biosynthesized nanoparticles are thought to have health benefits and are nontoxic also [4].

Many approaches, including bottom-up and top-down, have been used for the preparation of nanoparticles using chemical, biological, and physical methods. Many researchers are interested in green synthesis these days since a plant or one of its metabolites is employed to make nanoparticles [5]. Conventional way of nanoparticle synthesis is mostly toxic, unfriendly to the environment, and expensive, so naturally occurring sources are used for the synthesis of nanoparticles. Green synthesis is the use of microorganisms, plant and animal metabolites, or other natural resources to create nanoparticles [6]. Traditionally, microscale particles were used to reinforce composites, but inclusion sizes have shrunk to the nanoscale [7]. Because of their potential applications, nanocomposites can be used to manufacture glasses, ceramics, and other materials [8]. Many researchers are interested in polymer-based nanocomposites, which are used to improve reinforcement, flame resistance, electro-optical properties, bactericidal properties, and so on [9].

Polymer nanocomposites are generally made up of a polymer matrix and additives such as 1D, 2D, and 3D nanomaterials. Nanoadditives added to polymer-based nanocomposites have extremely high surface area-to-volume ratios, making them extremely elastic and strong [10]. Polymer nanocomposites are made up of different shaped nanoparticles, such as spheres and rods, that are dispersed in a polymer matrix which creates new properties and expands its applications [11]. These composites are environmentally friendly, biodegradable, and harmless, with high physical, chemical, and mechanical properties, as well as a superior compatible nature due to the presence of both counterparts in the composites [12]. However, the nature of these particles is heavily influenced by the scaling process [13]. Production of nanocomposites by polymer blending is another important process that improves the chemical, physical, and electrical properties of polymer substances because of their high flexibility, tunability, and wide temperature operating performance [14–16]. According to many publications, nanocomposites have many properties that allow them to be used in a variety of applications such as microelectronic devices, biomedical sciences, organic solar cells, piezoelectric sensors, electrolytes, and supercapacitors [17]. Various methods for synthesizing polymer-based nanocomposites using green synthesis, their characterization, and various applications are extensively discussed in this review.

8.2 Green-synthesized nanoparticles

Some of the disadvantages of physiochemical nanoparticle production include the generation of hazardous waste, the use of toxic chemicals, and the production of nanoparticles with imperfect surface structures. Nanoparticles synthesized using "Green synthesis" are typically produced using a bottom-up approach in which any plant extract or microbes are used to synthesize metal oxide or metal-based nanoparticles along with the use of a strong reducing agent [18]. Green-synthesized nanoparticles are nontoxic, environmentally friendly, and use safe reagents. Because they are created in a single step, these nanoparticles are more stable and have more precise dimensions [19]. The three main processes involved in green synthesis are the selection of a suitable solvent for the production of nanoparticles, the use of a bio-derived reducing agent, and the use of nontoxic materials to maintain its stability [20]. The production of nanoparticles from plant extracts is superior to the use of microorganisms as the nanoparticles are more stable, and the synthesis method is also faster [21]. Green synthesis is the best technique for producing nanoparticles on a large scale because it requires little energy and no high pressure, making it a highly energy-efficient method. It is based on the use of biological extracts and is heavily influenced by variables such as pressure, temperature, pH, and the solvent used [22]. Green synthesis, also known as biosynthesis, is the use of green chemistry to produce nanoparticles using unicellular and multicellular organisms such as bacteria, algae, fungi, yeast, plants, viruses, and actinomycetes [23]. Green nanoparticle synthesis has the advantage of producing nanoparticles with a homogeneous chemical composition and few defects. Plants and microorganisms are widely used in this process to produce nanoparticles. Some conditions, such as biological pathways, enzyme production, enzyme activities, and cell growth, must be optimal for effective nanoparticle production. This is determined by the type of plant extract and microorganisms used [24]. Plants and microorganisms are used to produce nanoparticles, which are elaborated in this chapter.

8.3 Production of nanoparticles from plants

Plants can be used to produce nanoparticles in green synthesis because they are widely distributed, easily available, contain an abundant amount of metabolites, and are much safer to handle [25]. Silver nanoparticles (AgNPs) were synthesized using an aqueous leaf extract of *Azadirachta indica* as both a reducing and capping agent. The particles were uniform and spherical in shape, and they were well dispersed, according to transmission electron microscopy analysis [26]. Plant extracts of *Origanum vulgare* L. were used as reducing agents in the synthesis of silver nanoparticles. Biosynthesized nanoparticles have also been shown to have high antibacterial activity [27]. In-vitro plant growth of *Salvia spinosa* extract was used to produce AgNPs. The particles had a spherical shape and were bactericidal against both gram-negative and gram-positive organisms [28]. Kumar et al. [29] reported that the silver nanoparticles were synthesized using the plant extract *Boerhaavia diffusa* as reducing agent. Nanoparticles synthesized were spherical, the average size being 25 nm, and it had face center cubic structure. The plant extracts from reetha and shikakai were used for the prodution of AgNPs which were in size of 50 nm and with good stability. The biosurfactants found in plant extracts were responsible for the formation of stable nanoparticles [30]. *Gloriosa superba* L. plant extract was used in the green synthesis of CuO nanoparticles. Transmission electron microscopy (TEM) images revealed that the size which were around 5–10 nm and had excellent antibacterial activities [31]. The leaf extract of *Cacumen Platycladi* was used to perform biological reduction of Au(III) and Pd(II) precursor. The produced nanoparticles were approximately 7 nm in size [32].

Gum karaya was used in the production of copper oxide (CuO) nanoparticles and the size was around 4.8, 5.5 and 7.8 nm when the $CuCl_2.H_2O$ was used in different molar concentrations like 1,

2 and 3 mM respectively [33]. CuO nanoparticles were synthesized using the leaf extract of *Aloe vera* plant and the size was around 20 nm. The synthesized nanoparticles were tested against fish pathogens like *Flavobacterium branchiophilum, Pseudomonas fluorescens*, and *Aeromonas hydrophila* and it was reported to be effective against all the fish pathogens even at minimum concentration [34]. Zinc oxide (ZnO) and CuO nanoparticles of size 34 nm were synthesized using aqueous extract of Oak fruit hull which showed that the ZnO and CuO nanoparticles were spherical and quasi spherical in shape, respectively [35]. The leaf extract of *Abutilon indicum* was used for the synthesis of copper oxide nanoparticles with high antioxidant activity and antibacterial activity [36]. *Carica papaya* leaf extracts were used for the biosynthesis of CuO nanoparticles, where the produced particles were crystalline and the size was around 140 nm [37].

ZnO nanoparticles were biosynthesized using *Aloe barbadensis miller* leaf extracts which had the size range between 25 and 40 nm and were spherical in shape [38]. *A. indica* leaf extract was used to make ZnO nanoparticles that were polycrystalline and porous in structure, and they had good antibacterial activity [39]. *Atlantia monophyla* leaf extracts were used for the green synthesis of ZnO nanoparticles with a size of around 30 nm and bactericidal activity against Gram-negative and Gram-positive microorganisms [40]. Aqueous extract of *Deverra tortuosa* was used to create zinc oxide nanoparticles that ranged in size from 9 to 32 nm and were crystalline in nature [41]. The aqueous seed extract of the plant *Piper nigrum* was used to create zinc ferrite nanoparticles. The nanoparticles synthesized were highly stable and pure, with sizes ranging from 60 to 80 nm [42].

Cobalt ferrite microparticles were synthesized using ginger root and cardamom seeds [43]. Tragacanth gum was used for the synthesis of copper ferrite nanoparticles which had average crystal size of 14 nm and it was cubic phase ferrite nanoparticles [44]. Zinc ferrite nanoparticles were synthesized by Sarala et al. [45] with the average crystal size of 17.12 nm and the particles were spherical and slightly agglomerated (Fig. 8.1). Table 8.1 shows the synthesis of various nanoparticles using various plant extracts. Silver nanoparticles were created using the gums of *Araucaria heterophylla, Prosopis chilensis*, and *A. indica* [46]. Plant-based green synthesis like utilizing *Capsicum annum* extracts is well reported [47].

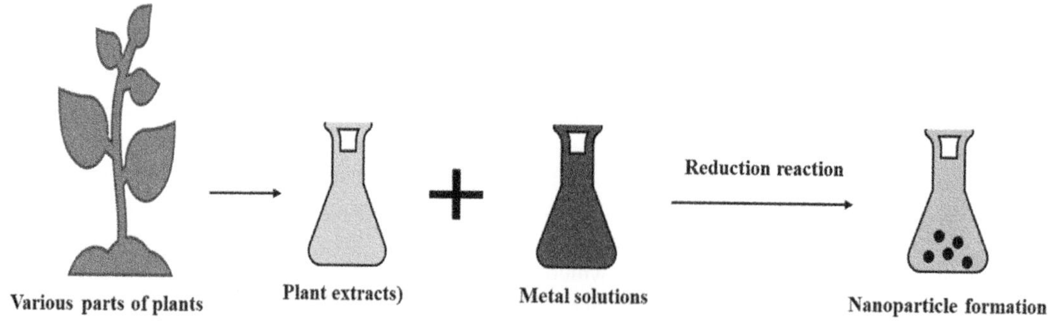

FIGURE 8.1

Green synthesis of nanoparticles using plant extract.

Table 8.1 Plant mediated synthesis of nanoparticles.

S. no.	Plant extract used	Nanoparticle produced	References
1	*A. heterophylla*, *P. chilensis*, and *A. indica* (Gums)	Silver nanoparticles	[46]
2	Ripened fruit of *C. annum*	Silver nanoparticles	[47]
3	Neem leaf extract	Silver nanoparticles	[48]
4	*A. heterophylla*	Silver nanoparticles	[49]
5	*Punica granatum*	Copper oxide nanoparticles	[50]
6	Black bean	Copper oxide nanoparticles	[51]
7	*Ixiro coccinea* leaf extract	Copper oxide nanoparticles	[52]
8	*Beta vulgaris*, *Cinnamomum tamala*, *Cinnamomum verum*, *Brassica oleracea* var. *italica*	Zinc oxide nanoparticles	[53]
9	*Hibiscus subdariffa* leaf extract	Zinc oxide nanoparticles	[54]
10	*C. guianensis* flower extract	Gold nanoparticles	[55]
11	Natural honey	Gold nanoparticles	[56]
12	*Salvia officinalis*, *P. granatum*, *Pelargonium graveolens*, and *Lippia citriodora*	Gold nanoparticles	[57]
13	*Agathosma betulina* leaf extract	Zinc oxide nanoparticles	[58]
14	*Cassia fistula* plant extract	Zinc oxide nanoparticles	[59]
15	*Calotropis gigantea* leaf extract	Zinc oxide nanoparticles	[60]
16	*Centella asiatica* leaf extract	Gold nanoparticles	[61]
17	*Murraya koenigii* leaf extract	Gold nanoparticles	[62]
18	Fenugreek seed extract	Selenium nanoparticles	[63]
19	*Clausena dentata* leaf extract	Selenium nanoparticles	[64]
20	*A. indica* leaf extract	Platinum nanoparticles	[65]
21	Orange peel	Platinum nanoparticles	[66]
22	Pomegranate crusts extract	Platinum nanoparticles	[67]
23	*Rosa* leaf extract	Aluminum nanoparticles	[68]
24	*Mulberry* leaf extract	Aluminum nanoparticles	[69]
25	*Jatropha curas* latex extract	Zinc sulfide nanoparticles	[70]
26	*Euphorbia hirta* aqueous extract	Silver nanoparticles	[71]
27	*Casia atlata* aqueous extract	Silver nanoparticles	[72]
28	*Thespesia populnea* aqueous extract	Silver nanoparticles	[73]
29	*Wrightia tinctoria* aqueous extract	Silver nanoparticles	[74]
30	Alkaloids, flavonoids, saponins and tannins	Silver nanoparticles	[75]
31	*Saraca indica*, *Annona retculata*, *Aegle marmelos*, and *Datura metel*	Silver nanoparticles	[76]

8.4 Green synthesis of nanoparticles using microorganism and algae

Microorganisms have been used to produce nanoparticles all over the world. Microorganisms such as algae, bacteria, and yeast are widely used for the green synthesis of nanoparticles because they are easy to cultivate and they grow quickly [77]. Many nanoparticles, including silver, gold, silica, selenium, platinum, palladium, titania, quantum dots, and magnetite are produced using microbial process [78]. Mixing the extract or culture of bacteria/fungi/algae with silver nitrate and incubating in dark yields silver nanoparticles (Fig. 8.2). Pugazhenthiran et al. [79] reported the synthesis of silver nanoparticles using *Bacillus* sp., size of which ranged between 5 and 15 nm. *Streptomyces glaucus* and *Spirulina platensis* were used for the production of silver nanoparticles with average size of around 13 nm which were spherical in structure [80]. The synthesis of lead sulfide nanoparticles was carried out by utilizing *Rhodobacter sphaeroides* [81]. *Bacillus cereus* was used for the production of silver nanoparticles with size of 5 − 7 nm which were spherical in structure [82]. Extracellular synthesis process was performed for the production of silver nanoparticles using the bacterium *Bacillus* strain CS 11. The synthesized nanoparticles were in the range between 45 and 92 nm [83]. Yeast strain MKY3 was also used for the extracellular production of very small silver nanoparticles with sizes ranging from 2 to 5 nm [84].

The silver nanoparticles were synthesized extracellularly using the microorganism *Fusarium solani*, and the nanoparticles which were approximately 16.23 nm in size [85]. *Morganella* sp. was used for the biosynthesis of AgNPs which were crystalline in structure and its average size was around 20 nm [86]. Culture supernatant of *Pseudomonas aeruginosa* was used for the synthesis of AgNPs which were around 13 nm [87]. *Pseudomonas* sp. THG-LS1.4 was used in the production of silver nanoparticles which was in between the size ranging from 10 and 40 nm which also showed great antimicrobial activity [88]. Psychrophilic bacteria including *Arthrobacter gangotriensis, Pseudomonas antarctica, Arthrobacter kerguelensis, Pseudomonas meridiana*, and *Pseudomonas proteolytica* and mesophilic bacteria like *Bacillus cecembensis* and *Bacillus indicus* were used for the synthesis of AgNPs, and it was reported that the synthesized silver nanoparticles had bactericidal effect [89]. Brown seaweed named *Sargassum wightii*, a marine algae was used for the biosynthesis of AgNPs and it was found to have good antibacterial activity against bacterias isolated from infected silkworms [90].

Fusarium oxysporum, a fungus was used for the production of gold nanoparticles (AuNPs) through the process of extracellular synthesis and the nanoparticles produced were in size ranging between 20 and 40 nm [91]. *Sargassum wightii* (Greville) was used for the extracellular biosynthesis of AuNPs with sizes ranging from 8 to 12 nm [92]. *Sporosarcina koreensis* DC4 was found to improve the antimicrobial activity of commercial antibiotics when used in the green synthesis of AgNPs and AuNPs [93]. *Streptomyces microflavus* was used in the extracellular synthesis of gold nanotriangles [94]. During the biosynthesis of AuNPs, *Paracoccus haeundaensis* was used and it was nontoxic to normal human cell lines, showed concentration dependent antiproliferative activity against cancer cell lines, and good antioxidant activity [95]. *Salmonella typhimurium* was used for the synthesis of copper nanoparticles and the mean size of the copper nanoparticles were 49 nm [96]. *Serratia nematodiphila* was used for the bioproduction of zinc oxide nanoparticles of 15–30 nm size which reported to have good antibacterial and antifungal activity [97]. *Thermoanaerobacter* sp. and *Shewanella* sp. were used for the production of magnetite nanoparticles [98]. Different fungi including *Penicillium* sp., *Aspergillus concius, Penicillium janthinellum* and *Phomosis* sp., *Penicillium italicum* etc. have been used for production of silver nanoparticles (Table 8.2).

8.5 Production of polymer nanocomposites

Polymer nanotechnology is based on the creation of polymer nanocomposites. The use of nanometric inorganic compounds improves the properties of the polymers, resulting in a variety of applications

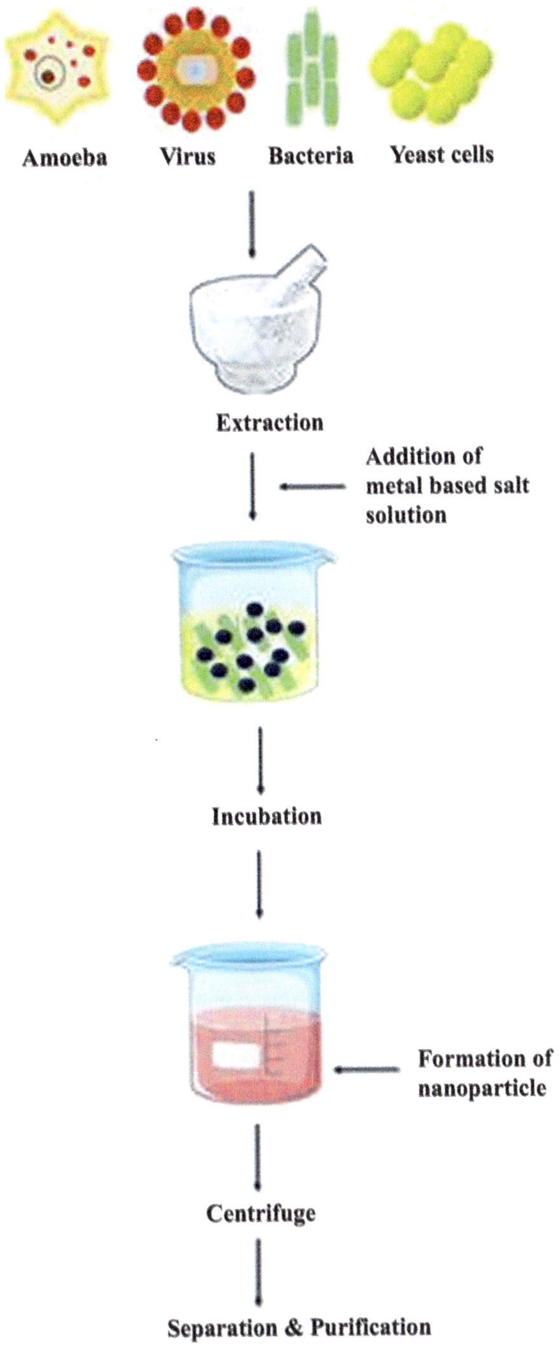

FIGURE 8.2

Green synthesis of nanoparticles using microorganism.

Table 8.2 Synthesis of nanoparticles using microbial routes.

S. no.	Microorganism	Type of nanoparticle	References
1	*Penicillium* sp.	Silver nanoparticles	[99]
2	*Aspergillus concius*, *Penicillium janthinellum* and *Phomosis* sp.	Silver nanoparticles	[100]
3	*Penicillium italicum*	Silver nanoparticles	[101]
4	*Scenedesmus* sp.	Silver nanoparticles	[102]
5	*Bacillus licheniformis*	Silver nanoparticles	[103]
6	*Penicillium brevicompactum*	Gold nanoparticles	[104]
7	*Hormoconis resinae*	Gold nanoparticles	[105]
8	*Bhargavaea indica* DC1	Gold nanoparticles	[105]
9	*Pseudomonas aeruginosa*	Gold nanoparticles	[106]
10	*Trichoderma koningii*	Gold nanoparticles	[107]
11	*Pseudomonas putida*	Zinc oxide nanoparticles	[108]
12	*A. hydrophila*	Zinc oxide nanoparticles	[109]
13	*Lactobacillus plantarum*	Zinc oxide nanoparticles	[110]
14	*Arthrospira platensis*	Zinc oxide nanoparticles	[111]
15	*Aspergillus aeneus*	Zinc oxide nanoparticles	[112]
16	*Shewanella loihica* PV-4	Copper nanoparticles	[113]
17	*Botryococcus braunii*	Copper and Silver nanoparticles	[114]
18	*Saccharomyces boulardii*	Platinum nanoparticles	[115]
19	*Schizosaccharomyces pombe* strain	Cadmium sulfide nanoparticles	[116]
20	*Shewanella oneidensis* MR-1 and *Bacillus subtilis*	Cadmium sulfide nanoparticles	[117]
21	*Penicillium oxalicum*	Cadmium oxide nanoparticles	[118]
22	*Bacillus licheniformis*	Cadmium sulfide nanoparticles	[119]
23	*Rhizopus oryzae*	Gold nanoparticles	[120]
24	*Pleurotus ostreatus*	Silver nanoparticles	[121]
25	*Vibrio natriegens*	Selenium nanoparticles	[122]

based on the polymer's inorganic, chemical-organic compound mixture [123]. The most important strategy for producing bulk nanocomposites is always after the production of nanoparticle suspensions or nanoparticle polymer blends [124]. Raza et al. [125] created silver nanoparticles from the fungus *Aspergillus fumigatus* where they combined it with the biopolymer chitosan to create silver-chitosan nanocomposites. Chitosan was dissolved in 1% glacial acetic acid and mixed with a solution of silver nanoparticles. This mixture was heated at 650 W for 1 min and 30 s at interval of 30 s each and then cooled to room temperature which was further stored at 4°C for characterization. *Cupriavidus necator* strain H16 was used to synthesize silver nanoparticles and a lab scale bioreactor scale up process was used to extract polymer from the culture flask using organic solvents. Polyhydroxybutyrate (PHB) was extracted and its purity was checked before being molded into 80 μm thick films using a hot plate hydraulic press and silver nanoparticles were fabricated onto them. Silver nanoparticle-PHB films demonstrated good antimicrobial properties [126]. Quince leaf extract was used to make silver nanoparticles, and polyethylene oxide (PEO) powder was used to make AgNPs-PEO nanocomposite solution using cast technique, in which 1 g of PEO (3 batches) was dissolved in 50 mL of acetonitrile and suspended silver nanoparticles were added in three different concentrations in each batch and labeled as three different PEO-AgNPs nanocomposites [127].

For the production of bacterial cellulose (BC), *Gluconacetobacter xylinum* was used, and later filtered Ti-BALDH was added and fermentation was carried out for 7 days at 30°C. BC-TiO_2 was harvested and immersed in NaOH solution to bring the pH to neutral, and was then stored at 4°C in deionized water for future use [128]. Silver nanoparticles were biosynthesized using *Bacillus* sp. SJ 14. Polycaprolactonate-silver nanoparticle nanocomposites were created by electrospinning 15% polycaprolactonate in acetone with silver nanoparticles and was characterized and reported that it can be used as a good wound dressing material [129]. Kiziltas et al. [130] prepared bacterial cellulose (BC) comprised of various nanoparticles such as cellulose nanofibrils, exfoliated graphite nanoplatelets, and nanoclay and loaded it with BC medium to produce nanocomposites, which were then characterized. The nanoparticles were discovered to be embedded on BC, resulting in the formation of a nanocomposite.

Chitosan-silver nanocomposite and chitosan gold nanocomposite were produced by dissolving chitosan in acetic acid and mixing in silver and gold nanoparticles. Later these were spread on a silicon coated petri dish and was dried for 24 h at room temperature and further characterized. It has been reported that it can be used as an effective packaging material [131]. To create nanocomposites, silver nanoparticles were synthesized from marine organisms (MER 1) and added to a polystyrene solution [132]. Starch polysaccharide-based nanocomposites were prepared after fabrication with copper nanoparticles synthesized using fungus and the results showed that the nanocomposites were 200 nm in size with excellent antimicrobial, antioxidant and anticancer activities [133]. Carboxy methylcellulose—AgNPs nanocomposites were prepared by Kanikireddy et al. [134] which showed excellent antimicrobial activities which can be used as an efficient packaging material when tested using strawberries.

Styrene maleic anhydride and 2-amino-5-mercapto-1,3,4-thiadiazole were combined to form a modified copolymer and ZnO nanoparticles were dispersed in this copolymer solution, heated, and dried to create nanocomposites [135]. To obtain nanocomposites, black tea extract was used to prepare silver nanoparticles, which were then injected into polyvinyl alcohol (PVA) solution, mixed to a homogenized mixture and casted onto a petri dish, cooled, and dried [136]. Silver nanoparticles were biosynthesized using *Ficus benghalensis* leaf extract. PVA film was made by dissolving PVA in deionized water, stirred it, and then biosynthesized silver nanoparticles were added to the mixture, which was poured onto a Teflon plate, cooled, dried, and then nanocomposites were obtained. [137]. *Jasminum subtriplinerve* leaves were used for the biosynthesis of silver nanoparticles for which 2-hydroxypropyl-β-cyclodextrin/alginate was used as a catalyst. Silver nanoparticles were capped on the aforementioned polymer, which was then poured onto a petriplates, cooled, and dried to produce a nanocomposite that was found to be useful in degrading pollutants from aqueous medium [138]. To create a polymer-based nanocomposite, nickel oxide and tungsten oxide nanoparticles were casted onto polyvinylchloride polymer using a casting technique inwhich the nanofillers were dispersed in dimethylformamide (DMF) and subjected to ultrasonication for 1 h while polyvinylchloride (PVC) was dissolved in DMF for upto 4 h at 60°C. Later the nanofillers were individually added to the PVC solution and sonicated each time. The composite solution was casted onto Teflon petri plates and dried for 6 h in a hot air oven at 60°C. The dried nanocomposites were removed from the petri dish and analyzed [139]. Polyvinyl alcohol and polyvinylidene fluoride nanocomposites were created by adding nafion and using DMF as a solvent, and were then characterized using various analytical techniques [140]. Solution casting method is widely used for the production for nanocomposites even for nonnanofiller based polymer nanocomposites (Fig. 8.3).

Pandey et al. [141] synthesized nanocomposites by using polyvinyl pyrrolidone and polyvinyl alcohol with the help of cadmium chloride using solvent casting method. The aforementioned method was used to create a polyvinylidenefuoride/graphene nanoplates/montmorillonite/nanoclay and titanium dioxide nanoparticle-based nanocomposite, and analysis revealed the adhesion of various nanoparticles to the polyvinylidenefuoride composite [142]. Nickel oxide and graphite oxide were used as nanofillers during the synthesis of polyvinyl alcohol and chitosan-based composites, and their thermal, structural, and mechanical properties were investigated [143]. Carbon dots nanocomposites can be made anywhere; in which a polymer, and the carbon dots acts as nanofillers, enhancing the properties. Its nature can be determined by the nature of the polymer and the mechanism by which the carbon dots are attached to the polymer, either covalently or physically blended (Fig. 8.4) [144]. *Datura sternum* leaf extract was

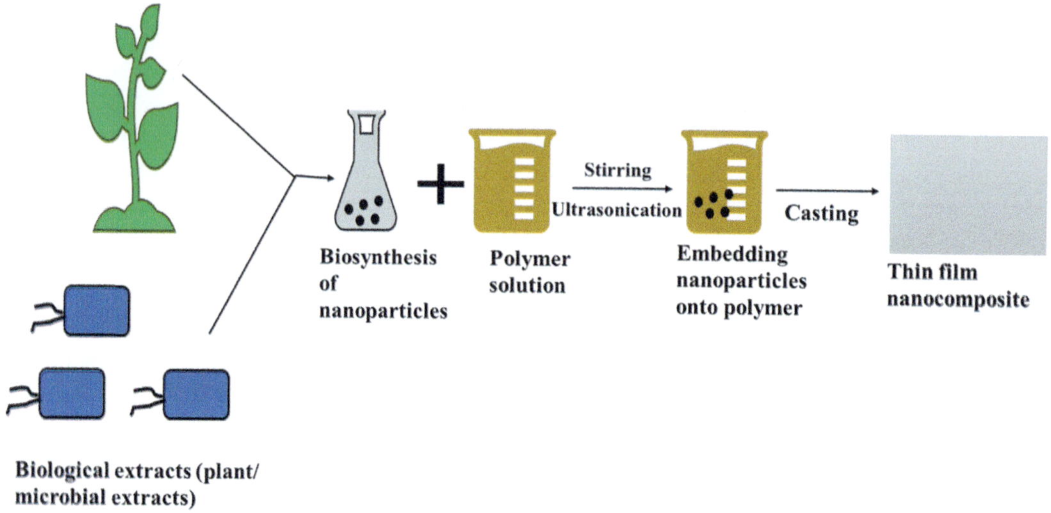

FIGURE 8.3

Green synthesis of nanoparticle-based polymer nanocomposites using solvent extraction method.

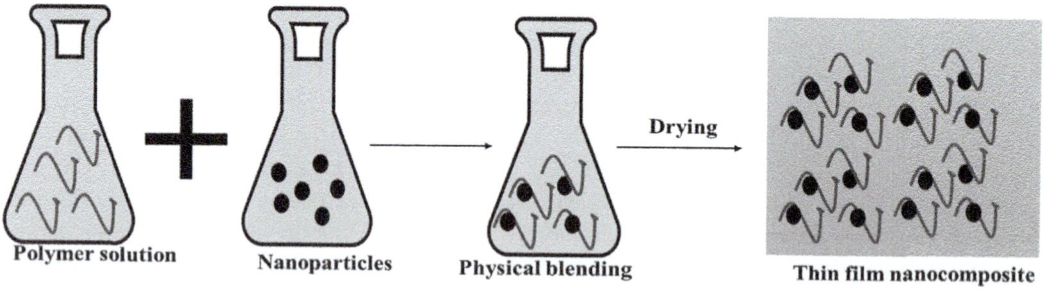

FIGURE 8.4

Green synthesis of nanoparticle-based polymer nanocomposites using physical blending method.

used in the synthesis of lead oxide nanoparticles, and it was then used as a nanofiller and reinforced into the polyvinylchloride polymer to create a nanocomposite [145]. Table 8.3 provides some of the green-synthesized nanoparticle-based polymer nanocomposites.

8.6 Characterization of green-synthesized nanoparticle-based polymer nanocomposites

Measuring and characterizing materials are critical for developing nanoformulations and improving their overall performance. Electron microscopy and spectroscopy analyses have been used extensively (Fig. 8.5) to study material of nanosize and are discussed below.

Table 8.3 Green-synthesized nanoparticle-based polymer nanocomposites.

S. no.	Nanoparticle used	Extract used for nanoparticle synthesis	Embedded surface	References
1	Fe_3O_4	*Gluconacetobacter xylinum*	Bacterial cellulose	[146]
2	Palladium	*Euphorbia larica* leaf	Sodium borosilicate	[147]
3	Silver	*Citrullus colocynthis* seed	Hydrogel beads based on *Kappa carrageenan*	[148]
4	Silver and magnesium oxide	*Olea cuspidate* leaf	–	[149]
5	Silver	*aqueous Rhubarb* stem	Chitosan	[150]
6	Palladium	*Gardenia taitensis* leaf	Bentonite	[151]
7	Silver	Pomegranate seed	Montmorilant, Xanthan	[152]
8	Zinc oxide	*S. lycopersicum* leaf	Chitosan	[153]
9	Platinum	*Padina gymnospora*	Polyvinyl pyrrolidone	[154]
10	Copper oxide	*Aspergillus terreus*	Poly(acrylic acid-acetonitrile)/*N*, *N'*-methyenediacrylamide	[155]

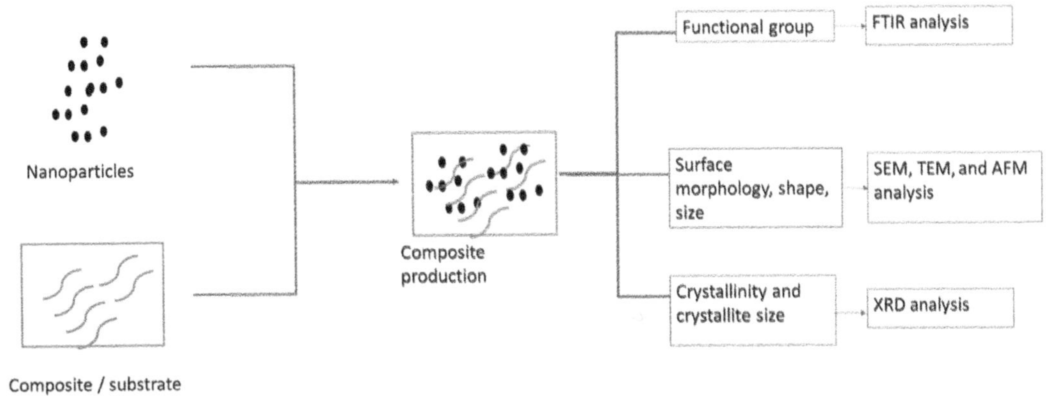

FIGURE 8.5

Characterization of nanoparticle-based polymer nanocomposites.

8.6.1 Fourier transform infrared spectroscopy

IR (Infrared) spectroscopy is based on the vibrations, stretching, or bending of molecule's atoms. Absorption of infrared ray corresponds to specific values, which can be used to identify functional groups in a molecule [156]. Polymer nanocomposites containing poly(hexamethylene isophthalamide) (APA) and montmorillonite nanoclay were characterized using Fourier transform infrared spectroscopy (FTIR) [157]. FTIR confirmed the presence of hydroxyl groups on the surface of the silica (or titania)-polymer interface. The water molecules were characterized by their O–H stretching and around 3700–3000 cm^{-1} and the H–O–H bending bands at 1640–1630 cm^{-1} [158]. Stretching of amide group and carboxylic group was observed at 1685 and 1775 cm^{-1} when chitosan-based silver nanoparticles were conjugated with amoxicillin to form a nanocomposites. Similarly, chitosan-based silver nanoparticles conjugated with cefixime and levofloxacin separately to form a nanocomposite displayed bands at 3260 cm^{-1} indicating presence of $-NH_2$ and –COO at 1760 cm^{-1} indicated the presence of cefixime onto the chitosan coated AgNPs. Peak at 2940 cm^{-1} indicated aromatic stretching of levofloxacin on chitosan coated AgNPs [159]. The bands at 1283–1083 cm^{-1} indicated the presence of stretching vibrations of hydroxyl group and C-O-C ether linkage, 1342 cm^{-1} and 1469 cm^{-1} indicated the presence of C-H bending vibrations in silver/chitosan/polyethylene glycol nanocomposites [160]. Graphene/silver nanoparticle-based nanocomposites were synthesized which showed bands near 3414, 1725, 1371, 1060 cm^{-1} indicating the OH group, C = O band on COOH groups of graphitic domains of graphene oxide and a band at 1570 cm^{-1} indicated the vibration of graphene sheets [161]. Hydroxyl groups were observed at 3441 cm^{-1} and 2929 cm^{-1} for silver nanoparticle and at 3433 cm^{-1} indicated the presence of hydroxy groups of chitosan. The CS peak shifts in the CS–Ag nanocomposite could be due to Ag–CS interaction in the nanocomposites. Similarly, the intensity of the hydroxyl peak had been reduced [162]. Thus, FTIR is found to be very useful in polymer-imposed metal nanoparticles.

8.6.2 Scanning electron microscopy

Despite having lower resolution than Atomic Force Microscopy (AFM) or Transmission Electron Microscopy (TEM), nanocomposites are most frequently examined structures using Scanning Electron Microscopy (SEM). Due to this limitation, a high resolution Field Emission Scanning Electron Microscope (FeSEM) that operates at extremely low voltages and allows the visualization of organic materials without the use of conductive coatings was developed [163]. Bio-composites derived from potato starch and reinforced with chestnut husks were visualized within 210–400 μm range using SEM analysis. The chestnut husks were found to be evenly distributed within the starch matrix [164]. When Silver/Chitosan/Polyethylene Glycol Nanocomposites were synthesized by stirring at different intervals (12, 24, and 48 h), flakes surface were observed, and SEM images revealed that more compatible nanocomposites can be produced after long term stirring [160]. Green synthesis of zinc sulfide nanoparticles used for embedding on to polyvinyl alcohol was observed using FeSEM to understand the morphology of its surface, and it was discovered that the surface was rough and slightly agglomerated, with good interaction and adhesion of zinc nanoparticles onto polyvinyl alcohol [165]. When biosynthesized nickel nanoparticles were used to create polyvinyl alcohol/Ni nanocomposite, the nickel nanoparticles were found to be rough, agglomerated, quasi shaped nanoparticles, and when nanocomposite was formed after adding nickel nanoparticles to polyvinyl alcohol, the surface became smooth and flat [166]. The morphology of this hybrid nanocomposite was compared to the chitosan/gelatin control, and the nanocomposite films were found to be translucent even after the addition of ZnS nanoparticle, and the nanocomposite was smooth, compact, and heterogeneous in nature when compared to the chitosan-gelatin control [167].

8.6.3 Transmission electron microscopy

Transmission Electron Microscopy (TEM) is transmission with high resolution Atomic-level resolution provided by electron microscopy. It is widely used in the field of nanocomposites for measuring microscopic structures because of its high precision and sensitivity [168]. It is necessary to prepare the sample using Ultramicrotomy before subjecting it to TEM analysis [163]. The preparation of ultrathin (<100 nm) polymer or cellulose nanocomposites involves three simple steps. Thermosetting in a resin, followed by curing in acetone or ethanol, sectioning with a microtome to obtain sections of thickness 50–70 nm on a metal mesh grid, and staining using electron dense stains, if necessary [169]. Furthermore, cryo-ultramicrotomy can be used to section soft polymers with glass transition temperatures below room temperature using liquid nitrogen [170]. Antibiotics such as amoxicillin, cefixime, and levofloxacin were conjugated with green-synthesized silver nanoparticle-based chitosan nanocomposite, and when characterized using a scanning electron microscope, the particles were found to be agglomerated in antibiotics conjugated nanocomposites rather than chitosan-silver nanoparticle-based nanocomposites [159]. The spherical bio-nanocomposite hydrogel beads based on Kappa Carrageenan had a smooth surface, but as the silver nanocomposite content increased, the surface became rough [148]. Biosynthesized nickel sulfide nanoparticles were synthesized, and when PVA-NiS was synthesized, the nickel nanoparticles were found to be uniformly distributed in the polymer matrix, and TEM imaging revealed good compatibility between nanofillers and the polymer matrix [166]. Cadmium oxide nanoparticles were prepared using a chemical precipitation method, with the samples being spherical in shape and agglomerated. This was further capped with polyethylene glycol to form a nanocomposite and it was in size range of 20–40 nm. After coating nanoparticles with polyethylene glycol, agglomeration was found to be reduced [171].

8.6.4 X-ray diffraction spectroscopy

X-ray diffraction (XRD) is an effective method for studying crystallographic materials such as crystalline, semicrystalline, and amorphous. A degree of crystallinity, which measures the degree of crystallization found in a material, is used to measure crystallinity. It is critical to understand the degree of crystallinity of a nanocomposite to predict its properties and determine its potential applications in various fields [172]. The XRD results of Kappa Carrageenan hydrogel and $-C/Ag_3$ bio-nanocomposite hydrogel was observed and 10–80° 2θ range was observed and two different peaks corresponding to silver and Kappa Carrageenan was observed and the peak for silver was similar to the peak of silver nanoparticle and peak of Kappa Carrageenan was observed at 2θ values of 21.35° [148]. Green-synthesized silver nanoparticle—carbon nanotube-reduced graphene oxide nanocomposite was performed and 2θ value at 10.8° proved the sharp peak of graphene oxide and peaks at 25.8 and 42.8° indicated the nanotube with plane of (002) and (100). The planes at (111), (200), (220), (311) indicated the peaks of silver and from these it was confirmed that the nanocomposite was successfully fabricated [173]. The presence of peaks corresponding to copper oxide nanoparticles and zinc oxide nanoparticles in the biofabricated nanocomposites confirmed their successful fabrication [174]. Green nanoparticle-based TiO_2/Graphene oxide nanocomposites were synthesized, and a strong peak at 10.2° was observed indicating graphene oxide, as well as corresponding peaks for TiO_2 nanoparticles. The diffraction peaks in the produced TiO_2/graphene oxide nanocomposites correspond to TiO_2 nanoparticles [175].

8.6.5 Thermogravimetric analysis

By observing the weight change that takes place as a sample is heated at a constant pace, Thermogravimetric Analysis (TGA) can be used to estimate a material's thermal stability and the amount of volatile components in it. Superparamagnetic iron oxide nanoparticles (Fe_3O_4) were synthesized and used as a nanofiller in conjunction with the polymer polyaniline, where their thermal properties were investigated using TGA analysis and thermograms were recorded. The first loss of weight was caused by vaporization due to water absorption by the surface at 115°C. The second weight loss occurred between 125°C and 300°C, which could be attributed to the loss of HCl to the polyaniline polymer [176]. Polymethyl methacrylate was used in conjunction with the carbon quantum dots, and its thermal stability was investigated using TGA, with the first weight loss discovered at 167.9°C with a 55% weight loss, which could be attributed to the carbon quantum dots. Weight loss continued from 322°C to 431°C [177]. The nanocomposites were created using nickel sulfide nanoparticles and polyviny alcohol. The weight loss of nickel sulfide nanoparticles was 36.06% overall. The weight loss of the polyvinyl alcohol/nickel sulfide nanocomposites occurred in three stages. The first decomposition occurred between 48°C and 134°C, which could be attributed to water removal from the surface of the nanocomposite. The second decomposition occurred between 147°C and 403°C, which could be attributed to the polymer's monomer splitting. The third decomposition occurred at 415°C, which could be attributed to structural changes in the polymer [166]. Polymethyl methacrylate and silver nanoparticles were tested for thermal stability, and it was discovered that the first decomposition occurred at 300°C due to water loss, followed by the second decomposition at 360–400°C due to polymer disintegration. Because of the addition of silver nanoparticles, the thermal stability of the nanocomposite was found to be high [178]. TGA was used to characterize polyvinyl alcohol reinforced with zirconia nanoparticles, and two decomposition stages were observed. The first weight loss occurred at 100–140°C due to water loss, and the second weight loss occurred at 385–530°C due to the decomposition of certain organic compounds [179].

8.7 Applications of green-synthesized nanoparticle-based polymer nanocomposites

The usage of metals, plastics and glasses as packaging materials created concerns as it causes environmental pollution and must be used in a safer quantities. Fig. 8.6 illustrates some of the applications of the nanocomposites. The section that follows goes over the various applications of green-synthesized nanoparticle-based polymer nanocomposites.

8.7.1 Food packaging

Every now and then, new alternatives for the use of packaging materials emerges, but polymer-based bio nanocomposites may be the best option (Fig. 8.7) [180]. Because both nanocomposite showed the best antimicrobial activity, chitosan embedded with silver and gold as a separate nanocomposite can be used as the best packaging material [131]. When polystyrene nanoparticles were synthesized with embedded silver nanoparticles, similar results were obtained [132]. Tripathi et al. [137] developed biodegradable, eco-friendly poly (vinyl alcohol) biogenic silver based nanocomposites that can be used as an excellent food packaging material. Guar gum silver nanocomposite

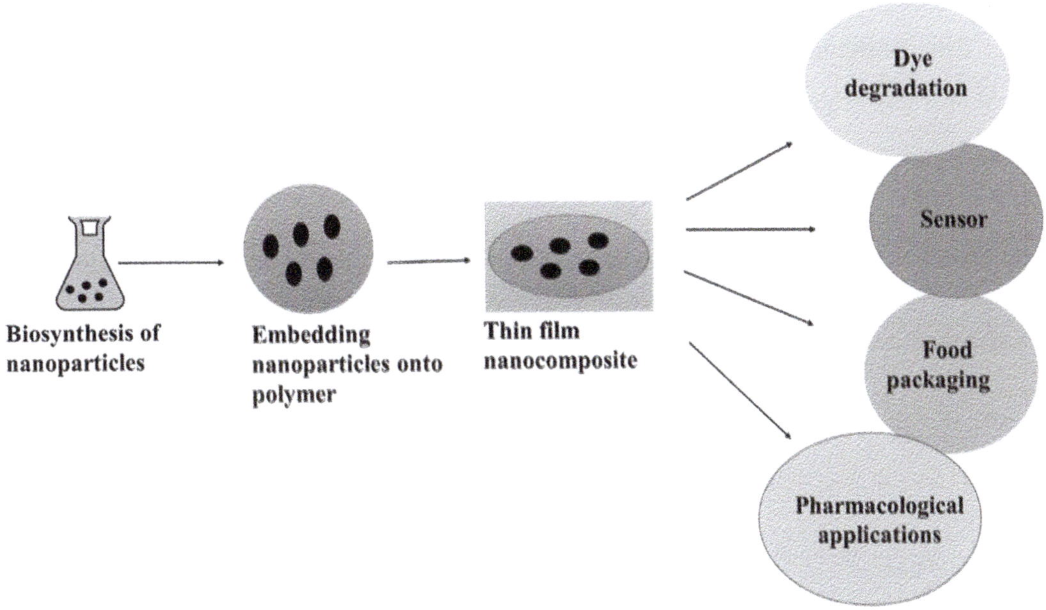

FIGURE 8.6

Applications of green-synthesized nanoparticle-based polymer nanocomposites.

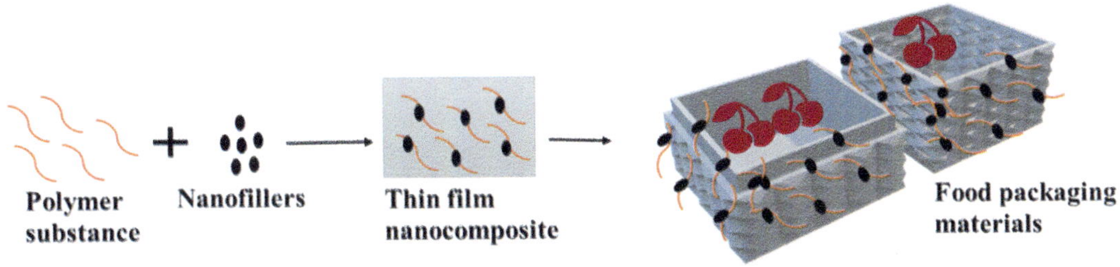

FIGURE 8.7

Applications of green-synthesized nanoparticle-based polymer nanocomposites in food packaging.

films were developed and found to be an excellent packaging material for strawberries, extending their shelf life and preserving the quality [134]. Silver nanoparticles were green-synthesized from pomegranate seed extract and used as nanofillers in the starch polymer to produce an AgNPs-starch nanocomposite with excellent antimicrobial applications and showed potential to be used as food packaging materials [181]. The nanocomposite was created using green-synthesized silver nanoparticles and cellulose acetate. It was discovered to have antibacterial activity against *Staphylococcus aureus*, *Escherichia coli*, *Klebsiella pneumoniae*, *Salmonella typhi*. As a result, it has been reported that silver nanoparticles in combination with cellulose acetate nanocomposite can be used as an active material in food packaging [182].

8.7.2 Environmental applications

Several nanocomposites have been used for dye removal [183]. Fig. 8.8 depicts the dye degradation process using nanocomposites. Copper oxide/zinc oxide nanocomposites were created and found to have photo degradation efficiency, with 97% of the methylene blue dye removed after 85 min [174]. Capping of silver nanoparticles onto the hybrid 2-hydroxypropyl-β-cyclodextrin/alginate nanocomposite has the application of degrading pollutants in aqueous medium [138]. Green-synthesized TiO_2 nanoparticle-based graphene oxide nanocomposites were created and found to degrade crystal violet dye with less reduction in its photocatalytic property, making it stable, easily scalable, and cost-effective process for water purification [175]. Co_3O_4/ZnO nanocomposites were synthesized and found to remove methylene blue and rhodamine B dyes with efficiency of 86% and 91%, respectively [184]. Green synthesized iron-aluminum nanocomposites were found to efficiently remove fluorides, with an adsorption capacity of approximately 42.95 mg/g and showed maximum removal of 82.1%, and followed the Langmuir adsorption pattern [185]. Silver nanoparticles were synthesized and it was fabricated onto δ-FeOOH/PUF to form a nanocomposite which had the ability

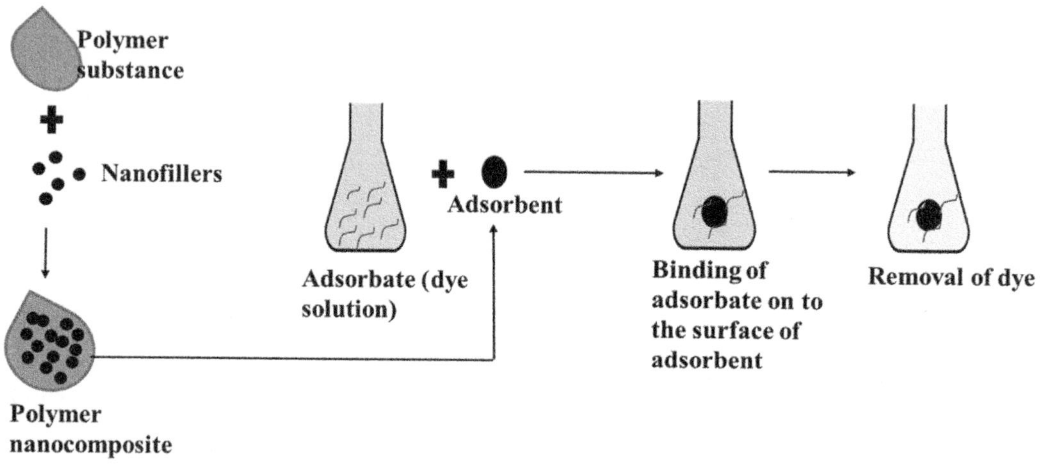

FIGURE 8.8

Applications of green-synthesized nanoparticle-based polymer nanocomposites in dye degradation.

to degrade both the congo red and methyl orange azo dyes at rate constant of 0.0357 and 0.0805 s^{-1} respectively [186]. Polypyrolle was embedded with zero valent iron nanocomposite and was used for the removal of chromium (VI). The absorption capacity of the nanocomposite was around 202.02 mg/g at 25°C. So, it was concluded that the nanocomposites can be used in waste water treatment [187].

8.7.3 Bioactivity and pharmacological applications

The production of polycaprolactone nanocomposites incorporated with green-synthesized silver nanoparticles can be used as an efficient wound dressing material because it has demonstrated effective bactericidal activity against *Staphylococcus epidermidis* and *Staphylococcus haemolyticus* [129]. Kappa Carrageenan hydrogel and $-C/Ag_3$ bio-nanocomposite demonstrated cytotoxicity less than 50%, which is an acceptable level of cytotoxicity, and as well as good antimicrobial activity, paving the way for their use in pharmaceutical applications (Fig. 8.9) [148]. The nanocomposite films of chitosan/gelatin/silver nanoparticles and agar/zinc oxide nanoparticles were found to have excellent antimicrobial activity and which was said to extend the shelf life of grapes from 10 to 15 days in ambient conditions. As a result, these can be used as a better food packaging alternatives [167].

Green synthesis of chitosan functionalized silver nanoparticle based nanocomposites with antimicrobial and anticoagulant activity was discovered. It was discovered to have low toxicity, paving the way for its use in the clinical sector for the development of nanomedicines [159]. The biosynthesis of silver nanoparticles and the production of nanocomposite by adding silver nanoparticles to graphene oxide were found to inhibit biofilm growth, to be nontoxic against HEK-293 cell lines, and to have promising therapeutic applications [188]. It was discovered that biosynthesis of nanozero valent copper nanoparticles supported with bentonite for the production of nanocomposites. It was found to remove tetracycline from natural water systems by about 90% and also had reusable application where it was used for around three times [189]. Green chitin nanoparticles synthesis were discovered to have numerous drug delivery applications [190]. Silver nanoparticles were biosynthesized from white tea extract and reinforced to the polymer montmorillonite to test their cytotoxic and antioxidant efficiency. It was discovered that they had very good antioxidant activity and could be used as a radical scavenger. In-vitro cytotoxic assays were performed on MOLT-4 cells, and the results were compared to anticancer drugs such as cisplatin and doxorubicin. The IC_{50} values of this nanocomposite were significantly lower than the IC_{50} values of anticancer drugs.

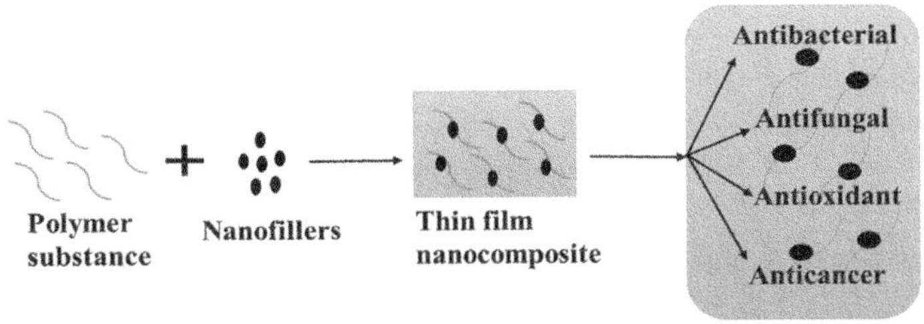

FIGURE 8.9

Green-synthesized nanoparticle-based polymer nanocomposites and its bioactivity.

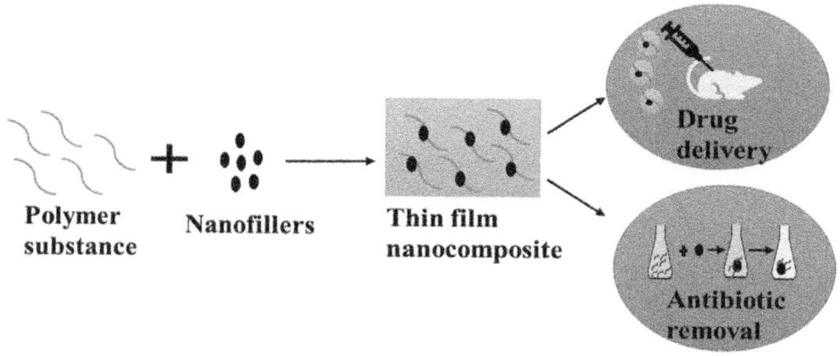

FIGURE 8.10

Applications of green-synthesized nanoparticle-based polymer nanocomposites in pharmaceutical industries.

As a result, it had a wide range of pharmaceutical applications [191]. Guar gum and gelatin polymer embedded with silver nanocomposites were used to produce nanocomposites and was found to have antibacterial activity against both gram-negative bacteria *E. coli* and *P. aeruginosa* and gram-positive bacteria *S. aureus* [192]. Fig. 8.10 illustrates the pharmacological applications of nanocomposites.

8.7.4 Other biosensing and supercapacitor applications

Aqueous ammonia sensing study was performed, and optical sensor's performance was observed by surface plasmon resonance and at room temperature, 1 ppm of ammonia solution was detected in 2–3 s using *Cyamopsis tetragonaloba* gum-mediated synthesis of silver nanoparticles and guar gum. This nanocomposite was used in biosensing application for the detection of ammonia [193].

Green-synthesized silver nanoparticles doped with graphene were used to make nanocomposites with great biosensing applications. It showed high catalytic activity for H_2O_2 reduction and a very good SERS (surface-enhanced Raman scattering) activity, which was detected in 2 s. Further on silver nanoparticles (AgNPs)-graphene biosensors, glucose oxidase was coated for glucose sensing applications where the detection limit of 100 μm was at noise ratio of 3. So, the synthesized nanocomposites showed excellent sensing application on SERS, H_2O_2 and glucose detection [194]. Rice powder was used as a reducing agent in the synthesis of graphene quantum dots before being doped with polyacrylonitrile and polyaniline to create polymer-based nanocomposites. The synthesized nanocomposite's supercapacitor application was tested practically on electronic and portable devices, and they concluded that it can be used as a power saving supercapacitor [195]. Mussel-inspired myoglobin/gold nanoparticles/polydopamine/graphene polymeric nanocomposite were synthesized and was found to have good electrocatalytic activity due to its high electron transfer capacity, making it suitable for biosensing applications [196]. Gold nanoparticles/ternary reduced graphene oxide/polyaniline nanocomposites were discovered to have supercapacitor applications. For 5000 cycles, the specific capacitance was around 86.9% [197]. Fig. 8.11 illustrates the biosensing applications of nanocomposites.

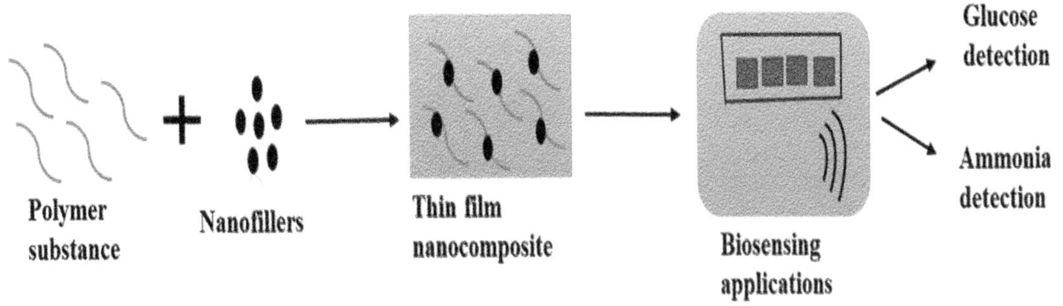

FIGURE 8.11

Applications of green-synthesized nanoparticle-based polymer nanocomposites in biosensing.

8.8 Conclusion

This review focuses primarily on the fabrication of polymer-based nanocomposites with green-synthesized nanoparticles. Green nanoparticle synthesis is an effective method for reducing or minimizing toxic effects, as the toxicity of nanoparticles is increased primarily due to the chemical composition of hazardous chemicals used. It can be used for a variety of applications by embedding these green-synthesized nanoparticles on the polymer substance. Because of their effective bactericidal effect, these nanocomposites have a wide range of applications in biomedical science, including food packaging materials. It is also used in dye degradation studies and have variety of applications.

References

[1] T. Singh, S. Shukla, P. Kumar, V. Wahla, V.K. Bajpai, I.A. Rather, Application of nanotechnology in food science: perception and overview, Front. Microbiol. 8 (2017). Available from: https://doi.org/10.3389/fmicb.2017.01501.

[2] X.Y. Wong, A. Sena-Torralba, R. Alvarez-Diduk, K. Muthoosamy, A. Merkoçi, Nanomaterials for nanotheranostics: tuning their properties according to disease needs, ACS Nano 14 (3) (2020) 2585–2627.

[3] F.A. Khan, Nanomaterials: types, classifications, and sources, Applications of Nanomaterials in Human Health, Springer, Singapore, 2020, pp. 1–13.

[4] C. Buzea, I.I. Pacheco, K. Robbie, Nanomaterials and nanoparticles: sources and toxicity, Biointerphases 2 (4) (2007) MR17–MR71.

[5] A. Gour, N.K. Jain, Advances in green synthesis of nanoparticles, Artif. Cells Nanomed. Biotechnol. 47 (1) (2019) 844–851.

[6] M. Rafique, I. Sadaf, M.S. Rafique, M.B. Tahir, A review on green synthesis of silver nanoparticles and their applications, Artif. Cells Nanomed. Biotechnol. 45 (7) (2017) 1272–1291.

[7] J. Jordan, K.I. Jacob, R. Tannenbaum, M.A. Sharaf, I. Jasiuk, Experimental trends in polymer nanocomposites—a review, Mater. Sci. Eng. A 393 (1–2) (2005) 1–11.

[8] S. Komarneni, Nanocomposites, J. Mater. Chem. 2 (12) (1992) 1219–1230.

[9] D.R. Paul, L.M. Robeson, Polymer nanotechnology: nanocomposites, Polymer 49 (15) (2008) 3187–3204.

[10] Y.W. Mai, Z.Z. Yu, Polymer Nanocomposites, first ed., Woodhead Publishing, 2006.
[11] K.I. Winey, R.A. Vaia, Polymer nanocomposites, MRS Bull. 32 (4) (2007) 314–322.
[12] G. Zhao, X. Huang, Z. Tang, Q. Huang, F. Niu, X. Wang, Polymer-based nanocomposites for heavy metal ions removal from aqueous solution: a review, Polym. Chem. 9 (26) (2018) 3562–3582.
[13] L. Peponi, D. Puglia, L. Torre, L. Valentini, J.M. Kenny, Processing of nanostructured polymers and advanced polymeric based nanocomposites, Mater. Sci. Eng. R Rep. 85 (2014) 1–46.
[14] M. Khutia, G.M. Joshi, K. Deshmukh, M. Pandey, Preparation of modified polymer blend and electrical performance, Comp. Interfaces 22 (3) (2015) 167–178.
[15] M. Khutia, G.M. Joshi, K. Deshmukh, M. Pandey, Optimization of dielectric constant of polycarbonate/polystyrene modified blend by ceramic metal oxide, Polym. Technol. Eng. 54 (4) (2015) 383–389.
[16] M. Pandey, G.M. Joshi, S. Bhattacharya, Influence of $K_2Ti_6O_{13}$ on dielectric and barrier properties of polymer, Ionics 22 (3) (2016) 445–449.
[17] T. Hanemann, D.V. Szabó, Polymer-nanoparticle composites: from synthesis to modern applications, Materials 3 (6) (2010) 3468–3517.
[18] I. Hussain, N.B. Singh, A. Singh, H. Singh, S.C. Singh, Green synthesis of nanoparticles and its potential application, Biotechnol. Lett. 38 (4) (2016) 545–560.
[19] K. Parveen, V. Banse, L. Ledwani, Green synthesis of nanoparticles: their advantages and disadvantages, AIP Conf. Proc. 1724 (1) (2016) 020048. AIP Publishing LLC.
[20] P. Raveendran, J. Fu, S.L. Wallen, Completely "green" synthesis and stabilization of metal nanoparticles, J. Am. Chem. Soc. 125 (46) (2003) 13940–13941.
[21] S. Iravani, Green synthesis of metal nanoparticles using plants, Green. Chem. 13 (10) (2011) 2638–2650.
[22] J. Singh, T. Dutta, K.H. Kim, M. Rawat, P. Samddar, P. Kumar, 'Green' synthesis of metals and their oxide nanoparticles: applications for environmental remediation, J. Nanobiotechnol. 16 (1) (2018) 1–24.
[23] M. Shah, D. Fawcett, S. Sharma, S.K. Tripathy, G.E.J. Poinern, Green synthesis of metallic nanoparticles via biological entities, Materials 8 (11) (2015) 7278–7308.
[24] K. Gopinath, V.K. Shanmugam, S. Gowri, V. Senthilkumar, S. Kumaresan, A. Arumugam, Antibacterial activity of ruthenium nanoparticles synthesized using *Gloriosa superba* L. leaf extract, J. Nanostruct. Chem. 4 (2014) 83.
[25] G. Geoprincy, B.V. Srri, U. Poonguzhali, N.N. Gandhi, S. Renganathan, A review on green synthesis of silver nanoparticles, Asian J. Pharm. Clin. Res. 6 (1) (2013) 8–12.
[26] S. Ahmed, Saifullah, M. Ahmad, B.L. Swami, S. Ikram, Green synthesis of silver nanoparticles using *Azadirachta indica* aqueous leaf extract, J. Radiat. Res. Appl. Sci. 9 (1) (2016) 1–7.
[27] M.R. Shaik, M. Khan, M. Kuniyil, A. Al-Warthan, H.Z. Alkhathlan, M.R.H. Siddiqui, et al., Plant-extract-assisted green synthesis of silver nanoparticles using *Origanum vulgare* L. extract and their microbicidal activities, Sustainability 10 (4) (2018) 913.
[28] S. Pirtarighat, M. Ghannadnia, S. Baghshahi, Green synthesis of silver nanoparticles using the plant extract of *Salvia spinosa* grown in vitro and their antibacterial activity assessment, J. Nanostruct. Chem. 9 (1) (2019) 1–9.
[29] P.V. Kumar, S.V.N. Pammi, P. Kollu, K.V.V. Satyanarayana, U. Shameem, Green synthesis and characterization of silver nanoparticles using *Boerhaavia diffusa* plant extract and their anti bacterial activity, Ind. Crop. Products 52 (2014) 562–566.
[30] U.K. Sur, B. Ankamwar, S. Karmakar, A. Halder, P. Das, Green synthesis of Silver nanoparticles using the plant extract of Shikakai and Reetha, Mater. Today Proc. 5 (1) (2018) 2321–2329.
[31] H.R. Naika, K. Lingaraju, K. Manjunath, D. Kumar, G. Nagaraju, D. Suresh, et al., Green synthesis of CuO nanoparticles using *Gloriosa superba* L. extract and their antibacterial activity, J. Taibah Univ. Sci. 9 (1) (2015) 7–12.
[32] G. Zhan, J. Huang, M. Du, I. Abdul-Rauf, Y. Ma, Q. Li, Green synthesis of Au–Pd bimetallic nanoparticles: single-step bioreduction method with plant extract, Mater. Lett. 65 (19–20) (2011) 2989–2991.

[33] V.V.T. Padil, M. Černík, Green synthesis of copper oxide nanoparticles using gum karaya as a biotemplate and their antibacterial application, Int. J. Nanomed. 8 (2013) 889.
[34] P.V. Kumar, U. Shameem, P. Kollu, R.L. Kalyani, S.V.N. Pammi, Green synthesis of copper oxide nanoparticles using Aloe vera leaf extract and its antibacterial activity against fish bacterial pathogens, BioNanoScience 5 (3) (2015) 135–139.
[35] M. Sorbiun, E.S. Mehr, A. Ramazani, S.T. Fardood, Green synthesis of zinc oxide and copper oxide nanoparticles using aqueous extract of oak fruit hull (jaft) and comparing their photocatalytic degradation of basic violet 3, Int. J. Environ. Res. 12 (1) (2018) 29–37.
[36] F. Ijaz, S. Shahid, S.A. Khan, W. Ahmad, S. Zaman, Green synthesis of copper oxide nanoparticles using *Abutilon indicum* leaf extract: antimicrobial, antioxidant and photocatalytic dye degradation activitie, Trop. J. Pharm. Res. 16 (4) (2017) 743–753.
[37] R. Sankar, P. Manikandan, V. Malarvizhi, T. Fathima, K.S. Shivashangari, V. Ravikumar, Green synthesis of colloidal copper oxide nanoparticles using *Carica papaya* and its application in photocatalytic dye degradation, Spectrochim. Acta A Mol. Biomol. Spectrosc. 121 (2014) 746–750.
[38] G. Sangeetha, S. Rajeshwari, R. Venckatesh, Green synthesis of zinc oxide nanoparticles by *Aloe barbadensis miller* leaf extract: structure and optical properties, Mater. Res. Bull. 46 (12) (2011) 2560–2566.
[39] K. Elumalai, S. Velmurugan, Green synthesis, characterization and antimicrobial activities of zinc oxide nanoparticles from the leaf extract of *Azadirachta indica* (L.), Appl. Surf. Sci. 345 (2015) 329–336.
[40] S. Vijayakumar, S. Mahadevan, P. Arulmozhi, S. Sriram, P.K. Praseetha, Green synthesis of zinc oxide nanoparticles using *Atalantia monophylla* leaf extracts: characterization and antimicrobial analysis, Mater. Sci. Semicond. Process. 82 (2018) 39–45.
[41] Y.A. Selim, M.A. Azb, I. Ragab, M.H. Abd El-Azim, Green synthesis of zinc oxide nanoparticles using aqueous extract of *Deverra tortuosa* and their cytotoxic activities, Sci. Rep. 10 (1) (2020) 1–9.
[42] M.I. Din, S. Jabbar, J. Najeeb, R. Khalid, T. Ghaffar, M. Arshad, et al., Green synthesis of zinc ferrite nanoparticles for photocatalysis of methylene blue, Int. J. Phytoremediat. 22 (13) (2020) 1440–1447.
[43] D. GingaŞu, I. Mindru, S. Preda, J.M. Calderon-Moreno, C.C. Daniela, L. Patron, et al., Green synthesis of cobalt ferrite nanoparticles using plant extracts, Rev. Roum. Chim. 62 (2017) (2017) 645–653.
[44] A. Ramazani, S. Taghavi Fardood, Z. Hosseinzadeh, F. Sadri, S.W. Joo, Green synthesis of magnetic copper ferrite nanoparticles using tragacanth gum as a biotemplate and their catalytic activity for the oxidation of alcohols, Iran. J. Catal. 7 (3) (2017) 181–185.
[45] E. Sarala, M.M. Naik, M. Vinuth, Y.R. Reddy, H.R. Sujatha, Green synthesis of Lawsonia inermis-mediated zinc ferrite nanoparticles for magnetic studies and anticancer activity against breast cancer (MCF-7) cell lines, J. Mater. Sci. Mater. Electron. 31 (11) (2020) 8589–8596.
[46] A.V. Samrot, J.L.A. Angalene, S.M. Roshini, P. Raji, S.M. Stefi, R. Preethi, et al., Bioactivity and heavy metal removal using plant gum mediated green synthesized silver nanoparticles, J. Clust. Sci. 30 (6) (2019) 1599–1610.
[47] A.V. Samrot, N. Shobana, R. Jenna, Antibacterial and antioxidant activity of different staged ripened fruit of *Capsicum annuum* and its green synthesized silver nanoparticles, BioNanoScience 8 (2) (2018) 632–646.
[48] P. Senthilkumar, S. Rashmitha, P. Veera, C.V. Ignatious, C. SaiPriya, A.V. Samrot, Antibacterial activity of neem extract and its green synthesized silver nanoparticles against *Pseudomonas aeruginosa*, J. Pure Appl. Microbiol. 12 (2) (2018) 969–974.
[49] A.V. Samrot, C. Saipriya, J.L.A. Angalene, S.M. Roshini, P.J. Cypriyana, S. Saigeetha, et al., Evaluation of nanotoxicity of *Araucaria heterophylla* gum derived green synthesized silver nanoparticles on *Eudrilus eugeniae* and *Danio rerio*, J. Clust. Sci. 30 (4) (2019) 1017–1024.
[50] A.Y. Ghidan, T.M. Al-Antary, A.M. Awwad, Green synthesis of copper oxide nanoparticles using *Punica granatum* peels extract: effect on green peach Aphid, Environ. Nanotechnol. Monit. Manag. 6 (2016) 95–98.

[51] P.C. Nagajyothi, P. Muthuraman, T.V.M. Sreekanth, D.H. Kim, J. Shim, Green synthesis: in-vitro anti-cancer activity of copper oxide nanoparticles against human cervical carcinoma cells, Arab. J. Chem. 10 (2) (2017) 215–225.
[52] K. Vishveshvar, M.A. Krishnan, K. Haribabu, S. Vishnuprasad, Green synthesis of copper oxide nanoparticles using *Ixiro coccinea* plant leaves and its characterization, BioNanoScience 8 (2) (2018) 554–558.
[53] A.M. Pillai, V.S. Sivasankarapillai, A. Rahdar, J. Joseph, F. Sadeghfar, K. Rajesh, et al., Green synthesis and characterization of zinc oxide nanoparticles with antibacterial and antifungal activity, J. Mol. Struct. 1211 (2020) 128107.
[54] N. Bala, S. Saha, M. Chakraborty, M. Maiti, S. Das, R. Basu, et al., Green synthesis of zinc oxide nanoparticles using *Hibiscus subdariffa* leaf extract: effect of temperature on synthesis, anti-bacterial activity and anti-diabetic activity, RSC Adv. 5 (7) (2015) 4993–5003.
[55] R. Geetha, T. Ashokkumar, S. Tamilselvan, K. Govindaraju, M. Sadiq, G. Singaravelu, Green synthesis of gold nanoparticles and their anticancer activity, Cancer Nanotechnol. 4 (4–5) (2013) 91–98.
[56] D. Philip, Honey mediated green synthesis of gold nanoparticles, Spectrochim. Acta A Mol. Biomol. Spectrosc. 73 (4) (2009) 650–653.
[57] P. Elia, R. Zach, S. Hazan, S. Kolusheva, Z.E. Porat, Y. Zeiri, Green synthesis of gold nanoparticles using plant extracts as reducing agents, Int. J. Nanomed. 9 (2014) 4007.
[58] F.T. Thema, E. Manikandan, M.S. Dhlamini, M. Maaza, Green synthesis of ZnO nanoparticles via *Agathosma betulina* natural extract, Mater. Lett. 161 (2015) 124–127.
[59] D. Suresh, P.C. Nethravathi, H. Rajanaika, H. Nagabhushana, S.C. Sharma, Green synthesis of multi-functional zinc oxide (ZnO) nanoparticles using Cassia fistula plant extract and their photodegradative, antioxidant and antibacterial activities, Mater. Sci. Semicond. Process. 31 (2015) 446–454.
[60] C. Vidya, S. Hiremath, M.N. Chandraprabha, M.L. Antonyraj, I.V. Gopal, A. Jain, et al., Green synthesis of ZnO nanoparticles by *Calotropis gigantea*, Int. J. Curr. Eng. Technol. 1 (1) (2013) 118–120.
[61] R.K. Das, B.B. Borthakur, U. Bora, Green synthesis of gold nanoparticles using ethanolic leaf extract of *Centella asiatica*, Mater. Lett. 64 (13) (2010) 1445–1447.
[62] D. Philip, C. Unni, S.A. Aromal, V.K. Vidhu, *Murraya koenigii* leaf-assisted rapid green synthesis of silver and gold nanoparticles, Spectrochim. Acta A Mol. Biomol. Spectrosc. 78 (2) (2011) 899–904.
[63] C.H. Ramamurthy, K.S. Sampath, P. Arunkumar, M.S. Kumar, V. Sujatha, K. Premkumar, et al., Green synthesis and characterization of selenium nanoparticles and its augmented cytotoxicity with doxorubicin on cancer cells, Bioprocess. Biosyst. Eng. 36 (8) (2013) 1131–1139.
[64] P. Sowndarya, G. Ramkumar, M.S. Shivakumar, Green synthesis of selenium nanoparticles conjugated *Clausena dentata* plant leaf extract and their insecticidal potential against mosquito vectors, Artif. Cells Nanomed. Biotechnol. 45 (8) (2017) 1490–1495.
[65] A. Thirumurugan, P. Aswitha, C. Kiruthika, S. Nagarajan, A.N. Christy, Green synthesis of platinum nanoparticles using *Azadirachta indica*—An eco-friendly approach, Mater. Lett. 170 (2016) 175–178.
[66] N.A. Karim, N.J. Rubinsin, M.A.A. Burukan, S.K. Kamarudin, Sustainable route of synthesis platinum nanoparticles using orange peel extract, Int. J. Green. Energy 16 (15) (2019) 1518–1526.
[67] B. Şahin, A. Aygün, H. Gündüz, K. Şahin, E. Demir, S. Akocak, et al., Cytotoxic effects of platinum nanoparticles obtained from pomegranate extract by the green synthesis method on the MCF-7 cell line, Colloids Surf. B Biointerfaces 163 (2018) 119–124.
[68] S.P. Goutam, S.K. Avinashi, M. Yadav, D. Roy, R. Shastri, Green synthesis and characterization of aluminium oxide nanoparticles using leaf extract of rosa, Adv. Sci. Eng. Med. 10 (7–8) (2018) 719–722.
[69] L. Xu, W. Li, Q. Shi, H. Li, Z. Yang, D. Liao, et al., Synthesis of mulberry leaf extract mediated gold nanoparticles and their ameliorative effect on Aluminium intoxicated and diabetic retinopathy in rats during perinatal life, J. Photochem. Photobiol. B Biol. 196 (2019) 111502.

[70] M. Hudlikar, S. Joglekar, M. Dhaygude, K. Kodam, Latex-mediated synthesis of ZnS nanoparticles: green synthesis approach, J. Nanopart. Res. 14 (5) (2012) 1–6.

[71] B.D. Rohan, P. Raji, D.M. Kumar, R.G. Gandham, K.V. Sharma, D. Keerthana, et al., Green Synthesis and antibacterial activity studies of silver nanoparticles from the aqueous extracts of *Euphorbia hirta*, J. Pure Appl. Microbiol. 14 (1) (2020) 301–306.

[72] R.G. Gandham, P. Raji, B.D. Rohan, D.M. Kumar, K.V. Sharma, D. Keerthana, et al., Green synthesis and antibacterial activity studies of silver nanoparticles from the aqueous extracts of *Cassia alata*, Lett. Appl. NanoBioSci. 9 (2) (2020) 1037–1041.

[73] D.M. Kumar, P. Raji, B.D. Rohan, R.G. Gandham, K.V. Sharma, D. Keerthana, et al., Green synthesis and antibacterial activity studies of silver nanoparticles from the aqueous extracts of *Thespesia populnea*, Lett. Appl. NanoBioSci. 9 (1) (2020) 931–934.

[74] V. Kripu Sharma, P. Raji, M. Divya Kumar, D. Bennet Rohan, Gandham Ratna Geethika, D. Keerthana, et al., Green synthesis and antibacterial activity studies of silver nanoparticles from the aqueous extracts of *Wrightia tinctoria*, Int. J. Adv. Res. Eng. Technol. (IJARET) 10 (6) (2019) 80–88.

[75] P. Raji, A.V. Samrot, D. Keerthana, S. Karishma, Antibacterial Activity of Alkaloids, Flavonoids, Saponins and Tannins Mediated Green Synthesised Silver Nanoparticles Against *Pseudomonas aeruginosa* and *Bacillus subtilis*, J. Clust. Sci. https://doi.org/10.1007/s10876-019-01547-2.

[76] A.V. Samrot, Silky, C.V. Ignatious, P. Raji, C. SaiPriya, J.A. Selvarani, Bioactivity studies of *Datura metel*, *Aegle marmelos*, *Annona retculata* and *Saraca indica* and their green synthesized silver nanoparticle, J. Pure Appl. Microbiol. 13 (1) (2019). Available from: https://doi.org/10.22207/JPAM.13.1.

[77] A. Fariq, T. Khan, A. Yasmin, Microbial synthesis of nanoparticles and their potential applications in biomedicine, J. Appl. Biomed. 15 (4) (2017) 241–248.

[78] K.B. Narayanan, N. Sakthivel, Biological synthesis of metal nanoparticles by microbes, Adv. Colloid Interface Sci. 156 (1-2) (2010) 1–13.

[79] N. Pugazhenthiran, S. Anandan, G. Kathiravan, N.K.U. Prakash, S. Crawford, M. Ashokkumar, Microbial synthesis of silver nanoparticles by Bacillus sp, J. Nanopart. Res. 11 (7) (2009) 1811–1815.

[80] N.Y. Tsibakhashvili, E.I. Kirkesali, D.T. Pataraya, M.A. Gurielidze, T.L. Kalabegishvili, D.N. Gvarjaladze, et al., Microbial synthesis of silver nanoparticles by *Streptomyces glaucus* and *Spirulina platensis*, Adv. Sci. Lett. 4 (11–12) (2011) 3408–3417.

[81] H.J. Bai, Z.M. Zhang, Microbial synthesis of semiconductor lead sulfide nanoparticles using immobilized *Rhodobacter sphaeroides*, Mater. Lett. 63 (9–10) (2009) 764–766.

[82] S. Ibrahim, Z. Ahmad, M.Z. Manzoor, M. Mujahid, Z. Faheem, A. Adnan, Optimization for biogenic microbial synthesis of silver nanoparticles through response surface methodology, characterization, their antimicrobial, antioxidant, and catalytic potential, Sci. Rep. 11 (1) (2021) 1–18.

[83] V.L. Das, R. Thomas, R.T. Varghese, E.V. Soniya, J. Mathew, E.K. Radhakrishnan, Extracellular synthesis of silver nanoparticles by the *Bacillus strain* CS 11 isolated from industrialized area, 3 Biotech. 4 (2) (2014) 121–126.

[84] M. Kowshik, S. Ashtaputre, S. Kharrazi, W. Vogel, J. Urban, S.K. Kulkarni, et al., Extracellular synthesis of silver nanoparticles by a silver-tolerant yeast strain MKY3, Nanotechnology 14 (1) (2002) 95.

[85] A. Ingle, M. Rai, A. Gade, M. Bawaskar, *Fusarium solani*: a novel biological agent for the extracellular synthesis of silver nanoparticles, J. Nanopart. Res. 11 (8) (2009) 2079–2085.

[86] R.Y. Parikh, S. Singh, B.L.V. Prasad, M.S. Patole, M. Sastry, Y.S. Shouche, Extracellular synthesis of crystalline silver nanoparticles and molecular evidence of silver resistance from Morganella sp.: towards understanding biochemical synthesis mechanism, ChemBioChem 9 (9) (2008) 1415–1422.

[87] C.G. Kumar, S.K. Mamidyala, Extracellular synthesis of silver nanoparticles using culture supernatant of *Pseudomonas aeruginosa*, Colloids Surf. B Biointerfaces 84 (2) (2011) 462–466.

[88] H. Singh, J. Du, P. Singh, T.H. Yi, Extracellular synthesis of silver nanoparticles by Pseudomonas sp. THG-LS1. 4 and their antimicrobial application, J. Pharm. Anal. 8 (4) (2018) 258–264.

[89] S. Shivaji, S. Madhu, S. Singh, Extracellular synthesis of antibacterial silver nanoparticles using psychrophilic bacteria, Process. Biochem. 46 (9) (2011) 1800–1807.

[90] K. Govindaraju, V. Kiruthiga, V.G. Kumar, G. Singaravelu, Extracellular synthesis of silver nanoparticles by a marine alga, *Sargassum wightii* Grevilli and their antibacterial effects, J. Nanosci. Nanotechnol. 9 (9) (2009) 5497–5501.

[91] P. Mukherjee, S. Senapati, D. Mandal, A. Ahmad, M.I. Khan, R. Kumar, et al., Extracellular synthesis of gold nanoparticles by the fungus *Fusarium oxysporum*, ChemBioChem 3 (5) (2002) 461–463.

[92] G. Singaravelu, J.S. Arockiamary, V.G. Kumar, K. Govindaraju, A novel extracellular synthesis of monodisperse gold nanoparticles using marine alga, *Sargassum wightii* Greville, Colloids Surf. B Biointerfaces 57 (1) (2007) 97–101.

[93] P. Singh, H. Singh, Y.J. Kim, R. Mathiyalagan, C. Wang, D.C. Yang, Extracellular synthesis of silver and gold nanoparticles by *Sporosarcina koreensis* DC4 and their biological applications, Enzyme Microb. Technol. 86 (2016) 75–83.

[94] M.S. Nejad, M. Khatami, G.H.S. Bonjar, Extracellular synthesis gold nanotriangles using biomass of *Streptomyces microflavus*, IET Nanobiotechnol. 10 (1) (2016) 33–38.

[95] M.P. Patil, M.J. Kang, I. Niyonizigiye, A. Singh, J.O. Kim, Y.B. Seo, et al., Extracellular synthesis of gold nanoparticles using the marine bacterium *Paracoccus haeundaensis* BC74171T and evaluation of their antioxidant activity and antiproliferative effect on normal and cancer cell lines, Colloids Surf. B Biointerfaces 183 (2019) 110455.

[96] H.R. Ghorbani, F.P. Mehr, A.K. Poor, Extracellular synthesis of copper nanoparticles using culture supernatants of *Salmonella typhimurium*, Orient. J. Chem. 31 (1) (2015) 527–529.

[97] D. Jain, A.A. Bhojiya, H. Singh, H.K. Daima, M. Singh, S.R. Mohanty, et al., Microbial fabrication of zinc oxide nanoparticles and evaluation of their antimicrobial and photocatalytic properties, Front. Chem. 8 (2020) 778.

[98] J.W. Moon, Y. Roh, R.J. Lauf, H. Vali, L.W. Yeary, T.J. Phelps, Microbial preparation of metal-substituted magnetite nanoparticles, J. Microbiol. Methods 70 (1) (2007) 150–158.

[99] K.S. Hemath Naveen, G. Kumar, L. Karthik, K.V. Bhaskara Rao, Extracellular biosynthesis of silver nanoparticles using the filamentous fungus Penicillium sp, Arch. Appl. Sci. Res. 2 (6) (2010) 161–167.

[100] R. Bharathidasan, A. Panneerselvam, Biosynthesis and characterization of silver nanoparticles using endophytic fungi *Aspergillus concius*, *Penicillium janthinellum* and Phomosis sp, Int. J. Pharm. Sci. Res. 3 (9) (2012) 3163.

[101] B.K. Nayak, A. Nanda, V. Prabhakar, Biogenic synthesis of silver nanoparticle from wasp nest soil fungus, *Penicillium italicum* and its analysis against multi drug resistance pathogens, Biocatalysis Agric. Biotechnol. 16 (2018) 412–418.

[102] J. Jena, N. Pradhan, R.R. Nayak, B.P. Dash, L.B. Sukla, P.K. Panda, et al., Microalga Scenedesmus sp.: a potential low-cost green machine for silver nanoparticle synthesis, J. Microbiol. Biotechnol. 24 (4) (2014) 522–533.

[103] S. Sarangadharan, S. Nallusamy, Biosynthesis and characterization of silver nanoparticles produced by *Bacillus licheniformis*, Int. J. Pharm. Med. Biol. Sci. 4 (4) (2015) 236.

[104] A. Mishra, S.K. Tripathy, R. Wahab, S.H. Jeong, I. Hwang, Y.B. Yang, et al., Microbial synthesis of gold nanoparticles using the fungus *Penicillium brevicompactum* and their cytotoxic effects against mouse mayo blast cancer C_2C_{12} cells, Appl. Microbiol. Biotechnol. 92 (3) (2011) 617–630.

[105] A.N. Mishra, S. Bhadauria, M.S. Gaur, R. Pasricha, Extracellular microbial synthesis of gold nanoparticles using fungus *Hormoconis resinae*, JOM 62 (11) (2010) 45–48.

[106] R.S. Tomar, S. Banerjee, S. Kaushik, S. Singh, A.S. Vidyarthi, Microbial synthesis of gold nanoparticles by biosurfactant producing *Pseudomonas aeruginosa*, Int. J. Adv. Life Sci. 8 (4) (2015) 520–527.

[107] I. Maliszewska, Microbial mediated synthesis of gold nanoparticles: preparation, characterization and cytotoxicity studies, Dig. J. Nanomater. Biostruct. (DJNB) 8 (3) (2013).

[108] J. Jayabalan, G. Mani, N. Krishnan, J. Pernabas, J.M. Devadoss, H.T. Jang, Green biogenic synthesis of zinc oxide nanoparticles using Pseudomonas putida culture and its In vitro antibacterial and anti-biofilm activity, Biocatalysis Agric. Biotechnol. 21 (2019) 101327.

[109] C. Jayaseelan, A.A. Rahuman, A.V. Kirthi, S. Marimuthu, T. Santhoshkumar, A. Bagavan, et al., Novel microbial route to synthesize ZnO nanoparticles using *Aeromonas hydrophila* and their activity against pathogenic bacteria and fungi, Spectrochim. Acta A Mol. Biomol. Spectrosc. 90 (2012) 78–84.

[110] H.M. Yusof, R. Mohamad, U.H. Zaidan, Sustainable microbial cell nanofactory for zinc oxide nanoparticles production by zinc-tolerant probiotic *Lactobacillus plantarum* strain TA4, Microb. Cell Factories 19 (1) (2020) 1–17.

[111] E.F. El-Belely, M. Farag, H.A. Said, A.S. Amin, E. Azab, A.A. Gobouri, et al., Green synthesis of zinc oxide nanoparticles (ZnO-NPs) using *Arthrospira platensis* (Class: Cyanophyceae) and evaluation of their biomedical activities, Nanomaterials 11 (1) (2021) 95.

[112] N. Jain, A. Bhargava, J.C. Tarafdar, S.K. Singh, J. Panwar, A biomimetic approach towards synthesis of zinc oxide nanoparticles, Appl. Microbiol. Biotechnol. 97 (2) (2013) 859–869.

[113] Q. Lv, B. Zhang, X. Xing, Y. Zhao, R. Cai, W. Wang, et al., Biosynthesis of copper nanoparticles using *Shewanella loihica* PV-4 with antibacterial activity: novel approach and mechanisms investigation, J. Hazard. Mater. 347 (2018) 141–149.

[114] A. Arya, K. Gupta, T.S. Chundawat, D. Vaya, Biogenic synthesis of copper and silver nanoparticles using green alga *Botryococcus braunii* and its antimicrobial activity, Bioinorg. Chem. Appl. (2018) 2018.

[115] V. Borse, A. Kaler, U.C. Banerjee, Microbial synthesis of platinum nanoparticles and evaluation of their anticancer activity, Int. J. Emerg. Trends Electr. Electron. 11 (2) (2015).

[116] M. Kowshik, N. Deshmukh, W. Vogel, J. Urban, S.K. Kulkarni, K.M. Paknikar, Microbial synthesis of semiconductor CdS nanoparticles, their characterization, and their use in the fabrication of an ideal diode, Biotechnol. Bioeng. 78 (5) (2002) 583–588.

[117] T.A. Voeikova, E.I. Kozhukhova, O.A. Zhuravliova, V.M. Retivov, E.A. Chigorina, V.S. Kuligin, et al., Microbial synthesis of cadmium sulfide nanoparticles: influence of bacteria of various species on the characteristics of biogenic nanoparticles, Nanotechnol. Russia 15 (2) (2020) 182–190.

[118] M. Asghar, S. Habib, W. Zaman, S. Hussain, H. Ali, S. Saqib, Synthesis and characterization of microbial mediated cadmium oxide nanoparticles, Microscopy Res. Tech. 83 (12) (2020) 1574–1584.

[119] A. Shivashankarappa, K.R. Sanjay, Study on biological synthesis of cadmium sulfide nanoparticles by *Bacillus licheniformis* and its antimicrobial properties against food borne pathogens, Nanosci. Nanotechnol. Res. 3 (1) (2015) 6–15.

[120] S.K. Das, A.R. Das, A.K. Guha, Gold nanoparticles: microbial synthesis and application in water hygiene management, Langmuir 25 (14) (2009) 8192–8199.

[121] R.S. Yehia, H. Al-Sheikh, Biosynthesis and characterization of silver nanoparticles produced by *Pleurotus ostreatus* and their anticandidal and anticancer activities, World J. Microbiol. Biotechnol. 30 (11) (2014) 2797–2803.

[122] H. Fernández-Llamosas, L. Castro, M.L. Blázquez, E. Díaz, M. Carmona, Speeding up bioproduction of selenium nanoparticles by using *Vibrio natriegens* as microbial factory, Sci. Rep. 7 (1) (2017) 1–9.

[123] A. Lagashetty, A. Venkataraman, Polymer nanocomposites, Resonance 10 (7) (2005) 49–57.

[124] Q.F. Guan, Z.M. Han, T.T. Luo, H.B. Yang, H.W. Liang, S.M. Chen, et al., A general aerosol-assisted biosynthesis of functional bulk nanocomposites, Natl Sci. Rev. 6 (1) (2019) 64–73.

[125] S. Raza, A. Ansari, N.N. Siddiqui, F. Ibrahim, M.I. Abro, A. Aman, Biosynthesis of silver nanoparticles for the fabrication of non cytotoxic and antibacterial metallic polymer based nanocomposite system, Sci. Rep. 11 (1) (2021) 1–15.

[126] J.L. Castro-Mayorga, F. Freitas, M.A.M. Reis, M.A. Prieto, J.M. Lagaron, Biosynthesis of silver nanoparticles and polyhydroxybutyrate nanocomposites of interest in antimicrobial applications, Int. J. Biol. Macromol. 108 (2018) 426–435.
[127] S.B. Aziz, R.B. Marif, M.A. Brza, A.N. Hassan, H.A. Ahmad, Y.A. Faidhalla, et al., Structural, thermal, morphological and optical properties of PEO filled with biosynthesized Ag nanoparticles: new insights to band gap study, Results Phys. 13 (2019) 102220.
[128] J. Geng, D. Yang, Y. Zhu, L. Cao, Z. Jiang, Y. Sun, One-pot biosynthesis of polymer–inorganic nanocomposites, J. Nanopart. Res. 13 (6) (2011) 2661–2670.
[129] R. Thomas, K.R. Soumya, J. Mathew, E.K. Radhakrishnan, Electrospun polycaprolactone membrane incorporated with biosynthesized silver nanoparticles as effective wound dressing material, Appl. Biochem. Biotechnol. 176 (8) (2015) 2213–2224.
[130] E.E. Kiziltas, A. Kiziltas, M. Blumentritt, D.J. Gardner, Biosynthesis of bacterial cellulose in the presence of different nanoparticles to create novel hybrid materials, Carbohydr. Polym. 129 (2015) 148–155.
[131] A.M. Youssef, M.S. Abdel-Aziz, S.M. El-Sayed, Chitosan nanocomposite films based on Ag-NP and Au-NP biosynthesis by *Bacillus subtilis* as packaging materials, Int. J. Biol. Macromol. 69 (2014) 185–191.
[132] A.M. Youssef, M.S. Abdel-Aziz, Preparation of polystyrene nanocomposites based on silver nanoparticles using marine bacterium for packaging, Polym. Technol. Eng. 52 (6) (2013) 607–613.
[133] M. Hasanin, M.A. Al Abboud, M.M. Alawlaqi, T.M. Abdelghany, A.H. Hashem, Ecofriendly synthesis of biosynthesized copper nanoparticles with starch-based nanocomposite: antimicrobial, antioxidant, and anticancer activities, Biol. Trace Elem. Res. (2021) 1–14.
[134] V. Kanikireddy, K. Varaprasad, M.S. Rani, P. Venkataswamy, B.J.M. Reddy, M. Vithal, Biosynthesis of CMC-Guar gum-Ag_0 nanocomposites for inactivation of food pathogenic microbes and its effect on the shelf life of strawberries, Carbohydr. Polym. 236 (2020) 116053.
[135] S. Chakraborty, N.L. Mary, Biocompatible supercapacitor electrodes using green synthesised ZnO/Polymer nanocomposites for efficient energy storage applications, J. Energy Storage 28 (2020) 101275.
[136] M.J. Uddin, B. Chaudhuri, K. Pramanik, T.R. Middya, B. Chaudhuri, Black tea leaf extract derived Ag nanoparticle-PVA composite film: structural and dielectric properties, Mater. Sci. Eng. B 177 (20) (2012) 1741–1747.
[137] R.M. Tripathi, R.N. Pudake, B.R. Shrivastav, A. Shrivastav, Antibacterial activity of poly (vinyl alcohol)—biogenic silver nanocomposite film for food packaging material, Adv. Nat. Sci. Nanosci. Nanotechnol. 9 (2) (2018) 025020.
[138] T.D. Nguyen, C.H. Dang, D.T. Mai, Biosynthesized AgNP capped on novel nanocomposite 2-hydroxypropyl-β-cyclodextrin/alginate as a catalyst for degradation of pollutants, Carbohydr. Polym. 197 (2018) 29–37.
[139] A. Muzaffar, M.B. Ahamed, K. Deshmukh, S.K. Pasha, Dielectric properties and electromagnetic interference shielding studies of nickel oxide and tungsten oxide reinforced polyvinylchloride nanocomposites, Polym. Technol. Mater. 59 (15) (2020) 1667–1678.
[140] M. Pandey, G.M. Joshi, K. Deshmukh, M. Khutia, N.N. Ghosh, Optimized AC conductivity correlated to structure, morphology and thermal properties of PVDF/PVA/Nafion composites, Ionics 20 (10) (2014) 1427–1433.
[141] M. Pandey, G.M. Joshi, K. Deshmukh, J. Ahmad, Impedance spectroscopy and conductivity studies of $CdCl_2$ doped polymer electrolyte, Adv. Mater. Lett. 6 (2) (2015).
[142] A.H. Kumar, M.B. Ahamed, K. Deshmukh, M.S. Sirajuddeen, Morphology, dielectric and EMI shielding characteristics of graphene nanoplatelets, montmorillonite nanoclay and titanium dioxide nanoparticles reinforced polyvinylidenefluoride nanocomposites, J. Inorg. Organomet. Polym. Mater 31 (2021) 2003–2016.

[143] P. Rani, M.B. Ahamed, K. Deshmukh, Structural, dielectric and EMI shielding properties of polyvinyl alcohol/chitosan blend nanocomposites integrated with graphite oxide and nickel oxide nanofillers, J. Mater. Sci. Mater. Electron 32 (2021) 764–779.

[144] Z. Feng, K.H. Adolfsson, Y. Xu, H. Fang, M. Hakkarainen, M. Wu, Carbon dot/polymer nanocomposites: from green synthesis to energy, environmental and biomedical applications, Sustain. Mater. Technol. 29 (2021) e00304.

[145] A. Hamid, M. Khan, A. Hayat, J. Raza, A. Zada, A. Ullah, et al., Probing the physio-chemical appraisal of green synthesized PbO nanoparticles in PbO-PVC nanocomposite polymer membranes, Spectrochim. Acta A Mol. Biomolecul. Spectrosc. 235 (2020) 118303.

[146] H. Zhu, S. Jia, T. Wan, Y. Jia, H. Yang, J. Li, et al., Biosynthesis of spherical Fe_3O_4/bacterial cellulose nanocomposites as adsorbents for heavy metal ions, Carbohydr. Polym. 86 (4) (2011) 1558–1564.

[147] M. Nasrollahzadeh, M. Sajjadi, M. Maham, S.M. Sajadi, A.A. Barzinjy, Biosynthesis of the palladium/sodium borosilicate nanocomposite using *Euphorbia milii* extract and evaluation of its catalytic activity in the reduction of chromium (VI), nitro compounds and organic dyes, Mater. Res. Bull. 102 (2018) 24–35.

[148] S. Azizi, R. Mohamad, R.A. Rahim, R. Mohammadinejad, A.B. Ariff, Hydrogel beads bio-nanocomposite based on Kappa-Carrageenan and green synthesized silver nanoparticles for biomedical applications, Int. J. Biol. Macromol. 104 (2017) 423–431.

[149] A.U. Khan, A.U. Khan, B. Li, M.H. Mahnashi, B.A. Alyami, Y.S. Alqahtani, et al., Biosynthesis of silver capped magnesium oxide nanocomposite using *Olea cuspidata* leaf extract and their photocatalytic, antioxidant and antibacterial activity, Photodiag. Photodyn. Ther. 33 (2021) 102153.

[150] R.R. Palem, S.D. Ganesh, Z. Kroneková, M. Sláviková, N. Saha, P. Sáha, Green synthesis of silver nanoparticles and biopolymer nanocomposites: a comparative study on physico-chemical, antimicrobial and anticancer activity, Bull. Mater. Sci. 41 (2) (2018) 1–11.

[151] M. Nasrollahzadeh, S.M. Sajadi, M. Maham, I. Kohsari, Biosynthesis, characterization and catalytic activity of the Pd/bentonite nanocomposite for base-and ligand-free oxidative hydroxylation of phenylboronic acid and reduction of chromium (VI) and nitro compounds, Microporous Mesoporous Mater. 271 (2018) 128–137.

[152] M.J. Nazarahari, A.K. Manshad, M. Ali, J.A. Ali, A. Shafiei, S.M. Sajadi, et al., Impact of a novel biosynthesized nanocomposite (SiO_2@ Montmorilant@ Xanthan) on wettability shift and interfacial tension: applications for enhanced oil recovery, Fuel 298 (2021) 120773.

[153] S. Preethi, K. Abarna, M. Nithyasri, P. Kishore, K. Deepika, R. Ranjithkumar, et al., Synthesis and characterization of chitosan/zinc oxide nanocomposite for antibacterial activity onto cotton fabrics and dye degradation applications, Int. J. Biol. Macromol. 164 (2020) 2779–2787.

[154] V.S. Ramkumar, A. Pugazhendhi, S. Prakash, N.K. Ahila, G. Vinoj, S. Selvam, et al., Synthesis of platinum nanoparticles using seaweed *Padina gymnospora* and their catalytic activity as PVP/PtNPs nanocomposite towards biological applications, Biomed. Pharmacother. 92 (2017) 479–490.

[155] A.M. Mousa, O.A.A. Aziz, O.E. Al-Hagar, M.A. Gizawy, K.F. Allan, M.F. Attallah, Biosynthetic new composite material containing CuO nanoparticles produced by *Aspergillus terreus* for 47Sc separation of cancer theranostics application from irradiated Ca target, Appl. Radiat. Isotopes 166 (2020) 109389.

[156] B. Jaleh, P. Fakhri, Chapter 5 - Infrared and Fourier transform infrared spectroscopy for nanofillers and their nanocomposites, Spectrosc. Polym. Nanocompos. (2016) 112–129.

[157] X. Zhang, S. Bhuvana, L.S. Loo, Characterization of layered silicate dispersion in polymer nanocomposites using Fourier transform infrared spectroscopy, J. Appl. Polym. Sci. 125 (2012) E175–E180.

[158] L. Bokobza, A.L. Diop, Reinforcement of silicone rubbers by sol-gel in situ generated particles, in: S. Thomas, R. Stephen (Eds.), Rubber Nanocomposites: Preparation, Properties, and Applications, John Wiley & Sons (Asia), Singapore, 2010.

[159] M.A. Asghar, R.I. Yousuf, M.H. Shoaib, M.A. Asghar, S. Ansar, M. Zehravi, et al., Synergistic nanocomposites of different antibiotics coupled with green synthesized chitosan-based silver nanoparticles: characterization, antibacterial, in vivo toxicological and biodistribution studies, Int. J. Nanomed. 15 (2020) 7841.

[160] M.B. Ahmad, M.Y. Tay, K. Shameli, M.Z. Hussein, J.J. Lim, Green synthesis and characterization of silver/chitosan/polyethylene glycol nanocomposites without any reducing agent, Int. J. Mol. Sci. 12 (8) (2011) 4872–4884.

[161] W. Yuan, Y. Gu, L. Li, Green synthesis of graphene/Ag nanocomposites, Appl. Surf. Sci. 261 (2012) 753–758.

[162] S. Govindan, E.A.K. Nivethaa, R. Saravanan, V. Narayanan, A. Stephen, Synthesis and characterization of chitosan–silver nanocomposite, Appl. Nanosci. 2 (3) (2012) 299–303.

[163] K. Oksman, R.J. Moon, Chapter 7: characterization of nanocomposites structure, Mater. Energy Handbook Green. Mater. (2014) 89–105.

[164] F.G. Torres, J.P. Mayorga, C. Vilca, et al., Preparation and characterization of a novel starch–chestnut husk biocomposite, SN Appl. Sci. 1 (2019) 1158. Available from: https://doi.org/10.1007/s42452-019-1204-y.

[165] P.L. Reddy, K. Deshmukh, K. Chidambaram, M.M.N. Ali, K.K. Sadasivuni, Y.R. Kumar, et al., Dielectric properties of polyvinyl alcohol (PVA) nanocomposites filled with green synthesized zinc sulphide (ZnS) nanoparticles, J. Mater. Sci. Mater. Electron. 30 (5) (2019) 4676–4687.

[166] P.L. Reddy, K. Deshmukh, T. Kovářík, D. Reiger, N.A. Nambiraj, R. Lakshmipathy, Enhanced dielectric properties of green synthesized Nickel Sulphide (NiS) nanoparticles integrated polyvinylalcohol nanocomposites, Mater. Res. Exp. 7 (6) (2020) 064007.

[167] S. Kumar, A. Mudai, B. Roy, I.B. Basumatary, A. Mukherjee, J. Dutta, Biodegradable hybrid nanocomposite of chitosan/gelatin and green synthesized zinc oxide nanoparticles for food packaging, Foods 9 (9) (2020) 1143.

[168] Y. Gao, L. Li, P. Tan, L. Liu, Z. Zhang, Application of Raman spectroscopy in carbon nanotube-based polymer composites, Chin. Sci. Bull. 55 (35) (2010) 3978–3988.

[169] T. Yaguchi, Y. Kuroda, M. Konno, T. Kamino, K. Kaji, M. Watanabe, A method for 3 dimensional structural and compositional imaging of nano-materials, Proc. Microsc. Microanal. 12 (Suppl. 2) (2006) 528–529.

[170] G. Michler, Preparation of thin sections: (cryo)ultramicrotomy and (cryo)microtomy, Electron Microscopy of Polymers (2008) 199–217. Available from: https://doi.org/10.1007/978-3-540-36352-1_11. Springer Laboratory. Springer, Berlin, Heidelberg.

[171] P.L. Reddy, K. Deshmukh, K. Chidambaram, B. Ahamed, K.K. Sadasivuni, D. Ponnamma, et al., Effect of poly ethylene glycol (PEG) on structural, thermal and photoluminescence properties of CdO nanoparticles for optoelectronic applications, Mater. Today Proc. 9 (2019) 175–183.

[172] A. Venkateshaiah, R. Nutenki, S.I. Kattimuttathu, Chapter 14 - X-ray diffraction spectroscopy of polymer nanocomposites, Spectrosc. Polym. Nanocompos. (2016) 410–451.

[173] F. Lorestani, Z. Shahnavaz, P. Mn, Y. Alias, N.S. Manan, One-step hydrothermal green synthesis of silver nanoparticle-carbon nanotube reduced-graphene oxide composite and its application as hydrogen peroxide sensor, Sens. Actuat. B Chem. 208 (2015) 389–398.

[174] A. Fouda, S.S. Salem, A.R. Wassel, M.F. Hamza, T.I. Shaheen, Optimization of green biosynthesized visible light active CuO/ZnO nano-photocatalysts for the degradation of organic methylene blue dye, Heliyon 6 (9) (2020) e04896.

[175] M. Sharma, K. Behl, S. Nigam, M. Joshi, TiO_2-GO nanocomposite for photocatalysis and environmental applications: a green synthesis approach, Vacuum 156 (2018) 434–439.

[176] F. Sadegh, A.R. Modarresi-Alam, M. Noroozifar, K. Kerman, A facile and green synthesis of superparamagnetic Fe_3O_4@ PANI nanocomposite with a core–shell structure to increase of triplet state population and efficiency of the solar cells, J. Environ. Chem. Eng. 9 (1) (2021) 104942.

[177] M. Sarno, W.A.A. Mustafa, A. Senatore, D. Scarpa, One-step "green" synthesis of dispersable carbon quantum dots/poly (methyl methacrylate) nanocomposites for tribological applications, Tribol. Int. 148 (2020) 106311.

[178] M.A. Awad, A.A. Hendi, K.M. Ortashi, A.B. Alanazi, B.A. ALZahrani, D.A. Soliman, Greener synthesis, characterization, and antimicrobiological effects of helba silver nanoparticle-PMMA nanocomposite, Int. J. Polym. Sci. (2019) 2019.

[179] F. Davar, A. Majedi, A. Mirzaei, Polyvinyl alcohol thin film reinforced by green synthesized zirconia nanoparticles, Ceram. Int. 44 (16) (2018) 19377–19382.

[180] A.M. Youssef, S.M. El-Sayed, Bionanocomposites materials for food packaging applications: concepts and future outlook, Carbohydr. Polym. 193 (2018) 19–27.

[181] M.S. Mohseni, M.A. Khalilzadeh, M. Mohseni, F.Z. Hargalani, M.I. Getso, V. Raissi, et al., Green synthesis of Ag nanoparticles from pomegranate seeds extract and synthesis of Ag-Starch nanocomposite and characterization of mechanical properties of the films, Biocatal. Agric. Biotechnol. 25 (2020) 101569.

[182] D.A. Marrez, A.E. Abdelhamid, O.M. Darwesh, Eco-friendly cellulose acetate green synthesized silver nano-composite as antibacterial packaging system for food safety, Food Packaging Shelf Life 20 (2019) 100302.

[183] L.M. Sanchez, R.P. Ollier, J.S. Gonzalez, V.A. Alvarez, Nanocomposite materials for dyes removal, Handbook of Nanomaterials for Industrial Applications Micro Nano Technol. (2018) 922–951. Available from: https://doi.org/10.1016/B978-0-12-813351-4.00053-5.

[184] M. Hassanpour, H. Safardoust-Hojaghan, M. Salavati-Niasari, Degradation of methylene blue and Rhodamine B as water pollutants via green synthesized Co_3O_4/ZnO nanocomposite, J. Mol. Liq. 229 (2017) 293–299.

[185] P. Mondal, M.K. Purkait, Preparation and characterization of novel green synthesized iron–aluminum nanocomposite and studying its efficiency in fluoride removal, Chemosphere 235 (2019) 391–402.

[186] M.E. Mahmoud, M.F. Amira, M.E. Abouelanwar, M.A. Salam, Green synthesis and surface decoration of silver nanoparticles onto δ-FeOOH-Polymeric nanocomposite as efficient nanocatalyst for dyes degradation, J. Environ. Chem. Eng. 9 (1) (2021) 104697.

[187] L.M. Katata-Seru, T.C. Maponya, E. Makhado, K.D. Modibane, M.J. Hato, S.M. Matomea, et al., Green synthesis of polypyrrole/nanoscale zero valent iron nanocomposite and use as an adsorbent for hexavalent chromium from aqueous solution, South. Afr. J. Chem. Eng. 34 (1) (2020) 1–10.

[188] S. Kulshrestha, S. Qayyum, A.U. Khan, Antibiofilm efficacy of green synthesized graphene oxide-silver nanocomposite using *Lagerstroemia speciosa* floral extract: a comparative study on inhibition of gram-positive and gram-negative biofilms, Microb. Pathogenesis 103 (2017) 167–177.

[189] G. Gopal, H. Sankar, C. Natarajan, A. Mukherjee, Tetracycline removal using green synthesized bimetallic nZVI-Cu and bentonite supported green nZVI-Cu nanocomposite: a comparative study, J. Environ. Manag. 254 (2020) 109812.

[190] I. Sivanesan, J. Gopal, M. Muthu, J. Shin, S. Mari, J. Oh, Green synthesized chitosan/chitosan nanoforms/nanocomposites for drug delivery applications, Polymers 13 (14) (2021) 2256.

[191] Z. Haghparasti, M.M. Shahri, Green synthesis of water-soluble nontoxic inorganic polymer nanocomposites containing silver nanoparticles using white tea extract and assessment of their in vitro antioxidant and cytotoxicity activities, Mater. Sci. Eng. C. 87 (2018) 139–148.

[192] N. Khan, D. Kumar, P. Kumar, Silver nanoparticles embedded guar gum/gelatin nanocomposite: green synthesis, characterization and antibacterial activity, Colloid Interface Sci. Commun. 35 (2020) 100242.

[193] S. Pandey, G.K. Goswami, K.K. Nanda, Green synthesis of biopolymer–silver nanoparticle nanocomposite: an optical sensor for ammonia detection, Int. J. Biol. Macromol. 51 (4) (2012) 583–589.

[194] Y. Zhang, S. Liu, L. Wang, X. Qin, J. Tian, W. Lu, et al., One-pot green synthesis of Ag nanoparticles-graphene nanocomposites and their applications in SERS, H_2O_2, and glucose sensing, RSC Adv. 2 (2) (2012) 538–545.

[195] D. Arthisree, W. Madhuri, Optically active polymer nanocomposite composed of polyaniline, polyacrylonitrile and green-synthesized graphene quantum dot for supercapacitor application, Int. J. Hydrog. Energy 45 (16) (2020) 9317–9327.

[196] P. Liu, F.Q. Bai, D.W. Lin, H.P. Peng, Y. Hu, Y.J. Zheng, et al., One-pot green synthesis of mussel-inspired myoglobin–gold nanoparticles–polydopamine–graphene polymeric bionanocomposite for biosensor application, J. Electroanalytical Chem. 764 (2016) 104–109.

[197] Z. Çıplak, A. Yıldız, N. Yıldız, Green preparation of ternary reduced graphene oxide-au@ polyaniline nanocomposite for supercapacitor application, J. Energy Storage 32 (2020) 101846.

CHAPTER

9 Biodegradable and biocompatible polymer nanocomposites for tissue engineering applications

Huda R.M. Rashdan[1] and Mehrez E. El-Naggar[2]

[1]*Chemistry of Natural and Microbial Products Department, Pharmaceutical and Drug Industries Research Institute, National Research Centre, Dokki, Cairo, Egypt* [2]*Textile Research and Technology Institute, National Research Centre, Dokki, Cairo, Egypt*

Chapter Outline

Biodegradable and Biocompatible Polymer Nanocomposites. DOI: https://doi.org/10.1016/B978-0-323-91696-7.00002-7
Copyright © 2023 Elsevier Inc. All rights reserved.

9.1 Introduction

Many concepts and definitions had to be discussed in order to understand what sustainable tissue engineering entails. Tissue engineering (TE) is a term used to describe the procedures used for the creation and regeneration of new tissue for the therapeutic reconstruction of lost or damaged human body tissues, which are accomplished through the controlled, systematic, and deliberate stimulation of specific selected cells via a systematic combination of mechanical and molecular signals [1]. The human body's tissues have an incredible ability to regenerate. Tissue regeneration, particularly for critical-sized body tissue defects, has piqued the interest of many researchers with the goal of replacing or repairing damaged tissues of various body organs. The central idea behind TE is designed to create living tissue grafts and can be used surgically to replace or repair congenitally damaged body tissues. TE consists of various body muscles that can be regenerated by seeding cell in a matrix scaffold, resulting in 3D engineered body tissues. It was implied indirectly that TE should include biomaterials, as mechanical and molecular signs cannot be delivered in vacuum and usually require a biomaterial to act as a carrier, which can be ascribed to scaffolds. The term "Template" is one of the platforms for regenerative medicine [2].

9.2 Biodegradable and biocompatible polymers for sustainable tissue engineering

The field of TE has enabled the reconstruction of body tissues for therapeutic purposes in order to treat various diseases within the human body associated with tissue damage or loss. Despite respectable progress in small animal and in vitro studies, commercial and clinical endpoints have been slow, particularly in the treatment of those who do not have acceptable conventional therapies. The selection of the biomaterial used as a scaffold to guide the tissue regeneration process is a critical issue in TE fields. For many years, the acceptable requirements of those biomaterials have not been well articulated; instead, the need for biodegradable and biocompatible polymers, as well as FDA approval for use in biomedical applications, has predominated the biomaterial selection process. Synthetic biomaterials including such synthetic bioabsorbable polymers, that lack biological activities and cannot induce target cells to form specific new tissues, will be ineffective. It has been reported that the use of traditional biomaterials for the fabrication of scaffolds in the tissue engineering approach is a poor selection. TE frameworks designed to stimulate the reproduction of a target cell's microenvironment have a better chance of success.

9.2.1 Natural polymers for tissue engineering

Chitosan (CS) is a biopolymer that has a wide range of applications in tissue engineering. Its chemical structure is composed of repeating units of D-glucosamine and *N*-acetyl D-glucosamine. CS is frequently used in TE owing to its high biocompatibility and biodegradability [3]. CS, because of its flexibility, can be combined with other materials to create a variety of tissues such as bone, cartilage, and skin. Because it possesses numerous desirable properties, CS is an excellent choice for scaffold construction in the field of bone tissue engineering (BTE) [4,5]. In vitro studies revealed that spheroidal clusters of osteoblasts were proliferating. In both in vitro and in vivo tests, the nano silica (nSi) composite scaffold improved biocompatibility and supported bone tissue regeneration. The commencement of mature bone growth utilizing this sort of scaffold needed 8 weeks, as seen in Fig. 9.1 [6].

Alginate is a naturally occurring anionic, hydrophilic polysaccharide composed of interchanging glucuronic and mannuronic acids. Seaweed is the primary source of alginate, which can be produced in a variety of forms, including alginic acid, sodium alginate, and propylene glycol alginate. Because of its inherent biocompatibility and immunogenicity, alginate has been formed into sponges, hydrogels, and fibers, and has found utility in tissue engineering modified via drug administration [8].

Gel is one of the most common structural proteins found in bones, skin, and ligaments (a natural polymer) [9]. The majority of the chemical make up of gel, which is comparable to that of its predecessor, is composed of glycine, proline, and hydroxyproline. The proteinaceous composition of the gel, which includes Arginylglycylaspartic acid (RGD) sequences that promote cell attachment. These RGD sequences interact with integrins in cells, promoting cell adhesion and improving host osteointegration [10]. Cross-linking, which can be done physically or chemically, can improve Gel's mechanical properties. The biostability and biocompatibility of Gel may be beneficial in maintaining bone function after implantation [11].

Cellulose appears to be the most common polymer on the planet, and it provides mechanical performance to plants while maintaining their established design [12]. It is a cellulase-degraded linear polymer composed of D glucose with 1,4 links. Regardless of whether or not mammals lack cellulase, mechanical erosion and acid hydrolysis are responsible for cellulase destruction. In comparison to its plant-derived counterpart, bacterial cellulose appears to have an exceptional tensile strength. As a result, it is the most common cellulose used in TE [13]. Furthermore, cellulose-based salts could be electrospunned in order to create unique nanofibrous frameworks, demonstrating cellulose's adaptability in any method of construction [14].

Fucoidan is a polysaccharide that is primarily derived from marine algae. It's primarily composed of sulfate groups, fucose, and uronic acid, with a trace of xylose, glucose, and other sugar fractions thrown in for good measure [15,16]. Fucoidan has excellent biodegradability, water absorption, and biocompatibility [16]. Fucoidan is also polymerized to a significant extent and can be combined with other polymers such as gels, alginate, and CS to improve scaffold properties. It has been shown that, fucoidan can be used to support the TE of the bone [16].

Hyaluronic acid (HA) is one of the primary glycosaminoglycans found in many tissues, including tendons, ligaments, synovial fluid, and the eye [17]. HA is involved in a wide range of physiological activities, including tissue morphogenesis, cell proliferation, migration, and tissue synthesis [18]. Because of its beneficial properties such as biocompatibility, biodegradability, and hydrophilicity, HA has a variety of applications in cosmetics and medicine, particularly TE [19]. However, due to the polymer's high hydrophilic nature, HA scaffolds have several drawbacks, including rapid breakdown and limited high stability [18].

To address these limitations, it appears that chemical or physical cross-linking of HA with functional groups such as methacrylate groups is necessary [20]. The interaction of HA through its carboxyl or hydroxyl groups is one method for producing the cross-linked scaffold [21]. Because of its various polar groups, HA can bind and interact with various growth factors, allowing them to be released continuously [19,22]. Variety of approaches have been used to create hybrid scaffolds

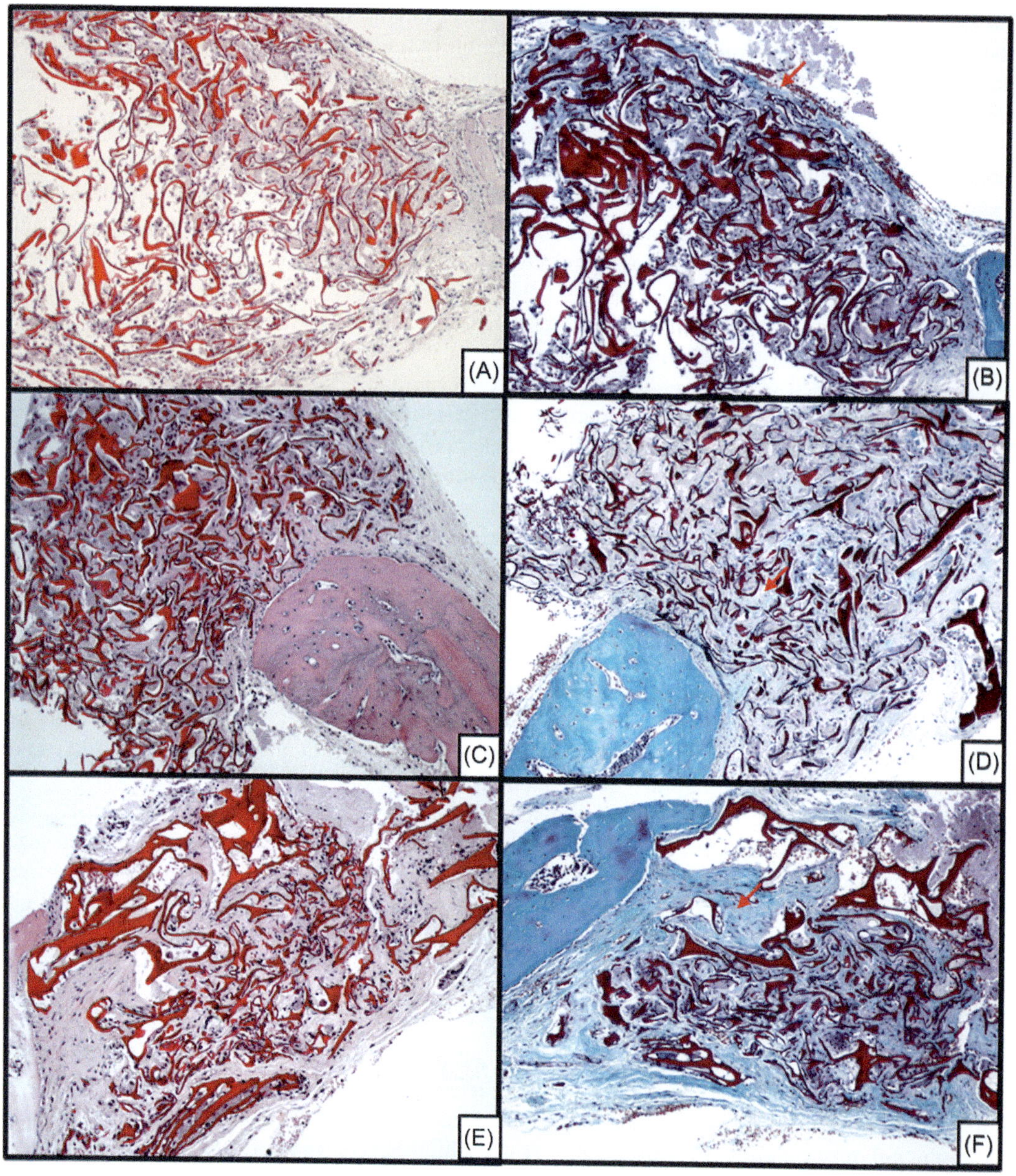

FIGURE 9.1

Histological images of nSi/CS with H&E staining after (A) 15 days, (C) 30 days, and (E) 45 days. Meanwhile, the histological images of nSi/CS with Mallory's trichrome staining after (B) 15 days, (D) 30 days, and (F) 45 days [7] https://s100.copyright.com/CustomerAdmin/PLF.jsp?ref = de460915-34c6-48f7-9de6-cee18ea40002.

Reproduced with permission from Elsevier.

depending on HA and silk fibroin, however the resulting scaffolds possess low mechanical characteristics. Burak Tavsanli et al., [23] used methacrylated HA (MeHA) rather than HA to prepare hydrogel of HA/silk fibroin (SF). MeHA's pendant vinyl groups serve as a macromer as well as a chemically cross-linking agent. The presence of SF during MeHA polymerization resulted in the formation of additional physical crosslinks in the sheet domains of molecules such as SF, which improved the mechanical properties of the hydrogels [19].

The responsiveness of an in situ chemically produced hybrid composite based on fucoidan and nHA to bone tissue regeneration was investigated [24]. In the case of human adipose-derived stem cells, biomineralization of the fucoidan/nHA/CS was superior to that of HA alone in the bone tissue regeneration environment; this affect was mainly owing to the presence of fucoidan in the nanocomposite. It was reported that the Gene expression for the fucoidan/nHA scaffolds enhanced on the seventh day after implantation. New bone development was observed in an in vivo rabbit model, although the bone formation was not thick, as illustrated in Fig. 9.2. nHA may have benefited osteoblastic activities, according to the increasement genes associated for Col Type-I, osteopontin, and osteocalcin in nHA-integrated composite [24].

Furthermore, bacterial cellulose (BC), a well-known bacterial-derived cellulose, is a simple to produce [25]. As a biocompatible natural polymer, BC has remarkable properties such as high tensile strength, high water-holding capacities, and crystallinity in the wet state [26]. It is widely used in TE when combined with many other compounds.

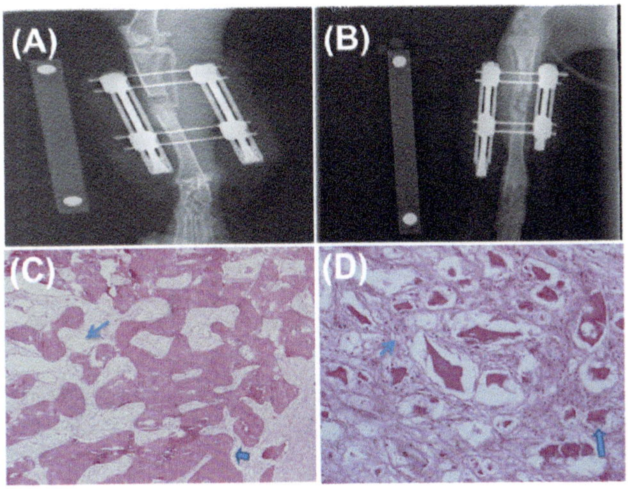

FIGURE 9.2

Radiographs of (A) nHA, and (B) Fucoidan/nHA nanocomposite scaffold. H&E staining shows (C) the filler of a bone deficiency through nanocomposite of fucoidan/nHA hybrid, and (D) the association of fat cells and aosteoblasts [24] https://s100.copyright.com/CustomerAdmin/PLF.jsp?ref = af660d17-1874-461b-8039-dd20c893fa58.

Reprinted with permission from Elsevier.

9.2.2 Biocompatible synthetic polymers for tissue engineering

Polycaprolactone (PCL) is an aliphatic polyester that is commonly used in bone tissue engineering (BTE). It is composed from repeating hexanoate structures. PCL has many advantages, including mechanical elasticity, biocompatibility, biodegradability, and thermal stability [27]. Because of its semicrystalline structure, PCL is frequently used in the electrospinning process to create nanofibers. To take the advantage of the biomaterials' exquisite properties in the composites and compensate for their shortcomings, PCL should be blended with certain other functional natural polymers in scaffold production [28]. Polylactic acid (PLA) has excellent thermal stability and mechanical properties, making it an excellent candidate for biomedical applications [29]. Because of its low stability, PLA can be combined with other materials for BTE. PLA/PCL mixture has been shown to aid in the growth of cranial bones [30]. Two lactide-based polymers are poly-l-lactic acid (PLLA) and poly D-lactic acid (PDLA). PLLA's biodegradability, mechanical properties, and biostimulatory properties make it a potential biomaterial [31]. PLLA is an excellent alternative for targeted drug administration in tissue engineering due to its proclivity to degrade in a nontoxic manner. To provide the controlled drug release, its fast breakdown rate is stabilized primarily through the development of composites with other biomaterials such as PCL. As a result, PLLA could be used to develop new drug delivery systems for TE [32]. Another water-soluble synthetic and biodegradable polymer is poly(vinyl alcohol) (PVA). It is created during the polymerization reaction via free radical interaction [33]. Chemical crosslinkers and associated physical procedures can be used to improve the mechanical performance of PVA. The use of 3D bio-composite scaffolds designed of PVA, Gel, and Collagen (Col) has been described for nonload-bearing bone healing [34].

Polyurethanes (PUs) are a class of segmental synthetic polymers manufactured by polymerizing polyols with diamine, diol or/and diisocyanates (low molecular weight) [35]. The soft domain of PUs is referred to diol, while the hard domain is referred to diisocyanate, and the chain extender is referred to a chain extender. The soft and hard categories of a polymer determine its elasticity and strength, respectively. Polyester, polyether, or polycarbonate soft domains, which are widely used in tissue engineering [36]. Bioink (water-based shape memory PU) was developed by Yu-Jen Wang et al., [37] for bone scaffolds printed in 3D. Soft segments fabricated from PCL diol and PLA diol can be used. They also used gelatin or polyethylene oxide (PEO) to increase printability, as well as superparamagnetic iron oxide nanoparticles (SPIO) to promote osteogenic activity. The proliferation of human mesenchymal stem cells was found to be greater in PU-gelatin and PU-gelatin-SPIO scaffolds than in PU-PEO and PU-PEO-SPIO scaffolds. Because of its versatility, PU can be used in cardiac engineering to provide appropriate Young's moduli. Biodegradable PUs can be made from bio-oils such as polyol and aliphatic diisocyanate, and their breakdown processes can take months. Because PUs are firmer than traditional hydrogels, they can be easily inserted into muscle cells [38]. Poly(lactic-co-glycolic acid) (PLGA) is one of the few synthetic polymers approved by the FDA for use in clinical trials. To assure angiogenic capacity, scaffolding were covered with an angiogenic peptide (AP) comprising collagen Type-I hydrogel (Fig. 9.3A) [39]. Following 3 days of development, in vitro angiogenesis assays utilizing rat endothelial cells (ECs) placed on the scaffolds revealed a vitality of more than 97% with living cells (Fig. 9.3B-a,b). Owing to the quick releasing of AP, there was also an increase in EC migrating on the extract of the scaffolding contained it (Fig. 9.3B-c,d).

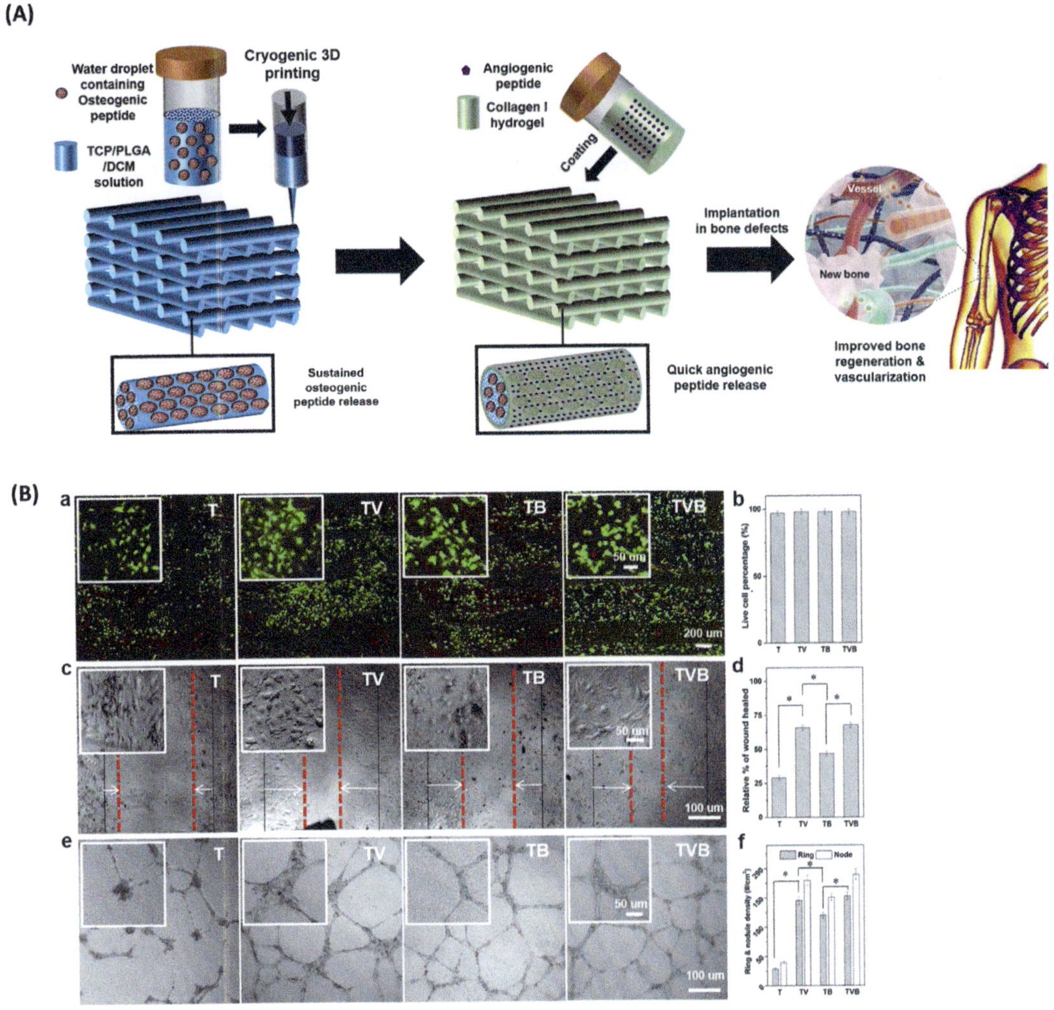

FIGURE 9.3

(A) diagrammatic representation of cryogenic 3D printing comprising osteogenic peptide (OP) and coated via angiogenic peptide (AP); (B) Angiogenesis investigate of the frameworks: (a, b), images of live and dead endothelial cells (ECs) on implants and survivability; (c, d), EC mobility of the scaffold extracts after 12 h of culture [40].

9.3 Hydrogels based on biodegradable and biocompatible polymer nanocomposites for tissue engineering

Smart hydrogels are biomaterials because of their sol–gel transition characteristics or volume phase changes that are reversible in response to changes in environmental variables. It was known that, the smart hydrogels can be used in biomedical domains including stem cell therapy with less invasive techniques. Smart hydrogels are capable of responding to a wide range of chemical (pH and ionic interactions) and physical (light, electric, temperature, and) stimuli [41]. pH-sensitive hydrogels are fabricated from basic or acidic polymeric compounds that receive or give protons as the pH of the environment changes. CS is a pH-responsive hydrogel capable of exhibiting sol–gel transition behavior at neutral pH 6.5. It was demonstrated that increasing the pH of deionized chitosan, 3D network was formed due to the effect of hydrogen bonds [42]. Overall, hydrogels have been used extensively in many biological activities, including medication delivery and tissue engineering. The porous structure of hydrogels allows for rapid nutrition and oxygen delivery, as well as a 3D permissive environment for growing cells [43]. Amiryaghoubi et al., [44] elaborated on hydrogel-based products and diverse hydrogel manufacturing processes, as well as their uses in cartilage and BTE (Fig. 9.4).

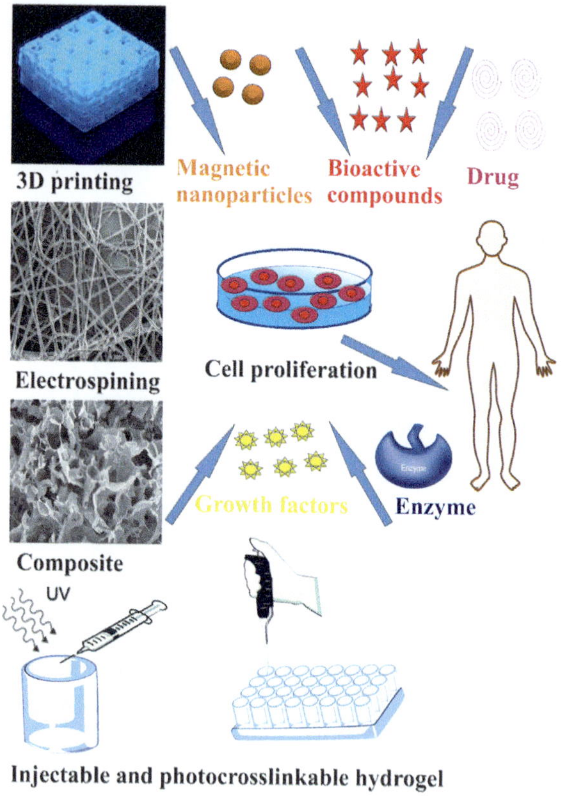

FIGURE 9.4

Hydrogel scaffolds for the regrowth of bone and cartilage tissue [44]. https://s100.copyright.com/CustomerAdmin/PLF.jsp?ref=978fba5b-1ecf-4bde-bf81-c1ae5052b9b2.

Reproduced with permission from Elsevier. Copyright 2022.

In order to prepare nanocomposite hydrogels, a cross-linking agent (physical or chemical cross-linking) blended with designated nanoparticles can be used [45]. Nanocomposite hydrogels have recently received a lot of interest as a substrate for TE due to their unique features such as electrical, mechanical, and magnetic properties [45]. Magnetic nanocomposite (MNPs) hydrogels have received a lot of attention in the field of TE [46]. For illustration, the utilization of the ultrasonic dispersion approach and freeze-thaw cross-linking decorating, a magnetic nanocomposite hydrogel was created by combining poly(vinyl alcohol) (PVA)-hydroxyapatite (HAP) with magnetic nanoparticles (MNPs; Fe_2O_3) (Fig. 9.5A). The addition of HAP improved the mechanical properties of hydrogels. Bone marrow mesenchymal stem cells (BMSCs) proliferated similarly and robustly in hydrogel scaffolds for bone and cartilage tissue regrowth. In addition, rising Fe_2O_3 concentration leads to growing the BMSC proliferation and chondrogenic differentiation (Fig. 9.5B) [47]. Another study was aimed the cyclic freeze-thaw process to create magnetic nano-HAP/PVA composite hydrogels.

For constructing a scaffold for cartilage tissue repair, a hydrogel with a magnetic hybrid composite (MagGel) containing HA, COL2, and poly(ethylene glycol) (PEG) mixed with MNPs was created. To demonstrate how the produced MagGel reacted, an outside magnet was used. Under controlled magnetic mediation, the hydrogel in biological fluids may travel forward into the defective tissue locations while retaining organizational integrity. The concentration of BMSCs on the

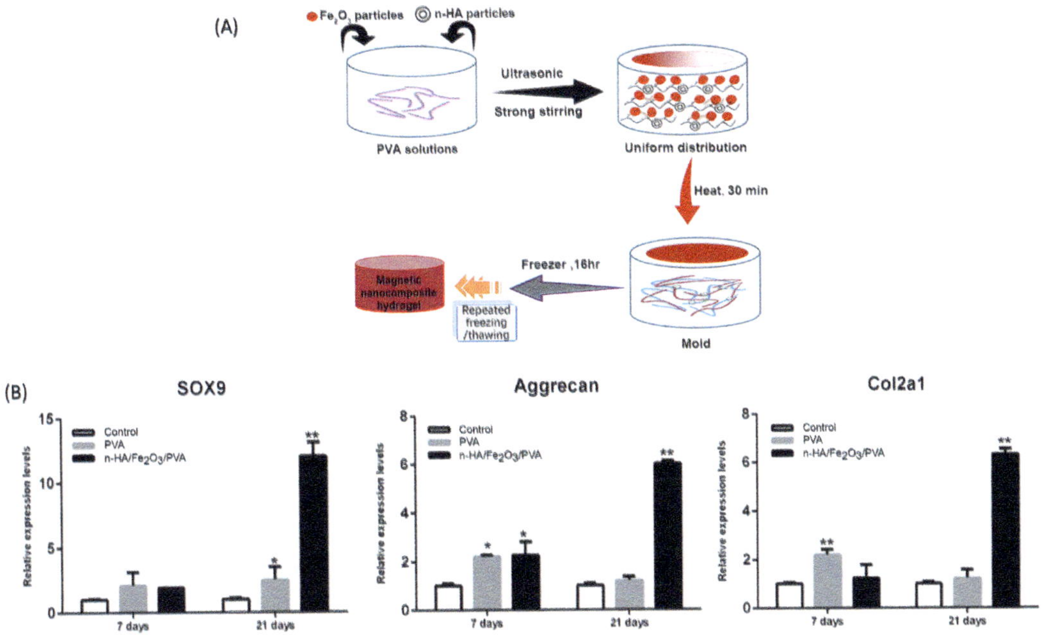

FIGURE 9.5

(A) The hydrogel preparation illustrated schematically for nHA/Fe_2O_3/PVA, (B) Quantitative real-time PCR analysis of genes grown on the constructed scaffolds [47] https://pubs.acs.org/page/rightslinkno.jsp.

Reproduced with permission from ACS.

MagGel specimen was significantly higher than in the baseline and magnetic-free groups. The microstructure of the MagGel group was comparable to that of hyaline cartilage, which has also been shown to be compatible with the growth of BMSC. Furthermore, the presence of MNPs had no effect on the survival of BMSCs when compared to the control group.

9.4 3D/4D printing of biodegradable and biocompatible polymer nanocomposites for tissue engineering

Necrosis, or the failure of body organs to function, is a serious threat to public health. Organ transplants are a highly effective treatment option for patients who have suffered organ damage. Meanwhile, a scarcity of donors and a low rate of perfectly matched success had already limited transplantation's therapeutic applications [1,48]. TE can solve this problem by inducing cell growth and tissue repair in body organs. Scaffolds can serve as a carrier for a combination of mechanical and molecular signals used to select target cells for therapeutic construction of human body tissues. As a result, high-quality scaffold fabrication is critical for achieving the goal of TE. Recently, 3D printing systems have been regarded as critical procedures for the development of promising frameworks in TE fields.

9.4.1 Aerogel scaffolds for 3D printing in tissue engineering field

Aerogels are a type of promising biomaterial with a perfectly preserved structural system having a volume of more than 90% containing air [49–51] which is suitable to be used in TE fields owing to their significant features of possessing high interconnected porous structure, highly water uptake along with their excellent permeability for oxygen, nutrients as well as the other water-soluble metabolites. These important characteristics make the use of aerogel in the fabrication of scaffolds used in TE fields. The three basic stages for the production of aerogel scaffolds are gel-aerogel transitions, solution-sol, and sol–gel. 3D printing and modeling are the two most common methods for sol–gel or solution—sol transitions. In comparison to modeling, 3D printing can create scaffolds of any specific dimensions by dumping the selected materials layer-by-layer. Aerogels have been created using a variety of 3D design techniques, including extrusion printing, inkjet printing (IJP), and stereolithography. Extrusion-based printing may be the most common method for producing a promising aerogel. Extrusion printing has many favorable characteristics, which would comprise versatility, the opportunity to construct complex structures, and affordability.

9.4.2 Tissue engineering aerogel scaffolds with adjustable macro/micropore structure

As known, aerogel is created by drying the wet gel and replacing the liquids within the pores with gas using methods such as supercritical drying, ambient pressure drying, or freeze drying techniques. An aerogel's micropore framework can have a significant impact on its properties [52,53]. Aerogel micropore structures, for example, could be used to improve cell proliferation, adhesion, and mechanical properties in TE fields. As a result, the micropore structure of the aerogels should be modified in order to obtain aerogels with significant properties that can be used to fabricate scaffolds with maximum TE effectiveness [54]. The ultrasound freeze-drying casting technique developed by Ogutcu et al., [55] involved ultrasound digital subtraction angiography (DSA) during the freezing process. The utilization of colloidal slurry to shape ultrasound waves into micro/nanoparticles in a circular ring design results in a porous pattern with alternating low and high levels of porosity. In another study, for adjusting the micropore morphologies, a model of high and low thermal conductivity areas on a material was developed by Tetik et al., [56] with using a one-direction freeze casting process. The thermal conductivity of the substrate was used to manipulate the location, shape, and mean size of the pores on the surface of the aerogel. Although these studies may achieve the regulation of pore diameters and dispersion in the aerogel scaffolds, the procedure for adjusting the macrostructure of the aerogels is complicated and thought-provoking. Recently, sacrificial materials such as Pluronic F127, a thermo-sensitive agent, have been used to regulate the microstructure of hydrogel scaffolds. It has shear recovery and thinning behavior, which are both required for 3D printing with extrusion. Pluronic F127 could not be used as a scaffold due to its decomposition at low temperatures and when immersed in the solution. Poly(ethylene glycol) diacrylate (PEGDA) is a photopolymerizable agent that is widely used in the production of aerogels for TE applications. Furthermore, PEGDA-Pluronic F127 has been reported to be used in the preparation of hydrogel scaffolds for extrusion-based 3D printing. Pluronic F127, due to its low thermal properties, can be used as a sacrificial substance that can be separated from the aerogel scaffolds. Wang Z. et al., [57] created micro/macro porous aerogel scaffolds for cell seeding using a 3D printing technique, resulting in porous macroscale specimens and sacrificial materials that controlled the microstructure of the formed aerogels.

As a material for aerogel scaffolds, they used a mixture of lithium phenyl-2,4,6-trimethylbenzoylphosphinate (LAP), pluronic F127, and PEGDA. They reported that PEGDA-Pluronic F127 hydrogels were immersed in phosphate buffer saline (PBS) at 4°C for 24 h after fabrication using 3D printing technique, and that the Pluronic F127 could then be removed. Aerogels with large micropores can be created after freeze drying, which aids in cell adhesion and nutrient transport. Fig. 9.6 illustrated the fabrication procedures of the 3D printing aerogels.

9.5 Modern tissue engineering concepts and applications: four-dimensional (4D) bioprinting for advanced tissue engineering applications

The additive manufacturing which known also by 3D printing technique was first investigated via Charles Hull in 1986 [58–60]. Since its inception, it has caused a revolution in the world by developing highly accurate, valuable, intricate, and sophisticated objects [60–62]. This technique has been extensively implemented in a wide range of different approaches, including biomedical, aviation, soft robotics, automotive, and healthcare systems, among others [63–65]. Many benefits of 3D printing include low cost, rapid prototyping, complex designs, environmental sustainability, and advances in human health. Layer-by-layer ceramics, metals, composites, and alloys processing methods were used to modify 3D objects using three-dimensional technology. Recently, 3D printing has captivated the medicinal and biomedical fields and has been extensively used in pharmaceutical, tissue engineering, and medicinal regenerative approaches [65,66]. Biomaterials created using 3D bioprinting technology, on the other hand, revealed a static nature, with no response to

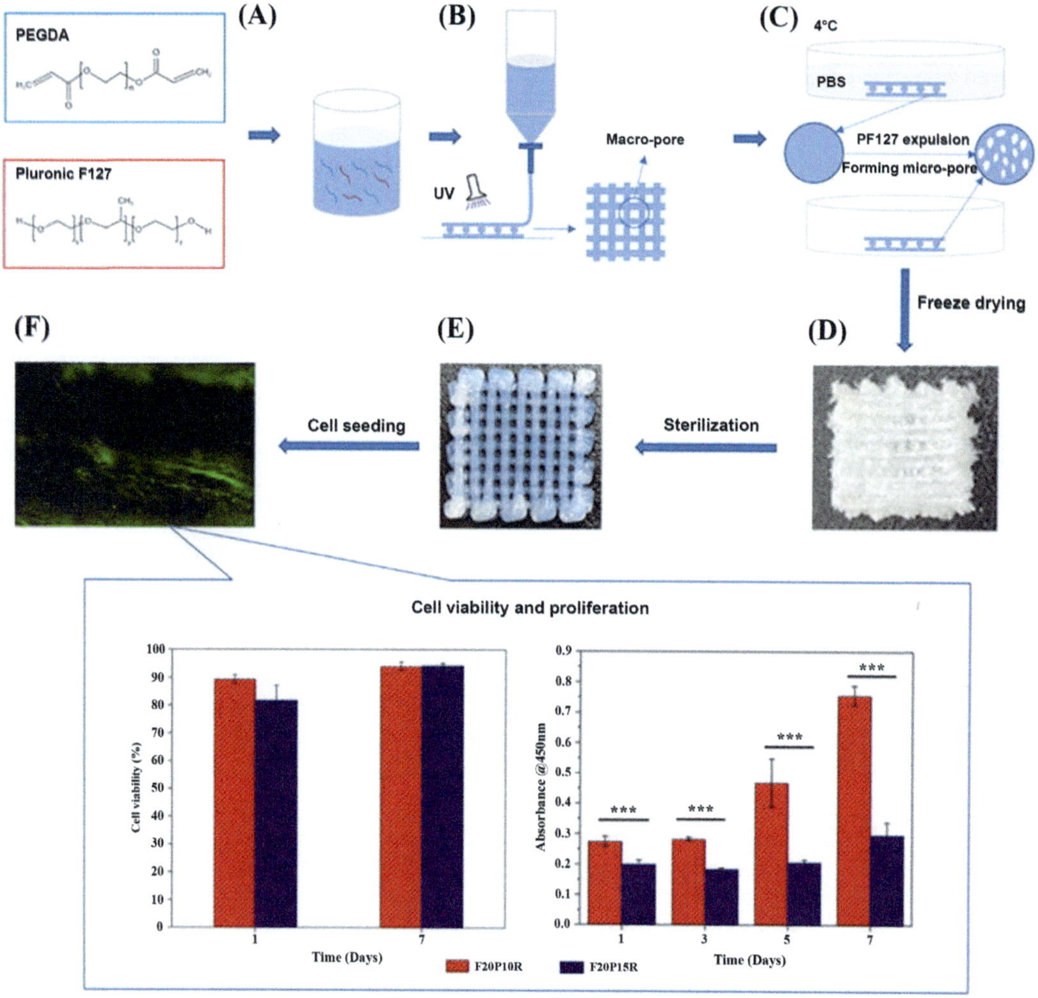

FIGURE 9.6

Schematic fabrication procedures of the scaffolds that made of aerogel with controlled porous structures. (A) the manufacturing of PEGDA-Pluronic F127 materials, (B) The 3D printing hydrogels with porous structures, (C) Formation of micropore thru dipping hydrogels in PBS overnight to omit the pluronic F127 at 4°C, (D) Freeze-dried aerogels, (E) Sterilized Aerogels, and (F) the shape of the aerogels after seeding the cell [57]. https://s100.copyright.com/CustomerAdmin/PLF.jsp?ref = c4f3390e-e1f1-4116-9562-6ec2db4e67bd.

Reproduced with permission from Elsevier. Copyright 2022.

dynamic changes in the human body's microenvironment. In the biomedical sector, a novel technique has recently emerged associated with the evolution of novel materials that respond to stimulus, specifically shape memory polymers (SMPs) or shape memory materials (SMMs). A modern technique has emerged and gained widespread attention in the pharmaceutical and biomedical sectors [67]. The 4D bioprinting technique required the 3D-printed tissues to mature and remodel. The primary goal of this technology is to promote advanced and sophisticated tissue structures with inherent significant characteristics of modifying and transitioning into new forms and features in response to various chemical, biological, or physical stimuli, as well as to aid in the repair or replacement of damaged organs, cells, and tissues [68]. 3D bioprinting is used to create bioengineered scaffolds by layering different biomaterials and incorporating viable cells to mimic the native tissues [68,69]. The bioprinting technology enabled the deposit of various factors, cells, drugs, and hormones in a significant and efficient systematic manner, allowing for relatively advanced tissue grafting [70]. In comparison to traditional 3D printing, 3D bioprinting is more complicated in terms of growth factors, biomaterial selection, cell types, and cell responses [71]. Furthermore, 3D bioprinting techniques were used to load cells and develop 3D printed tissue constructs with significant features [71]. In the meantime, unlike living tissues, engineered tissues created with 3D bioprinting technology are rigid and static. They do not have any shape transformation abilities. Furthermore, 3D bio-printed tissue composites have the capability of revealing the appropriate biological triggering in the human body micro-environment. Furthermore, their practical applications were limited due to tissue maturation and development shortages. As a result, further postprinting modifications involving self-organization and cell coating are required for the development and maturation of 3D bio-printed tissue constructions [72]. To address the challenges in the TE fields, 4D bioprinting has emerged. The 4D technology first appeared in 2013, when 3D printed parts were developed by promoting dynamic behavior [73]. Composites created with 4D printing technology can be adapted and transformed in terms of material features and shapes. These characteristics are common in liquid crystal elastomers (LCEs), SMMs/intelligent materials comprising shape memory hydrogels (SMHs), SMPs, and shape memory alloys (SMAs) as descried from Fig. 9.7.

The bioprinting technique produces biomedical composites by utilizing smart intelligent biomaterials that include hydrogels embedded with cells. The dynamic behavior is caused by energy inputs such as temperature, pH, magnetic, bioenzymes, electrics, light, or other external stimuli [75]. Living tissues are dynamic, vital structures that contain extracellular matrix (ECM) and dense cellular ensembles [75]. The extracellular matrix is a critical component of the cellular microenvironment that has evolved into a complex and realistic 3D matrix. As a response to environmental stimuli, the programmed biomaterials were transformed into specific defined shapes, sizes, or colors using 3D printing. The same 3D bioprinting processes were used in 4D bioprinting, including stereolithography (SLA), fused deposition modeling (FDM), digital light processing (DLP), direct ink writing (DIW), and IJP. These techniques made use of commercial biomaterials as well as self-developed materials that demonstrated promising compatibility with printers [76–78]. Furthermore, 4D bioprinting will open up new opportunities in a variety of biological and healthcare systems such as tissue engineering, cancer research, drug delivery, regenerative medicine, and biomedical devices [79–82]. 4D bioprinting techniques have recently revealed enormous potential in the healthcare and biomedical sectors, particularly in the tissue engineering approach. Table 9.1 compared the most well-known bioprinting processes by taking into account the following factors: cell density, cross-linking approaches, resolution, cell viability, material viscosity, printer cost, bioink materials, printing speed, and tissue engineering domains.

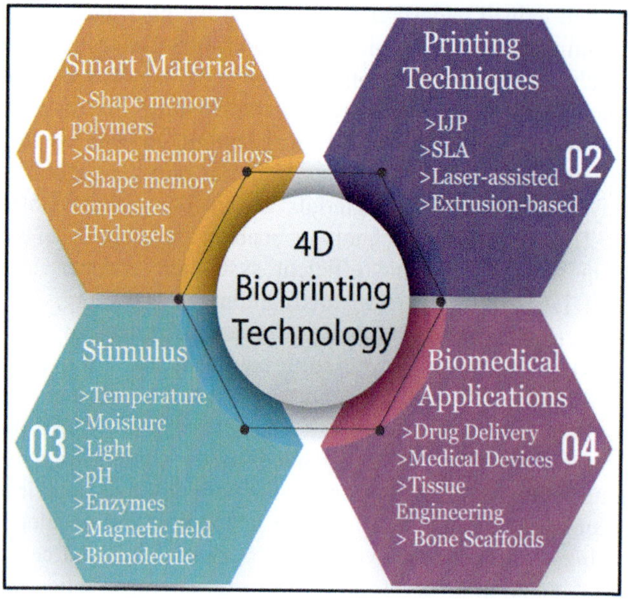

FIGURE 9.7

Various 4D bioprinting technologies smart materials, stimuli mechanisms, and biomedical applications [74]. https://s100.copyright.com/CustomerAdmin/PLF.jsp?ref = 80f6eb03-76ed-4765-bbf9-316caae6f72f.

Reproduced with permission from Elsevier. Copyright 2022.

9.5.1 Steps involved the shape transformation of the 3D printed objects into 4D printed responding to different stimuli and then recovery back to the primary shape

Because of their irreversible deformation, 3D printed structures are rigid and static, making them unsuitable for dynamic applications in emerging approaches. To address this issue, 4D printing manufacturing procedures were developed [89,90]. According to the Atlantic Council report, 4D printing is "the ability for material objects of programable matter to self-transform in form and function after interaction with appropriate stimuli, such as heat, light, moisture, humidity, stress, and other electromagnetic sources [91]. Fig. 9.8 illustrated the steps involved in the transformation process and the recovery back to the original shape.

9.5.2 Different stimuli-responsive techniques for 4D bioprinting applications in the field of tissue engineering

The stimuli-responsive materials (SRMs), also known as smart materials (SMs), have gained a lot of attention and interest in the field of TE. These materials have the potency to transfer their shapes and modify their properties responding to external biological [93], chemical [94], or physical stimuli [11]. For the 4D bioprinting of SMMs, various stimuli-responsive mechanisms were used.

Table 9.1 Bioprinting processes.

Parameters	Bioprinting techniques				References
	IJP	SLA	Extrusion-based	Laser-assisted	
Cross-linking approach	Chemical, photocuring	Chemical, photocuring	Chemical, thermal, photocuring	Chemical, photocuring	[82]
Speed of printing	Fast	Fast	Slow	Moderate	[83]
Cell density	Low (less than 106 cells/mL)	Medium	High	Medium	[84]
Materials viscosity (mPa/s)	3.5–12	No range	30 to >60 × 106	1–300	[85]
Viability of cell	>85%	>85%	40%–80%	>95%	[86,87]
cost of Printer	Low	Low	Low	High	[86]
Printing resolution	High	High	Medium	High	[86]
Materials for bioink	Alg, PEGDMA, Col	GelMA, PEGDA/GelMA	GelMA, Alg and Col	Col and matrigel matrix	[88]
Tissue engineering domains	Bone, blood vessels, neural, and cartilage,	Cartilage, and blood vessels	Neural, bone, blood vessels, muscle, and cartilage,	Skin, bone, adipose, and blood vessels	[82,88]

Reproduced with permission from Elsevier. Copyright 2022. https://s100.copyright.com/CustomerAdmin/PLF.jsp?ref=6737eca1-7c1c-4a53-807c-bf5e44143663.

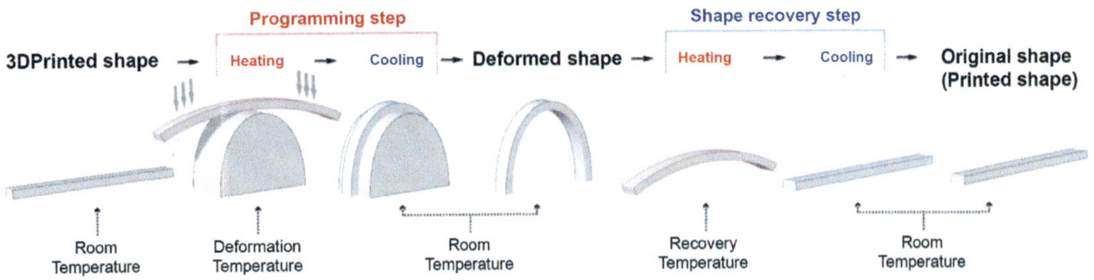

FIGURE 9.8

The steps of the transformation process from 3D to 4D printing and the recovery back to the primary shape [92]. https://s100.copyright.com/CustomerAdmin/PLF.jsp?ref=5f725872-e365-49e4-bb8e-0103f64b1a28.

Different environmental changes, known as cell traction forces (CTFs), were used to create 3D printed biomaterial scaffolds for cartilage, bone, and cardiac tissue regeneration, as well as the development of healthcare and biomedical devices [95–97]. CTFs have a wide range of applications in biology, including metastasis, inflammation, wound healing, embryogenesis, and angiogenesis. These forces can be used to improve communication with surrounding cells, maintain cell shapes, control migration within tissues, and recognize the extracellular matrix in a variety of biological situations. This technology can be used extensively to create the 4D printed cell origami technique, in which the cells use the folding process to transform a 2D element into a 3D predefined construct [98]. For instance, He et al., [99] developed a 3D structure by using the cell origami self-folding mechanism through incorporation of alginate lyase, fibroblasts (NIH/3T3) and heptatoma cells (HepG2). Fig. 9.9 described the production of 3D microstructure utilizing the CTF technology that can be used for drug release systems.

Among the electro-sensitive polymers-based hydrogels that can fold, swell, or shrink in response to external electric stimuli were polypyrroloe bioink, polythiophene, and polyaniline [100]. Different

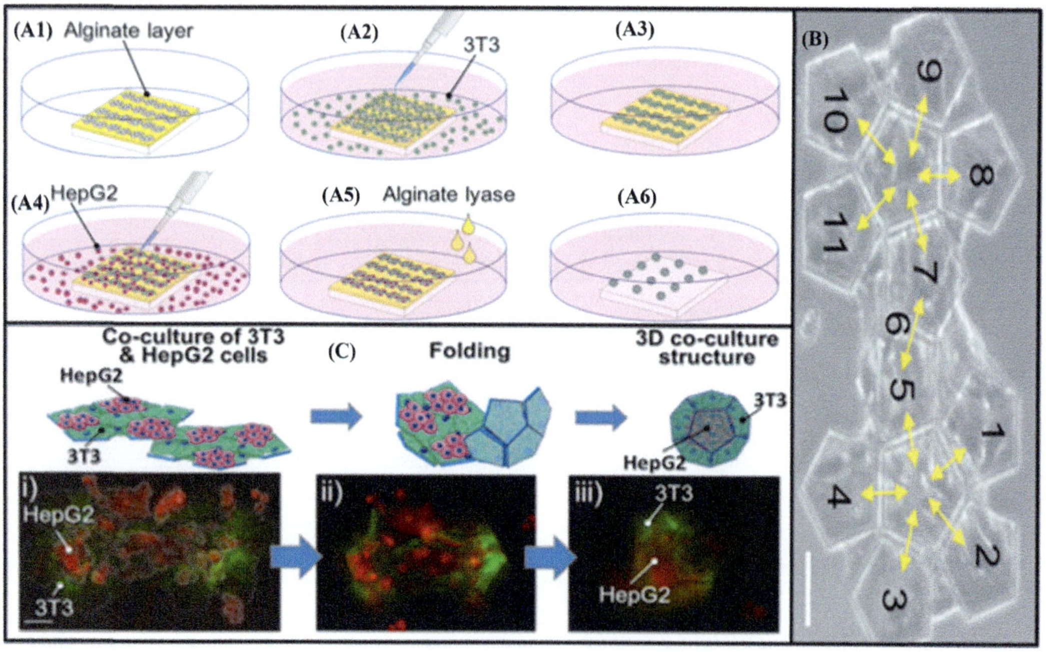

FIGURE 9.9

(A1–A6) diagrammatic description of culturing and seeding of NIH/3T3 and HepG2 cells on microplates; (B) Microplates micrographs for the generation of 3D dodecahedron composite; (C) fabrication of NIH/3T3 and HepG2 cells while cell origami self-folding mechanism by integration of alginate lyase in the cell-cultured microplates provide CTF which produced 3D microstructures [74]. https://s100.copyright.com/CustomerAdmin/PLF.jsp?ref=67f6f5b5-f299-4e30-9bbe-8553adadcaa6.

Reproduced with permission from Elsevier. Copyright 2022.

electric field strength and direction could be utilized to control and adjust the mechanical features of these materials. Because of their promising biocompatibility and printability, many conductive polymer-based hydrogels have an extraordinary capacity for the production of 4D bio-printed scaffolds [101]. Yang et al., [11] developed a cell-laden gelatin methacryloyl (GelMA) based microfibrl composite utilizing an electric-aided cross-linking technique. Moreover, the shape transforming potency (folding and swelling) of the 4D-conceptualized gelatin film is also elaborated in Fig. 9.10. These modified fibers have a wide range of applications in muscular skeletal tissue.

Light stimuli such as ultraviolet, infrared, and near infrared have been used to activate the light-responsive biomaterials. They are extensively used in a variety of clinical approaches, including drug delivery systems and TE applications [102,103]. In general, photoisomerization and photodegradation of polymeric chains are commonly used to create photo-responsive materials for 4D printing. Wei et al., [104] reported the development of poly(lactic acid)-based tubular composite using photo cross-linking materials. Furthermore, Luo et al., [105] designed poly(dopamine)/alginate-based and gelatin methacryloyl/alginate-based bioinks for the preparation of cell-laden structures that can transfer their shapes when stimulated with near infrared light, as shown in Fig. 9.11. While culturing for 15 days, the printed cell composites demonstrated promising extraordinary viability. As a result, these newly developed photo-sensitive bioinks can be used extensively in the dynamic micro-environment to mimic the general properties of the native extracellular matrix.

9.5.3 Bioprinting materials

The ability of 4D bio-printed materials to change their morphologies when stimulated is their most notable feature. Stimuli-responsive materials (SRMs) are the name given to these bioprinting materials [106]. By providing additional shape morphing character or shape memory, these stimulants induced the development of 4D dynamic products [107]. Shape morphing provided irreversible transformation in the products, whereas shape memory performs a transition between the programmed temporary shape and the permanent shape in response to external stimuli [108]. Shape-memory materials (SMMs) are generally transformed into a specific shape in response to external stimuli. Following the removal of the external stimulus, the SMMs are designed to transition into a secondary shape from the primary shape, which is then retained and replaced by the primary original shape after exposure to appropriate nonmechanical stimuli. SRMs such as shape memory polymers (SMPs), shape memory alloys (SMAs), shape memory hydrogels (SMHs), and liquid crystal elastomers (LCEs) are used to create 4D bio-printed items for a variety of biological and healthcare applications [108,109]. Because of their intricate nature, selecting bioprinting materials for use in 4D bioprinting technology is extremely difficult. Three fundamental components are required for the biorinting technique: the 3D printer, the polymer solution, and the viability of the cells [110]. In the ideal suitable polymer solution, low viscosity, low cross-linking mechanism, and superior permeability are required for cell movement, tissue formation, and oxygen transport [110–112]. The polymer used in 4D bioprinting should have superior structural and biomedical properties, as well as higher dimensional precision, in order to gelatinize quickly. Achieving a balance between the various properties of bioprinting substances is critical for the development of potential tissue composites [112].

SMPs are one of the most widely used types of 4D bio-printed materials. These polymers have the ability to recover and restore their original primary shape after being in the secondary shape for a short

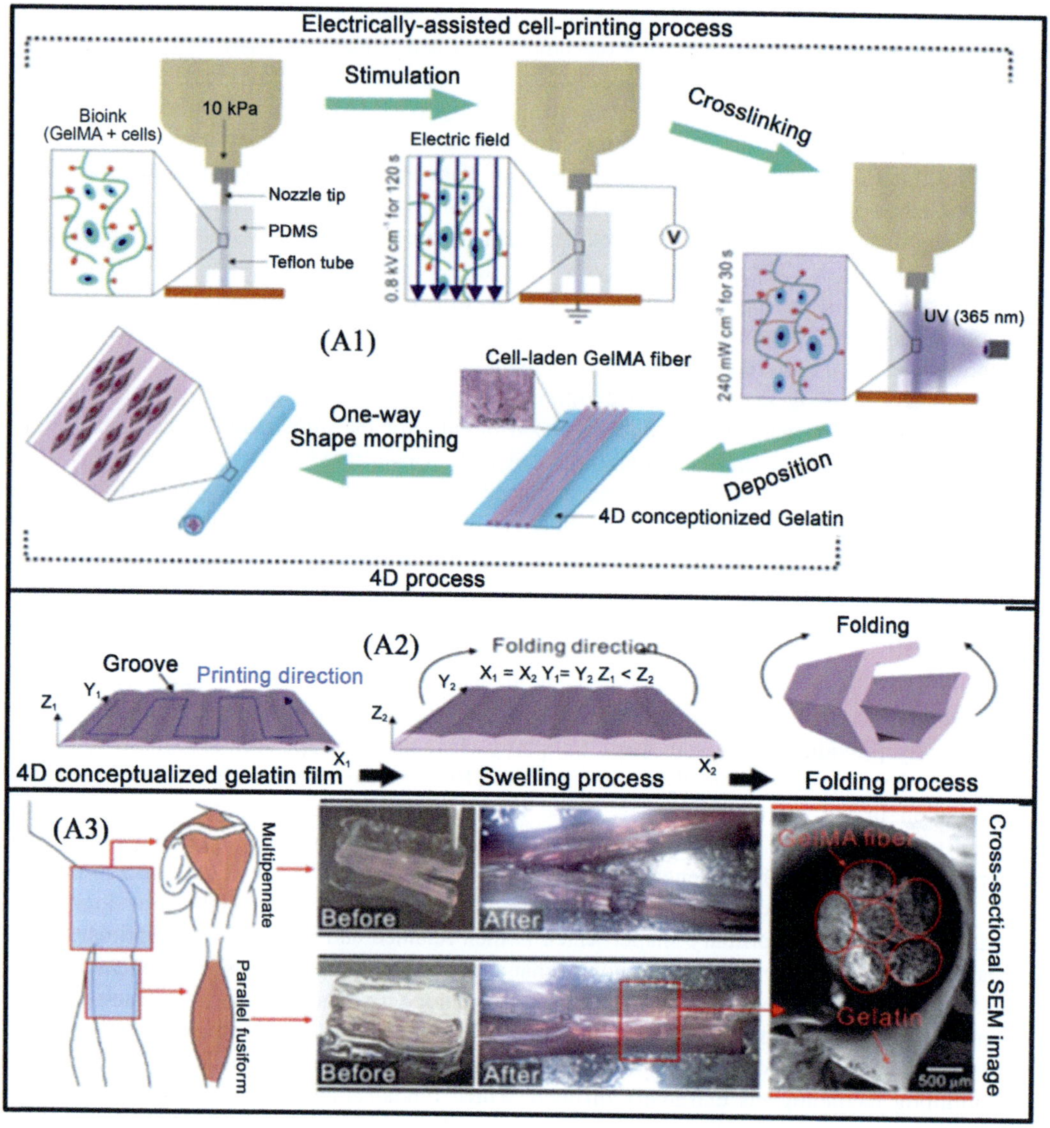

FIGURE 9.10

(A1) Diagrammatic scheme illustrated the electric field-assisted bioprinting technique used for fabrication of gelatin methacryloyl-based microfibers; (A2) diagrammatic description of the shape transformation procedures of gelatin film; (A3) SEM micrograph revealed the deposition of C2C12-laden gelatin methacryloyl fibers into the gelatin film [74]. https://s100.copyright.com/CustomerAdmin/PLF.jsp?ref = 67f6f5b5-f299-4e30-9bbe-8553adadcaa6.

Reproduced with permission from Elsevier. Copyright 2022.

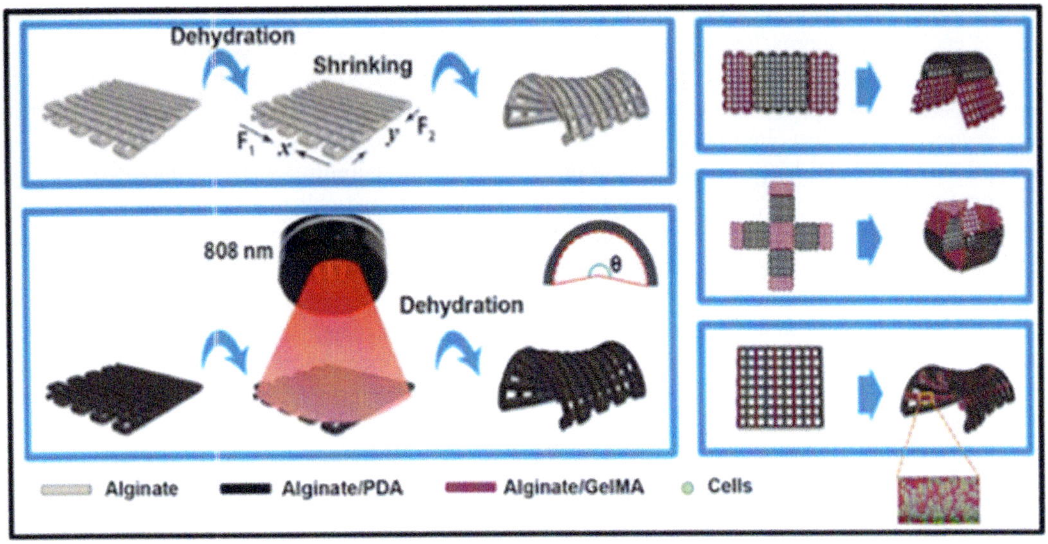

FIGURE 9.11

4D printed cell-laden bi-layered constructs which transformed into saddle-like scaffolds utilizing near infrared stimuli. The scaffolds were developed through poly(dopamine)/alginate-based and cell-laden gelatin methacryloyl/alginate-based layers, and poly(dopamine)/alginate instigated shape transforming potency using near infrared stimuli [74,105]. https://s100.copyright.com/CustomerAdmin/PLF.jsp?ref = 67f6f5b5-f299-4e30-9bbe-8553adadcaa6.

Reproduced with permission from Elsevier. Copyright 2022.

period of time. Furthermore, these polymers demonstrated excellent biodegradability, biocompatibility, regeneration and recoverable potency, as well as high sensitivity to a variety of external stimuli. SMMs demonstrated distinct properties; they can provide versatility and customizability via the 4D bioprinting technique. Because of their biological, mechanical, and physical properties, thermo-responsive SMPs have been reported to be the most modified 4D bio-printed biomaterials when compared to other stimuli-responsive SMPs [113]. Following specific development and modification, such polymers will be highly biocompatible and widely used in a variety of BTE applications [114]. Despite the fact that the cells seeded do not scatter uniformly and remain on the surface during the 4D bioprinting process, the synthetic polymers have received a lot of attention due to their excellent biochemical properties. The fabrication of these synthesized polymers aids in the achievement of promising and desirable mechanical properties that natural biopolymers cannot achieve. Despite concerns about their biocompatibility and biodegradability, these artificial polymers have so far demonstrated excellent prospects in biomedical and healthcare systems [115,116]. The most commonly synthesized 4D-printable SMPs are PGA, PCL, PVA, PE, polystyrene (PS), acrylonitrile butadiene styrene (ABS), polyamide (PA), and PLGA [117]. Similarly, a variety of acrylate monomers, including butyl acrylate, acrylic acid, methyl acrylate, iso-bornyl acrylate, benzyl methacrylate, lauryl acrylate, ethylene glycol phenyl ether acrylate, bisphenol a glycerolate dimethacrylate, acrylated epoxidized soybean oils (AESO), and 2-phenoxyethyl acrylate, are Furthermore, the cross-linking acrylates bisphenol an ethoxylate dimethacrylate, diethylene glycol diacrylate (DEGDA), and PEGDA are used in a variety of tissue engineering approaches [118,119].

Hydrogels are water-insoluble polymer composites that are cross-linked to retain more water than their weight [120]. Hydrogels resemble jelly, are pliable, and have low mechanical properties when compared to nonswollen polymer composites [121]. They are widely used in many different TE applications for cell bioprinting because of their promising stimuli-sensitivity, biodegradability, biocompatibility, low toxicity, and ability to mimic native external cell matrix and encapsulate living cells. Furthermore, these hydrogels are capable of making direct contact with cellular contractures by promoting cell adhesion, maturation, proliferation, and viability. They can also be used as scaffolds for cell differentiation, adhesion, and proliferation. Hydrogels, in general, have weak structures and cannot be used to sustain configurations including such bone, cartilage, and other TE applications because the extracellular matrix must have strong mechanical properties [122]. Many different polymers can be used to solve this problem. Unlike traditional 2D cell culture, hydrogel-based polymers can support cells or tissues by solving nutrients from all directions due to their high water content. Furthermore, stiffness can be adapted by the physiological environment based on the cell types. These advancements in hydrogel mechanical properties are mostly accomplished through blending and cross-linking strategies, or through the reinforcement of various nanoparticles [123,124]. Hydrogels must have certain important properties, such as high porosity, which allows for the exchange and filtration of nutrients and oxygen. Hydrogels should also have high biocompatibility, biodegradability, and printability, as well as a promising cross-linking mechanism [124]. Traditional techniques for developing hydrogels lack the ability to preserve cell viability while maintaining low resolution. The smart hydrogels have the ability to change shape in response to pH, temperature, light, pressure, liboenzymes, electric and magnetic stimuli via swelling mechanisms, which supports and allows the printing of scaffolds used in various tissue engineering applications [125]. The 4D printed hydrogel-based polymers hold numerous promises for use in the construction of various cells and tissues in a dynamic microenvironment. Cell maturation and infiltration into hydrogels are occasionally observed and documented [126]. Natural polymer-based hydrogels, synthetic polymer-based hydrogels, or a combination of both are widely used in the production of bioink materials [127,128].

Among the natural polymers-based hydrogels were alginate (Alg), agarose, CS, soy protein (SP), collagen (COL), gelatin (Gel), starch in the extracellular matrix of the cellular network, SF, and HA [129,130]. Fig. 9.12 illustrated the different polymers percentages applied for the bioink materials fabrication. In vitro and in vivo analysis were used to estimate the reaction of natural-polymer-based hydrogels with living cells [74,131]. COL, a protein found in large quantities in the extracellular matrix and mammalian tissues, has been used in the construction of cartilage. Meanwhile, biomaterial scaffolds made from COL are relatively weak; as a result, other bioactive polymers were used in conjunction with collagen. The helix-coil transition occurs as a result of partial hydrolysis of collagen, which produces gelatin biopolymer [132].

Gel-based hydrogels are commonly used in bioprinting and TE applications [133,134]. They have promising biomedical properties, such as low antigenicity and excellent mechanical properties. Gelatin is formed as a result of temperature changes. Gel was converted into a liquid at temperatures above 37°C, allowing it to be used as a sacrificial biomaterial scaffold for cellular structures to produce organ chips. After removing the gelatin after incubating the cells at a lower temperature, the cells remain undamaged. Synthetic photopolymerized hydrogels, including gelatin methacryloyl

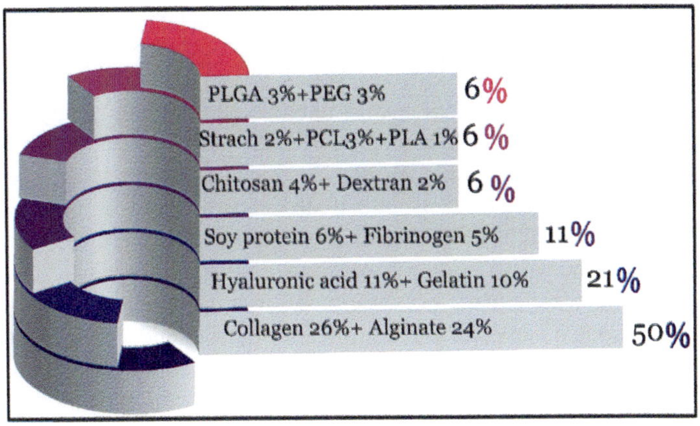

FIGURE 9.12

Schematic diagram described the different polymer (%) of both synthetic and natural biopolymers retained for bioink materials production [74]. https://s100.copyright.com/CustomerAdmin/PLF.jsp?ref=6737eca1-7c1c-4a53-807c-bf5e44143663.

Reproduced with permission from Elsevier. Copyright 2022.

(GelMA), are 4D products, whereas gelatin is not. GelMA-based acrylate polymers have been widely used in TE applications such as vascularization [59].

On the other hand, alginate and HA are widely used in the production and development of biomaterial scaffolds for skin tissue, cartilage, chondrocytes, muscular tissue, and vascular tissue contractures. HA aided in the maintenance of water in skin tissues, the regulation of blood vessel permeability, and the enhancement and acceleration of wound healing [135]. Meanwhile, HA revealed poor mechanical properties due to its high solubility, necessitating additional modifications and developments for bioprinting of biomaterial scaffolds. The chemical modifications of HA by PEG, thiolate, and supramolecular host-gust interactions that improve printability and stability. Furthermore, alginate requires further modification; it cannot be used alone due to a lack of cell adhesion [136]. Alginate was modified using cell adhesion motifs, resulting in promising biocompatibility with human pluripotent stem cells for the development of mini-liver. Furthermore, developing alginate-based hydrogels can improve cell adhesion. Bakarich et al., [137] created a 4D-printed biomedical apparatus using hydrogels created by incorporating poly(*N*-isopropylacrylamide) into alginate to increase the use of alginate in TE and bioengineering applications. The developed apparatuses demonstrated excellent promising swelling and deswelling mechanisms in response to temperature stimuli. Furthermore, fibrin, the main component of the extracellular matrix of normal skin, promotes gelling by enhancing cell differentiation and proliferation as well as encouraging the fibrinogen and thrombin reaction. In wound covering applications, these hydrogels were preferred. Chitin, a modified polysaccharide found in the extracellular matrix of invertebrates such as sponges, nematodes, fungi, mollusks, and arthropods, is similar. It revealed extremely promising characteristics that could be influenced by the tissue engineering approach [138]. Similarly, other natural polymer-based hydrogels such as matrigel, starch, and agarose are extensively used in a variety of tissue engineering applications such as liver, bone, and skin tissue constructs, as well as bioprinting of blood vascular contractures. Despite the widespread use of natural polymer-based hydrogels in the construction of various biological tissues, these natural-based hydrogels have several disadvantages, including low stability, mechanical strength, and immunogenicity [74,139,140].

Meanwhile, polyacrylamide, poly(ethylene glycol), and pluronics were identified as synthetic polymers-based hydrogels produced using various chemical synthetic techniques [141,142]. The incorporation of various synthetic polymers into hydrogels improved their biocompatibility, biochemical properties, biodegradability, and chemical properties. Furthermore, these advancements improved the cross-linkable networks required for tissue engineering applications. Intensive efforts have been made to improve the synthetic polymer-based hydrogels used in bioprinting. These synthetic polymer-based hydrogels demonstrated significant biomechanical properties that were superior to natural polymer-based hydrogels. Furthermore, the combination of natural and synthetic polymers via incorporation of natural polymers into synthetic polymers results in improved mechanical properties as well as excellent biocompatibility with native vital tissues. Saturated aliphatic polymers are extremely useful for bone and cartilage recovery and regeneration [143,144]. For example, Kao et al., [145] has deposited the poly(dopamine) onto the polylactic acid-based 3D printed scaffolds. The study found that surface modifications influenced the osteogenic and angiogenic differentiation of human adipose-derived stromal cells. Another instance, Kundu et al., [146] has printed polycarbolactone/alginate/chondrocyte-based hybrid hydrogels for regeneration of cartilage. These hydrogel scaffolds have increased and improved cartilage tissue and collagen fibrils type II formation. Similarly, Kim et al., [147] created dimethyl fumarate/poly(propylene fumarate)-based hydrogel frameworks with promising excellent biomechanical properties for BTE.

9.6 Electrospinning of tissue engineering scaffolds based on biodegradable and biocompatible polymer nanocomposites

The use of biodegradable and biocompatible nanofibers to provide target tissues is rapidly expanding [148]. These fiber scaffolds influenced cell-to-cell and cell-to-matrix interactions, and the fibers were identical to natural extracellular matrix, resulting in good growth factor transport [149]. For tissue engineering, scientists are turning to bio-natural polymers such as alginate, HA, COL, protein, silk, fibrinogen, starch, and chitosan due to their biodegradability and biocompatibility. Polymers are employed in cartilage tissue engineering [150], skin regeneration [151], and bone regeneration was also accomplished with the use of mixed SF and PCL [152]. Because of their smaller pore size, 2D structures of electrospun nanofibers have some limitations. The reduced pore size of 2D electrospun nanofiber architectures imposes several constraints. As a result of their outstanding biocompatibility and spatial geometry, 3D constructs of electrospun nanofibers have recently been used to substitute them for tissue regeneration, allowing them to be used in nerve regeneration [153]. Tao Xu et al., [154] used the biocompatible and biodegradable PCL as binding agent to create scaffold nanofibers. The development of tissue engineering scaffolds with morphological properties similar to natural extracellular matrices (ECMs) is a massive technological challenge. The resulting 3D scaffolds are soft and elastic, with interconnected and hierarchically organized holes; they are thus morphologically comparable to natural ECMs, making them suitable for cell functions and tissue formation. Various thermoplastic polymers could have been designed and manufactured into 3D nanofibrous scaffolds by first blending nanofibers with PCL, then processing induced via thermal (nanofiber) self-agglomeration method, and finally being thermally stabilized, with the end result electrospun 3D nanofibrous scaffolds potentially beneficial for a wide range of applications (Fig. 9.13).

The morphological characteristics of the 3D CA/PCL scaffolds were described using SEM. They examined the internal and external surfaces of 3D scaffolds, also known as the crack tip and the inner and outer surfaces of the sample, respectively. Such 3D scaffolds have interconnected, hierarchically constructed pores in both the inner and outer surfaces, as illustrated in Fig. 9.14 (A1–A2 and B1–B2). In contrast to the freeze-drying technique, which establishes the porous structure through the solidified solvents, the highly porous features in the obtained 3D CA/PCL scaffolds are produced by the randomized self-agglomeration of short nanofibers. Because the surface morphology of the scaffolds is similar to that of natural ECMs, they would be ideal for cellular functions and tissue formation. The CA/PCL blend nanofibers were blended together, as revealed in SEM images; Fig. 9.14 (A3 and B3), which significantly increased the mechanical properties.

While stabilized 3D scaffolds were as pliable as sponge, electrospun CA/PCL nanofibers were as smooth as silk. Prior to thermal stabilization, the CA/PCL scaffolds could not return to their original shape morphologies after being released from compression and/or bending pressures, as shown in Fig. 9.14 (C1 and C2). Nonetheless, as shown in Fig. 9.14, the 3D scaffolds may easily return to their original morphologies after thermal stabilization at high temperatures (D1 and D2). As a result, whenever compression and/or bending stresses were applied, the hierarchically constructed pores and partially fused CA/PCL mix nanofibers increased mechanical resilience significantly.

9.7 Foamed scaffolds based on biodegradable and biocompatible polymer nanocomposites

Supercritical fluid (SCF) has piqued the interest of many polymer applications, including polymerization, amendment, purifying and fractionation, extraction, coating, combining, composite materials formation, particulate manufacturing, and foam formation, and is now being used as an environmentally friendly processing solvent [155]. Carbon dioxide (CO_2) is the most widely used supercritical fluid due to its numerous advantageous properties, including its availability in high purity, being nonflammable, nontoxic, abundant, noncorrosive, affordable, and easily accessible supercritical states [155,156]. Because it combines a gas-like diffusivity with a liquid-like density and solvent power, $scCO_2$ is a unique medium for polymer synthesis and characterization. It is widely accepted that $scCO_2$ can be used to make porous polymer materials. To make polymer foams, gas bubbles must form and expand within the molten or plasticized polymer. Throughout this method, amorphous or semicrystalline polymer is first saturated with CO_2 at a specific pressure and temperature. When gas diffuses into the polymer matrix, single-phase polymer/CO_2 solutions form [155]. The first extremely porous biodegradable polymer scaffolds for possible application in tissue engineering were created by Mooney et al., [157] utilizing the high pressure CO_2 gas processing. Porous PLGA sponges with large holes (about 100 m) and porosities of 95% or more were thoroughly characterized. Poly(hydroxy acids) such as PCL, poly

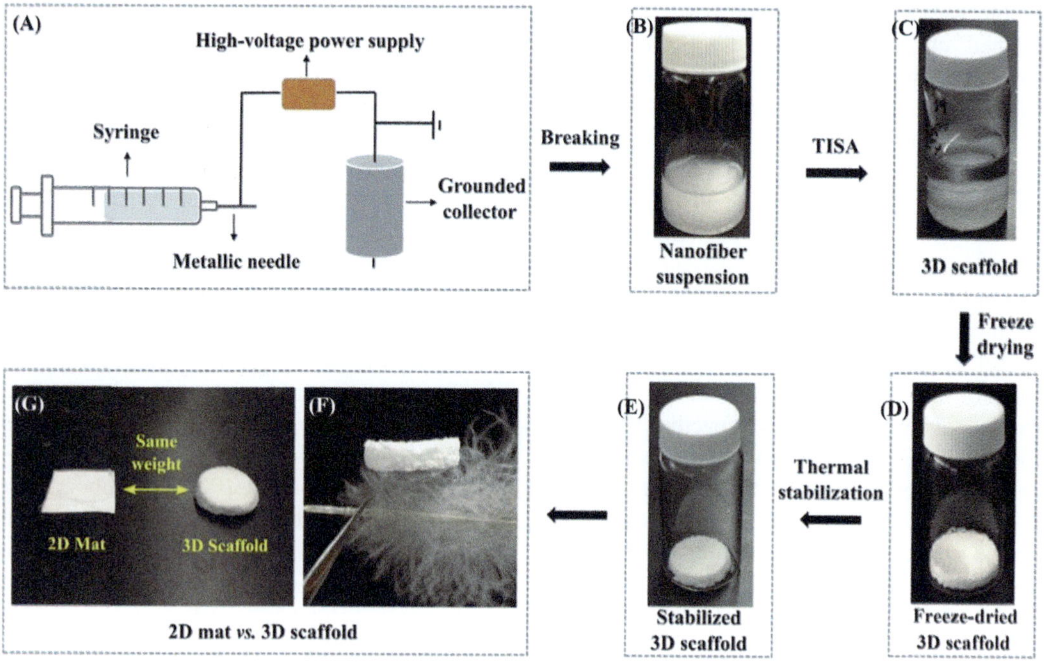

FIGURE 9.13

The steps for the construction of CA/PCL scaffold nanofibers. (A) Electrospinning setup, (B) suspension nanofibers, (C) formation of scaffold, (D) scaffold after freeze drying process, (E) 3D scaffold, and (F, G) 2D and 3D scaffolds [154]. https://s100.copyright.com/CustomerAdmin/PLF.jsp?ref = e60c3275-74ca-4b59-af86-fdfde3458d82.

(lactic acid) (PLA), poly(glycolic acid) (PGA), and the copolymer PLGA are receiving a lot of attention in this field of study because they are desirable possibilities for the creation of porous tissue engineering scaffolds [158–160]. Using the $scCO_2$ pressure quenching approach, Singh et al., [161] successfully created interconnected porous microcellular PLGA with pore diameters ranging from 30 to 100 μm and porosities as high as 90% of PLGA was thought to be the best renewable resource for making porous materials for TE applications due to its ability to convert harmful lactic and glycolic acid into non-harmful lactic and glycolic acid and its high foamability. The CO_2 foam technique has been shown to be a good potential technique for creating scaffolds laden with biological signals for more cell behavior stimulus and improved tissue regeneration. At room temperature and atmospheric pressures, Owen R. Davies et al., [162] successfully trapped protein in biodegradable and biocompatible polymers, including such PLA, PLGA, and PCL. Using spray drying and supercritical fluid foaming techniques, Nie H. et al., [163] fabricated bio-based polymer composite from both PLGA and chitosan composites as new carriers for DNA delivery.

The permeable biocompatible material used in tissue engineering must have a biocompatible surface and sufficient mechanical properties [164]. However, a single component polymer material cannot fully satisfy the functional criteria. It is possible to design and construct multiphase

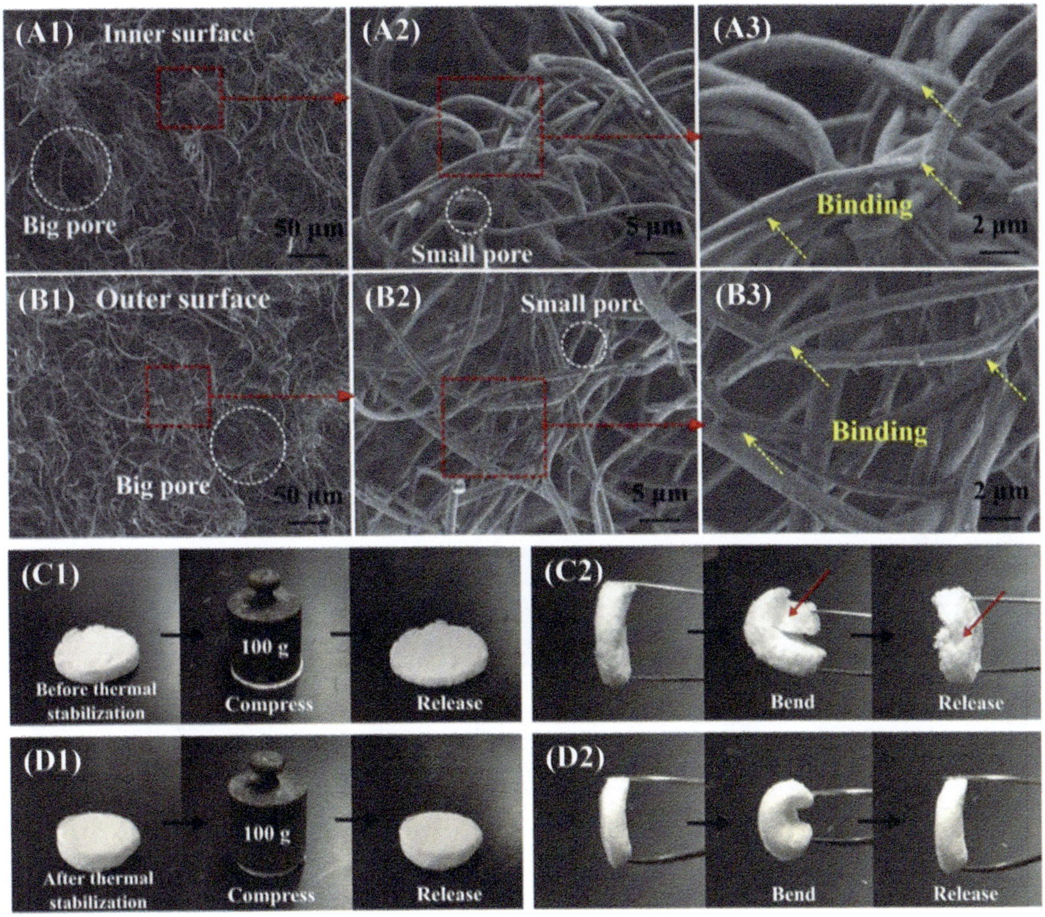

FIGURE 9.14

SEM images showing typical morphologies of CA/PCL nanofibrous scaffolds (A1–A3) inner surface and (B1–B3) outer surface. SEM pictures at high magnification (A3) and (B3) illustrating the binding sites images demonstrating the bending and compression of CA/PCL scaffolds (C1–C2) before and (D1–D2) after the thermal stabilization [154]. https://s100.copyright.com/CustomerAdmin/PLF.jsp?ref = e60c3275-74ca-4b59-af86-fdfde3458d82.

particulate matter in order to create multifunctional biodegradable scaffolds [165]. By incorporating nanostructures into a biodegradable polymer substrate, this study aims to create highly permeable nanocomposites with specific properties suitable for tissue engineering.

According to Liao et al., [166] many unique foam morphologies, including such layered, core-skin, and interconnected, may be produced by manipulating the CO_2 processing parameters and, therefore, the physical characteristics of PLLA via exposing to CO_2. Through minor adjustments to the processing settings, different pore geometries and diameters inside the center and also open and linked holes in the skin were generated (Fig. 9.15) [167]. The

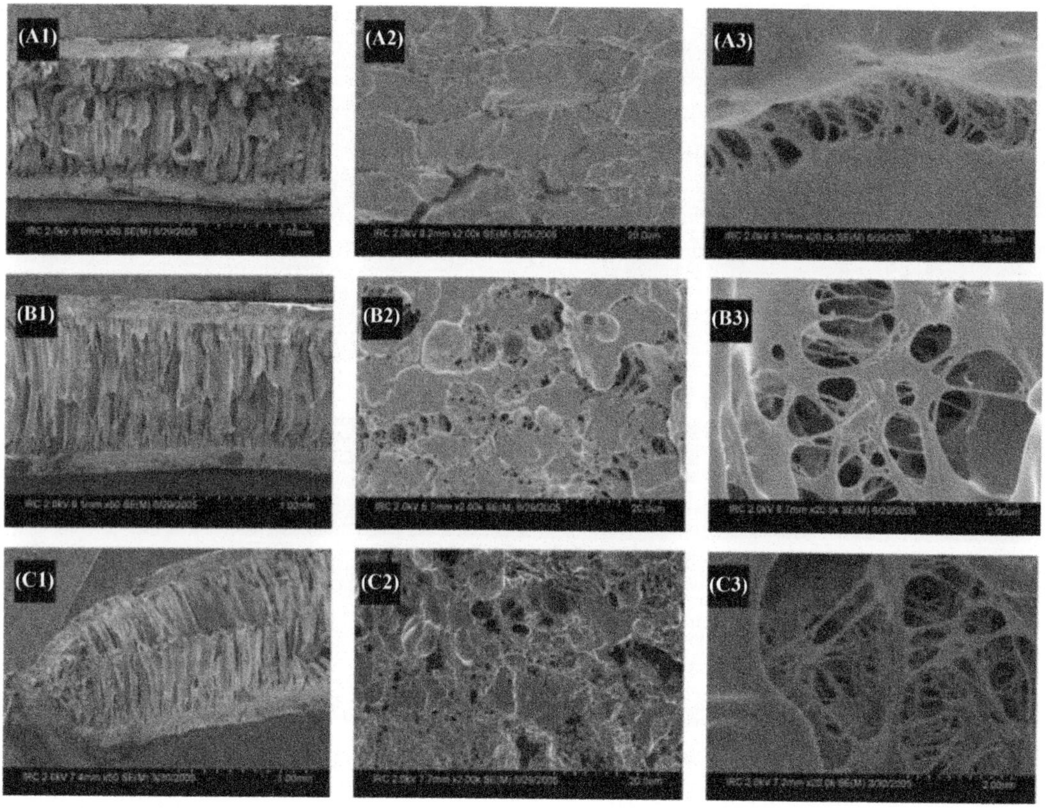

FIGURE 9.15

SEM of (a1–a3, b1–b3 and c1–c3) of CO_2-foamed PLLA samples saturated at 25°C and 5.5 MPa for 5 min and foamed [167].

epidermis with open and linked nanopores was thought to be an excellent conduit for nutrient flow, whereas the core with wider stretched holes was thought to be suitable for cell proliferation.

Fig. 9.16 shows the architecture of three distinct types of trabecular bones as well as three distinct polymer foams. The anisotropy varied depending on the density of the bubble nucleation and the cooling time. While a gradual cooling rate promoted pore elongation, a rapid cooling rate resulted in a significant number of spherical holes. Phosphate glass (PG) was used as a filling agent in PLLA foams made in super critical carbon dioxide ($scCO_2$) for bone regeneration. However, glass addition had no effect on compressive moduli or biocompatibility. The use of $scCO_2$ allows for the accessibility and interconnection of porous materials, as well as the incorporation of biological agents into nanocomposite matrix. The biocompatible and biodegradable ceramic- reinforced PLLA scaffolding produced through $scCO_2$ foaming is ideal for tissue engineering due to its remarkable mechanical properties and biocompatibility with main osteoblasts in vitro.

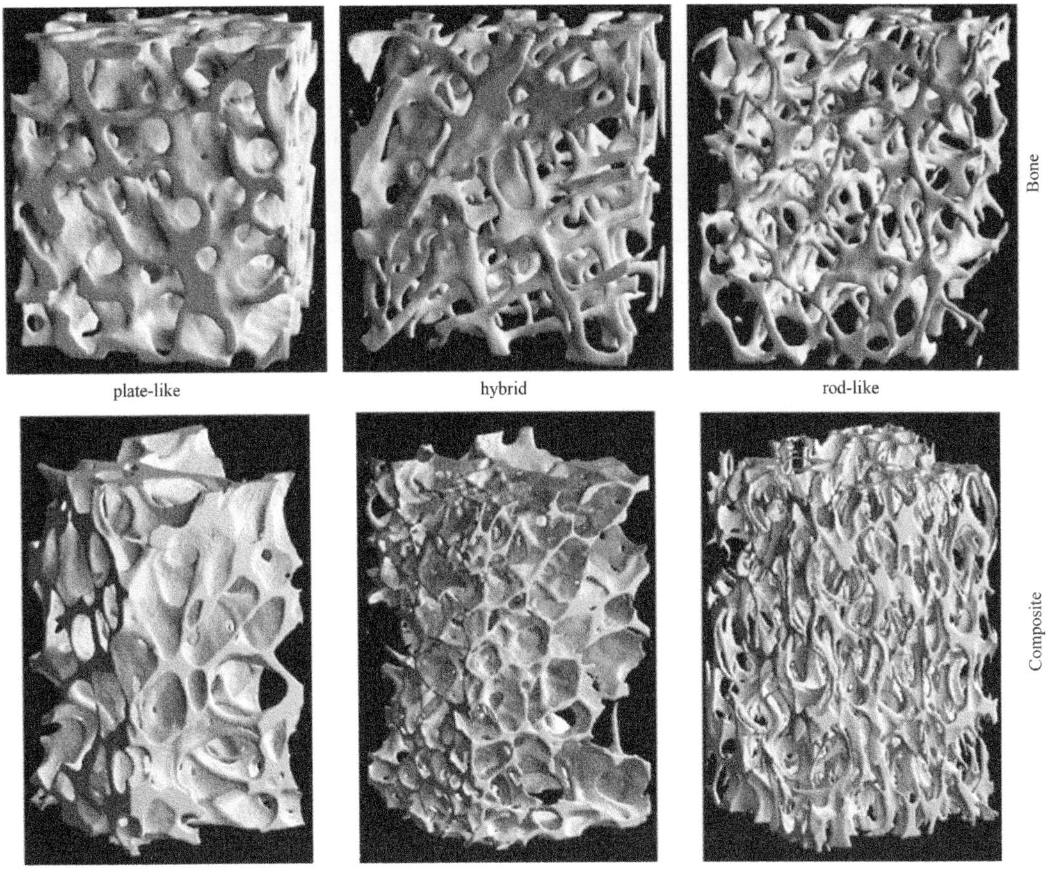

FIGURE 9.16

Similarities between the macrostructures of polymer foam and cancellous bone [168] https://s100.copyright.com/CustomerAdmin/PLF.jsp?ref = 126b09e1-26e9-4b61-beb0-cebcab4f8a4c.

With permission from Elsevier. Copyright (2006).

9.8 Other techniques of scaffold fabrication for tissue engineering

9.8.1 Sol–gel method

When calcium nitrate tetrahydrate, nonalkoxide, phosphoric acid, and ammonia are used as precursor materials, the interaction between calcium and phosphate influences particle size in the sol–gel procedure [169,170]. Raucci et al., [171] created a composite biomaterial composed of HA/PCL using the sol–gel process with di-phosphorous pentoxide and calcium nitrate tetrahydrate as substrates and ethanol and tetrahydrofuran as solvents. Furthermore, using salt as a porogen with a specific size range revealed an 88% porous structure. Following that, a CS (PCL/HA) was examined in simulated bodily fluid (SBF) with a regulated pH (6.5) [172]. The formation of apatite is caused by the dispersion of HA particles in the scaffold, and the presence of the coating was dyed with trypan blue. The composite scaffolding enhances the development of human stem cells (hMSCs) into the osteoblast phenotype shown in alkaline phosphatase (ALP) activity [173]. Costa et al., [174] employed calcium phosphate as a starting material in the sol–gel-hydrothermal technique to create HA nanorods. PCL/HA composite materials were created by combining HA powders with PCL, resulting in a scaffold of nanorods and nanowires. The incorporation of HA nanorods into PCL composites greatly improved the mechanical properties in terms of Young's and compressive modulus. Thus, HA nanowires and PCL composite can be used for bone regeneration.

9.8.2 Solvent casting and particulate leaching

The goal of the particulates leaching technique, which involves a polymer solution containing salt particles of a defined diameter, is to create the ideal porogen using paraffin beads, sucrose, and sodium bicarbonate as components. Fabbri et al., used salt chloride and sodium bicarbonate as a porogen to create a PCL/HA CS with 70–80% porosity, high interconnectivity, and superior mechanical properties [175]. PCL/HA scaffolds with huge porosity (greater than 90%) and five hundred μm pore in widths were created by Guarino et al. [176]. They were able to modulate porosity depending on the form and quantity of the porogens supplied.

The incorporation of HA nanoparticles into PCL composites increases cell differentiation while decreasing porosity and cell proliferation throughout the scaffolding [172,177]. The cell sensitivity assays MTTassay was used to assess the activation of ALP for cell differentiation and total protein content after 2 weeks of cell culture [178]. Guarino et al., [176] implanted Mesenchymal stem cells (MSCs) onto porous scaffolds of PCL/HA composite and cultured them for 1–5 weeks. MSCs can transfer and develop on composite surfaces in 3–4 weeks, and, used for evaluating the minor impact of signaling on biological function. In order to addressing this issue, a preosteoblastic cell line was used, which demonstrated improved cell adherence and greater preosteoblast response on PCL/HA 3D scaffolds created using this technique. Porous PCL/HA composites with a positive effect on structural features are suitable biomaterials for bone replacement [172,176,179]. Although the leaching process has a predetermined form, salt particles during scaffolding production cause poor connectivity, which may not provide the best scaffold permeability for cell dispersion in vitro.

9.8.3 Thermally induced phase separation (TIPS)

TIPS involves forcing liquid-liquid separation while cooling the polymer solution below the freezing point of the solvent [180]. By adding nanoparticles of HA to the polymer solution, composite scaffolds of PCL/HA were created in this procedure. The PCL/HA scaffold's compressive modulus increases after the addition of HA nanoparticles [181]. According to Guan et al., [182] decreasing the polymer solution's concentration results in the production of CS having smaller pores and less porosity. The HA/PCL/CSs demonstrated strong osteoconductive activity in applications requiring BTE [183].

9.9 Conclusions

To summarize, the primary goal of tissue engineering is to repair or replace damaged or diseased tissues. Some physical structures with the ability to connect cells, migrate cells, culture cells, differentiate cells, and eventually be replaced in tissues are required throughout this context. Such structures are created by biodegradable polymers used as an exterior framework in bioengineering. Scaffolds are constructed using a variety of methods, including gas foaming, solvent casting, and freeze drying. In addition, several studies on the formation and processing of polymer nanocomposite in tissue engineering have been discussed, with promising results. Nonetheless, there is no perfect biodegradable polymer that meets all of the tissue engineering parameters. This necessitates improved qualities, such as biocompatibility, which should be considered when selecting a scaffold. As a result, certain techniques for improving tissue engineering have been developed to improve the properties of biodegradable polymers, such as incorporating other components into a polymer to create a composite material. Finally, we looked at recent advances in biodegradable natural materials and synthetic polymers, as well as nanomedicine, that have been used to improve tissue engineering. Although there is no perfect material for heart tissue engineering, nanomedicine has the potential to be a valuable collaborator. Whatever the cause of the damage or illness, such nanomedicine outcomes provide significant hope for the construction of a healthy and functioning heart.

9.10 Challenges and future perspectives

Using optimal manufacturing techniques, it is still difficult to produce CSs with the appropriate tissue properties. Only by responding appropriately to the interaction between these essential properties and manufacturing technology will be possible to design acceptable biomaterials with hierarchical structures that are equivalent to those of natural bone tissue. Recently, technological constraints are to blame for structural and resolution flaws in other approaches. This strategy appears to be innovative for methods such as electrospinning. It is still difficult to manage fiber diameter across complex process-based interactions in electrospinning processes. Polymerization-based manufacturing techniques can address resolution and speed issues, but new materials must be developed while keeping the features listed below in mind. The combination of inorganic materials and polymers is thought to be the most promising of the various biomaterials used in scaffolding construction. Such composites have superior mechanical properties, appropriate biodegradability, and immunological function. To summarize, more work needs to be done on effective contribution and concurrent prototype development in order to produce bone tissue that mimics the original ECM. Because of the inherent constraints of the various manufacturing processes described, combination solutions hold promise. In the coming years, the combination of additive manufacturing and cutting-edge medical imaging methods will have the potential to significantly improve patient outcomes while bringing patient-specific treatments from the laboratory to the clinic. It is critical to understand the localized biomechanical environment as well as the kinetics of growth factor release. In this scenario, the development of scaffolds with tailored gradients may open up new avenues for tissue engineering. To improve in situ bone regeneration and achieve better results, biomimetic scaffold architecture with immunomodulatory features should be prioritized.

References

[1] D.F. Williams, Challenges with the development of biomaterials for sustainable tissue engineering, Frontiers in Bioengineering and Biotechnology 7 (2019) 127.

[2] C.A. Vacanti, The history of tissue engineering, Journal of Cellular and Molecular Medicine (2006).

[3] N.S. Sani, N.A.N.N. Malek, M.R.A. Kadir, H. Hamdan, In vitro bioactivity and osteoblast cell viability studies of hydroxyapatite-incorporated silica aerogel, Journal of Sol-Gel Science and Technology 96 (2020) 166–177.

[4] A. Muñoz-Ruíz, D.M. Escobar-García, M. Quintana, A. Pozos-Guillén, H. Flores, Synthesis and characterization of a new collagen-alginate aerogel for tissue engineering, Journal of Nanomaterials 2019 (2019) 1–5.

[5] R. Ghafari, M. Jonoobi, L.M. Amirabad, K. Oksman, A.R. Taheri, Fabrication and characterization of novel bilayer scaffold from nanocellulose based aerogel for skin tissue engineering applications, International Journal of Biological Macromolecules 136 (2019) 796–803.

[6] H. Maleki, L. Durães, C.A. García-González, P. Del Gaudio, A. Portugal, M. Mahmoudi, Synthesis and biomedical applications of aerogels: Possibilities and challenges, Advances in Colloid and Interface Science 236 (2016) 1–27.

[7] L. Keller, A. Regiel-Futyra, M. Gimeno, S. Eap, G. Mendoza, V. Andreu, M. Arruebo, N. Benkirane-Jessel, Chitosan-based nanocomposites for the repair of bone defects, Nanomedicine: Nanotechnology, Biology and Medicine 13 (2017) 2231–2240.

[8] H. Zhang, C. Junqiu, A.O. Qiang, Preparation of alginate-based biomaterials and their applications in biomedicine, Marine Drugs 19 (5) (2021) 264.

[9] J.R. Venugopal, S. Low, A.T. Choon, A.B. Kumar, S. Ramakrishna, Nanobioengineered electrospun composite nanofibers and osteoblasts for bone regeneration, Artificial Organs 32 (5) (2008) 388–397.

[10] L. Ziqian, L. Xiaoling, R. Seeram, Surface engineering of biomaterials in orthopedic and dental implants: Strategies to improve osteointegration, bacteriostatic and bactericidal activities, Biotechnology Journal 16 (7) (2021) 2000116.

[11] G.H. Yang, W. Kim, J. Kim, G. Kim, A skeleton muscle model using GelMA-based cell-aligned bioink processed with an electric-field assisted 3D/4D bioprinting, Theranostics 11 (2021) 48.

[12] G. Siqueira, B. Julien, D. Alain, Cellulosic bionanocomposites: a review of preparation, properties and applications, Polymers 2 (4) (2010) 728–765.

[13] R.J. Hickey, E.P. Andrew, Cellulose biomaterials for tissue engineering, Frontiers in Bioengineering and Biotechnology 7 (2019) 45.

[14] A. Frenot, M.W. Henriksson, P. Walkenström. Electrospinning of cellulose-based nanofibers. Journal of Applied Polymer Science 103 (2007) 1473–1482.

[15] N.M. Ponce, C.A. Pujol, E.B. Damonte, M.L. Flores, C.A. Stortz, Fucoidans from the brown seaweed *Adenocystis utricularis*: extraction methods, antiviral activity and structural studies, Carbohydrate Research 338 (2) (2003) 153–165.

[16] J. Venkatesan, S.K. Kim, I. Bhatnagar, Chitosan-alginate biocomposite containing fucoidan for bone tissue engineering, Marine Drugs 12 (1) (2014) 300–316.

[17] M. Dovedytis, Z.J. Liu, S. Bartlett, Hyaluronic acid and its biomedical applications: A review, Engineered Regeneration 1 (2020) 102–113.

[18] T. Kobayashi, T. Chanmee, N. Itano, Hyaluronan: Metabolism and function, Biomolecules 10 (11) (2020) 1525.

[19] H. Knopf-Marques, M. Pravda, L. Wolfova, V. Velebny, P. Schaaf, N.E. Vrana, P. Lavalle, Hyaluronic acid and its derivatives in coating and delivery systems: applications in tissue engineering, regenerative medicine and immunomodulation, Advanced Healthcare Materials 5 (22) (2016) 2841–2855.

[20] A. Kirschning, N. Dibbert, G. Dräger, Chemical functionalization of polysaccharides—Towards biocompatible hydrogels for biomedical applications, Chemistry–A European Journal 24 (6) (2018) 1231–1240.

[21] M.D. Raicopol, C. Andronescu, S.I. Voicu, C. Posarelli, A. Passani, M. Del Re, S. Fogli, M.D. Toro, A. Ferreras, M. Figus, Cross-linked hyaluronic acid as tear film substitute, Journal of Ocular Pharmacology and Therapeutics 35 (7) (2019) 381–387.

[22] J.A. Burdick, G.D. Prestwich, Hyaluronic acid hydrogels for biomedical applications, Advanced Materials 23 (12) (2011) H41–H56.

[23] B. Tavsanli, O. Okay, Mechanically robust and stretchable silk/hyaluronic acid hydrogels, Carbohydrate Polymers 208 (2019) 413–420.

[24] Y.A. Tae, H.K. Jeong, J.K. Dong, V. Jayachandran, K.C. Hee, B. Ira, C. Kwan-Young, H. Jae-Ho, S. Ziad, K. Se-Kwon, K. Hui Taek, K. Dong Gyu, Interaction of stem cells with nano hydroxyapatite-fucoidan bionanocomposites for bone tissue regeneration, International Journal of Biological Macromolecules 93 (2017) 1488–1491.

[25] J.C. Courtenay, M.A. Johns, F. Galembeck, C. Deneke, E.M. Lanzoni, C.A. Costa, et al., Surface modified cellulose scaffolds for tissue engineering, Cellulose 24 (2017) 253–267.

[26] B.V. Mohite, S.V. Patil, A novel biomaterial: bacterial cellulose and its new era applications, Biotechnology and Applied Biochemistry 61 (2) (2014) 101–110.

[27] C.L. Salgado, E.M. Sanchez, C.A. Zavaglia, P.L. Granja, Biocompatibility and biodegradation of polycaprolactone-sebacic acid blended gels, Journal of Biomedical Materials Research Part A 100 (1) (2012) 243–251.

[28] J.W.M. Jung, H.F. Lee, J.M.N. Hong, J.H. Park, J.H. Shim, T.H. Choi, D.W. Cho, A new method of fabricating a blend scaffold using an indirect three-dimensional printing technique, Biofabrication 7 (4) (2015) 045003.

[29] M.S. Singhvi, S.S. Zinjarde, D.V. Gokhale, Polylactic acid: synthesis and biomedical applications, Journal of Applied Microbiology 127 (6) (2019) 1612–1626.

[30] Q. Yao, J.G. Cosme, T. Xu, J.M. Miszuk, P.H. Picciani, H. Fong, H. Sun, Three dimensional electrospun PCL/PLA blend nanofibrous scaffolds with significantly improved stem cells osteogenic differentiation and cranial bone formation, Biomaterials 115 (2017) 115–127.

[31] S.G. Fitzpatrick, D.M. Cohen, A.N. Clark, Ulcerated lesions of the oral mucosa: clinical and histologic review, Head and Neck Pathology 13 (2019) 91–102.

[32] S. Liu, S. Qin, M. He, D. Zhou, Q. Qin, H. Wang, Current applications of poly(lactic acid) composites in tissue engineering and drug delivery, Composites Part B: Engineering 199 (2020) 108238.

[33] A.E. Ivanov, H. Larsson, I.Y. Galaev, B. Mattiasson, Synthesis of boronate-containing copolymers of N, N-dimethylacrylamide, their interaction with poly(vinyl alcohol) and rheological behavior of the gels, Polymer 45 (8) (2004) 2495–2505.

[34] I. Ielo, G. Calabrese, G. De Luca, S. Conoci, Recent advances in hydroxyapatite-based biocomposites for bone tissue regeneration in orthopedics, International Journal of Molecular Sciences 23 (17) (2022) 9721.

[35] M.A. Aristri, M.A.R. Lubis, S.M. Yadav, P. Antov, A.N. Papadopoulos, A. TurnPizzi, A. H Iswanto, Recent developments in lignin-and tannin-based non-isocyanate polyurethane resins for wood adhesives—A review, Applied Sciences 11 (9) (2021) 4242.

[36] J.P. Santerre, K. Woodhouse, G. Laroche, R.S. Labow, Understanding the biodegradation of polyurethanes: from classical implants to tissue engineering materials, Biomaterials 26 (35) (2005) 7457–7470.

[37] Y. Wang, W. Zhang, Q. Yao, Copper-based biomaterials for bone and cartilage tissue engineering, J. Orthop. Transl. 29 (2021) 60–71.

[38] F. Oveissi, S. Naficy, T.Y.L. Le, D.F. Fletcher, F. Dehghani, Tough hydrophilic polyurethane-based hydrogels with mechanical properties similar to human soft tissues, Journal of Materials Chemistry B 7 (22) (2019) 3512–3519.

[39] F. Donnaloja, E. Jacchetti, M. Soncini, M.T. Raimondi, et al., Natural and synthetic polymers for bone scaffolds optimization, Polymers 12 (4) (2020) 905.

[40] W. Chong, L. Jiahui, L. Kai, Z. Shaokui, L. Bingheng, L. Jia, T. Yujin, W. Yen, Cryogenic 3D printing of dual-delivery scaffolds for improved bone regeneration with enhanced vascularization, Bioactive Materials 6 (2021) 137–145.

[41] M.C. Koetting, J.T. Peters, S.D. Steichen, Stimulus-responsive hydrogels: Theory, modern advances, and applications, Materials Science and Engineering: Reports 93 (2015) 1–49.

[42] S.C. Chen, Y.C. Wu, F.L. Mi, Y.H. Lin, L.C. Yu, H.W. Sung, A novel pH-sensitive hydrogel composed of N, O-carboxymethyl chitosan and alginate cross-linked by genipin for protein drug delivery, Journal of Controlled Release 96 (2) (2004) 285–300.

[43] F. Zhou, Y. Hong, R. Liang, X. Zhang, Y. Liao, D. Jiang, H. Ouyang, Rapid printing of bio-inspired 3D tissue constructs for skin regeneration, Biomaterials 258 (2020) 120287.

[44] N. Amiryaghoubi, M. Fathi, J. Barar, Y. Omidi, Hydrogel-based scaffolds for bone and cartilage tissue engineering and regeneration, Reactive and Functional Polymers (2022) 105313.

[45] B. Xu, L. Wang, Y. Liu, H. Zhu, Q. Wang, Preparation of high strength and transparent nanocomposite hydrogels using alumina nanoparticles as cross-linking agents, Materials Letters 228 (2018) 104–107.

[46] A. Pardo, M. Gómez-Florit, S. Barbosa, P. Taboada, R.M. Domingues, M.E. Gomes, Magnetic nanocomposite hydrogels for tissue engineering: design concepts and remote actuation strategies to control cell fate, ACS Nano 15 (1) (2021) 175–209.

[47] J. Huang, L. Yujie, J. ZhaoFeng, C. Jielin, D. Li, L. Wei, Z. Feiyan, L. Qian, Z. Weimin, Y. Wei, X. Jianyi, W. Daping, Development of magnetic nanocomposite hydrogel with potential cartilage tissue engineering, ACS Omega 3 (2018) 6182–6189.

[48] Q. Yan, H. Dong, J. Su, J. Han, B. Song, Q. Wei, Y. Shi, A review of 3D printing technology for medical applications, Engineering 4 (5) (2018) 729–742.

[49] M. Mohammadian, T.S.J. Kashi, M. Erfan, F.P. Soorbaghi, Synthesis and characterization of silica aerogel as a promising drug carrier system, Journal of Drug Delivery Science and Technology 44 (2018) 205–212.

[50] P.C. Thapliyal, K. Singh, Aerogels as promising thermal insulating materials: An overview, Journal of Materials 2014 (2014) 14.

[51] L. Zheng, S. Zhang, Z. Ying, J. Liu, Y. Zhou, F. Chen, Engineering of aerogel-based biomaterials for biomedical applications, International Journal of Nanomedicine 15 (2363) (2020) 2363.

[52] Z. Ulker, I. Erucar, S. Keskin, C. Erkey, Novel nanostructured composites of silica aerogels with a metal organic framework, Microporous and Mesoporous Materials 170 (2013) 352–358.

[53] S. Liu, C. Zhou, S. Mou, L. Li, S. Xiang, S. Cao, J. Zhang, G. Ouyang, L. Chen, C.Y. Su, A synthetic route to ultralight hierarchically micro/mesoporous Al (III)-carboxylate metal-organic aerogels, Nature Communications 4 (1) (2013) 1–9.

[54] E.B. Yahya, A.A. Amirul, A.K. HPS, N.G. Olaiya, M.O. Iqbal, F. Jummaat, A.S. Adnan, Insights into the role of biopolymer aerogel scaffolds in tissue engineering and regenerative medicine, Polymers 13 (10) (2016) 1612.

[55] Y.E. Ogutcu M., N. Arifoglu, E. Dincer, Factors affecting tahini quality, Journal of International Scientific Publications: Agriculture & Food 5 (2017) 366–373.

[56] H. Tetik, D. Feng, S.W. Oxandale, G. Yang, K. Zhao, K. Feist, D. Lin, Bioinspired manufacturing of aerogels with precisely manipulated surface microstructure through controlled local temperature gradients, ACS Applied Materials & Interfaces 13 (2021) 924–931.

[57] Z. Wang, C. Huang, X. Han, S. Li, Z. Wang, J. Huang, Z. Chen, Fabrication of aerogel scaffolds with adjustable macro/micro-pore structure through 3D printing and sacrificial template method for tissue engineering, Materials & Design 217 (2022) 110662.

[58] C.W. Hull, Patent and Trademark Office., U.S. Patent No. 4,575,330. Washington, DC: U.S. (1986).

[59] W. Gao, Y. Zhang, D. Ramanujan, K. Ramani, Y. Chen, C.B. Williams, P.D. Zavattieri, The status, challenges, and future of additive manufacturing in engineering, Computer-Aided Design 69 (2015) 65–89.

[60] F. Murmura, L. Bravi, Additive manufacturing in the wood-furniture sector: Sustainability of the technology, benefits and limitations of adoption, Journal of Manufacturing Technology Management (2017).

[61] O.O. Spencer, O.T. Yusuf, T.C. Tofade, Additive manufacturing technology development: a trajectory towards Industrial Revolution, Journal of Mechanical and Industrial Engineering Research (2018) 80–90.

[62] O.G. Bhusnure, S.V. Gholve, B.K. Sugave, R.C. Dongre, S.A. Gore, P.S. Giram, 3D printing & pharmaceutical manufacturing: opportunities and challenges, International Journal of Bioassays 5 (1) (2016) 4723–4738.

[63] S.B. Kumar, S.D. Sekar, G. Sivakumar, J. Srinivas, R. Lavanya, G. Suresh, Modern concepts and application of soft robotics in 4D printing, Journal of Physics: Conference Series, IOP Publishing (2021), 12056.

[64] S.B. Kumar, J. Jeevamalar, P. Ramu, G. Suresh, K. Senthilnathan, Evaluation in 4D printing—a review, Materials Today: Proceedings 45 (2021) 1433–1437.

[65] J. Choi, O.-C. Kwon, W. Jo, H.J. Lee, M.-W. Moon, 4D printing technology: a review, 3D print, Addittive Manufacturing 2 (2015) 159–167.

[66] E.R. Ghomi, S.K. Eshkalak, S. Singh, A. Chinnappan, S. Ramakrishna, R. Narayan, Fused filament printing of specialized biomedical devices: a state-of-the art review of technological feasibilities with PEEK, Rapid Prototyping Journal (2021).

[67] V. Shanmugam, O. Das, K. Babu, U. Marimuthu, A. Veerasimman, D.J. Johnson, F. Berto, Fatigue behaviour of FDM-3D printed polymers, polymeric composites and architected cellular materials, International Journal of Fatigue 143 (2021) 106007.

[68] A.M. Bejoy, K.N. Makkithaya, B.B. Hunakunti, A. Hegde, K. Krishnamurthy, A. Sarkar, N. Mazumder, An insight on advances and applications of 3d bioprinting: A review., Bioprinting 24 (2021) e00176.

[69] T. Agarwal, I. Chiesa, D. Presutti, V. Irawan, K.Y. Vajanthri, M. Costantini, T.K. Maiti, Recent advances in bioprinting technologies for engineering different cartilage-based tissues, Materials Science and Engineering: C 123 (2021) 112005.

[70] N. Wang, S.T. Dheen, J.Y.H. Fuh, A.S. Kumar, A review of multi-functional ceramic nanoparticles in 3D printed bone tissue engineering, Bioprinting 23 (2021) e00146.

[71] D.A.S. Rodríguez, A.I. Ramos-Murillo, R.D. Godoy-Silva, Tissue engineering, 3D-Bioprinting, morphogenesis modelling and simulation of biostructures: relevance, underpinning biological principles and future trends, Bioprinting 24 (2021) e00171.

[72] I. Matai, G. Kaur, A. Seyedsalehi, A. McClinton, C.T. Laurencin, Progress in 3D bioprinting technology for tissue/organ regenerative engineering, Biomaterials 226 (2020) 119536.

[73] A. Selimis, V. Mironov, M. Farsari, Direct laser writing: principles and materials for scaffold 3D printing, Microelectronic Engineering 132 (2015) 83–89.

[74] Z.U. Arif, M.Y. Khalid, W. Ahmed, H. Arshad, A review on four-dimensional bioprinting in pursuit of advanced tissue engineering applications, Bioprinting (2022) e00203.

[75] Y.L. Tee, P. Tran, On bioinspired 4d printing: materials, design and potential applications, Australian Journal of Mechanical Engineering (2021) 1–11.

[76] M.Y. Khalid, A. Al Rashid, Z.U. Arif, W. Ahmed, H. Arshad, A.A. Zaidi, Natural fiber reinforced composites: sustainable materials for emerging applications, Results in Engineering 11 (2021) 100263.

[77] P. Rastogi, J. Njuguna, B. Kandasubramanian, Exploration of elastomeric and polymeric liquid crystals with photothermal actuation: a review, European Polymer Journal 121 (2019) 109287.
[78] G. Adam, A. Benouhiba, K. Rabenorosoa, C. Clévy, D.J. Cappelleri, 4D printing: enabling technology for microrobotics applications, Advanced Intelligent Systems 3 (2021) 2000216.
[79] C.M. O'Brien, B. Holmes, S. Faucett, L.G. Zhang, Three-dimensional printing of nanomaterial scaffolds for complex tissue regeneration, Tissue Engineering Part B: Reviews 21 (2015) 103–114.
[80] S. Amukarimi, M. Mozafari, 4D bioprinting of tissues and organs, Bioprinting 23 (2021) e00161.
[81] H. Chen, F. Zhang, Y. Sun, B. Sun, B. Gu, J. Leng, W. Zhang, Electrothermal shape memory behavior and recovery force of four-dimensional printed continuous carbon fiber/polylactic acid composite, Smart Materials and Structures 30 (2021) 25040.
[82] W. Aljohani, M.W. Ullah, X. Zhang, G. Yang, Bioprinting and its applications in tissue engineering and regenerative medicine, International Journal of Biological Macromolecules 107 (2018) 261–275.
[83] S. Joshi, K. Rawat, C. Karunakaran, V. Rajamohan, A.T. Mathew, K. Koziol, A.S.S. Balan, 4D printing of materials for the future: opportunities and challenges, Applied Materials Today 18 (2020) 100490.
[84] M.M. Zerankeshi, R. Bakhshi, R. Alizadeh, Polymer/metal composite 3D porous bone tissue engineering scaffolds fabricated by additive manufacturing techniques: a review, Bioprinting (2022) e00191.
[85] J. Zhou, S. Vijayavenkataraman, 3D-printable conductive materials for tissue engineering and biomedical applications, Bioprinting 24 (2021) e00166.
[86] S.H. Mahfouzi, S.H.S. Tali, G. Amoabediny, 3D bioprinting for lung and tracheal tissue engineering: criteria, advances, challenges, and future directions, Bioprinting 21 (2021) e00124.
[87] J.A. Semba, A.A. Mieloch, J.D. Rybka, Introduction to the state-of-the-art 3D bioprinting methods, design, and applications in orthopedics, Bioprinting 18 (2020) e00070.
[88] E. Pei, G.H. Loh, Technological considerations for 4D printing: an overview, Progress in Additive Manufacturing 3 (2018) 95–107.
[89] H.Z. Lu, C. Yang, X. Luo, H.W. Ma, B. Song, Y.Y. Li, L.C. Zhang, Ultrahigh-performance TiNi shape memory alloy by 4D printing, Materials Science and Engineering: A 763 (2019) 138166.
[90] E. Pei, G.H. Loh, S. Nam, Concepts and Terminologies in 4D Printing, Applied Sciences 10 (2020) 4443.
[91] T.A. Campbell, The next wave: 4D printing programming the material world, Atlantic Council., 2014.
[92] A. Ijaz, E.H. Mourad, E.M. Mohamed, L. Dimitri, Toward enabling manufacturing paradigm of 4D printing of shape memory materials: Open literature review, European Polymer Journal 168 (2022) 111106.
[93] S. Indermun, M. Govender, P. Kumar, Y.E. Choonara, V. Pillay, Stimuli-responsive polymers as smart drug delivery systems: classifications based on carrier type and triggered-release mechanism, Stimuli Responsive Polymeric Nanocarriers for Drug Delivery Applications - 1st Edition, Elsevier, 2018, pp. 43–58.
[94] M. Nadgorny, Z. Xiao, C. Chen, L.A. Connal, Three-dimensional printing of pH-responsive and functional polymers on an affordable desktop printer, ACS Applied Materials & Interfaces 8 (2016) 28946–28954.
[95] B. Jian, F. Demoly, Y. Zhang, H.J. Qi, J.-C. André, S. Gomes, Origami-based design for 4D printing of 3D support-free hollow structures, Engineering (2022).
[96] Y. Zhu, S. Deng, X. Zhao, G. Xia, R. Zhao, H.F. Chan, Deciphering and engineering tissue folding: a mechanical perspective, Acta Biomaterialia 134 (2021) 32–42.
[97] J.M. Taylor, H. Luan, J.A. Lewis, J.A. Rogers, R.G. Nuzzo, P.V. Braun, Biomimetic and biologically compliant soft architectures via 3D and 4D assembly methods: a perspective, Advanced Materials 34 (2022) 2108391.
[98] K. Kuribayashi-Shigetomi, H. Onoe, S. Takeuchi, Cell origami: self-folding of three-dimensional cell-laden microstructures driven by cell traction force, PLoS One 7 (2012) e51085.

[99] Q. He, T. Okajima, H. Onoe, A. Subagyo, K. Sueoka, K. Kuribayashi-Shigetomi, Origami-based self-folding of co-cultured NIH/3T3 and HepG2 cells into 3D microstructures, Scientific Reports 8 (2018) 1–7.

[100] R. Whenish, S. Ramakrishna, A.K. Jaiswal, G. Manivasagam, A framework for the sustainability implications of 3D bioprinting through nature-inspired materials and structures, Bio-Design and Manufacturing (2022) 1–12.

[101] Z.M. Khan, E. Wilts, E. Vlaisavljevich, T.E. Long, S.S. Verbridge, Electroresponsive hydrogels for therapeutic applications in the brain, Macromolecular Bioscience 22 (2022) 2100355.

[102] H. Tetsuka, S.R. Shin, Materials and technical innovations in 3D printing in biomedical applications, Journal of Materials Chemistry B 8 (2020) 2930–2950.

[103] X. Kuang, K. Chen, C.K. Dunn, J. Wu, V.C.F. Li, H.J. Qi, 3D printing of highly stretchable, shape-memory, and self-healing elastomer toward novel 4D printing, ACS Applied Materials & Interfaces 10 (2018) 7381–7388.

[104] H. Wei, Q. Zhang, Y Yao, Y. Liu, J. Leng, Direct-write fabrication of 4D active shape-changing structures based on a shape memory polymer and its nanocomposite, ACS Applied Materials & Interfaces 9 (1) (2017) 876–883.

[105] Y. Luo, X. Lin, B. Chen, X. Wei, Cell-laden four-dimensional bioprinting using near-infrared-triggered shape-morphing alginate/polydopamine bioinks, Biofabrication 11 (2019) 45019.

[106] A. Lendlein, M. Balk, N.A. Tarazona, O.E.C. Gould, Bioperspectives for shape-memory polymers as shape programmable, active materials, Biomacromolecules 20 (2019) 3627–3640.

[107] A. Melocchi, M. Uboldi, M. Cerea, A. Foppoli, A. Maroni, S. Moutaharrik, et al., Shape memory materials and 4D printing in pharmaceutics, Advanced Drug Delivery Reviews 173 (2021) 216–237.

[108] M.C. Biswas, S. Chakraborty, A. Bhattacharjee, Z. Mohammed, 4D printing of shape memory materials for textiles: mechanism, mathematical modeling, and challenges, Advanced Functional Materials 31 (2021) 2100257.

[109] R. Sarvari, P. Keyhanvar, S. Agbolaghi, M.S. Gholami Farashah, A. Sadrhaghighi, M. Nouri, et al., Shape-memory materials and their clinical applications, Journal of Polymeric Materials and Polymeric Biomaterials 71 (2022) 315–335.

[110] E. Koçak, A. Yıldız, F. Acartürk, Three dimensional bioprinting technology: applications in pharmaceutical and biomedical area, Colloids and Surfaces B: Biointerfaces 197 (2021) 111396.

[111] Z.U. Arif, M.Y. Khalid, W. Ahmed, H. Arshad, S. Ullah, Recycling of the glass/carbon fibre reinforced polymer composites: a step towards the circular economy, Polymer-Plastics Technology and Materials (2022) 1–28.

[112] D. Zhou, J. Chen, B. Liu, X. Zhang, X. Li, T. Xu, Bioinks for jet-based bioprinting, Bioprinting 16 (2019) e00060.

[113] Y. Wang, X. Yuan, B. Yao, S. Zhu, P. Zhu, S. Huang, Tailoring bioinks of extrusion-based bioprinting for cutaneous wound healing, Bioactive Materials (2022).

[114] S. Ghosh, E. Mostafavi, N. Thorat, T.J. Webster, Nanobiomaterials for three-dimensional bioprinting, Nanotechnology in Medicine and Biology, Elsevier, 2022, pp. 1–24.

[115] Q. Zhang, X. Kuang, S. Weng, L. Yue, D.J. Roach, D. Fang, et al., Shape-memory balloon structures by pneumatic multi-material 4D printing, Advanced Functional Materials 31 (2021) 2010872.

[116] B. Tan, S. Gan, X. Wang, W. Liu, X. Li, Applications of 3D bioprinting in tissue engineering: advantages, deficiencies, improvements, and future perspectives, Journal of Materials Chemistry B 9 (2021) 5385–5413.

[117] S.D. Eswaramoorthy, S. Ramakrishna, S.N. Rath, Recent advances in three-dimensional bioprinting of stem cells, Journal of Tissue Engineering and Regenerative Medicine 13 (2019) 908–924.

[118] S. Adepu, N. Dhiman, A. Laha, C.S. Sharma, S. Ramakrishna, M. Khandelwal, Three-dimensional bioprinting for bone tissue regeneration, Current Opinion in Biomedical Engineering 2 (2017) 22–28.

[119] H.K. Raut, R. Das, Z. Liu, X. Liu, S. Ramakrishna, Biocompatibility of biomaterials for tissue regeneration or replacement, Biotechnology Journal 15 (2020) 2000160.
[120] E. Kijeńska, S. Zhang, M.P. Prabhakaran, S. Ramakrishna, W. Swieszkowski, Nanoengineered biocomposite tricomponent polymer based matrices for bone tissue engineering, International Journal of Polymeric Materials and Polymeric Biomaterials 65 (2016) 807–815.
[121] P. Rastogi, B. Kandasubramanian, Breakthrough in the printing tactics for stimuli-responsive materials: 4D printing, Chemical Engineering Journal 366 (2019) 264–304.
[122] M.C. Biswas, B. Jony, P.K. Nandy, R.A. Chowdhury, S. Halder, D. Kumar, M.A. Imam, Recent advancement of biopolymers and their potential biomedical applications, Journal of Polymers and the Environment (2021) 1–24.
[123] P. Lavrador, M.R. Esteves, V.M. Gaspar, J.F. Mano, Stimuli-responsive nanocomposite hydrogels for biomedical applications, Advanced Functional Materials 31 (2021) 2005941.
[124] Chen, Z., Zhao, D., Liu, B., Nian, G., Li, X., Yin, J., ... & Yang, W. (2019). 3D printing of multifunctional hydrogels. *Advanced Functional Materials*, *29*(20), 1900971.
[125] C. Mota, S. Camarero-Espinosa, M.B. Baker, P. Wieringa, L. Moroni, Bioprinting: from tissue and organ development to in vitro models, Chemical Reviews 120 (2020) 10547–10607.
[126] A. Rayate, P.K. Jain, A review on 4D printing material composites and their applications, Materials Today: Proceedings 5 (2018) 20474–20484.
[127] A. Haleem, M. Javaid, R.P. Singh, R. Suman, Significant roles of 4D printing using smart materials in the field of manufacturing, Advanced Industrial and Engineering Polymer Research 4 (2021) 301–311.
[128] M. Maniruzzaman, 3D and 4D Printing in Biomedical Applications: Process Engineering and Additive Manufacturing, John Wiley & Sons, 2019.
[129] F. Pahlevanzadeh, R. Emadi, A. Valiani, M. Kharaziha, S.A. Poursamar, H.R. Bakhsheshi-Rad, et al., Three-dimensional printing constructs based on the chitosan for tissue regeneration: state of the art, developing directions and prospect trends, Materials (Basel)., 13, 2020, p. 2663.
[130] A. Indurkar, A. Pandit, R. Jain, P. Dandekar, Plant-based biomaterials in tissue engineering, Bioprinting 21 (2021) e00127.
[131] E.S. Ko, C. Kim, Y. Choi, K.Y. Lee, 3D printing of self-healing ferrogel prepared from glycol chitosan, oxidized hyaluronate, and iron oxide nanoparticles, Carbohydrate Polymers 245 (2020) 116496.
[132] Y. Bozkurt, E. Karayel, 3D printing technology; methods, biomedical applications, future opportunities and trends, Journal of Materials Research and Technology 14 (2021) 1430–1450.
[133] A. Mitchell, U. Lafont, M. Hołyńska, C. Semprimoschnig, Additive manufacturing—a review of 4D printing and future applications, Additive Manufacturing 24 (2018) 606–626.
[134] Z. Fu, L. Ouyang, R. Xu, Y. Yang, W. Sun, Responsive biomaterials for 3D bioprinting: a review, Materials Today (2022).
[135] P. Rastogi, B. Kandasubramanian, Review of alginate-based hydrogel bioprinting for application in tissue engineering, Biofabrication 11 (2019) 42001.
[136] R.A. Raus, W.M.F.W. Nawawi, R.R. Nasaruddin, Alginate and alginate composites for biomedical applications, Asian Journal of Pharmaceutical Sciences 16 (2021) 280–306.
[137] S.E. Bakarich, R. Gorkin III, M. In Het Panhuis, G.M. Spinks, 4D printing with mechanically robust, thermally actuating hydrogels, Macromolecular Rapid Communications 36 (2015) 1211–1217.
[138] S. Tandon, B. Kandasubramanian, S.M. Ibrahim, Silk-based composite scaffolds for tissue engineering applications, Industrial & Engineering Chemistry Research 59 (2020) 17593–17611.
[139] Y. He, F. Wang, X. Wang, J. Zhang, D. Wang, X. Huang, A photocurable hybrid chitosan/acrylamide bioink for DLP based 3D bioprinting, Materials & Design 202 (2021) 109588.
[140] M. Rajabi, M. McConnell, J. Cabral, M.A. Ali, Chitosan hydrogels in 3D printing for biomedical applications, Carbohydrate Polymers 260 (2021) 117768.

[141] M.Y. Khalid, Z.U. Arif, W. Ahmed, H. Arshad, Recent trends in recycling and reusing techniques of different plastic polymers and their composite materials, Sustainable Materials and Technologies (2021) e00382.

[142] Z. Wan, P. Zhang, Y. Liu, L. Lv, Y. Zhou, Four-dimensional bioprinting: current developments and applications in bone tissue engineering, Acta Biomaterialia 101 (2020) 26–42.

[143] C.A. Spiegel, M. Hackner, V.P. Bothe, J.P. Spatz, E. Blasco, 4D printing of shape memory polymers: from macro to micro, Advanced Functional Materials (2022) 2110580.

[144] D.A. Taylor, L.C. Sampaio, Z. Ferdous, A.S. Gobin, L.J. Taite, Decellularized matrices in regenerative medicine, Acta Biomaterialia 74 (2018) 74–89.

[145] C.-T. Kao, C.-C. Lin, Y.-W. Chen, C.-H. Yeh, H.-Y. Fang, M.-Y. Shie, Poly (dopamine) coating of 3D printed poly (lactic acid) scaffolds for bone tissue engineering, Materials Science and Engineering: C 56 (2015) 165–173.

[146] J. Kundu, J. Shim, J. Jang, S. Kim, D. Cho, An additive manufacturing-based PCL–alginate–chondrocyte bioprinted scaffold for cartilage tissue engineering, Journal of Tissue Engineering and Regenerative Medicine 9 (2015) 1286–1297.

[147] J.-H. Kim, J.W. Lee, W.-S. Yun, Fabrication and tissue engineering application of a 3D PPF/DEF scaffold using Blu-ray based 3D printing system, Journal of Mechanical Science and Technology 31 (2017) 2581–2587.

[148] B. Sun, Z. He, Q. Hou, Z. Liu, R. Cha, Y. Ni, Interaction of a spirooxazine dye with latex and its photochromic efficiency on cellulosic paper, Carbohydrate Polymers 95 (2013) 598–605.

[149] W. Li, L. Liu, H. Tian, X. Luo, S. Liu, Encapsulation of *Lactobacillus plantarum* in cellulose based microgel with controlled release behavior and increased long-term storage stability, Carbohydrate Polymers 223 (2019) 115065.

[150] N. Riahi-Noori, R. Sarraf-Mamoory, P. Alizadeh, A. Mehdikhani, Synthesis of ZnO nano powder by a gel combustion method, Journal of Ceramic Processing Research 9 (2008) 246–249.

[151] J. Chen, C. Glaus, R. Laforest, Q. Zhang, M. Yang, M. Gidding, et al., Gold nanocages as photothermal transducers for cancer treatment, Small 6 (2010) 811–817.

[152] D. Li, X. Xu, Z. Li, T. Wang, C. Wang, Detection methods of ammonia nitrogen in water: a review, TrAC Trends in Analytical Chemistry (2020) 115890.

[153] A.H. Moghaddam, M. Zare, Neuroprotective effect of hesperetin and nano-hesperetin on recognition memory impairment and the elevated oxygen stress in rat model of Alzheimer's disease, Biomedicine & Pharmacotherapy 97 (2018) 1096–1101.

[154] T. Xu, J.M. Miszuk, Z. Yong, S. Hongli, F. Hao, Electrospun polycaprolactone 3D nanofibrous scaffold with interconnected and hierarchically structured pores for bone tissue engineering, Advanced Healthcare Materials 4 (2015) 2238–2246.

[155] S.G. HossieKazarian, Polymer processing with supercritical fluids, Polymer Science Series CC/C of Vysokomolekuliarnye Soedineniia P42 (1) (2000) 78–101.

[156] J.L. Kendall, D.A. Canelas, J.L. Young, J.M. DeSimone, Polymerizations in supercritical carbon dioxide, Chemical Reviews 99 (1999) 543–564.

[157] D.J. Mooney, D.F. Baldwin, N.P. Suh, J.P. Vacanti, R. Langer, Novel approach to fabricate porous sponges of poly (D, L-lactic-co-glycolic acid) without the use of organic solvents, Biomaterials 17 (14) (1996) 1417–1422.

[158] P. Gentile, V. Chiono, I. Carmagnola, P.V. Hatton, An overview of poly (lactic-co-glycolic) acid (PLGA)-based biomaterials for bone tissue engineering, International Journal of Molecular Sciences 15 (3) (2014) 3640–3659.

[159] R.P. Pawar, S.U. Tekale, S.U. Shisodia, J.T. Totre, A.J. Domb, Biomedical applications of poly (lactic acid)., Recent Patents on Regenerative Medicine 4 (1) (2014) 40–51.

[160] S. Sharma, D. Srivastava, S. Grover, V. Sharma, Biomaterials in tooth tissue engineering: a review, Journal of Clinical and Diagnostic Research 8 (1) (2014) 309.
[161] L. Singh, V. Kumar, B.D. Ratner, Generation of porous microcellular 85/15 poly (DL-lactide-co-glycolide) foams for biomedical applications, Biomaterials 25 (13) (2004) 2611–2617.
[162] Owen R. Davies, Andrew L. Lewis, Martin J. Whitaker, Hongyun Tai, Kevin M. Shakesheff, Steven M. Howdle, Applications of supercritical CO_2 in the fabrication of polymer systems for drug delivery and tissue engineering, Advanced Drug Delivery Reviews 60 (3) (2008) 373–387.
[163] H. Nie, L.Y. Lee, H. Tong, C.H. Wang, PLGA/chitosan composites from a combination of spray drying and supercritical fluid foaming techniques: new carriers for DNA delivery, Journal of Controlled Release 129 (3) (2008) 207–214.
[164] B.S. Kim, D.J. Mooney, et al., Development of biocompatible synthetic extracellular matrices for tissue engineering, Trends in Biotechnology 16 (5) (1998) 224–230.
[165] M.P. Nikolova, M.S. Chavali, Recent advances in biomaterials for 3D scaffolds: A review, Bioactive Materials 4 (2019) 271–292.
[166] X. Liao, H. Zhang, T. He, Preparation of porous biodegradable polymer and its nanocomposites by supercritical CO_2 foaming for tissue engineering, Journal of Nanomaterials (2012)(2012).
[167] X. Liao, A.V. Nawaby, P. Whitfield, M. Day, M. Champagne, J. Denault, Layered open pore poly (L-lactic acid) nanomorphology, Biomacromolecules 7 (2006) 2937–2941.
[168] M. Mathieu, T. Mueller, P-E. Bourban, D. Pioletti, R. Müller, J-A Månson, Architecture and properties of anisotropic polymer compositescaffolds for bone tissue engineering, Biomaterials 27 (2006) 905–916.
[169] A. Rezaei, M.R. Mohammadi, Development of hydroxyapatite nanorods-polycaprolactone composites and scaffolds derived from a novel in-situ sol-gel process, Tissue Engineering and Regenerative Medicine 9 (6) (2012) 295–303.
[170] V.G. DileepKumar, M.S. Sridhar, P. Aramwit, V.K. Krut'ko, O.N. Musskaya, I.E. Glazov, N. Reddy, A review on the synthesis and properties of hydroxyapatite for biomedical applications, Journal of Biomaterials Science, Polymer Edition 33 (2) (2022) 229–261.
[171] M.G. Raucci, V. D'Antò, V. Guarinoa, E. Sardella, S. Zeppetelli, P. Favia, L. Ambrosio, Biomineralized porous composite scaffolds prepared by chemical synthesis for bone tissue regeneration, Acta Biomaterialia 6 (10) (2010) 4090–4099.
[172] S. Murugan, S.R. Parcha, Fabrication techniques involved in developing the composite scaffolds PCL/HA nanoparticles for bone tissue engineering applications, Journal of Materials Science: Materials in Medicine 32 (8) (2021) 1–10.
[173] A. Polini, D. Pisignano, M. Parodi, R. Quarto, S. Scaglione, Osteoinduction of human mesenchymal stem cells by bioactive composite scaffolds without supplemental osteogenic growth factors, PloS One 6 (10) (2011) e26211–1868.
[174] Daniel O. Costa, S. Jeffrey Dixon, Amin S. Rizkalla, One- and Three-Dimensional growth of hydroxyapatite nanowires during sol-gel-hydrothermal Synthesis, ACS Appl Mater Interfaces 4 (2012) 1490–9.
[175] P. Fabbri, F. Bondioli, M. Messori, C. Bartoli, D. Dinucci, F. Chiellini, Porous scaffolds of polycaprolactone reinforced with in situ generated hydroxyapatite for bone tissue engineering, Journal of Materials Science: Materials in Medicine 21 (2010) 343–351.
[176] V. Guarino, F. Causa, P.A. Netti, G. Ciapetti, S. Pagani, D. Martini, L. Ambrosio, The role of hydroxyapatite as solid signal on performance of PCL porous scaffolds for bone tissue regeneration, Journal of Biomedical Materials Research Part B: Applied Biomaterials: An Official Journal of The Society for Biomaterials, The Japanese Society for Biomaterials, and The Australian Society for Biomaterials and the Korean Society for Biomaterials 86 (2) (2008) 548–557.
[177] J. Venugopal, S. Low, A.T. Choon, S. Ramakrishna, Interaction of cells and nanofiber scaffolds in tissue engineering, Journal of Biomedical Materials Research Part B: Applied Biomaterials, 84 1 (2008) 34–48.

[178] N. Bloise, G. Ceccarelli, P. Minzioni, M. Vercellino, L. Benedetti, M.G.C. De Angelis, L. Visai, Investigation of low-level laser therapy potentiality on proliferation and differentiation of human osteoblast-like cells in the absence/presence of osteogenic factors, Journal of Biomedical Optics 18 (12) (2013) 128006.
[179] N. Ramesh, S.C. Moratti, G.J. Dias, Hydroxyapatite–polymer biocomposites for bone regeneration: A review of current trends, Journal of Biomedical Materials Research Part B: Applied Biomaterials 106 (5) (2018) 2046–2057.
[180] Y.S. Nam, T.G. Park, Porous biodegradable polymeric scaffolds prepared by thermally induced phase separation, Journal of Biomedical Materials Research: An Official Journal of The Society for Biomaterials, The Japanese Society for Biomaterials, and The Australian Society for Biomaterials and the Korean Society for Biomaterials 4147 (1) (1999) 8–17.
[181] S.I. Roohani-Esfahani, S. Nouri-Khorasani, Z. Lu, R. Appleyard, H. Zreiqat, The influence hydroxyapatite nanoparticle shape and size on the properties of biphasic calcium phosphate scaffolds coated with hydroxyapatite–PCL composites, Biomaterials 31 (21) (2010) 5498–5509.
[182] G. Guan, L. Bai, B. Zuo, M. Li, Z. Wu, Y. Li, L. Wang, Promoted dermis healing from full-thickness skin defect by porous silk fibroin scaffolds (PSFSs), Bio-Medical Materials and Engineering 20 (5) (2010) 295–308.
[183] N. Sultana, T. Hayat Khan, Polycaprolactone scaffolds and hydroxyapatite/polycaprolactone composite scaffolds for bone tissue engineering, Journal of Bionanoscience 7 (2013) 169–173.

CHAPTER

10

Biodegradable polymer nanocomposites as electrode materials for electrochemical double-layer capacitors and hybrid supercapacitor applications

Vraj Shah[1], Jaydip Bhaliya[1], Gautam M. Patel[1], Meera Popaliya[2], Arvnabh Mishra[2] and Parimal R. Patel[3]

[1]*Department of Applied Chemistry, School of Science, ITM SLS Baroda University, Vadodara, Gujarat, India* [2]*Department of Industrial Chemistry, Institute of Science & Technology for Advanced Studies & Research (ISTAR), CVM University, V.V. Nagar, Gujarat, India* [3]*Department of Chemistry, Government Science College, Pardi, Valsad, Gujarat, India*

Chapter Outline

Biodegradable and Biocompatible Polymer Nanocomposites. DOI: https://doi.org/10.1016/B978-0-323-91696-7.00019-2
Copyright © 2023 Elsevier Inc. All rights reserved.

10.1 Introduction

Due to the rapid increase in global energy demand and the environmental impact of traditional energy sources, research on clean and renewable energy sources has exploded over the last decade [1]. Indeed, energy technologies such as fuel cells, normal capacitors, batteries, and supercapacitors (SCs) have been widely used as energy-storage devices, all of which operate on the electrochemical energy conversion principle. SCs have received a lot of attention recently because of their properties such as long life cycle, high capacitance (Sc), high power density (P_d), quick charging, and discharge time. Due to the rapid increase in global energy demand and the environmental impact of traditional energy sources, research on clean and renewable energy sources has exploded over the last decade [2]. Furthermore, supercapacitors are advantageous for use in locations where the establishment of an electrical power station is not feasible, allowing the large budget required for the establishment of the power station to be avoided. Furthermore, because SCs are compact, flexible, and lightweight, they are used as a power supply for small portable devices such as cell phones, laptops, CPU, notepads, digitally operated cameras, and so on [3]. Although supercapacitors have numerous applications, their low energy densities in comparison to batteries have limited their use. For example, commercially available supercapacitors have energy densities of fewer than 10 Wh/kg, but batteries made up of Li-ion have densities greater than 180 Wh/kg. Nonetheless, supercapacitors outperform Li-ion batteries in terms of cycle life and power density [4]. Fig. 10.1 shows a comparison of various energy-storage devices, and it is clear that supercapacitors have been used as bridges between traditional batteries and capacitors, and they have higher power densities than other storage devices, but the energy density is still low [5]. As a result, in response to the ever-increasing demand for energy, researchers are looking for novel materials and technologies to develop vastly improved energy-storage supercapacitors [6].

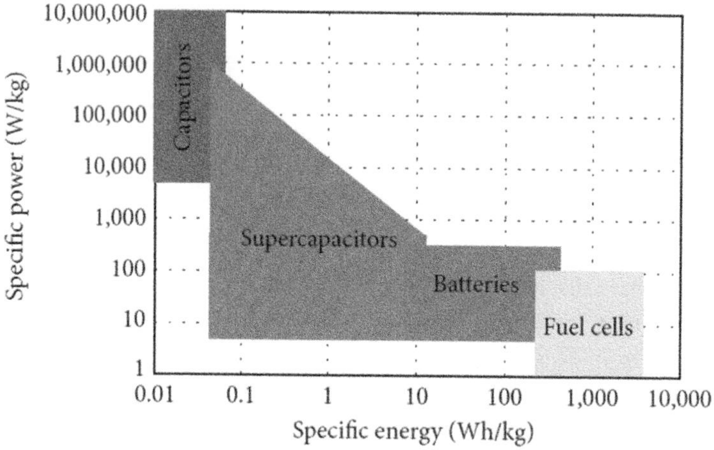

FIGURE 10.1

Various energy-storage systems are plotted on a ragone plot [6,7].

Reproduced with permission Hindawi. Copyright 2016.

Before we get into the most recent advances in supercapacitors, let's take a look at the history of the supercapacitor. In 1876, Fitzgerald et al. invented the dielectric capacitor, which was made of wax-impregnated paper with foil electrodes. Clarles Pollak invented the borax electrolyte aluminum electrolytic capacitor in 1897 [8]. In 1957, H. Becker developed a porous carbon electrode as a low voltage electrolytic capacitor [9]. In 1970, Donald L. Boos invented a carbon-activated electrolytic capacitor. In 1978, Nippon Electric Corporation commercialized Standard Oil Company of Ohio's patented double-layer capacitor concepts [10,11]. Furthermore, David A Evans created an electrolytic-hybrid electrochemical capacitor in 1994 by using the anode of a tantalum electrolytic capacitor with a voltage rating of 200 volts. Brian Evans Conway coined the term "supercapacitor" after discovering increased capacitance through surface redox processes using faradic charge transfer between ions and electrodes in 1994 [12]. VAG, a German public transportation company, was the first to use supercapacitors in the 21 century when they tested a bus with a diesel-electric drive system and supercapacitors [13].

The supercapacitor's energy storage is determined by the oxidation–reduction reaction or charge process. Based on storage criteria, supercapacitors are divided into three primary groups: EDLC, pseudocapacitor (PC), and hybrid supercapacitor. The hybrid supercapacitor, which is a cross between an EDLC and a PC, outperforms the combined components. In EDLC, energy is stored in the core area of the shell plus the length of the atomic charge partition [14]. In a PC, energy is stored by quick reproducible reduction–oxidation reactions between electro-active units that are placed on top of an electrode's active material and an electrolyte solution. The hybrid supercapacitor energy-storage technique is a combination of these two storage techniques [15]. One half of the hybrid supercapacitor functions as an EDLC, while the other half functions as a PC. In comparison to the standard EDLC and PC, the hybrid supercapacitor typically has a higher energy density and power density. As a result, they are better suited for use in energy-efficient systems than other energy-storing devices [16].

Furthermore, the primary components of supercapacitors include anode and cathode as electrodes, electrolytes, a current collector, sealants, and a separator, all of which contribute to the electrochemical performance of the devices. Because electrodes must have a large definite surface area, corrosion resistance, conductivity, and thermal and chemical stability, the materials used for electrodes, as well as their layout, are critical in improving the capacitance performance of supercapacitors. They must also be reasonably priced and environmentally friendly. They must also be able to transmit the faradic charge to increase the total capacitance value [17]. Furthermore, carbon-based materials such as carbon nanotubes (CNTs), graphene, and activated carbons (ACs). Transition-metal oxides (TMOs) are another type of material that contains MnO_2, RuO_2, NiO, and other elements. Third, conducting polymers such as polyaniline (PANI), polypyrrole (PPy), and polythiophene (PTh) are the three main categories of electrode materials for supercapacitors [18]. Traditional capacitor electrode materials are stiff and large, making them unsuitable for future applications. Multifunctional consumer electronics necessitate reedier, agiler, and visible SCs with a wide range of innovative types and functionalities. ACs and TMOs are the most commonly used electrode materials. ACs have amorphous morphologies, feebly graphene sheets frameworks, and broad pores, whereas TMOs have low electrical conductivity and cannot handle high rate energy-storage settings [19].

On the contrary, the lowest energy density (E_d) of SCs poses a significant challenge. Because the capacitance of the SC is proportional to the energy density and the Cs is the voltage squared.

This can be accomplished by using high-Cs electrode materials, electrolytes with large potential windows, and optimizing the topology of the integrated system. Although the production of independent SC units (such as electrode and electrolyte materials) is relatively simple, interoperability between the structure as well as pore size of the electrode material and the electrolyte ion size is critical for promoting their combined effect. A study of the pore size has an influence on carbide-derived carbon material specific capacitance, for example, demonstrated that increasing the hole size less than 2 nm can result in increased capacitance and thus better energy density, but increasing the pore size larger than 2 nm resulted in a fall in specific capacitance [20]. Furthermore, the solution used as an electrolyte is critical for properties such as power density, temperature range, and conductivity. Aside from that, the electrolyte for supercapacitors should have the following properties: a high potential window (PW), optimal ionic liquid concentration, high electrochemical stability, low equivalent series resistance (ESR), volatility, toxicity, low viscosity, small radius of ion, and low cost [21]. Ionic liquids (ILs) and liquid, polymeric, oxidative-reductive, crystalline, or quasi electrolytes are the electrolytes used in supercapacitors [22].

Aside from that, researchers have focused more on nanoscale materials to overcome the major challenges associated with conventional electrode materials. They are also widely used as an electrode in SC to improve electrochemical performance. Furthermore, nanomaterials are available in a variety of morphologies, such as from (0D to 3D). Nanomaterials have properties such as a high surface-to-volume ratio and surface energy, mechanical strength and microhardness, high specific heat, large thermal expansion, high catalytic activity, self-diffusion enhancement, high magnetic susceptibility, and a fast sintering rate. All of the nanomaterials' properties show a significant increase in capacitor capacitance [7,23–28]. Polymeric nanocomposites/nanomaterials (PNCs) are gaining global attention as an emerging technology in a variety of industries. First, polymers are great because of their simple production process, corrosion resistance, and low density; however, their strength and stiffness are far less than structural uses. Polymers can be used for structural uses metrics due to their low weight, ease of long-term stability, and processing [29].

Scientists are developing biodegradable electronic devices that are gaining popularity for use in gadgets, crop monitoring, healthcare, and environmental sensing to meet the requirements of environmentally friendly nature and sustainable development. Traditional polymer composites have issues such as low electrical conductivity, leakage, and incompatibility with the environment [30]. Biodegradable polymeric nanocomposites can be used to create an electrode that does not leak and is also suitable for solicitations in micro-, soft-, and deformable electronics [31]. Chemical bonds between distinct polymeric chains (chemical gels) can be used to create a three-dimensional polymeric matrix, though low-strength interchain contacts (physical gels), such as dipole-dipole or noncovalent interactions, can also be used. Furthermore, this matrix could be used for the insertion of nanomaterial to produce polymeric nanocomposites [32]. Furthermore, the use of biodegradable polymeric nanomaterials in electrodes allows for higher E_d, higher P_d, longer cycle life, improved security, and lower initial costs [33]. Despite this, the list of polymers that showcase the property of biodegradability has been depicted in Fig. 10.2 [34].

Various articles about supercapacitors have been published up to this point, but no brief chapter about biodegradable polymer nanocomposites as electrodes for supercapacitors applications has been published. As the global market for supercapacitors has exploded [36]. The demand for SCs appears to be high. As a result, the goal of this chapter is to provide a thorough investigation of various biodegradable polymer nanocomposites used as electrodes for EDLC/hybrid

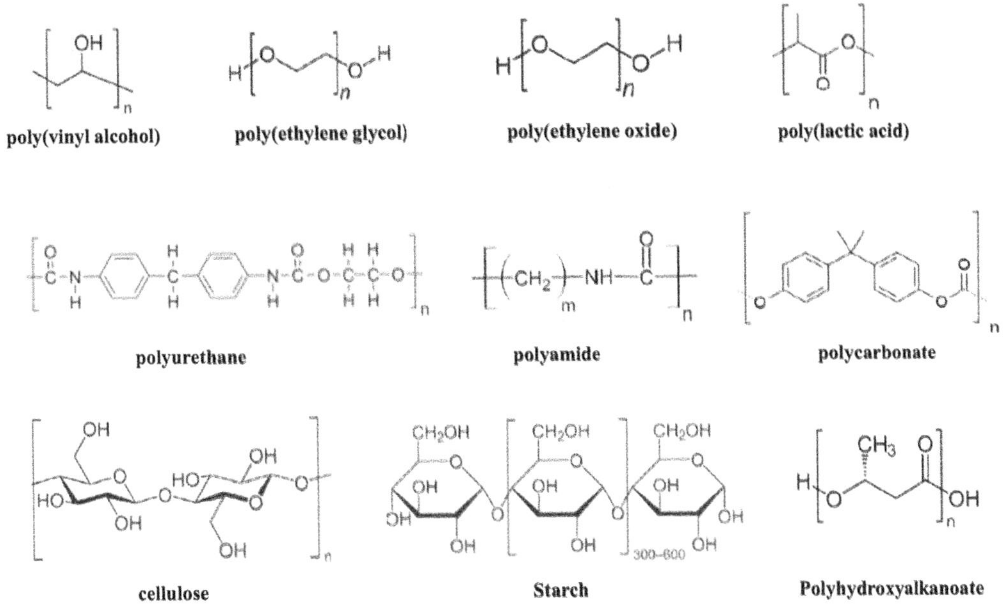

FIGURE 10.2

Examples of green and biodegradable polymers [34,35].

Reproduced with permission from MDPI. Copyright 2021.

supercapacitors, their mechanism, synthesis techniques, various examples, and some potential solutions for improving existing supercapacitors. Fig. 10.3 shows the overview of the chapter.

10.2 Classification and storage mechanism of supercapacitors

Nowadays, the market offers a wide range of supercapacitors based on their electrode materials and operating specifications. Fig. 10.4 depicts the taxonomy of supercapacitors classifications [5]. The development of supercapacitors has been accelerated as the transition from conventional capacitors (low voltage) to high energy supercapacitors occurs (high voltage) [37]. Let us first examine the mechanism of a standard capacitor before delving into the mechanism of a supercapacitor. A capacitor is an electrical component that can store a charge as an electrical potential. A capacitor is comprised of two parallel plates that are separated by a few millimeters. Capacitance is the amount of charge held in a capacitor and is determined by the size of the capacitor. When you connect a capacitor across the dielectric, $+Q$ (positive charge) accumulates on one plate and $-Q$ (negative charge) accumulates on the other. The capacitance changes as the size and materials used in the capacitor change ($C = A/d$) [38].

When compared to capacitors, supercapacitors (SC) have a much higher capacitance value and energy density. When confronted with electrochemical activity, the configuration of supercapacitors

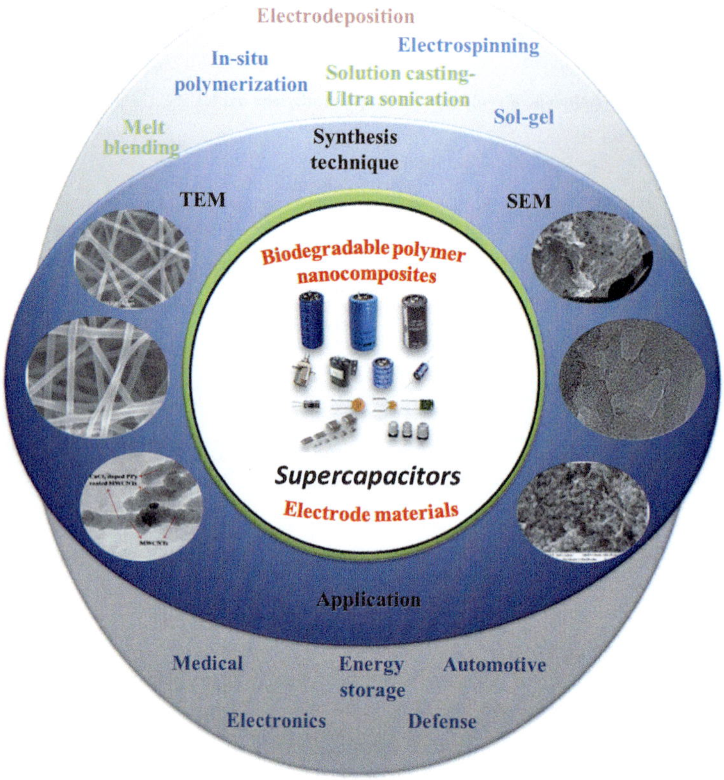

FIGURE 10.3

General overview of the described chapter.

is similar to that of a battery. A supercapacitor consists of two electrodes separated by an electrolyte-soaked separator. As illustrated in Fig. 10.5, the basic supercapacitor assembly consists of one cathode and one anode, electrolyte solutions, a current collector, and a separator partition between two electrodes. These elements serve as the foundation for supercapacitor properties [39].

A supercapacitor must have two porous electrodes and a membrane that separates positive and negative plated to function. This membrane is referred to as a separator. An electrolyte, which is an ionic liquid, connects the electrodes electrically. When a voltage is applied to the + Ve electrode, it attracts − Ve ions from the electrolyte. When a negative electrode is charged, positive ions are drawn from the electrolyte. These ions are deposited on the electrode surfaces, reducing the distance between them; as the distance between the electrodes decreases, the capacitance increases rapidly [40]. If a comparison is done for the storage capacity between the traditional capacitor and supercapacitor, the range exhibited by the capacitor is micro to millifarads whereas 100–1000 F, reveals by supercapacitors [41]. The EDL capacitance caused by Coulombian charge absorption near the electrode-electrolyte border, Pseudocapacitance caused by surface redox processes that are

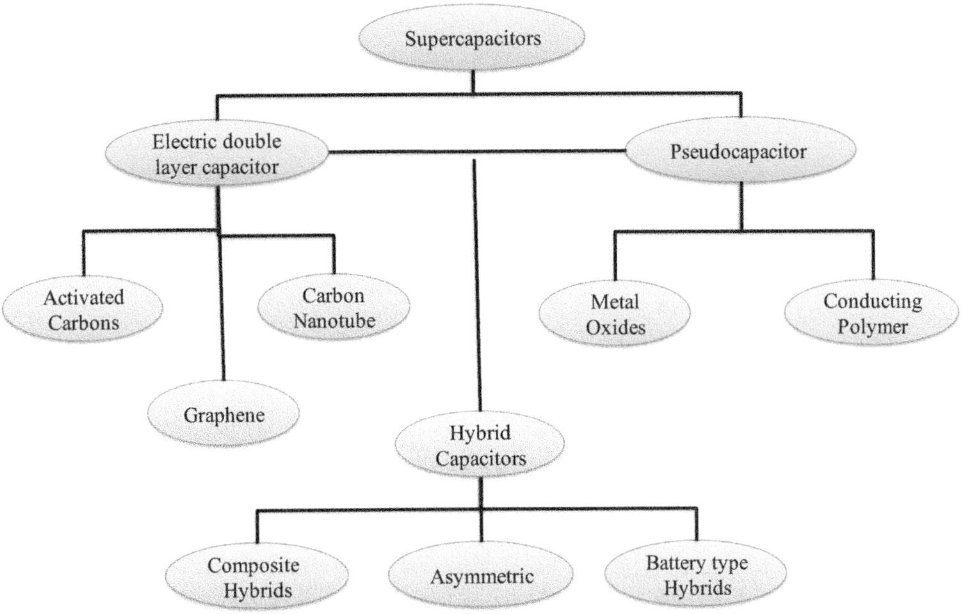

FIGURE 10.4

Types and classification of supercapacitors [5].

Reproduced with permission from Elsevier. Copyright 2018.

connected to their respective potential, and the hybrid supercapacitors storage concept are all associated with the supercapacitors mechanism [5]. Let us also look at the individual mechanisms of (1) electric-double-layer supercapacitors (EDLCs), (2) pseudosupercapacitors, and (3) hybrid supercapacitors.

10.2.1 Electric-double-layer supercapacitors

EDLCs are electrochemical materials in which charge storage occurs at the double-layer's electrode& electrolyte boundary. In the construction of EDLCs, two electrodes are used; one is the cathode (+Ve) and the other is the anode (−Ve), both of which are separated by a separator, which is usually a permeable membrane. Within the system, the electrolyte is filled [42,43]. As previously stated, charge is stored at the interface, so no charge transfer between electrode and electrolyte occurs, so the electrolyte remains constant, and charge storage occurs electrostatically or via a non-faradic process. When a voltage is applied, charge agglomeration occurs on the electrode surfaces. Similarly, the charge carried by an electrolyte's ions travels to their respective ion via the pores of a separator. If the surface has a negative charge, all negative ions attract to it, even if they are present at another interface, and vice versa if the surface has a positive charge. The layers generate as a result of this process, and the cycle continues until the time of optimum charge storage. Because ion recombination is not possible, EDLC is produced. The EDL increases the surface area of the

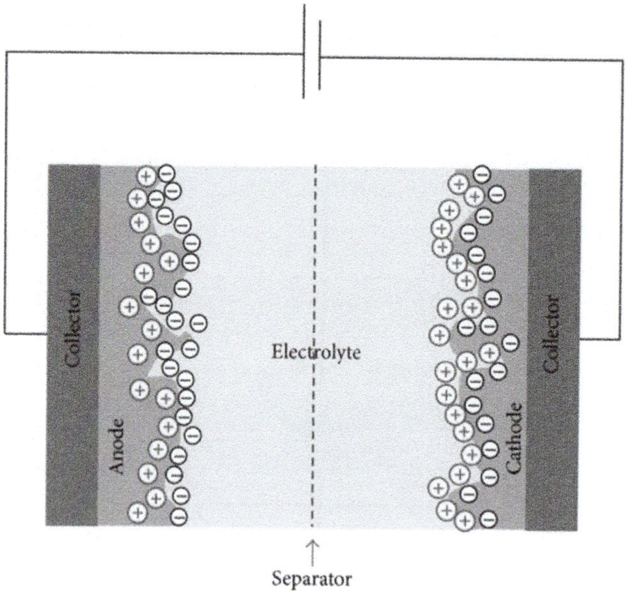

FIGURE 10.5

Construction and working of supercapacitor [7].

Reproduced with permission Hindawi. Copyright 2016.

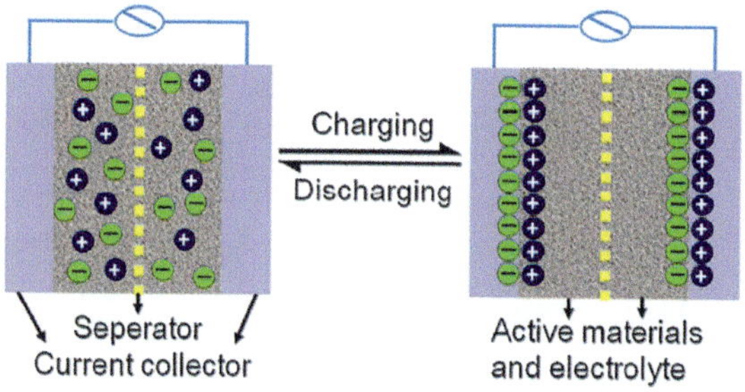

FIGURE 10.6

The schematic illustration of the charging and discharging process in EDLCs [47].

Reproduced with permission Royal Society of Chemistry. Copyright 2014.

electrode while decreasing the distance between the two electrodes. As a result, greater capacitance is obtained [44–46]. Furthermore, the charging and discharging process in EDLC is presented in Fig. 10.6 [47]. Also, the capacitance of the EDLCs is described in the Eq. (10.1) [48].

$$C = \frac{A \times \epsilon_0}{d} \tag{10.1}$$

In Eq. (10.1) C denotes the capacitance of the capacitor, measured in F, A denotes the surface area, E_0 denotes the permissiveness of open space, and d denotes the efficient thickness of the EDL or Debye length [48]. Some advantages of EDLCs include higher E_d than conventional capacitors, no change in chemical or structural composition due to the nonfaradic process, the ability to operate at high temperatures, and extremely high cycling stability. However, because of the limited surface area and the pore size of the electrode material, E_d is not as high as that produced by a battery. As a result, high capacitance may be overlooked [49,50]. To improve this, a material with a large surface area and pore size must be chosen or manufactured. In 2006, Kogasa and his colleagues discovered that the pore size of a material plays an important role in determining capacitance. They discovered that when the pore size is less than 1nm, the capacitance increases dramatically. This is because the pore size solvation shell of the ions reduces effective ion size and thus increases the number of ions that can be assaulted per surface area, so the normalized capacitance per surface area or the number of ions assaulted per surface area increases [51]. The use of biodegradable polymeric nanocomposites increases the surface area, which increases capacitance; an in-depth explanation of this has been provided later in this chapter with various examples.

10.2.2 Pseudocapacitors

Charge is stored in PCs via a faradaically process between electrolyte and electrode. The operation can be carried out using the reduction–oxidation reaction, the intercalation process, or under potential deposition. Because of these Faradaic processes, PCs may be able to achieve higher capacitances and energy densities than EDLCs. Furthermore, it includes a reduction–oxidation cycle, just like batteries, so it has a lack of cycling consistency as well as a lower power density [52–55]. PCs are constructed similarly to EDLCs, with the exception that the electrodes are made of metal oxides and polymers that transmit electric current. Fig. 10.7 shows the construction and operation of a personal computer [48,56]. Under potential deposition occurs when metal ions from an adsorbed monolayer are significantly above their redox potential at the surface of a different metal. Fig. 10.8A depicts a classic example of under potential lead deposition on a gold electrode's edge. Pseudocapacitance occurs in an oxidation–reduction reaction when electrochemically immobilized ions immobilize onto or around a material's surface, resulting in a Faradaic charge allocation (see Fig. 10.8B). Furthermore, as shown in Fig. 10.8C, pseudocapacitance occurs during the intercalation process as ions easily diffuse into the passages or layers of a redox-active structure, which is associated with a Faradaic charge transfer without changing the phase of the crystalline structure [56].

The pseudo electrodes are primarily composed of metal oxides, metal-doped carbon, and conductive polymers. As per Eq. (10.2), the charge derivative uptake (Δq) and altering potential (ΔV) yields the pseudocapacitance (C). The reaction that occurs during the process of charge storage is described in the reaction below.

$$C = \frac{d(\Delta q)}{d(\Delta V)} \tag{10.2}$$

$$O_{ad} + ne^{-} \rightarrow R_{ed}.$$

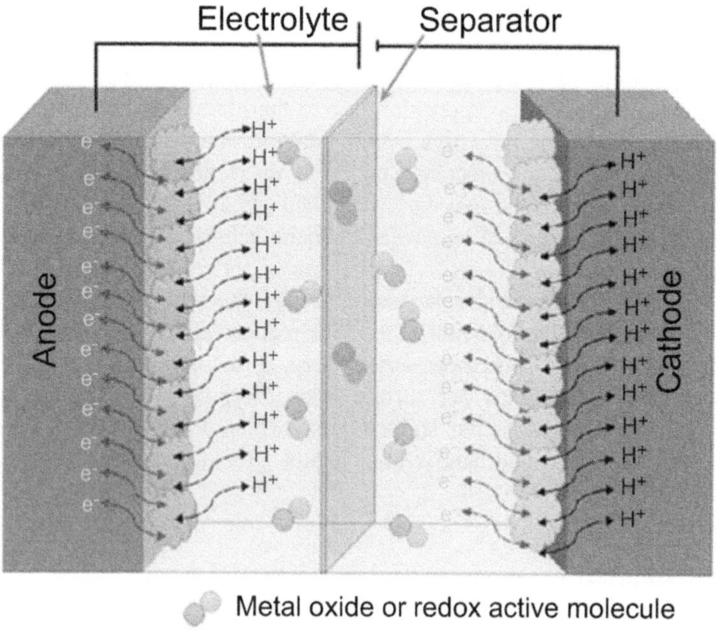

FIGURE 10.7

A diagrammatic depiction of construction and working of PCs [48].

Reprint from National Science Review. Copyright 2017.

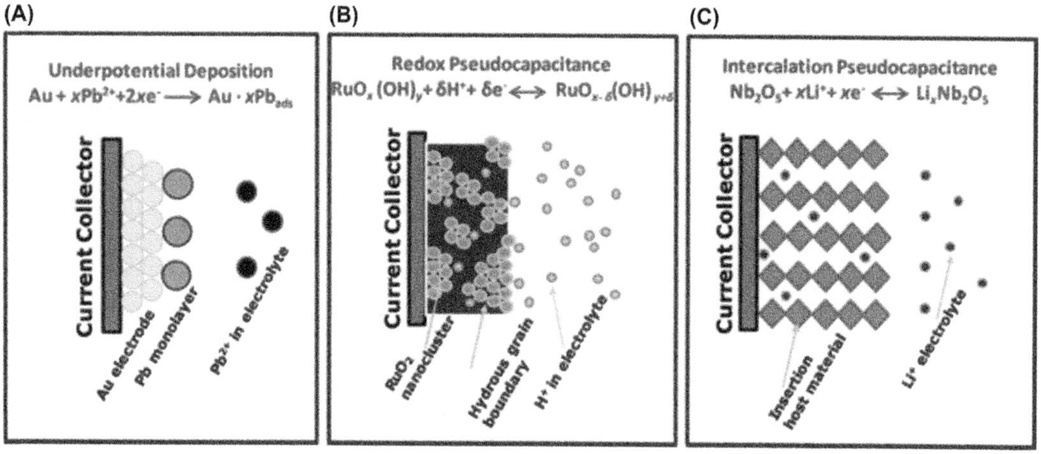

FIGURE 10.8

Typical reversible Faradaic processes that induce pseudocapacitance (A) under potential accumulation, (B) redox pseudocapacitance, (C) intercalation pseudocapacitance [56].

Reproduced with permission Elsevier. Copyright 2019.

PCs have few advantages and disadvantages. When hydrous ruthenium oxide doping provided excellent capacitance, E_d is greater in PCs than in EDLCs. Aside from that, the disadvantages are as follows: P_d is much less expensive than EDLCs, cycle life is limited due to the redox reaction, and negatively charged conducting polymers are inefficient; the finest metal-oxide electrodes are generally relatively expensive and require aqueous electrolytes, resulting in lower voltage; and negatively charged conducting polymers are inefficient [57]. The demerits could be eliminated using polymeric nanocomposite material.

10.2.3 Hybrid supercapacitors

The hybrid supercapacitor storage concept is guided by a combination of the principles of two supercapacitor types, PC and EDLC. Because the EDLC or PC is not capable of providing higher efficiency, their combination results in the merging components' restrictions being trumped, with the benefit of providing greater capacitance. Hybrid supercapacitor (HSC) is classified into two types based on the arrangement of the assembly: asymmetric and symmetric. Fig. 10.9 depicts a schematic representation of an HSC comprised of a lithium inclusion electrode and a carbon electrode [56]. The electrochemical behavior of a hybrid supercapacitor is superior to that of individual electrodes when two separate electrodes made of different materials are combined. Hybrid systems maintain cycle consistency and cost, but some specific variables limit the effectiveness of PCs [58].

Several + Ve and − Ve electrode amalgamations in liquid solution and inorganic electrolytes have previously been investigated. In general, the faradic electrode increases E_d at the expense of cyclic constancy, which is the primary disadvantage in comparison to EDLCs of hybrid devices; thus, avoiding converting an excellent SC into a regular battery is critical [59]. Canvassers are now focusing on three types of hybrid super capacitors, which can be distinguished by their electrode configuration, which includes battery type, asymmetric, and composite.

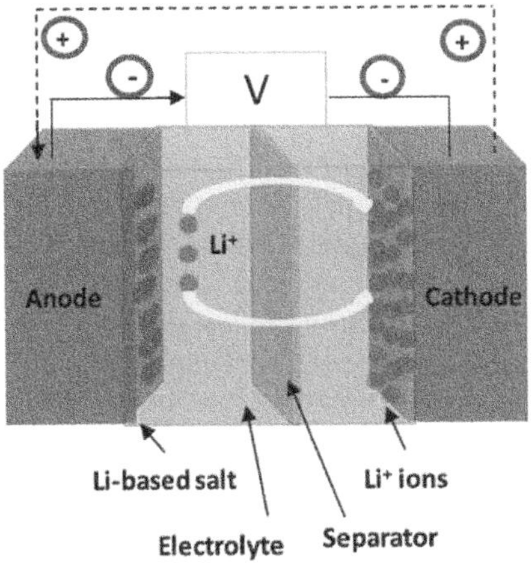

FIGURE 10.9

Schematic representation of hybrid supercapacitor [56].

Reproduced with permission Elsevier. Copyright 2019.

In a single electrode, carbon-based materials are combined with either metal oxide or conducting polymers, resulting in a single electrode that can store charge both physically and chemically. Carbon-based materials have a high capacitance and a large explicit surface area, allowing the electrolyte and the pseudocapacitive materials to make contact. By utilizing the Faradaic reaction, a pseudocapacitive material increases capacitance in composite electrodes [60]. Composites are currently divided into two types: binary composites and ternary composites. Binary composites use two different electrode materials, whereas ternary composites use three different anode and cathode materials to create a single electrode. Furthermore, by connecting EDLC to a PCs electrode, the asymmetric HSC consists of both faradic and nonfaradic progressions. They are designed in such a way that carbon serves as the − Ve electrode and a conducting metal oxide or polymer serves as the + Ve electrode [61]. Furthermore, battery type hybrids, like asymmetric hybrids, combine two separate electrodes; however, in this case, they are formed by combining a SC electrode with a battery electrode. This structure was created to take advantage of both supercapacitor and battery capabilities in a single cell [62].

10.3 Synthesis techniques of biodegradable polymer nanocomposites as an electrode

To produce material for specific applications, synthesis methods are extremely important. For example, in this case, the electrode materials must have a large surface area, a small pore size, a high density, and so on. The appropriate synthesis route can achieve these. Some methods for developing biodegradable polymeric nanocomposites have been discussed here. The methods listed are based on literature and surveys, but are not limited to them.

10.3.1 Solution casting-ultra sonication

This strategy is based on the law of stokes' principle. The polymer is first solubilized in a suitable solvent to form the polymeric matrix, and then the nanofillers or nanoparticles are thoroughly mixed into it using an ultrasonication process. The use of ultrasonic waves to generate energy greater than the interaction energy between nanofillers in aggregation, allowing small bundles to form, is referred to as ultrasonication in this context. As time passes, these bundles degrade into even smaller bundles, which exfoliate into a monodispersed phase in which all nanoparticles in the PNC are completely separated from one another. Allow the solvent to evaporate, leaving a thin layer of polymer nanocomposite behind. Furthermore, sonication time and power are critical for uniform distribution of nanofillers into the matrix. Bath sonicators and probe sonicators are two types of sonicators that may be used, and they provide 100 W to 100–150 W power, respectively [63–65]. Siva et al. developed $NiCo_2Fe_2O_4$ incorporated polyvinyl alcohol (PVA)/PVP nanocomposites for supercapacitor application using the solution casting method in 2022 [66].

10.3.2 In-situ polymerization

This polymer chemistry approach employs nanoparticles to create polymer nanocomposites "in the polymerization mixture." Several unstable oligomers (molecules) must be synthesized on-site for use in various processes (i.e., in the reaction mixture but not isolated on their own). The in-situ polymerization process begins with an initiation stage and then progresses through a series of polymerization stages, yielding a hybrid of nanoparticles and polymer molecules [67]. Initially, nanoparticles are dispersed in a liquid monomer or a low-molecular-weight precursor. After forming a homogeneous mixture, a suitable initiator is added, and the polymerization reaction is then subjected to heat, radiation, or another source of energy [68,69]. Following the completion of the polymerization procedure, a nanocomposite containing polymer molecules attached to nanoparticles is formed. Furthermore, certain requirements must be met, such as in-situ polymerization, which includes the use of low-viscosity prepolymers of precursor polymer molecules (typically less than 1 Pascal) for the formation of a polymer nanomaterial composite; short-term polymerization; the use of polymers with advantageous mechanical properties; and the absence of side products during polymerization [70]. In-situ polymerization has several advantages, including the use of low-cost ingredients, ease of automation, and the ability to combine with a variety of heating and curing procedures. However, several disadvantages of this method of preparation included a limited supply of usable ingredients, a shorter time to complete the reaction process, and the need for expensive equipment [70–72]. Tan et al. had developed Nano-Au@PANI core-shell nanoparticles for supercapacitor applications using in-situ polymerization [73].

10.3.3 Electrodeposition

The electrodeposition (EDP) method, which employs cathodic protection technology, is well-established and will improve further. The EDP process is the most dependable and cost effective of all application procedures because it is a simple, solvent-free method of coating intricate components. Its properties ensure that it will live for a long time. The monomer, doping agent (if necessary), and graphene oxide are electro-polymerized in an aqueous precursor solution to create a polymer/nanocomposite. The EDP cell typically consists of three electrodes: a working electrode for depositing the substrate, a counter electrode, and a platinum-based reference electrode. The working electrode is constructed from an electrically conductive material or substrate. When the required amount of charge is disseminated, the polymer is electrodeposited at a specific rate and stops. As a result, a composite layer forms on the conductive substance's surface [74,75]. Shahrokhian et al. created a V_2O_5/polypyrrole/graphene oxide nanocomposite for supercapacitor applications using an EDP technique [76].

10.3.4 Electrospinning

The electrospinning process results in the formation of a nonwoven network of micro- or nanofibers. The liquid solution is subjected to high-voltage electricity and a collector in this process, allowing the solution to eject from a nozzle and form a jet. During the drying process, the jet-shaped fibers formed and were then deposited in the collector. This method produces micro- or nanopolymer fibers quickly and easily. Inorganic fibers can also be created by electrospinning polymers mixed in a precursor solution [77,78]. Polymers can alter the viscosity, conductivity, surface tension, and other properties of the electrospinning solution, all of which are critical for this method. Furthermore, while this method may yield fibers with diameters ranging from 2 nm to more than 5 nm, most fiber sizes are in the tens to hundreds of nanometer range [79].

Electrospinning is a low-cost, easy-to-use method for producing nanofibers. Cooley was the first to patent electrospinning in 1900 [80,81]. Kalra et al. used electrospinning to create polyaniline and polyaniline-for supercapacitor applications [82].

10.3.5 Melt blending

When compared to solution mixing, the majority of industries use this method. There are two types of melt mixers used in this application: batch mixing and continuous mixing. Batch mixing is less effective because it cannot handle large capacities, but it is appropriate for laboratory processes. Furthermore, batch mixers manufactured by Brabender or Haake are referred to as internal mixers, whereas continuous mixers manufactured by Banbury or Moriyama are referred to as dispersion mixers [83]. Furthermore, this method can be used to make clay-polymer nanocomposites with either a thermoplastic or an elastomeric polymer matrix [84]. In general, an extruder or Banbury is used to melt the polymer with a certain amount of intercalated clay in the presence of noble gases such as argon, nitrogen, or neon. Furthermore, after being warmed in a mixer, the polymer was directly mixed with the intercalant, resulting in shear sufficient for the formation of required polymer nanocomposites. There are two types of melt intercalation: static and dynamic. Static melt intercalation occurs without mixing at temperatures 50°C above Tg, such as melting or glass transition, and is also known as melt annealing. Dynamic melt intercalation is another type of melt intercalation that is fairly common in extruders and internal mixers [85]. Deshmukh et al. had developed the ZnO/polymer nanocomposites through melt blending process [86].

10.3.6 Sol-gel method

Sol-gel is a simple method for producing materials with increased purity and uniformity. The sol-gel technique gets its name from the fact that, under controlled conditions, tiny particles in a solution (sol) clump together and connect to form a coherent matrix (gel). The colloidal approach and the polymeric or alkoxide approach appear to be two fundamental versions of the sol-gel approach that differ in the type of precursors used. The precursor is combined in a solution in both procedures and then activated by adding a basic or acidic solution. The activated reactant then reacts, forming a matrix that expands with temperature and time until it fills the container [87]. Sen et al. had developed the procedure of poly(3, 4-ethylenedioxythiophene)–$NiFe_2O_4$ nanocomposite by sol-gel method, for the material for electrode in SC [88].

10.4 Biodegradable polymer nanocomposites as electrode in supercapacitor; material synthesis method, characterization, and storage capacitance analysis

Contamination-free, long-term stability, and good energy-storing capabilities of SC with enhanced E_d without affecting P_d have recently been on the rise [5]. The efficient method for improving the creation of extremely porous or nanosized electrode materials is to increase the capacitance of the electrode materials [89]. In this regard, biodegradable polymer nanocomposites are the best choice for supercapacitor high performance in capacitance improvement as electrode materials. This section briefly discusses the recently reported biodegradable polymer nanocomposites as electrode materials, including their synthesis method, characterization, and high-performance storage capacitance capability in supercapacitors.

10.4.1 Electrode for supercapacitors

Polyaniline-based conducting polymer electrode materials have been widely used in energy-storage devices. Zin Htut et al. [90] create biodegradable polymer-modified graphene/polyaniline electrode materials for supercapacitors. As shown in Fig. 10.10A chitosan-modified graphene/polyaniline (CS-G/PANI) was synthesized via an in-situ polymerization process as a promising supercapacitor material. The composite material electrode has a high capacitance while preventing graphene aggregation and is a cost effective and simple method. Following the reduction procedure, the chitosan acts as a graphene oxide (GO) stabilizer. Its dispersion in water is good, so in the presence of the same polyaniline quantity, CS-G/PANI has higher PANI adsorption than graphene/PANI. Several researchers have discovered that HSCs made of graphene/PANI are effective in terms of capacitance [91–96], but Zin Htut et al. are the first to report on the improvement of electrochemical chitosan by in-situ polymerization of PANI using chitosan-stabilized graphene (CS-G) to synthesize CS-G/PANI for use in SCs. Because of the presence of two retelling entities formed by *N*-acetyl glucosamine (GlucNAc) and GlucN, the chitosan is a biodegradable and environmentally friendly polymer (glucosamine) [97].

Fig. 10.10B shows the morphological data of the composite material, it reveals that graphene are randomly aggregated with crumpled sheet. In addition, the transmission electron microscopic (TEM) image of the CS-G/PANI nanocomposite illustrated in Fig. 10.10C. The specific capacitive behavior of the CS-G/PANI also exanimate in a range of potential that is −0.2 to 0.8 V (Vs Ag/AgCl) at 10 mV/s rate of scan, as depicted in Fig. 10.10E. In comparison, CS-G/PANI 0.2 M, G/PANI 0.2 M, and graphene exhibits, the electrochemical surface area, among CS-G/PANI 0.2 M indicating the highest specific capacitance, Fig. 10.10E. Galvanostatic charge/discharge tests were also carried out to investigate the capacitance properties of the materials CS-G/PANI 0.2 M, G/PANI 0.2 M, and graphene, as shown in Fig. 10.10E. Graphene, G/PANI, and CS-G/PANI composite electrodes have discharge specific capacitances of 52, 234, and 340 F/g, respectively. In addition, the discharge & charge spell for the CS-G/PANI decline with growing in the density of current from 1 to 10 A/g as depicted in Fig. 10.10F. Another important parameter the electrochemical cycling performance for the determination of the effect CS modification on the electrode materials illustrated in Fig. 10.10G. Then, 1000 cycles were tested at a scan rate of 100 mV/s for the material. In which, the CS-G/PANI 0.2 M for that 74% drop is observed, whereas for the G/PANI 0.2 M nanocomposites at 53% drop is observed [90].

Plak et al. [98] created GO-Au@PANI and rGO-Au@PANI nanocomposites using polyaniline (PANI), GO, reduced graphene oxide (rGO), and gold (Au) nanoparticles. The green method of making the rGO-Au@PANI nanocomposite is shown in Fig. 10.11A. GO was initially dubbed rGO. After that, the rGO surface was coated with Au nanoparticles and aniline monomer. The composite was created at that precise moment. Because of in-situ polymerization, the polyaniline was constantly collected on the rGO nanosheet surface. There were electrostatic interactions between the nanosized Au particles and the rGO nanosheet. The PANI and the rGO had pie-in-the-sky

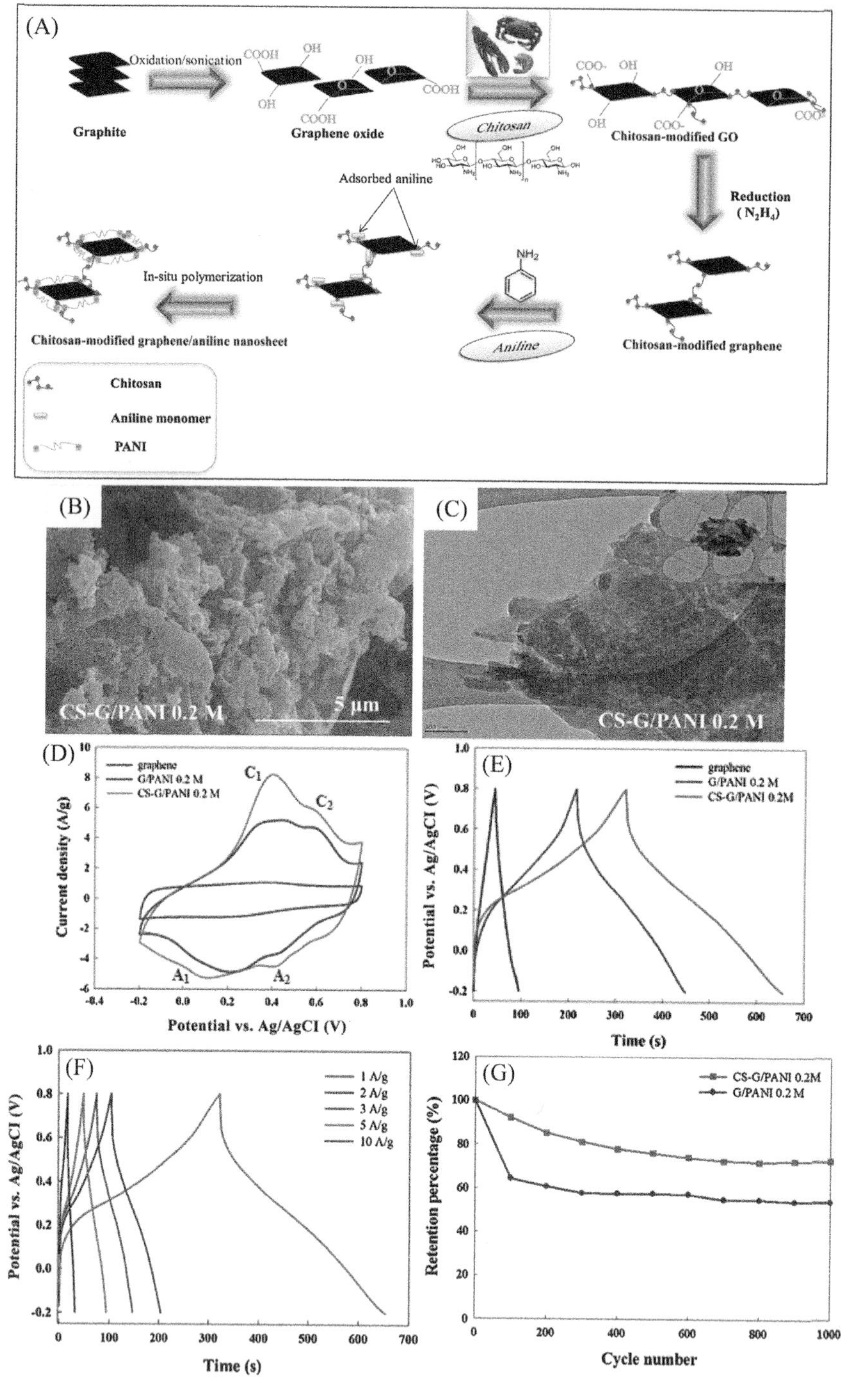

FIGURE 10.10

(*Continued*)

interactions. The pristine PANI, GO-Au@PANI, and rGO-Au@PANI nanocomposite electrodes had specific capacitances of 17.6%, 42.5%, and 63.5%, respectively. Fig. 10.11B shows the relationship between the specific capacitance of nanocomposites and their scan rate. At a current density of 1 A/g, the rGO-Au@PANI nanocomposite possesses a high specific capacitance of 212.8 F/g [35]. Sumanta et al. and his team have reported the production of graphene/polyaniline (GA/PANI) as an electrode material in SC. The GA/PANI has a specific capacitance of 242 F/g in 1 M KCl as an electrolyte. The cyclic stability of GA/PANI was investigated for 500 cycles at 1.5 A/g current density. The GA/PANI retained 86% capacitance after 500 cycles [99].

Murat et al. developed the SC electrode materials and used the electrospinning technique to demonstrate the simple fabrication of complex nanocomposites made of polypyrrole (PPy), reduced graphene (rGO), carbon black (CB), and PVA, such as GO/PPy/CB/PVA and rGO/PPy/CB/PVA [100]. Polypyrrole is a promising material for electrodes and SCs due to advantages such as low cost, higher electric conductivity, ease of synthesis, and low self-discharge, but it has poor long-term performance stability. Increasing the electrical conductivity of the materials and adding CB and reduced graphene oxide (rGO) may help to reduce this [101–104]. Fig. 10.12A and B show SEM images of nanocomposite materials GO/PPy/CB/PVA and rGO/PPy/CB/PVA. As shown in Fig. 10.12C, the rGO/PPy/CB/PVA cyclic voltammetry curves were monitored at various scan rates of 4, 6, 8, 10, and 75 mV/s between potential 0 and 0.8 V in 1 M H_2SO_4 as electrolyte solution. Fig. 10.12D shows the maximum capacitance, Csp = 951 F/g, for the rGO/PPy/CB/PVA nanofiber material. In addition, for the stability test y, the scan rate was set at 100 mV/s to achieve 1000 cycles. After 1000 cycles, the 99% retention capability of the rGO/PPy/CB/PVA nanofiber is detected, as shown in Fig. 10.12E.

The supercapacitor electrode material based on polypyrrole nanocomposites, Madhabi et al. [105] reported rGO nanosheets with nanotubes of polypyrrole (rGO/PPy NTs) as an electrode material, by in-situ reduction of GO in the existence of PPy NTs for enhancement in the capacitance of the supercapacitors application as illustrated in Fig. 10.13A. The rGO/PPy NTs are prepared with different weight % of rGO 5%, 10%, 20%, and 40%. In which, the most superior electrode material 40% weight of rGO to be found. It SEM and TEM morphological structures are confirmed as shown in Fig. 10.13B and C. The cyclic voltammetry measurement was carried out to investigate the electrochemical properties of the PPy NTs and rGO/PPy NTs composites within the −0.2 to 0.8 V potential range at variation in scan rate of 10, 25, 50, 75, and 100 mV/s in 1 M KCl aqueous solution as electrolyte as elicited in Fig. 10.13D. In addition to that, the SC of the material also measured with different scan rate as illustrated in Fig. 10.13E, the capacitance of the material increase with increase in the weight % of rGO in composites of rGO/PPy NTs. Moreover, the

◀ (A) CS-G/PANI 0.2 M composites preparation process, (B) Scanning electron microscopic (SEM) image of morphology CS-G/PANI 0.2 M composites, (C) High magnification TEM image of the CS-G/PANI 0.2 M composites, (D) Graphene, G/PANI 0.2 M, and CS-G/PANI 0.2 M cyclic voltammetry curves at 10 mV/s, (E) G, G/PANI 0.2 M, and CS-G/PANI 0.2 M galvanostatic charge/discharge curves at 1 A/g, (F) CS-G/PANI 0.2 M galvanostatic charge/discharge curves for various current densities: 1, 2, 3, 5, and 10 A/g, (G) The cycle stability of CS-G/PANI 0.2 M and G/PANI 0.2 M electrodes was evaluated after 1000 cycles [90].

Reproduced with permission Elsevier. Copyright 2017.

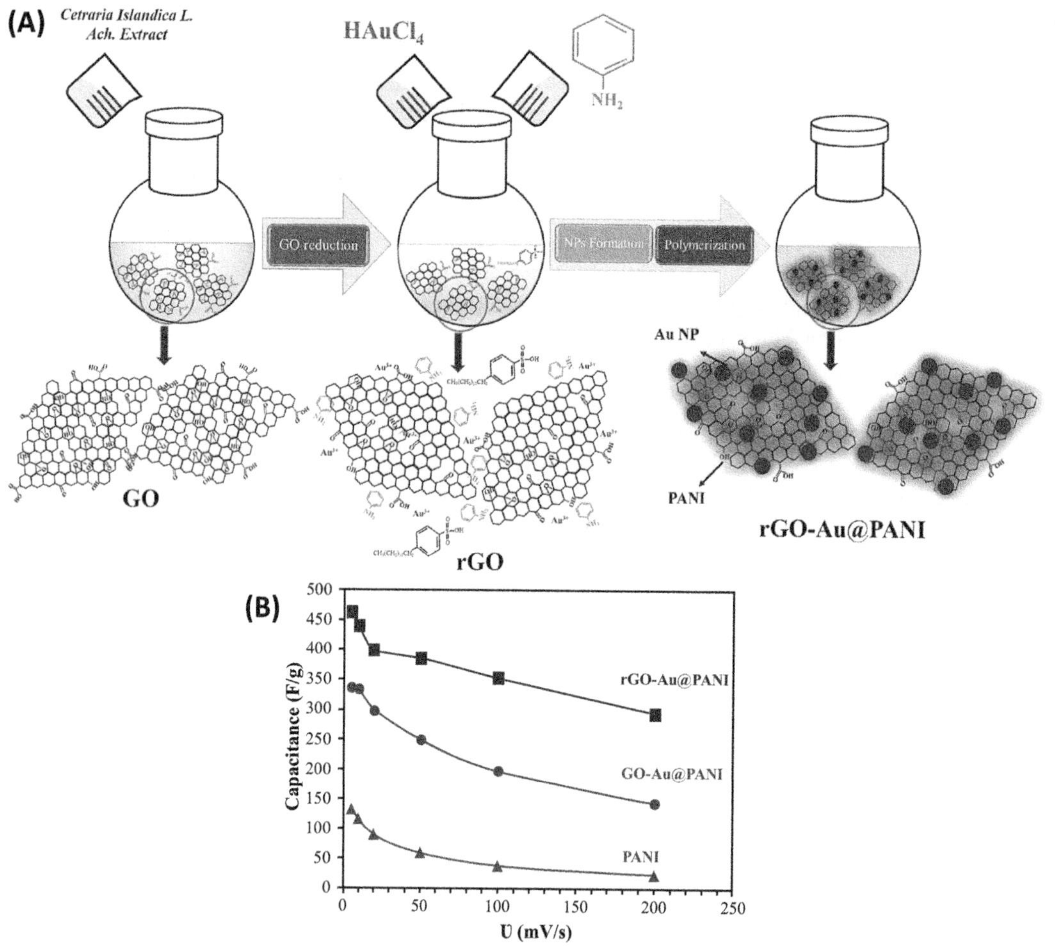

FIGURE 10.11

(A) The manufacturing of rGO-Au@PANI is depicted schematically, (B) the specific capacitance for GO-Au@PANI and rGO-Au@PANI at scan rates of 5–200 mV/s [35,98].

Reproduced with permission Elsevier. Copyright 2020.

stability test for the rGO/PPy NTs also conducted as shown in Fig. 10.13F. The 40 wt.% rGO/PPy NTs retains 77.28% after 1000 cycles. Whereas, the 5, 10, and 20 wt.% rGO/PPy NTs exhibit stability retention 64.39%, 67.19%, and 71.98%. The capacitance 141.79, 174.93, 220.83 and 288.57 F/g are observed for 5, 10, 20, and 40 wt.% rGO/PPy NTs, respectively.

Another work by Yun et al. [106], and his coworkers developed graphene/polypyrrole nanocomposites. With a capacitance of 650 F/g and a current density of 0.45 A/g, the GO/PPy offers the maximum cycle stability during the charge/discharge process. The maximum energy density was

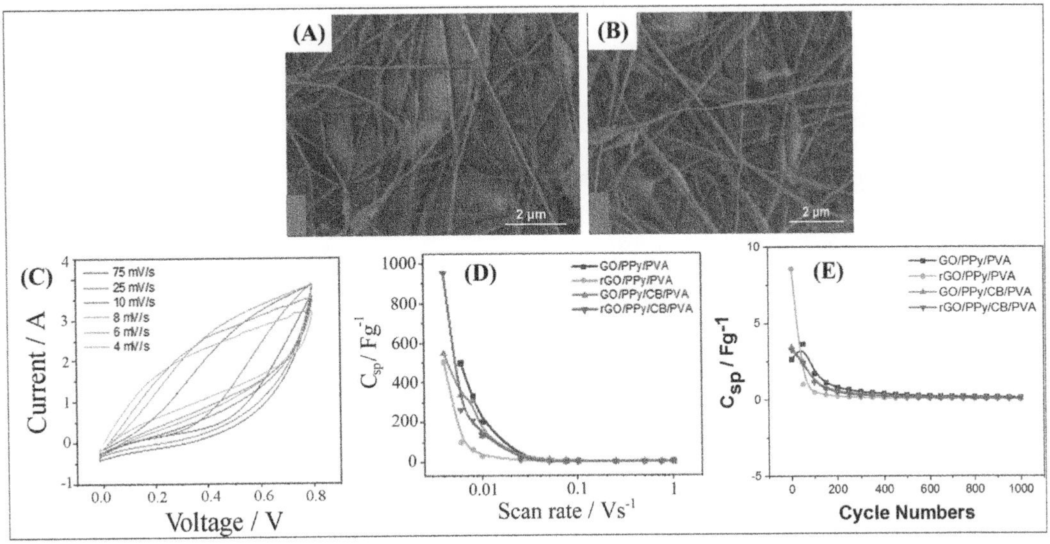

FIGURE 10.12

(A) SEM image of a GO/PPy/CB/PVA nanofiber; (B) SEM image of a rGO/PPy/CB/PVA nanofiber; (C) CV plots of rGO/PPy/CB/PVA nanofiber composites; (D) Csp vs scan rate graphs of GO/PPy/PVA, rGO/PPy/PVA, GO/PPy/CB/PVA, and rGO/PPy/CB/PVA fiber of nanocomposites, (E) Nanofiber composites using GO/PPy/PVA, rGO/PPy/PVA, GO/PPy/CB/PVA, and rGO/PPy/CB/PVA stability tests [100].

Reproduced with permission Springer Nature. Copyright 2021.

54 Wh/kg at 1 Ma current, while the highest P_d was 778.1 W/kg at 5 mA current. The GO/PPy investigated 96% specific capacitance after 5000 cycles.

Saptarshi et al. [107] created a biocompatible carbon nanotube-based composite, and PPy/multiwalled carbon nanotubes (MWCNTs) polymeric nanocomposites were created using an in-situ oxidative polymerization technique in an HCl medium with oxidant ammonium persulfate to determine the electrode material for a high capacitance supercapacitor. The morphological structure of the 2% CuCl2-doped polypyrrole/carbon nanotube synthesized (PPr C2 CNT). In addition, as illustrated in Fig. 10.14A and B, SEM and TEM examination pictures indicated the development of composite. Fig. 10.14C shows the electrochemical cyclic voltammetry characterization of the PPr C2 CNT from −0.8 to 0.8 at different scan rates of 10, 20, 50, 100, and 200 mV/s. The PPr C2 CNT had the maximum capacitance of 312 F/g at a scan rate of 10 mV/s, as illustrated in Fig. 10.14D. Moreover, the stability of the electrode materials PPr C2 CNT are conducted up to 1000 cycles at a 10 mV/s scan rate as illustrated in Fig. 10.14E. After 1000 cycles, the retention in capacitance dropped by 90%.

Karthikeyan et al. [108] reported the highest capacitance of 327 F/g for Zn^{2+}-doped PPy/MWCNT composites manufactured by in-situ oxidative polymerization at a 10 mV/s scan rate. Sun et al. [109] investigated electro polymerization method used fabrication PPy/F–MWCNT composites and found that the maximum capacitance was 243 F/g at a scanning rate of 10 mV/s. Hughes et al. [110] used an

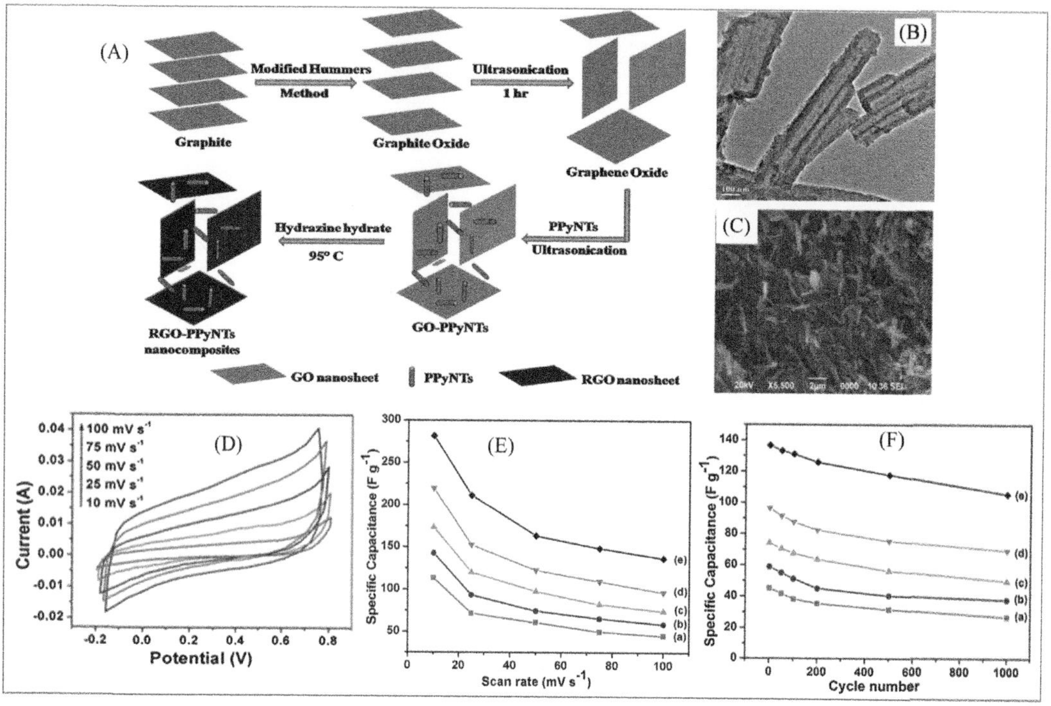

FIGURE 10.13

(A) Illustrate the process of fabrication of rGO/PPy NTs, (B) HR-TEM micrographs of 40 wt.% rGO-PPyNTs nanocomposite, (C) SEM images of 40 wt.% rGO-PPyNTs, (D) CV curves at varied voltage scan rates of 40% rGO-PPyNTs, (E)Specific capacitance of nanocomposites at varied scan rates of PPyNTs and rGO-PPyNTs at 5, 10, 20, and 40 wt.%, (F) stability test of PPyNTs and 5, 10, 20, and 40 wt.% rGO-PPyNTs in variation of capacitance for 1000 cycle at scan rate of 100 mV/s [105].

Reproduced with permission Elsevier. Copyright 2016.

electrochemical method to investigate the capacitive and structure characteristics of The maximum capacitance of 192 F/g was reported in composite films constructed from catalytically generated MWCNTs and PPy composites. At a scan rate of one millivolt per second. By studying PPy/MWCNT/CC composites made by chemical-oxidative polymerization processes, Paul et al. [111] created the SC that exhibits the maximum capacitance of 238 F/g. They also discovered that PPy/MWCNT composites had the highest capacitance of 167.2 F/g [112]. Jurewicz et al. [113] developed PPy/MWCNT composites with an extreme capacitance of 163 F/g and a scan rate of 2 mV/s using electrochemical polymerization techniques. Du et al. [114] synthesized PPy/CNT composites with a capacitance of 154.5 F/g.

Another material, polythiophene-based materials developed by Akram et al. [115], was used to create graphene oxide/polythiophene hybrid nanosheets. Polythiophene and its derivatives are high conductive polymers, and their high conductivity is responsible for high charge mobility [116].

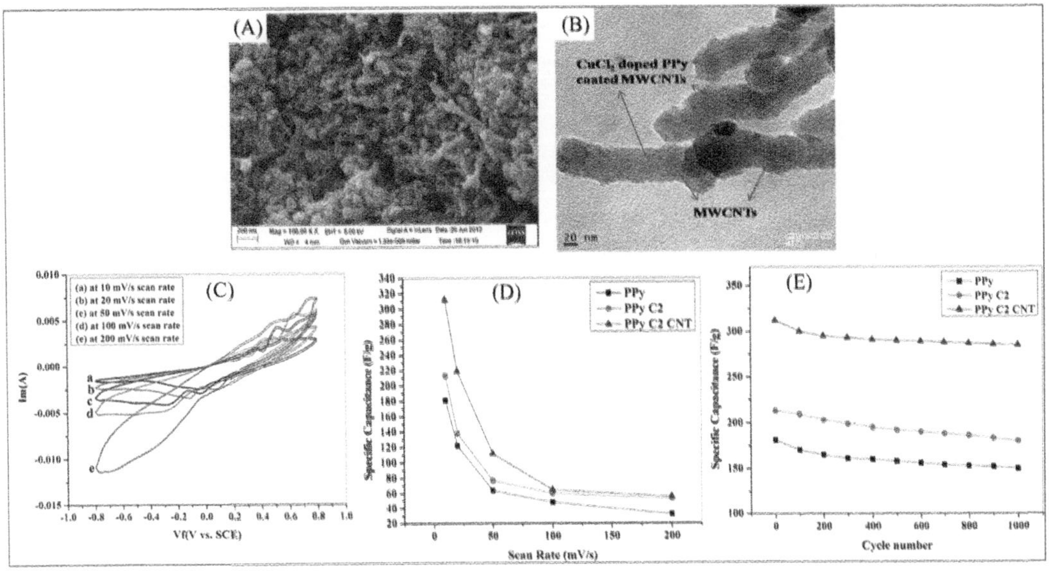

FIGURE 10.14

(A) High resolution FE-SEM image of the 2% Cu-doped PPy/CNTs, (B) High magnification TEM image of 2% Cu-doped PPy/CNTs, (C) Cyclic voltammetry of PPy C2 CNT at different scan rate, (D) the specific capacitance of the PPy, PPy C2, and PPY C2 CNT at various scan rate, (E) stability test as retention drop in capacitance after 1000 cycle for PPy, PPy C2 and PPy C2 CNT nanocomposite at 10 mV/s scan rate [107].

Reproduced with permission Wiley Online Library. Copyright 2013.

Graphene oxide/polythiophene (GO/TB) has the best electrochemical resistance, the lowest internal resistance, and the best cycle stability in capacitor retention drop. Fig. 10.15A and B show FE-SEM and HR-TEM images of GO/TB nanosheets, which are used to investigate electrode material manufacturing. In 2 M KOH solution as electrolyte, cyclic voltammetry charge curves of graphene/polythiophene (G/TB) and graphene oxide/polythiophene (GO/TB) are conducted at various scan speeds from 1–100 mV/s. As a result, the largest specific capacitances of G/TB and GO/TB at current density were found, 217 F/g and 296 F/g, respectively, in Fig. 10.15C and D. As shown in Fig. 10.15E, the electrode materials G/TB and GO/TB are tested for stability for 4000 cycles at 0.3 A/g current density. The retention rate for G/TB and GO/TB is 91.86% and 75.56%, respectively. In addition to that, various current density charge/discharge of the capacitance also measured as shown in Fig. 10.15F [115]. G-PEDOT and G-PTh nanocomposites with a maximum capacitance of 374 F/g and an 88% long-term stability in 2 M HCl after 100 cycles were produced by Farah Alvi et al. [117].

Supercapacitors are high-performance energy-storage devices. Rose et al. [118] described the electrospinning of PANI/PVA/GO and PANI/PVA nanofibers. Because of PVA's low mechanical and thermal properties, polyvinyl alcohol is highly soluble in water and has a high processability, allowing it to be easily and effectively strengthened into their structure [119]. The synthesized PANI/PVA and PANI/PVA/GO morphological SEM image are illustrated in Fig. 10.16A and B,

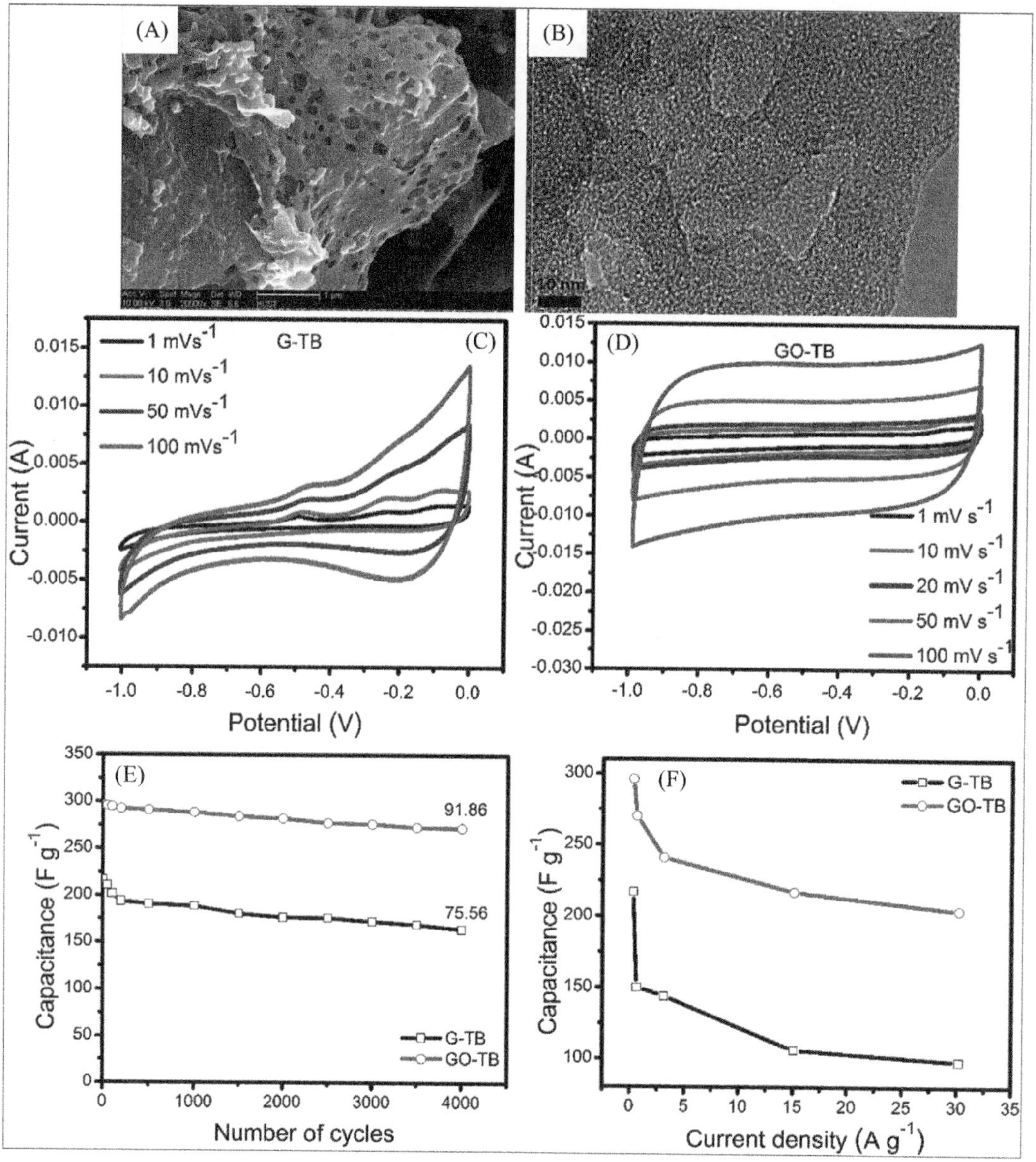

FIGURE 10.15

(A, B) FE-SEM and HR-TEM pictures of GO-TB, respectively, and graphene/polythiophene cyclic voltammetry curves (C) Different scan rates of graphene oxide/polythiophene (D) in 2 M KOH as electrolyte, (E) the capacitance retention of charge/discharge upto 4000 cycles at current density 0.3 A/g for G/TB and GO/TB, (F) At different current density for G/TB and GO/TB the specific capacitance retention in charge/discharge method [115].

Reproduced with permission Elsevier. Copyright 2015.

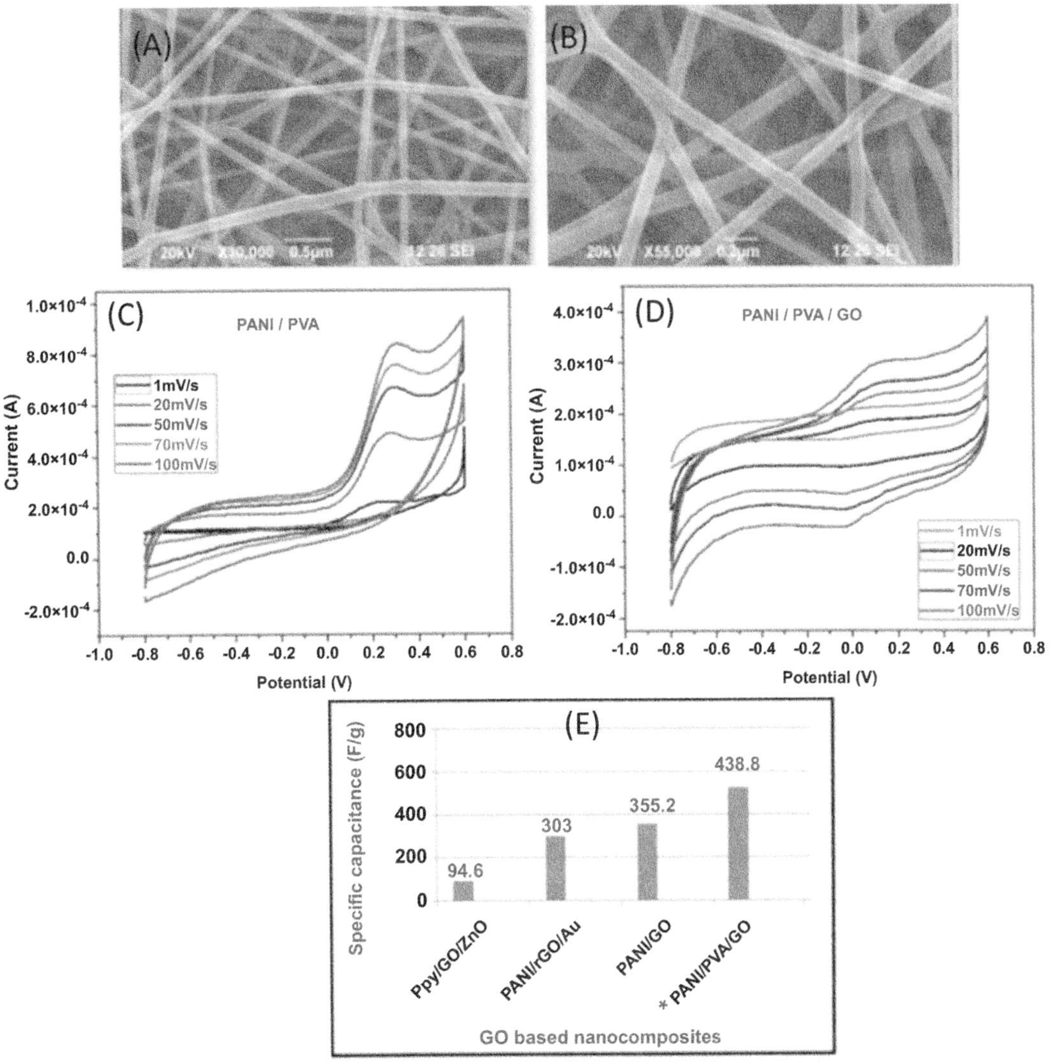

FIGURE 10.16

The morphological SEM image of (A) PANI/PVA and (B) PANI/PVA/GO nanofibers, (C) cyclic voltammetry curve for (C) PANI/PVA and (D) PANI/PVA/GO nanofibers at various scan rate, (E) plot of measured specific capacitance in comparison to GO based nanocomposites of PANI/PVA/GO nanofibers [118].

Reproduced with permission Elsevier. Copyright 2018.

respectively. These nanofibers having around 100 nm in diameter. The cyclic voltammetry process are used to checked capability as use of electrode materials by PANI/PVA and PANI/PVA/GO in supercapacitors application as shown in Fig. 10.16C and D respectively. The CV curve is monitored at various scan rates 1, 20, 50, 60, and 100 mV/s with PW −0.8 to 0.6 V in 1 M H_2SO_4 electrolyte, Fig. 10.16C and D for PANI/PVA and PANI/PVA/GO, correspondingly. The specific capacitance 143.3 and 438.8 F/g to be found for PANI/PVA and PANI/PVA/GO nanofibers to developed for electrode materials for supercapacitors.

Nayak et al. [120] used a green technique to synthesize Ag-Zirconia nanocomposites with biopolymer chitosan as a binder and Ag-Zirconia nanoparticles. In the electrolyte of 2 M KOH, Fig. 10.17A shows the construction of Ag-Zirconia nanocomposites as (+) ve electrodes with (−) ve electrode-AC. SEM structural morphology of Ag-Zirconia nanoparticles synthesized utilizing a biosynthetic technique is shown in Fig. 10.17B. Fig. 10.17C shows the most essential electrochemical parameters cyclic voltammetry (CV) curve for Ag-Zirconia with different scan rates ranging from 5 to 100 mV/s. Furthermore, the specific capacitance of Zirconia and Ag-Zirconia was compared. Fig. 10.17D indicates that the Ag-Zirconia has a higher capacitance than pure Zirconia. In

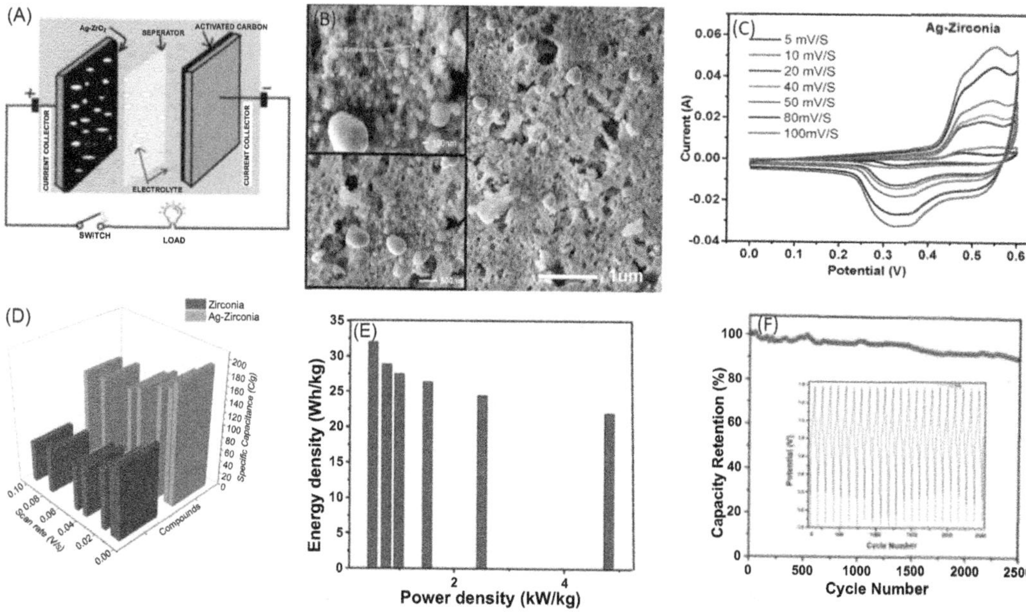

FIGURE 10.17

(A) The image illustrated supercapacitor construction, (B) SEM image of the Ag-Zirconia nanocomposite, (C) Cyclic Voltammetry curve of Ag-Zirconia at different scan rate, (D) Comparison of the specific capacitance between Zirconia and Ag-Zirconia, (E) Galvanostatic charge–discharge for specific capacitance at different current density of the Ag-Zirconia, (F) the capacitance retention of charge/discharge up to 2500 cycles at current density 1 A/g for Ag-Zirconia [120].

Reproduced with permission Elsevier. Copyright 2022.

addition to that, the Galvanostatic charge/discharge (GCD) (galvanostatic charge–discharge) curve measured at different current density as shown in Fig. 10.17E. The stability of the electrode also observed up to 2500 cycles at current density 1 A/g. The capacitance retention is observed 89.4% as shown in Fig. 10.17F [120].

As shown in Fig. 10.18A, Sun et al. [121] reported self-assembly and EDP methods for the fabrication of electrode materials such as polyacrylic acid/graphene gel (PAA-GG). The EDP process was used to deposit MnO_2 on the PAA-GG to form MnO_2/PAA-GG composites from the synthesized material. MnO_2/PAA-GG SEM images are given in Fig. 10.18B and C. The capacitance performance of the MnO_2/PAA-GG is investigated using the CV, GCD approach. The CV analysis of the MnO_2/PAA-GG and PAA-GG is shown in Fig. 10.18D at a scan rate of 5 mV/s. MnO_2/PAA-GG has outperformed PAA-GG in terms of CV curve performance. Furthermore, as shown in Fig. 10.18E with current density 1 A/g, the charge/discharge duration of MnO_2/PAA-GG is longer than that of electrode. When shown in Fig. 10.18F, the specific capacitance of the PAA-GG and MnO_2/PAA-GG decreases as the current density rises from 1 to 50 A/g. MnO_2/PAA-GG had a capacitance of 428 F/g, while PAA-GG had a capacitance of 252 F/g. In addition, the MnO_2/PAA-cyclic GG's performance was tested with 5000 cycles, as shown in Fig. 10.18E. MnO_2/PAA-GG loses just 13.8% of its specific capacitance after 5000 cycles at 50 A/g current density. The MnO_2/PAA-GG has also been employed as a biosensor for H_2O_2 detection [121].

Supercapacitors made of new materials are critical for future energy storage that is both sustainable and efficient. These goals inspired Chakraborty and his colleagues [122] to create electrode materials with biocompatible properties, such as ZnO/Polymer nanocomposites, for improved energy-storage applications. Styrene maleic anhydride (SMA) is a structural modification made with the thiadiazole moiety of the ZnO particle that is incorporated in the SMA. The SMA is structural modification done with thiadiazole moiety with that ZnO particle than incorporated in the SMA. The new materials synthesized ZnO/polymer nanocomposites (SMZ) are proven for the efficient capacitance storage materials up to 268.5 F/g at 0.1 A/g current density. The SEM and TEM morphological image of the SMZ electrode material shown in Fig. 10.19A and B, respectively. The electrochemical properties and CV curve of SMZ are depicted in Fig. 10.19C at various scan rates. The increase in the scan rate, it is increase current. Furthermore, Fig. 10.19D provides comparison of SMA-1 (bare polymer), SMZ and ZnO at 10 mV/s scan rate. In addition, Fig. 10.19E depicts the charge–discharge curves of ZnO, SMA-1, and SMZ. The stability of the SMZ electrode was also studied with 10,000 cycles, and the capacitance dropped by 27% as shown in Fig. 10.19G [122].

10.4.2 Other biodegradable polymer nanocomposite

PANI (polyaniline) was one of the most extensively researched conducting polymers in the field of supercapacitor electrode material development. Polyaniline's shape and dimensions were discovered to be in the nanoscale range to investigate the morphology-dependent PANI for electrochemical performance. As a result, Park et al. [123] used polyethylene glycol as a polymeric stabilizer to create three distinct PANI morphologies: nanorods, nanofibers (*N*-vinylpyrrolidone), and nanospheres. They created PANI with a variety of shapes by kinetically controlling the anisotropic development of polyaniline with the nanoscale scale as the presence of a stabilizer in varying concentrations. The electrochemical experiments showed that the size and shape of active components are critical for PANI-based electrodes to perform well electrochemically. The GCD curves at

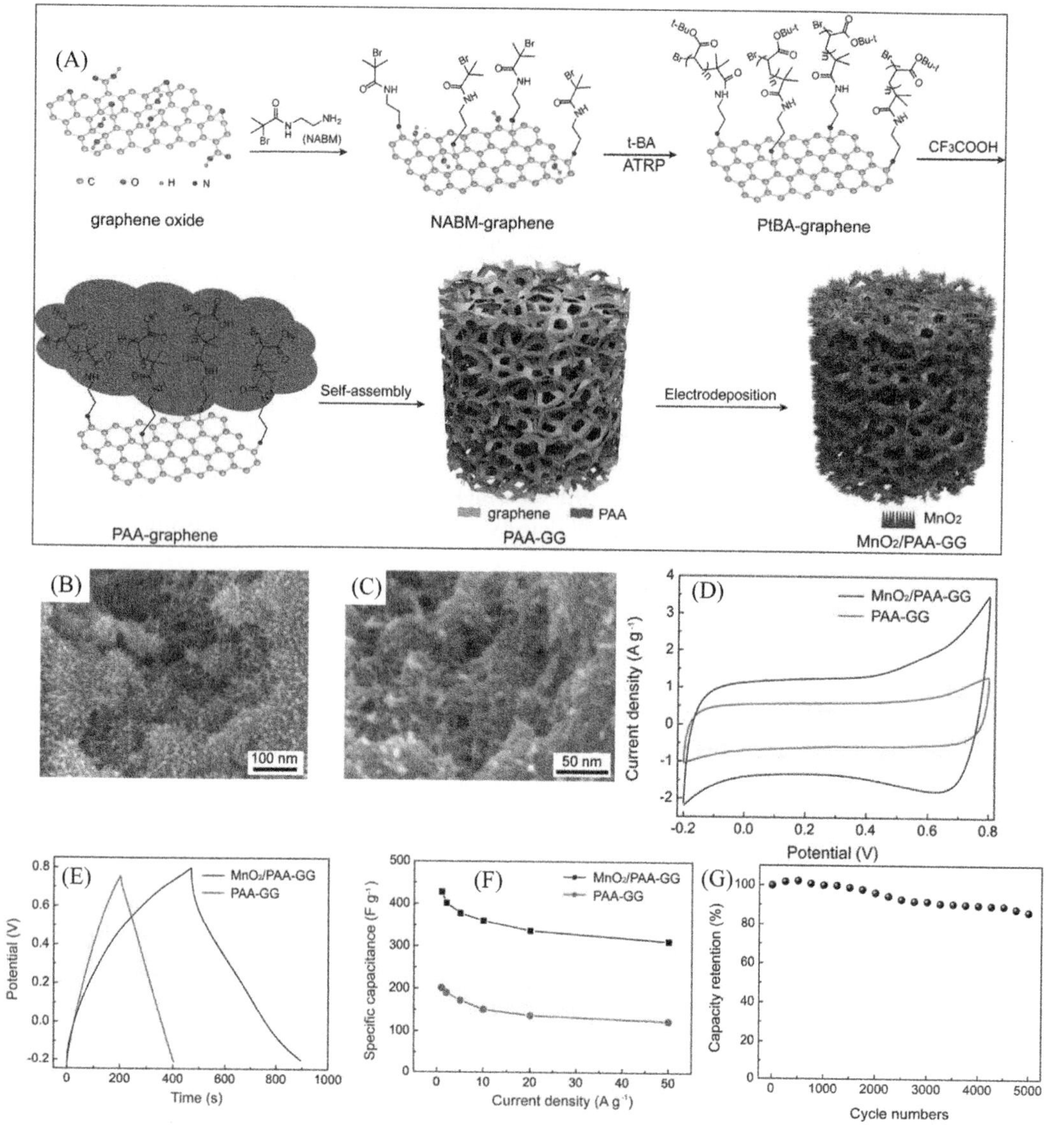

FIGURE 10.18

(A) The fabrication process of MnO_2/PAA-GG polymeric nanocomposite, (B, C) SEM image of MnO_2/PAA-GG PNCs in different magnification, (D) Cyclic voltammetry curves of PAA-GG and MnO_2/PAA-GG electrodes with 5 mV/s scan rate, (E) Galvanostatic charge–discharge curves of PAA-GG and MnO_2/PAA-GG electrodes with current density 1 A/g, (F) Galvanostatic charge–discharge for specific capacitance at different current density of the of PAA-GG and MnO_2/PAA-GG electrodes, (G) cycling performance up to 5000 cycles at 50 A/g current density of MnO_2/PAA-GG electrode [121].

Reproduced with permission Elsevier. Copyright 2018.

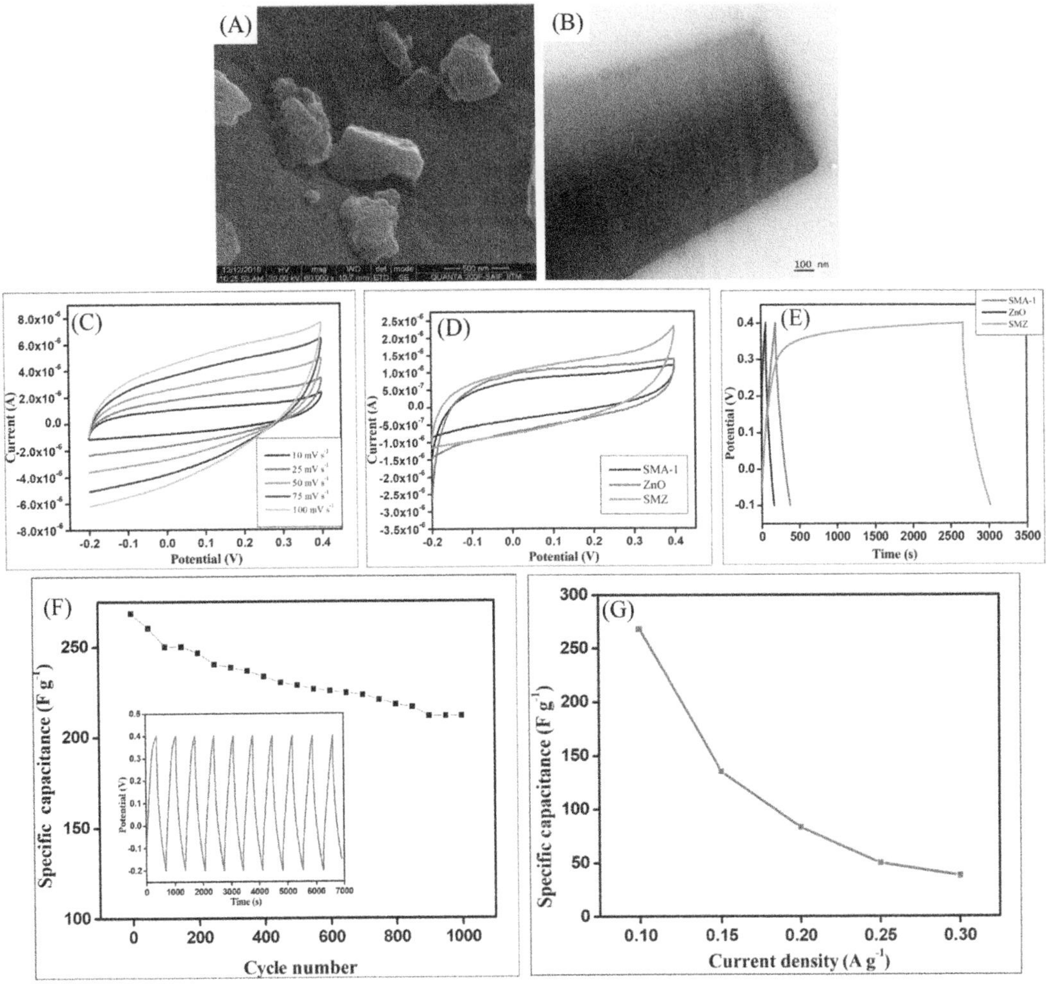

FIGURE 10.19

(A) SEM image of SMZ polymeric nanocomposite, (B) TEM image of SMZ polymeric nanocomposite, (C) Cyclic voltammogram of SMZ electrode at a10 mV/s scan rate in 0.5 M KCl electrolyte, (D) Cyclic voltammogram curves comparison of ZnO, SMZ and, SMA-1 nanoparticles at the a 10 mV/s scan rate in 0.5 M KCl electrolyte, (E) Charge–discharge curves comparison of ZnO, SMZ, and, SMA-1 at 0.1 A/g in 0.5 M KCl, (F) Cycle number vs specific capacitance of SMZ, (G) Specific capacitance vs current density of SMZ at 0.1 A/g [122].

Reproduced with permission Elsevier. Copyright 2020.

0.1 A/g current density were used to measure the capacitance of nanofibers (192 F/g), nanorods (133 F/g), and nanospheres (71 F/g) [123]. According to the findings, nanofiber-based electrodes demonstrated faster electrode kinetics and higher capacitance than nanospheres and nanorods.

Chen and his colleagues later published another paper [124]. They created PANI nanotubes, as well as nanofiber and nanosphere equivalents, using template-assisted chemical synthesis (Fig. 10.20). Charge–discharge and CV studies show the Faradaic reaction in aqueous H_2SO_4 electrolyte, just as SEM images show the formation of different shapes of PANI. When comparing nanospheres (345 F/g at current density 1 A/g), nanofibers (404 F/g at current density 1 A/g) with PANI nanotubes (502 F/g at current density 1 A/g), and the integral area of the charge–discharge time and CV curve vary as changes of shape, and the observed capacitance implies higher electrochemical performance (Fig. 10.21). Sharma et al. [125] created a hyper-cross-linked PANI nanoporous with a specific 1059 m^2/g surface area, and a capacitance 410 F/g for 1000 cycles in a report at a 3 mV/s scanning speed. The increased surface area of nanoporous PANI hyper-cross-linked aids in energy storage via the EDLC mechanism. At the same time, the redox sites accessible energy can be stored in the same electrode via the PC mechanism.

Cho et al. [126] proposed the self-stabilized dispersion polymerization (SSDP) method for fabricating a high-porosity PANI-based thin film for use as a electrode of supercapacitor. The superior extremely porous PANI-based prepared sample was coated on a stainless steel substrate to produce thick film type 10 μm electrode materials, which were then evaluated in electrolyte of 0.5 M H_2SO_4 as an electrochemical supercapacitor. It showed distinct oxidation–reduction peaks at 20 mV/s scan rate and generated a 361 F/g capacitance with a 0.25 A/g current density, which is substantially greater than a pure PANI electrode. When comparing the performance of a film formed of PANI fibers to that of PANI powders, electrospinning showed a significant improvement. The linked network of fibers in a film contributes to high capacitance via a quick and steady Faradaic reaction (230 F/g at 1000 cycle) [127].

Despite the fact that PANI has a higher specific capacitance (2000 F/g), it is difficult to achieve this in operation due to the material's limited availability of active sites on the surface for electrolyte accessibility [124]. Other disadvantages of PANI include its short cycle life, brittle nature at high current density with low-power density, limited rate capability, and excessive volume expansion during repeated charge–discharge processes. As a result, novel conducting polymers with good electrochemical performance have piqued the interest of many researchers. PPy (Polypyrrole) is one such potential that has piqued the interest of many in electrochemical research. PPy, like PANI, is semiconducting, but when protonated, it can simply transition from insulator to conductor.

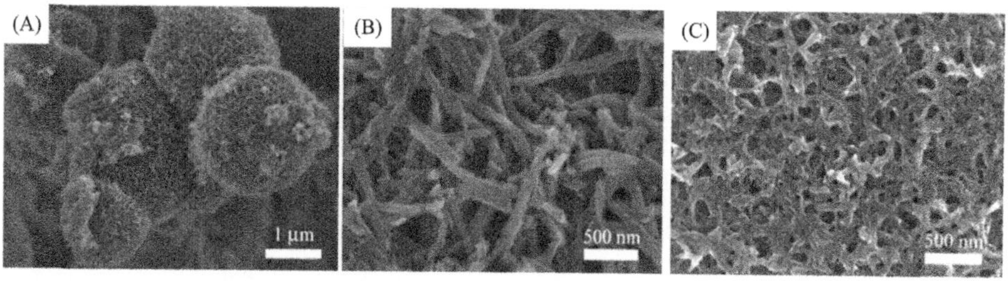

FIGURE 10.20

(A) SEM images of PANI nanospheres, (B) SEM images of PANI nanotubes, (C) SEM images of PANI nanofibers [124].

Reprint with permission American Chemical Society. Copyright 2013.

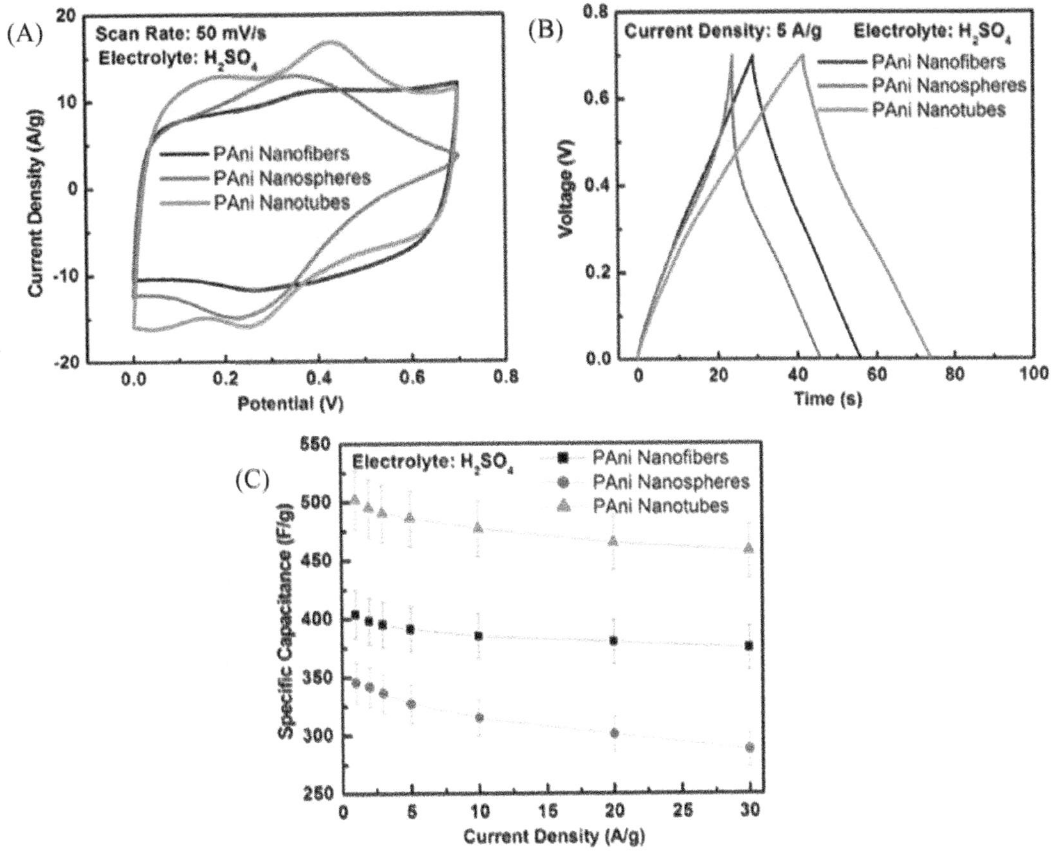

FIGURE 10.21

(A) CV at scan rate of 50 mV/s, (B) capacitance versus current density, (C) capacitance of all three PANI-based electrodes [124].

Reprint with permission American Chemical Society. Copyright 2013.

As a result, the PPy salt form is an excellent choice for electrical applications requiring higher conductivity. One of the most important uses of PPy in electronic applications is the development of electrodes for a supercapacitor. Sahoo et al. [128] used a simple chemical approach to create a PPy nanofiber-based electrode using a sodium alginate template. The highest capacitance of PPy nanofiber is 284 F/g, it is larger than PPy in its purest form with a normal a clump of particles.

The following PANI and PPy, PEDOT (poly(3,4-ethylenedioxythiophene)) is the most electrochemically investigated material, with 300–500 S/cm high conductivity, a larger range of potential (1.4 V), excellent mechanical strength and flexibility, and low oxidation potential. Chemical polymerization was used to create a thin film based PEDOT electrode on nonconducting substrate of plastic for electrochemical supercapacitor in a study [129]. To assess the super capacitive performance, the electrode film was submerged in the organic electrolyte liquid (0.1 M $LiClO_4$ in acetonitrile) and determinate in a three electrode setup with capacitive reactants being common-emitter amplifier and Ag/AgCl being a platinum foil in high surface [129]. The results showed that the response of CV is an outcome of both Faradaic redox and EDLC processes, and that it persists still at 200 mV/s scan rate. The voltage of cell can reach 0.8 volts after complete charging, and 1 to 4 Wh/kg ranges of energy density, depending on the discharge current densities utilized. However, at the greatest of 1 mA/cm current density, the power density exceeds 2.5 kW/kg [129]. Many more studies on supercapacitors based on PEDOT have been conducted in the search for high-performance, low-cost supercapacitors [130]. One such study [130] produced and electrochemically investigated symmetric and asymmetric supercapacitors based on PEDOT form AC (anode) and PEDOT (cathode). In the voltage range of 0–1 V, the curve of charge/discharge with 1000 cycles are measured for the both hybrid-type and redox supercapacitors.

PTh (polythiophene) is another major conducting polymer that has made an indelible mark on the energy areas. The capacitance of the PTh-based electrode is often smaller than that of PANI or PPy, but it has the advantage of being able to perform at a greater 1.2 V PW [131]. Laforgue and his group [132] developed a supercapacitors based on PTh with a260 F/g capacitance storage capacity. In another study, PTh was synthesized using a chemical-oxidative polymerization process in the absence and presence of three individual surfactants: anionic sodium dodecyl sulfate (SDS), cationic cetrimonium bromide (CTAB), and nonionic transistor (Triton X-100). The capacitance of PTh made with transistor nonionic surfactant was 117 F/g, The capacitance of PTh produced without surfactant (7 F/g) is 33% higher. The voltammogram was found to have a rectangular shape, indicating that PTh had capacitive behavior [133]. Gnanakanet et al. [134] reported PTh nanoparticles were produced using polymerization techniques with cationic surfactant dilute of $FeCl_3$ oxidant and the capacitance is 134 F/g. Fu et al. [135] used EDP to make a thin film of PTh with 110 F/g capacitance. PTh thin films were prepared utilizing a deposition process of chemical bath at ambient temperature around 300K by polymerization oxidative of the thiophene utilizing oxidizing agent as ammonium peroxydisulfate [136].

Attempts to develop such composites with superior electrochemical characteristics have also been made in a number of ways. Antiohos et al. [137] reported the 104 F/g capacitance at current density 0.2 A/g for freestanding PEDOT-PSS/SWCNTs. To achieve a capacitance of 383 F/g at 0.5 A/g, Dhibar et al. [138] employed in-situ chemical-oxidative polymerization to produce a homogenous coating of $CuCl_2$-doped poly(An-*co*-Py) over MWCNT surface [139]. Mao et al. [140] used in-situ polymerization to make Gr/PANI NFs with a of 526 F/g specific capacitance at 0.2 A/g and good cycle stability. Sahoo et al. [141] used simple chemical polymerization to make a poly(Anco-Py)/Gr fiber-like structure nanocomposite with a 351 F/g capacitance at 10 mV/s. Sahoo et al. [139] utilized a biopolymer, sodium alginate, to make Gr and PPy nanofiber. With the presence of Gr, a fiber-like morphology was formed. In the solution of 1 M KCl at room temperature, the composite had a high electrical conductivity of 1.45 S/cm and a maximum 466 F/g capacitance at 10 mV/s. Other biodegradable polymer nanocomposites material as electrode with excellent capacitance in the supercapacitors is briefly summarized in Table 10.1.

10.5 Application of supercapacitors

The ever-improving performance of supercapacitors has resulted in a significant increase in their applications. For the past decade, rechargeable batteries have dominated the energy-storage

Table 10.1 Biodegradable polymer nanocomposites as electrode materials in supercapacitors.

Electrode	Electrolyte	Specific capacitance (F/g)	Capacitance retention capability (cycle)	References
Chitosan-modified graphene/ polyaniline (CS-G/PANI)	Ag/AgCl	340 (1 A/g)	74% (1000 cycles)	[90]
Graphene/polyaniline	Ag/AgCl	234 (1 A/g)	53% (1000 cycles)	[90]
Reduced graphene oxide-gold nanoparticle Polyaniline (rGO-Au@PANI)	1 M H_2SO_4	212.8 (1 A/g)	86.9% (5000 cycles)	[35,98]
rGO/PPY/CB/PVA	1 M H_2SO_4	951 F/g (4 mV/s)	99% (1000 cycles)	[100]
rGO/PPy NTs	1 M KCl	288.57 F/g (20 mV/s)	77.28% after 1000 cycles	[105]
Graphene/polypyrrole	–	650 F/g (0.45 A/g)	96% after 5000 cycles	[106]
2% $CuCl_2$-doped polypyrrole/ carbon nanotube (PPy C2 CNT)	1 M KCl	312 F/g (10 mV/s)	90% after 1000 cycles	[107]
Graphene oxide-polythiophene	2 M KOH	296 F/g (0.3 A/g)	91.86% after 4000 cycles	[115]
PANI/PVA/GO nanofibers	1 M H_2SO_4	438.8 F/g	NA	[118]
Ag-Zirconia	2 M KOH	144 F/g (1 A/g)	89.4% after 2500 cycles	[120]
MnO_2/PAA-GG	1 M NaOH	428 F/g	73% after 5000 cycles	[121]
SMZ	0.5 M KCl	268.5 F/g (1 A/g)	73% after 1000 cycles	[122]
Chitosan-graphene-polyaniline (CGP)	1 M H_2SO_4	1519 F/g(5 mV/s)	88% (1000 cycles)	[142]
PANI-MOS_2	1 M H_2SO_4	390 (0.8 A/g)	86% (1000 cycles)	[143]
PANI-activated carbon	0.5 M H_2SO_4	115.2 (5 mV/s)	N/A	[144]
CeO_2-PANI	1 M H_2SO_4	180 F/g (2 A/g)	N/A	[145]
PANI-graphite	1 M H_2SO_4	180.5 (20 mV/s)	N/A	[146]
GO-Pt-PANI	1 M KOH	227.2 F/g (50 A/g)	N/A	[147]
PANI-Fe_3O_4	1 M H_2SO_4	228 F/g (1 mA/cm^2)	50.4% (1000 cycles)	[148]
Graphene/PPy	–	414 F/g	95% after 10000 cycles	[149]
PPy/MWCNT	–	374 F/g	88% after 1000 cycles	[150]

industry. The demand for improved energy storage for a wide range of applications, including electronic portability to hybrid electric vehicles, is rapidly increasing. In the energy-storage industry, performance optimization and cost reduction are critical [151]. Fig. 10.22 depicts the applications of supercapacitors.

The remote control was originally powered by two AAA batteries, but supercapacitor technology allowed for rapid charging [152]. The lifespan of a supercapacitor, Cresce et al. [153] claim that, exceeding the life expectancy of a capacitors of remote. Supercapacitors are utilized in digital cameras to generate flashes. The Sam Beck developed the portable speakers (Blue shift) with a

FIGURE 10.22

Application fields of supercapacitors [35].

Reproduced with permission MDPI. Copyright 2021.

supercapacitor assembly integrated are the most recent innovation. According to Denison [154], after charging for 5 minutes, on high volume the playback duration extends up to 6 hours. Supercapacitors had previously been primarily employed applications requiring memory backup and controllers [155]. They are used in renewable and diverse energy sources as wind stabilizers and power-line systems such as photovoltaic. Supercapacitors provide brief energy bursts to wind turbines due to their quick response to unpredictable weather fluctuations [156]. In a security application, hybrid supercapacitors are used when a short period of power supply is required. The hybrid supercapacitor serves a similar function in system security as a computer's uninterrupted power supply (UPS) [157].

Energy storage is a well-known requirement in the automobile industry for various tasks such as start-up, ignition, transmission, security, and lighting. The supercapacitors are coupled with the engine to provide a quick burst of energy during internal engine combustion. According to this, the thermal motor and supercapacitors are compatible in terms of durability. This feature reduces maintenance costs when compared to traditional lead acid batteries or accumulators [158]. PSA and Toyota's Yaris hybrid-R Peugeot Citroen use supercapacitors in the event of a power outage, beginning and stopping fuel savings, and faster preliminary speeding up. The use of supercapacitors in regenerative braking is well known. This is accomplished by bringing the vehicle to a complete stop while maintaining energy braking. This was demonstrated on a tramway in Switzerland that used a one-ton assembly of supercapacitors. With recharge duration of 20 seconds at each station, the trams were able to stop using regenerative braking [159].

Battery-powered navigators, sensors, and communication tools represent an untapped market for supercapacitor applications. Electromagnetic pulse weapons, torpedoes, radar systems, and other systems can all be powered by an appropriate supercapacitor assembly [5]. Powerful applications include phased array radar antennas, avionics display devices, missiles, air bag exploitation power, and GPS.

Among the most important applications of supercapacitors in electrical devices are complementary metal–oxide semiconductor (CMOS) otherwise microprocessors with semiconductor memory in memory backup [160]. In such cases, the supercapacitor serves as a memory backup rather than a source of energy storage during brief power outages. Supercapacitors are used to stabilize power supplies for random loads such as portable media players, laptop computers, and other electronic devices. They serve as a backup power supply for low-power components such as RAM, PC cards, and static random access memory (SRAM), as well as UPS. Using a supercapacitor protects memory from any significant loss in the primary power supply. Supercapacitors are used as a backup power supply in solid state drives (SSDs) that are designed to cache memory.

High-voltage pulses delivery applications can use supercapacitors in the field of medical. As a result, they are feed into defibrillators, it release a 500J blast of electricity to jolt the heart back to life [161]. Patients who have suffered from mental trauma are treated similarly. For example, the use of hybrid supercapacitors in conjunction with dental science could present a novel hybrid supercapacitors application. Li-ion ultra-supercapacitors (JSR) Micro developed hybrid supercapacitor equipment for medical imaging instruments [161], which is associated with backup power sources. This is the broadest range of hybrid supercapacitor applications, ranging from industrial electronics such as automated meter reading (AMR) to emergency power as backup sources to prevent catastrophic power outages until power is restored.

Supercapacitors could be used as an alternating storage of energy technology to improve the quality and reliability of power distribution. Each electrode material has its own set of advantages and disadvantages. The use of two distinct electrodes in a hybrid electrochemical supercapacitor [162,163] improves cycle life and energy density life. With rising fuel prices and environmental concerns, supercapacitors may become more appealing as ideal future energy sources.

10.6 Conclusion and future perspectives

To summarize, the energy crisis is worsening by the day due to a lack of fossil fuels, a rapid increase in population, and other factors. As a result, renewable energy sources are in high demand because they are environmentally friendly, cause no pollution, and require no major operations, saving money in the long run. Supercapacitors are popular in the field of renewable energy sources because they have properties such as high E_d, high P_d, long cycling stability, and affordability. Furthermore, supercapacitors serve as a link between the battery and the conventional capacitor. Apart from that, supercapacitors have been used in a wide range of applications, including mobile phones, sensors, buses, and cameras, to name a few. The capacitor's capacitance is greatly affected by its electrode and electrolyte, which means that if the surface area of the electrode material is large, the capacitance will be large, and similarly, if the electrolyte can remain constant throughout the process, the supercapacitor's life expectancy will be increased. EDLCs and hybrid supercapacitors are two types of supercapacitors. In terms of capacitance, the hybrid supercapacitors are quite good. The reason for this would be the use of two different electrodes.

Despite the overview of the supercapacitor, some major challenges are associated with existing supercapacitors, such as the energy density being lower than that of a battery due to surface area and pore size. These two terms are linked; for example, as pore size decreases, surface area increases due to large number charge accumulation. The use of electrode materials with the properties of narrow pore size and large surface area, for example, here in this chapter we have highlighted one approach that is the use of polymeric nanocomposites as an electrode material and by using this capacitance increased greatly. Aside from that, composite synthesis methods affect material morphology and thus deal with capacitance as well. Furthermore, according to the literature and surveys, it can be stated that much more research for the future energy source, supercapacitor, particularly its electrode materials, is required to obtain a large amount of energy. In the future, researchers in this field should concentrate on advanced techniques for synthesis of electrode materials, material selection, surface area, and pore size. Finally, students, scientists, and industrial delegates working in the fields of supercapacitors, advanced materials, and biodegradable polymeric nanocomposites will benefit from this chapter.

References

[1] B.K. Kim, S. Sy, A. Yu, J. Zhang, Electrochemical supercapacitors for energy storage and conversion, Handbook of Clean Energy Systems, Wiley, 2015, pp. 1–25.

[2] M. Adil, M.A. Abdelkareem, E.T. Sayed, C. Rodriguez, M. Ramadan, A.G. Olabi, Progress of metal chalcogenides in supercapacitors, Reference Module in Materials Science and Materials Engineering, Elsevier BV, 2021.

[3] M. Winter, R.J. Brodd, What are batteries, fuel cells, and supercapacitors? Chem. Rev. 104 (2004) 4245–4270.

[4] Z. Yu, L. Tetard, L. Zhai, J. Thomas, Supercapacitor electrode materials: nanostructures from 0 to 3 dimensions, Energy Environ. Sci. 8 (2015) 702–730.

[5] A. Muzaffar, M.B. Ahamed, K. Deshmukh, J. Thirumalai, A review on recent advances in hybrid supercapacitors: design, fabrication and applications, Renew. Sustain. Energy Rev. 101 (2019) 123–145.

[6] P. Simon, Y. Gogotsi, Materials for electrochemical capacitors, Nat. Mater. 7 (2008) 845–854.

[7] A.M. Saleem, V. Desmaris, P. Enoksson, Performance enhancement of carbon material for supercapacitors, J. Nanomater. (2016).

[8] J. Ho, T.R. Jow, S. Boggs, Historical introduction to capacitor technology, IEEE Electr. Insul. Mag. 26 (2010) 20–25.

[9] J. Both, Electrolytic capacitors, 1890 to 1925: early history and basic principle, IEEE Electr. Insul. Mag. 31 (2014) 22–29.
[10] D.L. Boos, Standard Oil Co, Electrolytic capacitor having carbon paste electrodes, 1970, U.S. Patent 3,536,963.
[11] B.E. Conway, Transition from “supercapacitor” to “battery” behavior in electrochemical energy storage, J. Electrochem. Soc. 138 (1991) 1539.
[12] M. Endo, T. Takeda, Y.J. Kim, K. Koshiba, K. Ishii, High power electric double layer capacitor (EDLC’s); from operating principle to pore size control in advanced activated carbons, Carbon Lett. 1 (2001) 117–128.
[13] P. Ball, Y. Gogotsi, Super-capacitors take charge in Germany, MRS Bull. 37 (2012) 802–803.
[14] Z. Lu, Z. Chang, W. Zhu, X. Sun, Beta-phased Ni $(OH)_2$ nanowall film with reversible capacitance higher than theoretical Faradic capacitance, Chem. Comm. 47 (2011) 9651–9653.
[15] J. Xie, X. Sun, N. Zhang, K. Xu, M. Zhou, Y. Xie, Layer-by-layer β-Ni $(OH)_2$/graphene nanohybrids for ultraflexible all-solid-state thin-film supercapacitors with high electrochemical performance, Nano Energy. 2 (2013) 65–74.
[16] M.D. Stoller, S. Park, Y. Zhu, J. An, R.S. Ruoff, Graphene-based ultracapacitors, Nano Lett. 8 (2008) 3498–3502.
[17] L. Xie, G. Sun, F. Su, X. Guo, Q. Kong, X. Li, et al., Hierarchical porous carbon microtubes derived from willow catkins for supercapacitor applications, J. Mater. Chem. A 4 (2016) 1637–1646.
[18] P. Forouzandeh, S.C. Pillai, Two-dimensional (2D) electrode materials for supercapacitors, Mater. Today Proc. 41 (2021) 498–505.
[19] Z.S. Wu, Y. Zheng, S. Zheng, S. Wang, C. Sun, K. Parvez, et al., Stacked-layer heterostructure films of 2D thiophene nanosheets and graphene for high-rate all-solid-state pseudocapacitors with enhanced volumetric capacitance, Adv. Mater. 29 (2017) 1602960.
[20] J. Chmiola, G. Yushin, R. Dash, Y. Gogotsi, Effect of pore size and surface area of carbide derived carbons on specific capacitance, J. Power Sour. 158 (2006) 765–772.
[21] G. Wang, L. Zhang, J. Zhang, A review of electrode materials for electrochemical supercapacitors, Chem. Soc. 41 (2012) 797–828.
[22] W.G. Pell, B.E. Conway, Voltammetry at a de Levie brush electrode as a model for electrochemical supercapacitor behaviour, J. Electroanal. Chem. 500 (2001) 121–133.
[23] S. Dsoke, K. Pfeifer, Z. Zhao, The role of nanomaterials for supercapacitors and hybrid devices, Frontiers of Nanoscience, Elsevier, 2021, pp. 99–136.
[24] G.M. Patel, V.R. Shah, G.J. Bhatt, P.T. Deota, Humidity nanosensors for smart manufacturing, Nanosensors for Smart Manufacturing, Elsevier, 2021, pp. 555–580.
[25] V. Shah, J. Bhaliya, G.M. Patel, P. Joshi, Room-temperature chemiresistive gas sensing of SnO_2 nanowires: a review, J. Inorg. Organomet. Polym. Mater. (2022) 1–32.
[26] V. Shah, J. Bhaliya, G.M. Patel, P. Joshi, Recent advancement in Pd-decorated nanostructures for its catalytic and chemiresistive gas sensing applications: a review, Top. Catal. (2022) 1–41.
[27] G. Patel, V. Pillai, M. Vora, Liquid phase exfoliation of two-dimensional materials for sensors and photocatalysis—a review, J. Nanosci. Nanotechnol. 19 (2019) 5054–5073.
[28] G. Patel, V. Pillai, P. Bhatt, S. Mohammad, Application of nanosensors in the food industry, Nanosensors for Smart Cities, Elsevier, 2020, pp. 355–368.
[29] K.A. Dubey, P.A. Hassan, Y.K. Bhardwaj, High performance polymer nanocomposites for structural applications, Materials under Extreme Conditions, Elsevier, 2017, pp. 159–194.
[30] K. Aschberger, H.J. Johnston, V. Stone, R.J. Aitken, S.M. Hankin, S.A. Peters, et al., Review of carbon nanotubes toxicity and exposure—appraisal of human health risk assessment based on open literature, Crit. Rev. Toxicol. 40 (2010) 759–790.

[31] D.P. Dubal, N.R. Chodankar, D.H. Kim, P. Gomez-Romero, Towards flexible solid-state supercapacitors for smart and wearable electronics, Chem. Soc. Rev. 47 (2018) 2065–2129.

[32] A.K. Nayak, B. Das, Introduction to polymeric gels, Polymeric Gels, Woodhead Publishing, Elsevier, 2018, pp. 3–27.

[33] L. Migliorini, T. Santaniello, F. Borghi, P. Saettone, M. Comes Franchini, G. Generali, et al., Eco-friendly supercapacitors based on biodegradable poly (3-hydroxy-butyrate) and ionic liquids, Nanomater. 10 (2020) 2062.

[34] M.D. Tabone, J.J. Cregg, E.J. Beckman, A.E. Landis, Sustainability metrics: life cycle assessment and green design in polymers, Environ. Sci. Technol. 44 (2010) 8264–8269.

[35] A. Kausar, Green nanocomposites for energy storage, J. Compos. Sci. 5 (2021) 202.

[36] Global Supercapacitor Market worth USD 2,096.16 Million by 2022, (n.d.). 188, https://www.zionmarketresearch.com/news/super-capacitor-market (accessed 15.02.22).

[37] C. Lian, D.E. Jiang, H. Liu, J. Wu, A generic model for electric double layers in porous electrodes, J. Phys. Chem. C. 120 (2016) 8704–8710.

[38] A. Nishino, Capacitors: operating principles, current market and technical trends, J. Power Sour. 60 (1996) 137–147.

[39] S.R.C. Vivekchand, C.S. Rout, K.S. Subrahmanyam, A. Govindaraj, C.N.R. Rao, Graphene-based electrochemical supercapacitors, J. Chem. Sci. 120 (2008) 9–13.

[40] L.L. Zhang, X.S. Zhao, Carbon-based materials as supercapacitor electrodes, Chem. Soc. Rev. 38 (2009) 2520–2531.

[41] J. Chmiola, G. Yushin, Y. Gogotsi, C. Portet, P. Simon, P.L. Taberna, Anomalous increase in carbon capacitance at pore sizes less than 1 nanometer, Science 313 (2006) 1760–1763.

[42] M.V. Kiamahalleh, S.H.S. Zein, G. Najafpour, S. Abdsata, S. Buniran, Multiwalled carbon nanotubes based nanocomposites for supercapacitors: a review of electrode materials, Nano 7 (2012) 1230002.

[43] M. Jayalakshmi, K. Balasubramanian, Simple capacitors to supercapacitors-an overview, Int. J. Electrochem. Sci. 3 (2008) 1196–1217.

[44] M.S. Halper, J.C. Ellenbogen, Ultra-capacitors: a brief overview. MITRE Nanosystems Group; 2006.

[45] H. Choi, H. Yoon, Nanostructured electrode materials for electrochemical capacitor applications, Nanomater. 5 (2015) 906–936.

[46] G. Guidi, T.M. Undeland, Y. Hori, An interface converter with reduced volt-ampere ratings for battery-supercapacitor mixed systems, IEEJ J. Ind. Appl. 128 (2008) 418–423.

[47] X. Zhang, H. Zhang, C. Li, K. Wang, X. Sun, Y. Ma, Recent advances in porous graphene materials for supercapacitor applications, RSC Adv. 4 (2014) 45862–45884.

[48] X. Chen, R. Paul, L. Dai, Carbon-based supercapacitors for efficient energy storage, Natl. Sci. Rev. 4 (2017) 453–489.

[49] P. Simon, Y. Gogotsi, Materials for electrochemical capacitors, Nat. Mater. (2010) 320–329.

[50] Y. Sun, M.S. Ata, I. Zhitomirsky, Electrophoretic deposition of linear polyethylenimine and composite films, Surf. Eng. 29 (2013) 495–499.

[51] P. Simon, Y. Gogotsi, Charge storage mechanism in nanoporous carbons and its consequence for electrical double layer capacitors, Philos. Trans. Royal Soc. 368 (2010) 3457–3467.

[52] S.M. Chen, R. Ramachandran, V. Mani, R. Saraswathi, Recent advancements in electrode materials for the high-performance electrochemical supercapacitors: a review, Int. J. Electrochem. Sci. 9 (2014) 4072–4085.

[53] T. Chen, L. Dai, Carbon nanomaterials for high-performance supercapacitors, Mater. Today. 16 (2013) 272–280.

[54] Y. Zhao, M.B. Zheng, J.M. Cao, X.F. Ke, J.S. Liu, Y.P. Chen, et al., Easy synthesis of ordered meso/macroporous carbon monolith for use as electrode in electrochemical capacitors, Mater. Lett. 62 (2008) 548–551.

[55] C.W. Huang, H. Teng, Influence of carbon nanotube grafting on the impedance behavior of activated carbon capacitors, J. Electrochem. Soc. 155 (2008) 739.
[56] M. Sarno, Nanotechnology in energy storage: the supercapacitors, Studies in Surface Science and Catalysis, Elsevier, 2020, pp. 431–458.
[57] M. Beidaghi, C. Wang, Micro-supercapacitors based on interdigital electrodes of reduced graphene oxide and carbon nanotube composites with ultrahigh power handling performance, Adv. Fun. Mater. 22 (2012) 4501–4510.
[58] K. Machida, S. Suematsu, S. Ishimoto, K. Tamamitsu, High-voltage asymmetric electrochemical capacitor based on polyfluorene nanocomposite and activated carbon, J. Electrochem. Soc. 155 (2008) 970.
[59] K. Naoi, 'Nanohybrid capacitor': the next generation electrochemical capacitors, Fuel Cells 10 (2010) 825–833.
[60] H. Wang, L. Zhang, X. Tan, C.M. Holt, B. Zahiri, B.C. Olsen, et al., Supercapacitive properties of hydrothermally synthesized Co_3O_4 nanostructures, J. Phys. Chem. C. 115 (2011) 17599–17605.
[61] G.G. Amatucci, F. Badway, A. Du Pasquier, T. Zheng, An asymmetric hybrid nonaqueous energy storage cell, J. Electrochem. Soc. 148 (2001) 930.
[62] J. Cherusseri, D. Pandey, J. Thomas, Symmetric, asymmetric, and battery-type supercapacitors using two-dimensional nanomaterials and composites, Batteries Supercaps 3 (2020) 860–875.
[63] E.I. Akpan, X. Shen, B. Wetzel, K. Friedrich, Design and synthesis of polymer nanocomposites, Polymer Composites with Functionalized Nanoparticles, Elsevier, 2019, pp. 47–83.
[64] M. Calisir, E. Stojanovska, A. Kilic, Polymer-Based Nanocomposites for Energy and Environmental Applications, Elsevier, 2018, pp. 361–396.
[65] F. Xu, H. Zhang, L. Jin, Y. Li, J. Li, G. Gan, et al., Controllably degradable transient electronic antennas based on water-soluble PVA/TiO_2 films, J. Mater. Sci. 53 (2018) 2638–2647.
[66] V. Siva, A. Murugan, A. Shameem, S. Thangarasu, S.A. Bahadur, A facile microwave-assisted combustion synthesis of $NiCoFe_2O_4$ anchored polymer nanocomposites as an efficient electrode material for asymmetric supercapacitor application, J. Energy Storage. 48 (2022) 103965.
[67] T. Hanemann, D.V. Szabó, Polymer-nanoparticle composites: from synthesis to modern applications, Materials 3 (2010) 3468–3517.
[68] W. Zheng, X. Lu, S.C. Wong, Electrical and mechanical properties of expanded graphite-reinforced high-density polyethylene, J. Appl. Poly. Sci. 91 (2004) 2781–2788.
[69] J. Liang, Y. Wang, Y. Huang, Y. Ma, Z. Liu, J. Cai, et al., Electromagnetic interference shielding of graphene/epoxy composites, Carbon N. Y. 47 (2009) 922–925.
[70] E.M. Sozer, P. Simacek, S.G. Advani, Resin transfer molding (RTM) in polymer matrix composites, Manufacturing Techniques for Polymer Matrix Composites (PMCs), Elsevier, Woodhead Publishing, 2012, pp. 245–309.
[71] S. Park, R.S. Ruoff, Chemical methods for the production of graphenes, Nat. Nanotechnol. 4 (2009) 217–224.
[72] Q. Guo, R. Ghadiri, T. Weigel, A. Aumann, E.L. Gurevich, C. Esen, et al., Comparison of in situ and ex situ methods for synthesis of two-photon polymerization polymer nanocomposites, Polymers 6 (2014) 2037–2050.
[73] Y. Tan, Y. Zhang, L. Kong, L. Kang, F. Ran, Nano-Au@ PANI core-shell nanoparticles via in-situ polymerization as electrode for supercapacitor, J. Alloys Compd. 722 (2017) 1–7.
[74] W.K. Chee, H.N. Lim, N.M. Huang, I. Harrison, Nanocomposites of graphene/polymers: a review, RSC Adv. 5 (2015) 68014–68051.
[75] J.S. Lee, D.H. Shin, W. Kim, J. Jang, Highly ordered, polypyrrole-coated Co $(OH)_2$ architectures for high-performance asymmetric supercapacitors, J. Mater. Chem. A 4 (2016) 6603–6609.
[76] P. Asen, S. Shahrokhian, One step electrodeposition of V_2O_5/polypyrrole/graphene oxide ternary nanocomposite for preparation of a high performance supercapacitor, Int. J. Hydrog. Energy. 42 (2017) 21073–21085.

[77] A.B. Reddy, G.S. M. Reddy, V. Sivanjineyulu, J. Jayaramudu, K. Varaprasad, E.R. Sadiku, Hydrophobic/hydrophilic nanostructured polymer blends, in: Design and Applications of Nanostructured Polymer Blends and Nanocomposite Systems, William Andrew Publishing, pp. 385–411.
[78] N. Bhardwaj, S.C. Kundu, Electrospinning: a fascinating fiber fabrication technique, Biotechnol. Adv. 28 (2010) 325–347.
[79] G. Verreck, I. Chun, J. Peeters, J. Rosenblatt, M.E. Brewster, Preparation and characterization of nanofibers containing amorphous drug dispersions generated by electrostatic spinning, Pharma. 20 (2003) 810–817.
[80] J.F. Cooley, Improved methods of and apparatus for electrically separating the relatively volatile liquid component from the component of relatively fixed substances of composite fluids. United Kingdom Patent, 6385, 1900.
[81] R. Goyal, K.M. Lauren, M.K. Hilton, K. Joachim, Nanoparticles and nanofibers for topical drug delivery, J. Control. Release 240 (2016) 77–92.
[82] S.K. Simotwo, C. DelRe, V. Kalra, Supercapacitor electrodes based on high-purity electrospun polyaniline and polyaniline–carbon nanotube nanofibers, ACS Appl. Mater. Interfaces 8 (2016) 21261–21269.
[83] L.A. Utracki, M.R. Kamal, Clay-containing polymeric nanocomposites, Arab. J. Sci. Eng. 27 (2002) 43–67.
[84] D. Verma, K.L. Goh, Functionalized graphene-based nanocomposites for energy applications, Functionalized Graphene Nanocomposites and Their Derivatives, Elsevier, 2019, pp. 219–243.
[85] A. VasudeoRane, K. Kanny, V.K. Abitha, S.S. Patil, S. Thomas, Clay–polymer composites: design of clay polymer nanocomposite by mixing, Clay-Polymer Nanocomposites, Elsevier, 2017, pp. 113–144.
[86] D. Ponnamma, J.J. Cabibihan, M. Rajan, S.S. Pethaiah, K. Deshmukh, J.P. Gogoi, et al., Synthesis, optimization and applications of ZnO/polymer nanocomposites, Mater. Sci. Eng. C. 98 (2019) 1210–1240.
[87] D.P. Dubal, P. Gomez-Romero, B.R. Sankapal, R. Holze, Nickel cobaltite as an emerging material for supercapacitors: an overview, Nano Energy 11 (2015) 377–399.
[88] P. Sen, A. De, Electrochemical performances of poly (3, 4-ethylenedioxythiophene)–NiFe2O4 nanocomposite as electrode for supercapacitor, Electrochim. Acta 55 (2010) 4677–4684.
[89] V. Gupta, S. Gupta, N. Miura, Potentiostatically deposited nanostructured Co_xNi_{1-x} layered double hydroxides as electrode materials for redox-supercapacitors, J. Power Sour. 175 (2008) 680–685.
[90] K.Z. Htut, M. Kim, E. Lee, G. Lee, S.H. Baeck, S.E. Shim, Biodegradable polymer-modified graphene/polyaniline electrodes for supercapacitors, Synth. Met. 227 (2017) 61–70.
[91] M.M. Sk, C.Y. Yue, R.K. Jena, Graphene/heparin template-controlled polyaniline nanofibers composite for high energy density supercapacitor electrode, Mater. Res. Express. 1 (2014) 045051.
[92] H.P. Cong, X.C. Ren, P. Wang, S.H. Yu, Flexible graphene–polyaniline composite paper for high-performance supercapacitor, Energy Environ. Sci. 6 (2013) 1185–1191.
[93] D.W. Wang, F. Li, J. Zhao, W. Ren, Z.G. Chen, J. Tan, Fabrication of graphene/polyaniline composite paper via in situ anodic electropolymerization for high-performance flexible electrode, ACS Nano 3 (2009) 1745–1752.
[94] Q. Wu, Y. Xu, Z. Yao, A. Liu, G. Shi, Supercapacitors based on flexible graphene/polyaniline nanofiber composite films, ACS Nano 4 (2010) 1963–1970.
[95] J. Li, H. Xie, Y. Li, J. Liu, Z. Li, Electrochemical properties of graphene nanosheets/polyaniline nanofibers composites as electrode for supercapacitors, J. Power Sour. 196 (2011) 10775–10781.
[96] J.E. Yang, I. Jang, M. Kim, S.H. Baeck, S. Hwang, S.E. Shim, Electrochemically polymerized vine-like nanostructured polyaniline on activated carbon nanofibers for supercapacitor, Electrochim. Acta 111 (2013) 136–143.
[97] R. Ramkumar, M. Minakshi, Fabrication of ultrathin CoMoO4 nanosheets modified with chitosan and their improved performance in energy storage device, Dalton Trans. 44 (2015) 6158–6168.

[98] Z. Çıplak, A. Yıldız, N. Yıldız, Green preparation of ternary reduced graphene oxide-au@ polyaniline nanocomposite for supercapacitor application, J. Energy Storage 32 (2020) 101846.
[99] S. Sahoo, G. Karthikeyan, G.C. Nayak, C.K. Das, Modified graphene/polyaniline nanocomposites for supercapacitor application, Macromol. Res. 20 (2012) 415–421.
[100] M. Ates, Y. Yuruk, Facile preparation of reduced graphene oxide, polypyrrole, carbon black, and polyvinyl alcohol nanocomposite by electrospinning: a low-cost and sustainable approach for supercapacitor application, Ionics 27 (2021) 2659–2672.
[101] X. Zhang, W. Yang, Y. Ma, Synthesis of polypyrrole-intercalated layered manganese oxide nanocomposite by a delamination/reassembling method and its electrochemical capacitance performance, Electrochem. Solid-State Lett. 12 (2009) A95.
[102] L. Li, H. Song, Q. Zhang, J. Yao, X. Chen, Effect of compounding process on the structure and electrochemical properties of ordered mesoporous carbon/polyaniline composites as electrodes for supercapacitors, J. Power Sour. 187 (2009) 268–274.
[103] M. Ates, M. El-Kady, R.B. Kaner, Three-dimensional design and fabrication of reduced graphene oxide/polyaniline composite hydrogel electrodes for high performance electrochemical supercapacitors, Nanotechnology 29 (2018) 175402.
[104] F.H. Hsu, T.M. Wu, In situ synthesis and characterization of conductive polypyrrole/graphene composites with improved solubility and conductivity, Synth. Met. 162 (2012) 682–687.
[105] M. Devi, A. Kumar, In-situ reduced graphene oxide nanosheets–polypyrrole nanotubes nanocomposites for supercapacitor applications, Synth. Met. 222 (2016) 318–329.
[106] Y. Liu, H. Wang, J. Zhou, L. Bian, E. Zhu, J. Hai, et al., Graphene/polypyrrole intercalating nanocomposites as supercapacitors electrode, Electrochim. Acta 112 (2013) 44–52.
[107] S. Dhibar, S. Sahoo, C.K. Das, Fabrication of transition-metal-doped polypyrrole/multiwalled carbon nanotubes nanocomposites for supercapacitor applications, J. Appl. Polym. Sci. 130 (2013) 554–562.
[108] G. Karthikeyan, C.K. Das, Transition metal salt doping on polypyrrole/MWCNT composites for supercapacitor applications, Macromolecular Symposia, WILEY-VCH Verlag, Weinheim, 2012, pp. 98–105.
[109] X. Sun, Y. Xu, J. Wang, S. Mao, The composite film of polypyrrole and functionalized multi-walled carbon nanotubes as an electrode material for supercapacitors, Int. J. Electrochem. 7 (2012) 3205–3214.
[110] M. Hughes, G.Z. Chen, M.S.P. Shaffer, D.J. Fray, A.H. Windle, Composites of carbon nanotubes and polypyrrole for electrochemical supercapacitors, Chem. Mater. 14 (2002) 1610–1613.
[111] S. Paul, J.H. Kim, D.W. Kim, Cycling performance of supercapacitors assembled with polypyrrole/multi-walled carbon nanotube/conductive carbon composite electrodes, J. Electrochem. Sci. Technol. 2 (2011) 91–96.
[112] S. Paul, Y.S. Lee, J. Choi, Y.C. Kang, D.W. Kim, Synthesis and electrochemical characterization of polypyrrole/multi-walled carbon nanotube composite electrodes for supercapacitor applications, B. Korean. Chem. Soc. 31 (2010) 1228–1232.
[113] K. Jurewicz, S. Delpeux, V. Bertagna, F. Beguin, E. Frackowiak, Supercapacitors from nanotubes/polypyrrole composites, Chem. Phys. Lett. 347 (2001) 36–40.
[114] B. Du, Q. Jiang, X.F. Zhao, B. Huang, Y. Zhao, Preparation of PPy/CNT composite applications for supercapacitor electrode material, Materials Science Forum, Trans Tech Publications Ltd, 2009, pp. 502–505.
[115] A. Alabadi, S. Razzaque, Z. Dong, W. Wang, B. Tan, Graphene oxide-polythiophene derivative hybrid nanosheet for enhancing performance of supercapacitor, J. Power Sour. 306 (2016) 241–247.
[116] R.B. Abdelkarim, A. Yahiaoui, M. Belbachir, A. Khelil, Synthesis and properties of polythiophène benzylidene and their photovoltaic applications, Int. J. Mater. Sci. Appl. 2 (2011) 1014.

[117] F. Alvi, M.K. Ram, P. Basnayaka, E. Stefanakos, Y. Goswami, A. Hoff, Electrochemical supercapacitors based on graphene-conducting polythiophenes nanocomposite, ECS Trans. 35 (2011) 167.

[118] A. Rose, K.G. Prasad, T. Sakthivel, V. Gunasekaran, T. Maiyalagan, T.J.A.S.S. Vijayakumar, Electrochemical analysis of graphene oxide/polyaniline/polyvinyl alcohol composite nanofibers for supercapacitor applications, Appl. Surf. Sci. 449 (2018) 551–557.

[119] S. Chuangchote, A. Sirivat, P. Supaphol, Mechanical and electro-rheological properties of electrospun poly (vinyl alcohol) nanofibre mats filled with carbon black nanoparticles, Nanotechnology 18 (2007) 145705.

[120] S. Nayak, A.A. Kittur, S. Nayak, Green synthesis of Silver-Zirconia composite using chitosan biopolymer binder for fabrication of electrode materials in supercapattery application for sustainable energy storage, CRGSC (2022) 100292.

[121] Y. Sun, W. Zeng, H. Sun, S. Luo, D. Chen, V. Chan, Inorganic/polymer-graphene hybrid gel as versatile electrochemical platform for electrochemical capacitor and biosensor, Carbon N. Y. 132 (2018) 589–597.

[122] S. Chakraborty, N.L. Mary, Biocompatible supercapacitor electrodes using green synthesised ZnO/Polymer nanocomposites for efficient energy storage applications, J. Energy Storage 28 (2020) 101275.

[123] H.W. Park, T. Kim, J. Huh, M. Kang, J.E. Lee, H. Yoon, Anisotropic growth control of polyaniline nanostructures and their morphology-dependent electrochemical characteristics, ACS Nano. 6 (2012) 7624–7633.

[124] W. Chen, R.B. Rakhi, H.N. Alshareef, Morphology-dependent enhancement of the pseudocapacitance of template-guided tunable polyaniline nanostructures, J. Phys. Chem. C. 117 (2013) 15009–15019.

[125] V. Sharma, A. Sahoo, Y. Sharma, P. Mohanty, Synthesis of nanoporous hypercrosslinked polyaniline (HCPANI) for gas sorption and electrochemical supercapacitor applications, RSC Adv. 5 (2015) 45749–45754.

[126] S. Cho, K.H. Shin, J. Jang, Enhanced electrochemical performance of highly porous supercapacitor electrodes based on solution processed polyaniline thin films, ACS Appl. Mater. Interfaces. 5 (2013) 9186–9193.

[127] S. Chaudhari, Y. Sharma, P.S. Archana, R. Jose, S. Ramakrishna, S. Mhaisalkar, et al., Electrospun polyaniline nanofibers web electrodes for supercapacitors, J. Appl. Polym. Sci. 129 (2013) 1660–1668.

[128] S. Sahoo, S. Dhibar, C.K. Das, Facile synthesis of polypyrrole nanofiber and its enhanced electrochemical performances in different electrolytes, Express Polym. Lett. 6 (2012).

[129] J.C. Carlberg, O. Inganäs, Poly (3, 4-ethylenedioxythiophene) as electrode material in electrochemical capacitors, J. Electrochem. Soc. 144 (1997) L61.

[130] A. Balducci, W.A. Henderson, M. Mastragostino, S. Passerini, P. Simon, F. Soavi, Cycling stability of a hybrid activated carbon//poly (3-methylthiophene) supercapacitor with N-butyl-N-methylpyrrolidinium bis (trifluoromethanesulfonyl) imide ionic liquid as electrolyte, Electrochim. Acta 50 (2005) 2233–2237.

[131] I. Shown, A. Ganguly, L.C. Chen, K.H. Chen, Conducting polymer-based flexible supercapacitor, Energy Sci. Eng. 3 (2015) 1–25.

[132] A. Laforgue, P. Simon, C. Sarrazin, J.F. Fauvarque, Polythiophene-based supercapacitors, J. Power Sour. 80 (1999) 142–148.

[133] B. Senthilkumar, P. Thenamirtham, R.K. Selvan, Structural and electrochemical properties of polythiophene, Appl. Surf. Sci. 257 (2011) 9063–9067.

[134] S.R.P. Gnanakan, M. Rajasekhar, A. Subramania, Synthesis of polythiophene nanoparticles by surfactant-assisted dilute polymerization method for high performance redox supercapacitors, Int. J. Electrochem. Sci. 4 (2009) 1289–1301.

[135] C. Fu, H. Zhou, R. Liu, Z. Huang, J. Chen, Y. Kuang, Supercapacitor based on electropolymerized polythiophene and multi-walled carbon nanotubes composites, Mater. Chem. Phys. 132 (2012) 596–600.

[136] M. Wesoły, X. Cetó, M. Del Valle, P. Ciosek, W. Wróblewski, Quantitative analysis of active pharmaceutical ingredients (APIs) using a potentiometric electronic tongue in a SIA flow system, Electroanalysis 28 (2016) 626–632.

[137] D. Antiohos, G. Folkes, P. Sherrell, S. Ashraf, G.G. Wallace, P. Aitchison, et al., Compositional effects of PEDOT-PSS/single walled carbon nanotube films on supercapacitor device performance, J. Mater. Chem. 21 (2011) 15987–15994.

[138] S. Dhibar, P. Bhattacharya, G. Hatui, C.K. Das, Transition metal doped poly (aniline-co-pyrrole)/multi-walled carbon nanotubes nanocomposite for high performance supercapacitor electrode materials, J. Alloys Compd. 625 (2015) 64–75.

[139] S. Sahoo, S. Dhibar, G. Hatui, P. Bhattacharya, C.K. Das, Graphene/polypyrrole nanofiber nanocomposite as electrode material for electrochemical supercapacitor, Polymer 54 (2013) 1033–1042.

[140] L. Mao, K. Zhang, H.S.O. Chan, J. Wu, Surfactant-stabilized graphene/polyaniline nanofiber composites for high performance supercapacitor electrode, J. Mater. Chem. 22 (2012) 80–85.

[141] S. Sahoo, P. Bhattacharya, S. Dhibar, G. Hatui, T. Das, C.K. Das, Graphene/poly (aniline-co-pyrrole) nanocomposite: potential candidate for supercapacitor and microwave absorbing applications, J. Nanosci. Nanotechnol. 15 (2015) 6931–6941.

[142] A. Torvi, S. Naik, M. Kariduraganavar, Development of supercapacitor systems based on binary and ternary nanocomposites using chitosan, graphene and polyaniline, Chem. Data Collect. 17 (2018) 459–471.

[143] J. Wang, Z. Wu, K. Hu, X. Chen, H. Yin, High conductivity graphene-like MoS_2/polyaniline nanocomposites and its application in supercapacitor, J. Alloys Compd. 619 (2015) 38–43.

[144] O. Misoon, K. Seok, Effect of dodecyl benzene sulfonic acid on the preparation of polyaniline/activated carbon composites by in situ emulsion polymerization, Electrochim. Acta 59 (2012) 196–201.

[145] R. Bortamuly, G. Konwar, P.K. Boruah, M.R. Das, D. Mahanta, P. Saikia, CeO_2-PANI-HCl and CeO_2-PANI-PTSA composites: synthesis, characterization, and utilization as supercapacitor electrode materials, Ionics 26 (2020) 5747–5756.

[146] X. Zhao, O. Grätz, J. Pionteck, Effect of dopant and oxidant on the electrochemical properties of polyaniline/graphite nanoplate composites, Polym. Inter. 67 (2018) 1429–1437.

[147] N.R. Dywili, A. Ntziouni, C. Ikpo, M. Ndipingwi, N.W. Hlongwa, A.L. Yonkeu, et al., Graphene oxide decorated nanometal-poly (anilino-dodecylbenzene sulfonic acid) for application in high performance supercapacitors, Micromachines 10 (2019) 115.

[148] S. Radhakrishnan, C.R. Rao, M. Vijayan, Performance of conducting polyaniline-DBSA and polyaniline-DBSA/Fe_3O_4 composites as electrode materials for aqueous redox supercapacitors, J. Appl. Polym. Sci. 122 (2011) 1510–1518.

[149] J. Chen, Y. Wang, J. Cao, L. Liao, Y. Liu, Y. Zhou, Pulsed electrochemical fabrication of graphene/polypyrrole composite gel films for high performance and flexible supercapacitors, Electrochim. Acta 361 (2020) 137036.

[150] X. Liu, L. Xue, Y. Lu, Y. Xia, Q. Li, Fabrication of polypyrrole/multi-walled carbon nanotubes composites as high performance electrodes for supercapacitors, J. Electroanal. Chem. 862 (2020) 114006.

[151] S. Akbulut, M. Yilmaz, S. Raina, S.H. Hsu, W.P. Kang, Advanced supercapacitor prototype using nanostructured double-sided MnO2/CNT electrodes on flexible graphite foil, J. Chem. 21 (2017) 15987–15994.

[152] C.G. Cameron, Cold temperature optimization of supercapacitors, ECS Trans. 41 (2012) 121.

[153] A. von Wald Cresce, O. Borodin, K. Xu, Correlating Li^+ solvation sheath structure with interphasial chemistry on graphite, J. Phys. Chem. C. 116 (2012) 26111–26117.

[154] C.R. Dennison, M. Beidaghi, K.B. Hatzell, J.W. Campos, Y. Gogotsi, E.C. Kumbur, Effects of flow cell design on charge percolation and storage in the carbon slurry electrodes of electrochemical flow capacitors, J. Power Sour. 247 (2014) 489–496.

[155] W. Münchgesang, P. Meisner, G. Yushin, Supercapacitors specialities-technology review, in: AIP Conference Proceedings, American Institute of Physics, 2014, pp. 196–20.

[156] J.P. Tey, M.A. Careem, M.A. Yarmo, A.K. Arof, Durian shell-based activated carbon electrode for EDLCs, Ionics 22 (2016) 1209–1216.

[157] A. Schneuwly, R. Gallay, Properties and applications of supercapacitors from the state-of-the-art to future trends, in: Proceeding PCIM, Citeseer, 2000.

[158] D.W. Park, N.A. Cañas, M. Schwan, B. Milow, L. Ratke, K.A. Friedrich, A dual mesopore C-aerogel electrode for a high energy density supercapacitor, Curr. Appl. Phys. 16 (2016) 658–664.

[159] J.R. Miller, A. Burke, Electrochemical capacitors: challenges and opportunities for real-world applications, Electrochem. Soc. Interface. 17 (2008) 53.

[160] M. Endo, T. Maeda, T. Takeda, Y.J. Kim, K. Koshiba, H. Hara, Capacitance and pore-size distribution in aqueous and nonaqueous electrolytes using various activated carbon electrodes, J. Electrochem. Soc. 148 (2001) A910.

[161] S.K.N. Kumar, P. Renuga, Optimal VAR planning using facts, Int. J. Electr. Power Energy Syst. 32 (2012) 49–56.

[162] G.J. Park, D. Kalpana, A.K. Thapa, H. Nakamura, Y.S. Lee, M. Yoshio, A novel hybrid supercapacitor using a graphite cathode and a niobium (V) oxide anode, B Korean Chem. Soc. 30 (2009) 817–820.

[163] Z. Fan, J. Yan, L. Zhi, Q. Zhang, T. Wei, J. Feng, A three-dimensional carbon nanotube/graphene sandwich and its application as electrode in supercapacitors, Adv. Mater. 22 (2010) 3723–3728.

CHAPTER

11 Advanced biodegradable polymer nanocomposites for rechargeable lithium and sodium-ion battery applications

Laila El Foujji[1], Abou el kacem Qaiss[2] and Rachid Bouhfid[2]

[1]*Moroccan Foundation for Advanced Science, Innovation and Research (MAScIR), Composites et Nanocomposites Center, Rabat Design Center, Madinat El Irfane, Rabat, Morocco* [2]*Mohammed VI Polytechnic University, Hay Moulay Rachid, Ben Guerir, Morocco*

Chapter Outline

11.1 Introduction

The world's reliance on fossil fuel energy is taking a serious and perilous turn. We will face a shortage of this energy in the not-too-distant future due to the depletion of its sources [1]. Furthermore, the use of this type of energy significantly increases environmental pollution, which has a negative impact on living species [2]. Thus, scientists and researchers are making broad efforts to reduce energy consumption and investigate new alternatives. Alternatives to traditional fossil fuel energy that are renewable and cleaner have emerged [3]. Renewable energy resources exist in nature in seven forms: solar, wind, wave, tidal, water, wood, and geothermal [4]. However, these energies are all intermittent in nature and cannot maintain a sustainable character in the face of our daily high energy demands. As a result, energy-storing assets are required to achieve a sustainable character, or to bridge the gap between intermittent and high peak energy demands [5]. To meet the various demands of several electric engines, the energy-storage systems (capacitors, batteries...) that will be used must be efficient. Batteries are a major component of electronic devices in our daily lives (portable cell phones and laptops, artificial pacemakers, hybrid and electric vehicles, etc.), and they are subject to new demands in terms of capacity, longevity, and safety,

Biodegradable and Biocompatible Polymer Nanocomposites. DOI: https://doi.org/10.1016/B978-0-323-91696-7.00014-3
Copyright © 2023 Elsevier Inc. All rights reserved.

posing a significant challenge to manufacturers [6]. The following framework (Fig. 11.1) shows the Ragone plot, used to compare the density of the different battery technology systems and can be used for other studies as comparison of analytical results and experimental ones... [7–9].

Among the new battery technologies, lithium-ion batteries (LIBs) and sodium-ion batteries (SIBs) stand out as promising energy-storage systems to replace the current ones [10]. LIBs have been widely used in a variety of applications due to properties such as high energy density, long cycling stability, high output potential, low self-discharge, and environmental protection [11,12]. SIBs have also been proposed as a valuable electrochemical energy-storage technology. They have interesting properties as LIB analogs, such as an abundance of sodium, promising electrochemical performance, and being environmentally benign [13,14]. Continuous research is being conducted to improve the properties of each battery component in order to improve the performance of both batteries [15–21]. Fig. 11.2 shows a general presentation of battery anatomy, showing the different components as well as the pathways of electrons and ions (M^+: here, it can be lithium ions or sodium ions), while the battery is operating (discharge and charge).

The anode and cathode (battery electrodes) are made of active materials (layers that are responsible for the first responsible criterion of a battery's performance). The electrode-specific capacity

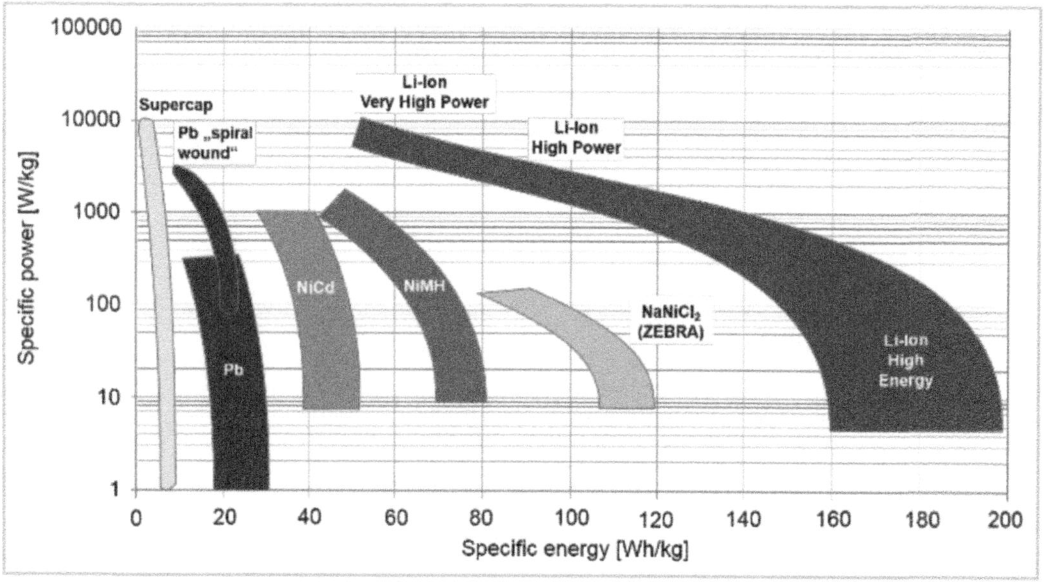

FIGURE 11.1

Ragone plot of various battery technologies with specification at the cell level for automotive applications without lithium-sulfur and metal-air batteries.

Reprinted from M. Richter, A. Sarram, C. Kaucher, H. Winkler, Investigation and analysis of accumulators for the use of electrochemical storage in hybrid shunting locomotives, Procedia CIRP. 81 (2019) 1010–1015. https://doi.org/10.1016/j.procir.2019.03.243 with permission from Elsevier. https://s100.copyright.com/MyAccount/viewLicenseDetails?ref= 51c2ad3d-4b5c-4b88-93f7-4d6d1729febf. Copyright 2019.

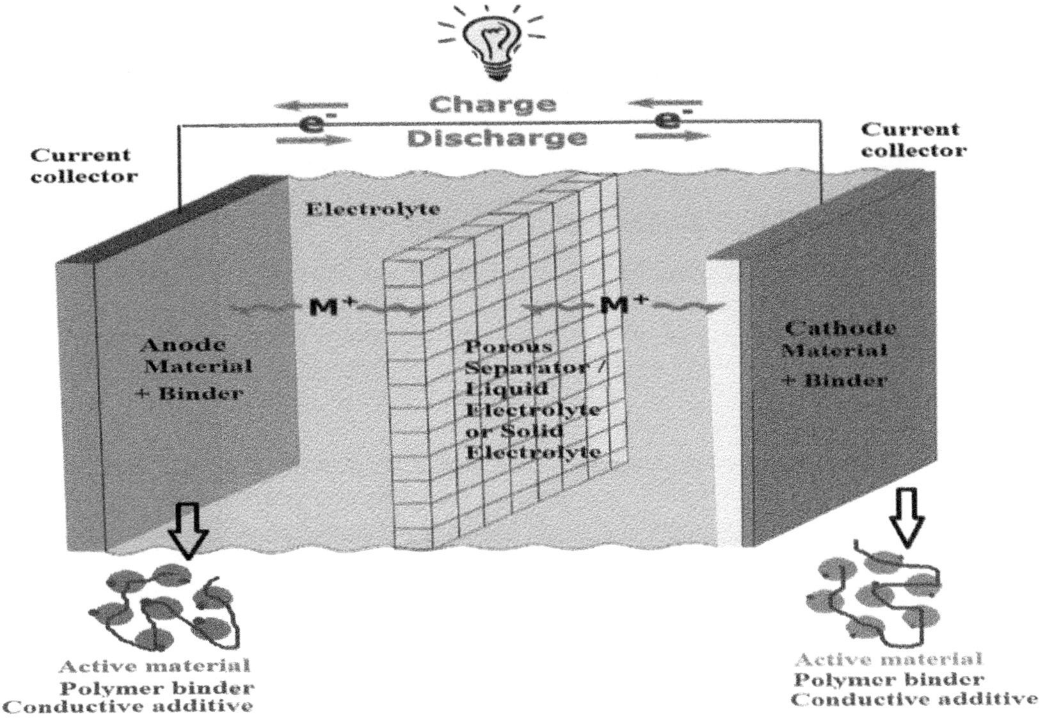

FIGURE 11.2

General battery anatomy.

and the differential potential between them determine the energy density of a battery system [22]. Conductive additives form percolating networks in the electrodes to ensure electronic transport between active material layers [23,24]. The binder is an important component of the electrodes. Polymer binder, which is typically made of polymers or a combination of polymers and nanocomposite modified polymers [25], serves the function of holding electrode materials together as well as providing flexibility and mechanical integrity of the current collector to the electrodes [26]. The separator and electrolyte can be thought of as complementary components. They maintain the electrical separation of electrodes by preventing electron movement while allowing ions to move freely [27–29]. The use of polymers in rechargeable batteries is gaining increasing interest [30], thanks to their intrinsic properties, specifically: lightweight [31], chemical resistance, mechanical flexibility, and easy process... Yet, the nonsustainability of these materials (resourced from fossil fuels already mentioned) and the harmful effects they got on the environment [32], is a strong motivation to come up with new materials with the same advantageous properties and lesser negative effects. Biodegradable polymers will be the foundation of polymer use in the future, particularly in batteries to make them safer. This book chapter will provide a small push toward this specific type of polymer. We will present a series of biodegradable polymers, both synthetically synthesized and natural, that will be integrated into the various components of the battery and used in conjunction with other materials to create nanocomposites functionalized materials. Electrodes, binders, separators, and polymer electrolytes are examples of battery components.

11.2 Biodegradable polymer classification

A polymer is considered biodegradable when it contains functional groups at its backbone chain that can undergo hydrolysis and oxidation reactions, eventually breaking down to smaller compounds. The biodegradation process is carried out solely by enzymes or in conjunction with a chemical deterioration of organic substances associated with microbial living organisms [33,34]. Biodegradable polymers are typically classified according to their synthesis, which includes biomass products, polymers derived from microorganisms, polymers synthesized via conventional routes from bio-derived monomers, and polymers derived from synthetic monomers obtained from petrochemicals [35]. As mentioned, the sources of biodegradable polymers are described as three origins: plants or vegetables, animals, and synthetic origins. The following list presents the most described biodegradable polymers in literature: poly(glycolic acid) (PGA), polylactic acid (PLA), poly(vinyl alcohol) (PVA), polyethylene oxide (PEO), polyacrylic acid (PAA), poly(lactide-*co*-glycolide) (PLGA), poly(ε-caprolactone) (PCL), polyhydroxyalkanoates (PHAs), polyethylene glycol (PEG), poly (butylene succinate) (PBS), poly (propylene fumarate) (PPF), polyesters, chitosan/chitin (CS/CT), collagen and sericin, lignin, alginate, cellulose acetate (CA), starch... [36]. Studies based on some of these biodegradable polymers will be presented in the following section. Fig. 11.3, shows the chemical structures of some synthetic and natural biodegradable polymers [37].

11.3 Biodegradable polymer nanocomposite in rechargeable LIBs and SIBs

The use of biodegradable polymers in battery components is an intriguing feature for improving a battery's overall properties. Both types of biodegradable polymers will be discussed separately in the following section. Furthermore, some recent research examples of biodegradable polymer nanocomposites will be presented for each battery component (electrode binder, separator, or polymer electrolyte).

11.3.1 Synthetic biodegradable polymer nanocomposites

Biodegradable polymeric materials are especially important in LIBs and SIBs because they are used as binders for electrode materials, but their role as a separator and solid electrolyte is more important. Their use as separators and solid electrolytes will be highlighted.

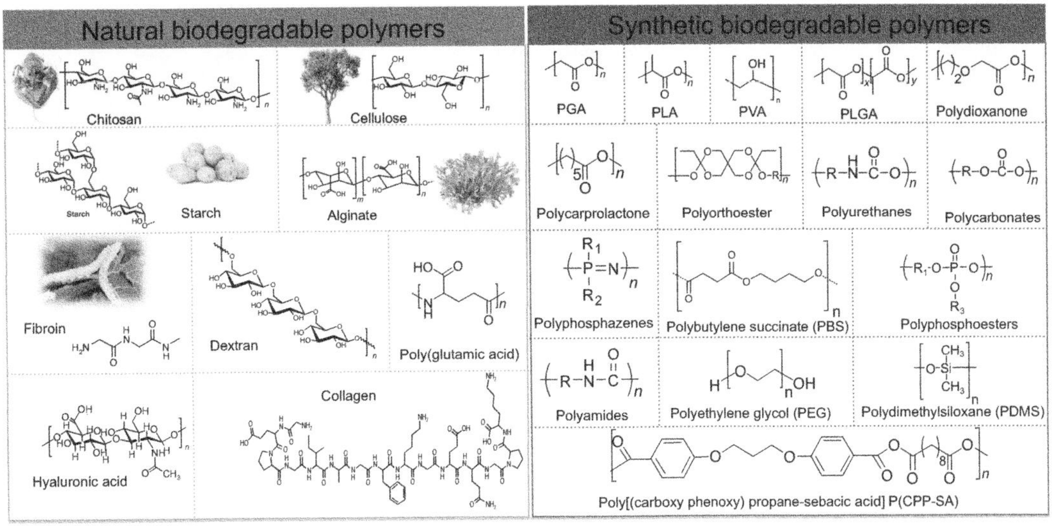

FIGURE 11.3

Chemical structures of some biodegradable polymers.

Adapted from E.S. Hosseini, S. Dervin, P. Ganguly, R. Dahiya, Biodegradable materials for sustainable health monitoring devices, ACS Appl. Bio Mater. 4 (2021) 163–194. https://doi.org/10.1021/acsabm.0c01139, 2021 Copyrights: This article is an open access article distributed under the terms and conditions of the Creative Commons Attribution (CC BY) license (http://creativecommons.org/licenses/by/4.0/).

11.3.1.1 Electrode binders

A binder's primary function in batteries is to ensure the interconnection of the electrode materials, specifically the active material and the conductive additive, while also ensuring perfect adherence of these materials to the current collectors. This critical decision results in the prevention of material disintegration [38]. A quick view of the interaction between the binder and other components of the electrode is shown in Fig. 11.4. Depending on the type of binders and their affinity for the other components, the binding mechanism can be viewed as a dot to surface contact, a segment to surface contact, or a network to surface contact [39,40].

An ideal binder should meet some stringent requirements, the illustration of these desired properties is shown in Fig. 11.5 [38,41].

The binder, which is typically made of polymers, has a significant impact on the battery's properties. The impact can be seen at both the macroscopic and microscopic levels Maurel et al. [42] developed a biodegradable PLA polymer and graphite nanocomposite material. The Fused Deposition Modeling technique was used to shape it in 3D-printable filament. This material was created with the intention of serving as the negative electrode in LIBs. Another biodegradable polymer (PEG) was used in this study to improve the material's flexibility, and mechanical property tests allowed the right amount of plasticizer to be added to avoid the brittle and very soft character of the slurry material. Carbon nanofibers were used as a conductive additive, increasing electrical conductivity and specific capacity. The increase in specific capacity indicates that the PLA matrix

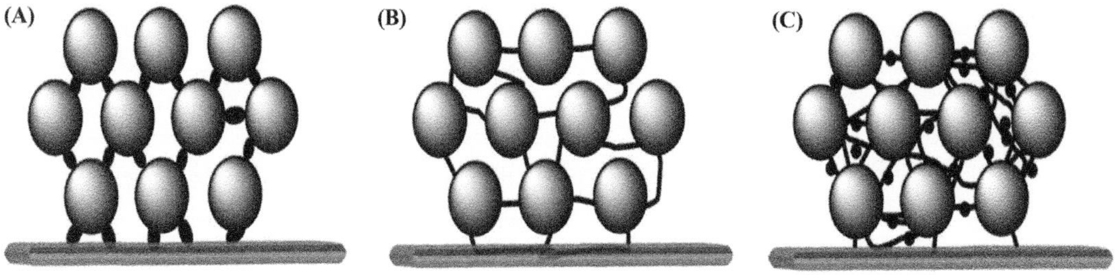

FIGURE 11.4

Illustration of interactions between binders and electrode components: (A) dot to surface contact, (B) segment to surface contact, and (C) network to surface contact.

Reprinted from P.S. Salini, S. V. Gopinadh, A. Kalpakasseri, B. John, M. Thelakkattu Devassy, Toward greener and sustainable Li-ion cells: an overview of aqueous-based binder systems, ACS Sustain. Chem. Eng. 8 (2020) 4003–4025. https://doi.org/10.1021/acssuschemeng.9b07478, with permission from American Chemical Society. Copyright 2020.

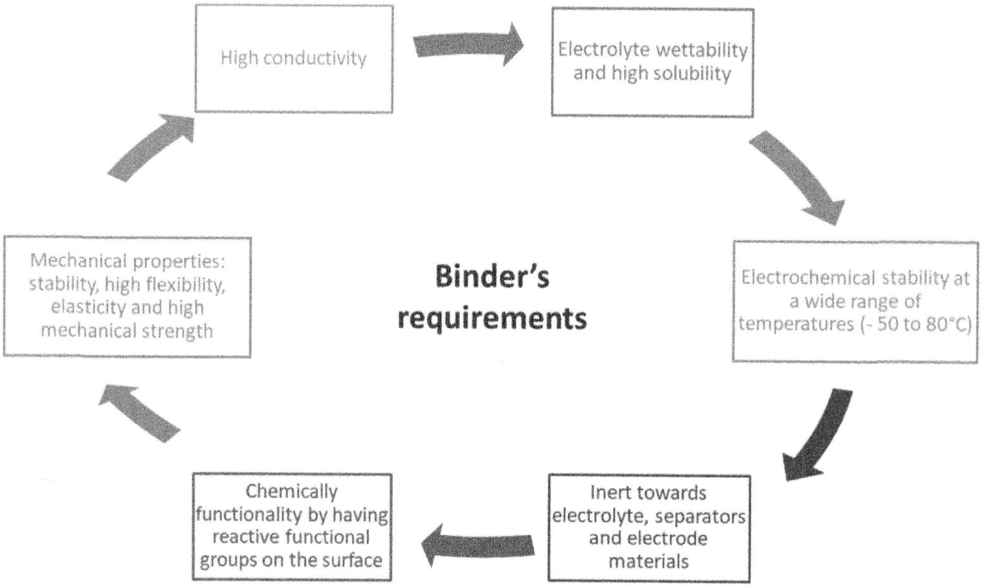

FIGURE 11.5

General requirements of a battery's binder.

has good electronic and ionic percolation. Gupta et al. [43], used PLA as well, and it was used as a binding agent. Carbon nanotubes are used to ensure the nanocomposite character, while lithium iron phosphate is used as the active material ($LiFePO_4$). The specific capacity of the PLA-elaborated cathode was increased to around 155 mAh/g. Fu et al. [44], reported the use of a nano-composite coating based on lithium fluoride (LiF) particles, embedded into the PEO polymer. The goal is to protect the electrode's surface from dendritic metal growth. This coating increased conductivity and the transfer of lithium ions. PEO was chosen because of its high donor number of lithium ions, as well as its ether functional groups, which have high reductive stability. A recent study by Phanikumar et al. [45], presented and for the first time, aqueous binders based on PVA and sodium alginate (natural biodegradable polymer). PVA's wettability by the electrolyte was superior to that of other polymers due to its hydroxyl groups. To investigate the electrochemical stability of both aqueous binders, cyclic voltammetry was used. The voltammograms after the first and third cycles demonstrate PVA's stability. Nguyen et al. [46] investigated the effect of using PVA as a binder on nanomechanical properties and adhesion using colloidal probe Atomic Force Microscopy. PVA was compared to other binders as well as a commercial polyvinylidene fluoride-based binder (PVDF). Fig. 11.6 depicts the results of a PVA-based binder in air and electrolyte.

Based on these findings, PVA appears to be more advantageous than commercial PVDF due to its relatively large elastic moduli, which is an important property for achieving better electrochemical performances. Another synthetic biodegradable polymer, PAA, was investigated as a matrix for holding composite materials of LIBs anode together. Parikh et al. [47] used a PPA binder for the control of the volume expansion of the silicon anode of LIBs. Fig. 11.7 shows the cycling performance of the half-cells. Fig. 11.7A shows the first cycle voltage profile for the electrode with the PAA as a binder, and to investigate the performances of this binder, a comparison to a carboxymethylcellulose (CMC) one is done.

The first discharge capacity is 3303 mAh/g and the charge capacity is 2540 mAh/g for the CMC binder, while the discharge capacity is 3650 mAh/g and the charge capacity is 3088 mAh/g for the PAA binder. Fig. 11.7B depicts the lithiation capacity; electrodes with the PAA binder have a greater delithiation capacity than electrodes with the CMC binder. Although both cells with different binders have a similar drop in capacity of about 500 mAh/g, the PAA binder allows the anode to retain 80% of its capacity compared to CMC's 73%. Hu et al. [48] investigated the effect of the polymer binder's molecular weight (MW) on the overall properties and performance of the battery cells. We know that the MW regulates physicochemical properties such as viscosity and adhesion, which influence the handling and performance of battery slurries. Several characterizations of the PAA binder were performed in this study, including rheological measurements and cycling tests, among others. Table 11.1 provides the values of the different parameters that were examined in this study. The number average MW (Mn) is presented for each PAA type.

The average-capacity results show that the higher Mn, the better the average-capacity. When these results are combined with the results of the average-capacity profiles, an optimal MW is observed. PAA binders with Mn less than 24 kDa performed worse than binders with kDa greater than this value. That same biodegradable polymer was used as an electrode binder in another study for SIBs. Ni et al. [49] demonstrated the ease and benefit of using an appropriate binder for electrode materials that change their volume dramatically during charging-discharging cycles (sodiation and desodiation process). PAA outperformed the commonly used PVDF in terms of mechanical properties. The cathode material is iron(II) fluoride-reduced graphene oxide (FeF_2-rGO). Several techniques were used to investigate the nanocomposite material. There was an improvement in rate and cycling performance. This time, we'll be looking at adhesion tests as another property that affects battery properties. The peeling test was used to investigate the adhesion performance in this case. A comparison was performed for the same active material, once with PVDF as a binder and once with PAA. Fig. 11.8 shows the results of the force-displacement curves of both electrode nanocomposites.

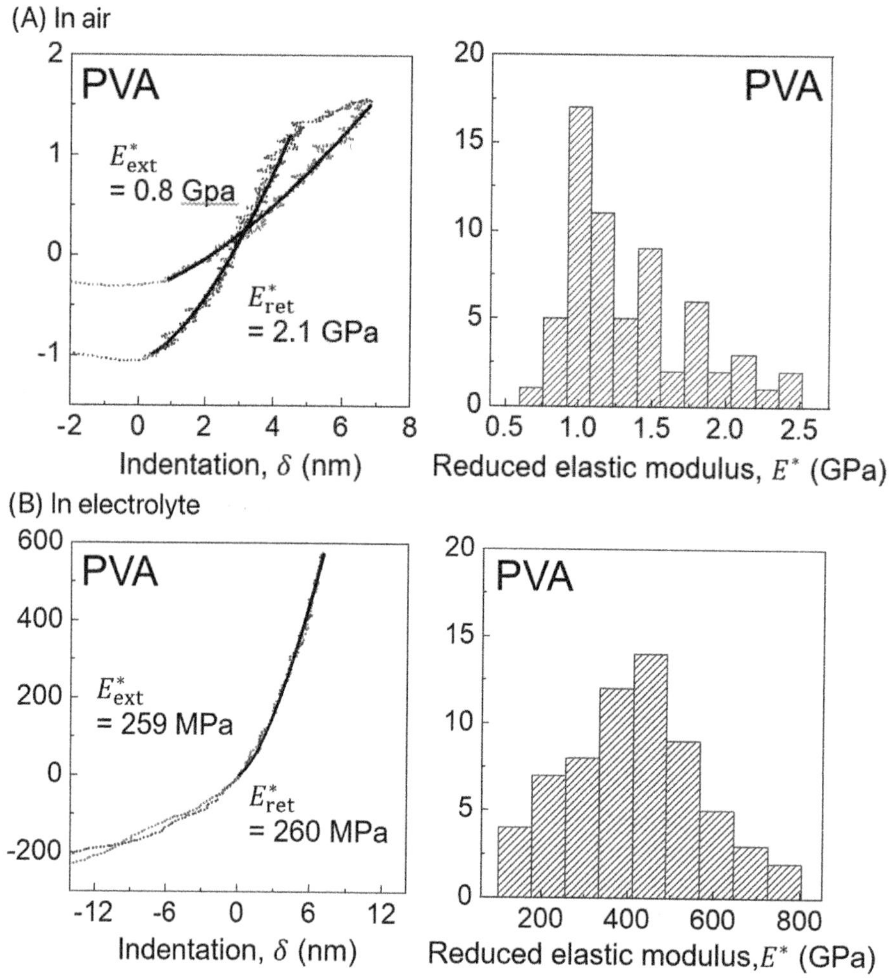

FIGURE 11.6

Force-indentation curves left (A) and left (B) in air and electrolyte, respectively. Histograms of the means of reduced elastic moduli determined from extension and retraction curves right (A) and right (B) in air and in electrolyte, respectively.

Reprinted from Q.D. Nguyen, E.S. Oh, K.H. Chung, Nanomechanical properties of polymer binders for Li-ion batteries probed with colloidal probe atomic force microscopy, Polym. Test. 76 (2019) 245–253. https://doi.org/10.1016/j.polymertesting.2019.03.025 with permission from Elsevier. https://s100.copyright.com/MyAccount/viewLicenseDetails?ref= a95a3f79-239c-4cbd-bea7-861f1d094669. Copyright 2019.

The results of the force-displacement showed a greater adhesion strength of the material made by the PAA binder which valued 108 N/cm, a nine times better value compared to the PVDF one giving only a value of 12 N/cm. In contrast to PAA, the optical images show a complete peel-off

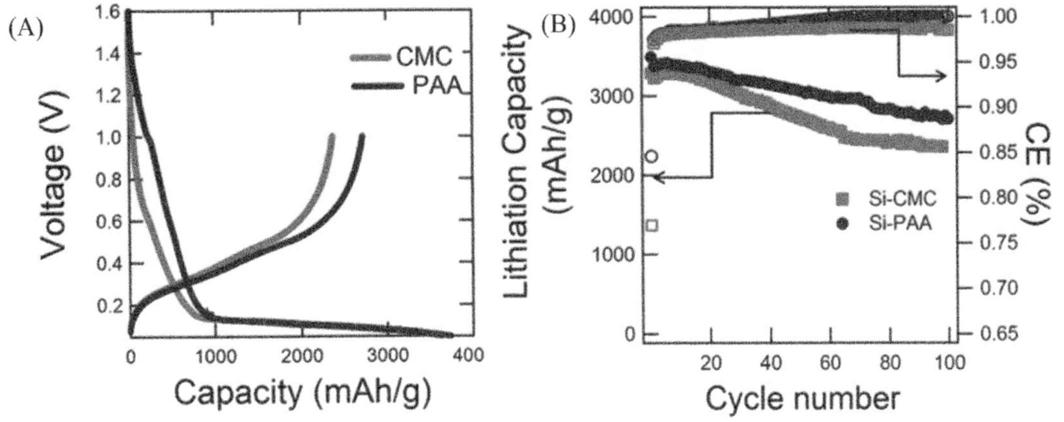

FIGURE 11.7

(A) 1st cycle voltage profile for electrodes with CMC and PAA binders. (B) Lithiation capacity and Coulombic efficiency as a function of cycle number.

Reprinted from P. Parikh, M. Sina, A. Banerjee, X. Wang, M.S. D'Souza, J.M. Doux, et al., Role of Polyacrylic Acid (PAA) Binder on the Solid Electrolyte Interphase in Silicon Anodes, Chem. Mater. 31 (2019) 2535–2544. https://doi.org/10.1021/acs.chemmater.8b05020 with permission from American Chemical Society. Copyright 2019.

Table 11.1 Porosity, initial specific capacity, capacity retention and the average-capacity data of Silicon/graphite electrodes fabricated using the corresponding lithiated-PAA binders.

Samples with different M_n (kDa)	Porosity (%)	Initial specific capacity (mAh/g)	Capacity retention (%)	Average capacity (mAh/g)
167	54	761	42.3	445
36	54	636	41.7	392
147	54	755	45.4	443
24	56	678	45	407
3.0	56	607	45.8	336
1.6	52	399	56.1	250

Adapted from B. Hu, I.A. Shkrob, S. Zhang, L. Zhang, J. Zhang, Y. Li, et al., The existence of optimal molecular weight for poly (acrylic acid) binders in silicon/graphite composite anode for lithium-ion batteries, J. Power Sources. 378 (2018) 671–676. https://doi.org/10.1016/j.jpowsour.2017.12.068 with permission of Elsevier. https://s100.copyright.com/MyAccount/viewLicenseDetails?ref=5b2a113b-2ace-4368-99c2-3373abcacd3b. Copyright 2018.

from the aluminum foil for the material with PVDF binder. This demonstrates that the stiffness and robustness of PAA improved adhesion between the active materials and the current collector. The influence of the binder was also investigated through analysis of the surface chemical composition. However, to avoid confusion caused by the use of reduced graphene oxide and the binder on the

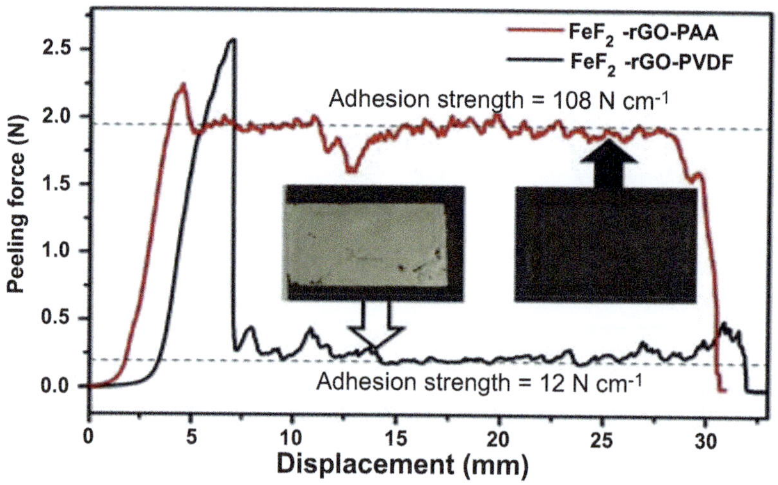

FIGURE 11.8

Force-displacement curves of FeF_2-rGO-PAA and FeF_2-rGO-PVDF electrodes. The insets are the optical images of electrodes after the peeling test.

Reprinted from D. Ni, W. Sun, C. Lu, Z. Wang, J. Qiao, H. Cai, et al., Improved rate and cycling performance of FeF_2-rGO hybrid cathode with poly (acrylic acid) binder for sodium ion batteries, J. Power Sources. 413 (2019) 449–458. https://doi.org/10.1016/j.jpowsour.2018.12.040 with permission from Elsevier. https://s100.copyright.com/MyAccount/viewLicenseDetails?ref= 849d067a-7589-41eb-94d8-1f83a8d9de4f. Copyright 2019.

chemical composition, electrodes made of iron fluoride FeF_2 alone were used. Results of X-ray photoelectron spectroscopy (XPS) are presented in Fig. 11.9.

Two intense peaks are seen at 712 and 725 eV corresponding to Fe^{2+}, yet the intensity in the electrode containing PAA is lower than the one with PVDF. The iron atomic percentage in the PAA-containing electrode and the PVDF electrode is 8.82 and 13.00, respectively. This can be explained by the uniform coverage and distribution of PAA on the electrode surface when compared to the PVDF electrode surface. Mandal et al. [50] used a PVA binder for nano silicon-based electrode LIBs and investigated the effect of PVA side group modification on battery performance. Nonetheless, they discovered that regardless of the type of functional modification introduced or how minor it is, it has a significant impact on the crystallinity of the PVA. As a result, different PVA binders were used to highlight the hypothesis of crystallinity affecting battery cycling. Because the crystallinity degrees of PVA binders are directly related to the number of OH groups on their surface, the main parameter to change for crystallinity change is the hydrolysis process. PVA crystallinity values of PVA-98, PVA-87, and PVA-80 were used, corresponding to degrees of hydrolysis of 98%, 87%–89%, and 80%, respectively. To confirm the hydrolysis effect on crystallinity, the melting temperature (T_m) and the values of the crystallinity were studied, the results are as follow: For the PVA-98 sample, T_m equals 211°C compared to 178°C for PVA-87 and 164°C for PVA-80. We can see a difference between melting temperatures that decrease as the hydrolysis process decreases; this can be explained by the size of the crystallites that decrease, which is

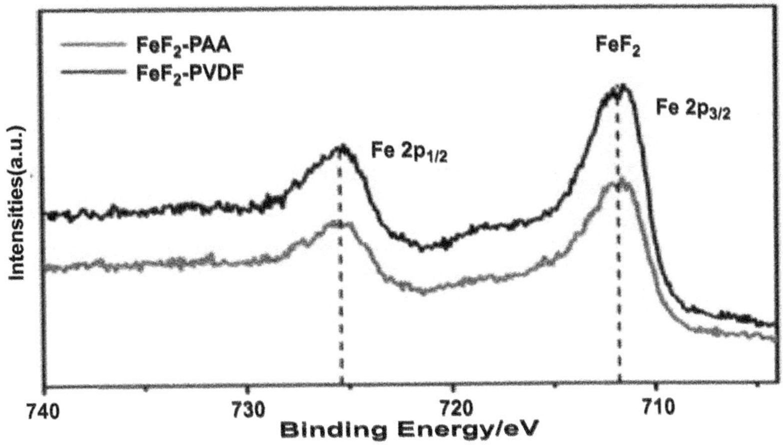

FIGURE 11.9

XPS spectra of pure FeF_2 electrodes prepared by PAA and by PVDF binders.

Reprinted from D. Ni, W. Sun, C. Lu, Z. Wang, J. Qiao, H. Cai, et al., Improved rate and cycling performance of FeF_2-rGO hybrid cathode with poly (acrylic acid) binder for sodium ion batteries, J. Power Sources. 413 (2019) 449–458. https://doi.org/10.1016/j.jpowsour.2018.12.040 with permission from Elsevier. https://s100.copyright.com/MyAccount/viewLicenseDetails?ref=849d067a-7589-41eb-94d8-1f83a8d9de4f. Copyright 2019.

confirmed by the crystallinity percentage, which shows a similar decrease. PVA-98 (the pristine form) has the highest value of 46%, while PVA-80 (the less hydrolyzed binder) has a value of 17%. The cycling tests are the next step in demonstrating the performance of the various binders. The tests are presented in Fig. 11.10.

The capacity value of the electrode with pristine PVA-98 binder shows a stable value at 869 mAh/g after the 100th cycle at C/5 rate. Yet, the value of this capacity for the other two binders is 472 and 498 mAh/g for PVA-87 and PVA-80, respectively. This study draws two conclusions: the first is that pristine PVA-98 performs well as a binder, and the second is that binder crystallinity degree has a positive effect on battery performance. The use of nanocomposite synthetic biodegradable polymers as battery binders demonstrates that they are superior to trivial blends of nonbiodegradable polymers. The battery binder is an important component of the electrodes because of its role in the electrochemical performance of the battery.

11.3.1.2 Battery separators

A barrier material called a separator is required to prevent physical contact between rechargeable battery electrodes, which leads to short circuits and thus battery failure. Along with ensuring ionic conductivity, this property is the separator's traditional and primary function (ions transport) [51]. Because not all battery separators are ionically conductive, the separator must perform an important function, which is electrolyte soaking and retention in significant amounts [52]. Generally, the requirements of battery separators for LIBs and SIBs are very similar and somehow we can say they are the same [53]. Because of the similarities discovered while comparing their properties, the

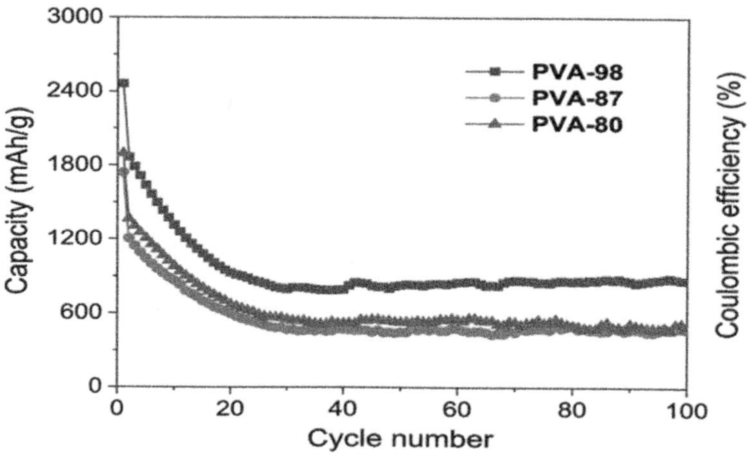

FIGURE 11.10

Specific capacity as a function of cycle number for the Si anode comprising 5 wt.% of PVA binders at C/20 (1st cycle) and at C/5 (subsequent cycles).

Adapted from P. Mandal, K. Stokes, G. Hernández, D. Brandell, J. Mindemark, Influence of Binder Crystallinity on the Performance of Si Electrodes with Poly(vinyl alcohol) Binders, ACS Appl. Energy Mater. 4 (2021) 3008–3016. https://doi.org/10.1021/acsaem.0c02051. Copyrights: This article is an open access article distributed under the terms and conditions of the Creative Commons Attribution (CC BY) license (http://creativecommons.org/licenses/by/4.0/).

Table 11.2 Comparison of some characteristics of Lithium and Sodium ions [54].

Properties	Lithium	Sodium
Potential (V vs SHE)	−3.04	−2.70
Cation radii (Å)	0.68	0.97
Atomic weight (g)	6.9	23
Capacity (mAh/g)	3860	1160

2015 Copyrights: This article is licensed under a Creative Commons Attribution 3.0 Unported Licence.

chemistry of the two ions is very similar. Table 11.2 shows a quick comparison of lithium-ion and sodium-ion [54]. The difference that can affect the choice of the appropriate separator in terms of porosity is the radii of both cations.

Although the separator is a nonactive material in the battery, its role is critical for battery performance, specifically energy densities, power densities, and safety [53]. The majority of synthetic polymers used in battery separators are nonbiodegradable, but they have interesting properties. Many examples of these synthetic polymers have been presented in a recent review by Costa et al. [55]. Some of those polymers are the following: polyacrylonitrile, polyethylene terephthalate, polyimide, poly(aryl ether ketone), polytetrafluoroethylene and poly(methyl methacrylate). The majority of the requirements of separators are the same either for LIBs or SIBs, some of those requirements are presented in the following Table 11.3 [51,56–58].

Recent LIB studies present some new developments in battery separators, which can be classified into four categories: polyolefin microporous separators, heterochain polymer microporous separators, polymer electrolyte separators, and nonwoven separators [59]. Similar to SIBs that include too: polyolefin microporous separators, nonwoven organic polymers along with glass-fibers films [29]. Many polymers were used as battery separators (polyvinylidene fluoride, polyol...) [29,60,61]. An example of the commercially used polymers or polymer blends are poly(ethylene) or poly(propylene). However, the drawbacks of these materials (nonbiodegradability, poor stability, poor wettability, nonrenewability...) [62] led to the exploration of new routes for battery separator materials' elaboration. Biodegradable polymers appear to be an intriguing alternative. Although their use in their original form is beneficial, new strategies are emerging to improve the performance of biodegradable polymer battery separators even further by incorporating either micro or nanometer-sized fillers into the polymeric matrix [55]. Xie et al. [63] used PVA as a synthetic biodegradable polymer to prepare a porous separator for LIBs. This new material was created by freeze-drying dehydrated PVA porous separator membranes with newly synthesized $CaCO_3$ particles in a spindle shape and using it as the pore-forming agent. The 10-PVA, 20-PVA, and 30-PVA separator materials are made of PVA with 10%, 20%, and 30% $CaCO_3$ by weight. The study used the poly(propylene) separator to compare the results of those separators to those of a commonly used commercial separator (PP). Many studies were conducted on the newly discovered material, including electrochemical performances, specifically ionic conductivity. This property of separators is highly dependent on the properties of the used material, specifically its porosity, tortuosity, and electrolyte uptake and retention abilities [64]. The electrochemical impedance spectroscopy (EIS) technique was used to investigate the ionic conductivity of this separator because it is a widely used nondestructive technique for characterizing the conductivities of various parts of batteries (both ionic and electronic conductivities) and specifically to inspect the properties of ionic conducting polymers in their various shapes (thin layers, surfaces, membranes, and films...) in general.

Table 11.3 General requirements of battery separators.

Parameters	Requirements
Chemical and electrochemical stabilities	Stable for a long period of time in harsh conditions (reducing and oxidizing)
Wettability	Wet out quickly and completely
Mechanical properties and thickness	Over >1000 kg/cm (98.06 MPa) 20–25 μm (to provide shorter)
Pore size, porosity, and tortuosity	<1 μm and 40%–60%, and the pore distribution should be uniform to favor homogeneous cations flux (to prevent dendrite growth)
Thermal stability	<5% shrinkage after 60 min at 90°C
Dimensional stability	No curl up and lay flat
Flame retardancy	Extinguish the fire when the electrolyte or even the battery is burning
Permeability	~25/mil using the air permeability method

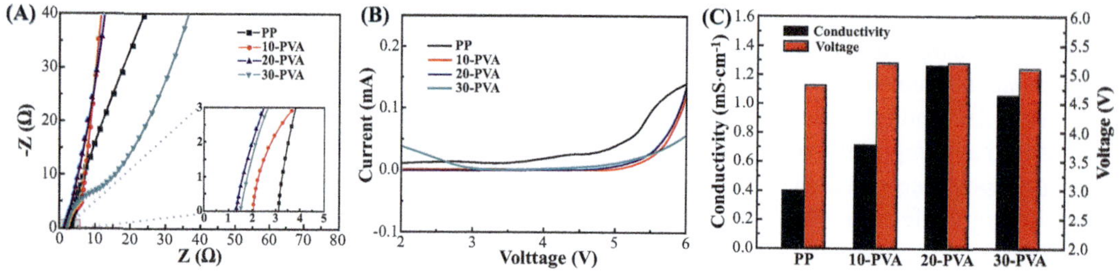

FIGURE 11.11

(A) Nyquist plots of the cells featuring the various elaborated separators and, (B) linear sweep voltammetry curves of Li/separators – electrolyte/SS cells, and (C) ionic conductivity and voltage of the PP separator along with elaborated separators.

Reprinted with permission from Y. Xie, Y. Pan, P. Cai, Novel PVA-based porous separators prepared via freeze-drying for enhancing performance of lithium-ion batteries, Ind. Eng. Chem. Res. 59 (2020) 15242–15254. https://doi.org/10.1021/acs.iecr.0c02182. Copyright 2020 American Chemical Society.

Fig. 11.11A shows the Nyquist plots of the battery cells that contain each elaborated type of separators (with different amounts of $CaCO_3$) in comparison with the one using the PP separator. In this figure, an inset shows the intercept of the Nyquist plot with the real Z' axis (Ohm (Ω)), this value corresponds to the ohmic resistance of the bulk material (R_b), in other terms the resistance of the separator. This value can be used to quantify the ionic conductivity (σ) of a separator by using the Eq. (11.1) shown afterwards:

$$\sigma = \frac{h}{R_b \times S} \qquad (11.1)$$

h is the thickness of the separator, and S is the surface area of the effective contact between the separator and the electrode.

The calculated ionic conductivities of the PP, 10-PVA, 20-PVA, and 30-PVA are the following: 0.40, 0.72, 1.27, and 1.06 cm^{-1}, respectively. Clearly, the elaborated separators exhibit a much better ionic conductivity compared to the one of PP (0.40 cm^{-1}), which has the lowest value. The 20-PVA separator presents the highest value (1.27 cm^{-1}). These results are due to several properties of these specific separators, including: first, the unique hierarchical 3D mazelike and porous structure, which allows for easy lithium-ion transport. Second, electrolyte uptake and retention were important criteria for improving ionic conductivity. The electrochemical stability of the separators is an important criterion. Fig. 11.11B shows the stability windows of the PP separator and the porous separators, these results are obtained by a linear sweep voltammetry analysis. As shown, the lowest initial oxidation voltage over 5.0 V is obtained by the elaborated separators (5.2 V for both 10-PVA and 20-PVA, and 5.1 V for 30-PVA), versus a relatively low value of 4.8 V for the PP separator (Fig. 11.11C). The EIS measurements after the first and hundredth cycles were used to investigate the compatibility in the interface between the separator and the electrodes. Fig. 11.12 shows the corresponding Nyquist plots. The intercept shows as previously discussed, R_S: the ohmic resistance, while the depressed semicircle in the relatively high-frequencies range, shows the interfacial resistance R_{ct}.

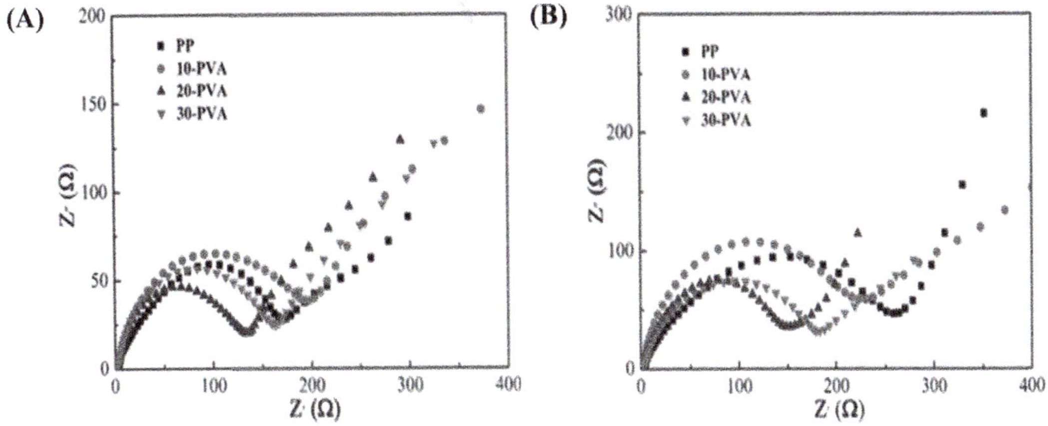

FIGURE 11.12

Electrochemical AC impedance spectra of the electrode material/Li half-cells with PP and the PVA separators (A) after activating two cycles at 0.1C rate and (B) after 100 cycles at 1.0C rate.

Adapted with permission from Y. Xie, Y. Pan, P. Cai, Novel PVA-based porous separators prepared via freeze-drying for enhancing performance of lithium-ion batteries, Ind. Eng. Chem. Res. 59 (2020) 15242–15254. https://doi.org/10.1021/acs.iecr.0c02182. Copyright 2020 American Chemical Society.

The obtained results for the PP during the 1st cycle are 2.05 and 170Ω for R_S and R_{ct}, respectively. Compared to lower values for the PVA separators and especially the 20-PVA one, that showed values of 1.62 and 134Ω for R_S and R_{ct}, respectively. After the 100th cycle, the same observations can be noted, for the PP 3.48 and 258Ω for R_S and R_{ct} while the 20-PVA showed 2.48 and 152Ω for R_S and R_{ct}. Resistances increased as the number of cycles increased, which could be due to pore clogging by trapped lithium ions during cycling or other products that decompose in the electrolyte and eventually get deposited on the pores and block them [65,66]. The results confirm that the elaborated PVA membranes outperform the commercial ones. Thermal protection is an important feature that batteries should have because it prevents casualties when their temperature rises significantly and causes a thermal runaway. Thermal shutdown features are provided in a multilayer composite separator material by using a readily melting layer (i.e., made of polymer microspheres...) that closes the pores of the other layer (which usually provides oxidation resistance as well as mechanical strength) near the critical temperature where thermal runaway occurs [65,67].

11.3.1.3 Polymeric electrolytes

The performance of SIBs and LIBs is highly dependent on the materials used. When using a liquid electrolyte, both types of batteries face significant latent risks and challenges. Leakage, short circuits, explosion, volatilization, and flammability are just a few examples... [68,69]. The substitution of a new class of materials known as polymer electrolytes for the electrolyte appears to be a brilliant step toward overcoming these drawbacks [70]. These materials are divided into three categories: solid polymer electrolytes (SPE), gel polymer electrolytes (GPE), and composite polymer electrolytes (CPE), which are a combination of the first two categories plus some inorganic nanofillers [55]. To some extent, the separator and the polymer electrolyte serve similar functions. The former is composed of a solid porous material or membrane, whereas the latter is composed of a polymeric matrix containing metal (lithium or sodium) salts, as well as ionic liquids and/or any other material yielding a composite while using a nonaqueous solvent. Fig. 11.13 illustrates a quick comparison of using a liquid electrolyte versus a polymeric electrolyte. The benefits are described in the caption; the polymeric electrolyte ensures the battery's flexibility while also preventing electrolyte leakage. Because of the mechanical properties of the polymer, dendritic growth is inhibited [71].

11.3.1.3.1 Solid polymer electrolyte

In the absence of an organic solvent, SPEs use the polymer host as a solid solvent for metal salts. PEO was the polymer that was used for the first time and for a long time [59], and since it is a

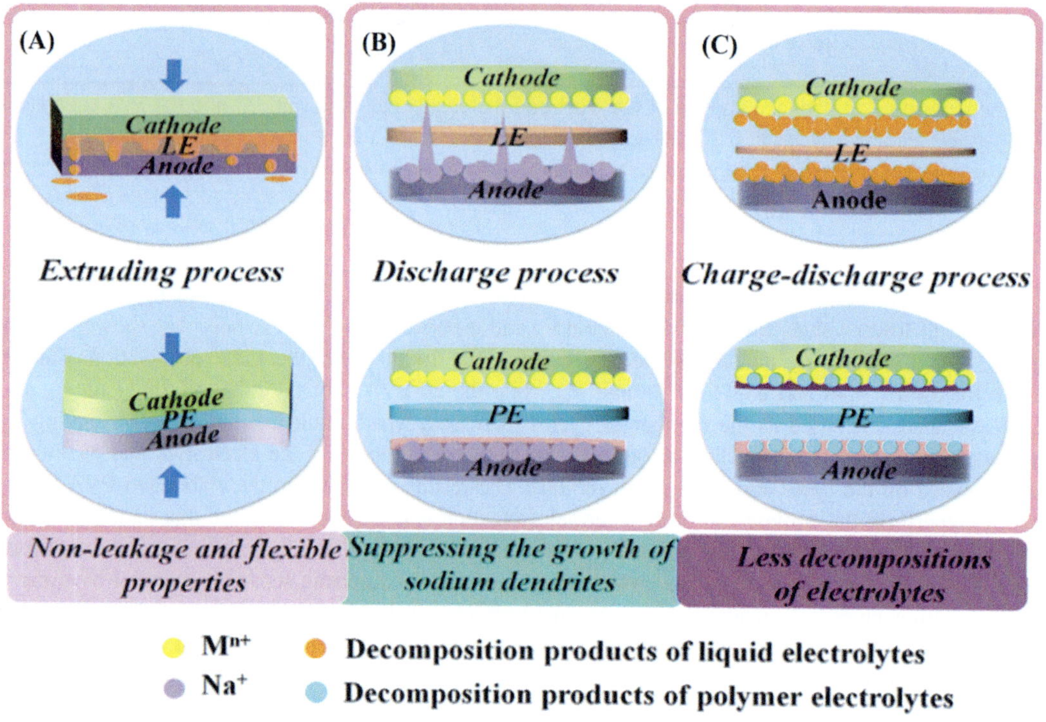

FIGURE 11.13

Illustration of the inherent advantages of polymer electrolytes for advanced sodium-ion batteries as compared with that of liquid electrolytes: (A) nonleakage and flexible properties, (B) suppressing the growth of sodium dendrites, and (C) less decompositions of electrolytes. Here, liquid electrolytes and polymer electrolytes are denoted as LE and PE, respectively.

Adapted with permission from J. Yang, H. Zhang, Q. Zhou, H. Qu, T. Dong, M. Zhang, et al., Safety-enhanced polymer electrolytes for sodium batteries: recent progress and perspectives, ACS Appl. Mater. Interfaces. 11 (2019) 17109–17127. https://doi.org/10.1021/acsami.9b01239. Copyright 2019 American Chemical Society.

synthetic biodegradable polymer, some studies have been conducted to improve its conductivity and its overall properties. Yang et al. [71] reviewed some interesting improvements conducted on PEO-based polymer matrices used in SIBs. The use of alkali salts in optimized concentrations improves the ions' conductivity by increasing the amorphous ratio, the value of 3.40×10^{-6} S/cm at 35°C is reached by utilizing the $NaClO_3$ saltUsing specific salt concentrations resulted in polymeric electrolytes with improved ionic conductivity, thermal stability, and oxidative stability. The addition of larger sodium salts results in more delocalized charge distribution, which improves ionic conductivity. The use of nanoparticles as fillers is also extensively discussed. For example, Ni'mah et al. [72] used TiO_2 nanoparticles in a blend polymer composite ($PEO/NaClO_4/TiO_2$) for SIBs, which allowed the conductivity to reach 2.62×10^{-4} S/cm at 60°C by adding TiO_2 nanoparticles with an average size of 3.4 nm at 5 wt.% (the ionic conductivity without TiO_2 addition reached 1.35×10^{-4} S/cm). Fig. 11.14 shows the effect of the temperature on the ionic conductivity.

The increase in temperature causes an increase in ionic conductivity, which is due to the increased mobility of Na + ions as the temperature rises. A significant increase in conductivity is observed at temperatures ranging from 40°C to 50°C and higher, which can be explained by the crystallinity transition of PEO from a semicrystal phase to an amorphous phase. The anatomic ratio of ethylene oxide to sodium also influences ionic conductivity (EO: Na). The ionic conductivity increases as the ratio rises from 0 to 20, but there is no discernible increase when the ratio reaches 30. This is attributed to ion-pairing at high salt concentrations. Another example [71] showed the

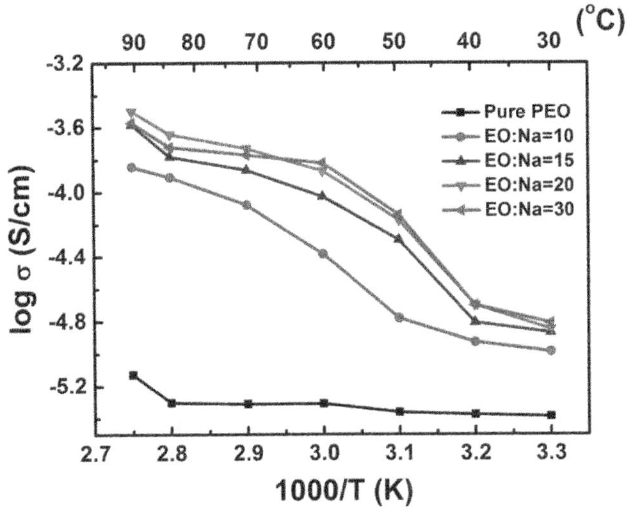

FIGURE 11.14

The conductivity plots of PEO and $PEO/NaClO_4$ polymer electrolyte films depending on temperature.

Reprinted from Y.L. Ni'Mah, M.Y. Cheng, J.H. Cheng, J. Rick, B.J. Hwang, Solid-state polymer nanocomposite electrolyte of TiO_2/$PEO/NaClO_4$ for sodium ion batteries, J. Power Sources. 278 (2015) 375–381. https://doi.org/10.1016/j.jpowsour.2014.11.047 with permission from Elsevier. https://s100.copyright.com/MyAccount/viewLicenseDetails?ref= e9f0677a-2462-4cea-99d7-27454a51110b. Copyright 2019.

use of SiO_2 nanoparticles in ($PEO/NaHCO_3/SiO_2$) polymeric blend where conductivity was 2.04 $\times$ 10^{-5} S/cm at 27°C. According to the authors, the interaction between the PEO and the SiO_2 created more hopping sites and thus more pathways for ion passage, resulting in high ionic conductivity. Sunandar et al. [73] presented a successful film elaboration of a solid polymeric electrolyte based on PCL, by using the casting method, where $LiClO_4$ was used as salt. This material presented an improved ionic conductivity that increased from 2.45 $\times$ 10^{-11} S/cm to 5.52 $\times$ 10^{-6} S/cm only by adding the lithium salt. This increase in ionic conductivity is explained by the strong base character of PCL's carbonyl groups, which are highly sensitive to ionic interactions. Solid-state batteries (with SPEs) present a critical problem due to their high interfacial resistance. Nkosi et al. [74] elaborated a new material that has the ability to address this problem, by integrating the merits of two different materials: A ceramic electrolyte and a copolymer electrolyte based on a biodegradable polyester PCL and a PCL-(CL)-trimethylene carbonate (PCL-PTMC) polymer, and by adding a lithium salt. The addition of this minor PCL-PTMC monomer, suppresses the crystallization that occurs in the PCL, thus improving ionic conductivity.

Fig. 11.15A shows a SEM image of the interface between the new composite SPE and the cathode material, which can be spotted by the dashed line. In this study, the new composite was also used as a cathode binder to form an integrated morphology, which improves interfacial resistance reduction and ions transport through the cathode. Fig. 11.15B shows the area-specific resistance (ASR) of the composite at several temperatures, 30, 40, and 60 degrees, obtained by EIS

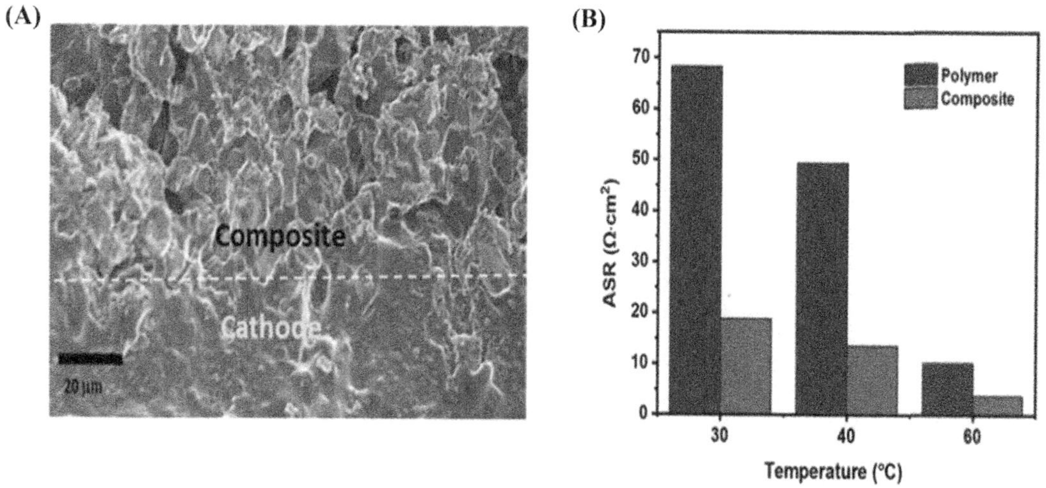

FIGURE 11.15

(A) Cross-sectional scanning electron microscope (SEM) image of the composite electrolyte – cathode interface, (B) area-specific resistance (ASR) vs temperature for Li/electrolyte/Li cells for the polymer.

Adapted from F.P. Nkosi, M. Valvo, J. Mindemark, N.A. Dzulkurnain, G. Hernández, A. Mahun, et al., Garnet-poly(ε-caprolactone-co-trimethylene carbonate) polymer-in-ceramic composite electrolyte for all-solid-state lithium-ion batteries, ACS Appl. Energy Mater. 4 (2021) 2531–2542. https://doi.org/10.1021/acsaem.0c03098 Copyrights: This article is an open access article distributed under the terms and conditions of the Creative Commons Attribution (CC BY) license (http://creativecommons.org/licenses/by/4.0/).

measurement after 2 hours calibration. When compared to the impedance of the pristine polymer, the composite impedance is clearly reduced (by about two factors). The use of ceramic particles improved the composite's stability towards the Li-metal and may have prevented the formation of the passivation layer. The increase in temperature improves the transportation of Li ions, which explains why the resistance decreases further.

11.3.1.3.2 Gel polymer electrolyte

GPE combines the benefits of both electrolyte types: liquid electrolytes, which are highly ionic conductive media, and solid electrolytes, which are nonleaking [59]. These materials are created by combining liquid electrolytes in which metal salts are dissolved in an organic solvent or ionic liquid and incorporating them into a polymer matrix. Ionic conductivity is aided by the retained solvent in the polymeric matrix [68]. Hybrid membranes of these materials are formed, whose ionic conductivity can be up to 6.0 mS/cm^2 [75]. Huy et al. [76], presented a study of a PVA GPE membrane elaboration, by dispersing PVA in 1M salt solution of ammonium thiocyanate (NH_5SCN) in dimethyl sulfoxide and different weight percentages (0, 2, 6, and 10 wt.%) of Al_2O_3 nanoparticles were added to the mixture, after syneresis and drying, the nano-Al_2O_3-filled PVA composite gel electrolyte was obtained. Fig. 11.16 shows the SEM images.

The PVA membrane shows a porous structure. Yet, the addition of Al_2O_3 nanoparticles (2 wt.%) reduced the porosity of the nanocomposite, this is due to the trapping of nanoparticles into the pores. After exceeding 6 wt.% the PVA is fully covered by the nanoparticles indicating the good dispersion of the nanofiller over the PVA chains. The cycling tests indicated the enhancement of the cycling stability that improved for (−1.3 to +1.4 V). $Li+$ ions conductivity was improved by adding 6 wt.% of Al_2O_3 nanoparticles. Liu et al. [77] recently presented a variety of polymer matrices used as GPE, including (PEO), poly(methyl methacrylate) (PMMA), poly(acrylonitrile) (PAN), poly(vinylidene fluoride) (PVDF), and others. The enhancement of these materials by incorporating conductive nanoparticles such as sodium conductive-alumina was carried out, and the results show that the material has a good enhanced ionic conductivity.

11.3.2 Natural biodegradable polymer nanocomposites

As previously stated, biodegradable polymers can be synthetic or natural. Natural biodegradable polymers are derived from living organisms and thus are renewable, sustainable, abundant, and can be regarded as low-cost raw materials [78]. We discussed some applications of synthetic biodegradable polymers in the previous section. Natural biodegradable polymers are an appealing alternative to synthetic ones due to their low cost and a variety of inherent properties. In the following section, we will look at some biodegradable polymers and nanocomposites that have been used as components of LIBs or SIBs.

11.3.2.1 Electrode materials and electrode binders

Batteries based on natural biodegradable polymers have emerged as a new area of study in the quest for long-term energy-storage devices. A variety of biodegradable bio-resourced polymers used as electrodes and binders will be presented in the following section. Lignin is the second most abundant biopolymer on the planet [79], and its use as a functional material in batteries is an exciting prospect. Luo et al. [80] developed a new material based on lignin as a biodegradable polymer,

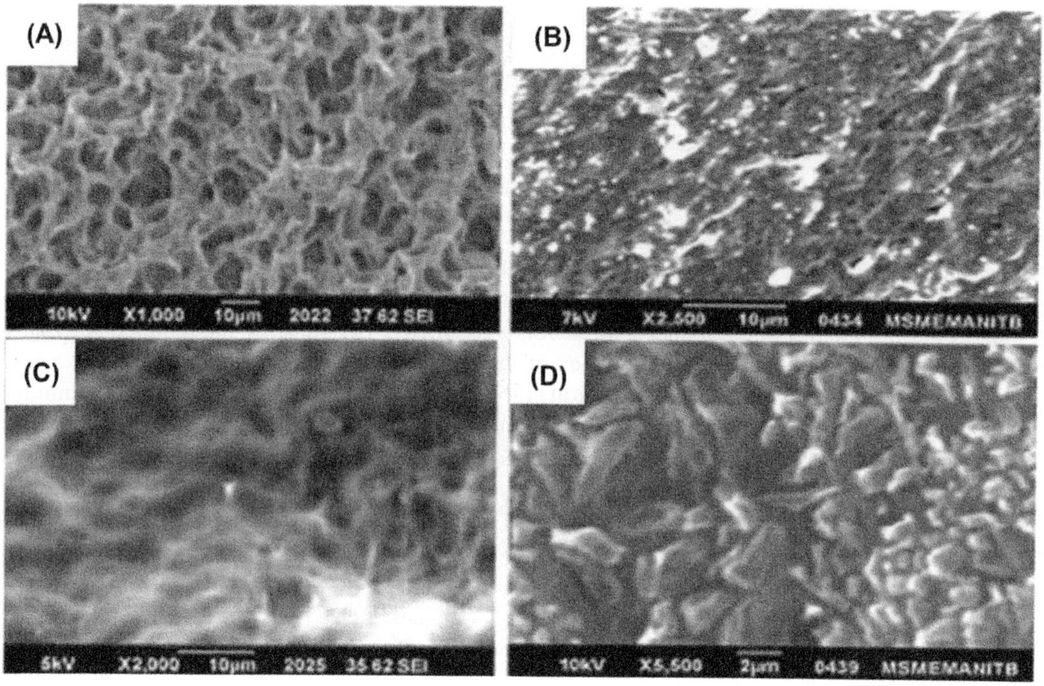

FIGURE 11.16

SEM images of the (A) PVA:NH_4SCN/DMSO gel polymer electrolyte containing (B) 2, (C) 6, and (D) 10 wt.% Al_2O_3 nanoparticles.

Reprinted from V.P.H. Huy, S. So, J. Hur, Inorganic fillers in composite gel polymer electrolytes for high-performance lithium and non-lithium polymer batteries, Nanomaterials. 11 (2021) 1–40. https://doi.org/10.3390/nano11030614 Copyrights: This article is an open access article distributed under the terms and conditions of the Creative Commons Attribution (CC BY) license (http://creativecommons.org/licenses/by/4.0/).

which was the first water-soluble lignin-derived binder (its treatment provided: purified alkali lignin, PAL). The used anode is a microsilicon-based material that performs poorly when coupled to trivial binders. This study proposed a microsilicon-based anode for LIBs. The binder's surface, made of sodium polyacrylate-grafted lignin (PAL-NaPAA), has a high number of hydroxyl and carboxyl groups that can easily interact with silicon particles and the current collector (made of copper) via hydrogen or ester bonding. Fourier transform infrared spectroscopy (FTIR) tests were done to confirm the grafting (a peak appears at 2246 cm^{-1} confirming the obtention of the C≡N stretching). To evaluate the performances of this elaborated binder, it was compared to CMC commercial binder, and they are used for cycling tests. Fig. 11.17 shows the cycling performance of the electrodes featuring both binders.

A limitation of lithiation capacity to 800 mAh/g was used to record lithiation for each electrode. The PAL-NaPAA binder proves better cycling performances, about 940 cycles without any capacity fade contrarily, to the CMC that cycled only 250 times, and then its capacity of cycling faded. The

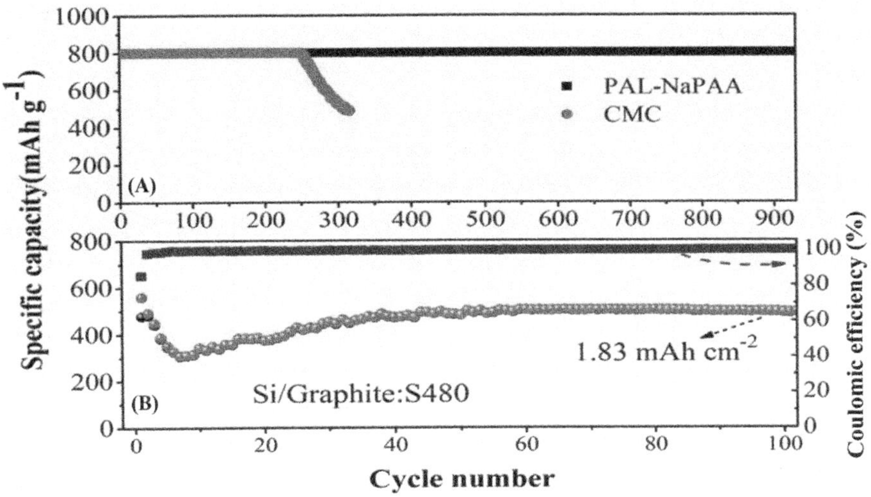

FIGURE 11.17

(A) Cycling performance with the capacity limited to 800 mAh/g at 840 mA/g over the potential window of 0.01–1.0 V (vs Li/Li^+); (B) cycling performance of Si/graphite mixture electrode with PAL-NaPAA binder in a cycling voltage range of 0.01–2.0 V at 50 mA/g.

Reprinted with permission from C. Luo, L. Du, W. Wu, H. Xu, G. Zhang, S. Li, et al., Novel lignin-derived water-soluble binder for micro silicon anode in lithium-ion batteries, ACS Sustain. Chem. Eng. 6 (2018) 12621–12629. https://doi.org/10.1021/acssuschemeng.8b01161. Copyright 2018 American Chemical Society.

use of lignin as an electrode and a binder material have been reviewed by Nirmale et al. [81]. Thanks to its aromatic rings, the lignin forms a quite similar network to the one of graphite. Fig. 11.18 depicts the isotropic nature of lignin-based carbon fibers, which makes them an excellent candidate for lithium insertion. The SEM images show that the fabrication process can be used to control the fusion and mat densities of fibers. One disadvantage of lignin-based carbon fibers is their lower flexibility when compared to graphite.

The use of lignin as a binder for LIBs was reported in the same study; the lignin was used here for both the anode (Graphite) and the cathode ($LiFePO_4$). Acetone was used to dissolve the lignin, and a 5 wt.% PEG plasticizer was added to help improve the elasticity towards the electrodes. In that study, leaching lignin before using it as an electrode helped remove the low MW soluble fractions. Lignin can be used as a raw material for the synthesis and fabrication of mesoporous carbon hybrid nanospheres, which will eventually be used to make LIB electrode materials [82]. Molybdenum disulfide (MoS_2) was synthesized on the surface of the nanospheres in this study. Its application was intriguing because of its structure; when accumulated in the ABAB stacking sequence, its interlayer distance is 0.62 nm which can aid in the insertion/extraction of lithium ions during charge and discharge. It is worth noting that the original source of the lignin used as a precursor has a significant impact on the hard carbon properties, and thus on its performance as an electrode (in this study the hard carbon was used for SIBs) [83]. Lignin is the second biodegradable polymer on the planet, after cellulose, which is the first. It is a renewable polymer and the most

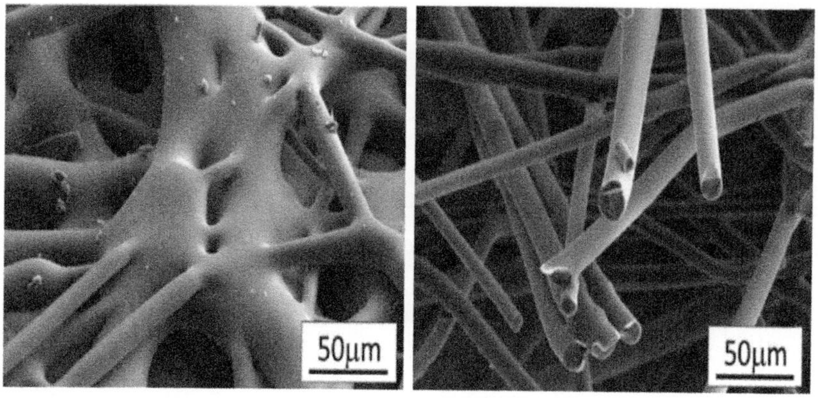

FIGURE 11.18

Scanning electron images of varying levels of fiber–fiber fusion in lignin-based carbon fibers mats.

Reprinted from T.C. Nirmale, B.B. Kale, A.J. Varma, A review on cellulose and lignin based binders and electrodes: Small steps towards a sustainable lithium ion battery, Int. J. Biol. Macromol. 103 (2017) 1032–1043. https://doi.org/10.1016/j.ijbiomac.2017.05.155, with permission from Elsevier. https://s100.copyright.com/MyAccount/viewLicenseDetails?ref= aff55445-6b3c-43a0-8857-46447838bd2b. Copyright 2017.

abundant material on the planet [84]. Because of its solvent retention capacity and rate capability demonstrated by electrochemical results, CMC was the first biodegradable candidate to replace traditional PVDF binders in rechargeable batteries [81]. When the volume expands during charge/discharge cycles, the CMC prevents cracks from forming in the electrode. Here, silicon has a high theoretical specific capacity, making it a very interesting and promising material for the next generation of anodes. However, the large volume change during cycling prevents it from being used in high-capacity LIBs. New research and extensive exploration are being conducted to address the issue of massive volume expansion. The development of active materials is not only difficult but also costly. There appears to be an effective solution for the development of binders with high intrinsic performances. Wang et al. [85], elaborated an improved binder based on dopamine-modified CMC (DOP-CMC). To ultimately evaluate the performances of the new binder, electrochemical performance studies are done, and the results are presented in Fig. 11.19.

The cycling performances presented in Fig. 11.19A of silicon-based electrodes using different binders (PVDF, CMC, and CMC-DOP) show sharp decrease of capacity after 200 cycles for silicon-PVDF and silicon-CMC electrodes that remain 816.6 and 628.4 mAh/g, respectively. While the new elaborated binder delivered an initial specific capacity of 3418.2 mAh/g and maintained an excellent one (1650.6 mAh/g) even after 200 cycles, with a capacity fade ratio of only 0.19% for each cycle. These results are in accordance with rate performance on Fig. 11.19B, which show high specific capacity for CMC-DOP binders contrarily to the ones of PVDF. A study of the binder's effect on electrochemical properties of SIBs' cathode was done by Zhang et al. [19]. Guar gum, sodium alginate, and xanthan gum were the biodegradable polymers used as binders. These biopolymers are all water-soluble and contain OH and COO functional groups, which are expected to improve the binder's adhesion to the active material nanoparticles. They are also expected to

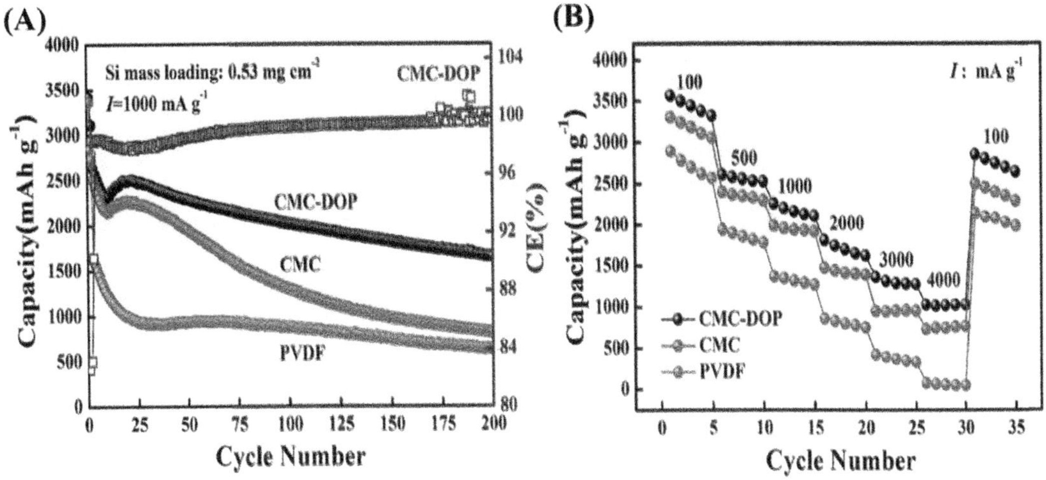

FIGURE 11.19

(A) The cycling performance of different binders for 200 cycles in the potential range of 0.01 – 1.20 V, and their corresponding (B) rate performance at different current densities (100, 500, 1000, 2000, 3000, 4000 and 100 mA/g).

Reprinted from Z. Wang, T. Huang, Z. Liu, A. Yu, Dopamine-modified carboxymethyl cellulose as an improved aqueous binder for silicon anodes in lithium-ion batteries, Electrochim. Acta. 389 (2021) 138806. https://doi.org/10.1016/j.electacta.2021.138806, with permission from Elsevier. https://s100.copyright.com/MyAccount/viewLicenseDetails?ref= d39bc5f6-0726-4cf5-aad0-0b374f81dbb1. Copyright 2021.

prevent metal dissolution, which is one of the main causes of poor cycling. The XPS analysis of the electrodes was used to verify this property; the results show manganese (Mn) dissolution with PVDF-based-binder electrodes, and an inhibition of this undesirable phenomenon was observed when water-soluble binders were used. Cellulose nanofibrils (CNF) combine to form new light-weight nanocomposite materials that could be used as battery binders. The use of this nanosized cellulose as binder was mentioned by El Baradai et al. [86], in both cathode and anode materials. CNF content should be tuned to meet the requirements of binder. Since CNF is promoting high viscosity because of its thickening ability. The used amount of CNF is then fixed to 2 wt.%. CMC biodegradable polymer was also used in this study as a dispersing agent to prevent CNF flocculation. It should be used sparingly because it drastically reduces the viscosity of the ink. Fig. 11.20 displays the rheological properties of the electrode inks. Each viscosity is given depending on the rate of shearing.

The results are in good accordance with the screen-printing process viscosity requirements that range between 10 and 150 Pa/s at a shear rate corresponding to 1 s^{-1}. A recent study using a dual anionic polysaccharides as a binder in LIBs, was investigated by Li et al. [87]. Hyaluronic acid and soybean polysaccharides were used to create a 3D network binder by weaving these two polysaccharides via supramolecular interactions. The material developed can significantly reduce silicon electrode disintegration while also ensuring continuous ion transport through the 3D circuit.

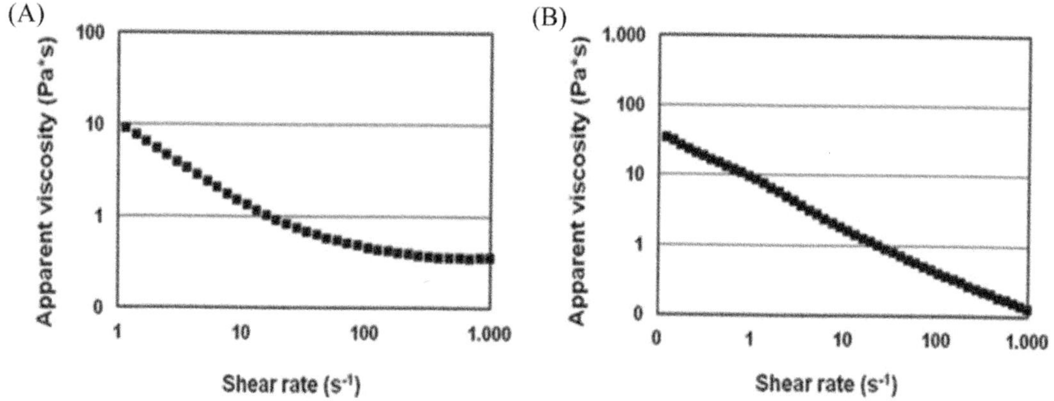

FIGURE 11.20

Rheological properties of the inks in (A): the negative electrode made of graphite powder ink and (B): the positive electrode made of lithium iron phosphate ink.

Adapted from O. El Baradai, D. Beneventi, F. Alloin, Y. Bultel, D. Chaussy, Use of cellulose nanofibers as an electrode binder for lithium ion battery screen printing on a paper separator, Nanomaterials. 8 (2018) 982. https://doi.org/10.3390/nano8120982. 2018 Copyrights: This article is an open access article distributed under the terms and conditions of the Creative Commons Attribution (CC BY) license (http://creativecommons.org/licenses/by/4.0/).

Fig. 11.21 shows the electrochemical performance of the electrodes with the different binders, namely: CMC as a common binder, a hyaluronic acid (HA) binder, solid soybean polysaccharide (SSPS) binder, and the 3D network elaborated binder (SHA). The used binders were tester at two different discharge current densities (4 A/g Fig. 11.21A and 2 A/g Fig. 11.21B).

The electrode binder made of SHA showed the highest discharge capacity (3256 mAh/g) at 4 A/g and, even after 100 cycles, the SHA discharge capacity remains the highest (1407 mAh/g). As the current density decreases to 2 A/g better cyclability was also displayed by the SHA electrode binder. Cellulose nanocrystals (CNC) were used in several studies as a green and low-cost carbon source, for example Zhang et al. [88], processed to the carbonization of CNC into short-range ordered amorphous carbon. The first material was mixed with a MoS_2 with enhanced crystallinity to create a nanocomposite material. The final nanocomposite exhibited a 3D hierarchical structure with not only a high specific capacity but also excellent stability as an anode for SIBs. Electrochemical measurements revealed that this new material was promising. Zhu et al. [89] also presented CNC carbonization; the advances made here are the low carbonization temperature (1000°C). The carbonized CNC-derived carbon was used in SIBs and demonstrated excellent performance, including the highest capacity carbon used in SIBs, better rate capability, and good cycling stability. The interlayer spacing in the porous structure of CNC-based carbon was responsible for the good electrochemical performances. The percolated structure of carbon nanofibers, as well as the more ordered carbon rings in a hexagonal shape shown in Fig. 11.22, resulted in a high electrical conductivity. This image is obtained by the Transmission Electron Microscopy (TEM) and High-resolution transmission electron microscope (HRTEM). In the inset Fourier transform we can observe a very highly ordered region that displays two sets of reflections at 0.24 and 0.14 nm,

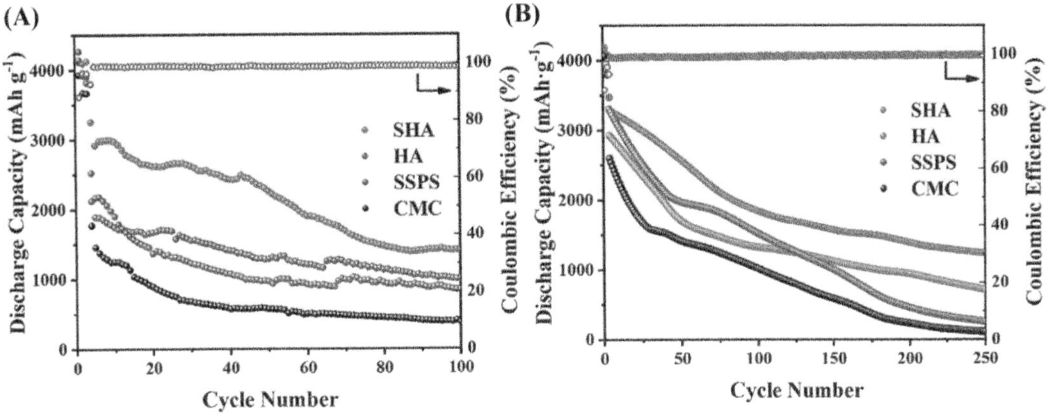

FIGURE 11.21

(A) Cycling performance of electrodes with various binders at 4 A/g and (B) at 2 A/g.

Reprinted from Y. Li, B. Jin, K. Wang, L. Song, L. Ren, Y. Hou, et al., Coordinatively-intertwined dual anionic polysaccharides as binder with 3D network conducive for stable SEI formation in advanced silicon-based anodes, Chem. Eng. J. 429 (2021) 132235. https://doi.org/10.1016/j.cej.2021.132235, with permission from Elsevier. https://s100.copyright.com/MyAccount/viewLicenseDetails?ref= 11c1e343-cba4-4a02-87fb-aa8a84dc7d7f. Copyright 2021.

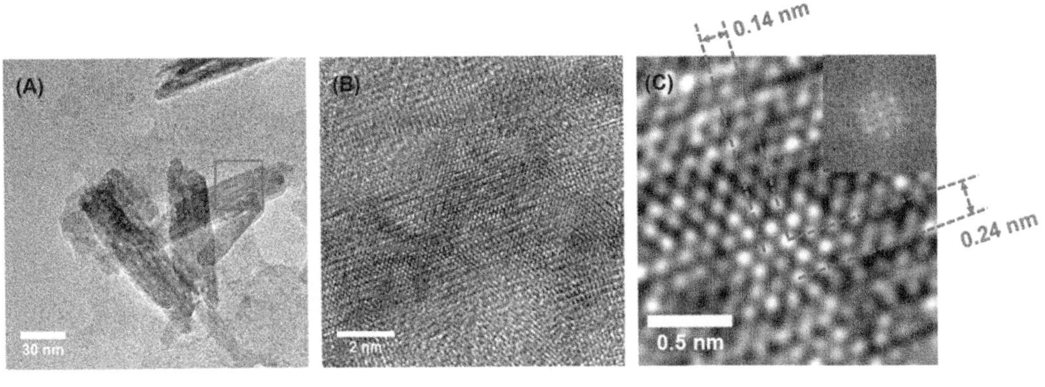

FIGURE 11.22

TEM image of the CNC derived carbon, (B) Higher magnification HRTEM image of the area contained in the red squared in (A); (C) higher magnification HRTEM image of the area contained in the red squared in (B) with the Fourier transform shown in the inset.

Reprinted from H. Zhu, F. Shen, W. Luo, S. Zhu, M. Zhao, B. Natarajan, et al., Low temperature carbonization of cellulose nanocrystals for high performance carbon anode of sodium-ion batteries, Nano Energy. 33 (2017) 37–44. https://doi.org/10.1016/j.nanoen.2017.01.021, with permission from Elsevier. https://s100.copyright.com/MyAccount/viewLicenseDetails?ref= dcff15f7-40a6-4eef-ae24-3024e79f9e86. Copyright 2017.

that correspond to the lattice constant of the hexagonal stacking and to the carbon-carbon nearest neighbor distance. The effect of temperature on the carbon radicals' spatial density and on the formation of graphene carbon network was investigated based on molecular dynamics as shown in Fig. 11.23.

Temperature had a clear effect on the number of generated molecule H_2O, CO_2, CO, and carbon radicals. At 1000°C, the CNC bundles were completely depolymerized into carbon radicals, increasing the spatial density of carbon radicals until graphene-like carbon networks formed (Fig. 11.23A–D). To further prove the crystalline carbon formation, in-situ TEM image is given in Fig. 11.23E. Kim et al. [90], on the other hand, demonstrated that the variation of the structure of the CN_2C-based carbon affects the mechanism of sodiation, and that sodium ions adsorption is more easily favorable in the less ordered structural carbon than sodium ions intercalate the more ordered ones. Many other studies investigated the use of natural biodegradable polymers, such as alginate, as battery binders [45,91]. Hu et al. [92] used both natural (sodium alginate) and synthetic

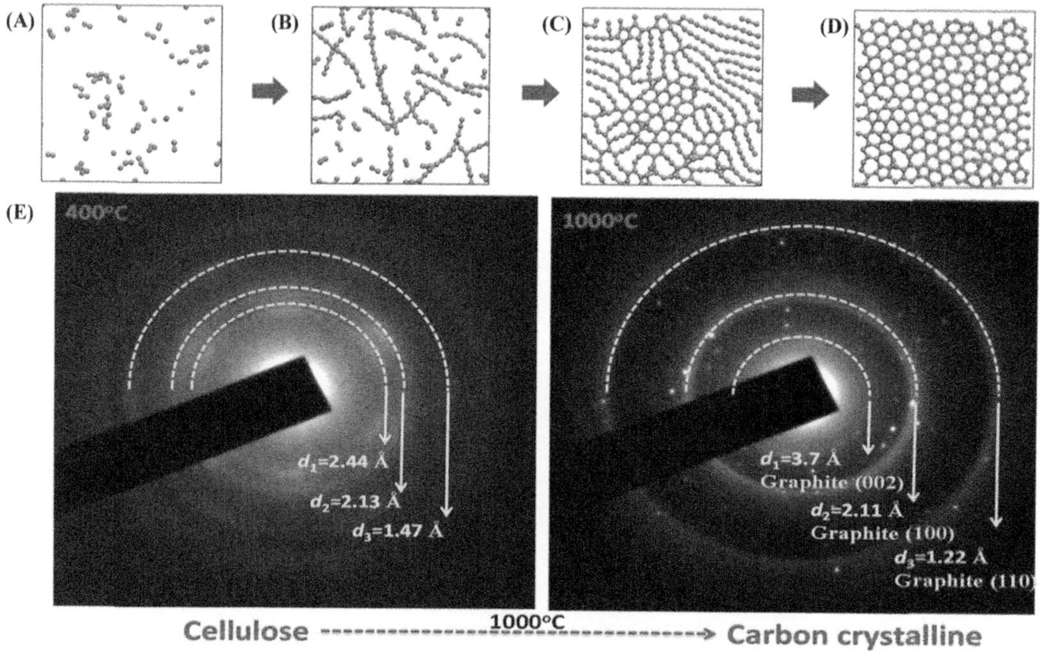

FIGURE 11.23

(A)–(D) show the formation of a graphene-like structure as the as function of spatial density of carbon radicals increases. (E) In-situ TEM diffraction patterns at 400°C and 1000°C to show the transformation from cellulose to crystalline carbon.

Reprinted from H. Zhu, F. Shen, W. Luo, S. Zhu, M. Zhao, B. Natarajan, et al., Low temperature carbonization of cellulose nanocrystals for high performance carbon anode of sodium-ion batteries, Nano Energy. 33 (2017) 37–44. https://doi.org/10.1016/j.nanoen.2017.01.021, with permission from Elsevier. https://s100.copyright.com/MyAccount/viewLicenseDetails?ref= dcff15f7-40a6-4eef-ae24-3024e79f9e86. Copyright 2017.

(PAA) biodegradable polymers as a battery binder, but the polymers were not only homogeneously mixed, but the pores of the hydroxy-rich sodium alginate, which is presented as a solid network, were filled up with a solution of lithiated carboxyl-rich PAA to form a three-dimensional composite binder. Because of the extensive ester bindings, the PAA is tightly bound to alginate. This new binder showed good cycling performances (1584.3 mAh/g after 150 cycles at charge-discharge current density of 420 mA/g). To investigate the electrochemical stability of the binders, the study of the cyclic voltammetry is shown in Fig. 11.24, where the copper electrode coated with pure alginate (m-Si/SA) is shown in Fig. 11.24A, and the calcinated mixture using PAA (of a lithiation degree of 0.5) and alginate (m-Si/b-$Li_{0.5}$PAA@SA) is shown in Fig. 11.24B.

A small peak can be seen around 1.2 V, which can be attributed to the reduction of the fluoroethylene chloride additive, which was used as an electrolyte, indicating that pure alginate alone was unable to suppress the formation of the surface electrolyte interphase. The cyclic voltammogram of the binder, which contains a mixture of calcinated lithiated-PAA and alginate, shows a clear reduced intensity of this peak, confirming the binder's ability to suppress SEI formation. Jin et al. [93] used chitosan, another natural biodegradable polymer, as an alkaline solution for silicon nanoparticle phase-etching. This process added a SiO_2 layer to the surface of silicon nanoparticles, and experimental analysis confirmed that these particles were incorporated into the carbonized chitosan matrix. The designed material is used as a material in LIB anodes. Chitosan was also used in SIBs as a source of nitrogen doped hard carbon via hydrothermal carbonization, and the finished material was coated with SnO_2 to form a nanocomposite anode [94]. Starch, another biodegradable abundant polymer, was used by Hapuarachchi et al. [95] as a binder for silicon nanoparticles of the electrode. The elaborated material was PEG-modified, resulting in a water-soluble material that provided new properties to the separator. SEM images of the electrode were examined to investigate the effect of the binder on the surface morphology of the electrode after cycling. Fig. 11.25

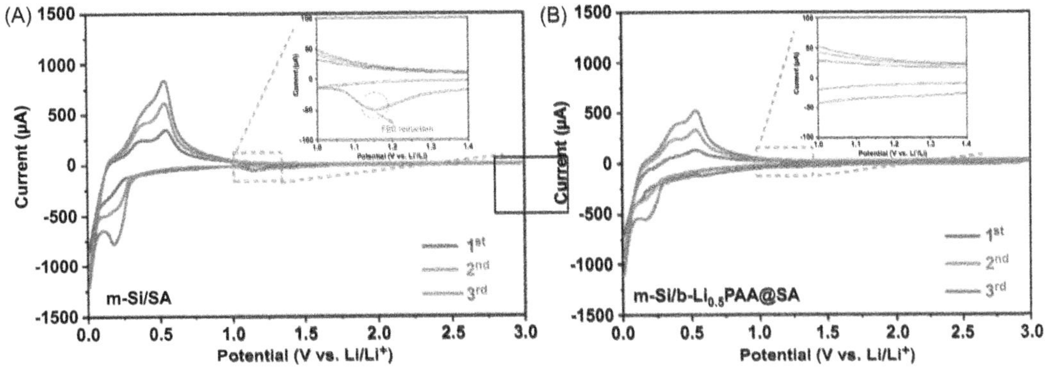

FIGURE 11.24

The cyclic voltammetry of the cells with the (A) m-Si/SA or (B) m-Si/b-Li 0.5 PAA@SA electrodes.

Reprinted from Y.Y. Hu, J.H. You, S.J. Zhang, H. Lin, W.F. Ren, L. Deng, et al., Li0.5PAA domains filled in porous sodium alginate skeleton: A 3D bicontinuous composite network binder to stabilize micro-silicon anode for high-performance lithium ion battery, Electrochim. Acta. 386 (2021) 138361. https://doi.org/10.1016/j.electacta.2021.138361 with permission from Elsevier. https://s100.copyright.com/MyAccount/viewLicenseDetails?ref= 36503f91-5527-4fce-9466-b14337ca017b. Copyright 2021.

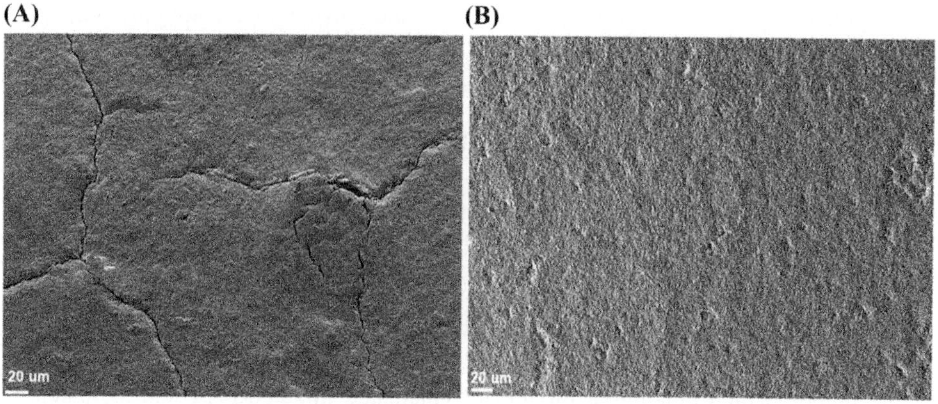

FIGURE 11.25

(A) SEM image of Si@PVDF electrode and (B) Si@Starch/PEG electrode after 3 cycles.

Adapted with permission from S.N.S. Hapuarachchi, K.C. Wasalathilake, J.Y. Nerkar, E. Jaatinen, A.P. O'Mullane, C. Yan, Mechanically robust tapioca starch composite binder with improved ionic conductivity for sustainable lithium-ion batteries, ACS Sustain. Chem. Eng. 8 (2020) 9857–9865. https://doi.org/10.1021/acssuschemeng.0c02843. Copyright 2020 American Chemical Society.

shows a water-soluble, environmentally friendly tapioca starch binder modified with PEG groups as a novel binder for Si anodes electrodes in Li-ion batteries in this study.

As seen in the SEM images, the surface of the electrode with the PVDF binder has some clear cracks after three discharge/charge cycles, whereas the surface of the electrode with starch and PEG as binder biodegradable polymers has a few tiny cracks that are insignificant when compared to the PVDF ones. To further investigate the effect of these binders on the electrodes, a cross section image was observed by SEM as well (Fig. 11.26) after 10 electrochemical cycles.

The thickness of the electrodes postcycling for both binders showed an increase (for PVDF, the section increased from 15 to 22.2 μm postcycling, and for the starch/PEG binder the value increased from 15 to 17.77 μm). The electrode with the PVDF binder, on the other hand, showed a greater increase. This finding confirms that the biodegradable polymer is more adaptable to volume expansion changes in silicon electrodes. Jin et al. [96] used fluorinated corn starch emulsion for silicon oxide anodes as well as silicon nanoparticles. Although only a few examples of natural biodegradable polymers were presented in this section, research is still ongoing to find new ways to value these natural materials and use them to improve LIBs and SIBs [38].

11.3.2.2 Battery separators and polymeric electrolytes

The separator is an important component in LIBs and SIBs, and it should be improved. Using biodegradable polymers, particularly those derived from natural resources, is a promising step toward improving battery sustainability and environmental friendliness. The first natural biodegradable polymer we'll look at is the most abundant polymer on the planet: cellulose. This biopolymer is widely used in batteries, particularly as a separator. Lv et al. [97] presented the use of cellulose in LIBs in its two forms. First in its natural state (as cellulose paper), then in CNF form. CNF was

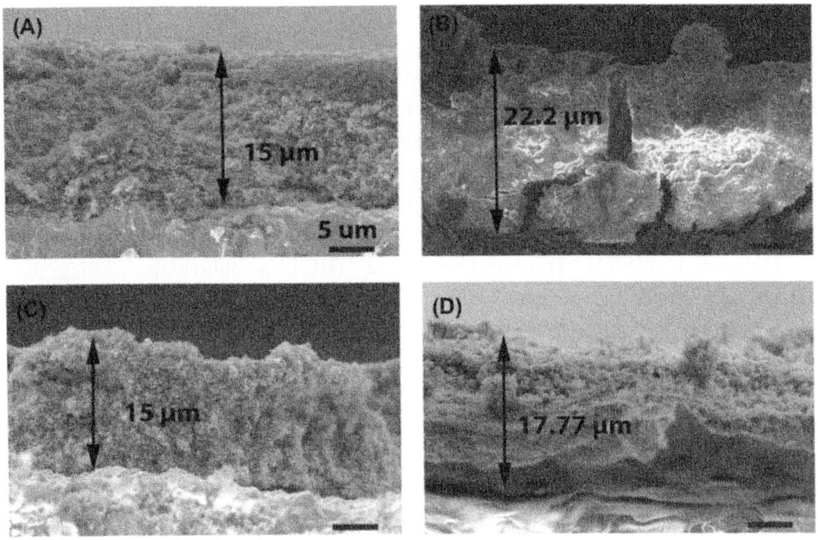

FIGURE 11.26

Cross section SEM images of Si@PVDF electrode (A) before cycling, (B) after 10 cycles, Si@Starch/PEG electrode (C) before cycling, and (D) after 10 cycles.

Reprinted with permission from S.N.S. Hapuarachchi, K.C. Wasalathilake, J.Y. Nerkar, E. Jaatinen, A.P. O'Mullane, C. Yan, Mechanically robust tapioca starch composite binder with improved ionic conductivity for sustainable lithium-ion batteries, ACS Sustain. Chem. Eng. 8 (2020) 9857–9865. https://doi.org/10.1021/acssuschemeng.0c02843. Copyright 2020 American Chemical Society.

used not only as a reinforcing agent to improve the mechanical properties of the separator, but also as a mediating tuner to optimize the cellulose membrane pore size. The porosity of the separators was measured in this study. Each cellulose paper was treated with a different percentage of CNF (10%, 20%, and 30%). The separator made with 10% CNF had a high porosity value and a large mean pore size. The electrolyte uptake for the 10% CNF separator was also the highest (142.3%), with an ionic conductivity of 1.09 S/cm. Higher Gurley values, on the other hand, were observed, indicating reduced porosity due to the fine size of CNF that blocks in some way the large pores of the cellulose paper. Despite this, the electrochemical properties were promising (the separator based on cellulose paper with 20% CNF outperformed the commercial one based on polypropylene). These materials are biodegradable and present high tensile strength (49 MPa for cellulose paper improved with 20% CNF). This material, along with its other properties, makes it a promising candidate for LIB separators. Gou et al. [98] also used cellulose in its purest form, but a novel method was used to prepare the highly porous separator. As previously stated, the capillarity during the drying process causes the CNF to form a denser network. The employment of the 5 wt.% polystyrene as colloidal monodispersed spheres of 1 µm diameter assured the tackling of this problem. After casting the solution into glass models to prepare the membranes from aqueous CNF/polystyrene in a 2:3 ratio, the transfer to an ethanol bath helped remove water and gave mechanically stable gels. These later were dried in a 50°C oven for 5 hours which were immersed afterwards into toluene bath to dissolve the polystyrene embedded in the CNC network, and then dried for 12 hours in 50°C oven. This method was less time consuming (30 hours) compared to other methods that take approximately 1–2 weeks. The ionic conductivity of these separators showed a good performance, with a value of 1.24 mS/cm compared to the previous research with only a maximum value of 1.09 μS/cm. Casas et al. [62] used CMC and hydroxyl cellulose (HEC) as battery separators for SIBs by nonsolvent induced phase separation. And the results show some promising aspects in terms of long-term high galvanostatic cycling, after 10 cycles at a C/10 rate, this separator delivers a capacity of 74 mAh/g with 100% Coulombic efficiency compared to 61 mAh/g and 96% for a commercial separator (Whatman GF/D). Bacterial cellulose (BC), a nanoscale biodegradable material available in a labyrinthic morphology, was used in LIBs as presented by Zhu et al. [99]. This material was incorporated into a polyphenylene sulfide matrix, which has been widely used as a battery separator but has many flaws, including mediocre wettability and thermal stability. Because of its unique three-dimensional network morphology and by tailoring and optimizing the pore size of this polymer, the use of BC heals these defects. Lizundia et al. [100] reviewed the majority of cellulose derivatives and their applications as LIBs separators. Alginate another biodegradable was presented by Song et al. [101]. Attapulgite, a nanofiber mineral with high thermal and structural stabilities up to 800°C, was used to incorporate the polymer host: alginate. Because of its hollow morphology, it can be used to improve electrolyte adsorption in this application. A phase inversion process was used to prepare the separator. Flame retardancy is an important property that separators should have to ensure safety in the event of a short circuit. In this study, the made separators were tested by hanging them over to a flame, the following images in Fig. 11.27, show their combustion behavior.

A commercial separator (Celgard 2325) was used as a reference, the presented separator was made of sodium alginate in 1 weight ratio to which added the attapulgite nanofibers in a 4-weight ratio (SA/

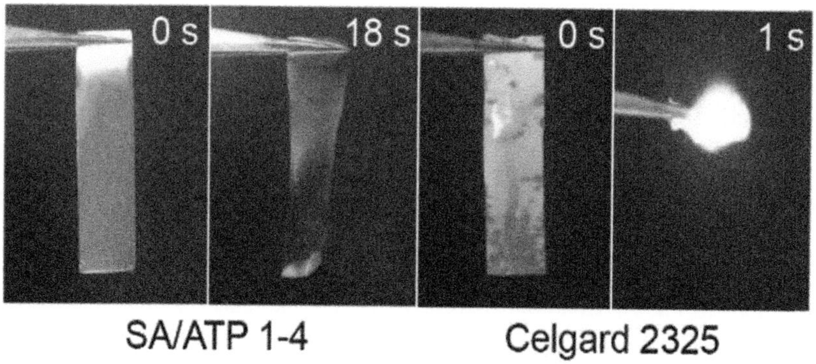

FIGURE 11.27

Combustion test of Celgard commercial separator and SA/ATP 1–4 separator wetted with commercial carbonate electrolyte.

Reprinted from Q. Song, A. Li, L. Shi, C. Qian, T.G. Feric, Y. Fu, et al., Thermally stable, nano-porous and eco-friendly sodium alginate/attapulgite separator for lithium-ion batteries, Energy Storage Mater. 22 (2019) 48–56. https://doi.org/10.1016/j.ensm.2019.06.033 with permission from Elsevier. https://s100.copyright.com/MyAccount/viewLicenseDetails?ref= 627d45ec-56f7-4d13-9d1c-8badbe641609. Copyright 2019.

ATP 1–4). The results show that even after the alginate was burned, the attapulgite fibers provided significant retardancy to the composite, while the commercial separator shrank completely when it came into contact with the flame. The as-prepared separator has excellent electrochemical performances due to its high porosity, electrolyte wettability, and uptake that exceeds 400%. Because the ingredients used are all nontoxic, the separator membranes are soil degradable. Because of its mesoporous morphology, agarose, a marine-based biodegradable polysaccharide, is also used in SIBs as a water-soluble separator, ensuring a stable electrodeposition of sodium onto the electrodes [61]. As previously stated, the electrolyte and separator in solid-state batteries are replaced by a single material: solid electrolyte. There are two types of this material: SPE and GPE [102]. Torres et al. [103] reviewed the use of Algal-based polysaccharides as a SPE in electrochemical devices, and its use in LIBs was promising due to its beneficial parameter, namely, the amorphous property that promotes charge carrier transport. Ben Youcef et al. [104] presented a SPE based on a functionalized cellulose and used it in both LIBs and SIBs. This SPE was created by functionalizing ethyl cellulose with lithium and sodium salt anions (lithium/sodium fluorosulfonylimide groups) to create quasi single-ion conductors within the PEO matrix. The degree of functionalization by both elements was measured by elemental analysis, and the values are as follows: Li anion (48%), and sodium salts (53%). The following are the transfer numbers: 0.9 for lithium-ion transfer and 0.6 for sodium-ion transfer The obtained results were compared to the ambipolar PEO using nonmodified commercial salts and proved higher electrochemical stability and better ionic conductivity of both materials (approximately 10^{-4} S/cm at 80°C). Chatterjee et al. [70] elaborated a material combining both types of biodegradable polymers: natural (starch) and synthetic (PVA) as a SPE. A polymeric matrix of these two polymers in 60:40 starch and PVA ratio was prepared by solution casting method, to which fumed nanosilica (14 nm) is added in different amounts (1, 2, 3, and 4 wt.%). Fig. 11.28 shows the effect of the fumed silica on ionic conductivity of bulk nanocomposite.

As a salt, lithium bromide is incorporated into the used polymer. The variation in ionic conductivity shows an increase while the amount of silica increases, followed by a decrease. This is a common finding in the study of nanocomposites used in SPE. The maximum conductivity measured was 1.0×10^{-3} S/cm. The increase is due to the nanofillers' large surface area, which induces several pathways, promoting ion motion. The increase in ionic conductivity could also be due to interactions between the polar groups of PVA and starch and the electrolyte's lithium ions. When the content exceeds 3 wt.% (4 wt.%), ionic conductivity decreases, which could be a result of aggregation of silica nanoparticles in the starch/PVA composite [105]. The combination of two biodegradable polymers is interesting, Ai et al. [106] investigated too, the use of a natural biodegradable polymer: chitosan and a synthetic one: PEO, to elaborate a SPE. The other type of polymeric electrolytes is the gel polymeric electrolyte. These materials exist in an intermediate state between liquid and solid states and are formed here by the swelling of a biodegradable polymer matrix with a liquid electrolyte [107]. Because of their much softer physical character, these materials provide better interfacial contact than solid ones and can serve as both an electrolyte and a separator. Zhu et al. [108] used a mixture of polymers: Soy protein isolate polymer (SPI) and PVA via electrospinning to elaborate a skeleton GPE in LIBs. The composite nanofiber membrane GPE based on SPI/PVA was investigated using the TEM (Fig. 11.29) to prove a previous observation on the diameter distribution of the PVA nanofiber.

The use of increased amounts of SPI, induces a descending trend on the diameter of the composite nanofibers. While the average diameter was about 160 nm, this value decreased to about 100 nm when the amount of SPI to PVA was raised to 3:1. No further results can be acquired because of the brittleness of the SPI chains. The study of electrochemical performances presented

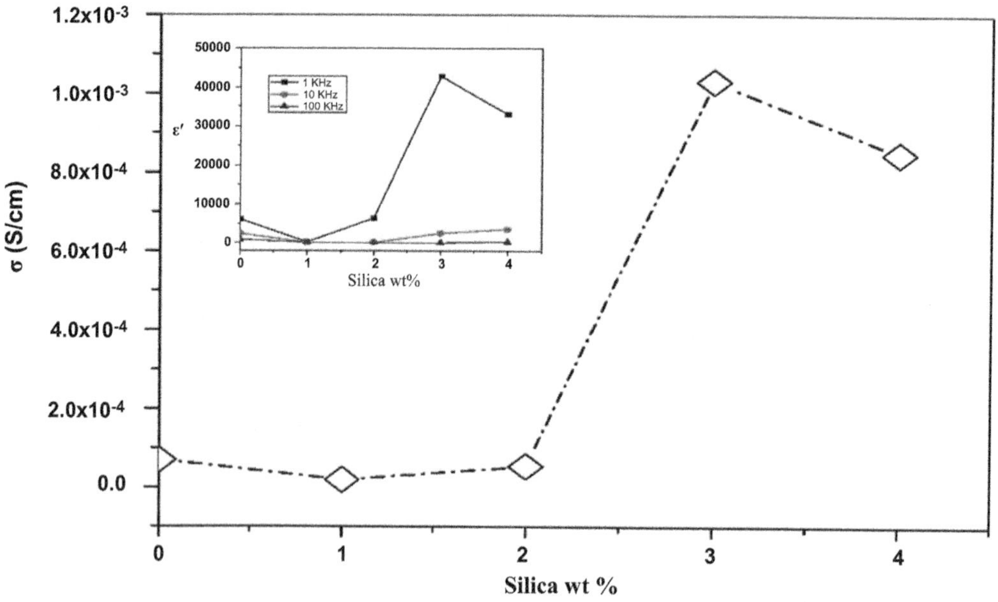

FIGURE 11.28

Bulk conductivity of starch/PVA nanocomposite polymer electrolyte with different contents of silica.

Reprinted from B. Chatterjee, N. Kulshrestha, P.N. Gupta, Nano composite solid polymer electrolytes based on biodegradable polymers starch and poly vinyl alcohol, Measurement. 82 (2016) 490–499. https://doi.org/10.1016/j.measurement.2016.01.022 with permission from Elsevier. https://s100.copyright.com/MyAccount/viewLicenseDetails?ref= f31c0b38-4e45-486e-a788-0f1357a33f69. Copyright 2016.

interesting results in terms of ionic conductivity (reached 3.8×10^{-3} S), and the excellent compatibility with the electrodes. Fig. 11.30 shows some of the electrochemical properties. Fig. 11.30A shows linear sweep voltammogram of the GPE used in the cell by using $LiCoO_2$ as a cathode and the Li as the counter reference electrode. A comparison to the cell featuring Celgard 2500 commercial separator took place. All the composite materials discomposed over 5.5 V which corresponds to the value of the electrochemical oxidation limit of the $LiPF_6$ electrolyte (5.1 V), higher values were observed then the one of Celgard 2500 (5.2 V). The obtained result supports the suitability of using these GPE in high voltage LIBs. The cycling performance presented in Fig. 11.30B of the separator made of (3:1) SPI/PVA ratio showed the higher reversible discharge capacity 118.2 mAh/g compared to Celgard 2500 with only 103.9 mAh/g. The observed superior performances can be ascribed to the high saturated electrolyte uptake and the good electrochemically interface stability.

A recent study by Han et al. [109] presented the use of both natural biodegradable polymers, chitosan, and lignocellulose. The material shows some very interesting properties in terms of electrolyte uptake, which reaches 749.1 wt.%, and ionic conductivity that surpassed 2.89 mS/cm. For the elaborated materials, EIS was used to study ionic conductivity. The range of frequencies that have been used are 0^{-2} to 10^5 Hz, the tests were done while controlling temperature as well (from 20°C to 90°C). The obtained values of ionic conductivity depend on the percentage of chitosan added. At first,

FIGURE 11.29

TEM images of (A) pure PVA, (B) SPI/PVA (1:1), (C) SPI/PVA (2:1), (D) SPI/PVA (3:1) nanofiber membranes.

Reprinted with permission from M. Zhu, C. Tan, Q. Fang, L. Gao, G. Sui, X. Yang, High performance and biodegradable skeleton material based on soy protein isolate for gel polymer electrolyte, ACS Sustain. Chem. Eng. 4 (2016) 4498–4505. https://doi.org/10.1021/acssuschemeng.6b01218. Copyright 2020 American Chemical Society.

the use of 5 wt.% of chitosan induced a high decrease of ionic conductivity (0.99 mS/cm) compared to pure lignocellulose matrix (1.79 mS/cm). But while increasing the amount beyond 10 wt.%, the ionic conductivity started to increase until it reaches its maximum (2.89 mS/cm) for 15 wt.% and decreases again to 2.53 mS/cm for 20 wt.%. This indicates that the optimal value and content to use is 15 wt.%. These results can be attributed to the best formation of the gelation phase, and it is in good agreement with the optimal electrolyte uptake that nears 750% obtained by the composite membrane of 15 wt.% of chitosan. Because of drawbacks such as weak mechanical properties, insufficient electrolyte absorption, and low ionic conductivity, the use of an excellent lignocellulose-based GPE alone is insufficient for use in LIBs.Luo et al. [110] proposed the use of a hydroxy propyl methyl cellulose (HPMC), as a novel biodegradable polymer with remarkable swelling effect. The phase-transfer method was used, which helped for obtaining extra free volume allowing better electrolyte absorption (1168 wt.%), and thus better electrochemical performances. For the lithium-ion transference number (t_{Li+}) for GPE, the calculation is done using the Eq. (11.2) shown afterwards:

$$t_{Li^+} = \frac{I_s \quad (\Delta V - \quad I_0 R_0)}{I_0 \quad (\Delta V - \quad I_s R_s)} \tag{11.2}$$

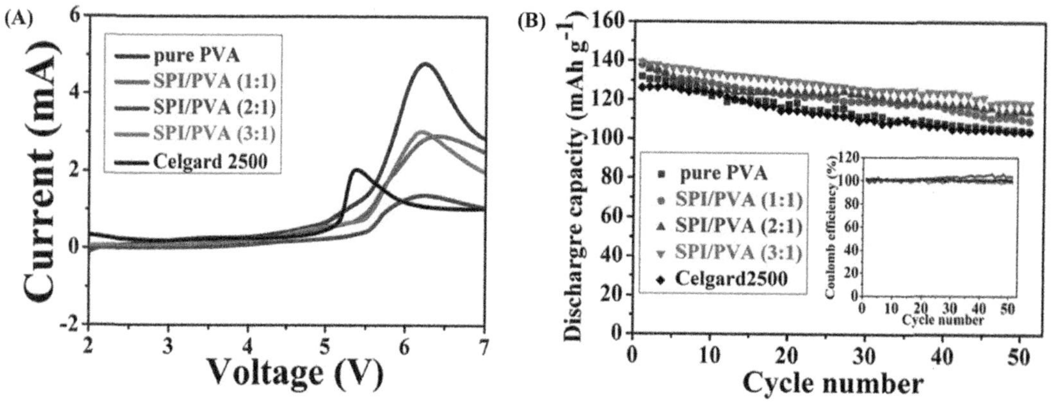

FIGURE 11.30

(A) Linear sweep voltammogram of Li/GPE/SS cells, (B) Cycling behavior up to 50 cycles.

Reprinted with permission from M. Zhu, C. Tan, Q. Fang, L. Gao, G. Sui, X. Yang, High performance and biodegradable skeleton material based on soy protein isolate for gel polymer electrolyte, ACS Sustain. Chem. Eng. 4 (2016) 4498–4505. https://doi.org/10.1021/acssuschemeng.6b01218. Copyright 2020 American Chemical Society.

where ΔV is the direct current (DC) polarization voltage and it equals to 10 mV, and I_0 and I_s are the initial and steady current, respectively. R_0 and R_s are the responses of interface impedance before and after polarization, respectively. The obtained value was 0.80 for the GPE containing only lignocellulose, and 0.75 for the GPE containing HPMC modified lignocellulose. Even though the number of transference is not as high as with lignocellulose alone, it is sufficient to meet the requirement for use in LIBs. As shown in Fig. 11.31, the use of HPMC improved other properties, specifically the plating and stripping performance of Li.

The galvanostatic charge/discharge tests evaluated the effect of HPMC on the deposition of Li on the surface of the electrodes. Long-term cycling shown in Fig. 11.30 indicates that the initial overpotential of unmodified lignocellulose is 12.4 mV and keeps increasing after a 1000 hours to reach approximately 70 mV, indicating an uneven deposition of Li, whereas the HPMC modified one showed an overpotential of 9.3 mV during the initial cycle and maintain a more stable value of 9.3 mV even after 1000 hours. These findings point to very stable plating and stripping of Li ions. Gou et al. [111] present a recent study on the use of cellulose in GPE. The abundance of hydroxyl groups on cellulose provides the GPE with very good ionic conductivity, but also results in poor interfacial resistance. The combination of a cross-linked cellulose network and polyethylene glycol diacrylate tailored this conflict (PEGDA). These GPE were created using a green two-step process. These GPE were elaborated via a green two-step method. Fig. 11.32 shows the investigation of the interfacial compatibility between the electrodes and the elaborated GPE by Nyquist plots. The used cellulose types are the following: double network hydrogel (GP), cellulose hydrogel (GC) and PEGDA modified hydrogel (GPC). The equivalent circuit was given, and it was used to fit the plots and to calculate the corresponding resistances. The R_f and R_{ct} correspond to the resistances of the Solid electrolyte interphase and the ionic transfer resistance of the interface between the electrode and the GPE, respectively. The obtained values are 206.4, 257.6 and 64.8Ω for GP, GC, and GPC, respectively.

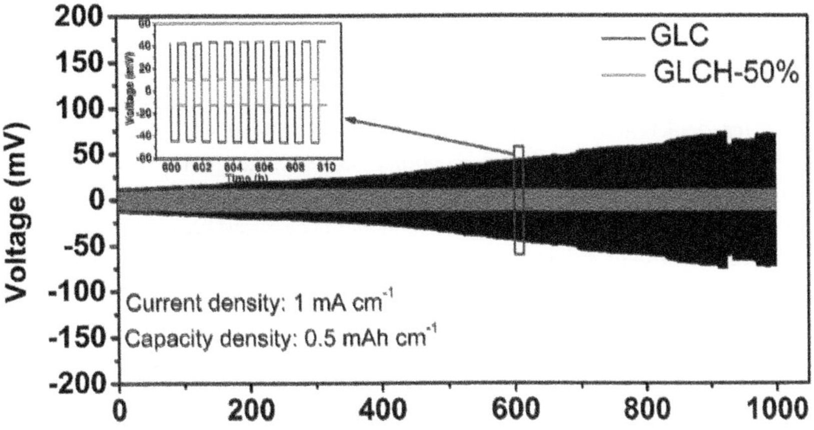

FIGURE 11.31

Long-term cycling of symmetric cells of Li-electrodes using lignocellulosic GPE (GLC in black shade) and the 50 wt.% HPMC modified lignocellulose GPE (GLCH-50%) with current densities of 1 mA/cm^2 and a deposition/stripping capacity of 1 mAh/cm^2.

Reprinted from C. Luo, Y. Huang, Z. Yin, H. Xu, X. Qin, X. Li, et al., A universal natural hydroxy propyl methyl cellulose polymer additive for modifying lignocellulose-based gel polymer electrolytes and stabilizing lithium metal anodes, Mater. Chem. Phys. 250 (2020) 123174. https://doi.org/10.1016/j.matchemphys.2020.123174, with permission from Elsevier. https://s100.copyright.com/MyAccount/viewLicenseDetails?ref= 9c5bdf9c-436f-4b1f-bb5d-9e8f513f4ccf. Copyright 2020.

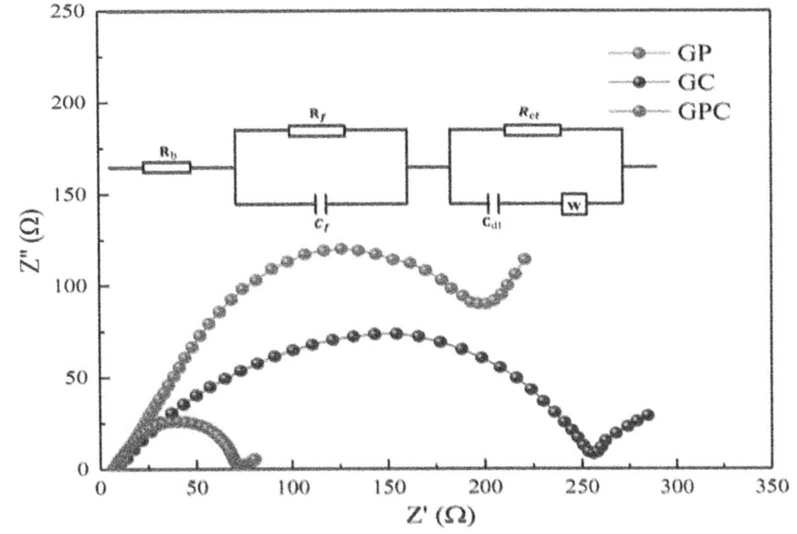

FIGURE 11.32

The AC impedance spectra of Li/GPE/Li and the equivalent circuits to fit the Nyquist plots.

Reprinted from J. Gou, W. Liu, A. Tang, To improve the interfacial compatibility of cellulose-based gel polymer electrolytes: a cellulose/PEGDA double network-based gel membrane designed for lithium ion batteries, Appl. Surf. Sci. 568 (2021) 150963. https://doi.org/10.1016/j.apsusc.2021.150963, with permission from Elsevier. https://s100.copyright.com/MyAccount/viewLicenseDetails?ref= 415e08b9-c1c7-4048-9799-0e271135398e. Copyright 2021.

11.4 Conclusion

We have begun to face serious energy concerns in recent decades as a result of high energy demands that contrast with the depletion of its fossil fuel resources. Renewable energies arose as a solution to the previous pitfalls in order to make a balanced choice between energy needs for the growing industry and the environmental aspect. However, because this energy production is intermittent and fluctuating, its storage is at the heart of current issues. Concerns about access and energy-storage device optimization led to the use of more appropriate systems, namely LIBs and SIBs. Many new battery materials have emerged, and the use of polymeric materials based on petroleum resources in these energy-storage devices has been extensively researched. The replacement of these materials with more sustainable and benign materials has received a lot of attention. Biodegradable polymers are a diverse class of materials that are widely used in daily life. Their importance has grown as a result of their application in a variety of domains. There are two types of biodegradable polymers: natural biodegradable polymers and synthetic biodegradable polymers.The use of biodegradable polymer composites and nanocomposites in the energy field was thoroughly reviewed in this book chapter. We were interested in two types of energy-storage systems: LIBs and SIBs. It was demonstrated how to use biodegradable materials in lithium and SIBs as nanocomposite materials for positive and negative electrodes, binders, separators, and SPEs. Some recent advances in improving battery properties, such as the use of various types and formulations, were presented. New synthesis and elaboration methods for these materials were mentioned. Characterization techniques were used to better understand the structure-properties relationship. Biodegradable polymers were used as starting materials for carbon-based nanocomposites, which played an important role in improving electrochemical performances for both battery types. Some elaboration techniques were developed for the use of biodegradable nanocomposites as binders, providing binders with good volume expansion control for silicon-based electrodes. Their nanoscale range has a lot of benefits in terms of ionic conductivity and high-rate cycling performances for the biodegradable composite and nanocomposite used in the other nonactive components of batteries, namely separator and polymer electrolytes. We noticed a large number of research papers about LIBs compared to SIBs, which were far from sufficient, while writing this book chapter. In this manuscript, we concentrated on providing recent research items, particularly those from the last 5 years. Suggestions for testing the use of the exact same formulations of polymer composites and nanocomposite previously used in LIBs are highly recommended for the elaboration of SIBs counterparts.

References

[1] F. Wei, Q. Zhang, P. Zhang, W. Tian, K. Dai, L. Zhang, et al., Review—research progress on layered transition metal oxide cathode materials for sodium ion batteries, J. Electrochem. Soc. 168 (2021). Available from: https://doi.org/10.1149/1945-7111/abf9bf.

[2] S. Li, Z. Gu, J. Guo, X. Hou, X. Yang, B. Zhao, et al., Enhanced electrode kinetics and electrochemical properties of low-cost $NaFe_2PO_4(SO_4)_2$ via Ca^{2+} doping as cathode material for sodium-ion batteries, J. Mater. Sci. Technol. 78 (2020) 176–182. Available from: https://doi.org/10.1016/j.jmst.2020.10.047.

[3] A. Qayoom, A. Ali, A. Ahmed, Meritorious spatially on hierarchically Co_3O_4/MoS_2 phase nanocomposite synergistically a highefficient electrocatalyst for hydrogen evolution reaction performance: recent advances & future perspectives, Int. J. Hydrog. Energy. 46 (2021) 22707–22718. Available from: https://doi.org/10.1016/j.ijhydene.2021.04.122.

[4] A. Kalair, N. Abas, M.S. Saleem, A. Raza, K. Nasrullah, Role of energy storage systems in energy transition from fossil fuels to renewables, Energy Storage 3 (2020) 1−27. Available from: https://doi.org/10.1002/est2.135.

[5] M.A. Morales-Mora, J.J.H. Pijpers, A.C. Antonio, J. de la, C. Soto, A.M.A. Calderón, Life cycle assessment of a novel bipolar electrodialysis-based flow battery concept and its potential use to mitigate the intermittency of renewable energy generation, J. Energy Storage. 35 (2021). Available from: https://doi.org/10.1016/j.est.2021.102339.

[6] Q. Zhang, Z. Gao, X. Shi, C. Zhang, K. Liu, J. Zhang, Recent advances on rare earths in solid lithium ion conductors, J. Rare Earths. 39 (2020) 1−10. Available from: https://doi.org/10.1016/j.jre.2020.04.001.

[7] M. Richter, A. Sarram, C. Kaucher, H. Winkler, Investigation and analysis of accumulators for the use of electrochemical storage in hybrid shunting locomotives, Procedia CIRP 81 (2019) 1010−1015. Available from: https://doi.org/10.1016/j.procir.2019.03.243.

[8] H. Budde-Meiwes, J. Drillkens, B. Lunz, J. Muennix, S. Rothgang, J. Kowal, et al., A review of current automotive battery technology and future prospects, Proc. Inst. Mech. Eng. Part. D. J. Automob. Eng. 227 (2013) 761−776. Available from: https://doi.org/10.1177/0954407013485567.

[9] E. Catenaro, D.M. Rizzo, S. Onori, Experimental analysis and analytical modeling of enhanced-ragone plot, Appl. Energy. 291 (2021) 116473. Available from: https://doi.org/10.1016/j.apenergy.2021.116473.

[10] Y. Liang, C.-Z. Zhao, H. Yuan, Y. Chen, W. Zhang, J.-Q. Huang, et al., A review of rechargeable batteries for portable electronic devices, InfoMat Wiley Online Libr. 1 (2019) 1−27. Available from: https://doi.org/10.1002/inf2.12000.

[11] J.C. Zheng, Y.Y. Yao, G.Q. Mao, H.Z. Chen, H. Li, L. Cao, et al., Iron-zinc sulfide $Fe_2Zn_3S_5/Fe_{1-x}S$@C derived from a metal-organic framework as a high performance anode material for lithium-ion batteries, J. Mater. Chem. A 7 (2019) 16479−16487. Available from: https://doi.org/10.1039/c9ta03271a.

[12] G. Liang, V.K. Peterson, K.W. See, Z. Guo, W.K. Pang, Developing high-voltage spinel $LiNi_{0.5}Mn_{1.5}O_4$ cathodes for high-energy-density lithium-ion batteries: current achievements and future prospects, J. Mater. Chem. A 8 (2020) 15373−15398. Available from: https://doi.org/10.1039/d0ta02812f.

[13] K. Chayambuka, G. Mulder, D.L. Danilov, P.H.L. Notten, Sodium-ion battery materials and electrochemical properties reviewed, Adv. Energy Mater. 8 (2018) 1−49. Available from: https://doi.org/10.1002/aenm.201800079.

[14] H.S. Hirsh, Y. Li, D.H.S. Tan, M. Zhang, E. Zhao, Y.S. Meng, Sodium-ion batteries paving the way for grid energy storage, Adv. Energy Mater. 10 (2020) 2001274. Available from: https://doi.org/10.1002/aenm.202001274.

[15] S. Zhu, C. Hu, Y. Xu, Y. Jin, J. Shui, Performance improvement of lithium-ion battery by pulse current, J. Energy Chem. 46 (2020) 208−214. Available from: https://doi.org/10.1016/j.jechem.2019.11.007.

[16] H. Shen, Y. Huang, Y. Chang, R. Hao, Z. Ma, K. Wu, et al., Narrowing working voltage window to improve layered GeP anode cycling performance for lithium-ion batteries, ACS Appl. Mater. Interfaces. 12 (2020) 17466−17473. Available from: https://doi.org/10.1021/acsami.9b23515.

[17] Z. Zou, H. Xu, H. Zhang, Y. Tang, G. Cui, Electrolyte therapy for improving the performance of $LiNi_{0.5}Mn_{1.5}O_4$ cathodes assembled lithium-ion batteries, ACS Appl. Mater. Interfaces. 12 (2020) 21368−21385. Available from: https://doi.org/10.1021/acsami.0c02516.

[18] F. Li, Z. Zhou, Micro/nanostructured materials for sodium ion batteries and capacitors, Small 14 (2018) 1−25. Available from: https://doi.org/10.1002/smll.201702961.

[19] Y.Y. Zhang, S.J. Zhang, J.T. Li, K. Wang, Y.C. Zhang, Q. Liu, et al., Improvement of electrochemical properties of P2-type N2/3 Mn2/3 Ni1/3 O2 sodium ion battery cathode material by water-soluble binders, Electrochim. Acta. 298 (2019) 496−504. Available from: https://doi.org/10.1016/j.electacta.2018.12.089.

[20] Y. Zhang, J. Qin, S.E. Lowe, W. Li, Y. Zhu, M. Al-Mamun, et al., Enhanced electrochemical production and facile modification of graphite oxide for cost-effective sodium ion battery anodes, Carbon N. Y. 177 (2021) 71−78. Available from: https://doi.org/10.1016/j.carbon.2021.02.067.

[21] X. Li, X. Sun, X. Hu, F. Fan, S. Cai, C. Zheng, et al., Review on comprehending and enhancing the initial Coulombic efficiency of anode materials in lithium-ion/sodium-ion batteries, Nano Energy 77 (2020) 105143. Available from: https://doi.org/10.1016/j.nanoen.2020.105143.
[22] C. Liu, Z.G. Neale, G. Cao, Understanding electrochemical potentials of cathode materials in rechargeable batteries, Mater. Today. 19 (2016) 109–123. Available from: https://doi.org/10.1016/j.mattod.2015.10.009.
[23] R. Wang, H. Mo, S. Li, Y. Gong, B. He, H. Wang, Influence of conductive additives on the stability of red phosphorus-carbon anodes for sodium-ion batteries, Sci. Rep. 9 (2019) 1–6. Available from: https://doi.org/10.1038/s41598-018-36797-z.
[24] S. Hein, T. Danner, D. Westhoff, B. Prifling, R. Scurtu, L. Kremer, et al., Influence of conductive additives and binder on the impedance of lithium-ion battery electrodes: effect of morphology, J. Electrochem. Soc. 167 (2020) 013546. Available from: https://doi.org/10.1149/1945-7111/ab6b1d.
[25] F.E. Bouharras, M. Raihane, M. Baccour, N. Louvain, B. Ameduri, Evaluation of core–shell poly(vinylidene fluoride)-grafted-Barium titanate (PVDF-g-$BaTiO_3$) nanocomposites as a cathode binder in batteries, Solid. State Ion. 356 (2020) 115441. Available from: https://doi.org/10.1016/j.ssi.2020.115441.
[26] K. Peng, Y. He, H. Hu, S. Li, B. Tao, Mechanical integrity of conductive carbon-black-filled aqueous polymer binder in composite electrode for lithium-ion battery, Polym. (Basel) 12 (2020) 1–16. Available from: https://doi.org/10.3390/polym12071460.
[27] K. Kim, H. Ma, S. Park, N.-S. Choi, Electrolyte-additive-driven interfacial engineering for high-capacity electrodes in lithium-ion batteries: promise and challenges, ACS Energy Lett. 5 (2020) 1537–1553. Available from: https://doi.org/10.1021/acsenergylett.0c00468.
[28] J.C. Barbosa, D.M. Correia, R. Gonçalves, V. de Zea Bermudez, M.M. Silva, S. Lanceros-Mendez, et al., Enhanced ionic conductivity in poly(vinylidene fluoride) electrospun separator membranes blended with different ionic liquids for lithium ion batteries, J. Colloid Interface Sci. 582 (2021) 376–386. Available from: https://doi.org/10.1016/j.jcis.2020.08.046.
[29] X. Ma, F. Qiao, M. Qian, Y. Ye, X. Cao, Y. Wei, et al., Facile fabrication of flexible electrodes with poly(vinylidene fluoride)/Si_3N_4 composite separator prepared by electrospinning for sodium-ion batteries, Scr. Mater. 190 (2021) 153–157. Available from: https://doi.org/10.1016/j.scriptamat.2020.08.053.
[30] A. Saal, T. Hagemann, U.S. Schubert, Polymers for battery applications—active materials, membranes, and binders, Adv. Energy Mater. 2001984 (2020) 1–40. Available from: https://doi.org/10.1002/aenm.202001984.
[31] Y. Cui, J. Wan, Y. Ye, K. Liu, L.-Y. Chou, Y. Cui, A fireproof, lightweight, polymer – polymer solid-state electrolyte, Nano Lett. 20 (2020) 1686–1692.
[32] G.-Z. Yin, X.-M. Yang, Biodegradable polymers : a cure for the planet, but a long way to go, J. Poymer Res. 27 (2020). Available from: https://doi.org/10.1007/s10965-020-2004-1.
[33] I. Vroman, L. Tighzert, Biodegradable polymers, Mater. (Basel) 2 (2009) 307–344. Available from: https://doi.org/10.3390/ma2020307.
[34] N.R. Nair, V.C. Sekhar, K.M. Nampoothiri, A. Pandey, Biodegradation of biopolymers, Curr. Dev. Biotechnol. Bioeng. Prod. Isol. Purif. Ind. Prod. (2016) 739–755. Available from: https://doi.org/10.1016/B978-0-444-63662-1.00032-4. Elsevier B.V.
[35] L. Avérous, E. Pollet, Environmental silicate nano-biocomposites, Green Energy Technol (2012) 13–39. Available from: https://doi.org/10.1007/978-1-4471-4108-2.
[36] S. Bandehali, H. Sanaeepur, A. Ebadi Amooghin, S. Shirazian, S. Ramakrishna, Biodegradable polymers for membrane separation, Sep. Purif. Technol. 269 (2021) 118731. Available from: https://doi.org/10.1016/j.seppur.2021.118731.
[37] E.S. Hosseini, S. Dervin, P. Ganguly, R. Dahiya, Biodegradable materials for sustainable health monitoring devices, ACS Appl. Bio Mater. 4 (2021) 163–194. Available from: https://doi.org/10.1021/acsabm.0c01139.
[38] N. Lingappan, L. Kong, M. Pecht, The significance of aqueous binders in lithium-ion batteries, Renew. Sustain. Energy Rev. 147 (2021) 111227. Available from: https://doi.org/10.1016/j.rser.2021.111227.

[39] P.S. Salini, S.V. Gopinadh, A. Kalpakasseri, B. John, M. Thelakkattu Devassy, Toward greener and sustainable li-ion cells: an overview of aqueous-based binder systems, ACS Sustain. Chem. Eng. 8 (2020) 4003–4025. Available from: https://doi.org/10.1021/acssuschemeng.9b07478.

[40] Y. Ma, J. Ma, G. Cui, Small things make big deal: powerful binders of lithium batteries and post-lithium batteries, Energy Storage Mater. 20 (2019) 146–175. Available from: https://doi.org/10.1016/j.ensm.2018.11.013.

[41] F. Zou, A. Manthiram, A review of the design of advanced binders for high-performance batteries, Adv. Energy Mater. 10 (2020) 1–28. Available from: https://doi.org/10.1002/aenm.202002508.

[42] A. Maurel, M. Courty, B. Fleutot, H. Tortajada, K. Prashantha, M. Armand, et al., Highly loaded graphite-polylactic acid composite-based filaments for lithium-ion battery three-dimensional printing, Chem. Mater. 30 (2018) 7484–7493. Available from: https://doi.org/10.1021/acs.chemmater.8b02062.

[43] V. Gupta, F. Alam, P. Verma, A.M. Kannan, S. Kumar, Additive manufacturing enabled, microarchitected, hierarchically porous polylactic-acid / lithium iron phosphate / carbon nanotube nanocomposite electrodes for high performance Li-Ion batteries, J. Power Sources. 494 (2021) 229625. Available from: https://doi.org/10.1016/j.jpowsour.2021.229625.

[44] C. Fu, C. Battaglia, Polymer – inorganic nanocomposite coating with high ionic conductivity and transference number for a stable lithium metal anode, Appl. Mater. Interfaces. 12 (2020) 41620–41626. Available from: https://doi.org/10.1021/acsami.0c13485.

[45] V.V.N. Phanikumar, V.R. Rikka, B. Das, R. Gopalan, B.A. Rao, Raju Prakash, Investigation on polyvinyl alcohol and sodium alginate as aqueous binders for lithium-titanium oxide anode in lithium-ion batteries, Ion. (Kiel.) 25 (2019) 2549–2561.

[46] Q.D. Nguyen, E.S. Oh, K.H. Chung, Nanomechanical properties of polymer binders for Li-ion batteries probed with colloidal probe atomic force microscopy, Polym. Test. 76 (2019) 245–253. Available from: https://doi.org/10.1016/j.polymertesting.2019.03.025.

[47] P. Parikh, M. Sina, A. Banerjee, X. Wang, M.S. D'Souza, J.M. Doux, et al., Role of polyacrylic acid (PAA) binder on the solid electrolyte interphase in silicon anodes, Chem. Mater. 31 (2019) 2535–2544. Available from: https://doi.org/10.1021/acs.chemmater.8b05020.

[48] B. Hu, I.A. Shkrob, S. Zhang, L. Zhang, J. Zhang, Y. Li, et al., The existence of optimal molecular weight for poly(acrylic acid) binders in silicon/graphite composite anode for lithium-ion batteries, J. Power Sources 378 (2018) 671–676. Available from: https://doi.org/10.1016/j.jpowsour.2017.12.068.

[49] D. Ni, W. Sun, C. Lu, Z. Wang, J. Qiao, H. Cai, et al., Improved rate and cycling performance of FeF_2-rGO hybrid cathode with poly (acrylic acid) binder for sodium ion batteries, J. Power Sources 413 (2019) 449–458. Available from: https://doi.org/10.1016/j.jpowsour.2018.12.040.

[50] P. Mandal, K. Stokes, G. Hernández, D. Brandell, J. Mindemark, Influence of binder crystallinity on the performance of Si electrodes with poly(vinyl alcohol) binders, ACS Appl. Energy Mater. 4 (2021) 3008–3016. Available from: https://doi.org/10.1021/acsaem.0c02051.

[51] A. Li, A. Chun, Y. Yuen, W. Wang, I. Miguel, D.C. Cordeiro, et al., A review on lithium-ion battery separators towards enhanced safety performances and modelling approaches, Molecules 26 (2021) 478. Available from: https://doi.org/10.3390/molecules26020478.

[52] Y. Li, Q. Li, Z. Tan, A review of electrospun nanofiber-based separators for rechargeable lithium-ion batteries, J. Power Sources 443 (2019) 227262. Available from: https://doi.org/10.1016/j.jpowsour.2019.227262.

[53] S. Janakiraman, M. Khalifa, R. Biswal, S. Ghosh, S. Anandhan, A. Venimadhav, High performance electrospun nanofiber coated polypropylene membrane as a separator for sodium ion batteries, J. Power Sources 460 (2020) 228060. Available from: https://doi.org/10.1016/j.jpowsour.2020.228060.

[54] L.P. Wang, L. Yu, M. Srinivasan, Z.J. Xu, X. Wang, Recent developments in electrode materials for sodium-ion batteries, J. Mater. Chem. A. 3 (2015) 9353–9378. Available from: https://doi.org/10.1039/c4ta06467d.

[55] C.M. Costa, E. Lizundia, S. Lanceros-méndez, Polymers for advanced lithium-ion batteries: state of the art and future needs on polymers for the different battery components, Prog. Energy Combust. Sci. 79 (2020) 100846. Available from: https://doi.org/10.1016/j.pecs.2020.100846.
[56] L. Zhang, X. Li, M. Yang, W. Chen, High-safety separators for lithium-ion batteries and sodium-ion batteries: advances and perspective, Energy Storage Mater. (2021). Available from: https://doi.org/10.1016/j.ensm.2021.06.033.
[57] P. Zhai, K. Liu, Z. Wang, L. Shi, S. Yuan, Multifunctional separators for high-performance lithium ion batteries, J. Power Sources 499 (2021) 229973. Available from: https://doi.org/10.1016/j.jpowsour.2021.229973.
[58] J. Nunes-Pereira, C.M. Costa, S. Lanceros-Méndez, Polymer composites and blends for battery separators: state of the art, challenges and future trends, J. Power Sources 281 (2015) 378–398. Available from: https://doi.org/10.1016/j.jpowsour.2015.02.010.
[59] H. Zhang, M.-Y. Zhou, C.-E. Lin, B. Zhu, Progress in polymeric separators for lithium ion batteries, RSC Adv. 5 (2015) 89448–89860. Available from: https://doi.org/10.1039/c5ra14087k.
[60] H. Zhang, H. Zhao, M.A. Khan, W. Zou, J. Xu, L. Zhang, et al., Recent progress in advanced electrode materials, separators and electrolytes for lithium batteries, J. Mater. Chem. A. 6 (2018) 20564–20620. Available from: https://doi.org/10.1039/C8TA05336G.
[61] A. Ojanguren, N. Mittal, E. Lizundia, M. Niederberger, Stable Na electrodeposition enabled by agarose-based water-soluble sodium ion battery separators, ACS Appl. Mater. Interfaces. 13 (2021) 21250–21260. Available from: https://doi.org/10.1021/acsami.1c02135.
[62] X. Casas, M. Niederberger, E. Lizundia, A sodium-ion battery separator with reversible voltage response based on water-soluble cellulose derivatives, ACS Appl. Mater. Interfaces. 12 (2020) 29264–29274. Available from: https://doi.org/10.1021/acsami.0c05262.
[63] Y. Xie, Y. Pan, P. Cai, Novel PVA-based porous separators prepared via freeze-drying for enhancing performance of lithium-ion batteries, Ind. Eng. Chem. Res. 59 (2020) 15242–15254. Available from: https://doi.org/10.1021/acs.iecr.0c02182.
[64] C.F.J. Francis, I.L. Kyratzis, A.S. Best, Lithium-ion battery separators for ionic-liquid electrolytes: a review, Adv. Mater. 32 (2020) 1–22. Available from: https://doi.org/10.1002/adma.201904205.
[65] D.R. Rajagopalan Kannan, P.K. Terala, P.L. Moss, M.H. Weatherspoon, Analysis of the separator thickness and porosity on the performance of lithium-ion batteries, Int. J. Electrochem. 2018 (2018) 1–7. Available from: https://doi.org/10.1155/2018/1925708.
[66] X. Zhang, J. Zhu, E. Sahraei, Degradation of battery separators under charge-discharge cycles, RSC Adv. 7 (2017) 56099–56107. Available from: https://doi.org/10.1039/c7ra11585g.
[67] C. Zhang, H. Li, S. Wang, Y. Cao, H. Yang, X. Ai, et al., A polyethylene microsphere-coated separator with rapid thermal shutdown function for lithium-ion batteries, J. Energy Chem. 44 (2020) 33–40. Available from: https://doi.org/10.1016/j.jechem.2019.09.017.
[68] Y. Wang, S. Song, C. Xu, N. Hu, J. Molenda, L. Lu, Development of solid-state electrolytes for sodium-ion battery-A short review, Elements 1 (2021) 91–100. Available from: https://doi.org/10.1016/j.nanoms.2019.02.007.
[69] L. Li, Y. Shan, X. Yang, New insights for constructing solid polymer electrolytes with ideal lithium-ion transfer channels by using inorganic filler, Mater. Today Commun. 26 (2021) 101910. Available from: https://doi.org/10.1016/j.mtcomm.2020.101910.
[70] B. Chatterjee, N. Kulshrestha, P.N. Gupta, Nano composite solid polymer electrolytes based on biodegradable polymers starch and poly vinyl alcohol, Measurement 82 (2016) 490–499. Available from: https://doi.org/10.1016/j.measurement.2016.01.022.
[71] J. Yang, H. Zhang, Q. Zhou, H. Qu, T. Dong, M. Zhang, et al., Safety-enhanced polymer electrolytes for sodium batteries: recent progress and perspectives, ACS Appl. Mater. Interfaces. 11 (2019) 17109–17127. Available from: https://doi.org/10.1021/acsami.9b01239.

[72] Y.L. Ni'Mah, M.Y. Cheng, J.H. Cheng, J. Rick, B.J. Hwang, Solid-state polymer nanocomposite electrolyte of TiO_2/PEO/$NaClO_4$ for sodium ion batteries, J. Power Sources 278 (2015) 375–381. Available from: https://doi.org/10.1016/j.jpowsour.2014.11.047.

[73] M. Sunandar, E. Yulianti, D. Deswita, S. Sudaryanto, Study of solid polymer electrolyte based on biodegradable polymer polycaprolactone, Malaysian J. Fundam. Appl. Sci. 15 (2019) 467–471. Available from: https://doi.org/10.11113/mjfas.v15n3.1185.

[74] F.P. Nkosi, M. Valvo, J. Mindemark, N.A. Dzulkurnain, G. Hernández, A. Mahun, et al., Garnet-poly (ϵ-caprolactone-co-trimethylene carbonate) polymer-in-ceramic composite electrolyte for all-solid-state lithium-ion batteries, ACS Appl. Energy Mater. 4 (2021) 2531–2542. Available from: https://doi.org/10.1021/acsaem.0c03098.

[75] C.M. Costa, Y.H. Lee, J.H. Kim, S.Y. Lee, S. Lanceros-Méndez, Recent advances on separator membranes for lithium-ion battery applications: from porous membranes to solid electrolytes, Energy Storage Mater. 22 (2019) 346–375. Available from: https://doi.org/10.1016/j.ensm.2019.07.024.

[76] V.P.H. Huy, S. So, J. Hur, Inorganic fillers in composite gel polymer electrolytes for high-performance lithium and non-lithium polymer batteries, Nanomaterials 11 (2021) 1–40. Available from: https://doi.org/10.3390/nano11030614.

[77] Z. Liu, X. Wang, J. Chen, Y. Tang, Z. Mao, D. Wang, Gel polymer electrolyte membranes boosted with sodium-conductive β-alumina nanoparticles: application for Na-ion batteries, ACS Appl. Energy Mater. 4 (2021) 623–632. Available from: https://doi.org/10.1021/acsaem.0c02513.

[78] E. Lizundia, D. Kundu, Advances in natural biopolymer-based electrolytes and separators for battery applications, Adv. Funct. Mater. 31 (2021) 1–29. Available from: https://doi.org/10.1002/adfm.202005646.

[79] A.L. Yaguchi, S.J. Lee, M.A. Blenner, Synthetic biology towards engineering microbial lignin biotransformation, Trends Biotechnol. (2021). Available from: https://doi.org/10.1016/j.tibtech.2021.02.003.

[80] C. Luo, L. Du, W. Wu, H. Xu, G. Zhang, S. Li, et al., Novel lignin-derived water-soluble binder for micro silicon anode in lithium-ion batteries, ACS Sustain. Chem. Eng. 6 (2018) 12621–12629. Available from: https://doi.org/10.1021/acssuschemeng.8b01161.

[81] T.C. Nirmale, B.B. Kale, A.J. Varma, A review on cellulose and lignin based binders and electrodes: small steps towards a sustainable lithium ion battery, Int. J. Biol. Macromol. 103 (2017) 1032–1043. Available from: https://doi.org/10.1016/j.ijbiomac.2017.05.155.

[82] F. Chen, L. Wu, Z. Zhou, J. Ju, Z. Zhao, M. Zhong, et al., MoS_2 decorated lignin-derived hierarchical mesoporous carbon hybrid nanospheres with exceptional Li-ion battery cycle stability, Chin. Chem. Lett. 30 (2019) 197–202. Available from: https://doi.org/10.1016/j.cclet.2018.10.007.

[83] X. Dou, I. Hasa, M. Hekmatfar, T. Diemant, R.J. Behm, D. Buchholz, et al., Pectin, hemicellulose, or lignin? Impact of the biowaste source on the performance of hard carbons for sodium-ion batteries, ChemSusChem 10 (2017) 2668–2676. Available from: https://doi.org/10.1002/cssc.201700628.

[84] S. Zhou, L. Nyholm, M. Strømme, Z. Wang, Cladophora cellulose: unique biopolymer nanofibrils for emerging energy, environmental, and life science applications, Acc. Chem. Res. 52 (2019) 2232–2243. Available from: https://doi.org/10.1021/acs.accounts.9b00215.

[85] Z. Wang, T. Huang, Z. Liu, A. Yu, Dopamine-modified carboxymethyl cellulose as an improved aqueous binder for silicon anodes in lithium-ion batteries, Electrochim. Acta. 389 (2021) 138806. Available from: https://doi.org/10.1016/j.electacta.2021.138806.

[86] O. El Baradai, D. Beneventi, F. Alloin, Y. Bultel, D. Chaussy, Use of cellulose nanofibers as an electrode binder for lithium ion battery screen printing on a paper separator, Nanomaterials 8 (2018) 982. Available from: https://doi.org/10.3390/nano8120982.

[87] Y. Li, B. Jin, K. Wang, L. Song, L. Ren, Y. Hou, et al., Coordinatively-intertwined dual anionic polysaccharides as binder with 3D network conducive for stable SEI formation in advanced silicon-based anodes, Chem. Eng. J. 429 (2021) 132235. Available from: https://doi.org/10.1016/j.cej.2021.132235.

[88] W. Zhang, H. Zhou, Z. Huang, S. Li, C. Wang, H. Li, et al., 3D hierarchical microspheres constructed by ultrathin MoS_2-C nanosheets as high-performance anode material for sodium-ion batteries, J. Energy Chem. 49 (2020) 307–315. Available from: https://doi.org/10.1016/j.jechem.2020.03.001.

[89] H. Zhu, F. Shen, W. Luo, S. Zhu, M. Zhao, B. Natarajan, et al., Low temperature carbonization of cellulose nanocrystals for high performance carbon anode of sodium-ion batteries, Nano Energy 33 (2017) 37–44. Available from: https://doi.org/10.1016/j.nanoen.2017.01.021.

[90] Y.E. Kim, S.J. Yeom, J.E. Lee, S. Kang, H. Kang, G.H. Lee, et al., Structure-dependent sodium ion storage mechanism of cellulose nanocrystal-based carbon anodes for highly efficient and stable batteries, J. Power Sources 468 (2020) 228371. Available from: https://doi.org/10.1016/j.jpowsour.2020.228371.

[91] P.R. Kumar, P. Kollu, C. Santhosh, K. Eswara Varaprasada Rao, D.K. Kim, A.N. Grace, Enhanced properties of porous $CoFe_2O_4$-reduced graphene oxide composites with alginate binders for Li-ion battery applications, N. J. Chem. 38 (2014) 3654–3661. Available from: https://doi.org/10.1039/c4nj00419a.

[92] Y.Y. Hu, J.H. You, S.J. Zhang, H. Lin, W.F. Ren, L. Deng, et al., $Li_{0.5}$PAA domains filled in porous sodium alginate skeleton: a 3D bicontinuous composite network binder to stabilize micro-silicon anode for high-performance lithium ion battery, Electrochim. Acta. 386 (2021) 138361. Available from: https://doi.org/10.1016/j.electacta.2021.138361.

[93] H. Jin, M. Zhu, J. Liu, L. Gan, Z. Gong, M. Long, Alkaline chitosan solution as etching phase to design Si@SiO_2@N-carbon anode for Lithium-ion battery, Appl. Surf. Sci. 541 (2021) 148436. Available from: https://doi.org/10.1016/j.apsusc.2020.148436.

[94] M. Aydin, E. Demir, B. Unal, B. Dursun, A.S. Ahsen, R. Demir-Cakan, Chitosan derived N-doped carbon coated SnO_2 nanocomposite anodes for Na-ion batteries, Solid. State Ion. 341 (2019) 115035. Available from: https://doi.org/10.1016/j.ssi.2019.115035.

[95] S.N.S. Hapuarachchi, K.C. Wasalathilake, J.Y. Nerkar, E. Jaatinen, A.P. O'Mullane, C. Yan, Mechanically robust tapioca starch composite binder with improved ionic conductivity for sustainable lithium-ion batteries, ACS Sustain. Chem. Eng. 8 (2020) 9857–9865. Available from: https://doi.org/10.1021/acssuschemeng.0c02843.

[96] B. Jin, D. Wang, L. Song, Y. Cai, A. Ali, Y. Hou, et al., Biomass-derived fluorinated corn starch emulsion as binder for silicon and silicon oxide based anodes in lithium-ion batteries, Electrochim. Acta. 365 (2021). Available from: https://doi.org/10.1016/j.electacta.2020.137359.

[97] D. Lv, J. Chai, P. Wang, L. Zhu, C. Liu, S. Nie, et al., Pure cellulose lithium-ion battery separator with tunable pore size and improved working stability by cellulose nanofibrils, Carbohydr. Polym. 251 (2021) 116975. Available from: https://doi.org/10.1016/j.carbpol.2020.116975.

[98] J. Gou, W. Liu, A. Tang, A novel method to prepare a highly porous separator based on nanocellulose with multi-scale pore structures and its application for rechargeable lithium ion batteries, J. Memb. Sci. 639 (2021) 119750. Available from: https://doi.org/10.1016/j.memsci.2021.119750.

[99] C. Zhu, J. Zhang, S. Qiu, Y. Jia, L. Wang, H. Wang, Tailoring the pore size of polyphenylene sulfide nonwoven with bacterial cellulose (BC) for heat-resistant and high-wettability separator in lithium-ion battery, Compos. Commun. 24 (2021) 100659. Available from: https://doi.org/10.1016/j.coco.2021.100659.

[100] E. Lizundia, C.M. Costa, R. Alves, S. Lanceros-Méndez, Cellulose and its derivatives for lithium ion battery separators: a review on the processing methods and properties, Carbohydr. Polym. Technol. Appl. 1 (2020) 100001. Available from: https://doi.org/10.1016/j.carpta.2020.100001.

[101] Q. Song, A. Li, L. Shi, C. Qian, T.G. Feric, Y. Fu, et al., Thermally stable, nano-porous and eco-friendly sodium alginate/attapulgite separator for lithium-ion batteries, Energy Storage Mater. 22 (2019) 48–56. Available from: https://doi.org/10.1016/j.ensm.2019.06.033.

[102] F. Gebert, J. Knott, R. Gorkin, S.-L. Chou, S.-X. Dou, Polymer electrolytes for sodium-ion batteries, Energy Storage Mater. 36 (2021) 10–30. Available from: https://doi.org/10.1016/j.ensm.2020.11.030.

[103] F.G. Torres, G.E. De-la-Torre, Algal-based polysaccharides as polymer electrolytes in modern electrochemical energy conversion and storage systems: a review, Carbohydr. Polym. Technol. Appl. 2 (2021) 100023. Available from: https://doi.org/10.1016/j.carpta.2020.100023.

[104] H. Ben Youcef, B. Orayech, J.M.L. Del Amo, F. Bonilla, D. Shanmukaraj, M. Armand, Functionalized cellulose as quasi single-ion conductors in polymer electrolyte for all-solid−state Li/Na and Li[sbnd]S batteries, Solid. State Ion. 345 (2020) 115168. Available from: https://doi.org/10.1016/j.ssi.2019.115168.

[105] S.K. Patla, R. Ray, S. Karmakar, S. Das, S. Tarafdar, Nanofiller-induced ionic conductivity enhancement and relaxation property analysis of the blend polymer electrolyte using non-debye electric field relaxation function, J. Phys. Chem. C. 123 (2019) 5188−5197. Available from: https://doi.org/10.1021/acs.jpcc.8b10460.

[106] S. Ai, S. Mazumdar, H. Li, Y. Cao, T. Li, Nano-silica doped composite polymer chitosan/poly(ethylene oxide)-based electrolyte with high electrochemical stability suitable for quasi solid-state lithium metal batteries, J. Electroanal. Chem. 895 (2021) 115464. Available from: https://doi.org/10.1016/j.jelechem.2021.115464.

[107] Q. Yang, N. Deng, J. Chen, B. Cheng, W. Kang, The recent research progress and prospect of gel polymer electrolytes in lithium-sulfur batteries, Chem. Eng. J. 413 (2021) 127427. Available from: https://doi.org/10.1016/j.cej.2020.127427.

[108] M. Zhu, C. Tan, Q. Fang, L. Gao, G. Sui, X. Yang, High performance and biodegradable skeleton material based on soy protein isolate for gel polymer electrolyte, ACS Sustain. Chem. Eng. 4 (2016) 4498−4505. Available from: https://doi.org/10.1021/acssuschemeng.6b01218.

[109] J.Y. Han, Y. Huang, Y. Chen, A.M. Song, X.H. Deng, B. Liu, et al., High-performance gel polymer electrolyte based on chitosan−lignocellulose for lithium-ion batteries, ChemElectroChem 7 (2020) 1213−1224. Available from: https://doi.org/10.1002/celc.202000007.

[110] C. Luo, Y. Huang, Z. Yin, H. Xu, X. Qin, X. Li, et al., A universal natural hydroxy propyl methyl cellulose polymer additive for modifying lignocellulose-based gel polymer electrolytes and stabilizing lithium metal anodes, Mater. Chem. Phys. 250 (2020) 123174. Available from: https://doi.org/10.1016/j.matchemphys.2020.123174.

[111] J. Gou, W. Liu, A. Tang, To improve the interfacial compatibility of cellulose-based gel polymer electrolytes: a cellulose/PEGDA double network-based gel membrane designed for lithium ion batteries, Appl. Surf. Sci. 568 (2021) 150963. Available from: https://doi.org/10.1016/j.apsusc.2021.150963.

CHAPTER

12 Advances in biodegradable polymer nanocomposites for solar cell applications: current status and future directions

Sreekanth M.S.[1], Ariful Rahaman[1] and Sumangala T.P.[2]

[1]Department of Manufacturing Engineering, School of Mechanical Engineering, Vellore Institute of Technology, Vellore, Tamil Nadu, India [2]Department of Physics, School of Advanced Sciences, Vellore Institute of Technology, Vellore, Tamil Nadu, India

Chapter Outline

Biodegradable and Biocompatible Polymer Nanocomposites. DOI: https://doi.org/10.1016/B978-0-323-91696-7.00001-5
Copyright © 2023 Elsevier Inc. All rights reserved.

12.1 Introduction

Polymer composites are a type of engineering material that has a wide range of applications [1–11]. Biodegradable polymers (BP) and their composites are a viable alternative to the flexible, non-BP derived from petroleum, because they are nontoxic, they are green, low-cost, easily processed, and environmentally friendly. Despite their excellent nature, engineering applications of BPC are limited due to the need for improved mechanical and electrical properties, as well as less sensitivity to humidity. There is a lot of work being done to improve the properties of biodegradable polymers/composites so that they can be used in a variety of applications. Photovoltaics and energy harvesting are two of the most effective and inspiring applications of BP. Solar energy is clean energy and is one of the most important renewable energy sources. It can be used to create efficient and long-lasting solar cells. Solar cells are an excellent substitute for nonrenewable energy sources. However, the most efficient and commercially available solar cells on the market are based on silicon (Si), and the degradability of the electronic components is extremely poor, posing an environmental threat. As a result, to maximize the utility of solar cells, the emphasis should be on transient electronic devices. Biodegradable materials are being researched extensively to develop eco-friendly solar cells due to their unique properties such as renewable, abundant, sustainable, scalable, and eco-friendly energy production. They serve as a flexible and transparent substrate for the construction of transient solar cells. Furthermore, BP are being researched for their potential use as a substrate in perovskite and OSC. Biodegradable polymers are widely utilized as electrolytes in solid and gel form in dye-sensitized solar cell (DSSC). Biodegradable polymers are explored for their usefulness in barrier protection and encapsulation of solar cells.

12.2 BP composites used in solar cell application

Traditional energy sources, such as fossil fuels, are rapidly depleting. Renewable energy has drawn the attention of researchers as a solution to the constant supply of energy required for a thriving human society. Solar energy is the most abundant renewable energy source. Harvesting and converting solar radiation energy to other forms, particularly electrical energy, could solve our energy crisis. As a result, dedicated devices must be developed. This conversion is carried out by the devices known as solar cells (SC) or photovoltaics. A great deal of research has gone into developing efficient solar cells. Si-based solar cells currently achieve a power conversion efficiency of around 25%. However, because Si-based cells are not biodegradable, they may result in white waste. The use of biodegradable components could result in a successful conversion of clean energy. Biodegradable polymers (BP) are those that degrade to monomer stages in response to external stimuli. By incorporating a variety of filler materials, the properties of these BPs could be modified or improved. Poly(lactic acid) (PLA), polycaprolactone (PCL), poly(L-lactic-*co*-glycolide) (PLGA), poly(vinyl alcohol) (PVA), cellulose, starch from various sources such as rice, potato, and other biomass, and silk fibroin are all biodegradable polymers that can be used in solar cells. Fillers added to BP result in Biodegradable polymer composites (BPC). Biodegradable polymer composites are used as organic solar cell (OSC) substrates, encapsulation layers, and electrolytes in DSSC applications. Examples of different polymer composites used as solar cell materials are

Table 12.1 Examples of biodegradable polymer composites for solar cell applications.

	Composites			
Sl. no.	Filler	Polymer	Applications	References
1	Nanoclay	Poly-L-lactic acid (PLLA)	Substrate	[12]
2	Potassium iodide	Agarose biopolymer matrix	Electrolyte in DSSC	[13]
3	Potassium iodide	Arrowroot biopolymer	Electrolyte in DSSC	[14]
4	Sodium iodide (NaI) and 1-methyl-3-propylimidazolium (MPII) ionic liquid	Rice starch	Electrolyte in DSSC	[15]
5	Sucrose or gelatin	Polyvinyl alcohol	Transparent transient substrates of OSC	[16]
6	Chemically modified using hexamethyl disilazane (HMDS)	Cellulose nanocrystal	Substrate	[17]
7	Co_3O_4-NMP-LiI-I_2, NiO-NMP-LiI-I_2	Agarose biopolymer matrix	Electrolyte DSSC	[18]
8	NH_4I as doping salt	Chitosan	Electrolyte	[19]
9	Nanoporous TiO_2 film	Polysaccharide (j-carrageenan or agarose)	Solidification of electrolyte	[20]
10	Indium tin oxide, TiO_2 paste, k-carrageenan	Polyethylene naphthalate (PEN/ITO) film	Electrolyte for DSSC	[21]
11	NH_4I, ethylene carbonate, 1-butyl-3-methylimidazolium iodide	Chitosan	Electrolyte for DSSC	[22]
12	NH_4I, and I_2 salts	Agar polymer host	Gel electrolyte for DSSC	[23]
13		Nanocellulose paper	Substrate	[24]

presented in Table 12.1. A schematic representation of various BPC and their application in SC is given in Fig. 12.1.

12.3 Synthesis, processing and fabrication of biodegradable polymer-based SC

Solar cell efficiency is primarily determined by the properties of the materials used, as well as their processing/fabrication methodology and recyclability and/or biodegradability. Biodegradable polymer composite substrates for the fabrication of transient OSC are gaining popularity due to their biodegradability. PVA, PLGA, PLA, PCL, rice paper, cellulose, and silk are the most common polymer matrix-based OSC substrates. Because of their high optical transparency (89%–91% for

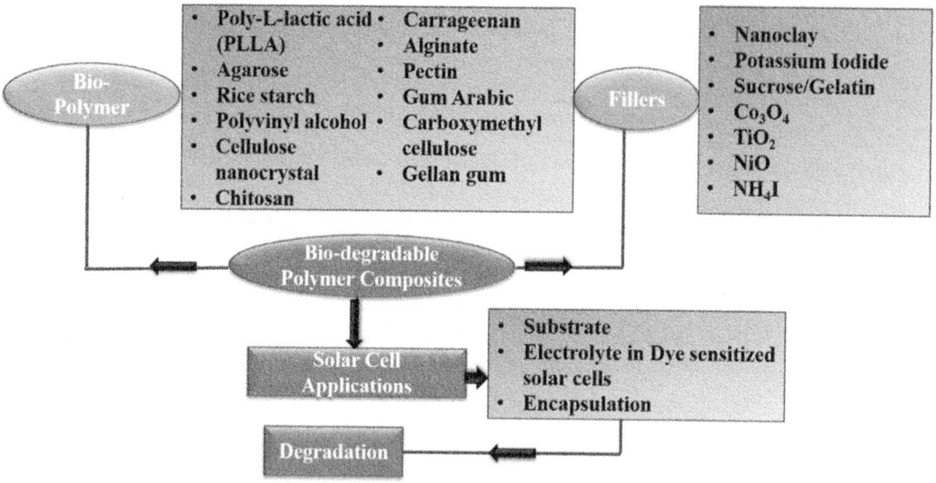

FIGURE 12.1

Schematic illustration of biodegradable polymer composites used for solar cell application.

PVA-based composites), water solubility, and desired physical properties, PVA composite substrates were developed for OSCs [16]. The efficiency of SC is primarily determined by the properties of the materials used, as well as their processing/fabrication methodology and recyclability and/or biodegradability. Biodegradable polymer composite substrates for the fabrication of transient OSC are becoming more popular due to their biodegradability. PVA, PLGA, PLA, PCL, rice paper, cellulose, and silk are the most commonly used polymer matrix-based OSC substrates. PVA composite substrates were developed for OSCs due to their high optical transparency (89%–91% for PVA-based composites), water solubility, and desired physical properties. The optimum thickness was taken as 11 nm and was deposited on PVA transparent electrode to function as the bottom transparent electrode. In this case, the Ag layer was deposited by thermal evaporation on a PVA composite electrode to serve as the bottom electrode. Next, a layer of molybdenum oxide (MoO_3) was deposited as a hole transporting layer to construct OSCs with planar heterojunction.

Polymer-based photovoltaic devices were investigated by incorporating nanoclay into the biodegradable polymeric substrate PLLA to improve thermal stability, mechanical strength, and gas permeability. The compounding of PLLA/clay (2 wt.%) was carried out by melt-mixing technique using twin-screw extruder and is shown in Fig. 12.2A. The compounded granules were again fed to a single screw extruder with a film die unit to make a uniform film thickness and width of 200 μm and 12 cm, respectively [12]. The film extrusion is shown in Fig. 12.2B. As prepared film may deform when temperature increase above 40°C, so the film is laminated between polyethylene terephthalate (PET) sheets (200 μm) to serve as a smooth surface. The device construction is done by the following steps. Initially, 100 nm thick aluminum (Al) layer is deposited on substrate by thermal evaporation. A photonegative photoresist is applied to the PLLA-based aluminum coated electrode. Then the photoresist is illuminated for 45 seconds using an anode grid photonegative mask. A 50% geometric fill factor anode was used.

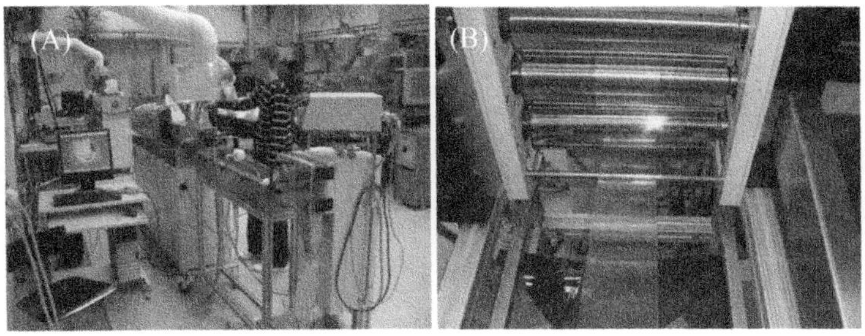

FIGURE 12.2

(A) Melt compounding using a twin-screw extruder and (B) film extrusion of PLLA substrate [12].

Reproduced with permission from Elsevier. Copyright 2008.

Recently cellulose-based flexible substrates received lots of attention in organic photovoltaics. Gao et al. [25] have demonstrated the preparation of nanocellulose paper (NCP)-based PSC. Fabrication of NCP substrate and NCP-based PSC is represented in Fig. 12.3. The process is divided into six steps. The first step was to extract nanocellulose from naturally occurring cotton. This nanocellulose was then transformed into a viscous liquid in the second step. The prepared viscous liquid was used to create an NCP substrate. In the third step, an acrylic resin-coated NCP is coated with a waterproof layer. In the following two steps, flexible PSC was created on an NCP substrate. A doped PEDOT: PSS layer is used as the bottom anode. The hole transport layer PEDOT: PSS was spin-coated above this anode layer. The perovskite layer ($CH_3NH_3PbI_3$) was deposited using an antisolvent, one-step, low-temperature method. [6,6]-phenyl C61-butyric acid methylester (PCBM), the electron transport layer, was spin-coated onto the perovskite layer. Vacuum evaporation was used to deposit metal aluminum on top of the electron transport layer. The final step, step 6, is the flame burning disposal of the flexible PSC.

In addition to being used as a substrate for PSC, BP is also used as a substrate for OSC. Lin et al. created a flexible, conducting substrate-based on cellulose nanofiber (CNF) by incorporating silver nanowires (AgNWs) [26]. The following is the procedure for creating a conductive substrate-based on CNF/AgNWs. The first step is to spray AgNWs from suspension onto a preheated silicon (Si) substrate. To enable film separation, the substrate surface is modified with a self-assembled monolayer of octadecyltrichlorosilane (ODTS). Drying the coated AgNWs resulted in a flexible conducting film. Then, the CNFs aqueous solution of 0.5 wt.% concentration was cast onto the AgNWs film. It is then annealed at 80°C for 4.5 hours followed by drying at 60°C to remove traces of water. The dried CNF/AGNWs film was removed from the silicon substrate and flattened by hot pressing at 60°C for 30 minutes.

The DSSC was created by combining a gel polymer electrolyte based on hydroxyethyl cellulose (HEC) with nanocrystalline TiO_2-coated conducting glass fluorine-doped tin oxide (FTO) [50]. Gel polymer electrolytes were synthesized with the following formulation; HEC (500 mg), KI (30 mg), I_2 (10 mg) and organic compound TPIPA (Tris(4-(1H-phenanthro[9,10-d]imidazol-2-yl)phenyl) amine) (10 mg) added in dimethylformamide (DMF) were heated to 80°C–100°C to form a uniform mixture of gel. The TiO_2 layered electrodes are dipped in methanol solution of cis-bis(isothiocyanato)bis(2,2-bipyridyl-4,4′-dicarboxylato)ruthenium(II) N_3 dye for 24 hours and was used as the

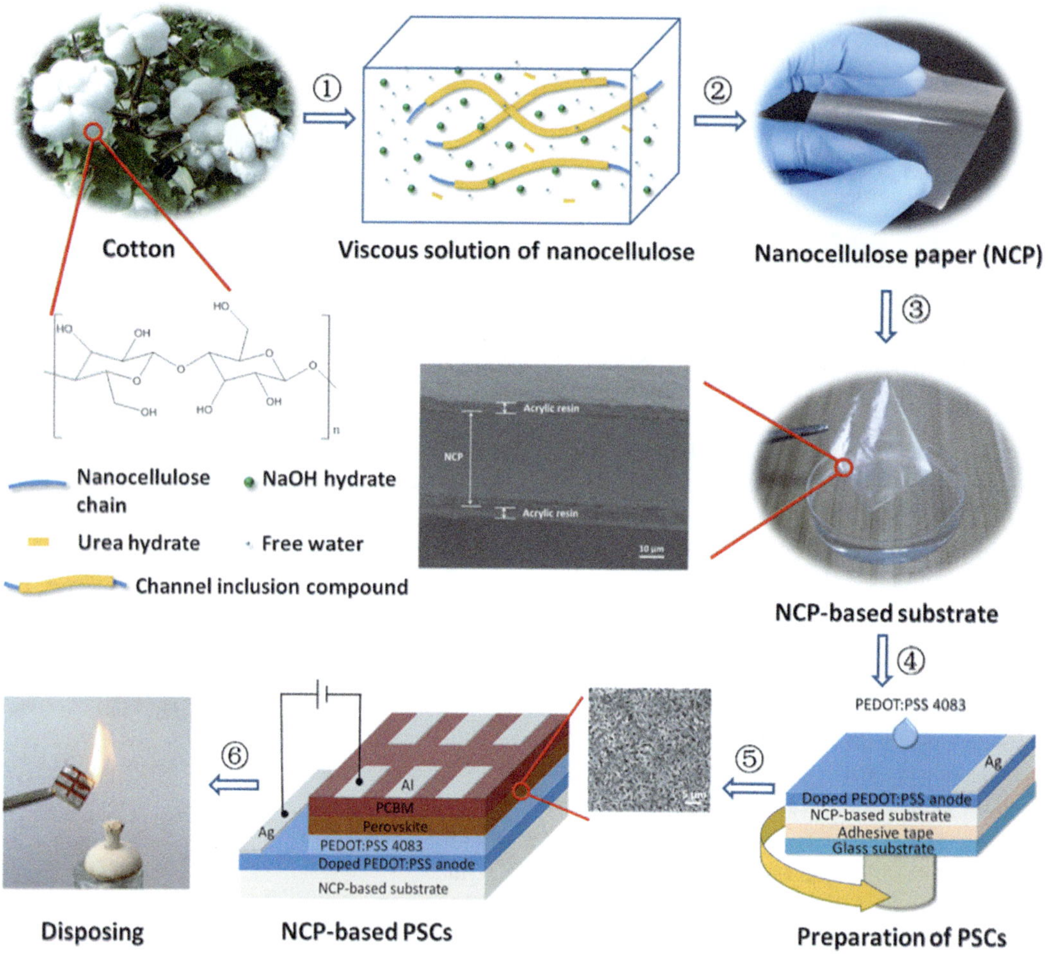

FIGURE 12.3

Fabricating flexible eco-friendly nanocellulose paper (NCP)-based substrate and NCP-based perovskite solar cell [25].

Reproduced under creative common license from Springer Nature. Copyright 2019.

DSSC working electrode. The synthesized gel polymer electrolyte with triphenylamine core was sandwiched between TiO_2 and coated FTO plate, which was used as a counter electrode. Finally, the electrodes in DSSC were held in place with alligator clips.

The steady-state increase in OSC growth over the last few years reveals significant potential for use as an economically viable renewable energy source. Efficiency and stability are the two most difficult challenges in OSC for commercial applications. Because of the sensitivity to external atmospheric factors such as humidity, oxygen, and temperature, efficient encapsulation is required for commercial development. To improve the stability and performance of OSC, such an encapsulating layer must be developed without affecting the existing properties. Preparation of such a multilayer structure using PVA nanocomposites with 5–10 wt.% sodium montmorillonite (MMt-Na^+) for OSC (OSC) encapsulation is demonstrated by Gaume et al. [27]. The solution intercalation film casting technique was used to create the nanocomposites. Except for the top silver electrode, all of the layers are created using a spin coating technique on a plastic substrate made of indium tin oxide (ITO)-coated polyethylene terephthalate (PET) film. The top silver (Ag) electrode is made by PVD. A pressure-sensitive adhesive is applied to the barrier film, PET coated with nanocomposite. It was further dried under vacuum at 50°C. A similar barrier is used on both sides of the cell and is laminated with barrier foil using a hot laminator using a glove box.

Solar cell encapsulation studies using ethylene-vinyl acetate copolymer (EVA) matrix and bacterial cellulose (BC) nanofibers were similar [28]. EVA composites were created using a twin-screw extruder and the melt-mixing technique. Before melt-mixing with the EVA matrix, the BC nanofibers were treated with propionic anhydride. Encapsulation made from modified BC nanofiber-based EVA composites is expected to outperform unmodified encapsulation in terms of mechanical and barrier performance due to improved interaction between matrix and reinforcement.

In the case of PSC, the stability can be improved by pacifying the perovskite layer. In this regard, Qiu et al. [29], have reported the use of biopolymer PLL for the passivation of PSC prepared by blade coating. They observed that the introduction of PLL to the solar cell resulted in a more uniform and defect free perovskite layer of Methylammonium lead iodide ($MAPbI_3$). They have observed a reduced grain size of these layers. They have noted that the reduction in defects can be attributed to the presence of ions like $-COO^-$ and $-NH_2^+$. These ions in turn act as chelating agents for Pb^{2+} which otherwise leads to Pb metallic state. These metal Pb are capable of forming trap states and hence need to be removed. The use of PLL-based passivated perovskite has led to an increase in V_{OC} by 100 mV, and they have also obtained a PCE value of 19.45%. The structure of the prepared PSC (Fig. 12.4A), the fabrication method (Fig. 12.4B), and the cross sectional SEM (Fig. 12.4C), with various concentrations of PLL is given.

12.4 Properties of biodegradable polymer-based solar cells

BP serve a variety of functions in solar cell applications. It consists of a substrate for perovskite/OSC, an electrolyte for DSSC, and an organic SC encapsulant. The suitability of a particular polymer for these applications is determined by a variety of factors such as hydrophobicity, surface roughness, tensile strength, and radiation transmittance. As a result, it is critical to comprehend the properties of BP for solar cell applications.

12.4.1 Microstructural characterization

Optical microscopy, atomic force microscopy, SEM, and transmission electron microscopy are used to investigate the microstructures of BP.

The optical micrographs of biodegradable polymer PLLA with the dispersion of nanoclay [12], have been reported. The optical micrographs revealed information about the optical range's dispersion as well as the absence of agglomerates.

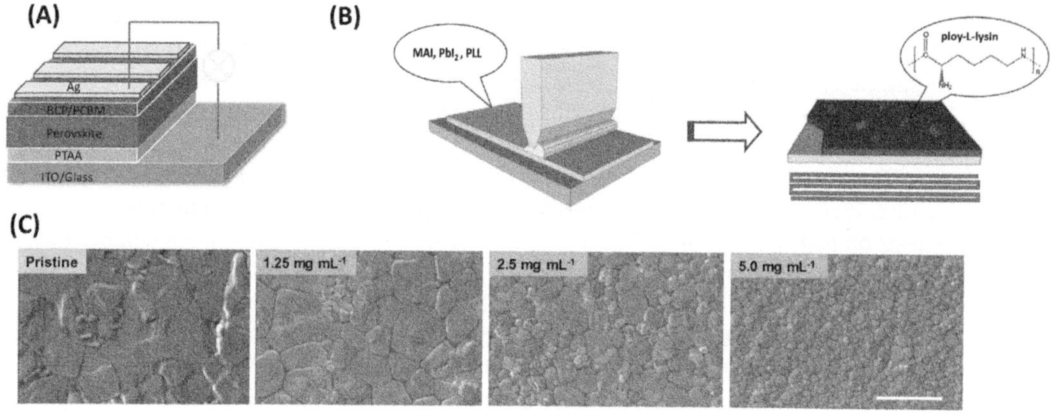

FIGURE 12.4

(A) Device structure, (B) fabrication and (C) cross sectional scanning electron microscopy (SEM) of PSC with passivation with poly-L-lysine (PLL) [29].

Reproduced with permission from Elsevier. Copyright 2021.

Another microscopic technique used in the micron and submicron range is SEM. On the other hand, atomic force microscopy (AFM) is used to probe the surface of a sample at the atomic level and provides information on surface roughness. The root mean square (RMS) roughness, R_q is defined by Eq. (12.1).

$$R_q = \sqrt{\frac{\Sigma(z_t - z_{ave})^2}{N}} \tag{12.1}$$

where z_{ave} is the average z value for the area under consideration, z_t is the z value at a particular point and N is the total number of points studied [27].

12.4.2 Optical properties

The transmission properties of the biodegradable polymer-based substrate are an important parameter in determining the device's efficiency. A transparent substrate allows radiation to pass through to the subsequent layers, improving efficiency. The UV-Visible/Near Infrared spectrometer is used in this case. UV-Visible spectra can also be used to calculate sample figures of merit (FoM) [26]. FoM is calculated using Eq. (12.2) below.

$$T(\lambda) = \left(1 + \frac{188.5}{R_s}\frac{\sigma_{op}(\lambda)}{\sigma_{DC}}\right)^{-2} \tag{12.2}$$

where $T(\lambda)$ is the transmittance at a particular wavelength, generally chosen as 550 nm corresponding to the maximum intensity of sunlight and human eye sensitivity, R_s is the sheet resistance, σ_{op} is the optical conductivity and σ_{DC} is the direct current conductivity [19].

UV-Visible transmission spectra of ultrathin cellophane paper substrate useful for PSC are depicted in Fig. 12.5A [30]. By varying the thickness of a sandwiched Ag layer, the spectra were obtained. The transmission showed a decrease with increased Ag thickness for the lower wavelength range (300–400 nm). However, in this visible range (400–800 nm), the transmittance showed an increase with Ag layer thickness. This provided information regarding the usefulness of the substrate in the visible range. Gao et al. [25] also used UV-Visible spectroscopy to understand the transmittance of nanocellulose paper (NCP)-based substrate. The transmittance of NCP substrate, acrylic resin and acrylic-coated NCP is shown in Fig. 12.5B [25]. Acrylic and acrylic-coated NCP had higher transmittance than NCP substrate alone. The inset contains digital photographs of each of these substrates. The comparable transparency of each of the samples is evident from Fig. 12.5B.

12.4.3 Hydrophobicity

BP are extremely environmentally sensitive. As a result, understanding their interaction with water is critical for achieving device stability. Hydrophobicity must be measured in this regard. This is accomplished by measuring the contact angle with a Goniometer. Using a syringe, water drops are released onto the surface under consideration [25]. Photographs are captured and processed using dedicated software to calculate the contact angle following Young's relation given by Eq. (12.3).

$$\gamma^{sv} = \gamma^{sl} + \gamma^{lv}\cos\theta \tag{12.3}$$

where γ^{sv} is the solid surface free energy, γ^{sl} is the solid/liquid interfacial energy, γ^{lv} is the liquid surface free energy and θ is the contact angle [18].

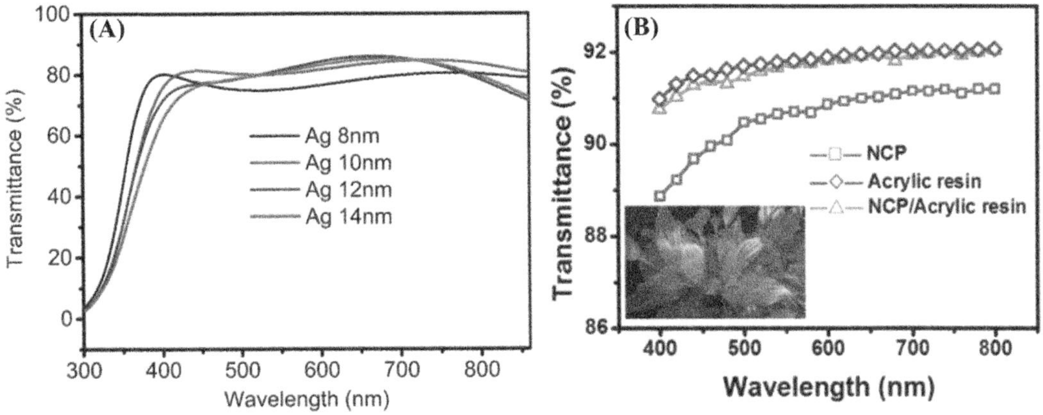

FIGURE 12.5

(A) UV-Visible spectra of cellophane paper with various Ag wire thickness [30] and (B) transmittance of nancocellulose paper, acrylic resin and NCP/acrylic resin [25].

For (A) reprinted with permission from Elsevier, (B) reproduced under creative common license from Springer Nature. Copyright 2019.

FIGURE 12.6

Optical photograph of NCP (left) and acrylic-coated NCP (right) before dropping water, after dropping and after drying [25].

Reproduced under creative common license from Springer Nature. Copyright 2019.

The hydrophobic nature of NCP and acrylic-coated NCP measured through contact angle measurement is represented in Fig. 12.6 [25].

Before the water was dropped, the substrates were transparent. The NCP substrate changed significantly after the water was dried. However, when exposed to water, the acrylic-coated NCP substrate demonstrated exceptional stability. They have reported a contact angle of 60 degrees for NCP which further decreased with time to about 40 degrees in a minute. However, the acrylic-coated NCP exhibited a contact angle of 80 degrees and remained within 70 degrees up to 50 seconds.

12.4.4 Mechanical properties

Only if the prepared devices have sufficient mechanical properties to withstand the strain undergone by the device can the replacement of non-BP with paper-based or other BP be fruitful. As a result, it is critical to investigate the mechanical properties of biodegradable polymer-based substrates. Dynamic mechanical analysis, tensile measurements, and comparing electrical properties after multiple bending cycles are used to investigate mechanical properties. The mechanical properties of composites of *O*-(2,3-dihydroxypropyl) cellulose (DHPC) and various concentrations of tunicate cellulose nanocrystal (TCNC) are shown in Fig. 12.7 [31]. Temperature causes a decrease in storage modulus for samples of varying concentrations. However, with an increase in the concentration of TCNC, an increase in the storage modulus is observed at a particular temperature (say 45°C, Fig. 12.7B).

The experimental data on storage modulus was compared to a percolation model as given in Eqs. (12.4) and (12.5) [32].

$$E' = \frac{(1 - 2\Psi + \Psi X_r)E'_s E'_r + (1 - X_r)\Psi E'^2_r}{(1 - X_r)E_r + (X_r - \Psi)E'_s} \tag{12.4}$$

$$\Psi = X_r\left(\frac{X_r - X_c}{1 - X_c}\right)^{0.4} \tag{12.5}$$

where Ψ is the volume fraction of the percolation network, X_r is the volume fraction of TCNC, X_c is the critical percolation volume. X_c is calculated from the aspect ratio of TCNC using 0.7/aspect ratio. E_s' is the storage modulus of DHPC which exhibits soft nature and E_r' is that of the filler TCNC which exhibits a rigid nature.

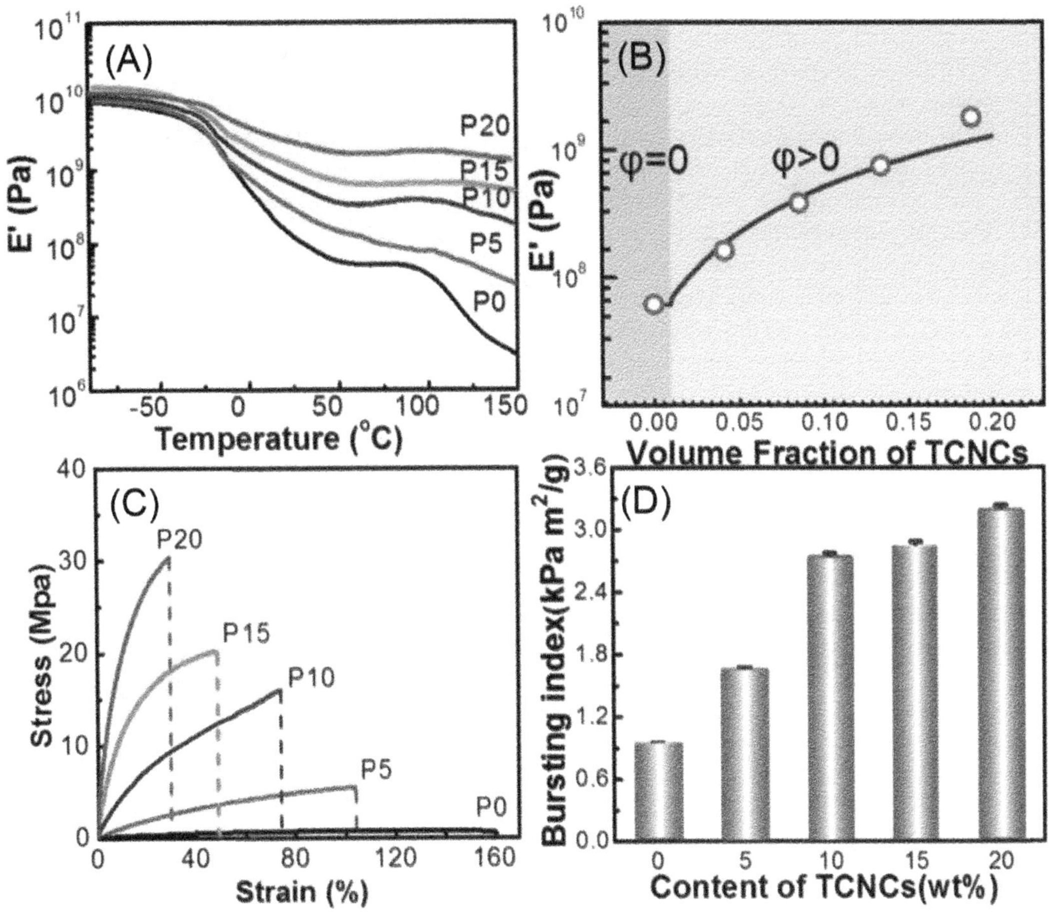

FIGURE 12.7

Tensile storage modulus as a function of temperature E′ (A), as a function of cellulose nanocrystal concentration (B), stress-strain curve (C) and bursting index (D) of *O*-(2,3-Dihydroxypropyl) cellulose (DHPC) with varying concentration of tunicate cellulose nanocrystals (TCNCs) [31].

Reproduced with permission from American Chemical Society. Copyright 2018.

The tensile properties studied using a Universal Testing Machine are shown in Fig. 12.7C. The stress versus strain curve demonstrates that the tensile strength of the composite increases as the TCNC content increases. The impact resistance, tested using a burst tester and the bursting index obtained is represented in Fig. 12.7D. An improvement in the bursting index with the incorporation of TCNC is clear from Fig. 12.7D.

A thermomechanical analyzer was used to investigate the thermomechanical properties of an acrylic-coated NCP substrate [25]. The test was performed at a heating rate of 5°C/min in a nitrogen atmosphere applying 29 mN. The coefficient of thermal expansion (CTE) was calculated from the mean value of three measurements at temperatures ranging from 20°C to 150°C. A CTE of 12.7 ppm/K was reported for NCP substrate and 26.7 ppm/K for acrylic-coated NCP substrates. The CTE for pure acrylic resin, on the other hand, was much higher, at 217 ppm/K.

The glass transition temperature of various biodegradable electrolytes has been studied using dynamic mechanical analysis (DMA). In a typical system of electrolytes based on the blend of derivatives of k-carrageenan and cellulose [33] the experiment was carried over in dual cantilever mode within a temperature of range −40°C to 100°C, with a heating rate of 2°C/min under nitrogen liquid flow. The glass transition temperature of carboxymethyl kappa-carrageenan (CMKC)/carboxymethyl cellulose (CMCE) blend with various quantities of NH_4I was obtained from the peak of tan δ measured as a function of temperature. An increase in T_g from −13.5°C to −5.7°C is observed in this case. The increase in T_g is attributed to the increase in transient cross-link happening between the polymer chains and NH_4I.

12.4.5 Stability measurement

The stability of devices prepared with biodegradable substrates is important for product translation. This is usually accomplished by measuring the properties after several bending and peeling cycles. The stability of the prepared TOCN/AgNWs substrate in comparison with commercial PEN/ITO is depicted in Fig. 12.8A and B for bending and peeling cycles [26].

Even after 500 cycles, the relative resistance of TOCN/AgNWs remains constant. However, despite having the highest tensile strength and lowest elastic modulus, commercial PEN/ITO exhibited an increase in resistance, which is attributed to the cracks formed, indicating its inferior mechanical stability. Similarly, the effect of device fabrication method on peeling properties of TOCN/AgNWs was investigated. Another property to consider is the effect of different fabrication routes on device stability. Lin et al. used a fabrication method involving spray coating and

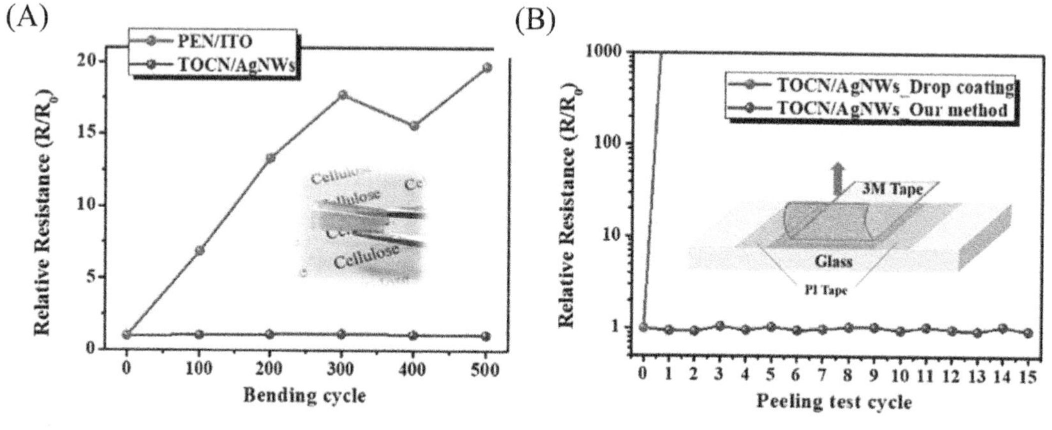

FIGURE 12.8

Stability of TOCN/AgNWs substrates with (A) bending cycle and (B) peeling cycles [26].

Reproduced with permission from American Chemical Society. Copyright 2021.

subsequent hot pressing in their work. They demonstrated that this method outperformed drop cast substrates in terms of peeling off stability [26].

Another important factor in device stability is the strain induced by bending. The effect of bending radius on strain in PSC as a function of substrate thickness has been reported [30]. The concept of neutral axis tuning, where no strain exists, is used here to reduce strain on the device [34]. The neutral axis in turn is calculated using Eq. (12.6).

$$z_{NA} = \frac{\sum_{k=1}^{n} E_k^* t_k z_k}{\sum_{k=1}^{n} E_k^* t_k} \tag{12.6}$$

$$\text{Where } E_k^* = \frac{E_k}{1 - v_k^2} \tag{12.7}$$

Here E_k is Young's modulus, v_k is the Poisson's ratio, t_k is the thickness of the layer and z_k is the middle point of each layer along z-axis.

The strain can be calculated for each radius of curvature at various points in the device following Eq. (12.8) [30],

$$\varepsilon(z, R) = \frac{z - z_{NA}}{R} \tag{12.8}$$

Here ε is the strain at position z with a radius of curvature R.

This effect of bending radius on the strain and the distribution of strain is represented in Fig. 12.9A and B, respectively. Fig. 12.9A represents a continuous increase in strain, with a decrease in the radius of curvature. For the cellophane thickness of 150 μm, the strain is around 65% for a radius of 0.1 mm. However, for a radius of 1 mm, the strain is 10%. An abrupt decrease in the strain is seen with the decrease in the cellophane thickness. The strain distribution in PSC for a cellophane substrate of thickness 25 μm as a function of bending radius is represented in Fig. 12.9B. The decrease in strain as the

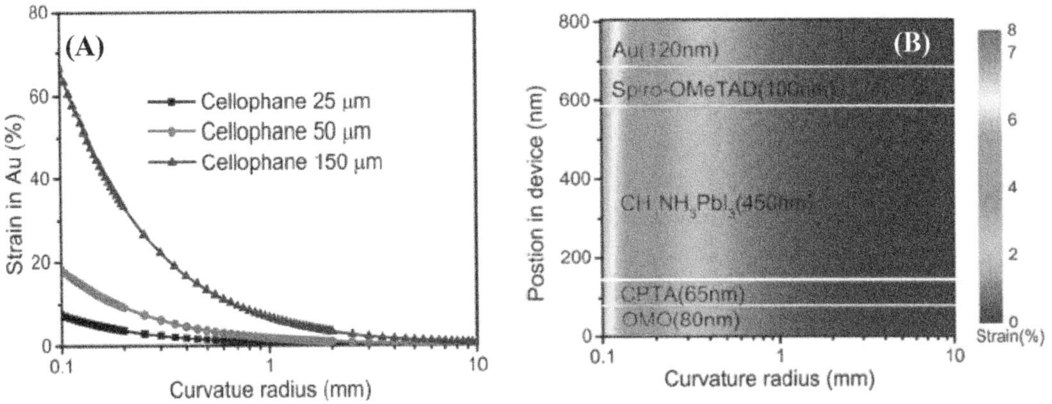

FIGURE 12.9

(A) Effect of bending radius on the Strain on Au layer in a PSC and (B) Strain distribution in PSC on 25 μm cellophane substrate [30].

Copyright 2019. Reproduced with permission from Elsevier.

curvature radius increases is obvious. When the position was moved away from the substrate, the strain increased slightly.

12.4.6 Thermal analysis

Thermal analyses such as thermogravimetric and differential scanning calorimetry can be used to understand the thermal stability of polymer samples. Verma et al. investigated the stability of biopolymers such as agarose substrate using derivative thermogravimetric (DTG). The DTG data revealed no significant differences between the samples heated at various temperatures. This demonstrated the agarose sample's thermal stability [35].

The effect of nanoparticles (TiO_2, Co_3O_4, and NiO) addition on the thermal properties of agarose polymer electrolyte has been studied utilizing differential scanning calorimetry [18]. The change in melting temperature of the polymer electrolyte with additives was sorted from the DSC plot. A change in melting point from 280°C to 227°C–240°C was observed with the addition of nanoparticles which was attributed to the cross-linking interactions between the agarose polymer host and added nanoparticles.

TGA is another technique used to investigate the thermal degradation properties of BPC. The effect of various salts on the thermal behavior of rice starch based electrolyte studied using TGA has been studied by Khanmirzaei et al. [36]. The TGA plot was obtained for pure rice starch and rice starch with NH_4I (35%, 40%, and 45% in samples RN-7, 8, and 9, respectively). The change in the decomposition temperature in the presence of salt was studied using this technique. The decomposition temperature for pristine starch was 292°C and was high in contrast to composite samples which exhibited temperatures from 184°C to 187°C. A minor decrease in decomposition temperature was also observed as salt concentration increased. They have expanded the research to include LiI and NaI. According to their findings, the decomposition temperature of LiI and NaI at 35% concentration was 245°C, and these samples had better properties than NH_4I.

12.4.7 Structural analysis

The structure and composition of the biodegradable polymer are investigated in order to comprehend the impact of various components or additives on the structure. Fourier Transform Infrared spectroscopy (FTRI) and X-ray diffraction are the most commonly used techniques for probing structural characteristics (XRD). The presence of silylation reaction in their agarose substrate prepared by functionalizing agarose chains with HMDS is demonstrated using FTIR [35]. The presence of $-Si-CH_3$ peaks at 1253 and 843 cm^{-1} was used to identify this. As a control experiment, FTIR was obtained for agarose without functionalization and a physical mixture of agarose, HMDS and Toluene was also demonstrated. A similar missing of peaks was observed on swelling the substrate for 6 hours. This was consistent with the work to develop a substrate that does not absorb moisture during processing but can absorb moisture later for decomposition. FTIR spectroscopy was used to explain this.

The ion migration scheme of a gel electrolyte based on esterified starch was studied using FTIR spectroscopy [37]. There has been a reported shift toward lower wavelength numbers for hydroxyl, ester carbonyl, and ether compounds. This observed shift during the addition of LiI was most noticeable for ester carbonyl and ether groups. This indicated their involvement as mobile Li ion migration sites. Another study used FTIR to confirm the composite nature of chitosan doped with ionic liquid [38]. Chitosan, ionic liquid, and ionic liquid doped chitosan film FTIR spectra were obtained. All of the prominent chitosan and ionic liquid vibrations were visible in the composite film, confirming its composite nature.

X-ray diffraction techniques are typically applied to crystalline materials rather than amorphous materials. In studies involving polymer electrolytes, however, XRD is used to understand the interaction between the polymers and the ionic liquid. Such a study was performed by Khanmirzaei et al. [15]. The samples studied include MPII (iodide ionic liquid), rice starch (RS) with sodium iodide (55:45 wt.%) and RS:NaI:X MPII ($X = 5$, 10, 15 and 20 wt.%). The complexation occurring between the rice starch/NaI and the ionic liquid was indicated by a change in the amorphous nature and peak width observed in the composite samples.

Gaume et al. [27] have reported the use of XRD for the analysis of their PVA-MMt-Na^+ (Clay) composite. They have studied the XRD of pristine PVA, MMt-Na^+ and their composite materials. They have reported an intercalated hybrid structure for all composite samples. A parallel layer of MMt-Na^+ along with an ultrathin film of PVA was depicted for intercalated composites. In the presence of clay, they discovered a change in the PVA structure. They have calculated the domain size along a particular [hkl] direction using Scherrer's formula given in Eq. (12.9).

$$D = \frac{k\lambda}{\beta \cos\theta} \tag{12.9}$$

where D is the domain size, k is the Scherrer's constant (taken as 1), β is the full width at half maxima for a particular angle θ represented in radian.

With increasing clay content, the size of the coherent domain was observed to increase along the [10-1] direction. The structure of a nanocellulose-based substrate and PSC fabricated on this flexible paper-based substrate has also been studied using XRD [25].

12.4.8 Conductivity

LCR meters can be used to investigate the ionic conductivity of polymer-based electrolyte systems. The addition of iodine salts increased the conductivity of a rice starch-based polymer electrolyte system [36]. Among various iodine salts used, viz. LiI, NH_4I, and NaI, the highest conductivity was observed in NaI samples. The increase in ionic conductivity of NaI added samples in comparison to others was attributed to the higher mass of sodium ion following the Rice and Roth model represented in Eq. (12.10).

$$\sigma = \frac{2}{3}\left[\frac{(Ze)^2}{kTm}\right] nE_a\tau_0 \exp\left(\frac{-E_a}{kT}\right) \tag{12.10}$$

where Ze represents the charge carried by an ion with valency Z, n is the number density of mobile ions, τ_0 is the lifetime of free ion, E_a the activation energy and k the Boltzmann constant.

The increase in ionic conductivity is followed by a decrease in electrical bulk resistance. Cole-cole plot obtained for samples with the addition of each of the three salts (LiI—35 wt.%), NH_4I (45 wt.%) and NaI (45%) is represented in Fig. 12.10A–C. The equivalent circuit model for RS:NaI is given in the inset of Fig. 12.10C. They investigated temperature-dependent ionic conductivity and discovered

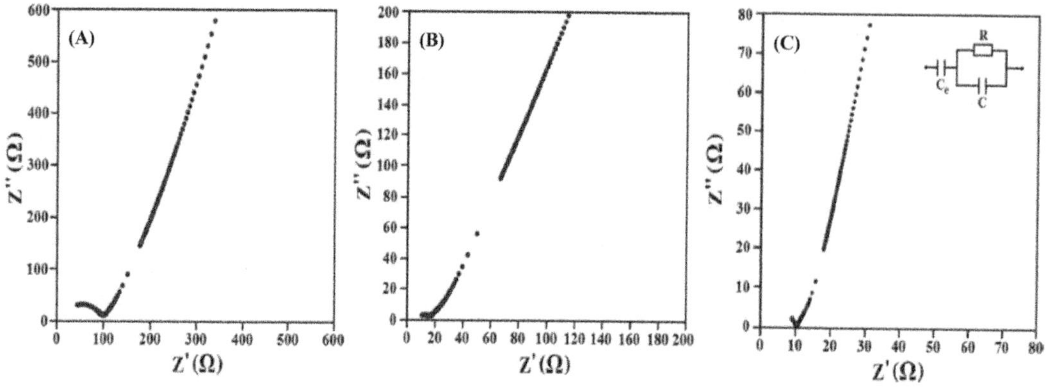

FIGURE 12.10

Cole-cole plot for RS:X electrolytes with (A) X = LiI (35 wt.%), (B) X = NH_4I (45 wt.%) and (C) X = NaI (45 wt.%); The equivalent circuit model for RS:NaI is given in inset of (C) [36].

Reproduced with permission from Springer Nature. Copyright 2015.

that it follows the Arrhenius equation. The activation energy was calculated using the Arrhenius equation represented by Eq. (12.11).

$$\sigma = \sigma_0 \exp\left[\frac{-E_a}{kT}\right] \quad (12.11)$$

where σ_0 is the preexponential factor, E_a is the activation energy and k is the Boltzmann constant.

It has also been reported that the ionic conductivity of biopolymer electrolytes based on a carboxymethyl kappa-carrageenan/carboxymethyl cellulose blend is high [33]. The ionic conductivity at room temperature, as well as the variation in ionic conductivity with temperature was measured. The Vogel–Tammann–Fulcher (VTF) model was used to explain the observed increase in ionic conductivity with temperature. The corresponding equation is given by Eq. (12.12).

$$\sigma = \sigma_0 T^{\frac{-1}{2}} \exp\left[\frac{-E_a/k_B}{(T - T_g)}\right] \quad (12.12)$$

where σ_0 is the preexponential factor, E_a is the activation energy and k_B is the Boltzmann constant and T and T_g are the absolute temperature and glass transition temperature. The activation energy was calculated using this equation and found to decrease as ionic conductivity increased. This lower activation energy value represents a decrease in the energy required for ion migration.

Singh et al. [13] have reported the use of an electrochemical workstation in the frequency range 100 Hz to 1 MHz for gel electrolyte for DSSC. They have used platinum electrodes and measured ionic conductivity using Eq. (12.13).

$$\sigma = R_b\left(\frac{l}{A}\right) \quad (12.13)$$

where R_b is the bulk resistance, l is the thickness of the sample and A is the area of the sample.

Singh et al. [38] have demonstrated the use of complex impedance spectroscopy for finding the ionic conductivity of ionic liquid doped chitosan polymer. They looked into the electrochemical performance of a chitosan:ionic liquid electrolyte system. Measurements were performed in 0.1M phosphate buffer solution, maintaining a pH of 7. A three-electrode system comprising of Ag/AgCl reference electrode, a platinum counter electrode and printed carbon with drop coating of electrolyte as the working electrode was used for measurement. Cyclic voltammograms of chitosan + NaI/I_2 and chitosan + NaI/I_2 + 150 wt.% Ionic liquid are represented in Fig. 12.11. The inset shows measurements taken at faster scan rates. They used voltammograms to investigate the electrochemical performance of chitosan in the presence and absence of ionic liquid. In the absence of ionic liquid, they observed a good interaction between chitosan and the iodide redox couple. In the presence of ionic liquid, however, a competition between the iodide redox couple and the ionic liquid was observed in the case of chitosan interaction.

12.4.9 Solar cell performance

The impact of BP on solar cell performance is at the heart of the transition from conventional to biodegradable materials. A solar cell's performance is typically measured using a white light solar simulator with illumination from a Xenon or halogen lamp. For most measurements, the intensity referred to as sun intensity is between 0.5 and 2. A source meter or electrochemical instrument is used to investigate the current density versus voltage (J-V) characteristics further. The fill factor and efficiency of the prepared SC are the most important parameters obtained from this J-V curve. These are given by Eqs. (12.14) and (12.15) respectively.

$$FF = \frac{V_{max} \times J_{max}}{V_{oc} \times J_{oc}} \tag{12.14}$$

$$\eta = \frac{P_{max}}{P_{min}} = \frac{V_{oc} \times J_{oc} \times FF}{P_{min}} \tag{12.15}$$

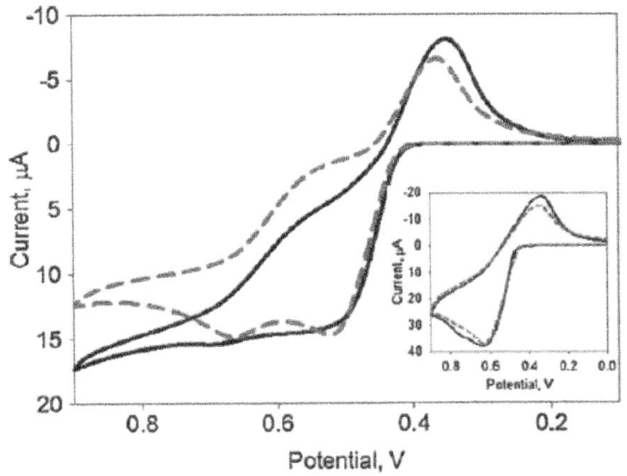

FIGURE 12.11

Cyclic voltammogram of chitosan with NaI/I_2 (black line) and NaI/I_2 + 150 wt.% ionic liquid (red line) [38].

Reproduced with permission from Elsevier. Copyright 2010.

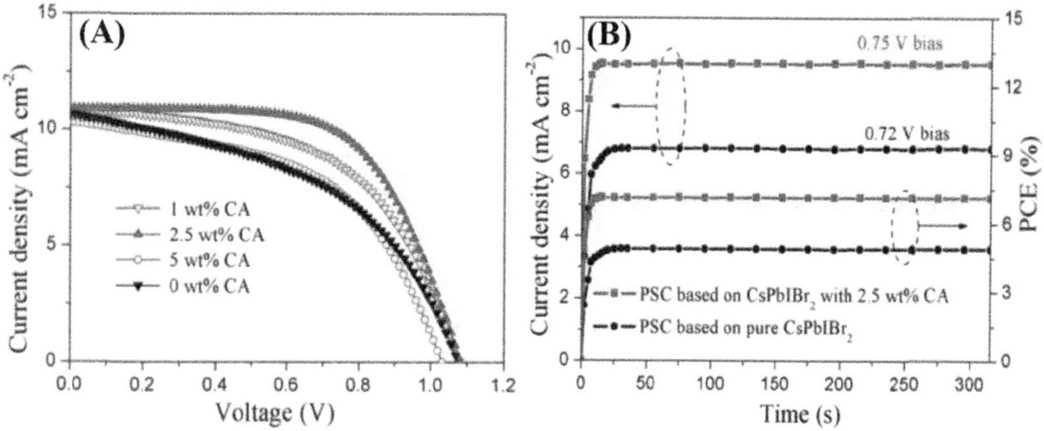

FIGURE 12.12

(A) Current density versus voltage for $CsPbIBr_2$ perovskite film with and without chitosan and (B) variation in current density and PCE with time [39].

Reproduced with permission from Elsevier. Copyright 2021.

Here, FF is the fill factor, V_{max} and J_{max} are the voltage and current density corresponding to maximum power, V_{OC} is the open-circuit voltage, I_{sc} is the short-circuit current and P_{min} is the input power. The J-V of $CsPbIBr_2$ perovskite film with and without chitosan is represented in Fig. 12.12A. The change in the current density and the photoconversion efficiency (PCE) with time is represented in Fig. 12.12B [39]. The stability of biodegradable-based SC is investigated by tracking efficiency over time. This is reported by Gaume et al. [27] when they were studying the stability of OSC having PET/PVA films for encapsulation. The change in the normalized power conversion efficiency using PET film (A), PET/PVA (B) and PET/PVA-MMt (5 wt.%) is given in Fig. 12.13. The stability of PET with modified PVA is reported using the PCE change over time.

12.5 Biodegradable polymer/composites as substrate for solar cells

12.5.1 Substrate for transient solar cells

Xi et al. [16] have reported the use of PVA composites as substrates for transient OSC. PVA blends were created by filling them with either sucrose or gelatin. The solution technique blends were drop cast to produce polymer films. The schematic of the structure is represented in Fig. 12.14A. The layered structure construction follow from bottom to top with PVA composite substrate/Ag/MoO_3/P3HT(poly(3-hexylthiophene)):$PC_{61}BM$([6,6]-phenyl-C61-butyric acid methyl ester)/Ca/Ag arrangement. Glass substrates are commonly used as supporting substrates during OSC fabrication. The optical micrograph of the OSC prepared on PVA-based substrate is shown in Fig. 12.14B. Using a PVA/sucrose substrate, a maximum power conversion efficiency of 2.05% was obtained. External stimuli such as deionized water triggered the disintegration of the prepared devices. The

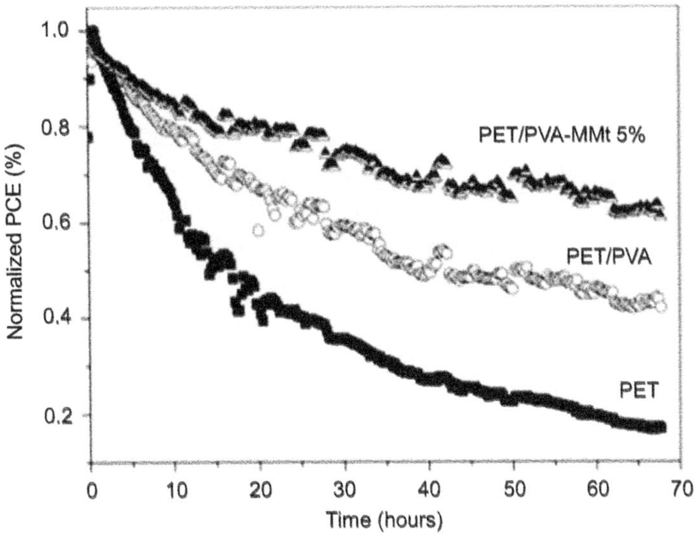

FIGURE 12.13

Change in the photoconversion efficiency with various encapsulated barrier film for OSC [27].

Reproduced with permission from Elsevier. Copyright 2012.

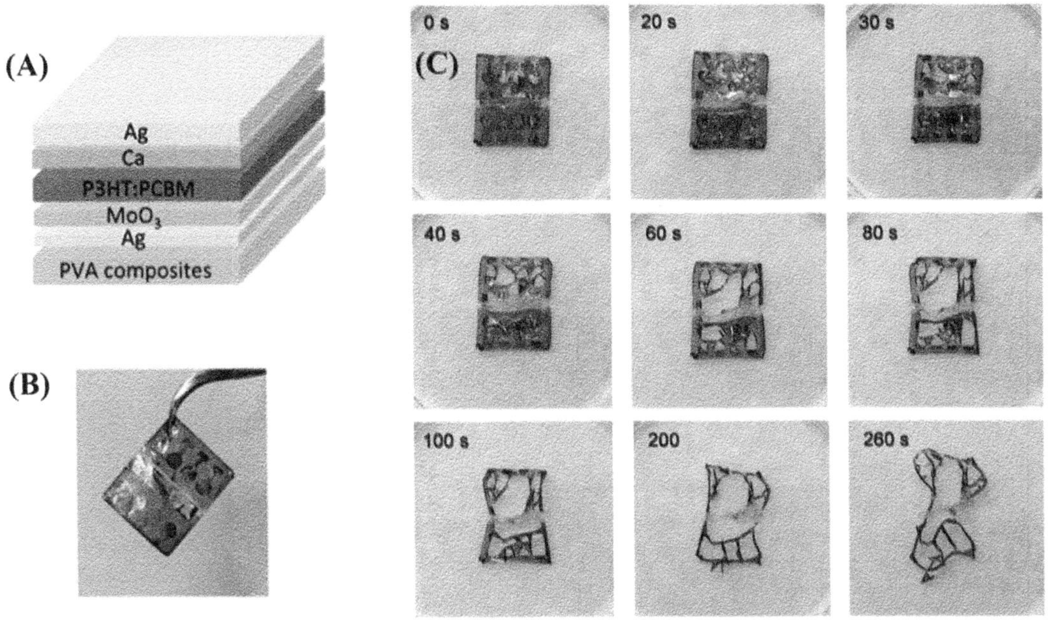

FIGURE 12.14

(A) Schematic of the device structure of OSC with PVA substrate, (B) Optical image of the device, (C) Optical image of dissolution of OSC with time [16].

Reproduced under creative common license from Springer Nature. Copyright 2017.

time for disintegration was about 10 seconds in the case of PVA/sucrose substrate whereas it was more for PVA/gelatin substrate. Fig. 12.14C, gives the optical illustration of the disintegration PVA-based OSC under exposure to water.

Zhou et al. [17] have reported the study on the recyclable OSC using cellulose nanocrystal (CNC) substrate. Cellulose nanocrystals were obtained from sulfuric acid hydrolysis of softwood pulp. Fig. 12.15A shows the TEM image of isolated CNC which is used to prepare CNC films. A CNC suspension and glycerol solution was used to make the films. Films were prepared by pouring the solution in a Petri dish of diameter 80 mm and drying under ambient conditions. Dried films

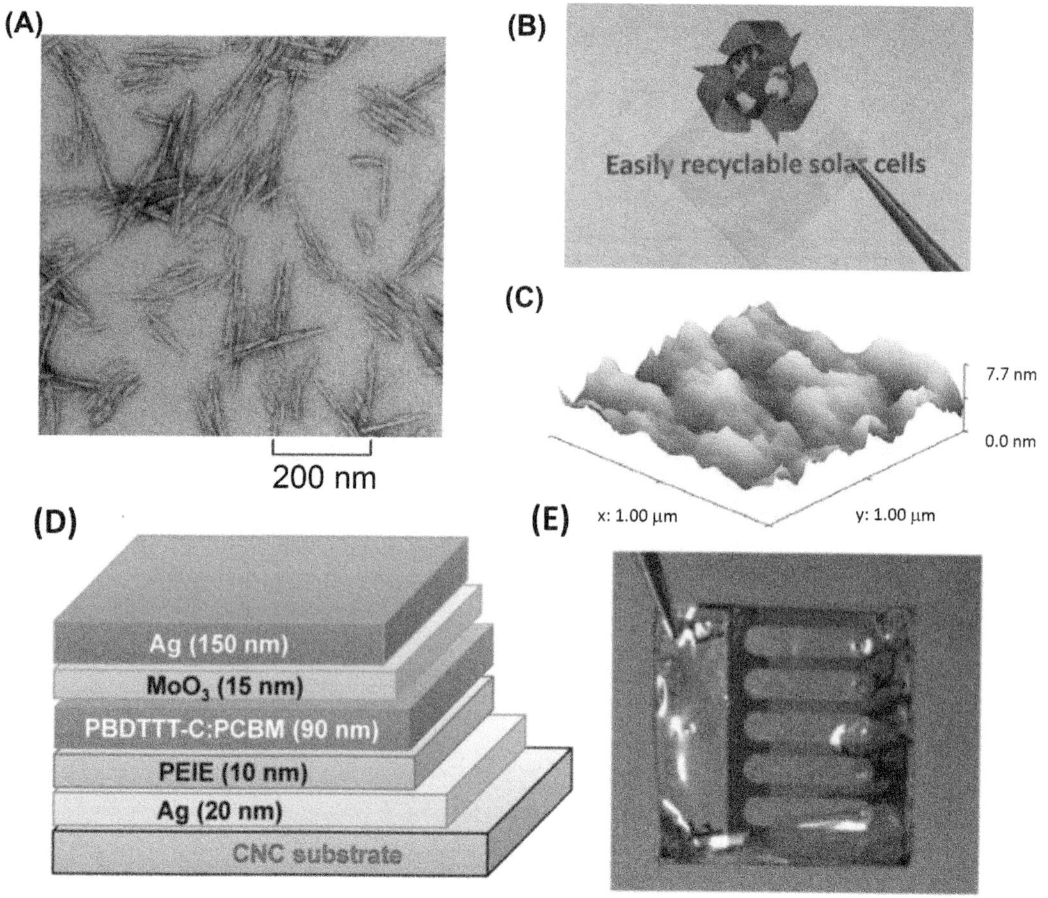

FIGURE 12.15

(A) TEM image of isolated CNC, (B) photograph of CNC film showing the transparency, (C) AFM image of CNC surface, (D) schematic of the OSC device fabricated on CNC substrate and (E) the picture of the prepared device [17].

Reproduced under creative common license from Springer Nature. Copyright 2013.

were peeled off and cut to 2.5 cm × 2.5 cm for use as substrate. The transparency of the obtained substrate is shown in Fig. 12.15B and the surface roughness of CNC surface can be seen from AFM is shown in Fig. 12.15C. Polydimethylsiloxane (PDMS) was used to secure the substrate to the glass. To create an OSC, additional layers were spin-coated onto a dried substrate. The schematic of the OSC structure architecture on CNC substrate and the picture of the fabricated solar cell is represented in Fig. 12.15D and E respectively [17].

A PCE value of $2.7 \pm 0.1\%$ was obtained using this solar cell. They have obtained $V_{OC} = 0.065 \pm 0.01$ V, $J_{SC} = 7.5 \pm 0.1$ mA/cm^2, FF value of 0.54. However, the value of PCE obtained using CNC was lower compared to their earlier reports using glass/ITO substrate. This was attributed to the lower J_{SC} value. The FF and V_{OC} of CNC substrate and glass substrates were comparable. The recyclability and combustion properties were thoroughly investigated. CNC substrates and SC were immersed in distilled water. Following CNC dispersion, the solution was filtered through P5 filter paper and the residuals were washed with chlorobenzene. The photoactive layer was dispersed in chlorobenzene after the first washing. The residuals were separated after a second washing with the same solvent. On CNC/solar cell, a combustion experiment was carried out. The burning took place in less than 2 seconds [17].

Consequently, Varma et al. [35], have demonstrated the suitability of agarose as a substrate for transient electronics. In their research, they used HMDS to modify agarose, converting it from hydrophilic to hydrophobic. It is thus distinct from other biodegradable hydrophilic substrates. AFM image of agarose substrate before and after treatment with KOH is represented in Fig. 12.16A and B, respectively. They demonstrated that the produced substrates are compatible with both water-based UV lithography and screen printing. The prepared devices were dissolved in 85% formic acid, and the silver was collected after the device and encapsulation layer were dissolved. This is shown in Fig. 12.16C [35].

12.5.2 Cellulose substrate-based solar cells

Leonat et al. [40], showed the use of paper substrate for an inverted structure to obtain an efficiency of 4%. They used vacuum thermal evaporation to deposit top contact using a shadow mask. Yusoff et al. [41] demonstrated the use of cellulose nanocrystals as substrates for perovskite based SC. They have reported a current density of 9.94 mA/cm^3 and a fill factor of 0.63. The PCE for this flexible, recyclable substrate was 5.4%, which was significantly lower than the efficiency obtained for ITO glass substrate-based SC (10.44%). The cellulose nanocrystal achieved a high transparency of about 80%, with the added benefit of not requiring a high annealing temperature. As a result, it is compatible with solution processing techniques for device fabrication, which are generally temperature sensitive. Cheng et al. [31] demonstrated the synthesis and application of *O*-(2,3-Dihydroxypropyl) cellulose (DHPC) by homogeneous etherification. The prepared cellulose exhibited exceptional properties such as ductility, transparency, and adhesiveness. The main disadvantage of this material was its poor mechanical properties. Biodegradable composites were created by using cellulose nanocrystals as fillers. Tin-doped indium oxide electrodes were used to create inverted polymer SC. They reported a PCE of 4.98%. Notte et al. [42], have demonstrated the use of a unique top electrode composed of conductive cellulose substrate. This was achieved by spray coating GnP on cellulose. This was then hot-pressed to additional layers to serve as the top electrode. As a result, cellulose was used to create a top electrode laminate. Spray coating was used to prepare the entire solar cell, resulting in a flexible solar cell. The schematic of the prepared solar

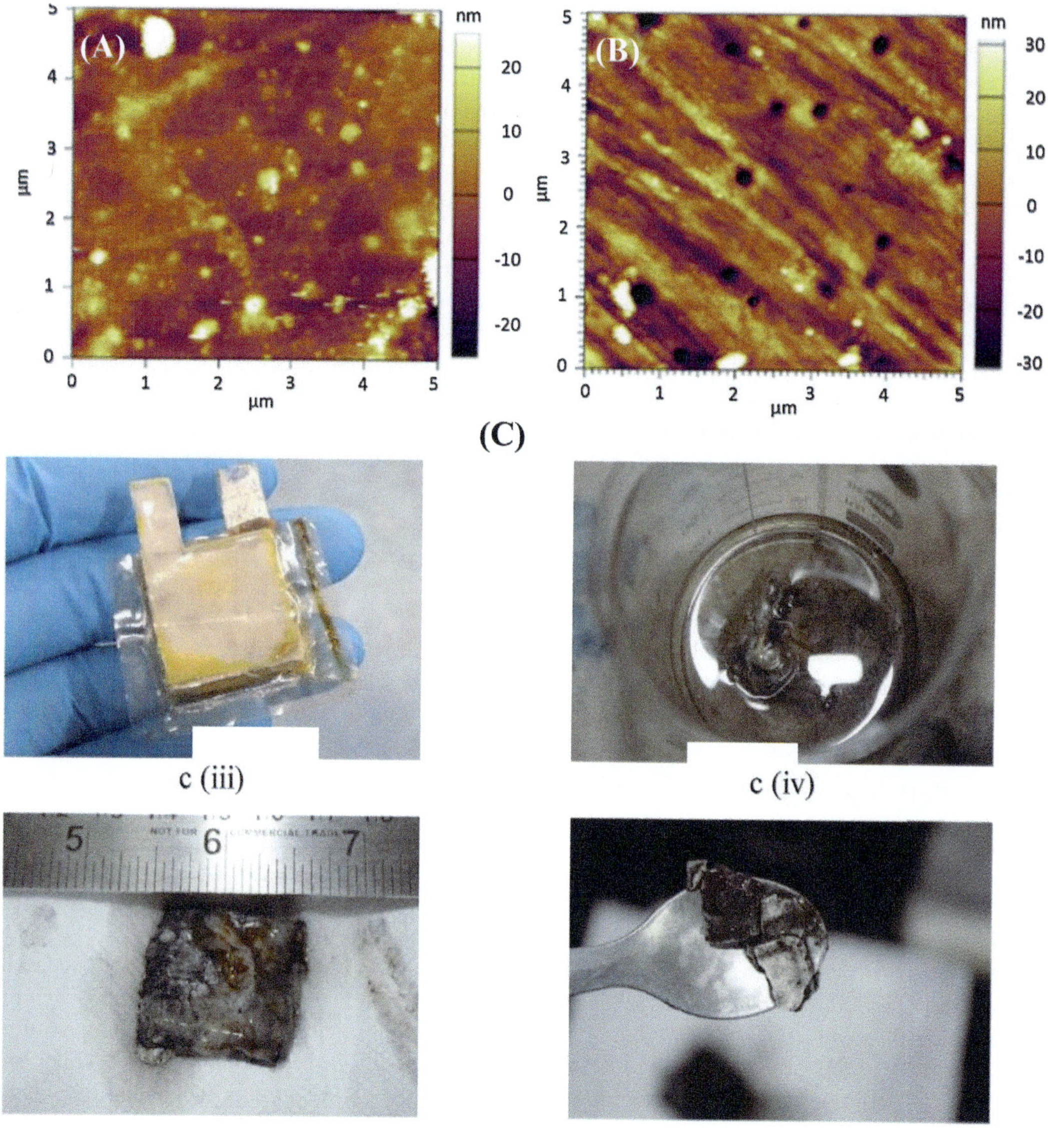

FIGURE 12.16

AFM image of agarose substrate (A) without treatment, (B) with KOH treatment. (C) Steps involved in the dissolution of substrates made of agarose using 85% formic acid [35].

Reproduced with permission from Elsevier. Copyright 2018.

cell is shown in Fig. 12.17A. The J-V characteristics of the large-area device are given in Fig. 12.17B. A photograph of the flexible device on PET substrate with conductive cellulose laminate is shown in the inset of Fig. 12.17 [42].

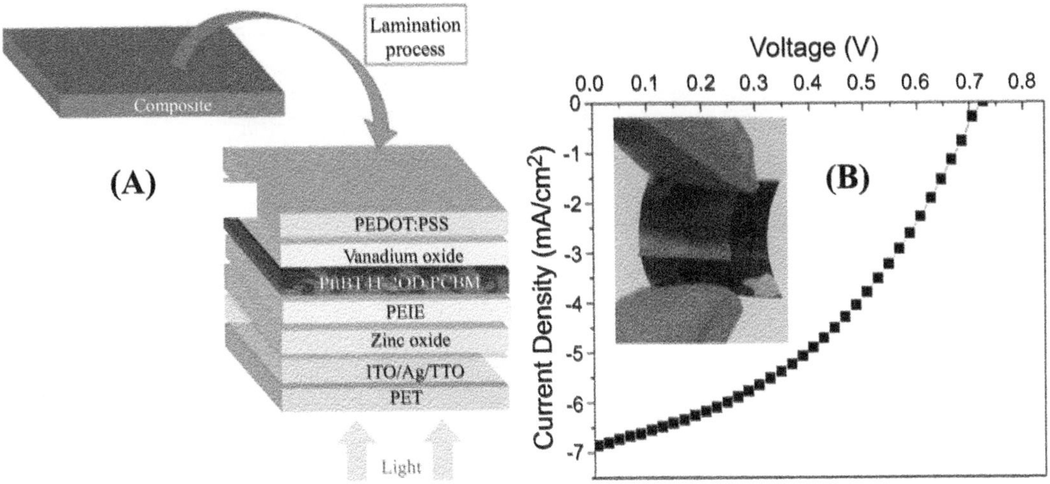

FIGURE 12.17

(A) Schematic of the OSC with laminated conductive cellulose, (B) J-V characteristics of OSC device. A photograph of spray-coated flexible solar cell with conducting cellulose laminate is shown in the inset of (B) [42].

Reproduced with permission from Elsevier. Copyright 2018.

Gao et al. [25] has reported the fabrication of nanocellulose paper (NCP) with acrylic resin coating. Nanocellulose was extracted from cotton to create a viscous solution in this study. This solution was then used to make NCP. A waterproof NCP was created by coating it with acrylic resin. They achieved a maximum PCE of 4.25% and an FF of 34.65%. The prepared substrates demonstrated excellent stability, with >80% property retention even after 50 bending cycles. The power per weight of these substrates was about 0.56 W/g. The main component in these devices was NCP constituting about 66 wt.%. Hence the disposal of this device is easy and can be done by flame burning for less than 3 seconds. The use of minimal electrode materials reduces the impact of position. Carbon was the most abundant component after incineration. However, the lead used in PSC was used in these cells, and more work is needed to eliminate it.

Lin et al. [26] showed a greater improvement in the PCE of biodegradable, flexible, conducting substrate. They demonstrated the preparation and performance of TEMPO oxidized (TO) cellulose nanofiber (CNF) incorporated with silver nanowires (AgNWs) for improved conductivity in their work. Furthermore, they prepared conductive substrate using unmodified CNF/AgNWs and compared the performance to that of commercially available PEN/ITO substrate. Fig. 12.18 shows the surface (A) and cross sectional SEM (B) of TEMPO-oxidized CNF (TOCN)/AgNWs. The surface micrograph depicts the sample's surface roughness caused by the presence of AgNWs. They used AFM to investigate the roughness of the samples. The AFM micrograph of ToCN/AgNWs is represented in Fig. 12.18C. From AFM, they have obtained an RMS value of 28 for TOCN-based substrate whereas the same for CNF-based substrate was 35.5 nm. This was consistent with their observation from SEM micrographs that the TOCN-based sample had more surface roughness than the CNF-based sample.

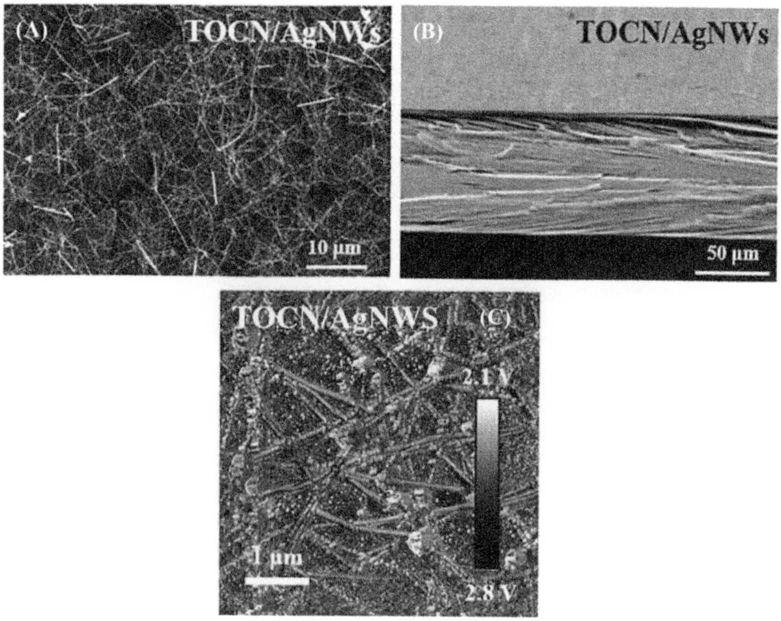

FIGURE 12.18

SEM micrograph (A) surface and (B) cross sectional view of TOCN/AgNWs composites and (C) AFM image of TOCN/AgNWs based substrate [26].

Copyright 2021. Reproduced with permission from Elsevier.

They have obtained very high transparency of 78% using TOCN/AgNWs. A PCE of 7.47% was reported with a high retainability of 43.5% even after 200 bending cycles. Fig. 12.19A shows the J-V curve for TOCN/AgNWs self-standing substrates. A comparison of PCE with other reported biomaterial-based substrate is shown in Fig. 12.19B. The prepared conductive substrate exhibited excellent mechanical stability, withstanding up to 15 peeling cycles and 500 bending cycles. They described a novel method for creating a high-performance biomaterial-based flexible conductive substrate. In their work, Liu et al. [39] have demonstrated the use of cellulose acetate (CA), a biodegradable polymer, to improve the stability of $CsPbIBr_2$-based perovskite films. They have reported an increase in PCE of up to 40% using 2.5 wt.% CA. The obtained efficiency was 7.52%.

12.5.3 Other biodegradable polymer substrates

Stranage et al. [12], studied the use of biodegradable poly-L-lactic acid (PPLA) loaded with nanoclay prepared by compounding via. extrusion as the substrate for SC. Transparency of up to 80% in PLLA and 60% in nanoclay loaded PLLA was demonstrated in the study. However, due to their lower thermal and mechanical properties, the prepared substrates have limited use in a solar cell. It precludes using it at high processing temperatures and dealing with mechanical stress caused by extrusion. Furthermore, the soluble nature of these biopolymers limited the solvents that could be

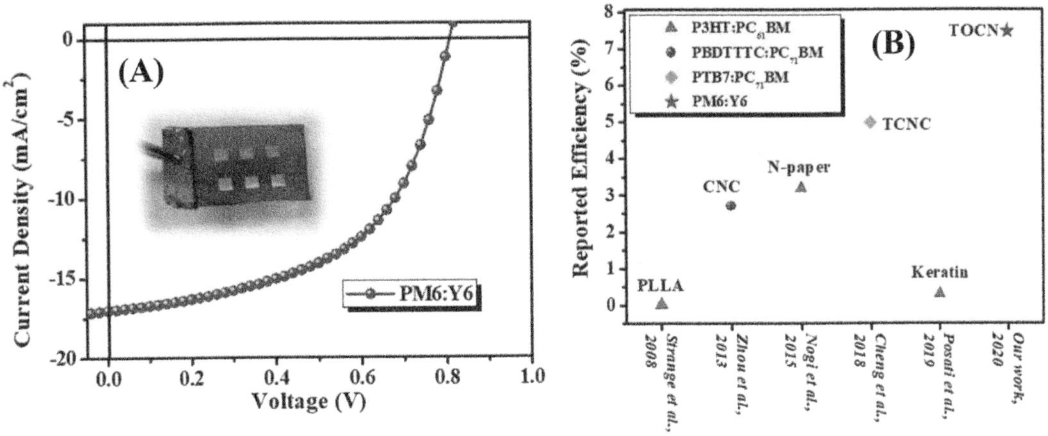

FIGURE 12.19

(A) J-V curve of TOCN/AgNW based flexible OSC and (B) Comparison of PCE with TOCN/AgNW and other reported self-standing biomaterial substrates [26].

Reproduced with permission from Elsevier. Copyright 2021.

used to water and ethanol. Lam et al. [43], have demonstrated the use of biomass derived polyethylene furonate (PEF) as a promising material for OSC. They used PEF in conjunction with silver nanowire to create flexible substrates with high conductivity, mechanical properties, and 90% transparency. The PCE value obtained was 6.7%, which was higher than the PCE value obtained from ITO/PEN substrates.

12.6 BP for solar cell encapsulation/passivation

The main issue with organic and perovskite SC is their lower thermal and mechanical stability, as well as their performance degradation when exposed to the environment. Encapsulating photovoltaic cells could be a solution to this problem. It acts as a barrier layer in the solar cell, limiting moisture and oxygen diffusion and assisting in the protection of the cathode interface and active layer from deterioration. The encapsulation material should have low water absorption, good processability, and permeability to protect the solar cell from moisture and oxygen. It should also be light transmitting, have a high dielectric constant, and be resistant to UV degradation as well as thermal oxidation. Chemical inertness, mechanical strength, and adhesion must all be improved. In addition to their use as solar cell substrates, cellulose-based materials have been investigated for applications such as encapsulation [44]. Onder et al. [45] have reported the use of natural polymers, gelatin and gum arabic for encapsulation of paraffin wax. This is a phase change material that is used in the textile industry. These materials have heat storage and release capabilities that can be applied to the encapsulation of SC.

Gaume et al. [46] provided a detailed analysis of the stability and degradation property of PVA for its application for multilayer stacking for OSC. They thoroughly investigated the degradation mechanism in the presence and absence of oxygen. Their studies revealed that the absence of oxygen gave higher stability in terms of the retention time of the volatile compounds. In their further work, Gaume et al. [27] demonstrated the optimization of PVA—MMt-Na^+ (a naturally occurring phyllosilicate) composite to improve the barrier properties. Even with 10 wt.% MMt-Na^+, it was seen that the transparency is not affected in the visible region. Improved intercalation and exfoliation were reported for composites with 2 and 5 wt.% of clay. For the integration to a multilayer stack of encapsulates, 5 wt.% of clay was suggested based on the compromise between wettability and surface roughness. The coating of PET + PVA-MMt-Na^+ (5 wt.%) showed a PCE value of 65% after 70 hours in comparison with PET + PVA coating which showed a value of 40% after the same time. They reported a reduction in PCE of up to 80% when using only PET coating. This increase in OSC stability could be attributed to better gas barrier protection for water vapor and oxygen.

Yuwawech et al. [28], reported the study of enhanced mechanical, thermal and barrier properties of EVA copolymer. They used BC nanofiber to modify EVA. BC was modified with propionic anhydride and combined with EVA in a twin-screw extruder to create composites. These films with significant barrier protection properties achieved a visible light transparency of approximately 75%. As a result of the functional groups present in the composite, they can be used as an encapsulation material for SC because the deacetylation of EVA is delayed.

Ansari et al. [47], have reported the use of biomass of polysaccharide Polygalacturonic acid as the template for the growth of the 3D superstructure of ZnO. The photovoltaic conversion property is determined using a specific electron injection dye D1. In comparison to ZnO nanorods, which had a PCE of 3.48%, the 3D ZnO structure had a PCE of 5.37%. The biopolymer acts as a crystal growth modifier. It also aids in the separation of ZnO nanorods for arrangement in the 3D superstructure.

Khan et al. [48] have demonstrated the use of ZnO/PVA nanocomposite films prepared by solution casting to improve the efficiency of OSC. They created composite films by varying the ratio of ZnO to water-soluble PVA. These films were stacked on top of OSC using the [Carbon Fiber/ (ZnO/Epoxy Resin)/(CuO/Epoxy Resin)/Carbon Fiber] architecture. The efficiency of such a design was 10.07%. However, with the addition of ZnO/PVA films with 14.25 wt.% of ZnO, an efficiency of 13.57% was obtained. They discovered that as ZnO concentration increases, the optical band gap decreases, as does efficiency. The efficiency decreases as the thickness of the ZnO/PVA layer increases.

12.7 Biodegradable polymer as electrolyte for dye-sensitized SC

O'Regan and Grätzel invented DSSC, also known as third-generation photovoltaics, in 1991. They differ from traditional p-n heterojunctions in their ability to separate photo-generated charge carriers. In the case of DSSC, the electrolyte plays a larger role in regulating stability and efficiency. An electrolyte usually controls the short-circuit current density (J_{sc}), open-circuit voltage (V_{OC}), and *FF* to improve efficiency. Electrolytes are classified into three types: solid, liquid, and gel. There has been a lot of research into the use of biodegradable polymer electrolytes in DSSC. The basic construction, mechanism, and working principle of DSSC are shown in Fig. 12.20 [49]. It is

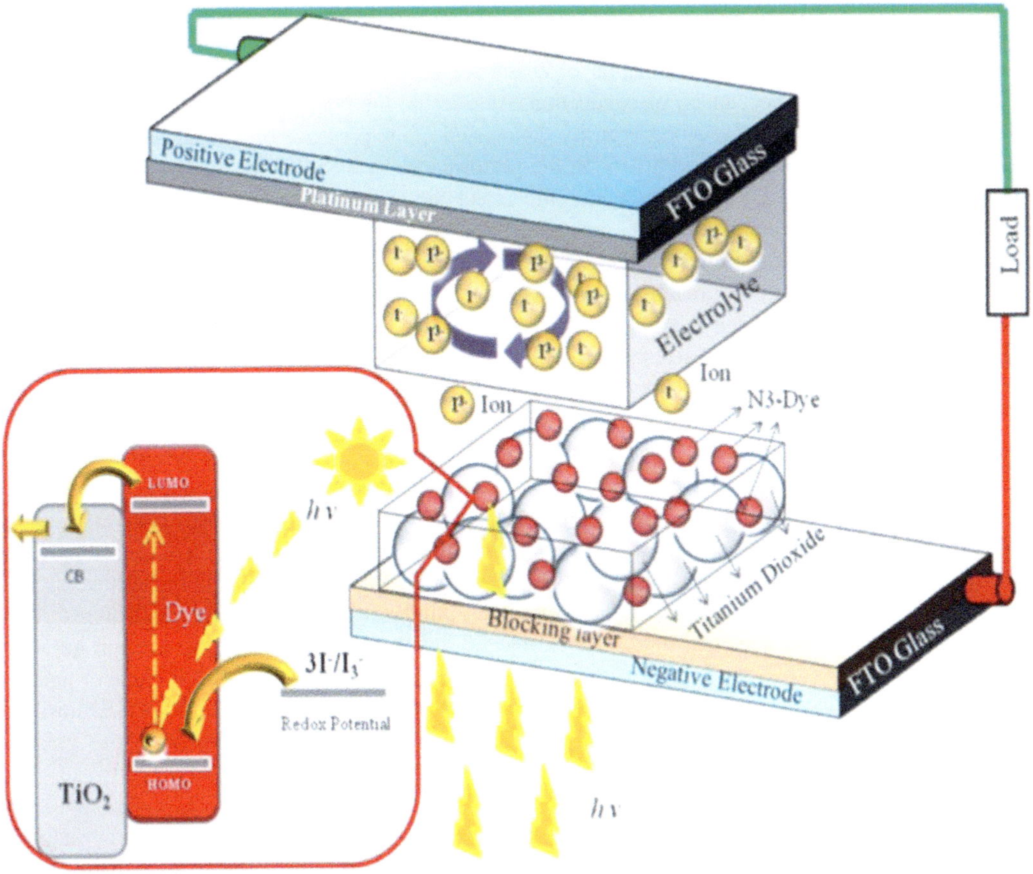

FIGURE 12.20

Schematic and working principle of DSSC [49].

Reproduced with permission from Elsevier. Copyright 2016.

made up of an anode, a cathode, and an electrolyte sandwiched together. The anode is made of mesoporous metal oxides like titanium oxide (TiO_2), while the cathode is made of conductive layers like platinum coated transparent conductive oxide. The anode TiO_2 was dipped in the dye solution and left overnight before filling the space between the anode and cathode with electrolyte. The cell's illumination causes current to flow through an external circuit load connected to the electrodes. The absorption of light in DSSC takes place due to the dye molecule and the charge separation by electron injection takes place from dye to TiO_2 at the semiconductor electrolyte interface.

The chemical and physical structure and composition of the electrolyte have a significant impact on the energy conversion efficiency and stability of SC. Polysaccharides are one of the most extensively researched biopolymers in the development of efficient electrochemical devices. Singh et al. [13] reported a novel agarose-based biopolymer electrolyte for DSSC. By incorporating potassium iodide into the agarose biopolymer matrix, the polymer electrolyte was created. The ionic conductivity increase with the addition of KI and 60:40 (agarose: KI, in wt.%) composition has shown the highest ionic conductivity of 9.02×10^{-3} S/cm. The coordination of ether oxygen of agarose, potassium cation, and iodine anion from KI salt is the mechanism of conduction. As a result, the potassium cation is tightly bound to the matrix, while the weak iodine anion is easily dissociated in the presence of a small DC electric field.

Fig. 12.21 shows the effect of additives on the performance of DSSC. Fig. 12.21A shows the electrolyte without additive in DSSC and which could affect the performance due to recombination reaction. The improved performance of DSSC with triphenylamine core organic additives absorbed on TiO_2 is shown in Fig. 12.21B. The Lewis basicity of a semiconductor near the Fermi level resulted in better performance. This is due to the availability of electrons within excited dye molecules, which reduces recombination reactions gradually. With N_3 dye as a sensitizer, the presence of triphenylamine core organic additive in HEC results in an efficient PCE of 5.72%. The presence of a highly conjugated system and heteroatom-based additives aided in improving mobility transport and reducing the recombination reaction between TiO_2 and I_3^-, thereby improving photovoltaic performance.

The use of BP is critical for SC, particularly DSSC. Polymers are used in DSSC as a significant improvement over ionic liquid electrolyte-based cells, which had leakage and gas permeability issues. To overcome the problems associated with ionic liquid electrolytes, polymers are used to create a quasi-solid or gel-based electrolyte. Many studies have been conducted to compare the PCE of gel-based and liquid-based electrolytes. They found that PCE has a comparable value to liquid electrolytes. Much research has been conducted on the use of BP as gel electrolytes. They are being investigated for the solidification of ionic liquids in DSSC while being used in gel electrolytes. One such work is reported by Bidikoudi et al. [51]. They have reported the use of agarose natural polysaccharides for the solidification of ionic liquid comprised of $[C_nC_1im]I/EMimDCA$ double salt. Polyethylene oxide (PEO) and natural agarose polymer were used in the gelation

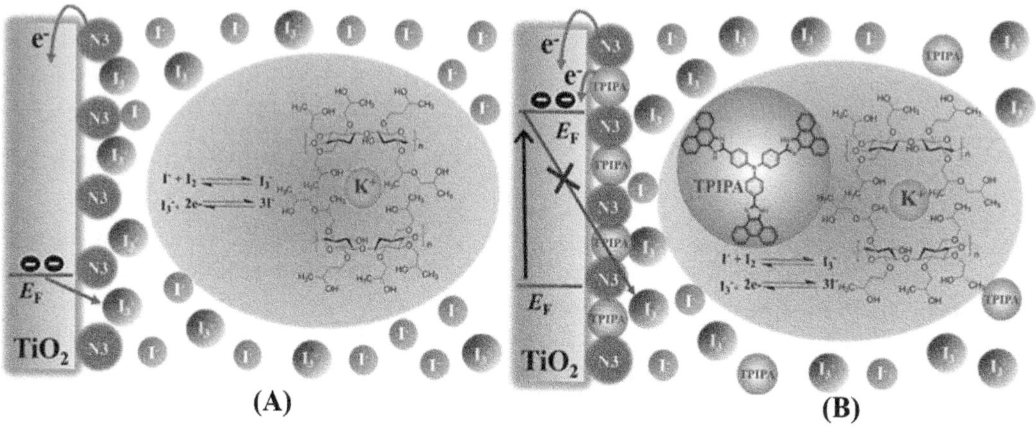

FIGURE 12.21

Interactions at TiO_2/dye/electrolyte interface in DSSC (A) without additive and (B) with triphenylamine core organic additive [50].

Reproduced with permission from American Chemical Society. Copyright 2021.

process. Agarose gelation involves the formation of aggregated double helices structures. Seaweeds provide the agarose, linear agarobiose disaccharide. In solution, they form single coils. Some of these coils form a double-helical structure when cooled. This intermediate state is made up of a mix of double helices and single coils. Further cooling results in the formation of a helical aggregate [52]. The device with PEO showed a PCE value of 4.41% against agarose which exhibited 3.98%. The reference electrode consisted of 0.1M LiI and 0.2M I_2 dissolved in [C_4C_1im]I/EMimDCA. The reference electrode showed a PCE value of 3.84%. This demonstrated the utility of agarose polymer as an effective gelator for the DSSC electrolyte.

Wang et al. [53] have studied the agarose-based polymer electrolyte. They investigated the effect of LiI concentration on the electrochemical properties of polymer electrolytes based on agarose. The study found that as LiI concentration increased, electron lifetime and resistance to charge recombination and transfer decreased. Gao et al. [54] varied the concentration of agarose in novel polysaccharide agarose-based electrolytes for DSSC. The concentration was systematically varied from 1 to 5 wt.%. Photoelectric performance of these DSSC revealed that relative high conductivity and low T_g was observed at lower concentration (1–2 wt.%). They reported a maximum conductivity with 1.5 wt.% agarose which was 3.94×10^{-4} S/cm. A PCE of 4.14% was observed for 2 wt.% agarose. Alias et al. [23], have demonstrated the use of agar gel polymer with NH_4I and I_2 additives as potential electrolyte candidate for DSSC. They have observed an Arrhenius behavior and activation energy as low as 0.25 eV was obtained. Hsu et al. [55] investigated the effect of TiO_2 pore size on DSSC performance in an agarose-based gel electrolyte. They discovered that porous TiO_2 improves performance by increasing redox couple diffusion. Guo et al. [56] investigated the effect of surfactants on the electrochemical performance of a novel agarose-based magnetic polymer electrolyte. They investigated the effects of four surfactants: sodium dodecyl sulfate (SDS), polyvinylpyrrolidone (PVP), polyethylene glycol (PEG200), and polysorbate 80 (TW-80). They have added Fe_3O_4 to the polymer electrolyte to induce magnetic properties. They discovered that the performance varies depending on the surfactant used. Using TW-80 surfactant, a maximum PCE of 1.83% was obtained. In another investigation, Khanmirzaei et al. [15], have reported the use of biodegradable RS based solid polymer electrolyte in DSSC. The polymer electrolyte was treated with the ionic liquid 1-methyl-3-propylimidazolium iodide (MPII). The achievement of larger ionic conductivity of 1.20×10^{-3} S/cm is observed by adding 20 wt.% of MPII. With the same composition, they have reported an energy conversion efficiency of up to 2.09%.

In 2017, Singh et al. [57] have reported the use of agarose-based electrolytes for DSSC application. They discovered that the viscosity of ionic liquids and their dispersion have a significant impact on the electronic conductivity of these polymers. They compared the effects of two solvents for DSSC application, DMF and DMSO, and discovered that DMF is the best solvent for fabricating DSSC using agarose biopolymers. Gupta et al. [58] have reported photovoltaic properties of lapnotite nanoclay dispersed in [poly (ethylene oxide)-succinonitrile]-LiI-I_2 solid polymer electrolytes. The photovoltaic measurement shows that the J_{SC} value increased 123% with doping lapnotite clay in the biopolymer. The cell efficiency was improved by 54%, which is due to the better concentration of I^-/I^{3-} ionic species in presence of nanoclay at the interface of dye-sensitized TiO_2 and solid polymer electrolyte. Selvanathan [37] has also developed DSSC based on esterified starch as quasi-solid biopolymer electrolyte. Due to the presence of LiI, ionic conductivity reaches 4.82×10^{-3} S/cm, resulting in 0.7 fill factor in DSSC. The photoconversion efficiency of various biodegradable polymer-based systems used as electrolytes in DSSC along with the dye used is summarized in Table 12.2.

Table 12.2 Various biodegradable polymer-based electrolytes used for DSSC with the type of electrolyte, the dye used and photoconversion efficiency obtained.

Sl. no.	Type	Electrolyte	DYE	PCE	Year	References
1	Gel	Triphenylamine added HEC	N3	5.72%	2021	[50]
2	Liquid	Chitosan dissolved in formic acid	N719	1.6%	2021	[59]
3	Solid	Chitosan with NaI (30 wt.%)	N719	0.06	2021	[60]
4	Gel	Phthaloyl starch	N719	3.02%	2021	[37]
5	Ionic liquid	Biopolymer rice starch; RS:LiI: MPII: TiO_2	N3	0.17%	2014	[61]
6	Solid	$PEO/Pr_4N + I^- + I_2$		0.01%	2008	[62]
7	Solid	Chitosan doped with low viscosity IL, 1-ethyl 3-methylimidazolium thiocyanate (EMImSCN)		0.73%	2010	[38]
8	Gel	Rice starch + LiI	N719	0.35%	2017	[63]
9	Gel	Agarose + KI + DMF + IL + I_2	N3	1.36%	2017	[57]
10	Gel	*N*-Phthaloylchitosan + Succinonitrile	N3	4.82%	2020	[64]
11	Aqueous	Preoxidized kraft lignin with poly(ethylene glycol) diglycidyl ether (PEGDGE)	D131	1.5%	2021	[65]
12	Gel	Potato starch + 1-glycidyl-3-methylimidazolium chloride (GMIC)/KI/I2	N719 and chenodeoxycholic acid	0.51%	2019	[66]
13	Gel	Agarose + KI	535-bisTBA	0.54%	2013	[13]
14	Solid	Arrowroot biopolymer doped with potassium iodide	N3	0.63%	2014	[14]
15	Gel	Rice starch with LiI, NH_4I, NaI	N719	0.78%	2014	[18]
16	Solid	Rice starch + Sodium iodide + MPII	N719	2.09	2014	[15]
17	Gel	Guar gum	MK-2	4.96%	2020	[67]
18	Gel	Phthaloylchitosan + tetrapropylammonium iodide (TPAI) and 1-butyl-3-methylimidazolium iodide (BMII)	N3	9.61%	2017	[68]
19	Gel	Grafted cellulose with acrylic acid	$RuL_2(NCS)_2$	5.51%	2011	[69]
20	Gel	Citric acid cross-linked potato starches	N719	1.4%	2017	[70]
21	Gel	Nonedible oil-based Polyurethane acrylate	N719	1.97%	2020	[71]
22	Gel	HEC; tris(1,10-phenanthroline) cobalt(II)/(III)bis/tris (hexafluorophosphate) and synthesized tetramethyl benzidine-based thiourea	N719	9.1%	2019	[72]

(Continued)

Table 12.2 Various biodegradable polymer-based electrolytes used for DSSC with the type of electrolyte, the dye used and photoconversion efficiency obtained. ***Continued***

Sl. no.	Type	Electrolyte	DYE	PCE	Year	References
23	Gel	Cellulose; (MPII), 1-ethyl-3-methylimidazolium thiocyanate (EMISCN),LiI, 4-tert-butylpyridine	N719	3.3%	2014	[73]
24	Gel	Hydroxy ethyl cellulose (natrosol); 2,5bis(1(nanofibers-2-ylmethyl)1Hbenzo[d] imidazole-2-yl)thiophene	N3	4.5%	2020	[74]
25	Gel	Hydroxyl propyl cellulose, sodium iodide and MPII	N719	5.79%	2015	[75]
26	Gel	Cyanoethylated hydroxypropyl cellulose, LiI/I_2 and $MHII/I_2$	N719	7.55%	2012	[76]
27	Solid	Chitosan-PEO, NH_4I, Ag to TiO_2	N3	1.13%	2016	[77]
28	Gel	Chitosan, I^- and I_2 in acetonitrile	N719	2%	2017	[78]
29		LiI: I_2, TiO_2 filler, ethylene carbonate (plasticizer)	Anthocyanin	1.8%	2019	[79]
30	Solid-film	Blend of carboxymethyl kappa-carrageenan/carboxymethyl cellulose, NH_4I-I_2	N719	0.13%	2015	[33]
31	Gel	Cellulose powder, MPII, EMISCN, iodine (I_2), lithium iodide (LiI) and 4-tert-butylpyridine (TBP)	N719	3.33%	2014	[80]
32	Gel	CMC and microfibrillated cellulose (MFC), sodium iodide (NaI), MPII, 4-tert-butylpyridine (TBP) and iodine (I_2)	N719	5.18% (CMC) & 7.03% (MFC)	2014, 2013	[81,82]
33	Gel	Agarose, 1-allyl-3-ethylimidadolium iodide (AEII)	N719	5.89%	2013	[83]
34	Gel	Electro spun cellulose acetate (CA) nanofibers, tetrapropyl ammonium iodide (Pr_4NI), TBA = tertrabutylamonium	N719	4.0%	2017	[84]
35	Gel	Agarose, 1-allyl-3-ethylimidazolium iodide (AEII).	N719	7.43%	2014	[55]
36	Gel	Magnetic agarose, LiI and I_2	N719	1.83	2013	[56]
37	Gel	Co_3O_4-modified agarose, LiI and I_2	N719	2.97	2014	[85]

The majority of the electrolytes investigated based on BP are gel or quasi-solid in nature. Ionic liquid and solid states are also investigated in a few studies. Various BPs are being investigated, including cellulose, agarose, chitosan, RS, guar gum, and nonedible oil-based BPs. The gel electrolyte is made by combining BP and ionic liquids. Various ionic liquids including LiI, I_2, Pr4NI, TBA, AEII, MHII, NaI, EMISCN, LiI, 4-tert-butyl pyridine, and BMII in combination with redox couples are explored in these SC. N719 is the most commonly used dye. However, there have been reports of N_3 use in a few cases. In this case, the photoconversion efficiency ranges from 0.01% to 9.61%. The main work on biodegradable polymer-based DSSC began around 2008 and saw a significant increase in 2013. Throughout these years, there has also been a resurgence of interest in this field. However, there is no evidence of a consistent rise in PCE over time. This could be due to the wide range of materials investigated for use as electrolytes.

12.8 Conclusion and future direction

Over the last two decades, there has been tremendous growth in research based on BP and composites used in solar cell applications. A wide variety of materials, including biomass, such as RS, nonedible oil, potato starch, agar gum, algae, and cellulose, have been investigated for potential applications. BPC is most commonly used as a transparent substrate in organic and perovskite SC. Furthermore, they have been used to create a quasi-solid or gel electrolyte for DSSC with the addition of ionic liquid. BPC has a less-explored application in environmental barrier protection and encapsulation. With the incorporation of BPC, there is a significant change in the PCE in each of the areas. Several strategies have been implemented to improve PCE. One of them was to use the BP in a variety of forms, including nanocrystals, nanofibers, and grafted agarose/cellulose. In a second method, the BP was functionalized, primarily to increase hydrophobicity and stability. The final method was the incorporation of filler materials including nanoclay, cellulose nanocrystal, metal oxides like Co_3O_4, Nio, and TiO_2 and silver nanowire. The use of these BPC ensured the prepared SC' biodegradability and flexibility. However, the main issues with BPC-based SC are their stability, mechanical properties, poor performance when compared to nonbiodegradable counterparts, hydrophilic nature, thermal stability, and transparency. A lot of effort has gone into addressing these issues, such as functionalization to make hydrophobic, manipulating the processing technique to improve stability and mechanical properties, and using various redox couples to improve PCE. Coating with other synthetic materials, such as acrylic resin, is also attempted to improve transparency. However, there is still work to be done to improve the properties of BPC-based SC such as perovskite, organic, and DSSC.

To enable the use of robust, low-cost, flexible BPC-based SC, the future direction could be to achieve each of these properties with an emphasis on the eco-friendly processing technique. Better processing techniques, such as hot pressing, stamping, and lamination of casted films, are some avenues to pursue. A baking cycle prior to processing could improve the thermal properties of the BPC. The use of BP blends rather than the current single polymer-filler approach will help to improve mechanical properties and thermal stability. Multilayer architecture with rough/smooth layer stacking, in addition to BP blends, could improve stability. Another approach is to use magnetic orientation of nanoparticles to align the filler for increased strength. The research on

BPC-based SC is critical as a renewable, green, and sustainable solution to the world's ever-increasing energy demand.

Acknowledgement

The authors thank Vellore Institute of Technology, Vellore for providing 'VIT SEED Grant - RGEMS fund (SG20220118, SG20220116)' for carrying out this research work.

References

[1] H. Kumar, M.B. Ahamed, K. Deshmukh, M.S. Sirajuddeen, Morphology, dielectric and EMI shielding characteristics of graphene nanoplatelets, montmorillonite nanoclay and titanium dioxide nanoparticles reinforced polyvinylidenefluoride nanocomposites, J. Inorg. Organomet. Polym. Mater. 31 (2021) 2003–2016.

[2] P. Rani, M.B. Ahamed, K. Deshmukh, Structural, dielectric and EMI shielding properties of polyvinyl alcohol/chitosan blend nanocomposites integrated with graphite oxide and nickel oxide nanofillers, J. Mater. Sci. Mater. Electron. 32 (2021) 764–779.

[3] P. Rani, M.B. Ahamed, K. Deshmukh, Dielectric and electromagnetic interference shielding properties of zeolite 13X and carbon black nanoparticles based PVDF nanocomposites, J. Appl. Polym. Sci. 138 (2021) 50107.

[4] Muzaffar, M.B. Ahamed, K. Deshmukh, S.K.K. Pasha, Dielectric properties and electromagnetic interference shielding studies of nickel oxide and tungsten oxide reinforced polyvinylchloride nanocomposites, Polym. Plast. Technol. Mater. 59 (2020) 1667–1678.

[5] P.L. Reddy, K. Deshmukh, K. Chidambaran, M.B. Ahamed, K.K. Sadasivuni, D. Ponnamma, et al., Effect of poly ethylene glycol (PEG) on structural, thermal and photoluminescence properties of CdO nanoparticles for optoelectronic applications, Mater. Today Proceed 9 (2019) 175–183. 6.

[6] Pandey, G.M. Joshi, K. Deshmukh, M. Khutia, N.N. Ghosh, Optimized AC conductivity correlated to structure, morphology and thermal properties of PVDF/PVA/Nafion composites, Ionics 20 (2014) 1427–1433.

[7] M. Pandey, G.M. Joshi, K. Deshmukh, J. Ahmad, Impedance spectroscopy and conductivity studies of $CdCl_2$ doped polymer electrolyte, Adv. Mater. Lett. 6 (2015) 165–171.

[8] M. Khutia, G.M. Joshi, K. Deshmukh, M. Pandey, Preparation of modified polymer blend and electrical performance, Comp. Interf. 22 (2015) 167–178.

[9] M. Khutia, G.M. Joshi, K. Deshmukh, M. Pandey, Optimization of dielectric constant of polycarbonate/polystyrene modified blend by ceramic metal oxide, Polym. Plast. Technol. Eng. 54 (2015) 383–389.

[10] M. Pandey, G.M. Joshi, S. Bhattacharya, Influence of $K_2Ti_6O_{13}$ on dielectric and barrier properties of polymer, Ionics 22 (2016) 445–449.

[11] T. Remis, P. Belski, T. Kovarik, J. Kadlec, M.G. Azar, R. Medlin, et al., Study on structure, thermal Behavior, and viscoelastic properties of nanodiamond-reinforced poly (vinyl alcohol) nanocomposites, Polymers. 13 (2021) 1426.

[12] M. Strange, D. Plackett, M. Kaasgaard, F.C. Krebs, Biodegradable polymer solar cells, Sol. Energy Mater. Sol. Cell 92 (2008) 805–813.

[13] R. Singh, N.A. Jadhav, S. Majumder, B. Bhattacharya, P.K. Singh, Novel biopolymer gel electrolyte for dye-sensitized solar cell application, Carbohydr. Polym. 91 (2013) 682–685.

[14] R. Singh, B. Bhattacharya, H.W. Rhee, P.K. Singh, New biodegradable polymer electrolyte for dye sensitized solar cell, Int. J. Electrochem. Sci. 9 (2014) 2620–2630.

[15] M.H. Khanmirzaei, S. Ramesh, K. Ramesh, Polymer electrolyte based dye-sensitized solar cell with rice starch and 1-methyl-3-propylimidazolium iodide ionic liquid, Mater. Des. 85 (2015) 833–837.

[16] H. Xi, D. Chen, L. Lv, P. Zhong, Z. Lin, J. Chang, et al., High performance transient organic solar cells on biodegradable polyvinyl alcohol composite substrates, RSC Adv. 7 (2017) 52930–52937.

[17] Y. Zhou, C.F. Hernandez, T.M. Khan, J.C. Liu, J. Hsu, J.W. Shim, et al., Recyclable organic solar cells on cellulose nanocrystal substrates, Scr. Rep. 3 (1536) (2013) 1–5.

[18] Y. Yang, J. Cui, P. Yi, X. Zheng, X. Guo, W. Wang, Effects of nanoparticle additives on the properties of agarose polymer electrolytes, J. Power Sources 248 (2014) 988–993.

[19] M.H. Buraidah, L.P. Teo, S.R. Majid, R. Yahya, R.M. Taha, A.K. Arof, Characterizations of chitosan-based polymer electrolyte photovoltaic cells, Int. J. Photoenergy 2010 (2010) 1–7.

[20] Masao Kaneko, Takayuki Hoshi, Yuuki Kaburagi, Hirohito Ueno, Solid type dye-sensitized solar cell using polysaccharide containing redox electrolyte solution, J. Electroanal. Chem. 572 (1) (2004) 21–27.

[21] Junichi Nemoto, Mami Sakata, Takayuki Hoshi, Hirohito Ueno, Masao Kaneko, All-plastic dye-sensitized solar cell using a polysaccharide film containing excess redox electrolyte solution, J. Electroanal. Chem. 599 (1) (2007) 23–30.

[22] M.H. Buraidah, L.P. Teo, S.R. Majid, A.K. Arof, Characteristics of TiO_2/solid electrolyte junction solar cells with I^-/I_3^- redox couple, Opt. Mater. 32 (6) (2010) 723–728.

[23] S.S. Alias, A.A. Mohamad, Effect of NH_4I and I_2 concentration on agar gel polymer electrolyte properties for a dye-sensitized solar cell, Ionics. 19 (2013) 1185–1194.

[24] Yei Hwan Jung, Tzu-Hsuan Chang, Huilong Zhang, Chunhua Yao, Qifeng Zheng, Vina W. Yang, et al., High-performance green flexible electronics based on biodegradable cellulose nanofibril paper, Nat. Commun. 6 (1) (2015) 1–11.

[25] L. Gao, L. Chao, M. Hou, J. Liang, Y. Chen, H.D. Yu, et al., Flexible, transparent nanocellulose paper-based perovskite solar cells, Npj Flex. Electron. 3 (2019) 4.

[26] P.C. Lin, C.T. Hsieh, X. Liu, F.C. Chang, W.C. Chen, J. Yu, et al., Fabricating efficient flexible organic photovoltaics using an eco-friendly cellulose nanofibers/silver nanowires conductive substrate, Chem. Eng. J. 405 (2021) 126996.

[27] J. Gaume, C. Taviot-Gueho, S. Cros, A. Rivaton, S. Therias, J.L. Gardette, Optimization of PVA clay nanocomposite for ultra-barrier multilayer Encapsulation of organic solar cells, Sol. Energy Mater. Sol. Cells. 99 (2012) 240–249.

[28] K. Yuwawech, J. Wootthikanokkhan, S. Tanpichai, Enhancement of thermal, mechanical and barrier properties of EVA solar cell encapsulating films by reinforcing with esterified cellulose nanofibers, Polym. Test. 48 (2015) 12–22.

[29] S. Qiu, X. Xu, L. Zeng, Z. Wang, Y. Chen, C. Zhang, et al., Biopolymer passivation for high-performance perovskite solar cells by blade coating, J. Energy Chem. 54 (2021) 45–52.

[30] H. Li, X. Li, W. Wang, J. Huang, J. Li, S. Huang, et al., Ultraflexible and biodegradable perovskite solar cells utilizing ultrathin cellophane paper substrates and $TiO_2/Ag/TiO_2$ transparent electrodes, Sol. Energy. 188 (2019) 158–163.

[31] Q. Cheng, D. Ye, W. Yang, S. Zhang, H. Chen, C. Chang, et al., Construction of transparent cellulose-based nanocomposite papers and potential application in flexible solar cells, ACS Sustain. Chem. Eng. 6 (2018) 8040–8047.

[32] J.R. Capadona, K. Shanmuganathan, D.J. Tyler, S.J. Rowan, C. Weder, Stimuli-responsive polymer nanocomposites inspired by the sea cucumber dermis, Science. 319 (5868) (2008) 1370.

[33] S. Rudhziah, A. Ahmad, I. Ahmad, N.S. Mohamed, Biopolymer electrolytes based on blend of kappa-carrageenan and cellulose derivatives for potential application in dye sensitized solar cell, Electrochim. Acta. 175 (2015) 162–168.

[34] A. Gerthoffer, C. Poulain, F. Roux, F. Emieux, L. Grenet, S. Perraud, CIGS solar cells on ultra-thin glass substrates: determination of mechanical properties by nanoindentation and application to bending-induced strain calculation, Sol. Energy Mater. Sol. Cell 166 (2017) 254–261.

[35] S. Verma, Gaganjot, J. Tripathi, M. Katiyar, V. Verma, Biodegradable photolithography compatible substrate for transparent transientelectronics and flexible energy storage devices, Appl. Mater. Today. 13 (2018) 83–90.

[36] M.H. Khanmirzaei, S. Ramesh, K. Ramesh, Effect of different iodide salts on ionic conductivity and structural and thermal behavior of rice-starch-based polymer electrolyte for dye-sensitized solar cell application, Ionics. 21 (2015) 2383–2391.

[37] V. Selvanathan, M.H. Ruslan, A. Ahmed, N. Alkahtani, N. Amin, K. Sopian, et al., Organosoluble, esterified starch as quasi-solid biopolymer electrolyte in dye-sensitized solar cell, J. Mater. Res. Technol. 12 (2021) 1638–1648.

[38] K. Pramod, B. Singh, R.K. Bhattacharya, Nagarale, Kang Wook Kim, Hee-Woo Rhee, Synthesis, characterization and application of biopolymer-ionic liquid composite membranes, Synth. Met. 160 (1-2) (2010) 139–142.

[39] J. Liu, Q. He, J. Bi, M. Lei, W. Zhang, G. Wang, Remarkable quality improvement of $CsPbIBr_2$ perovskite film by cellulose acetate addition for efficient and stable carbon-based inorganic perovskite solar cells, Chem. Eng. J. 424 (2021) 130324.

[40] L. Leonat, M.S. White, E.D. Głowacki, M.C. Scharber, T. Zillger, J. Rühling, et al., 4% Efficient polymer solar cells on paper substrates, J. Phys. Chem. C. 118 (2014) 16813–16817.

[41] R. Bin Mohd Yusoff, M.K. Nazeeruddin, Perovskite solar cells based on recyclable and biodegradable substrates, in: International Renewable and Sustainable Energy Conference (IRSEC), 2017, 1–4.

[42] L. Notte, P. Cataldi, L. Ceseracciu, I.S. Bayer, A. Athanassiou, S. Marras, et al., Fully-sprayed flexible polymer solar cells with a cellulose-graphene electrode, Mater. Today Energy 7 (2018) 105–112.

[43] J.Y. Lam, C.C. Shih, W.Y. Lee, C.C. Chueh, G.W. Jang, C.J. Huang, et al., Bio-based transparent conductive film consisting of polyethylene furanoate and silver nanowires for flexible optoelectronic devices macromol, Rapid Commun. 39 (2018) 1800271.

[44] A. Uddin, M.B. Upama, H. Yi, L. Duan, Encapsulation of organic and perovskite solar cells: a review, Coatings 9 (2) (2019) 65.

[45] E. Onder, N. Sarier, E. Cimen, Encapsulation of phase change materials by complex coacervation to improve thermal performances of woven fabrics, Thermochim. Acta. 467 (1-2) (2008) 63–72.

[46] J. Gaume, P.W. Chung, A. Rivaton, S. Thérias, J.L. Gardette, Photochemical behavior of PVA as an oxygen-barrier polymer for solar cell encapsulation, RSC Adv. 1 (2011) 1471–1481.

[47] M.S. Ansari, R. Maragani, A. Banik, R. Misra, M. Qureshi, Enhanced photovoltaic performance using biomass derived nano 3D ZnO hierarchical superstructures and a D – A type CS-Symmetric triphenylamine linked bisthiazole, Electrochim. Acta. 259 (2018) 262–275.

[48] S.A. Khan, Z.M. Zain, M. Mansoor, M.M.H. Mahfuz, A. Rahman, M.A.N. Rashid, et al., Performance investigation of ZnO/PVA nanocomposite film for organic solar cell, Mater. Today Proc. (2021). Available from: https://doi.org/10.1016/j.matpr.2021.05.197.

[49] R. Singh, A.R. Polu, B. Bhattacharyya, H.W. Ree, C. Varlikli, P.K. Singh, Perspectives for solid biopolymer electrolytes in dye sensitized solar, cell and battery application, Renew. Sustain. Energy Rev. 65 (2016) 1098–1117.

[50] D. Raja, B. Selvaraj, G. Shanmugam, A. Maruthapillai, D. Sundaramurthy, Improving the efficiency of dye-sensitized solar cells via the impact of triphenylamine-based inventive organic additives on biodegradable cellulose polymer gel electrolytes, Energy Fuels 35 (2021) 4273–4282.

[51] M. Bidikoudi, D. Perganti, C.S. Karagianni, P. Falaras, Solidification of ionic liquid redox electrolytes using agarose biopolymer for highly performing dye-sensitized solar cells, Electrochim. Acta. 179 (2015) 228–236.

[52] S. Arnott, A. Fulmer, W.E. Scott, I.C.M. Dea, R. Moorhouse, D.A. Rees, The agarose double helix and its function in agarose gel structure, J. Mol. Biol. 90 (2) (1974) 269–284.

[53] W. Wang, X. Guo, Y. Yang, Lithium iodide effect on the electrochemical behavior of agarose based polymer electrolyte for dye-sensitized solar cell, Electrochim. Acta 56 (21) (2011) 7347–7351.

[54] Y. Yang, X.Y. Guo, X.Z. Zhao, Influence of polymer concentration on poly-saccharide electrolyte for quasi-solid-state dye-sensitized solar cell, Mater. Sci. Forum. 685 (2011) 76–81.

[55] H.L. Hsu, C.F. Tien, J. Leu, Effect of pore size / distribution in TiO_2 films on agarose gel electrolyte-based dye-sensitized solar cells, J. Solid. State Electrochem. 18 (2014) 1665–1671.

[56] X. Guo, P. Yi, Y. Yang, J. Cui, S. Xiao, W. Wang, Effects of surfactants on agarose-based magnetic polymer electrolyte for dye-sensitized solar cells, Electrochim. Acta. 90 (2013) 524–529.

[57] R. Singh, B. Bhattacharya, S.K. Tomar, V. Singh, P.K. Singh, Electrical, optical and electrophotochemical studies on agarose based biopolymer electrolyte towards dye sensitized solar cell application, Measurement 102 (2017) 214–219.

[58] R.V. Gupta, H.W. Rhee, I. Bedja, A.N. AlHazaa, A. Khan, Effect of Laponite® nanoclay dispersion on electrical, structural, and photovoltaic properties of dispersed [Poly (Ethylene oxide)-succinonitrile]-LiI-I_2 solid polymer electrolyte, J. Power Sources 490 (2021) 229509.

[59] E. Praveen, J. Peter, A. Muthu kumar, K. Ramachandran, K. Jayakumar, Performance of ZnO/ZnS nanocomposite based dye-sensitized solar cell with chitosan-polymer electrolyte, Mater. Today Proc. 35 (1) (2021) 27–30.

[60] N.A. Rahman, S.A. Hanifah, N.N. Mobarak, A. Ahmad, N.A. Ludin, F. Bella, et al., Chitosan as a paradigm for biopolymer electrolytes in solid-state dye-sensitised solar cells, Polymer 230 (2021) 124092.

[61] M.H. Khanmirzaei, S. Ramesh, Nanocomposite polymer electrolyte based on rice starch/ionic liquid/ TiO_2 nanoparticles for solar cell application, Measurement 58 (2014) 68–72.

[62] T.M.W.J. Bandara, M.A.K.L. Dissanayake, O.A. Ileperuma, et al., Polyethyleneoxide (PEO)-based, anion conducting solid polymer electrolyte for PEC solar cells, J. Solid. State Electrochem. 12 (2008) 913–917.

[63] K.C. Yogananda, E. Ramasamy, S. Kumar, S. Vasantha Kumar, M. Navya Rani, D. Rangappa, Novel rice starch based aqueous gel electrolyte for dye sensitized solar cell application, Mater. Today Proc. 4 (11) (2017) 12238–12244.

[64] A. Ratan, M. Buraidah, P.L. Teo, P.K. Singh, A. Arof, Enhanced photo-current conversion efficiency by incorporation of succinonitrile in N-Phthaloylchitosan based bio-polymer electrolyte for dye sensitized solar cell, Optik. 222 (2020) 165467.

[65] J.C. De Haro, E. Tatsi, L. Fagiolari, M. Bonomo, C. Barolo, S. Turri, et al., Lignin-based polymer electrolyte membranes for sustainable aqueous dye-sensitized solar cells, ACS Sustain. Chem. Eng. 9 (2021) 8550–8560.

[66] M.O.S. Lobregas, D.H. Camacho, Gel polymer electrolyte system based on starch grafted with ionic liquid: synthesis, characterization and its application in dye-sensitized solar cell, Electrochim. Acta. 298 (2019) 219–228.

[67] Ahalya Gunasekaran, A. Sorrentino, A. M.Asiri, S. Anandan, Guar gum-based polymer gel electrolyte for dye-sensitized solar cell applications, Sol. Energy. 208 (2020) 160–165.

[68] M.H. Buraidah, S. Shah, L.P. Teo, F.I. Chowdhury, M.A. Careem, I. Albinsson, et al., High efficient dye sensitized solar cells using phthaloylchitosan based gel polymer electrolytes, Electrochim. Acta. 245 (2017) 846–853.

[69] P. Li, Y. Zhang, W. Fa, Y. Zhang, B. Huang, Synthesis of a grafted cellulose gel electrolyte in an ionic liquid ([BMIM]I) for dye-sensitized solar cells, Carbohydr. Polym. 86 (2011) 1216–1220.

[70] P. Nagaraj, A. Sasidharan, V. David, A. Sambandam, Effect of cross-linking on the performances of starch-based biopolymer as gel electrolyte for dye-sensitized solar cell applications, Polymers. 9 (12) (2017) 667.

[71] K.L. Chai, I.M. Noor, M.M. Aung, L.C. Abdullah, M.Z. Kufian, Non-edible oil based polyurethane acrylate with tetrabutylammonium iodide gel polymer electrolytes for dye-sensitized solar cells, Sol. Energy. 208 (2020) 457–468.

[72] P. Karthika, S. Ganesan, A. Thomas, T.M. Sheeba Rani, M. Prakash, Influence of synthesized thiourea derivatives as a prolific additive with tris(1,10-phenanthroline)cobalt(II/III)bis/tris(hexafluorophosphate)/hydroxypropyl cellulose gel polymer electrolytes on dye-sensitized solar cells, Electrochim. Acta. 298 (2019) 237–247.

[73] G.P. Salvador, D. Pugliese, F. Bella, A. Chiappone, A. Sacco, S. Bianco, et al., New insights in long-term photovoltaic performance characterization of cellulose-based gel electrolytes for stable dye-sensitized solar cells, Electrochim. Acta. 146 (2014) 44–51.

[74] S. Balamurugan, S. Ganeshan, Novel cobalt redox materials admitted in natrosol polymer with a thiophene based additive as a gel polymer electrolyte to tune up the efficiency of dye sensitized solar cells, Electrochim. Acta. 329 (2020) 135169.

[75] M.H. Khanmirzaei, et al., Hydroxypropyl cellulose based non-volatile gel polymer electrolytes for dye-sensitized solar cell applications using 1-methyl-3-propylimidazolium iodide ionic liquid, Sci. Rep. 5 (2015) 18056.

[76] X. Huang, Y. Liu, J. Deng, B. Yi, X. Yu, P. Shen, et al., A novel polymer gel electrolyte based on cyanoethylated cellulose for dye-sensitized solar cells, Electrochim. Acta. 80 (2012) 219–226.

[77] M.H. Buraidah, L.P. Teo, C.M.A. Yong, S. Shah, A.K. Arof, Performance of polymer electrolyte based on chitosan blended with poly(ethylene oxide) for plasmonic dye-sensitized solar cell, Opt. Mater. 57 (2016) 202–211.

[78] M. Khalili, M. Abedi, H.S. Amoli, S.A. Mozaffari, Comparison of chitosan and chitosan nanoparticles on the performance and charge recombination of water-based gel electrolyte in dye sensitized solar cells, Carbohydr. Polym. 175 (2017) 1–6.

[79] P. Chawla, A. Srivastava, M. Tripathi, Performance of chitosan based polymer electrolyte for natural dye sensitized solar cell, Environ. Prog. Sustain. Energy 38 (2) (2019) 630–634.

[80] G. Paolo, D. Pugliese, F. Bella, A. Chiappone, A. Sacco, S. Bianco, M. Quaglio, New insights in long-term photovoltaic performance characterization of cellulose-based gel electrolytes for stable dye-sensitized solar cells, Electrochim. Acta. 146 (2014) 44–51.

[81] F. Bella, A. Chiappone, J.R. Nair, G. Meligrana, C. Gerbaldi, Effect of different green cellulosic matrices on the performance of polymeric dye-sensitized solar cells, Italian Assoc. Chem. Eng. 41 (2014) 211–216.

[82] F. Bella, J.R. Nair, C. Gerbaldi, Towards green: efficient and durable quasi-solid dye-sensitized solar cells integrated with a cellulose-based gel-polymer electrolyte optimized by a chemometric DoE approach, RSC Adv. 3 (2013) 15993–16001.

[83] H.L. Hsu, C.F. Tien, Y.T. Yang, J. Leu, Dye-sensitized solar cells based on agarose gel electrolytes using allylimidazolium iodides and environmentally benign solvents, Electrochimca Acta 91 (2013) 208–213.

[84] A.M.J.S. Weerasinghe, M.A.K.L. Dissanayake, G.K.R. Senadeera, V.A. Senaviratne, C.A. Thotawatthage, J. M.K.W. Kumari, Application of electrospun cellulose acetate nanofibre membrane based quasi-solid state electrolyte for dye sensitized solar cells, Ceylon J. Sci. 46 (2) (2017) 93–98.

[85] Si Xiao, Jiarui Cui, Pengfei Yi, Ying Yang, Xueyi Guo, Insight into electrochemical properties of Co_3O_4–modified magnetic polymer electrolyte, Electrochim. Acta. 144 (2014) 221–227.

CHAPTER

13 Biodegradable polymer nanocomposite for electromagnetic interference shielding applications

Krishnendu Nath and Narayan Ch. Das

Rubber Technology Centre, Indian Institute of Technology Kharagpur, Kharagpur, West Bengal, India

Chapter Outline

13.1 Introduction

Nanoscience and nanotechnology have seen tremendous technological advancement in the last decade. Composite materials involving polymer matrix and different carbonaceous nanoparticles (e.g., multiwalled carbon nanotube, conductive carbon black, graphene oxide (GO), etc.) [1–4] as conductive inclusion to transform the insulating polymer into a conductive one. As a result, polymer nanocomposites have been exploited to innovate different sensing devices [5], EMI shielding materials [6–17] light-emitting devices [18,19], etc. polymer nanocomposites with conductive fillers are lightweight, flexible, easily processable as well as electrically conductive. As a result, in terms of applications, they have effectively replaced heavy-weight nonflexible metal devices. Metal encasing of various electronic devices, which were previously very heavy-weight and unattractive, was seen by the older generation. Metallic electronic devices are heavy-weight and have high processing temperature requirements, making them unportable, expensive, and visually unappealing. We have now discussed the general application of polymer nanocomposites. Among all applications, the development of EMI shielding materials is the most difficult due to two major development challenges. One, while maintaining a high level of electrical conductivity, the material should be lightweight and flexible.

Biodegradable and Biocompatible Polymer Nanocomposites. DOI: https://doi.org/10.1016/B978-0-323-91696-7.00009-X
Copyright © 2023 Elsevier Inc. All rights reserved.

Second, the material's processing should be simple and inexpensive. These days, a larger area of nanotechnology research is devoted to developing EMI shielding materials by transforming lightweight and electrically insulating polymeric materials into conductive materials by incorporating electrically conductive and lightweight carbon nanoparticles via simple and cost-effective processing mechanisms (e.g., melt-blending [20] or solution blending and casting [21]). In the melt-blending process [22], the polymers are blended with fillers at the melting/processing temperature of the polymer in a melt blender or Brabender then compression-molded to get the final shaped composite material. In the case of solution mixing process [23], the polymers are first solubilized in a solvent and a predetermined amount of filler is suspended in the same solvent with the help of an ultra-sonicator/stirrer. Following the mixing of the polymer solution and the filler suspension, the composite solution is cast on a petri dish for air/oven drying. The dried composite is then compressed molded to achieve a well-shaped and optimized filler dispersion on the polymer nanocomposite. Electromagnetic radiation is also harmful to one's health [24–26]. Due to exposure to the electromagnetic wave, several health hazards such as headache [27], neurological disorders [28], numbness [29], skin cancer [30] prevail and can affect human health quite possibly. The need for EMI shield materials drives new research, and then another problem arises. The present era of electronics faces some new challenges regarding the excessive electronic wastages known as "e-wastes" [31,32]. After use, electronic devices are discarded. Because they are made of biologically inert polymers, the polymeric materials of dumped electronic wastes are now indestructible by the environment. Excessive "e-wastes" gradually fill up the sea and lands and affects the marine life impactfully [33]. So, what is left next? We have to make "e-wastes" biodegradable so that as soon as there are dumped, they would be degraded as time passes. How to make "e-wastes biodegradable?" The answer is—by replacing the polymer matrix of nanocomposite with biologically degradable polymers, for example, PLA [34], PBAT [35], poly(ε-caprolactone) (PCL) [36], etc. Now the obvious question is "what is biodegradability?" as per ASTM D 5488-94d and European standard EN 13432, "biodegradability" of a polymer has been decided by its capability to undergo gradual decomposition to give methane, carbon dioxide, and other inorganic or organic chemical derivatives under normal biological condition (environment or in presence of specific enzyme as a catalyst) [37]. The mechanism of biodegradation [38–40] is given in Fig. 13.1.

People have been looking for an effective replacement for ceramics, metals, and other nonflexible materials since World War II, and plastics and polymers have come into the light of science and technology. The problem of "e-waste" and nonbiodegradable plastic waste that the modern world is dealing with was not faced by people during World War II. Someone correctly stated that the modern world is confronted with modern problems. The word "biodegradable plastic" and "biopolymers" [41] were first introduced during the era of 1980s. The biodegradability reaction of biopolymers is catalyzed by enzymes used by microorganisms for degradation, according to European standard EN 13432-2000. Biodegradation, more precisely, is the degradation of organic materials caused by biological activity (catalytical biodegradation) by microorganisms [42], resulting in carbon dioxide, water, methane, and other end products as shown in Fig. 13.1. Another important consideration is the nature of the end product after the degradation process has been completed. Generally, water is produced if the biodegradation is done in the aerobic condition [43] and anaerobic condition [44] produces methane as the main end product due to the degradation. There are two types of ways [45] depending upon which biodegradable polymers are degraded, one is physical degradation/decomposition [46] associated with hydrolysis reaction, another way is photo-degradation process [47] where the degradation reaction is aided by light energy.

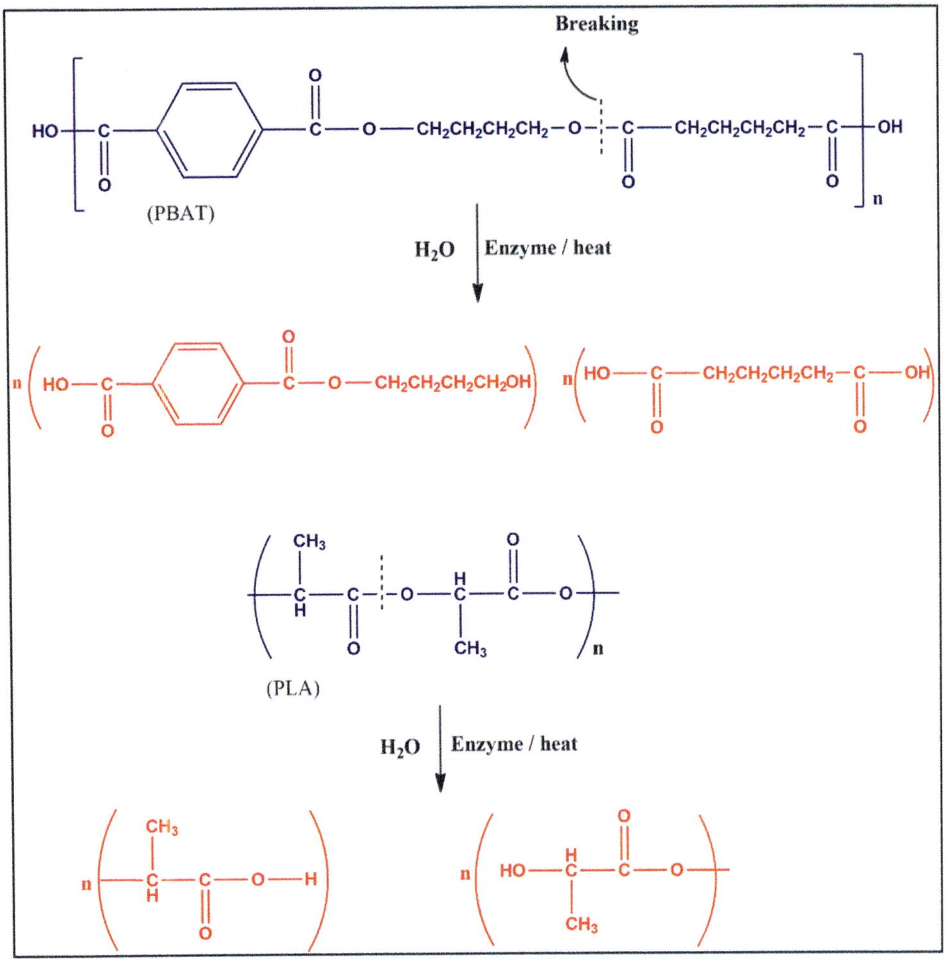

FIGURE 13.1

Mechanism of physical biodegradation of some biodegradable polymers.

There are two types of biodegradable polymer sources: biological (renewable) sources (e.g., natural biodegradable/biopolymer) and petroleum derivatives (e.g., synthetic biodegradable polymer). Polypeptides, proteins, polysaccharides, lipids, and other natural biopolymers are the most common. Before being used as a commercially viable polymer, natural biopolymers must undergo structural modification to improve mechanical, electrical, or thermal properties. Hydrolyzable esters, amides, urethanes, and other compounds are found in synthetic biopolymers. In the case of synthetic biopolymers, their properties can be improved by blending them with another polymer, incorporating fillers to improve conductivity, and so on. Importantly, biodegradable polymer degradation involves two steps: Mineralization follows fragmentation [48]. Before mineralization, the large chain biopolymer molecules are fragmented into small fragments. Let us now look at how microorganisms

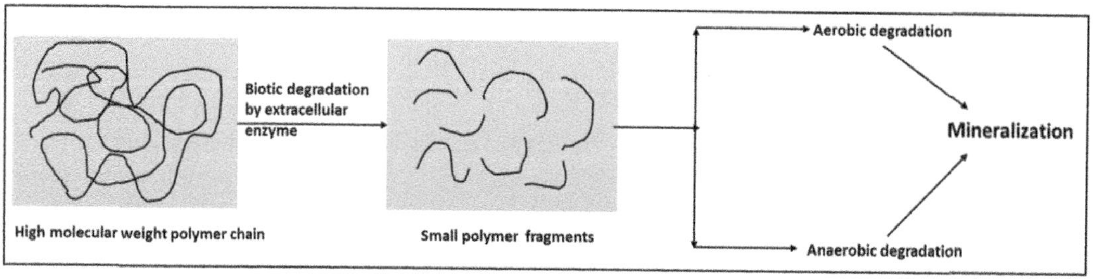

FIGURE 13.2

Mechanism of biotic degradation of some biodegradable polymers by microorganisms.

in nature degrade biopolymers. The long-chain biopolymers are first degraded into small fragments by an extracellular enzyme, which is known as an abiotic reaction. Smaller fragments are easily carried to the cell, where microorganisms deposit and mineralize them. Now, microbial fragmentation of biopolymers can occur in two ways: aerobic degradation or in the presence of air (oxygen) and anaerobic degradation or in the absence of oxygen. The evolution of gas or the formation of salts indicates the complete degradation of biopolymers. A probable mechanism of biodegradation by microorganisms has been graphically represented in Fig. 13.2.

Before discussing the application of some popular biodegradable polymer nanocomposites as EMI shield materials, we must first understand the mechanism of EMI shield materials.

13.2 Basic theory and mechanism of electromagnetic inference shielding

Materials for electromagnetic interference (EMI) shielding are naturally electrically conductive [2]. As shown in Fig. 13.3 an EMI shield material cross-section effectively absorbs, reflects, or transmits the incident electromagnetic wave. Now, if we divide the fractions of waves, we can broadly divide the incident wave as reflection component (SE_R), absorption component (SE_A), and a small portion of the incident wave as internal reflection loss component (SE_M), which is negligible. Then, if total shielding efficiency (SE_T) of the shielding material is given as Eq. (13.1).

$$SE_T = SE_R + SE_A \tag{13.1}$$

The unit of shielding efficiency is decibel (dB) and if the value of SET is less than -15 dB, the term internal reflection loss component can be ignored when calculating the overall shielding efficiency of the material.

13.3 Poly(lactic acid)-related polymeric electromagnetic inference shielding materials

Concerns about the widespread industrial and commercial use of petroleum-derived nonbiodegradable plastics, as well as the nondecomposability of e-waste, prompted the development of biodegradable

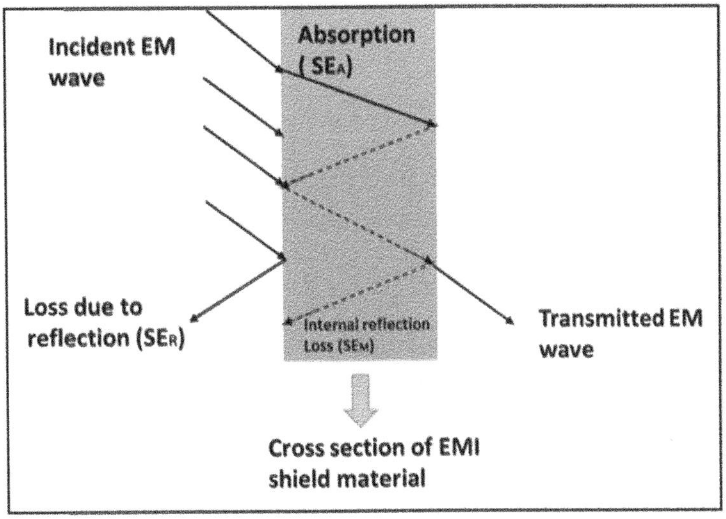

FIGURE 13.3

Mechanism of electromagnetic inference (EMI) shielding phenomena across a cross-section of EMI shield material.

plastics. There was a need to achieve biodegradability as well as other utility on the polymeric materials. Bioplastics have several advantages, including (1) environmental friendliness and (2) the rising cost of manufacturing petroleum-based polymers. The first-ever known bioplastic was poly(glycolic acid) (PGA). PLA was popularized later as a bioplastic [49] which was first produced by Carothers in 1932 by heat treatment of lactic acid in absence of air. But the molecular weight to low so that procedure was later discarded. Later, PLA was generated, based on ring-opening polymerization (ROP) of lactide where metal oxides (e.g., tin octoate) are used as catalyst [48]. The general reaction mechanism [50] is given in Fig. 13.4.

Aside from all of its advantages, PLA has a significant disadvantage: it is highly flammable. Poly (lactic acid) is highly flammable, and this disadvantage must be overcome. Polyphosphoramide, a flame retardant phosphorus derivative, could be used as a flame retardant for PLA without compromising its mechanical properties [51]. The reaction for the synthesis [51] of polyphosphoramide/polyphosphonate is suggested in Fig. 13.5. Polyphosphonates are found to be a great flame retardant for PLA.

In general, phosphorus and nitrogen-containing flame retardants are very effective because they decompose during combustion to produce incombustible ammonia gas and other phosphorus-containing gaseous materials. As a result, the ammonia produced dilutes the concentration of oxygen in the air surrounding the polymer, slowing the overall combustion process. The decomposition of polyphosphoramide is shown in Fig. 13.6.

The breakdown of polyphosphoramide produces PO, a radical scavenger that can easily neutralize O. and HO. PLA's decomposition is the fault of radicals. The limiting oxidation index (LOI) measures the effect of polyphosphoramide on the flame retardancy of a PLA/polyphosphoramide blend. For the pristine PLA the LOI value was found to be around 20% and with the addition of 0.5 wt.% of the flame

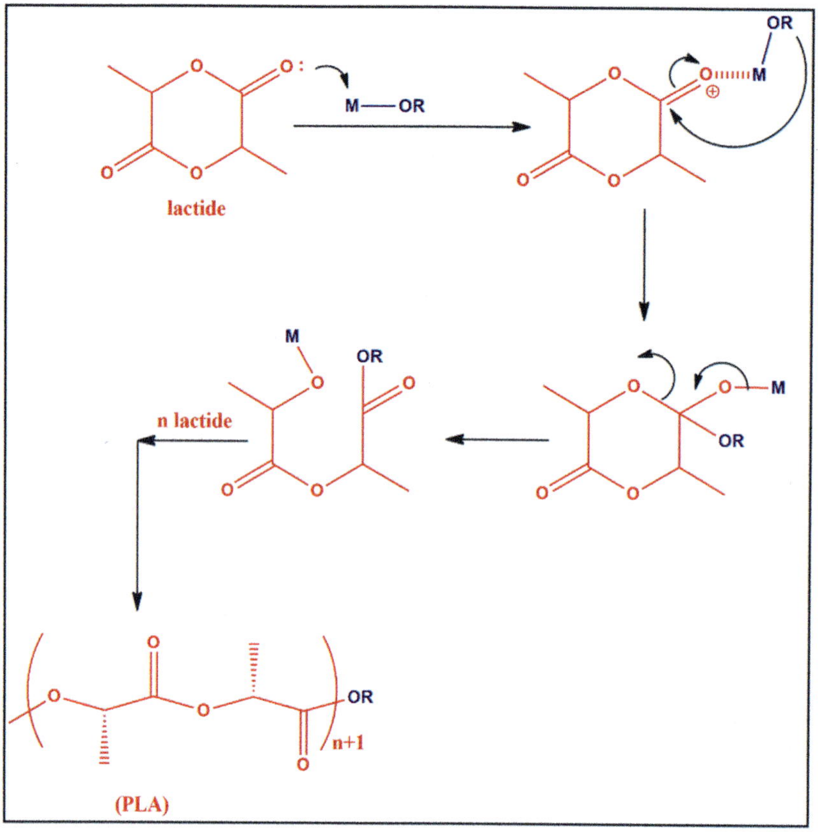

FIGURE 13.4

A general mechanism of synthesis of poly(lactic acid) (PLA) following on ring-opening polymerization (ROP) in presence of metal oxide (MOR) as a catalyst.

retardant had increased the value around 27%. In general, for a polymer, an LOI value above 26% is considered to be very good concerning flame retardancy. Let's talk about the EMI shielding performance of PLA-based polymer composites now. In this context, Shi et al. [52] has developed PLA-based 3-dimensional cellular polymer nanocomposite that contains a hybrid filler graphite nanoparticle (GNP)/ carbon nanotube (CNT). Firstly, the filler blends in different GNP/CNT ratios (GNP/CNT = 2:0, 2:0.5, 2:1, 2:2, and 2:4) are dispersed welly in distilled water keeping the suspension under mechanical stirrer (~600 rpm) followed by ultrasonication for about 3 hours. Then the PLA particles are added to the distilled water suspension of hybrid filler and slowly stirred for about 1 hour. so that a uniform coating of hybrid filler is achieved throughout the surface area of the PLA particles. Following filtration and drying, the hybrid filler coated PLA particles are extruded to produce the composite in the form of filament for 3D printing. The process is discussed in Fig. 13.7.

The 3-D printed parts (length of 40 mm and thickness of 4 mm) are produced in different shapes of honey-comb like structures (hexagon, square, and triangle) and EMI shielding values (between

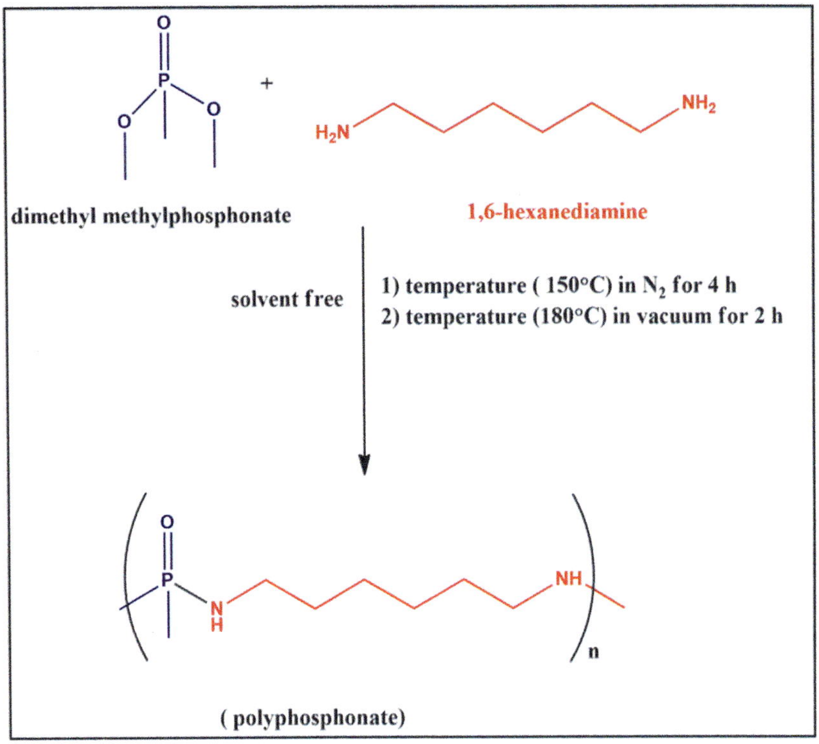

FIGURE 13.5

Synthesis of polyphosphoramide.

8.2 and 12.6 GHz) are checked. The EMI shielding values would vary depending on the cell's structural geometry and size. The structural geometry was discovered to have very little effect on the EMI shielding values. It was also discovered that the EMI shielding values of 3-D composites decrease as the cell size decreases. The graphical representations are given in Fig. 13.8.

The size of the cells (24–50 nm) is equivalent to the X-band frequency as a result the former has a huge impact on the latter. In another attempt, Shi et al. [53] had prepared only GNPs/polylactic acid-based 3D nanocomposite following the same procedure as mentioned in their previous job. With 10 wt.% of filler content, the EMI shielding value achieved was around 29 dB for 2.5 mm of 3D composite. The value of EMI shielding is positive depending upon Eq. (13.2).

$$SE_T = 10 \log \left(P_{In}/P_{out}\right) \tag{13.2}$$

where P_{in} is the incoming power and P_{out} is the outgoing power on a vector network analyzer (VNA). Xu et al. [54] has prepared poly(L-lactide) (PLLA)/PLA stereo complex crystallites (PLAsc)/CNT-based polymer nanocomposites where CNT particles are specifically distributed to amorphous PLLA phase to build a conductive network of CNTs to reduce the electrical conductivity percolation threshold. The idea behind the work is basically to selectively distribute CNT in the

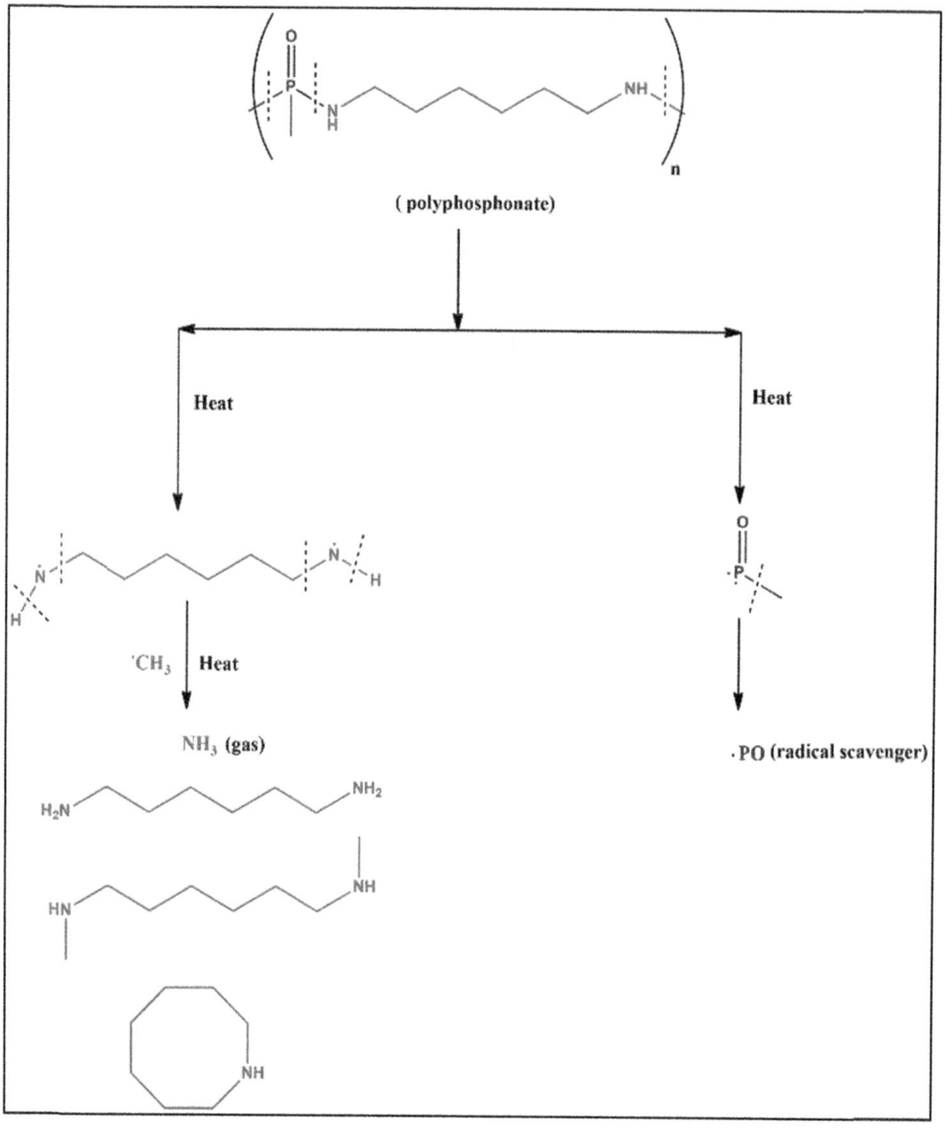

FIGURE 13.6

Decomposition of polyphosphoramide.

amorphous PLLA phase by exploiting the melting point difference between PLLA and PLAsc. In a typical process, PLLA granules are solubilized in dichloromethane (DCM) and CNT of the required amount is suspended in ethanol. The CNT/ethanol suspension is mixed with PLLA/DCM solution and ultrasonicated before the solvent casting. After solvent casting and after subsequent drying the

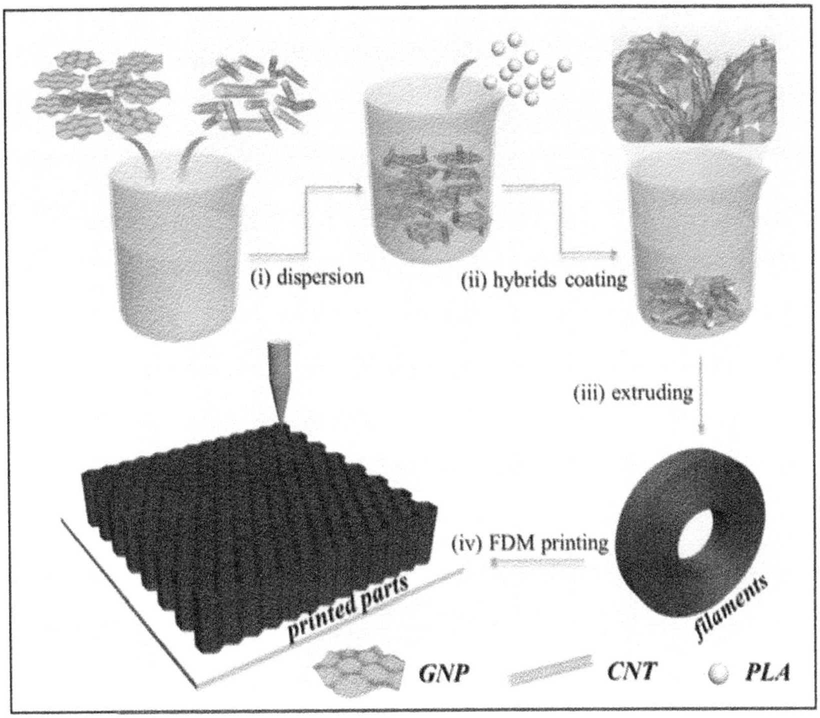

FIGURE 13.7

Graphical representation of the formation of poly(lactic acid), poly(ε-caprolactone)/carbon nanotube (PLAPLA/GNP/CNT) nanocomposite.

Reproduced with the permission from S. Shi, Z. Peng, J. Jing, L. Yang, Y. Chen, 3D printing of delicately controllable cellular nanocomposites based on polylactic acid incorporating graphene/carbon nanotube hybrids for efficient electromagnetic interference shielding. ACS Sustain. Chem. Eng. 8 (2020) 7962–7972.

solid PLLA/CNT disc composite is pulverized to make composite granules. The PLLA/CNT granules are then mixed with PLAsc granules with the aid of a functional grinder. Then, PLLA/CNT/PLAsc composite granules are then compression-molded under a pressure of 10 MPa and 210°C to get the PLLA/CNT/PLAsc composite with different wt.% of CNT (0.2%, 0.3%, 0.5%, 1.0%, and 2.0%). The overall procedure has been described in Fig. 13.9.

It was found out that, PLLA/CNT/PLAsc composite with only 2 wt.% of CNT gave EMI shielding around 30 dB. From Fig. 13.10, we can understand the above fact.

Wang et al. [55] created a microporous or foamed PLA/MWCNT-based polymer nanocomposite MWCNT has formed a 3-D network to increase electrical conductivity and lower the percolation threshold as well as the EMI shielding property. A bead foaming process was inducted to produce the composite. In a typical process, 60–120 mesh-sized PLA particles are subjected to a bead foaming process to generate expanded PLA particles (EPLA) where carbon dioxide is used as the blowing agent. A mixed solvent is prepared by mixing 95 vol.% of waters with 5 vol.% of

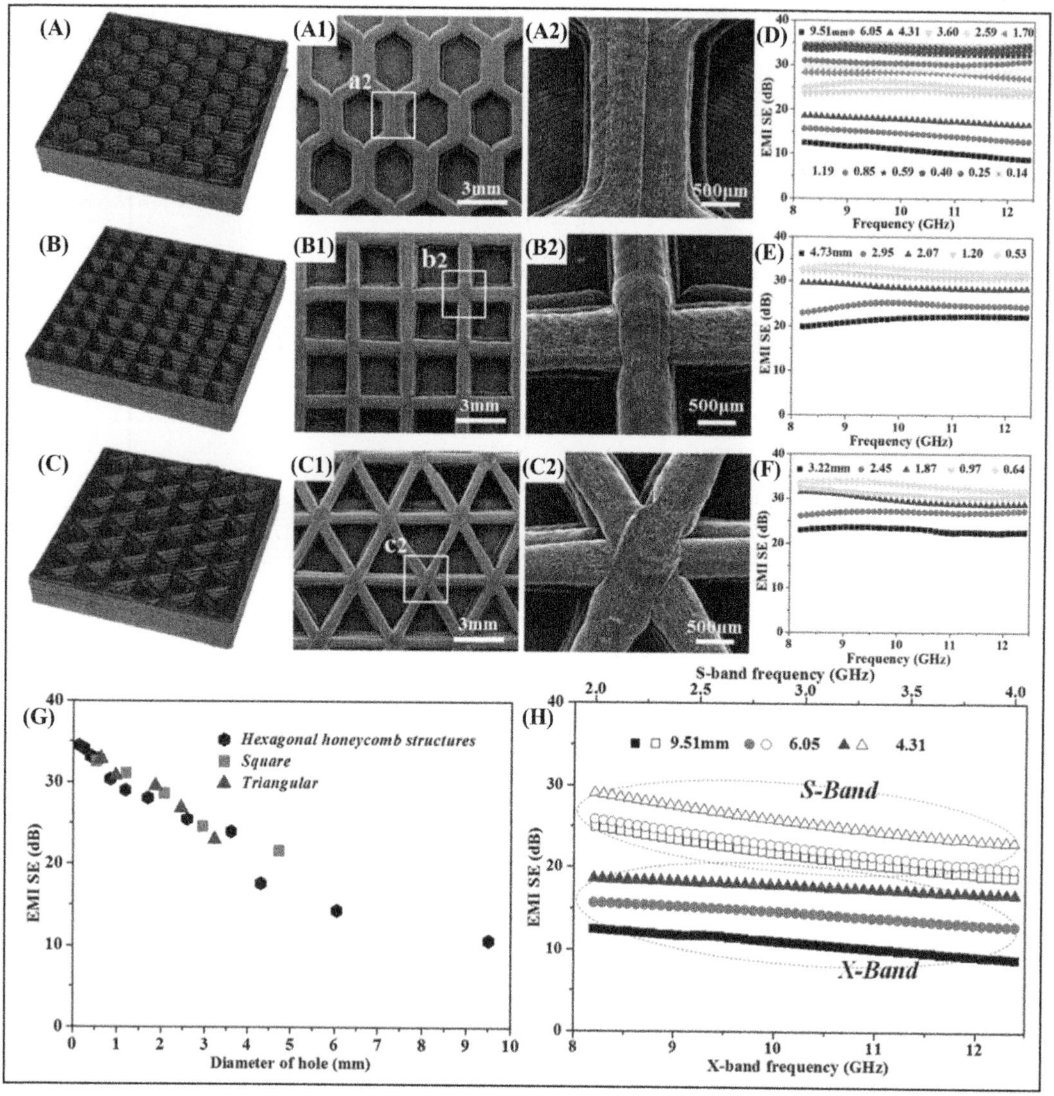

FIGURE 13.8

Graphical representations along with SEM images of three dimensionally printed hexagonal structural composites with different cell morphology: hexagonal (A, A2), square structures (B, B2), and triangular (C, C2); EMI shielding performance of three dimensionally printed composites with various structural geometries of different cell size: hexagonal (D), square structures (E), and triangular (F) in the frequency range 8.4–12.6 GHz; average EMI SE of three dimensionally printed composites with various geometrical structures or cells as a function of the size of it (G); EMI shielding activities of three dimensionally printed parts with hexagonal structures of various sizes within the X- and S-band (H).

Reproduced with the permission from S. Shi, Z. Peng, J. Jing, L. Yang, Y. Chen, 3D printing of delicately controllable cellular nanocomposites based on polylactic acid incorporating graphene/carbon nanotube hybrids for efficient electromagnetic interference shielding. ACS Sustain. Chem. Eng. 8 (2020) 7962–7972.

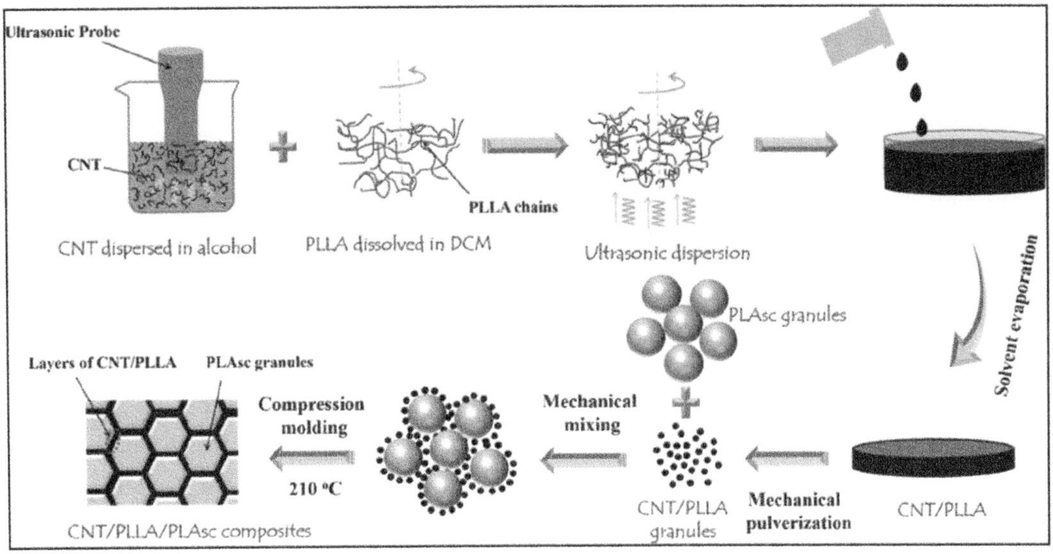

FIGURE 13.9

Graphical representation of the formation of poly(L-lactide)/carbon nanotube/PLA stereo complex (PLLA/CNT/PLAsc) nanocomposites.

Reproduced with the permission from L. Xu, X.-P. Zhang, C.-H. Cui, P.-G. Ren, D.-X. Yan, Z.-M. Li, Enhanced mechanical performance of segregated carbon nanotube/poly (lactic acid) composite for efficient electromagnetic interference shielding Ind. Eng. Chem. Res.58 (2019) 4454–4461.

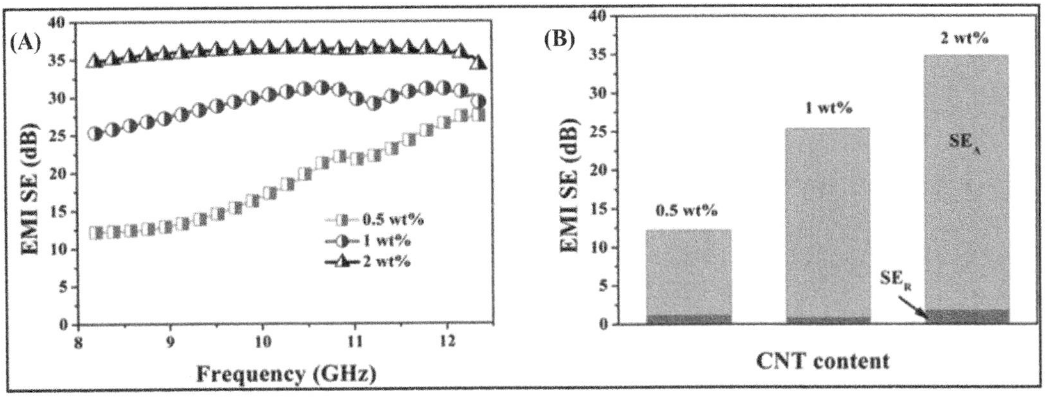

FIGURE 13.10

(A) EMI shielding efficiency of the isolated CNT/PLLA/PLAsc nanocomposites with different CNT loadings, within the X-band. (B) Comparative study of SE_A and SE_R of the isolated composites at a particular frequency (8.2 GHz).

Reproduced with the permission from L. Xu, X.-P. Zhang, C.-H. Cui, P.-G. Ren, D.-X. Yan, Z.-M. Li, Enhanced mechanical performance of segregated carbon nanotube/poly (lactic acid) composite for efficient electromagnetic interference shielding Ind. Eng. Chem. Res.58 (2019) 4454–4461.

dimethylformamide in which the required amount of MWCNT is suspended by ultrasonication method in the water bath at 50°C. The EPLA beads are then suspended in the MWCNT suspension to coat the surface of the EPLA beads with MWCNT. The MWCNT coated EPLA beads are now sintered using a steam chest molding process to produce EPLA/MWCNT composite foams with varying MWCNT volume percentages. The detailed process has been described in Fig. 13.11.

The EMI shielding measurement was conducted between the frequency range of X-band using a VNA following Eq. (13.2). The graphical representation is given in Fig. 13.12.

As we can see from Fig. 13.12, a total electromagnetic interference (EMI) shielding of around 45 dB could be easily achievable. Cui et al. [56] have prepared stereo-complexed poly(lactic acid) (PLA) or PLA stereo complex (scPLA)/carbon nanotube (CNT)-based conductive foam nanocomposite material. The schematic representation is given in Scheme 13.1.

Stereo-complexed PLA is formed by stereo complex crystallization of a mixture of enantiomeric poly(L-lactide) (PLLA) and poly(D-lactide) (PDLA). Fig. 13.13A shows with an increase in filler content foam density decreases and electrical conductivity increases. As the CNT content is increased, the gelation process increases which indulges the phase separation causing a decrease in the density. Fig. 13.13B shows an increase in both EMI SE and specific EMI SE with the increase in filler concentration.

Wang et al. [57] discovered a way to replace indium tin oxide (ITO) with silver nanowire (AgNW) to create a pliable and transparent EMI shielding material by sandwiching the AgNWs between biodegradable gelatin-based films using the rod coating method, and the composites are known as G/Ag NW/G. (G signifies gelatin-based film). The schematic representation is given in Scheme 13.2.

The EMI shielding effectiveness (EMI SE_T) of G/Ag NW/G is shown in Fig. 13.14. It could be seen that with the increase in AgNW concentration (in mg/mm^2) EMI shielding effectiveness increases in the range between 35 and 40 dB for G/Ag NW/G with AgNW concentration at around 250.0 mg/m^2 (Fig. 13.8A) both EMI SE_R and EMI SE_A showed a similar trend (Fig. 13.14C and D). Fig. 13.14B shows that theoretical EMI shielding values changed almost similarly with an increase in sheet resistance (Ω/s).

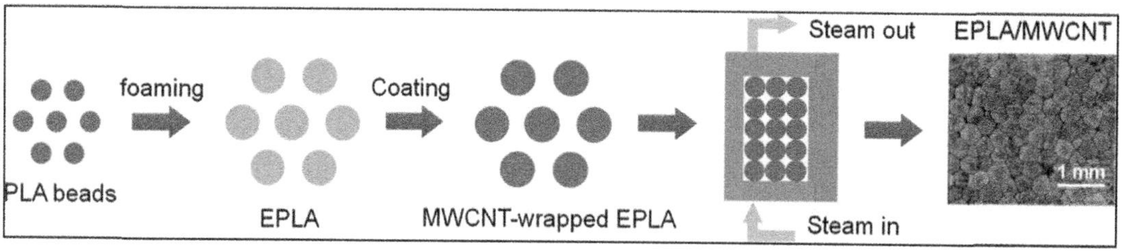

FIGURE 13.11

Stepwise production of expanded PLA particles/multiwalled carbon nanotubes (EPLA/MWCNT) composite foam.

Reproduced with the permission from G. Wang, L. Wang, L. Howe Mark, et al., Ultralow-threshold and lightweight biodegradable porous PLA/MWCNT with segregated conductive networks for high-performance thermal insulation and electromagnetic interference shielding applications. ACS Appl. Mater. Interfaces 10 (2018) 1195–1203.

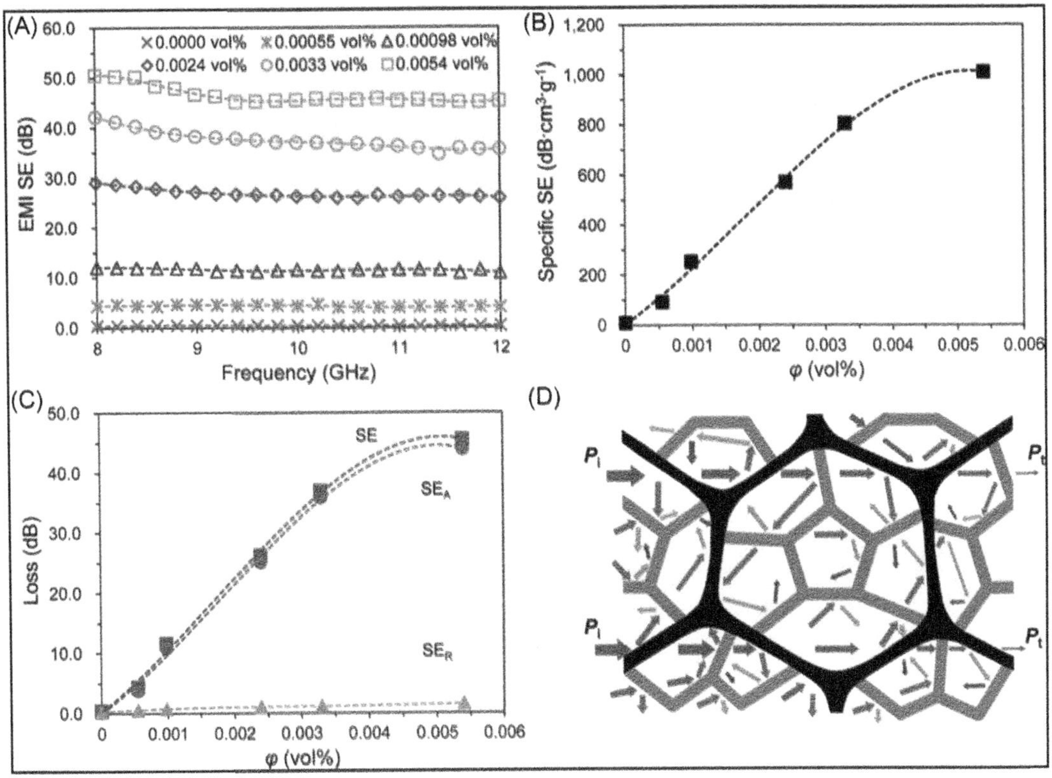

FIGURE 13.12

(A) variation of electromagnetic interference (EMI) shielding activities of the poly(lactic acid) (PLA)/multiwalled carbon nanotube (MWCNT) composites at various MWCNT contents. (B) Variation of the specific shielding activity of poly(lactic acid) (PLA)/MWCNT composites with the variation of MWCNT amount. (C) Variation of SE_T, SE_A, and SE_R on MWCNT amount at a particular frequency, 10 GHz. (D) Mechanism of EMI shielding through the foam composite.

Reproduced with the permission from G. Wang, L. Wang, L. Howe Mark, et al., Ultralow-threshold and lightweight biodegradable porous PLA/MWCNT with segregated conductive networks for high-performance thermal insulation and electromagnetic interference shielding applications. ACS Appl. Mater. Interfaces 10 (2018) 1195–1203.

Fang et al. [58] have developed poly(L-lactide) (PLLA)/MWCNT-based polymer nanocomposite in presence of in situ crosslinked poly(ethylene glycol) diacrylate (CPEGDA) as an impact modifier. The MWCNT particles were preferably dispersed in the PLLA phase, resulting in a CPEGDA domain that was significantly smaller than the PLLA domain. The production method of the composite is described in Scheme 13.3.

The composites are designated as PCM-x where x stands for the content of MWCNT (vol.%) in the composite. Fig. 13.15A gives the EMI SE of the composites and the percolation threshold was achieved with just 0.19 vol.% of MWCNT and a mere 2.4 vol.% of MWCNT had given EMI SE of around

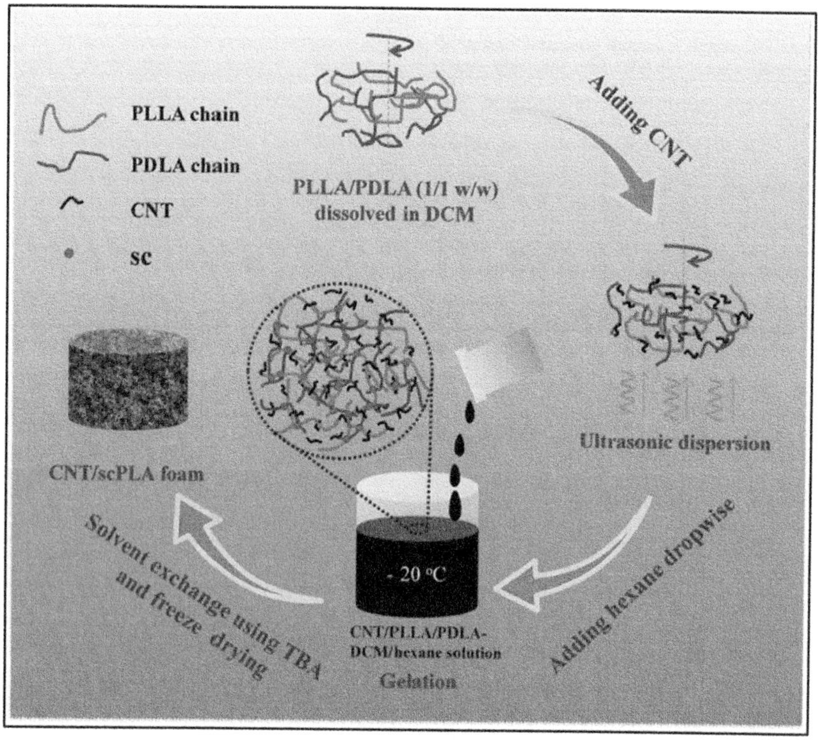

SCHEME 13.1

Graphical representation of production of carbon nanotube/Sc poly(lactic acid) (CNT/Sc PLA) foam composite.

Reproduced with the permission from C.H. Cui, D.-X. Yan, H. Pang, et al., A high heat-resistance bioplastic foam with efficient electromagnetic interference shielding. Chem. Eng. J. 323 (2017) 29–36.

26.4 dB. PCM-3.6 had shown very little improvement in EMI SE compared to that of PCM-2.4. Absorption dominancy of EMI SE becomes more prominent with the increment in MWCNT content (Fig. 13.15B).

Zhang et al. [59] used a straightforward method to create a PLA/Ag composite by coating PLA fibers with silver prior to compression molding the silver coated PLA fibers. PLA/Ag composites are denoted as S-x, where x denotes the time (minutes) of Ag deposition on the PLA fibers' surface. The composite fabrication procedure is depicted in Scheme 13.4.

From Fig. 13.16A it is clear that with an increase in Ag deposition time on the Ag fibers the EMI shielding effectiveness (SE_{Total}) of S-x increases and S-7 has given the SE_{Total} of around 50 dB in the X-band region and absorption dominant (Fig. 13.16B).

PLLA/MWCNT foam-based composite was prepared by Kuang et al. [60]. The "brick and muds" foaming structure manufactured first by mechanical mixing of PLLA and MWCNT followed by hot-pressing and pressure induced flow (PIF) processing is shown below in Scheme 13.5.

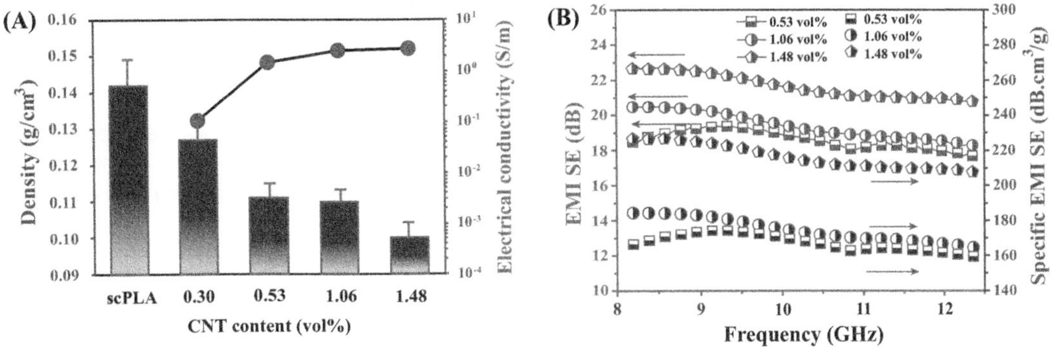

FIGURE 13.13

(A) Change in electrical conductivity and foam density of carbon nanotube/sc poly(lactic acid) (CNT/scPLA) foams with the variation in CNT loadings. (B) Variation in electromagnetic interference shielding effectiveness (EMI SE) and specific EMI SE of CNT/sc poly(lactic acid) (PLAsc (scPLA)) foams with the variation in filler loading.

Reproduced with the permission from C.H. Cui, D.-X. Yan, H. Pang, et al., A high heat-resistance bioplastic foam with efficient electromagnetic interference shielding. Chem. Eng. J. 323 (2017) 29–36.

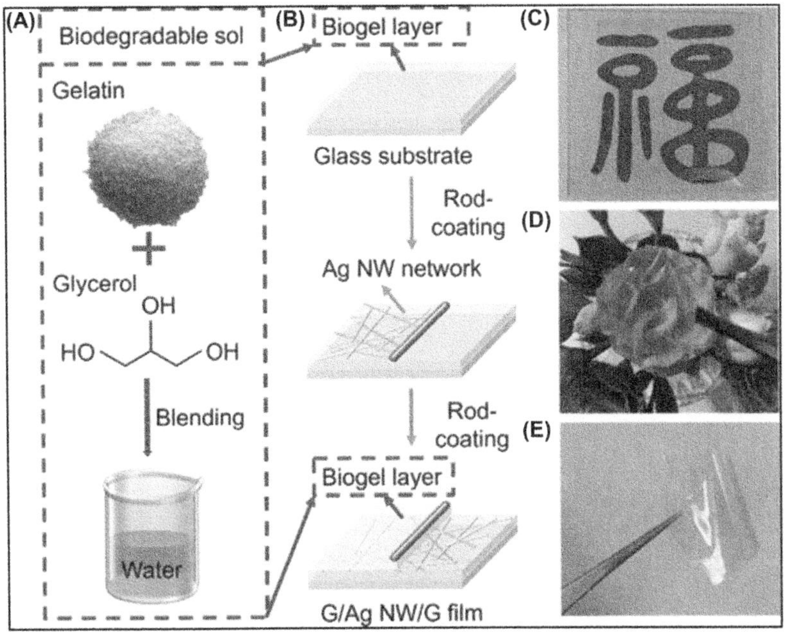

SCHEME 13.2

Graphical representation of (A) biodegradable sol formation, (B) biogel layer formation over rod coated substrate with AgNW, (C) clear digital image could be seen while keeping the G/Ag NW/G above, (D) picturization of G/Ag NW/G film, and (E) flexibility of the G/Ag NW/G film.

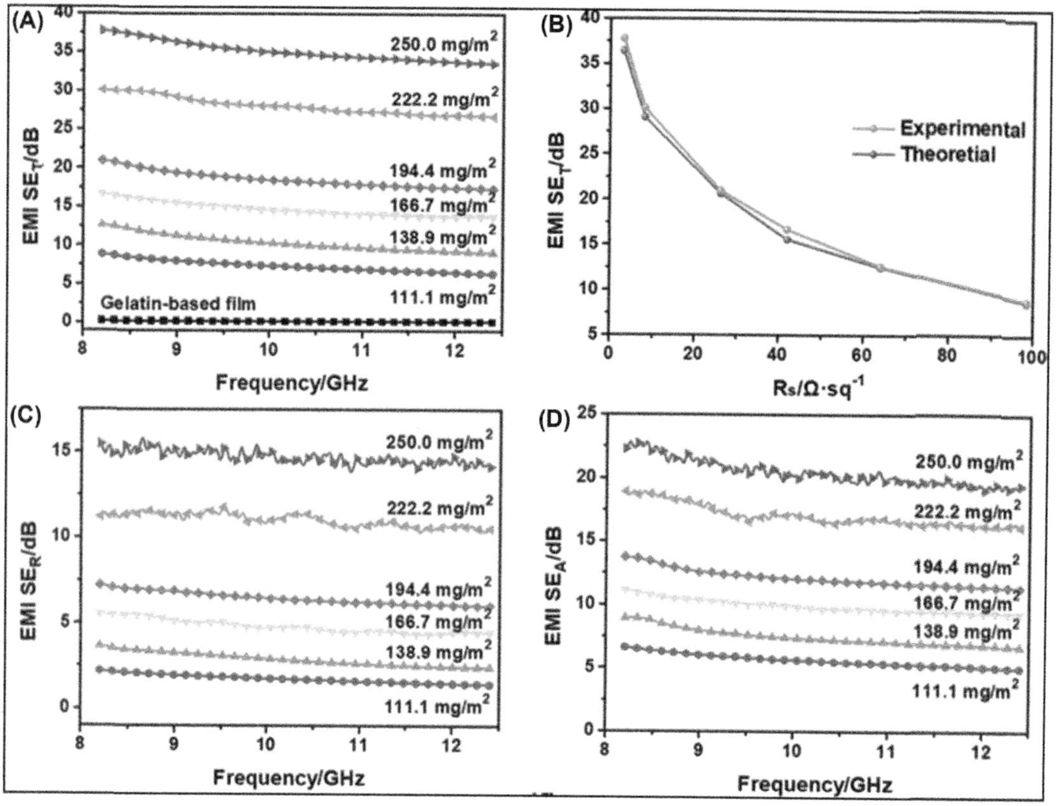

FIGURE 13.14

(A) Electromagnetic interference (EMI) shielding efficiency of gelatin-based film and G/Ag NW/G films. (B) The theoretical calculated and experimental SE value versus sheet resistance. (C) EMI SE_R and (D) EMI SE_A values of different G/Ag NW/G films in the X-band.

Reproduced with the permission from G. Wang, L. Hao, X. Zhang, S. Tan, M. Zhou, W. Gu, et al., Flexible and transparent silver nanowires/biopolymer film for high-efficient electromagnetic interference shielding. J. Colloid Interface Sci. 607 (2022) 89–99.

The EMI shielding effectiveness is depicted in Fig. 13.17A. The foam composite with 10 wt.% of MWCNT content had given shielding effectiveness in the range of 20–25 dB. Absorption dominates over reflection in the contribution to the total shielding effectiveness (Fig. 13.17B).

13.4 Poly(butylene adipate-*co*-terephthalate)-related polymeric electromagnetic inference shielding materials

Synthesis of PBAT involves polycondensation reaction in which adipic acid (AA), butanediol (BDO), and terephthalic acid (PTA) [61] are used as reactants. The procedure used is similar to

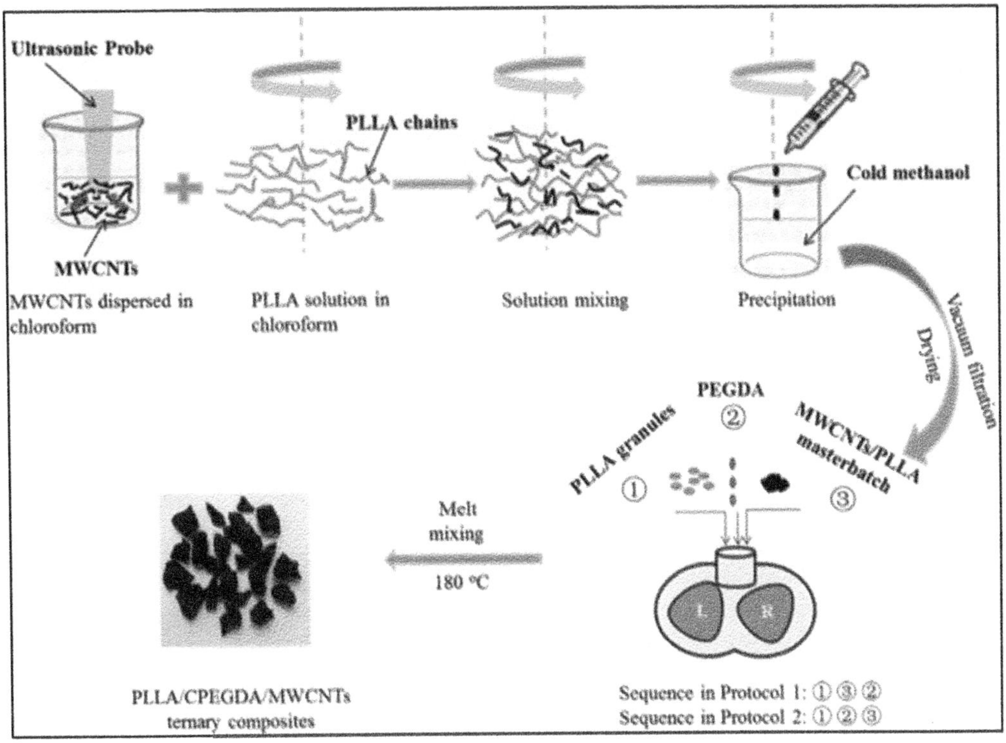

SCHEME 13.3

Graphical representation of the production of the composite poly(L-lactide)(PLLA)/crosslinked poly(ethylene glycol) diacrylate (CPEGDA)/multiwalled carbon nanotubes (MWCNTs).

Reproduced with the permission from H. Fang, S. Wang, W. Ye, X. Chen, X. Wang, P. Xu, et al., Simultaneous improvement of mechanical properties and electromagnetic interference shielding performance in eco-friendly polylactide composites via reactive blending and MWCNTs induced morphological optimization. Compos. B. Eng. 178 (2019) 107452.

traditional polyester production technology. As polycondensation reaction catalysts, organometallic compounds could be used. The polycondensation probably follows the pathway as Fig. 13.18.

The synthetic pathway can be divided into three stages: premixing, prepolymerization, and final polymerization. Commercial production of PBAT is very time-consuming as well as the process requires high temperature (200°C) and vacuum. Water (a byproduct of the reaction) is removed from the reaction mixture by the high temperature. By adding nucleating agents (talc, chalk, mica, etc.) during the polymerization process, the crystallization behavior of PBAT could be improved. Because of the presence of long aliphatic chains, PBAT has much better mechanical properties than other biodegradable polymers such as PLA, and its flexibility is comparable to low-density polyethylene (LDPE) [62,63]. However, the mechanical properties of the polymer are heavily influenced by its monomer composition, that is, PTA content and AA or BDO content. It was discovered that as PTA content in the chain increases, young modulus increases but elongation at break

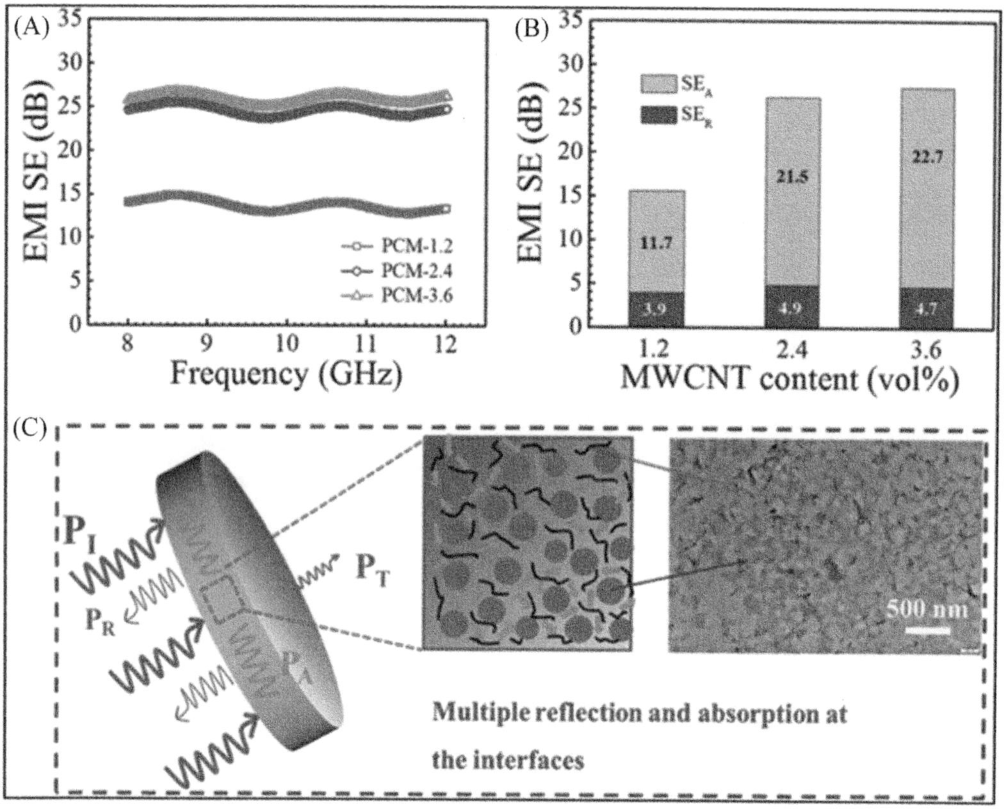

FIGURE 13.15

(A) Change in electromagnetic interference (EMI) SE with frequency (X-band) for the composite PCM-x. (B) Comparative study of SE_A and SE_R at a particular frequency (8.6 GHz). (C) Graphical representation of the EMI SE of PCM-x.

Reproduced with the permission from H. Fang, S. Wang, W. Ye, X. Chen, X. Wang, P. Xu, et al., Simultaneous improvement of mechanical properties and electromagnetic interference shielding performance in eco-friendly polylactide composites via reactive blending and MWCNTs induced morphological optimization. Compos. B. Eng. 178 (2019) 107452.

decreases [64]. Kashi et al. [65] discuss the DC conductivity and EMI shielding activities of PBAT/carbonaceous nanoparticle-based composites. had prepared GNP/PBAT-based nanocomposites with varying filler contents and compared them to GNP/PLLA-based nanocomposites with similar filler compositions as the former using a simple melt-blending process in an internal mixer at the respective processing temperatures (140°C for PLA and 180°C for PBAT). The blending process was continued for 10 minutes and compression-molded in square-shaped sheets with a thickness of 1.0 mm. The EMI shielding results are tabulated below in Figs. 13.19 and 13.20.

In Fig. 13.20, we can see that the total shielding efficiency can move up to 15 dB for PLA/GNP composite with GNP content up to 15% but for PBAT/GNP composite with 15% filler content

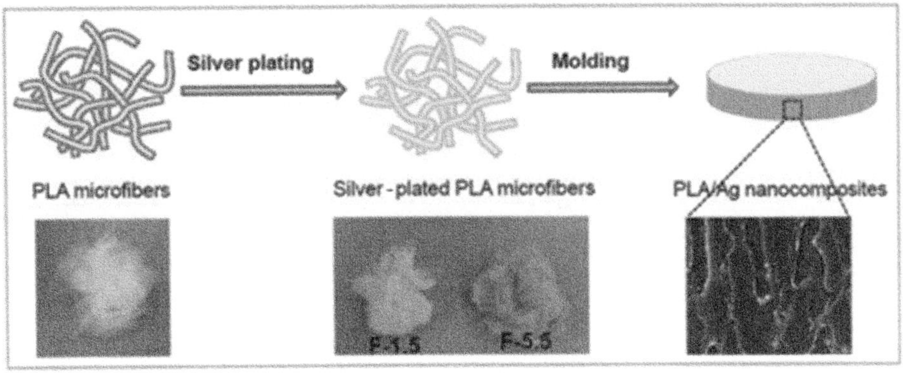

SCHEME 13.4

Graphical representation of the fabrication of poly(lactic acid) (PLA)/Ag composite.

Reproduced with the permission from K. Zhang, H.-O. Yu, K.-X. Yu, Y. Gao, M. Wang, J. Li, et al., A facile approach to constructing efficiently segregated conductive networks in poly (lactic acid)/silver nanocomposites via silver plating on microfibers for electromagnetic interference shielding. Compos. Sci. Technol. 156 (2018) 136–143.

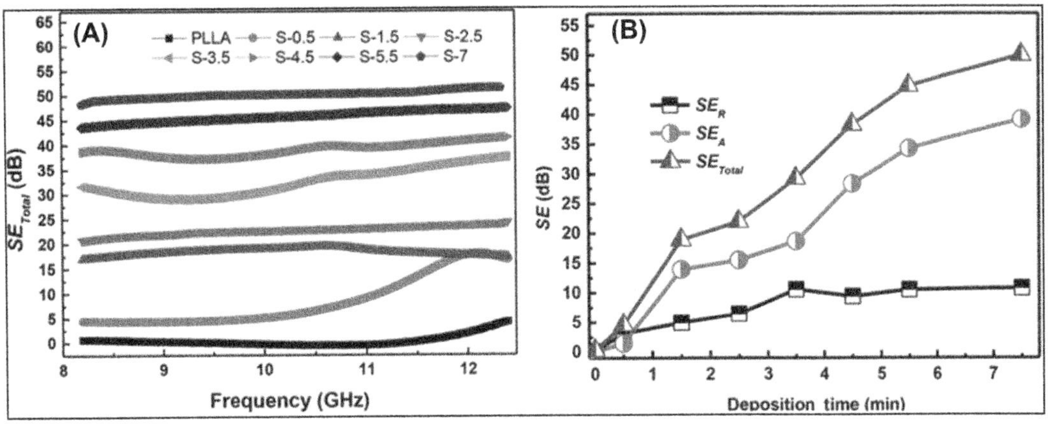

FIGURE 13.16

(A) Electromagnetic interference (EMI) shielding activity of the poly(lactic acid) (PLA)/Ag composites with different Ag accumulation time at different frequency in the X-band region, and (B) comparative study of SE_{Total}, SE_A, and SE_R of the PLA/Ag composites with different silver accumulation time at 9 GHz.

Reproduced with the permission from K. Zhang, H.-O. Yu, K.-X. Yu, Y. Gao, M. Wang, J. Li, et al., A facile approach to constructing efficiently segregated conductive networks in poly (lactic acid)/silver nanocomposites via silver plating on microfibers for electromagnetic interference shielding. Compos. Sci. Technol. 156 (2018) 136–143.

gives near 14 dB. Here the measurement of EMI shielding followed Eq. (13.2) and the thickness was 2.5 mm for all of the test samples. Fig. 13.10 represents the reflection component (SE_R) and absorption component (SE_A) of total EMI shielding efficiency (SE_T). PLA/GNP composites absorb

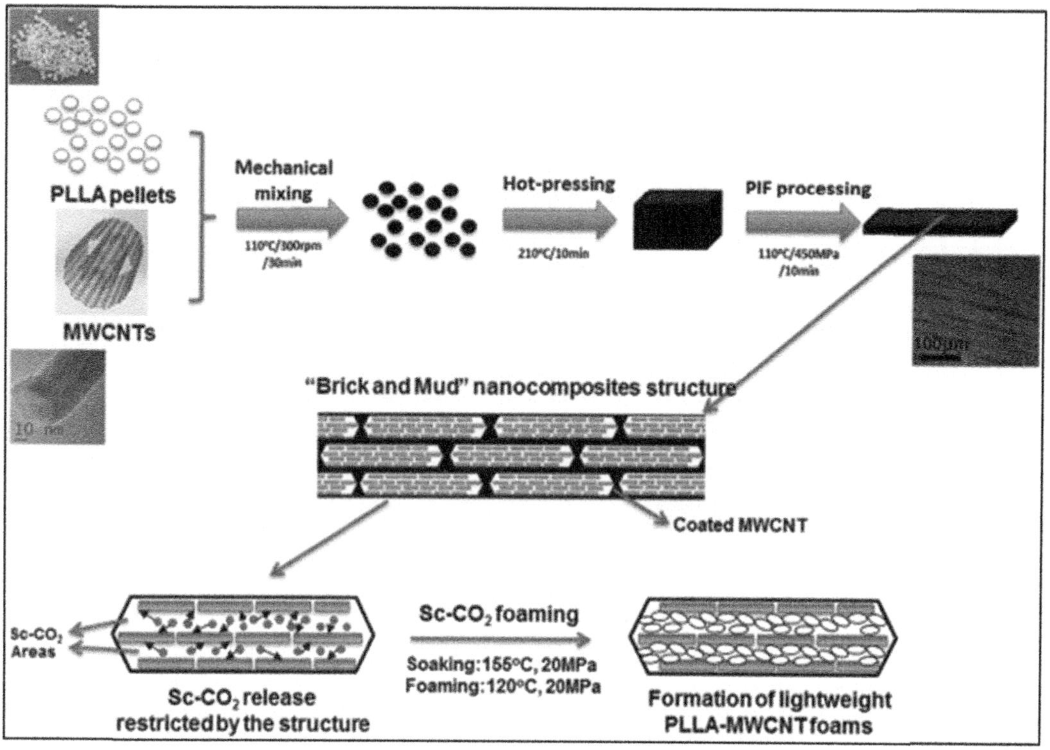

SCHEME 13.5

Graphical representation of poly(L-lactide) (PLLA)/multiwalled carbon nanotube (MWCNT) composite foam.

Reproduced with the permission from T. Kuang, L. Chang, F. Chen, Y. Sheng, D. Fu, X. Peng, Facile preparation of lightweight high-strength biodegradable polymer/multi-walled carbon nanotubes nanocomposite foams for electromagnetic interference shielding. Carbon N. Y.105 (2016) 305–313.

more electromagnetic waves and reflect fewer of them, whereas PBAT/GNP composites are better reflectors of electromagnetic waves than PLA/GNP composites.

13.5 Poly(ε-caprolactone)-related polymeric electromagnetic inference shielding materials

PCL is a kind of aliphatic polyester that is derived from ε-caprolactone by ROP of the former or polycondensation reaction of 6-hydroxyhexanoic acid. The PCL's skeleton is made up of hexanoate repeating units. PCL is a semicrystalline [66] (~70% crystallinity) polymer with excellent mechanical properties, and biodegradability. PCL solvents at room temperature include common organic solvents such as DCM, carbon tetrachloride, benzene, toluene, cyclohexanone, and others. The most common synthetic procedure used previously was enzyme catalyzed condensation of

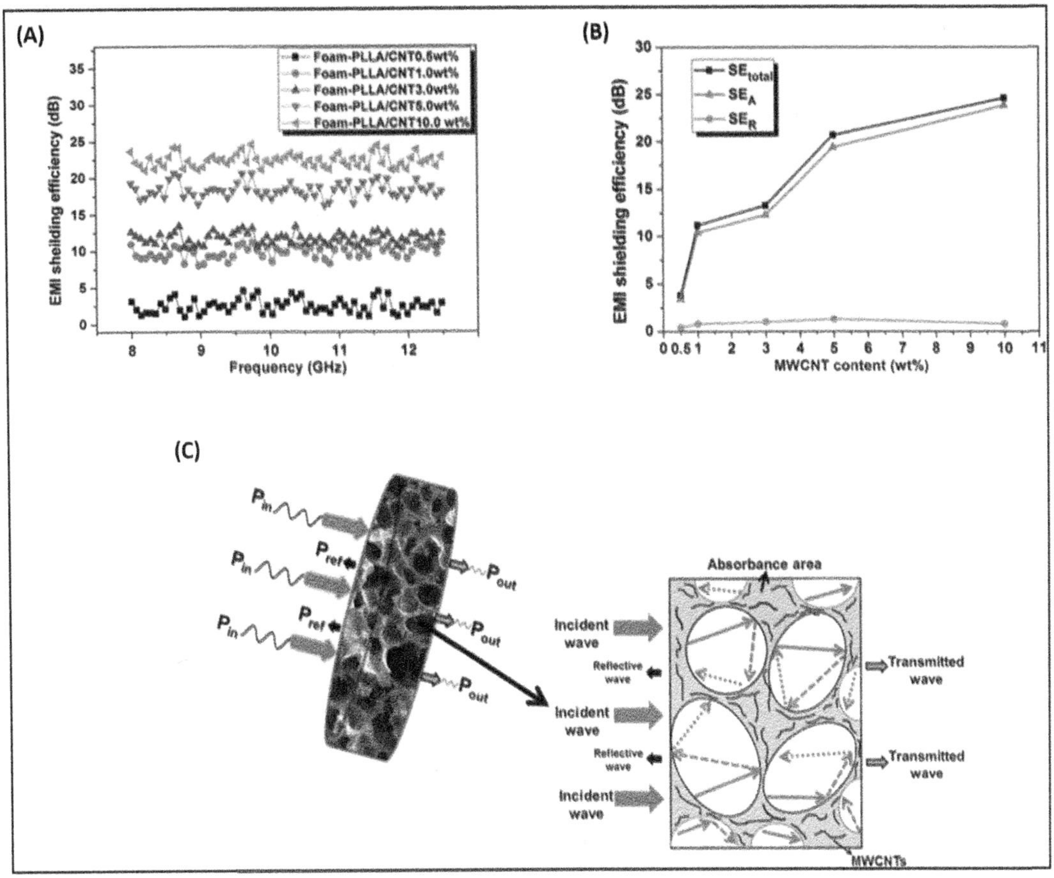

FIGURE 13.17

(A) SE_{total} of poly(L-lactide) (PLLA)/multiwalled carbon nanotubes (MWCNT) composites with various MWCNT wt.% measured in the X-band region; (B) SE_{total}, SE_R, and SE_A of PLLA/MWCNT composites at a frequency of 9.75 GHz; (C) graphical representation of attenuation of electromagnetic wave in the PLLA/MWCNT composite foams.

Reproduced with the permission from T. Kuang, L. Chang, F. Chen, Y. Sheng, D. Fu, X. Peng, Facile preparation of lightweight high-strength biodegradable polymer/multi-walled carbon nanotubes nanocomposite foams for electromagnetic interference shielding. Carbon N. Y. 105 (2016) 305–313.

6-hydroxycaproic acid [67]. In a typical process, a bulk polymerization of 6-hydroxy acid catalyzed by 10% w/w lipase B enzyme (from candida Antarctica) concerning monomer at 90°C gives PCL with an average molecular weight around 9000 g/mol [68]. But the most interesting fact is ROP of ε-caprolactone gives high molecular weight PCL and the polymerization technique follows four different pathways: (1) Anionic ROP, (2) Cationic ROP, (3) Monomer-activated ROP, and (4) Coordination insertion ROP [51]. The anionic ROP entails the attack of anionic species on the carbonyl carbon of the monomer followed by ring-opening, the cationic ROP entails a bimolecular

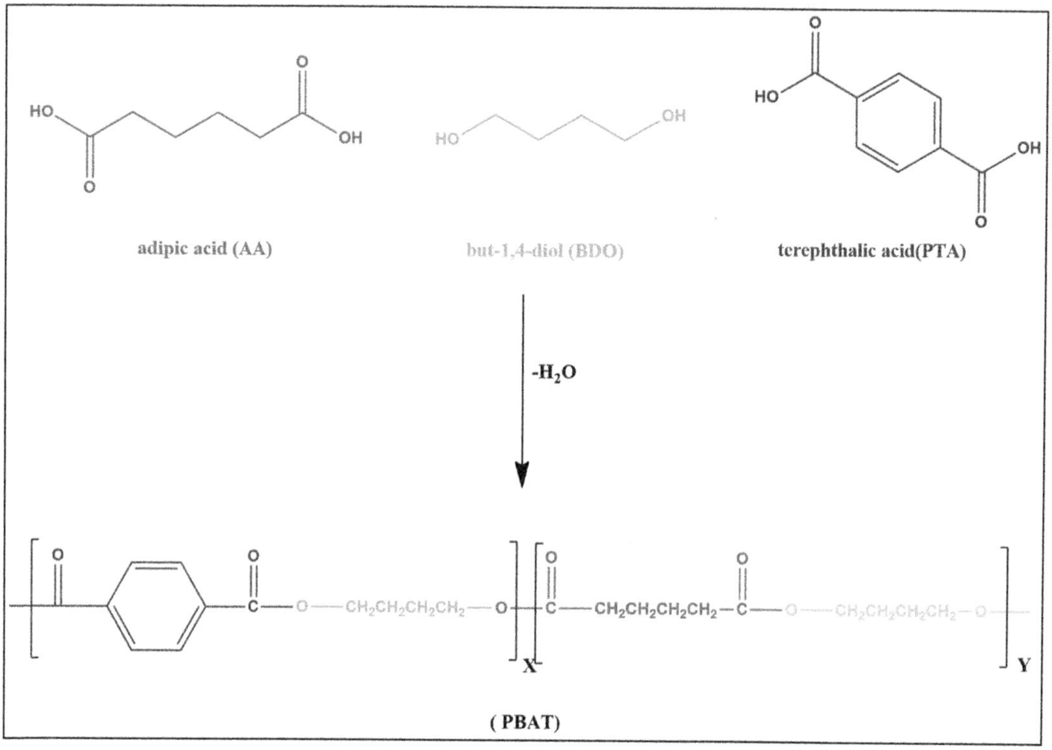

FIGURE 13.18

Synthesis of poly(butylene adipate-*co*-terephthalate) (PBAT).

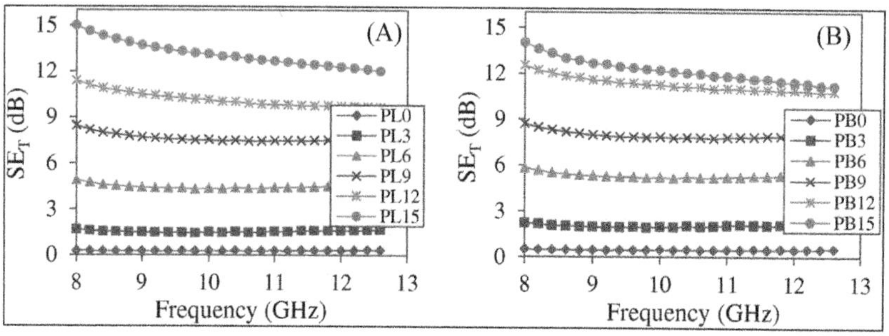

FIGURE 13.19

SE_T of (A) Poly(lactic acid) (PLA)/graphite nanoparticle (GNPO composites (1 mm – thick)) with different GNP contents (0%, 3%, 6%, 9%, 12%, and 15% referred to as PL0, PL3, PL6, PL12, and PL15) and (B) PBAT/GNP composites (thickness = 1.00 mm) with different GNP contents (0%, 3%, 6%, 9%, 12%, and 15% referred to as PB0, PB3, PB6, PB12, and PB15) over X-band.

Reproduced with the permission from S. Kashi, R.K. Gupta, T. Baum, N. Kao, S.N. Bhattacharya, Dielectric properties and electromagnetic interference shielding effectiveness of graphene-based biodegradable nanocomposites. Mater. Des.109 (2016) 68–78.

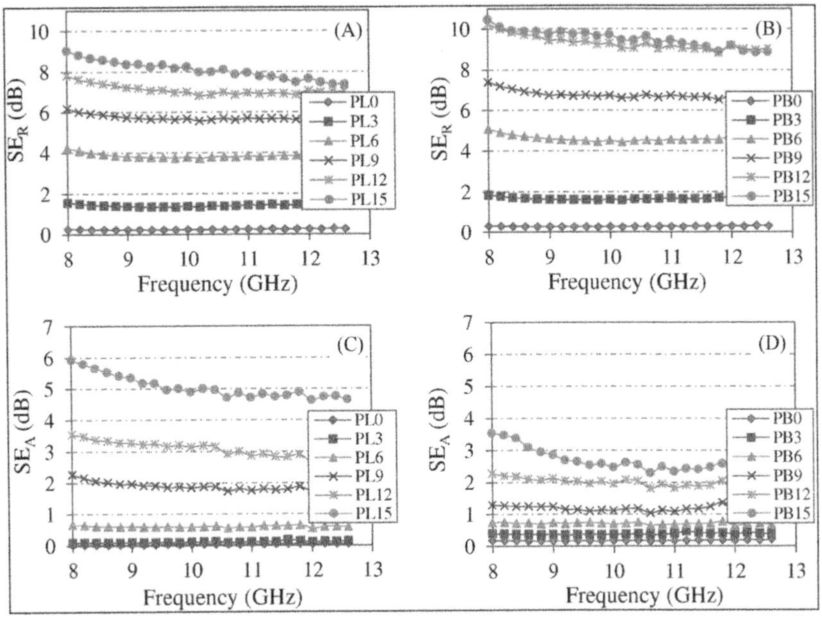

FIGURE 13.20

SE_R of (A) poly(lactic acid) (PLA)/graphite nanoparticle (GNP) composites (1 mm – thick) with different GNP contents (0%, 3%, 6%, 9%, 12%, and 15% referred to as PL0, PL3, PL6, PL12, and PL15) and (B) Poly (butylene adipate-*co*-terephthalate) (PBAT)/graphite nanoparticle (GNP) nanocomposites (1 mm thick) with different GNP contents (0%, 3%, 6%, 9%, 12%, and 15% referred to as PB0, PB3, PB6, PB12, and PB15) and SE_A of (C) PLA/GNP composites (1 mm – thick) with different GNP contents (0%, 3%, 6%, 9%, 12%, and 15% referred to as PL0, PL3, PL6, PL12, and PL15) and (D) PBAT/GNP nanocomposites (1 mm thick) with different GNP contents (0%, 3%, 6%, 9%, 12%, and 15% referred to as PB0, PB3, PB6, PB12, and PB15) over X-band.

Reproduced with the permission from S. Kashi, R.K. Gupta, T. Baum, N. Kao, S.N. Bhattacharya, Dielectric properties and electromagnetic interference shielding effectiveness of graphene-based biodegradable nanocomposites. Mater. Des.109 (2016) 68–78.

substitution nucleophilic reaction in which the carbonyl oxygen of the monomer is attacked by a cationic species, the monomer-activated ROP entails the attack of a monomer-activated catalyst on the polymer chain end, and the coordination insertion R The overall reaction mechanisms are shown in Fig. 13.21.

PCL could be well used as an EMI shield material as recommended by Thomassin et al. [69]. The PCL-based composites are made by inclusion of two types of MWCNTs (thick MWCNT (outer diameter = 25 nm) and thin MWCNT (outer diameter = 10 nm)) into the PCL as the polymer matrix by following: (1) melt mixing method and (2) coprecipitation method. In the melt mixing method, the PCL polymer is blended with the filler MWCNT in a lab-scale microextruder at 200 RPM and 80°C. In the coprecipitation technique, PCL is first dissolved in tetrahydrofuran (THF) and the required amount of MWCNT is dispersed in the polymer solution and dispersion is precipitated as PCL/MWCNT composite after 30 minutes of ultrasonication by a PCL nonsolvent, heptane. The composites have shown promising mechanical properties as depicted by Fig. 13.22.

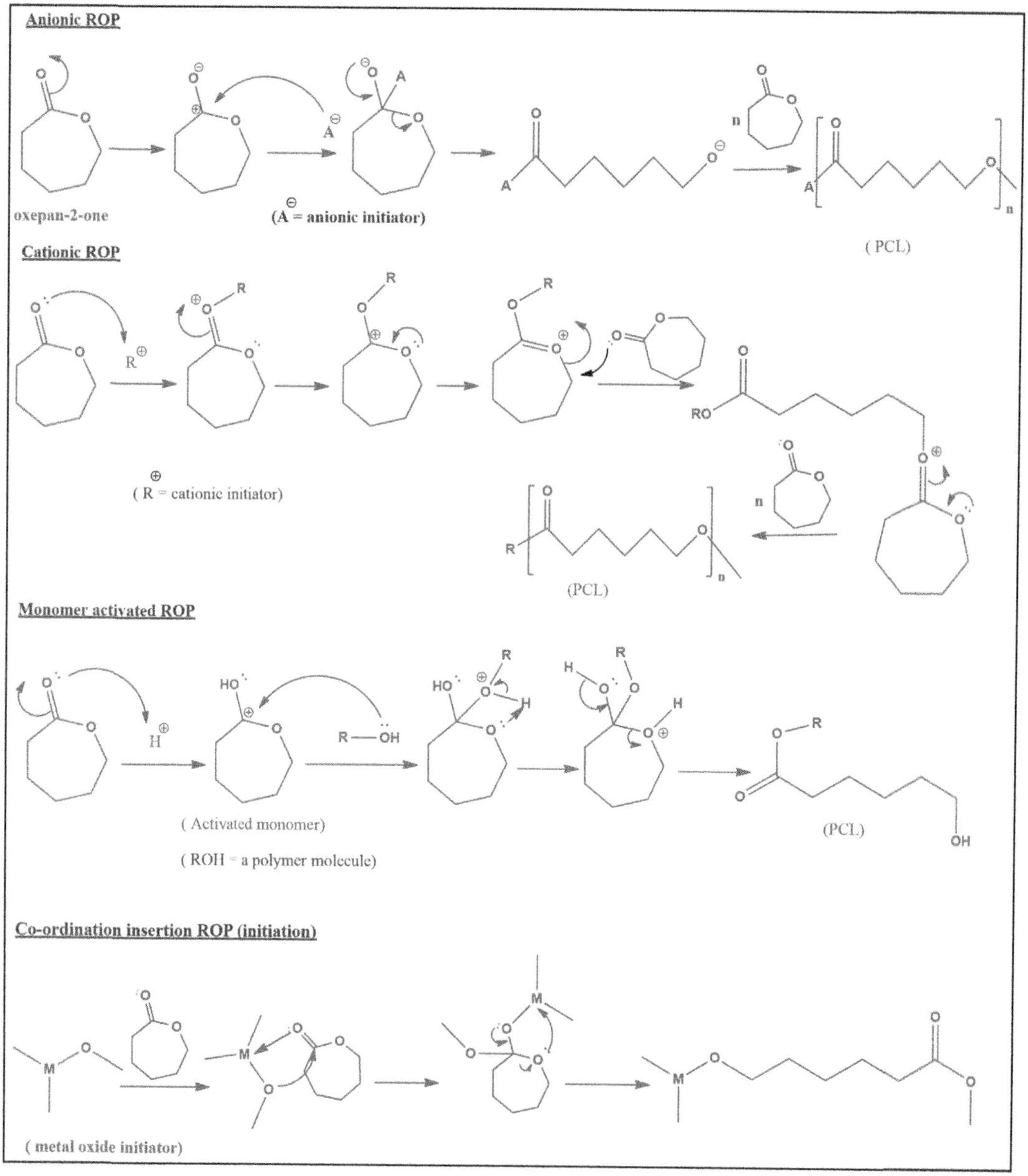

FIGURE 13.21

Different synthetic procedures of poly(ε-caprolactone) (PCL).

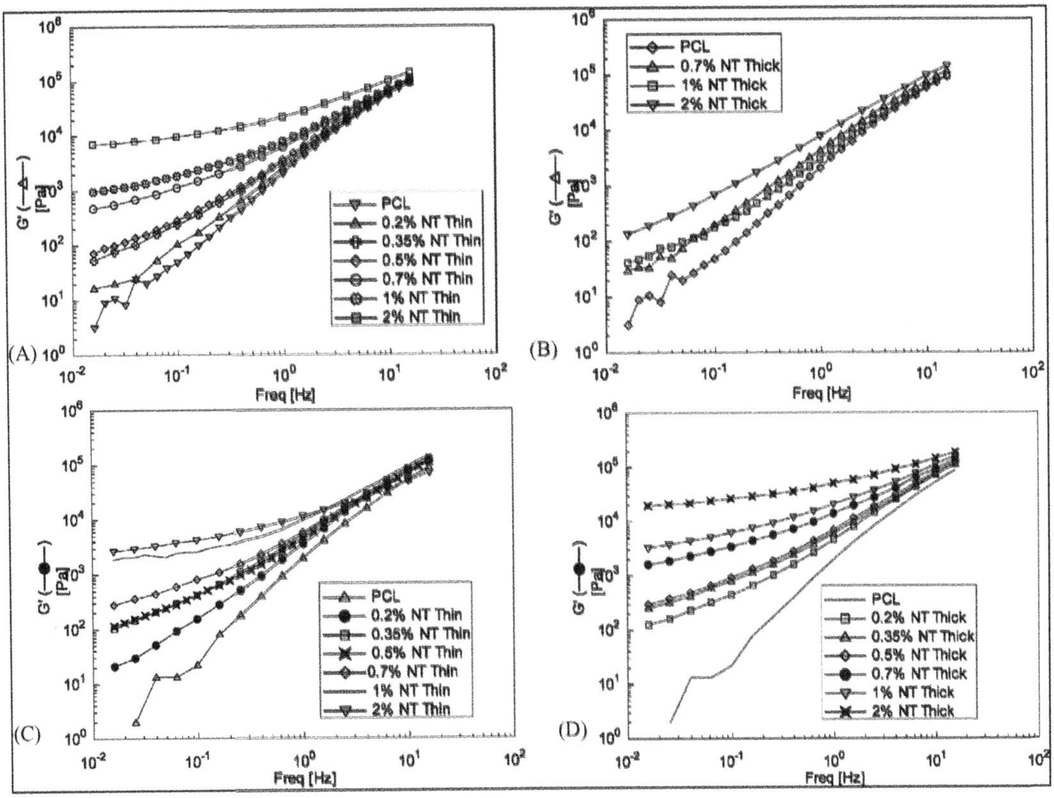

FIGURE 13.22

Change in storage modulus with various multiwalled carbon nanotubes (MWCNT) contents: (A) for thin MWCNTs and (B) for thick MWCNTs incorporated by melt-blending technique, (C) thin MWCNTs incorporated by coprecipitation technique, (D) thick MWCNTs incorporated by coprecipitation technique.

Reproduced with the permission from J.M. Thomassin, X. Lou, C. Pagnoulle, A. Saib, L. Bednarz, I. Huynen, et al., Multi-walled carbon nanotube/poly (ε-caprolactone) nanocomposites with exceptional electromagnetic interference shielding properties. J. Phys. Chem. C. 111(2007) 11186–11192.

As shown in the figure above, PCL/MWCNT-based composites with both thick and thin MWCNTs have a high storage modulus at very high frequencies (~100 Hz). The composites' DC conductivity validates their good EMI shielding properties as well (Fig. 13.23).

The EMI shielding efficiency (range approximately 40 MHz to 40 GHz) of the composites and 1–2 wt.% of both kinds of MWCNT incorporated in the polymer matrix shows electrical conductivity higher than 1 S/m at the higher frequency range. The experiment also confirms the coprecipitation method's superiority over the melt-blending process in terms of the mechanical and electrical properties of the composites. The EMI shielding measurement result shows a larger contribution of absorption component than reflection (frequency range is approximately 25–40 GHz) (Fig. 13.24).

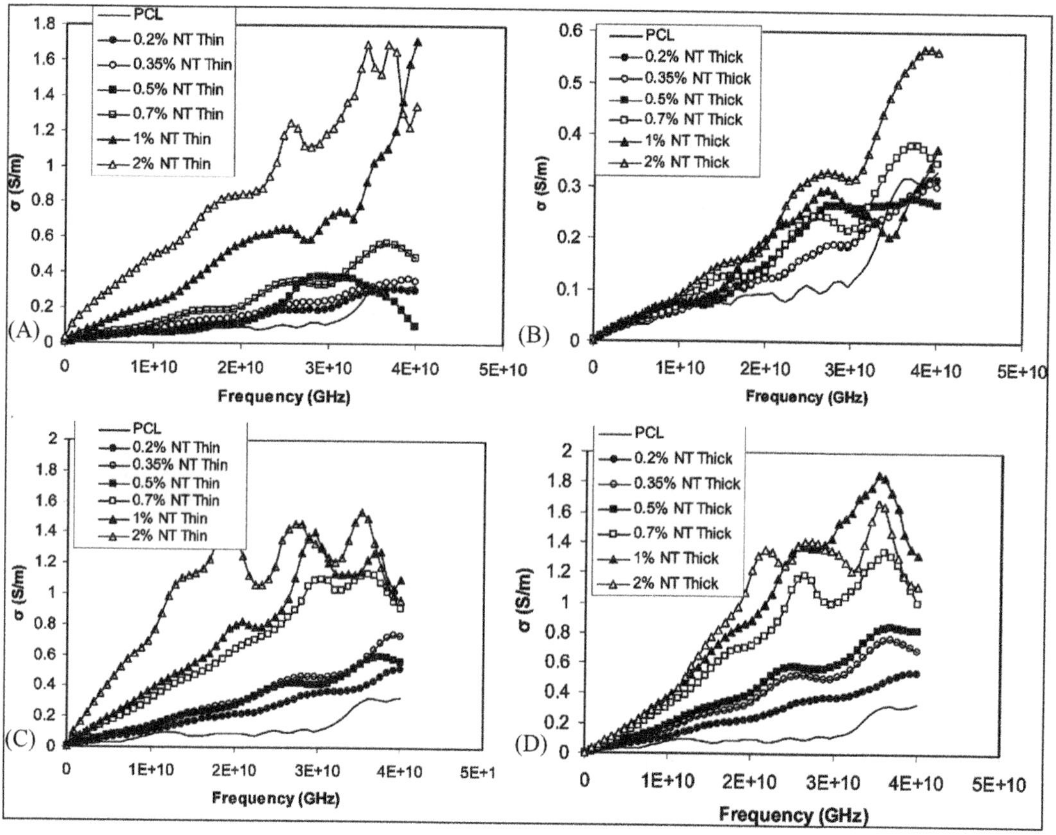

FIGURE 13.23

Change in electrical conductivity with the change in frequency for poly(ε-caprolactone) (PCL)/multiwalled carbon nanotubes (MWCNTs) nanocomposites: (A) thin MWCNTs incorporated by melt-blending technique, (B) thick MWCNTs incorporated by melt-blending technique, (C) thin MWCNTs incorporated by coprecipitation technique, (D) thick MWCNTs incorporated by coprecipitation technique.

Reproduced with the permission from J.M. Thomassin, X. Lou, C. Pagnoulle, A. Saib, L. Bednarz, I. Huynen, et al., Multi-walled carbon nanotube/poly (ε-caprolactone) nanocomposites with exceptional electromagnetic interference shielding properties. J. Phys. Chem. C. 111(2007) 11186–11192.

13.6 Biodegradable epoxy-related polymeric electromagnetic inference shielding materials

Despite their high brittleness, biodegradable epoxy-based polymer composites (e.g., DGEBA) have impressed the scientific community with their good mechanical properties and toughness (i.e., low impact strength). The natural impact modifier (e.g., castor oil) could compensate for the polymer blend's shortcomings prior to composite formation [70]. Park et al. had used epoxidized castor oil as an impact modifier in B-epoxy resin (Fig. 13.25).

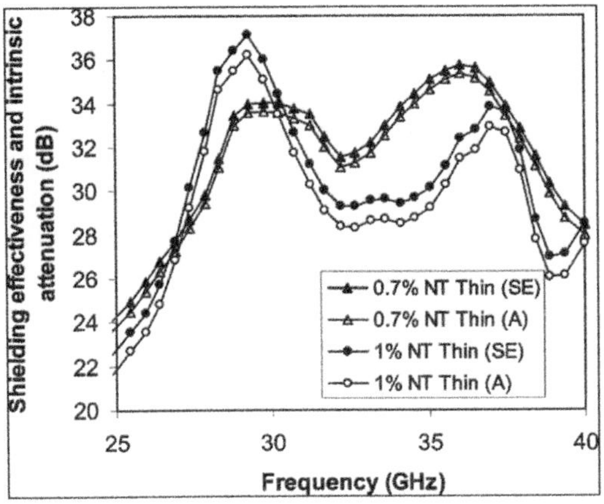

FIGURE 13.24

Electromagnetic interference (EMI) SE and intrinsic absorption A for poly(ε-caprolactone) (PCL) filled with multiwalled carbon nanotubes (MWCNTs).

Reproduced with the permission from J.M. Thomassin, X. Lou, C. Pagnoulle, A. Saib, L. Bednarz, I. Huynen, et al., Multi-walled carbon nanotube/poly (ε-caprolactone) nanocomposites with exceptional electromagnetic interference shielding properties. J. Phys. Chem. C. 111(2007) 11186–11192.

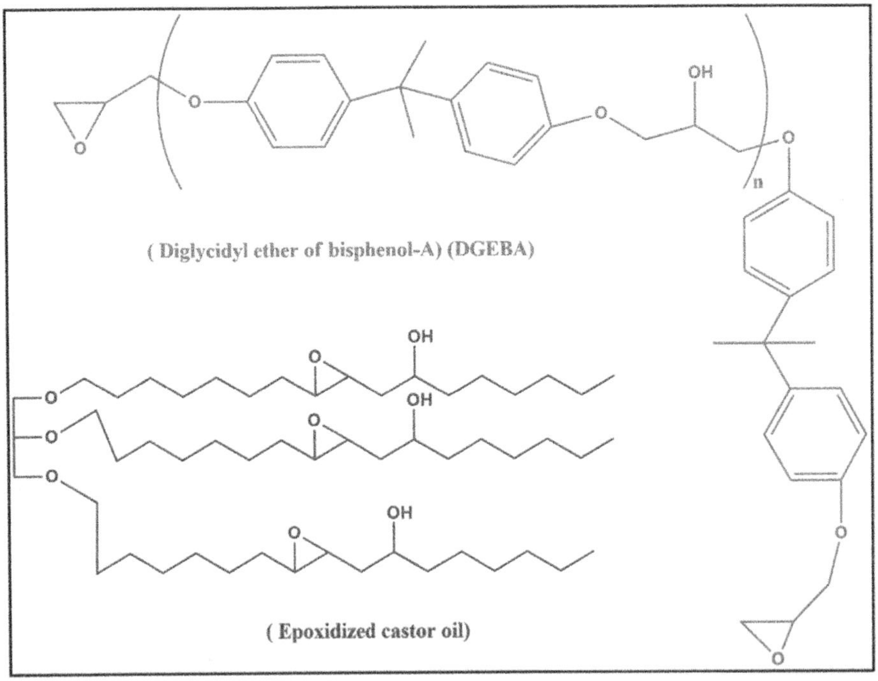

FIGURE 13.25

Bisphenol A diglycidyl ether (DGEBA) and epoxidized castor oil.

Gao et al. has used functionalized MWCNT (f-MWCNT) as a conducting nanoparticle to incorporate into the DGEBA polymer matrix [71]. A polyether amine and organosilane chain mixture was used to functionalize oxidized MWCNT. The primary goal of organically modifying oxidized MWCNT is to improve dispersion and plasticization in the epoxy matrix. In a typical procedure, first, the pristine MWCNT is treated with a mixture of H_2SO_4/HNO_3 at 50°C for 5 hours to get the purified MWCNT. The acidified MWCNT is then treated with a mixture of 3 wt.% polyether amine (M2070) and 0.3 wt.% of organosilane compound (KH560) at 45°C for 24 hours. The f-MWCNT is then dried in the absence of air at 60°C. To obtain the foamed epoxy matrix, a batch foaming process in the presence of supercritical CO_2 is now used. A predetermined amount of epoxy resin (DGEBA) with a curing agent (E4MZ) and a variable amount of off-MWCNT is dissolved in DCM in a typical process. The mixture was then powdered and subjected to compression molding at 160°C and 10 MPa to get the composite material epoxy/f-MWCNT. Now, the epoxy/f-MWCNT composite materials are subjected to batch foaming process by placing them inside a high-pressure autoclave at 50°C and 5 MPa for 48 hours. Then, the composites are foamed at 110°C at an oil bath. The overall procedure has been represented in Fig. 13.26.

The f-MWCNT/epoxy and acid-modified MWCNT/epoxy composites with 20 wt.% of MWCNT content have given EMI shielding effectiveness in the range of 15–20 dB and 20–25 dB, respectively.

FIGURE 13.26

Schematic representation of fabrication of epoxy/f-multiwalled carbon nanotubes (f-MWCNTs) composite foam material.

Reproduced with the permission from Q. Gao, G. Zhang, X. Fan, H. Zhang, Y. Zhang, F. Huang, et al., Enhancements of foamability, electromagnetic interference shielding and mechanical property of epoxy microcellular composite foam with well-dispersed f-MWCNTs. Compos. Part. A Appl. Sci. 138 (2020) 106060.

Whereas, -MWCNT/epoxy and acid-modified MWCNT/epoxy foams have also given EMI SE (measured in 8–12 GHz) 15–20 dB and 20–25 dB, respectively. The frequency range for the measurement was between 8 and 12 GHz. It was discovered that functionalizing MWCNT improves electromagnetic wave absorbability when compared to reflection. Metal carbides or carbonitrides (MXenes) have advanced significantly in terms of electrical conductivity and EMI shielding [72,73]. Wang et al. [74] have produced $Ti_3C_2T_x$ MXene foam/C (MCF) by sol-gel method in which Ti_3AlC_2 powder is added to the solution containing LiF and 9 M HCl and the reaction is maintained at 35°C for 24 hours. followed by a few cycles of ultrasonication, centrifugation, and decantation and finally freeze drying to get $Ti_3C_2T_x$ MXene which was then dispersed in distilled water and mixed with a predetermined amount of resorcinol and formaldehyde in presence of sodium carbonate as a catalyst and the mixture is sonicated. The mixture is then cured at 90°C for 5 hours. under the environment of nitrogen. And the resulting gel is then heated to and kept in 400°C for 2 hours. under a mixture of Ar and 5% H_2 environment to obtain $Ti_3C_2T_x$ MXene foam/C (MCF). The epoxy along with its curing agent are mixed at 70°C for 1 hour. The mixture is then poured into the MCF foam by employing a vacuum-assisted impregnation procedure and cured at 120°C for 5 hours. to get MCF/epoxy nanocomposites. The overall procedure is shown in Fig. 13.27.

The MCFs with different compositions of MXenes were prepared (MCF-0 (2.61 wt.% CF), MCF-1 (2.61 wt.% CF + 0.36 MXene), MCF-2 (2.61 wt.% CF + 0.67 MXene), MCF-3 (2.61 wt.%

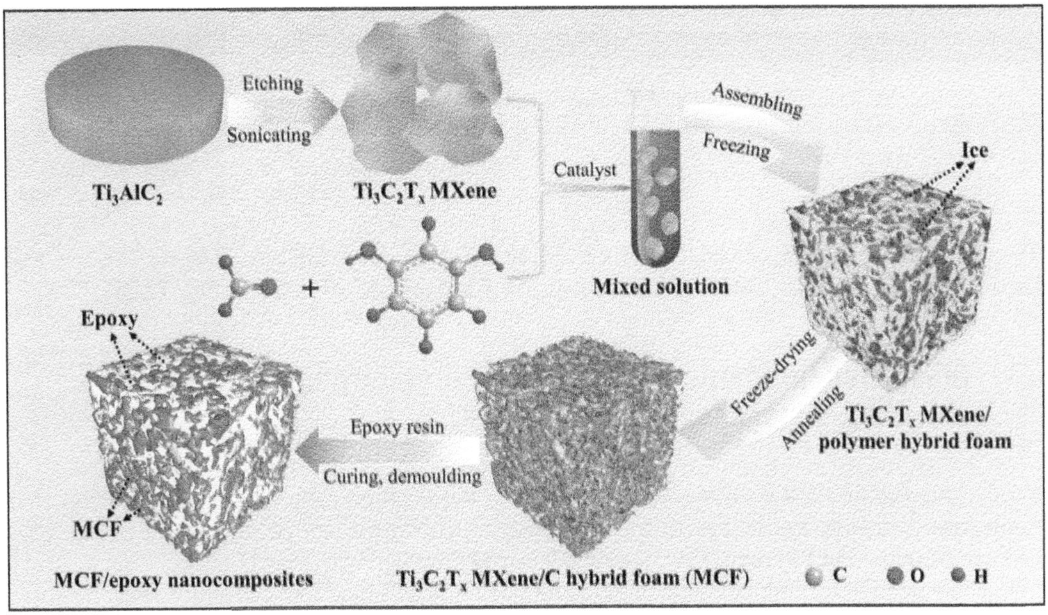

FIGURE 13.27

Schematic representation of construction procedure for MXene foam/C (MCF)/epoxy electromagnetic interference (EMI) shielding nanocomposites.

Reproduced with the permission from L. Wang, H. Qiu, P. Song, Y. Zhang, Y. Lu, C. Liang, et al., 3D Ti3C2Tx MXene/C hybrid foam/epoxy nanocomposites with superior electromagnetic interference shielding performances and robust mechanical properties. Compos. Part. A Appl. Sci. 123 (2019) 293–300.

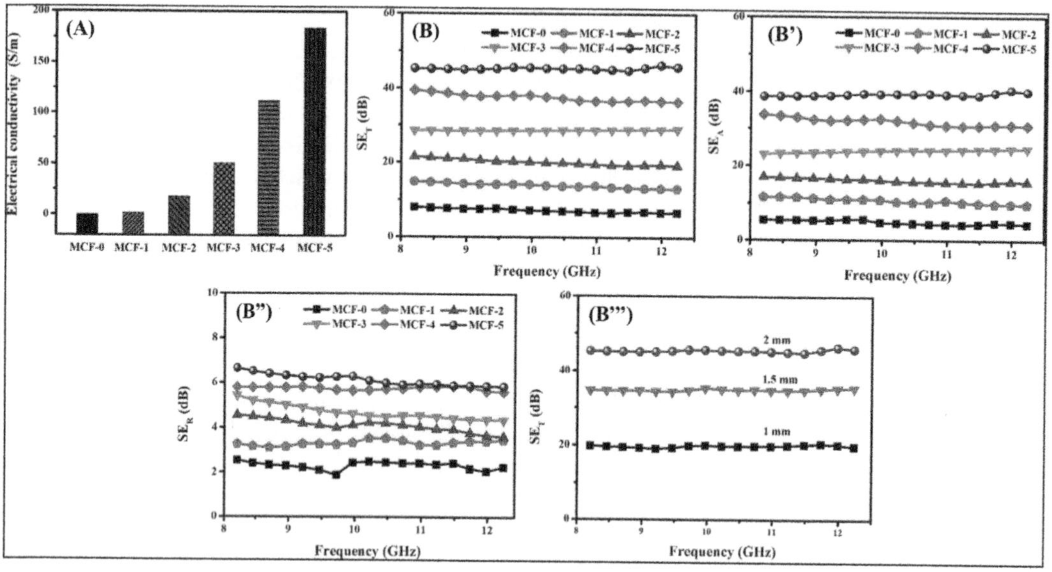

FIGURE 13.28

(A) Electrical properties of the MXene foam/C (MCF)/epoxy composites; (B) SE_T, (B′) SE_A, (B″) SE_R for MCF/epoxy EMI shielding nanocomposites and (B‴) electromagnetic interference (EMI) SE of the MCF-5/epoxy composites at X-band with various thickness, respectively.

Reproduced with the permission from L. Wang, H. Qiu, P. Song, Y. Zhang, Y. Lu, C. Liang, et al., 3D Ti3C2Tx MXene/C hybrid foam/epoxy nanocomposites with superior electromagnetic interference shielding performances and robust mechanical properties. Compos. Part. A Appl. Sci. 123 (2019) 293–300.

CF + 0.98 MXene), MCF-4 (2.61 wt.% CF + 1.34 MXene), and MCF-5 (2.61 wt.% CF + 1.64 MXene), respectively) and EMI shielding values are measured and depicted in Fig. 13.28.

13.7 Other biodegradable polymer-related electromagnetic inference shielding materials

As we discuss the technological advancements in composite manufacturing on behalf of MXenes as the filler material, we will continue to do so in the current section to get a sense of the current state of composite research. Recent research in composite science has been progressing at a good pace thanks to MXenes and Liu et al. [75]. has developed MXene/chitosan-based polymer composites to achieve an effective EMI shield material. Here the $Ti_3C_2T_x$ MXenes are prepared by etching Al on the Ti_3AlC_2 precursor using a solution of LiF in HCl solution. 1 wt.% chitosan (CS) solution is prepared by adding 1.0 g of CS in 100 mL of 2 wt.% acetic acid solution as an initial step to prepare the CS/MXene films. Following that, CS/MXene films are created by combining different weight ratios of CS with MXene while keeping the total solid content of the composites constant.

The EMI SE activities (in X-band) of the composites are measured and the results are presented in Fig. 13.29.

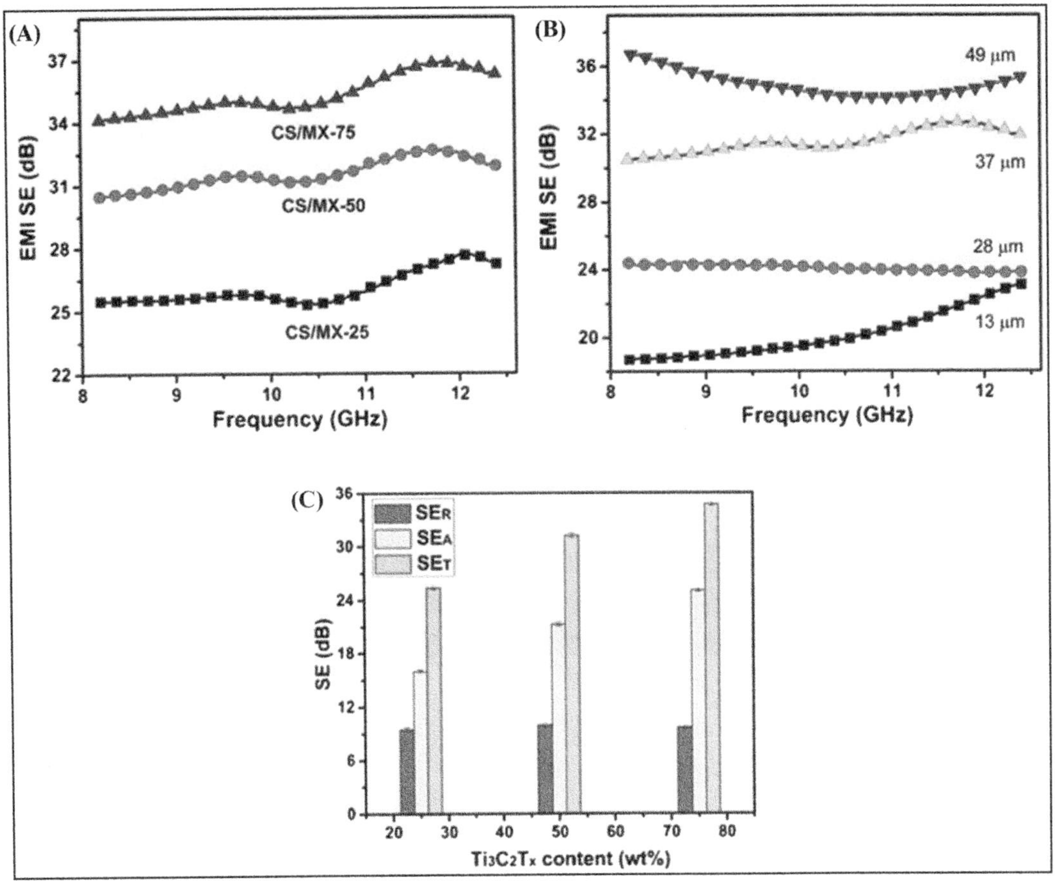

FIGURE 13.29

(A) Electromagnetic interference (EMI) SE activities of the CS/MXene composite films with change in Ti_3C_2Tx content (film thickness = 37 μm), (B) EMI SE activities of the CS/MX-50 film with the change in thickness (C) SE_R, SE_A, and SET of CS/MXene films with the change in Ti_3C_2Tx content at 10.3 GHz.

Reproduced with the permission from F. Liu, Y. Li, S. Hao, Y. Cheng, Y. Zhan, C. Zhang, et al., Well-aligned MXene/chitosan films with humidity response for high-performance electromagnetic interference shielding. Carbohydr. Polym. 243 (2020) 116467.

As we can see from Fig. 13.29C that the EMI shielding mechanism is basically absorption which is directly proportional to MXene content in the composite films. Yin et al. [76] has developed polyvinyl alcohol (PVA)/MXene-based composite, that is, Ti_2CT_x MXene/poly(vinyl alcohol) composite foams produced by layer structured Ti_2CT_x (f-Ti_2CT_x) MXene and poly(vinyl alcohol) (PVA) are constructed via a flexible freeze-drying method. The EMI shielding value reaches about 5136 dB cm^2/g. Hong et al. [77] have reinforced polybutylene succinate (PBS) by the inclusion of bamboo fibers which were modified with Fe_3O_4 nanoparticles and polypyrrole to biocomposite termed as "EMS." The synthetic procedure for the PDA-Fe_3O_4 coating is shown below in Scheme 13.6.

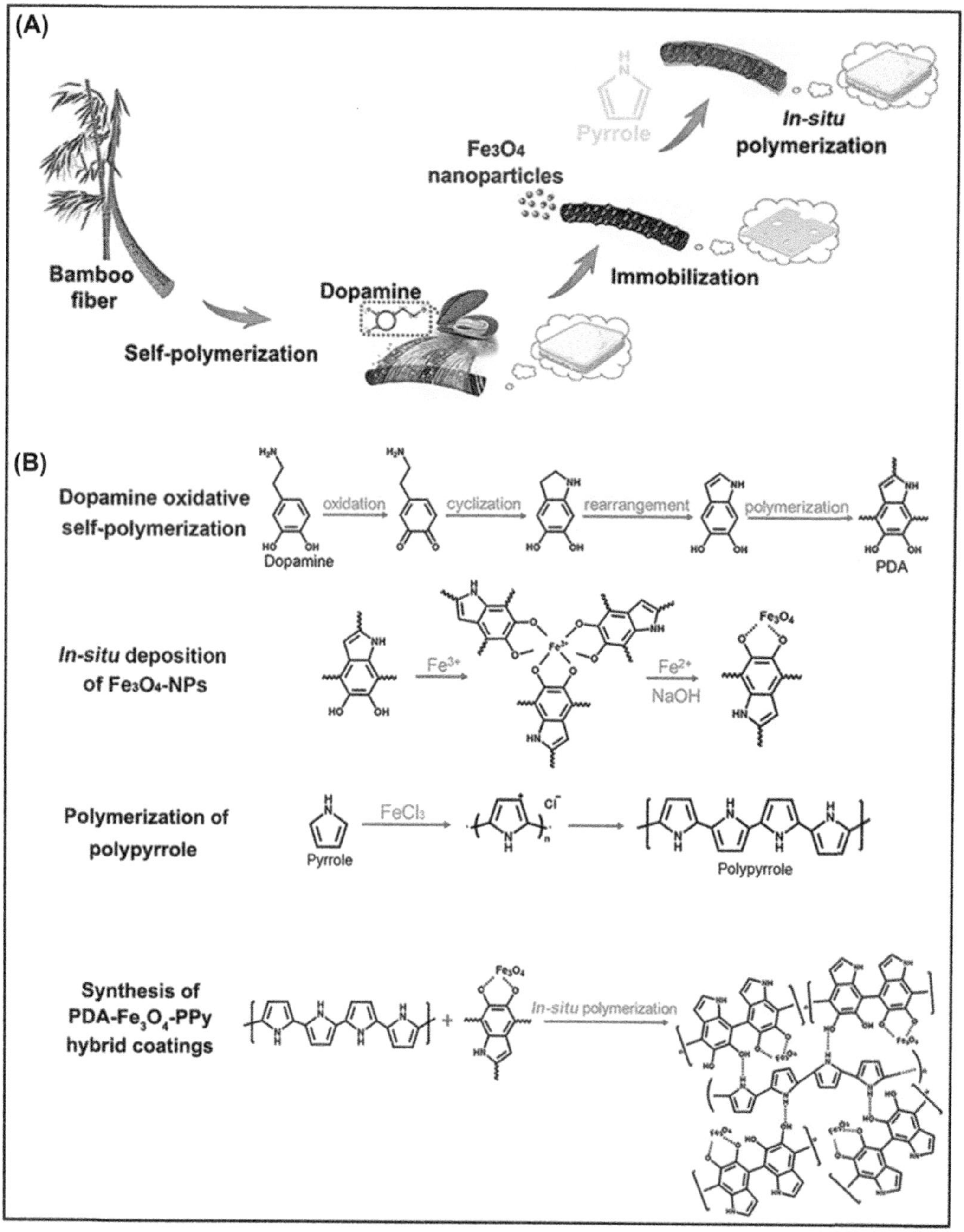

SCHEME 13.6

(A) In situ coating formation diagram on bamboo fibers (B) chemical synthesis of PDA-Fe_3O_4-Ppy coating.

Reproduced with the permission from G. Hong, H. Cheng, S. Zhang, OJ Rojas, Polydopamine-treated hierarchical cellulosic fibers as versatile reinforcement of polybutylene succinate biocomposites for electromagnetic shielding. Carbohydr. Polym. 277 (2022) 118818.

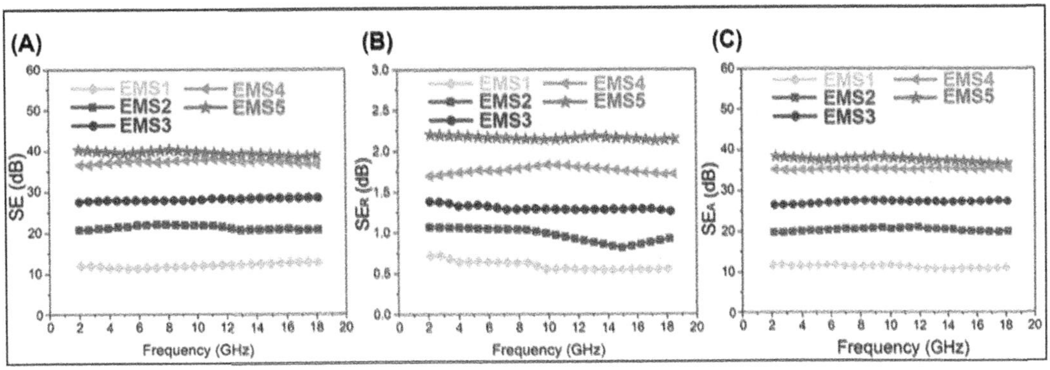

FIGURE 13.30

(A) Shielding effectiveness of PBS-based biocomposite; (B) SE$_R$ of PBS-based biocomposite; and (C) SE$_A$ of PBS-based biocomposite.

Reproduced with the permission from G. Hong, H. Cheng, S. Zhang, OJ Rojas, Polydopamine-treated hierarchical cellulosic fibers as versatile reinforcement of polybutylene succinate biocomposites for electromagnetic shielding. Carbohydr. Polym. 277 (2022) 118818.

The results are shown in Fig. 13.30. As shown in the figure, as the amount of Fe_3O_4 nanoparticles and polypyrrole in the bamboo fiber increases, the electromagnetic radiation absorption increases (the highest amount of Fe_3O_4 nanoparticles and polypyrrole is referred to as EMS5), contributing the most to the overall EMI shielding effectiveness.

Qian et al. [78] have prepared "egg-box" type structural composite films in which carbonized cellulose microspheres (CCM) or "eggs" are encapsulated within MXene or $Ti_3C_2T_x$ layer or "box" and different composites were prepared by varying MXene contents keeping CCM wt.% constant by annealing (termed as "CVMF"), and the composite film before annealing is termed as "CMF." The scheme for the formation of "egg-box" is shown below in Scheme 13.7.

The composite films having the designation of CMF$_X$ and CVMF$_X$ denote the ratio of CCM wt.% to MXene wt.% being 1:X without or with annealing treatment respectively. Pure MXene films after annealing and before annealing are designated as aMF and pMF, respectively. The electrical and EMI shielding properties are shown in Fig. 13.31.

The EMI shielding efficiency of CVMF composites was demonstrated to be significantly higher than that of CMF film composites. The significant improvement in electrical conductivity for annealed MXene and composite films may be attributed to increased carbon content, decreased fluorine, and increased miscellaneous hydrophilic groups on the surface as a result of the annealing process. Wan et al. [79] used a solution mixing process to create hybrid aerogels with GO and cellulose to create lightweight nanocomposites. The chemical interaction of a cellulose molecule with GO-nanosheet and schematic procedure to produce GO/CA composite is shown in Scheme 13.8.

If the GO in the aerogels could be reduced followed by pyrolysis of the aerogels under an inert environment, increase the EMI shielding efficiency of the pyrolyzed composite up to 58.4 dB. In Fig. 13.32 where pyrrolized carbon aerogels (PCNA), graphene nanosheets (GN), and CNA signify pyrolyzed pure cellulose aerogel, graphene nanosheet, and carbon aerogel respectively. and PCNA

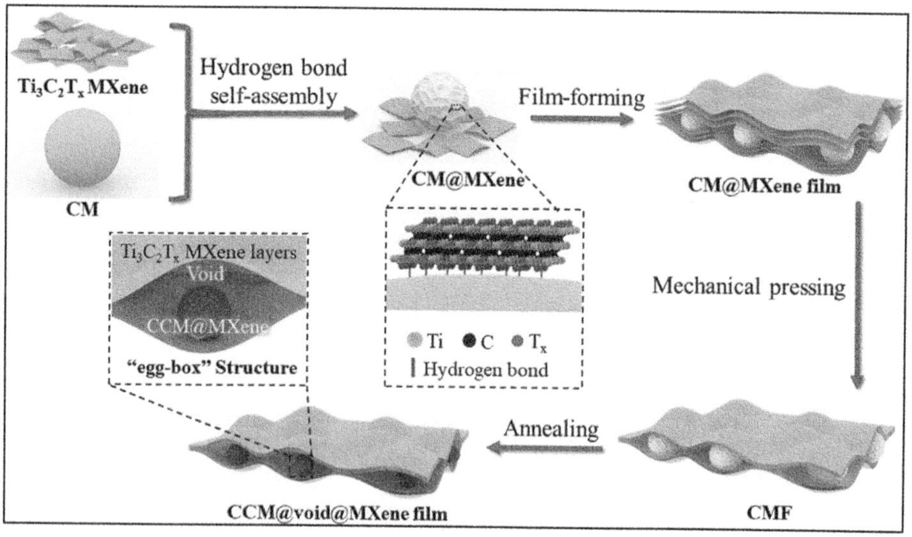

SCHEME 13.7

Graphical representation of the formation of $CVMF_X$ "egg-box" composite.

Reproduced with the permission from K. Qian, Q. Zhou, H. Wu, J. Fang, M. Miao, Y. Yang, et al., Carbonized cellulose microsphere@ void@ MXene composite films with egg-box structure for electromagnetic interference shielding. Compos. Part. A Appl. Sci. 141 (2021) 106229.

shows the lowest shielding effectiveness and GN/CNA-5 containing 5 wt.% of GN to 100 wt.% cellulose shows the highest EMI shielding efficiency (X-band region) being a composite with the highest graphene nanosheet content. The SE_A and SE_R values of GN/CNA-5 are also highest among others.

Zhu et al. [80] have prepared bilayered MXene (f-$Ti_3C_2T_x$)/cellulose-based paper sheet composites. As shown in Fig. 13.33, five different composites with MXene/cellulose ratios 1:4, 1:2, 1:1, 3:2, and 2:1 were prepared, and it was found out that electrical conductivity increases with the increment in MXene content. MXene/cellulose 2:1 had shown the highest EMI SE (in the X-band region) in the range of 60–65 dB and SE/t values (t = sample thickness, SE = shielding efficiency) in the range of 600–650 dB/mm^2.

Du et al. [81] have prepared PLA (P)/MXene (M)-based composite by electrospinning PLA and then incorporating MXene into the electrospun PLA nanofiber mats alternatively in a layered format and the composites are designated as PMPM where P stands for layer of PLA and M stands for adjacent MXene layer to the P layer. EMI shielding properties of a different number of alternating P and M layers showed that the six-layered composite (i.e., PMPMPM) has the highest EMI shielding efficiency (Fig. 13.34A). The EMI shielding efficiency of a bilayer composite MMMPPPP (3M4P) was compared to that of a sample PMPMPMP. Although the thickness of both composites was kept constant for comparison, the EMI shielding efficiency of PMPMPMP was found to be greater than that of bilayer MMMPPP. In a comparison with the bilayer composite membrane 3M4P, PMPMPMP with an alternating layer of P and M shows 28.2% (43.2–55.4 dB), 24.8% (31.8–39.7 dB), and 37.7% (11.4–15.7 dB) enlargement in the SE_T, SE_A, and SE_R, respectively

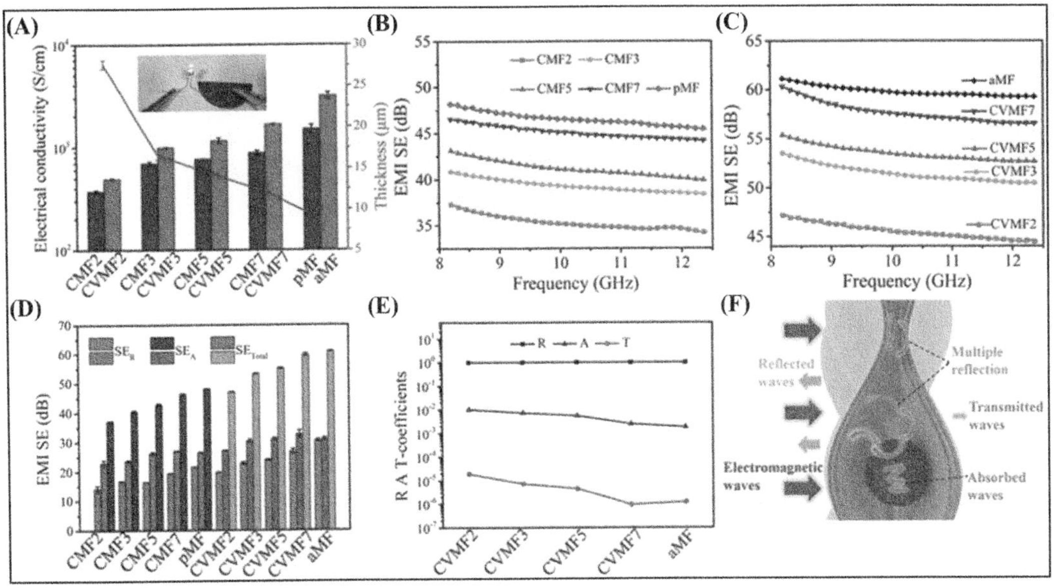

FIGURE 13.31

(A) Electrical conductivity and thickness of the composite films and the inset describe CVMF5 when used as an electrical conductor inside of a circuit where the LED bulb is illuminated. (B) EMI SE of CMF2, CMF3, CMF5, CMF7, and pMF within X-band region. (C) EMI SE of CVMF2, CVMF3, CVM5, CVM7, and aMF in the X-band region. (D) Comparison of average SE_A, SE_R, and SE_T total for CMF, pMF, CVMF, and aMF. (E) Average R, A, and T coefficients of aMFT and aMF in X-band. (F) Diagram of electromagnetic waves (EMW) through CVMF.

Reproduced with the permission from K. Qian, Q. Zhou, H. Wu, J. Fang, M. Miao, Y. Yang, et al., Carbonized cellulose microsphere@ void@ MXene composite films with egg-box structure for electromagnetic interference shielding. Compos. Part. A Appl. Sci. 141 (2021) 106229.

(Fig. 13.34B), despite having the same material composition. This phenomenon demonstrates that an alternating structure is a more effective electromagnetic (EM) radiation suppressant than a bilayer structure. As shown in Fig. 13.34C, R is mostly greater than 0.9 for all layered composites, portraying a commendable reflection of an electromagnetic wave due to impedance misconfiguration between incident electromagnetic wave and MXene layer. The *R*-value is enlarged by 4.0%, whereas the A and T values are decreased by 5.6% and 99.5%, respectively, when the number of MXene interlayers are increased (from 1 to 3).

Pei et al. [82] used a highly efficient mechanical ball milling method to increase the dispersibility of CNT into the CS matrix, resulting in CNT/CS-based 3D printing ink. It was demonstrated that a 3D porous CNT/CS composite is an extremely effective EM wave absorber. The sample designation CNTs/CS-50/8h signifies the composite with CNT concentration of 50 wt.% and ball milling time of 8hrs. Keeping the same CNT concentration, if the milling time is increased the EMI shielding effectiveness is also increased up to 8 hours. of milling time and the composite where ball milling process is not employed (shown in the inset of Fig. 13.35A) have shown the EMI

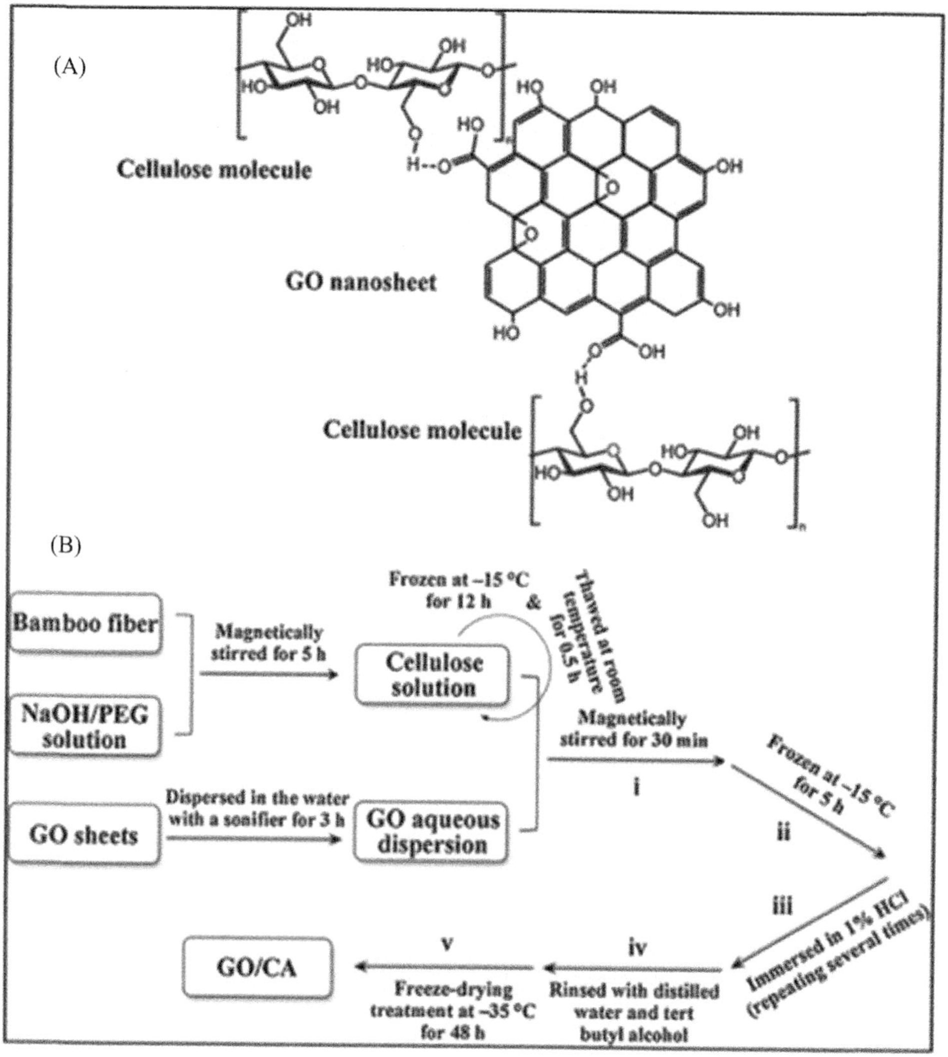

SCHEME 13.8

(A) Graphene oxide (GO)-nanosheet/cellulose chemical interaction (B) GO/CA formation procedure.

Reproduced with the permission from C. Wan, J. Li, Graphene oxide/cellulose aerogels nanocomposite: preparation, pyrolysis, and application for electromagnetic interference shielding. Carbohydr. Polym. 150 (2016) 172–179.

shielding value around 6 dB. When the milling time is increased to 10 hours, the results contradict the convention, which states that increasing the milling time improves EMI shielding performance. Because of the shear and peeling forces of the zirconia ball and stirring rods, the zirconia spheres in the ball mill affect CNTs/CS composites. When the ball milling time is so long that it causes a significant structural change in the CNT particles, destroying the microstructure of the CNTs. As a

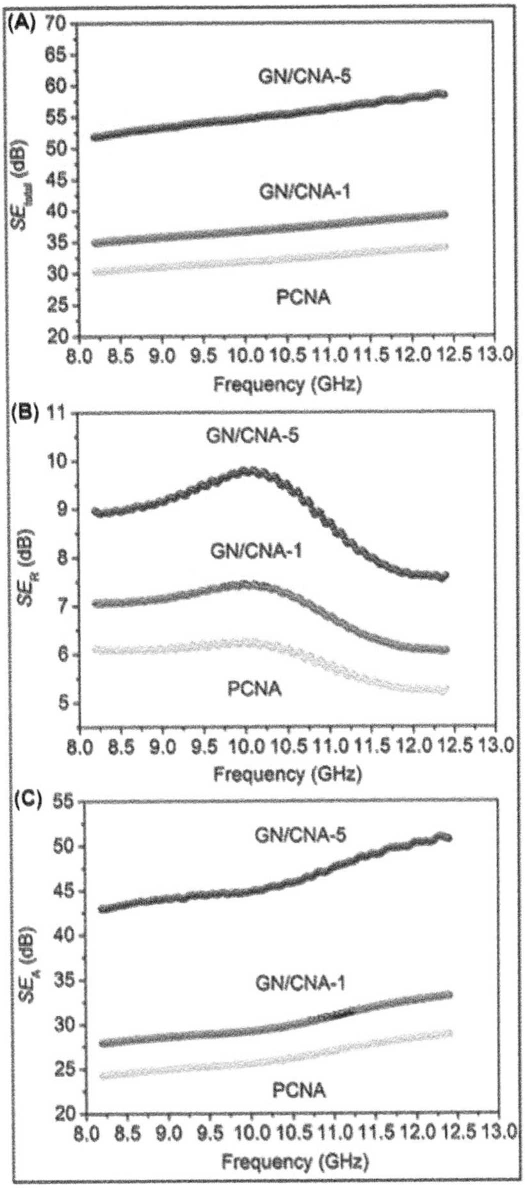

FIGURE 13.32

Study of electromagnetic interference (EMI) shielding properties of GN/CNA-1, GN/CNA-5, and PCNA: (A) SE_{total}, (B) SE_R, (C) SE_A.

Reproduced with the permission from C. Wan, J. Li, Graphene oxide/cellulose aerogels nanocomposite: preparation, pyrolysis, and application for electromagnetic interference shielding. Carbohydr. Polym. 150 (2016) 172–179.

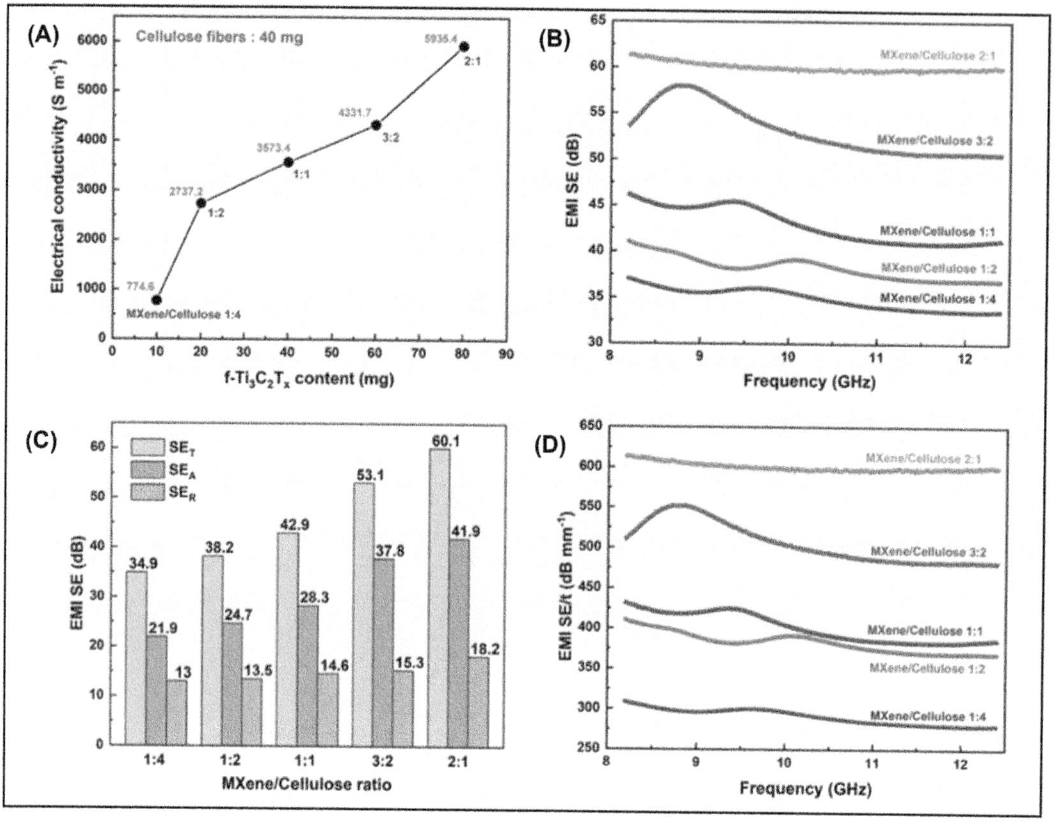

FIGURE 13.33

(A) Electrical conductivity, (B) Net electromagnetic interference (EMI) SE, (C) average EMI shielding effectiveness, and (D) SE/t values of the MXene/cellulose paper composite sheets.

Reproduced with the permission from M. Zhu, X. Yan, H. Xu, Y. Xu, L. Kong, Highly conductive and flexible bilayered MXene/cellulose paper sheet for efficient electromagnetic interference shielding applications. Ceram. Int. 47 (2021) 17234–17244.

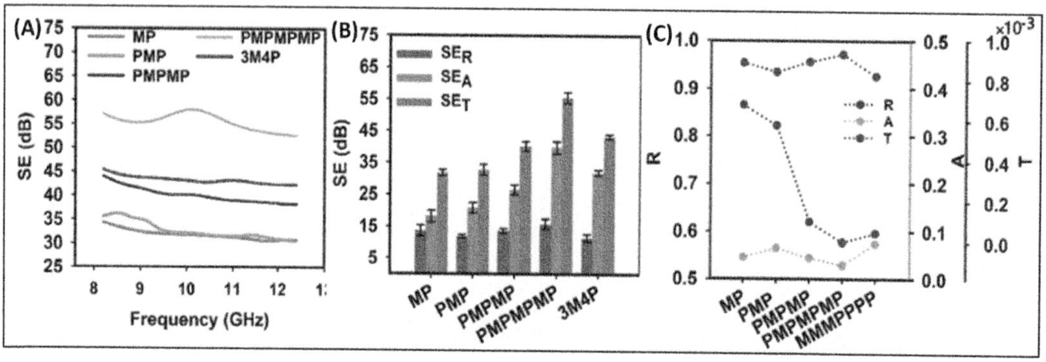

FIGURE 13.34

Electromagnetic interference (EMI) shielding efficiency of multilayered MXene/poly(lactic acid) (PLA) composite films. (A) EMI SE as a function of frequency; (B) SE_T, SE_A, SE_R, and (C) R, A, T values.

Reproduced with the permission from Z. Du, K. Chen, Y. Zhang, Y. Wang, P. He, H.-Y. Mi, et al., Engineering multilayered MXene/electrospun poly (lactic acid) membrane with increscent electromagnetic interference (EMI) shielding for integrated Joule heating and energy generating. Compos. Commun. 26 (2021) 100770.

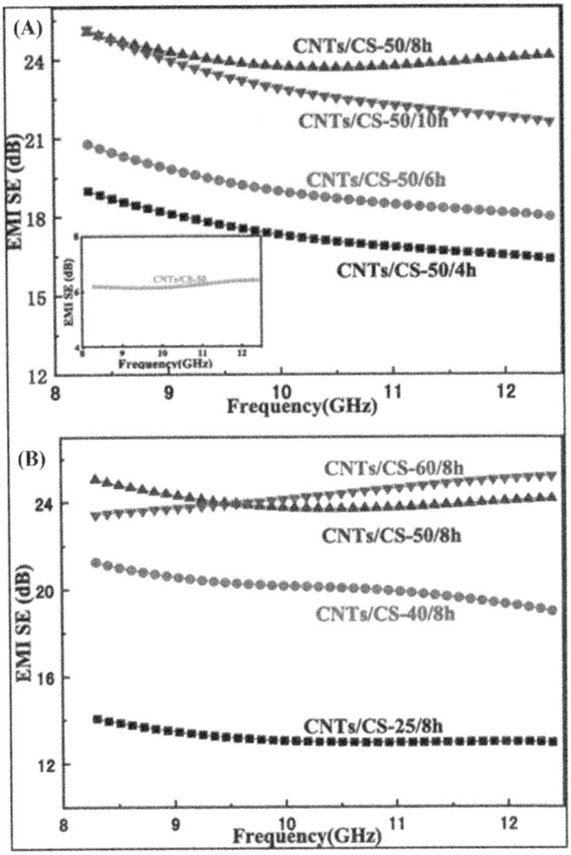

FIGURE 13.35

Electromagnetic interference (EMI) shielding values of carbon nanotube (CNTs)/CS composites: (A) with the variation in milling time and (B) with the variation in CNT concentrations.

Reproduced with the permission from X. Pei, M. Zhao, R. Li, H. Lu, R. Yu, Z. Xu, et al., Porous network carbon nanotubes/chitosan 3D printed composites based on ball milling for electromagnetic shielding. Compos. Part. A Appl. Sci. 145 (2021) 106363.

result, it has been demonstrated that an 8-hours ball milling time is optimal for achieving effective EMI shielding performance. If the CNT concentration is increased from 50 to 60 wt.% the EMI shielding performance is not effectively increased (Fig. 13.35B) suggesting optimum electrical conductivity is achieved with 50 wt.% of filler as they have just connected to form an electrically conductive channel.

Liu et al. [83] have used PLA/PCL blend as a polymer matrix in which the addition of SiO_2 to the PLA phase gave the cocontinuous morphology with a mere 30 wt.% of PCL concerning 70 wt.% of PLA. To achieve effective conductivity and EMI shielding, the conductive inclusion, MWCNT, was preferentially distributed into the PCL phase. The graphical representation of the fabrication of PLA/PCL/SiO_2 composites are shown below in the Scheme 13.9.

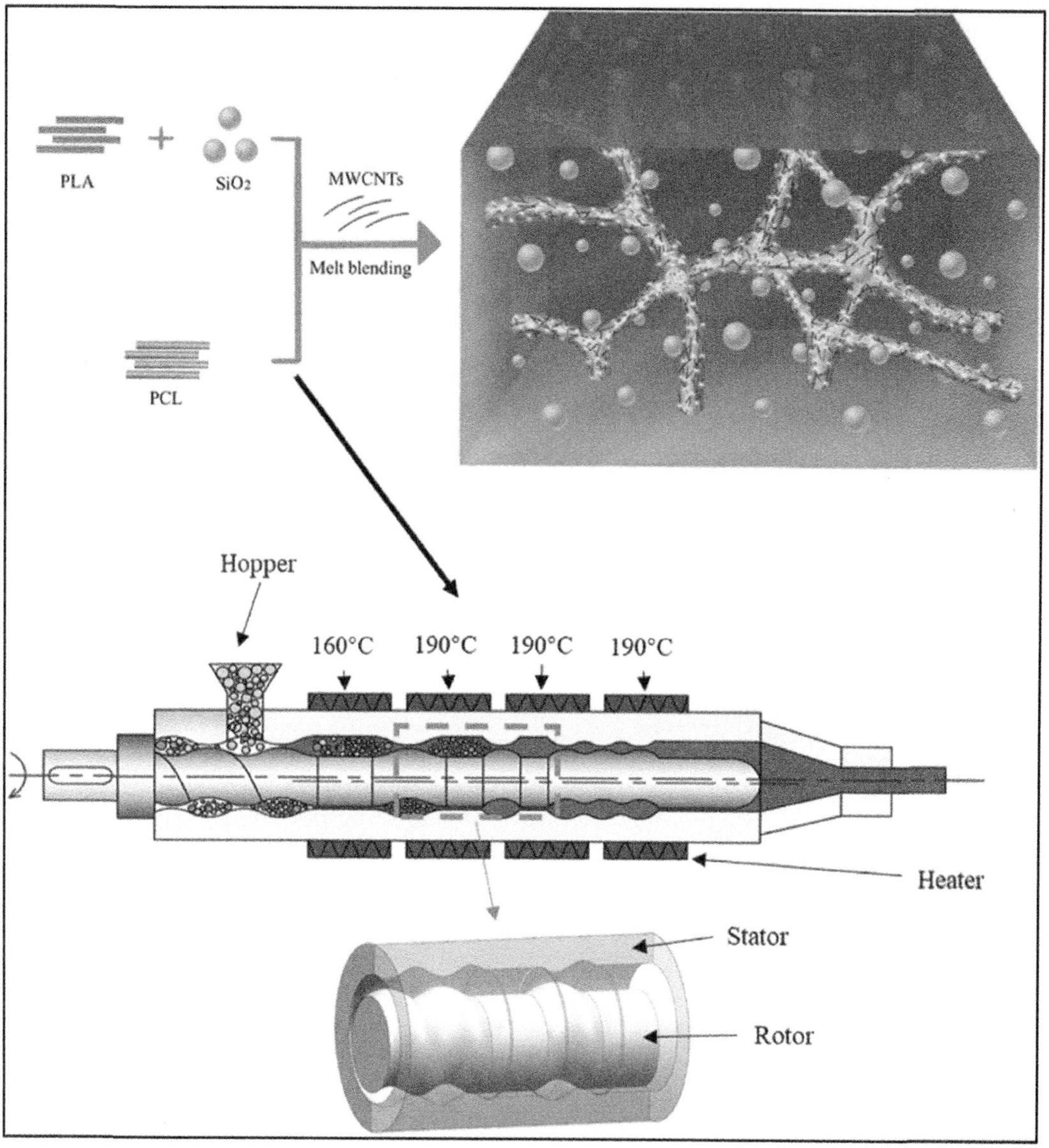

SCHEME 13.9

Graphical representation of the formation of poly(lactic acid) (PLA)/poly(ε-caprolactone) (PCL)/SiO_2/multiwalled carbon nanotube (MWCNT) composite.

Reproduced with the permission from Y. Liu, H. He, G. Tian, Y. Wang, J. Gao, C. Wang, et al., Morphology evolution to form double percolation polylactide/polycaprolactone/MWCNTs nanocomposites with ultralow percolation threshold and excellent EMI shielding. Compos. Sci. Technol. 214 (2021) 108956.

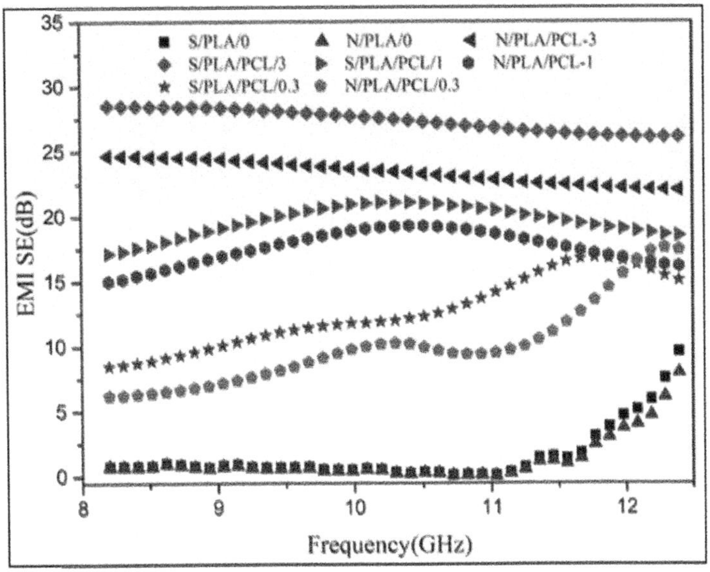

FIGURE 13.36

Electromagnetic interference (EMI) shielding efficiency of different composites in the X-band region.

Reproduced with the permission from Y. Liu, H. He, G. Tian, Y. Wang, J. Gao, C. Wang, et al., Morphology evolution to form double percolation polylactide/polycaprolactone/MWCNTs nanocomposites with ultralow percolation threshold and excellent EMI shielding. Compos. Sci. Technol. 214 (2021) 108956.

The nanocomposites are designated as S/PLA/PCL/x or N/PLA/PCL/x, here S signifies the nanocomposites with 6.0 wt.% SiO_2, N means the nanocomposites are deprived of 6.0 wt.% SiO_2, x describes the weight fraction of MWCNTs included. The description of S/PLA/PCL/1 is given as an example, it suggests that the PLA/PCL polymer blend with 6.0 wt.% SiO_2 incorporated in the PLA phase and 1.0 wt.% MWCNT included in the PCL phase. The EMI shielding measurements are given in Fig. 13.36. It was found out that SiO_2 particles have no significant effect on the composite on its EMI shielding performance as both N/PLA and S/PLA marched together. However, at the high-frequency range, the EMI shielding performance of both N/PLA and S/PLA increased up to 2–8 dB. which is the inherent EMI shielding performance of neat PLA. The EMI shielding efficiency of all composites increased with increasing MWCNT concentration in the composite, which was attributed to the fact that more MWCNT forms a conductive network that can effectively absorb or reflect EMI waves. S/PLA/PCL have indicated better EMI shielding efficiency compared to N/PLA/PCL due to constrained and continuous conductive network due to the presence of SiO_2 particles on the former.

Hu et al. [84] used a deep coating method followed by a coating polydimethylsiloxane (PDMS) polymer to create an MXene ($Ti_3C_2T_x$)/cellulose-based highly flexible sheet composite material. 0.2 mm thick composite sheet has shown EMI shielding efficiency of 43 dB. The composite sheet samples are designated as PM-x where x = number of dip cycles of the cellulose in

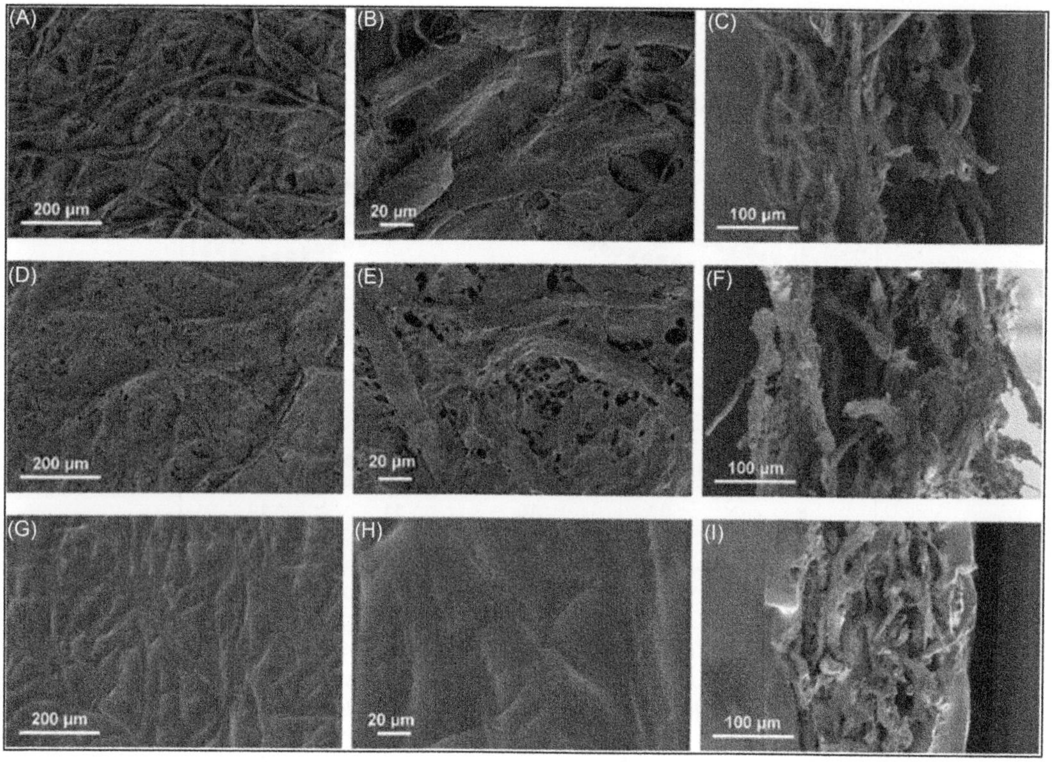

FIGURE 13.37

Scanning electron microscopic (SEM) images of surface morphology of (A, B) cellulose sheet, (D, E) M-7, and (G, H) PM-7. Cross-sectional image of (C) cellulose sheet, (F) M-7, and (I) PM-7.

Reproduced with the permission from D. Hu, X. Huang, S. Li, P. Jiang, Flexible and durable cellulose/MXene nanocomposite paper for efficient electromagnetic interference shielding. Compos. Sci. Technol. 188 (2020) 107995.

the MXene latex. The SEM (Scanning Electron Microscopic) images of the composite sheet samples (both MXene treated cellulose sheet or M-7 and PM-7) depicts the structural morphology which has been shown in Fig. 13.37 where (A) to (F) gives the morphology of M-7 and (G) to (I) gives the morphology of PM-7. The porous microstructure of cellulose sheet with cell structures ensures good electrical conductivity, as shown in (A) to (C). From (D) to (F), the MXene coating is formed throughout the cellulose sheet as well as the pores. From (G) to (I) we can observe that the PDMS layer is covered throughout the M-7 microstructures. Fig. 13.38 depicts the electrical properties and EMI shielding performance of the PM-*x* samples. Fig. 13.38A shows that after 7 dip cycles the conductivity reached the maximum for the PM-7 and the inset picture shows that the cellulose sheet after 7 cycles of dipping in MXene latex (M-7) became electrically conductive whereas cellulose sheets are electrically insulating

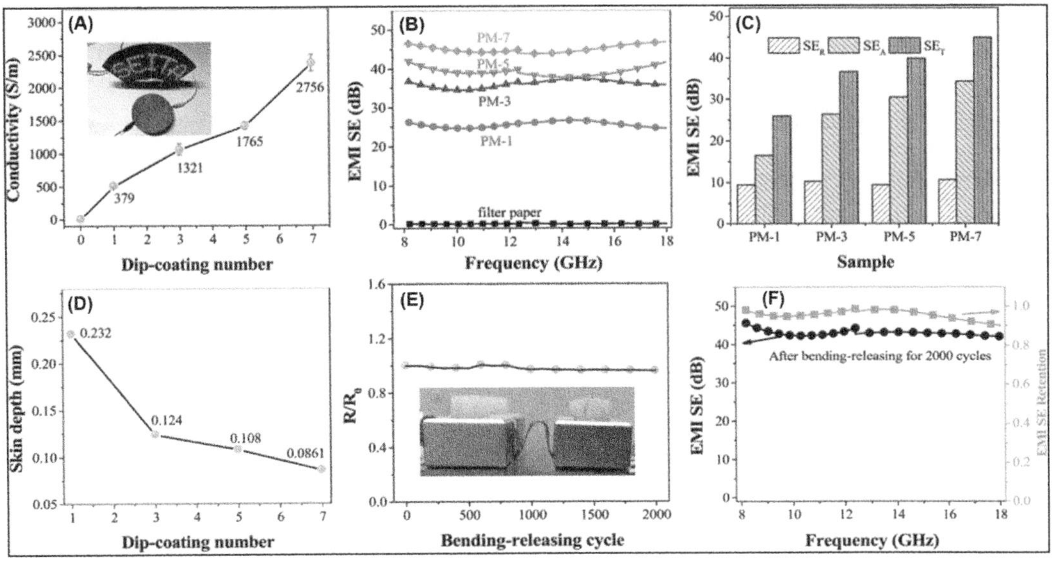

FIGURE 13.38

(A) Electrical conductivity of MXene-coated cellulose sheet (M-filter) as a function of dipping cycles. (B) electromagnetic interference (EMI) shielding effectiveness of the cellulose sheet and the polydimethylsiloxane (PDMS)-layered M-filter in the X and Ku band. (C) Comparative study of SE_T, SE_R, and SE_A values of the PDMS-layered M-filter with several dipping cycles at 12.4 GHz. (D) Variation of skin depth w.r.t the number of dipping cycles at 12.4 GHz. (E) variation of the ratio R/R_0 of PM-7 w.r.t the bending-releasing cycles. (F) electromagnetic interference (EMI) shielding effectiveness of the PM-7 post-2000 bending-releasing cycles.

Reproduced with the permission from D. Hu, X. Huang, S. Li, P. Jiang, Flexible and durable cellulose/MXene nanocomposite paper for efficient electromagnetic interference shielding. Compos. Sci. Technol. 188 (2020) 107995.

material. Fig. 13.38D portrays the variation of skin depth (δ) (mm) with an increase in dipping cycles where skin depth gradually decreases with increment in dipping cycles. Fig. 13.38E shows the variation of the ratio of electrical resistance (R) after a certain number of bending cycles with initial electrical resistance (R_0) and after 2000 cycles there is a very small decrease in the electrical resistance and EMI shielding performance remains almost same with the value of 42.1 dB (Fig. 13.38F).

Kashi et al. [85] has incorporated graphite nanoplatelets (GNP) separately in PBAT (polybutylene adipate-*co*-terephthalate) and PLA (polylactide) polymer matrix and showed that 1.5 mm thick PBAT/GNP and PLA/GNP composites do not necessarily give increased total shielding efficiency (SE_T) with an increase in GNP wt.% (Fig. 13.39) and reflection is the dominant contributor to the total shielding effectiveness. 2.8 mm thick composites (both PBAT/GNP and PLA/GNP) showed the increment in the contribution of absorption compared to the reflection after crossing the GNP loading of 9 wt.%.

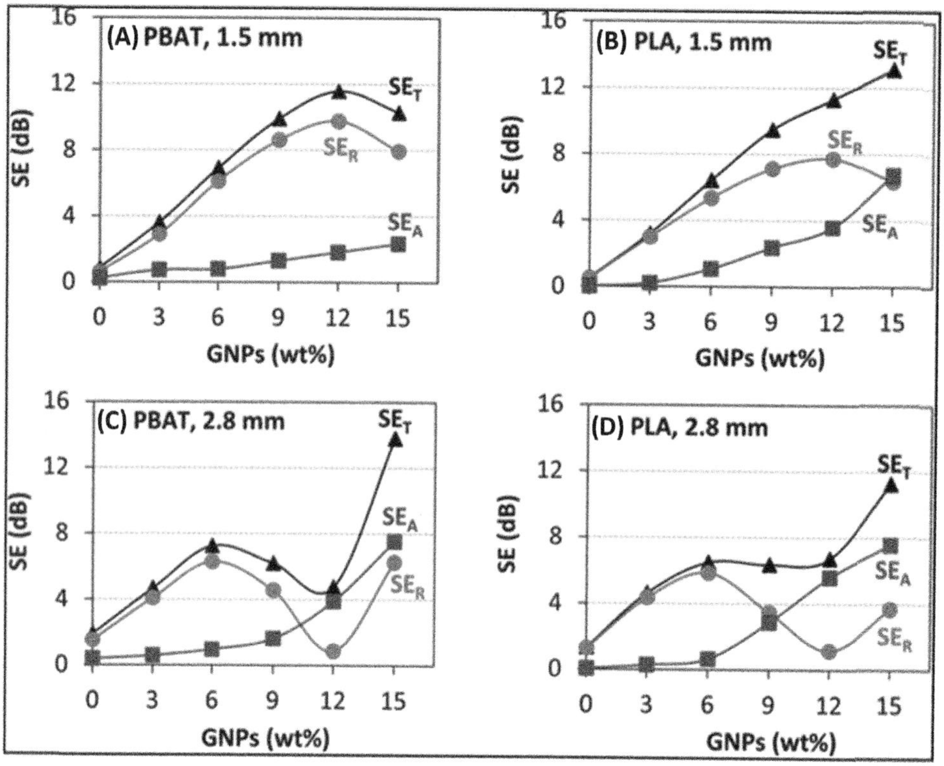

FIGURE 13.39

SE_T, SE_R, and SE_A of (A and C) poly(butylene adipate-*co*-terephthalate) (PBAT) and (B and D) poly(lactic acid) (PLA) nanocomposites versus graphite nanoparticle (GNP) loading at 10 GHz for 1.5 and 2.8 mm thick samples.

Reproduced with the permission from S. Kashi, R.K. Gupta, S.N. Bhattacharya, R.J. Varley, Experimental and simulation study of effect of thickness on performance of (butylene adipate-co-terephthalate) and poly lactide nanocomposites incorporated with graphene as stand-alone electromagnetic interference shielding and metal-backed microwave absorbers. Compos. Sci. Technol. 195 (2020) 108186.

13.8 Conclusion

To summarize, manufacturing biodegradable polymer nanocomposite as EMI shield material is an endeavor to reduce the ever-increasing "e-waste" on the planet. The composites are versatile, efficient EMI radiation absorbers with low production costs and ease of processing. In addition, the last decade has seen a remarkable shift in the direction of technology with a green and sustainable approach. The chapter discusses various synthetic approaches, as well as the electrical conductivity and EMI shielding properties of various polymer nanocomposites based on biodegradable polymers such as PLA, PBAT, biodegradable epoxy polymer, PCL, and others. This chapter also discusses the significance of quantitative inclusion of various conductive fillers such as MWCNT, CNT, conductive carbon blacks, and silica (to improve the mechanical properties of the composite) in the biodegradable polymer matrix to achieve optimal electrical conductivity and EMI shielding efficiency. This chapter also discusses innovative foaming technology processes for making light weight biodegradable polymer composites. The discovery of MXenes as a new type of conductive agent is also a critical topic to discuss. The chapter delves into various synthetic methods for producing layered MXenes. This chapter discusses more innovative MXenes research, such as the use of MXenes as a coating material to achieve desired conductivity. The importance of biodegradability in addition to excellent electrical and EMI shielding properties on modern polymer nanocomposites is a trend, and this chapter strongly supports this apology.

Acknowledgements

I thank my PhD supervisor Dr. Narayan Chandra Das for supporting me with his knowledge and guidance to eradicate any mistakes while writing the chapter and I am also thankful to Indian Institute of Technology, Kharagpur for providing me internet facilities to search related research papers while writing the chapter.

References

[1] F. Su, M. Miao, Effect of MWCNT dimension on the electrical percolation and mechanical properties of poly (vinylidenefluoride-hexafluoropropylene) based nanocomposites, Synth. Met. 191 (2014) 99–103.

[2] K. Nath, S. Ghosh, S.K. Ghosh, P. Das, N.C. Das, Facile preparation of light-weight biodegradable and electrically conductive polymer based nanocomposites for superior electromagnetic interference shielding effectiveness, J. Appl. Polym. Sci. 138 (2021) 50514.

[3] L. Lingamdinne, J.R. Prasanna, Koduru, R.R. Karri, A comprehensive review of applications of magnetic graphene oxide based nanocomposites for sustainable water purification, J. Environ. Manage. 231 (2019) 622–634.

[4] F.V. Ferreira, F.S. Brito, W. Franceschi, E.A.N. Simonetti, L.S. Cividanes, M. Chipara, et al., Functionalized graphene oxide as reinforcement in epoxy based nanocomposites, Surf. Interfaces 10 (2018) 100–109.

[5] B. Sakthivel, G. Nammalvar, Selective ammonia sensor based on copper oxide/reduced graphene oxide nanocomposite, J. Alloy. Compd. 788 (2019) 422–428.

[6] P. Sawai, P.P. Chattopadhaya, S. Banerjee, Synthesized reduce Graphene Oxide (rGO) filled Polyetherimide based nanocomposites for EMI Shielding applications, Mater. Today Proc. 5 (2018) 9989–9999.

[7] W. Tao, Z. Sun, H. Da, Z. Yang, Q. Ji, N. Hu, et al., Studies on NH_3 gas sensing by zinc oxide nanowire-reduced graphene oxide nanocomposites, Sens. Actuators B: Chem. 252 (2017) 284–294.

[8] P. Rani, M.B. Ahamed, K. Deshmukh, Dielectric and electromagnetic interference shielding performance of graphene nanoplatelets and copper oxide nanoparticles reinforced polyvinylidenefluoride/poly(3,4-ethylenedioxythiophene)-block-poly (ethylene glycol) blend nanocomposites, Synth. Met. 282 (2021) 116923.

[9] J. Joseph, K. Deshmukh, A.N. Raj, S.K.K. Pasha, Electromagnetic interference shielding characteristics of $SrTiO_3$ nanoparticles induced polyvinyl chloride and polyvinylidene fluoride blend nanocomposites, J. Inorg. Organomet. Polym. Mater. 31 (2021) 1–15.

[10] H. Kumar A, M.B. Ahamed, K. Deshmukh, M.S. Sirajuddeen, Morphology, dielectric and EMI shielding characteristics of graphene nanoplatelets, montmorillonite nanoclay and titanium dioxide nanoparticles reinforced polyvinylidenefluoride nanocomposites, J. Inorg. Organomet. Polym. Mater. 31 (2021) 2003–2016.

[11] P. Rani, M.B. Ahamed, K. Deshmukh, Structural, dielectric and EMI shielding properties of polyvinyl alcohol/chitosan blend nanocomposites integrated with graphite oxide and nickel oxide nanofillers, J. Mater. Sci. Mater Electron. 32 (2021) 764–779.

[12] P. Rani, M.B. Ahamed, K. Deshmukh, Dielectric and electromagnetic interference shielding properties of zeolite 13X and carbon black nanoparticles based PVDF nanocomposites, J. Appl. Polym. Sci. 138 (2021) 50107.

[13] P. Rani, M.B. Ahamed, K. Deshmukh, Significantly enhanced electromagnetic interference shielding effectiveness of montmorillonite nanoclay and copper oxide nanoparticles based polyvinylchloride nanocomposites, Polym. Test. 91 (2020) 106744.

[14] A. Muzaffar, M.B. Ahamed, K. Deshmukh, S.K.K. Pasha, Dielectric properties and electromagnetic interference shielding studies of nickel oxide and tungsten oxide reinforced polyvinylchloride nanocomposites, Polym. Technol. Mater. 59 (2020) 1667–1678.

[15] P. Rani, M.B. Ahamed, K. Deshmukh, Dielectric and electromagnetic interference shielding properties of carbon black nanoparticles reinforced PVA/PEG blend nanocomposite films, Mater. Res. Express 7 (2020) 064008.

[16] P. Rani, M.B. Ahamed, K. Deshmukh, Electromagnetic interference shielding properties of graphene quantum-dots reinforced poly(vinyl alcohol)/polypyrrole blend nanocomposites, J. Appl. Polym. Sci. 45 (2020) 49392.

[17] M.K. Mohanapriya, K. Deshmukh, J. Kadlec, K.K. Sadasivuni, M. Faisal, N.A.N. Raj, et al., Dynamic mechanical analysis and broadband electromagnetic interference shielding characteristics of poly (vinyl alcohol)-poly (4-styrenesulfonic acid)-titanium dioxide nanoparticles based tertiary nanocomposites, Polym-Plast Tech. Mat. 59 (2020) 847–863.

[18] X. Tang, Z. Zu, Z. Zang, et al., $CsPbBr_3$/reduced graphene oxide nanocomposites and their enhanced photoelectric detection application, Sens. Actuators B Chem. 245 (2017) 435–440.

[19] H. Hu, X. Wang, F. Liu, J. Wang, C. Xu, Rapid microwave-assisted synthesis of graphene nanosheets–zinc sulfide nanocomposites: optical and photocatalytic properties, Synth. Met. 161 (2011) 404–410.

[20] P. Bhawal, S. Ganguly, T.K. Das, S. Mondal, S. Choudhury, N.C. Das, Superior electromagnetic interference shielding effectiveness and electro-mechanical properties of EMA-IRGO nanocomposites through the in-situ reduction of GO from melt blended EMA-GO composites, Compos. B. Eng. 134 (2018) 46–60.

[21] R. Ravindren, S. Mondal, K. Nath, N.C. Das, Investigation of electrical conductivity and electromagnetic interference shielding effectiveness of preferentially distributed conductive filler in highly flexible polymer blends nanocomposites, Compos. Part. A Appl. Sci. 118 (2019) 75–89.

[22] P. Bhawal, S. Ganguly, T.K. Chaki, N.C. Das, Synthesis and characterization of graphene oxide filled ethylene methyl acrylate hybrid nanocomposites, RSC Adv. 6 (2016) 20781–20790.

[23] R. Ravindren, S. Mondal, K. Nath, N.C. Das, Prediction of electrical conductivity, double percolation limit and electromagnetic interference shielding effectiveness of copper nanowire filled flexible polymer blend nanocomposites, Compos. B. Eng. 164 (2019) 559–569.

[24] A. Agarwal, N.R. Desai, K. Makker, A. Varghese, R. Mouradi, E. Sabanegh, et al., Effects of radiofrequency electromagnetic waves (RF-EMW) from cellular phones on human ejaculated semen: an in vitro pilot study, Fertil. Steril. 92 (2009) 1318–1325.

[25] A. Obajuluwa, A.J. Olukayode, O.B. Akinyemi, K. Afolabi, J.O. Adekoya, A.O.Ishola Sanya, Exposure to radio-frequency electromagnetic waves alters acetylcholinesterase gene expression, exploratory and motor coordination-linked behaviour in male rats, Toxicol. Rep. 4 (2017) 530–534.

[26] S. Banik, S. Bandyopadhyay, S. Ganguly, Bioeffects of microwave—a brief review, Bioresour. Technol. 87 (2003) 155–159.

[27] A. Agarwal, F. Deepinder, R.K. Sharma, G. Ranga, J. Li, Effect of cell phone usage on semen analysis in men attending infertility clinic: an observational study, Fertil. Steril. 89 (2008) 124–128.

[28] M. Terzi, B. Ozberk, O.G. Deniz, S. Kaplan, The role of electromagnetic fields in neurological disorders, J. Chem. Neuroanat. 75 (2016) 77–84.

[29] V.I.T. Ahamed, N.G. Karthick, P.K. Joseph, Effect of mobile phone radiation on heart rate variability, Comput. Biol. Med. 38 (2008) 709–712.

[30] S. Szmigielski, Cancer morbidity in subjects occupationally exposed to high frequency (radiofrequency and microwave) electromagnetic radiation, Sci. Total. Env. 180 (1996) 9–17.

[31] B. Tansel, From electronic consumer products to e-wastes: global outlook, waste quantities, recycling challenges, Env. Int. 98 (2017) 35–45.

[32] Z. Jun-hui, M. Hang, Eco-toxicity and metal contamination of paddy soil in an e-wastes recycling area, J. Hazard. Mater. 165 (2009) 744–750.

[33] Y. Qu, Q. Zhu, J. Sarkis, Y. Geng, Y. Zhong, A review of developing an e-wastes collection system in Dalian, China. J. Clean. Prod. 52 (2013) 176–184.

[34] S.M. Lebedev, O.S. Gefle, E.T. Amitov, D.Y. Berchuk, D.V. Zhuravlev, Poly(lactic acid)-based polymer composites with high electric and thermal conductivity and their characterization, Polym. Test. 58 (2017) 241–248.

[35] K. Fukushima, M.H. Wu, S. Bocchini, A. Rasyida, M.C. Yang, PBAT based nanocomposites for medical and industrial applications, Mater. Sci. Eng. C. 32 (2012) 1331–1351.

[36] A.-L. Goffin, J.-M. Raquez, E. Duquesne, G. Siqueira, Y. Habibi, A. Dufresne, et al., Poly (ε-caprolactone) based nanocomposites reinforced by surface-grafted cellulose nanowhiskers via extrusion processing: morphology, rheology, and thermo-mechanical properties, Polymer. 52 (2011) 1532–1538.

[37] A. Romero, A. Santos, J. Tojo, A. Rodríguez, Toxicity and biodegradability of imidazolium ionic liquids, J. Hazard. Mater. 151 (2008) 268–273.

[38] L.M.G. Gonçalves, T.R. Rigolin, B.M. Frenhe, S.H.P. Bettini, On the recycling of a biodegradable polymer: multiple extrusion of poly (lactic acid), Mater. Res. 23 (2020).

[39] Y. Fu, G. Wu, X. Bian, J. Zeng, Y. Weng, Biodegradation behavior of poly (butylene adipate-co-terephthalate) (PBAT), poly (lactic acid) (PLA), and their blend in freshwater with sediment, Molecules 25 (2020) 3946.

[40] Filipe V. Ferreira, M. Mariano, I.F. Pinheiro, E.M. Cazalini, D.H.S. Souza, L.S.S. Lepesqueur, et al., Cellulose nanocrystal-based poly (butylene adipate-co-terephthalate) nanocomposites covered with antimicrobial silver thin films, Polym. Eng. Sci. 59 (2019) E356–E365.

[41] D.A. Wood, Biodegradable drug delivery systems, Int. J. Pharm. 7 (1980) 1–18.

[42] A. Kumar, K. Anjana, M. Hinduja, K. Sujitha, G. Dharani, Review on plastic wastes in marine environment – Biodegradation and biotechnological solutions, Mar. Pollut. Bull. 150 (2020) 110733.

[43] V. Massardier-Nageotte, C. Pestre, T. Cruard-Pradet, R. Bayard, Aerobic and anaerobic biodegradability of polymer films and physico-chemical characterization, Polym. Degrad. Stab. 91 (2006) 620–627.

[44] H. Yagi, F. Ninomiya, M. Funabashi, M. Kunioka, Mesophilic anaerobic biodegradation test and analysis of eubacteria and archaea involved in anaerobic biodegradation of four specified biodegradable polyesters, Polym. Degrad. Stab. 110 (2014) 278–283.

[45] T. Ishigaki, W. Sugano, A. Nakanishi, M. Tateda, M. Ike, M. Fujita, The degradability of biodegradable plastics in aerobic and anaerobic waste landfill model reactors, Chemosphere. 54 (2004) 225–233.

[46] G. Kale, T. Kijchavengkul, R. Auras, M. Rubino, S.E. Selke, S.P. Singh, Compostability of bioplastic packaging materials: an overview, Macromol. Biosci. 7 (2007) 255–277.

[47] L. Wei, A.G. McDonald, Accelerated weathering studies on the bioplastic, poly(3-hydroxybutyrate-co-3-hydroxyvalerate), Polym. Degrad. Stab. 126 (2016) 93–100.

[48] K. Nath, S.K. Bhattacharyya, N.C. Das, Biodegradable polymeric materials for EMI shielding, Mater. Potential. EMI Shield. Appl. (2020) 165–178.

[49] R. Mehta, V. Kumar, H. Bhunia, S.N. Upadhyay, Synthesis of poly (lactic acid): a review, J. Macromol. Sci. Polym. Rev. 45 (2005) 325–349.

[50] G. Xiao, B. Yan, R. Ma, W.J. Jin, X.Q. Lü, L.Q Ding, et al., Bulk ring-opening polymerization (ROP) of L-lactide catalyzed by Ni (ii) and Ni (ii)-Sm (iii) complexes based on a salen-type schiff-base ligand, Polym. Chem. 2 (2011) 659–664.

[51] J. Feng, X. Xu, Z. Xu, H. Xie, P. Song, L. Li, et al., One-Pot, solvent-and catalyst-free synthesis of polyphosphoramide as an eco-benign and effective flame retardant for poly (lactic acid), ACS Sustain. Chem. Eng. 8 (2020) 16612–16623.

[52] S. Shi, Z. Peng, J. Jing, L. Yang, Y. Chen, 3D printing of delicately controllable cellular nanocomposites based on polylactic acid incorporating graphene/carbon nanotube hybrids for efficient electromagnetic interference shielding, ACS Sustain. Chem. Eng. 8 (2020) 7962–7972.

[53] S. Shi, Z. Peng, J. Jing, L. Yang, Y. Chen, R. Kotsilkova, et al., Preparation of highly efficient electromagnetic interference shielding polylactic acid/graphene nanocomposites for fused deposition modeling three-dimensional printing, Ind. Eng. Chem. Res. 59 (2020) 15565–15575.

[54] L. Xu, X.-P. Zhang, C.-H. Cui, P.-G. Ren, D.-X. Yan, Z.-M. Li, Enhanced mechanical performance of segregated carbon nanotube/poly (lactic acid) composite for efficient electromagnetic interference shielding, Ind. Eng. Chem. Res. 58 (2019) 4454–4461.

[55] G. Wang, L. Wang, L.H. Mark, et al., Ultralow-threshold and lightweight biodegradable porous PLA/MWCNT with segregated conductive networks for high-performance thermal insulation and electromagnetic interference shielding applications, ACS Appl. Mater. Interfaces 10 (2018) 1195–1203.

[56] C.-H. Cui, D.-X. Yan, H. Pang, et al., A high heat-resistance bioplastic foam with efficient electromagnetic interference shielding, Chem. Eng. J. 323 (2017) 29–36.

[57] G. Wang, L. Hao, X. Zhang, S. Tan, M. Zhou, W. Gu, et al., Flexible and transparent silver nanowires/biopolymer film for high-efficient electromagnetic interference shielding, J. Colloid Interface Sci. 607 (2022) 89–99.

[58] H. Fang, S. Wang, W. Ye, X. Chen, X. Wang, P. Xu, et al., Simultaneous improvement of mechanical properties and electromagnetic interference shielding performance in eco-friendly polylactide composites via reactive blending and MWCNTs induced morphological optimization, Compos. B. Eng. 178 (2019) 107452.

[59] K. Zhang, H.-O. Yu, K.-X. Yu, Y. Gao, M. Wang, J. Li, et al., A facile approach to constructing efficiently segregated conductive networks in poly (lactic acid)/silver nanocomposites via silver plating on microfibers for electromagnetic interference shielding, Compos. Sci. Technol. 156 (2018) 136–143.

[60] T. Kuang, L. Chang, F. Chen, Y. Sheng, D. Fu, X. Peng, Facile preparation of lightweight high-strength biodegradable polymer/multi-walled carbon nanotubes nanocomposite foams for electromagnetic interference shielding, Carbon N. Y. 105 (2016) 305–313.

[61] J. Jian, Z. Xiangbin, H. Xianbo, An overview on synthesis, properties and applications of poly(butylene-adipate-co-terephthalate)–PBAT, Adv. Ind. Eng. Polym. Res. 3 (2020) 19–26.

[62] P. Bordes, E. Pollet, L. Avérous, Nano-biocomposites: biodegradable polyester/nanoclay systems, Prog. Polym. Sci. 34 (2009) 125–155.

[63] V. Nagarajan, M. Misra, A.K. Mohanty, New engineered biocomposites from poly(3-hydroxybutyrate-co-3-hydroxyvalerate) (PHBV)/poly(butylene adipate-co-terephthalate) (PBAT) blends and switchgrass: fabrication and performance evaluation, Ind. Crop. Prod. 42 (2013) 461–468.

[64] R. Herrera, L. Franco, A. Rodríguez-Galán, J. Puiggalí, Characterization and degradation behavior of poly (butylene adipate-co-terephthalate), J. Polym. Sci. A: Polym. Chem. 40 (2002) 4141–4157.

[65] S. Kashi, R.K. Gupta, T. Baum, N. Kao, S.N. Bhattacharya, Dielectric properties and electromagnetic interference shielding effectiveness of graphene-based biodegradable nanocomposites, Mater. Des. 109 (2016) 68–78.

[66] M. Labet, W. Thielemans, Synthesis of polycaprolactone: a review, Chem. Soc. Rev. 38 (2009) 3484–3504.

[67] M. Bartnikowski, T.R. Dargaville, S. Ivanovski, D.W. Hutmacher, Degradation mechanisms of polycaprolactone in the context of chemistry, geometry and environment, Prog. Polym. Sci. 96 (2019) 1–20.

[68] A. Mahapatro, A. Kumar, R.A. Gross, Mild, solvent-free ω-hydroxy acid polycondensations catalyzed by *Candida antarctica* lipase B, Biomacromolecules. 5 (2004) 62–68.

[69] J.-M. Thomassin, X. Lou, C. Pagnoulle, A. Saib, L. Bednarz, I. Huynen, et al., Multiwalled carbon nanotube/poly (ε-caprolactone) nanocomposites with exceptional electromagnetic interference shielding properties, J. Phys. Chem. C. 111 (2007) 11186–11192.

[70] S.J. Park, F.L. Jin, J.R. Lee, Effect of biodegradable epoxidized castor oil on physicochemical and mechanical properties of epoxy resins, Macromol. Chem. Phys. 205 (2004) 2048–2054.

[71] Q. Gao, G. Zhang, X. Fan, H. Zhang, Y. Zhang, F. Huang, et al., Enhancements of foamability, electromagnetic interference shielding and mechanical property of epoxy microcellular composite foam with well-dispersed f-MWCNTs. Compos, Part. A Appl. Sci. 138 (2020) 106060.

[72] G. Choi, F. Shahzad, Y.M. Bahk, Y.M. Jhon, H. Park, M. Alhabeb, et al., Enhanced terahertz shielding of MXenes with nano-metamaterials, Adv. Opt. Mater. 6 (2018) 1701076.

[73] G.M. Weng, J. Li, M. Alhabeb, C. Karpovich, H. Wang, J. Lipton, et al., Layer-by-layer assembly of cross-functional semi-transparent MXene-carbon nanotubes composite films for next-generation electromagnetic interference shielding, Adv. Funct. Mater. 28 (2018) 1803360.

[74] L. Wang, H. Qiu, P. Song, Y. Zhang, Y. Lu, C. Liang, et al., 3D $Ti_3C_2T_x$ MXene/C hybrid foam/epoxy nanocomposites with superior electromagnetic interference shielding performances and robust mechanical properties, Compos. Part. A Appl. Sci. 123 (2019) 293–300.

[75] F. Liu, Y. Li, S. Hao, Y. Cheng, Y. Zhan, C. Zhang, et al., Well-aligned MXene/chitosan films with humidity response for high-performance electromagnetic interference shielding, Carbohydr. Polym. 243 (2020) 116467.

[76] H. Xu, X. Yin, X. Li, M. Li, S. Liang, L. Zhang, et al., Lightweight Ti_2CT_x MXene/poly (vinyl alcohol) composite foams for electromagnetic wave shielding with absorption-dominated feature, ACS Appl. Mater. Interfaces. 11 (2019) 10198–10207.

[77] G. Hong, H. Cheng, S. Zhang, O.J. Rojas, Polydopamine-treated hierarchical cellulosic fibers as versatile reinforcement of polybutylene succinate biocomposites for electromagnetic shielding, Carbohydr. Polym. 277 (2022) 118818.

[78] K. Qian, Q. Zhou, H. Wu, J. Fang, M. Miao, Y. Yang, et al., Carbonized cellulose microsphere@ void@ MXene composite films with egg-box structure for electromagnetic interference shielding, Compos. Part. A Appl. Sci. 141 (2021) 106229.

[79] C. Wan, J. Li, Graphene oxide/cellulose aerogels nanocomposite: preparation, pyrolysis, and application for electromagnetic interference shielding, Carbohydr. Polym. 150 (2016) 172–179.

[80] M. Zhu, X. Yan, H. Xu, Y. Xu, L. Kong, Highly conductive and flexible bilayered MXene/cellulose paper sheet for efficient electromagnetic interference shielding applications, Ceram. Int. 47 (2021) 17234–17244.

[81] Z. Du, K. Chen, Y. Zhang, Y. Wang, P. He, H.-Y. Mi, et al., Engineering multilayered MXene/electrospun poly (lactic acid) membrane with increscent electromagnetic interference (EMI) shielding for integrated Joule heating and energy generating, Compos. Commun. 26 (2021) 100770.

[82] X. Pei, M. Zhao, R. Li, H. Lu, R. Yu, Z. Xu, et al., Porous network carbon nanotubes/chitosan 3D printed composites based on ball milling for electromagnetic shielding, Compos. Part. A Appl. Sci. 145 (2021) 106363.

[83] Y. Liu, H. He, G. Tian, Y. Wang, J. Gao, C. Wang, et al., Morphology evolution to form double percolation polylactide/polycaprolactone/MWCNTs nanocomposites with ultralow percolation threshold and excellent EMI shielding, Compos. Sci. Technol. 214 (2021) 108956.
[84] D. Hu, X. Huang, S. Li, P. Jiang, Flexible and durable cellulose/MXene nanocomposite paper for efficient electromagnetic interference shielding, Compos. Sci. Technol. 188 (2020) 107995.
[85] S. Kashi, R.K. Gupta, S.N. Bhattacharya, R.J. Varley, Experimental and simulation study of effect of thickness on performance of (butylene adipate-co-terephthalate) and poly lactide nanocomposites incorporated with graphene as stand-alone electromagnetic interference shielding and metal-backed microwave absorbers, Compos. Sci. Technol. 195 (2020) 108186.

CHAPTER

14 Biodegradable polymer nanocomposites for gas-sensing and bio-sensing applications: prospects and challenges

Rashmi Singh[1], A.K. Shrivastava[2] and A.K. Bajpai[3]

[1]Department of Physics, Starex University, Gurugram, Haryana, India [2]School of Studies in Physics, Jiwaji University, Gwalior, Madhya Pradesh, India [3]Bose Memorial Research Laboratory, Department of Chemistry, Govt. Autonomous Science College, Jabalpur, Madhya Pradesh, India

Chapter Outline

Biodegradable and Biocompatible Polymer Nanocomposites. DOI: https://doi.org/10.1016/B978-0-323-91696-7.00005-2
Copyright © 2023 Elsevier Inc. All rights reserved.

14.1 Introduction

The entire world is currently fighting a COVID-19 pandemic that is threatening human life. Combating an invisible enemy that causes serious health problems is always difficult. Various environmental and medical reports have indicated that biosensors will be extremely useful in combating this pandemic [1]. The evolution of smart gas-sensing materials has increased the demand for and growth of biosensors and gas sensors. In this field of study, nanocomposites based on biocompatible polymers have been widely used [2–6]. Many biodegradable polymer-based biosensors and gas sensors have been designed and developed for monitoring environmental disturbances, biological disturbances, and disease diagnosis [7,8]. The current peak of the second corona virus wave has caused serious human breath issues, which has sharply increased the daily mortality rate. Biosensors and gas sensors can detect cancer, diabetes, and other diseases by monitoring human breath [9]. The advantage of biodegradable polymers is that they can be recycled, and as green materials, they have no negative impact on our environment. The demand for biosensors and gas sensors made from these materials has resulted in a more secure human life. We can reduce treatment costs and improve our economy if we can detect viruses and other diseases in their early stages using sensors [10,11]. Aside from that, global warming is one of the most serious problems in the modern world, causing massive environmental pollution. Most pollutants are invisible to human eyes, such as pipeline leaks, natural gas extraction, and methane gas leakage. These gases have an impact on the earth's atmosphere if they leak into the environment. Global warming is one of the most serious challenges facing our planet, and it is caused by the presence of toxic gases in our environment, such as methane and carbon dioxide (CO_2). These harmful gases are produced as more air conditioners, refrigerators, cars, and factory smoke are used. The demand for portable, lightweight devices that can detect these harmful gases has grown. Material science is a rapidly expanding field as a result of the increasing use of biodegradable polymer nanocomposites as smart and green gas-sensing materials [12].

14.2 Current need of gas sensors and biosensors

In the current scenario, humans are constantly polluting the air in order to make their lives more comfortable, such as smoke produced by factories, smoke produced by vehicles, hazardous gases produced by some home appliances, and so on [13]. It will be fatal to human health if we do not control it in time. If we can detect these dangerous gases, we can protect our environment and control industrial production, which is critical for saving lives. These gases pollute the environment and pose serious health risks. As a result, our skin, eyes, and respiratory system may suffer serious consequences [14].

14.2.1 Gas sensors

The name "gas sensors" implies that they are tools for providing qualitative and quantitative presence of vapors or gases of environmental interest such as volatile organic molecules of alcohols, ketones, and oxides of carbon, sulfur, nitrogen, and so on. Absorption of light, change in electrical conductivity, electrochemical behavior, and other methods are commonly used to detect these toxic vapors and gases. Other gas sensors, however, use physical properties of gases such as thermal conductivity, pressure, temperature, super compressibility, flow, and octane number (for gaseous fuels), among others. When gas sensors provide additional outputs such as mass spectrometry, differential thermal analysis, nuclear magnetic resonance, gas chromatography, and so on, they are referred to as gas analyzers. To achieve sensing properties, gas sensors typically use materials such as polymers, nanocomposites, oxide nanomaterials, and graphene-based materials. Whether it is a gas sensor or a gas analyzer, the end performance of these tools is best described by parameters such as minimum detection limit (MDL), signal to noise (S/N) ratio, response time, power consumption, dimensions and weight, and so on.

14.2.2 Types of gas sensors

Depending on the approach used in sensing, gas sensors can be broadly classified as semiconductor-based sensors, ion selective electrode-based sensors, metal oxides-semiconductors-based sensors, micro machined semiconductor-based sensors, solid state gas sensors, and so on.

14.2.3 What are biosensors?

Biosensors, also known as bio-recognition devices, are a type of analytical tool that can convert biological and/or chemical signals to electrical signals [15]. They are primarily used in quantitative mode to determine the presence of biologically or chemically active species in a given sample. A biosensor should ideally have some intrinsic properties such as high reproducibility, robust stability over a wide range of experimental and environmental conditions, economic viability, eco- and user-friendliness, and resistance to interference from foreign elements. Another significant point is that it must produce a linear response over the concentration range of analytes, resulting in a linear relationship between the amount of signal produced and the concentration of unknown biological or biochemical species to be sensed [16,17]. From compositional viewpoint, a biosensor is essentially composed of following parts:

1. *Biocatalyst:* It has a property to transform a substrate to product.
2. *Transducer:* The function of a transducer is to converts the reaction to electrical signals. Greater the extent of reaction, larger would be the intensity of the signal.
3. *Amplifier:* It simply amplifies the output signal received from the transducer so that it can be measured and recorded.
4. *Processor:* It processes the amplified signals.
5. *Display:* The signal received from processor is displayed on the monitor. A schematic presentation of a typical biosensor may be depicted as shown in Fig. 14.1.

14.2.4 Working of a biosensor

It is worth noting that the role of the transducer is critical in determining the efficiency and type of a biosensor [18]. The biosensor is referred to as a thermal biosensor if the transducer can monitor the heat absorbed or evolved in the reaction (thermal transducer). If the transducer has the capability of measuring the quantum of light absorbed or emitted as a result of the reaction (optical

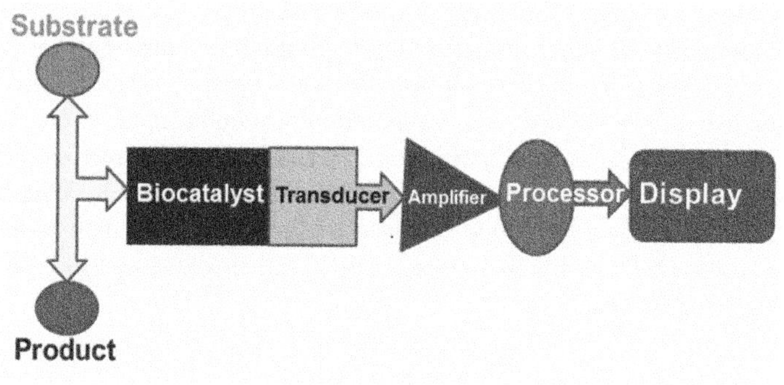

FIGURE 14.1

A schematic presentation of a typical biosensor.

transducer), the biosensors are referred to as optical biosensors. Potentiometric biosensors are used when it is convenient to measure the potential developed as a result of unequal charge distribution (electrical transducers). However, if the reaction is accompanied by electron transformation (such as a redox reaction), the biosensors are referred to as amperometric biosensors [19].

14.2.5 Classification of biosensors

In broad sense, the biosensors may also be classified into first, second and third generation biosensors [20]. In the first generation of biosensors, the reaction products diffuse into the transducer and produce a signal. In the second generation of biosensors, a mediator is placed between the reaction and the transducer solely to improve the transducer's response. The reaction itself produces a response in the third type of biosensor, and no mediator is required.

14.2.6 Mechanism of biosensors

Certain enzyme-catalyzed reactions are exothermic in nature and may be best studied with a thermal biosensor [21]. Certain reactions, for example, glucose oxidase catalyzed reaction of glucose, amylase catalyzed reaction of starch, urease catalyzed reaction of urea, uricase catalyzed reaction of uric acid, cholesterol oxidase catalyzed reaction of cholesterol, etc., have molar heat enthalpy ranging from 10 to 100 kJ/mole. Thermistors are commonly used in thermal sensors to register the temperature developed during the reaction. Potentiometer biosensors are devices that use a specific type of electrode known as ion selective electrodes (ISE) to convert biological reactions to electrical signals [22]. Glass electrodes for cations are commonly used, as are glass electrodes coated with a gas-permeable membrane for gases such as carbon dioxide, ammonia, hydrogen sulfide, and others, and solid state electrodes coated with a thin membrane of silver sulfide and another silver halide. Amperometric biosensors, on the other hand, use a current generated when a potential is applied between two electrodes [23]. They are almost identical to potentiometric biosensors in terms of sensitivities, response time, and dynamic ranges. Optical biosensors measure the optical energy of reactants and products in a reaction or the emission of light in a luminescent process.

Colored disposable strips for measuring blood sugar levels are a common example of the former type of optical biosensor. They are composed of a cellulosic bad containing relevant enzymes such as glucose oxidase and horseradish peroxydase (HRP) as well as appropriate reagents such as o-toludine. Such sensors are best suited for analysis such as whole blood monitoring in diabetes control, among other things [24]. What actually happens is that the aerobic oxidation of glucose produces hydrogen peroxide which oxidizes light colored o-toludine to highly colored dye. The reactions involved may be depicted as below:

$$\text{D} - \text{glucose} + \text{Oxygen} \rightarrow \text{D} - \text{glucono} - 1,\ 5 - \text{lactone} + \text{hydrogen peroxide perosidase}$$

$$\text{Chromogen} + \text{Hydrogen peroxide} \rightarrow \text{Dye} + \text{Water}$$

The color of the dye stripes is best read with a colorimeter, but the developed colors are frequently compared to a standard color chart. There are numerous commercially available strips on the market that contain various types of enzymes depending on the nature of the biosensor. For example, a biosensor that makes use of luminescence produced by firefly contains an enzyme luciferase (Photinus–luciferin 4-monooxygenase, ATP hydrolyzing, EC 1.1312.7) can detect the presence of bacteria in food samples as well as clinical samples. What happens is that bacteria are lysed and the ATP is released in the process which is almost equal to the number of bacteria. The released ATP reacts with the D-lucinferin (a protein) and results in production of yellow light. The involved reaction may be represented as shown below,

$$\text{Luciferease ATP} + \text{D} - \text{luciferin} + O_2 \rightarrow \text{Oxyluciferin} + \text{AMP} + \text{pyrophosphate} + CO_2 + \text{light}$$

The yellow glow produced can be easily quantified using either photomultiplier tube or spectrophotometer my measuring the absorbance at 562 nm. The biosensors have several advantages like economic viability, specificity, ease of operate, excellent reproducibility etc. Furthermore, they can be used comfortably over a wide range of analyte concentrations. Although many materials are used in the design of various types of biosensors, the material used is very important in determining the efficiency of the biosensor. Bio-nanocomposites have recently become an integral part of biosensors, with numerous applications in sensing various biomolecules and gases.

14.3 Biodegradable polymers as gas sensors and biosensors

Biodegradable polymers were discovered in 1980 as reported by Ghanbarzadeh and Almasi [25,26]. Prior to the discovery of biodegradable polymers, the first polymer used in industry by humans was a natural polymer. In the 18th century, natural rubber was discovered [27]. First technical biodegradable polymer launched in the market was Rilsan (or Polyamide 11) in the year 1947. This is the first patent, commercialized by a small French firm called Organico [28]. The demand for biodegradable polymers grows by 30% each year. According to available data, Europe used the greatest proportion of these polymers. This discovery has ushered in a new era in polymer and material sciences. The first polymer reported to investigate the unique electrical properties of polymeric materials was polysulfurnitride [29].

The movement of enzymes and the degradation of living organisms can aid in the biodegradation process. Polymer fragmentation, such as oxidation, photo-degradation, or hydrolysis, as well as biotic reactions, can occur during this process. These are known as microorganism degradation processes [30–33]. Chemical structure of polymer and the degradation of surrounding molecule also depends for biodegradability [34,35]. Fig. 14.2 shows the structure and sources of various naturally occurring polymers (biopolymers).

14.3.1 Properties

Biodegradable polymers have unique electrical, mechanical, structural, thermal properties [36]. The following properties are given

1. *Physical Properties:* Biodegradable polymers can be formed in any shape due to their excellent physical properties such as flexibility, softness, light weight etc.
2. *Chemical Properties:* They are simple to make and can react with any molecular group. Many molecular groups can be attached to their derivatives, which have the properties of their parent polymer. It also has the properties of processibility and reactivity.

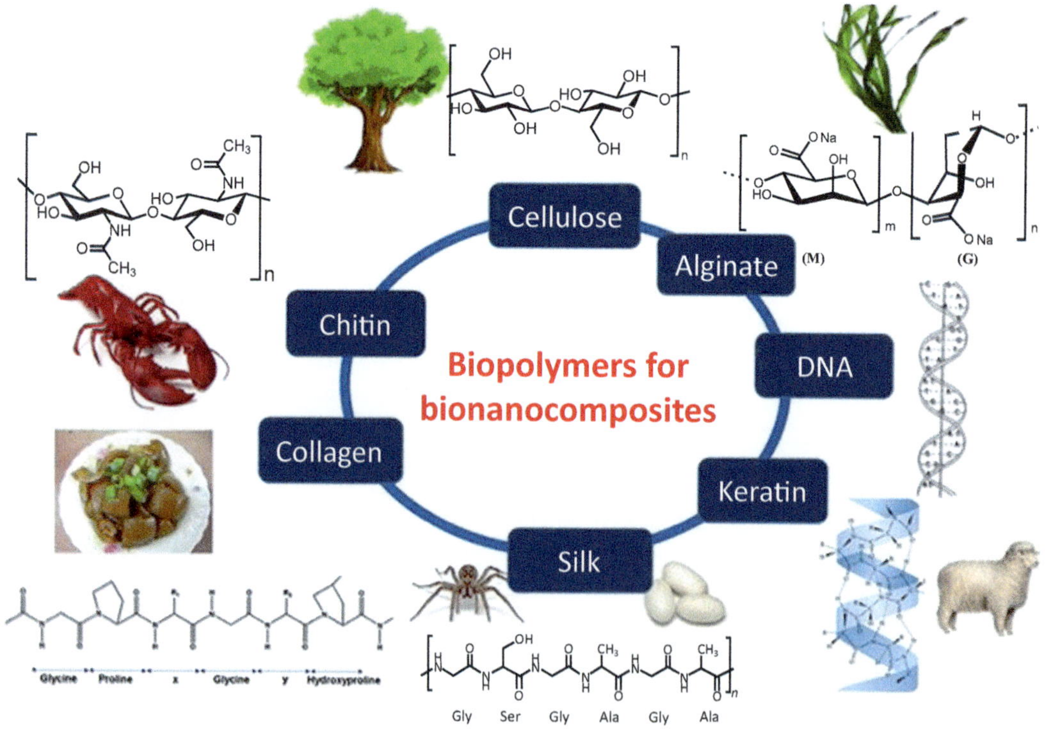

FIGURE 14.2

The structure and source of various naturally occurring polymers (biopolymer) [6].

Copyright 2018 reproduced with permission from Elsevier.

3. *Electrical Conduction and dielectric Property:* A redox reaction occurs when we add or remove p-type or n-type material from a polymer chain. Electrical conduction of charge carriers in the form of free electrons or holes is possible in these polymers.
4. *Biocompatibility:* It is the property of the material to being compatible with living tissue and surrounding. These materials can easily interact with living tissue in the human body with no negative consequences. It can be used in drug delivery devices as well as a variety of other medical devices [37–41].
5. *Biodegradability:* Because of its biodegradability, polymer becomes biodegradable polymer. When enzyms pass through microorganisms, they break down organic substances into simpler materials. After the completion of this whole reaction organic substance transforms to simpler inorganic substance such as water CO_2 and CH_4. Biodegradable polymers have been used in a variety of applications [34,42,43].
6. *Recyclability:* These polymers have good recyclability property. Their disposal can fuel and recycle after use. It is not hazardous for our nature and ecosystem [44].
7. *Mechanical properties:* These polymers have good tensile strength, flexibility, stiffness. These are the parameters which are important for making devices and different industrial application [45].
8. *Toxicity and Biological Properties:* Biodegradable polymers and their products are nontoxic and endotoxin, and they do not produce undesirable foreign bodies.
9. *Thermal Properties:* The time of degradation ought to be matched with the reconstruction time. Aside from these characteristics, they have high water vapor permeability, a good oxygen barrier, low thermal stability, and so on [46].

14.3.2 Applications

Biodegradable polymers have a wide range of applications due to their unique properties such as biocompatibility, flexibility, lightweight, low cost, and good electrical, thermal, and structural properties. This is because biodegradable polymers are used in different areas as agriculture and environmental monitoring, drug delivery devices and other biomedical applications, chemical, biological and gas sensors, artificial muscles, nerves, electronic tongues, noses, gas sensors, bio-imaging, food packaging, electronic waste management, display devices such as light emitting diodes, energy storage systems, rechargeable batteries, electromechanical batteries, and electromechanical actuators [44,47–51].

Guan et al. in 2020 developed NH_3 gas sensor which is highly sensitive and sensitivity increases with increasing NH_3 concentration. They designed flexible nonwoven fabric substrates by using zinc oxide (ZnO) sheets and polyaniline (PANI). The ZnO/PANI/nonwoven composite material possesses appropriate mechanical, thermal, and sensing properties such as flexibility, thermostability, and quick response time. The composites exhibit good response at room temperature, and the value of response increases with increasing NH_3 concentration in the range of 0.5–10 ppm. The NH_3 gas sensor is used in a variety of applications, including environmental monitoring, the creation of flexible devices, and so on [52]. Fig. 14.3A and B show the thermal analysis of nonwoven, ZnO/nonwoven, PANI/nonwoven and ZnO/PANI/nonwoven which exhibit the thermal stability of the composite materials. Fig. 14.4 shows dynamic responses curve of PANI/nonwoven and ZnO/PANI/nonwoven toward NH_3 in the concentration range of 0.5–10 ppm.

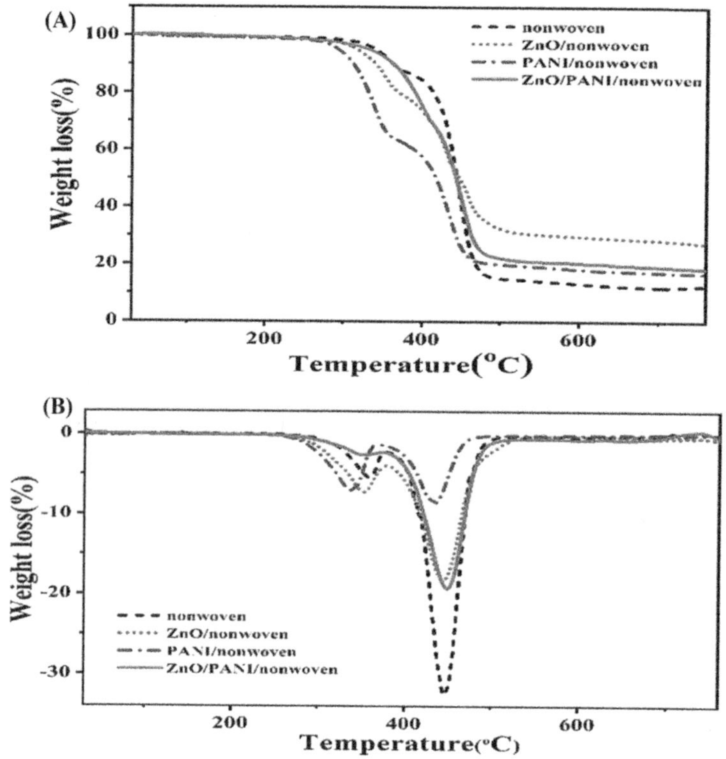

FIGURE 14.3

(A) Thermogravimetric analysis (TG). (B) Differential thermogravimetric analysis (DTG) curves of nonwoven, ZnO/nonwoven, PANI/nonwoven and ZnO/PANI/nonwoven [52].

Copyright 2020 reproduced with permission from Materials Science for Energy Technologies (Science Direct).

Fig. 14.5 shows the responses curve of ZnO/PANI/nonwoven and PANI/nonwoven to the NH_3 concentration of (0.5–10 ppm).

Lou et al. in 2020 have synthesized porous nickel molybdate nanosheets/chitosan ($NiMoO_4$/CHIT) nanocomposite by sonochemical method. They investigated both voltammetric and electrochemical sensors. They investigated the electrocatalytic performance of $NiMoO_4$/CHIT nanocomposites in the detection of amlodipine. Volammetry studies show that this nanocomposites material is highly sensitive and can be used to make electrochemical sensors, biosensors, and other devices. Fig. 14.6A depicts the DPV responses of $NiMoO_4$/CHIT/GCE in phosphate buffer with different concentration of amlodipine and 6(B) Calibration plot of peak current/μA vs [amlodipine]/μM. Fig. 14.7A shows the reproducibility analysis of $NiMoO_4$/CHIT/GCE s(n = 3) and 7(B) Repeatability analysis of $NiMoO_4$/CHIT/GCE [53].

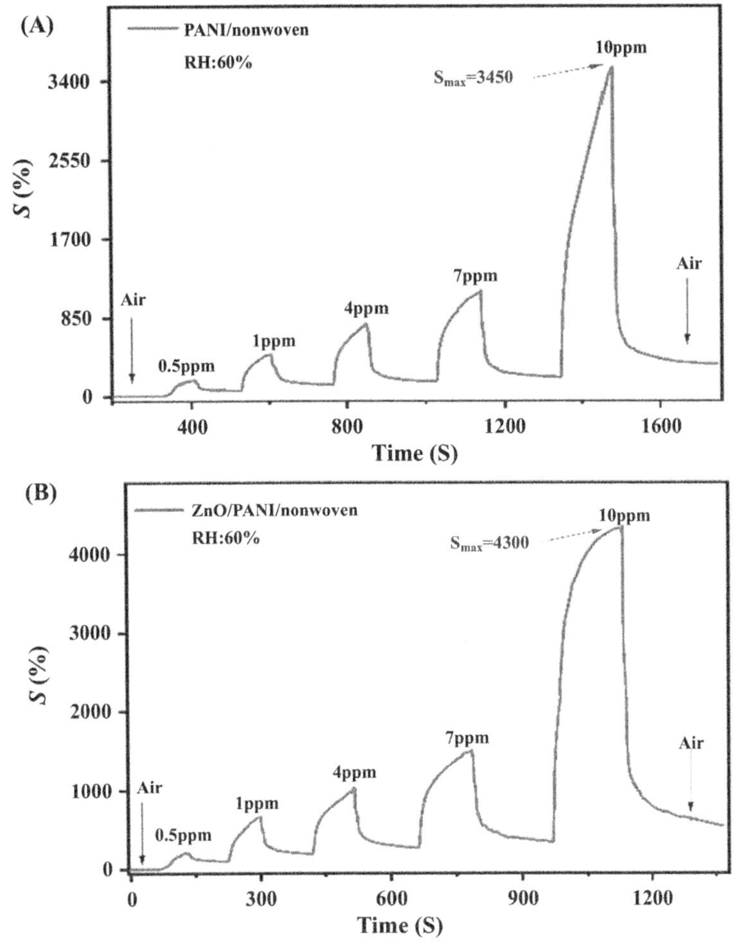

FIGURE 14.4

Dynamic responses of PANI/nonwoven and ZnO/PANI/nonwoven toward NH_3 in the concentration range of 0.5–10 ppm at room temperature [52].

Copyright 2020 reproduced with permission from Materials Science for Energy Technologies (Science Direct).

14.4 Bio-nanocomposites

When one filler material is reinforced into another host biopolymer, a bionanocomposite is formed. The final product has improved in terms of physical, chemical, and electrical properties [54]. Previously, the scientists created bulk and macromolecular materials. Because of technological advancements, a wide range of analytical instruments such as transmission electron microscopes

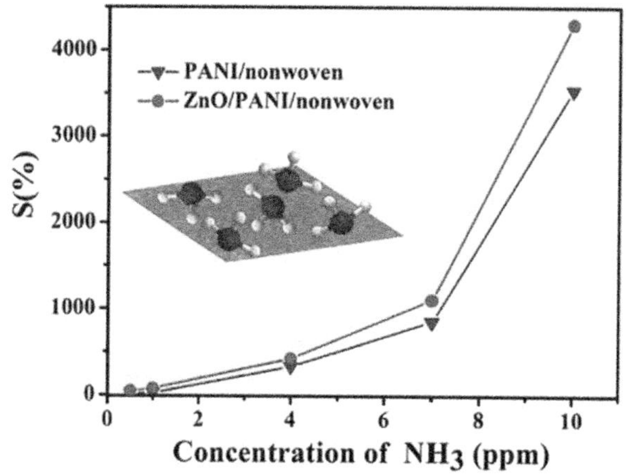

FIGURE 14.5

Responses of ZnO/PANI/nonwoven and PANI/nonwoven to the NH_3 concentration of (0.5–10 ppm) [52].

Copyright 2020 reproduced with permission from Materials Science for Energy Technologies (Science Direct).

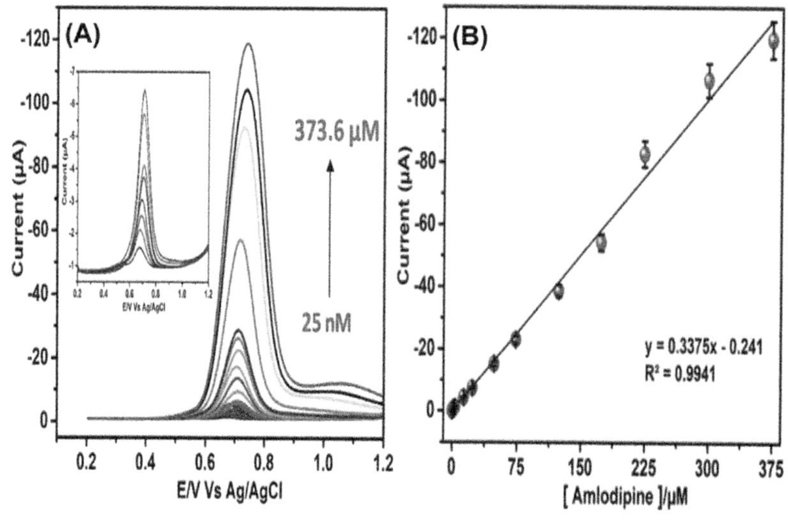

FIGURE 14.6

(A) DPV responses of $NiMoO_4$/CHIT/GCE in phosphate buffer with different concentration of amlodipine. (B) Calibration plot of peak current/μA vs [amlodipine]/μM [53].

Copyright 2020 reproduced with permission from Elsevier.

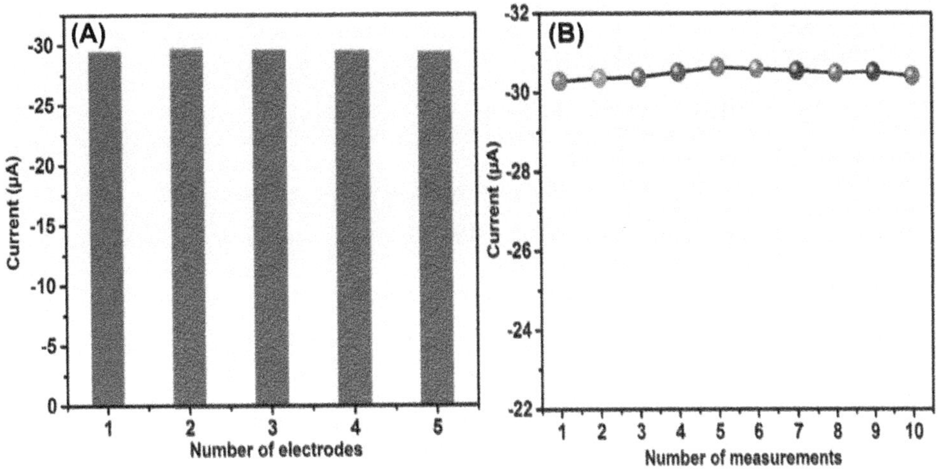

FIGURE 14.7

(A) Reproducibility analysis of $NiMoO_4$/CHIT/GCE s(n = 3) and (B) Repeatability analysis of $NiMoO_4$/CHIT/GCE [53].

Copyright 2020 reproduced with permission from Elsevier.

(TEM) and scanning electron microscopes (SEM) are now available. Nanocomposites are typically made with 1D-, 2D-, or 3D-structured nanoparticles. We can see the structure of materials at the nanoscale level. A nanocomposite is a composite in which at least one of the ingredients is on the nanometer scale. A nanocomposite with novel properties of both materials has been created. It has mechanical strength, light weight, ductility, electrical conductivity, and so on [36,55–58]. Electrical conductivity, mechanical properties such as strength, modulus, stability, thermal degradation, chemical resistance, surface appearance, decreased gas, water, and hydrocarbon permeability, and reduced energy band gap are some of the enhanced properties of nanocomposites. Nanocomposites are used in a variety of fields due to their superior improved properties, including sports, food, electronics, construction, agriculture, transportation, and aeronautics engineering [57,59]. There are several synthetic routes to prepare bio-nanocomposites as depicted in Fig. 14.8.

14.5 Biodegradable polymer nanocomposites for gas-sensing applications

There are many different types of toxic gases that are hazardous to our environment and livelihoods, as well as causing problems in agriculture, so we must detect these harmful gases. During World War II, Germany developed chemical weapons and poisonous gases for use against Japan, resulting in serious health problems such as cancer, skin and genetic disorders, environmental issues, and so on. The demand for gas sensors devices has gradually increased in order to save human lives. Scientists have created a wide range of gas sensors for a variety of applications,

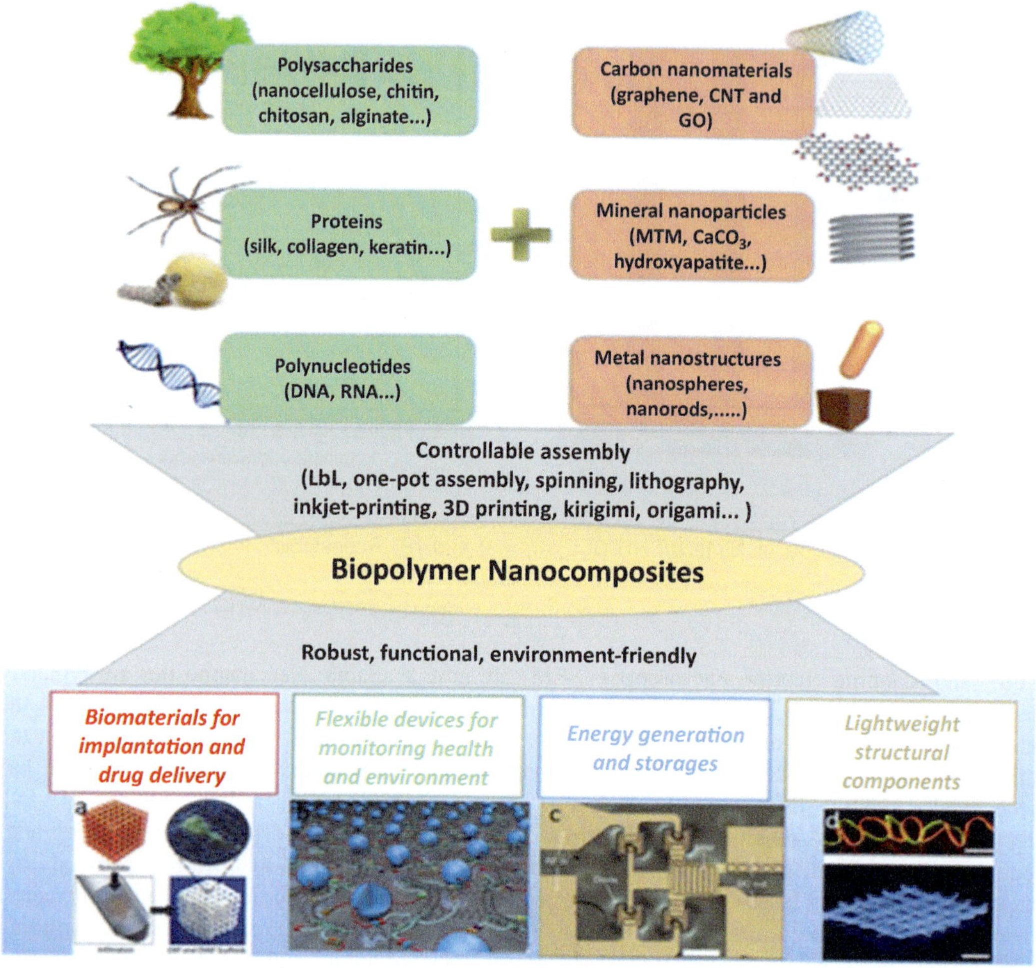

FIGURE 14.8

Several synthetic routes to prepare bio-nanocomposites [6].

Copyright 2018 reproduced with permission from Elsevier.

including optical gas sensors, metal oxide-based gas sensors, acoustic-based gas sensors, electrochemical sensors, and so on. Because a synthetic polymer does not degrade, green materials are used to make gas sensor devices. Many industrial and environmental applications require these polymers' environmentally friendly behavior [60].

Polysaccharide is one of the most abundant groups of biopolymers, derived from a variety of natural resources such as plants (cellulose and xylan), sea-animal shells (chitin and chitosan), and algae (alginate and carrageenam) [61,62]. Non-renewable energy resources degrade day by day, polysaccharide is one of the renewable biopolymers which is less toxic, biocompatible, stable, economic etc.

[63]. This biopolymer and nanocellulose based nanocomposites are used in a variety of sensor and biosensor devices [64]. Chitin is the second most abundant biopolymer found in nature after polysaccharides. The presence of chitin in nature demonstrates its crystalline nature [65]. Chitin's structural properties demonstrate that this biopolymer and its components provide structural integrity and protection to arthropod exoskeletons as well as fungi and yeast cell walls [6].

Fig. 14.9 depicts a biopolymer composite fabrication technique. When biopolymer reacts with a solvent, carbon-black or carbon nanotubes are formed. Carbon-black (CB)-based biopolymers and composites have a high surface area, excellent physical, electronic, structural, and chemical properties, and a small particle size [34]. CB and its nanocomposites are mainly used in the production of variety of gas sensors, biosensors, sensor devices, catalysts etc. [66] (Fig. 14.10).

Carbon nanotube (CNT)-based biopolymers and composites have superior electronic, dielectric, structural, and gas-sensing properties. CNT-based materials also have a large surface area, making it easy to detect many hazardous toxic gases [68,69]. Cellulose, chitosan, polylactic acid (PLA), k-carrageenan, and a 40-mer polypeptide are five biopolymers which can be incorporated with CNTs to design different gas-sensing, vapor sensing, and bio-sensing devices. These devices are used in a variety of applications such as medical, military, industrial, and electrical. Ammonia is a toxic gas found in the environment, and a single-walled carbon nanotube (SWCNT)-cellulose based biopolymer and its composite are used to detect it. SWCNT-based ammonia gas sensors have a fast response time and good repeatability at 10–100 ppm [67,70].

In many studies researcher have found that silver nanoparticles (AgNPs) have shown excellent optical, electronic and chemical properties [71,72]. Because of their high surface energy and other fluorescence activity, size, shape, inter-particle spacing they is suitable for gas-sensing applications [73]. The linear and non-linear optical behavior of AgNPs can be controlled. These novel properties

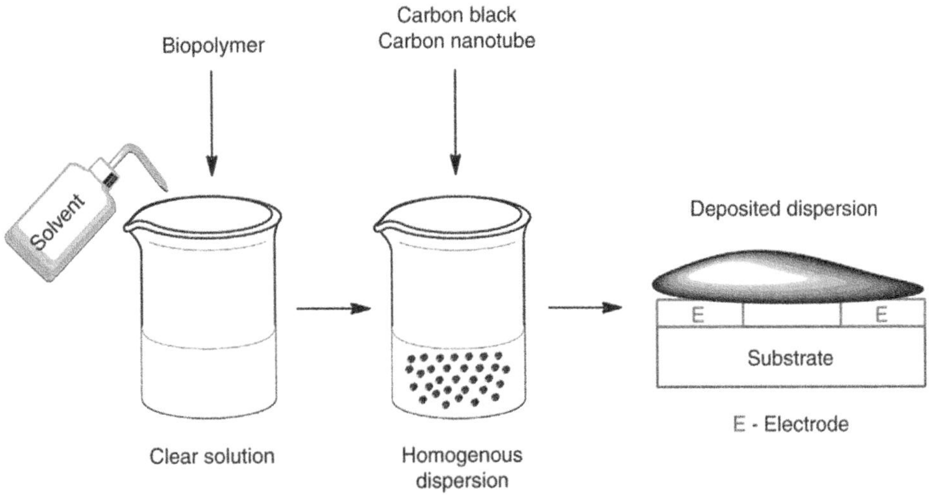

FIGURE 14.9

Fabrication of biopolymer composite [67].

Copyright 2017 reproduced with permission from Elsevier.

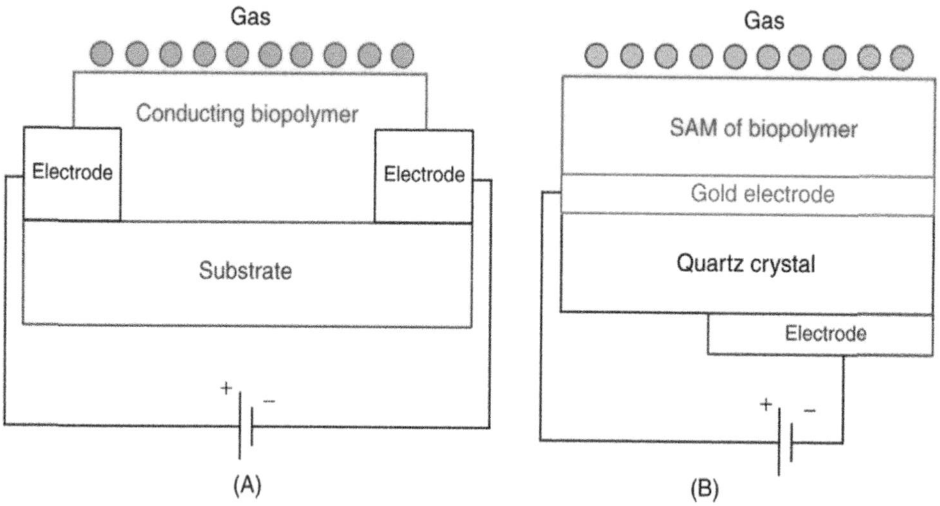

FIGURE 14.10

Device configuration for chemiresistor. (A) Piezoelectric crystal. (B) Type gas sensor devices [67].

Copyright 2017 reproduced with permission from Elsevier.

enable AgNPs to be used in many imaging methods. Essentially it can also be used in gas and biosensing applications like detection of human serum antibody [74], laser desorption/ionization mass spectrometry of peptides [75], colorimetric sensor for histidine [76,77].

PLA has been extensively available in the form biopolymer [78,79]. The ring-opening polymerization method can be used to create PLA. The fermentation of sugar feed stocks yields monomers of lactides and lactic acid in this method [80]. In nature, these polymers are not completely crystalline. Polymers are amorphous or non-crystalline in nature, but some are semi-crystalline. Polymer structural analysis reveals whether the polymer is semi-crystalline or amorphous. The XRD technique detects both crystalline and amorphous regions. XRD provides information such as phase, average particle or crystallite size, lattice constant, and so on [81,82]. The structural analysis of PLA shows semi-crystalline nature of this biopolymer [79,80]. When composites of PLA are prepared with nanofillers their crystallinity and electrical conductivity increase [83]. PLA is used in many industrial and medical applications like e-nose, critical disease detection etc. [84,85]. PLA/CNT-based nanocomposites are smart materials which are used for gas and vapor sensing [86] (Figs. 14.11 and 14.12).

Kausik et al. used polymer nanocomposites-based gas sensors to detect toxic gases in the environment. With the use of organic-inorganic hybrid nanocomposites, they can also monitor other hazardous gases such as hydrochloric acid (HCl), ammonia (NH_3), hydrogen disulfide (H_2S), nitrogen oxides (NO), carbon dioxide (CO_2), and carbon monoxide (CO) [12]. To detect toxic gases, gas sensors are now designed using biodegradable polymers and their nanocomposite materials. According to various structural analyses, biodegradable polymers have a porous structure that allows gaseous molecules from the environment to penetrate and become trapped within the polymer. Their sensing behavior is also affected by surface area, size, and volume [60,87]. Biodegradable polymers and their nanocomposites are also used in the production of electronic

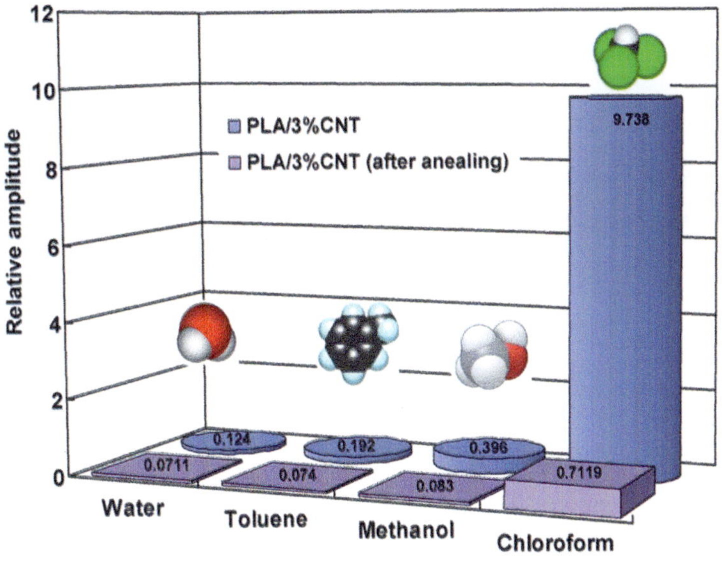

FIGURE 14.11

Effect of annealing on sensitivity of PLA/3%CNT toward different solvents [86].

Copyright 2012 reproduced with permission from Elsevier.

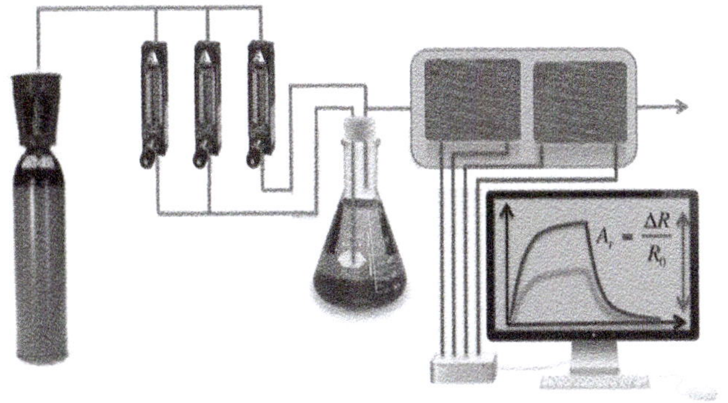

FIGURE 14.12

Scheme of chemo-resistive vapor sensing device [84].

Copyright 2012 reproduced with permission from Elsevier.

devices such as rechargeable batteries, photovoltaic cells, LEDs, molecular sensors, electromagnetic shielding, gas-separation membranes, and so on [87–89] (Fig. 14.13).

The use of gas sensors has increased globally due to their user-friendliness, quick response time, ease of processing, and economic viability. These sensors use less energy and thus play an important role in detecting hazardous gases. There are currently numerous toxic and flammable

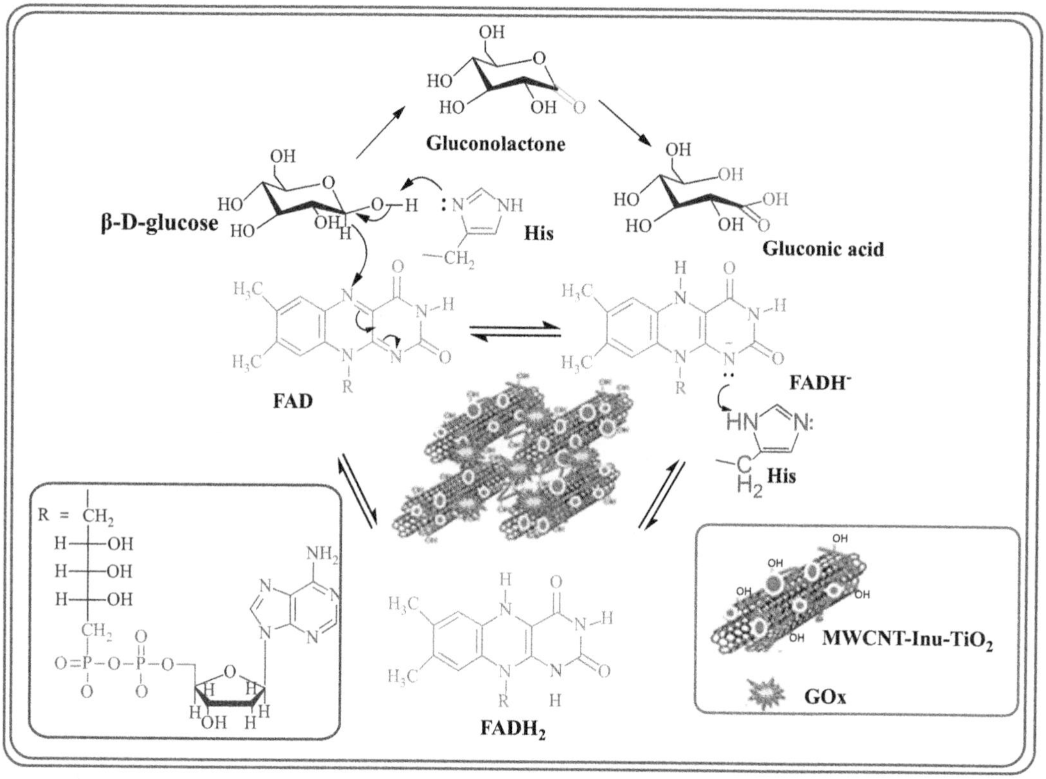

FIGURE 14.13

Mechanism of glucose oxidation at the CP/MWCNT-Inu-TiO_2/GOx electrode [90].

Copyright 2019 reproduced with permission from Elsevier.

gases such as H_2S, NO_2, CO, SO_2, NH_3, O_3 and VOCs that are hazardous to both the environment and human health. It is therefore critical to detect these gases in order to protect ourselves from them. Skin cancer, asthma, lungs infection, and other serious diseases are caused by exposure to these toxic gases. To detect toxic gases, gas sensors are used in many devices for safety reasons. They are used in liquid petroleum gas (LPG), home appliances, mining, automotive industries, electronic industries, military etc. [69]. In addition to these systems, there are so many bionanocomposite-based systems which are summarized in Table 14.1 [67].

14.6 Biodegradable polymer nanocomposites for bio-sensing applications

A biosensor is a device that converts biological reactions into data. Because of its superior properties, biodegradable polymers are used to create biosensors. It employs, investigates, and diagnoses

Table 14.1 Applications of biopolymer-based gas/vapor sensors [67].

Analytes	Biopolymer composites	Transducer	References
Ammonia	SWCNT–cellulose	Chemiresistor	[91]
Water, methanol, ethanol, acetone, chloroform, THF, toluene, *n*-hexane	MWCNT–cellulose	Aerogel; Chemiresistor	[92]
Ammonia	MWCNT–polypeptide	Chemiresistor	[93]
Chloroform, methanol, toluene, water	MWCNT–PLA	Chemiresistor	[86]
Ethyl acetate, carbon tetrachloride, methanol, ethanol, xylene, *n*-hexane	CB–poly(lactic acid)	Chemiresistor	[94]
Water, methanol, toluene	CB–chitosan	Chemiresistor	[95]
Formaldehyde	Polyethyleneimine-chitosan	PZ(QCM)–gold	[96]
Methylamine, dimethylamine, ethanol	Polyaniline–chitosan	PZ(QCM)–gold	[97]
Hydrogen	Polyaniline–chitosan	Chemiresistor	[98]
Water	SWCNT–carrageenan	Chemiresistor	[99]
Water	MWCNT–carrageenan	Chemiresistor	[99]
Water, methanol, ethanol, isopropanol, toluene, Chloroform	MWCNT–chitosan	Chemiresistor	[100,101]
Water, methanol, ethanol, isopropyl alcohol	MWCNT-chitosan-polyvinyl Alcohol	Nanohybrid, Chemiresistor	[102,103]
Water, chloroform, ethanol, isopropyl alcohol	MWCNT–chitosan–polyethylene oxide	Nanohybrid, Chemiresistor	[103]
Methylamine, dimethylamine, diethylamine, methanol, ethanol, isopropyl alcohol	Chitosan	PZ(QCM)–gold	[104]
Chloroform	Chitosan	Chemiresistor	[105]
Water	Chitosan	Optical sensor	[106]
Ammonia	Chitosan–carrageenan	Optical sensor	[106]
Ammonia, hydrochloric acid	Chitosan–carrageenan	Optical sensor	[107]
Water, methanol, toluene	Amylose	Chemisensor	[108]
Water, methanol, toluene	Starch	Chemisensor	[108]
Methanol, propionic acid, trimethylamine, dinitrotoluene, dimethyl methylphosphonate	SWCNT–ss-DNA	FET	[109,110]
Toluene, acetonitrile, *n*-hexane, *n*-heptane, chloroform, ethanol, methanol, cyclohexane, 2-propanol, dichloromethane, diethylether, tetrahydrofuran, cyclohexanone, triethylamine, anisole 1-propanol, 1-butanol, 1-pentanol, butyraldehyde, styrene	DNA	PZ(QCM)–gold	[111]
Toluene, gasoline	Polyglycolic acid	PZ(silicon)	[112]
Aceticacid	Polypeptide SAM	PZ(QCM)–gold	[113]
1-Hexanol, 1-pentanol	Polypeptide SAM	PZ(QCM)–gold	[114]

(Continued)

Table 14.1 Applications of biopolymer-based gas/vapor sensors [67]. ***Continued***

Analytes	Biopolymer composites	Transducer	References
Aceticacid, butyricacid, ammonia, dimethylamine, benzene, chlorobenzene, their mixtures	Oligopeptide SAM	PZ(QCM)−gold	[115]
Triethylamine ammonia, acetone, methanol, ethyl acetate, o-xylene, p-xylene, Toluene	Oligopeptide SAM	PZ(QCM)−gold	[116]
Acetone, 2-propanol	Poly-l-proline SAM	PZ(QCM)	[117]

SAM, *Self assembled monolayer;* PZ, *piezoelectric;* QCM, *quartz crystal microbalance;* CB, *carbon-black;* MWCNT, *multiple-wall carbon nanotube;* SWCNT, *single-walled carbon nanotube;* FET, *field effect transistor;* PLA, *poly(lactic acid).*
Copyright 2017 reproduced with permission from Elsevier.

various diseases such as cancer, diabetes, virus and fungal infection, and so on. Biosensors are used in a wide range of medical applications [38,118,119]. Diabetes is now one of the most common health problems, causing a slew of other issues such as lungs and kidney infections. In everyday life, the demand for glucose sensing devices and sugar-free products has increased. To that end, researchers created sensors and sugar-free products that are now available on the market [90].

Biodegradable polymers have properties that are ideal for creating a gas sensor with high sensitivity, selectivity, and a short response time. Biodegradable polymers and their nanocomposites are long-lasting, reversible, low cost, low-maintenance, and easy to use. Because they are biocompatible with the human body, biodegradable polymer materials are ideal for detecting serious diseases. After use, these green materials pose no threat to our environment. In today's world, the demand for biodegradable polymers is increasing. When used as sensing materials in gas-sensing, biodegradable polymers or their nanocomposites show good agreement. At room temperature, MnO_2, SnO_2, and ZnO exhibit high sensitivity and stability for sensing various toxic gases [120]. Biosensors are the result of the recombination of biological or biocompatible materials, and they are also known as bio-receptor to transducers. Biosensors have been used in a variety of applications in recent years, including biomedical, food, electrical and electronic industries, pharma, agriculture, metrological measurement, and security [120].

14.7 Literature survey of recent advancement and prospects of biopolymer nanocomposites for gas-sensing and bio-sensing applications

Zare et al. succeeded in expanding a nanobiosensor using biodegradable polymers (poly (lactic acid)) (PLA) and poly (ethylene oxide) (PEO) in 2021. They calculated the conductivity of the material using a simple model advancement. All parameters such as degradation fraction, interphase depth, and tunneling distance were analyzed and compared with experimental data using this model. The nanobiosensor's conductivity was increased by a short tunneling distance and a thick interface. This review article discusses recent advances in the field of biosensors [121].

Inderan et al. developed an ethanol gas-based gas sensor in 2020. They created biodegradable nanocomposites from nickel and palladium SnO_2 nanorods, PANI, and polyhydroxy-3-butyrate (P3HB). The hydrothermal and electrospinning methods were used to create the bio-nanocomposites. For ethanol gas-sensing, the response and recovery times were tested at various temperatures and concentrations. The structural and chemical analysis of the bio-nanocomposites demonstrates that they can be used to fabricate P-N junction diodes and gas-sensing applications [122].

Palanisamy et al. developed a highly stable and sensitive biosensor based on H_2O_2 gas-sensing in 2020. They created this biosensor by first sonochemically preparing composites of multiwall carbon nanotubes (MWCNT)/cellulose nanocrystals (CNC)/zinc oxide (ZnO). The electrochemical analysis revealed that bio-nanocomposites have superior electrocatalytic properties with a lower potential value for H_2O_2 sensing. The results show that this biosensor has a good recovery time for H_2O_2 gas sensing. It has a wide range of potential applications [123].

14.8 Some significant literature survey on gas sensors and biosensors

Thangamani et al. used solution casting method to prepare polyvinyl alcohol (PVA)/polypyrrole (WPPy)/vanadium pentoxide (V_2O_5) nanocomposite films. They discovered that mechanical properties such as tensile strength and flexibility of nanocomposites films are improved. They tested the sensitivity, selectivity, and response time of chemiresistive sensors by detecting LPG. Surface morphology is critical for detecting LPG sensing or other sensing properties. In this study, the morphology of V_2O_5 was changed from spherical to nanorods in nanocomposites, indicating that LPG sensing was improved. Thermal testing revealed that nanocomposites were also thermally stable. At room temperature, they discovered a maximum sensitivity of 1.16% for LPG. The demand for LPG gas detectors, which detect harmful gas leaks, has increased. Most people in rural areas are unaware of LPG leakage. As a result, in terms of safety, LPG sensors can warn people before a leak occurs, allowing them to avoid unanticipated events [124] (Figs. 14.14 and 14.15).

Deshmukh et al. used solution casting to create thin films of polyvinylidene fluoride (PVDF)/cerium oxide (CeO_2)/graphene oxide (GO) nanocomposites. They investigated the nanocomposites' gas-sensing properties for NH_3 gas as well as volatile organic compounds (VOCs) such as ethanol, acetone, formaldehyde, and toluene. They discovered that NH_3 gas produces better results than others. When GO is combined with composites, it exhibits excellent response time and sensing properties [91]. Netto et al. developed an electronic nose that is built when a chemo-resistive gas sensor is connected to a computational system. Iono-gels were incorporated into metallic interdigitized electrodes to create chemo-resistive gas sensors. They tested this nose at room temperature with five organic volatile compounds (methanol, ethanol, acetone, ethyl acetate, and hexane) as evidence. Machine learning and two cross-validation processes were used to ensure that this nose was perfect. They had achieved a 96% hit rate [125].

Okechukwu et al. developed nanocomposites-based sensors for detecting and observing the sensing behavior of 2-methylcyclopentanone vapors prepared with cellulose acetate (CA), carbon nanoparticles (CNPs), and manganese dioxide (MnO_2) nanorods. They investigated the material's sensing behavior using deep oxidation decomposition of CO_2. When metal oxide was incorporated in composites (MnO_2/CNPs/CA) in different mass ratios such as 1:1.5:3, 2:1.5:3, and 2.5:1.5:3, the

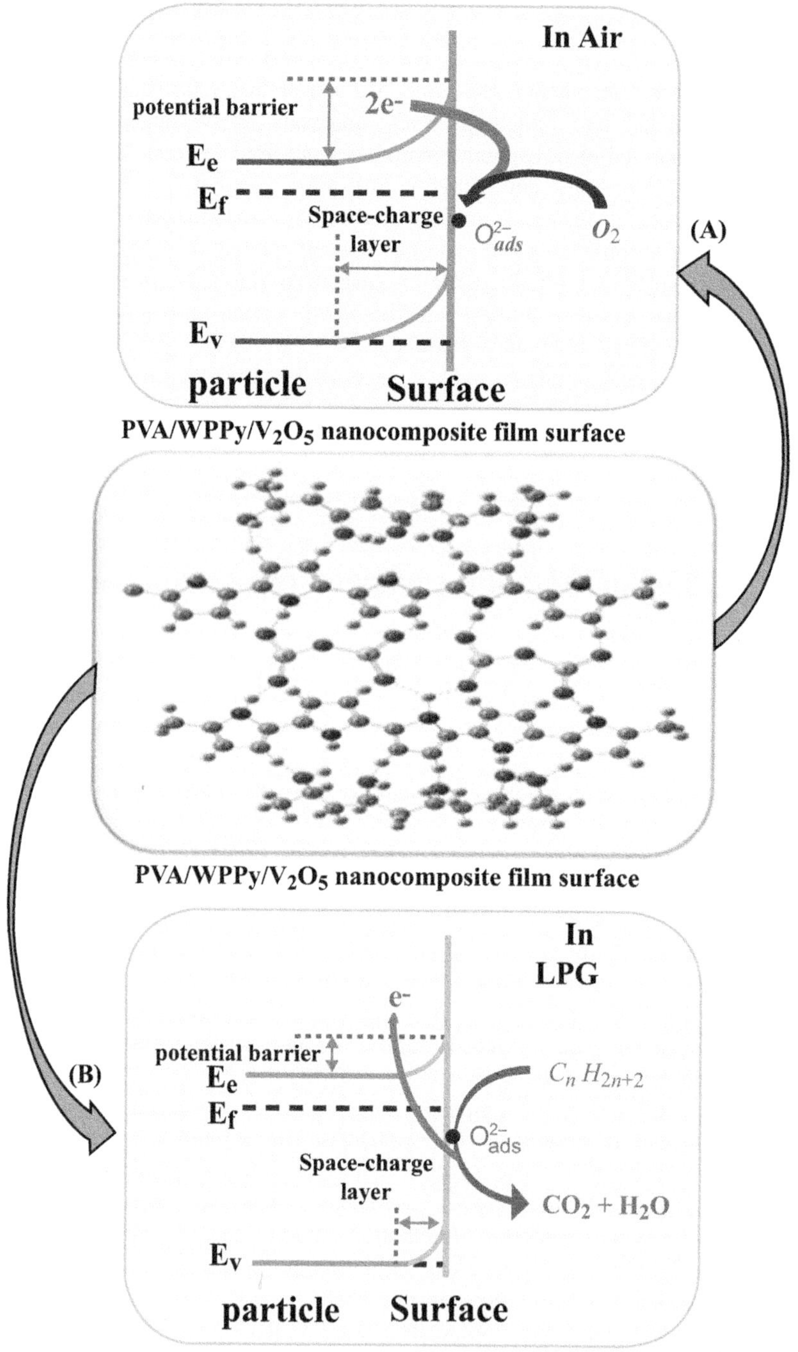

FIGURE 14.14

A schematic diagram of LPG sensing mechanisms of $PVA/WPPy/V_2O_5$ nanocomposite sensors (A) In presence of air (B) In presence of gas [124].

Copyright 2021 reproduced with permission from Elsevier.

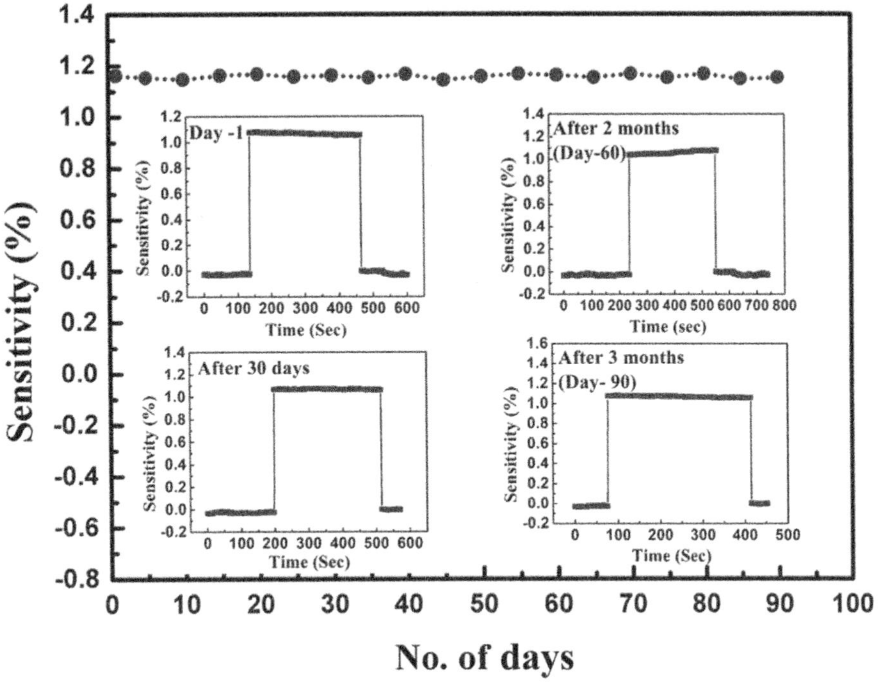

FIGURE 14.15

The stability of chemiresistive sensor based on PVA/WPPy/V_2O_5 nanocomposite film with 15 wt.% V_2O_5 loading for the detection of 600 ppm LPG concentration at RT [124].

Copyright 2021 reproduced with permission from Elesvier.

sensing properties of prepared nanocomposites were improved. In comparison to others, the second sensor composition (2:1.5:3) provides a faster response time and better sensing results [120]. Kalaivani and Suja created a sensor out of multiwalled carbon nanotubes (MWCNT). MWCNT-Isulin-TiO_2 composites disable the body's glucose. A CP/MWCNT-Inulin-TiO_2/GOx electrode was used to detect glucose in food products. In comparison to carbon-based electrodes, this bio-sensing device is extremely stable [90]. Chang et al. created photonic crystal (PC) sensors to detect human health and environmental problems due to their high sensitivity, colorimetric measurements, and low cost. They developed a PC-based VOC sensor that was ultra-fast and was used in the packaging industry. The nanoscale easy tear (NET) method was used to create the sensor in this case. This novel method has never been studied before. They measured the vapor-phase VOCs used in multiplexed selective sensing and nanomolding templates [126].

Thangamani et al. created nanocomposites by reinforcing PVA/WPPy with hexagonal boron nitride nanoparticles (hBNNP). They studied the nanocomposites' sensing behavior in the presence of hBNNP nanoparticles in composite materials. The impact of hBNNP loading on sensing performance was studied. At room temperature maximum sensitivity of composites material was obtained S = 0.25% at 600 ppm concentration. The recovery time, sensitivity, and other sensing parameters of hBNNP nanoparticles reinforced with PVA/WPPy composites are altered. hBNNP nanoparticles with hexagonal structures are referred to as "white graphene." When hBNNP nanoparticles were added to the nanocomposites, their thermal conductivity increased. Mechanical, structural, and morphological observations indicate that this material is suitable for LPG detection. Polypyrrole (PPy) is a conducting polymer with gas-sensing and bio-sensing properties. It's also used in polymeric batteries, electronic devices, and a variety of other things [127]. Pandey et al. investigated methods for producing biopolymer nanoparticles. Silver nanoparticles were used as a sensing material for optical sensors in their study. These sensors recovered in 2–3 seconds. They tested the ammonia at room temperature. This type of sensor is known as an optical ammonia sensor, and it is used in medical science to test plasma, saliva, and cerebrospinal fluid in the human body [77].

14.9 Current challenges and future perspectives

Despite the fact that the fabrication of various types of gas sensors and biosensors is increasing on a daily basis, there are some challenges such as practical difficulties, processing issues, sensor device efficiency, and so on that limit their wide acceptability and utility. In addition, it reacts with other analytic or materials present in the environment at the time of synthesis. Everyone is responsible for maintaining a clean and green environment. Because of the ever-increasing volume of e-waste, the demand for green electronic materials has increased in recent years. This e-waste has serious environmental and ecological implications [44]. One of the most pressing issues today is the scarcity of fossil fuels, which are rapidly depleting. Recent biodegradable polymer research indicates that these polymers can be obtained from renewable resources [128]. Because of global environmental concerns and the degradation of fossil fuels, the demand for and production of biodegradable polymers has increased in recent years. There are numerous reasons for the development of these polymers, including environmental concerns, the degradation of fossil fuels, pollution concerns, electronic waste, and so on [1,34]. The current chapter investigates some new biodegradable polymer nanocomposites research directions for the future. The first is that biodegradable polymers, because they are lighter materials, can serve as low-weight carriers for electronic devices. Sensor devices made from these biodegradable materials will be competent, light in weight, easy to miniaturize, and low in cost. The surface morphology of biodegradable polymer-based nanocomposites, on the other hand, indicates that the present nanocomposites can be efficiently used for gas sensor device applications for various sensing gases. Third, some biodegradable polymers may be used in the development of aptasensors for use in astrobiology. Fourth, the current nanocomposites can be used for super capacitor applications, and their specific capacitance will be measured. The current hybrid nanocomposites, on the other hand, will be used to make Schottky diodes.

14.10 Conclusions

Nanocomposites of biodegradable polymers have emerged as one of the most promising materials for use as gas sensors and biosensors in fields ranging from industrial to biomedical and environmental. A typical biosensor is made up of five parts: a biocatalyst, a transducer, an amplifier, a processor, and a display, all of which work together to convert a biological or chemical reaction into a recordable signal. Although a variety of materials are used in the fabrication of various types of biosensors, biodegradable polymer nanocomposites have risen to prominence as constituent materials in the design of biosensors and gas sensors. While biodegradable polymers are mostly naturally occurring degradable polymers, their nanocomposites include a diverse range of nanomaterials, including semiconductor materials, graphene quantum dots, titanium dioxide, and so on. The dispersion of nanomaterials into biodegradable polymers results in a variety of nanocomposites, which have a wide range of applications as sensors for hazardous gases and toxic biomolecules. As biosensors and gas sensors, a wide range of biodegradable polymer nanocomposites containing various biodegradable polymers and nanostructure materials are used.

References

[1] R. Bhatt, A. Mishra, A.K. Bajpai, Role of diaminonaphthalene based polymers as sensors in detection of biomolecules: a review, Results Mater. 9 (2021) 100174.

[2] M. Bilal, H.M.N. Iqbal, Microbial-derived biosensors for monitoring environmental contaminants: recent advances and future outlook, Process. Saf. Environ. 124 (2019) 8–17.

[3] D.G. Prajapati, B. Kandasubramanian, Progress in the development of intrinsically conducting polymer composites as biosensors, Macromol. Chem. Phys. 220 (10) (2019) 1800561.

[4] S. Kumar, S.D. Bukkitgar, S. Singh, Pratibha, V. Singh, K.R. Reddy, et al., Electrochemical sensors and biosensors based on graphene functionalized with metal oxide nanostructures for healthcare applications, Chemistry 4 (2019) 5322–5337.

[5] W. Silva, M.E. da, Ghica, C.M.A. Brett, Biotoxic trace metal ion detection by enzymatic inhibition of a glucose biosensor based on a poly(brilliant green)-deepeutectic solvent/carbon nanotubes modified electrode, Talanta 208 (2020) 120427.

[6] A. Khan, E.S. Salama, Z. Chen, H. Ni, S. Zhao, T. Zhou, et al., A novel biosensor for zinc detection based on microbial fuel cell system, Biosens. Bioelectron. 147 (2020) 111763.

[7] D.F. Katowah, M.A. Hussein, M.M. Alam, M.A. Gabal, T.R. Sobahi, A.M. Asiri, et al., Selective fabrication of an electrochemical sensor for Pb^2 based on poly(pyrrole-co-o-toluidine)/$CoFe_2O_4$ nanocomposites, Chemistry Select 4 (35) (2019) 10609–10619.

[8] T. Kubo, K. Otsuka, Recent progress in molecularly imprinted media by new preparation concepts and methodological approaches for selective separation of targeting compounds, Trends Anal. Chem. 81 (2016) 102–109.

[9] N. Nasiri, C. Clarke, Nanostructured gas sensors for medical and health applications: low to high dimensional materials, Biosensors 9 (1) (2019) 43.

[10] L. Uzun, A.P.F. Turner, Molecularly-imprinted polymer sensors: realising their potential, Biosens. Bioelectron. 76 (2016) 131–144.

[11] Y.Y. Qiu, Q.Q. Wen, L. Zhang, P.H. Yang, Label-free and dynamic evaluation of cell-surface epidermal growth factor receptor expression via an electro-chemi luminescence cytosensor, Talanta 150 (2016) 643–648.

[12] A. Kausik, R. Kumar, S.K. Arya, M. Nair, B.D. Malhotra, S. Bhansali, Organic-inorganic hybrid nanocomposites based gas sensors for environmental monitoring, Chem. Rev. 115 (2015) 4571–4606.

[13] S.M.C. Davison, M.P. White, S. Pahl, T. Taylor, K. Fielding, B.R. Roberts, et al., Public concern about, and desire for research into, the human health effects of marine plastic pollution: results from a 15-country survey across Europe and Australia, Global Environ. Change 69 (2021) 102309.

[14] M.J.S. Fernandesa, C.O. Carletti, L.F. Sierra de Araújo, R.C. Santos, J. Rei, Respiratory gases, air pollution and epilepsy, Rev. Neurologique 175 (10) (2019) 604–613.
[15] H.D. Alan Lindquist, Microbial biosensors for recreational and source water, J. Microbiol. Methods 177 (2020) 106059.
[16] M. Imani, N. Mohajeri, M. Rastegar, N. Zarghamia, Recent advances in FRET-based biosensors for biomedical applications, Anal. Biochem. 630 (2021) 114323.
[17] A. Sharma, M. Majdinasab, R. Khan, Z. Lie, J.L. Hayat a, Martyd, Nanomaterials in fluorescence-based biosensors: defining key role, Nanostru Nano-Objects 27 (2021) 100774.
[18] C.S. Pundir, B. Nohwal, R. Chaudhary, A comprehensive review of methods for determination of l-lysine with detailed description of biosensors, Int. J. Biol. Macromol. 186 (2021) 445–461.
[19] X. Chen, Y. Lv, R. Wu, Current technologies of synthetic biosensors for disease detection: design, classification and future perspectives, Chin. Med. Sci. J. 33 (4) (2018) 240–251.
[20] Z. Wang, L. Jinlong, Z. An, M. Kimura, T. Ono, Enzyme immobilization in completely packaged freestanding SU-8 microfluidic channel by electro click chemistry for compact thermal biosensor, Process. Biochem. 79 (2019) 57–64.
[21] T. Hyun, C. Bipro, R. Dhar, Paper-based platforms for microbial electrochemical cell-based biosensors: a review, Biosens. Bioelectron. 192 (2021) 113485.
[22] Y.H. Cheong, L. Gec, G. Lisak, Highly reproducible solid contact ion selective electrodes: emerging opportunities for potentiometry – a review, Analytica Chim. Acta 1162 (2021) 338304.
[23] M.N.P. Vittorio, D. Fracensco, C. Biagio, D.L. Francesco, Sensors 7 (2007) 508.
[24] I. Vroman, L. Tighzert, Biodegradable polymer, Materials 2 (2009) 307–344.
[25] B. Ghanbarzedeh, H. Almas, Biodegradable polymer, Biodegradation: Life Sci. (2020) 141–185.
[26] A. Barrett, History of bioplastics in the automotive industry, Bioplastics N. (2019).
[27] L.M. Sherman, One of the first biobased nylon turns 70, Biopolymers (Plastic Technol.) (2017).
[28] M. Kolybaba, L.G. Tabil, S. Panigrahi, W.J. Crerar, T. Powell, B. Wang, Biodegradable polymers: past, present and future, The Society for Engineering in Agriculture, Food and Biological System 2003 CSAE/ASAE Annual International Meeting, ASABE, 2003.
[29] I. Armentano, M. Dottori, E. Fortunati, S. Mattioli, J.M. Kenny, Biodegradable polymer matrix nanocomposites for tissue engineering: a review, Polym. Degrad. Stab. 95 (2010) 2126–2146.
[30] Y.N. Li, Y. Guo, W. Niu, M. Chen, Y.M. Xue, J. Ge, et al., Biodegradable multifunctional bioactive glass-based nanocomposite elastomers with controlled biomineralization activity, real-time bioimaging tracking, and decreased inflammatory response, ACS Appl. Mater. Interfaces 10 (2018) 17722–17731.
[31] L. Sha, Z. Chen, Z. Chen, A. Zhang, Z. Yang, Polylactic acid based nanocomposites: promising safe and biodegradable materials in biomedical field, Int. J. Polym. Sci. (2016) 1–11.
[32] V.R. Feig, H. Tran, Z. Bao, Biodegradable polymeric materials in degradable electronic devices, ACS Cent. Sci. 4 (2018) 337–348.
[33] R. Smith, Biodegradable Polymers for Industrial Application, Woodhead Publishing Limited, Cambridge, England, 2005.
[34] J.W. Lee, Y.H. Kim, K.D. Park, K.S. Jee, J.W. Shin, S.B. Hahn, Importance of integrin β1-mediated cell adhesion on biodegradable polymers under serum depletion in mesenchymal stem cells and chondrocytes, Biomaterials 25 (10) (2004) 1901–1909.
[35] S.J.A. Mohamed, Synthesis and Characterization of Conductive Polymers Nanocomposite Thin Films, Department of Physics, Panjab University, Chandigarh, 2014.
[36] J.M. Anderson, Polymer in biology and medicine, Polym. Sci. A Compr. Ref. (2012).
[37] S. Ramanvicius, A. Ramanvicius, Conducting polymer in the design of biosensors and biofuel, Polymers 13 (2021) 49.

[38] A. Heller, Implanted electrochemical glucose sensors for the management of diabetes, Annu. Rev. Biomed. Eng. 1 (1999) 153–175.

[39] C.S. Park, C. Lee, O.S. Kwon, Conducting polymer based nanobiosensors, Polymers 8 (2016) 249.

[40] S. Ullah, F. Hammade, J. Engblom, A. Ramanvicius, U. Bubniene, A. Ramanaviciene, et al., In-vitro model for assessing glucose diffusion through skin, Biosens. Bioelectron. 110 (2018) 175–179.

[41] H. Idrees, S.Z.J. Zaidi, A. Sabir, R.U. Khan, X. Zhang, S.ul Hassan, A review of biodegradable natural polymer-based nanoparticles for drug delivery applications, Nanomaterials 10 (2020) 1970.

[42] S.K. Prajapati, A. Jain, A. Jain, Biodegradable polymers and constructs: a novel approach in drug delivery, Eur. Polym. J. 120 (2019) 109191.

[43] H. Liu, R. Jian, H. Chen, X. Tian, C. Sun, J. Zhu, et al., Application of biodegradable and biocompatible nanocomposites in electronics: current status and future direction, Nanomaterials 9 (2019) 950.

[44] A. Gleadall, Mechanical Properties of Biodegradable Polymers for Medical Applications. Woodhead Publishing Series in Biomaterials, Elsevier, 2015, pp. 163–199.

[45] A. Lendlein, A. Sisson, Handbook of Biodegradable Polymers: Synthesis, Characterization and Applications, Wiley-VCH, 2010.

[46] D. Zhang, Y. Wang, Synthesis and applications of one-dimensional nanostructured polyaniline: an overview, Mater. Sci. Eng. B 134 (2006) 9–19.

[47] P.C. Innis, G.G. Wallace, Inherently conducting polymeric nanostructures, J. Nano Sci. Nanotech. 2 (2002) 441–451.

[48] T. Ahuja, I.A. Mir, D. Kumar, Rajesh, Recent progress in the development of nanostructured conducting polymers/nanocomposites for sensor applications, Sens. Actuat. B 136 (2009) 275–286.

[49] M. Zhu, G. Zhang, Z. Hu, W. Gong, M.P. Aldred, Z. Huang, Biodegradable polymer nanoparticles with photo switchable fluorescene for super-resolution bioimagimg, Optics in the Life Science, OSA Technical Digest (Online), Optical Society of America, 2013.

[50] B. Arkema, Natureworks, novamont, plantic, science & education from big market research, in: Application of Biodegradable Polymer, Global Biodegradable Polymer Market, 2020.

[51] M. Angelopoulos, A. Ray, A.G. MacDiarmid, A.J. Epstein, Polyaniline: processability from aqueous solutions and effect of water vapor on conductivity, Synth. Met. 21 (1–3) (1987) 21–30.

[52] Y. Guan, C. Wang, H. Yu, Z. Zou, Y. Zhou, G. Cao, J. Yao, Flexible ZnO/PANI/nonwoven nanocomposites based high-sensitive NH_3 gas sensor via vapor phase polymerization method, Mater. Sci. Energy Technol. 3 (2020) 862–867.

[53] B.S. Lou, U. Rajaji, S.M. Chen, T.W. Chen, A simple sonochemical assisted synthesis of $NiMoO_4$/chitosan nanocomposites for electrochemical sensing of amlodipine in pharmaceutical and serum samples, Ultrasonic – Sonochem. 64 (2020) 104827.

[54] Z. Li, Synthesis and Characterization of Conducting Polymer Nanostructures and Their Application in Sensor (Doctoral Dissertation 1906), Department: Material Science and Engineering, Missouri University of Science and Technology, 2010.

[55] A.G. Green, A.E. Woodhead, Aniline-black and allied compounds: P-1, J. Chem. Soc. Trans. 97 (1910) 2388–2403.

[56] D. Shrivastava, Investigation on polyimide nanocomposites comprising in-situ generated silica or siloxanes derives nanoparticles, RDVV Jabalpur (2009).

[57] J.A. Schlueter, Y. Orihashi, M.G. Kanatzidis, W. Liang, T.J. Marks, D.C. Groot, et al., Charge transport and optical properties of the first highly conductive one-dimensional molecular metal containing fluoride counterions: $(TSeT)F_{0.25}Cl_{0.50}$, Chem. Mater. 3 (1991) 1013–1015.

[58] M. Mousavi, A. Kompany, N. Shahtahmasebi, M.M. Baghari-Mohagheghi, Study of structural, electrical and optical properties of vanadium oxide condensed films deposited by spray pyrolysis technique, Adv. Manuf. 1 (2013) 320–328.

[59] L. Yesappa, M. Niranjana, S. Ashok kummar, H. Vijetha, S. Raghu, H. Devendrappa, Characterization, electrical conductivity and electrochemical performance of poly−aniline-LiClO4-CuO nano composite for energy storage applications, Polym. Plastic Technol. Eng. 58 (2018) 193−205.
[60] J. Liu, S. Willför, C. Xu, A review of bioactive plant polysaccharides: biological activities, functionalization, and biomedical applications, Bioact. Carbohydr. Diet. Fibre 5 (2015) 31−61.
[61] S.E. Harding, M.P. Tombs, G.G. Adams, B.S. Paulsen, K.T. Inngjerdingen, H. Barsett, An Introduction to Polysaccharide Biotechnology, CRC Press, 2017.
[62] N.B. Shelke, R. James, C.T. Laurencin, S.G. Kumbar, Polysaccharide biomaterials for drug delivery and regenerative engineering, Polym. Adv. Technol. 25 (2014) 448−460.
[63] D. Klemm, F. Kramer, S. Moritz, T. Lindström, M. Ankerfors, D. Gray, et al., Nanocelluloses: A new family of nature-based materials, Angew. Chem. Int. Ed. 50 (2011) 5438−5466.
[64] M. Rinaudo, Chitin and chitosan: properties and applications, Prog. Polym. Sci. 31 (2006) 603−632.
[65] R. Xiong, A.M. Grant, R. Ma, S. Zhang, V.V. Tsukruk, Naturally-derived biopolymer nanocomposites: interfacial design, properties and emerging applications, Mater. Sci. Eng. R 125 (2018) 1−4.
[66] E. Llobet, Gas sensors using carbon nanomaterials: a review, Sens. Actuators B 179 (2013) 32−45.
[67] S. Schultz, D. Smith, J. Mock, D. Schultz, Single-target molecule detection with nonbleaching multicolor optical immunolabels, Proc. Natl. Acad. Sci. U. S. A. 97 (3) (2000) 996−1001.
[68] D.R. Kauffman, A. Star, Carbon nanotube gas and vapor sensors, Angew. Chem. Int. Ed. 47 (2008) 6550−6570.
[69] J.W. Han, B. Kim, J. Li, M. Meyyappan, A carbon nanotube based ammonia sensor on cellulose paper, RSC Adv. 4 (2014) 549−553.
[70] M. Krebsz, T. Pasinszki, T.T. Tung, D. Losic, Development of vapor/gas sensors from biopolymer composites, Biopolym. Compos. Electron. (2017) 385−403.
[71] T. Taton, C. Mirkin, R. Letsinger, D.N.A. Scanometric, Array detection with nanoparticle, Probes. Sci. 289 (2000) 1757−1760.
[72] C. Rao, G. Kulkarni, P. Thomas, P. Edwards, Size-dependent chemistry: properties of nanocrystals, Chem. − A Eur. J. 8 (1) (2002) 28−35.
[73] M. Otamiri, K.G.I. Nilsson, Analysis of human serum antibody−carbohydrate interaction using biosensor based on surface plasmonresonance, Int. J. Biol. Macromol. 26 (1999) 263−268.
[74] L. Hua, J. Chen, Ta. S.N. GeL, Silver nanoparticles as matrix for laser desorption/ionization mass spectrometry of peptides, J. Nanopart. Res. 9 (2007) 1133−1138.
[75] D. Xiong, M. Chen, H. Li, Synthesis of para-sulfonatocalix[4]arene-modified silver nanoparticles as colorimetric histidine probes, Chem. Commun. 7 (2008) 880−882.
[76] S. Pandey, G.K. Goswami, K.N. Karuna, Green synthesis of biopolymer−silver nanoparticle nanocomposite: an optical sensor for ammonia detection, Int. J. Biol. Macromol. 51 (2012) 583−589.
[77] O. Martin, L. Avérous, Poly(lactic acid): plasticization and properties of biodegradable multiphase systems, Polymer 42 (2001) 6209−6219.
[78] M. Agarwal, K.W. Koelling, J.J. Chalmers, Characterization of the degradation of polylactic acid polymer in a solid substrate environment, Biotechnol. Prog. 14 (3) (1998) 517−526.
[79] L. Yu, H. Liu, F. Xie, L. Chen, X. Li, Effect of annealing and orientation on microstructures and mechanical properties of polylactic acid, Polym. Eng. Sci. 48 (4) (2008) 634−641.
[80] R. Singh, A.K. Bajpai, A.K. Shrivastava, Structural, morphological and electrical studies of poly (1, 8-diaminonaphthalene)/CdSe(S) nanocomposites for charge storage applications, Mater. Res. Express 6 (2019) 045033.
[81] R.B. Kale, S.D. Sartale, B.K. Chougule, C.D. Lokhande, Growth and characterization of nanocrystalline CdSe thin films deposited by the successive ionic layer adsorption and reaction method, Semicond. Sci. Technol. 19 (8) (2004) 980−986.

[82] K. Kobashi, T. Villmow, T. Andres, P. Pötschke, Liquid sensing of melt-processed poly(lactic acid)/multi-walled carbon nanotube composite films, Sens. Actuators B: Chem. 134 (2008) 787–795.
[83] K. Persaud, G. Dodd, Analysis of discrimination mechanisms in the mammalian olfactory system using a model nose, Nature 5881299 (1982) 352–355.
[84] M.S. Freund, N.S. Lewis, A chemically diverse conducting polymer-based electronic nose, Proc. Natl. Acad. Sci. U. S. A. 92 (1995) 2652–2656.
[85] B. Kumar, M. Castro, J.F. Feller, Poly(lactic acid)–multi-wall carbon nanotube conductive biopolymer nanocomposite vapour sensors, Sens. Actuators B 161 (2012) 621–628.
[86] K.C. Sajjan, M. Fiasal, B.S. Khened, S. Khasim, Electrical conductivity, dielectric behavior and humidity sensing properties of polyaniline-graphite oxide composites, Int. J. Electr. Electron. Eng. (IJEEE) 2 (2) (2013) 67–76.
[87] R. Singh, A.K. Bajpai, A.K. Shrivastava, CdSe reinforced polyaniline nanocomposites as superior material for future applications as gas sensor and diodes, Mater. Res. Express 6 (2019) 1250a9.
[88] S. Abusalemadnabirqudri, M.M. Mohiuddin, A.M. Farhan, J. Husain, Thermal and sensing properties polyaniline/lead oxide nanocomposites, J. Eng. Res. Application 8 (2018) 45–47.
[89] K. Deshmukh, S.K. Khadheer Pasha, Room temperature ammonia sensing based on graphene oxide integrated flexible polyvinylidenefluoride/cerium oxide nanocomposite films, Polym. Plast. Technol. Mater. 59 (13) (2020) 1429–1446.
[90] V.O. Okechukwu, V. Mavumengwana, I.A. Hümmelgen, M.A. Mamo, Concomitant in situ FTIR and impedance measurements to address the 2-methylcyclopentanone vapor-sensing mechanism in MnO_2 – polymer nanocomposites, ACS Omega 4 (2019) 8324–8333.
[91] H. Qia, J. Liua, J. Piontecka, P. Pötschkea, E. Mäder, Carbon nanotube–cellulose composite aerogels-forvapour sensing, Sens. Actuators B 213 (2015) 20–26.
[92] L.C. Wang, T.H. Su, C.L. Ho, S.R. Yang, S.W. Chiu, KuoHW, et al., A bio-inspired two-layer sensing-structure of polypeptide and multiple-walled carbon nanotube to sense small molecular gases, Sensors 15 (2015) 5390–5401.
[93] K. Li, K. Dai, X. Xu, G. Zheng, C. Liu, J. Chen, et al., Organic vapor sensing behaviors of carbon-black/poly(lacticacid) conductive biopolymer composite, Colloid Polym. Sci. 291 (2013) 2871–2878.
[94] A. Bouvree, J.F. Feller, M. Castro, Y. Grohens, M. Rinaudo, Conductive polymer nano-biocomposites (CPC): chitosan-carbon nanoparticle a good candidate to design polar vapour sensors, Sens. Actuators B 138 (2009) 138–147.
[95] N. Wang, X. Wang, Y. Jia, X. Li, J. Yu, B. Ding, Electrospun nanofibrous chitosan membranes modified with polyethyleneimine for formaldehyde detection, Carbohydr. Polym. 108 (2014) 192–199.
[96] M.M. Ayad, N.A. Salahuddin, I.M. Minisy, W.A. Amer, Chitosan/polyaniline nanofibers coating on thequartz crystal microbalance electrode for gas sensing, Sens. Actuators B 202 (2014) 144–153.
[97] W. Li, D.M. Jang, S.Y. An, D. Kim, S.K. Hong, K. Kim, Polyaniline–chitosan nanocomposite: high performance hydrogen sensor from new principle, Sens. Actuators B 160 (2011) 1020–1025.
[98] A. Aldalbahi, J. Chu, P. Feng, M.H. Panhuis, Conducting composite materials from the biopolymer kappa-carrageenan and carbon nanotubes, Beilstein J. Nanotechnol. 3 (2012) 415–427.
[99] B. Kumar, J.F. Feller, M. Castro, J. Lu, Conductive bio-polymer nano-composites (CPC): chitosan-carbon nanotube transducers assembled via spray layer-by-layer for volatile organic compound sensing, Talanta 81 (3) (2010) 908–915.
[100] B. Kumar, M. Castro, J.F. Feller, Controlled conductive junction gap for chitosan–carbon nanotube-quantum resistive vapour sensors, J. Mater. Chem. 22 (2012) 10656–10664.
[101] P. Molla-Abbasi, S.R. Ghaffarian, E. Dashtimoghadam, Wrapping carbon nanotubes by biopolymer chains: role of nano interfaces in detection of vapors in conductive polymer composite transducers, Polym. Compos. (2015) 1–8.

[102] P. Molla-Abbasi, S.R. Ghaffarian, Decoration of carbon nanotubes by chitosan in a nanohybrid conductive polymer composite for detection of polar vapours, RSC Adv. 4 (2014) 30906–30913.
[103] M.M. Ayad, N. Salahuddin, I.M. Minisy, Detection of some volatile organic compounds with chitosan-coated quartz crystal microbalance, Des. Monomers Polym. 17 (2014) 795–802.
[104] D.S. Chandrasakaran, I. Nainggolan, N.D.T. Ikhsan, Chloroform gas sensor based on chitosan biopolymer, Appl. Mech. Mater. 679 (2014) 45–49.
[105] A. Mironenko, A. Sergeev, S. Voznesensky, S. Bratskaya, Thin chitosan films for optical gas sensors, Key Eng. Mater. 605 (2014) 536–539.
[106] A.Y. Mironenko, A.A. Sergeev, S.S. Voznesenskiy, D.V. Marinin, S.Y. Bratskaya, P-indicators doped polysaccharide LbL coatings for hazardous gases optical sensing, Carbohydr. Polym. 92 (2013) 769–774.
[107] B. Kumar, M. Castro, J.F. Feller, Tailoring the chemo-resistive response of self-assembled polysaccharide-CNT sensors by chain conformation at tunnel junctions, Carbon 50 (2012) 3627–3634.
[108] C. Staii, A.T. Johnson Jr., M. Chen, A. Gelperin, DNA-decorated carbon nanotubes for chemical sensing, Nano Lett. 5 (9) (2005) 1774–1778.
[109] S.M. Khamis, R.R. Johnson, Z. Luo, A.T.C. Johnson, Homo-DNA functionalized carbon nanotube chemical sensors, J. Phys. Chem. Solids 71 (2010) 476–479.
[110] K. Fu, B.G. Willis, Characterization of DNA as a solid-state sorptive vapour sensing material, Sens. Actuators B 220 (2015) 1023–1032.
[111] A. Hajjam, S. Pourkamali, Fabrication and characterization of MEMS-based resonant organic gas sensors, IEEE Sens. J. 12 (2012) 1958–1964.
[112] S. Panigrahi, S. Sankaran, S. Mallik, B. Gaddam, A.A. Hanson, Olfactory receptor-based polypeptides ensor for acetic acid VOC detection, Mater. Sci. Eng. C 32 (2012) 1307–1313.
[113] S. Sankaran, S. Panigrahi, S. Mallik, Olfactory receptor based piezoelectric biosensors for detection ofalcohols related to food safety applications, Sens. Actuators B 155 (2011) 8–18.
[114] H.H. Lu, Wu. T.Z. RaoYK, Y.M. Tzeng, Direct characterization and quantification of volatile organic compounds by piezoelectric module chips sensor, Sens. Actuators B 137 (2009) 741–746.
[115] T.Z. Wu, Y.R. Lo, E.C. Chan, Exploring the recognized bio-mimicrymaterials for gas sensing, Biosens. Bioelectron. 16 (2001) 945–953.
[116] M.M. Salleh, A.A. Umar, M. Yahaya, Quartz crystal microbalance coated with poly-L-proline self-assembly layer to detect organic gases, Proc. SPIE 5276 (2004) 107–112.
[117] J. Bai, B. Zhou, Titanium dioxide nanomaterials for sensor applications, Chem. Rev. 114 (19) (2014) 10131–10176.
[118] S. Singh, V. Kumar, D.B. Singh, S. Datta, R. Prasad, J. Singh, Biological for monitoring and diagnosis, Microb. Biotechnol. Basic Res. Appl. (2020) 317–335.
[119] G.J. Kalaivani, S.K. Suja, Nanomolar level sensing of glucose in food samples using glucose oxidase confined MWCNT-Inulin-TiO_2 bio-nanocomposite, Food Chem. 298 (2019) 124981.
[120] Z. Nezafat, M. Nasrollahzadeh, T. Baran, N. Shafiei, Biopolymer-based metal nanoparticles for biosensing, Chem. Sustain. Appl. 2 (2021) 573–608.
[121] Y. Zare, S. SooRhim, Kyong, Y. Rhee, The interface degradation in a nanobiosensor including biopolymers and carbon nanotubes, Sens. Acutator A: Phys. 331 (2021) 112967.
[122] V. Inderan, M.M. Arafat, A.S.M.A. Haseeb, K. Sudesh, H.L. Lee, Electrospun (Nickel and palladium) tin(IV) oxide/polyaniline/polyhydroxy-3-butyrate biodegradable nanocomposite fibers for low temperature ethanol gas sensing, Nanotechnology 31 (2020) 425503.
[123] S. Palanisamy, V. Velusamy, S. Balu, S. Velmurugan, C.K. Yang, S.W. Thomas, Chen, Sonochemical synthesis and anchoring of zinc oxide on hemin-mediated multi walled carbon nanotubes-cellulose nanocomposites for ultra-sensitive biosensing of H_2O_2, Ultrasonic-Sonochemistry 63 (2020) 104917.

[124] G.J. Thangamani, K. Deshmukh, N.A. Nambiraj, S.K. Khadheer Pasha, Chemiresistive gas sensors based on vanadium pentoxide reinforced polyvinyl alcohol/polypyrrole blend nanocomposites for room temperature LPG sensing, Synth. Met. 273 (2021) 116687.

[125] M.O.M. Netto, B.W. Gonçalves, W.C.R. Li, J. Gruber, Biopolymer based ionogels as active layers in low-cost gas sensors for electronic noses, Sens. Actuators B: Chem. 315 (2020) 128025.

[126] H.K. Chang, G.T. Chang, A.K. Thokchom, T. Kim, J. Park, Ultra-fast responsive colloidal–polymer composite-based volatile organic compounds (VOC) sensor using nanoscale easy tear process, Sci. Rep. 8 (2018) 5291.

[127] J.G. Thangamani, K. Deshmukh, K.K. Sadasivuni, D. Ponnamma, S. Goutham, K. Venkateswara Rao, et al., White graphene reinforced polypyrrole and poly(vinyl alcohol) blend nanocomposites as chemiresistive sensors for room temperature detection of liquid petroleum gases, Microchem. Acta 184 (2017) 3977–3987.

[128] S. Isah, Advanced materials for energy storage devices, Asian J. Nanosci. Mater. 1 (2) (2018) 90–103.

CHAPTER

Applications of biodegradable polymer nanocomposites in water and wastewater treatment

15

Momina[1], Kafeel Ahmad[1] and Mohd Rafatullah[2]

[1]*Civil Engineering Department, Jamia Millia Islamia, New Delhi, India* [2]*School of Industrial Technology, Universiti Sains Malaysia, Penang, Malaysia*

Chapter Outline

15.1 Introduction

In recent years, there has been a surge in research interest in the development of sustainable nanocomposites for use in environmental remediation. Polymer nanocomposite is a class of nanostructured hybrid materials made up of polymers and nanofillers. These nanocomposites have a broad spectrum of applications such as food packaging, paramedical, electronics, wastewater treatment, and antimicrobial performance (Fig. 15.1). Because of their superior properties, polymer nanocomposites can

Biodegradable and Biocompatible Polymer Nanocomposites. DOI: https://doi.org/10.1016/B978-0-323-91696-7.00003-9
Copyright © 2023 Elsevier Inc. All rights reserved.

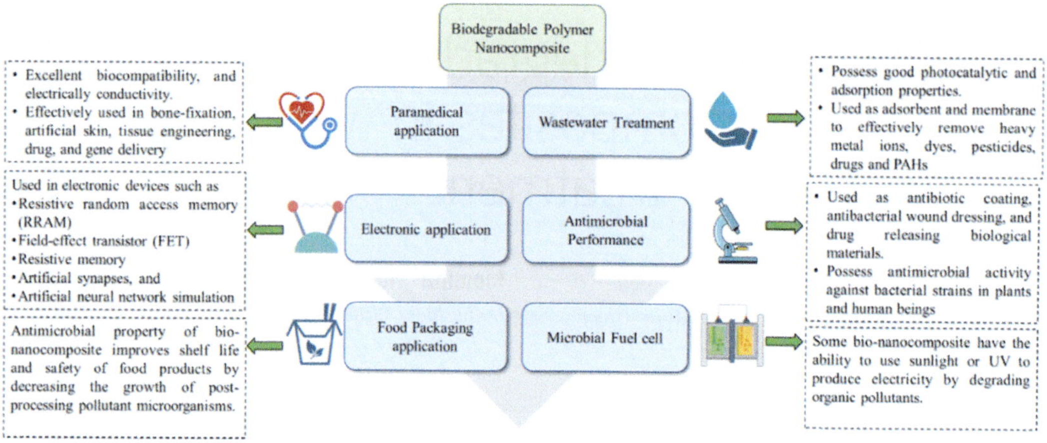

FIGURE 15.1

Various applications of biodegradable polymer nanocomposite.

overcome the limitations of monolithic materials' performance. The particle size of these nanocomposites is in the range of 1–100 nm with excellent mechanical, thermal, good recyclability, low weight, and transparency [1]. Polyvinyl alcohol (PVA), polystyrene, polypyrrole, aliphatic polyesters, and other synthetic polymers have been extensively used in the development of polymer nanocomposite. These polymers, however, present a number of challenges in terms of sustainability, uneconomical cost, toxicity, and recycling [2]. These issues give rise to the development of such polymers that can overcome these obstacles while retaining their properties for a variety of applications, such as water and wastewater treatment. As a result, biodegradable polymers are preferable to synthetic polymers to address environmental issues and reduce waste generated by synthetic polymers [3]. The most exciting characteristics of biodegradable polymer nanocomposites are their eco-friendliness, biodegradability, and sustainability. Because of the high surface-to-volume ratio of the nanofiller, the combination of biodegradable polymers and nanofillers can provide stability to the polymer matrix, improving the overall performance of biodegradable polymer nanocomposite [4]. The primary requirement for biodegradable polymer nanocomposites is that the polymer be green (renewable). As a result, they are also referred to as "green composites," "bio-nano composites," "biohybrids," and "biocomposite" [5].

Water is regarded as one of the most important resources for the survival of all living organisms, including flora and fauna. However, the freshwater water system is directly threatened by humans, with less than 1% of fresh water available for human consumption. Furthermore, the rapid acceleration of climate change, industrialization, population growth, and urbanization raises the demand for fresh water supply. According to a World Bank report, 4.6 billion people will be affected by a lack of access to safe drinking water by 2080 [6,7]. As a result, many researchers and government agencies are becoming increasingly interested in wastewater purification and treatment. Numerous methods and techniques, including advanced nanotechnology-based approaches, have been used to treat wastewater discharged by industries and agricultural practices. Due to their high surface area, good adsorbability, superb mechanical stability, good chemical reactivity and selectivity, and economic feasibility in treating emerging pollutants in water such as dyes, heavy metal ions, and pesticides, biodegradable polymer nanocomposite was effective in treating wastewater. This chapter examines the adsorptive capability of various types of biodegradable polymer nanocomposite, as well as their synthetic route and application in the removal of organic and inorganic pollutants from contaminated water.

15.2 Types of biodegradable polymers

Biodegradable plastics can be degraded in at least one step of the degradation process by naturally occurring organisms through digestion. At appropriate environmental conditions such as temperature, moisture, and oxygen availability, this degradation leads to plastic disintegration and fragmentation, biodegradation with less harmful residue, and no toxicity. Furthermore, regardless of raw material or manufacturing process, the chemical and physical microstructure of a polymer influences its degradability [8]. In general, the rate of biopolymer degradation is inversely proportional to the degree of chemical modification. Different classifications of biodegradable polymers based on their source or raw materials are described in Fig. 15.2. Some authors proposed categorizing biodegradable polymers extracted from biomass, such as polysaccharides, proteins, and lipids; polymers derived from biotechnology-based products; chemically synthesized polymers derived from monomers, such as polylactic acid (PLA); and polymers derived from microorganism extraction, such as polyhydroxybutryte, xanthan, pullulan, and polyhydroxyalkanoates [1,3,9].

15.2.1 Biomass-derived biodegradable polymers

Biodegradable polymers derived from biomass are polymeric organic molecules that offer a promising route to the production of renewable materials. Their distinct physical and chemical properties,

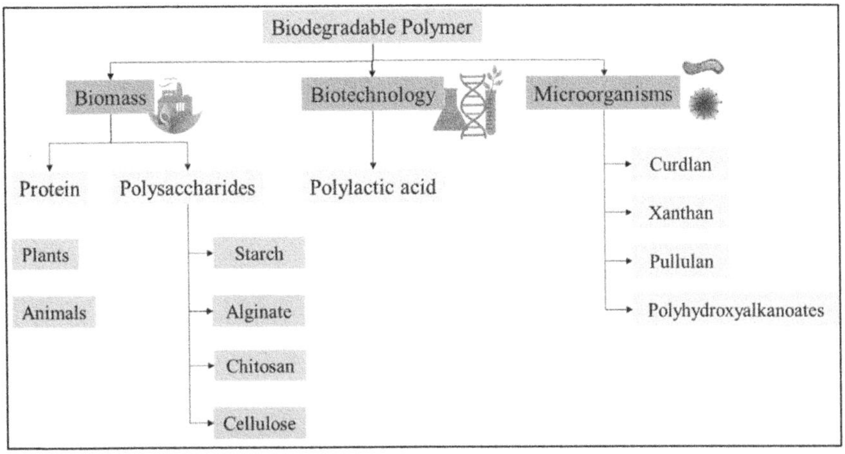

FIGURE 15.2

Classification of biodegradable polymers.

availability, biocompatibility, and non-toxic nature make them suitable nano-adsorbents and nano-catalysts for the removal of various pollutants from contaminated water [10–12]. Polysaccharides are biomass-derived polymers that, due to their diversity in structure, size, and molecular chains, can be stabilized and immobilized, enhancing catalytic property and stability [13,14]. The usage of biomass-based biodegradable polymers is not only confined to the removal of organic and inorganic compounds but also utilized in the removal of hydrocarbons [15], pharmaceutical [16], phosphates [17], pesticides [18], nitrites [19], perchlorates [20], and fluorides [21]. Polysaccharides consist of a repeated unit of monosaccharides bound together by glycosidic linkages having a molecular formula $Cn(H_2O)_n$. Their structure varies from highly branched to linear branches.

Cellulose (CS) is one of the bountiful biopolymers on earth that is widely employed to treat wastewater and is mainly extracted from wood sources, as shown in Fig. 15.3. Because of its high surface area to volume ratio, improved mechanical properties, biocompatibility, low density, and low cost, CS is an appealing candidate for use in a variety of fields, including water and wastewater treatment. However, the environmental effects of their toxic-free and biodegradable properties have yet to be studied. Because of the presence of strong hydrogen bonds with polymer chains, cellulosic materials are hydrophilic in nature and insoluble in water and organic solvents [22]. As a result, they are widely used in the field of filtration and distillation as a membrane CS nano-material integrated into a polymer matrix. The combination of cellulosic nanomaterials with polymer matrices such as CS triacetate, polypyrrole, poly(acrylonitrile), poly(ether sulfone), poly (ethylene oxide), poly(vinylidene fluoride), poly(vinyl alcohol), and poly(3-hydroxybutyrate) improves hydrophilicity, selectivity, permeability, and biofouling resistance [23,24]. CS can be converted to its various derivatives via two methods: the first is metal chelation to functionalize CS, and the second is monomers grafting to lose the backbone of CS [25]. These functionalization methods improve CS's

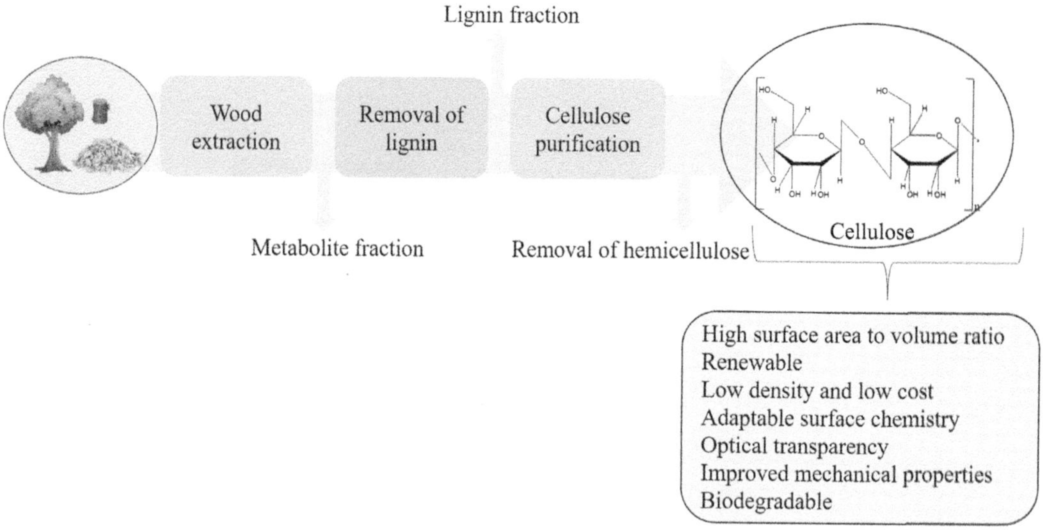

FIGURE 15.3

Extraction process of CS biopolymers and their physical and chemical properties.

adsorptive properties while also improving its elasticity, microbial activity, hydrophobicity, hydrophilicity, and mechanical resistance.

Chitosan is the most abundant biodegradable N-deacetylated chitin in nature, a renewable polymer derived from crustaceous shells and insect cuticles. They are formed by glucosamine which is linear amino polysaccharides, as shown in Fig. 15.4. Enzymatic N-deacetylation of chitin takes the conversion of chitin to chitosan. The deacetylation process exposes the hydroxyl and amino groups of chitosan, which are responsible for its cationic nature and suitability for pollutant adsorption [26]. Aside from their cationic behavior, chitosan has good adsorption performance, is easily modified, safe, and cost-effective. Chitosan's solubility is described by the presence of an amine group, which makes it insoluble in water, alkali, and organic solvents. Aside from their cationic behavior, chitosan has good adsorption performance, is easily modified, safe, and cost-effective. Chitosan's solubility is described by the presence of an amine group, which makes it insoluble in water, alkali, and organic solvents. They can be used in different forms, such as nanofiber [27], nanopowder [28], microbeads [29], micro membrane [28], hydrogel [30]. Chitosan can be modified by grafting functional groups into it or by cross-linking reactions. These functional groups bind to

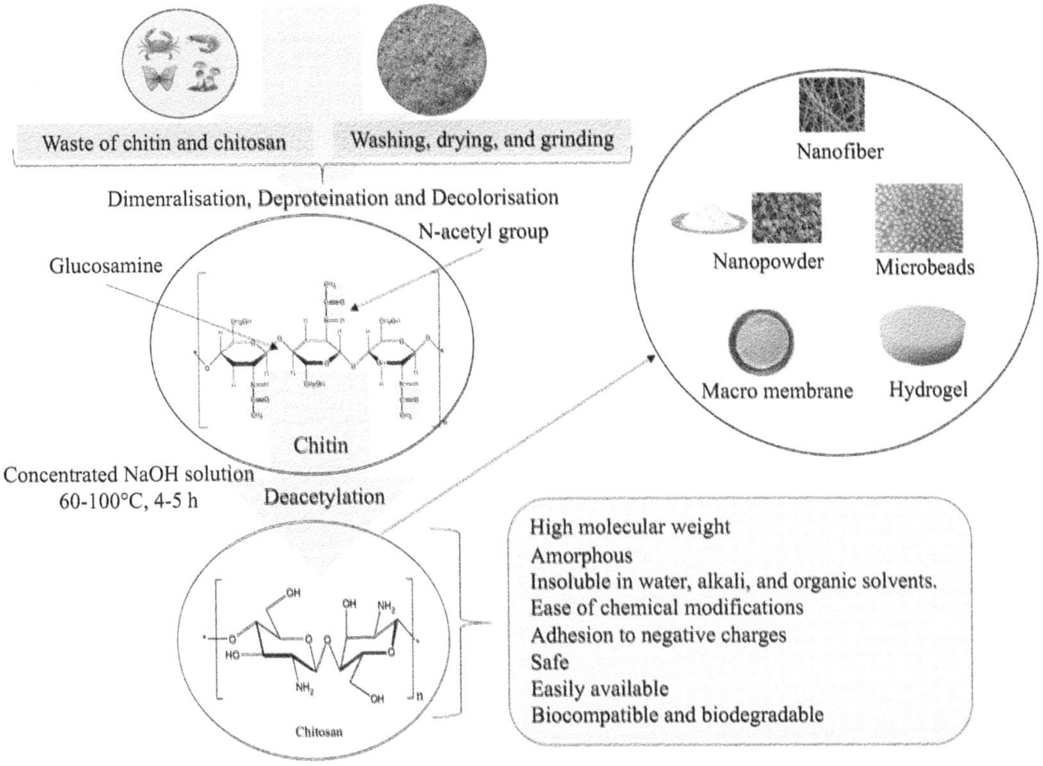

FIGURE 15.4

Extraction process of chitosan biopolymer and their physical and chemical characteristics.

macromolecular chains, resulting in the formation of chitosan derivatives with improved properties over unmodified chitosan [25].

Starch is also a biodegradable polymer, renewable, consists of 15%–45% crystallinity, and abundant derived from stalks, roots, and seeds of various plants such as potato, corn, wheat, and rice. Starch granules have a 3D architecture and contain D-glucose units with biomolecules such as amylopectin, 1-4 glucoside linkage, amylose, 1-6 glucoside linkage [31]. The extraction process of starch biopolymers with their chemical and physical properties is illustrated in Fig. 15.5. Because starch alone does not provide good adsorption efficiency, a combination of starch biopolymer and clay nanoparticles was developed to create a sustainable, eco-friendly composite [32]. Chivrac et al. claim that combining sepiolite with starch biopolymer improves its physical properties such as Young's modulus [33].

Alginate biopolymer is one of the low-cost biomaterials, and its chemical structure consists of a linear polymer of two uronic acids, e.g., β (1–4) linked D-mannuronic and β-(1–4) linked α-L-guluronic acid by glycosidic bonds [34]. Alginate is an anionic polysaccharide derived primarily from brown algae, i.e., kelps. Alginic acid is a significant alginate component; however, sodium alginate is the sodium salt of alginic acid that contains a high concentration of hydroxyl and carboxyl groups dispersed along the polymer's backbone chain [35]. Because of its strong gelation and complexing properties, sodium alginate is environmentally stable. A thermo-irreversible gel formation occurs after reacting with polyvalent cations (except Mg^{2+}) by forming cross-linking bonds. For example, Ca^{2+} reacts with sodium alginate solution; it forms calcium alginate gel by displacing H^+ and Na^+ ions [35]. Furthermore, the availability of carboxyl and hydroxyl groups on the surface of alginate lends itself to a variety of environmental applications, such as ion-exchange adsorption of metallic or cationic ions. Alginate has some limitations when compared to other polysaccharides, including low elasticity, high rigidity, and low mechanical stability. As a result, they were combined with organic and inorganic nanomaterials to improve mechanical and thermal stability [36].

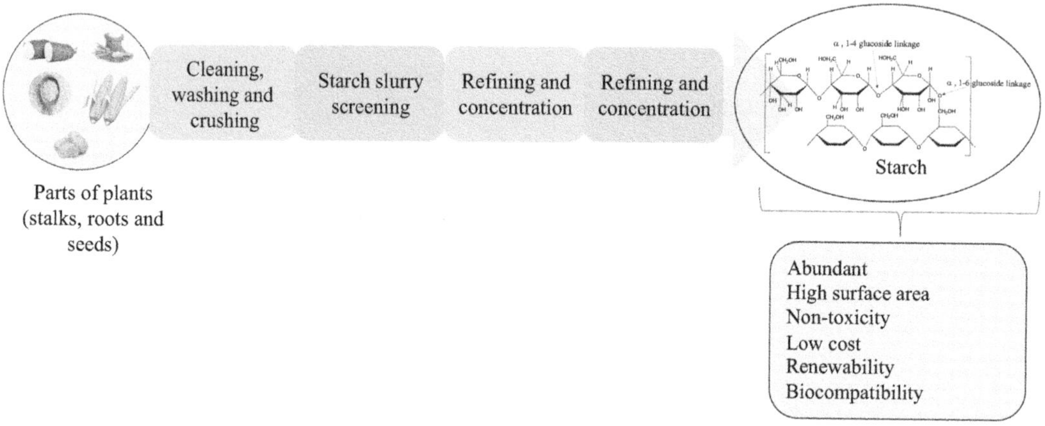

FIGURE 15.5

Extraction process of starch biodegradable polymer and its physical and chemical properties.

15.2.2 Biotechnology-derived biodegradable polymers

PLA possesses high strength and compatibility, derived from the fragmentation of sugar and starch in corn starch, sugarcane, and tapioca (as shown in Fig. 15.6). It is made of thermoplastic aliphatic polyester, which can be used as an adsorbent to clean contaminated water. The ratio of two types of lactic acid monomer, L or D, is related to the PLA's properties. L-PLA is crystalline and has a high melting point, whereas a combination of D- and L-PLA is amorphous and has a low glass transition temperature [3]. However, the effect of PLA alone in the removal of pollutants from wastewater is insufficient; thus, a variety of nanoparticles are incorporated into the polymer matrix to enhance its pollutant removal properties [1].

15.2.3 Microorganisms derived biodegradable polymers

Recently, there has been increased interest in obtaining biopolymers from third generation feedstocks, i.e., microbes that do not compete with human food, animal feed, land, or freshwater. These biopolymers are formed in the cytoplasm of some cells as water-insoluble "inclusion bodies" that are completely biodegradable and natural [37]. Xanthan gum (extracellular heteropolysaccharide) is a corn-derived microbial-derived biopolymer. It is produced under aerobic conditions by *Xanthomonas campestris* fermentation. Because of the accessibility of carbonyl and hydroxyl groups on the surface,

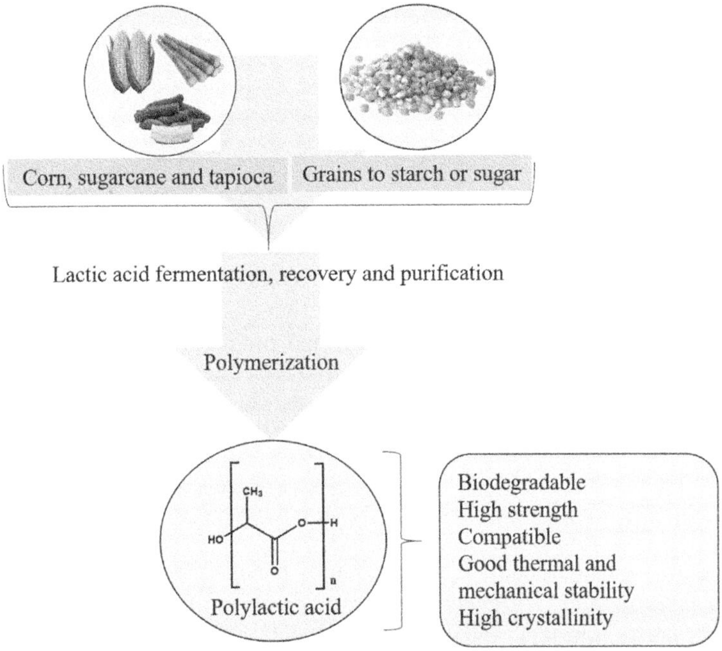

FIGURE 15.6

Extraction process of PLA and their physical and chemical properties.

xanthan is an excellent adsorbent capable of chelating heavy metal ions [38]. However, because xanthan adsorbents are soluble in water at room temperature, it is difficult to separate them after the adsorption process [39]. This challenge can be overcome by modifying xantham and incorporating nanoparticles. *Dextran* is also a microbial-derived biopolymer, biocompatible, and water-soluble. Its chemical structure consists pendant of α-1, 6-linked D-glucopyranose with α-2, α-3, α-4 linked side chains. It has pendant groups of hydroxyl in backbone, which can be modified to remove pollutants [38].

15.3 Synthesis of biodegradable polymers composites

Polymers are supporting materials or host matrices for composite synthesis. The lightweight and excellent properties of biodegradable polymers have piqued the interest of researchers who want to combine them with nanoparticles to serve as a host matrix. Hydrothermal, sol-gel, electrochemical polymerization, and chemical reduction methods were used to create biodegradable polymer nanocomposites. The polymers, on the other hand, are synthesized either ex situ, with the nanoparticles incorporated in a polymer matrix, or in situ, with in situ polymerization of monomer or in situ synthesis of nanoparticles in the presence of the nanoparticles or polymers, spin-coating, drop casting, and dip coating [40,41].

15.3.1 Sol-gel

The "Sol-gel method" is one of the easiest, most cost-effective, and simplest methods for producing high-quality metal oxide nanoparticles from their respective aqueous salt solutions [42]. This method can be used to control the surface and texture properties of nanoparticles. Sols are converted to gel using five key steps in this method: hydrolysis, condensation, aging and drying, and calcination [43].

15.3.2 Hydrothermal

This method can be used to control the surface and texture properties of nanoparticles. Sols are converted to gel using five key steps in this method: hydrolysis, condensation, aging and drying, and calcination [44]. Nanoparticles are formed in a Teflon-coated stainless steel vessel heated to a specific time and temperature in a furnace. This method produces nanomaterials that are stable at high temperatures and pressures with minimal material loss [44]. The Solvo-thermal method is a modified version of the hydrothermal method that allows for the preparation of nanomaterials in a variety of solvents. This method monitors and controls the crystallinity and particle size [45].

15.3.3 Chemical reduction

The chemical reduction method entails the reduction of nanoparticle ionic salt solutions by reducing agents in the presence of a surfactant for the preparation of metal nanoparticles [46]. After reacting with other atoms, clusters, and metal ions, metal salts are reduced to zero-valent metal atoms, which then diffuse into the solution to form nuclei [47]. Some common reducing agents used in the preparation of nanoparticles include hydrazine, sodium borohydride, dimethylformamide, and ammonium formate [47,48].

15.3.4 In situ polymerization

In this method, nanofiller is incorporated into the monomer of a polymeric material while the polymerization reaction occurs in the presence of an oxidant, initiator, thermal, or photocatalyst [49,50]. In comparison to other techniques, it is a simple method for distributing nanoparticles uniformly in polymeric materials. Furthermore, the size of nanoparticles in the polymer matrix can be tuned by designing the polymer structure [51]. This method can produce one-dimensional, two-dimensional, and bulk composites.

15.3.5 Melt blending

Nanoparticles are dispersed in the molten state of a thermoplastic polymer by injection molding and extrusion without the use of solvents in the melt blending method [52]. In polymeric chains, an intercalated network of polymer and nanoparticles forms to form nanocomposites. Furthermore, melt mixing of polymer and nanoparticle is possible after applying shear force greater than the melting point of the host polymer [51].

15.3.6 Solution intercalation

This method is based on Stoke's law, in which pre-polymer and polymer are merged equally and are soluble in suitable solvents [53]. The colloidal suspension of nanoparticles was then mixed in the polymer matrix using magnetic stirring and simple shear mixing, while the solvent in which the pre-polymer was mixed was evaporated. Some common solvents used in the solution casting method for the preparation of biopolymer nanocomposite include acetone, deionized water, cyclohexanone, dimethylformamide, and ethanol [51].

15.4 Application of biodegradable polymer nanocomposite in treatment of wastewater

15.4.1 Adsorbents for removal of heavy metal ion and dyes

The presence of heavy metal ions discharged into freshwater from various industries raises concerns about ecological and public health because of their toxicity and other negative effects. They have the potential to enter the human food chain due to their non-biodegradable nature. In this regard, polymer-containing nanoparticles demonstrate excellent performance in removing or rejecting toxic pollutants from industrial discharge. Organic dyes, such as cationic, anionic, or nonionic dyes, are complex organic pollutants emitted or produced by industries such as textile, paper and pulp, leather, printing, and cosmetics [54,55]. The water consumed in the average textile mill for producing 8000 kg of fabric is approximately 1.6 million liters. The dyeing process uses 10%–15% of the water and contributes 15%–20% of the wastewater, resulting in water pollution and human illness [56]. Various biodegradable polymer nanocomposites have been used to efficiently remove and recover a wide variety of heavy metal ions and dyes from wastewater (Table 15.1) through a process called biosorption.

Table 15.1 List of biodegradable nanocomposite used for the removal of heavy metal ions.

Biodegradable polymer nanocomposite	Pollutant	Adsorption capacity/efficiency	References
CS-based composite			
SCNC	Cd^{2+}/Pb^{2+}	367.6 and 259.7 mg/g	[57]
NaSCNC	Cd^{2+}/Pb^{2+}	465.1 and 344.8 mg/g	[57]
Fe_3O_4/ATP@(BCNs/CS)	Pb^{2+}/Cu^{2+}/CR/Cr^{6+}	67.8, 70.5, 230, and 91 mg/g	[58]
Nano-magnetic CS (MCGT)	$Hg^{2+}/Cu^{2+}/Ag^{+}$	2, 1.5, and 1.2 mmol/g	[59]
MSC-GMA-EDA	Cr^{6+}	171.5 mg/g	[60]
APS-CS-MFC	$Ni^{2+}/Cu^{2+}/Cd^{2+}$	2.734, 3.150, and 4.195 mmol/g	[61]
Aminated magnetite-BC nanocomposites	As^{5+}	90 mg/g	[62]
Mercerized nanocellulose with succinic anhydride	$Zn^{2+}/Ni^{2+}/Cu^{2+}/Co^{2+}/Cd^{2+}$	1.338, 0.744, 1.900, 1.610, and 2.062 mmol/g	[63]
Itaconic acid/magnetite CS/2-mercaptobenzamide	Hg^{2+}	240 mg/g	[64]
Bisphosphonate nanoCS	Vanadium	1.98 mmol/g	[65]
Chitosan-based composite			
Cs-TPP	Pb^{2+}	398 mg/g	[66]
Chitosan	Cr^{4+}	323.6 mg/g	[67]
Nanochistosan	Cd^{2+}	358 mg/g	[68]
Hexanediamine/magnetite/chitosan	Cd^{2+}/Cu^{2+}, $/Pb^{2+}$	95%, 89%, and 92%	[69]
CS/PAA-MCM	Cu^{2+}	108.0 mg/g	[70]
CMNs	Fe^{3+}	35.98 mg/g	[71]
α−KA−CCMNPs	Cu^{2+}	60.61	[72]
Thiourea−modified Fe_3O_4−imprinted chitosan/TiO_2	Cr^{4+}	256.4 mg/g	[73]
Nano-TiO_2-enabled CS beads	As^{3+}	71.9%	[74]
PEG-chitosan composite	Ni^{2+}	50.68 mg/g	[19]
Carboxymethyl chitosan	Cd^{2+} and Cr^{6+}	81 and 31 mg/g	[75]
Goethite/chitosan nanocomposites	Pb^{2+}	98.26%	[76]
Polypyrrole(PPy)/CS/CNT	Cu^{2+}	16.83 mg/g	[77]
Starch-based nanocomposites			
St-PEG-AcAH	Hg^{2+}	188.67 mg/g	[78]
AgNPs-St-PEG-AcANCH	Hg^{2+}	263.15 mg/g	[78]
mCVP	Pb^{2+}, Cd^{2+}, and Cu^{2+}	65, 53.2, and 83.6 mg/g	[79]
Starch stabilized Fe^0 NPs	Cr^{4+}	99.25%	[80]
Magnetic dialdehyde/starch	Hg^{2+}	312.5 mg/g	[81]
Superparamagnetic starch functionalized maghemite NPs	Cr^{6+}	24.76 mg/g	[82]

(Continued)

Table 15.1 List of biodegradable nanocomposite used for the removal of heavy metal ions. *Continued*

Biodegradable polymer nanocomposite	Pollutant	Adsorption capacity/efficiency	References
Alginate-based nanocomposites			
Mesoporous calcium alginate/titania hybrid	Cr^{6+}	8.4 mg/g	[83]
PVA/alginate/zeolite	Ni^{2+} and Cu^{2+}	81.5 and 79.6 mg/g	[84]
Fe^0-alginate beads	Cr^{6+}	381 ± 17.1 mg/g	[85]
Glycine/magnetic NPs/calcium alginate	Cu^{2+}	120 mg/g	[86]
AC-Cu^0-hydroxyapatite-alginate	As^{3+}	39.06 mg/g	[87]
CdSQD	Hg^{2+}	88.33 μg/g	[88]
Fe_3O_4/nano-hydroxyapatite/alginate	Fluoride	4.05	[89]
PVA/alginate/Fe^0	$Cu^{2+}/Cr^{6+}/Zn^{2+}/As^{5+}$	84.2%, 70.8%, 31.2%, and 39.8%	[90]
NCBs	Co^{2+}/Ni^{2+}	9.04 and 10.93 mg/g	[91]
Alg-MGO	Pb^{2+} and Cu^{2+}	96.13 and 103.09 mg/g	[92]
PLA-based nanocomposites			
Boehmite@ Fe_3O_4@PLA@SiO_2	Hg^{2+}	36.94 mg/g	[93]
PLA/chitosan nanofibers	Cu^{2+}	111.66 ± 3.22 mg/g	[94]
Xanthan gum (XG)-based nanocomposites			
Xanthan gum (XG) with EDA (EDA)	Cu^{2+}	46.95 mg/g	[95]
Hybrid NaA-zeolite/xanthan gum (XG)-alginate	Co^{2+} and Ni^{2+}	25.58 ± 0.45 and 45.87 ± 1.13 mg/g	[96]
Fe_3O_4/silica–XG	Pb^{2+}	21.32 mg/g	[97]
Silica modified calcium alginate–XG	Pb^{2+}	18.9 mg/g	[98]
XG-g-PAM/SiO_2bionanocomposite	Pb^{2+}	537.634 mg/g	[39]
XG-glutathione/Zeolite nanocomposite	Ni^{2+} and Pb^{2+}	42.91 and 47.98 mg/g	[99]
Poly(XG-g-AAM-g-AA)	Cu^{2+}	130.31 ± 2.97 mg/g	[100]
MgO_2/PPy	Fluoride	4328 mg/F/kg	[101]

15.4.1.1 CS-based nanocomposites

Removal of Cd^{2+} and Pb^{2+} metal ions was reported using CS nanocrystal (CNCs) that are modified by succinic anhydride (SCNC) and $NaHCO_3$ (NaSCNC). The adsorption capacities of metal ions strongly rely on the solution pH and shows increment with rise in pH value. The complexion mechanism was the primary mechanism for metal ion adsorption on SCNC, whereas NaSCNC was an ion-exchange process. NaCl was used to regenerate both adsorbents, resulting in negligible metal ion desorption due to the destruction of active functional groups in the SCNC adsorbent. However, regeneration had little effect on the active sites of NaSCNC, and adsorption of both metal ions remained constant after two cycles [57]. Bacterial cellulose (BC) nanofibers derived from *Acetobacter xylinum* blended with magnetic ATP composite (Fe_3O_4/ATP@(BCNs/CS)) were reported to remove Cr^{6+}, Pb^{2+}, Cu^{2+}, and CR [58]. The oxidized BC nanofibers allow subsequent reaction with chitosan amino groups due to the transformation of − OH groups of BC to −COO groups. Then, BCNs/CS coated alternately on the surface of Fe_3O_4/ATP by electrostatic attraction and form Fe_3O_4/ATP@(BCNs/CS) composite. The adsorption of Cu^{2+}, CR, and Cr^{6+} on biocomposite was much better than Pb^{2+}. The uniform dispersion of ATP on BCNs, as well as the presence of −COO and amino groups on the adsorbent's surface, resulted in good adsorption performance. Five adsorption-desorption cycles can be performed on the biocomposite.

The graft co-polymerization method for functionalizing biopolymers improves polymerization, increases graft density and uniformity, and introduces appropriate functional groups needed to absorb dyes and heavy metal ions [38]. For example, grafting of CS with glycidyl methacrylate (GMA) using ceric ammonium nitrate (CAN) initiator was carried out for the adsorption of Hg^{2+}, Cu^{2+}, and Ag^{+} [59]. The free radical ions formed by the ceric ion initiator act as active sites for grafting polymerization. The functionalization of cellulose acetate (CA) using grafting offers active sites to adsorb heavy metal ions by complexion mechanisms due to the presence of NH_2, SO_3H, and COOH functional groups. Similar results were reported for CA grafted with poly (methacrylic acid) (PMAA) using CAN initiator for the removal of Hg^{2+}, Cd^{2+} and Cu^{2+} [102]. The FTIR results confirmed that the carbonyl group −COOR had been changed to −COOH as a result of PMMA grafting on the surface of the CA fibers. Functionalized CA is thermally and chemically stable in terms of metal ion removal.

Magnetic CS composite have been explored to adsorp Cr^{6+} ions. CS coated with magnetic silica was prepared first, then it is functionalized with an amino group by the grating of GMA and ethylenediamine (EDA) [60]. The TEM images of the composite after grafting with GMA show a thin homogenous layer of magnetic nanoparticles, are given in Fig. 15.7. The Q_{max} of composite for the adsorption of Cr^{6+} was 171.5 mg/g. Coupling agents such as silanes (aminopropyltriethoxysilane, APTS) are used to chemically modify CS, forming a chemical bridge between matrix and reinforcement [61]. The additional coupling agents generate amine groups with high adsorption capacity for heavy metal ions.

The mercerization of succinic anhydride CS improves its surface area and intensify the number of −OH groups that can be used to remove Zn^{2+}, Ni^{2+}, Cu^{2+}, Co^{2+}. Magnetic CS nanocomposites have been broadly explored to remove broad array of pollutants. The bacterial CS-based composite modified with 1,6−hexane diamine and shows adsorption of As^{5+} 3 times better performance than magnetic CS with an adsorption capacity of 90 mg/g [61]. Graft co-polymerization of itaconic acid on magnetic CS composite using cross-linking agent Ethylene glycol dimethylacrylate (EGDMA) was performed by Anirudhan and Shainy to remove Hg^{2+} ion [64]. A multilayer adsorption process can remove the mercury ion with a Q_{max} of 240 mg/g. The ion-exchange process, accompanied by the complexion reaction, was responsible for the adsorption of mercury ions.

The adsorption behavior of dyes was investigated using various types of CS-based adsorbents and their composites. Functionalization of CS with carboxyl groups is one of the most studied procedures for improving adsorption performance.

COO-modified CS, for example, is created by ammonium persulphate (APS) hydrolysis of microcrystalline CS carboxylation on the adsorbent's surface [103]. Using COO-modified CS, where carboxyl groups bind with positively charged dyes, the removal of methylene blue (MB) dye was investigated. The Q_{max} of COO-modified CS was 127 mg/g at 22°C. Also, the adsorbent was efficiently regenerated up to seven cycles with 90% of dye removal even after seven cycles. Another study in carboxylated CS nanocomposite for MB dye adsorption was developed by hydrolysis of microcrystalline CS with

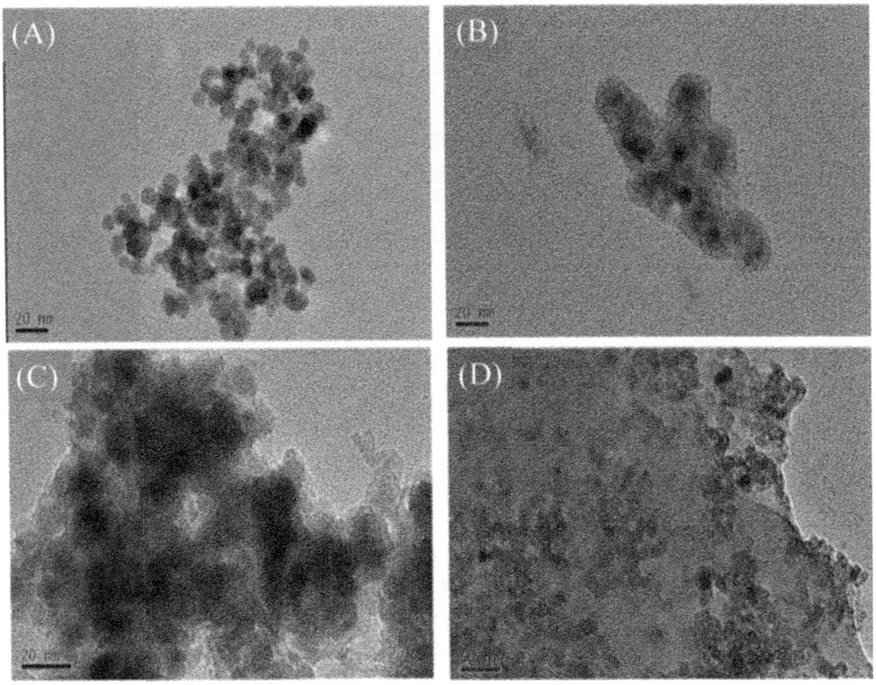

FIGURE 15.7

TEM images of (A) magnetic nanoparticles, (B) Magnetic silica nanoparticles (Ms), (C) Ms-CS composite (MSC), and (D) MSC-GMA-EDA (MSCGE).

Reprinted from S. Xitong, Y. Liangrong, L. Qian, Z. Junmei, W. Xiaoqin, L. Huizhou. Amino-functionalized magnetic cellulose nanocomposite asadsorbent for removal of Cr(VI): synthesis and adsorption studies, Chemical Engineering Journal 241 (2014) 175–183. Copyright 2014, Elsevier. https://s100.copyright.com/CustomerAdmin/PLF.jsp?ref= 2e9a8e20-3123-4229-9546-60db3c8b02a1.

HCl/citric acid [104]. COO-modified CS outperforms other modified CS (modified with sulfuric acid and formic acid), with complete MB degradation after 4 hours. Amination of CS improves the catalytic performance for removing anionic dyes by strengthening the bio-active functional groups. Jin et al., functionalized CS with EDA increase the amino groups and uptake of yellow K-4G, acid red, and CR4BS dyes [105]. The Q_{max} of the adsorbent for AR was turn out to be 555.6 mg/g, which follows chemisorption process and follows pseudo-second-order kinetics. In alkaline conditions, dye-adsorbent interaction was reduced due to a decrease in positively charged ions on the adsorbent's surface, resulting in electrostatic repulsion between dye and adsorbent. The removal of MO and MB dyes was investigated using activated carbon and magnetic nanoparticles combined in a CS matrix with a spherical structure. The adsorbent equilibrium was reached in 180 min with Q_{max} of 6.5×10^{-3} and 4.4×10^{-3} mmol/g for both dyes [106].

Polyacrylamide-CS nanocrystal (HPAM/CNC) was prepared using a simple thermal treatment and was used to investigate the removal of MB [107]. The anionic nature of HPM, increase in the concentration of CNC (up to 20 wt.%), and less pH boost the adsorption capability of the nanocomposite. Decoration of cobalt phthalocyanine (CoPc) on BC nanofibers to fabricate novel nanocatalyst CoPc@BC was used to remove RhB dye. Greater than 90% of RhB dye was degraded in the presence of H_2O_2 in 3 h [108]. Yang et al. incorporate and stabilizes Cd nanoparticles through coordination effect with CS nanofibers by hydrothermal reactions [109]. This nanocomposite shows 82% degradation of MO in 90 min of exposure to UV light with recycle of composite for five consecutive cycles without a decrease in catalytic properties. In other studies, 3D porous interconnected blocks of BC were immersed in titanium solution (BC@TiO_2) to study the degradation of MO [110,111]. A hybrid polyamine (PDA)/BNC membrane was developed to remove RhB, MB, and MO [112]. The separated membrane catalyst was reused ten times with no loss of catalytic activity.

15.4.1.2 Chitosan-based nanocomposites

The chitosan and tripolyphosphate (CS-TPP) ionic gelation has been studied to remove Pb^{2+} ions with a Q_{max} of 398 mg/g. The incorporation of cross-linking agents in the form of phosphate groups into chitosan improves intraparticle diffusion and adsorption capacity [66]. Other studies explored the adsorption capability of chitosan nanorods [67] and nanochitosan [68] for the removal of Cr^{6+} and Cd^{2+} with Q_{max} of 323.6 and 358 mg/g.

The amine groups on CS are responsible for the pollutant's electrostatic attraction. To increase amine groups, various mutation methods such as EDA, diethylenetriamine, and hexane diamine have been developed. For example, hexane diamine was used to modify magnetic nanoparticles to synthesize magnetic CS nanocomposites for the adsorption of Pb^{2+}, Cu^{2+} and Cd^{2+}. The removal efficiency of metal ions was reported to be 95%, 89%, and 92%, respectively [70]. CS was also modified with nanoparticles such as carboxylation, magnetic nanoparticle incorporation, carbon nanotubes, and clay-based nanoparticles for wastewater purification. Many literature and review articles are available that discussed the adsorption of heavy metal ions and dyes using CS nanocomposites (some of them are given in Table 15.1). In a study, the chemical co-precipitation method synthesizes Fe_3O_4/polyacrylic acid grafted CS (CS-MCM) composite to remove Cu^{2+} ions. The Q_{max} of 108 mg/g was reported with monolayer adsorption of Cu^{2+} ions. The adsorbent can be recycle by 0.1 mol/dm^3 EDTA solution for six consecutive cycles without any reduction of adsorption properties [113]. According to Yan et al., if a large amount of solvent is required for metal ion desorption, the overall cost of the process will rise, resulting in secondary pollution. As a result, they proposed that the adsorbent, with its altered surface structure, can be used to adsorb other pollutants under certain conditions [113]. PEG-CS composites have been studied to adsorb Ni^{2+} ions from aqueous solution and achieved an Q_{max} of 50.68 mg/g, while Goethite-CS nanocomposites [76] carboxymethyl CS [76] and carboxymethyl CS [75] were found to be suitable for complexion of Cd^{2+} and Cr^{6+}.

CS nanoparticles are projected to be excellent adsorbents to achieve superior adsorption of dyes in light of their large surface area and availability of -OH and amine groups. Hu et al. developed nanochitosan of 180 nm by gelation reaction between amino groups (positively charged) and sodium TPP (negatively charged) for the removing acid green 27 [114]. The adsorption of dye at equilibrium is mainly occurring due to availability of nanoparticles and it follows Langmuir model with monolayer adsorption capacity of 2103.6 mg/g. The principal mechanism for dye adsorption on CS was an electrostatic attraction between dye and amino groups of CS. Another study used nanochitosan acetylation to increase the amino groups for acid green 25 adsorption [115].

ZnO nanoparticles dispersed on CS biopolymer for the adsorptive removal of AB26 and DB78 and attained maximum adsorption of 34.5 and 52.6 mg/g [116]. Adsorption was primarily caused by electrostatic interaction between dyes and positively charged active sites, according to the isotherm model best fitted with Langmuir and Temkin. Gopi and colleagues created a CS nanofibril-chitin nanocrystal hybrid aerogel and investigated its application to remove dye as well as its anti-oxidant and antibacterial properties [117]. The aerogels were made by freeze-drying chitin nanocrystals at concentrations of 1% (AR1) and 2% (AR2) and decorated on CNFs. CNCs were derived from shrimp shells and corn husk. The AR2 hydrogel shows better morphology, alignment, orientation, higher crystallinity, and good adsorption performance towards MB and RhB with Q_{max} of 8 and 9.2 mg/g, respectively.

Because clay nanoparticles have a high surface area and good cation exchange capacity, CS/clay adsorbent has been extensively studied. However, their ability to adsorb anionic dyes is ineffective. Marrakchi et al. develop a CS/sepiolite nanocomposite for removing both cationic (MB) and anionic dyes using sepiolite clay (Orange 16) [118]. Sepiolite clay was dispersed in CS containing 50% v/v acetic acid and stirred continuously to achieve a homogeneous mixture. The epichlorohydrin (EPH) is used as a crosslinker for CS/sepiolite composite and achieved Q_{max} of 40.99 and 190.97 mg/g for MB and RO16. In previous studies with MB dye, the cross-linked composite demonstrated good adsorption performance. The cross-linking reaction improves the physiochemical stability and selectivity of biopolymers towards specific ions or molecules; it also prevents CS dissolution in acidic media [38]. However, excessive cross-linking can reduce mechanical stability; thus, it must be carefully controlled. In a recent study, a hybrid composite of cross-linked EPH/TiO_2/CS is synthesized (synthesis steps are shown in Fig. 15.8) to remove RR120 dye [119]. The optimum conditions achieved were temperature 300K, pH of solution 3, and adsorbent dosage 0.2 g/100 mL for removal of dyes. Because of the presence of amine and –OH groups, dye adsorption on composite occurred. These groups interact in a variety of ways, including dipole-dipole hydrogen bonding, electrostatic interaction, Yoshida H-bonding, and n-π interaction. Kadam and Lee investigated the removal of AR2 by using glutaraldehyde (GLD) as a crosslinker for magnetic CS nanocomposite [120]. The XPS analysis revealed that hydroxyl and amino groups cross-link chitosan with GLD and EPH, forming a new structure with different functional groups [121]. GLD cross-linked magnetic chitosan nanocomposite removed acid red dye with a removal efficiency of 91.6%. The application of chitosan nanocomposite in removing dyes and heavy metal ions is listed in Tables 15.1 and 15.2.

15.4.1.3 Starch-based nanocomposites

Because of their tunable mechanical properties, high water retention, self-degradation, and wide environmental applications, hydrogels have received a lot of attention for their applications in wastewater treatment. Nanoparticles such as magnetic nanoparticles and silver nanoparticles are incorporated and nanodispersed to form nanocomposite, which has a large surface area to volume ratio and mechanical and chemical stability. Some of the studies have been executed on the synthesis of starch hydrogel-based nanocomposites, for example, starch-based/PEG-ACA hydrogel (St-PEG-AcAH) and AgNPs-starch/PEG-ACA nanocomposite hydrogel (AgNPs-St-PEG-AcANCH) to study the removal of Hg^{2+} and antimicrobial activity [78]. The adsorption of Hg^{2+} was found to be monolayer adsorption for both adsorbents with a Q_{max} of 188.67 and 263.15 mg/g. PVA/carboxymethyl starch-g-poly(vinyl imidazole) (mCVP) hydrogel modified with magnetic nanoparticles was investigated for the removal

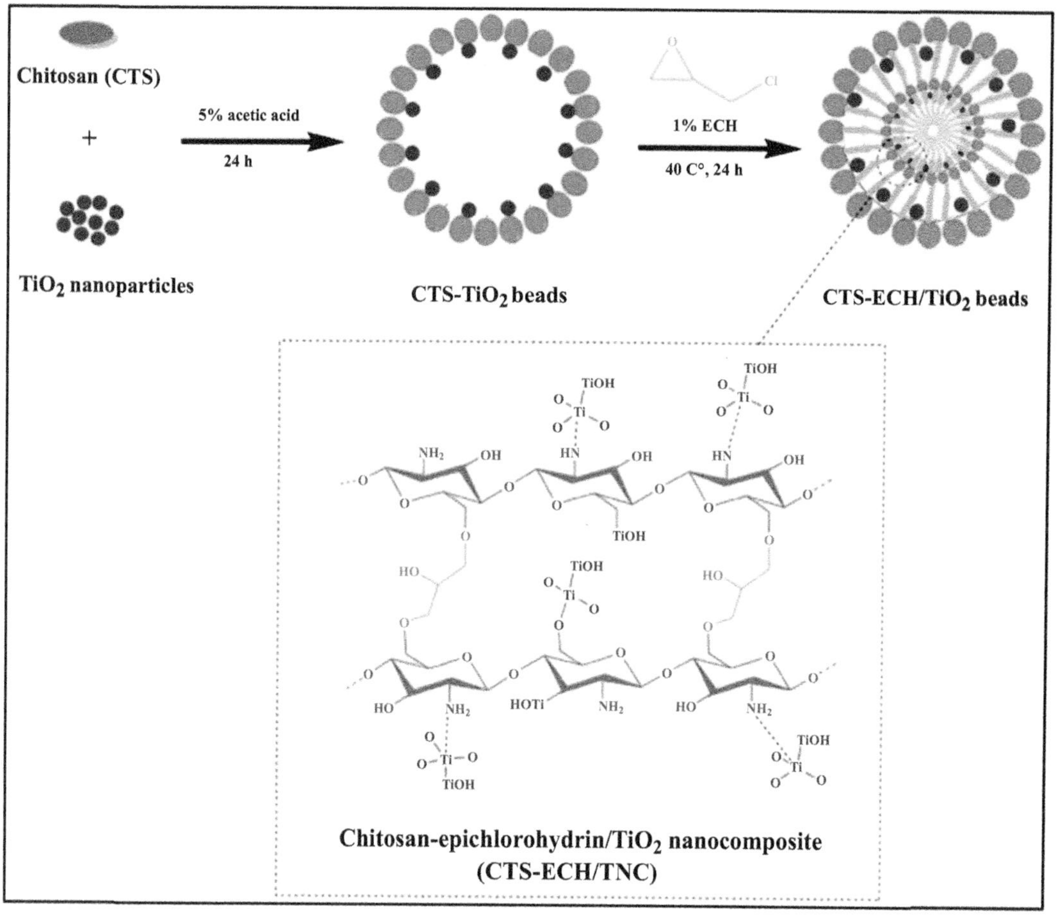

FIGURE 15.8

Synthesis steps of hybrid composite of CTS-ECH/TNC [119].

of Pb^{2+}, Cd^{2+}, and Cu^{2+} [79]. The Q_{max} of all metal ions was found to be 65, 53.2, and 83.6 mg/g. some other reported starch-based nanocomposites are listed in Table 15.1.

Diverse active groups that can form strong bonds with the –OH groups on the surface of starch can be used to modify starch. The abundance of hydroxyl groups aids in the formation of hydrogels via etherification, cross-linking, or grafting. Etherification involves the substitution of ether groups for a few hydroxyl groups, whereas grafting is accomplished through the use of vinyl monomers (acrylic and acrylamide acid) that can be grafted to starch. Guo and colleagues created cross-linked cationic starch from corn starch, using EPH as a cross-linking agent and 3-chloro-2-hydroxypropyl trimethylammonium chloride as an etherification agent, to investigate the elimination of golden yellow SNE dye [135]. The process results in providing Q_{max} around 208.77 mg/g at 308K. Magnetic cross-linked starch/PVA was synthesized for the removal of MG and MB dyes

Table 15.2 List of biodegradable nanocomposite used for the removal of dyes.

Biodegradable polymer nanocomposite	Pollutant	Adsorption capacity (mg/g)/efficiency (%)	References
CS-based biodegradable nanocomposite			
Fe_3O_4/ATP@(BCNs/CS)	CR	230 mg/g	[58]
COO-modified CS	MB	127 mg/g	[103]
ANCC	Acid red (AR)	555.6 mg/g	[122]
Polyvinylamine–modified CS nanocomposite	LY-K-4G/CR-4BS/ AR-GR	1250.9, 869.1, and 1469.7 mg/g	[123]
R-POSS	Reactive blue B–RN/ Reactive yellow B–4RFN	22.3 and 17.4 mg/g	[124]
Polyacrylamide-CS nanocrystal (HPAM/CNC)	MB	224.8 mg/g	[107]
Carboxymethyl CS (CMC)/MMT	CR	161.1 mg/g	[107]
CoPc@BC	RhB	≥90%	[108]
CdS nanoparticles/BC	MO	80%	[109]
PDA/BNC	R6G	8.7 mg/g	[112]
BNC/TiO_2	MO	—	[125]
Silver/TiO_2-CS CNF	MB	70%	[126]
Gold/TiO_2-CS CNF	MB	75%	[126]
Chitosan-based biodegradable nanocomposite			
Nanochitosan	AG17	2103.6 mg/g	[114]
Nanochitosan	AR73/AO10/AO12// R18	645.1, 922.9, 973.3, 693.2, and 728.2 mg/ g	[127]
Nanochitosan	Eosin Y	3.333 g/g	[128]
Chitosan–zinc oxide nanoparticle	AB26 /DB78	34.5 and 52.6 mg/g	[116]
CNFs–CNCs aerogels	MB/RhB	8 and 9.2 mg/g	[117]
Chitosan/sepiolite	MB/O16	40.99 and 190.97 mg/g	[118]
Glutaraldehyde/magnetic /chitosan	AR2	91.6%	[120]
Chitosan-4-nitroacetophenone/CuO-CeO_2/ Al_2O_3 and chitosan-4-nitroacetophenon/CuO-CeO_2/Fe_2O_3	AR60	92.8 and 88.5 mg/g	[129]
PVDF/CS&DA	MB/OG	96.8% and 92.7%	[130]
Chitosan/ZnO and Ce-ZnO	MG	87%	[131]
$MnFe_2O_4$/chitosan	MB	371.7 mg/g	[132]
Magnetic/graphene/chitosan	AO7	42.7 mg/g	[133]
CC–Fe_3O_4-NPs	Dr16	99%	[134]
Ag_2O/TiO_2 modified chitosan	MO	96%	[134]

(Continued)

Table 15.2 List of biodegradable nanocomposite used for the removal of dyes. ***Continued***

Biodegradable polymer nanocomposite	Pollutant	Adsorption capacity (mg/g)/efficiency (%)	References
Starch-based biodegradable nanocomposite			
Cross-linked cationic starch	Golden yellow SNE dye	208.77 mg/g	[135]
MNPs@Starch-g-poly(vinyl sulfate) nanocomposite	MG and MB	567 and 621 mg/g	[136]
CCMS/TiO_2	Cationic yellow 28	84.98%	[137]
CMS/poly(vinyl alcohol)	MB	23.53 mg/g	[138]
Starch/nanoscale zero-valent iron	Acid Blue-25	100%	[139]
Fe_3O_4/starch/poly (acrylic acid)	MV and CR	96.7% and 97.5%	[140]
Alginate-based biodegradable nanocomposite			
AgClNPs	BCB	95%	[141]
GOx/$MnFe_2O_4$/Calcium alginate nanocomposite	MB	73.68%	[142]
TiO_2/Ca-alginate	MO	98.14%	[143]
CNC–ALG	MB	97%	[144]
Fe_3O_4/alginate microspheres	MG	93.9%	[145]
Fe_3O_4/Na-alginate/H_2O_2	Reactive Blue 222	90.39%	[146]
Fe_3O_4/AC/CD/Alg	MB	99.53%	[147]
Pd NPs@CNCC–AHB	MB	99%	[148]
PLA-based biodegradable nanocomposite			
Zeolite (FAU)/PLA composite	MB	90%	[149]
Fe3O4@PLA@SiO_2	MB	10.17 mg/g	[93]
3D PLA@GO/chitosan sponge bionic filter	CV	97.8% ± 0.5%	[122]
XG-based biodegradable nanocomposite			
XG–poly(AA-co-AAM)-Fe_3O_4	MG	497.15 mg/g	[150]
XG-cl-PAA/Fe_3O_4	MV	642 mg/g	[151]
XG/Methionine-bentonite nanocomposite	CR	530.540 mg/g	[152]
XG/SiO_2 nanocomposite	MB and BB	432.90 and 448.43 mg/g	[153]
XG-cl-PAA	MB	510.9 mg/g	[154]
XG-cl-PAA/reduced GO	MB and MV	793.65 and 1052.63 mg/g	[155]
XG-cl-(PAA-co-PICA)	EBT	90.53%	[156]
XG-TiO_2 composite	MO	89%	[157]

[136]. The modifications performed with vinyl acetate to cross-link with starch (substitution of sulfur groups with hydroxyl groups) in the presence of magnetic NPs produced MNPs@Starch-g-poly (vinyl sulfate) nanocomposite. The Q_{max} for MG and MB dyes was 567 and 621 mg/g and can reuse for five successive regeneration cycles.

Metal oxide/metal acid salts of n type's semiconductors, for instance, ZnO, Fe_2O_3, TiO_2 have been comprehensively deliberated as a photocatalyst to degrade organic pollutants. Photocatalyst is relatively inexpensive, stable, and non-toxic. For example, Guo et al. studied photodegradation of cationic yellow 28 by cross-linked carboxymethyl starch (CCMS)/TiO_2, synthesized via sol-gel method [137]. The degradation of dye was found to be 84.98% and can be reused for four adsorption-desorption cycles. Sharma et al. utilizes Fe nanoparticles with amine and starch to develop versatile Pd NPs@Fe_3O_4-NH_2/Starch and reduces the amination of nitroarenes EtOH: H_2O at room temperature [158]. In another study, magnetic CMS/PVA hydrogel is studied to adsorb MB [138]. The Q_{max} of the composite reaches 23.53 mg/g with eight successive regeneration cycles with minimal loss of adsorption efficiency (not more than 15%).

15.4.1.4 Alginate-based nanocomposites

Because of the large number of functional groups and high surface area, an alginate-based adsorbent could be used as a catalyst and could form stable bio-hydrogel beads. Mesoporous calcium alginate/titania hybrid was employed to remove Cr^{6+}, Cd^{2+}, Cr^{3+}, Cu^{2+}, and Co^{2+} from the simulated wastewater [83]. The tendency of the adsorption rate of heavy metal ions follows in the order of $Cu^{2+} > Co^{2+} > Cd^{2+} > Cr^{6+} > Cr^{3+}$ with an adsorption capability of 8.4 mg/g for Cr^{6+}. These Cr^{6+} are negatively charged and adsorbed on alginate hybrid via electrostatic attraction due to rich cationic surface functionalities on the adsorbent. In a recent study, PVA/alginate/zeolite nanohybrid adsorbent was synthesized by casting method to adsorb Ni^{2+} and Cu^{2+} [84]. The monolayer and homogenous adsorption of metal ions has occurred at surface of adsorbent with a Q_{max} of 81.5 and 79.6 mg/g. Ravikumar et al. performed the 60 days continuous flow system to remove Cr^{6+} using Zr(VI)-alginate beads [85]. The system was tested for industrial wastewater and shows the removal of Cr^{6+} with 381 ± 17.1 mg/g of adsorption capacity for an initial concentration of 10 mg/L. Calcium alginate microbeads sensor was developed which consist of immobilized Cadmium sulfide (CdS) quantum dots (QD) to detect Hg^{2+} ions [88]. Synthesized cadmium sulfide quantum dot particles (CdSQD) possesses a high specific surface area which favors the removal of metal ions with Q_{max} of 88.33 μg/g. Some of the associated studies on the application of alginate-based adsorbent are listed in Table 15.1.

To investigate the adsorption of brilliant cresyl blue, silver chloride nanoparticles were incorporated in the polymer via the co-precipitation method by co-polymerization of acrylic acid (AA) and hydroxyethyl methacrylate (HEMA) in the presence of alginate (BCB) [141]. The dye removal efficiency in the batch column was greater than 95%, whereas in the column study, it was reduced (20%–35%), and the adsorbent can be recycled for five successive regeneration cycles without any structural degradation. Zolfaghari et al. created a heterogeneous GOx/$MnFe_2O_4$/Calcium alginate nanocomposite to investigate and optimize the adsorptive removal or degradation of MB dye [142]. Optimization for degradation of MB and modeling were investigated by varying $MnFe_2$ dose, glucose concentration, and pH value. At optimum conditions, the degradation efficiency of dye was found to be 73.68% after 2 h. Zhao et al. developed TiO_2/calcium alginate (T/CA) membrane by cross-linking of sodium alginate and TiO_2 in aqueous solution with supporting matrix ethylenetri (β-methoxy)ethoxy silane (KH-570) (as a monomer) and γ-amidopropyltriethyoxysilane (KH-550) (as a cross-linking agent) to investigate the degradation of MO dye [143]. Because of the stable nature of molecularly imprinted polysiloxane and TiO_2, the membrane's degradation of MO reached 98.14% and it can be reused six times with minimal change in MO degradation.

15.4.1.5 PLA-based nanocomposites

PLA is a renewable polymer that can be used in composites as a promoter to remove dyes and heavy metal ions. Alinezhad et al. created magnetic boehmite from SiO_2 and PLA for Hg^{2+} ion adsorption [93]. Boehmite is a low-cost hydroxyl-rich material that can be used for catalysis and cation adsorption from aqueous solution. Magnetic boehmite's adsorption performance is improved by coating it with SiO_2 and PLA. The Q_{max} of adsorbent for Hg^{2+} was 36.94 mg/g and followed the Langmuir isotherm model with reuse of adsorbent for six successive cycles. In another study, porous poly(L−lactic acid)/chitosan nanofibers membrane was synthesized by direct immersion coating method to remove Cu^{2+} ions [94]. At pH 7, the Q_{max} of composite for Cu^{2+} approached 111.66 ± 3.22 mg/g. The large surface area provided by PLA nanofibers and aminic and hydroxide groups on chitosan of composite material enhance the ion adsorption capacity.

Mascarenhas et al. used electrospinning to create faujasite zeolite (FAU)/poly(lactic acid) (PLA) composite fibers to investigate MO, RB, and MB dye removal from aqueous solution [149]. More uniform and bead-free fibers were formed as the polymer concentration in the solution increased. The nanofibers successfully adsorb cationic dye with 90% removal efficiency, but not anionic dyes. Alinezhad et al. use boehmite@ Fe_3O_4@PLA@SiO_2 to remove cationic dye MB and obtain a Q_{max} of 10.17 mg/g. More uniform and bead-free fibers were formed as the polymer concentration in the solution increased [93]. Zhou et al. synthesized 3D PLA@graphene oxide (GO)/chitosan sponge bionic filter using 3D printing and double freeze-drying technology [122]. This 3D filter was effectively utilized for the adsorption of CV dye with a removal efficiency of 97.8% ± 0.5% at pH 8, 35°C, 9 wt.% of GO, and contact time of 46 minutes, respectively.

15.4.1.6 Xanthan gum (XG)-based nanocomposites

Some of the molecules on the surface of XG, such as acetic and pyruvic acid moieties, carboxyl and hydroxyl groups, have a high adsorption capacity due to chelation between these multiple groups and heavy metal ions. However, there is a disadvantage to using XG composite for wastewater applications because of its solubility in water, making it difficult to separate XG from the water after adsorption. To overcome the aforementioned challenge, XG is modified by either combining with inorganic compounds or introducing functional groups. Qiu et al. modified the XG biopolymer with EDA to remove Cu^{2+} ions from an aqueous solution [95]. The adsorption of metal ions follows the Langmuir model with a Q_{max} of 46.95 mg/g. Moreover, the adsorption of Cu^{2+} has shown slightly adverse effects in the coexisting salts; also, the adsorbent was not ideal after four successive regeneration cycles. Zhang et al. synthesized NaA/XG−alginate to adsorb Co^{2+} and Ni^{2+}, where XG biopolymer is utilized to control suspension of highly concentrated zeolite NaA particles [96]. To interact with zeolite NaA particles, the natural XG was first denatured in the partial unwound from under suitable temperature, pH, and salinity conditions. After re-naturing in wound form, it is mixed with alginate to form more complex hydrogel networks. The Langmuir model turns out to be the best fitting for the adsorption of Co^{2+} and Ni^{2+} with an adsorption capacity of 25.58 ± 0.45 and 45.87 ± 1.13 mg/g at 293K. Magnetic Fe_3O_4/silica/XG composite was applied for examining the adsorptive behavior of Pb^{2+} ions [97]. The mechanism of lead ions was primarily dependent on the interaction of metal ions with adsorbent hydroxyl groups via the formation of meta-OH complexes. The poly(XG-g-AAM-g-AA) synthesis was investigated for the adsorption of Cu^{2+} ions, synthesized by a crosslinker TTGE and an initiator KPS [100]. The adsorption capacity for Cu^{2+} ions was 130.31 ± 2.97 mg/g with 5 pH and ambient temperature. The amine and carboxyl groups introduced by the monomers AAM and AA improve thermal stability, provide a regular arrangement of the XG molecules, and provide a porous surface morphology.

To remove CV dye, XG was modified with trimethylolpropane triglycidyl ether (TTE) (cross-linking agent) and synthetic polymers such as acrylamide (AM) with KPS [159]. CV adsorption follows the Freundlich and pseudo-second-order models with a Q_{max} of 183 ± 12 mg/g at pH 7. Because of the accessibility of active sites at the adsorbent's surface, dye removal increases with adsorbent dosage. The incorporation of Fe_3O_4 nanoparticles in XG–poly(AA-co-AAM), prepared by ultrasonication, facilitated synthetic methodology for the adsorption of MG dye [150]. Magnetic nanoparticles grafted onto copolymer provide a coherent surface for dye molecules to efficiently adsorb on. Mittal and coworkers synthesized an XG/Fe_3O_4 hydrogel by a cross-linking agent polyacrylic acid for the adsorption of MV [151]. Because of the addition of a cross-linking agent, the surface morphology of the nanocomposite is smooth and homogeneous. The adsorption of MV follows the Langmuir model with a Q_{max} of 642 mg/g at pH 6.5. The reusability of the adsorbent was accomplished by 0.1 M HCl solution without any decrement up to three consecutive cycles; however, after three cycles, efficiency reduces to 5% at the last two-cycle. XG's methionine-bentonite nanocomposite was investigated for CR elimination [152]. The Q_{max} of the composite was 530.54 mg/g at pH 3. The nanocomposite synthesized by incorporating SiO_2 in XG possesses a high surface area and specific pore volume to eliminate MB and BB [153]. The adsorption capacities of an adsorbent for MB and BB at pH 9 and 7 were 432.90 and 448.43 mg/g, respectively. Other XG-based nanocomposites are summarized in Table 15.2.

15.4.2 Membrane filtration

Membrane technology is an appealing technology for wastewater treatment and desalination because of its ease of operation, good performance in separating mineral components from saline water, low cost, and high efficiency. The categorization of different types of the membrane is depend on their pores sizes such as reverse osmosis (0.0001–0.001 μm) ultrafiltration (0.01–0.1 μm), microfiltration (0.1–5 μm), and nanofiltration (0.001–0.01 μm) membranes as shown in Fig. 15.9 [160]. Since a decade, there has been a clear increase in research development in the field of membrane technology. Polymeric and ceramic-based membranes are the most commonly used because they are thermally and chemically stable and achieve good performance with fast separation. However, the high cost of these membranes, as well as their poor tolerance to high temperatures, organic solvents, and brittleness, limited their use in wastewater treatment applications. As a result, biodegradable polymer-based membranes have received a lot of attention in recent years and have been widely used to treat wastewater because they are non-hazardous and sustainable. High water flux, high pollutant removal performance (organic, inorganic, and antifouling), good mechanical properties, exceptional physical and chemical properties such as high surface area, and good interfacial activity of nanofillers all contribute to the excellent performance of biodegradable polymer nanocomposite membrane [161,162]. Because of their metal chelating, good

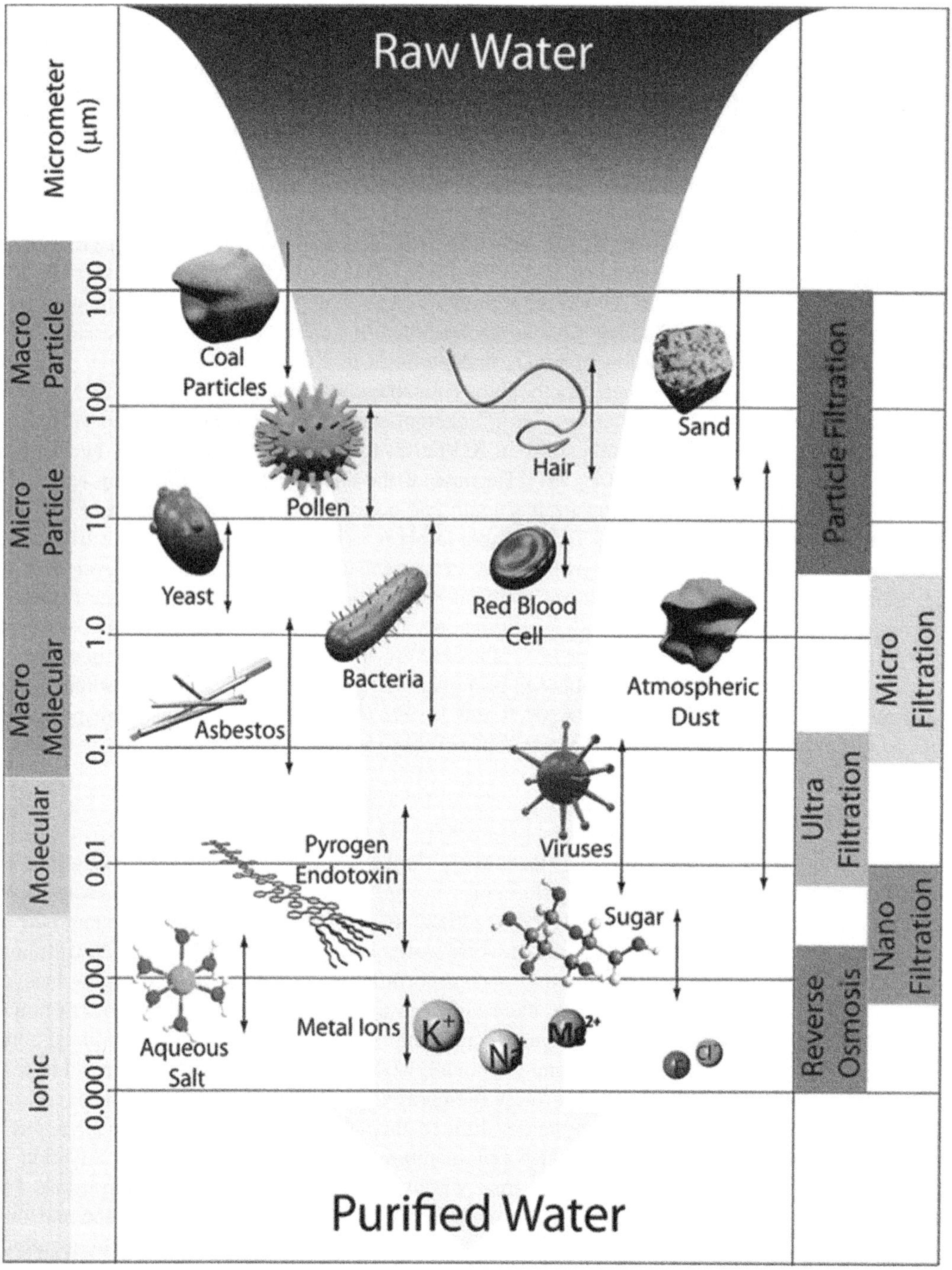

FIGURE 15.9

Schematic illustration spectrum of membrane filtration.

Republished with permission of Royal Society of Chemistry from A. Lee, J.W. Elam, S.B. Darling, Membrane materials for water purification: design, development, and application, Environmental Science: Water Research &Technology 2 (2016) 17–42.

flocculating, and availability of amino and hydroxyl groups, CS and chitosan biopolymer-based membranes have recently received a lot of attention for wastewater treatment. The combination of these biodegradable polymers with various organic and inorganic nanomaterials improves membrane flux and pollutant removal performance. In general, two methods are used to synthesize biopolymer-based nanocomposite membranes: the first is the deposition/incorporation of nanofillers on the surface of the biopolymer membrane, and the second is that a simple frame of nanoparticles in polymeric mesh can be used for the development of nanocomposite membranes [163,164]. Incorporating nanomaterials into the polymeric matrix membrane can be accomplished through a variety of methods, including physical and chemical deposition, layer-by-layer assembly, chemical grafting, self-assembly, and corona plasma coating or vacuum assisted assembly [165].

Abolhassani et al. create a membrane (CS-GO) made of CS and GO to remove MB and MO [166]. The chemical composition and surface morphology of the membrane suggested that the GO size influences the membrane's structure, internal morphology, and chemical composition. The rejection of MB using CS-GO membrane was 95%, and the adsorption capacity of GO and CS-GO composite was 139.29 and 52.40 mg/g. The decrease in composite adsorption capacity was caused primarily by electrostatic and hydrogen bonding interactions between the amine groups of both CS and GO. Adsorption was the primary removal mechanism for MO dyes, which is affected by GO particle size and time. Sile-Yuksel et al. used three types of polymers, polyethersulfone (PES), polysulfone (PS), and CA (CA), to form nanocomposite membranes using silver nanoparticles to study anti-biofouling [167]. To investigate the location of silver nanoparticles and their effect on the membrane matrix, different polymer/silver nanoparticle ratios were investigated. The SEM and EDS analysis show that silver nanoparticles are located at the skin layer and sublayer of PES and PS matrix, while in the CA matrix, silver nanoparticles accumulate inside big pores and locate at the bottom surface, as shown in Fig. 15.10. Water could permeate more into polysulfone by increasing the silver-nanoparticle/polymer ratio; however, the opposite effect was observed when polyethersulphone and CA polymers were used.

By dispersing silver particles in CA matrix and chemically attaching silver particles in CA matrix, the biofouling resistance of silver/CA membrane is improved [168]. Casein's thiol group has a higher affinity for silver nanoparticles and reduces accumulation due to proper dispersion in the CA matrix. Silver nanoparticles combined with CA membrane are also applied to remove pesticides and salts from water, such as triazines (prometryn, propazine, and atrazine) and $MgSO_4$ salts [169]. When compared to a pure CA membrane, the nanocomposite membrane exhibits high rejection of $MgSO_4$ salts and low flux. When compared to other triazines, the membrane has a high removal of prometryn, and removal performance improves with increasing feed concentration and pressure. Furthermore, as particle loadings increased, salt rejection increased.

Other nanofillers, such as TiO_2, have been widely investigated as a photocatalyst and have found their way into a variety of membranes due to their ease of fabrication and solidity [170]. Because of their Bronsted and Lewis acidity, high surface area, and layered structure, clay nanomaterials have been functionalized and widely used for water purification. However, due to clay surface hydration, organic molecule accessibility in the clay interlayer is primarily low. As a result, the development of clay-polymer nanocomposite membranes has overcome this disadvantage. Melt intercalation, solution casting, and in situ polymerization can all be used to create clay-polymer nanocomposite materials [171]. Further, modification and functionalization of the clay-based membrane such as surfactant modification [172], coupling agent modification [173], intercalaction modification [174], and

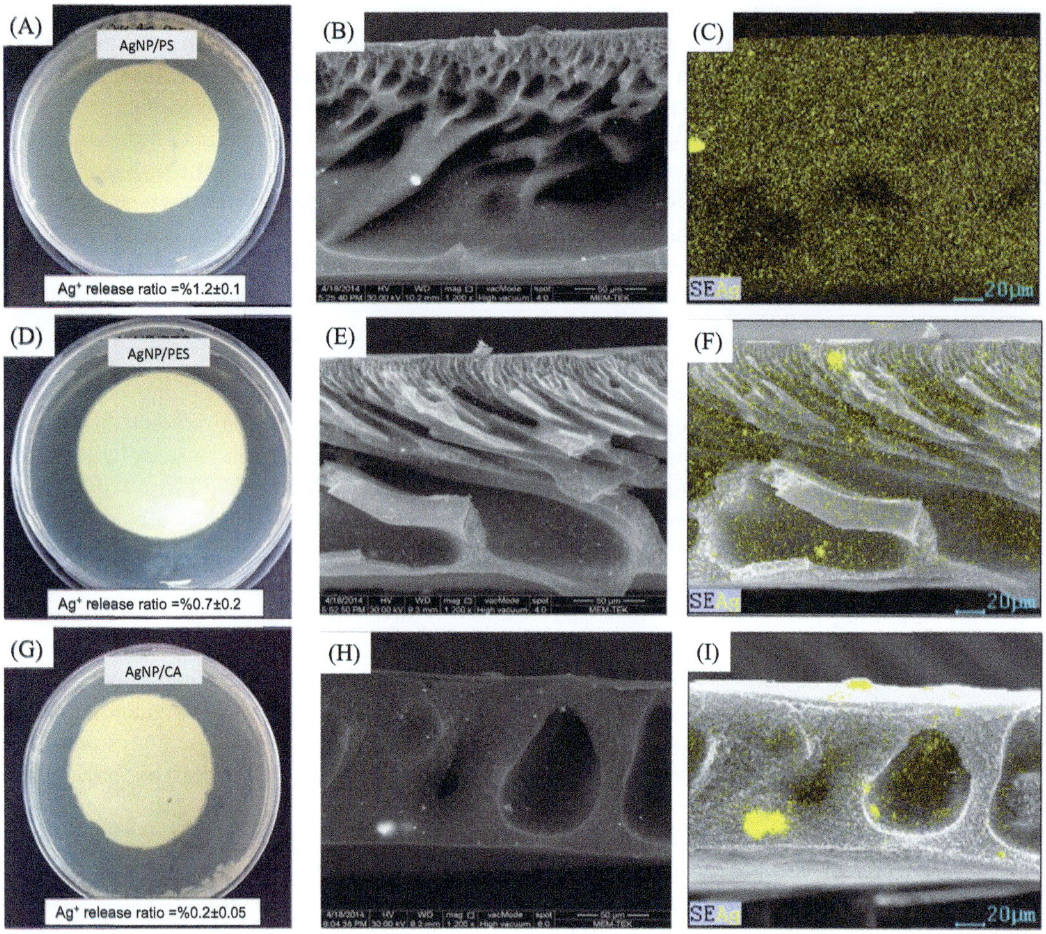

FIGURE 15.10

(A) Effect of PS/silver nanoparticles on *Escherichia coli* (B) SEM images of PS/silver membrane (C) EDS mapping of PS/silver membrane (D) effect of PES/silver nanoparticles on *E. coli* (E) SEM images of PES/silver membrane (C) EDS mapping of PES/silver membrane (G) effect of CA/silver nanoparticles on *E. coli* (H) SEM images of CA/silver membrane (I) EDS mapping of CA/silver membrane.

Reprinted from S.Y. Merve, T. Bahar, Y.K.I. Derya, K. Ismail, Effect of silver-nanoparticle (AgNP) location in nanocomposite membrane matrix fabricated with different polymer type onantibacterial mechanism, Desalination 347 (2014) 120–130. Copyright 2014, Elsevier https://s100.copyright.com/CustomerAdmin/PLF.jsp?ref= 1c4fc8d0-83e6-4d62-ab7b-805eb3b37ff4.

modification with polymers [175], advances the performance of clay-polymer nanocomposite membrane. Chitosan/cloisite clay (10 Å) polymer composite membrane utilized to remove hexavalent Cr^{3+} from wastewater. The chitosan/clay nanocomposite membrane fabricated from solvent casting The Qe of membrane was 357.14 mg/g [176]. The successful removal of Cr^{3+} and Fe^{3+} ion from both single and binary systems have been performed by chitosan/attapulgite membrane [177]. The adsorption percentage of Fe^{3+} increases from 69.4% to 91.1% and for Cr^{3+} it reaches to 95% for 3 mg/L ion concentration. Various membranes reported in literature consisting of chitosan biopolymer with clay-based nanofiller to remove heavy metal ions [178], dyes [179] and shows good antimicrobial and antifouling properties [180].

A biopolymer-based membrane other than CS and chitosan was created by assembling exfoliated silk nanofibrils from native Bombyx mori silk fibers. The membrane exhibit separation of dyes with at least a 64% rejection of RhB and pure water fluxes of 13,000 LHM/bar [181]. Nonwoven CA nanofiber membrane prepared by electrospinning and hot pressing also has a high rejection rate and water permeability [182]. In another study, CA/polybenzoxazine (CA/PB) was electrospun and cross-linked with citric acid [183]. The addition of a crosslinker to CA stabilizes the membrane's thermal and mechanical properties, resulting in 98.5% pollutant rejection from wastewater.

Forward osmosis (FO) combined with a thermal membrane reduces energy demand, improves water flux, hydrophilicity, chemical and thermal stabilities, antifouling and antibacterial properties, and long-term resistance [165]. Different fabrications methods have been performed on the FO membrane, such as interfacial polymerization, phase inversion, layer by layer and electro-spinning method as shown in Fig. 15.11. The selection of suitable fabrication methods depends upon the choice of polymer and desired structure of membrane.

15.4.3 Microbial fuel cell

Bioelectrochemical systems have emerged in recent years as a new and promising technology in which microbes can treat wastewater with minimal energy expenditure, as reported in numerous studies in the literature [185]. Electrochemically active microbes reduce the electron donor before transferring the electron generated in the process from anode to cathode, resulting in electric current. Microbial fuel cell (MFC), microbial desalination cell (MDC), microbial electrolysis cell (MEC), and microbial solar cell are the four types of bioelectrochemical systems (MSC). In MFC, wastewater is used to generate electricity as a final product; in MDC, saline water is used to generate electricity through the desalination process; in MEC, hydrogen is primarily generated at the cathode; and in MSC, electricity is generated using solar energy. MFC-based bioelectrochemical systems have recently found use in wastewater and provide several operational, economic, and environmental benefits. The foremost constituents of MFC are anaerobic cathode, anaerobic anode, and proton exchange membrane (PEM) coupled with an external circuit and separating electrodes (as shown in Fig. 15.12). The anode is typically fed various types of wastewater, while the cathode generates beneficial chemicals while removing pollutants. The overall performance of MFC is primarily determined by the performance of the electrodes, which includes effective electron generation and proton transfer [186]. As a result, several materials were investigated to improve the performance of electrodes by improving features such as surface area, conductivity, and stability. Furthermore, the selection of appropriate PEM is critical for the overall performance of MFC because the efficiency of PEM in proton cathode transfer can be reduced due to substrate loss, biofouling, and oxygen crossover. Ideally, the membrane should be biofouling resistant, inexpensive, reduce oxygen crossovers, and allow unhindered proton-cathode channeling [185].

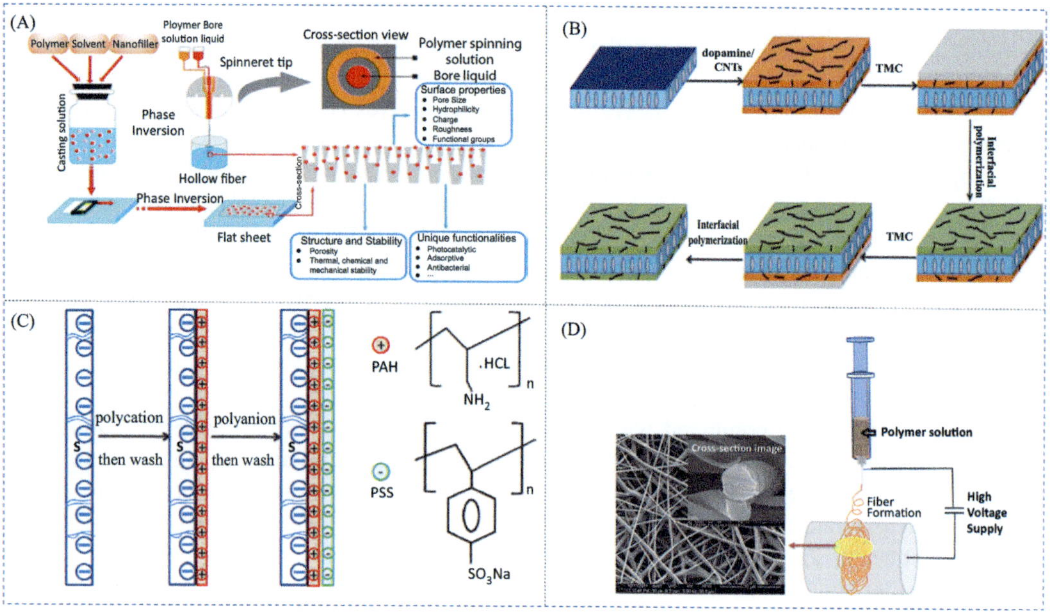

FIGURE 15.11

Modification methods of membrane. (A) Phase inversion. (B) Interfacial polymerization [184]. (C) Layer-by-layer. (D) Electrospinning.

(A) Reprinted with permission from W. Jianlong, L. Xiaojing, Forward osmosis technology for water treatment: recent advances and future perspectives, Journal of Cleaner Production 280 (2021) 124354. Copyright 2021, Elsevier https://s100.copyright.com/CustomerAdmin/PLF.jsp?ref= d2127a10-bb08-49a4-912c-2ebcb6b12a70. (B) Reprinted with permission from Xiangju, W. Li, Y.T. Chuyang, W. Zhining, G. Congjie, Fabrication of carbon nanotubes incorporated double-skinned thin film nanocomposite membranes for enhanced separation performance and antifouling capability in forward osmosis process, Desalination 369 (2015) 1–9. Copyright 2015, Elsevier https://s100.copyright.com/CustomerAdmin/PLF.jsp?ref= 4947c6b3-4d22-4bd8-bd23-2f12676c9c1d. (C) Reprinted with permission from S. Qi, Q.Q. Chang, Y.T. Chuyang, Synthesis and characterization of novel forward osmosis membranes based on layer-by-layer assembly, Environmental Science & Technology (2011). Copyright 2011, American Chemical Society. (D) Reprinted with permission from S.L. Boor, K. Victor, H. Raed, H. Nidal, A review on membrane fabrication: structure, properties and performance relationship, 326 (2013) 77–95. Copyright 2013, Elsevier. https://s100.copyright.com/CustomerAdmin/PLF.jsp?ref= 79e6f473-e991-405e-962d-af4201afe0dd.

Biopolymer nanocomposites have been used as a replacement for PEM in MFC technology, which reduces biofouling and improves biofouling colonization while decreasing PEM degradation, resulting in a better ion-exchange process due to the formation of micropores [185]. For example, in MFC, the modification of CS biopolymer with GO to form CS-GO membrane has been studied for chemical oxygen demand (COD) removal and power intensity enhancement [187]. Because of the formation of an ionic cluster and improved mechanical and thermal stability, the CS-GO is cross-linked with phosphoric acid (CS-GO-P), which increases COD's power density and removal efficiency to 89.52%.

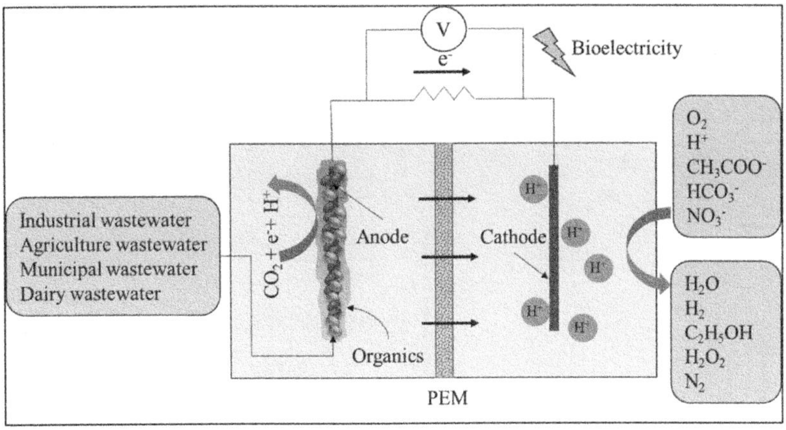

FIGURE 15.12

Microbial fuel cell for the treatment of wastewater.

Similarly, BC was modified with sulfur and cross-linked with polyaniline (PANI) to form SBC/PANI membranes, which were then used as gel polymer electrolytes in MFCs [188]. Biochemical properties and ion-exchange conductivity are improved by the nanocomposite membrane. The results showed that 41.8% of the degree of sulfonation (DS) enhances the ionic conductivity to 5.2×10^{-3} S/cm and ion-exchange capacity to 3.92 mequiv/g. Conductive polymers are also used and combined with biopolymers to improve their stability due to their ion-exchange and doped processes. For example, polypyrrole/chitosan/CNT nanoelectrode was used to absorb Cu^{2+} ions from an aqueous solution by capacitive deionization (CDI) process [77]. Polypyrrole and chitosan were mixed in different mass ratios and synthesized using in situ polymerizations. The specific capacitance of the composite as determined by the CV test decreases by only 3.4% after 50 cycles, indicating that the composite is stable after 50 cycles.

Ceramic and clay materials demonstrated excellent performance in MFC membranes, particularly for large-scale and wastewater treatment applications. However, the performance of MFC electricity generation is hampered by high oxygen levels and the rate of substrate diffusion through a porous membrane. Combining clay or ceramic material with biopolymer thus improves ionic conductivity while decreasing substrate diffusion and oxygen. Yousefi et al. assembled CS/MMT membrane as layer-by-layer (LBL) over the ceramic surface to utilize nanocomposite film for membrane in MFC technology, as shown in Fig. 15.13 [189]. When compared to blank ceramic, the nanocomposite with 1/2% wt./wt. performs best and generates twice as much electricity. COD and BOD removal from wastewater have been studied, and all experiments achieved 85% removal efficiency.

Furthermore, the columbic efficiency and power generated more than triple, while the oxygen coefficient decreases by two orders of magnitude. In another study, polyhydroxyalkanoates (PHA) biopolymer combined with poly-(R)-3-hydroxybutyrate (PHB) composite was used for PEM for electricity generation and wastewater treatment [185]. The composite showed 70% removal efficiency of COD, 912 mW/m^2 power density, and 958 mV voltage potential.

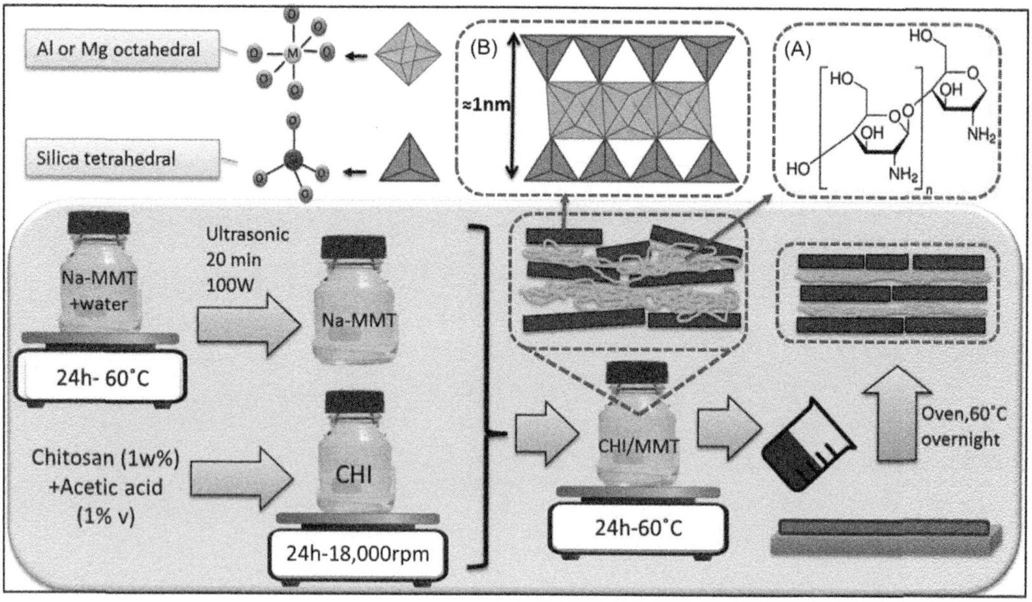

FIGURE 15.13

Procedure for the preparation of the nanocomposite; (A) the chemical structure of chitosan. (B) Structure of MMT sheets [189].

Reprinted with permission from Y. Vajihe, M.K. Davod, S. Abdolreza. Start-up investigation of the self-assembled chitosan/ montmorillonite nanocomposite over the ceramic support as a low-cost membrane for microbial fuel cell application, International Journal of Hydrogen Energy 45 (2020) 4804–4820. Copyright 2020, Elsevier. https://s100.copyright.com/CustomerAdmin/PLF.jsp? ref= d6d3652b-5ed3-4da9-8178-acc24a4b4a95.

15.5 Conclusion and future perspective

The chapter summarizes various types of biodegradable polymer nanocomposites, synthesis methods, and wastewater purification applications. Because of their low cost, tunable structure, abundance availability, physicochemical and biochemical properties, and ease of processability, biodegradable polymers and nanocomposite are in high demand. Furthermore, it has been discovered that there are no standard rules for theoretically predicting the outcome of the catalytic activity, adsorption/degradation processes of bio-nanocomposites. The amount of nanomaterials, their types, surface area, morphology, and porosity, loading of nanomaterials on the surface of biopolymer, temperature and pH of pollutant solution, and contact time between pollutant and bionanocomposite are all factors that influence adsorption performance and catalytic activity.

Many knowledge gaps must be filled before biodegradable polymer nanocomposite can be used effectively in wastewater treatment. If one wishes to create smart polymer nanocomposites for specific applications, one must first understand the mechanism to be used for removal, the constituents of polymer nanocomposites, and their properties, particularly the structure-related properties.

Furthermore, it is critical to investigate the relationship between various properties, removal mechanisms, and properties of polymer nanocomposites. The chemical and physical stability of bio-nanocomposite in aqueous media can also be improved by using a suitable crosslinker/polymer or developing new modification/synthesis approaches. As a result, more research is required to develop bionanocomposite via grafting or cross-linking methods to maximize the benefit of these methods in scaling up the process and balancing the cost of the process. Nonetheless, much more work is required to decipher the laws that govern the various limiting behaviors of polymer nanocomposites, such as reusability, recycling ability, negative environmental effects, and life cycle. Since biodegradable polymer nanocomposite exhibits interesting properties in membrane separation, more research is needed to address some other issues such as biodegradability rates of biodegradable polymers, poor mechanical properties, and infeasibility to industrial applications due to poor performance. Of course, all of the above factors are strongly related to the overall cost of the polymer nanocomposites-based water treatment solution, as cost is a major driving factor in the selection of the water treatment method in industrial applications. Because most studies investigate performance on a batch scale, the main limitation of bionanocomposite is the implementation of experimental procedures in practical application. Involving MFC also makes wastewater treatment methods more sustainable. Over the last decade, researchers have significantly improved power densities; however, some issues require immediate attention before MFC-based systems can be economically used in industries on a large scale. For cost-effective, industry-compatible, pragmatic solutions, efforts should be focused on increasing electrode surface areas while using low-cost materials. The prototypes developed in research must use actual wastewater discharge from industries rather than a laboratory-made alternative.

Acknowledgment

The authors acknowledge the Department of Civil Engineering, Jamia Millia Islamia (JMI), and Department of Chemistry, Faculty of Science, Aligarh Muslim University (AMU) in India for providing research facilities. The authors are grateful to Prime Minister Research Fellowship (PMRF) for its financial support.

References

[1] S.S. Bari, A. Chatterjee, S. Mishra, Biodegradable polymer nanocomposites: an overview, Polym. Rev. 56 (2016) 287–328.

[2] S.O. Adeosun, G.I. Lawal, S.A. Balogun, E.I. Akpan, Review of green polymer nanocomposites, J. Min. Mater. Charact. Eng. 11 (2012) 385–416.

[3] A. Sorrentino, G. Gorrasi, V. Vittoria, Potential perspectives of bio-nanocomposites for food packaging applications, Trends Food Sci. Technol. 18 (2007) 84–95.

[4] B. Joseph, S. Krishnan, V.K. Sagarika, A. Tharayil, N. Kalarikkal, S. Thomas, Bionanocomposites as industrial materials, current and future perspectives: a review, Emer Mat. 3 (2020) 711–725.

[5] O.A. Dar, M.A. Malik, M.I.A. Talukdar, A.A. Hashmi, Bionanocomposites in water treatment, Bionanocomposites (2020) 505–518.

[6] A.K. Momina, Study of different polymer nanocomposites and their pollutant removal efficiency: review, Polymers (Guildf.) 217 (2021) 123453.

[7] World Bank. Drinking Water – Overview, 2020.
[8] A.D. La Rosa, Life cycle assessment of biopolymers, Biopolymers and Biotech Admixtures for Eco-Efficient Construction Materials, Elsevier, 2016, pp. 57–78.
[9] T.J. Gutiérrez, Polymers for food applications: news, Polymers for Food Applications, Springer, 2018, pp. 1–4.
[10] T. Baran, M. Nasrollahzadeh, Cyanation of aryl halides and Suzuki-Miyaura coupling reaction using palladium nanoparticles anchored on developed biodegradable microbeads, Int. J. Biol. Macromol. 148 (2020) 565–573.
[11] W. Den, V.K. Sharma, M. Lee, G. Nadadur, R.S. Varma, Lignocellulosic biomass transformations via greener oxidative pretreatment processes: access to energy and value added chemicals, Front. Chem. 6 (2018).
[12] S. Iravani, R.S. Varma, Greener synthesis of lignin nanoparticles and their applications, Green Chem. 22 (2020) 612–636.
[13] N.A. Oladoja, R.O.A. Adelagun, A.L. Ahmad, E.I. Unuabonah, H.A. Bello, Preparation of magnetic, macro-reticulated cross-linked chitosan for tetracycline removal from aquatic systems, Colloids Surf. B Biointerfaces 117 (2014) 51–59.
[14] G. Crini, Recent developments in polysaccharide-based materials used as adsorbents in wastewater treatment, Prog. Polym. Sci. 30 (2005) 38–70.
[15] R. Xu, L.C. Yong, Y.G. Lim, J.P. Obbard, Use of slow-release fertilizer and biopolymers for stimulating hydrocarbon biodegradation in oil-contaminated beach sediments, Mar. Pollut. Bull. 51 (2005) 1101–1110. Pergamon.
[16] P. Amouzgar, B. Salamatinia, A short review on presence of pharmaceuticals in water bodies and the potential of chitosan and chitosan derivatives for elimination of pharmaceuticals, Mol. Genet. Med. S4 (2015) 001.
[17] B. An, K.Y. Jung, D. Zhao, S.H. Lee, J.W. Choi, Preparation and characterization of polymeric ligand exchanger based on chitosan hydrogel for selective removal of phosphate, React. Funct. Polym. 85 (2014) 45–53.
[18] K. Sahithya, D. Das, N. Das, Effective removal of dichlorvos from aqueous solution using biopolymer modified MMT-CuO composites: equilibrium, kinetic and thermodynamic studies, J. Mol. Liq. 211 (2015) 821–830.
[19] A. Rajeswari, A. Amalraj, A. Pius, Adsorption studies for the removal of nitrate using chitosan/PEG and chitosan/PVA polymer composites, J. Water Process. Eng. 9 (2016) 123–134.
[20] S. Sayed, A. Jardine, Chitosan derivatives as important biorefinery intermediates. Quaternary tetraalkylammonium chitosan derivatives utilized in anion exchange chromatography for perchlorate removal, Int. J. Mol. Sci. 16 (2015) 9064–9077.
[21] S. Jagtap, M.K.N. Yenkie, N. Labhsetwar, S. Rayalu, Defluoridation of drinking water using chitosan based mesoporous alumina, Microporous Mesoporous Mater. 142 (2011) 454–463.
[22] S. Doppalapudi, A. Jain, W. Khan, A.J. Domb, Biodegradable polymers—an overview, Polym. Adv. Technol. 25 (2014) 427–435.
[23] T. Russo, P. Fucile, R. Giacometti, F. Sannino, Sustainable removal of contaminants by biopolymers: and future perspectives, Processes 9 (2021) 719.
[24] J. Yin, B. Deng, Polymer-matrix nanocomposite membranes for water treatment, J. Memb. Sci. 479 (2015) 256–275.
[25] J. Pandey, Biopolymers and their application in wastewater treatment, Emerging Eco-Friendly Green Technologies for Wastewater Treatment, Vol. 18, Springer Nature, Singapore, 2020.
[26] J. Wang, S. Zhuang, Removal of various pollutants from water and wastewater by modified chitosan adsorbents, Crit. Rev. Environ. Sci. Technol. 47 (2017) 2331–2386.

[27] U. Habiba, A.M. Afifi, A. Salleh, B.C. Ang, Chitosan/(polyvinyl alcohol)/zeolite electrospun composite nanofibrous membrane for adsorption of Cr6 + , Fe3 + and Ni2 + , J. Hazard. Mater. 322 (2017) 182–194.
[28] Y.C. Chang, S.W. Chang, D.H. Chen, Magnetic chitosan nanoparticles: studies on chitosan binding and adsorption of Co(II) ions, React. Funct. Polym. 66 (2006) 335–341.
[29] L. Ren, J. Xu, Y. Zhang, J. Zhou, D. Chen, Z. Chang, Preparation and characterization of porous chitosan microspheres and adsorption performance for hexavalent chromium, Int. J. Biol. Macromol. 135 (2019) 898–906.
[30] J. Nie, Z. Wang, Q. Hu, Difference between chitosan hydrogels via alkaline and acidic solvent systems open, Sci. Rep. 6 (2016) 36053.
[31] P.M. Visakh, A.P. Mathew, K. Oksman, S. Thomas, Starch-based bionanocomposites: processing and properties, Polysaccharide Building Blocks: A Sustainable Approach to the Development of Renewable Biomaterials, John Wiley and Sons, 2012, pp. 287–306.
[32] A. Li, R. Liu, A. Wang, Preparation of starch-graft-poly(acrylamide)/attapulgite superabsorbent composite, J. Appl. Polym. Sci. 98 (2005) 1351–1357.
[33] F. Chivrac, E. Pollet, M. Schmutz, L. Avérous, Starch nano-biocomposites based on needle-like sepiolite clays, Carbohydr. Polym. 80 (2010) 145–153.
[34] A.E.S.I. Ahmed, H.Y. Moustafa, A.M. El-Masry, S.A. Hassan, Natural and synthetic polymers for water treatment against dissolved pharmaceuticals, J. Appl. Polym. Sci. 131 (2014).
[35] B. Wang, Y. Wan, Y. Zheng, X. Lee, T. Liu, Z. Yu, et al., Alginate-based composites for environmental applications: a critical review, Crit. Rev. Environ. Sci. Technol. 49 (2019) 318–356.
[36] S. Thakur, S. Pandey, O.A. Arotiba, Development of a sodium alginate-based organic/inorganic superabsorbent composite hydrogel for adsorption of methylene blue, Carbohydr. Polym. 153 (2016) 34–46.
[37] A.A. Amadu, S. Qiu, S. Ge, G.N.D. Addico, G.K. Ameka, Z. Yu, et al., A review of biopolymer (Poly-β-hydroxybutyrate) synthesis in microbes cultivated on wastewater, Sci. Total Environ. 756 (2021) 143729.
[38] E. Khademian, E. Salehi, H. Sanaeepur, F. Galiano, A. Figoli, A systematic review on carbohydrate biopolymers for adsorptive remediation of copper ions from aqueous environments-part A: classification and modification strategies, Sci. Total Environ. (2020) 738.
[39] S. Ghorai, A. Sinhamahpatra, A. Sarkar, A.B. Panda, S. Pal, Novel biodegradable nanocomposite based on XG-g-PAM/SiO_2: application of an efficient adsorbent for Pb2 + ions from aqueous solution, Bioresour. Technol. 119 (2012) 181–190.
[40] J.L. Wilson, P. Poddar, N.A. Frey, H. Srikanth, K. Mohomed, J.P. Harmon, et al., Synthesis and magnetic properties of polymer nanocomposites with embedded iron nanoparticles, J. Appl. Phys. 95 (2004) 1439–1443.
[41] A. Sellinger, P.M. Weiss, A. Nguyen, Y. Lu, R.A. Assink, W. Gong, et al., Continuous self-assembly of organic-inorganic nanocomposite coatings that mimic nacre, Nature 394 (1998).
[42] S. Sultana, Rafiuddin, M. Zain Khan, K. Umar, Synthesisand characterization of copper ferrite nanoparticles doped polyaniline, J. Alloy. Compd. 535 (2012) 44–49.
[43] S.M. Gupta, M. Tripathi, A review on the synthesis of TiO_2 nanoparticles by solution route, Cent. Eur. J. Chem. 10 (2012) 279–294.
[44] Y.X. Gan, A.H. Jayatissa, Z. Yu, X. Chen, M. Li, Hydrothermal synthesis of nanomaterials, J. Nanomater. 2020 (2020).
[45] G. Demazeau, Solvothermal processes: definition, key factors governing the involved chemical reactions and new trends, J. Chem. Sci. 65 (2010) 999–1006.
[46] M. Shahadat, M.Z. Khan, P.F. Rupani, A. Embrandiri, S. Sultana, S.Z. Ahammad, et al., A critical review on the prospect of polyaniline-grafted biodegradable nanocomposite, Adv. Colloid Interface Sci. 249 (2017) 2–16.

[47] R. Ferrando, Synthesis and experimental characterization of nanoalloy structures, Front. Nanosci. 10 (2016) 47–74. Elsevier.
[48] Z. Khan, S.A. Al-Thabaiti, E.H. El-Mossalamy, A.Y. Obaid, Studies on the kinetics of growth of silver nanoparticles in different surfactant solutions, Coll. Surf. B Biointerfaces 73 (2009) 284–288.
[49] K.J. Yao, M. Song, D.J. Hourston, D.Z. Luo, Polymer/layered clay nanocomposites: 2 polyurethane nanocomposites, Polymers (Guildf.) 43 (2001) 1017–1020.
[50] M. Okamoto, S. Morita, Y.H. Kim, T. Kotaka, H. Tateyama, Synthesis and structure of smectic clay/poly(methyl methacrylate) and clay/polystyrene nanocomposites via in situ intercalative polymerization, Polymers (Guildf.) 41 (2000) 3887–3890.
[51] B. Sharma, P. Malik, P. Jain, Biopolymer reinforced nanocomposites: a comprehensive review, Mater. Today Commun. 16 (2018) 353–363.
[52] S. Kim, I. Do, L.T. Drzal, Thermal stability and dynamic mechanical behavior of exfoliated graphite nanoplatelets-LLDPE nanocomposites, Poly Comp. 31 (2010) 755–761.
[53] R. Das, A.J. Pattanayak, S.K. Swain, Polymer nanocomposites for sensor devices, Polymer-Based Nanocomposites for Energy and Environmental Applications A Volume Woodhead Publishing Series in Composites Science and Engineering, Elsevier, 2018, pp. 206–216.
[54] M. Nasrollahzadeh, M. Sajjadi, S. Iravani, R.S. Varma, Starch, cellulose, pectin, gum, alginate, chitin and chitosan derived (nano)materials for sustainable water treatment: a review, Carbohydr. Polym. 251 (2021) 116986.
[55] B. Thomas, E.P. Shilpa, L.K. Alexander, Role of functional groups and morphology on the pH-dependent adsorption of a cationic dye using banana peel, orange peel, and neem leaf bio-adsorbents, Emergent Mater. 2021 (2021) 1–9.
[56] R. Kant, Textile Dyeing Industry an Environmental Hazard, Vol. 2012, Scientific Research Publishing, 2011.
[57] X. Yu, S. Tong, M. Ge, L. Wu, J. Zuo, C. Cao, et al., Adsorption of heavy metal ions from aqueous solution by carboxylated cellulose nanocrystals, J. Environ. Sci. (China) 25 (2013) 933–943.
[58] X. Chen, J. Cui, X. Xu, B. Sun, L. Zhang, W. Dong, et al., Bacterial cellulose/attapulgite magnetic composites as an efficient adsorbent for heavy metal ions and dye treatment, Carbohydr. Polym. (2020) 229.
[59] A.M. Donia, A.A. Atia, F.I. Abouzayed, Preparation and characterization of nano-magnetic cellulose with fast kinetic properties towards the adsorption of some metal ions, Chem. Eng. J. 191 (2012) 22–30.
[60] X. Sun, L. Yang, Q. Li, J. Zhao, X. Li, X. Wang, et al., Amino-functionalized magnetic cellulose nanocomposite as adsorbent for removal of Cr(VI): synthesis and adsorption studies, Chem. Eng. J. 241 (2014) 175–183.
[61] S. Hokkanen, E. Repo, T. Suopajärvi, H. Liimatainen, J. Niinimaa, M. Sillanpää, Adsorption of Ni(II), Cu(II) and Cd(II) from aqueous solutions by amino modified nanostructured microfibrillated cellulose, Cellulose 21 (2014) 1471–1487.
[62] I.F. Nata, M. Sureshkumar, C.K. Lee, One-pot preparation of amine-rich magnetite/bacterial cellulose nanocomposite and its application for arsenate removal, RSC Adv. 1 (2011) 625–631.
[63] S. Hokkanen, E. Repo, M. Sillanpää, Removal of heavy metals from aqueous solutions by succinic anhydride modified mercerized nanocellulose, Chem. Eng. J. 223 (2013) 40–47.
[64] T.S. Anirudhan, F. Shainy, Effective removal of mercury(II) ions from chlor-alkali industrial wastewater using 2-mercaptobenzamide modified itaconic acid-grafted-magnetite nanocellulose composite, J. Colloid Interface Sci. 456 (2015) 22–31.
[65] J. Antti, S. Tapani, H. Tiina, L. Henrikki, L. Osmo Hormi, Bisphosphonate nanocellulose in the removal of vanadium(V) from water, Cellulose 23 (2016) 689–697.
[66] L. Qi, Z. Xu, Lead sorption from aqueous solutions on chitosan nanoparticles, Colloids Surf. A Physicochem. Eng. Asp. 251 (2004) 183–190.

[67] M.S. Sivakami, T. Gomathi, J. Venkatesan, H.S. Jeong, S.K. Kim, P.N. Sudha, Preparation and characterization of nano chitosan for treatment wastewaters, Int. J. Biol. Macromol. 57 (2013) 204–212.

[68] S.M. Seyedi, B. Anvaripour, M. Motavassel, N. Jadidi, Comparative cadmium adsorption from water by nanochitosan and chitosan, Int. J. Eng. Innov. Technol. 2 (2013) 145–148.

[69] X. Liu, Q. Hu, Z. Fang, X. Zhang, B. Zhang, Magnetic chitosan nanocomposites: a useful recyclable tool for heavy metal ion removal, Langmuir 25 (2009) 3–8.

[70] S. Hu, H. Wang, J. Cao, J. Liu, B. Fang, M. Zheng, et al., Synthesis of mesostructure anatase TiO_2 particles in room-temperature ionic liquids, Mater. Lett. 62 (2008) 2954–2956.

[71] M. Namdeo, S.K. Bajpai, Chitosan-magnetite nanocomposites (CMNs) as magnetic carrier particles for removal of Fe(III) from aqueous solutions, Colloids Surf. A Physicochem. Eng. Asp. 320 (2008) 161–168.

[72] Y.T. Zhou, H.L. Nie, C. Branford-White, Z.Y. He, L.M. Zhu, Removal of Cu^{2+} from aqueous solution by chitosan-coated magnetic nanoparticles modified with α-ketoglutaric acid, J. Colloid Interface Sci. 330 (2009) 29–37.

[73] A. Chen, G. Zeng, G. Chen, X. Hu, M. Yan, S. Guan, et al., Novel thiourea-modified magnetic ion-imprinted chitosan/TiO_2 composite for simultaneous removal of cadmium and 2,4-dichlorophenol, Chem. Eng. J. 191 (2012) 85–94.

[74] L.N. Pincus, F. Melnikov, J.S. Yamani, J.B. Zimmerman, Multifunctional photoactive and selective adsorbent for arsenite and arsenate: evaluation of nano titanium dioxide-enabled chitosan cross-linked with copper, J. Hazard. Mater. 358 (2018) 145–154.

[75] F.G.L. Medeiros Borsagli, A.A.P. Mansur, P. Chagas, L.C.A. Oliveira, H.S. Mansur, O-carboxymethyl functionalization of chitosan: complexation and adsorption of Cd (II) and Cr (VI) as heavy metal pollutant ions, React. Funct. Polym. 97 (2015) 37–47.

[76] S. Rahimi, R.M. Moattari, L. Rajabi, A.A. Derakhshan, Optimization of lead removal from aqueous solution using goethite/chitosan nanocomposite by response surface methodology, Colloids Surf. A Physicochem. Eng. Asp. 484 (2015) 216–225.

[77] Y.J. Zhang, J.Q. Xue, F. Li, J.Z. Dai, X.Z.Y. Zhang, Preparation of polypyrrole/chitosan/carbon nanotube composite nano-electrode and application to capacitive deionization process for removing Cu^{2+}, Chem. Eng. Process. – Process Intensif. 139 (2019) 121–129.

[78] A. Saberi, M. Sadeghi, E. Alipour, Design of AgNPs-base starch/PEG-poly (acrylic acid) hydrogel for removal of mercury (II), J. Polym. Environ. 28 (2020) 906–917.

[79] Z. Sekhavat Pour, M. Ghaemy, Removal of dyes and heavy metal ions from water by magnetic hydrogel beads based on poly(vinyl alcohol)/carboxymethyl starch-g-poly(vinyl imidazole), RSC Adv. 5 (2015) 64106–64118.

[80] L. Alidokht, A.R. Khataee, A. Reyhanitabar, S. Oustan, Reductive removal of Cr(VI) by starch-stabilized Fe0 nanoparticles in aqueous solution, Desalination 270 (2011) 105–110.

[81] Y. Wang, Y. Zhang, C. Hou, Z. Qi, X. He, Y. Li, Facile synthesis of monodisperse functional magnetic dialdehyde starch nano-composite and used for highly effective recovery of Hg(II), Chemosphere 141 (2015) 26–33.

[82] P.N. Singh, D. Tiwary, I. Sinha, Chromium removal from aqueous media by superparamagnetic starch functionalized maghemite nanoparticles, J. Chem. Sci. 127 (2015) 1967–1976.

[83] N. Wu, H. Wei, L. Zhang, Efficient removal of heavy metal ions with biopolymer template synthesized mesoporous titania beads of hundreds of micrometers size, Environ. Sci. Technol. 46 (2012) 419–425.

[84] A. Tabatabaeefar, A.R. Keshtkar, M. Talebi, H. Abolghasemi, Polyvinyl alcohol/alginate/zeolite nanohybrid for removal of metals, Chem. Eng. Technol. 43 (2020) 343–354.

[85] K.V.G. Ravikumar, S.V. Sudakaran, M. Pulimi, C. Natarajan, A. Mukherjee, Removal of hexavalent chromium using nano zero valent iron and bacterial consortium immobilized alginate beads in a continuous flow reactor, Environ. Tech. Innov. 12 (2018) 104–114.

[86] A. Asthana, R. Verma, A.K. Singh, M.A.B.H. Susan, Glycine functionalized magnetic nanoparticle entrapped calcium alginate beads: a promising adsorbent for removal of Cu(II) ions, J. Environ. Chem. Eng. 4 (2016) 1985–1995.
[87] J. Iqbal, N.S. Shah, M. Sayed, M. Imran, N. Muhammad, F.M. Howari, et al., Synergistic effects of activated carbon and nano-zerovalent copper on the performance of hydroxyapatite-alginate beads for the removal of As^{3+} from aqueous solution, J. Clean. Prod. 235 (2019) 875–886.
[88] A. Kumar, R.K. Dutta, CdS quantum dots immobilized on calcium alginate microbeads for rapid and selective detection of Hg^{2+} ions, RSC Adv. 5 (2015) 76275–76284.
[89] K. Pandi, N. Viswanathan, Enhanced defluoridation and facile separation of magnetic nano-hydroxyapatite/alginate composite, Int. J. Biol. Macromol. 80 (2015) 341–349.
[90] Y. Sun, C. Lei, E. Khan, S.S. Chen, D.C.W. Tsang, Y.S. Ok, et al., Aging effects on chemical transformation and metal(loid) removal by entrapped nanoscale zero-valent iron for hydraulic fracturing wastewater treatment, Sci. Total Environ. 615 (2018) 498–507.
[91] W. Jung, B.H. Jeon, D.W. Cho, H.S. Roh, Y. Cho, S.J. Kim, et al., Sorptive removal of heavy metals with nano-sized carbon immobilized alginate beads, J. Ind. Eng. Chem. 26 (2015) 364–369.
[92] A. Sharif, M. Khorasani, F. Shemirani, Nanocomposite bead (NCB) based on bio-polymer alginate caged magnetic graphene oxide synthesized for adsorption and preconcentration of lead(II) and copper (II) ions from urine, saliva and water samples, J. Inorg. Organomet. Polym. Mater. 28 (2018) 2375–2387.
[93] H. Alinezhad, M. Zabihi, D. Kahfroushan, Design and fabrication the novel polymeric magnetic boehmite nanocomposite (boehmite@ Fe_3O_4@PLA@SiO_2) for the remarkable competitive adsorption of methylene blue and mercury ions, J. Phys. Chem. Solids 144 (2020) 109515.
[94] Q. Zia, M. Tabassum, Z. Lu, M.T. Khawar, J. Song, H. Gong, et al., Porous poly(L–lactic acid)/chitosan nanofibres for copper ion adsorption, Carbohydr. Polym. 227 (2020) 115343.
[95] H. Qiu, J. Yan, G. Lan, Y. Liu, X. Song, W. Peng, et al., Removal of Cu^{2+} from wastewater by modified xanthan gum (XG) with ethylenediamine (EDA), RSC Adv. 6 (2016) 83226–83233.
[96] W. Zhang, F. Xu, Y. Wang, M. Luo, D. Wang, Facile control of zeolite NaA dispersion into xanthan gum-alginate binary biopolymer network in improving hybrid composites for adsorptive removal of Co^{2+} and Ni^{2+}, Chem. Eng. J. 255 (2014) 316–326.
[97] X. Peng, F. Xu, W. Zhang, J. Wang, C. Zeng, M. Niu, et al., Magnetic Fe_3O_4 @ silica-xanthan gum composites for aqueous removal and recovery of Pb^{2+}, Colloids Surf. A Physicochem. Eng. Asp. 443 (2014) 27–36.
[98] S. Zhang, F. Xu, Y. Wang, W. Zhang, X. Peng, F. Pepe, Silica modified calcium alginate-xanthan gum hybrid bead composites for the removal and recovery of Pb(II) from aqueous solution, Chem. Eng. J. 234 (2013) 33–42.
[99] R. Ahmad, A. Mirza, Adsorptive removal of heavy metals and anionic dye from aqueous solution using novel Xanthan gum-Glutathione/Zeolite bionanocomposite, Groundw. Sustain. Dev. 7 (2018) 305–312.
[100] M. Zheng, F. Lian, Y. Zhu, Y. Zhang, B. Liu, L. Zhang, et al., pH-responsive poly (xanthan gum-g-acrylamide-g-acrylic acid) hydrogel: preparation, characterization, and application, Carbohydr. Polym. 210 (2019) 38–46.
[101] A. Nagaraj, D. Govindaraj, M. Rajan, Magnesium oxide entrapped polypyrrole hybrid nanocomposite as an efficient selective scavenger for fluoride ion in drinking water, Emergent Mater. 1 (2018) 25–33.
[102] Y. Tian, M. Wu, R. Liu, Y. Li, D. Wang, J. Tan, et al., Electrospun membrane of cellulose acetate for heavy metal ion adsorption in water treatment, Carbohydr. Polym. 83 (2011) 743–748.
[103] X. He, K.B. Male, P.N. Nesterenko, D. Brabazon, B. Paull, J.H.T. Luong, Adsorption and desorption of methylene blue on porous carbon monoliths and nanocrystalline cellulose, ACS Appl. Mater. Interfaces 5 (2013) 8796–8804.

[104] H.Y. Yu, D.Z. Zhang, F.F. Lu, J. Yao, New approach for single-step extraction of carboxylated cellulose nanocrystals for their use as adsorbents and flocculants, ACS Sustain. Chem. Eng. 4 (2016) 2632–2643.

[105] L. Jin, W. Li, Q. Xu, Q. Sun, Amino-functionalized nanocrystalline cellulose as an adsorbent for anionic dyes, Cellulose 22 (2015) 2443–2456.

[106] X. Luo, L. Zhang, High effective adsorption of organic dyes on magnetic cellulose beads entrapping activated carbon, J. Hazard. Mater. 171 (2009) 340–347.

[107] C. Zhou, Q. Wu, T. Lei, Negulescu II. Adsorption kinetic and equilibrium studies for methylene blue dye by partially hydrolyzed polyacrylamide/cellulose nanocrystal nanocomposite hydrogels, Chem. Eng. J. 251 (2014) 17–24.

[108] S. Chen, Y. Huang, Bacterial cellulose nanofibers decorated with phthalocyanine: preparation, characterization and dye removal performance, Mater. Lett. 142 (2015) 235–237.

[109] J. Yang, J. Yu, J. Fan, D. Sun, W. Tang, X. Yang, Biotemplated preparation of CdS nanoparticles/bacterial cellulose hybrid nanofibers for photocatalysis application, J. Hazard. Mater. 189 (2011) 377–383.

[110] B.X. Zhang, H. Yu, Y. Zhang, Z. Luo, W. Han, W. Qiu, et al., Bacterial cellulose derived monolithic titania aerogel consisting of 3D reticulate titania nanofibers, Cellulose 25 (2018) 7189–7196.

[111] P. Zhang, D. Hou, D. O'Connor, X. Li, S. Pehkonen, R.S. Varma, et al., Green and size-specific synthesis of stable Fe-Cu oxides as earth-abundant adsorbents for malachite green removal, ACS Sustain. Chem. Eng. 6 (2018) 9229–9236.

[112] H. Gholami Derami, Q. Jiang, D. Ghim, S. Cao, Y.J. Chandar, J.J. Morrissey, et al., A robust and scalable polydopamine/bacterial nanocellulose hybrid membrane for efficient wastewater treatment, ACS Appl. Nano Mater. 2 (2019) 1092–1101.

[113] H. Yan, L. Yang, Z. Yang, H. Yang, A. Li, R. Cheng, Preparation of chitosan/poly(acrylic acid) magnetic composite microspheres and applications in the removal of copper(II) ions from aqueous solutions, J. Hazard. Mater. 229–230 (2012) 371–380.

[114] Z.G. Hu, J. Zhang, W.L. Chan, Y.S. Szeto, The sorption of acid dye onto chitosan nanoparticles, Polymers (Guildf.) 47 (2006) 5838–5842.

[115] G. Gibbs, J.M. Tobin, E. Guibal, Sorption of acid green 25 on chitosan: influence of experimental parameters on uptake kinetics and sorption isotherms, J. Appl. Polym. Sci. 90 (2003) 1073–1080.

[116] R. Salehi, M. Arami, N.M. Mahmoodi, H. Bahrami, S. Khorramfar, Novel biocompatible composite (Chitosan-zinc oxide nanoparticle): preparation, characterization and dye adsorption properties, Colloids Surf. B Biointerfaces 80 (2010) 86–93.

[117] S. Gopi, P. Balakrishnan, C. Divya, S. Valic, E. Govorcin Bajsic, A. Pius, et al., Facile synthesis of chitin nanocrystals decorated on 3D cellulose aerogels as a new multi-functional material for waste water treatment with enhanced anti-bacterial and anti-oxidant properties, N. J. Chem. 41 (2017) 12746–12755.

[118] F. Marrakchi, W.A. Khanday, M. Asif, B.H. Hameed, Cross-linked chitosan/sepiolite composite for the adsorption of methylene blue and reactive orange 16, Int. J. Biol. Macromol. 93 (2016) 1231–1239.

[119] A.H. Jawad, N.S.A. Mubarak, A.S. Abdulhameed, Hybrid crosslinked chitosan-epichlorohydrin/TiO_2 nanocomposite for reactive red 120 dye adsorption: kinetic, isotherm, thermodynamic, and mechanism study, J. Polym. Envion. 28 (2019) 624–637.

[120] A.A. Kadam, D.S. Lee, Glutaraldehyde cross-linked magnetic chitosan nanocomposites: reduction precipitation synthesis, characterization, and application for removal of hazardous textile dyes, Bio Tech. 193 (2015) 563–567.

[121] R.S. Vieira, M.L.M. Oliveira, E. Guibal, E. Rodríguez-Castellón, M.M. Beppu, Copper, mercury and chromium adsorption on natural and crosslinked chitosan films: an XPS investigation of mechanism, Colloids Surf. A Physicochem. Eng. Asp. 374 (2011) 108–114.

[122] G. Zhou, K.P. Wang, H.W. Liu, L. Wang, X.F. Xiao, D.D. Dou, et al., Three-dimensional polylactic acid@graphene oxide/chitosan sponge bionic filter: highly efficient adsorption of crystal violet dye, Int. J. Biol. Macromol. 113 (2018) 792–803.

[123] L. Jin, Q. Sun, Q. Xu, Y. Xu, Adsorptive removal of anionic dyes from aqueous solutions using microgel based on nanocellulose and polyvinylamine, BioTech 197 (2015) 348–355.

[124] K. Xie, X. Gao, W. Zhao, Thermal degradation of nano-cellulose hybrid materials containing reactive polyhedral oligomeric silsesquioxane, Carbohydr. Polym. 81 (2010) 300–304.

[125] D. Sun, J. Yang, X. Wang, Bacterial cellulose/TiO_2 hybrid nanofibers prepared by the surface hydrolysis method with molecular precision, Nanoscale 2 (2010) 287–292.

[126] A. Snyder, Z. Bo, R. Moon, J.C. Rochet, L. Stanciu, Reusable photocatalytic titanium dioxide-cellulose nanofiber films, J. Colloid Interface Sci. 399 (2013) 92–98.

[127] Y.C. Wong, Y.S. Szeto, W.H. Cheung, G. McKay, Adsorption of acid dyes on chitosan - Equilibrium isotherm analyses, Process. Biochem. 39 (2004) 695–704.

[128] W.L. Du, Z.R. Xu, X.Y. Han, Y.L. Xu, Z.G. Miao, Preparation, characterization and adsorption properties of chitosan nanoparticles for eosin Y as a model anionic dye, J. Hazard. Mater. 153 (2008) 152–156.

[129] H.S. Mohamed, N.K. Soliman, A.F. Moustafa, O.F. Abdel-Gawad, R.R. Taha, S.A. Ahmed, Nano metal oxide impregnated chitosan-4-nitroacetophenone for industrial dye removal, Int. J. Environ. Anal. Chem. (2019).

[130] G. Zhang, Y. Li, A. Gao, Q. Zhang, J. Cui, S. Zhao, et al., Bio-inspired underwater superoleophobic PVDF membranes for highly-efficient simultaneous removal of insoluble emulsified oils and soluble anionic dyes, Chem. Eng. J. 369 (2019) 576–587.

[131] A.M. Saad, M.R. Abukhadra, S. Abdel-Kader Ahmed, A.M. Elzanaty, A.H. Mady, M.A. Betiha, et al., Photocatalytic degradation of malachite green dye using chitosan supported ZnO and Ce–ZnO nanoflowers under visible light, J. Environ. Manage 258 (2020) 110043.

[132] M.S. Jyothi, V.J. Angadi, T.V. Kanakalakshmi, M. Padaki, B.R. Geetha, K. Soontarapa, Magnetic nanoparticles impregnated, cross-linked, porous chitosan microspheres for efficient adsorption of methylene blue from pharmaceutical waste water, J. Poly Environ 27 (2019) 2408–2418.

[133] S. Sheshmani, A. Ashori, S. Hasanzadeh, Removal of Acid Orange 7 from aqueous solution using magnetic graphene/chitosan: a promising nano-adsorbent, Int. J. Biol. Macromol. 68 (2014) 218–224.

[134] S. Zinadini, A.A. Zinatizadeh, M. Rahimi, V. Vatanpour, H. Zangeneh, M. Beygzadeh, Novel high flux antifouling nanofiltration membranes for dye removal containing carboxymethyl chitosan coated Fe3O4 nanoparticles, Desalination 349 (2014) 145–154.

[135] J. Guo, J. Wang, G. Zheng, X. Jiang, Optimization of the removal of reactive golden yellow SNE dye by cross-linked cationic starch and its adsorption properties, J. Eng. Fiber Fabr. (2019) 14.

[136] A. Pourjavadi, A. Abedin-Moghanaki, A. Tavakoli, Efficient removal of cationic dyes using a new magnetic nanocomposite based on starch-g-poly(vinylalcohol) and functionalized with sulfate groups, RSC Adv. 6 (2016) 38042–38051.

[137] J. Guo, J. Wang, G. Zheng, X. Jiang, A TiO_2/crosslinked carboxymethyl starch composite for high-efficiency adsorption and photodegradation of cationic golden yellow X-GL dye, Environ. Sci. Pollut. Res. 26 (2019) 24395–24406.

[138] G. Gong, F. Zhang, Z. Cheng, L. Zhou, Facile fabrication of magnetic carboxymethyl starch/poly(vinyl alcohol) composite gel for methylene blue removal, Int. J. Biol. Macromol. 81 (2015) 205–211.

[139] M. Elkady, H. Shokry, A. El-Sharkawy, G. El-Subruiti, H. Hamad, New insights into the activity of green supported nanoscale zero-valent iron composites for enhanced acid blue-25 dye synergistic decolorization from aqueous medium, J. Mol. Liq. 294 (2019) 111628.

[140] A. Saberi, E. Alipour, M. Sadeghi, Superabsorbent magnetic Fe_3O_4-based starch-poly (acrylic acid) nanocomposite hydrogel for efficient removal of dyes and heavy metal ions from water, J. Polym. Res. 26 (2019) 1–14.

[141] B.K. Bhangi, S.K. Ray, Nano silver chloride and alginate incorporated composite copolymer adsorbent for adsorption of a synthetic dye from water in a fixed bed column and its photocatalytic reduction, Int. J. Biol. Macromol. 144 (2020) 801–812.

[142] P. Zolfaghari, R. Shojaat, A. Karimi, N. Saadatjoo, Application of fluidized bed reactor containing GOx/$MnFe_2O_4$/calcium alginate nano-composite in degradation of a model pollutant, J. Environ. Chem. Eng. 6 (2018) 6414–6420.

[143] K. Zhao, L. Feng, H. Lin, Y. Fu, B. Lin, W. Cui, et al., Adsorption and photocatalytic degradation of methyl orange imprinted composite membranes using TiO_2/calcium alginate hydrogel as matrix, Catal. Today 236 (2014) 127–134.

[144] N. Mohammed, N. Grishkewich, R.M. Berry, K.C. Tam, Cellulose nanocrystal–alginate hydrogel beads as novel adsorbents for organic dyes in aqueous solutions, Cellulose 22 (2015) 3725–3738.

[145] A. Soni, A. Tiwari, A.K. Bajpai, Removal of malachite green from aqueous solution using nano-iron oxide-loaded alginate microspheres: batch and column studies, Res. Chem. Intermed. 40 (2014) 913–930.

[146] R. Shokoohi, Z. Torkshavand, M.M. Mahmoudi, A.M. Behgoo, E. Ghaedrahmati, F.M. Hosseini, Effective removal of azo dye reactive blue 222 from aqueous solutions using modified magnetic nanoparticles with sodium alginate/hydrogen peroxide, Environ. Prog. Sustain. Energy 38 (2019) 205–213.

[147] S. Yadav, A. Asthana, R. Chakraborty, B. Jain, A.K. Singh, S.A.C. Carabineiro, et al., Cationic dye removal using novel magnetic/activated charcoal/β-cyclodextrin/alginate polymer nanocomposite, Nanomaterials 10 (2020) 170.

[148] Ni.B. Wang Miao Ran Guigan Fang Ting Wu Qingwen Tian Linqiang Zheng Laura Romero-Zerón Yonghao, Á. Wang, Á.L. Zheng, Á.Y. Romero-Zerón, B.L. Ni, Á. Ran, et al., Palladium nano-catalyst supported on cationic nanocellulose-alginate hydrogel for effective catalytic reactions, Cellulose 27 (2020) 6995–7008.

[149] B.C. Mascarenhas, F.A. Tavares, E.C. Paris, Functionalized faujasite zeolite immobilized on poly(lactic acid) composite fibers to remove dyes from aqueous media, J. Appl. Polym. Sci. 137 (2020) 1–12.

[150] H. Mittal, V. Parashar, S.B. Mishra, A.K. Mishra, Fe_3O_4 MNPs and gum xanthan based hydrogels nanocomposites for the efficient capture of malachite green from aqueous solution, Chem. Eng. J. 255 (2014) 471–482.

[151] H. Mittal, V. Kumar, Saruchi, S.S. Ray, Adsorption of methyl violet from aqueous solution using gum xanthan/Fe_3O_4 based nanocomposite hydrogel, Int. J. Biol. Macromol. 89 (2016) 1–11.

[152] R. Ahmad, A. Mirza, Green synthesis of Xanthan gum/Methionine-bentonite nanocomposite for sequestering toxic anionic dye, Surf. Interfaces 8 (2017) 65–72.

[153] S. Thakur, S. Pandey, O.A. Arotiba, Sol-gel derived xanthan gum/silica nanocomposite—a highly efficient cationic dyes adsorbent in aqueous system, Int. J. Biol. Macromol. 103 (2017) 596–604.

[154] E. Makhado, S. Pandey, P.N. Nomngongo, J. Ramontja, Preparation and characterization of xanthan gum-cl-poly(acrylic acid)/o-MWCNTs hydrogel nanocomposite as highly effective re-usable adsorbent for removal of methylene blue from aqueous solutions, J. Colloid Interface Sci. 513 (2018) 700–714.

[155] E. Makhado, S. Pandey, J. Ramontja, Microwave assisted synthesis of xanthan gum-cl-poly (acrylic acid) based-reduced graphene oxide hydrogel composite for adsorption of methylene blue and methyl violet from aqueous solution, Int. J. Biol. Macromol. 119 (2018) 255–269.

[156] S. Chaudhary, J. Sharma, B.S. Kaith, S. Yadav, A.K. Sharma, A. Goel, Gum xanthan-psyllium-cl-poly (acrylic acid-co-itaconic acid) based adsorbent for effective removal of cationic and anionic dyes:

adsorption isotherms, kinetics and thermodynamic studies, Ecotoxicol. Environ. Saf. 149 (2018) 150–158.

[157] Inamuddin, Xanthan gum/titanium dioxide nanocompositefor photocatalyt, Int. J. Biol. Macromol. 211 (2018) 1046–1053.

[158] H. Sharma, M. Bhardwaj, M. Kour, S. Paul, Highly efficient magnetic Pd(0) nanoparticles stabilized by amine functionalized starch for organic transformations under mild conditions, Mol. Catal. 435 (2017) 58–68.

[159] M. Zheng, Y. Zhu, B. Liu, Z. Chen, F. Lian, Y. Zhang, et al., Modified xanthan gum for crystal violet uptake: kinetic, isotherm, and thermodynamic behaviors, Water Sci. Technol. 79 (2019) 165–174.

[160] A. Lee, J.W. Elam, S.B. Darling, Membrane materials for water purification: design, development, and application, Environ. Sci. Water Res. Tech. 2 (2016) 17–42.

[161] R.S. Dongre, K.K. Sadasivuni, K. Deshmukh, A. Mehta, S. Basu, J.S. Meshram, et al., Natural polymer based composite membranes for water purification: a review, Polym. Technol. Mater. 58 (2019) 1295–1310.

[162] A. Hussain, A. Janson, J.M. Matar, S. Adham, Membrane distillation: recent technological developments and advancements in membrane materials, Emergent Mater. 2021 (2021) 1–21.

[163] O. Agboola, O. Sunday Isaac Fayomi, A. Ayodeji, A. Omoniyi Ayeni, E.E. Alagbe, S.E. Sanni, et al., A review on polymer nanocomposites and their effective applications in membranes and adsorbents for water treatment and gas separation, Membrane (Basel) 11 (2021) 139.

[164] J.H. Jhaveri, Z.V.P. Murthy, A comprehensive review on anti-fouling nanocomposite membranes for pressure driven membrane separation processes, Desalination 379 (2016) 137–154.

[165] H. Jain, M.C. Garg, Fabrication of polymeric nanocomposite forward osmosis membranes for water desalination—a review, Environ. Technol. Innov. 23 (2021) 101561.

[166] M. Abolhassani, C.S. Griggs, L.A. Gurtowski, J.A. Mattei-Sosa, M. Nevins, V.F. Medina, et al., Scalable chitosan-graphene oxide membranes: the effect of GO size on properties and cross-flow filtration performance, ACS Omega 2 (2017) 8751–8759.

[167] M. Sile-Yuksel, B. Tas, D.Y. Koseoglu-Imer, I. Koyuncu, Effect of silver nanoparticle (AgNP) location in nanocomposite membrane matrix fabricated with different polymer type on antibacterial mechanism, Desalination 347 (2014) 120–130.

[168] C. Sprick, S. Chede, V. Oyanedel-Craver, I.C. Escobar, Bio-inspired immobilization of casein-coated silver nanoparticles on cellulose acetate membranes for biofouling control, J. Environ. Chem. Eng. 6 (2018) 2480–2491.

[169] N. Rakhshan, M. Pakizeh, The effect of functionalized SiO_2 nanoparticles on the morphology and triazines separation properties of cellulose acetate membranes, J. Ind. Eng. Chem. 34 (2016) 51–60.

[170] H.D. Beyene, T.G. Ambaye, Application of sustainable nanocomposites for water purification process, Sustain. Polym. Compos. Nanocompos. (2019) 387–412.

[171] K. Buruga, H. Song, J. Shang, N. Bolan, T.K. Jagannathan, K.H. Kim, A review on functional polymer-clay based nanocomposite membranes for treatment of water, J. Hazard. Mater. (2019) 379.

[172] L. Ma, Q. Chen, J. Zhu, Y. Xi, H. He, R. Zhu, et al., Adsorption of phenol and Cu(II) onto cationic and zwitterionic surfactant modified montmorillonite in single and binary systems, Chem. Eng. J. 283 (2016) 880–888.

[173] B. Sharma, R. Chhibber, R. Mehta, Effect of surface treatment of nanoclay on the mechanical properties of epoxy/glass fiber/clay nanocomposites, Composite Interfaces 23 (2016) 623–640.

[174] Y. Tang, S. Deng, L. Ye, C. Yang, Q. Yuan, J. Zhang, et al., Effects of unfolded and intercalated halloysites on mechanical properties of halloysite–epoxy nanocomposites, Compos. Part A Appl. Sci. Manuf. 42 (2011) 345–354.

[175] S. Xie, S. Zhang, F. Wang, Synthesis and characterization of poly(propylene)/montmorillonite nanocomposites by simultaneous grafting-intercalation, J. Appl. Polym. Sci. 94 (2004) 1018–1023.

[176] S. Pandey, S.B. Mishra, Organic-inorganic hybrid of chitosan/organoclay bionanocomposites for hexavalent chromium uptake, J. Colloid Interface Sci. 361 (2011) 509–520.

[177] X. Zou, J. Pan, H. Ou, X. Wang, W. Guan, C. Li, et al., Adsorptive removal of Cr(III) and Fe(III) from aqueous solution by chitosan/attapulgite composites: equilibrium, thermodynamics and kinetics, Chem. Eng. J. 167 (2011) 112–121.

[178] X. Wang, A. Wang, Adsorption characteristics of chitosan-g-poly(acrylic acid)/attapulgite hydrogel composite for Hg(II) ions from aqueous solution, Sep. Sci. Technol. 45 (2010) 2086–2094.

[179] J. Zhu, M. Tian, Y. Zhang, H. Zhang, J. Liu, Fabrication of a novel "loose" nanofiltration membrane by facile blending with Chitosan–Montmorillonite nanosheets for dyes purification, Chem. Eng. J. 265 (2015) 184–193.

[180] A. Rehab, In vitro antimicrobial activity of intercalated modified chitosan-clay nanocomposite against different microorgamnisms, Chem. Mater. Res. 6 (2014).

[181] S. Ling, K. Jin, D.L. Kaplan, M.J. Buehler, Ultrathin free-standing *Bombyx mori* silk nanofibril membranes, Nano Lett. 16 (2016) 3795–3800.

[182] Z. Zhou, W. Lin, X.F. Wu, Electrospinning ultrathin continuous cellulose acetate fibers for high-flux water filtration, Colloids Surf. A Physicochem. Eng. Asp. 494 (2016) 21–29.

[183] Y. Ertas, T. Uyar, Fabrication of cellulose acetate/polybenzoxazine cross-linked electrospun nanofibrous membrane for water treatment, Carbohydr. Polym. 177 (2017) 378–387.

[184] X. Song, L. Wang, C.Y. Tang, Z. Wang, C. Gao, Fabrication of carbon nanotubes incorporated double-skinned thin film nanocomposite membranes for enhanced separation performance and antifouling capability in forward osmosis process, Desalination 369 (2015) 1–9.

[185] A.A. Olayiwola Sirajudeen, M.S. Mohamad Annuar, K.A. Ishak, H. Yusuf, R. Subramaniam, Innovative application of biopolymer composite as proton exchange membrane in microbial fuel cell utilizing real wastewater for electricity generation, J. Clean. Prod. 278 (2021) 123449.

[186] A.A. Yaqoob, M.N.M. Ibrahim, M. Rafatullah, Y.S. Chua, A. Ahmad, K. Umar, Recent advances in anodes for microbial fuel cells: an overview, Materials (Basel) 13 (2020) 2078.

[187] S.L. Holder, C.-H. Lee, S.R. Popuri, Simultaneous wastewater treatment and bioelectricity production in microbial fuel cells using cross-linked chitosan-graphene oxide mixed-matrix membranes, Environ. Sci. Pollut. Res. 24 (2017) 13782–13796.

[188] L. Yue, Y. Xie, Y. Zheng, W. He, S. Guo, Y. Sun, et al., Sulfonated bacterial cellulose/polyaniline composite membrane for use as gel polymer electrolyte, Compos. Sci. Technol. 145 (2017) 122–131.

[189] V. Yousefi, D. Mohebbi-Kalhori, A. Samimi, Start-up investigation of the self-assembled chitosan/montmorillonite nanocomposite over the ceramic support as a low-cost membrane for microbial fuel cell application, Int. J. Hydrog. Energy 45 (2020) 4804–4820.

CHAPTER

16 Membranes based on biodegradable polymer nanocomposite for gas separation applications

Mostafa Khodakarami and Hongbo Zeng

Department of Chemical and Materials Engineering, University of Alberta, Edmonton, AB, Canada

Chapter Outline

16.1 Introduction

The demands for natural gas and individual industrial gases such as hydrogen, nitrogen, oxygen, noble gases, and so on have been steadily increasing over the past decades and are expected to continue to rise due to their critical role in human life. Natural gas is an important energy source that is primarily used forheating, transportation, and electricity generation for residential, industrial, and commercial sections. Global natural gas consumption and production have both grown by ~35% within the period of 2009–19 [1]. In 2019, global energy consumption increased by 1.3%, in which the natural gas contributed 36% of the growth [2], and based on the estimations made by the U.S. According to the Energy Information Administration (EIA), natural gas demand will rise to 201.8 exajoules in 2040, which is 40% higher than natural gas consumption in 2019 [1,3]. Aside from

Biodegradable and Biocompatible Polymer Nanocomposites. DOI: https://doi.org/10.1016/B978-0-323-91696-7.00008-8
Copyright © 2023 Elsevier Inc. All rights reserved.

natural gas, industrial gases such as oxygen, hydrogen, nitrogen, noble gases, carbon dioxide, and acetylene have critical applications in a variety of industries such as electronics, lighting, medicine, steel and metal manufacturing, automotive industry, and so on. Because of their diverse applications, the production and consumption of these gases has increased globally in various sectors. The global market size of industrial gases was ~ \$92.4 billion in 2019 and is expected to increase at an annual growth rate of 5.5% during the period of 2020–30 [4]. Gas separation technologies are critical from a variety of perspectives, including economics, environmental and safety concerns, process efficiency, product purity, and technological challenges. Aside from the aforementioned factors that are critical for gas production industries, the development of gas separation technologies is critical for addressing global warming and capturing greenhouse gases emitted from various sources. The main greenhouse gases including carbon dioxide (CO_2), nitrous oxide (N_2O), methane (CH_4), and fluorine-containing gases (F-gases) are emitted by various human activities such as power generation based on fossil fuel combustion, transportation, enteric fermentation, wastewater treatment, decomposition in landfills, and manure management. These gases can alter the earth's radiative balance by absorbing infrared radiation and/or chemically converting other atmosphere gases to greenhouse gases. Carbon dioxide contributes the most to total greenhouse gas emissions (74.4%), followed by methane (17.3%), nitrous oxide (6.2%), and F-gases (2.1%) [5]. Energy sector and fossil fuel combustion for electricity generation and transportation accounted for 73.2% of CO_2 emission in 2016 [6]. Despite the Paris Agreement in 2015 to reduce the fossil fuel production by 6% per year during the period of 2020–30 to limit warming to 1.5°C or below 2°C [7], countries continue to use fossil fuel-based energy much more than the level required to comply with 1.5°C limit [8]. Because it is difficult for many sectors to comply with the required regulations and significantly reduce fossil fuel production, the development of technologies for efficient gas separation and capture can play a significant role in decarbonizing different sectors. The current state of gas separation technologies, as well as the associated challenges and limitations, are briefly reviewed in the following section.

16.1.1 Conventional gas separation technologies

Gas separation technologies are employed for a variety of processes, including natural gas treatment, industrial gas production, and dealing with greenhouse gas emissions. Several separation steps are involved in the natural gas production process, regardless of the type of gas reservoir, for gas treatment and recovery of its components. In Table 16.1, the composition of some typical natural gases is presented, and in Fig. 16.1, the schematic diagram of a natural gas treatment plant is shown.

Although the process required for gas treatment is dictated by the end product, the process is generally composed of purification (i.e., removal of materials that hinder the application of gas as fuel or industrial gas) and separation (i.e., splitting the components into individual industrial gases (e.g., helium) or fuels (e.g., propane)). In addition to gas separation processes, gas treatment operations may include dehydration and water removal, solids removal, and hydrocarbon liquid separation. Removal of H_2S and CO_2 are usually the first separation steps required for the natural gases containing $CO_2 \geq 2\%$ and $H_2S \geq 4$ ppmv [10]. The maximum allowed concentration of CO_2 and H_2S gases is determined by the following stages of the manufacturing process, environmental standard limits, and transportation requirements. The most common methods for removing these acid gases are solvent-based absorption, physical absorption, cryogenic fractionation, molecular sieves, and H_2S scavengers. Amines are one of the most commonly used absorption materials in water solutions for

Table 16.1 Composition of some natural gas resources [9].

Natural gas source	CO_2	H_2S	N_2	H_2	C1	C2	C3	C4	C5+
Alberta (Canada)	1.7	3.3	3.2	0	77.1	6.6	3.1	2	3
Beach Ho Fielda (Vietnam)	0.06	0	0.21	0	7.85	13.41	7.5	4.02	2.64
Miskar Field (Tunisia)	13.58	0.09	16.9	0	63.9	3.35	0.96	0.544	0.63
Southwest Kansan (USA)	0	0	14.65	0.45	72.89	6.24	3.74	1.38	0.62
Rio Arriba County (USA)	0	0	25.6	1.8	65.8	3.8	1.7	0.8	0.5
Western Colorado (USA)	42.66	0	26.1	0	29.98	0.55	0.28	0.21	0.25

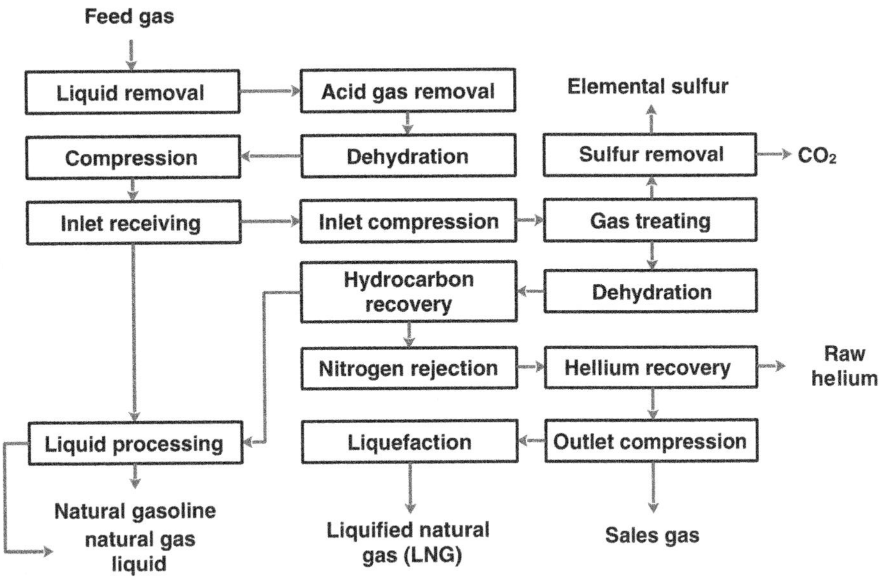

FIGURE 16.1

A simplified schematic of a natural gas processing plant [10].

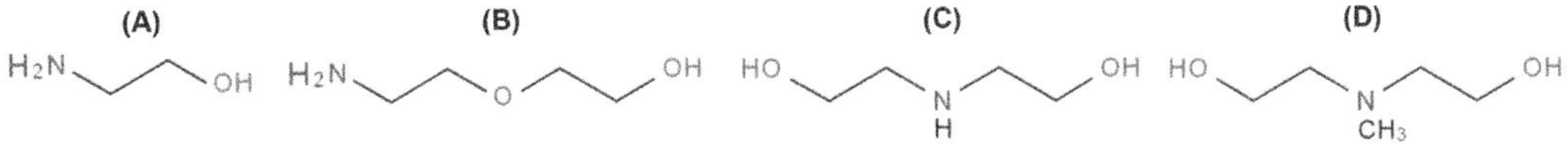

FIGURE 16.2

Chemical structure of (A) monoethanolamine, (B) diglycolamine, (C) diethanolamine, (D) methyl diethanolamine.

removing CO_2 and H_2S gases via acid-base reaction and salt formation. Monoethanolamine [11], diglycolamine [12], diethanolamine [13], and methyl diethanolamine [14] are some of the common amines used for acid gas removals (Fig. 16.2).

The major drawbacks of using amines for acid gas removal are the formation of corrosive materials, solution foaming, the formation of heat-stable salts, and low reclaimability. Using alkali salts such as hot potassium carbonate is another method developed in 1950s for removing CO_2 [15], by the way, H_2S can also be removed using this absorbent via the following reaction:

$$K_2CO_3 + CO_2 + H_2O \Leftrightarrow 2KHCO_3$$

$$K_2CO_3 + H_2S \Leftrightarrow KHS + KHCO_3$$

The energy required for regeneration is relatively low because the heat of reaction for carbon dioxide absorption in potassium carbonate is much lower than that in amines. Selexol, made of the dimethyl ether of polyethylene glycol, is another type of absorbent [16] that is used to remove acid gases solely based on physical absorption, whereas removal of acid gases using amines and alkaline salts engages both physical absorption and chemical reaction. This reduces the energy required for solvent regeneration, which is required in chemical processes. Working at ambient temperature and a partial gas dehydration are the other advantages of the separation by Selexol. To overcome the weaknesses of these absorbents and take advantage of their individual constituents, hybrid systems composed of physical solvents and amines are also used. Sulfinol is a hybrid absorbent composed of sulfolane, as a physical solvent, and methyl diethanolamine that is commercially used for removal of acid gases like H_2S, COS, and CO_2 [17]. In addition to absorbents, molecular sieves, such as synthetic zeolites, can reduce the amount of acid gases through physical adsorption; however, their application is limited because water can also co-adsorb on the adsorption bed and displace the gas molecules. Furthermore, the regeneration of the adsorption bed necessitates a high temperature, which may result in the formation of COS.

Distillation is another popular method for separating liquids; however, it can also be used to separate gases such as H_2S and CO_2 from natural gas as these gases have different vapor pressures. Separation of CO_2 from gases such as H_2S, methane, and ethane, on the other hand, is fraught with difficulties; for example, CO_2 and ethane typically form an azeotrope, making complete separation impossible [18]. The Ryan/Holmes process is an extractive distillation method that is widely used to solve some of the problems associated with conventional distillation methods, such as the high freezing point of CO_2 [19]. A series of distillation columns operate at low temperatures with a solvent added to the system to modify the volatility of components with close boiling points, break the CO_2/C2 azeotrope, and avoid CO_2 freeze up. Biological processes have also been reported to be used in the separation and removal of unwanted gases in the natural gas production process [20]. In this process, H_2S-containing natural gas is absorbed in an alkaline solution, and the elemental sulfur is removed via oxidation of the gas's sulfur content by microorganisms. It has been used commercially in a gas production plant in Canada, and according to reports, it has a 95% sulfur removal efficiency and a sulfur production capacity of 1 ton/day [21]. In addition to H_2S removal, which is critical for downstream processes and meeting pipeline specifications, hydrocarbon separation and recovery is required to reduce heating value and produce valuable byproducts. Refrigeration is the most common method for cooling the gas stream and recovering hydrocarbons. Propane vapor is a key working fluid and refrigerant in the compression refrigeration process [10]. Another simple method for cooling gas is Joule-Thomson expansion, which uses a pressure drop and a control valve to expand the gas stream [22]. Turboexpanders, which are centrifugal compressors that allow condensation to occur inside the expander, have replaced Joule-Thomson expanders. Meshed screens are typically used prior to turboexpanders because the gas entering these expanders must be free of solids and liquids. Demethanizers are distillation columns that function as cold separators to recover methane and natural gas liquid [23]. The temperature gradient is large with operating temperature about −110°C at the top of the column. Twister is another device in which droplet nucleation is caused by a drop in temperature and pressure as the gas enters through a nozzle and expands at sonic velocity [24]. The gas/liquid separation then occurs as a result of the centrifugal force created by the wing, which creates a swirl. Twisters have no moving parts, making them simple devices; however, an inlet filter is usually required because this device requires a gas free of fine solid particles. Vortex tubes are another type of equipment that uses pressure drop to cool the gas [25]. It generates two cold and warm streams, which separate as the gas expands after passing through several nozzles and spirals down the tube at near-sonic velocity. Vortex tubes are comparable with Joule-Thomson expanders, but both gas expansion and separation in vortex tubes take place in a single step.

Nitrogen rejection is an additional step in gas treatment processes, particularly for low-quality gas feedstocks. The nitrogen rejection process employs cryogenic methods, as well as adsorbents and membrane technology. Cryogenic distillation is a popular and efficient method for nitrogen rejection; however, membranes and adsorption methods are more cost effective in smaller scale operations. Depending on the operating pressure of the cryogenic distillation column, the temperature is around −150°C to ~ −190°C [26]. "Pressure swing adsorption" is the adsorption method used to remove nitrogen, which is different from the adsorption process employed in gas dehydration which is based on thermal swing adsorption [27,28]. Pressure swinge adsorption is useful for bulk separation of one component from another when their partial pressures differ. Another contaminant that may not occur naturally but may enter the stream via valves and piping is oxygen. The oxygen concentrations up to 1% may also be okay for sales [29], but presence of 50 ppmv oxygen can result in sever problems such as corrosion of pipeline if water is present, and possible reaction with hydrocarbons, glycol, and amine which can lead to a reduction in production capacity [10]. Scavengers are used to remove small concentrations of oxygen [30], and in higher concentrations, oxygen is catalytically converted to water and then removed by dehydration methods, if the gas stream is free of sulfur. In some cases, oxygen is removed alongside nitrogen rejection; however, nitrogen rejection units are not present in all gas processing plants.

Although helium recovery is not common in gas treatment facilities, it is usually performed when the helium content exceeds 0.5% [31]. Although helium recovery from natural gas is difficult, its recovery is appealing since it is a valuable and expensive product. Recovery of helium from air may not be economically feasible since the helium concentration is very low (~5 ppm) but its recovery from natural gas is viable in a couple of gas fields. Cryogenic methods, natural gas cooling and condensation, and then distillation columns are the first steps in producing a raw stream of helium. The traditional method of refining the crude helium stream is adsorption on activated charcoal; however, pressure swing adsorption and membranes have recently been developed to refine and purify the crude helium stream [32,33]. In Fig. 16.3, a schematic flowsheet of helium recovery process of Ladder Creek Plant is presented.

Radon is another noble gas that may be present in natural gas and cause health problems as a radioactive material. Radon can degrade into lead, bismuth, and polonium, which can accumulate on inlet filters or condense on pipeline walls before being discarded as sludge in appropriate disposal wells [35]. Except for helium, which can be recovered from some natural gas sources, all other noble gases are separated and produced from air. Neon, krypton, and xenon concentrations in air are 18, 1.14, and 0.086 ppmv, respectively. To extract a valuable quantity of these gases, a large amount of air must be processed. To separate and purify the pre-enriched crude krypton/xenon mixture, a crude mixture of concentrated

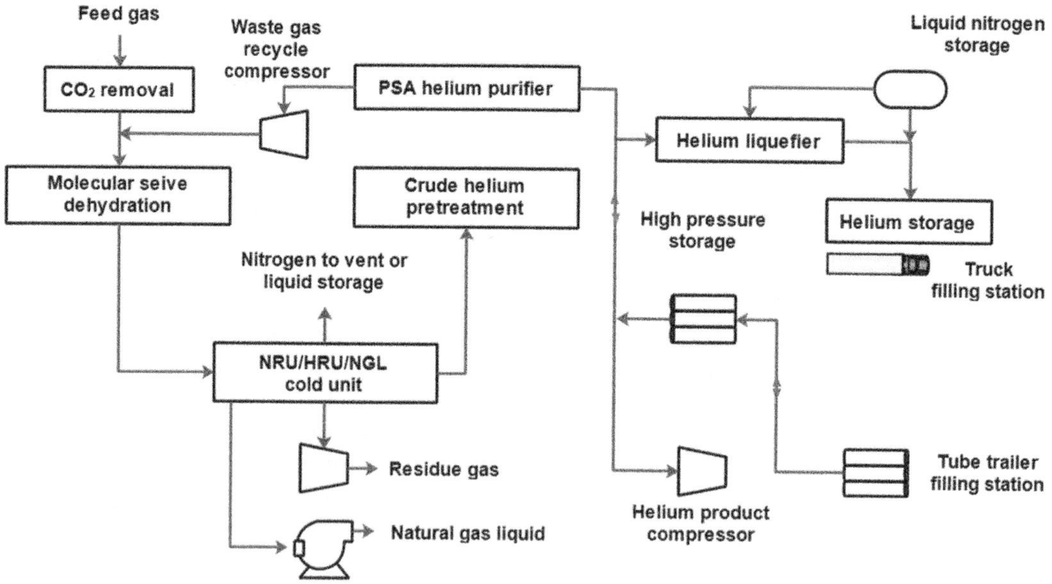

FIGURE 16.3

Schematic flowsheet of helium recovery process (Ladder Creek Plant) [34].

krypton (~4000 ppmv) and xenon (~400 ppmv) is produced in a liquid oxygen column integrated into the air separation unit, and then a variety of separation technologies including cryogenic methods, membranes, adsorption, and chemisorption technologies are used in combination with each other [36]. The main impurities to be removed in krypton/xenon processing are hydrocarbons, nitrous oxide, and fluorine-based greenhouse gases. Because hydrocarbons can form flammable and explosive materials, they are usually removed first through catalytic combustion with Pd- or Pt-based catalysts. Chemisorption is used to remove fluorocarbons, and cryogenic methods are used to separate krypton and xenon. Another important noble gas is neon, which is obtained primarily from air via pre-enrichment via rectification and condensation before being purified and recovered via adsorption, catalytic oxidation, partial condensation, and cryogenic methods. Because the boiling point of argon (−185.9°C) is between the boiling points of oxygen (−183.0°C) and nitrogen (−195.8°C), recovering argon requires a complete separation of nitrogen and oxygen, which makes argon recovery more energy-consuming. Extra refrigeration using nitrogen as a refrigerant is usually required to increase argon recovery. Pressure columns and molecular sieves are critical components of argon preconcentration processes. In the argon refining process, an argon stream is injected alongside hydrogen and then oxygen is removed in a combustion tower [36].

16.1.2 Limitations and challenges of current technologies

The cryogenic process, which separates gases based on their boiling points, is one of the most widely and commercially used methods in gas treatment. In cryogenic operations, the gas must be cooled to extremely low temperatures. Although cryogenic methods have some advantages, such as not requiring chemicals or additives, the major disadvantages and challenges of cryogenic processes are high operating and capital costs, significant energy requirements for the refrigeration process, and the requirement of removing the frozen components during refrigeration [37]. Another unit required in cryogenic separations is compression. In addition to the risks associated with high-pressure operation, working with compressors raises concerns about noise and exhaust emissions.

Adsorption, another major gas separation method, has some drawbacks that make its application difficult [38]. As one of the most commonly used adsorbents in gas processing, amines react with contaminants in the gas stream to form organic acids and heat-stable salts, which can accumulate and cause problems in the process. Corrosion and foaming solution generation are two other major issues with amines that can lead to low process efficiency. Furthermore, amines have an affinity for organic molecules such as benzene, toluene, and xylene, which can be released into the environment during the amine regeneration process. In addition to amines, most solid scavengers are irritating to the eyes and lungs and must be handled with extreme caution when using or disposing of them. Using pressure swing adsorption necessitates the use of switch valves, which must operate at close cycles, require more maintenance, and pose operational issues. Gas velocity through the adsorption bed is also limited to avoid particle abrasion; on the other hand, slow switching cycle operation to reduce particle erosion necessitates larger pressure vessels, which can result in higher material costs. In summary, the major challenges and reasons for researchers to develop new technologies with higher efficiency and lower environmental issues are operational and technical drawbacks, separation process efficiency, and environmental issues.

16.1.3 Recent developments in gas separation technologies

16.1.3.1 Novel materials for gas separation

Intensive efforts have been made to develop technologies and new materials to improve gas separation performance and address challenges in the gas production industry. Many efforts have been directed toward the design and synthesis of novel materials as adsorbents in order to address issues such as cost, secondary pollution and environmental problems, efficiency, and recyclability. The main parameters considered in the design and development of novel adsorbents are stability, regenerability, selectivity, separation efficiency, adsorption capacity, mass and heat transfer properties, cost, and environmental factors. Metal–organic frameworks (MOFs) are a new class of hybrid porous materials with uniform structures made up of inorganic nodes linked by organic struts that have piqued the interest of researchers interested in using them for separation and purification applications. MOFs have demonstrated great potential in selectively separation of CO_2 from other gases as its kinetic diameter (3.3 Å) is smaller than that of other gases such as H, He, and Ne [39]. Size exclusion or differences in the interactions between the frameworks and gas molecules can explain the selective adsorption toward CO_2. Different gases can be separated by designing MOFs with specific pore or cage sizes; however, synthesis of MOFs with cage sizes that are exactly the same as the predetermined size is difficult. In some studies, although the actual pore sizes were larger than all gas molecules, separation could still occur due to the kinetic molecular sieving effect [40]. Because the quadrupole moment of other gases differs from that of CO_2, the quadrupole moment of CO_2 can also be used to aid in the adsorption process. Another technique for selectively separating different gases is the functionalization of pores in MOFs. Fig. 16.4 presents various functionalized MOFs with varying pore structures that are used for selective recognition and binding of gas molecules. Despite the fact that MOFs have several advantages for selectively separating different gases, they do not appear to be efficient for CO_2 adsorption due to their low affinity toward CO_2 molecules.

Aside from MOFs, several porous polymers have been developed for gas separation applications [41]. Unlike MOFs, which have metal nodes interconnected by organic linkers, porous polymers are metal-free organic frameworks connected by covalent bonding. These materials' structural properties and functionalities can be more easily tuned by changing the monomers and their composition, for example. In Fig. 16.5, the structure of several polymers of intrinsic microporosity (PIM) used for gas separation is shown.

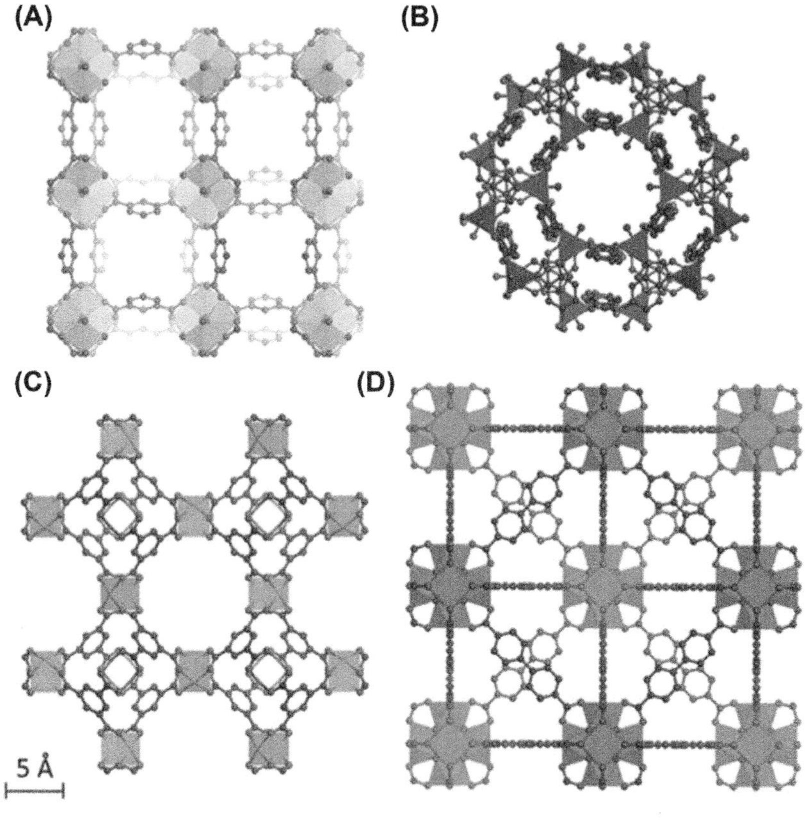

FIGURE 16.4

(A) CAU-10, an aluminum MOF-based on isophthalate ions; (B) Zeolite imidazolate framework-300 (ZIF-300); (C) HKUST-1 ($Cu_3(btc)_2$; btc = 1,3,5-benzenetricarboxylate); (D) UiO-67, a Zr-based MOF with 4,4′-biphenyl dicarboxylate linkers [40].

Copyright 2021. Reproduced with permission from American Chemical Society.

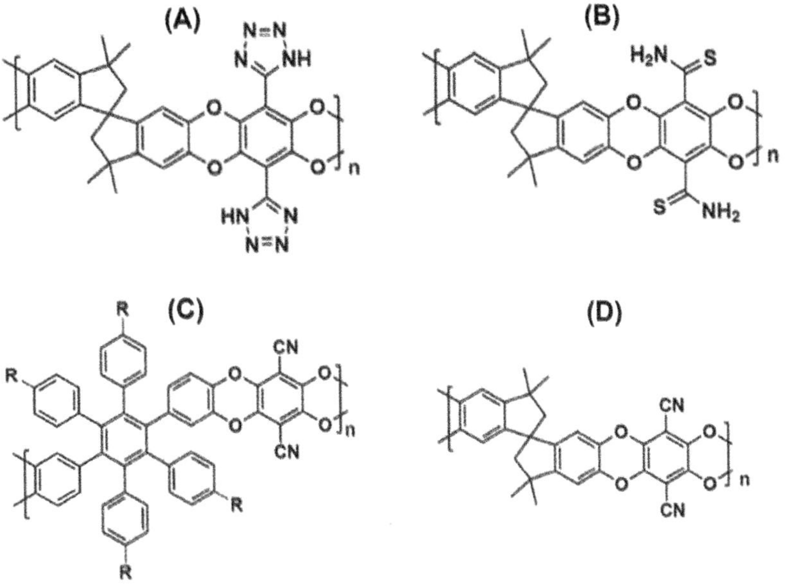

FIGURE 16.5

Polymers of intrinsic microporosity used for gas separation; (A) Tetrazole-PIM, (B) Thio-PIM-1, (C) PIM-R-Hexaphenylbenzene (R can be replaced by H, Br, CN, or metal ions), (D) PIM-1.

However, although the porous polymers are promising materials for gas separation, application of these materials for CO_2 separation is still in early stages and need more work to improve the CO_2 adsorption/separation performance. Carbonaceous porous materials have also been investigated as potential gas separation materials [42]. Activated carbons do not exhibit a suitable CO_2 adsorption performance and CO_2/N_2 or CO_2/CH_4 selectivity at low pressures, however, various strategies have been used to improve their separation performance by grafting functional groups on the surface, doping of heteroatoms, and modification of the pore structures. Both N_2 and CH_4 molecules are nonpolar as they are linear symmetric molecules, therefore functionalization of carbonaceous materials with polar groups can increase the selectivity toward CO_2 molecules due to quadrupole moment of these molecules and polarity of C=O bonds. Doping of heteroatoms such as nitrogen is one of the methods used to improve the selectivity toward CO_2 molecules and increase the adsorption capacity [43]. In addition to dopant and surface group functionalization, engineering of porous structure and modification of pore size, geometry, volume, and surface area can improve gas separation efficiency. Ionic liquids (ILs) are another new material that has shown promise in gas separation [44]. ILs are defined as ionic compounds that have melting point below 100°C and possess a myriad of features such as negligible vapor pressure and tunability [45]. Because CO_2 is soluble in several ILs, it has shown promising results for capturing CO_2 from waste gases. By changing the polarity of ILs, they have also demonstrated good performance in the separation and removal of acidic gases such as H_2S, CO_2, SO_2, and NO_2 from industrial waste gases. The structure of some ILs used for gas separation purposes is presented in Fig. 16.6.

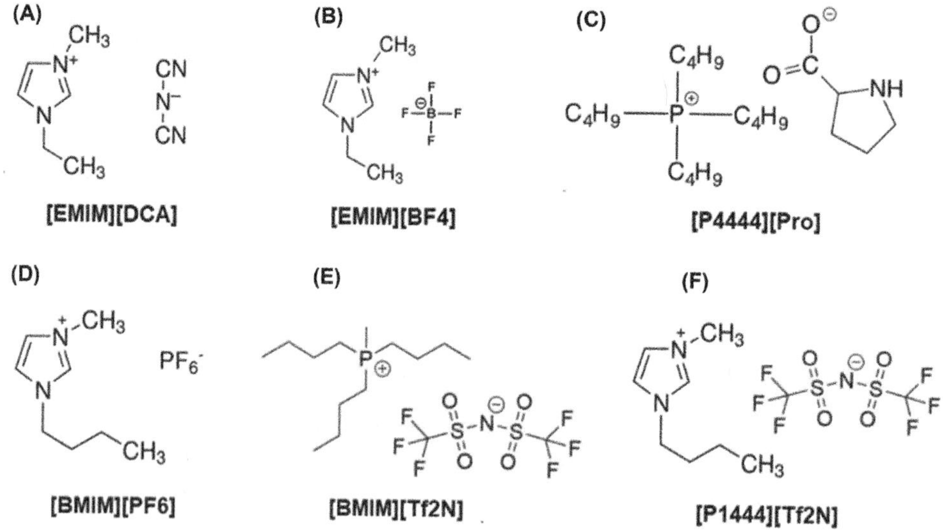

FIGURE 16.6

Chemical structure of (A) 1-Ethyl-3-methylimidazolium dicyanamide, (B) 1-Ethyl-3-methylimidazolium tetrafluoroborate, (C) tetrabutylphosphonium *L*-prolinate, (D) 1-Butyl-3-methylimidazolium hexafluorophosphate, (E) tributylmethylphosphonium bis(triflouromethylsulfonyl)imide, (F) 1-Butyl-3-methylimidazolium bis(trifluoromethylsulfonyl)imide.

16.1.3.2 Development of gas separation membranes

In recent years, the use of membrane-based systems for gas separation processes has increased due to their appealing advantages such as operational simplicity, no phase change requirement, ease of scaling-up, and higher energy efficiency when compared to traditional techniques such as adsorption and distillation. A membrane is a semipermeable barrier that allows a specific gas component in a gas mixture to dissolve into or pass through it while preventing other gas species from passing through. Gas species are transported under the influence of a driving force, such as a difference in pressure, concentration, or temperature between the two sides of the membrane, or a difference in the permeability or diffusion coefficients of the gases. Since the 1980s, when the first major membranes were commercially used for hydrogen separation, membrane technology has advanced dramatically for a wide range of gas separation applications, including O_2/N_2 separation, methane separation from different gases such as CO_2, H_2S, and He in natural gas treatment processes, and hydrogen separation from CO, N_2, and CH_4 [46]. To reduce costs or solve problems associated with traditional methods, hybrid systems combining membrane processes with other techniques have been widely reported for gas separation applications. Membranes combined with the Ryan-Holmes process have been designed and applied to the problem of separating CO_2 from light hydrocarbons in the conventional amine process [47]. Membranes are also used in a wide variety of small-scale applications. In commercial airplanes, polymeric membranes are used to enrich nitrogen gas as an inert gas for fuel tanks [48]. Despite advancements in membrane technology, the use of membranes for gas separation has been slow, owing to the inverse relationship between selectivity and permeability. In order to address the challenges associated with the permeability-selectivity tradeoff and make membrane technology more competitive, intensive research on membrane materials, structure, and configuration is still ongoing. As depicted in Fig. 16.7, hollow fibers and spiral wound modules are two major and common configurations of membranes that have been commercially used in different arrangements for gas separation applications. These geometries have a higher surface-to-volume ratio, which reduces the size of the pressure vessels and the entire membrane plant. The search for new highly selective and permeable materials has resulted in a massive effort to improve the purity and productivity of gas separation membranes. For gas separation membranes, both organic and inorganic materials, porous and non-porous, have been developed. Non-porous membranes, also known as dense membranes, are typically made of inorganic polycrystalline materials such as ceramics and are used in gas separation applications where diffusivity and solubility differences between gas molecules are driving forces. The pore size, shape, and distribution of inorganic materials can be altered to improve membrane permeability and selectivity. For the fabrication of microporous gas separation membranes, various inorganic materials such as zeolite, carbon, silica, and metals have been investigated. Zeolites are crystalline porous materials that have recently gained popularity for gas separation applications due to their well-defined and uniform porous structure, as well as their high mechanical and chemical stability and selectivity [50]. Adsorption and diffusion processes are used to separate gases through a zeolitic membrane. The affinity of a gas molecule toward zeolite pores depends on the quadrupole and dipole moments of gas molecules and their polarizability (Table 16.2). Till now, zeolite-based membranes have been explored for various gas separation applications including air separation, H_2 recovery, CO_2 capture, separation of noble gases, H_2S capture, etc. [51].

Membranes based on microporous silica are also extensively studied to improve the selectivity of the gas separation due to their small pore sizes (<1 nm) [52], however, the low diffusivity and

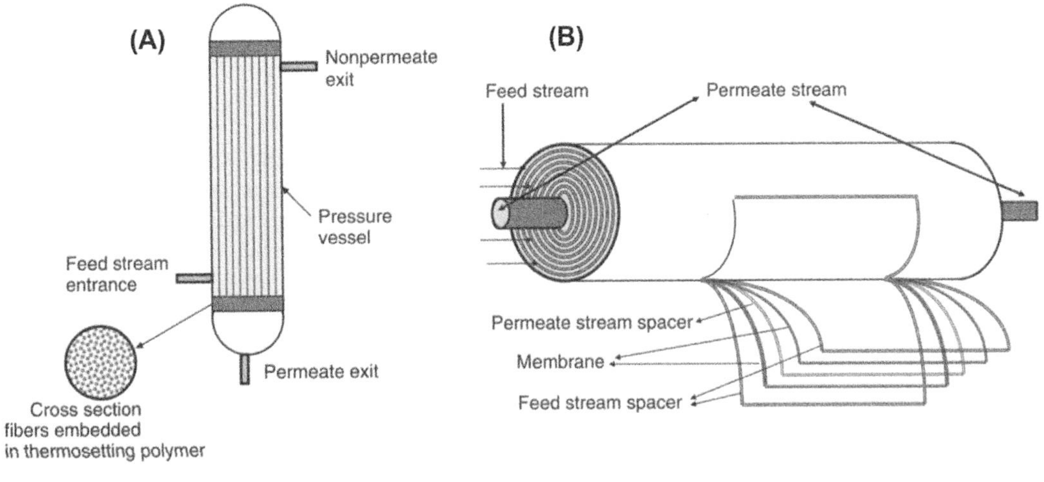

FIGURE 16.7

Common configurations of membranes: (A) hollow fibers and (B) spiral wound modules [49].

Copyright 2021. Reproduced with permission from Elsevier.

Table 16.2 Polarizability, kinetic diameter, quadrupole and dipole moments of some gas molecules [51].

Gas molecule	Polarizability (Å^3)	Dipole moment (D)	Quadrupole moment (D Å)	Kinetic diameter (Å)
O_2	1.600	0.000	0.39	3.47
N_2	1.760	0.000	1.52	3.64
H_2	0.80	0.000	0.66	2.89
CO_2	2.650	0.000	4.30	3.30
CO	1.95	0.112	2.50	3.69
H_2O	1.450	1.87	2.30	2.65
CH_4	2.600	0.000	0.02	3.76

lower hydrothermal stability are the main drawbacks of these materials that need to be addressed. Chemical vapor deposition and surface grafting are two methods used to address these flaws [53]. Carbon-based membranes are other types of membranes used for separation of different gases such as CO_2, H_2, etc. [54,55]. These materials can be created by pyrolyzing various polymers such as cellulose. To increase the rate of gas permeation, ultrathin graphene oxide membranes with nano-sized thickness have recently been developed [56]. However, the brittleness of these membranes and the difficulty in controlling the fabrication process parameters are major drawbacks that limit their applicability.

Metal–organic frameworks are another emerging class of materials that have recently received a lot of attention for gas separation applications due to their tunability and designability. MOF synthesis uses less energy than zeolites because it can be done at lower temperatures. Despite the promising potential of MOF-based membranes for gas separation, their application is still limited and difficult due to MOFs' low stability and poor interactions with common substrates such as silica and alumina [57]. To improve the binding between the substrate and the MOFs, surface modification and support conditioning were used. Polymers are the most commonly used materials for the preparation of gas separation membranes among all the materials investigated for gas separation membranes [58,59]. Almost all gas separation membranes used on an industrial scale are polymeric membranes. Polymeric membranes have been used commercially for biogas purification, hydrogen recovery in ammonia production, CO_2 separation in oil recovery, nitrogen enrichment, and CO_2/CH_4 separation in natural gas treatment. The nature of polymers has a significant impact on the gas permeation properties of membranes. The permeability and selectivity of gas separation membranes can be affected by physical, chemical, and structural properties such as the chemical structure of the repeat units, chain flexibility, degree of crystallinity, glass transition temperature, and so on. Rubbery polymers typically have higher permeability than glassy polymers due to their lower solubility coefficient. The chemical structure of the repeating units has a greater impact on gas permeation through glassy polymers [58]. Permeability and transport properties of polymers with aromatic rings can be improved by changing the connecting groups or by symmetrical or asymmetrical substitution of phenylene rings [60]. The two major classes of polymers investigated for gas separation membrane fabrication are intrinsic microporous polymers and conjugated microporous polymers [61]. Despite the fact that hundreds of polymers have been synthesized and investigated for membrane fabrication, only a few have been commercially used for gas separation applications. The main factors preventing commercial applications of these polymeric membranes are the permeability-selectivity tradeoff, as well as membrane stability and cost. Polymer nanocomposites are a new class of advanced materials being researched and offered for the fabrication of gas separation membranes.

16.2 Polymer nanocomposites for gas separation membranes

Polymer nanocomposite membranes were proposed to improve the efficiency and overcome the shortcomings of both polymeric and inorganic membranes in terms of gas permeability, selectivity, mechanical and thermal stability, and so on [62]. Polymer nanocomposites are created through the dispersion of nanoscale inorganic materials (nanofillers) in a polymeric phase (continuous matrix). The effects of nanofiller and polymeric matrix on the performance of polymer nanocomposite membranes are significant. Robeson introduced a theoretical upper bound for the performance of membranes in separation of various gas pairs to address the selectivity/permeability tradeoff (Fig. 16.8) [63]. Many researchers who design gas separation membranes hope to exceed or come close to the Robeson upper limit.

Many different types of inorganic nanoparticles have been incorporated into various polymers to create polymer nanocomposite membranes for efficient gas separation. Metals [64], clays [65], and carbonaceous inorganic materials [66] have been studied as nanofiller in various synthetic [67] and natural polymeric [68] matrices. The polymeric phase chosen is an important factor in determining the gas separation and transportation behavior of the membranes. When selecting or designing a polymeric phase, important parameters to consider include physicochemical properties, gas solubility, and diffusivity. Among all properties, biodegradability has received a lot of attention in order to address environmental concerns. When discussing biodegradable polymers, natural biopolymers have received a lot of attention. However, despite the link between biopolymers and biodegradability, biodegradation is not always associated with the origin of the polymer. The biodegradability of polymers is determined by their

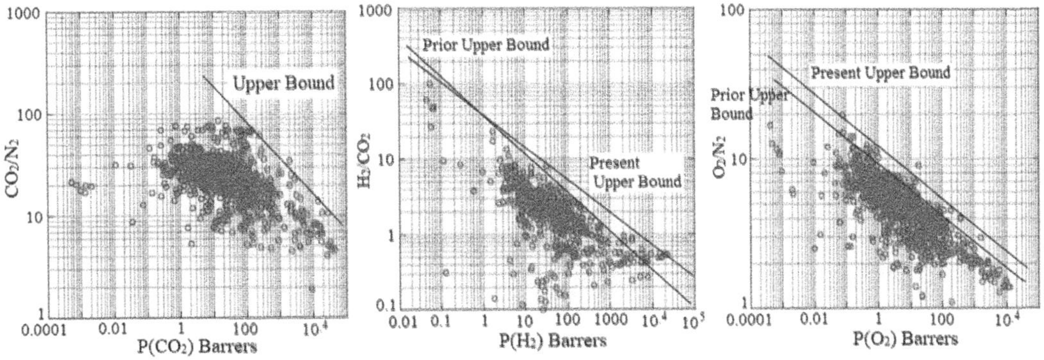

FIGURE 16.8

Robeson upper bound for different gas pairs [63].

Copyright 2021. Reproduced with permission from Elsevier.

chemical structure and composition, and it reflects their end-of-life [69]. Although natural polymers such as cellulose have been widely used in the preparation of gas separation nanocomposite membranes, the synthesis of biodegradable polymers has the potential to expand their use in the fabrication of biodegradable polymer nanocomposite membranes for gas separation. In addition to natural biopolymers (e.g., cellulose, chitosan, lignin, amylose), synthetic polymers such as aliphatic polyesters (e.g., polyglycolide, polylactide, polycaprolactone, poly(butylene succinate), poly(p-dioxanone), and polycarbonates), aromatic co-polyesters (e.g., poly(butylene adipate-co-terephtalate)), polyamides and poly (ester-amide)s, polyurethanes, polyanhydrides (e.g., poly(sebacic anhydride)) are synthetic polymers with some sort of biodegradability properties that can be employed for fabrication of polymer nanocomposites for gas separation applications. In Fig. 16.9, the chemical structure of several synthetic biodegradable polymers has been demonstrated.

16.2.1 Gas separation mechanism and transport phenomena through polymer nanocomposite membranes

Membrane separation of gas molecules can occur due to diffusivity differences, molecule size and/or charge differences, carrier-facilitated transportation, and molecular sieving. The difference in concentration of gas molecules on the two sides of the membrane drives gas flow through the membrane. The pressure difference between the sides of the membrane causes the concentration variation. The membrane material as well as the membrane structure influence separation efficiency, selectivity, permeability, and transport mechanism. The mechanism of gas separation differs between porous and non-porous membranes. Porous membranes have voided structures with a network of interconnected pores that are distributed throughout the membrane. In addition to the size of gas molecules, the size and distribution of pores are generally regarded as the most important factors in porous membrane separation. Gas flow through porous membranes is typically high, resulting in lower selectivity. The mechanism of gas transportation through porous membranes depends on the mean free path of gas molecules (i.e., the mean distance between collisions),

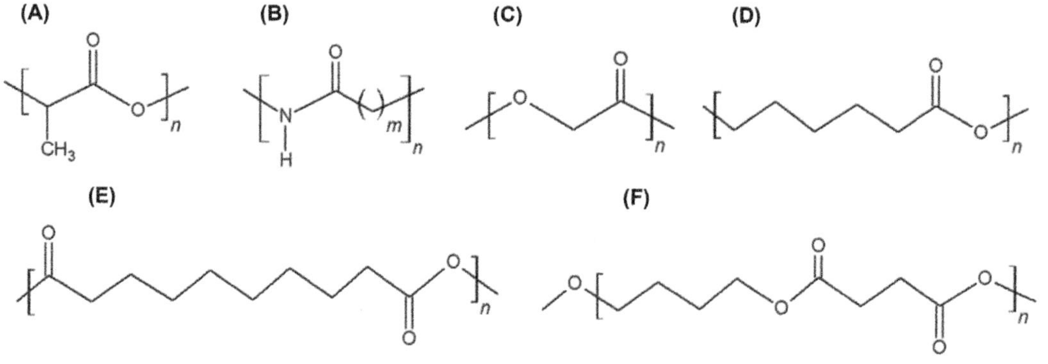

FIGURE 16.9

Chemical structure of synthetic biodegradable polymers: (A) poly(lactic acid), (B) polyamides, (C) poly(glycolide acid), (D) polycaprolactone, (E) poly(sebacic anhydride), (F) poly(butylene succinate).

membrane porosity, pore sizes, and the interactions between gas molecules and the membrane surface. As shown in Fig. 16.10, the gas permeation can occur through four main mechanisms namely molecular sieving, capillary condensation, Knudsen flow, and surface diffusion. In some cases, however, the gas molecules pass through the membrane via more than one mechanism [70]. Different types of gas molecules diffuse independently of each other in the Knudsen mechanism because the interaction between gas molecules and the membrane surface is greater than the interaction between gas molecules themselves. This mechanism is more pronounced when the pore radius of the membrane is smaller than the length of the mean free path of gas molecules [71]. The Knudsen number is the ratio of the mean free path length of gas molecules (λ) to the radius of pore (r), and can help to explain the flow of gas molecules through the membrane:

$$K_n = \frac{\lambda}{r} \tag{16.1}$$

Viscosity (η) and molecular weight of gas molecules (M_W) as well as the temperature (T) and pressure (P) of the system are the factors determining the mean free path of gas molecules:

$$\lambda = \left(\frac{\eta}{P}\right) \times \sqrt{\frac{\pi RT}{2M_W}} = \left(\frac{\eta}{P}\right) \times \sqrt{\frac{\pi K_B T}{2m}} \tag{16.2}$$

where R is the gas constant, m is the molecular mass, and K_B is the Boltzmann constant.

Knudsen diffusivity (D_K) of a gas molecule can be obtained by

$$D_K = \left(\frac{2r}{3}\right) \times \sqrt{\frac{8RT}{\pi M_W}} \tag{16.3}$$

As it can be seen from Eq. (16.3), because a gas molecule's diffusivity is inversely proportional to its molecular weight, the separation factor for two different types of gas molecules can be estimated using the ratio of one type of gas molecule to another. If the pore radius is greater than the mean free path of gas molecules, the gas molecules may collide and cause convective flow across the membrane pores, obstructing gas separation.

Separation occurs in the molecular sieving mechanism based on the size of gas molecules. Because the pores are so small and close to the size of gas molecules, transportation of gas molecules is more

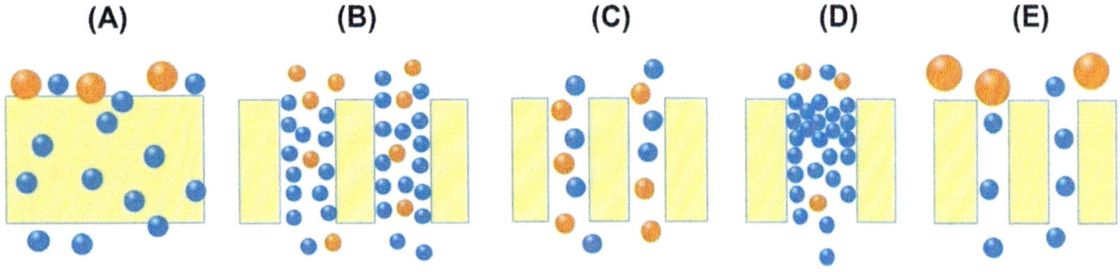

FIGURE 16.10

Different gas permeation mechanisms: (A) solution-diffusion, (B) surface diffusion, (C) Knudsen diffusion, (D) capillary condensation, (E) molecular sieving.

selective, resulting in higher permselectivity of this molecular sieving mechanism [72]. Pore sizes are also small in the capillary condensation mechanism, which occurs when a gas mixture contains both condensable and non-condensable molecules [73]. Adsorption of condensable gas in this mechanism prevents non-condensable molecules from permeating.

In surface diffusion mechanism, gas molecules move along the surface of membrane pores after being adsorbed on the pore walls. Adsorption of gas molecules on pore walls reduces effective pore diameter, making non-adsorbing molecules difficult to pass through. This improves the selectivity and permeability of the target gas molecules. The surface diffusion mechanism and Knudsen diffusion can occur concurrently; however, surface diffusion only occurs in certain temperature and pore size ranges.

The material of the membrane has a large impact on gas separation and permeation through non-porous membranes. The solution-diffusion mechanism is the most commonly used model to explain the movement of gas molecules across dense non-porous membranes. In this mechanism, permeable gas molecules first adsorb on the membrane's surface upstream, then dissolve in the polymeric matrix and diffuse across the membrane due to a concentration gradient and the presence of a driving force such as pressure. They are eventually released from the membrane surface downstream. The permeability of a certain gas molecule (P_i) depends on the solubility (S_i) and diffusivity (D_i) of that gas component in polymeric matrix:

$$P_i = S_i.D_i \tag{16.4}$$

Barrer is the common unit used for gas permeability in membranes. To obtain the permeance of a gas component (P) through a membrane, the membrane thickness (δ) needs to be considered, and to obtain the molar flux of a gas component (J_i), the pressure difference between the two sides of the membrane ($\Delta p = p_{i,u} - p_{i,d}$) should be considered:

$$P = \frac{P_i}{\delta} \tag{16.5}$$

$$J_i = \frac{K_i.D_i.\Delta p}{\delta} \tag{16.6}$$

where D_i is the diffusion coefficient for the gas species i, and K_i is its sorption coefficient. GPU is the main unit used to express the gas permeance. While the diffusion coefficient of a specific gas molecule is strongly dependent on the type of polymer, the sorption coefficient is nearly identical for different types of polymers and is more correlated with the gas's saturation vapor pressure and boiling temperature. However, the diffusion and sorption coefficients of gas molecules are the major parameters determining the separation factor (α_{ij}) between two different gas species i and j:

$$\alpha_{ij} = \left(\frac{D_i}{D_j}\right)\left(\frac{K_i}{K_j}\right) \tag{16.7}$$

The terms $\left(\frac{K_i}{K_j}\right)$ and $\left(\frac{D_i}{D_j}\right)$ represent the solubility selectivity and diffusivity selectivity, respectively. Dense non-porous membranes typically exhibit good selectivity while having a low permeation rate. Several studies and modifications to operational conditions and membrane materials have been carried out in order to improve permeability in non-porous membranes [74]. Increased free volume should improve permeability and gas transport properties [75]. The void space between polymer chains is defined as the free volume in polymers, and it is used to describe the motion of gas molecules through polymeric membranes. The definition of free volume and occupied volume may be different for different materials, however, the free volume (FV) for membranes can be obtained by

$$FV = \frac{V_s - V_o}{V_s} \tag{16.8}$$

where V_o is the volume occupied by polymer chains and V_s is the specific volume (i.e., reciprocal of density). The total concentration of a gas in a membrane with glassy polymer matrix (C_t) is the summation of the gas present in the free volume (C_{FV}) and the gas dissolved in polymeric matrix (C_D):

$$C_t = C_{FV} + C_D \tag{16.9}$$

According to the Henry's law, the concentration of a gas dissolved in polymer can be obtained by

$$C_D = K_D f \tag{16.10}$$

where f is the fugacity and K_D is the Henry's solubility coefficient. Also, the Langmuir isotherm can be used to obtain C_H and explain the gas present in free volume:

$$C_H = \frac{q_m bf}{1 + bf} \tag{16.11}$$

where q_m is the maximum adsorption of gas molecules in free volume, and b is the Langmuir affinity constant, which represents the interaction between the gas molecules and the polymer-free volume. Contrary to glassy polymers, the gas diffusion through rubbery polymers is often independent of gas concentration. As shown in Fig. 16.11, the permeability of rubbery polymers is higher than that of glassy polymers.

The free volume contributes significantly to gas diffusion and permeability through the membrane and is affected by various parameters such as crystallinity, cross-linking, and side groups in the polymer chain. Because of the increase in free volume, using bulky side groups in polymer chains has a positive impact on gas molecule permeability and mobility [77]. The incorporation of inorganic particles such as silica into polymeric matrix has been shown to increase free volume and thus gas permeability [67]. However, the gas transportation and separation mechanisms in polymer nanocomposite membranes are distinct. The majority of the models proposed for describing gas transport in polymer nanocomposite membranes are based on approximations made on the medium and interfacial reactions, and the variation of inorganic phase is not adequately considered. Maxwell's model, which was developed primarily to estimate the dielectric properties of composites, is widely used to predict the permeability of membranes made of polymeric matrix containing impermeable spherical particles. The permeability of polymer nanocomposite membranes can be calculated by Maxwell's model according to the following equation:

$$P_{NC} = P_p \left(\frac{1 - \varphi_f}{1 + \frac{\varphi_f}{2}} \right) \tag{16.12}$$

where P_{NC} is the permeability of polymer nanocomposite, P_p is the permeability of pure polymeric matrix, and φ_f is the nanofiller volume fraction. Based on Maxwell's model, permeability of polymer nanocomposite membranes will be lower since any increase in particle volume fraction can result in smaller numerator and greater denominator in Eq. (16.12). Although this model explains the loss of permeability in some polymer nanocomposite membranes, it neglects the interactions

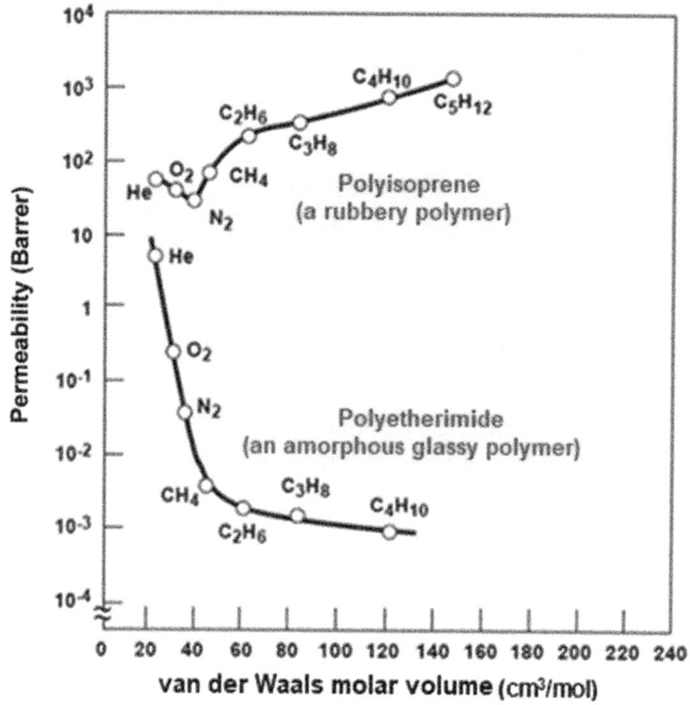

FIGURE 16.11

Gas permeation properties of rubbery vs glassy polymers; permeability as a function of molar volume [76].

Copyright 2021. Reproduced with permission from MTR Inc.

between polymer nanocomposite with gas molecules as well the interactions between polymer matrix and nanofiller. Several modified models based on Maxwell's model have been proposed to include the role of the penetrant molecule and the nanofiller properties in gas transportation through the polymer nanocomposite membranes [78]. The defects formed as a result of void formation and interfacial rigidification have a significant impact on gas diffusion across polymer nanocomposites. The nanofiller can disrupt the packing of polymer chains, increasing the free volume as well as the permeability and diffusivity of permeants. To describe the free volume increase mechanism and estimate the gas diffusion coefficient in polymer nanocomposite membranes, statistical-mechanical models have been proposed. According to the theory developed by Cohen and Turnbull, the diffusion coefficient of a molecule is correlated to the likelihood of finding a free volume by that molecule. This model is formulated as:

$$D = A\,exp\left[\frac{-\gamma V^*}{V_f}\right] \tag{16.13}$$

where D is the self-diffusion coefficient, A is an empirical constant, V^* is the minimum hole size that can be occupied by a gas molecule, V_f is the average free volume per molecule, and γ is the overlap factor as the same free volumes can be shared by different molecules. This model contributes to a better understanding of the interactions between the polymeric matrix and the nanofiller. Aside from the disruption of polymer chain packing caused by the addition of nanofiller, which results in voids and free volume in polymer nanocomposite, the poor compatibility of polymeric matrix and nanofiller can also result in the formation of nanogaps surrounding the nanofiller particles through which gas molecules can permeate. However, in addition to the polymer-nanofiller interactions, the interaction between gas molecules and nanofiller has an impact on gas solubility and permeability. The Arrhenius equation considers the effect of these interactions on the gas permeability by involving the enthalpy of gas sorption:

$$P = P_0\, exp\left[\frac{-E_p}{RT}\right] \tag{16.14}$$

where P is the gas permeability, P_0 is the pre-exponential factor, T is absolute temperature, R is the ideal gas constant, and E_p is the activation energy of permeation which can be obtained from

$$E_p = E_d + \Delta H_s \tag{16.15}$$

where ΔH_s is the enthalpy of gas sorption and E_d is the activation energy of diffusion. As evident, the incorporation of different nanoparticles in polymeric matrix can result in a change in sorption enthalpy since the gas-nanofiller interactions changes if the type of nanofiller changes.

16.2.2 Preparation and characterization of polymer nanocomposite membranes

The method and technologies used to fabricate membranes are determined by the membrane materials and properties, as well as the application for which the membrane will be used. As a result, the methods used to make polymeric or inorganic membranes may not be applicable to the fabrication of polymer nanocomposite membranes. Regarding the morphology and structure, there are two major categories of polymer nanocomposite membranes namely symmetric and asymmetric (Fig. 16.12) [79]. Symmetric or isotropic membranes can be both dense and porous and are considered to have uniform structure and composition over the cross-section.

Solution blending is a popular method for creating symmetric polymer nanocomposite membranes. The polymer is first dissolved in a solvent in this method, and then the inorganic particles are

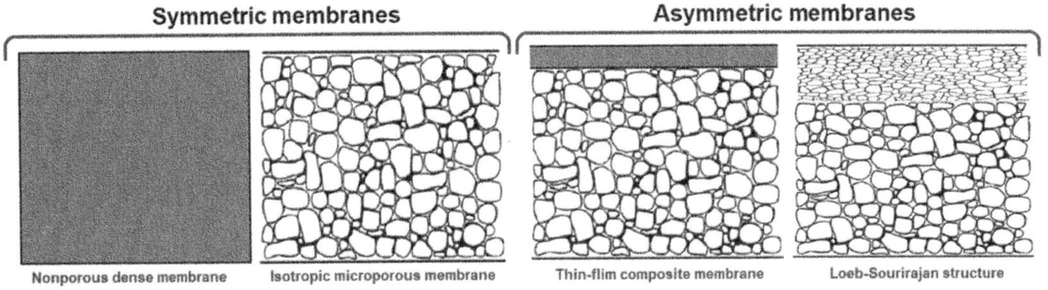

FIGURE 16.12

Schematic cross-sectional view of symmetric and asymmetric membranes.

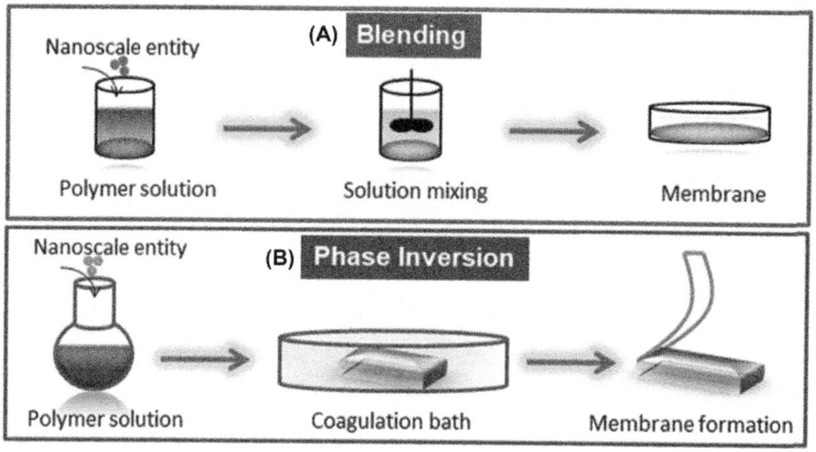

FIGURE 16.13

Method for fabrication of polymer nanocomposite membranes: (A) solution blending, (B) phase inversion [80].

Copyright 2021. Reproduced with permission from Elsevier.

dispersed by mechanical mixing followed by solvent evaporation (Fig. 16.13A). Various solvents such as dimethylformamide, toluene, chloroform, methanol, acetone, tetrahydrofuran, etc. have been employed in solution-blending method [81]. It is important to use a solvent compatible with both polymer and nanofiller; however, it is possible to dissolve polymer and nanofiller in different solvents and then mix them with each other if the solvents are miscible. Sun and Sue, for example, created a PMMA/ZnO nanocomposite by dispersing ZnO nanoparticles in methanol and dissolving polymethyl methacrylate in acetone, then mixing the ZnO/methanol and PMMA/acetone mixtures and evaporating the solvents under vacuum [82]. The removal of solvent is essential with respect to both environmental considerations and membrane quality and efficiency. Suer et al. prepared polyethersulfone/zeolite membrane for gas separation and used three procedures for drying the membrane after casting: (1) drying at room temperature for one day followed by annealing at 80°C–100°C for 6–8 hours; (2) drying at barometric pressure with nitrogen circulation followed by water quenching and annealing in vacuum at 80°C–100°C for 16–24 hours; and (3) drying in partial vacuum with nitrogen circulation followed by water quenching and annealing at barometric pressure at 80°C–100°C for 16–24 hours [83]. The drying procedure, as discovered by the authors, has a significant effect on the permeability of the membrane. Despite its simplicity, the solution-blending method has some environmental issues and challenges due to the high consumption of organic solvents.

Phase inversion is a widely used method for preparation of gas separation membranes (Fig. 16.13B). In this process, the polymer in liquid phase is transformed to a solid state. There are several techniques for phase inversion including nonsolvent-induced phase separation (NIPS), vapor induced phase separation (VIPS), thermally induced phase separation (TIPS), and evaporation induced phase separation (EIPS). The polymer-containing solution is cast on a suitable surface using the NIPS technique, also known as the Loeb-Sourirajan method, and then immersed in a non-solvent or coagulation bath to solidify the polymer film after partial solvent evaporation. For polymer dissolution, a volatile solvent that is easily evaporable is typically used. The initial solvent evaporation creates a thin solid layer, also known as the skin layer, on top of the membrane that can control the flux and selectivity. After immersion in a coagulation bath and during the solvent-nonsolvent exchange process, porous sublayers form. Although the phase inversion method produces an asymmetric membrane, the polymer concentration and solvent properties can be modified or changed during the casting stage to make the skin layer porous. Fig. 16.14 shows an asymmetric polymer nanocomposite membrane prepared by phase inversion and NIPS method for gas separation application.

The VIPS method is similar to the NIPS method except that the nonsolvent used in VIPS is a vapor phase such as water vapor. In VIPS, the solution is placed in a nonsolvent environment with a vapor phase, and phase inversion occurs as the vapor flows into the polymer film. Despite their similarities, the difference in nonsolvent results in different pore characteristics and membrane structure. In EIPS, also known as the solution casting method, the nonsolvent is present in the polymer-containing solution alongside the volatile solvent. As the nonsolvent evaporates and flows out of the polymer film, phase separation occurs. The evaporation process and the selection of solvents with different boiling points can control the morphology and porosity of the membrane.

TIPS is commonly used for polymers that are difficult to dissolve at room temperature, such as polypropylene and polyvinylidene fluoride. The polymer is first dissolved at high temperatures, then casted on a support, and finally cooled to solidify the polymer. Because the solvent is typically removed via sublimation or freeze-drying, the crystallization temperature of the solvent is an

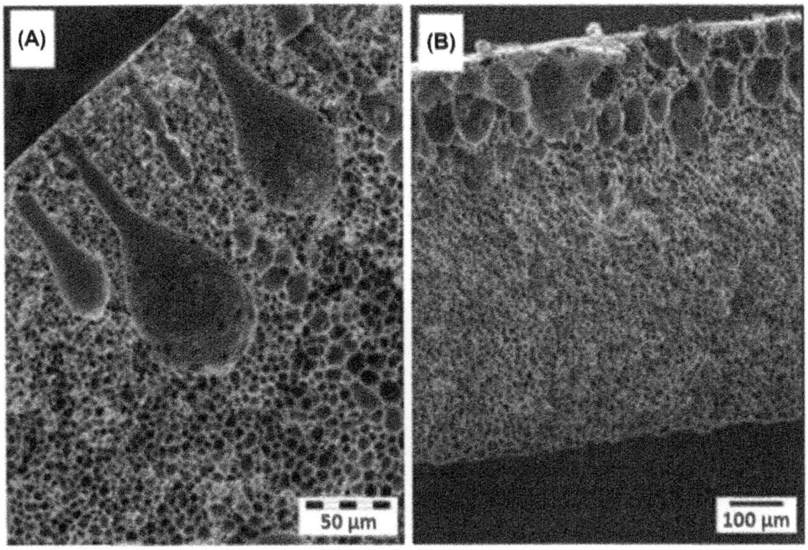

FIGURE 16.14

Morphology of polyetherimide/palladium nanocomposite asymmetric membranes prepared by phase inversion for gas separation [84].

Copyright 2021. Reproduced with permission from Elsevier.

important factor in the TIPS process. In this process, you can also control the porosity and morphology of the membrane by using different solvents with different crystallization temperatures. In Fig. 16.15, scanning electron microscopy (SEM) images of SiO_2/polybenzimidazole nanocomposite, prepared by TIPS, is shown. Sadeghi et al. fabricated this membrane for separation of a gas mixture containing CO_2, N_2, and CH_4 [85].

The incorporation of nanofiller into the polymeric matrix is a necessary step in the membrane fabrication process. The method used to introduce nanoparticles into a polymeric matrix is determined by the polymer type as well as the desired membrane properties. Direct blending is a straightforward method that involves dispersing and stirring nanofiller in polymer solution. The nanoparticles can be mixed directly into the polymer solution or first dissolved in a separate solvent before being mixed in. The agglomeration of nanoparticles is a significant limitation of this method. To address the agglomeration issue, surfactants and dispersing agents such as block copolymers can be used [86]. Other techniques that are commonly used to distribute the nanoparticles and homogenize the nanocomposite include high-speed stirring and ultrasonication. The incorporation of nanofillers such as graphene oxide which have functional groups that can form hydrogen bonds with the functional groups in polymeric matrix is less difficult [87]. The addition of nanofiller can have a significant impact on the morphological properties of the membrane during the phase inversion process. As shown in Fig. 16.16, introducing nanoparticles resulted in two different morphologies namely sponge-like and finger-like morphologies of membranes, depending on the rate of solvent-nonsolvent exchange.

To improve the performance of a polymer nanocomposite membrane, the nanofiller loading must be optimized in relation to the polymer/nanofiller interfacial properties. It has been reported that high loading can cause nanoparticle aggregation and the formation of undesirable voids, resulting in lower permeability [89]. Fig. 16.17 shows how the change of loading percentage can lead to change in membrane separation efficiency. In gas separation membranes, the loading can be up to 50% of the total weight of the membrane [91], depending on the membrane application and the properties of polymer and inorganic phases. Aside from agglomeration, another challenge in the preparation of polymer nanocomposite membranes is the settling of nanoparticles by gravity, which not only affects membrane separation efficiency but also reduces membrane reproducibility. This challenge can be addressed through the in situ formation of nanofiller.

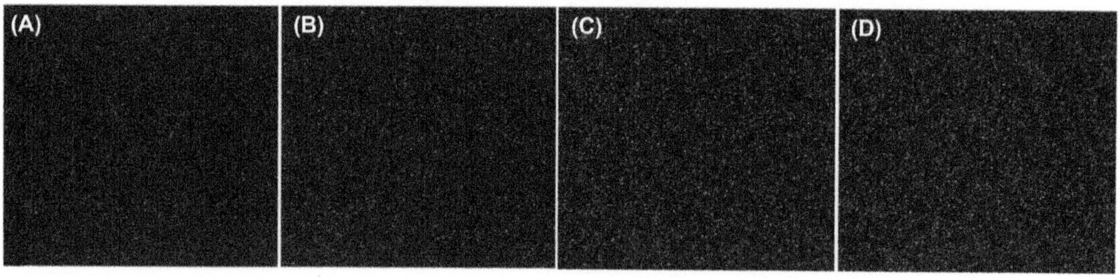

FIGURE 16.15

SEM mapping of silica nanoparticles distribution in membrane cross-section; silica contents (A) 2.8 wt.%, (B) 9.4 wt.%, (C) 16 wt.%, (D) 20 wt.% [85].

Copyright 2021. Reproduced with permission from Elsevier.

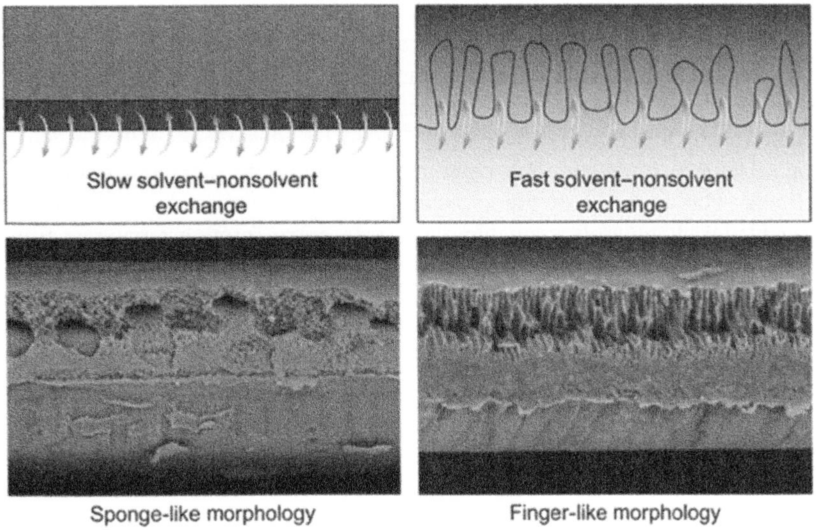

FIGURE 16.16

Effect of the rate of solvent-nonsolvent exchange on the morphology of membrane [88].

Copyright 2021. Reproduced with permission from Elsevier.

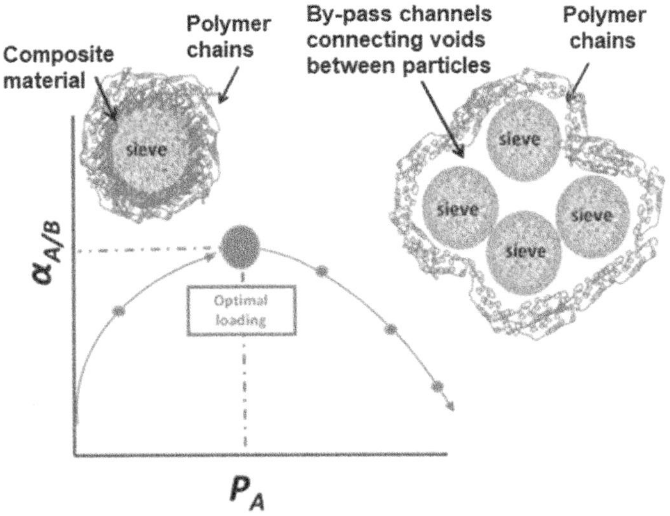

FIGURE 16.17

Effect of loading percentage on membrane separation efficiency [90].

Copyright 2021. Reproduced with permission from Elsevier.

The sol-gel method employs a reduction reaction with reducing agents to form nanofiller within the polymer solution in the coagulation bath. In addition to the sol-gel method, nanofiller can be grown in situ by dispersing the nanoparticles in monomers before polymerization. In situ nanoparticle formation in polymeric matrix can prevent interfacial void formation and produce a monodispersed nanofiller in polymeric matrix, which appear to be advantages of the in situ method over direct blending. The two most common types of nanocomposite membranes are flat membranes and hollow fibers. Flat membranes are typically created using a solution casting method that involves doctor blades slipping on a suitable glass surface and forming a polymer film that is then transferred to a coagulation bath. The thickness of the flat membranes can be adjusted by adjusting the gap between the blades and the glass surface. The main fabrication factors that influence the properties of flat membranes are casting rate, solvent evaporation, solvent-nonsolvent exchange, and polymer, nanofiller, solvent, and nonsolvent properties. The primary method for producing hollow-fiber membranes is dry-jet wet spinning. Fig. 16.18 schematically shows the set-up of a dry-jet wet spinning method and its different components including polymer solution, bore fluid, spinneret, coagulation bath, and take-up drum. After being extruded through an air gap, the polymer solution enters the coagulation bath. The main factors influencing spinning operation and efficiency are flow rate, extrusion rate, spinneret type and structure, air gap, and bore fluid properties.

16.2.3 Characterization of polymer nanocomposite membranes

Membrane characterization and a thorough understanding of membrane properties enable us to create novel membrane materials and structures that improve separation efficiency. Mechanical

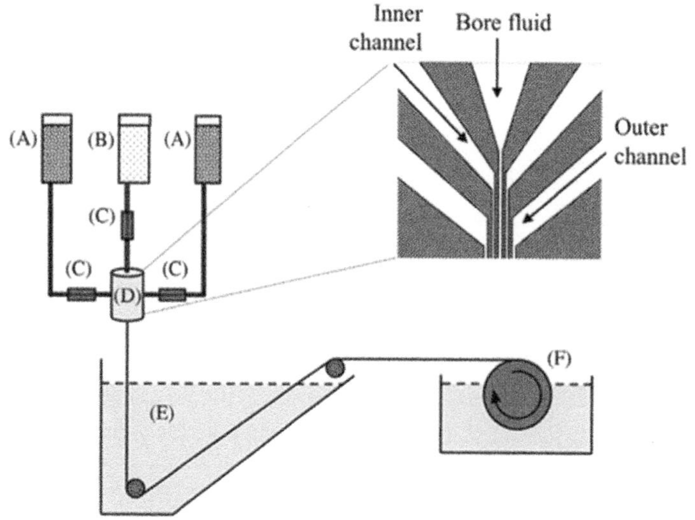

FIGURE 16.18

Schematic view of a dry-jet wet spinning set-up; (A) bore fluid, (B) polymer solution, (C) filters, (D) spinneret, (E) coagulation bath, (F) take-up drum [91].

Copyright 2021. Reproduced with permission from Elsevier.

properties, thermal properties and membrane stability, mass transport properties, surface morphology, porosity and free volume, purity and content of membrane, molecular structure, surface chemistry, distribution of nanofiller and interfacial properties are the major membrane characteristics that need to be identified.

Mechanical property retention is critical for membranes to avoid brittleness and maintain integrity over long periods of operation at different temperatures. Tensile strength, Young's modulus, and elongation are the most common parameters used to study membrane mechanical properties. Tensile strength is a material's resistance to breaking apart when subjected to a pulling force. A pulling force is applied to a sample of membrane material with a specific cross-sectional area in this measurement, and it is then elongated until it fails or breaks apart. In Fig. 16.19A, a tensiometer used for the measurement of tensile strength and elongation is shown. Tensile strength is calculated by dividing the magnitude of force at rupture by the cross-sectional area of the sample before elongation in standard tests based on ASTM specifications. During a tensile test, both the applied force and the sample elongation values are recorded and plotted as a stress-strain curve, which provides important information about the mechanical properties of materials (Fig. 16.19B). Young's modulus (E) is another important mechanical property that shows the stiffness of a material and its ability to withstand against permanent deformation under tension. As shown in Fig. 16.18B, Young's modulus is the slope of the linear region of stress-strain curve and can be obtained by

$$E = \frac{\sigma}{\varepsilon} = \frac{F/A}{\Delta L/L_0} \tag{16.16}$$

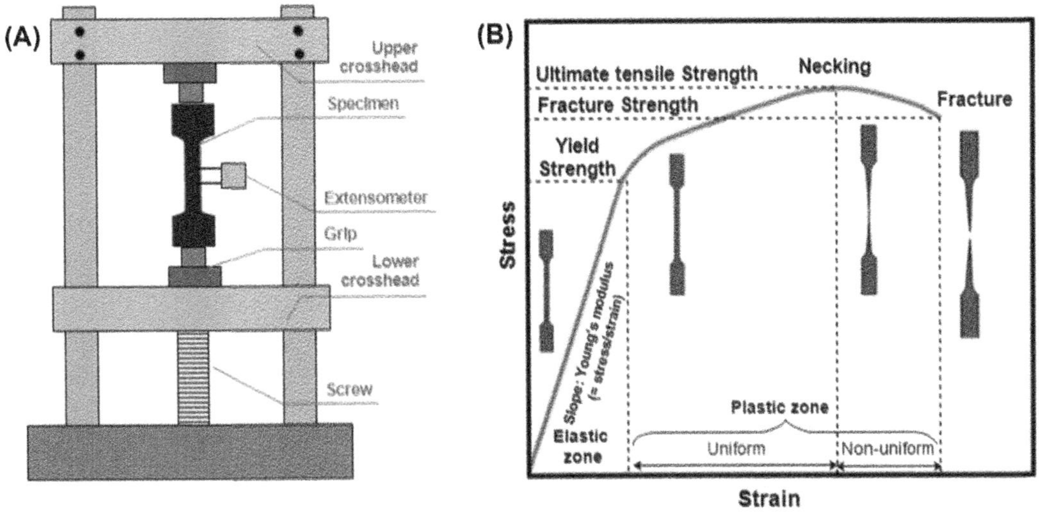

FIGURE 16.19

(A) Tensiometer used for the measurement of tensile strength and elongation, (B) stress-strain curve explaining mechanical properties.

where σ is stress, i.e., the applied force (F) per unit area (A), and ε is the strain, i.e., the change in length (ΔL) divided by the original length before deformation (L_0). Although the modulus can be obtained from the tensile test, there are difficulties in obtaining data from the linear region of the stress-strain curve, and precise measurement requires a separate experimental set-up to focus on the initial part of the stress-strain curve where the elastic deformation occurs. ASTM D638 and ISO 527 are two standard methods commonly used for studying the tensile properties of plastics and plastic composites, respectively. ASTM D882 is another method for determination of tensile properties of polymer films.

Several studies have demonstrated that incorporating nanofillers into polymeric matrices has a significant impact on the thermal properties of polymer nanocomposites [92,93]. Thermogravimetric analysis (TGA) and differential scanning calorimetry (DSC) are two major thermal analysis techniques commonly used for polymer nanocomposite membranes. TGA analysis continuously measures weight loss while heating the sample at a constant rate in a controlled environment with an inert gas. Material dehydration, oxidation, degradation, evaporation, sublimation, and decomposition can all cause weight loss. TGA is required when using the membrane in high-temperature gas separation processes. The amount of heat absorbed or released by the sample is measured over time and over a specific temperature range in DSC analysis. DSC analysis can determine important parameters such as specific heat capacity, thermal stability, oxidation behavior, thermal phase transitions (e.g., glass transition temperature, melting point, and crystallization), relaxation behavior, and segmental mobility in polymer nanocomposites. In Fig. 16.20, TGA and DSC graphs for two polymer nanocomposites are presented, showing the effect of nanofiller concentration on the thermal properties of polymer nanocomposite.

Because gas permeation is strongly influenced by the structural, physical, and chemical properties of membrane materials, gas transport properties of membranes must be characterized in order to design

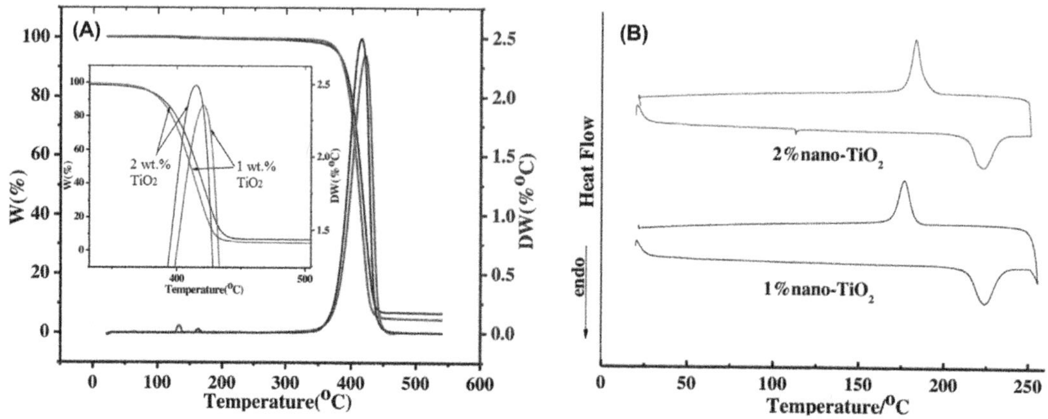

FIGURE 16.20

(A) TGA graphs for polymerized cyclic butylene terephthalate (pCBT) sample with different content of nano-TiO_2, (B) DSC graphs of pCBT with various content of nano-TiO_2 during the first heating and cooling [94].

Copyright 2021. Reproduced with permission from Elsevier.

materials with high permeability. Integral permeation and differential permeation are two techniques for estimating the rate of gas permeation through membranes experimentally. The permeants rate is dynamically monitored as a function of time in the differential permeation method. In this method, the gas molecule passes through the membrane into a flowing gas stream, and the permeation rate is calculated by multiplying the rate of flowing gas by the measured concentration of gas molecule. The membrane is first degassed in the integral permeation technique, also known as the time-lag method, and then the gas molecules pass through the degassed membrane and accumulate in an evacuated chamber at the receiving side of the membrane. The change in volume at constant pressure or pressure at constant volume is then measured over time. D1434-82 is an ASTM standard test for determining the rate of gas permeation using barometric and volumetric methods. Dow and Linde permeation cells are commonly used for measuring gas permeation using barometric and volumetric methods, respectively. In Linde cell, the fluid in capillary tube is displaced by the penetrating gas and the velocity of displaced fluid is measured and used to estimate the gas permeability. In Dow cell, the change in pressure in receiving chamber is measured by a mercury manometer and then the permeability coefficient as well as the number of gas moles permeated through the membrane can be calculated by the following equations:

$$n = \frac{P_c V}{RT} \tag{16.17}$$

$$P = \frac{nL}{AtP_d} \tag{16.18}$$

where n is the number of gas moles passed across the membrane, P_c is the change of pressure in receiving side, V is the volume of receiving chamber, R is gas constant, T is the temperature, P is the permeability coefficient, L is the thickness of membrane, A is the membrane area, P_d is the pressure difference between two sides of the membrane, and t is the time of operation.

The permeability of a polymer nanocomposite is affected by several properties, including free volume and membrane morphology. The main technique commonly used to study the free volume and voids in the structure of polymer nanocomposites is positron annihilation lifetime spectroscopy (PALS). In this technique, a positron is injected into the material from a positron source such as ^{22}Na and annihilates through interaction with electrons in the material. Positron annihilation produces gamma radiation, which scintillator detectors can detect. The free volume and the size of voids are estimated based on the fact that the lifetime and annihilation rapidity of a positron in material are different from those in voids.

Crystallinity is another important factor in determining the permeability of polymer nanocomposite membranes. The matrix phase of most polymeric membranes used for gas separation is semi-crystalline or amorphous polymers. The incorporation of inorganic nanofillers can disrupt the arrangement of polymer chains, resulting in a decrease in membrane crystallinity [95]. X-ray diffraction analysis is the main technique widely used to probe the crystallinity in polymer nanocomposite membranes. In this method, a beam of X-ray is radiated toward the sample, and the diffraction pattern is obtained by measuring the angles and intensities of reflected X-ray. Brag's law is used to calculate the d-spacing (i.e., distance between adjacent polymer chains), which is an indication of crystal structure as well as free volume.

$$d = \frac{\lambda}{2\, sin\theta} \tag{16.19}$$

where d represents d-spacing, and λ is the wavelength of X-ray, which is 1.5406 Å when Cu Kα is used as X-ray source. Ahmadizadegan et al. investigated the change in crystallinity of a polyimide matrix caused by the incorporation of TiO_2 nanoparticles and observed a broadening of diffraction peaks as the nanofiller content increased [96]. The disruption in the order of polymer chain packing was reported to be the cause of the decrease in crystallinity. Amini et al. incorporated nanoclay particles into Pebax 1657, which is composed of amorphous polyethylene oxide and crystalline polyamide, and discovered that the addition of nanofiller increased d-spacing and thus free volume, which could lead to higher gas permeability [97].

Microscopic methods are typically used to examine the surface and cross-sectional morphology of membranes in order to provide detailed information on the structure of the polymeric matrix and nanofiller. The most common and convenient method for studying the morphology of polymer nanocomposite membranes is electron microscopy (EM). The two main EM techniques for studying membrane morphology and effective parameters are SEM and transmission electron microscopy (TEM). Both techniques rely on an electron source to generate a stream of electrons and electromagnetic lenses to focus the electron beam on a sample in a vacuum. SEM can provide sample surface composition and images by capturing X-rays and secondary electrons, also known as backscattered electrons, produced by the interaction of primary high-energy electrons with the sample surface. In TEM, the electron beam passes through the specimen and is captured by a detector below after transmission. As a result, TEM necessitates more sample preparation and handling to obtain an electron-transparent sample, increasing the cost of TEM. The thickness of sample required for TEM analysis is very small ($\sim$150 nm). While SEM can reveal information about the sample's surface, TEM can reveal information about the sample's internal structure as well. Both methods have distinct advantages and applications. SEM can generate 3D images, whereas TEM can generate 2D images with higher resolution (10–50 million times). Focused ion beam scanning electron microscopy (FIB-SEM) is used to learn more about the spatial distribution of nanofillers and the interaction between nanofiller and polymer matrix. In the FIB-SEM technique, an ion beam is directed toward the sample, removing a nanosized block of sample from the surface before imaging the exposed block with a SEM. In Fig. 16.21, the images taken from a polymer nanocomposite by SEM and TEM are shown for comparison.

Atomic force microscopy (AFM) is another widely used microscopic method for studying surface topology and roughness, which provide indications of the presence and distribution of nanoparticles in polymeric matrix. The AFM technique analyzes surface properties by measuring the adhesion force between the sample surface and the tip by determining the cantilever deflection (Fig. 16.22A). In Fig. 16.22B, the AFM image of polyether block amide/nanoclay/polyacrylonitrile/polyester nanocomposite (PEBA/NC/PAN/PE) is shown and the effect of clay nanoparticles on the surface topology of membrane is observable.

Spectroscopic techniques are commonly used to characterize the functional groups and chemical structure of the membrane surface. Fourier transform infrared (FTIR) and Raman spectroscopy are powerful tools for identifying chemical bonds on the membrane surface. FTIR spectroscopy is based on analyzing the sample's absorption of infrared radiation. The absorption spectrum is a sample's molecular fingerprint. A list of chemical bonds as a function of FTIR absorbance is presented in Table 16.3. Raman spectroscopy has been used to investigate the molecular conformation and orientation of polymer chains as they are affected by nanofiller incorporation, mechanical and thermal stresses, and interactions with penetrant molecules. Wanger has discussed the application of

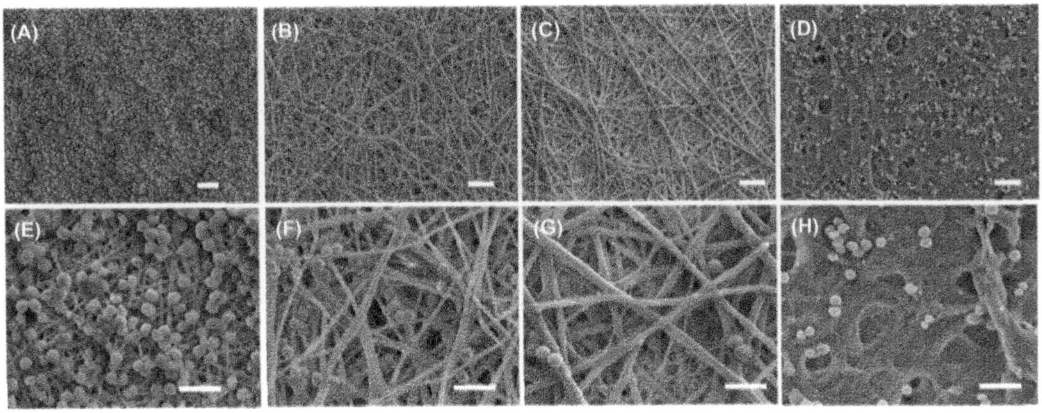

FIGURE 16.21

SEM images of nanofibrous membranes with (A) neat polycaprolactone (PCL), (B) 75% PCL and 25% chitosan, (C) 50% PCL and 50% chitosan, and (D) 25% PCL and 75% chitosan; TEM images of nanofibrous membranes with (E) PCL only, (F) 75% PCL and 25% chitosan, (G) 50% PCL and 50% chitosan, and (H) 25% PCL and 75% chitosan (the scale bars in SEM and TEM images represent 2 μm and 400 nm, respectively) [98].

Copyright 2021. Reproduced with permission from Elsevier.

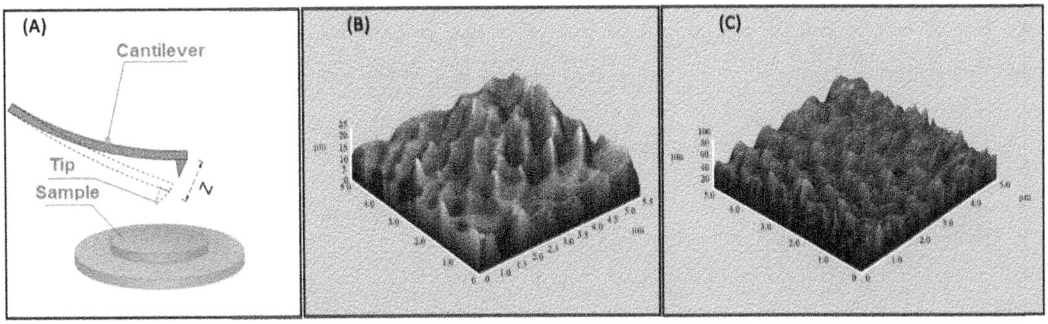

FIGURE 16.22

(A) schematic of a cantilever deflection; AFM images: (B) PEBA/PAN/PE (C) PEBA/NC/PAN/PE membranes [97,99].

Copyright 2021. Reproduced with permission from American Chemical Society and Elsevier.

Raman spectroscopy to the study of interactions between carbon nanotubes and polymeric matrix [100]. Such interactions are reflected in the peak shift or change in peak width. IR and Raman spectroscopies complement each other, allowing for a more in-depth understanding of the interactions of molecules and functional groups with their surroundings and nanofiller.

Table 16.3 Chemical bonds as a function of FTIR absorbance.

Functional group		Absorbance (cm^{-1})
Alkanes	C–H stretch	3000–2850
	=C–H stretch	3100–3000
Aromatic compounds	C–C stretch	1600–1585
	C=C	800, 1610, 1500
	Amine	1350, 1250
Alcohols	O–H	3500–3200
	C–O	1260–1050
Aldehyde	H–C=O	2830–2695
Ketones	C=O stretch	1715
Organic nitrogen compounds	N–O asymmetric stretch	1550–1475
	N–O symmetric stretch	1360–1290
Ether	C–O–C	1150, 1070, 1062
Amines	N–H	3350–3300
Inorganic compounds	CO_3^{2-}	880–800
	SO_4^{2-}	680–610
	NO_3^-	860–800
	PO_4^{3-}	1100–950

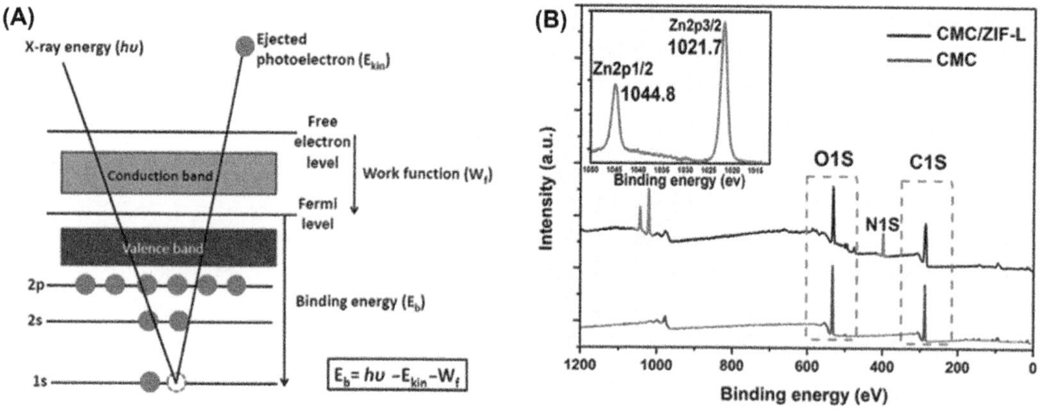

FIGURE 16.23

(A) schematic view of XPS method, (B) XPS spectrum of a polymer nanocomposite composed of zeolitic imidazolate framework incorporated into carboxymethyl cellulose for gas separation applications [101,102].

Copyright 2021. Reproduced with permission from Elsevier and MDPI.

Elemental composition as well as the types of bonds and oxidation state can be probed by X-ray photoelectron spectroscopy (XPS) to a depth of ~10 nm from the membrane surface. XPS technique involves the emission of an X-ray beam from a soft X-ray source (e.g., Mg Kα and Al Kα) toward the sample. The X-ray photons interact with the electrons in the inner shells of the atoms on the surface. The energy of photons enables the photoelectrons to escape from the subshells (s, p, d, etc.) and surface of the sample (Fig. 16.23A). The kinetic energy of the electrons escaped from the sample surface are recorded and analyzed to identify the elements and structure of molecules on the surface. In the XPS spectrum, the peaks are presented as a function of binding energy which is an element characteristic and can be calculated by

$$E_b = h\upsilon - K_E - \varphi \tag{16.20}$$

where E_b is the binding energy, K_E is the kinetic energy of escaped electrons, $h\upsilon$ represents the energy of X-ray photon, and φ is the spectrometer work function. This method detects all elements except helium and hydrogen, making it difficult to characterize and distinguish carboxylic acid-containing materials from esters. Fig. 16.23B shows the XPS spectrum of a polymer nanocomposite composed of zeolitic imidazolate framework incorporated into carboxymethyl cellulose for gas separation applications. There are two sharp binding energy peaks of C 1s and O 1s, which are indication of polysaccharide carboxymethyl cellulose. In C 1s region, there are a series of peaks observed at 284.4 eV (C–H and C–C bonds), 286.3 eV (C–O bond), 287.8 eV (C–O and O–C–O bonds), and 290.5 eV (O–C–OR bonds). In O 1s spectrum, the binding energy peaks located at 531.3 and 533.1 eV represent O–C, and COO– bonds, respectively. Two Zn 2p peaks were observed at 1044.8 and 1021.7 eV after incorporation of zeolitic imidazolate framework nanosheets.

16.3 Applications of biodegradable polymer nanocomposite membranes for gas separation

16.3.1 Carbon dioxide separation

Carbon dioxide separation and capture is critical because it is the primary greenhouse gas emitted by a wide range of industrial and non-industrial processes. To reduce CO_2 emissions, several polymer nanocomposite membranes have recently been developed. Because of their biodegradability and biocompatibility, nanocomposite membranes based on biopolymers have received a lot of attention for such an environmental goal. Cellulose-based biopolymers have been employed for the fabrication of polymer nanocomposite membranes for CO_2 separation. Cellulose is a polysaccharide composed of β linked D-glucose units (Fig. 16.24). Najafi et al. incorporated SiO_2 nanoparticles

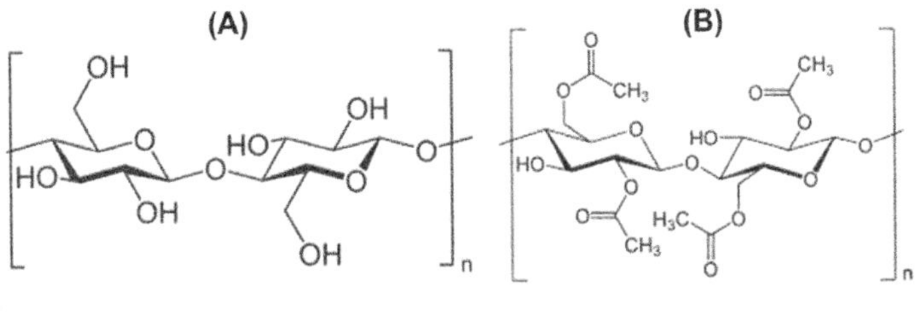

FIGURE 16.24

Chemical structure of (A) cellulose (B) cellulose acetate.

into cellulose acetate membrane and used for separation of CO_2 from O_2 and N_2 molecules [103]. The compatibility of silica nanoparticles and cellulose acetate was reported to be good; however, morphological analysis by SEM revealed that nanoparticle agglomeration was unavoidable. Increasing the concentration of SiO_2 filler resulted in the formation of more active sites in the polymeric matrix, which increased the solubility of polar CO_2 gas molecules. The CO_2 permeability was increased from 6.32 barrer to 7.3 barrer as the silica content increased from 0 to 20 wt.%, while the N_2 and O_2 permeabilities decreased from 0.186 and 0.95 barrer to 0.092 and 0.489 barrer, respectively. The CO_2/N_2 selectivity was found to be increased from 32 to 80 due to the enhancement in CO_2 permeability. Various types of nanocellulose, such as cellulose nanofibrils, bacterial nanocellulose, and cellulose nanocrystals, have been synthesized and used as membranes for gas separation applications using mechanical and/or chemical techniques. However, due to cellulose's low permeability, these membranes have mostly been used as gas barriers. To improve permeability, these nanomaterials have been incorporated into other polymeric matrices. Ansaloni et al. incorporated cellulose nanofibrils into a hydrophilic polyvinylamine and studied the CO_2 permeability and selectivity of the prepared polymer nanocomposite and compared them with those of pure cellulose nanofibrils film [104]. The CO_2/CH_4 and CO_2/N_2 selectivity were as high as 350 and 500, respectively, however, the CO_2 permeability never exceeded 25 barrer. The addition of cellulose nanofibrils to polyvinylamine increased CO_2 permeability by up to one order of magnitude while decreasing selectivity by the same amount. Torstensen et al. combined cellulose nanocrystals with polyvinyl alcohol (PVA), another biodegradable polymer, and investigated the efficiency of the resulting polymer nanocomposite membrane for CO_2 separation [105]. The experimental results indicated that the CO_2 permeance (pressure-normalized flux) increased from 105.5 ± 1.9 GPU to 127.8 ± 5.5 GPU as cellulose nanocrystals were incorporated into PVA. According to the authors, the homogeneous dispersion of cellulose nanocrystals created a continuous network in the PVA matrix, providing a favorable path for CO_2 molecules to diffuse across the active layer. Other effective factors that improved CO_2 permeability were the charge and hydrophilicity of nanocellulose. The increase in CO_2 permeance was most likely due to the accumulation of water at the nanocellulose surface, which resulted in the absorption of polar CO_2 molecules rather than nonpolar N_2 molecules. This also led to higher CO_2/N_2 separation factor for cellulose nanocrystal/PVA nanocomposite membrane (39 ± 0.4) compared to pure PVA membrane (36 ± 0.5). Dilshad et al. investigated the CO_2 permeation properties of PVA/PEG cross-linked membrane tethered with multiwalled carbon nanotubes (MWCNT) [106]. The surface of MWCNTs were engineered through functionalization with −OH groups. It was found that the CO_2 permeability increased from 22.6 ± 0.45 barrer to 115.57 ± 2.89 barrer as the concentration of MWCNT in polymeric matrix increased from 0 to 0.5 wt.%. The increased CO_2 permeability was reported to be due to the increased surface area and pores created by MWCNTs, which provided suitable channels for CO_2 adsorption and transportation. Also, the CO_2/N_2 separation factor increased from 20.92 ± 0.42 to 22.75 ± 5.07, probably due to the interactions between polar CO_2 molecules and the carbonyl group present on the surface of MWCNTs. Both CO_2 permeability and CO_2/N_2 selectivity were observed to decrease as MWCNT concentration was increased above 0.5 wt.%. According to SEM analysis, they occurred as a result of MWCNT agglomeration and a change in nanotube orientation from vertical to irregular positions. The decrease in N_2 permeability occurred as MWCNT concentration increased above 0.25 wt.%, most likely due to blockage of nanotubes by polymer chains. In another study, Borgohain et al. incorporated MWCNTs into carboxymethyl chitosan and prepared a membrane for CO_2 separation [107]. The humidity of separation environment and the membrane's moisture holding capacity were found to be crucial to facilitate the CO_2 transportation. The optimum values for CO_2 permeance and CO_2/N_2 separation were ~43 GPU and ~45, respectively, which achieved with 1 wt.% of MWCNTs at 80°C. Hydrophilicity and the presence of functional groups make biopolymers good candidates for fabrication of nanocomposite membranes for CO_2 separation processes.

16.3.2 Air separation

The development of air separation technologies is critical because air is a rich source of nitrogen and oxygen, two of the most demanding chemicals. Cryogenic distillation, solvent absorption, and adsorption techniques are the most common industrial air separation processes. Membrane technology has several advantages, particularly in small-scale applications, because of its portability and compact size, lower energy consumption, and potential to solve specific problems. However, because N_2 and O_2 have similar sizes, they cannot be effectively separated from air because their permselectivity is low when the difference in diameter is the only driving force of separation. Several polymer nanocomposite membranes have been developed and tested in order to improve N_2/O_2 separation efficiency while optimizing the permeability/selectivity tradeoff. Cobalt Schiff's bases have been used as carriers for O_2 facilitated transfer in polymeric membranes. Samarasinghe et al. incorporated cobalt(II) phthalocyanine microparticles into the Matrimid 528 polyimide matrix [108]. The authors used PluronicF-127 as compatibilizer to avoid undesired defects which could lead to lower selectivity. Pluronic is a polyethylene-b-polypropylene block copolymer. To test the gas permeation, a constant pressure-variable volume set-up was used. The O_2/N_2 selectivity increased from 14.5% to 31.6% as the filler concentration increased, according to the permeability analysis. The increase in selectivity came at the expense of a significant decrease in O_2 permeability. The use of Pluronic compatibilizer, on the other hand, resulted in a 64% increase in O_2 permeability and a 34% increase in O_2/N_2 selectivity. The amphiphilic nature of Pluronic compatibilizer resulted in a homogeneous dispersion of filler in polymeric matrix, leading to an effective permeation of O_2 molecules. Han et al. incorporated tetra(p-methoxylphenyl)porphyrin cobalt chloride into poly(amide-12-b-ethylene oxide) (also called Pebax-2533) and prepared a thin membrane for O_2/N_2 separation [109]. Because of the increase in free volume, the permeability of N_2 molecules was found to increase with increasing filler concentration. The O_2 permeability was 2.6 times higher when the filler content of membrane was 1.2 wt.% compared to pure polymeric membrane. However, the O_2 permeability decreased as the filler concentration increased more probably due to the agglomeration of filler particles. At the feed pressure of 0.1 MPa and 0.6 wt.% filler concentration, the O_2/N_2 selectivity was 4.2 times that of neat Pebax-2533. It was found that decreasing the feed pressure to 0.035 MPa could significantly increase the O_2 permeability and O_2/N_2 selectivity to 12.2 barrer and 7.6, which is close to Robeson upper bound. Ogbole et al. prepared SiO_2-polydimethylsiloxane nanocomposite membrane through solution casting method and studied the transportation behavior of N_2 and O_2 molecules [110]. The authors used fumed silica nanoparticles, pretreated with polydimethylsiloxane, to improve the compatibility and the interaction between polymeric matrix and nanofiller. The pretreatment of SiO_2 nanoparticles lead to optimal loading and homogeneous dispersion of filler, which resulted in an increase in O_2 permeability from 590 to 640 Barrer at 0.20 MPa. It was found that the O_2 permeability increased over time whereas the N_2

permeability decreased and then reached steady state. The O_2/N_2 selectivity increased from 2.43 to 3.46 as the SiO_2 concentration increased from 0 to 10 wt.%. Ahmad et al. synthesized TiO_2-polyvinyl alcohol (TiO_2-PVA) membrane via solution blending and tested for separation of several gas pairs including O_2/N_2 [111]. According to the authors, when the neat PVA membrane was used, low permeability was observed for both gases, most likely due to the semicrystalline nature of PVA. Both gases' permeability decreased as TiO_2 was added; however, the decrease in permeability was greater for N_2, most likely due to its larger kinetic diameter. The O_2/N_2 selectivity increased by 59% while their permeability kept decreasing as the TiO_2 content increased to 20 wt.%. In another study, Ghalei et al. synthesized silica particles using a micelle-templating method in the presence of PVA and Pluronic and then incorporated them into a polyurethane (PU) matrix [112]. The OH groups in PVA were considered to increase the gas separation efficiency and the amphiphilic Pluronic block copolymer was used as templating agents for synthesis of modified silica particles. The permeability of O_2 was found to be higher than that of N_2. It has been reported that the permeability coefficient of gas molecules is directly related to their critical temperature. The gas permeability was found to decrease as the concentration of silica particles increased. According to the glass transition temperature of the composite, the majority of the silica particles were found in the soft regions of the polymeric matrix. This was probably the reason for the reduction in permeability of O_2 and N_2 gas molecules. The permselectivity of O_2/N_2 pairs increased by 28%, compared to the pure PU. Koh et al. prepared a biodegradable poly(lactic acid)/layered silicate nanocomposite membrane and studied the permeation properties of N_2, O_2, and CO_2 gases [113]. Organoclays namely Closite15A, Closite20A, and Closite30B were used as layered silicate nanofiller in polylactic acid (PLA) matrix (Fig. 16.25A and B). The permeability of all gases decreased with increasing the nanofiller content, however, the d-spacing (i.e., the gap between

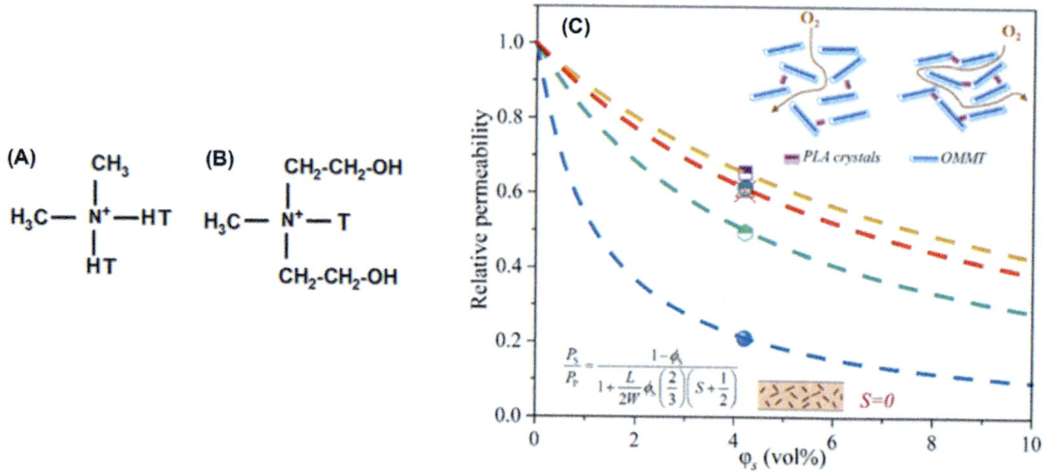

FIGURE 16.25

Chemical structure of organoclays (A) Closite15A, Closite20A, and (B) Closite30B; (C) effect of organoclay incorporation on O_2 permeability of PLA [114].

Copyright 2021. Reproduced with permission from American Chemical Society.

silicate layers) was found to be greater for the membranes containing Closite30A, which resulted in higher gas barrier properties. In another investigation, Li et al. studied the effect of the incorporation of an organically modified montmorillonite (OMMT) into PLA matrix on the barrier performance of polymer nanocomposite membrane [114]. In comparison with PLA film, the O_2 permeability decreased by 79% when ethylene glycol diglycidyl ether was used to regulate the d-spacing of organoclays and increase the polymer crystallinity (Fig. 16.25C).

16.3.3 Hydrogen recovery

Hydrogen is a plentiful nonpolar gas with important applications in a variety of industries, including oil refining, metallurgy, chemistry, electronics, and food processing. Hydrogen purification and recovery is critical for long-term energy development because hydrogen is expected to be one of the primary energy sources in the coming decades due to its high-energy content (120 MJ/kg). For hydrogen enrichment and recovery, a variety of technologies (e.g., cryogenic distillation, pressure swinge adsorption, membranes) have been used. Among various types of membranes (i.e., inorganic, metallic, and polymeric), membranes based on palladium and its alloys have been applied to hydrogen separation applications due to their permeability and selectivity toward H_2 molecules [115]. However, due to the limitations that exist for the application of Pd, such as high cost and phase change at high temperatures, other inorganic membranes as well as polymeric membranes have been developed for hydrogen separation. Membranes based on metal-organic frameworks [116], zeolite [117], carbon [118], silica [119], and polymers such as polysulfones [120] and polycarbonates [121] have been prepared and used for hydrogen separation, purification, and recovery. Recently, biopolymers have gained attention for fabrication of membranes, however, few nanocomposite membranes based on biopolymers have been fabricated and investigated for hydrogen separation in particular.

El-Azzami and Grulke studied the CO_2/H_2 separation by a chitosan membrane modified by arginine–sodium salt [122]. Arginine–Na is an amino acid salt that plays the role of a carrier and facilitates the CO_2 transportation. It was observed that the addition of arginine salt resulted in an increase in both permeability and CO_2/H_2 selectivity. Increasing the feed stream pressure from ~1.5 to ~5 bar decreased the CO_2/H_2 separation factor from 76 to 48. The hydrophilicity of chitosan biopolymer is an important factor affecting the CO_2 transportation, and the increase in feed pressure can result in lower water content in the membrane, reducing the CO_2 permeability. Compared to CO_2/N_2 selectivity, the CO_2/H_2 selectivity was less decreased due to higher diffusivity of H_2 molecules.

Prasad and Mandal studied the separation of H_2 from CO_2 by a nanocomposite membrane composed of chitosan, silk fibroin, and graphene nanoparticles [123]. Due to the presence of amines such as glycine and alanine in the SF molecules, chitosan was used as a matrix and silk fibroin (SF) as a carrier for CO_2 transportation. The incorporation of graphene nanoparticles (GN) prevented pore blockage caused by low-viscosity polymer in the active layer. Swelling of the active layer and penetration into the membrane support is another phenomenon that could affect CO_2 transportation and lower the permeability due to pore clogging. It was reported that adding GNs into chitosan matrix could positively control the swelling affect. The temperature had a positive effect on CO_2/H_2 selectivity of chitosan/SF/GN membrane and increased it from 35 to 52 as temperature increased from 60°C to 90°C. However, a further temperature increase to 120°C led to the drop of CO_2/H_2 selectivity to ~37. Also, the increase in the feed absolute pressure from 2 to

5.9 bar resulted in a decrease in CO_2/H_2 selectivity from 52 to 24 due to the membrane compression and reduction in its ability to hold water. As authors reported, the CO_2/H_2 selectivity achieved by chitosan/SF/GN membrane was sufficient for hydrogen extraction for fuel cell applications. In another investigation, Okada et al. added cesium carbonate (Cs_2CO_3), as CO_2 carrier, to polyvinyl alcohol/polyacrylic acid (PVA/PAA) membrane to selectively remove CO_2 from CO shift reactor [124]. In hydrogen production process, hydrocarbons are first converted to carbon monoxide (CO) and hydrogen, then the CO reacts with H_2O steam through a reaction called CO shift reaction:

$$CO + H_2O \Leftrightarrow CO_2 + H_2$$

Employing a membrane being able to selectively remove CO_2 from CO shift reactor can shift the chemical equilibrium to the right side of the CO shift reaction, achieving higher H_2 conversion values. Since the performance of catalysts used in CO shift reaction deteriorates at lower temperatures, it is a requisite to have a membrane with capability of working and being selective at temperatures $\geq$100°C. Based on the experimental results, the Cs_2CO_3/PVA/PAA membrane made it possible to achieve CO_2 permeance and CO_2/H_2 selectivity of 60 GPU and 50 at 100°C, respectively. As Okada et al. discussed in their patent, by incorporation of rubidium carbonate (Rb_2CO_3) into PVA/PAA, it will be possible to obtain CO_2 permeance and CO_2/H_2 selectivity as high as 60 GPU and 90–100 at 100°C, respectively, due to higher solubility of Rb_2CO_3 in water which can facilitate the CO_2 transportation.

16.3.4 Natural gas treatment

As discussed earlier, natural gas is an important energy source composed of hydrocarbon gases, mainly methane (CH_4), and a variety of contaminants such as H_2S and CO_2 which need to be removed before transporting to the pipelines for distribution. In addition to the removal of contaminants and separation of methane as the main product, other valuable components such as ethane, butane, and propane are also separated as by-products in natural gas treatment processes. Various nanocomposite membranes based on biodegradable polymers have been prepared and studied for separating different gases in natural gas treatment. Cellulose acetate (CA) membranes have been commercially employed in gas treatment facilities for CO_2 separation. Kim et al. incorporated AMH-3, as a nanoporous layered silicate with crystallographic pore size of 3.4 Å, into CA membrane and studied the efficiency of the membrane for separation of CO_2 from CH_4 (Fig. 16.26A) [70]. A high-shear mixer with a rotating speed of 7000 rpm was used to effectively exfoliate the layered silicates in polymeric matrix. Compared to bare CA membrane, the CO_2 permeability increased by 54% to 11.6 barrer as the nanofiller loading percentage increased to 6 wt.%, while the CO_2/CH_4 selectivity remained at the same level (~30). Based on SAX and TEM analysis, the exfoliated flakes were stacked in 4–8 layers with the interlayer distance of 2 nm. The authors used the Cussler model to explain the permeation properties of the membrane because conventional gas permeability models could not describe the improvement in CO_2 permeability achieved by adding a small amount of nano flakes. The Cussler model is an extended version of Maxwell's model that calculates the flux of a species through a membrane containing particles with high aspect ratios. The comparison of the experimental results and the values predicted by the Cussler model, on the other hand, demonstrated that this model, too, cannot explain the permeation properties of the AMH-3/cellulose nanocomposite membrane. As authors proposed, the gas transportation through

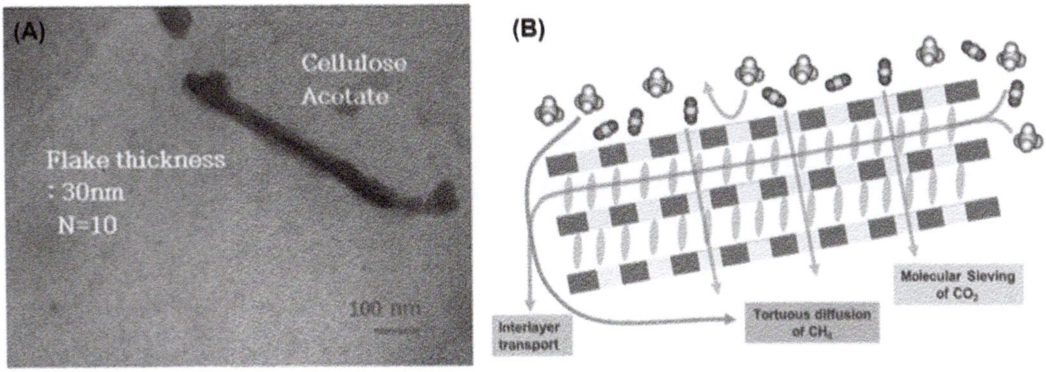

FIGURE 16.26

(A) SEM image of silicate flakes in CA matrix, (B) possible mechanisms of gas transportation through AMH-3/cellulose nanocomposite membrane [70].

Copyright 2021. Reproduced with permission from Elsevier.

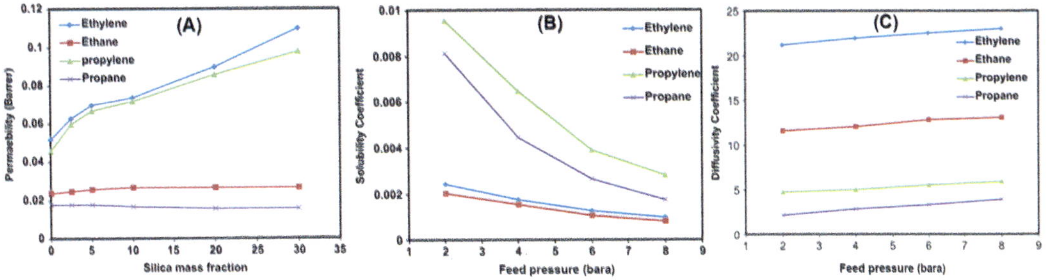

FIGURE 16.27

(A) Effect of silica content on the permeability of olefines and paraffines; Effect of feed pressure on (B) solubility and (C) diffusivity of olefines and paraffines [126].

Copyright 2021. Reproduced with permission from Elsevier.

the nanoflake/polymer nanocomposite membrane is governed by a complex mechanism depicted in Fig. 16.26B.

In another study, Farrukh et al. incorporated TiO_2 nanoparticles into CA and studied the CO_2/CH_4 selectivity of the membrane [125]. The CO_2 permeability was found to be increased by incorporation of TiO_2 nanoparticles while the CH_4 permeability of TiO_2/CA nanocomposite was lower than the bare CA membrane at all titania loading levels. This could result in an increase in CO_2/CH_4 selectivity to 23.3. As authors reported, the permeability of both CO_2 and CH_4 through bare CA membrane were close to each other and the small difference was due to the difference in the kinetic diameters of CH_4 (3.8 Å) and CO_2 (3.3 Å) molecules.

Naghsh et al., incorporated SiO_2 nanoparticle into CA matrix and fabricated a nanocomposite membrane for ethylene (C_2H_4)/ethane (C_2H_6) and propylene (C_3H_6)/propane (C_3H_8) separation [126]. The permeability of C_2H_4 and C_2H_6 increased from 0.052 and 0.046 barrer to 0.11 and 0.098

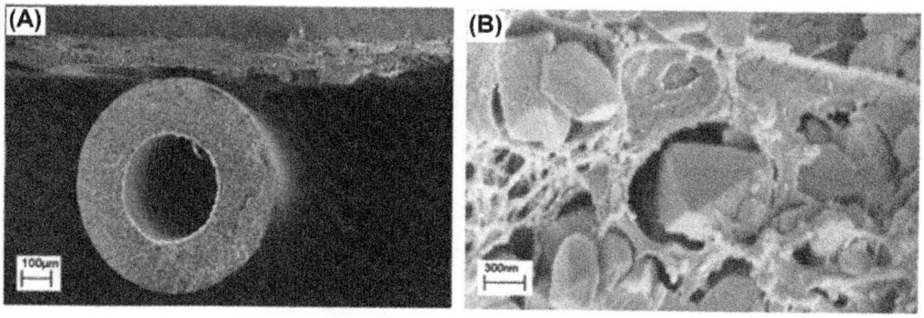

FIGURE 16.28

(A) SEM image of zeolite NaY/CA fiber showing no macro-voids, (B) sieve-in-a-cage morphology of NaY filler in CA matrix [127].

Copyright 2021. Reproduced with permission from American Chemical Society.

barrer, respectively, and the C_2H_4/C_2H_6 and C_3H_6/C_3H_8 selectivity increased from 2.16 and 2.55 to 4.07 and 6.12, respectively, as the SiO_2 concentration in CA increased from 0 to 30 wt.%. As shown in Fig. 16.26, the permeability of olefins was higher than that of paraffins, probably due to the smaller size of olefins molecules. As it is evident from Fig. 16.27, the permeability of olefins increases with the increase of SiO_2 content. The authors attributed this observation to changes in solubility and diffusion coefficients, as well as structural changes caused by silica nanoparticles.

Although a variety of membranes have been investigated for the removal of H_2S from natural gas streams, studies on the use of nanocomposite membranes based on biopolymers for the removal of acid gases are uncommon. Using cellulose acetate as a binder and zeolite NaY as a filler, Bhandari et al. created hollow-fiber membrane adsorbers. The hollow fiber was used for natural gas desulfurization [127]. The results indicated that the hollow fiber had a sieve-in-a-cage structure with a desirable permeation rate (Fig. 16.28). The hollow-fiber sorbent showed high H_2S removal efficiency and ability to be completely regenerated by application of heat or decrease in pressure.

16.4 Concluding remarks and future prospects

Membrane-based systems have grown in popularity for gas separation processes in recent years due to their appealing advantages such as operational simplicity, no phase change requirement, ease of scaling-up, and higher energy efficiency when compared to traditional techniques such as distillation and adsorption. Polymer nanocomposites offer a great opportunity to create materials with on-demand properties to address permeability-selectivity tradeoff issues in polymeric and inorganic membranes. In order to improve the overall performance of the membranes, intensive research is being conducted on the constituents of polymer nanocomposites, namely the polymeric matrix and nanofiller. The use of biodegradable polymers has grown in popularity as environmental concerns have grown. Biopolymers like cellulose have been widely used in the fabrication of polymer nanocomposite membranes for natural gas treatment, air separation, and the production of industrial gases. Biopolymers can separate polar and nonpolar gases due to the presence of functional groups in their structure. Although several studies on biodegradable polymer-based gas separation membranes have been conducted, the majority of those biodegradable polymers are natural polymers. To broaden their application for gas separation applications, more research on synthetic biodegradable polymers is required. Because the incorporation of inorganic nanofillers into biodegradable polymers may disrupt the biodegradability of the entire polymer nanocomposite, a life-cycle assessment appears necessary to ensure the biodegradability of these polymers and their respective nanocomposite.

References

[1] B.S.R.O.W. Energy, Statistical Review of World Energy, 2020.

[2] Global Gas Report 2020, Snam, International Gas Union, BloombergNEF, 2020.

[3] Global Natural Gas Markets Overview, Leidos, Inc., U.S. Energy Information Administration, Washington, DC, 2014.

[4] Industrial Gases Market Research Report: By Type, Delivery, End User – Global Industry Analysis and Demand Forecast to 2030, P&S Intelligence, 2020.

[5] H. Ritchie, M. Roser, CO_2 and greenhouse gas emissions, Our World Data (2020).

[6] H. Ritchie, Sector by sector: where do global greenhouse gas emissions come from? Ourworldindata.org (2020).

[7] Paris Agreement (Articles), United Nations, 2015.

[8] P. Erickson, et al., The Production Gap Report: 2020 Special Report, 2020.

[9] L.E. Cardwell, Analyses of natural gases, 1971, US Department of the Interior, 8554, Bureau of Mines, 1972.

[10] A.J. Kidnay, W. Parrish, Fundamentals of Natural Gas Processing, CRC, New York, 2006.

[11] B. Lv, et al., Mechanisms of CO_2 capture into monoethanolamine solution with different CO_2 loading during the absorption/desorption processes, Environ. Sci. Technol. 49 (17) (2015) 10728–10735.

[12] M. Al-Juaied, G.T. Rochelle, Absorption of CO_2 in aqueous diglycolamine, Ind. Eng. Chem. Res. 45 (8) (2006) 2473–2482.

[13] T. Prosochkina, A. Nikitina, E. Kantor, Hydrogen sulfide removal from hydrocarbon gas mixtures by diethanolamine (computer simulation), Pet. Chem. 56 (7) (2016) 616–622.

[14] U. Shoukat, D.D. Pinto, H.K. Knuutila, Study of various aqueous and non-aqueous amine blends for hydrogen sulfide removal from natural gas, Processes 7 (3) (2019) 160.

[15] K. Smith, N. Nicholas, G. Stevens, Inorganic salt solutions for post-combustion capture, Absorption-Based Post-Combustion Capture of Carbon Dioxide, Elsevier, 2016, pp. 145–166.

[16] B. Miller, 8-Greenhouse gas-carbon dioxide emissions reduction technologies, Foss. Fuel Emiss. Control. Technol. (2015) 367–438.

[17] A. Kohl, R. Nielsen, Physical solvents for acid gas removal, Gas Purification, Gulf Professional Publishing, Houston, 1997, pp. 1188–1234.

[18] W.L. Luyben, Control of an extractive distillation system for the separation of CO_2 and ethane in enhanced oil recovery processes, Ind. Eng. Chem. Res. 52 (31) (2013) 10780–10787.

[19] A.S. Holmes, J.M. Ryan, Cryogenic distillative separation of acid gases from methane, Google Patents, 1982.

[20] N.J. Kraakman, J. Rocha-Rios, M.C. van Loosdrecht, Review of mass transfer aspects for biological gas treatment, Appl. Microbiol. Biotechnol. 91 (4) (2011) 873–886.

[21] C. Cline, et al., Biological process for H∼ 2S removal from gas streams: the shell-paques/THIOPAQ™ gas desulfurization process, in: Proceedings of the Laurance Reid Gas Conditioning Conference, 2003.

[22] S.F. Interlenghi, J.L. de Medeiros, F.A. Ofélia de Queiroz, On small-scale liquefaction of natural gas with supersonic separator: energy and second law analyses, Energy Convers. Manage. 221 (2020) 113117.

[23] A. Dutta, I.A. Karimi, S. Farooq, Adjusting the heating value of LNG using a superstructure for hydrocarbon recovery, Computer Aided Chemical Engineering, Elsevier, 2018, pp. 1105–1110.

[24] J. Wilson, J. Yuvancic, Process selection for dehydrating Gulf of Mexico offshore platform gas, in: Proceedings of the Laurance Reid Gas Conditioning Conference, 2004.

[25] G. Agarwal, Z.P. McConkey, J. Hassard, Optimisation of vortex tubes and the potential for use in atmospheric separation, J. Phys. D: Appl. Phys. 54 (1) (2020) 015502.

[26] G. De Guido, F. Messinetti, E. Spatolisano, Cryogenic nitrogen rejection schemes: analysis of their tolerance to CO_2, Ind. Eng. Chem. Res. 58 (37) (2019) 17475–17488.

[27] F. Brandani, P. Pullumbi, C. Monereau, Nitrogen rejection from landfill gas using pressure swing adsorption, Adsorption 27 (4) (2021) 603–617.

[28] M. Netusil, P. Ditl, Natural gas dehydration, Natural Gas-Extraction to End Use, IntechOpen, 2012.

[29] Product Specifications, Engineering Data Book, Gas Processors Supply Association, Tulsa, OK, 2004.

[30] R. Jones, K. McIntush, C. Wallace, Oxygen removal in natural gas systems. Gas Processors Association Research Report RR-201, GPA Res. (073) (2010).

[31] E. Grynia, P.J. Griffin, Helium in natural gas-occurrence and production, J. Nat. Gas. Eng. 1 (2) (2016) 163–215.

[32] C.A. Scholes, U.K. Gosh, M.T. Ho, The economics of helium separation and purification by gas separation membranes, Ind. Eng. Chem. Res. 56 (17) (2017) 5014–5020.

[33] N.K. Das, et al., Purification of helium from natural gas by pressure swing adsorption, Curr. Sci. (2008) 1684–1687.

[34] D. Johnson, K. Rydjord, Unique NGL/NRU/HRU plant processes new gas in eastern Colorado, in: Proceedings of the Eightieth Annual Convention of the Gas Processors Association, Tulsa, OK, 2001.

[35] S. Chanyotha, et al., Determination of radon in natural gas pipelines, J. Radioanalytical Nucl. Chem. 307 (3) (2016) 2095–2099.

[36] A. Belloni, Industrial Gases Processing, John Wiley & Sons, 2008.

[37] N. Ghasem, CO_2 removal from natural gas, Advances in Carbon Capture, Elsevier, 2020, pp. 479–501.

[38] M.T. Ravanchi, S. Sahebdelfar, Carbon dioxide capture and utilization in petrochemical industry: potentials and challenges, Appl. Petrochem. Res. 4 (1) (2014) 63–77.

[39] J. Yu, et al., CO_2 capture and separations using MOFs: computational and experimental studies, Chem. Rev. 117 (14) (2017) 9674–9754.

[40] Q. Qian, et al., MOF-based membranes for gas separations, Chem. Rev. 120 (16) (2020) 8161–8266.

[41] Z. Chang, et al., Microporous organic polymers for gas storage and separation applications, Phys. Chem. Chem. Phys. 15 (15) (2013) 5430–5442.

[42] R.L. Siegelman, E.J. Kim, J.R. Long, Porous materials for carbon dioxide separations, Nat. Mater. 20 (8) (2021) 1060–1072.

[43] W. Xing, et al., Superior CO_2 uptake of N-doped activated carbon through hydrogen-bonding interaction, Energy Environ. Sci. 5 (6) (2012) 7323–7327.

[44] T.U. Rashid, Ionic liquids: innovative fluids for sustainable gas separation from industrial waste stream, J. Mol. Liq. (2020) 114916.

[45] M. Khodakarami, L. Alagha, Separation and recovery of rare earth elements using novel ammonium-based task-specific ionic liquids with bidentate and tridentate O-donor functional groups, Sep. Purif. Technol. 232 (2020) 115952.

[46] Y. Alqaheem, et al., Polymeric gas-separation membranes for petroleum refining, Int. J. Polym. Sci. 2017 (2017).

[47] F. Lastari, Ryan-Holmes and Modified Ryan-Holmes Processes for LNG Production, Curtin University, 2009.

[48] R.S. Murali, T. Sankarshana, S. Sridhar, Air separation by polymer-based membrane technology, Sep. Purif. Rev. 42 (2) (2013) 130–186.

[49] L.M. Robeson, Polymer membranes, in: M.M. Krzysztof Matyjaszewski (Ed.), Polymer Science: A Comprehensive Reference, Elsevier, 2012, pp. 325–347.

[50] T. De Baerdemaeker, D. De Vos, Trapdoors in zeolites, Nat. Chem. 5 (2) (2013) 89–90.

[51] N. Kosinov, et al., Recent developments in zeolite membranes for gas separation, J. Membr. Sci. 499 (2016) 65–79.

[52] T. Peters, et al., Hollow fibre microporous silica membranes for gas separation and pervaporation: synthesis, performance and stability, J. Membr. Sci. 248 (1–2) (2005) 73–80.

[53] H. Kato, et al., Gas separation silica membranes prepared by chemical vapor deposition of methyl-substituted silanes, Membranes 9 (11) (2019) 144.

[54] L. Lei, et al., Carbon membranes for CO_2 removal: status and perspectives from materials to processes, Chem. Eng. J. (2020) 126084.

[55] N. Sazali, A review of the application of carbon-based membranes to hydrogen separation, J. Mater. Sci. 55 (2020) 11052–11070.

[56] S.K. Alen, S. Nam, S.A. Dastgheib, Recent advances in graphene oxide membranes for gas separation applications, Int. J. Mol. Sci. 20 (22) (2019) 5609.

[57] M. Shah, et al., Current status of metal–organic framework membranes for gas separations: promises and challenges, Ind. Eng. Chem. Res. 51 (5) (2012) 2179–2199.

[58] Y. Yampolskii, Polymeric gas separation membranes, Macromolecules 45 (8) (2012) 3298–3311.

[59] L.M. Robeson, Polymer membranes for gas separation, Curr. Opin. Solid State Mater. Sci. 4 (6) (1999) 549–552.

[60] D.R. Paul, Polymeric Gas Separation Membranes, CRC press, 2018.

[61] Y. Wang, et al., Polymers of intrinsic microporosity for energy-intensive membrane-based gas separations, Mater. Today Nano 3 (2018) 69–95.

[62] M. Khodakarami, M. Bagheri, Recent advances in synthesis and application of polymer nanocomposites for water and wastewater treatment, J. Clean. Prod. (2021) 126404.

[63] L.M. Robeson, The upper bound revisited, J. Membr. Sci. 320 (1–2) (2008) 390–400.

[64] H.D. Ying Kong, Y. Jinrong, S. Deqing, W. Yunfang, Z. Yuanyuan, X. Wei, Study on polyimide/TiO_2 nanocomposite membranes for gas separation, Desalination 146 (1–3) (2002) 49–55.

[65] S.K. Yanbin Cui, R.K. Balakantha, H. Daniel van, Gas barrier properties of polymer/clay nanocomposites, RSC Adv. 5 (2015) 63669–63690.

[66] S. Singh, et al., Graphene-based membranes for carbon dioxide separation, J. CO_2 Utilization 49 (2021) 101544.

[67] C. Gao, et al., The effect of nanoparticles on gas permeability with polyimide membranes and network hybrid membranes: a review, Rev. Inorg. Chem. 41 (1) (2021) 1–20.

[68] D. Zhongde, O. Vegar, D. Jing, R.M. Lilleby Helberg, D. Liyuan, A brief review of nanocellulose based hybrid membranes for CO_2 separation, Fibers 7 (2019) 40.

[69] A.K.N. Bernhard Rieger, G.W. Coates, R. Reichardt, E. Dinjus, T.A. Zevaco, Synthetic Biodegradable Polymers, Verlag Berlin Heidelberg Springer, 2012.

[70] W.-g Kim, et al., Nanoporous layered silicate AMH-3/cellulose acetate nanocomposite membranes for gas separations, J. Membr. Sci. 441 (2013) 129–136.

[71] S. Ted Oyama, M. Yamada, T. Sugawara, A. Takagaki, R. Kikuchi, Review on mechanisms of gas permeation through inorganic membranes, J. Jpn. Pet. Inst. 54 (5) (2011) 298–309.

[72] E. Nagy, Mass transport through a membrane layer, in: E. Nagy (Ed.), Basic Equations of Mass Transport Through a Membrane Layer, second ed., Elsevier, Massachusetts, 2019, pp. 21–68.

[73] M.B. Karimi, S. Hassanajili, G. Khanbabaei, Capillary condensation mechanism for gas transport in fiber reinforced poly (ether-b-amide) membranes, Chem. Eng. Res. Des. 148 (2019) 180–190.
[74] S. Zhao, et al., Design and fabrication of highly selective and permeable polymer membranes, J. Appl. Phys. 128 (13) (2020) 131102.
[75] G. Choudalakis, A. Gotsis, Free volume and mass transport in polymer nanocomposites, Curr. Opin. Colloid Interface Sci. 17 (3) (2012) 132–140.
[76] R.W. Baker, Membranes for vapor/gas separation, Membr. Technol. Res. Inc. (2006) 25.
[77] C.M. Zimmerman, W.J. Koros, Polypyrrolones for membrane gas separations. I. Structural comparison of gas transport and sorption properties, J. Polym. Sci. Part B: Polym. Phys. 37 (12) (1999) 1235–1249.
[78] R. Bouma, et al., Permeation through a heterogeneous membrane: the effect of the dispersed phase, J. Membr. Sci. 128 (2) (1997) 141–149.
[79] B.V.D. Bruggen, Membrane technology, Kirk-Othmer Encyclopedia of Chemical Technology, John Wiley & Sons, 2017.
[80] A. Nasir, et al., Progress in polymeric nanocomposite membranes for wastewater treatment: preparation, properties and applications, J. Ind. Eng. Chem. 79 (2019) 29–40.
[81] L.-C. Tang, et al., Mechanical properties of rubber nanocomposites containing carbon nanofillers, Carbon-Based Nanofillers and Their Rubber Nanocomposites, Elsevier, 2019, pp. 367–423.
[82] H.-J.S. Dazhi Sun, Multifunctional polymer/ZnO nanocomposites: controlled dispersion and physical properties, Multifunctionality of Polymer Composites, Elsevier Inc, 2015, pp. 858–874.
[83] M.G. Süer, N. BaC, L. Yilmaz, Gas permeation characteristics of polymer-zeolite mixed matrix membranes, J. Membr. Sci. 91 (1–2) (1994) 77–86.
[84] S. Simon, E. Espuche, Effect of different metal in situ growing routes on the morphology and gas separation properties of polyetherimide/palladium nanocomposite asymmetric membranes, Sep. Purif. Technol. 129 (2014) 41–49.
[85] M. Sadeghi, M.A. Semsarzadeh, H. Moadel, Enhancement of the gas separation properties of polybenzimidazole (PBI) membrane by incorporation of silica nano particles, J. Membr. Sci. 331 (1–2) (2009) 21–30.
[86] S. Fu, et al., Some basic aspects of polymer nanocomposites: a critical review, Nano Mater. Sci. 1 (1) (2019) 2–30.
[87] A.T. Smith, et al., Synthesis, properties, and applications of graphene oxide/reduced graphene oxide and their nanocomposites, Nano Mater. Sci. 1 (1) (2019) 31–47.
[88] S. Mohsenpour, et al., Comparison of the membrane morphology based on the phase diagram using PVP as an organic additive and TiO_2 as an inorganic additive, Polymer 97 (2016) 559–568.
[89] M.S. Islam, R. Masoodi, H. Rostami, The effect of nanoparticles percentage on mechanical behavior of silica-epoxy nanocomposites, J. Nanosci. 2013 (2013).
[90] B. Zornoza, et al., Metal organic framework based mixed matrix membranes: an increasingly important field of research with a large application potential, Microporous Mesoporous Mater. 166 (2013) 67–78.
[91] A.F.I. Pei Sean Goh, Nanocomposite membrane fabrication, in: A.F.I. Pei Sean Goh (Ed.), Nanocomposite Membranes for Gas Separation, Micro and Nano Technologies, Elsevier, 2020, pp. 125–162.
[92] A. Rapacz-Kmita, et al., Thermal, structural and mechanical analysis of polymer/clay nanocomposites with controlled degradation, J. Therm. Anal. Calorim. 127 (1) (2017) 389–398.
[93] C.E. Corcione, M. Frigione, Characterization of nanocomposites by thermal analysis, Materials 5 (12) (2012) 2960–2980.
[94] B. Yang, J. Zhang, Effect of nano-fillers on the thermal performance of pCBT resin, Procedia Eng. 102 (2015) 773–778.
[95] G. George, et al., Effect of nanofillers on the crystalline and mechanical properties of EVACO polymer nanocomposites, Materials Today: Proceedings (2021).

[96] H. Ahmadizadegan, et al., Polyimide-TiO_2 nanocomposites and their corresponding membranes: synthesis, characterization, and gas separation applications, Solid State Sci. 89 (2019) 25–36.

[97] Z. Amini, M. Asghari, Preparation and characterization of ultra-thin poly ether block amide/nanoclay nanocomposite membrane for gas separation, Appl. Clay Sci. 166 (2018) 230–241.

[98] A. Cooper, et al., Chitosan-based nanofibrous membranes for antibacterial filter applications, Carbohydr. Polym. 92 (1) (2013) 254–259.

[99] M. Khodakarami, L. Alagha, D.J. Burnett, Probing surface characteristics of rare earth minerals using contact angle measurements, atomic force microscopy, and inverse gas chromatography, ACS Omega 4 (8) (2019) 13319–13329.

[100] H. Wagner, Raman spectroscopy of polymer–carbon nanotube composites, Polymer–Carbon Nanotube Composites, Elsevier, 2011, pp. 400–427.

[101] A. Mane, V. Patil, X-ray photoelectron spectroscopy of nanofillers and their polymer nanocomposites, Spectroscopy of Polymer Nanocomposites, Elsevier, 2016, pp. 452–467.

[102] F. Zhang, J. Dou, H. Zhang, Mixed membranes comprising carboxymethyl cellulose (as capping agent and gas barrier matrix) and nanoporous ZIF-L nanosheets for gas separation applications, Polymers 10 (12) (2018) 1340.

[103] M. Najafi, et al., Gas permeation properties of cellulose acetate/silica nanocomposite membrane, Adv. Polym. Technol. 37 (6) (2018) 2043–2052.

[104] L. Ansaloni, et al., Nanocellulose-based membranes for CO_2 capture, J. Membr. Sci. 522 (2017) 216–225.

[105] J.Ø. Torstensen, et al., PVA/nanocellulose nanocomposite membranes for CO_2 separation from flue gas, Int. J. Greenh. Gas. Control. 81 (2019) 93–102.

[106] M.R. Dilshad, et al., Novel PVA/PEG nano-composite membranes tethered with surface engineered multi-walled carbon nanotubes for carbon dioxide separation, Microporous Mesoporous Mater. 308 (2020) 110545.

[107] R. Borgohain, et al., Carboxymethyl chitosan/carbon nanotubes mixed matrix membranes for CO_2 separation, Reactive Funct. Polym. 143 (2019) 104331.

[108] S.A.S.C. Samarasinghe, et al., Enhanced O_2/N_2 separation of mixed-matrix membrane filled with pluronic-compatibilized cobalt phthalocyanine particles, Membranes 10 (4) (2020) 75.

[109] J. Han, et al., Highly selective oxygen/nitrogen separation membrane engineered using a porphyrin-based oxygen carrier, Membranes 9 (9) (2019) 115.

[110] E.O. Ogbole, et al., Influence of surface-treated SiO_2 on the transport behavior of O_2 and N_2 through polydimethylsiloxane nanocomposite membrane, Sep. Purif. Technol. 175 (2017) 358–364.

[111] J. Ahmad, K. Deshmukh, M.B. Hägg, Influence of TiO_2 on the chemical, mechanical, and gas separation properties of polyvinyl alcohol-titanium dioxide (PVA-TiO_2) nanocomposite membranes, Int. J. Polym. Anal. Charact. 18 (4) (2013) 287–296.

[112] B. Ghalei, et al., Effect of polyvinyl alcohol modified silica particles on the physical and gas separation properties of the polyurethane mixed matrix membranes, Silicon 11 (3) (2019) 1451–1460.

[113] H.C. Koh, et al., Preparation and gas permeation properties of biodegradable polymer/layered silicate nanocomposite membranes, Desalination 233 (1–3) (2008) 201–209.

[114] F. Li, C. Zhang, Y. Weng, Improvement of the gas barrier properties of PLA/OMMT films by regulating the interlayer spacing of OMMT and the crystallinity of PLA, ACS Omega 5 (30) (2020) 18675–18684.

[115] M. Rahimpour, et al., Palladium membranes applications in reaction systems for hydrogen separation and purification: a review, Chem. Eng. Processing: Process. Intensif. 121 (2017) 24–49.

[116] B. Ghalei, et al., Rational tuning of zirconium metal–organic framework membranes for hydrogen purification, Angew. Chem. Int. Ed. 58 (52) (2019) 19034–19040.

[117] H. Verweij, Y. Lin, J. Dong, Microporous silica and zeolite membranes for hydrogen purification, MRS Bull. 31 (10) (2006) 756–764.

[118] X. He, Techno-economic feasibility analysis on carbon membranes for hydrogen purification, Sep. Purif. Technol. 186 (2017) 117–124.

[119] S.J. Khatib, S.T. Oyama, Silica membranes for hydrogen separation prepared by chemical vapor deposition (CVD), Sep. Purif. Technol. 111 (2013) 20–42.

[120] I.N. Mohamad, et al., Permeation properties of polymeric membranes for biohydrogen purification, Int. J. Hydrog. Energy 41 (7) (2016) 4474–4488.

[121] L. Rodríguez-Jardón, et al., Effect of porous organic polymers in gas separation properties of polycarbonate based mixed matrix membranes, J. Membr. Sci. 619 (2021) 118795.

[122] L.A. El-Azzami, E.A. Grulke, Carbon dioxide separation from hydrogen and nitrogen: facilitated transport in arginine salt–chitosan membranes, J. Membr. Sci. 328 (1–2) (2009) 15–22.

[123] B. Prasad, B. Mandal, Graphene-incorporated biopolymeric mixed-matrix membrane for enhanced CO_2 separation by regulating the support pore filling, ACS Appl. Mater. Interfaces 10 (33) (2018) 27810–27820.

[124] O. Osamu, T. Massaki, Y. Reza, M. Hideto, S. Keiko, M. Kaori, CO_2-facilitated transport membrane and method for producing the same, U.S. Patent, Editor, USA, 2013.

[125] S. Farrukh, et al., Blending of TiO_2 nanoparticles with cellulose acetate polymer: to study the effect on morphology and gas permeation of blended membranes, Asia-Pacific J. Chem. Eng. 9 (4) (2014) 543–551.

[126] M. Naghsh, et al., Separation of ethylene/ethane and propylene/propane by cellulose acetate–silica nanocomposite membranes, J. Membr. Sci. 423 (2012) 97–106.

[127] D.A. Bhandari, N. Bessho, W.J. Koros, Hollow-fiber sorbents for desulfurization of natural gas, Ind. Eng. Chem. Res. 49 (23) (2010) 12038–12050.

CHAPTER

17

Advances and prospects of biodegradable polymer nanocomposites for fuel cell applications

Mohammad Mizanur Rahman Khan[1], Md. Khairul Amin[2,3] and Nilave Chakraborty[4]

[1]Department of Chemistry, Shahjalal University of Science and Technology, Sylhet, Bangladesh [2]Chemistry Discipline, Khulna University, Khulna, Bangladesh [3]School of Chemistry, University of Edinburgh, Edinburgh, United Kingdom [4]Department of Chemistry & Biochemistry, Auburn University, AL, United States

Chapter Outline

17.1 Introduction

Due to rapid urbanization and population growth, global energy consumption is expected to rise by 56% by 2040. Furthermore, the asymmetric geographical distribution of fuel, ecological aspects, health, and the economy have compelled researchers to pursue renewable energy sources. Numerous energy technologies have been developed in response to these issues, with the fuel cell being recognized as a green, efficient, sustainable, and cost-effective energy source [1–3]. Polymer electrolyte membrane fuel cell or proton exchange membrane fuel cell (PEMFC) is one of the most commonly used fuel cells. Because polymeric membranes are the "nucleus" of PEMFC technology,

Biodegradable and Biocompatible Polymer Nanocomposites. DOI: https://doi.org/10.1016/B978-0-323-91696-7.00018-0
Copyright © 2023 Elsevier Inc. All rights reserved.

researchers are constantly working to develop new materials that are suitable for this technology. Nafion, Acipex, and Flemion, which are based on perfluorosulfonic acid, are currently the most widely used commercial membranes for PEMFC. These membranes have high stability and excellent proton conductivity. However, their diverse applications are limited by their high cost, production complexity, high methanol permeability, and environmental incompatibility [1,4]. To address these shortcomings, considerable effort has been expended in developing alternative cost-effective biodegradable polymer composites for PEMFC technology. As a result, this chapter focuses on the different types, syntheses, properties, and future prospects of biodegradable polymer composites for fuel cell applications.

17.2 Fuel cell

Fuel cell is an electrochemical cell which transforms chemical energy of fuel (H_2, CH_4, CH_3OH etc.) and oxidant (usually O_2) into electric energy coupled with heat and water as by-products [2]. Unlike batteries, it provides continuous electricity until inputs (fuel and oxidant) are provided. These cells are using in various sectors, including portable power systems (e.g., mobile, laptop, detector, camera), stationary applications (e.g., domestic, official, and industrial), and transportations (e.g., car, aircraft, forklift, motorcycle, train, and submarine) [5–7]. The numerous applications of fuel cells are presented in Fig. 17.1. The COVID-19 pandemic is a devastating blow to the global economy; however, the global market for fuel cells increased by 14% during the pandemic, reaching $2.62 billion in 2020. The market

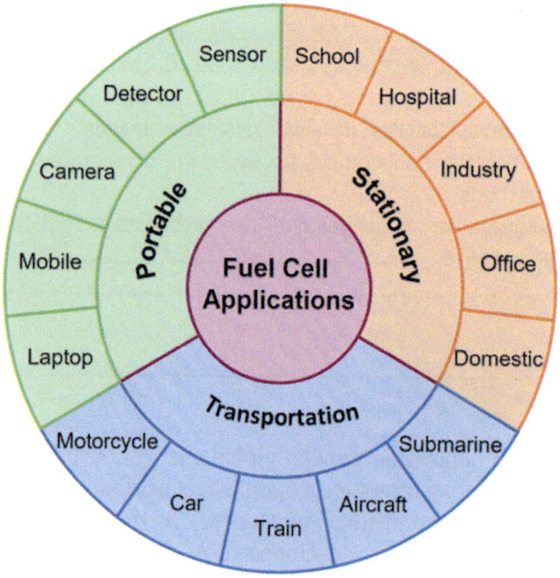

FIGURE 17.1

The different applications of fuel cells.

will grow from \$3.36 billion in 2021 to \$28.95 billion in 2028 [8]. In recent years, the stationary fuel cell market has seen an increase in technology transfer. This technology transfer reached 395,000 units in 2014, and it is expected to reach 575,000 units by 2018 [9].

17.2.1 History of fuel cell

Fuel cell concerned the researchers for about 180 years (Fig. 17.2). For the first time, the concept of fuel cell was successfully described by Sir Humphry Davy in 1801. This was followed by a revolutionary work published in 1838 at "The Philosophical Magazine," describing the electrochemical reaction between H_2 and O_2, by German-Swiss chemist Christian Friedrich Schönbein [10]. These earlier advancements motivated Sir William Robert Grove, a British scientist and lawyer, to construct the history's first fuel cell in 1842, consisted of platinum electrodes placed in two glass vessels containing H_2 and O_2, respectively, both electrodes submerged in dilute sulfuric acid electrolyte. He referred to it as a

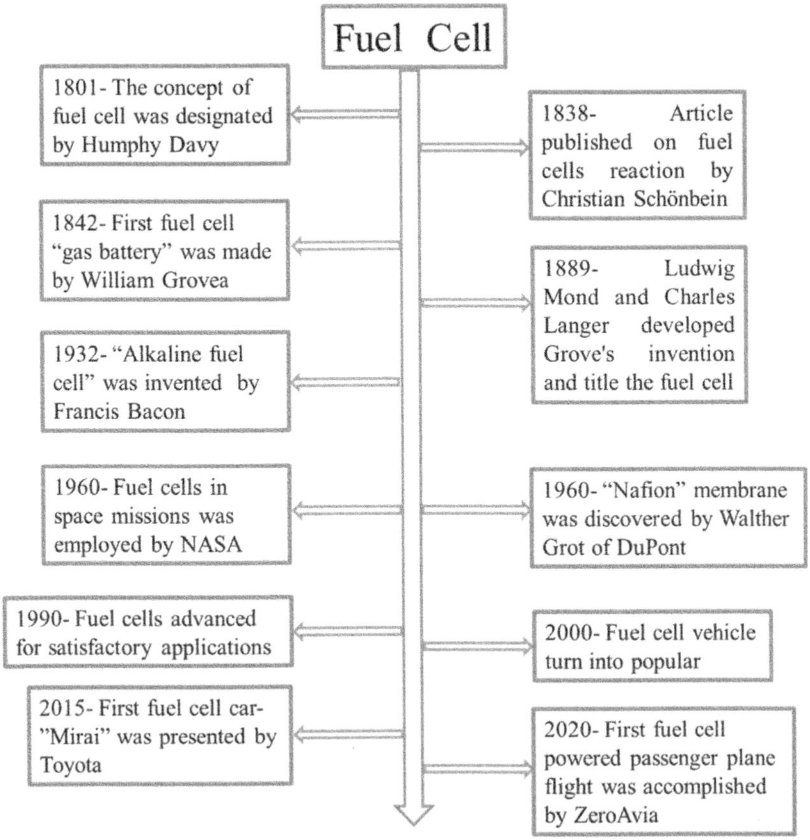

FIGURE 17.2

Fuel cell history [10,11].

"gas battery," also known as a "Grove battery." He is known as the "Father of the Fuel Cell" for his contributions. Grove's cell produced unfeasible results due to electrode corrosion and low material stability [9,12]. Despite previous accomplishments, the term "fuel cell" was first used in 1889 by chemist Ludwig Mond and his assistant Charles Langer in reference to their air-coal gas-based fuel cell construction [13]. Due to the development of the heat engine, further technological progress was obscured for more than 50 years [10]. In 1932, British Professor Francis T Bacon modified Mond and Langer's concept to create the first fully operational alkaline fuel cell [14,15]. Here, instead of expensive platinum electrodes and corrosive sulfuric acid, he introduced cheap nickel electrodes, less corrosive potassium hydroxide electrolyte, and fueled by H_2 and O_2 [13]. It took almost 30 years until Bacon exhibited a 5×10^3 W fuel cell capable of running a welding device in 1959 [5]. The same year, W. T. Grubb and L. Niedrach from General Electric (GE) company invented first PEMFC following Bacon's protocol [10]. NASA employed this PEMFCs fuel cells in Gemini V mission and were still used in the 90's space missions [12]. W. Grot of DuPont discovered "Nafion"—a sulfonated tetrafluoroethylene-based membrane that is still widely used in fuel cell technology—at the end of the 1960s. Researchers concentrated on stationary applications of fuel cell technology in the 1990s [10]. The 2000s experienced an increasing concern on energy security, energy efficiency, and CO_2 emission by a group of governments, companies, and consumers. Consequently, fuel cell vehicles are still promising because of their superior efficiency, zero-emissions, and easy refueling [14]. In August 2011, there were around one hundred operational buses worldwide. Majority of them were produced by Toyota, UTC Power, Proton Motor, Hydrogenics, and Ballard [16]. In 2015, world's first fuel cell car named Mirai (signifies "the future") launched by Toyota. It initiated a new era in automobile technology [10]. ZeroAvia successfully completed the world's first passenger plane flight powered by hydrogen fuel cell in September 2020 [11]. Fuel cell research has recently expanded dramatically, covering a wide range of applications in portable, transportation, and stationary fields [9].

17.2.2 Types of fuel cell

Based on the electrolyte used, fuel cells can be categorized into (1) alkaline fuel cells (AFCs), (2) PEMFCs, (3) direct methanol fuel cells (DMFCs), (4) microbial fuel cells (MFCs), (5) phosphoric acid fuel cells (PAFCs), (6) molten carbonate fuel cells (MCFCs), (7) solid oxide fuel cells (SOFCs), (8) enzymatic fuel cells (EFCs) etc. [2,17]. Furthermore, it can be categorized into two classes based on operational temperature: (1) low temperature cells—which operate below 250°C, and (2) high temperature cells—which operate above 600°C [2,12]. The electrical efficiency of these fuel cells are ranging from 60% to 65% except the PAFCs and MCFCs where the efficiency are 40% and 40%–50%, respectively. Table 17.1 represents the comparison of basic parameters of various fuel cells [2].

PEMFCs are the most effective and encouraging renewable energy source, receiving increasing global attention because they can reduce energy consumption, contaminant emissions, and reliance on fossil fuels. The PEMFC transfers ions selectively by using a solid polymer electrolyte. As a result, it aids in the elimination of the need for liquid electrolytes and toxic acid-base solutions [10]. Fig. 17.3 shows total number of papers on PEMFCs over the last 60 years. It is evident that the research on PEMFCs has tremendously increased in the last 15 years.

Table 17.1 Different fuel cells, characteristic electrolytes, fuel, and functioning temperature are listed.

Fuel cells	Characteristic electrolytes	Fuel	Functioning temperature (°C)
AFCs	Aqueous KOH solution	H_2	50–200
PEMFCs	Polymer membrane	H_2	50–80
DMFCs	Polymer membrane	CH_3OH/H_2O	90–120
MFCs	Ion-exchange membrane	Wastewater, $C_6H_{12}O_6$, $C_2H_3O^{2-}$	20–60
PAFCs	H_3PO_4 in SiC	H_2	160–220
MCFCs	Solution of (Li_2CO_3, Na_2CO_3, K_2CO_3) in $LiAlO_2$	CH_4	600–700
SOFCs	Yttria-stabilized zirconia (YSZ)	H_2	800–1000

Modified from K. Pourzare, Y. Mansourpanah, S. Farhadi, Advanced nanocomposite membranes for fuel cell applications: a comprehensive review, Biofuel Res. J. 3 (2016) 496–513.

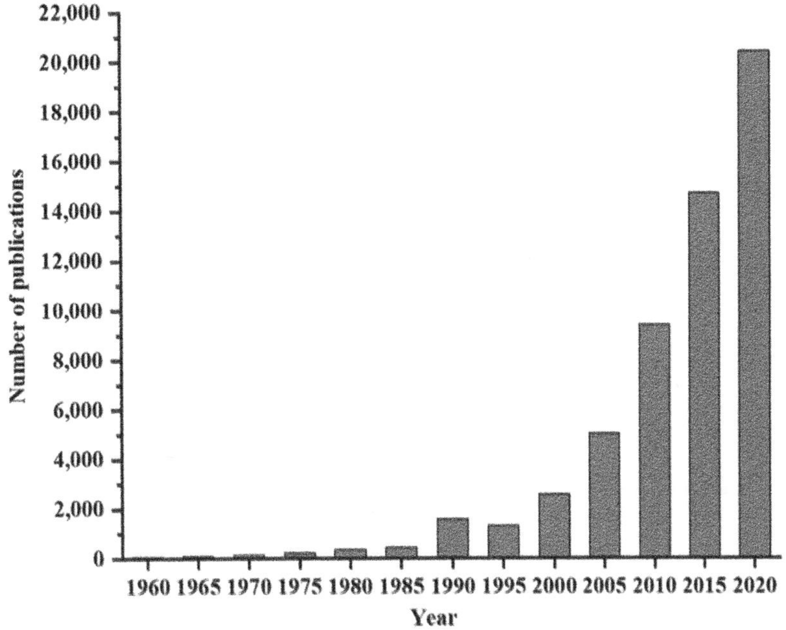

FIGURE 17.3

Number of publications in the period of 1960–2020 based on Google Scholar with the key words "polymer electrolyte membrane fuel cell"; (data acquired on 13 July 2021).

17.2.3 Fuel cell mechanism

A fuel cell is made up of two electrodes (anode and cathode) that are filled with a polymer electrolyte. A fuel cell typically uses hydrogen, methane, methanol, ethanol, and other fuels, as well as an oxygen oxidant. In a hydrogen fuel cell (Fig. 17.4), H_2 and O_2 are provided constantly to the anode and cathode, respectively. In anode, H_2 interacts with the catalyst present in electrolyte membrane's surface, generating protons (H^+) and electrons (e^-). These protons travel through the polymer electrolyte, while electrons travel through the outer circuit to the cathode and generate electricity. Water and heat are produced at the cathode by combining protons, oxygen, and electrons [10,17,18]. The PEMFCs use polymer membrane as an electrolyte (Table 17.1). The electrolyte membrane has two key functions: (1) it isolates the H_2 and O_2 from direct contact, and (2) it allows only H^+ for anode-cathode transportation. Hence, polymer electrolytes must have "H^+ conduction" properties [17]. Thus, very few membranes are suitable for this application. The essential properties of a suitable electrolyte membrane are: (1) excellent H^+ conductivity, (2) lower e^- conductivity, (3) lower fuel and oxidant permeability, (4) high chemical, electrochemical, mechanical and thermal stability, (5) appreciable oxidative and hydrolytic stability, (6) suitable H_2O transport, (7) satisfactory morphological and dimensional stability, (8) long-lasting, and (9) low cost [2,4,17,18].

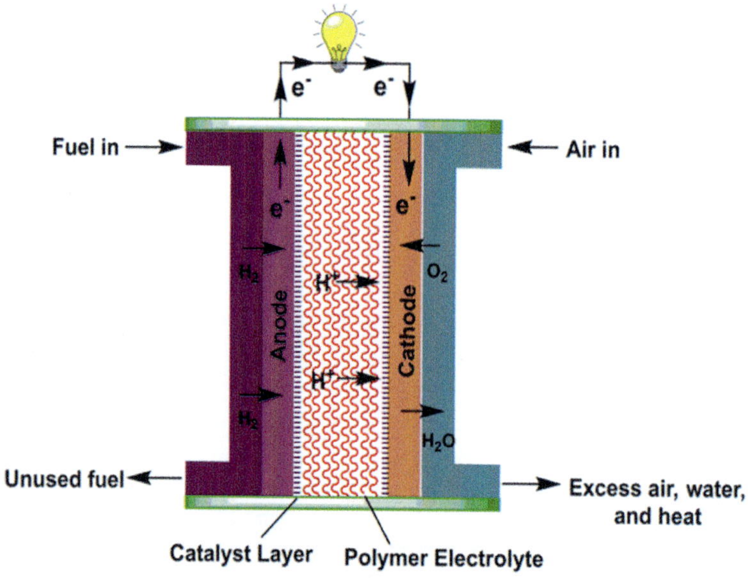

Anode: $H_2 \rightarrow 2H^+ + 2e^-$

Cathode: $1/2\ O_2 + 2H^+ + 2e^- \rightarrow H_2O + \text{Heat}$

FIGURE 17.4

Schematic depiction of a hydrogen fuel cell.

17.2.4 Assembly of fuel cell

The heart of a single PEMFC is membrane-electrode assembly (MEA). It contains polymer electrolyte membrane, electro-catalyst layers, and gas diffusion layers (GDLs), crammed between bipolar plates [19]. Usually, a single cell has the output of around 0.7 V, which is very low for practical applications. Thus, several single cells are coupled in series to build a fuel cell stack for enhanced output (Figs. 17.5 and 17.6). The power of a PEMFC stack governs by total number of connected

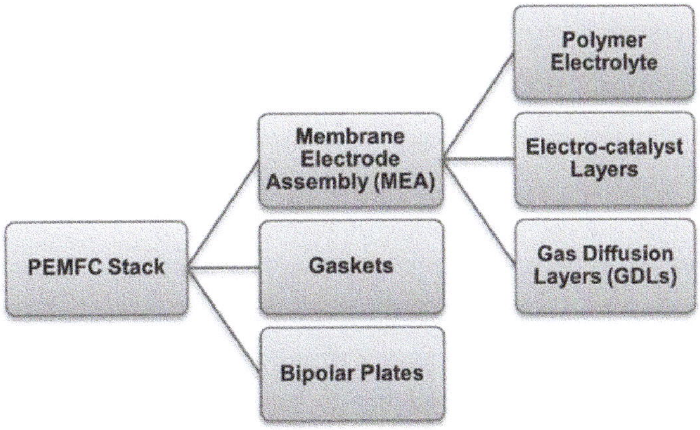

FIGURE 17.5

Key components of a PEMFC stack.

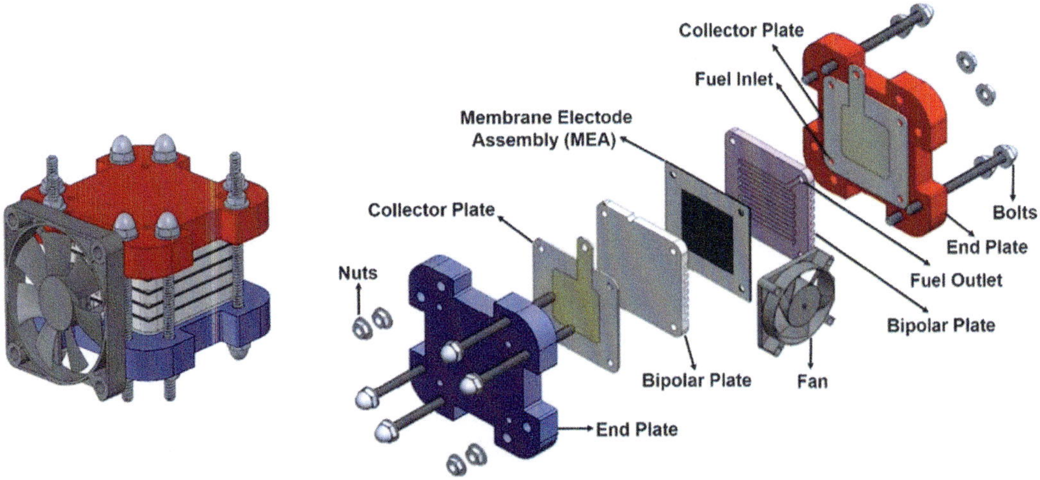

FIGURE 17.6

A five-stack fuel cell and its simplified exploded view [20].

single cells [21]. Because of its multiple functions, such as receiving output and managing oxidant, fuel, temperature, water, humidity, structure, and so on, a stack is much more complex than a single cell. To meet these requirements, the stack includes, in addition to MEA, output collectors, end plates, coolants, gaskets, bipolar plates, and bolts [22].

17.2.5 Fuel cell efficiency

Fuel cell efficiency is defined as the ratio of produced electricity to consumed hydrogen. The units for the above parameters are W or kW [23,24].

$$\eta = \frac{W_{el}}{W_{H2}} \tag{17.1}$$

The generated electricity can be expressed as a product of voltage (V) and current (I):

$$W_{el} = I \times V \tag{17.2}$$

The units of current and cell potential are ampere (A) and volt (V), respectively. Based on Faraday's law, the H_2 consumed is directly proportional to the current:

$$N_{H2} = \frac{I}{nF} \tag{17.3}$$

here the unit of N_{H2} is mol/s, and

$$W_{H2} = \Delta H \frac{I}{nF} \tag{17.4}$$

where W_{H2} denotes the energy (J/s) of hydrogen consumed and ΔH is the higher heating value of hydrogen (286.0 kJ/mol).

It is mentioned that the dimension of $\Delta H/nF$ is volt (V), and for $\Delta H = 286.0$ kJ/mol, the obtained value 1.482 V is addressed as thermo-neutral potential.

By adding two Eqs. (17.1) and (17.4), the efficiency (η) is directly proportional to cell potential (V):

$$\eta = \frac{V}{1.482} \tag{17.5}$$

The η can also be stated considering hydrogen's lower heating value $\Delta H_{LHV} = 241.0$ kJ/mol. The difference between ΔH_{LHV} and ΔH_{HHV} is the heat of water condensation. From the fuel cell, the produced water can escape either as liquid or vapor. Thus, it is important to specify which type (ΔH_{LHV} or ΔH_{HHV}) is being used for η measurement. Considering ΔH_{LHV}, the efficiency is expressed as follows:

$$\eta_{LHV} = \frac{V}{1.254} \tag{17.6}$$

If few hydrogens are lost due to internal current or hydrogen diffusion through the membrane, then the consumption of hydrogen will be greater compared to produced current (according to Eq. 17.3). Therefore, the efficiency ought to be fairly lesser than assumed by the Eq. (17.5). Usually, this hydrogen loss is very small, and thus it alters the efficiency only at very lower current density. Thus, the efficiency is now a product of voltage and current efficiency [23–25]:

$$\eta = \frac{V}{1.482} \frac{i}{(i + i_{loss})} \tag{17.7}$$

If the supplied H_2 exceeds the amount required to maintain stoichiometry, the excess will escape from the fuel and can be reused by recirculating if it remains pure. But, if it is impure, the excess hydrogen escape fuel cell without participating in electrochemical reaction. Then, the efficiency is:

$$\eta = \frac{V}{1.482}\eta_{fu} \tag{17.8}$$

here η_{fu} indicates fuel utilization ($= 1/S_{H2}$), where S_{H2} is called H_2 stoichiometric ratio—a ratio of the actually provided and consumed hydrogen's amount:

$$S_{H2} = \frac{N_{H2,act}}{N_{H2,theor}} = \frac{nF}{I}N_{H2,act} \tag{17.9}$$

A well-designed fuel cell operating with reformate may achieve 83%–85% fuel utilization, whereas it exceeds 90% when operating with pure hydrogen [23,24].

17.2.6 Properties of fuel cell

17.2.6.1 Mechanical properties

The electrolyte membranes must be mechanically stable both during fuel cell assembly and operation. An ideal membrane has high mechanical stability to stretch and shear in both dry and wet conditions [26]. Dynamic mechanical analysis (DMA) can be used to determine the mechanical stability of a membrane. Membranes in DMA are subjected to a periodic mechanical strain or stress in relation to temperature. The storage modulus E′ (elastic) and loss modulus E″ (viscous) of sample are calculated as a function of temperature, T. The E″/E′ ratio, also called tan δ, is linked to mechanical damping [2]. Below T_g, the polymeric materials show glassy behavior (higher elastic modulus and brittle). Whereas, above glass transition or melting temperature ($>T_g$ or T_m), the modulus falls with varying rates based on the structure of polymer. Thus, by tuning of temperature carefully, T_g, α-transition, and other relaxations can be studied by DMA. These transitions and relaxations can also be studied by differential scanning calorimetry (DSC), but DMA provides stronger signals [27]. The tensile strength can be measured by universal testing machine (UTM). To measure tensile strength, the membranes cut into a dumbbell shape and then put in instrument's sample gauze. The membrane is then expanded with a rate of 1 mm/min and the tensile strength is calculated using the following equation [1,27,28]:

$$\text{Tensile strength} = \frac{\text{Maximum load}}{\text{Cross} - \text{sectional area}} \tag{17.10}$$

17.2.6.2 Thermal stability

During their potential applications, electrolyte membranes are exposed to a hoarse environment. As a result, the thermal stability of these materials must be higher in order for them to be durable [18]. The thermogravimetric analyzer is used to test thermal stability (TGA). The TGA profile is affected by several factors, including heating rate, gas flow rate, and sample preparation technique. Membranes typically decompose in three stages: (1) moisture release, (2) functional group decomposition, and (3) loss of membrane core fragments. DSC or DMA are also used to investigate thermal transition behavior (glass transition, T_g) [2,27,29].

17.2.6.3 Ion-exchange capacity

Ion-exchange capacity (IEC) denotes the number of charges per unit mass (in milliequivalent/g) of the membrane [1,2]. The IEC can be calculated by titration. The weighted membrane is dripped in sodium chloride solution (2 M) for 2 d. Later, the mixture is titrated using sodium hydroxide solution (0.01 M) in presence of phenolphthalein indicator [28,30]. The IEC is calculated using following equation:

$$IEC\left(m_{eq}/g\right) = \frac{V_{NaOH} \times C_{NaOH}}{W_{dry}} \times 100 \quad (17.11)$$

where V_{NaOH}, C_{NaOH}, and W_{dry} indicate the volume of NaOH solution used, strength of NaOH solution, and weight of the dry membrane, respectively.

17.2.6.4 Proton conductivity

A desirable proton conductivity of the membrane is critical for cell fabrication. Electrochemical impedance spectroscopy (EIS) is used to examine it. The EIS can be evaluated in two directions, namely in-plane and through-plane, using either a four-probe or a two-probe approach [2]. Proton conductivity (σ) is expressed as follows:

$$\sigma = \frac{L}{RA} \quad (17.12)$$

In the case of an in-plane test, L represents the gap between electrodes and A represents the cross-sectional area of the membrane; in the case of a through-plane test, L represents the thickness of the membrane and A represents the overlapping area of electrodes. R is calculated from the frequency at the minimum imaginary response in the in-plane test and the electrochemical impedance curve in the through-plane approach [2,30,31].

17.2.6.5 Water uptake

The weight ratio of water absorbed and dry membrane is used to calculate water uptake [2]. The presence of water or moisture in the membrane facilitates proton transport. As a result, the water uptake and retention capability of the membrane are critical in PEMFCs [18]. Before analysis, the membrane is dried at 80°C for 12 hours, weighed, and then dipped in deionized water for 1 d. Then it is mopped with tissue and immediately weighted [2,30,31]. A higher water content membrane usually ensures excellent proton conductivity. High water uptake, however, has some disadvantages, including low mechanical strength, low hydrolytic and dimensional stability, and high methanol permeability [2]. The water uptake (WU) and swelling ratio (SR) of the membrane are calculated as follows [18,28,32].

$$WU\,(\%) = \frac{W_{wet} - W_{dry}}{W_{dry}} \times 100 \quad (17.13)$$

$$SR\,(\%) = \frac{L_{wet} - L_{dry}}{L_{dry}} \times 100 \quad (17.14)$$

here W_{dry} and W_{wet} indicate the weights of dry and wet membranes, respectively. L_{dry} and L_{wet} denote the thicknesses of the dry and the wet membranes, respectively.

17.3 Biodegradable polymer nanocomposites: developments and scenarios

Human civilization enters a technologically dominated period. The demand for energy is increasing every day as technology advances. Because they are a cheap and reliable source of energy, fossil fuels have been used as a primary source of energy for a long time to meet the astronomical demand. The extensive and improper use of fossil fuels has resulted in the depletion of fossil fuel resources while also posing negative threats to our environment. As a result, developing alternative, renewable energy sources to mitigate the negative effects on the environment and/or improving fuel efficiency becomes critical. Fuel cells have recently received a lot of attention from the research community because of their potential application as green, sustainable, and productive energy conversion systems. They have little or no negative impact on the environment [10,33]. However, the commercialization of fuel cell technologies is currently hampered by a number of factors, including the high cost of catalysts (e.g., platinum-based electrodes), synthetic and nearly non-biodegradable membranes (e.g., Nafion), insufficient MEA stability, water handling intricacies in the systems, and so on [30,32,34]. One of the crucial challenges, aiming to limit or eliminate environmental pollution, could be overcome by developing biodegradable membranes for fuel cell systems.

Fuel cell technologies, emerging power generation alternatives, have aimed to eliminate fossil fuel pollution over the last few decades [35]. PEMFCs, MFCs, and DMFCs, among others, have advanced significantly because they are inexpensive, have a high power density, and are environmentally friendly [1,10,30,32,35]. The efficiency and speed of operation of a fuel cell are dictated by proton transfer through the membrane; thus, polymer-based membranes, specifically proton-conducting polymer electrolytes, play a critical role in overall performance and lifetime. There are two mechanisms for proton transport across the membrane: Grotthuss' mechanism (or hopping mechanism) and vehicle mechanism [36,37]. In Grotthuss mechanism, it assumes that the presence of H-bond networks carries protons from one place to another with synchronous forming and breaking of OH bonds. Contrarily, in vehicle mechanism, small molecules, such as H_2O or NH_3, convey protons as H_3O^+ or NH^{4+} ions, through the membrane. Schematic illustration of the two mechanisms is exhibited in Fig. 17.7.

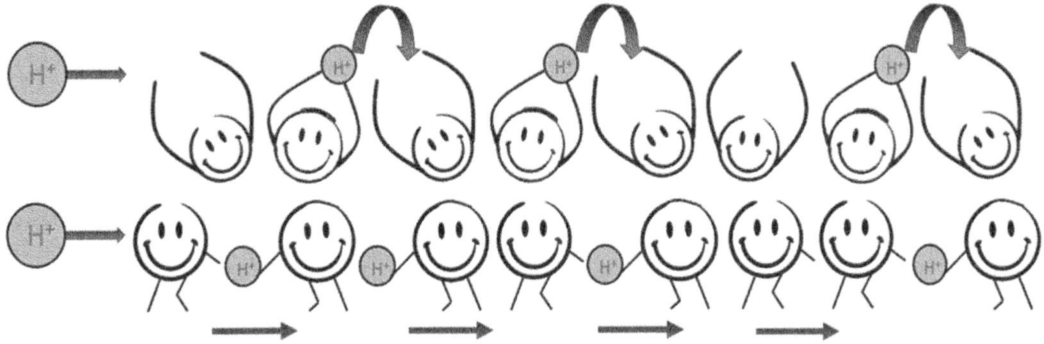

FIGURE 17.7

Schematic representation of the Grotthuss mechanism (top) and the vehicle mechanism (bottom) [38].

Because of its excellent proton conductivity and long lifetime, Nafion is a widely used commercially available perfluorinated ionomer membrane [39]. However, Nafion has a number of drawbacks, including high manufacturing costs, a difficult recycling process, and relatively high methanol permeability, which limit its widespread application [34]. Besides, polymer blend composites with Nafion (5–15 wt.% loadings) can also be useful for fuel cell applications but hazardous with regard to biodegradability and processing method which requires organic solvent [40]. As a result, new polymer electrolyte membranes with low electronic but high proton conductivity, low fuel permeability, excellent stability, optimum lifetime, low electro-osmotic drag coefficient, cost-effectiveness, biodegradability, and favorable mechanical properties are required. Many studies have recently focused on developing environmentally compatible and commercially viable novel proton-conducting membranes to improve the performance of fuel cells. Because of their biodegradability, sustainability, tunable properties, higher oxidative stability, and lower methanol permeability, biodegradable polymer nanocomposites are the best candidates [1,2,28].

17.3.1 Biodegradable polymer nanocomposites based on incorporated filler materials

Biodegradable polymers have been extensively researched in recent years as bio-inspired polymer membranes for fuel cells [34,41]. Biodegradable polymers such as chitosan, cellulose, starch, pectin, gelatine, agar, alginate, and poly(lactic acid) are of particular interest for the production of polymer electrolyte membranes for fuel cells due to their low cost, abundance in nature, good film-forming capability, ease of chemical functionalization, and biocompatibility [1,10,34]. However, biopolymers and biodegradable polymers cannot fully meet the performance requirements of fuel cells due to their low proton conductivity, insufficient stability, and fuel permeability, all of which have a negative impact on the overall performance of the fuel cells [1,30]. A myriad of modification methods have been proposed to negate the drawbacks of the biodegradable polymers in fuel cell applications including sulfonation [42], phosphorylation [43], chemical crosslinking [28], blending with other polymers [1,32], incorporating inorganic or organic fillers etc. [44]. Recently, research has focused on improving the desirable properties of bio-based membranes (e.g., thermochemical stability, lower fuel permeability, and so on) by incorporating nanofillers (organic or inorganic environmentally safe materials) with excellent properties into biopolymer matrix, i.e., nanocomposites. As a result, the properties of nanocomposite are determined by the chemistry of the polymer moiety and the filling materials incorporated. It is also thought that the relative amounts of the nanocomposite's components (for example, polymer and filler) as well as homogeneous dispersion, shape, size, and alignment of the fillers within the polymer moiety influence the properties of the nanocomposite membranes [45]. Nanocomposites are hybrid materials composed of a polymer matrix reinforced with nanometer-sized (at least in one dimension) fiber, platelet, or particle reinforcement [46]. In the following sub-section, such nanocomposite systems are classified based on the type of filler materials to provide a better understanding of how nanocomposites are currently modified to promote industrial applications of fuel cells.

17.3.1.1 Organic materials

Organic fillers in biopolymers are intriguing issues that have received a lot of attention in recent years due to their incredible biocompatibility and biodegradability. Polysaccharide nanocrystals, such as cellulose, chitin, and starch, have piqued the interest of researchers as reinforcing nanofillers in polymer nanocomposites due to their inherent biocompatibility and biodegradability, as well as other appealing properties such as nontoxicity, renewability, low density, high surface area, and ease of functionalization [47–49]. Polysaccharides such as cellulose, chitin, and starch are semi-crystalline, with both amorphous and crystalline domains. Interestingly, the crystalline domains with higher modulus can be separated on the nanometer scale, making polysaccharides potential candidates for organic nanofillers in biodegradable polymer nanocomposites systems [48]. Such polysaccharides as a nanofiller are described below:

17.3.1.1.1 Nanocellulose (NCC)

Nanocellulose (NCC)-based biodegradable polymers have sparked enormous interest due to their superior biological properties. Cellulose, one of the most abundant naturally occurring materials, is formed by plant-forming micro-fibrils that aggregate to form fibers. Using various physical and chemical methods, cellulose fibers can be disintegrated into micro-fibrils or nanofibers [50]. NCC are being prepared by acid hydrolysis, with H_2SO_4, HCl, or H_3PO_4 in controlled conditions [47]. The hydrolysis treatment of cellulose with H_2SO_4 to prepare NCC is appeared to be the most used and effective method [50–52]. According to the literature, nanocellulose (NCC) can be manufactured as cellulose-nanofibers (CNFs), cellulose-nanocrystals (CNCs), cellulose nanowhiskers (CNWs), and so on, and used as a reinforcement material due to its exceptional properties such as rich resources, biodegradability, low cost, nanosize, good stability, and reusability [50,52–54]. Furthermore, because of its large aspect ratio and ability to form an interconnected network structure with the polymer matrix via hydrogen bonding, NCC can significantly improve the mechanical properties and dimensional stability of biopolymers [54]. Zhao et al. [30] demonstrated that incorporation of NCC with chitosan-based membrane materials enhances the mechanical properties of membranes. Furthermore, the incorporation of NCC into the membrane reduces porosity, increases hydrogen bonding force, tightly connects molecules, decreases SR, and thus ensures the membrane's dimensional stability. Similarly, Hou et al. [52] achieved high proton conductivity by incorporating an suitable amount of NCC into sulfonated poly(arylene ether ketone)s (SPAEKs) copolymers with $-SO_3H$ and $-COOH$ groups. Furthermore, NCC insertion significantly improved the mechanical strength and dimensional stability of the copolymers. Another study examined the performance of proton-conducting mixed-matrix membranes synthesized by inserting amino-acid-functionalized cellulose whiskers into sulfonated polysulfone. The results show that incorporating amino-acid-functionalized cellulose whiskers into sulfonated polysulfone significantly improves proton conductivity while decreasing methanol permeability, WU, and single-cell performance. As a result, NCC is an excellent enhancer for polymer composite proton exchange membranes.

17.3.1.1.2 Chitin nanowhiskers (CNWs)

Chitin is a second most abundant naturally occurring polysaccharide after cellulose extracted from exoskeletons of crustaceans as well as from the cell walls of fungi and insects [48]. Chitin has natural biocompatibility, biodegradability, nontoxicity, and adsorption properties as a component of living organisms [48]. Furthermore, chitin has been shown to form highly crystalline micro-fibrils from nanofibers [55]. Acid hydrolysis, wet crushing, (2,2,6,6-tetramethylpiperidin-1-yl) oxyl (TEMPO)-facilitated oxidation, and ultrasonication can all be used to isolate and convert chitin nanofibers into rod-like nanowhiskers [56]. Like NCC, CNWs are encouraging reinforcing nanofillers for PEMFCs. Nasirinezhad et al. [57] prepared and characterized polymer membranes for fuel cell by inclusion of sulfonic acid-functionalized CNWs into the chitosan polymer skeleton. The nanocomposite membranes produced demonstrated increased proton conductivity and decreased methanol permeability. Likewise, Zhang et al. [58] formulated chitin nanowhisker-supported nanocomposite membranes by incorporating CNWs obtained from TEMPO method with an average length and diameter of 317 and 33 nm, respectively, into sulfonated poly(ether-sulfone) (SPES). The nanocomposites were tested further as proton exchange membranes. The nanocomposite membranes outperformed SPES films in terms of proton conductivity, WU, thermal stability, and mechanical properties due to the strong interaction between SPES molecules and CNWs via H-bonding. These results suggest that CNWs are promising fillers for the development of biopolymers for fuel cell applications.

17.3.1.1.3 Chitosan nanofibers (CSNFs)

Chitosan, a naturally occurring polymer and the most important chitin derivative, has excellent chitin properties such as biocompatibility, biodegradability, nontoxicity, absorption properties, and many more, and can thus be used as a nanocomposite's material to develop PEMFCs [48]. Chitosan is formed by deacetylation of chitin, which contains a reactive amino group on its structural units [48]. The chemical structure of chitin and chitosan is revealed in Fig. 17.8. The presence of amino groups makes chitosan a better polyelectrolyte by facilitating the proton transfer in an acidic matrix [59]. Chitosan nanofibers (CSNFs) with many amino groups can improve water absorption and proton conductivity in nanocomposite membranes by forming proton transport sites in the membranes. Biopolymers have several drawbacks as proton exchange membranes, which can be overcome by incorporating CSNFs into the polymer skeleton. By forming nanocomposites with CSNFs, the proton conductivity of synthetic polymers can be improved. For example, Wang et al. [60] employed the electrospinning method to prepare CSNFs followed by modifying with the amino acids, and then nanocomposite membranes were fabricated with sulfonated polysulfone (SPSF) through the casting method. When compared to pristine SPSF membranes, the nanocomposite membranes demonstrated improved proton conductivity, reduced methanol crossover, increased WU, and tensile strength. As a result, CSNFs as filler materials improve the performance of pure

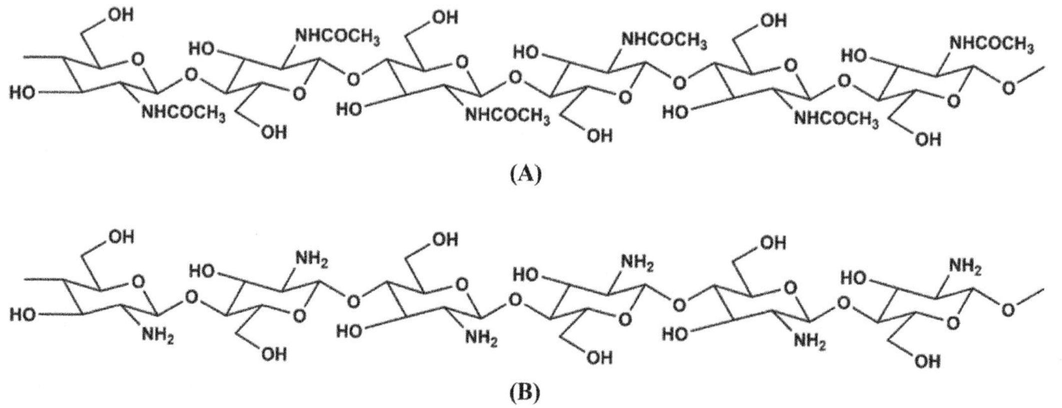

FIGURE 17.8

The chemical structures of (A) chitin and (B) chitosan [48].

polymers, and the resulting nanocomposites can be potential replacements for conventional membranes in fuel cell applications.

17.3.1.1.4 Starch nanoparticles (SNPs)

Starch is a naturally occurring, sustainable, and environmentally friendly polymer that is typically produced by plants to store the energy obtained from photosynthesis. After being extracted from plants, starch appears as a flour-like white powder with microscopic granules [61]. These granules are made up of nano-sized semi-crystalline blockets [62]. Acid hydrolysis could be used to separate the nano-blockets from the starch [63,64]. Starch nanoparticles (SNPs) with platelet-like morphology, prepared by Putaux et al. [64], can increase the methanol molecule migration pathways and consequently limit permeability when incorporated into biopolymer matrix. Furthermore, the incorporation of SNPs increases the nanocomposites' tensile strength and elastic modulus [49]. SNPs' reinforcing effect can be attributed to the formation of increased H-bonding between filler and matrix, and/or between fillers [65]. Furthermore, the WU of SNPs filled nanocomposites can be increased as the filler (SNPs) content increases [66]. At this point, it is certain that SNPs have all of the necessary properties to serve as a potential filler material in the fabrication of PEMFCs.

17.3.1.2 Inorganic materials

Organic-inorganic nanocomposite membranes are typically organic polymeric composites made up of inorganic nanoscale filler materials. These nanocomposite membranes have the potential for fuel cell applications because they combine the advantages of inorganic materials such as rigidity and stability with the benefits of organic polymers such as flexibility, dielectric, ductility, and processability [67]. High proton conductivity through the membrane, as well as outstanding chemical and mechanical stability, are required for the development of proton exchange membranes for fuel cells. Incorporating environmentally friendly inorganic nanoparticles into biopolymer membranes improves mechanical properties and WU while limiting SR and fuel crossover in pure polymer membranes [39]. The interface interaction between organic-inorganic components via H-bonding, Van der Wall forces, and/or electrostatic forces modifies the polymers' original properties [2,41]. Based on the proton-conducting ability of the respective system components, the organic-inorganic nanocomposite membranes could be composed of proton conductive polymers and less-proton conductive inorganic particles, or proton conductive inorganic particles and less-proton conductive organic polymers [68]. Furthermore, inorganic fillers can be porous or nonporous [2]. There are numerous inorganic filler materials that can improve the properties of biopolymers and make them suitable for fuel cells in a variety of ways. Because it is nearly impossible to describe all of the inorganic fillers used to develop biopolymers that have been reported thus far, only the most common types will be discussed in this context.

17.3.1.2.1 Metal oxides (MO)

Hygroscopic oxides like SiO_2, ZrO_2, and TiO_2 are widely used inorganic fillers to fabricate nanocomposite membranes because of their cost-effectiveness, water retention capacity at relatively low humid conditions, and easy accessibility [28,45,69]. Embedding these metal oxides with the proper structure and pore size into the polymer matrix improves chemical and thermal stability, decreases fuel permeability, and thus improves fuel cell performance [70]. Though, the oxides (e.g., SiO_2, ZrO_2, and TiO_2) do not possess any novel inherent proton conductivity, and sometime, proton conductivity reduces with increasing filler content [41,45]. Furthermore, unmodified inorganic nanoparticles are incompatible with the polymer moiety, which may result in phase separation of the nanocomposite [44]. As a result, inorganic fillers are typically functionalized prior to incorporation into polymers [1]. Surface modification reduces the agglomeration of nanoparticles, resulting in quasi-uniform dispersion of the nanofillers throughout the composite membranes. For example, the morphological analysis of uncapped and polyethylene glycol (PEG) capped CdO nanoparticles by transmission electron microscopy (TEM) explicitly shows the effect of surface modification (Fig. 17.9) as the capped nanoparticles (Fig. 17.9A) were exhibiting lesser agglomeration compared to the uncapped nanoparticles (Fig. 17.9B) [71]. The TEM images divulged the size and shape of the CdO nanoparticles (both uncapped and PEG capped) which were nearly spherical shape and 20–40 nm in range, respectively.

Furthermore, surface modification of metal oxides with organic compounds containing amines and sulfonic acid functional groups not only improves nanofiller dispersibility into the polymer matrix but also increases proton conductivity [28,44]. Recently, Kalaiselvimary et al. [72] modified sulfonated-chitosan and poly (ethylene oxide) (PEO) blends with sulfonated silica dioxide (s-SiO_2) nanoparticles. The as-prepared sulfonated-chitosan/PEO-based membranes containing 7.5 wt.% of s-SiO_2 nanoparticles displayed the superior IEC, proton conductivity, little SR, and remarkable thermochemical stability. The enhanced performance was attributed to s-SiO_2 nanoparticles with greater proton conductivity and WU. Similar phenomena were also reported by Mohanapriya et al. [1], when sulfonated titanium dioxide (s-TiO_2) nanoparticles were embedded into pectin (PC)-polyvinyl alcohol (PVA) blend. Flexible polymeric network of PC-PVA blend that is suitable to disperse rigid s-TiO_2 nanoparticles causing in enhanced proton conductivity and hindered methanol permeability. Furthermore, incorporating phosphorylated titanate nanotubes (PTNTs) into chitosan (CS) membranes significantly reduced methanol crossover through the membranes while also improving proton conductivity by easing proton transfer by creating continuous conductive channels with the help of P-OH groups and absorbed water

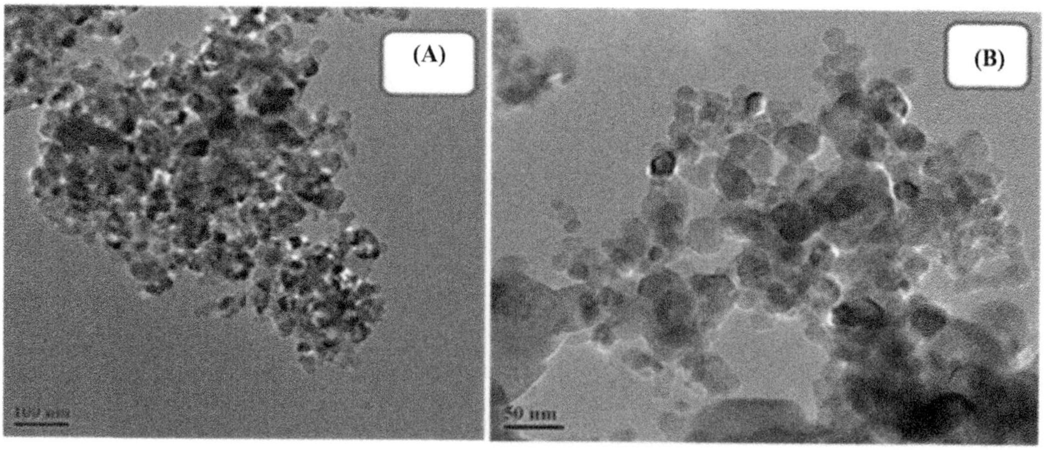

FIGURE 17.9

TEM images of (A) uncapped and (B) PEG capped CdO nanoparticles [71].

molecules [73]. It is worth mentioning that other low-cost metal oxides such as nickel oxide (NiO) [74,75], aluminum oxide (Al_2O_3) [76–78], potassium hexatitanate ($K_2Ti_6O_{13}$) [79], etc. are also being used as inorganic filler materials to develop structural and functional properties of polymeric membranes. In this connection, the SEM image and the EDS data of the PVA-CaO-CuO composites are presented in Figs. 17.10 and 17.11, respectively. Moreover, several studies have been reported that perovskite-type oxide (e.g., $BaZrO_3$, $SrCeO_3$, etc.) nanoparticles, owing to their higher chemical stability, superb thermal and mechanical stability, cheap, and hygroscopic nature, can be used to develop proton-conducting membranes for high temperature PEMFCs [80,81].

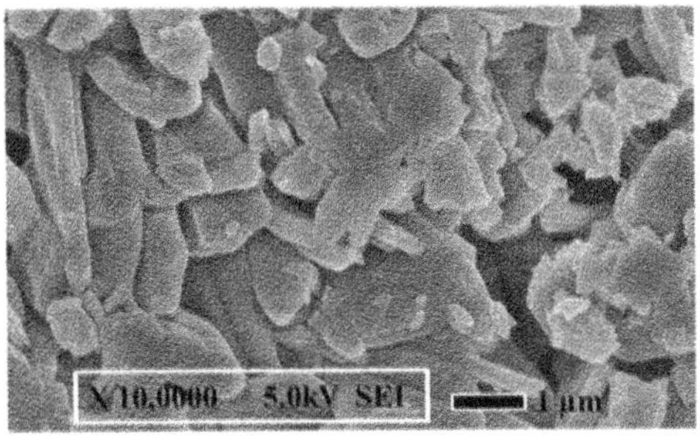

FIGURE 17.10

SEM micrograph of PVA-CaO-CuO composites.

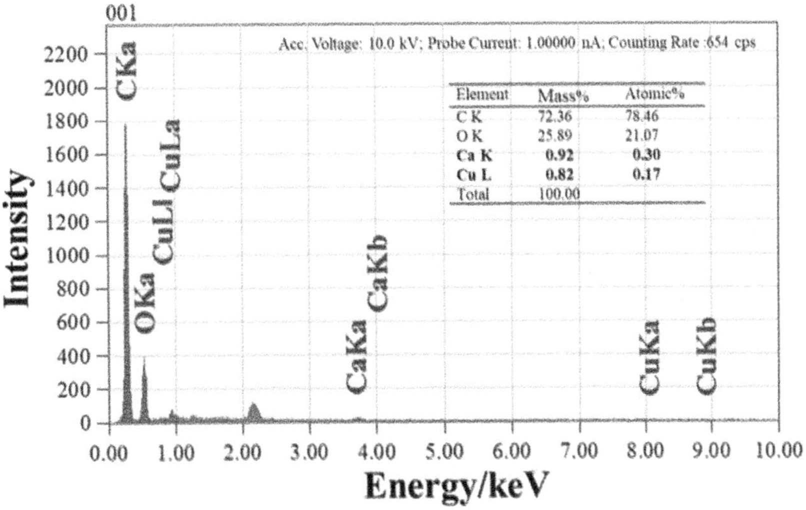

FIGURE 17.11

EDS spectrum of PVA-CaO-CuO composites.

17.3.1.2.2 Zeolites

Zeolites are crystalline aluminosilicates with a rigid framework structure that encloses cavities that can be occupied by various ions or molecules and have a high cation exchange capacity [34,45]. The incorporation of zeolite particles into polymers results in significant changes in the properties of nanocomposite membranes depending on pore size, content, and hydrophilic/hydrophobic nature [82,83]. For example, the appropriate addition of zeolite content improved the mechanical strength [84], on a contrary, excessive addition triggered the diminution in mechanical strength of the membranes because of forming too many interfacial voids within polymer and zeolite [82]. Furthermore, studies revealed that the addition of hydrophobic zeolites increased methanol diffusion resistance and thus decreased methanol permeability, whereas hydrophilic zeolites had the opposite effect [82,85]. By incorporating plasticizers, such as sorbitol, and controlling the membrane formation temperature, the interfacial morphology of polymer/zeolite composites could be developed [86]. Furthermore, surface modification of zeolites, such as with a silane coupling agent, is another technique for promoting interfacial morphology of nanocomposites while limiting methanol crossover [84,87]. Although zeolite-incorporated nanocomposite membranes demonstrated improved methanol resistance, they had lower proton conductivity when compared to pristine polymer membranes; this may be due to the inclusion of zeolites, which reduces WU [87,88]. The schematic representation of the transient manner of proton and CH_3OH in the polymer-zeolite composites membrane is shown in Fig. 17.12. The proposed approaches to enhance the ionic conductivity of polymer/zeolite composites are introducing either $-SO_3H$ groups by sulfonation or triphosphate groups by phosphorylation onto the surface of zeolite particles [87,89].

17.3.1.2.3 Layered nanoclays

Nanoclays, including montmorillonite (MMT) and laponite (LP), represent a type of materials usually composed of layered silicates, are the most widely used inorganic fillers utilized to fabricate PEMFCs because they are cheap and bear ion-exchange properties as well as excellent barrier properties [45,90]. The thickness of the layer is usually around 1 nm and is 30-several μm, causing platelets with a high aspect ratio [26]. Poor dispersion and phase separation in nanoclays/polymer (hydrophilic/hydrophobic) nanocomposites can reduce the proton conductivity of the nanocomposite membranes. The compatibility between nanoclays and polymer can be improved by surface modification of nanoclays through hydrophilic functional groups, like $-NH_2$, $-OH$, $-SO_3H$, etc. [44]. He et al. [91] investigated the effects in three types of commercially available clays, the

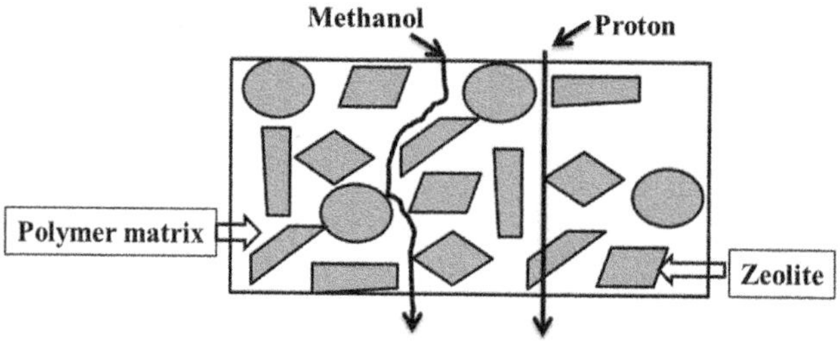

FIGURE 17.12

Passing way of proton and methanol in the polymer-zeolite composite membrane [88].

sodium montmorillonite (IC), long alkyl chains (OC) containing hydrophobic organo-clay, and organo-clay having −COOH end groups (HC), on the structure and properties of sulfonated poly (ether ether ketone) (SPEEK)/clay nanocomposites. This study revealed that the SPEEK/HC hybrid membranes exhibited superior overall performance, considering clay dispersion, proton conductivity, and selectivity (because of the interaction between −COOH groups in HC and $-SO_3H$ groups in SPEEK) when the HC loading was below 10 wt.%. Furthermore, Purwanto et al. [92] pre-treated pristine montmorillonite (MMT) filler using 3-glicidoxy propyltrimethoxysilane (GPTMS) to produce surface-modified montmorillonite (G-MMT) then incorporated into biopolymer chitosan at different wt.%, ranging from 5 to 25 wt.%. The composite membrane with 5 wt.% loadings of G-MMT filler demonstrated superior compatibility, improved physicochemical properties, best proton conductivity, and least methanol permeability among chitosan/G-MMT membranes. They claimed that the phase interaction between polymer and fillers was obtained owing to interaction between −OH groups of G-MMT (fillers) and −OH or $-CONH_2$ groups of chitosan (biopolymer).

17.3.1.2.4 Metal phosphates

As inorganic fillers in polymer nanocomposite membranes for fuel cell applications, numerous metal phosphates have been used. Of them, hydroxyapatite ($Ca_{10}(PO_4)_6(OH)_2$) is encouraging ion conductive compound and emerged as a potential inorganic filler for PEMFCs [41]. For instance, Zhao et al. [93] prepared hybrid membranes by incorporating biomimetic mineralized hydroxyapatite (BMHA) into chitosan (CS) polymer matrices. The thermal stability of the CS/BMHA hybrid membrane was significantly improved due to strong H-bonding between the BMHA and CS surfaces. Furthermore, CS/BMHA hybrid membranes outperformed pure CS membranes in terms of methanol resistance and proton conductivity. In another study, Gholizadeh et al. [94] investigated the consequence of hydroxyapatite nanoparticles (HA NPs) towards antibacterial activity, physical, and mechanical properties of sodium alginate (SA) film and found that the incorporation of HA NPs improved the physical and mechanical properties while declining the water solubility and water vapor permeability of the biodegradable SA/HA nanocomposites than pure alginate. As a result, hydroxyapatite fillers may be a viable material for fuel cell applications.

17.3.1.3 Organic-inorganic materials

Carbon-based fillers, like other filler materials, have been widely investigated as promising materials for PEMFCs due to their high aspect ratios, high surface areas, low densities, and outstanding mechanical properties. Carbon nanotubes (CNTs), cylindrical graphene tubes with a nanometer-size diameter consisting of single or several graphene layers, have been widely investigated as promising materials for PEMFCs [2]. CNTs, on the other hand, are difficult to disperse in polymeric matrices because they agglomerate during processing due to the low polymer/CNT interaction [95]. The processing conditions may affect the dispersion of CNTs in the nanocomposites. Besides, the properties and performance of the CNTs can be considerably improved via modification with different functionalized groups such as sulfonated CNTs [96], phosphonated CNTs [97], carboxylic acid-functionalized CNTs [98], imidazole functionalized CNTs [99], etc. For instance, Tohidian et al. [99] prepared and characterized nanocomposite polyelectrolyte membranes based on biopolymer (e.g., chitosan) and both hydroxyl and imidazole functionalized multi-walled CNTs for fuel cell applications. The membrane containing imidazole functionalized multi-walled CNTs demonstrated increased proton conductivity and decreased methanol permeability. Furthermore, sulfonated functionalized multi-walled CNTs in poly (ether sulfone) demonstrated excellent proton conductivity and low methanol permeability [100].

Graphene oxides (GO), like CNTs, are very promising fillers for designing nanocomposite membranes. GO is a two-dimensional structured material with –OH and epoxy groups on the basal plane and –COOH groups on the sheet edges [2]. These functional groups allow GO to hold more water, improving proton conductivity. GO is also well known for its inherent mechanical and chemical stability. All of GO's excellent features have piqued the interest of fuel cell developers. In a recent study, polybenzimidazole (PBI)/GO nanocomposite membranes doped with H_3PO_4 were fabricated for high temperature PEMFCs, and the performance of the as-prepared membrane demonstrated that GO additive boosted the performance of PEMFC compared to the bare PBI [101]. Similarly, another carbon-based filler, nanodiamond (ND) particles, is being used primarily to improve the stability and mechanical properties of nanocomposite membranes, and thus can be used to develop proton exchange membranes that can withstand harsh environments [102].

17.3.2 Preparation methods

The method of preparation is critical in the development of nanocomposite membranes for fuel cell applications because it affects the dispersion of the filler materials within the polymer moiety and thus the properties. Nanocomposite membranes can be made in a variety of ways. Among them, blending, sol-gel, and infiltration (also known as "in situ method") have received tremendous attention as general preparative methodologies of nanocomposites as shown in Fig. 17.13 [103], due to their wide applicability and versatility which are described in the following sub-sections.

17.3.2.1 Blending method

This is the most basic and widely used method for creating nanocomposites, involving the direct mixing of NPs into the polymer moiety. The schematic illustration of blending method is presented in Fig. 17.14. The mixing can be achieved by either solution blending or melt blending [103]. The solvent

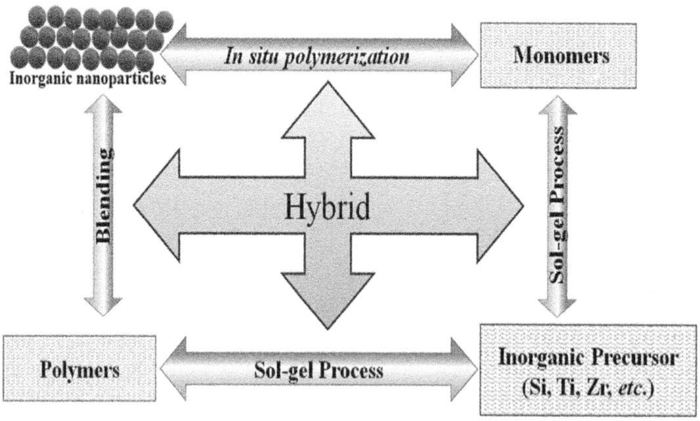

FIGURE 17.13

Different synthesis routes for polymer electrolyte membranes are depicted schematically (PEMs) [103].

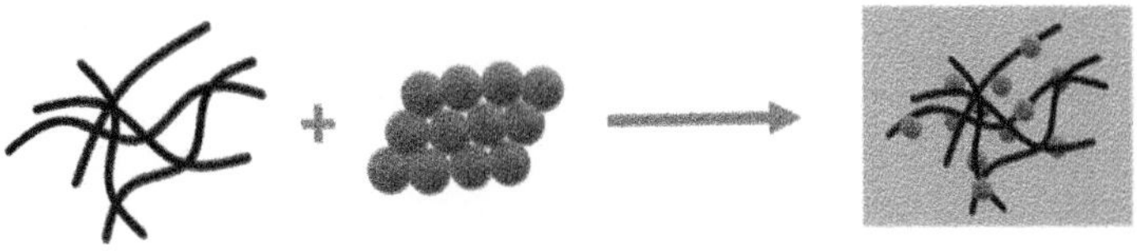

FIGURE 17.14

Schematic illustration of blending method.

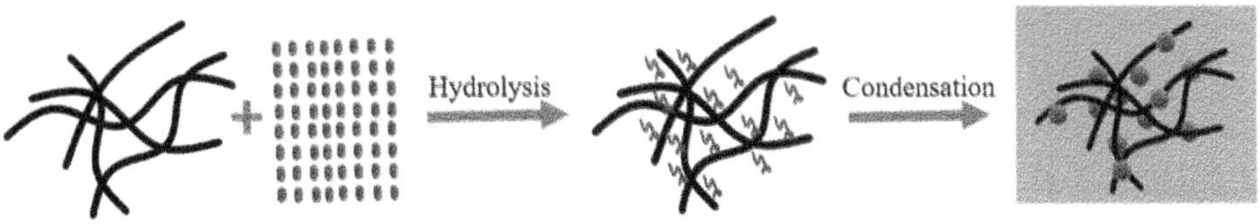

FIGURE 17.15

Schematic representation of sol-gel method.

used in solution blending is important because it dictates the dissolution of the selected polymer while also providing a good dispersion of the nanofillers in the polymer moiety. Furthermore, the polymer and solvent concentrations, as well as the external conditions (e.g., pressure, temperature, and humidity) during solvent evaporation, are thought to be determining factors of nanocomposites properties and surface morphology. Melt blending, a solvent-free nanocomposites processing technique, is considered more industrially and economically viable because it is a simple, continuous, quick, and versatile operation to obtain finished products from raw ingredients [50]. Filler agglomeration, on the other hand, may cause a problem in the blending method, which can be solved by surface modification or functionalization of the filler materials [1,44].

17.3.2.2 Sol-gel method

Because of its eco-friendliness, the sol-gel method, a low temperature synthetic protocol, has been widely used to fabricate nanocomposite membranes. This process is typically conducted by two consecutive reactions, hydrolysis, and condensation reactions, respectively (Fig. 17.15). Here, the of metal alkoxides, $M(OR)_n$ ($M = Zn$, Al, Si, Ti, Zr, etc. and $R = C_xH_{2x+1}$), inside a polymer dissolved in aqueous or non-aqueous solution. The reactions can be expressed as follows (for alkoxides $n = 4$) [104]:

$$M(OR)_4 + 4H_2O \rightarrow M(OH)_4 + 4\,ROH \tag{17.15}$$

$$mM(OH)_4 \rightarrow (MO_2)_m + 2\,mH_2O \tag{17.16}$$

It is worth noting that silicon alkoxides are not very sensitive to hydrolysis and thus require acid or base catalysts, whereas hydrolysis and condensation of non-silicate metal alkoxides do not require

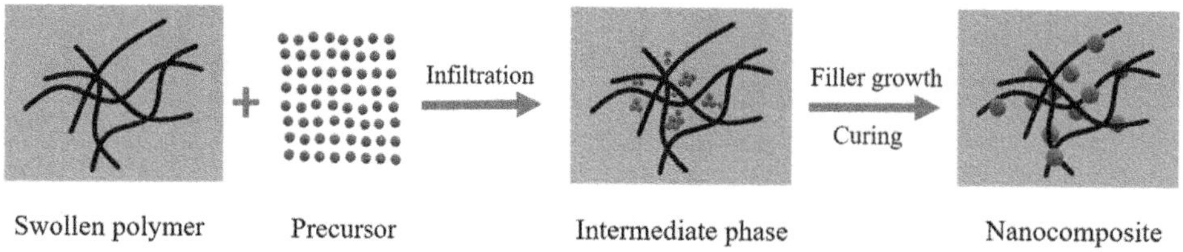

FIGURE 17.16

Schematic illustration of infiltration method.

catalysts [103]. Furthermore, some factors, such as water/silane ratio, pH of the solution (if a catalyst is used), temperature, and solvent nature, can influence the kinetics of hydrolysis and condensation reaction, and thus the structure, morphology, and properties of the final product [103,105].

17.3.2.3 Infiltration method

Infiltration or in situ methods have been widely used to prepare nanocomposite membranes, specifically to alter the membranes' properties by initiating NPs or precursors [103]. Precursors of filler materials infiltrate into the swollen membrane, which has an increased pore or void volume, during this process. After the impregnation, the nanocomposite membranes are achieved through filler growth, removing the impurities, and polymer curing (Fig. 17.16). Because the size and distribution of the nanoparticles are governed by the pore size of the polymer matrices, the infiltration method can avoid undesirable agglomeration of the nanoparticles, resulting in controlled size and uniform distribution of the nanoparticles within the nanocomposite membranes [2].

17.3.3 Types of biodegradable composites and their prospects as fuel cell application

Fuel cells are promising high-efficiency power generation alternatives that are gaining attention due to their potential to reduce energy consumption, reliance on fossil fuels, and pollutant emissions. Because of its electron nonconductive nature, the proton exchange membrane is the interior part of most commonly used fuel cells (e.g., PEMFCs, DMFCs, etc.) that separates the anode from the cathode and efficiently transfers generated protons from the anode to the cathode side of the cell. Furthermore, to be suitable for fuel cell applications, proton exchange membranes must meet a number of criteria, including increased proton conductivity, limited fuel permeability, and long-term mechanical, chemical, and dimensional stability under heated and humidified conditions [41]. The biodegradability of these membranes is an additional requirement that has become an unavoidable requirement of the present time. As a result, research into biodegradable proton exchange membranes is currently intensifying. The prime goal of this section is to present recent approaches to develop biodegradable PEMFCs. Table 17.2 summarizes the fuel cell performance of some reported biodegradable membranes.

17.3.3.1 Chitosan-based composites

Chitosan-based membrane materials are being extensively researched as low-cost and environmentally friendly alternatives for proton exchange membranes in order to potentially produce economical fuel

Table 17.2 List of literature reported biodegradable membranes for fuel cell with their performances.

Membrane	Ion conductivity (mS/cm)	Power density (mW/cm^2)	References
AMP-CS	186.7 (100°C)	—	[30]
AMP-CS-NCC-2	116.3 (100°C)	—	[30]
PIC	42.0	—	[117]
CS-sChW nanowhisker	22.1 (65°C)	—	[57]
NCC-PVA-T5	13.0 (120°C)	—	[108]
5 wt.% G-MMT	4.66	—	[92]
CNC	4.6 (120°C)	—	[110]
Agar-glycerol (40 wt.%)-NH_4NO_3	1.44	—	[129]
AP-GA	1.10 (25°C)	—	[122]
Agar-HCHO-glycerol-50 wt.% AcOH	0.96 (80°C)	—	[125]
Gelatine-HCHO-glycerol-26.3 wt.% AcOH	0.36 (80°C)	—	[133]
CS/0.5 wt.% MWCNT-Im	0.34 (25°C)	—	[99]
Gelatine-HCHO-glycerol-0.1 M HCl	0.05 (25°C)	—	[133]
CNF	0.05 (100°C)	—	[110]
5.9% NaCl-treated Cassava starch	—	1068.5	[112]
PC-PVA/s-TiO_2 nanocomposite	—	27.0 (70°C)	[1]
Am-sCNC	210.0 (100°C)	227.0 (75°C)	[138]
s-NBF-qAPEEK	119.0 (80°C)	930.0 (60°C)	[139]
BC-PSSA	100.0 (94°C)	0.06 (25°C)	[140]
CS-PMA-HPA	92.1 (80°C)	88.7 (30°C)	[31]
CS-sGO-SiO_2	92.0 (80°C)	87.18 (80°C)	[141]
qCNC/qPPO	60.0 (80°C)	392.0 (60°C)	[142]
BC-Nafion	58.0 (30°C)	106.0 (25°C)	[143]
CS-CS@CNT	34.6 (80°C)	47.5 (30°C)	[144]
CS-sZr@CNT	34.0 (80°C)	64.6 (80°C)	[145]
CS-PTA-MMT	30.0 (25°C)	49.7 (70°C)	[146]
smpCS-PTA	29.0 (80°C)	16.0 (80°C)	[147]
PVA-CS	27.0 (80°C)	67.5 (60°C)	[148]
KOH doped qPVA-qCS	21.0 (60°C)	73.0 (60°C)	[149]
CS-PTA	18.0 (25°C)	350.0 (25°C)	[150]
CS-EMImC-Co-EP	11.0 (80°C)	21.7 (25°C)	[151]
CS-sGO	10.9 (120°C)	146.7 (80°C)	[152]
CS-PANI-SiO_2	8.39 (80°C)	56.27 (80°C)	[153]
CNF	0.05 (100°C)	0.79 (80°C)	[110]

cells. These are excellent biomaterials with promising properties like biodegradability, biocompatibility, hydrophilicity, and low cost. So far, many studies on the development and fabrication of chitosan-based nanocomposites as high-performance alternative membranes for fuel cell applications have been published. Recently, Nasirinezhad et al. [57] fabricated and characterized CS based polymeric membranes using sulfonic acid-functionalized chitin nanowhiskers (sChW) as nanofillers (at different wt.%) and CS as a matrix for application in DMFCs. Besides, the manufactured membranes were compared with pure CS membrane and membrane made of 5 wt.% unmodified CNWs to demonstrate not only the enhancement of desired properties in the nanocomposites but also the effectiveness of functionalization of nanofillers before inserting into the polymer matrix. The selectivity parameter, calculated from the ratio of the proton conductivity to methanol permeability at room temperature, showed a sharp rise from 3900 for pure chitosan to 26,888 Ss/cm^3 for a membrane with 5 wt.% sChW, whereas it was 6207 Ss/cm^3 for membrane with ChW. The attractive interactions between the sulfonic acid groups on the surface of nanofillers (i.e., sChW) and amine groups in the polymer (i.e., CS) chain were attributed to the improved proton conductivity and reduced methanol permeability. Even though the obtained results suggest that the synthesized CS/sChW nanocomposite membrane could be a proton exchange membrane alternative, it is always better to compare the as-prepared nanocomposites with Nafion membranes. Tohidian et al. [99] reported a comparison study of chitosan-based nanocomposite polyelectrolyte membrane and commercial Nafion 117 membrane. The membrane was made by incorporating imidazole surface-modified multi-walled nanotubes (MWCNT-Im) into a chitosan (CS) polymer matrix. The characterization results showed that introducing imidazole groups into the membrane structure can improve proton conductivity by increasing proton-conducting sites and decrease methanol permeability by providing a better nanofiller/matrix interface, resulting in a higher selectivity parameter. The obtained selectivity parameter of CS/0.5 wt.% MWCNT-Im at room temperature was 1.18 times of that in Nafion 117. In another study, Purwanto et al. [92] prepared nanocomposite membranes using surface-modified montmorillonite (G-MMT) fillers and chitosan polymer. The composite membrane with unmodified MMT exhibits severe agglomeration of the MMT particles on the prepared membrane's surface. The membrane with 5 wt.% G-MMT loadings exhibited best characteristics in terms of proton conductivity (4.66 mS/cm) and methanol permeability (3.03×10^{-7} cm^2/s). Higher G-MMT loading, on the other hand, reduced WU efficiency and, as a result, proton conductivity of the membranes. To achieve the best performance of the nanocomposite membrane, an optimal amount of nanofillers should be incorporated with the polymer.

It is worth noting that all of the membranes discussed thus far have used surface-modified nanofillers to achieve a homogeneous dispersion of the fillers into the polymer matrix, significant proton-conducting ability, and excellent mechanical and chemical stability. Interestingly, Zhao et al. [30] developed an efficient strategy for fabricating chitosan-based composites by modifying the polymer with purine groups such as adenine (A) and adenosine monophosphate (AMP). In addition, nanocomposite membranes were created by combining nanocellulose (NCC) in various weight ratios with the modified chitosan. Purine groups generate numerous proton-hopping sites within the membrane, while NCC improves the membrane's mechanical properties and dimensional stability. The AMP modified chitosan (AMP-CS) membrane materials showed highest proton conductivity (0.1867 S/cm) at 100°C, which is higher than Nafion at (100°C). However, the NCC-doped membranes were more compatible for operation conditions of fuel cell because of their improved mechanical properties and comparable proton conductivity at 100°C (0.1041 S/cm for A-CS-NCC-2 and 0.1163 S/cm for AMP-CS-NCC-2).

17.3.3.2 Cellulose-based composites

Cellulose is a well-known natural biopolymer due to its nearly limitless resources, biocompatibility, high specific strength and modulus, easily modifiable surface, high hydrophilicity, and resistance to many organic solvents [106]. From the structural point of view, cellulose is comprised of glucose units containing three hydroxyl groups on each of them, linked together by β-1–4 glucosidic bonds [106]. Cellulose has been extensively researched as a potential electrolyte membrane for fuel cell applications, whether neat, blended with other polymers, or chemically modified [107]. To begin with, sulfonated nanocrystalline cellulose (NCC)-PVA composite membrane doped with 1,2,4-triazole was synthesized by Etuk et al. [108] The synergistic effect of sulfonic acid and 1,2,4-triazole on nanocrystalline cellulose composites was demonstrated by a high proton conductivity of 13 mS/cm for NCC-poly vinyl alcohol-triazole5 (NCC-PVA-T5) composite membrane at 120°C under anhydrous conditions, which is a significant improvement for the proton conductivity of nanocrystalline cellulose-based membranes for fuel cell applications. Thermal stability was compromised, however, due to the interaction of sulfonic groups and imidazole. Additionally, Sriruangrungkamol et al. [109] modified cellulose membrane with sulfosuccinic acid (SSA) using the impregnation method. The concentrations of SSA were varied from 0.1 to 10%w/v. Although nanocellulose membrane became more rigid with increasing SSA concentration, the stability of the membrane was lost when modified with 10%w/v SSA. Besides, the proton conductivity values of the as-prepared membranes were lower than Nafion 117 (1.07×10^{-2} S/cm) but exhibited better methanol barrier properties. Bayer et al. [110] presented research on an environmentally friendly and biodegradable nanocellulose membrane for fuel cells that increases proton conductivity up to 120°C. The researchers compared two types of nanocellulose, CNFs and CNCs, and discovered that maximum conductivity for CNF paper membranes was 0.05 mS/cm at 100°C and 4.6 mS/cm for CNC paper membranes at 120°C (both at 100% relative humidity). Furthermore, the results showed that fuel cells with nanocellulose membranes and hydrogen fuel can operate successfully at 80°C and 95% relative humidity. As a result of this research, nanocellulose paper membranes can be used as eco-friendly and biodegradable proton exchange membranes in fuel cell applications.

17.3.3.3 Starch-based composites

Starch, a low-cost biopolymer that can be obtained from maize, wheat, potatoes, and cassava, can form biodegradable films [61]. However, due to its high hydrophilic nature, starch is water sensitive. Furthermore, starch has poor mechanical and ion-conducting properties, making it unsuitable for certain applications [34]. To make starch-based composites suitable for fuel cell applications, their mechanical properties and ion-conducting ability must be improved. Various physical and chemical strategies can be used to improve the physical and functional properties of starch. For example, Dai et al. [111] prepared starch-based nanocomposite films reinforced by cross-linked starch nanocrystals (CSNCs) with different content (6%–12%). The study of the characteristics revealed that the self-reinforcing effect improved the physicochemical properties of starch-based films. In comparison to the others, the nanocomposite films with 10% CSNCs content demonstrated the best mechanical and water barrier properties. Although starch polymers have a wide range of applications, only a few have been reported on their use as polymer electrolyte membranes. Obasi et al. [112] utilized cassava starch as a proton exchange membrane (PEM) to study the viability of power generation in a mediatorless dual chambered MFCs. Three different PEMs (gelatinized cassava starch) were prepared with different degrees of modifications, which are, (A) untreated pristine starch, (B) starch treated with 5.9% sodium chloride, and (C) starch modified with SA and treated with activated carbon and $CaCl_2$ in NaOH solution. The highest power outputs from the constructed MFCs were 945.69, 1068.54, and 570.83 mW/m^2 for A, B, and C, respectively, which were less than that already reported in the literature (3600 mW/m^2) from a MFC system involving mediator. Finally, the cost-effectiveness and the feasibility of MFC operation utilizing the cassava starch-based composite would augment the commercial potential of MFCs.

17.3.3.4 Alginate-based composites

Alginate is a linear, natural, and water-soluble polysaccharide that can be extracted from vegetable sources like brown algae, seaweeds, etc. [113,114]. It is the only known polysaccharide that contains carboxyl groups naturally in each constituent residue, as its structure is mainly comprised of (1–4)-β-D-mannuronic acid and (1–4)-α-L-guluronic acid units. The alginate has many excellent properties including biocompatibility, nontoxicity, cost-efficiency, biodegradability, and can form gel with divalent cations [115]. These characteristics of alginate have led to its use in many areas such as food, pharmaceuticals, tissue engineering, etc. [116], and encouraged studies on using alginate to fabricate electrolyte membranes for fuel cell applications. Smitha et al. [117] reported the preparation of a polyion complex (PIC) membrane consisting of chitosan and SA for DMFC applications. The cost-effectiveness and simple fabrication technique along with low methanol permeability (4.6×10^{-8} cm^2/s at 50 vol.% methanol concentration), remarkable physicomechanical properties, and relatively high proton conductivity (0.042 S/cm) of the PIC membranes rendered the blend quite attractive for DMFC applications. However, co-biopolymer membranes are relatively frail in most cases due to their low mechanical strength, and thus do not provide significant advantages over Nafion. These disadvantages can be overcome by incorporating inorganic and/or organic fillers, i.e., forming nanocomposites. To this end, SA (Al)/GO nancomposite films were prepared with different wt.% loadings of graphene oxides [118]. The quality of the prepared films was investigated by SEM. Fig. 17.17 illustrates both the surface (A, B) and cross-sectional (C, D) SEM micrographs of pure alginate (Al) and Al/GO composite films embedded with 2.5 wt.% GO. From the SEM images (Fig. 17.17A and B) it is evident that the surface morphology of the film changes from smooth (for Al only film) to rough when GO sheets incorporated into Al. Besides, the cross-section image of the Al/GO composite film (Fig. 17.17D) indicates well dispersion and clear stratification of the GO in the Al polymer matrix. Based on the morphological analysis, Ionita et al. [118] proposed "brick-and-mortar" structure (Fig. 17.18) for the prepared Al/GO composite films. Finally, the obtained results demonstrated significant improvements in thermal stability and mechanical properties of the nanocomposites. Moreover, Abdollahi et al. [114] fabricated alginate-based nanocomposites by adding montmorillonite (MMT) and cellulose nanoparticles (CNT) to alginate biopolymer. The addition of both nanoreinforcements resulted in reduced water vapor permeability, improved tensile strength, and tensile modulus (at ≤ 1 wt.% MMT). Furthermore, Gholizadeh et al. [94] investigated the effect of hydroxyapatite nanoparticles (HA NPs) on different properties of SA films. The researchers discovered that HA NPs additives improved the elongation, tensile strength, and moisture content of the nanocomposite films. Furthermore, the nanocomposites' water solubility and water vapor permeability were reduced by 50% (at 5% HP NPs concentration).

17.3.3.5 Pectin-based composites

Pectin is a heterogeneous grouping of acidic polysaccharides that regulates the mechanical behavior of plant cell walls and can be extracted from the outer cells of most plants. Pectin fits into a family of pectic polysaccharides, in terms of structural resemblances, containing 1,4-linked α-D-galacturonic residues with partial methyl ester groups of carboxylic acid residues and is recognized for its widespread industrial applications [119–121]. Pectin is a naturally abundant, biodegradable,

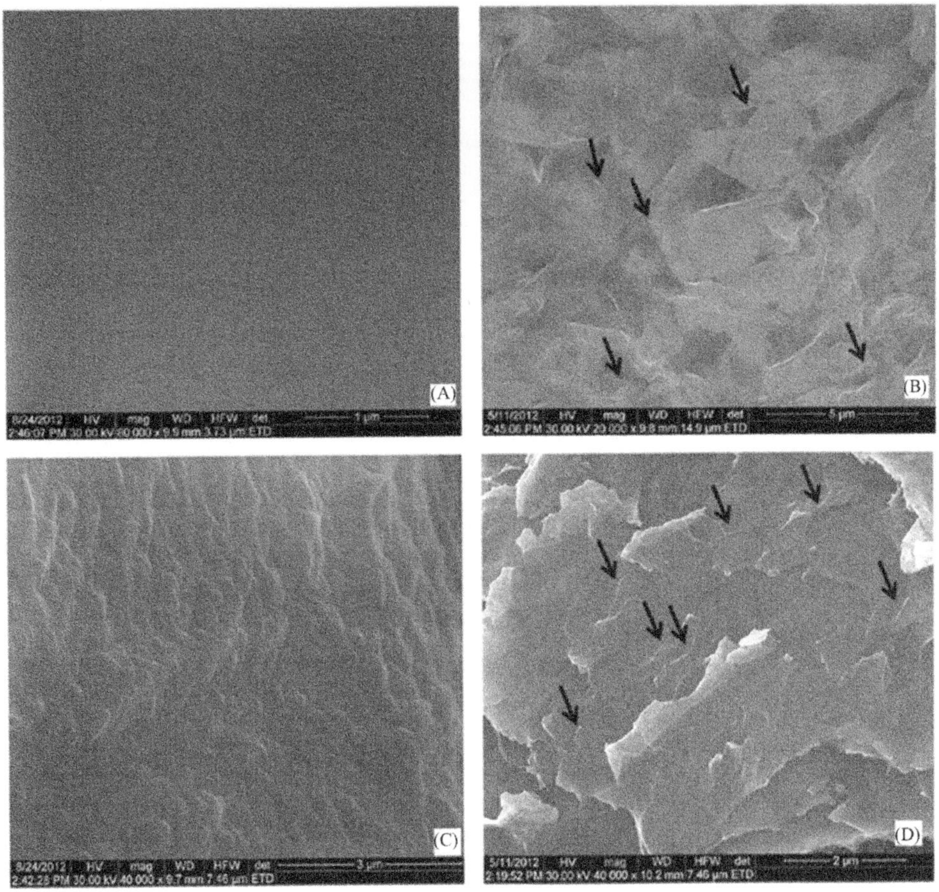

FIGURE 17.17

SEM images of surface (A, B) and cross section (C, D) of pure Al (A, C) and Al/GO (B, D) composite films [118].

economical, hydrophilic, and anionic biopolymer with excellent gelling and stabilizing properties, film-forming capabilities, and low methanol permeability, making it a suitable material for developing polymer electrolyte membranes. However, when compared to chitosan, cellulose, or alginate, PC is the least investigated polysaccharide for fuel cell applications. The utilization of PC-based membranes was first reported by Mishra et al. [122]. They reported the synthesis and characterization of amidated pectin (AP) based polymer electrolyte membranes cross-linked with glutaraldehyde (GA) and achieved a maximum proton conductivity of 1.098×10^{-3} S/cm at room temperature. Recently, Mohanapriya et al. [1] prepared PC-based membranes by blending PC with PVA, which is also a biodegradable polymer, followed by the addition of sulfonated titanium oxide (s-TiO_2) up to 9.75 wt.%. However, the addition of filler (s-TiO_2) amounting beyond 10 wt.% could develop fragility and brittleness in the prepared membranes. The flexible polymeric network of PC-PVA blends is appropriate for dispersing rigid s-TiO_2 nanoparticles, resulting in enhanced proton conductivity and restricted methanol permeability of the nanocomposite

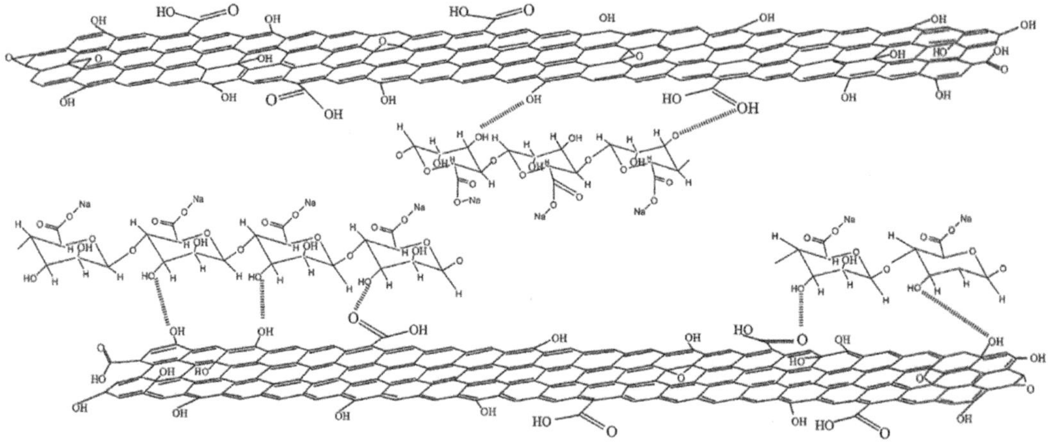

FIGURE 17.18

Schematic depiction of Al/GO nanocomposite structure [118].

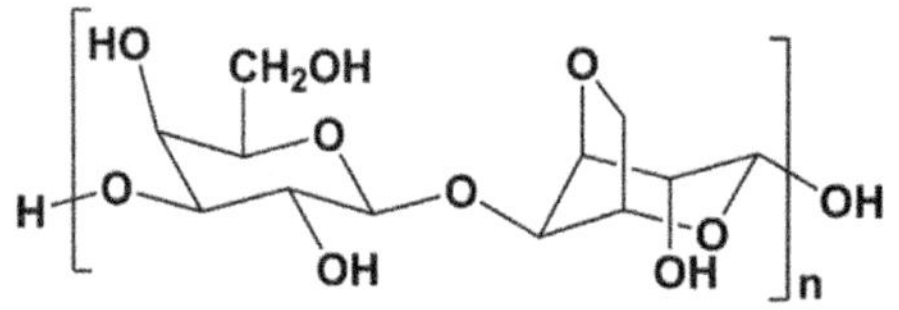

FIGURE 17.19

Chemical structure of agarose.

membranes, according to this study. In DMFCs, the fabricated PC-PVA/s-TiO_2 nanocomposite showed a maximum power density of 27 mW/cm^2 at 70°C. Pectin-based nanocomposite membranes were also prepared using pectin-chitosan blends containing an optimal amount of s-TiO_2, and adequate thermo-mechanical stabilities as well as high charge-density per unit volume were discovered [123].

17.3.3.6 Agar-based composites

Another viable candidate as a matrix for polymer electrolyte membranes is agar, also known as agar-agar. Agar is a strong gelling polysaccharide obtained from certain species of red algae or seaweeds. It is a hydrophilic, biodegradable, low-cost, non-toxic, and chemically stable natural polymer composed of alternating 1,3-linked D-galactose and 1,4-linked 3,6 anhydro-L-galactose unit [124]. Agar, which like starch is a heterogeneous mixture of two polysaccharides: agarose (Fig. 17.19) and agaropectin. Despite having the same galactose backbone, agaropectin is heavily modified with acidic side groups such as sulfate, pyruvate, and so on, whereas agarose is a low-sulfate, neutral gelling fraction of agar [125]. The presence of two types of cations coordinating sites, i.e., hydroxyl (–OH) and ether (–C–O–C–) in the structure and high chains flexibility makes agar capable of trapping a large number of ionic species and facilitating their migration as well [126]. As a result, agar has recently been investigated as a possible biopolymer in the development of polymer electrolyte membranes for fuel cell applications.

Hernández-Flores et al. [127] presented a comparison study between low-cost agar-based membranes, i.e., alternative membranes (AMs), and high-cost Nafion 117 membrane in MFCs. In this work, the performances of the membranes were evaluated by fitting MFCs with either hybrid membrane made of agar and Nafion MH, agar membrane M6 (6% agar), agar membrane M2 (2% agar), or the control Nafion 117 membrane (NF-117). Despite the powers delivered by MFC fitted with AMs were significantly lower (less than half) compared with the power of the cell fitted with Nafion 117, the power/cost ratios were found to be higher (0.9–4.4 mW/US$) than that of the NF-117 membrane (0.23 mW/US$). Therefore, agar-based AMs provide an economical and eco-friendly pathway to develop polymer electrolyte membranes. Besides, Raphael et al. [125] prepared agar-based composites based on agar cross-linked with formaldehyde, plasticized with glycerol, and doped with different amounts of acetic acid (6.3–66.7 wt.%). The obtained membranes were transparent (70%–92%) (Fig. 17.20) and thermally stable (up to 150°C). The acetic acid concentration, temperature, and humidity greatly influence the ion conductivity of the prepared membranes. The highest ionic conductivity values varied from 1.1×10^{-4} S/cm at room temperature and 9.6×10^{-4} S/cm at 80°C were found for samples containing 50 wt.% acetic acids. Moreover, Boopathi et al. [128] fabricated agar-based solid polymer electrolytes membrane doped with ammonium nitrate (NH_4NO_3) by solvent casting method. The amorphous nature of the membrane increases with increasing NH_4NO_3 concentration and the highest ionic conductivity of 6.57×10^{-4} S/cm at room temperature was obtained for a sample with 60 wt.% NH_4NO_3. However, in a recent study, Boopathi et al. [129] demonstrated

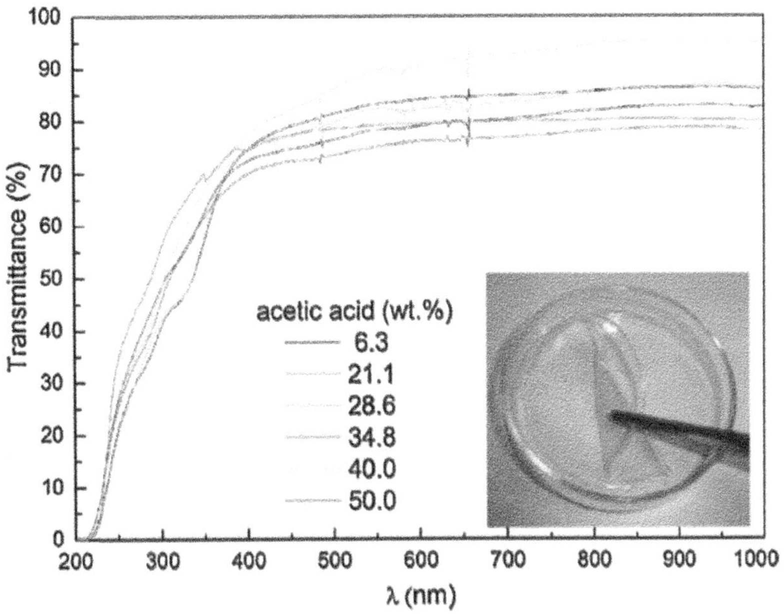

FIGURE 17.20

Optical transmittance measurements of agar-based composites doped with different amounts of acetic acid (6.3–66.7 wt.%) content and membrane image (inset) [125].

that the introduction of plasticizers (glycerol) in agar/NH_4NO_3 system, reported previously [128], led to increased ionic conductivity. The addition of 40 wt.% glycerol hiked the ionic conductivity value to a maximum of 1.44×10^{-3} S/cm for the sample containing 60 wt.% NH_4NO_3 and 40 wt.% agar.

17.3.3.7 Gelatine-based composites

Gelatine is a natural polymer and fibrous protein obtained primarily through the hydrolysis of collagens [130]. It possesses numerous excellent properties, including good film-forming properties, biodegradability, nontoxicity, cost efficiency, and the ability to form a transparent solution with a high viscosity [131]. Because of these properties, gelatine is a very promising material for polymer electrolyte membrane applications. Gelatine-based composites have been studied as fuel cell alternatives over the last few decades. The conductivity of gelatine-based protonic solid polymeric electrolyte films was studied by Vieira et al. [132]. The films were made from uncolored commercial gelatine, glycerol (as a plasticizer), formaldehyde, and varying amounts of acetic acid (17.5%–36 wt.%). This study discovered that the amount of acetic acid in the samples (i.e., films) influences proton conduction, and the film with 26.3% acetic acid content had high ionic conductivity values of 4.5×10^{-5} S/cm at room temperature and 3.6×10^{-4} S/cm at 80°C. Furthermore, the gelatine-based films were both mechanically stable and transparent. Furthermore, the use of hydrochloric acid instead of acetic acid has been reported [133]. The membrane with 0.1 M of HCl showed conductivity of 4.5×10^{-5} S/cm at room temperature. As a result, gelatin-based composites with high ionic conductivity have proven to be extremely promising materials for fuel cell applications.

17.3.3.8 Poly (lactic acid) based composite

Poly (lactic acid) (PLA) is a biodegradable thermoplastic derived from L- and D-lactic acid fermentation of corn starch [134]. It's a semi-crystalline polymer with excellent mechanical properties, biocompatibility, and absorbability [113]. However, it has several drawbacks, including brittleness, low impact strength, and poor resistance to thermal deformation, which limit its widespread application [135]. Besides, the PLA matrix demonstrates poor compatibility with hydrophilic nanofillers due to their hydrophobic nature [136]. It is worth noting that the application of polymer electrolyte membranes necessitates a certain level of water retention ability within the membrane; thus, at least one component of the composite should be hydrophilic. However, hydrophobic polymers are typically incompatible with hydrophilic polymers or filler materials. Qu et al. [137] reported the reduction of mechanical properties of PLA with the addition of cellulose nanofibers. Nonetheless, the addition of poly (ethylene glycol) (PEG) as a compatibilizer to the PLA matrix allowed for the homogeneous dispersion of nanofillers and improved mechanical strength. The preparation method used a hazardous organic solvent, and PEG is a non-biodegradable polymer, so it is not completely environmentally friendly. To the best of our knowledge, no report has been published on the use of PLA-based composites as polymer electrolyte membranes in fuel cells.

17.4 Summary and future prospective

Biodegradable polymer nanocomposites are gaining popularity as materials for various fuel cells due to their low cost and high efficiency. Furthermore, fuel cells such as polymer-based electrolyte membrane based on biodegradable polymer nanocomposites have advanced due to their increased power density and environmental friendliness. Various nanocomposites have been investigated for fuel cell applications to date. In terms of low cost, efficiency, and eco-friendliness, such nanocomposites typically do not provide significant advantages over biodegradable polymer nanocomposites, and, more importantly, the power density of associated fuel cells is lower than that of biodegradable polymer nanocomposites. As a result, not only must power density be improved further, but also cost, efficiency, and environmental pollution be considered for fuel cell applications. To develop the properties of nanocomposites for fuel cell applications, numerous efforts have been prepared, such as modification of synthesis methods, use of various filler materials with the polymer matrix, use of different types of degradable nature polymers, preparation approaches, and so on. These trials aim to improve the properties of degradable polymers in nanocomposites, with or without sacrificing the others. Various biopolymer modifications, such as chemical reforms for the formation of efficient nanocomposites, may be able to improve the properties of fuel cells. However, during such changes, the solubility nature of composite materials must be considered in order to maintain the proper mechanical strength of the nanocomposites.

Biopolymers alone have several drawbacks, including low proton conductivity, insufficient mechanical and thermal stability, and fuel permeability, all of which reduce fuel cell efficiency. Thus, methods of modification such as sulfonation, phosphorylation, chemical crosslinking, blending with other polymers, and incorporating fillers are effective ways of improving fuel cell efficiency. Such performances are determined by factors such as the chemistry of the polymer moiety and the types of filler materials used. Under these conditions, nanocomposites are currently being modified to support industrial applications of fuel cells. Furthermore, in terms of fuel cell performance and environmentally compatible and commercially viable membrane, biodegradable polymer nanocomposites may be an excellent candidate. Furthermore, preparation methods such as blending, sol-gel, and infiltration are critical steps in developing nanocomposite membranes to improve fuel cell performance.

The primary goal at the moment is to develop biodegradable PEMs for fuel cell applications, as the biodegradability of such membranes is a major concern in terms of environmental issues. Biodegradable polymer membranes, such as chitosan, cellulose, alginate, and others, are capable of enhancing the desired properties as well as the effectiveness of functionalization of nanofillers in nanocomposites due to their diverse properties, making them very promising materials for fuel cell applications. However, the number of issues that must be resolved remains relatively large. Overcoming such obstacles would allow us to meet global power demand. Thus, biodegradable polymer nanocomposites in fuel cell applications enable an industrial, economic, and sustainable use of natural resources.

References

[1] S. Mohanapriya, G. Rambabu, S.D. Bhat, V. Raj, Pectin based nanocomposite membranes as green electrolytes for direct methanol fuel cells, Arab. J. Chem. 13 (2020) 2024–2040.

[2] K. Pourzare, Y. Mansourpanah, S. Farhadi, Advanced nanocomposite membranes for fuel cell applications: a comprehensive review, Biofuel Res. J. 3 (2016) 496–513.

[3] X. Xiao, H. Xia, R. Wu, L. Bai, L. Yan, E. Magner, et al., Tackling the challenges of enzymatic (bio) fuel cells, Chem. Rev. 119 (2019) 9509–9558.
[4] Y.-S. Ye, J. Rick, B.-J. Hwang, Water soluble polymers as proton exchange membranes for fuel cells, Polymers 4 (2012) 913–963.
[5] J. Garche, L. Ju Rissen, Applications of fuel cell technology: status and perspectives, Interface Mag. 24 (2015) 39–43.
[6] S. Bhadra, M. Rahaman, P. Noorunnisa Khanam, Electrical and electronic application of polymer–carbon composites, in: M. Rahaman, D. Khastgir, A.K. Aldalbahi (Eds.), Carbon-Containing Polymer Composites, Springer Singapore, Singapore, 2019, pp. 397–455.
[7] I. Staffell, D. Scamman, A. Velazquez Abad, P. Balcombe, P.E. Dodds, P. Ekins, et al., The role of hydrogen and fuel cells in the global energy system, Energy Environ. Sci. 12 (2019) 463–491.
[8] Fuel cell market size, share, and COVID-19 impact analysis. Fortune Business Insights.
[9] R.-A. Felseghi, E. Carcadea, M.S. Raboaca, C.N. Trufin, C. Filote, Hydrogen fuel cell technology for the sustainable future of stationary applications, Energies 12 (2019) 4593.
[10] J. Walkowiak-Kulikowska, J. Wolska, H. Koroniak, Polymers application in proton exchange membranes for fuel cells (PEMFCs), Phys. Sci. Rev. 2 (2017).
[11] ZeroAvia completes world first hydrogen-electric passenger plane flight, ZeroAvia Ltd. https://www.zeroavia.com/press-release-25-09-2020 (accessed July 11, 2021).
[12] A.G. Olabi, T. Wilberforce, M.A. Abdelkareem, Fuel cell application in the automotive industry and future perspective, Energy 214 (2021) 118955.
[13] M. Bellis, Hydrogen fuel cells innovation for the 21st century. https://www.thoughtco.com/hydrogen-fuel-cells-1991799.
[14] Fuel cells works – fuel cells, Energy and Hydrogen News. https://fuelcellsworks.com/knowledge/history/.
[15] Fuel cell history. http://www.esru.strath.ac.uk/EandE/Web_sites/03–04/hydrogen/history.htm.
[16] Fuel cell, Wikipedia. https://en.wikipedia.org/wiki/Fuel_cell (accessed July 11, 2021).
[17] N. Mahato, H. Jang, A. Dhyani, S. Cho, Recent progress in conducting polymers for hydrogen storage and fuel cell applications, Polymers 12 (2020) 2480.
[18] A.K. Mishra, M.C. Valodkar, Polymer nanocomposites for energy and fuel cell applications, in: D.K. Tripathy, B.P. Sahoo (Eds.), Properties and Applications of Polymer Nanocomposites, Springer Berlin Heidelberg, Berlin, Heidelberg, 2017, pp. 107–137.
[19] A. Baroutaji, J.G. Carton, J. Stokes, A.G. Olabi, Design and development of proton exchange membrane fuel cell using open pore cellular foam as flow plate material, J. Energy Chall. Mech. 1 (2014) 95–102.
[20] T. Wilberforce, Z. El-Hassan, F.N. Khatib, A. Al Makky, A. Baroutaji, J.G. Carton, et al., Modelling and simulation of proton exchange membrane fuel cell with serpentine bipolar plate using MATLAB, Int. J. Hydrog. Energy 42 (2017) 25639–25662.
[21] F. de Bruijn, The current status of fuel cell technology for mobile and stationary applications, Green Chem. 7 (2005) 132.
[22] O.Z. Sharaf, M.F. Orhan, An overview of fuel cell technology: fundamentals and applications, Renew. Sustain. Energy Rev. 32 (2014) 810–853.
[23] F. Barbir, Fuel cell electrochemistry, PEM Fuel Cells, Elsevier, 2005, pp. 33–72.
[24] F. Barbir, Fuel cell electrochemistry, PEM Fuel Cells, Elsevier, 2013, pp. 33–72.
[25] F. Barbir, Fuel cells – exploratory fuel cells | Regenerative fuel cells, Encyclopedia of Electrochemical Power Sources, Elsevier, 2009, pp. 224–237.
[26] C. Laberty-Robert, K. Vallé, F. Pereira, C. Sanchez, Design and properties of functional hybrid organic–inorganic membranes for fuel cells, Chem. Soc. Rev. 40 (2011) 961.
[27] Z.S.M. Javaid, T. Matsuura (Eds.), Polymer Membranes for Fuel Cells, Springer US, Boston, MA, 2009.

[28] M. Gouda, A.H. Konsowa, H.A. Farag, N.A. Elessawy, T.M. Tamer, M.S. Mohy Eldin, Novel nanocomposite membranes based on cross-linked eco-friendly polymers doped with sulfated titania nanotubes for direct methanol fuel cell application, Nanomater. Nanotechnol. 10 (2020). Available from: https://doi.org/10.1177/1847980420964368.

[29] Z. Zakaria, S.K. Kamarudin, S.N. Timmiati, Membranes for direct ethanol fuel cells: an overview, Appl. Energy 163 (2016) 334–342.

[30] G. Zhao, Y. Chen, X.-F. Li, S. Zhang, Y. Situ, Fabrication of highly proton-conductive chitosan whole-bio-membrane materials functionalized with adenine and adenosine monophosphate, Green Chem. 22 (2020) 2426–2433.

[31] J. Kalaiselvimary, M. Ramesh Prabhu, Fabrications and investigation of physicochemical and electrochemical properties of heteropoly acid-doped sulfonated chitosan-based polymer electrolyte membranes for fuel cell applications, Polym. Bull. 76 (2019) 1401–1422.

[32] M.J. González-Pabón, F. Figueredo, D.C. Martínez-Casillas, E. Cortón, Characterization of a new composite membrane for point of need paper-based micro-scale microbial fuel cell analytical devices, PLoS One 14 (2019) e0222538.

[33] E.T. Sayed, T. Eisa, H.O. Mohamed, M.A. Abdelkareem, A. Allagui, H. Alawadhi, et al., Direct urea fuel cells: challenges and opportunities, J. Power Sources 417 (2019) 159–175.

[34] M. Agostini de Moraes, C.F. Silva, R.S. da, Vieira, Biopolymer Membranes and Films Health, Food, Environment, and Energy Applications, Elsevier, Amsterdam, 2020.

[35] I. Merino-Jiménez, C. Ponce de León, A.A. Shah, F.C. Walsh, Developments in direct borohydride fuel cells and remaining challenges, J. Power Sources 219 (2012) 339–357.

[36] N. Agmon, The Grotthuss mechanism, Chem. Phys. Lett. 244 (1995) 456–462.

[37] K.-D. Kreuer, A. Rabenau, W. Weppner, Vehicle mechanism, a new model for the interpretation of the conductivity of fast proton conductors, Angew. Chem. Int. Ed. Engl. 21 (1982) 208–209.

[38] Y. Ren, G.H. Chia, Z. Gao, Metal–organic frameworks in fuel cell technologies, Nano Today 8 (2013) 577–597.

[39] S.J. Peighambardoust, S. Rowshanzamir, M. Amjadi, Review of the proton exchange membranes for fuel cell applications, Int. J. Hydrog. Energy 35 (2010) 9349–9384.

[40] M. Pandey, G.M. Joshi, K. Deshmukh, M. Khutia, N.N. Ghosh, Optimized AC conductivity correlated to structure, morphology and thermal properties of PVDF/PVA/nafion composites, Ionics 20 (2014) 1427–1433.

[41] J. Ma, Y. Sahai, Chitosan biopolymer for fuel cell applications, Carbohydr. Polym. 92 (2013) 955–975.

[42] B. Smitha, S. Sridhar, A.A. Khan, Chitosan–poly(vinyl pyrrolidone) blends as membranes for direct methanol fuel cell applications, J. Power Sources 159 (2006) 846–854.

[43] N.A. Choudhury, J. Ma, Y. Sahai, High performance and eco-friendly chitosan hydrogel membrane electrolytes for direct borohydride fuel cells, J. Power Sources 210 (2012) 358–365.

[44] D.J. Kim, M.J. Jo, S.Y. Nam, A review of polymer–nanocomposite electrolyte membranes for fuel cell application, J. Ind. Eng. Chem. 21 (2015) 36–52.

[45] A.-C. Dupuis, Proton exchange membranes for fuel cells operated at medium temperatures: materials and experimental techniques, Prog. Mater. Sci. 56 (2011) 289–327.

[46] J.K. Pandey, A.P. Kumar, M. Misra, A.K. Mohanty, L.T. Drzal, R. Palsingh, Recent advances in biodegradable nanocomposites, J. Nanosci. Nanotechnol. 5 (2005) 497–526.

[47] C. Zinge, B. Kandasubramanian, Nanocellulose based biodegradable polymers, Eur. Polym. J. 133 (2020) 109758.

[48] J.-B. Zeng, Y.-S. He, S.-L. Li, Y.-Z. Wang, Chitin whiskers: an overview, Biomacromolecules 13 (2012) 1–11.

[49] H.-Y. Kim, S.S. Park, S.-T. Lim, Preparation, characterization and utilization of starch nanoparticles, Colloids Surf. B Biointerfaces 126 (2015) 607–620.
[50] I. Armentano, D. Puglia, F. Luzi, C. Arciola, F. Morena, S. Martino, et al., Nanocomposites based on biodegradable polymers, Materials 11 (2018) 795.
[51] T.C. Mokhena, A.S. Luyt, Investigation of polyethylene/sisal whiskers nanocomposites prepared under different conditions, Polym. Compos. 35 (2014) 2221–2233.
[52] X. Hou, Z. Liu, Y. Wei, Q. Zhao, J. Dong, B. Liu, et al., Proton conducting nanocomposite membranes of nanocellulose reinforced poly(arylene ether ketone)s containing sulfonic/carboxylic groups, Solid State Ion. 311 (2017) 31–40.
[53] X. Xu, G. Zhao, H. Wang, X. Li, X. Feng, B. Cheng, et al., Bio-inspired amino-acid-functionalized cellulose whiskers incorporated into sulfonated polysulfone for proton exchange membrane, J. Power Sources 409 (2019) 123–131.
[54] Y. Aitomäki, K. Oksman, Reinforcing efficiency of nanocellulose in polymers, React. Funct. Polym. 85 (2014) 151–156.
[55] M.-C. Li, Q. Wu, K. Song, H.N. Cheng, S. Suzuki, T. Lei, Chitin nanofibers as reinforcing and antimicrobial agents in carboxymethyl cellulose films: influence of partial deacetylation, ACS Sustain. Chem. Eng. 4 (2016) 4385–4395.
[56] A.M. Salaberria, J. Labidi, S.C.M. Fernandes, Different routes to turn chitin into stunning nano-objects, Eur. Polym. J. 68 (2015) 503–515.
[57] M. Nasirinezhad, S.R. Ghaffarian, M. Tohidian, Eco-friendly polyelectrolyte nanocomposite membranes based on chitosan and sulfonated chitin nanowhiskers for fuel cell applications, Iran. Polym. J. 30 (2021) 355–367.
[58] C. Zhang, X. Zhuang, X. Li, W. Wang, B. Cheng, W. Kang, et al., Chitin nanowhisker-supported sulfonated poly(ether sulfone) proton exchange for fuel cell applications, Carbohydr. Polym. 140 (2016) 195–201.
[59] P. Mukoma, B.R. Jooste, H.C.M. Vosloo, Synthesis and characterization of cross-linked chitosan membranes for application as alternative proton exchange membrane materials in fuel cells, J. Power Sources 136 (2004) 16–23.
[60] S. Wang, L. Shi, S. Zhang, H. Wang, B. Cheng, X. Zhuang, et al., Proton-conducting amino acid-modified chitosan nanofibers for nanocomposite proton exchange membranes, Eur. Polym. J. 119 (2019) 327–334.
[61] D. Le Corre, J. Bras, A. Dufresne, Starch nanoparticles: a review, Biomacromolecules 11 (2010) 1139–1153.
[62] H. Tang, T. Mitsunaga, Y. Kawamura, Molecular arrangement in blocklets and starch granule architecture, Carbohydr. Polym. 63 (2006) 555–560.
[63] A. Dufresne, Polysaccharide nano crystal reinforced nanocomposites, Can. J. Chem. 86 (2008) 484–494.
[64] J.-L. Putaux, S. Molina-Boisseau, T. Momaur, A. Dufresne, Platelet nanocrystals resulting from the disruption of waxy maize starch granules by acid hydrolysis, Biomacromolecules 4 (2003) 1198–1202.
[65] H. Angellier, S. Molina-Boisseau, P. Dole, A. Dufresne, Thermoplastic starch – waxy maize starch nanocrystals nanocomposites, Biomacromolecules 7 (2006) 531–539.
[66] A. Dufresne, J.-Y. Cavaille, Clustering and percolation effects in microcrystalline starch-reinforced thermoplastic, J. Polym. Sci. Part B Polym. Phys. 36 (1998) 2211–2224.
[67] H. Zou, S. Wu, J. Shen, Polymer/silica nanocomposites: preparation, characterization, properties, and applications, Chem. Rev. 108 (2008) 3893–3957.
[68] L. Zhang, S.-R. Chae, Z. Hendren, J.-S. Park, M.R. Wiesner, Recent advances in proton exchange membranes for fuel cell applications, Chem. Eng. J. 204–206 (2012) 87–97.
[69] A. Harish Kumar, M.B. Ahamed, K. Deshmukh, M.S. Sirajuddeen, Morphology, dielectric and emi shielding characteristics of graphene nanoplatelets, montmorillonite nanoclay and titanium dioxide

nanoparticles reinforced polyvinylidenefluoride nanocomposites, J. Inorg. Organomet. Polym. Mater. 31 (2021) 2003–2016.
[70] E. Bakangura, L. Wu, L. Ge, Z. Yang, T. Xu, Mixed matrix proton exchange membranes for fuel cells: state of the art and perspectives, Prog. Polym. Sci. 57 (2016) 103–152.
[71] P. Lokanatha Reddy, K. Deshmukh, K. Chidambaram, B. Ahamed, K. Kumar Sadasivuni, D. Ponnamma, et al., Effect of poly ethylene glycol (peg) on structural, thermal and photoluminescence properties of Cdo nanoparticles for optoelectronic applications, Mater. Today Proc. 9 (2019) 175–183.
[72] J. Kalaiselvimary, M. Sundararajan, M.R. Prabhu, Preparation and characterization of chitosan-based nanocomposite hybrid polymer electrolyte membranes for fuel cell application, Ionics 24 (2018) 3555–3571.
[73] J. Wang, Y. Zhao, W. Hou, J. Geng, L. Xiao, H. Wu, et al., Simultaneously enhanced methanol barrier and proton conductive properties of phosphorylated titanate nanotubes embedded nanocomposite membranes, J. Power Sources 195 (2010) 1015–1023.
[74] A. Muzaffar, M.B. Ahamed, K. Deshmukh, S.K.K. Pasha, Dielectric properties and electromagnetic interference shielding studies of nickel oxide and tungsten oxide reinforced polyvinylchloride nanocomposites, Polym-Plast Technol. Mater. 59 (2020) 1667–1678.
[75] P. Rani, M.B. Ahamed, K. Deshmukh, Structural, dielectric and EMI shielding properties of polyvinyl alcohol/chitosan blend nanocomposites integrated with graphite oxide and nickel oxide nanofillers, J. Mater. Sci. Mater Electron. 32 (2021) 764–779.
[76] M. Khutia, G.M. Joshi, K. Deshmukh, M. Pandey, Optimization of dielectric constant of polycarbonate/polystyrene modified blend by ceramic metal oxide, Polym-Plast Technol. Eng. 54 (2015) 383–389.
[77] M. Khutia, G.M. Joshi, K. Deshmukh, M. Pandey, Preparation of modified polymer blend and electrical performance, Compos. Interfaces 22 (2015) 167–178.
[78] M.M. Rahman Khan, M. Akter, Md.K. Amin, M. Younus, N. Chakraborty, Synthesis, luminescence and thermal properties of PVA–ZnO–Al_2O_3 composite films: towards fabrication of sunlight-induced catalyst for organic dye removal, J. Polym. Environ. 26 (2018) 3371–3381.
[79] M. Pandey, G.M. Joshi, S. Bhattacharya, Influence of $K_2Ti_6O_{13}$ on dielectric and barrier properties of polymer, Ionics 22 (2016) 445–449.
[80] A. Shabanikia, M. Javanbakht, H.S. Amoli, K. Hooshyari, M. Enhessari, Polybenzimidazole/strontium cerate nanocomposites with enhanced proton conductivity for proton exchange membrane fuel cells operating at high temperature, Electrochim. Acta 154 (2015) 370–378.
[81] K. Hooshyari, M. Javanbakht, A. Shabanikia, M. Enhessari, Fabrication $BaZrO_3$/PBI-based nanocomposite as a new proton conducting membrane for high temperature proton exchange membrane fuel cells, J. Power Sources 276 (2015) 62–72.
[82] J. Wang, X. Zheng, H. Wu, B. Zheng, Z. Jiang, X. Hao, et al., Effect of zeolites on chitosan/zeolite hybrid membranes for direct methanol fuel cell, J. Power Sources 178 (2008) 9–19.
[83] P. Rani, B. Ahamed, K. Deshmukh, Dielectric and electromagnetic interference shielding properties of zeolite 13X and carbon black nanoparticles based PVDF nanocomposites, J. Appl. Polym. Sci. 138 (2021) 50107.
[84] H. Wu, B. Zheng, X. Zheng, J. Wang, W. Yuan, Z. Jiang, Surface-modified Y zeolite-filled chitosan membrane for direct methanol fuel cell, J. Power Sources 173 (2007) 842–852.
[85] Y. Wang, D. Yang, X. Zheng, Z. Jiang, J. Li, Zeolite beta-filled chitosan membrane with low methanol permeability for direct methanol fuel cell, J. Power Sources 183 (2008) 454–463.
[86] W. Yuan, H. Wu, B. Zheng, X. Zheng, Z. Jiang, X. Hao, et al., Sorbitol-plasticized chitosan/zeolite hybrid membrane for direct methanol fuel cell, J. Power Sources 172 (2007) 604–612.
[87] Y. Wang, Z. Jiang, H. Li, D. Yang, Chitosan membranes filled by GPTMS-modified zeolite beta particles with low methanol permeability for DMFC, Chem. Eng. Process. Process Intensif. 49 (2010) 278–285.

[88] B. Libby, W.H. Smyrl, E.L. Cussler, Polymer-zeolite composite membranes for direct methanol fuel cells, AIChE J. 49 (2003) 991–1001.

[89] S. Wafiroh, A. Abdulloh, Phosphorylated zeolite-A/chitosan composites as proton exchange membrane fuel cell, Chem. Chem Technol. 12 (2018) 229–235.

[90] A.K. Mishra, S. Bose, T. Kuila, N.H. Kim, J.H. Lee, Silicate-based polymer-nanocomposite membranes for polymer electrolyte membrane fuel cells, Prog. Polym. Sci. 37 (2012) 842–869.

[91] S. He, H. Jia, Y. Lin, H. Qian, J. Lin, Effect of clay modification on the structure and properties of sulfonated poly(ether ether ketone)/clay nanocomposites, Polym. Compos. 37 (2016) 2632–2638.

[92] M. Purwanto, L. Atmaja, M.A. Mohamed, M.T. Salleh, J. Jaafar, A.F. Ismail, et al., Biopolymer-based electrolyte membranes from chitosan incorporated with montmorillonite-crosslinked GPTMS for direct methanol fuel cells, RSC Adv. 6 (2016) 2314–2322.

[93] Y. Zhao, Z. Jiang, L. Xiao, T. Xu, S. Qiao, H. Wu, Chitosan membranes filled with biomimetic mineralized hydroxyapatite for enhanced proton conductivity, Solid State Ion. 187 (2011) 33–38.

[94] B.S. Gholizadeh, F. Buazar, S.M. Hosseini, S.M. Mousavi, Enhanced antibacterial activity, mechanical and physical properties of alginate/hydroxyapatite bionanocomposite film, Int. J. Biol. Macromol. 116 (2018) 786–792.

[95] A.D. Oliveira, C.A.G. Beatrice, F.R. Passador, L.A. Pessan, Polyetherimide-Based Nanocomposites Materials for Hydrogen Storage (2016) 040006. Graz, Austria.

[96] D.M. Yu, I.H. Sung, Y.J. Yoon, T.-H. Kim, J.Y. Lee, Y.T. Hong, Properties of sulfonated poly(arylene ether sulfone)/functionalized carbon nanotube composite membrane for high temperature PEMFCs, Fuel Cell 15 (2013) 843–850.

[97] R. Kannan, H.N. Kagalwala, H.D. Chaudhari, U.K. Kharul, S. Kurungot, V.K. Pillai, Improved performance of phosphonated carbon nanotube–polybenzimidazole composite membranes in proton exchange membrane fuel cells, J. Mater. Chem. 21 (2011) 7223.

[98] J.-M. Thomassin, J. Kollar, G. Caldarella, A. Germain, R. Jérôme, C. Detrembleur, Beneficial effect of carbon nanotubes on the performances of nafion membranes in fuel cell applications, J. Membr. Sci. 303 (2007) 252–257.

[99] M. Tohidian, S.R. Ghaffarian, Polyelectrolyte nanocomposite membranes based on chitosan and surface modified multi-walled carbon nanotubes for use in fuel cell applications, J. Macromol. Sci. Part A (2021) 1–14.

[100] S. Yun, Y. Heo, H. Im, J. Kim, Sulfonated multiwalled carbon nanotube/sulfonated poly(ether sulfone) composite membrane with low methanol permeability for direct methanol fuel cells, J. Appl. Polym. Sci. 126 (2012) E513–E521.

[101] N. Üregen, K. Pehlivanoğlu, Y. Özdemir, Y. Devrim, Development of polybenzimidazole/graphene oxide composite membranes for high temperature PEM fuel cells, Int. J. Hydrog. Energy 42 (2017) 2636–2647.

[102] T. Remiš, P. Bělský, T. Kovářík, J. Kadlec, M. Ghafouri Azar, R. Medlín, et al., Study on structure, thermal behavior, and viscoelastic properties of nanodiamond-reinforced poly (vinyl alcohol) nanocomposites, Polymers 13 (2021) 1426.

[103] B.P. Tripathi, V.K. Shahi, Organic–inorganic nanocomposite polymer electrolyte membranes for fuel cell applications, Prog. Polym. Sci. 36 (2011) 945–979.

[104] A.D. Pomogailo, Polymer sol-gel synthesis of hybrid nanocomposites, Colloid J. 67 (2005) 658–677.

[105] V. Bounor-Legaré, P. Cassagnau, In situ synthesis of organic–inorganic hybrids or nanocomposites from sol–gel chemistry in molten polymers, Prog. Polym. Sci. 39 (2014) 1473–1497.

[106] I. Jankowska, R. Pankiewicz, K. Pogorzelec-Glaser, P. Ławniczak, A. Łapiński, J. Tritt-Goc, Comparison of structural, thermal and proton conductivity properties of micro- and nanocelluloses, Carbohydr. Polym. 200 (2018) 536–542.

[107] Z. Ghassemi, G. Slaughter, Biological fuel cells and membranes, Membranes 7 (2017) 3.

[108] S.S. Etuk, I. Lawan, W. Zhou, Y. Jiang, Q. Zhang, X. Wei, et al., Synthesis and characterization of triazole based sulfonated nanocrystalline cellulose proton conductor, Cellulose 27 (2020) 3197–3209.
[109] A. Sriruangrungkamol, W. Chonkaew, Modification of nanocellulose membrane by impregnation method with sulfosuccinic acid for direct methanol fuel cell applications, Polym. Bull. 78 (2021) 3705–3728.
[110] T. Bayer, B.V. Cunning, R. Selyanchyn, M. Nishihara, S. Fujikawa, K. Sasaki, et al., High temperature proton conduction in nanocellulose membranes: paper fuel cells, Chem. Mater. 28 (2016) 4805–4814.
[111] L. Dai, H. Yu, J. Zhang, F. Cheng, Preparation and characterization of cross-linked starch nanocrystals and self-reinforced starch-based nanocomposite films, Int. J. Biol. Macromol. 181 (2021) 868–876.
[112] L.A. Obasi, C.C. Opara, A. Oji, Performance of cassava starch as a proton exchange membrane in a dual chambered microbial fuel cell, Int. J. Eng. Sci. Technol. 4 (2012) 227–238.
[113] S. Bandehali, H. Sanaeepur, A. Ebadi Amooghin, S. Shirazian, S. Ramakrishna, Biodegradable polymers for membrane separation, Sep. Purif. Technol. 269 (2021) 118731.
[114] M. Abdollahi, M. Alboofetileh, M. Rezaei, R. Behrooz, Comparing physico-mechanical and thermal properties of alginate nanocomposite films reinforced with organic and/or inorganic nanofillers, Food Hydrocoll. 32 (2013) 416–424.
[115] S.N. Pawar, K.J. Edgar, Alginate derivatization: a review of chemistry, properties and applications, Biomaterials 33 (2012) 3279–3305.
[116] L. Yang, G. Liang, Z. Zhang, S. He, J. Wang, Sodium alginate/Na^+-rectorite composite films: preparation, characterization, and properties, J. Appl. Polym. Sci. 114 (2009) 1235–1240.
[117] B. Smitha, S. Sridhar, A.A. Khan, Chitosan–sodium alginate polyion complexes as fuel cell membranes, Eur. Polym. J. 41 (2005) 1859–1866.
[118] M. Ionita, M.A. Pandele, H. Iovu, Sodium alginate/graphene oxide composite films with enhanced thermal and mechanical properties, Carbohydr. Polym. 94 (2013) 339–344.
[119] B.R. Thakur, R.K. Singh, A.K. Handa, M.A. Rao, Chemistry and uses of pectin—a review, Crit. Rev. Food Sci. Nutr. 37 (1997) 47–73.
[120] C.D. May, Industrial pectins: sources, production and applications, Carbohydr. Polym. 12 (1990) 79–99.
[121] P.J.P. Espitia, W.-X. Du, R.de J. Avena-Bustillos, N.de F.F. Soares, T.H. McHugh, Edible films from pectin: physical-mechanical and antimicrobial properties—a review, Food Hydrocoll. 35 (2014) 287–296.
[122] R.K. Mishra, A. Anis, S. Mondal, M. Dutt, A.K. Banthia, Preparation and characterization of amidated pectin based polymer electrolyte membranes, Chin. J. Polym. Sci. 27 (2009) 639–646.
[123] M. Subramanian, V. Raj, Preparation and characterization of pectin based proton exchange membranes derived by solution casting method for direct methanol fuel cells, Int. J. Chem. Mol. Eng. 11 (2017) 639–644.
[124] P. Guerrero, A. Etxabide, I. Leceta, M. Peñalba, K. de la Caba, Extraction of agar from Gelidium sesquipedale (Rodhopyta) and surface characterization of agar based films, Carbohydr. Polym. 99 (2014) 491–498.
[125] E. Raphael, C.O. Avellaneda, B. Manzolli, A. Pawlicka, Agar-based films for application as polymer electrolytes, Electrochim. Acta 55 (2010) 1455–1459.
[126] S.C. Nunes, R.F.P. Pereira, N. Sousa, M.M. Silva, P. Almeida, F.M.L. Figueiredo, et al., Eco-friendly red seaweed-derived electrolytes for electrochemical devices, Adv. Sustain. Syst. 1 (2017) 1700070.
[127] G. Hernández-Flores, H.M. Poggi-Varaldo, O. Solorza-Feria, Comparison of alternative membranes to replace high cost Nafion ones in microbial fuel cells, Int. J. Hydrog. Energy 41 (2016) 23354–23362.
[128] G. Boopathi, S. Pugalendhi, S. Selvasekarapandian, M. Premalatha, S. Monisha, G. Aristatil, Development of proton conducting biopolymer membrane based on agar–agar for fuel cell, Ionics 23 (2017) 2781–2790.

[129] G. Boopathi, S. Pugalendhi, P. Subramanian, D. Manohar Jesudas, K.S. Subramanian, Performance evaluation of single PEM fuel cell (PEMFC) with novel bio-degradable polymer membrane based on agar-agar, Environ. Ecol. 37 (2019) 198–203.

[130] T. Furuike, T. Chaochai, T. Okubo, T. Mori, H. Tamura, Fabrication of nonwoven fabrics consisting of gelatin nanofibers cross-linked by glutaraldehyde or N-acetyl-d-glucosamine by aqueous method, Int. J. Biol. Macromol. 93 (2016) 1530–1538.

[131] N.S. Hettiarachchy, G.R. Ziegler, Protein Functionality in Food Systems, CRC Press, 1994.

[132] D.F. Vieira, C.O. Avellaneda, A. Pawlicka, Conductivity study of a gelatin-based polymer electrolyte, Electrochim. Acta 53 (2007) 1404–1408.

[133] A. Pawlicka, R.I. Mattos, J.F. Lima, C.E. Tambelli, C.J. Magon, J.P. Donoso, Magnetic resonance and conductivity study of a gelatin-based polymer gel electrolyte, Electrochim. Acta 57 (2011) 187–191.

[134] J. Lunt, Large-scale production, properties and commercial applications of polylactic acid polymers, Polym. Degrad. Stab. 59 (1998) 145–152.

[135] C.-S. Wu, Characterizing biodegradation of PLA and PLA-g-AA/starch films using a phosphate-solubilizingbacillus species, Macromol. Biosci. 8 (2008) 560–567.

[136] J. Lu, P. Askeland, L.T. Drzal, Surface modification of microfibrillated cellulose for epoxy composite applications, Polymer 49 (2008) 1285–1296.

[137] P. Qu, Y. Gao, G. Wu, L. Zhang, Nanocomposites of poly(lactic acid) reinforced with cellulose nanofibrils, BioResources 5 (2010) 1811–1823.

[138] Q. Zhao, Y. Wei, C. Ni, L. Wang, B. Liu, J. Liu, et al., Effect of aminated nanocrystal cellulose on proton conductivity and dimensional stability of proton exchange membranes, Appl. Surf. Sci. 466 (2019) 691–702.

[139] Y. Peng, Y. Wang, X. Wei, J. Zhou, H. Peng, L. Xiao, et al., Sulfonated nanobamboo fiber-reinforced quaternary ammonia poly(ether ether ketone) membranes for alkaline polymer electrolyte fuel cells, ACS Appl. Mater. Interfaces 10 (2018) 33581–33588.

[140] T.D.O. Gadim, A.G.P.R. Figueiredo, N.C. Rosero-Navarro, C. Vilela, J.A.F. Gamelas, A. Barros-Timmons, et al., Nanostructured bacterial cellulose–poly(4-styrene sulfonic acid) composite membranes with high storage modulus and protonic conductivity, ACS Appl. Mater. Interfaces 6 (2014) 7864–7875.

[141] M. Ranjani, M. Pannipara, A.G. Al-Sehemi, A. Vignesh, G.G. kumar, Chitosan/sulfonated graphene oxide/silica nanocomposite membranes for direct methanol fuel cells, Solid State Ion. 338 (2019) 153–160.

[142] X. Cheng, J. Wang, Y. Liao, C. Li, Z. Wei, Enhanced conductivity of anion-exchange membrane by incorporation of quaternized cellulose nanocrystal, ACS Appl. Mater. Interfaces 10 (2018) 23774–23782.

[143] G. Jiang, J. Zhang, J. Qiao, Y. Jiang, H. Zarrin, Z. Chen, et al., Bacterial nanocellulose/Nafion composite membranes for low temperature polymer electrolyte fuel cells, J. Power Sources 273 (2015) 697–706.

[144] Y. Ou, W.-C. Tsen, C. Gong, J. Wang, H. Liu, G. Zheng, et al., Chitosan-based composite membranes containing chitosan-coated carbon nanotubes for polymer electrolyte membranes, Polym. Adv. Technol. 29 (2018) 612–622.

[145] Y. Ou, W.-C. Tsen, S.-C. Jang, F.-S. Chuang, J. Wang, H. Liu, et al., Novel composite polymer electrolyte membrane using solid superacidic sulfated zirconia-functionalized carbon nanotube modified chitosan, Electrochim. Acta 264 (2018) 251–259.

[146] S.E. Shakeri, S.R. Ghaffarian, M. Tohidian, G. Bahlakeh, S. Taranejoo, Polyelectrolyte nanocomposite membranes, based on chitosan-phosphotungstic acid complex and montmorillonite for fuel cells applications, J. Macromol. Sci. Part B 52 (2013) 1226–1241.

[147] Q. Wu, H. Wang, S. Lu, X. Xu, D. Liang, Y. Xiang, Novel methanol-blocking proton exchange membrane achieved via self-anchoring phosphotungstic acid into chitosan membrane with submicro-pores, J. Membr. Sci. 500 (2016) 203–210.

[148] P.-C. Li, G. Liao, S.R. Kumar, C.-M. Shih, C.-C. Yang, D.-M. Wang, et al., Fabrication and characterization of chitosan nanoparticle-incorporated quaternized poly(vinyl alcohol) composite membranes as solid electrolytes for direct methanol alkaline fuel cells, Electrochim. Acta 187 (2016) 616–628.

[149] G.-M. Liao, C.-C. Yang, C.-C. Hu, Y.-L. Pai, S.J. Lue, Novel quaternized polyvinyl alcohol/quaternized chitosan nano-composite as an effective hydroxide-conducting electrolyte, J. Membr. Sci. 485 (2015) 17–29.

[150] M. Santamaria, C.M. Pecoraro, F. Di Quarto, P. Bocchetta, Chitosan–phosphotungstic acid complex as membranes for low temperature H_2-O_2 fuel cell, J. Power Sources 276 (2015) 189–194.

[151] F. Song, Y. Fu, Y. Gao, J. Li, J. Qiao, X.-D. Zhou, et al., Novel alkaline anion-exchange membranes based on chitosan/ethenylmethylimidazoliumchloride polymer with ethenylpyrrolidone composites for low temperature polymer electrolyte fuel cells, Electrochim. Acta 177 (2015) 137–144.

[152] Y. Liu, J. Wang, H. Zhang, C. Ma, J. Liu, S. Cao, et al., Enhancement of proton conductivity of chitosan membrane enabled by sulfonated graphene oxide under both hydrated and anhydrous conditions, J. Power Sources 269 (2014) 898–911.

[153] V. Vijayakumar, D. Khastgir, Hybrid composite membranes of chitosan/sulfonated polyaniline/silica as polymer electrolyte membrane for fuel cells, Carbohydr. Polym. 179 (2018) 152–163.

CHAPTER

18 Biodegradable polymer nanocomposites for food packaging applications

Fatih Tornuk and Perihan Kubra Akman

Food Engineering Department, Chemical and Metallurgical Engineering Faculty, Yildiz Technical University, Esenler, Istanbul, Turkey

Chapter Outline

18.1 Introduction

A packaging material's primary function is to protect food from environmental effects such as air, humidity, and microbial organisms by providing a physical barrier while also retaining its initial nutritional properties. For centuries, traditional food packaging materials such as plastics, glass, metals, wrapping paper, and board have been used. Plastics have increasingly dominated packaging applications since the synthesis and commercialization of polyethylene in the middle of the twentieth century, owing to their versatility, good mechanical barrier properties, and low cost [1]. Annual

Biodegradable and Biocompatible Polymer Nanocomposites. DOI: https://doi.org/10.1016/B978-0-323-91696-7.00006-4
Copyright © 2023 Elsevier Inc. All rights reserved.

global plastic production has increased dramatically since the 1950s, reaching nearly 368 million metric tons in 2019 [2]. In Europe, an estimated 8 million tons of plastic are used for food products, with plastic output increasing by 4% each year [3]. Packaging is the largest consumer of plastics, and plastic materials are the second most common packaging fabric used in packaging applications [4].

Plastic packaging materials are primarily synthetic and petroleum-based. The major petroleum-based plastic packaging materials are polyethylene, polypropylene, polyvinyl chloride, polyethylene terephthalate, and polystyrene, with polyethylene being the most common and used for both rigid and flexible packaging materials production [5]. Polypropylene has several advantages, including high chemical stability and mechanical properties, as well as a high melting point, which makes it suitable for hot filling processes. The most common material used for packaging of both drinking water and carbonated beverages is polyethylene terephthalate in the form of a bottle [6].

The vast majority of plastic packaging materials are nonrenewable and nonbiodegradable. When they reach the end of their useful life, they can be recycled or recovered, or they can be discarded into the environment. Plastic-based environmental wastes cause a serious plastic pollution problem in ecosystems such as lands and oceans because biodegradation takes a long time (>100 years). It is estimated that one-third of the plastic waste generated in oceans and landfills each year is the result of petroleum-derived plastics used as food packaging materials [7]. Another concern of plastic pollution that threatens living organisms and even human life is the accumulation of microplastics, plastic particles smaller than 5 mm in size formed by fragmentation and degradation of waste plastics [8,9]. Based on these global environmental problems caused by increasing plastic accumulation, there has been an increase in public concern to reduce plastic waste and use eco-friendly alternatives to petrochemical-based plastics.

Bio-based and/or biodegradable polymers are the most promising alternatives of nonbiodegradable synthetic plastics for food packaging applications. A wide range of bioplastics with varying properties (hydrophilic, hydrophilic, etc.) are available, including starch, cellulose, polyhydroxyalkanoates (PHAs), polylactic acid (PLA), polycaprolactone (PCL), alginates, zein, and chitosan derived from biomass and microorganisms. Despite the fact that different biopolymers exhibit different polymeric properties, they are still far from being a complete replacement for petrochemical-based plastics due to several drawbacks such as high brittleness, poor processability, and high gas permeability. Several solutions, such as blending with synthetic polymers, chemical/physical modification, and nanoparticle incorporation, have been proposed to overcome these drawbacks [10,11]. One of the most promising and cost-effective routes for improving biopolymer properties is the fabrication of bionanocomposites, which are multiphase materials composed of at least two components, biopolymer and nanofillers [12]. Overall, the goal of this chapter is to provide a synopsis of recent discoveries pertaining to the properties of bionanocomposites as food packaging materials.

18.2 Biodegradable polymers: their classification and properties

Biodegradable polymers are defined as biopolymers which can be decomposed to simple molecules like CO_2 and H_2O or biomass, which are nontoxic for environment, through enzymatic reactions realized by naturally-occurring organisms [13]. In the early 1970s, scientists and several industries began to pay attention to biopolymers in order to solve the "landfill crisis" by biodegrading waste plastic materials. Almost all plastic materials are not recyclable and eventually end up in landfills due to their short lifespan. This results in an annual economic loss of \$80–120 billion [14]. As a result of their biodegradable, renewable, and abundant properties, biopolymers have established themselves in the industry as an alternative material to petroleum-based plastics [15]. Fig. 18.1 shows categorization of biodegradable polymers based on their origins and manufacturing process [16].

18.2.1 Biodegradable polymers from bioresources

Natural polymers have been used since the dawn of time and are made entirely of renewable resources. Protein (gluten, zein, gelatin, soy protein, and so on), starch, cellulose, chitosan, and wax are the most commonly used bioresource-based polymers in the production of biodegradable products [17].

Proteins are complex compounds composed of monomeric amino acids linked together by covalent bonds [15]. Although both plant and animal protein sources are commonly used, waste-derived proteins are more appealing for commercial applications due to their low cost [18]. Because of their heterogenic structure, proteins' thermal resistance and nonhydrophobic nature play an important role in the emergence of functional features of biodegradable packaging [19]. Several plant-based (wheat gluten, corn zein, soy protein, etc.) and animal-based (milk proteins, whey protein, gelatin, etc.) proteins are used in the food packaging industry as edible packaging materials that can be consumed

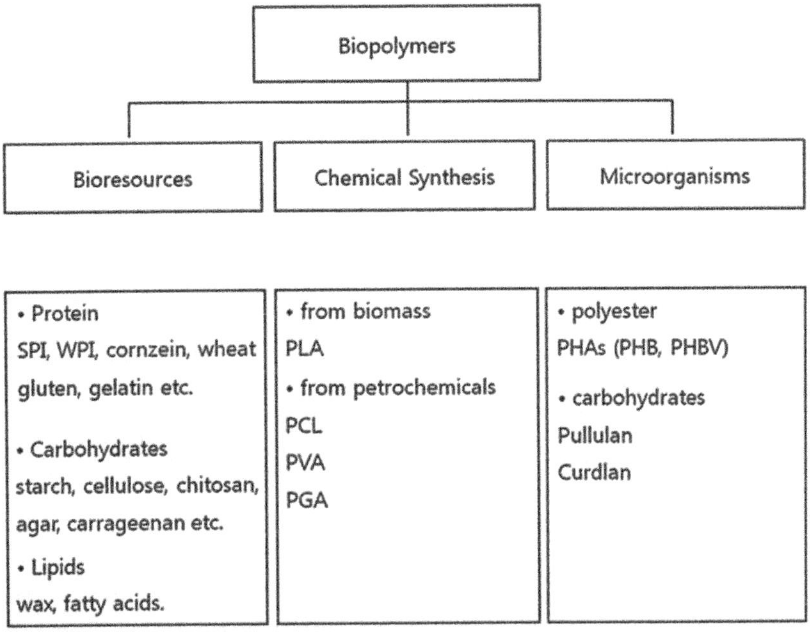

FIGURE 18.1

Classification of biopolymers [16].

Reproduced with permission from Elsevier. Copyright 2021.

with food [20]. Protein-based polymers, on the other hand, can be blended with other polymers such as starch, pectin, polypropylene, and so on. These applications offer the opportunity to improve the oxygen-water barrier and mechanic properties of packaging systems while also effectively incorporating bioactive materials [21].

Starch, chitosan, cellulose derivatives, pectin, and agar are the most commonly used biopolymers in packaging systems [22]. Because of its low cost, easy availability, and degradability, starch is extremely important [23]. It can, however, be used as a thermoplastic polymer with plasticizers, heat, and shear pressures [19]. Furthermore, its high water permeability limits its application [20]. Cellulose is the most abundant biopolymer on the planet, accounting for 33% of all plant materials. Furthermore, some bacteria can synthesize cellulose [24]. The most important limiting factors when using cellulose fibers are high water absorption, low processing temperature, and low microbial resistance. Green composites made from cellulose and other biopolymers, on the other hand, are the most promising applications in packaging systems [25]. Chitosan, another common biopolymer, is made from deacetylated chitin. Chitosan films are transparent, durable, and elastic, with a good oxygen barrier. As a packaging material, chitosan-based films can protect foods from fungal rottenness [1,26].

In recent years, lipid-based biopolymers such as wax and fatty acids have attracted attention due to their favorable water vapor barrier properties and ability to give foods, particularly vegetables and fruits, a bright and glossy appearance [27]. Moreover, they are excellent copolymers for protein or polysaccharide-based biopolymers to enhance barrier properties as a packaging material [28].

18.2.2 Biodegradable polymers from microorganisms

PHAs are biodegradable polyesters that are produced through fermentation of renewable substrates such as sugar and lipids by various microorganisms from [29]. The molecular weight and chain length of PHAs vary from tens to thousands depending on factors such as producer strain, growing environment, and carbon source used [30]. However, only a few PHA derivatives, such as medium chain length PHA (mcl-PHA), polyhydroxybutyrate (PHB), and poly(hydroxy butyrate-*co*-valerate) (PHBV), are commercially available today [31]. PHAs are widely used in the packaging industry to make bottles, vessels, plates, films, membranes, and coatings [20]. However, there are some restrictions on how PHAs can be used. PHB's performance as a packaging material, for example, is hampered by its brittleness, slow crystallization, poor heat stability, and poor melt processability. However, copolymerization with hydorxyvalerate (HV) can improve it, resulting in PHBV. The resulting copolymer is softer and more resistant. Furthermore, the copolymers biodegrade faster than the homopolymer PHAs [32].

PHAs are frequently used in shopping bags, disposable items, food and cosmetic packaging, and some medical products [33,34]. Research on future developments of PHAs is increasing day-by-day.

18.2.3 Biodegradable polymers from biotechnology

PLA is one of the most used polymers in the food packaging industry. It is an aliphatic polyester produced by controlled polymerization of lactic acid produced by bacterial strains from sugar feedstock, corn, and other sources [35]. PLA has received a lot of attention in recent years because of its low cost, 100% biodegradability, compatibility, and excellent mechanical properties [18]. PLA's performance varies according to its stereochemical composition (L-lactic acid, D-lactic acid or L–D-lactic acid). PLA can be processed using injection molding, thermoforming, stretch blow molding, extrusion blown film, foaming, fiber spinning, and sheet and film extrusion to achieve the desired result. It is an excellent alternative packaging material for short shelf-life products such as shopping bags and trash bags [36]. PLA's use, however, is limited due to its low decomposition temperature and poor stroke strength. As a result, additional research is required to address the aforementioned limitations and increase its availability as a food packaging material [37,38].

18.2.4 Biodegradable polymers from petrochemical sources

Biodegradable polymers can be made from both petrochemical and renewable bioresources. For many years, various nonnatural, namely synthetic polymers, have been well-known and used in a variety of applications. Petroleum-based biodegradable polyesters are created through polycondensation reactions between aliphatic diacids and aliphatic diols or ring-opening polymerization of lactones. Polycaprolactone (PCL), polyethylene adipate (PEA), polybutylene succinate and polybutylene succinate-co-butylene adipate) (PBS and PBSA), and polybutylene adipate-*co*-terephalate (PBAT) are the most commonly used petroleum-based polymers [17].

PCL is a semicrystalline biodegradable polymer. The ring-opening polymerization technique is used to create it from caprolactone. It is widely used in biomedical applications and has recently been used for food packaging [39]. This polyester has low glass transition temperature, low melting point, high crystallinity, high permeability, high break elongation, low modulus, and exceptional flexibility. PCL, on the other hand, is frequently mixed with other polymers (cellulose, chitosan, etc.,) due to its high cost [40,41]. PBS, another polyester in this group, is gaining popularity due to its excellent biodegradability, melt workability, and thermal and chemical strength. Polycondensation synthesis of 1,4-butanediol and succinic acid yields it [19]. Copolymerization of succinic acid, 1,4-butanediol, and adipic acid monomers results in PBSA. PBS and PBSA are both easily processed using conventional methods for food packaging applications [42,43]. The most common method of producing PEA is a polycondensation reaction between ethylene glycol and adipic acid. It is commonly combined with other polymers to increase its plasticity and flexibility [44]. PBAT is produced by a polycondensation process that yields an aliphatic–aromatic copolyester from 1,4-butanediol, terephthalic acid, and adipic acid monomers. PBAT has well elasticity and it has mechanical properties are quite comparable to low-density polyethylene (LDPE). On the other hand, it has a lower oxygen barrier property as compared to LDPE [43]. PBAT has focused on products with short shelf lives, such as films and packaging. Its use is limited due to its low tensile strength, low modulus, poor water vapor permeability and high cost. These negative effects can be avoided by combining various polymers [18].

18.3 Biodegradable polymer nanocomposites (bionanocomposites)

Polymer composites are made by combining polymers with organic or inorganic fillers in various geometries such as spherical, tubular, platelet, and fibrous [45]. When the filling agent is on the nanoscale, the resulting filler and composite are referred to as nanofiller and nanocomposite, respectively. The composite is bionanocomposite if the matrix is bio-based or biodegradable.

Nanomaterials have at least one nanoscale (1–100 nm) dimension and their physical and chemical properties can be vastly different from those of their bulk equivalents [46]. Furthermore, due to their extremely high surface-area-to-volume and aspect ratios, nanosized systems are regarded as distinct structures that could be used in polymer nanocomposites. As a result, nanocomposite structures with improved properties created by incorporating nanoparticles into polymer matrix have found many applications in fields such as engineering and biomedical [47,48]. However, several obstacles must be overcome in order to create a true polymer nanocomposite structure: (1) The nanofiller and polymer must be compatible, and the appropriate nanofiller-polymer pair must be chosen; (2) Nanofillers must be dispersed properly and without aggregation in the polymer matrix; and (3) Polymer nanocomposite must be cost-effective [49].

To overcome these constraints, various types of nanomaterials have been used in the food industry. Nanomaterials used as reinforcement agents in food packaging materials can have a variety of geometrical structures, including nanoparticles, nanofibers, nanoplatelet-type nanomaterials, nanotubes, and nanowires [50]. Another classification of nanomaterials is based on their chemical origin, which can be organic, inorganic, or metallic, with each having distinct properties. Nanomaterials are incorporated into conventional biodegradable food packaging materials to improve functional properties such as water vapor permeability, oxygen permeability, mechanical properties, and thermal resistance, as well as to gain antioxidant and/or antimicrobial activity to extend food shelf life [51].

18.3.1 Preparation of bionanocomposites

For the preparation of bionanocomposites on a laboratory and industrial scale, several common processing methods are available. The most important goal of processing is to achieve a uniform distribution of the bionanocomposite constituents. Several pretreatment processes are applied to nanofillers in order to make their surface properties compatible with the biopolymer matrix, as discussed in the following sections of this chapter.

The most common and simplest bionanocomposite preparation method is solvent casting. The polymer is first dissolved in an organic solvent, and then the nanofiller particles are added to the polymer solution at a suitable temperature and mixed together. The nanocomposite solution is then cast onto a plate as a thin layer. A thin bionanocomposite film is formed when the solvent is evaporated during the drying process [52]. The ideal moisture content and drying conditions are critical parameters for achieving the desired film [53]. The major challenge of this process is the potential need for toxic chemicals to solve the biopolymer problem [54].

The most widely used industrialized thin film preparation method is extrusion molding (melt mixing). It is also eco-friendly because no toxic chemicals or solvents are required [55]. This method is also advantageous due to its low cost [56]. This method employs single, twin, or multiple extruders. In general, batch extrusion systems are used for laboratory-scale mixing of nanofillers with biopolymers, whereas continuous methods involve continuously feeding the film constituents into the extruder and are used for scaling up [57]. Although biopolymers have become more widely used for extrusion processing in recent years for the preparation of food packaging materials, extrusion blowing of many biopolymers is difficult due to their low melting points and possible degradation/decomposition upon heating [53].

18.4 Organic nanofillers

To improve their physical, mechanical, and barrier properties, biopolymers such as chitosan, PLA, starch, and PHAs are frequently combined with natural organic nanofillers. Cellulose is a common organic filling agent used in the production of bionanocomposites. It is the most abundant natural polymer on the planet [58]. The cell walls of green plants are the most abundant source of cellulose, but it can also be found in algae, fungus, bacteria, and even animals. Plant-derived cellulose contains hemicellulose and lignin, whereas bacterial cellulose is pure [59].

Cellulose is a linear homopolysaccharide that include several hundred to many thousands of cellobiose (Fig. 18.2), a disaccharide comprised from D-glucose units covalently bonded by β-1,4-glycosidic linkages. Van der Waals and hydrogen bond interactions between cellulose chains allow for parallel aggregation, resulting in cellulose fibrils with the length of 5–50 nm in diameter and several microns [60]. Strong intermolecular interactions cause crystalline regions to form, but amorphous regions are also present in the cellulose fibrous structure.

For the extraction of nanoscale elements from lignocellulosic rawmaterials, two main processes are used: acid hydrolysis and mechanical treatment. Strong acid hydrolysis generally separates the crystalline region from the amorphous domain, resulting in the formation of cellulose nanocrystals (CNCs) [61]. Many factors, including acidity, hydrolysis temperature and time, acid type, and cellulose origin, influence the CNC's characteristic properties [61]. In general, sulfuric acid is used for extraction of CNC, enabling sulfate-grafted groups on the CNC surface [62] while other acids like phosphoric, maleic and formic acids have also been tested [63–65]. Fig. 18.3 shows the visual appearance and transmission electron microscopy (TEM) images of 100× and 1000× times water-diluted CNCs obtained from cotton pulp [66].

Microfibrillated cellulose (MFC) is another nanoscale cellulose-based lignocellulosic resource derivative. To make MFC from cellulose fibers, a strong mechanical treatment is required. Production of

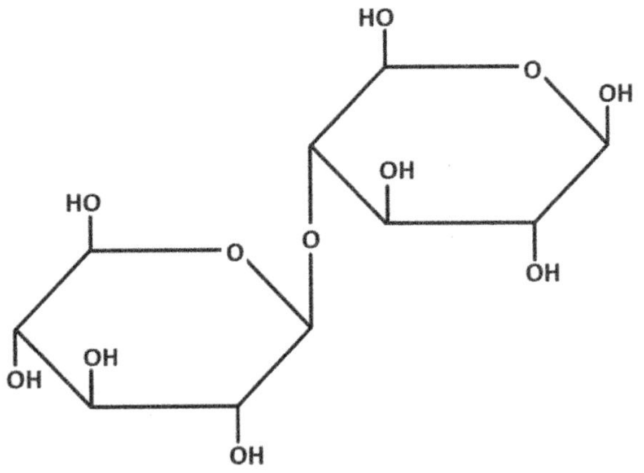

FIGURE 18.2

Chemical structure of cellobiose.

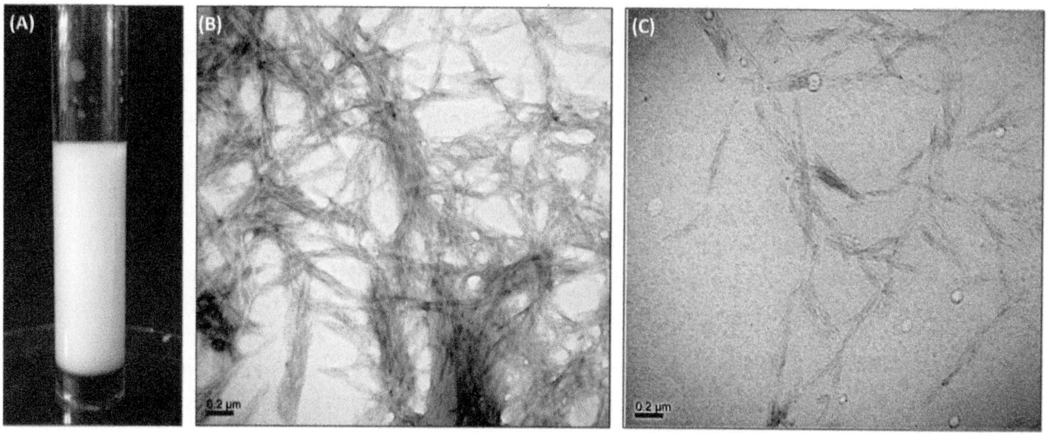

FIGURE 18.3

Visual appearance (A) and TEM images of 100× (B) and 1000× times (C) diluted CNC samples from cotton pulp [66].

Reproduced with permission from Elsevier. Copyright 2021.

MFCs by mechanical process was first reported by Herrick et al. [67] and Turbak et al. [68]. Several mechanical actions are used to reduce cellulose fibers to smaller particles with more uniform diameters, which is then followed by a filtration step to separate the MFCs from unfibrillated cellulose [60]. Enzymatic pretreatment before mechanical process has also been demonstrated to remarkably reduce the energy requirement and increase the product yield while obtainment of MFCs with larger diameters and higher tensile properties [69–71]. Fig. 18.4 shows the scanning electron micrographs (SEM) of (acacia craft pulp origin) MFC fibers which were pretreated with carboxyethyl at different concentrations [72]. Untreated MFC had a generally flat and smooth surface, as shown in the figure, whereas carboxyethyl pretreatment resulted in the formation of rod-like fibers.

A type of cellulose, namely bacterial nanocellulose (BNC), is synthetized by a variety of bacteria as a linear exopolysaccharide formed of D-glucose monomers covalently connected by β-1,4-glycosidic bonds likewise plant-derived cellulose [73]. However, BNC differs by its purity from plant-based cellulose that is present in the lignocellulose structure together with lignin and hemicellulose, making its purification very simple and easy [74]. A number of bacteria belonging to different genera such as *Komagataeibacter* (previously named as *Gluconacetobacter*), *Acetobacter*, *Achromobacter*, *Alcaligenes*, *Aerobacter*, *Agrobacterium*, *Medusomyces*, *Azotobacter*, *Pseudomonas*, *Rhizobium*, *Sarcina*, *Dickeya*, and *Rhodobacter* have been identified to produce extracellular BNC [75–83]. Following the biosynthesis of the cellulose by the bacteria, it can be finely extracted from the extracellular biomass in the nanoscale with the values of width less than 100 nm and length up to 2–4 nm [84,85]. The conversion of BNC to bacterial CNCs is also similar to the process applied to vegetal MFC [84]. Physical and morphological properties of BNC is mainly dependent to species used for production and the cultivation conditions such as medium and temperature, air supply etc. [86].

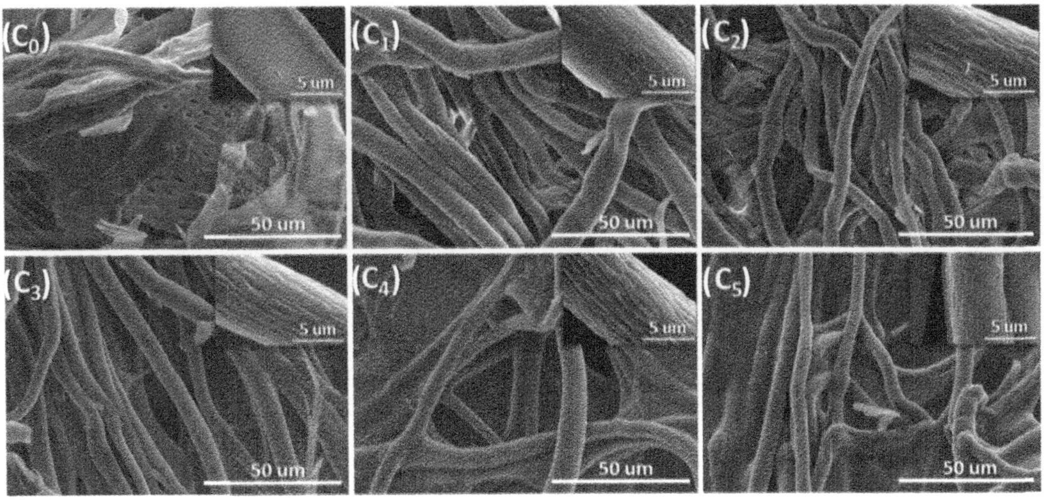

FIGURE 18.4

SEM micrographs of MFC fibers before (C0) and after carboxyethyl treatment at 0.5 (C1), 1 (C2), 2 (C3), 3 (C4), and 4 g/g (C5) concentrations [72].

Reproduced with permission from Elsevier. Copyright 2021.

18.4.1 Modification of nanocellulose

Along with the numerous findings indicating improved biopolymer properties, there are several drawbacks when incorporating with polymers, which are primarily due to the hydrophilicity of nanocellulose. Due to the incompatibility of nanocellulose as a nanofiller, the main issue during the preparation of bionanocomposites using hydrophobic matrices is the lack of good dispersion of the nanoscale fillers in the combined substance [87]. Several approaches have been developed to overcome this difficulty and improve nanocellulose dispersibility in polymer matrix, including surface modification, biopolishing, covalent and noncovalent binding with biopolymers, and the use of compatibilizers.

Because it contains many reactive hydroxyl compounds, nanocellulose is a very hydrophilic material. Surface modification is a common method of reducing nanocellulose's hydrophilicity by introducing hydrophobic moieties capable of amination, acetylation, esterification, urethanization, oxidation, silylation, carboxymethylation, and epoxidation, among other things. Several physical methods have also been used to modify the surface. Table 18.1 shows a number of current studies on cellulose surface modification and its impact on material qualities.

As seen in Table 18.1, Nanocellulose samples from various sources could be chemically modified to be useful for a variety of applications such as food packaging, biomedical engineering, and tissue engineering.

18.4.2 Nanocellulose reinforced bionanocomposites

Because of its numerous advantages, such as biodegradability and renewability, affordability, high abundancy, low density, desirable surface properties, and ease of modification, the use of

Table 18.1 Several papers conducted on surface modification of nanocellulose and key findings related to food packaging and other applications.

Surface modifier or modification method	Nanocellulose type	Biopolymer	Results	References
EPTMAC	CNC	CMC	The degree of modification (substitution) of mCNCs has a significant impact on their reinforcing potential. Obtaining mCNC/CMC films with uniform dispersion and outstanding mechanical characteristics required a high modification degree.	Li et al. [88]
Castor oil-dopamine	CNC		With a contact angle of up to 95.6 degrees, the hydrophobicity of modified CNC was improved. Protection of morphology and crystalline structure of CNC was provided after modification	Shang et al. [89]
Lactic acid	CNC	PLLA	PLLA crystallization was improved. CNC was evenly distributed throughout the polymer matrix. The mechanical characteristics of nanocomposites have significantly improved.	Wu et al. [90]
n-Octadecyl isocyanate	CNC	PLA	The grafted modified CNCs might be very well distribute in the polymer matrix. Mechanical properties were improved while oxygen barrier properties were not affected.	Espino-Pérez et al. [91]
Tannic acid—decylamin	CNC		The modified product was easily redistributable in organic solvents such as toluene. The particle hydrophobicity increased by measurement of contact angle shifting from 21 to 74 degrees.	Hu et al. [92]
Acetic anhydride	CNC	PLA	Tensile characteristics, thermal resistance, dimensional stability, and dynamic mechanical properties of the PLA composites were all improved by surface modified CNC reinforcing. The crystallinity of PLA/CNC composites increased as a result of the improved CNC.	Lee et al. [93]
Succinic anhydride	CNC		By the controlling of succinylation parameters, CNCs with desired sizes, extent of esterification or preserved thermal stability could be manufactured.	Leszczyńska et al. [94]
Cationic-modified polyacrylate latex	MFC	CNF based thin paper	The high contact angles and low WVTR value indicated that MFC was extremely hydrophobic and had enhanced barrier properties.	Pan et al. [95]

(*Continued*)

Table 18.1 Several papers conducted on surface modification of nanocellulose and key findings related to food packaging and other applications. ***Continued***

Surface modifier or modification method	Nanocellulose type	Biopolymer	Results	References
Acetic anhydride	MFC	PLA	The hydrophobicity of the nanocellulose was improved. Increased compatibility of modified cellulose with PLA was obtained.	Jamaluddin et al. [96]
Fatty acid chlorides	MFC	MFC-based film	The maximum mechanical strength was preserved after surface modification. Water vapor permeability was considerably decreased after bulk modification.	Balasubramaniam et al. [97]
Layer-by-layer deposition of chitosan and lignosulfonates	MFC		Multilayered MFCs with antioxidant and antimicrobial activity were manufactured.	Li and Peng [98]
Oxygen plasma and silanization with TCMS	BNC		Hydrophobicity was maximized. Even after 6 months under dry circumstances BNC remained hydrophobic as well as a month submerged in distilled water.	Leal et al. [99]
Acetic acid, hexanoic acid, dodecanoic acid	BNC	PLLA	Between PLLA and BNC, better compatibility (interfacial adhesion) was achieved.	Lee et al. [100]
Nitrogen-containing plasma	BNC		After the nitrogen plasma treatment, wettability of the material did not improve, but its porosity and surface chemistry improved significantly. It was showed that adhesion and proliferation of HMEC-1 and N1E-115 cells was remarkably increased.	Pertile et al. [101]
Acetylation	BNC	TCDDMA	Acetylation changed the surface properties and reduced the moisture content, although excessive acetylation increased hygroscopicity.	Ifuku et al. [102]

CNC, *cellulose nanocrystal;* EPTMAC, *2,3-epoxypropyl trimethylammonium chloride;* CMC, *carboxymethyl cellulose;* PLLA, *poly-L-lactic acid;* MFC, *microfibrillated cellulose;* TCMS, *trichloromethyl silane;* TCDDMA, *tricyclodecane dimethanol dimethacrylate.*

nanocellulose as a reinforcing material for bionanocomposites has received a lot of attention in recent years [87]. Improved bionanocomposites containing nanocellulose have been employed in a variety of applications such as food packaging, electronics, biomedical, automotive, construction etc. [103]. Many researchers have looked into the mechanical, barrier, thermal, and other properties of bio-based packaging materials reinforced with nanocellulose particles in the food packaging field. Table 18.2 summarizes the results of some relevant recent papers.

Table 18.2 shows that nanocellulose used in bionanocomposites is one of the most promising and cost-effective nanofillers. According to the findings in the literature listed in Table 18.2, as long as they

Table 18.2 Use of nanocellulose in bionanocomposites for food packaging applications.

Biopolymer type	Nanocellulose type and ratio (wt.%)	Results	References
Sodium alginate	CNC (1–9)	Silver nitrate was also incorporated into the nanocomposite films. Tensile strength of the composite films improved by 39%–57% depending on the kind of nanofillers used, while water vapor permeability reduced by 17%–36%.	Yadav et al. [104]
PVA/CMC	CNC (0.5–10)	Sugarcane bagasse was used to extract CNC. The addition of 5% CNC enhanced mechanical and barrier characteristics while having little effect on transparency.	El Achaby et al. [105]
Sodium alginate	MCC (1–10)	When the CN concentration was increased to 10%, the water solubility and water vapor permeability of the nanocomposites fell by around 40% and 17%, respectively. With increasing NC content from 0% to 5%, the tensile strength of the composite films raised from 18.03 to 22.4 MPa; but it reduced as the filler content increased.	Abdollahi et al. [106]
κ-Carrageenan	CNC (0–9)	CNC incorporation increased mechanical and barrier properties. Contact angle increased and water vapor permeability decreased by CNC addition.	Yadav and Chiu [107]
CMC/Starch	CNC (0.5–5)	Sugarcane bagasse was used to extract CNC on a nanometric scale. Due to CNC dispersion at the nanoscale, bionanocomposite films remained clear. With the addition of CNC, the WVP was greatly lowered, and the elastic modulus and tensile strength were gradually raised.	El Miri et al. [108]
Gelatin	CNC (1–10)	Tensile strength reduced once CNC was added to the gelatin matrix. CNC provided an increase strong oxygen gas barrier characteristics of gelatin.	Mondragon et al. [109]
Agar	CNC (1–10)	The mechanical and water vapor barrier qualities of agar film were greatly improved by blending with the CNC. In the composite film with 5 wt.% of CNC, the tensile modulus and tensile strength of agar film improved by 40% and 25%, respectively, and after forming a nanocomposite with 3 wt.% CNC, the WVP of agar film reduced by 25%.	Reddy and Rhim [110]
Wheat gluten	CNC (2.5–12.5)	Gluten, cellulose nanocrystals, and titanium dioxide nanoparticles were combined to create a new nanocomposite. Gluten films with CNCs and titanium dioxide nanoparticles provide better mechanical and water barrier qualities. Antimicrobial activity of bionanocomposite-coated paper sheets against G (+) and G () bacteria and yeast was good.	El-Wakil et al. [111]

(Continued)

Table 18.2 Use of nanocellulose in bionanocomposites for food packaging applications.
Continued

Biopolymer type	Nanocellulose type and ratio (wt.%)	Results	References
CMC	CNC (1–7)	CNCs were placed equally throughout the polymer to produce homogenous films. Depending on the type and concentration of CNCs, the mechanical and water vapor barrier qualities varied. Tensile strength (TS) raised by 45.7%, 25.2%, and 42.6%, and the WVP reduced by 26.3%, 19.1%, and 20.4% after composing form with 5 wt.% of CNCs originated from RS, WS, and BS, respectively.	Oun and Rhim [112]
Chitosan/PVP	CNF (1–5)	The addition of 5% CNF to manufactured bionanocomposite films increased their thermal stability. The inclusion of CNFs improved the mechanical properties of bionanocomposite films substantially. Furthermore, when the concentration of CNFs increased, the rate of water and oxygen transmission of the bionanocomposite film reduced.	Kumar et al. [113]
PVA/chitosan	CNC (1–5)	The tensile strength and thermal stability of the films were increased with addition of CNC. CNCs were scattered on the nanoscale. Good antifungal and antibacterial activity was observed in the bionanocomposite films.	Perumal et al. [114]
Pullulan	MFC (0.1–1)	The mechanical properties of the final films improved in direct proportion to the filler loading. The addition of MFC had no effect on the amount of water that the samples absorbed. The films had high relative humidity values, but their oxygen barrier performance was outstanding in dry conditions.	Cozzolino et al. [115]
Gelatin/PVA	BNC (1%–10%)	Up to 7.5 wt.% BNC good miscibility was achieved between gelatin/PVA blend film matrix and BNC. Water vapor permeability was decreased by up to 7.5% by including BNC into the film, while tensile strength, elongation at break, and elastic modulus raised by 21.5%, 41%, and 19%, respectively.	Haghighi et al. [116]
Pullulan	BNC (5–60)	When compared to unfilled pullulan films, all bionanocomposites demonstrated significant improvements in thermal stability and mechanical properties.	Trovatti et al. [117]
PVA	BNC (0.17, 0.33, 0.5)	The addition of nanofiber to PVA film raised its tensile strength, thermal stability, water vapor impermeability, and moisture resistance, but not its strain at break. Increased nanofiber content resulted in a small reduction in transparency.	Abral et al. [118]

(Continued)

Table 18.2 Use of nanocellulose in bionanocomposites for food packaging applications. ***Continued***

Biopolymer type	Nanocellulose type and ratio (wt.%)	Results	References
PHB	BNC (50:50 BNC-PHB)	PHB molecules filled the gaps between the BC nanofibrils. The incorporation of BNC into the PHB matrix resulted in an increase in mechanical characteristics.	Zhijiang and Guang [119]
Agar	BNC (0–10)	SEM revealed a good dispersion of BNC at low concentrations (3%–5%) in the films. BNC (10%) considerably reduced moisture content, water solubility, and WVP by 60.4%, 13.3%, and 25.7%, respectively. After adding BNC (0%–10%), the tensile strength of the films raised from 22.10 to 44.51 MPa, whereas the elongation at break increased at first with raising BNC concentrations (0%–5%), then reduced with more BC (8%–10%).	Wang et al. [120]

CNC, *cellulose nanocrystal;* PVA, *polyvinyl alcohol;* CMC, *carboxymethyl cellulose;* MCC, *microcrystalline cellulose;* WVP, *water vapor permeability;* CNF, *cellulose nanofiber;* PVP, *polyvinylpyrrolidone;* MFC, *microfibrillated cellulose;* BNC, *bacterial nanocellulose;* PHB, *polyhydroxybutyrate.*

are used in appropriate amounts in the biopolymer matrix and there is good biocompatibility between the nanocellulose and biopolymer, film properties, particularly mechanical, barrier, and thermal attributes, are effectively improved, owing to its high crystallinity. In addition to the literature listed in the table, many scientists have investigated the use of nanocellulose in biomaterials for active packaging. Fortunati et al. [121] used melt extrusion method to obtain multifunctional bionanocomposite films by combining CNCs, surfactant modified CNC, and silver (Ag) nanoparticles in a PLA matrix. The surfactant improved CNC dissolution in the biopolymer matrix, and the nanocomposite films were antibacterial against *Staphylococcus aureus* and *Escherichia coli* cells, according to the findings. Another study used clove essential oil and MFC to create antimicrobial and antioxidant nanocomposite films based on soy protein isolate. MFCs appeared to promote the release of bioactive components of clove essential oil, possibly due to increased clove essential oil dissolution in nanocomposites due to a greater number of tiny droplets [122]. Lavoine et al. [123], who produced a new antibacterial bionanocomposite material using carvacrol grafted β-cyclodextrin and MFC, reported that the β-cyclodextrin—carvacrol was gradually released 21 hours later in grafted samples, with a release kinetic 24% slower in water. Salari et al. [124] showed that the incorporation of BNC or silver nanoparticles into chitosan decreased its sensibility to water and the addition of nanoparticles and BNC increased the mechanical and thermal properties of nanocomposite films.

18.5 Inorganic nanofillers

Inorganic nanofillers incorporated into biopolymers have been successfully used to prepare bionanocomposites with improved properties. Nanoclays are the most important of the inorganic nanofillers.

Although clay has been known since ancient times, its use in nanotechnology is a relatively new field. Clays are abundant in nature and are inexpensive materials to use as polymer reinforcement agents [125]. Polymer-nanoclay nanocomposites have attracted great attention since 1990s especially in automotive and packaging industry. This type of nanocomposites was firstly used in the form of a timing belt cover for the Toyota Camry cars in 1993.

Nanoclays are a class of minerals that occur naturally and are frequently combined with trace amounts of metal oxides. Based on their morphological and chemical properties, they are classified as bentonite, illite, halloysite, montmorillonite (MMT), and hectorite [126]. MMT, saponite and hectorite are the most commonly studied layered silicates while halloysite is a relatively new class of nanoclays [127,128]. Halloysite, belonging to kaolin group, is also an aluminasilicate which has a nanotubular structure with a 15 nm lumen, 50 nm external diameter, and length of 800 ± 300 nm [129].

18.5.1 Layered silicates

Among the nanoclays, layered silicates have drawn the greatest attention. Layered silicates like MMT consist of hundreds of very thin layers (approximately 1 nm of thickness) which are stacked in approximately 10 μm-sized multilayer stacks [130,131]. Each layer is combined of two tetrahedral sheets formed by Si and O atoms and one octahedral sheet containing Al and Mg atoms, generating 2:1 TOT (tetrahedral-octahedral-tetrahedral) structures (Fig. 18.5) [132,133]. Thanks to exchangeable ions present on the surface of the layers, MMT has a strong cation exchange capacity which enables its chemical modification with organic and inorganic cations and molecules [125].

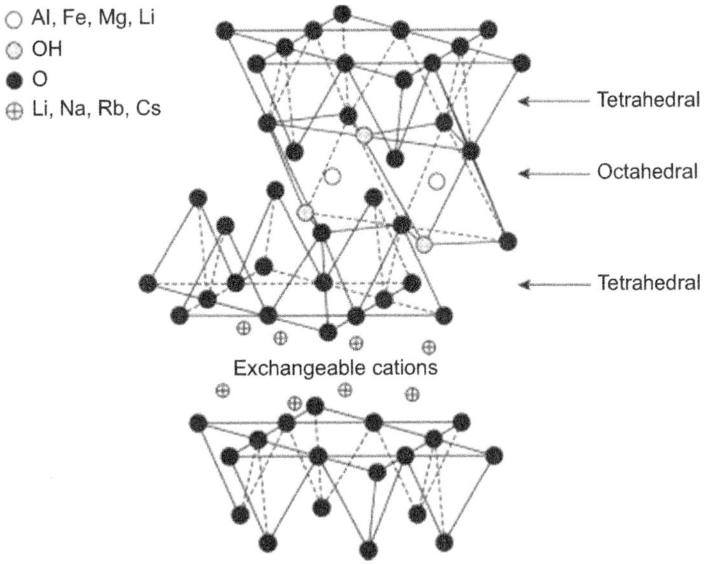

FIGURE 18.5

The crystal structure of 2:1 type clay minerals [134].

Reproduced with permission from Elsevier. Copyright 2021.

MMT is the most common natural layered aluminasilicate-based nanoclay. MMT's main advantages are its abundance, high aspect ratio, stiffness, and low cost. However, due to MMT's high hydrophilicity, it can only be dispersed succesfully in hydrophilic polymers such as starch, gelatin, and polyacrylamide (ethylene glycol). It has been clearly demonstrated that the incorporation of MMT into polymer matrix at appropriate concentrations can develop mechanical, barrier, and thermal properties [135]. In order to make hydrophilic MMT more hydrophobic and enable it to produce a homogeneous dispersion with hydrophobic polymers, the exchangeable cations (Fig. 18.4) found in the interlayers can be readily replaced with organic cationic surfactants such as ammonium or phosphonium [134,136]. This process is called organophilization or organomodification. By a successful ion-exchange reaction, due to the weakening cohesive energy, the gallery between the layers expands (intercalation) or basal spacing increases or the layers are completely separated (exfoliation). Although exfoliation is the best, it is very difficult to achieve [137]. Nanoclay layers can be effectively dispersed in polymer matrix via intercalation or exfoliation, resulting in nanocomposite structures. However, unless the gallery between the layers is expanded, homogeneous dispersion of the silicate layers and polymer phases cannot be achieved, resulting in the formation of a true nanocomposite [6,134]. The polymer/nanoclay nanocomposite or microcomposite structures can be visualized in Fig. 18.6.

Improved barrier, mechanical, and thermal properties, among other things, in food packaging materials are achieved through the fabrication of intercalated or exfoliated nanocomposites. Several approaches have been proposed to explain the contribution of layered silicates to the limitation of gas permeation from nanocomposite films. According to Nielsen's theory, good dispersion of clay layers in the polymer matrix creates a tortuous path for gas permeation, forcing the permeant to travel a greater distance to diffuse throughout the nanocomposite film than the neat one [139]. Incorporation of layered silicates into biopolymers have also been demonstrated to enhance their mechanical stability. Table 18.3 summarizes the results of improved biopolymer properties by the layered silicate addition by citing some recent papers. As can be seen in Table 18.3, modified or unmodified layered silicates are successfully incorporated into the bionanocomposites as nanofillers. The biocompatibility

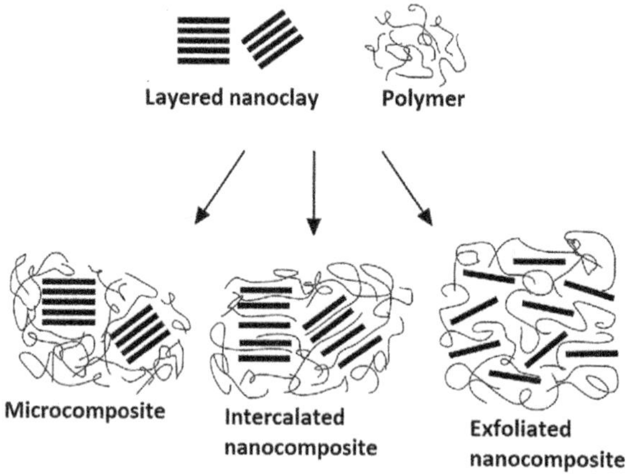

FIGURE 18.6

Polymer/clay microcomposites and nanocomposites [138].

Table 18.3 Use of platelet type nanoclays in bionanocomposites for food packaging applications.

Biopolymer type	Nanoclay type and ratio (wt.%)	Results	References
Chitosan	MMT-K10 (0.5–5)	As a result of the increase in filler concentration, the water vapor and gas permeability values of the composite films reduced significantly. With the addition of clay, the tensile strength of the composites increased significantly and the elongation at break decreased as a result of the increase in clay concentration. The dispersed clay enhanced the thermal resistance and improved the strength and elastic modulus of the matrix systematically with the increased loading of clay.	Kasirga et al. [140]
Chitosan	MMT-Na + and Cloisite 30B (1–5)	According to XRD and TEM results, with the addition of little amounts of MMT-Na + to the chitosan matrix, an exfoliated structure was obtained. Addition of Cloisite 30B to the chitosan matrix provided microscale composite (tactoids) formation. Surface roughness increased as a result of adding a small amount of clay. With the addition of clay to the chitosan matrix, the tensile strength of a chitosan film was increased while the elongation at break was reduced. The addition of clays had no effect on melting properties or thermal resistance.	Xu et al. [141]
Chitosan	KSF-MMT (1–5)	The FTIR and SEM data showed that an intercalated-and-exfoliated structure was formed at low KSF level while an intercalated-and-flocculated structure was formed at high KSF level. The dispersed clay improves the thermal stability of the matrix systematically with the increase of clay loading. The thermal resistance of the matrix improved systematically as the distributed clay loading increased.	Altınışık et al. [142]
PHBV/PBAT	Nanomer I.30T (0.6–1.8)	When compared to compression molded films, cast extruded PHBV/PBAT nanocomposite films had better barrier characteristics (oxygen and water vapor). TEM confirmed the intercalation/exfoliation of nanoclay in PHBV, as well as at the PHBV/PBAT blend matrix interface. The rheological characteristics favored the better dispersion of the nanoclay and the interaction between the matrix and the filler.	Pal et al. [143]
κ-carrageenan	Hydrophilic bentonite (0–15)	Between BT and -carrageenan, hydrogen bonds and/or electrostatic interactions may have formed. The ability to absorb water and sorption of water reduced as the clay concentration increased. The water contact angle decreased when the nanoclay content was increased. If the reinforcing components are uniformly disseminated in the matrix, water absorption can be greatly reduced.	Dogaru et al. [144]

(Continued)

Table 18.3 Use of platelet type nanoclays in bionanocomposites for food packaging applications. ***Continued***

Biopolymer type	Nanoclay type and ratio (wt.%)	Results	References
Fenugreek seed gum	Na–MMT and Nanomer ® I.44P (0–7.5)	The maximum TS was obtained by incorporating 5% MMT, however there was no significant difference in the TS values of nanocomposite films containing nanomer. The addition of nanoclay reduced the elongation at break. Oxygen permeability decreased by incorporation of nanoclays up to 5%.	Memiş et al. [145]
PCL/chitosan	Nanomer I.34TCN (0.4–3)	0.4 wt.% of organo nanoclay, exhibited antimicrobial activity on *Candida albicans.* The addition of clay improved the thermal degradation properties of the PCL by raising the degradation temperature. Small levels of inorganic material additions slow down soil biodegradation.	Cesur et al. [146]
Starch	Cloisite 30 (0–10)	When biofilm was filled with 2–3 wt.% nanoclay, ~22% reduced water absorption, 40% reduced moisture uptake, 30% reduced oxygen permeation and 31% reduced swelling were observed. Nanoclay addition delays the biodegradation.	Mohan et al. [147]
Starch	Na–MMT (0–5)	Na–MMT increased TS gently. Hydrophilicity of the films decreased utilizing Na–MMT.	Heydari et al. [148]
TPS	Organic activated MMT (0–8)	Crystallization behavior of starch was restrained in the presence of MMT. Lower MMT amounts dispersed well in the TPS matrix.	Ren et al. [149]
TPS-PLA	Na–MMT (0.5–1)	The TPS matrix was well exfoliated with unmodified Na–MMT. The inclusion of PLA resulted in an increase in mechanical and thermomechanical properties, which were further impacted by the presence of clay.	Suin and Khatua [150]
Carrageenan/ pectin	Nanomer I.34TCN (1–10)	At a concentration of 10% nanoclay, WVP lost around 35% of its original value. The carbon dioxide permeability of the films was similarly lowered by 50% when 1% nanoclay was added.	Alves et al. [151]
Methyl cellulose	MMT (10–60)	The film thickness and opacity values of MC/MMT nanocomposite films increased as the MMT concentration increased, but their water adsorption and water solubility decreased.	Tunç and Duman [152]
PCL	OMMT1 and OMMT2 (3)	When compared to clean PCL, the degradation process of PCL nanocomposites occurs at a lower temperature. Antibacterial activity was detected when 3% of OMMT1 or OMMT2 was added to the PCL matrix. WVP of PCL matrix was considerably reduced when OMMT1 or OMMT2 nanoclays were added.	Yahiaoui et al. [153]

(*Continued*)

Table 18.3 Use of platelet type nanoclays in bionanocomposites for food packaging applications. ***Continued***

Biopolymer type	Nanoclay type and ratio (wt.%)	Results	References
Fish myofibrillar protein	MMT (1–5)	To increase the functional features of the fish myofibrillar protein film, MMT and transglutaminase were added. Addition of MMT and transglutaminase into film enhanced water gain, WVP and solution ability of the film. Clay and transglutaminase, which showed combined effect, enhanced the TS and elongation of nanocomposites.	Rostamzad et al. [154]
Whey protein isolate	MMT (0–3)	Whey protein isolate/MMT/citric acid nanocomposite films were developed. Interaction between MMT and citric acid showed lower WVP. Possible dispersion findings for films containing 1% MMT as well as 5% and 10% citric acid.	Azevedo et al. [155]

MMT, *montmorillonite;* PHBV, *polyhydroxybutyrate valerate;* PBAT, *poly-butylene adipate terephthalate;* TS, *tensile strength;* PCL, *polycaprolactone;* WVP, *water vapor permeability;* TPS, *thermoplastic starch;* OMMT1, *hexadecyl trimethylammonium chloride modified MMT;* OMMT2, *hexadecyl pyridinium chloride modified MMT.*

FIGURE 18.7

TEM (A) and SEM (B) micrographs of halloysite [172].

of the nanoclays with the biopolymer, as well as the amount of nanoclays, are critical factors in the production of bionanocomposite based food packaging materials.

18.5.2 Halloysite

Although halloysite is also an aluminasilicate, it has a completely different morphology than layered silicates such as MMT. Halloysite is a tubular nanoclay of the kaolin group with a 1:1 sheet structure consisting of one octahedral and one tetrahedral layer [156]. It is naturally found in hydrated form and drying of kaolin gives halloysite. Internal of external diameters of halloysite nanotubes are 10–15 nm and 50–70 nm, respectively while the average length of the tubes is 1 μm (Fig. 18.7) [157]. Halloysite is a green substance that has been designated by the US Food and

Drug Administration (FDA) as generally regarded as safe (GRAS) [158,159]. Halloysite has been proven as a versatile material to be utilized in food industry for different aims such as reinforcement of food packaging materials [128,145,157,160–163], controlled release/protection of bioactive compounds [127,129,164–167] and ethylene scavenging [168–171].

As previously stated, biopolymers have a few significant drawbacks when compared to petroleum-derived polymers as food packaging materials, including high brittleness and hydrophilicity, as well as weak mechanical and barrier properties. In the literature, halloysite has been tested for the elimination of these drawbacks of biopolymers by using it as a nanofiller. In Table 18.4, the improvements in the biopolymer properties by halloysite nanofiller incorporation has been summarized.

Table 18.4 Use of halloysite in bionanocomposites for food packaging applications.

Biopolymer type	Halloysite ratio (wt.%)	Results	References
PLA	0–9	The best mechanical and oxygen barrier qualities were observed when PLA mixed with 3 wt.% nanoclay concentration.	Othman et al. [173]
PBAT-PLA	0–2	80PBAT/20PLA/HNT filled blends have better mechanical properties than nonfilled ones.	Sancheti et al. [174]
Chitosan	2–10	Mechanical properties were improved and glass transition temperature was increased by halloysite addition.	Liu et al. [175]
Fenugreek seed gum	0–7.5	Incorporation of up to 5% nanoclay increased tensile strength but significantly reduced elongation at break. When the amount of nanoclay increased, the oxygen barrier and thermal properties were significantly improved.	Memiş et al. [145]
Pectin	2.5–30	Two types of halloysite were used. The encapsulation of salicylic acid into the lumen of shorter nanotubes (Matauri Bay) was more successful, whereas the patchy structure and longer tubes of patch halloysite nanotubes made the process more challenging. Halloysite-reinforced films have obtained exceptional thermal, tensile, and contact angle characteristics.	Makaremi et al. [160]
PLA	0–6	The concentration and dispersion of HNTs affect the film properties. In PLA films, the optimal concentration of halloysite was determined as 3.0 wt.%.The increased characteristics of PLA films were due to the good dispersion of halloysite. Cherry tomatoes which packaged with PLA/halloysite bionanocomposite had a longer shelf life.	Risyon et al. [161]
Potato starch	0–5	The mechanical and barrier properties of films were improved by increasing the nanoclay concentration.	Sadegh-Hassani and Mohammadi Nafchi [176]
PVA and starch	0.25–5	With increased halloysite content, water absorption capacity and solubility both reduced noticeably. The presence of halloysite lowered migration rates overall.	Abdullah and Dong [177]
PLA	3–9	The addition of 3% halloysite to the lactic acid oligomer-including PLA films delayed thermal deterioration by 7°C and lowered permeabilities to water and limonene vapors by about 8% and 47%, respectively. The unfilled PLA film plasticized with 10 wt.% lactic acid oligomer achieved the maximum barrier performance.	Rojas-Lema et al. [178]

(Continued)

Table 18.4 Use of halloysite in bionanocomposites for food packaging applications. ***Continued***

Biopolymer type	Halloysite ratio (wt.%)	Results	References
Bovine gelatin	0–5	The films were reinforced with halloysite or nano-SiO_2. Both nanocomposite films exhibited better mechanical and water solubility compared to the pristine gelatin films. Lower seal strength was obtained by halloysite incorporation.	Voon et al. [179]
Chitosan/PCL	2–6	Chitosan/PCL electrospun nanofibers were obtained using chlorogenic acid-loaded halloysite nanotubes. Addition of halloysite enhanced thermal resistance of the nanofibers. The integration of chlorogenic acid-loaded halloysite resulted in a hydrophilic surface with improved water vapor barrier qualities, although mechanical properties were compromised.	Zou et al. [180]

PLA, *polylactic acid;* PBAT, *poly-butylene adipate terephthalate;* PVA, *polyvinyl alcohol;* PCL, *polycaprolactone.*

18.5.3 Graphene and its derivatives

Graphene is essentially the main building block of carbon-derived graphitic materials. It has a two-dimensional honeycomb structure made up of a single atomic layer of carbon [181]. Its main application in electronics and optoelectronics is due to its excellent electrical, optical, and thermal conductivity [182,183]. It also stands out with its outstanding mechanical properties, having approximately 100 times higher strength than that of steel [184]. It is known as the thinnest material existing with a thickness value of 0.34 nm, which was firstly discovered by Geim and coworkers in 2004 [185]. This property enables it many advantages such as outstanding mechanical properties [186]. Antibacterial activity of graphene and its derivatives against different bacterial strains and their mechanism of antimicrobial action has also been clarified by several researchers [187–189]. In another study, lateral dependence of antibacterial activity of GO was highlighted [190].

Since the 1990s, graphene and graphene-derived materials have been used to meet specific requirements in various application fields. Graphene can be modified using a variety of methods, including physical and chemical techniques, to improve its properties or tailor new functionalities. The most studied graphene derivative, GO, is a partially oxidized form of graphite with oxygenated functional groups [191]. The presence of oxygen-containing hydrophilic groups such as hydroxyl, epoxy, carbonyl, and carboxyl groups, which can easily form hydrogen linkages, contributes to the hydrophilic nature of GO [192]. It is typically made in two steps: To begin, graphite is oxidized with a variety of acidic-based oxidants, yielding stacked graphite oxide. The material is then exfoliated using a gentle mechanical process to produce single GO layers [193]. GO not only has properties similar to graphene, but it also has advantages such as ease of processing, low cost and large-scale production, and ease of incorporation with various polymers and solvents [194,195]. Cytocompatibility of GO was also reported by Wojtoniszak et al. [196].

One of the most common applications of graphene and its derivatives is as nanofiller agents in biopolymers to produce bionanocomposites. In this context, as with other nanofillers, homogeneous dispersion of graphene-based fillers in biopolymers is required to achieve a true bionanocomposite [186]. As a result, exfoliated graphene derivatives such as GO may be able to meet this requirement. The incorporation of GO into various biopolymers has been shown to improve their mechanical, barrier, and thermal properties [192,197–200]. The incorporation of very low levels (0.2%) of GO, for example, significantly improved the oxygen permeability of pullulan-based films. The GO addition significantly improved the mechanical properties of the nanocomposite films [201]. Thanks to its excellent electrical conductivity, it could be used for fabrication of conductive materials to be utilized instead of metals [182,202–204]. In addition, antimicrobial nanocomposite films have also been developed by using GO as nanofiller in biodegradable polymers. Grande et al. [205] reported that cross linking of chitosan with GO increased the antibacterial activity of the film against *E. coli* and *Bacillus subtilis.* Starch reduced GO which was doped with iodine showed remarkable antimicrobial activity against Gram-positive (*S. aureus*) and Gram-negative (*E. coli*) strains [206].

18.5.4 Metal nanoparticles

Antimicrobial packaging is a type of active packaging that aims to prevent or eliminate foodborne pathogen survival/growth. Antimicrobial packaging materials can be created by incorporating active agents into sachets, direct incorporation into polymers, the use of innate antimicrobial polymers (such as chitosan), and immobilization on polymer surfaces [207]. Because antimicrobial agents have a selective effect on undesirable microorganisms found in food, true antimicrobials that affect the target microorganisms should be used when designing antimicrobial packaging materials.

Metals can be manufactured as nanoparticles. Smaller metal nanoparticles may exhibit biocidal activity at lower concentrations due to their higher surface area compared to their bulk counterparts [208]. Metal nanoparticles are antimicrobial agenta used in the development of antimicrobial food packagings. Metals' biocidal activity has long been known, so they have been used in a variety of fields such as agriculture, medicine, and food packaging [209]. The incorporation of metal nanoparticles into a biopolymer matrix has also been shown to improve the barrier and mechanical properties of bionanocomposites. Although a number of metals, including zinc, silver, gold, titanium, and copper, have been extensively researched for packaging applications, silver, a GRAS inorganic compound, has received the most attention due to its biological activity against a broad range of microorganisms [210,211]. Mode of antimicrobial action of silver against different bacterial and fungal strains has been studied by many researchers [212–216]. Although metallic silver is inactive, ionic silver (Ag^+) which is formed by reaction with moisture becomes active to inhibit the bacteria by making interaction with their cell wall [211]. The bacterial toxicity of silver is dependent on the Ag + concentration released, while particle size and shape of the silver particles may have an indirect effect on antibacterial activity [217,218]. Bionanocomposite films containing silver nanoparticles prepared using different biopolymers such as PVA, PLA, pullulan, chitosan, gelatin, and sodium alginate have been widely demonstrated to show remarkable antimicrobial activity against many foodborne pathogenic bacteria including *Salmonella*, *S. aureus*, *E. coli* O157:H7, *Listeria monocytogenes*, *Bacillus cereus*, *Pseudomonas aeruginosa*, *Klebsiella pneumonia*, *Micrococcus luteus*, and fungi including *Fusarium solani*, *Fusarium oxysporum*, *Aspergillus niger*, *Alternaria solani*, and *Botrytis cinerea* [104,121,124,219–221]. Shelf-life extension of foods including fresh fruits and vegetables, meat and dairy products has been acquired by packaging with bionanocomposite films enriched with silver nanoparticles by studies conducted by different researchers [222–230].

Aside from silver, metallic and metal oxide nanoparticles such as titanium dioxide (TiO_2), zinc oxide (ZnO), and copper oxide (CuO) have shown promise as antimicrobial agents in bionanocomposite films. TiO_2 is a multifunctional material with antimicrobial and ethylene scavenging activity that could be used in bionanocomposite films for a variety of applications [231,232]. It can be photo-activated by ultraviolet light and has many advantages to have low cost, low toxicity, biocompatibility and high chemical stabilization [233–235]. In recent years, TiO_2 has been used as a nanofiller and/or antimicrobial agent in biodegradable films primarily made of polysaccharide/protein-based materials and tested for extending the shelf-life of fresh foods such as tomato, pear, mushroom, and meat [111,236–246]. ZnO, an inorganic compound classified as GRAS, has been used to enable antimicrobial activity in bionanocomposite films. ZnO nanoparticles have been shown to have antibacterial and antifungal activity against a variety of Gram-negative and Gram-positive bacteria, as well as fungi [247]. As a result, it is not only added to food supplements, drugs, cosmetics, and foods such as breakfast cereals, but it is also incorporated into biopolymers as antimicrobial agents for food packaging applications [248]. ZnO, like other metal nanoparticles, can develop polymer properties such as mechanical and barrier properties when added to biopolymers [249–252].

18.6 Conclusion

Petroleum-derived plastics continue to pose serious global issues, such as nonbiodegradability and low recycling rates, resulting in a global accumulation of plastics. One of the most promising solutions to these problems is the use of renewable and/or biodegradable plastics. In the natural environment, biodegradable plastics can be rapidly degraded into H_2O, CO_2, and biomass. However, due to shortcomings such as high brittleness and poor barrier properties, biodegradable plastics are still far from being widely used as alternatives to petroleum-based plastics. One of the most versatile approaches to overcoming these issues is the use of nanoscale reinforcement agents in biodegradable packaging materials. Bionanocomposites made by incorporating organic or inorganic nanofillers into biopolymers in appropriate amounts frequently have better mechanical and barrier properties than the neat polymers. However, some disadvantages of bionanocomposites as food packaging materials, such as low water resistance, should be addressed before considering them as alternatives to petroleum-based packaging materials. In conclusion, the use of biodegradable polymers as alternatives to petroleum-derived plastics in all fields, including food packaging, appears to be the most promising route to eliminating the growing global plastic problem that threatens the future of humans and other living beings, with the help of solutions provided by intensive research in nanocomposites.

References

[1] X.Z. Tang, P. Kumar, S. Alavi, K.P. Sandeep, Recent advances in biopolymers and biopolymer-based nanocomposites for food packaging materials, Crit. Rev. Food Sci. Nutr. 52 (5) (2012) 426–442.

[2] I. Tiseo, Global plastic production 1950–2019; 2021. [05 May 2021]; Available from: https://www.statista.com/statistics/282732/global-production-of-plastics-since-1950/.
[3] T. Geijer, Plastic packaging in the food sector: six ways to tackle the plastic puzzle, 2019 [cited 2021 01.05.2021]; pp. 1–21. Available from: https://think.ing.com/uploads/reports/ING_-_The_plastic_puzzle_-_December_2019_%28003%29.pdf.
[4] Y.T. Kim, B. Min, K.W. Kim, General characteristics of packaging materials for food system, in: J.H. Han (Ed.), Innovations in Food Packaging, second ed., Academic Press, 2014, pp. 13–35.
[5] Y. Zhong, P. Godwin, Y. Jin, H. Xiao, Biodegradable polymers and greenbased antimicrobial packaging materials: a mini-review, Adv. Ind. Eng. Polym. Res. 3 (1) (2020) 27–35.
[6] A. Arora, G.W. Padua, Review: nanocomposites in food packaging, J. Food Sci. 75 (1) (2010) R43–R49.
[7] A. Nesic, C. Castillo, P. Castano, G. Cabrera-Barjas, J. Serrano, Bio-based packaing materials, in: C.M. Galanakis (Ed.), Biobased Products and Industries, Elsevier, 2020, pp. 279–309.
[8] M. Padervand, E. Lichtfouse, D. Robert, C. Wang, Removal of microplastics from the environment. A review, Env. Chem. Lett. 18 (3) (2020) 807–828.
[9] K. Zhang, H. Shi, J. Peng, Y. Wang, X. Xiong, C. Wu, et al., Microplastic pollution in China's inland water systems: a review of findings, methods, characteristics, effects, and management, Sci. Total. Env. 630 (2018) 1641–1653.
[10] Z.W. Abdullah, Y. Dong, I.J. Davies, S.P.V.A. Barbhuiya, PVA blends, and their nanocomposites for biodegradable packaging application, Polym-Plastics Technol. Eng. 56 (12) (2017) 1307–1344.
[11] F. Garavand, M. Rouhi, S.H. Razavi, I. Cacciotti, R. Mohammadi, Improving the integrity of natural biopolymer films used in food packaging by crosslinking approach: a review, Int. J. Biol. Macromol. 104 (2017) 687–707.
[12] R.M.S. Cruz, Food Packaging: Innovations and Shelf-Life, first ed., CRC Press, Boca Raton, 2019.
[13] S. Shankar, J.W. Rhim, Bionanocomposite films for food packaging applications, Ref. Module Food Sci. 1 (2018) 1–10.
[14] F. Wu, M. Misra, A.K. Mohanty, Challenges and new opportunities on barrier performance of biodegradable polymers for sustainable packaging, Prog. Polym. Sci. 117 (2021) 101395.
[15] J. Wroblewska-Krepsztul, T. Rydzkowski, G. Borowski, M. Szczypinski, T. Klepka, V. Thakur, Recent progress in biodegradable polymers and nanocomposite-based packaging materials for sustainable environment, Int. J. Polym. Anal. Charact. 23 (4) (2018) 383–395.
[16] J.W. Rhim, H.M. Park, C.S. Ha, Bio-nanocomposites for food packaging applications, Prog. Polym. Sci. 38 (10–11) (2013) 1629–1652.
[17] L. Jiang, J. Zhang, Biodegradable and biobased polymers, in: M. Kutz (Ed.), Applied Plastics Engineering Handbook, William Andrew Publishing, 2011, pp. 145–158.
[18] R. Muthuraj, M. Misra, A.K. Mohanty, Biodegradable compatibilized polymer blends for packaging applications: a literature review, J. Appl. Polym. Sci. 135 (24) (2018) 45726.
[19] S.S. Bari, A. Chatterjee, S. Mishra, Biodegradable polymer nanocomposites: an overview, Polym. Rev. 56 (2) (2016) 287–328.
[20] S. Mangaraj, A. Yadav, L.M. Bal, S.K. Dash, N.K. Mahanti, Application of biodegradable polymers in food packaging industry: a comprehensive review, J. Packag. Technol. Res. 3 (1) (2019) 77–96.
[21] P.L. Nayak, Biodegradable polymers: opportunities and challenges, J. Macromol. Sci. Part. C. 39 (3) (1999) 481–505.
[22] P. Cazon, G. Velazquez, J.A. Ramirez, M. Vazquez, Polysaccharide-based films and coatings for food packaging: a review, Food Hydrocoll. 68 (2017) 136–148.
[23] A. Valdes, A.C. Mellinas, M. Ramos, M.C. Garrigos, A. Jimenez, Natural additives and agricultural wastes in biopolymer formulations for food packaging, Front. Chem. 2 (6) (2014).

[24] L. Jiang, J. Zhang, Biodegradable and biobased polymers, in: M. Kutz (Ed.), Applied Plastics Engineering Handbook, second ed., William Andrew Publishing, 2017, pp. 127–143.

[25] D. Plackett, T.L. Andersen, W.B. Pedersen, L. Nielsen, Biodegradable composites based on l-polylactide and jute fibres, Compos. Sci. Technol. 63 (9) (2003) 1287–1296.

[26] J.M. Krochta, Edible protein films and coatings, in: S. Damodaran, A. Paraf (Eds.), Food Proteins and Their Applications, CRC Press, New York, 1997, p. 22.

[27] F. Debeaufort, A. Voilley, Lipid-based edible films and coatings, in: K.C. Huber, M.E. Embuscado (Eds.), Edible Films and Coatings for Food Applications, Springer, New York, 2009, pp. 135–168.

[28] M.B. Perez-Gago, J.W. Rhim, Edible coating and film materials: lipid bilayers and lipid emulsions, in: J.H. Han (Ed.), Innovations in Food Packaging, second ed., Academic Press, San Diego, 2014, pp. 325–350.

[29] Z.A. Raza, S. Abid, I.M. Banat, Polyhydroxyalkanoates: characteristics, production, recent developments and applications, Int. Biodeterior. Biodegrad. 126 (2018) 45–56.

[30] J. Chesterman, Z. Zhang, O. Ortiz, R. Goyal, J. Kohn, Biodegradable polymers, in: R. Lanza, R. Langer, J. Vacanti, A. Atala (Eds.), Principles of Tissue Engineering, fifth ed., Academic Press, 2020, pp. 317–342.

[31] G.Q. Chen, A microbial polyhydroxyalkanoates (PHA) based bio- and materials industry, Chem. Soc. Rev. 38 (8) (2009) 2434–2446.

[32] M.M. Reddy, S. Vivekanandhan, M. Misra, S.K. Bhatia, A. Mohanty, Biobased plastics and bionanocomposites: current status and future opportunities, Prog. Polym. Sci. 38 (10) (2013) 1653–1689.

[33] A.M. Clarinval, J. Halleux, Classification of biodegradable polymers, in: R. Smith (Ed.), Biodegradable Polymers for Industrial Applications, Woodhead Publishing, 2005, pp. 3–31.

[34] S. Philip, T. Keshavarz, I. Roy, Polyhydroxyalkanoates: biodegradable polymers with a range of applications, J. Chem. Technol. Biotechnol. 82 (3) (2007) 233–247.

[35] D. Garlotta, A literature review of poly(lactic acid), J. Polym. Env. 9 (2) (2001) 63–84.

[36] L.T. Lim, R. Auras, M. Rubino, Processing technologies for poly(lactic acid), Prog. Polym. Sci. 33 (8) (2008) 820–852.

[37] J.B. Zeng, K.A. Li, A.K. Du, Compatibilization strategies in poly(lactic acid)-based blends, RSC Adv. 5 (41) (2015) 32546–32565.

[38] V. Nagarajan, A.K. Mohanty, M. Misra, Perspective on polylactic acid (PLA) based sustainable materials for durable applications: focus on toughness and heat resistance, ACS Sustain. Chem. Eng. 4 (6) (2016) 2899–2916.

[39] M.A. Woodruff, D.W. Hutmacher, The return of a forgotten polymer-polycaprolactone in the 21st century, Prog. Polym. Sci. 35 (10) (2010) 1217–1256.

[40] S. Ali Akbari Ghavimi, M.H. Ebrahimzadeh, M. Solati-Hashjin, N.A. Abu Osman, Polycaprolactone/starch composite: fabrication, structure, properties, and applications, J. Biomed. Mater. Res. A 103 (7) (2015) 2482–2498.

[41] R. Ortega-Toro, J. Contreras, P. Talens, A. Chiralt, Physical and structural properties and thermal behaviour of starch-poly(ε-caprolactone) blend films for food packaging, Food Packag. Shelf Life 5 (2015) 10–20.

[42] J. Xu, B.H. Guo, Microbial succinic acid, its polymer poly(butylene succinate), and applications, in: G. G.Q. Chen (Ed.), Plastics from Bacteria: Natural Functions and Applications, Springer, Springer Berlin Heidelberg, 2010, pp. 347–388.

[43] R. Muthuraj, M. Misra, A.K. Mohanty, Injection molded sustainable biocomposites from poly(butylene succinate) bioplastic and perennial grass, ACS Sustain. Chem. Eng. 3 (11) (2015) 2767–2776.

[44] J. Liu, P. Chen, J. Li, S.H. Jiang, Z.Q. Jiang, Q. Gu, Synthesis of poly(ethylene adipate-co-l-lactic acid) copolymers via ring opening polymerization, Polym. Bull. 66 (2) (2011) 187–197.

[45] A. Sorrentino, G. Gorrasi, V. Vittoria, Potential perspectives of bio-nanocomposites for food packaging applications, Trends Food Sci. Technol. 18 (2) (2007) 84–95.

[46] J. Njuguna, F. Ansari, S. Sachse, H. Zhu, V.M. Rodriguez, Nanomaterials, nanofillers, and nanocomposites: types and properties, in: J. Njuguna, K. Pielichowski, H. Zhu (Eds.), Health and Environmental Safety of Nanomaterials, Woodhead Publishing, 2014, pp. 3–27.

[47] R.A. Hule, D.J. Pochan, Polymer nanocomposites for biomedical applications, MRS Bull. 32 (4) (2007) 354–358.

[48] H. Fischer, Polymer nanocomposites: from fundamental research to specific applications, Mater. Sci. Eng. C. 23 (6) (2003) 763–772.

[49] J.H. Koo, Polymer Nanocomposites: Processing, Characterization, and Applications., second ed., McGraw-Hill Education, New York, 2019.

[50] A.M. Youssef, S.M. El-Sayed, Bionanocomposites materials for food packaging applications: concepts and future outlook, Carbohydr. Polym. 193 (2018) 19–27.

[51] Y. Huang, L. Mei, X. Chen, Q. Wang, Recent developments in food packaging based on nanomaterials, Nanomaterials 8 (10) (2018) 830.

[52] J.W. Rhim, A.K. Mohanty, S.P. Singh, P.K.W. Ng, Effect of the processing methods on the performance of polylactide films: thermocompression versus solvent casting, J. Appl. Polym. Sci. 101 (6) (2006) 3736–3742.

[53] R.N. Tharanathan, Biodegradable films and composite coatings: past, present and future, Trends Food Sci. Technol. 14 (3) (2003) 71–78.

[54] M.U. Akbar, F. Rehman, M. Ibrahim, M. Barikani, M. Mohammadi, H. Sobhani, et al., Processing methods of bionanocomposites, in: K.M. Zia, F. Jabeen, M.N. Anjum, S. Ikram (Eds.), Bionanocomposites, Elsevier, 2020, pp. 87–104.

[55] A.M. Youssef, Polymer nanocomposites as a new trend for packaging applications, Polym. Plastics Technol. Eng. 52 (7) (2013) 635–660.

[56] D. Bondeson, Biopolymer-based nanocomposites: processing and properties, in: Department of Engineering Design and Materials. Norwegian University of Science and Technology: Trondheim, 2007. p. 114.

[57] K. Oksman, Y. Aitomäki, A.P. Mathew, G. Siqueira, Q. Zhou, S. Butylina, Review of the recent developments in cellulose nanocomposite processing, Compos. A Appl. Sci. Manuf. 83 (2016) 2–18.

[58] G. Siqueira, J. Bras, A. Dufresne, Cellulosic bionanocomposites: a review of preparation, properties and applications, Polymers 2 (4) (2010) 728–765.

[59] D. Klemm, B. Heublein, H.P. Fink, A. Bohn, Cellulose: fascinating biopolymer and sustainable raw material, Angew. Chem. Int. Ed. 44 (22) (2005) 3358–3393.

[60] R.J. Moon, A. Martini, J. Nairn, J. Simonsen, J. Youngblood, Cellulose nanomaterials review: structure, properties and nanocomposites, Chem. Soc. Rev. 40 (7) (2011) 3941–3994.

[61] F.V. Ferreira, A. Dufresne, I.F. Pinheiro, D.H. Souza, R.F. Gouveia, L. Mei, et al., How do cellulose nanocrystals affect the overall properties of biodegradable polymer nanocomposites: a comprehensive review, Eur. Polym. J. 108 (2018) 274–285.

[62] L. Couret, M. Irle, C. Belloncle, B. Cathala, Extraction and characterization of cellulose nanocrystals from post-consumer wood fiberboard waste, Cellulose 24 (5) (2017) 2125–2137.

[63] S. Camarero Espinosa, T. Kuhnt, E.J. Foster, C. Weder, Isolation of thermally stable cellulose nanocrystals by phosphoric acid hydrolysis, Biomacromolecules 14 (4) (2013) 1223–1230.

[64] B. Li, W. Xu, D. Kronlund, A. Määttänen, J. Liu, J.H. Smått, et al., Cellulose nanocrystals prepared via formic acid hydrolysis followed by TEMPO-mediated oxidation, Carbohydr. Polym. 133 (2015) 605–612.

[65] F.T. Seta, X. An, L. Liu, H. Zhang, J. Yang, W. Zhang, et al., Preparation and characterization of high yield cellulose nanocrystals (CNC) derived from ball mill pretreatment and maleic acid hydrolysis, Carbohydr. Polym. 234 (2020) 115942.

[66] L. Di Giorgio, L. Martin, P.R. Salgado, A.N. Mauri, Synthesis and conservation of cellulose nanocrystals, Carbohydr. Polym. 238 (2020) 116187.
[67] F.W. Herrick, R.L. Casebier, J.K. Hamilton, K.R. Sandberg, Microfibrillated cellulose: morphology and accessibility, J. Appl. Polym. Sci. (1983) 797–813. Appl. Polym. Symp.; (United States); Journal Volume: 37; Conference: 9. Cellulose Conference, Syracuse, NY, USA, 24 May 1982.
[68] A.F. Turbak, F.W. Snyder, K.R. Sandberg, Microfibrillated cellulose, a new cellulose product: properties, uses, and commercial potential, J. Appl. Polym. Sci. Appl. Polym. Symp. (1983).
[69] M. Pääkkö, M. Ankerfors, H. Kosonen, A. Nykänen, S. Ahola, M. Osterberg, et al., Enzymatic hydrolysis combined with mechanical shearing and high-pressure homogenization for nanoscale cellulose fibrils and strong gels, Biomacromolecules 8 (6) (2007) 1934–1941.
[70] M. Henriksson, G. Henriksson, L.A. Berglund, T. Lindström, An environmentally friendly method for enzyme-assisted preparation of microfibrillated cellulose (MFC) nanofibers, Eur. Polym. J. 43 (8) (2007) 3434–3441.
[71] M.L. Hassan, J. Bras, E.A. Hassan, C. Silard, E. Mauret, Enzyme-assisted isolation of microfibrillated cellulose from date palm fruit stalks, Ind. Crop. Prod. 55 (2014) 102–108.
[72] J.H. Chen, J.G. Liu, Y.Q. Su, Z.H. Xu, M.C. Li, R.F. Ying, et al., Preparation and properties of microfibrillated cellulose with different carboxyethyl content, Carbohydr. Polym. 206 (2019) 616–624.
[73] P. Cazón, M. Vázquez, Bacterial cellulose as a biodegradable food packaging material: a review, Food Hydrocoll. 113 (2021) 106530.
[74] O. Shezad, S. Khan, T. Khan, J.K. Park, Physicochemical and mechanical characterization of bacterial cellulose produced with an excellent productivity in static conditions using a simple fed-batch cultivation strategy, Carbohydr. Polym. 82 (1) (2010) 173–180.
[75] M. Matsuoka, T. Tsuchida, K. Matsushita, O. Adachi, F. Yoshinaga, A synthetic medium for bacterial cellulose production by acetobacter xylinum subsp. sucrofermentans, Biosci. Biotechnol. Biochem. 60 (4) (1996) 575–579.
[76] D. Mikkelsen, B.M. Flanagan, G.A. Dykes, M.J. Gidley, Influence of different carbon sources on bacterial cellulose production by *Gluconacetobacter xylinus* strain ATCC 53524, J. Appl. Microbiol. 107 (2) (2009) 576–583.
[77] E. Tsouko, C. Kourmentza, D. Ladakis, N. Kopsahelis, I. Mandala, S. Papnikolaou, et al., Bacterial cellulose production from industrial waste and by-product streams, Int. J. Mol. Sci. 16 (7) (2015) 14832–14849.
[78] E.K. Gladysheva, E.A. Skiba, V.N. Zolotukhin, G.V. Sakovich, Study of the conditions for the biosynthesis of bacterial cellulose by the producer *Medusomyces gisevii* Sa-12, Appl. Biochem. Microbiol. 54 (2) (2018) 179–187.
[79] A. Okiyama, H. Shirae, H. Kano, S. Yamanaka, Bacterial cellulose I. Two-stage fermentation process for cellulose production by *Acetobacter aceti*, Food Hydrocoll. 6 (5) (1992) 471–477.
[80] T. Ishida, Y. Sugano, T. Nakai, M. Shoda, Effects of acetan on production of bacterial cellulose by *Acetobacter xylinum*, Biosci. Biotechnol. Biochem. 66 (8) (2002) 1677–1681.
[81] Y. Yang, J. Jia, J. Xing, J. Chen, S. Lu, Isolation and characteristics analysis of a novel high bacterial cellulose producing strain *Gluconacetobacter intermedius* CIs26, Carbohydr. Polym. 92 (2) (2013) 2012–2017.
[82] M. Shoda, Y. Sugano, Recent advances in bacterial cellulose production, Biotechnol. Bioproc. Eng. 10 (1) (2005) 1–8.
[83] P. Ross, R. Mayer, M. Benziman, Cellulose biosynthesis and function in bacteria, Microbiol. Rev. 1 (1991) 35–58.
[84] A. Sharma, M. Thakur, M. Bhattacharya, T. Mandal, S. Goswami, Commercial application of cellulose nano-composites - a review, Biotechnol. Rep. 21 (2019) e00316.

[85] E.E. Brown, M.P.G. Laborie, Bioengineering bacterial cellulose/poly(ethylene oxide) nanocomposites, Biomacromolecules 8 (10) (2007) 3074–3081.
[86] H.M.C. Azeredo, H. Barud, C.S. Farinas, V.M. Vasconcellos, A.M. Claro, Bacterial cellulose as a raw material for food and food packaging applications, Front. Sustain. Food Syst. 3 (7) (2019) 1–14.
[87] H. Kargarzadeh, M. Mariano, J. Huang, N. Lin, I. Ahmad, A. Dufresne, et al., Recent developments on nanocellulose reinforced polymer nanocomposites: a review, Polymer 132 (2017) 368–393.
[88] M.-C. Li, C. Mei, X. Xu, S. Lee, Q. Wu, Cationic surface modification of cellulose nanocrystals: toward tailoring dispersion and interface in carboxymethyl cellulose films, Polymer 107 (2016) 200–210.
[89] Q. Shang, C. Liu, Y. Hu, P. Jia, L. Hu, Y. Zhou, Bio-inspired hydrophobic modification of cellulose nanocrystals with castor oil, Carbohydr. Polym. 191 (2018) 168–175.
[90] H. Wu, S. Nagarajan, J. Shu, T. Zhang, L. Zhou, Y. Duan, et al., Green and facile surface modification of cellulose nanocrystal as the route to produce poly(lactic acid) nanocomposites with improved properties, Carbohydr. Polym. 197 (2018) 204–214.
[91] E. Espino-Pérez, J. Bras, V. Ducruet, A. Guinault, A. Dufresne, S. Domenek, Influence of chemical surface modification of cellulose nanowhiskers on thermal, mechanical, and barrier properties of poly (lactide) based bionanocomposites, Eur. Polym. J. 49 (10) (2013) 3144–3154.
[92] Z. Hu, R.M. Berry, R. Pelton, E.D. Cranston, One-pot water-based hydrophobic surface modification of cellulose nanocrystals using plant polyphenols, ACS Sustain. Chem. Eng. 5 (6) (2017) 5018–5026.
[93] J.H. Lee, S.H. Park, S.H. Kim, Surface modification of cellulose nanowhiskers and their reinforcing effect in polylactide, Macromol. Res. 22 (4) (2014) 424–430.
[94] A. Leszczyńska, P. Radzik, E. Szefer, M. Mičušík, M. Omastová, K. Pielichowski, Surface modification of cellulose nanocrystals with succinic anhydride, Polymers 11 (5) (2019) 866.
[95] Y. Pan, H. Xiao, Z. Song, Hydrophobic modification of cellulose fibres by cationic-modified polyacrylate latex with core-shell structure, Cellulose 20 (1) (2013) 485–494.
[96] N. Jamaluddin, T. Kanno, T.A. Asoh, H. Uyama, Surface modification of cellulose nanofiber using acid anhydride for poly(lactic acid) reinforcement, Mater. Today Commun. 21 (2019) 100587.
[97] S.L. Balasubramaniam, A.S. Patel, B. Nayak, Surface modification of cellulose nanofiber film with fatty acids for developing renewable hydrophobic food packaging, Food Packag. Shelf Life 26 (2020) 100587.
[98] H. Li, L. Peng, Antimicrobial and antioxidant surface modification of cellulose fibers using layer-by-layer deposition of chitosan and lignosulfonates, Carbohydr. Polym. 124 (2015) 35–42.
[99] S. Leal, C. Cristelo, S. Silvestre, E. Fortunato, A. Sousa, A. Alves, Hydrophobic modifiation of bacterial cellulose using oxygen plasma treatment and chemical vapor deposition, Cellulose 27 (18) (2020) 10733–10746.
[100] K.Y. Lee, J.J. Blaker, A. Bismarck, Surface functionalisation of bacterial cellulose as the route to produce green polylactide nanocomposites with improved properties, Compos. Sci. Technol. 69 (15) (2009) 2724–2733.
[101] R.A.N. Pertile, F.K. Andrade, C. AlveS, M. Gama, Surface modification of bacterial cellulose by nitrogen-containing plasma for improved interaction with cells, Carbohydr. Polym. 82 (3) (2010) 692–698.
[102] S. Ifuku, M. Nogi, K. Abe, K. Handa, F. Nakatsubo, H. Yano, Surface modification of bacterial cellulose nanofibers for property enhancement of optically transparent composites: dependence on acetyl-group DS, Biomacromolecules 8 (6) (2007) 1973–1978.
[103] S. Kalia, A. Dufresne, B.M. Cherian, B.S. Kaith, L. Avérous, J. Njuguna, et al., Cellulose-based bio- and nanocomposites: a review, Int. J. Polym. Sci. (2011) 1–35.
[104] M. Yadav, Y.K. Liu, F.C. Chiu, Fabrication of cellulose nanocrystal/silver/alginate bionanocomposite films with enhanced mechanical and barrier properties for food packaging application, Nanomaterials 9 (11) (2019) 1523.

[105] M. El Achaby, N. El Miri, A. Aboulkas, M. Zahouily, E. Bilal, A. Barakat, et al., Processing and properties of eco-friendly bio-nanocomposite films filled with cellulose nanocrystals from sugarcane bagasse, Int. J. Biol. Macromol. 96 (2017) 340–352.

[106] M. Abdollahi, M. Alboofetileh, R. Behrooz, M. Rezaei, R. Miraki, Reducing water sensitivity of alginate bio-nanocomposite film using cellulose nanoparticles, Int. J. Biol. Macromol. 54 (2013) 166–173.

[107] M. Yadav, F.C. Chiu, Cellulose nanocrystals reinforced κ-carrageenan based UV resistant transparent bionanocomposite films for sustainable packaging applications, Carbohydr. Polym. 211 (2019) 181–194.

[108] N. El Miri, K. Abdelouahdi, A. Barakat, M. Zahouily, A. Fihri, A. Solhy, et al., Bio-nanocomposite films reinforced with cellulose nanocrystals: rheology of film-forming solutions, transparency, water vapor barrier and tensile properties of films, Carbohydr. Polym. 129 (2015) 156–167.

[109] G. Mondragon, C. Peña, A. González, A. Eceiza, A. Arbelaiz, Bionanocomposites based on gelatin matrix and nanocellulose, Eur. Polym. J. 62 (2015) 1–9.

[110] J.P. Reddy, J.W. Rhim, Characterization of bionanocomposite films prepared with agar and paper-mulberry pulp nanocellulose, Carbohydr. Polym. 110 (2014) 480–488.

[111] N.A. El-Wakil, E.A. Hassan, R.E. Abou-Zeid, A. Dufresne, Development of wheat gluten/nanocellulose/titanium dioxide nanocomposites for active food packaging, Carbohydr. Polym. 124 (2015) 337–346.

[112] A.A. Oun, J.W. Rhim, Isolation of cellulose nanocrystals from grain straws and their use for the preparation of carboxymethyl cellulose-based nanocomposite films, Carbohydr. Polym. 150 (2016) 187–200.

[113] R. Kumar, B. Rai, G. Kumar, A simple approach for the synthesis of cellulose nanofiber reinforced chitosan/PVP bio nanocomposite film for packaging, J. Polym. Env. 27 (12) (2019) 2963–2973.

[114] A.B. Perumal, P.S. Sellamuthu, R.B. Nambiar, E.R. Sadiku, Development of polyvinyl alcohol/chitosan bio-nanocomposite films reinforced with cellulose nanocrystals isolated from rice straw, Appl. Surf. Sci. 449 (2018) 591–602.

[115] C.A. Cozzolino, G. Cerri, A. Brundu, S. Farris, Microfibrillated cellulose (MFC): Pullulan bionanocomposite films, Cellulose 21 (6) (2014) 4323–4335.

[116] H. Haghighi, M. Gullo, S. La China, F. Pfeifer, H.W. Siesler, F. Licciardello, et al., Characterization of bio-nanocomposite films based on gelatin/polyvinyl alcohol blend reinforced with bacterial cellulose nanowhiskers for food packaging applications, Food Hydrocoll. 113 (2021) 106454.

[117] E. Trovatti, S.C.M. Fernandes, L. Rubatat, C.S.R. Freire, A.J.D. Silvestre, C.P. Neto, Sustainable nanocomposite films based on bacterial cellulose and pullulan, Cellulose 19 (3) (2012) 729–737.

[118] H. Abral, Kadriadi, M. Mahardika, D. Handayani, E. Sugiarti, A.N. Muslimin, Characterization of disintegrated bacterial cellulose nanofibers/PVA bionanocomposites prepared via ultrasonication, Int. J. Biol. Macromol. 135 (2019) 591–599.

[119] C. Zhijiang, Y. Guang, Optical nanocomposites prepared by incorporating bacterial cellulose nanofibrils into poly(3-hydroxybutyrate), Mater. Lett. 65 (2) (2011) 182–184.

[120] X. Wang, C. Guo, W. Hao, N. Ullah, L. Chen, Z. Li, et al., Development and characterization of agar-based edible films reinforced with nano-bacterial cellulose, Int. J. Biol. Macromol. 118 (2018) 722–730.

[121] E. Fortunati, I. Armentano, Q. Zhou, A. Iannoni, E. Saino, L. Visai, et al., Multifunctional bionanocomposite films of poly(lactic acid), cellulose nanocrystals and silver nanoparticles, Carbohydr. Polym. 87 (2) (2012) 1596–1605.

[122] C.M. Ortiz, P.R. Salgado, A. Dufresne, A.N. Mauri, Microfibrillated cellulose addition improved the physicochemical and bioactive properties of biodegradable films based on soy protein and clove essential oil, Food Hydrocoll. 79 (2018) 416–427.

[123] N. Lavoine, C. Givord, N. Tabary, I. Desloges, B. Martel, J. Bras, Elaboration of a new antibacterial bio-nano-material for food-packaging by synergistic action of cyclodextrin and microfibrillated cellulose, Innov. Food Sci. Emerg. Technol. 26 (2014) 330–340.
[124] M. Salari, M. Sowti Khiabani, R. Rezaei Mokarram, B. Ghanbarzadeh, Samadi, H. Kafil, Development and evaluation of chitosan based active nanocomposite films containing bacterial cellulose nanocrystals and silver nanoparticles, Food Hydrocoll. 84 (2018) 414–423.
[125] S. Ganguly, S. Dana, T.K. Mukhopadhyay, T.K. Parya, S. Ghatak, Organophilic nano clay: a comprehensive review, Trans. Indian. Ceram. Soc. 70 (4) (2011) 189–206.
[126] N. Saba, M. Jawaid, M. Asim, Recent advances in nanoclay/natural fibers hybrid composites, in: M. Jawaid, A.E.K. Qaiss, R. Bouhfid (Eds.), Nanoclay Reinforced Polymer Composites: Natural Fibre/Nanoclay Hybrid Composites, Springer, Singapore, 2016, pp. 1–28.
[127] G. Gorrasi, Dispersion of halloysite loaded with natural antimicrobials into pectins: characterization and controlled release analysis, Carbohydr. Polym. 127 (2015) 47–53.
[128] F. Tornuk, O. Sagdic, M. Hancer, H. Yetim, Development of LLDPE based active nanocomposite films with nanoclays impregnated with volatile compounds, Food Res. Int. 107 (2018) 337–345.
[129] Y.M. Lvov, D.G. Shchukin, H. Möhwald, R.R. Price, Halloysite clay nanotubes for controlled release of protective agents, ACS Nano 2 (5) (2008) 814–820.
[130] E.P. Giannelis, Polymer layered silicate nanocomposites, Adv. Mater. 8 (1) (1996) 29–35.
[131] F. Guo, S. Aryana, Y. Han, Y. Jiao, A review of the synthesis and applications of polymer–nanoclay composites, Appl. Sci. 8 (9) (2018) 1696.
[132] F. Uddin, Montmorillonite: An Introduction to Properties and Utilization, IntechOpen, London, 2018.
[133] M. Jorda-Beneyto, N. Ortuño, A. Devis, S. Aucejo, M. Puerto, D. Gutiérrez-Praena, et al., Use of nanoclay platelets in food packaging materials: technical and cytotoxicity approach, Food Addit. Contam. Part. A 31 (3) (2014) 354–363.
[134] M. Alexandre, P. Dubois, Polymer-layered silicate nanocomposites: preparation, properties and uses of a new class of materials, Mater. Sci. Eng. R. Rep. 28 (1) (2000) 1–63.
[135] A. Okada, A. Usuki, Twenty years of polymer-clay nanocomposites, Macromol. Mater. Eng. 291 (12) (2006) 1449–1476.
[136] K. Zandsalimi, B. Akbari, F. Mehrnejad, R. Bagheri, Compatibilization of clays and hydrophobic polymers: the case of montmorillonite and polyetheretherketone, Polym. Bull. 77 (10) (2020) 5505–5527.
[137] S. Ray, S.Y. Quek, A. Easteal, X.D. Chen, The potential use of polymer-clay nanocomposites in food packaging, Int. J. Food Eng. 2 (4) (2006).
[138] H.M.C. de Azeredo, L.H.C. Mattoso, T.H. McHugh, Nanocomposites in food packaging: a review, in: B. Reddy (Ed.), Advances in Diverse Industrial Applications of Nanocomposites, IntechOpen, Rijeka, Crotia, 2012.
[139] L.E. Nielsen, Models for the permeability of filled polymer systems, J. Macromol. Sci. A 1 (5) (1967) 929–942.
[140] Y. Kasirga, A. Oral, C. Caner, Preparation and characterization of chitosan/montmorillonite-K10 nanocomposites films for food packaging applications, Polym. Compos. 33 (11) (2012) 1874–1882.
[141] Y. Xu, X. Ren, M.A. Hanna, Chitosan/clay nanocomposite film preparation and characterization, J. Appl. Polym. Sci. 99 (4) (2006) 1684–1691.
[142] A. Altınışık, Y. Seki, K. Yurdakoç, Preparation an characterization of chitosan/KSF biocomposite film, Polym. Compos. 30 (8) (2009) 1035–1042.
[143] A.K. Pal, F. Wu, M. Misra, A.K. Mohanty, Reactive extrusion of sustainable PHBV/PBAT-based nanocomposite films with organically modified nanoclay for packaging applications: compression moulding vs. cast film extrusion, Compos. B Eng. 198 (2020) 108141.

[144] B.I. Dogaru, B. Simionescu, M.C. Popescu, Synthesis and characterization of κ-carrageenan bio-nanocomposite films reinforced with bentonite nanoclay, Int. J. Biol. Macromol. 154 (2020) 9–17.

[145] S. Memiş, F. Tornuk, F. Bozkurt, M.Z. Durak, Production and characterization of a new biodegradable fenugreek seed gum based active nanocomposite film reinforced with nanoclays, Int. J. Biol. Macromol. 103 (2017) 669–675.

[146] S. Cesur, C. Köroğlu, H.T. Yalçın, Antimicrobial and biodegradable food packaging applications of polycaprolactone/organo nanoclay/chitosan polymeric composite films, J. Vinyl Addit. Technol. 24 (4) (2018) 376–387.

[147] T. Mohan, K. Devchand, K. Kanny, Barrier and biodegradable properties of corn starch-derived biopolymer film filled with nanoclay fillers, J. Plast. Film. Sheeting 33 (3) (2017) 309–336.

[148] A. Heydari, I. Alemzadeh, M. Vossoughi, Functional properties of biodegradable corn starch nanocomposites for food packaging applications, Mater. Des. 50 (2013) 954–961.

[149] P. Ren, T. Shen, F. Wang, X. Wang, Z. Zhang, Study on Biodegradable Starch/OMMT Nanocomposites for Packaging Applications, J. Polym. Env. 17 (3) (2009) 203–207.

[150] B.A.S. Suin, B.B. Khatua, Highly exfoliated eco-friendly thermoplastic starch (TPS)/poly (lactic acid) (PLA)/clay nanocomposites using unmodified nanoclay, Carbohydr. Polym. 110 (2014) 430–439.

[151] V. Alves, R. Castelló, A. Ferreira, N. Costa, I. Fonseca, I. Coelhoso, Barrier properties of carrageenan/pectin biodegradable composite films, Procedia Food Sci. 1 (2011) 240–245.

[152] S. Tunç, O. Duman, Preparation and characterization of biodegradable methyl cellulose/montmorillonite nanocomposite films, Appl. Clay Sci. 48 (3) (2010) 414–424.

[153] F. Yahiaoui, F. Benhacine, H. Ferfera-Harrar, A. Habi, A.S. Hadj-Hamou, Y. Grohens, Development of antimicrobial PCL/nanoclay nanocomposite films with enhanced mechanical and water vapor barrier properties for packaging applications, Polym. Bull. 72 (2) (2015) 235–254.

[154] H. Rostamzad, S.Y. Paighambari, B. Shabanpour, S.M. Ojagh, S.M. Mousavi, Improvement of fish protein film with nanoclay and transglutaminase for food packaging. Food Packag Shelf, Life 7 (2016) 1–7.

[155] V.M. Azevedo, M.V. Dias, S.V. Borges, R.V. Fernandes, B. de, E.K. Silva, et al., Optical and structural properties of biodegradable whey protein isolate nanocomposite films for active packaging, Int. J. Food Prop. 20 (sup2) (2017) 1869–1878.

[156] J. Zhang, Z. Jia, D. Jia, D. Zhang, A. Zhang, Properties of halloysite nanotube–epoxy resin hybrids and the interfacial reactions in the systems, Nanotechnology 18 (45) (2007) 455703.

[157] Y. Lvov, E. Abdullayev, Functional polymer–clay nanotube composites with sustained release of chemical agents, Prog. Polym. Sci. 38 (10) (2013) 1690–1719.

[158] M.H. Lee, H.S. Seo, H.J. Park, Thyme oil encapsulated in halloysite nanotubes for antimicrobial packaging system, J. Food Sci. 82 (4) (2017) 922–932.

[159] M.H. Lee, H.J. Park, Preparation of halloysite nanotubes coated with Eudragit for a controlled release of thyme essential oil, J. Appl. Polym. Sci. 132 (46) (2015).

[160] M. Makaremi, P. Pasbakhsh, G. Cavallaro, G. Lazzara, Y.K. Aw, S.M. Lee, et al., Effect of morphology and size of halloysite nanotubes on functional pectin bionanocomposites for food packaging applications, ACS Appl. Mater. Interfaces 9 (20) (2017) 17476–17488.

[161] N.P. Risyon, S.H. Othman, R.K. Basha, R.A. Talib, Characterization of polylactic acid/halloysite nanotubes bionanocomposite films for food packaging, Food Packag. Shelf Life 23 (2020) 100450.

[162] Q. Li, T. Ren, P. Perkins, X. Hu, X. Wang, Applications of halloysite nanotubes in food packaging for improving film performance and food preservation, Food Control. 124 (2021) 107876.

[163] N. Alipoormazandarani, S. Ghazihoseini, Mohammadi, A. Nafchi, Preparation and characterization of novel bionanocomposite based on soluble soybean polysaccharide and halloysite nanoclay, Carbohydr. Polym. 134 (2015) 745–751.

[164] B.A. Tas, E. Sehit, C.E. Tas, S. Unal, F. Cebeci, Y. Menceloglu, et al., Carvacrol loaded halloysite coatings for antimicrobial food packaging applications, Food Packag. Shelf Life 20 (2019) 100300.

[165] E. Abdullayev, Y. Lvov, Halloysite clay nanotubes for controlled release of protective agents, J. Nanosci. Nanotechnol. 11 (11) (2011) 10007–10026.

[166] J. Hári, M. Sárközi, E. Földes, B. Pukánszky, Long term stabilization of PE by the controlled release of a natural antioxidant from halloysite nanotubes, Polym. Degrad. Stab. 147 (2018) 229–236.

[167] L. Buendía – Moreno, M.J. Sánchez – Martínez, V. Antolinos, M. Ros – Chumillas, L. Navarro – Segura, S. Soto – Jover, et al., Active cardboard box with a coating including essential oils entrapped within cyclodextrins and/or halloysite nanotubes. A case study for fresh tomato storage, Food Control. 107 (2020) 106763.

[168] C.E. Tas, S. Hendessi, M. Baysal, S. Unal, F.C. Cebeci, Y.Z. Menceloglu, et al., Halloysite nanotubes/polyethylene nanocomposites for active food packaging materials with ethylene scavenging and gas barrier properties, Food Bioproc Tech. 10 (4) (2017) 789–798.

[169] K.K. Gaikwad, S. Singh, Y.S. Lee, High adsorption of ethylene by alkali-treated halloysite nanotubes for food-packaging applications, Env. Chem. Lett. 16 (3) (2018) 1055–1062.

[170] J. Joung, A. Boonsiriwit, M. Kim, Y.S. Lee, Application of ethylene scavenging nanocomposite film prepared by loading potassium permanganate-impregnated halloysite nanotubes into low-density polyethylene as active packaging material for fresh produce, LWT 145 (2021) 111309.

[171] A. Boonsiriwit, Y. Xiao, J. Joung, M. Kim, S. Singh, Y.S. Lee, Alkaline halloysite nanotubes/low density polyethylene nanocomposite films with increased ethylene absorption capacity: applications in cherry tomato packaging, Food Packag. Shelf Life 25 (2020) 100533.

[172] V.A. Vinokurov, A.V. Stavitskaya, Y.A. Chudakov, E.V. Ivanov, L.K. Shrestha, K. Ariga, et al., Formation of metal clusters in halloysite clay nanotubes, Sci. Technol. Adv. Mater. 18 (1) (2017) 147–151.

[173] S.H. Othman, H.N. Ling, R.A. Talib, M.N. Naim, N.P. Risyon, M.D. Saifullah, PLA/MMT and PLA/Halloysite bio-nanocomposite films: mechanical, barrier, and transparency, J. Nano Res. 59 (2019) 77–93.

[174] I. Sancheti, P. Katkar, S. Thorat, S. Radhakrishnan, M.B. Kulkarni, PBAT/PLA/HNT filled blends for active food packaging, Mater. Res. Express 6 (12) (2020) 125378.

[175] M. Liu, Y. Zhang, C. Wu, S. Xiong, C. Zhou, Chitosan/halloysite nanotubes bionanocomposites: structure, mechanical properties and biocompatibility, Int. J. Biol. Macromol. 51 (4) (2012) 566–575.

[176] F. Sadegh-Hassani, Mohammadi, A. Nafchi, Preparation and characterization of bionanocomposite films based on potato starch/halloysite nanoclay, Int. J. Biol. Macromol. 67 (2014) 458–462.

[177] Z.W. Abdullah, Y. Dong, Biodegradable and water resistant poly(vinyl) alcohol (PVA)/starch (ST)/glycerol (GL)/halloysite nanotube (HNT) nanocomposite films for sustainable food packaging, Front. Mater. 6 (58) (2019).

[178] S. Rojas-Lema, L. Quiles-Carrillo, D. Garcia-Garcia, B. Melendez-Rodriguez, R. Balart, S. Torres-Giner, Tailoring the properties of thermo-compressed polylactide films for food packaging applications by individual and combined additions of lactic acid oligomer and halloysite nanotubes, Molecules 25 (8) (2020) 1976.

[179] H.C. Voon, R. Bhat, A. Easa, M. Liong, A. Karim, Effect of addition of halloysite nanoclay and SiO2 nanoparticles on barrier and mechanical properties of bovine gelatin films, Food Bioproc Tech. 5 (5) (2012) 1766–1774.

[180] Y. Zou, C. Zhang, P. Wang, Y. Zhang, H. Zhang, Electrospun chitosan/polycaprolactone nanofibers containing chlorogenic acid-loaded halloysite nanotube for active food packaging, Carbohydr. Polym. 247 (2020) 116711.

[181] R.J. Young, I.A. Kinloch, L. Gong, K.S. Novoselov, The mechanics of graphene nanocomposites: a review, Compos. Sci. Technol. 72 (12) (2012) 1459–1476.

[182] A.H. Kumar, M. Ahamed, K. Deshmukh, M.S. Sirajuddeen, Morphology, dielectric and EMI shielding characteristics of graphene nanoplatelets, montmorillonite nanoclay and titanium dioxide nanoparticles reinforced polyvinylidenefluoride nanocomposites, J. Inorg. Organom. Polym. Mater. 31 (5) (2021) 2003–2016.

[183] X. Li, L. Tao, Z. Chen, H. Fang, X. Li, X. Wang, et al., Graphene and related two-dimensional materials: structure-property relationships for electronics and optoelectronics, Appl. Phys. Rev. 4 (2) (2017) 021306.

[184] T.M. Radadiya, A properties of graphene, Eur. J. Mater. Sci. 2 (1) (2015) 6–18.

[185] K. Novoselov, A.K. Geim, S. Morozov, D. Jiang, Y. Zhang, S. Dubonos, et al., Electric field effect in atomically thin carbon films, Science 306 (5696) (2004) 666–669.

[186] V.B. Mohan, K. Lau, D. Hui, D. Bhattacharyya, Graphene-based materials and their composites: a review on production, applications and product limitations, Compos. B Eng. 142 (2018) 200–220.

[187] S. Liu, T.H. Zeng, M. Hofmann, E. Burcombe, J. Wei, R. Jiang, et al., Antibacterial activity of graphite, graphite oxide, graphene oxide, and reduced graphene oxide: membrane and oxidative stress, ACS Nano 5 (9) (2011) 6971–6980.

[188] S. Gurunathan, J.W. Han, A.A. Dayem, V. Eppakayala, J.H. Kim, Oxidative stress-mediated antibacterial activity of graphene oxide and reduced graphene oxide in *Pseudomonas aeruginosa*, Int. J. Nanomed. 7 (2012) 5901–5914.

[189] K. Krishnamoorthy, N. Umasuthan, R. Mohan, J. Lee, S.J. Kim, Antibacterial activity of graphene oxide nanosheets, Sci. Adv. Mater. 4 (11) (2012) 1111–1117.

[190] S. Liu, M. Hu, T.H. Zeng, R. Wu, R. Jiang, J. Wei, et al., Lateral dimension-dependent antibacterial activity of graphene oxide sheets, Langmuir 28 (33) (2012) 12364–12372.

[191] K. Li, S. Jin, J. Li, H. Chen, Improvement in antibacterial and functional properties of mussel-inspired cellulose nanofibrils/gelatin nanocomposites incorporated with graphene oxide for active packaging, Ind. Crop. Prod. 132 (2019) 197–212.

[192] S. Afshar, H. Baniasadi, Investigation the effect of graphene oxide and gelatin/starch weight ratio on the properties of starch/gelatin/GO nanocomposite films: the RSM study, Int. J. Biol. Macromol. 109 (2018) 1019–1028.

[193] Y. Zhu, S. Murali, W. Cai, X. Li, J. Suk, J.R. Potts, et al., Graphene and graphene oxide: synthesis, properties, and applications, Adv. Mater. 22 (35) (2010) 3906–3924.

[194] Y. Wang, L. Sisi, Y. Haiyan, L. Jie, Progress in the functional modification of graphene/graphene oxide: a review, RSC Adv. 10 (26) (2020) 15328–15345.

[195] A. Barra, J.D.C. Santos, M.R.F. Silva, C. Nunes, E. Ruiz-Hitzky, I. Gonçalves, et al., Graphene derivatives in biopolymer-based composites for food packaging applications, Nanomaterials 10 (10) (2020) 2077.

[196] M. Wojtoniszak, X. Chen, R.J. Kalenczuk, A. Wajda, J. Łapczuk, M. Kurzewski, et al., Synthesis, dispersion, and cytocompatibility of graphene oxide and reduced graphene oxide, Colloids Surf. B Biointerfaces 89 (2012) 79–85.

[197] P. Rani, M.B. Ahamed, K. Deshmukh, Structural, dielectric and EMI shielding properties of polyvinyl alcohol/chitosan blend nanocomposites integrated with graphite oxide and nickel oxide nanofillers, J. Mater. Sci. Mater Electron. 32 (1) (2021) 764–779.

[198] M. Aqlil, A. Moussemba Nzenguet, Y. Essamlali, A. Snik, M. Larzek, M. Zahouily, Graphene oxide filled lignin/starch polymer bionanocomposite: structural, physical, and mechanical studies, J. Agric. Food Chem. 65 (48) (2017) 10571–10581.

[199] A. Venkateshaiah, J.Y. Cheong, C. Habel, S. Wacławek, T. Lederer, M. Cernik, et al., Tree gum–graphene oxide nanocomposite films as gas barriers, ACS Appl. Nano Mater. 3 (1) (2020) 633–640.

[200] R. Zhang, Y. Wang, D. Ma, S. Ahmed, W. Qin, Y. Liu, Effects of ultrasonication duration and graphene oxide and nano-zinc oxide contents on the properties of polyvinyl alcohol nanocomposites, Ultrason. Sonochem. 59 (2019) 104731.

[201] I.U. Unalan, C. Wan, Ł. Figiel, R.T. Olsson, S. Trabattoni, S. Farris, Exceptional oxygen barrier performance of pullulan nanocomposites with ultra-low loading of graphene oxide, Nanotechnology 26 (27) (2015) 275703.
[202] R. Justin, B. Chen, Strong and conductive chitosan–reduced graphene oxide nanocomposites for transdermal drug delivery, J. Mater. Chem. B 2 (24) (2014) 3759–3770.
[203] O.K. Park, M.G. Hahm, S. Lee, H.I. Joh, S.I. Na, R. Vajtai, et al., In situ synthesis of thermochemically reduced graphene oxide conducting nanocomposites, Nano Lett. 12 (4) (2012) 1789–1793.
[204] A. Barra, N.M. Ferreira, M.A. Martins, O. Lazar, A. Pantazi, A.A. Jderu, et al., Eco-friendly preparation of electrically conductive chitosan - reduced graphene oxide flexible bionanocomposites for food packaging and biological applications, Compos. Sci. Technol. 173 (2019) 53–60.
[205] C. Grande, J. Mangadlao, J. Fan, A. Leon, J. Delgado-Ospina, J.G. Rojas, et al., Chitosan cross-cinked graphene oxide nanocomposite films with antimicrobial activity for application in food industry, Macromol. Symposia 374 (1) (2017) 1600114.
[206] K.B. Narayanan, G.T. Park, S.S. Han, Antibacterial properties of starch-reduced graphene oxide–polyiodide nanocomposite, Food Chem. 342 (2021) 128385.
[207] P. Appendini, J.H. Hotchkiss, Review of antimicrobial food packaging, Innov. Food Sci. Emerg. Technol. 3 (2) (2002) 113–126.
[208] F. Martinez-Gutierrez, P.L. Olive, A. Banuelos, E. Orrantia, N. Nino, E.M. Sanchez, et al., Synthesis, characterization, and evaluation of antimicrobial and cytotoxic effect of silver and titanium nanoparticles, Nanomedicine. 6 (5) (2010) 681–688.
[209] J.W. Alexander, History of the medical use of silver, Surg. Infect. 10 (3) (2009) 289–292.
[210] R.D. Toker, N. Kayaman-Apohan, M.V. Kahraman, UV-curable nano-silver containing polyurethane based organic–inorganic hybrid coatings, Prog. Org. Coat. 76 (9) (2013) 1243–1250.
[211] M. Rai, A. Yadav, A. Gade, Silver nanoparticles as a new generation of antimicrobials, Biotechnol. Adv. 27 (1) (2009) 76–83.
[212] A.D. Russell, W.B. Hugo, 7 Antimicrobial activity and action of silver, in: G.P. Ellis, D.K. Luscombe (Eds.), Progress in Medicinal Chemistry, Elsevier, 1994, pp. 351–370.
[213] P. Dibrov, J. Dzioba, K.K. Gosink, C.C. Häse, Chemiosmotic mechanism of antimicrobial activity of Ag(+) in Vibrio cholerae, Antimicrob. Agents Chemother. 46 (8) (2002) 2668–2670.
[214] L. Sintubin, B. De Gusseme, P. Van der Meeren, B.F. Pycke, W. Verstraete, N. Boon, The antibacterial activity of biogenic silver and its mode of action, Appl. Microbiol. Biotechnol. 91 (1) (2011) 153–162.
[215] K.J. Kim, W.S. Sung, B.K. Suh, S.K. Moon, J.S. Choi, J.G. Kim, et al., Antifungal activity and mode of action of silver nano-particles on *Candida albicans*, Biometals. 22 (2) (2009) 235–242.
[216] C.P. Randall, L.B. Oyama, J.M. Bostock, I. Chopra, A.J. O'Neill, The silver cation (Ag +): antistaphylococcal activity, mode of action and resistance studies, J. Antimicrob. Chemother. 68 (1) (2013) 131–138.
[217] Z.M. Xiu, Q.B. Zhang, H.L. Puppala, V.L. Colvin, P.J. Alvarez, Negligible particle-specific antibacterial activity of silver nanoparticles, Nano Lett. 12 (8) (2012) 4271–4275.
[218] G.A. Martínez-Castañón, N. Niño-Martínez, F. Martínez-Gutierrez, J.R. Martínez-Mendoza, F. Ruiz, Synthesis and antibacterial activity of silver nanoparticles with different sizes, J. Nanopart. Res. 10 (8) (2008) 1343–1348.
[219] S. Mathew, J. Mathew, E.K. Radhakrishnan, Polyvinyl alcohol/silver nanocomposite films fabricated under the influence of solar radiation as effective antimicrobial food packaging material, J. Polym. Res. 26 (9) (2019) 223.
[220] J.H. Lee, D. Jeong, P. Kanmani, Study on physical and mechanical properties of the biopolymer/silver based active nanocomposite films with antimicrobial activity. Carbohydr. Polym. 201;224:115159.
[221] P. Kanmani, J.W. Rhim, Physical, mechanical and antimicrobial properties of gelatin based active nanocomposite films containing AgNPs and nanoclay, Food Hydrocoll. 35 (2014) 644–652.

[222] S. Kumar, A. Shukla, P.P. Baul, A. Mitra, D. Halder, Biodegradable hybrid nanocomposites of chitosan/gelatin and silver nanoparticles for active food packaging applications, Food Packag. Shelf Life 16 (2018) 178–184.

[223] S. Mathew, J.M. Nil, S.S. Babu, Biodegradable and active nanocomposite pouches reinforced with silver nanoparticles for improved packaging of chicken sausages, Food Packag. Shelf Life 19 (2019) 155–166.

[224] B.S. Sooch, M.K. Mann, Nanoreinforced biodegradable gelatin based active food packaging film for the enhancement of shelf life of tomatoes (*Solanum lycopersicum* L.), Food Control. 130 (2021) 108322.

[225] S. Ediyilyam, B. George, S.S. Shankar, T.T. Dennis, S. Wacławek, M. Černík, et al., Chitosan/gelatin/silver nanoparticles composites films for biodegradable food packaging applications, Polymers 13 (11) (2021) 1680.

[226] M.P. Indumathi, K. Saral Sarojini, G.R. Rajarajeswari, Antimicrobial and biodegradable chitosan/cellulose acetate phthalate/ZnO nano composite films with optimal oxygen permeability and hydrophobicity for extending the shelf life of black grape fruits, Int. J. Biol. Macromol. 132 (2019) 1112–1120.

[227] S. Girdthep, P. Worajittiphon, R. Molloy, S. Lumyong, T. Leejarkpai, W. Punyodom, Biodegradable nanocomposite blown films based on poly(lactic acid) containing silver-loaded kaolinite: a route to controlling moisture barrier property and silver ion release with a prediction of extended shelf life of dried longan, Polymer 55 (26) (2014) 6776–6788.

[228] V.K. Pandey, S.N. Upadhyay, K. Niranjan, P.K. Mishra, Antimicrobial biodegradable chitosan-based composite Nano-layers for food packaging, Int. J. Biol. Macromol. 157 (2020) 212–219.

[229] A.L. Incoronato, A. Conte, G. Buonocore, M.D. Nobile, Agar hydrogel with silver nanoparticles toprolong the shelf life of Fior di Latte cheese, J. Dairy. Sci. 94 (4) (2011). 1697-04.

[230] A.C.F. Vieira, J. de Matos Fonseca, N.M.C. Menezes, A.R. Monteiro, G.A. Valencia, Active coatings based on hydroxypropyl methylcellulose and silver nanoparticles to extend the papaya (*Carica papaya* L.) shelf life, Int. J. Biol. Macromol. 164 (2020) 489–498.

[231] K.K. Gaikwad, S. Singh, Y.S. Negi, Ethylene scavengers for active packaging of fresh food produce, Env. Chem. Lett. 18 (2) (2020) 269–284.

[232] H. Wei, F. Seidi, T. Zhang, Y. Jin, H. Xiao, Ethylene scavengers for the preservation of fruits and vegetables: a review, Food Chem. 337 (2021) 127750.

[233] B. Lin, Y. Luo, Z. Teng, B. Zhang, B. Zhou, Q. Wang, Development of silver/titanium dioxide/chitosan adipate nanocomposite as an antibacterial coating for fruit storage, LWT - Food Sci. Technol. 63 (2) (2015) 1206–1213.

[234] T. Tong, C.T. Binh, J.J. Kelly, J.F. Gaillard, K.A. Gray, Cytotoxicity of commercial nano-TiO2 to *Escherichia coli* assessed by high-throughput screening: effects of environmental factors, Water Res. 47 (7) (2013) 2352–2362.

[235] T.A. Egerton, I.R. Tooley, Effect of changes in TiO_2 dispersion on its measured photocatalytic activity, J. Phys. Chem. B 108 (16) (2004) 5066–5072.

[236] P. Kaewklin, U. Siripatrawan, A. Suwanagul, Y.S. Lee, Active packaging from chitosan-titanium dioxide nanocomposite film for prolonging storage life of tomato fruit, Int. J. Biol. Macromol. 112 (2018) 523–529.

[237] X. Zhang, G. Xiao, Y. Wang, Y. Zhao, H. Su, T. Tan, Preparation of chitosan-TiO_2 composite film with efficient antimicrobial activities under visible light for food packaging applications, Carbohydr. Polym. 169 (2017) 101–107.

[238] W. Zhang, J. Chen, Y. Chen, W. Xia, Y.L. Xiong, H. Wang, Enhanced physicochemical properties of chitosan/whey protein isolate composite film by sodium laurate-modified TiO2 nanoparticles, Carbohydr. Polym. 138 (2016) 59–65.

[239] A. Vejdan, S.M. Ojagh, A. Adeli, M. Abdollahi, Effect of TiO_2 nanoparticles on the physico-mechanical and ultraviolet light barrier properties of fish gelatin/agar bilayer film, LWT - Food Sci. Technol. 71 (2016) 88–95.

[240] S.A. Oleyaei, Y. Zahedi, B. Ghanbarzadeh, A.A. Moayedi, Modification of physicochemical and thermal properties of starch films by incorporation of TiO_2 nanoparticles, Int. J. Biol. Macromol. 89 (2016) 256–264.

[241] Q. He, Y. Zhang, X. Cai, S. Wang, Fabrication of gelatin-TiO2 nanocomposite film and its structural, antibacterial and physical properties, Int. J. Biol. Macromol. 84 (2016) 153–160.

[242] R. Ahmadi, A. Tanomand, F. Kazeminava, F.S. Kamounah, A. Ayaseh, K. Ganbarov, et al., Fabrication and characterization of a titanium dioxide (TiO_2) nanoparticles reinforced bio-nanocomposite containing *Miswak* (*Salvadora persica* L.) extract – the antimicrobial, thermo-physical and barrier properties, Int. J. Nanomed. 14 (2019) 3439–3454.

[243] H. Bodaghi, Y. Mostofi, A. Oromiehie, Z. Zamani, B. Ghanbarzadeh, C. Costa, et al., Evaluation of the photocatalytic antimicrobial effects of a TiO2 nanocomposite food packaging film by in vitro and in vivo tests, LWT - Food Sci. Technol. 50 (2) (2013) 702–706.

[244] N. Baek, Y.T. Kim, J.E. Marcy, S.E. Duncan, S.F. O'Keefe, Physical properties of nanocomposite polylactic acid films prepared with oleic acid modified titanium dioxide, Food Packag. Shelf Life 17 (2018) 30–38.

[245] R. Sami, A. Elhakem, M. Alharbi, N. Benajiba, M. Almatrafi, J. Jing, et al., Effect of titanium dioxide nanocomposite material and antimicrobial agents on mushrooms shelf-life preservation, Processes 8 (12) (2020) 1632.

[246] M. Alizadeh Sani, A. Ehsani, M. Hashemi, Whey protein isolate/cellulose nanofibre/TiO2 nanoparticle/rosemary essential oil nanocomposite film: its effect on microbial and sensory quality of lamb meat and growth of common foodborne pathogenic bacteria during refrigeration, Int. J. Food Microbiol. 251 (2017) 8–14.

[247] P.J.P. Espitia, C.G. Otoni, N.F.F. Soares, Zinc oxide nanoparticles for food packaging applications, in: J. Barros-Velázquez (Ed.), Antimicrobial Food Packaging, Academic Press, San Diego, 2016, pp. 425–431.

[248] P.J.P. Espitia, N.F.F. Soares, J.S.R. Coimbra, N. de Andrade, R.S. Cruz, E.A.A. Medeiros, Zinc oxide nanoparticles: synthesis, antimicrobial activity and food packaging applications, Food Bioproc. Tech. 5 (5) (2012) 1447–1464.

[249] P. Kanmani, J.W. Rhim, Properties and characterization of bionanocomposite films prepared with various biopolymers and ZnO nanoparticles, Carbohydr. Polym. 106 (2014) 190–199.

[250] M. Zare, K. Namratha, S. Ilyas, A. Hezam, S. Mathur, K. Byrappa, Smart fortified PHBV-CS biopolymer with Zno-Ag nanocomposites for enhanced shelf life of food packaging, ACS Appl. Mater. Interfaces 11 (51) (2019) 48309–48320.

[251] G. Guan, L. Zhang, J. Zhu, H. Wu, W. Li, Q. Sun, Antibacterial properties and mechanism of biopolymer-based films functionalized by CuO/ZnO nanoparticles against *Escherichia coli* and *Staphylococcus aureus*, J. Hazard. Mater. 402 (2021) 123542.

[252] S. Sherafatkhah Azari, A. Alizadeh, L. Roufegarinejad, N. Asefi, H. Hamishehkar, Preparation and characterization of gelatin/β-glucan nanocomposite film incorporated with ZnO nanoparticles as an active food packaging system, J. Polym. Env. 29 (4) (2021) 1143–1152.

CHAPTER

19

Biodegradable polymer nanocomposites for catalytic and photocatalytic applications

Radha D. Pyarasani[1], John Amalraj[2], Poonam Negi[3], Saravanan Chandrasekaran[4] and Swarnalatha Kalayar[5]

[1]Centro de Investigación de Estudios Avanzados del Maule (CIEAM), Vicerrectoria de Investigacion y Postgrado, Universidad Católica del Maule, Talca, Maule, Chile [2]Laboratory of Materials Science, Instituto de Química de Recursos Naturales, Universidad de Talca, Talca, Maule, Chile [3]School of Pharmaceutical Sciences, Shoolini University of Biotechnology and Management Sciences, Solan, Himachal Pradesh, India [4]Department of Chemistry, School of Engineering, Presidency University, Bangalore, Karnataka, India [5]Department of Chemistry, Manonmaniam Sundaranar University, Tirunelveli, Tamil Nadu, India

Chapter Outline

19.1 Introduction

In recent years, there has been a lot of interest in the development of environmentally friendly biopolymer-based nanomaterials derived primarily from natural resources such as plant-based feedstocks or bacterial fermentation. These materials address the need for replacements for fossil-fuel feedstock while also addressing growing concerns about end-of-life disposal issues and the need for landfill alternatives. It is believed that the massive amount of synthetic polymer production can be

Biodegradable and Biocompatible Polymer Nanocomposites. DOI: https://doi.org/10.1016/B978-0-323-91696-7.00020-9

Copyright © 2023 Elsevier Inc. All rights reserved.

reduced by using biodegradable polymer materials, which are less expensive and have positive environmental and economic effects.

Furthermore, biodegradable polymers are nonhazardous to the environment and are used in a variety of applications, including biomedical and tissue engineering. Because of their low or acceptable toxicity, biodegradable polymers are widely used in drug delivery applications, sutures, and disposable gloves [1–3]. There are some obvious applications in agriculture, such as compostable materials in plant pots, mulching film, and yarn [4,5]. Biopolymeric materials are primarily used in the catering industry for the packaging of food and convenience items such as disposable cutlery, plates, and cups [6,7]. Furthermore, these biodegradable polymer matrices have received a lot of attention as catalyst immobilization substrates [8]. These polymer-supported materials are popular for catalytic applications due to the presence of polar groups in the polymer chain, which have a strong affinity for holding inorganic semiconductors on its surface and allow reuse of these materials without loss of catalytic efficiency. However, it is fascinating to note that the positive properties of biodegradable polymers can be a good alternative to synthetic polymers, but some of their limitations, such as their poor technical performance, particularly optical, mechanical, and barrier properties, make them unfavorable when compared to many conventional polymers [9]. To address these shortcomings, nanocomposite technology has been applied to biodegradable polymers.

Before we get into the nanocomposites section, let's go over some of the fundamentals and definitions of biodegradable polymers, including their classification and sources. Biodegradation is defined as "a process in which degradation occurs as a result of microorganisms such as fungi, bacteria, or algae." Biodegradable polymers can be either natural or synthetic, and they can be broken down by microorganism enzymes, producing CO_2 in aerobic environments and CH_4 in anaerobic environments. In addition to producing hummus, the polymer substrate serves as a carbon source for microorganism metabolism during this process. Biodegradable polymers can be broadly classified into two broad categories based on their origin. One of these groups is natural biopolymers, which are derived from biomass or natural resources. The other category includes biopolymers derived from fossil fuels, where the monomer is derived from fossil fuel [9]. As shown in Fig. 19.1 among the natural biopolymers, most of the biopolymers are either derived from plants, animals, microorganisms, or bio-derived polymers. Proteins (collagen, gelatin, silk, albumin, fibrin, etc.) and polysaccharides (cellulose, alginates, starch, chitin/chitosan, hyaluronic acid derivatives, etc.) are two types of natural biopolymers. Biopolymers based on gums (Xanthan gum, gum arabica, gellan gum, etc.) and polyesters (Polyhydroxy alkonates). Biodegradable polymers typically have ester, amide, or ether bonds in their structure, making them a suitable material for a wide range of applications. Starch, cellulose, and lignin, for example, are carbon-based compounds that are hydrophilic, can absorb water, swell, and are easily biodegradable. As a result, they are increasingly used in biomedical and tissue engineering applications [3].

19.2 Natural biopolymers

Starch is a biopolymer composed of polysaccharides that is found only in plants and is stored in their tissues. Starch can be obtained naturally from a variety of sources, including corn, tapioca, wheat, and potatoes [10,11]. However, the main component of starch is glucose, which can be

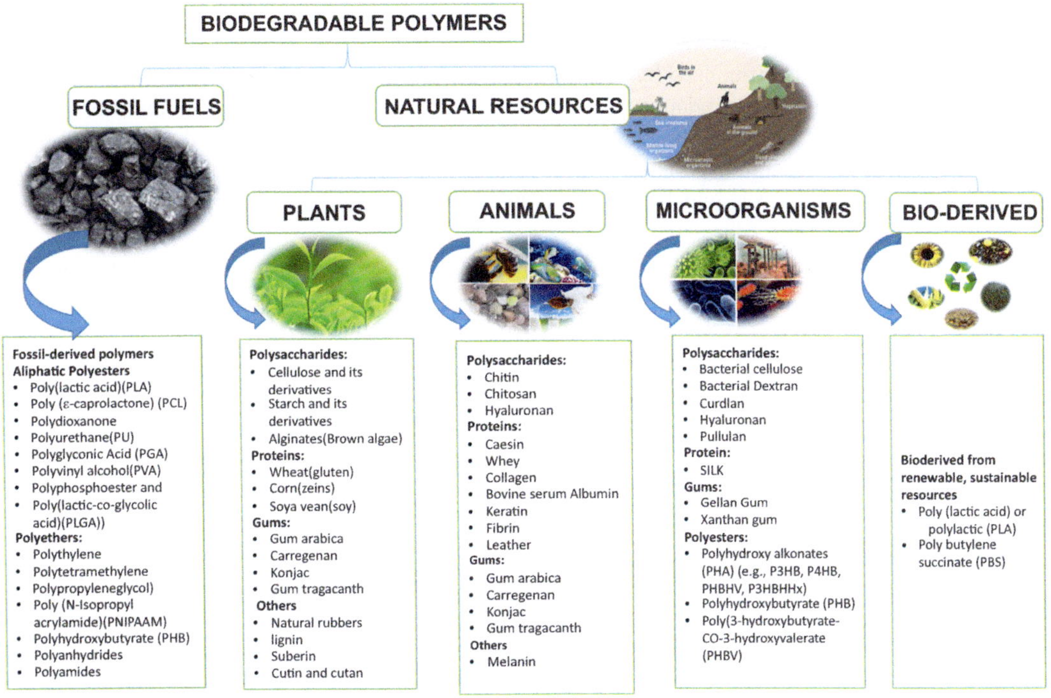

FIGURE 19.1

Classification of biodegradable biopolymers.

converted to glucose via thermal treatment or other chemical stimulants. Many starch derivatives are already being used in tissue engineering, food industry applications, and nanomaterial fabrication. Cellulose, on the other hand, is the most common biopolymer found in plant cell walls, and glucose is its primary constituent. Cellulose has a long molecular chain that is insoluble and infusible in most solvents but highly soluble in strong polar solvents. As a result, it is frequently converted into derivatives to improve solubility, which increases matrix adhesion even further [12]. Natural cellulose and its derivatives are commonly used as fillers or plasticizers in conjunction with other biopolymers to improve mechanical properties. Furthermore, cellulose is easily attacked by a wide range of microorganisms, which use a variety of enzymes to degrade it. However, in some cases, cellulose biodegradation is complicated because cellulose coexists with lignin (for example, in wood cell walls), which delays its degradation. Pure cellulose, on the other hand, decomposes quickly [12]. Recently, a wide range of biodegradable polymers have been developed, as have microbes and enzymes capable of decomposing them [13].

Chitin is a polysaccharide-based biopolymer found primarily in shrimp, lobsters, crab shells, and insects. Chitosan, a partially deacetylated form of chitin, is water-soluble in its native form. This biopolymer is biocompatible and has good antimicrobial activity as well as the ability to absorb heavy metal ions. to enhance therapeutic properties chitosan is functionalized into derivatives such as *N*-carboxymethyl chitosan and *N*-carboxybutyl chitosan, which have superior biological properties to chitosan and are more commonly used in wound healing [14]. The rate of degradation of chitosan is influenced by its residual acetyl content, and the rate of polymer deacetylation influences the rate of degradation, which can disintegrate rapidly in vivo. Furthermore, chitosan is highly resistant to oxidation and has excellent mechanical properties that favor catalyst support as well as the separation of catalyst components after their use by burning the organic support [15]. Although chitosan is widely used in many industries, there is currently very little data on its use in catalysis.

Poly(hydroxybutyrate) (PHB) is another important biodegradable polymer produced by microorganisms through carbon breakdown from various forms of glucose. When the microorganism's nutrition supply is limited, it is used as an energy storage molecule. PHB biopolymers have gotten a lot of attention in the scientific community because of their unique properties at low concentrations and weight percentages. Previously, PHB was never used directly as a catalyst via photoactivation, but advances in biomaterials research have led to the use of PHB polymer film to support catalytic systems. As a result, it can boost the production of oxidative free radicals while also being biodegradable, allowing for effluent treatment and eliminating the need for catalyst recovery and reutilization [8].

There is a group of biopolymers (aliphatic polyesters) in the biodegradable polymers classification that are obtained from bio-derived monomers using biotechnology. Poly(lactic acid) (PLA) is a good example of this category; the monomer lactic acid is derived from lactose, a milk sugar that will be used to make PLA. Poly(lactic acid) is of particular interest to researchers due to its high biodegradability, compatibility, and strength. The fact that bio-derived polymers are mostly similar to biological tissues is a positive feature. They have excellent biological affinity, which may help cells adhere and perform better, but the only limitation is that they have poor mechanical functions. PLA is being enhanced with nanoparticles to improve its thermal and mechanical properties, which will lead to improved properties and expand its application. However, many of these bio-derived biopolymers are in short supply and require numerous production steps, making them expensive. It is important to note that in the previous classification, we discussed natural and bio-derived polymers. These are simply a group of biopolymers derived from renewable or sustainable resources; a definition that excludes fossil fuels such as coal and crude oil. Poly(caprolactones) is a classic example in this group (e.g., PCL).

19.3 Biopolymers based on fossil fuels

Synthetic biopolymers are a type of biopolymer produced through a chemical reaction involving monomers derived from a petroleum feedstock. These biopolymers can be copolyesters, aliphatic or aromatic polyesters, or a combination of the two. Some of the most commonly used synthetic biopolymers are poly(caprolactone) (PCL), poly(glycolide) (PGA), PLA, and poly(3-hydroxybutyrate-*co*-3-hydroxy valerate) (PHBV). Two well-known aromatic copolyesters derived chemically from fossil fuels are poly (butylene adipate-*co*-terephthalate) (PBAT) and poly(butylene succinate-*co*-terephthalate) (PBST) [16]. Although these synthetic biopolymers have excellent mechanical and physicochemical properties similar to biological tissues, their surfaces are hydrophobic and lack cell recognition signals. These synthetic biopolymers are hydrolyzed with or without enzymes and absorbed by the body through the metabolic pathway. However, the longevity of these biodegradable polymers in the body varies depending on the type of biopolymer [1]. Furthermore, when compared to aromatic ester, high molecular weight compounds like aliphatic ester were shown to be biodegradable due to the presence of hydrolyzable functional groups on the backbone of these polyesters [17]. Polyesters derived from diacids of medium-sized monomers (C6-C12) were found to be more degradable by fungi (*Aspergillus niger* and *Aspergillus flavus*) than polyesters derived from short or long monomers (C20-C4). Synthetic biopolymers have recently received a lot of attention as a potential replacement for natural biopolymers like peptides, nucleic acids, and polysaccharides, which are expensive to make and purify. Because synthetic biopolymers have excellent mechanical and barrier properties, they have potential applications in the biomedical and food packaging sectors.

Taking into account all aspects of biodegradable biopolymers, there are numerous advantages and disadvantages that distinguish these two types of biopolymers. The main impediment to the use of biodegradable polymers is their poor technical performance in comparison to many conventional polymers, particularly their mechanical and barrier properties [18]. As a result, nanocomposite technology has been applied to biodegradable polymers to improve mechanical strength, chemical resistance, thermal stability, barrier properties, water vapor, and gas (carbon dioxide and oxygen).

19.4 Polymer nanocomposites

Polymer nanocomposites are materials made by incorporating nanosized particles into a polymer matrix or a standard matrix material. These nanosized particles are generally nanoclusters, fullerenes, clays, metal o metal oxide, or semiconductors on a nanometer scale [19] in other words they are also called organic /inorganic fillers. The addition of these nanosized particles into the polymer matrix will result in an interaction between the nanostructures/nanoparticles and the polymer matrix, resulting in the formation of new sustainable polymeric materials with significantly improved mechanical strength, toughness, antimicrobial, electrical or thermal conductivity properties [9,20]. Fig. 19.2 shows the formation of polymer nanocomposite material. The effectiveness of the nanoparticles is so great that the amount of material added is normally only between 0.5% and 5% by weight, but it produces a better nanocomposite than conventional micro composites. Furthermore, nanoparticles have an extremely high surface-to-volume ratio, which significantly alters their properties when compared to bulk-sized particles. It's worth noting that the way nanoparticles interact with polymer materials can

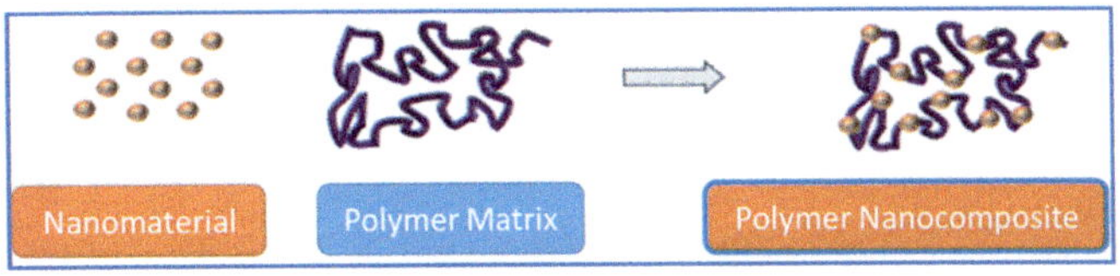

FIGURE 19.2

Formation of polymer nanocomposite.

enhance the properties of nanocomposite materials intrinsically, such as optical, catalytic, and electrical properties, as well as extrinsically, flexibility, and transparency [21,22]. Polymer nanocomposites have been recommended for a variety of potential applications in medicine, photonics, catalysis, and electronics due to these advantages.

It is important to mention that nanocomposite will have a multiphase structure in which at least one of the phases of the polymer nanocomposite should have dimensions in the range of 10–100 nm. In some cases, the resulting nanocomposite will have different properties due to differences in structure and chemistry, whereas in others, the nanoreinforcement in the polymer matrix can be 1000 times tougher than the bulk component materials. As a result, any advancements in nanocomposites are dependent on the type of nanoparticle used for the application. Numerous studies have been conducted to create nanoreinforced biodegradable polymer nanocomposite materials with mechanical and physical properties comparable to pristine polymer matrices. Biodegradable polymer-based nanomaterials are the most advanced nanocomposites in current research and development. Biopolymer-based nanocomposites will benefit greatly from properties such as film-forming ability, activated functionalities, and dimensional diversity [23]. Immobilizing inorganic nanostructured semiconductor materials in a polymeric matrix can also result in photocatalytically active polymeric nanocomposites. When light and polluted aqueous solutions interact with these materials, a redox medium is formed that acts as a photocatalyst to eliminate organic pollutants [24]. The main goal of this chapter is to highlight new research that has been published in the literature that uses biodegradable polymer nanocomposites as a catalyst or for photocatalytic applications. The purpose of this chapter is to provide a thorough understanding of the methods used in the preparation of biodegradable polymer nanocomposites, their properties, and the evaluation of their catalytic and photocatalytic applications in recent studies.

19.5 Biodegradable polymer nanocomposites as photocatalyst and catalyst

The semiconductor metal oxides, such as titanium dioxide (TiO_2) and zinc oxide (ZnO), have received the most attention as photocatalysis catalysts [25,26]. However, these inorganic nanomaterials have a proclivity to accumulate in aqueous media, making catalyst recovery and reuse difficult. Potential solutions to these problems include the development of polymer nanocomposites materials that can act as supports or directly as photocatalysts (conjugated polymers) [8,27].

19.5.1 Cellulose-based catalysts for organic transformations

Because of their structural properties, biomass has been considered a good support for catalysts in recent years. They are porous, containing numerous functional groups that are critical for the functionalization with selective functional groups and immobilization of metal nanoparticles or other smart materials. Among the biomass, cellulose is a plentiful, inexpensive, renewable, organic raw resource on the planet. It is a carbohydrate polymer with three hydroxyl groups per anhydro glucose unit composed up of repeating β-D-glucopyranose units. Simple reactions can easily functionalize them. Because of environmental awareness and necessity, cellulose-based smart materials are appealing, and their applications have been evaluated in all possible domains. Cellulose is an appealing material for catalyst support.

Pei et al., have reported immobilization of Ag-Fe_3O_4 on cellulose microspheres and used this biodegradable composite as a catalyst to prepare 4-aminophenol (4-AP) from 4-nitrophenol (4-NP) [28]. In this work.an in-situ redox reaction was used to immobilize AgFe_3O_4 NPs onto cellulose microspheres, resulting in a magnetic nanocatalyst. The particle size of the $Ag-Fe_3O_4$ NPs was found to be around 24–38 nm which shows that cellulose is functioning as a very good capping agent. The pore size and volume of these materials were determined using N_2 adsorption/desorption, which indicated pore sizes ranging from 5 to 40 nm. These catalysts have thermal stability of up to 200°C, so they can be used as a catalyst for many industrial processes. Magnetic cellulose spheres have been demonstrated to be effective catalysts for the conversion of 4-NP to 4-AP, as well as for cyclic and reusable applications. The catalyst can be separated and reused for up to nine cycles with relatively high catalytic activity by using an external magnetic field. Samaneh et al. have prepared a similar composite but with a small modification where a magnetically separable composites including nano-Fe_3O_4@Cellulose-NH_2-CuI can be obtained and can be used as a heterogeneous catalyst to manufacture 1,4-disubstituted 1,2,3-triazoles with increased yields and regioselectivity [29]. Amine group functionalized cellulose was prepared by reaction of cellulose with (3-aminopropyl) triethoxysilane (APTES) and then Fe_3O_4 and CuI were anchored by simple coprecipitation method. The composition of the catalyst was optimized by changing the amount of cellulose, Fe_3O_4, and CuI, and it was discovered that a magnetic composite with a weight ratio of 2:1:1 of celluloseNH_2: nanoFe_3O_4:CuI has greater catalytic activity, yielding 95%. Magnetic catalysts provide several advantages, including easy separation and reusability in subsequent cycles. In this aspect, nano Fe_3O_4@CelluloseNH_2CuI magnetic catalyst was found to be recyclable for four cycles with no significant differences in reaction time or yield.

In another research work, nanocellulose-$TiCl_3$ nanocomposite was prepared by the simple physical mixing method, FTIR analysis revealed that –OH group of cellulose reacted with $TiCl_3$ [30]. This composite was used as a biodegradable and economical nanobio catalyst for the production of 2,3′-Dihydroquinazolin-4-(1H)-ones. Changing the solvent, reaction temperature, and catalyst amount optimized the reaction conditions. It was found that ethanol solvent at the reflexing condition with 0.72 mol% of catalyst offered the best result with a 90% yield.

Many metals and metal oxides NPs have been anchored in cellulose backbone, in this series of work, novel paper matrices coated with CuO nanoflakes (CP) were prepared by microwave irradiation method and The synthesized material was used as a catalyst in the production of cyclic carbonates and epoxides from CO_2 [31]. FESEM analyses showed that uniform distribution of CuO nanoflakes with 50–100 nm width and around 200 nm length; XPS analysis confirmed the formation of exclusive CuO by the absence of signature peak of Cu_2O at 933 eV. In this research, it was discovered that when blank paper was employed as a catalyst, only 8% of the product was obtained, but pristine CuO nanoparticles only gave 40% of the product. However, under identical experimental settings, the CP produced 100% of the product. Immobilizing CuO nanoflakes on the surface of cellulose fibers in paper matrices has a synergistic effect on product development, according to the findings. This method has a number of appealing advantages, including biodegradable composites, a green manufacturing process, easy separation of the catalyst from the reaction mixture, and the ability to reuse the catalyst for multiple cycles without losing catalytic activity. Maleki et al. created a pumice-modified cellulose fiber nanocomposite (Cellulose-pumice).

Pumice powder has a natural, porous structure, high stability, and remarkable magnetic properties [32]. SEM-EDX analysis revealed the presence of several metals such as Na (1.24%), Mg (0.67%), Al (1.06%), Si (2.33%), K (0.16%), Ca (1.19%) and Fe (1.36%) in Cellulose-pumice. This material was used as a catalyst in the production of 2,4,5-triarylimidazole derivatives and offered an excellent yield (98%) (Scheme 19.1).

19.5.2 Other biopolymer-based catalysts for organic transformations

Aside from cellulose, other biopolymers such as starch, chitosan, alginate, wool, and silk have been used as catalyst supports. Safari et al. have prepared sulfonated starch NPs (HO_3S-SNPs) from starch by treating with chlorosulfonic acid and used as acidic nanocatalyst for the production of 14-aryl-14-H-dibenzo[a,j]xanthenes under solvent-free conditions, This method has several benefits, including simple reaction conditions, low manufacturing costs, a biodegradable catalyst, and no lengthy workup or purification [33]. This process offered the product is very good yield (92%), the versatility of the reaction was checked with a variety of substituted aldehydes with improved product (90%–97%) (Scheme 19.2).

The possible explanation for the catalytic activity of HO_3S-SNPs to produce 14-aryl-14-H-dibenzo [a,j]xanthenes is given in Scheme 19.3, in which the catalyst is acting as a Lewis acid to activate the aldehyde molecule and then it undergoes Knoevenagel addition and then Michael addition to offer the final product.

Similarly, Shaabani et al. created Manganese (II) tetrasulfophthalocyanine (MnTSPc), which was covalently anchored on natural silk material and tested as a catalyst for the synthesis of benzoxazepine derivatives. Natural white silk is a readily available renewable, biodegradable, and biocompatible natural resource. It was functionalized in this study with a Manganese (II) complex containing phthalocyanine moieties as a metal-ligand. TGA analysis shows that the catalyst is stable up to 250°C [30].

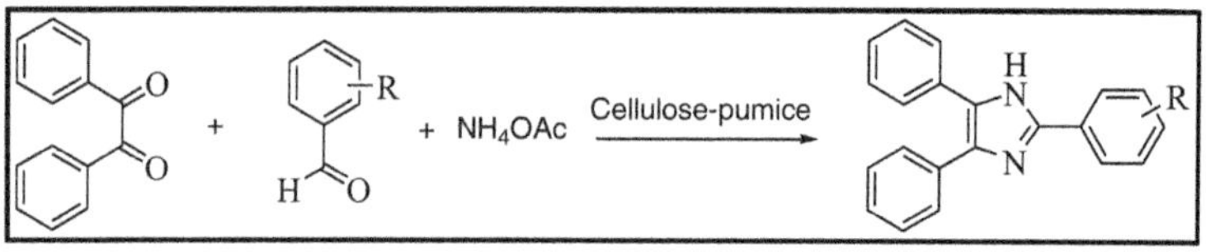

SCHEME 19.1

Synthesis of 2,4,5-triarylimidazole [32].

Reproduced with permission from Elsevier. Copyright 2020.

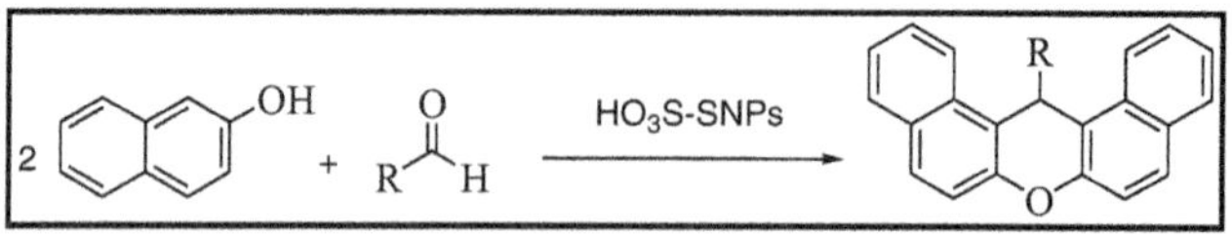

SCHEME 19.2

Synthesis of 14-aryl-14-H-dibenzo[a,j]xanthenes [33].

Reproduced with the permission from Elsevier. Copyright 2017.

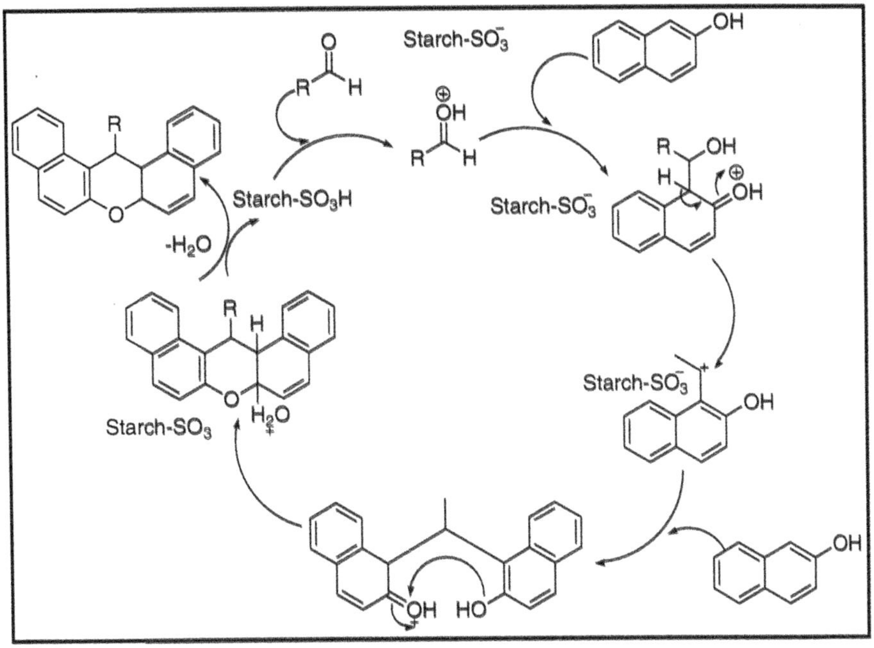

SCHEME 19.3

Tentative reaction mechanism of the catalytic activity of HO_3S-SNPs to synthesis 14-aryl-14-H-dibenzo[a,j] xanthenes [33].

Reproduced with permission from Elsevier. Copyright 2017.

The catalyst was characterized using spectral and thermal techniques. The serine amino acid group of the fibroin protein, which contains many hydroxyl functional groups, was discovered to covalently interact with MnTSPc. Metal phthalocyanine molecules are prone to aggregation, which significantly reduces their catalytic properties. Aggregation was prevented, however, when they were immobilized on silk due to the special structure of silk. In 2016, Shaabani et al., have prepared novel wool-supported catalysts, in this category, they have prepared wool-SO_3H and wool-Fe_3O_4 catalysts and they have been used as catalysts to prepare various one-pot multicomponent reactions to produce polyfunctional heterocyclic compounds [31]. Wool is a chiral, natural, insoluble natural polymer made up of repeating units of amino acids connected by S–S bonds, which can be oxidized to form the $-SO_3H$ group and act as a solid acid catalyst. Wool, on the other hand, is an excellent chelating agent for the preparation of metal nanocomposites due to the presence of several peptide bonds. Both catalysts offered very good yields in the synthesis of 3,4-dihydroquinoxalin-2-amine,1,6-dihydropyrazine-2,3-icarbonitrile,(cyanophenylamino)aceta-mide, tetrahydro-1H-1,5-benzodiazepine-2-carboxamide, and 4,5,6,7-tetrahydro-1H-1,4-diazepine-5-carboxamide derivatives under mild reaction conditions (Scheme 19.4). When comparing the performance wool-SO_3H and wool-Fe_3O_4 catalysts. Wool-Fe_3O_4 can be separated easily using a simple external magnet.

Alginate is a water-soluble natural polysaccharide derived from the cell walls of various brown algae species, including Laminaria, Macrocystis, and Ascophyllum. Alginate is a linear block polymer

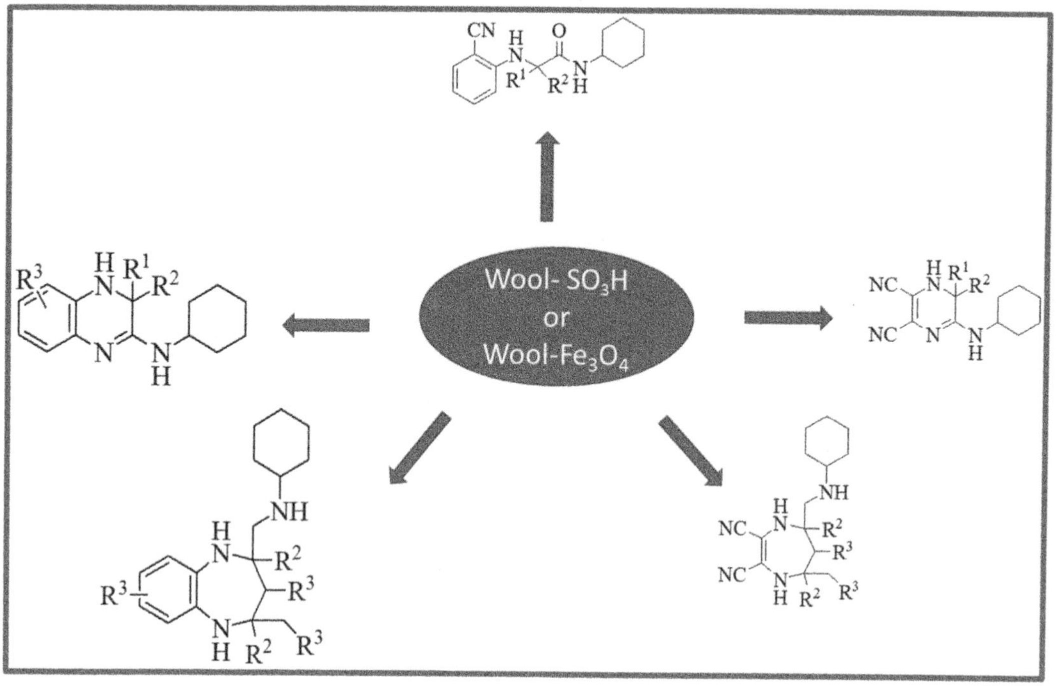

SCHEME 19.4

Synthesis of polyfunctional heterocyclic compounds.

comprising of 1β-D-mannuronic acid (M) and α-L-guluronic acid (G) units; the content of M and G units defines its physical properties and applications. Water-soluble alginate can be converted into water-insoluble hydrogels by cross-linking with Ca^{2+} ions. Calcium alginate-TiO_2-SiO_2 was prepared from calcium alginate by attaching TiO_2-SiO_2 NPs onto calcium alginate, and used as a catalyst for the conversion of glucose molecule to 5-hydroxymethylfurfural [32]. Primarily, TiO_2-SiO_2 nanocomposite was obtained by sol-gel method and later it was added to the solution of sodium alginate and cross-linked by $CaCl_2$. The better conversion (97%) was achieved for this reaction at 140°C for 4 hours reaction time.

19.5.3 Graphene nanocomposite-based catalysts for organic transformations

Carbon-based smart materials such as graphene, graphene oxide (GO), multiwalled carbon nanotubes (MWCNTs), and functionalized GOs are important materials in modern technological applications due to their electrical conductivity, thermal stability, chemical stability, and biodegradability. They are well-known materials in electronics, sensors, and biomedical applications, but few reports on their catalytic applications are available (Table 19.1).

Kaya et al. have developed functionalized MWCNT (f-MWCNT) based heterogeneous catalysts for the synthesis of acridinedione derivatives (Scheme 19.5) [33] these (f-MWCNT) are efficient,

Table 19.1 Biodegradable polymer composite as catalysts for the organic transformations.

Sl. no.	Catalyst system	Product	References
1	Cellulose-Ag-Fe_3O_4	4-Aminophenol	[27]
2	Cellulose-Fe_3O_4-CuI	1,4-Disubstituted 1,2,3-triazoles	[28]
3	Nanocellulose-$TiCl_3$	2,3′-Dihydroquinazolin-4-(1H)-ones	[29]
4	Cellulose-CuO	Cyclic carbonates	[30]
5	Cellulose-pumice	2,4,5-Triarylimidazole	[31]
6	HO_3S-SNPs	14-Aryl-14-H-dibenzo[a,j]xanthenes	[32]
7	Silk-Mn	Benzoxazepine	[33]
8	Wool-SO_3H	Polyfunctional heterocyclic derivatives	[34]
9	Wool-Fe_3O_4	Polyfunctional heterocyclic derivatives	[35]
10	Ca-Al-TiO_2-SiO_2	5-Hydroxymethylfurfural	[36]
11	f-MWCNT	Acridinedione	[37]
12	PFGN	Aza-Michael addition	[38]
13	CS-Pd	4-Aminophenol	[39]

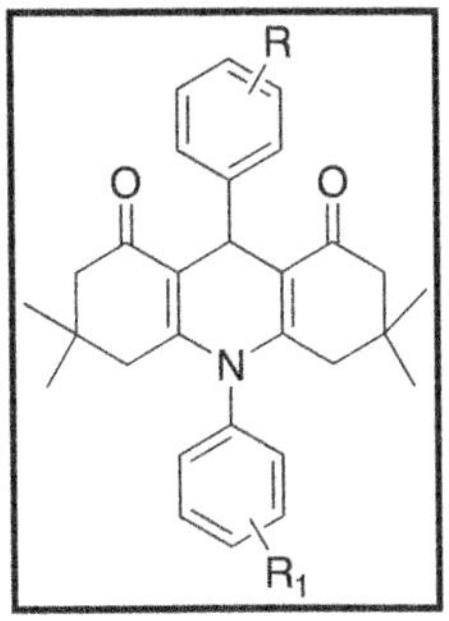

SCHEME 19.5

Structure of acridinedione.

reusable, and biodegradable. For the functionalization, carboxylic acid groups were immobilized onto MWCNT by a simple chemical reaction. Raman analysis confirmed the functionalization of MWCNT by checking the I_D/I_G value; TEM analysis revealed the tubular structure of the catalyst with a diameter of 20–30 nm with a length of several hundred nm. In addition, f-MWCNTs catalysts were separated from the reaction mixture easily and reused for seven cycles without compromising its catalytic performance.

Karak et al. published eco-friendly functionalized GO nanocomposites with a waterborne polyester component in 2018 [34]. Because of their inherent mechanical strength, high aspect ratio, very good thermal and electrical conductivity, and large surface area, graphene-based materials are well-known in the nanotechnology field. However, In comparison between graphene and GO; GO contains a numerous oxygen-based functional groups such as carboxylic acid, epoxy and hydroxyl groups; which will facilitate the interaction with other materials. As a result, GO can be used to create composite materials. Because of the abundance of oxygen-based functional groups, GO has a very robust intermolecular interaction, which limits its interaction with other polymeric materials. The authors propose that functionalization with appropriate reactants will improve composite material formation. In this study, GO was simply modified with toluene diisocyanate and butanediol before being incorporated into waterborne PEG and citric acid-based polyester. The nanocomposite obtained has the ability to disperse common solvents such as water, ethanol, DMAc, DMSO, and THF. Finally, the PFGN nanocomposite thermoset was created by treating it with glycerol-based hyperbranched epoxy and polymer (amidoamine) When we loaded a smaller amount of nanomaterial, the resulting nanocomposite improved significantly in thermal and mechanical properties. Furthermore, the nanocomposite was observed to be biodegradable in the presence of the *Pseudomonas aeruginosa* bacterial strain and exhibited a strong catalytic property for the Aza-Michael addition reaction (almost 100% yield within 3 minutes at the ambient condition). Hence, PFGN can be used as a versatile material in a variety of modern applications.

Dhanavel et al. created a CS-Pd nanocomposite with varying Pd metal concentrations (5%, 10%, and 15%) and used it as a catalyst to convert 4-nitrophenol to 4-aminophenol [35]. As the amount of Pd increased, so did the catalytic performance. The authors tested the antimicrobial properties of these composites and discovered that they were more effective against all types of bacteria. The hemolytic study was carried out to determine the toxicity of these composites, and it was discovered to be a safe material.

19.5.4 Degradable polymer composites as photocatalysts

Photocatalytic removal/degradation of toxic organic compounds such as pollutants from the dye, pharmaceutical, chemical, leather, paper, plastic, food, and fertilizer industries, as well as other inorganic and heavy metal pollutants, is the most researched research area; however, more research is needed to provide better technology to address this issue. Adsorption and photocatalysis are two common methods for dealing with this problem. Adsorption only removes toxic compounds and does not degrade them into nontoxic compounds. One of the most promising technologies for environmental cleanup and solar light usage is semiconductor-based photocatalysis. Although TiO_2, CeO_2, ZnS, and ZnO are photocatalysts for toxic compound breakdown, they only respond to UV light, which accounts for a very small fraction of total sunlight (around 4%). Furthermore, photocatalytic activity is dependent on particle size; NPs have higher efficiency than macroparticles, so there is a greater chance of contamination than purification. To address this issue, we require environmentally friendly materials to act as a support for these NPs.

Sharma et al. prepared cellulose from fruit waste and used it to make cellulose-bismuth vanadate (Cellulose-$BiVO_4$) nanocomposites that were thoroughly characterized using thermal, physical, and morphological analysis [36]. TEM analysis revealed that $BiVO_4$ had a size of 24–35 nm in the composite. XPS analysis also confirmed the formation of $BiVO_4$ and its interaction with the –OH group of cellulose. Similar experiments were performed with pristine $BiVO_4$, cellulose, and commercial cellulose to compare the photocatalytic activity of Cel-$BiVO_4$ nanocomposite for the decomposition of methyl orange (MO) dye under visible light. Cellulose and commercial cellulose have very low photocatalytic activity levels of 4% and 8%, respectively, and pristine $BiVO_4$ has around 67%, whereas Cel-$BiVO_4$ has a better performance of around 80%. It is possible that this is because cellulose acts as a substrate for additional dye adsorption on its surface, delaying electron and hole recombination and thus increasing photocatalytic activity. Similarly, Konstantin et al. have prepared novel poly(lactic acid)-iron oxide (PLA-Fe_2O_3) nanocomposite and has been as excellent UV filters in cosmetic and sunscreen products with better sun protecting factor (SPF) [37]. In general, commercial sunscreen formulations contain ZnO or TiO_2 nanoparticles as the main component, which are excellent photocatalysts with the ability to pass through the skin and are thus considered hazardous to health. PLA-Fe_2O_3 nanocomposites with different weight percentages, such as 5%, 10%, and 20%, were prepared in this study, and their SPF and protection factor-UVA (PF-UVA) were measured. Cosmetic products containing PLA with 10–20 wt.% Fe_2O_3 exhibited better SPF values in comparison to commercial ZnO based products, showing SPF up to 47 $\pm$ 4; without compromising other physical properties of the PLA-Fe_2O_3 and retaining its original color of the skin. Geetha et al. developed novel chitosan-PEG stabilized Ag-doped TiO_2 by simple one-pot reaction for photochemical degradation; in this method, chitosan and PEG biopolymers also act as reducing agents [38]. The composition of the composites is varied by changing the amount of TiO_2 (0.03, 0.06, and 0.09 mM) and the obtained nanocomposites were characterized by spectral, thermal methods and has been employed as a photocatalyst for the decomposition of MB dye using a UV-Vis spectrophotometer technique. The photocatalytic activity of the nanocomposite was performed by sunlight irradiation (Fig. 19.3). In this research, it was explained that due to the large bandgap energy (3.2 eV), nano-TiO_2 photocatalyst was used at shorter wavelengths (285–320 nm) hence TiO_2 nanostructure do not have preferable efficiency under visible light. Whereas, pristine Ag NPs are excellent photocatalysts under visible light. It was also discovered that the photocatalytic activity of the Ag/TiO_2 nanocomposite is superior to that of pure Ag nanoparticles. It is possible that the presence of Ag nanoparticles increases light absorbance in the UV and visible regions, while the presence of TiO_2 increases photocatalytic activity. Because this degradation process produces no toxic byproducts, it can be considered a clean wastewater treatment process.

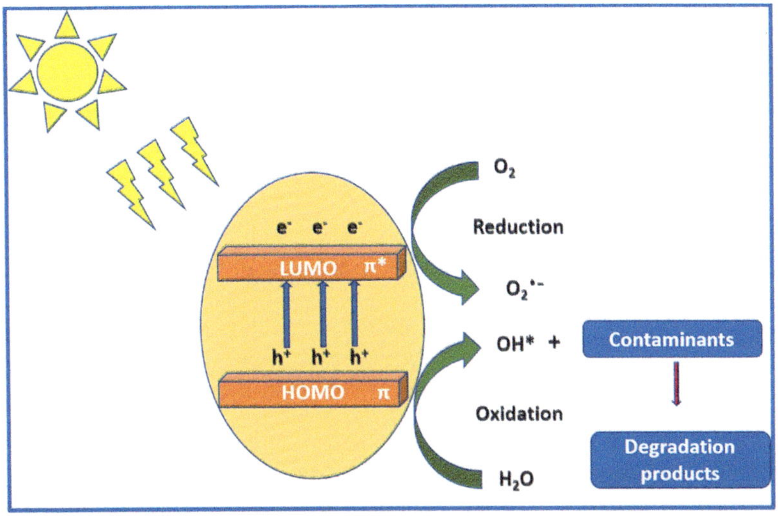

FIGURE 19.3

Photochemical degradation of contaminants.

Tripathy et al. have synthesized Ag-Au bimetallic nanocomposites using hydroxyethyl starch-g-poly (acrylamide-*co*-acrylic acid) (HES-g-poly(AM-*co*-AA)) as stabilizing as well as a reducing agent [39]. $K_2S_2O_8$ was used as a radical initiator in a one-pot free radical polymerization technique to create the graft polymer. The unreacted monomer, polyacrylamide, and polyacrylic acid homopolymers were then extracted using solvent. The HES-g-poly(AM-*co*-AA) was then analyzed using FTIR, XRD, and SEM. The degradation of azo dyes such as Congo red, MO, and Sudan-1, as well as the reduction of 4-nitrophenol to 4-aminophenol, were used to test the catalytic activity of this composite (Fig. 19.4). The reduced products were obtained within 10 minutes at ambient conditions, indicating that this composite has very good catalytic activity.

In another study, carboxymethyl cellulose (CMC) films functionalized with gelatin and Ag-TiO_2 were created, and the composition of the nanocomposite was altered by varying the amounts of gelatin (0.5% and 1%) and TiO_2-Ag nanoparticles (0.1%, 0.2%, 0.5% (W/W)) [40]. CMC/Gel/TiO_2-Ag films were completely characterized by spectral and mechanical methods. The SEM images of CMC/Gel/TiO_2-Ag films clearly show that nanoparticles size of 50–100 nm were homogeneously spread in polymer nanocomposites. The thickness and water content of the CMC/Gel/TiO_2-Ag films increased as the amount of gelatin increased; however, as the amount of TiO_2-Ag nanoparticles increased, the moisture content decreased significantly. Water vapor permeability (WVP) is a unique property of films used in food packaging applications. The WVP analysis revealed that gelatin has a positive impact on WVP, whereas increasing the TiO_2-Ag percentage results in poor water vapor permeability. Furthermore, gelatin increases the flexibility of the film, and the uniform distribution of TiO_2-Ag nanoparticles increases the tensile strength; however, as the concentration increases from 0.2% to 0.4%, the tensile strength decreases due to TiO_2-Ag nanoparticle agglomeration. The degradation of benzene, ammonia, and ethanol was used to test the photocatalytic activity of CMC/Gel/TiO_2-Ag films. When TiO_2-Ag nanoparticles are added to films, their photocatalytic activity against NH_3, ethanol, and benzene is increased. In a very similar work, Zahedi and Achachlouei have prepared CMC reinforced with sodium montmorillonite (Na-MMT) and TiO_2 nanoparticles and the composition of the composites was varied by changing the amount of Na-MMT (1, 3 and 5 wt.%) and TiO_2 (1, 3 and 5 wt.%) [41]. The results showed that the addition of Na-MMT, and Na-MMT + TiO_2 NPs has reduced the WVP of the films up to 39% and 50% respectively while the moisture content, density, and T_g have increased considerably.

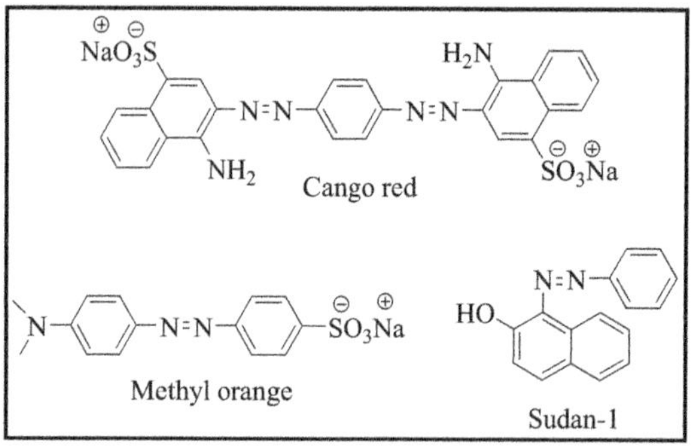

FIGURE 19.4

Structure of azo dyes.

This composite absorbed almost 99% of the UV radiation, which favors as a potential material in food packing and photocatalytic applications.

Helmiyati have demonstrated the synthesis of sodium alginate-TiO_2-bentonite nanocomposite (STB nanocomposite) where bentonite was used as a template for the TiO_2 NPs and sodium alginate (NaAlg) was used as catalyst support [42]. Finally, synthesized STB nanocomposite was used as photocatalyst for the elimination of MB, and the efficiency was compared with NaAlg-TiO_2 and bentonite-TiO_2 composite. It was found that STB nanocomposite had better photocatalytic activity (90.5%) than the NaAlg-TiO_2 (69%), and bentonite-TiO_2 (79%). It could be because the STB nanocomposite, which is made up of sodium alginate and bentonite, has a high adsorption capacity and a combination of TiO_2 has a high degrading capacity.

In another interesting work, Kalaivani and Suja extracted inulin, water-soluble carbohydrates from garlic and prepared an inulin-TiO_2 composite [43]. The photocatalytic activity of Inulin-TiO_2 was investigated using MB decomposition under UV irradiation. Changing the concentration of catalyst, dye, and irradiation time optimized the process. Surprisingly, the inulin-TiO_2 composite has nearly 100% higher photocatalytic activity than pure TiO_2. It could be because inulin helps to absorb more dye molecules in the composite and by preventing TiO_2 particle agglomeration, it increases the surface area of the photocatalysts. Furthermore, the inulin composite stabilizes the photo-generated electrons and prevents recombination with the hole, increasing photocatalytic activity.

In a very similar work, Rehim et al., have prepared sodium alginate-TiO_2 (SAT) composite and then mixed this SAT with paper to prepare a modified paper sheet [44]. By varying the amount of SAT (3%, 7%, 10%, 15%, and 20%), the composition of the paper sheet was altered. The photocatalytic activity of this composite was determined by calculating the chemical oxygen demand (COD) of a dichromate solution. It was discovered that paper sheets with 15% SAT remove COD better (42%), whereas paper sheets with 20% SAT remove only 18%. The authors explained that an excess of SAT may affect the interaction of TiO_2 and cellulose, resulting in poor photocatalytic activity. A similar pattern was discovered in antimicrobial activity; the paper sheet containing 155 of SAT is more effective against all bacteria.

There are very few articles on zirconium-based materials available in this research area. Pd impregnated ZrO_2-chitosan nanocomposite (Pd/ZrO_2-CS), ZrO_2-chitosan nanocomposite (ZrO_2-CS), Pd/ZrO_2, and pristine ZrO_2 were prepared by Abdelaal and Mohamed, and their photocatalytic activity was tested by the degradation of thiophene under visible light irradiation [45]. TEM analysis revealed that CS controls the agglomeration of the NPS and provides uniform size distribution. Since ZrO_2 absorbs only in the UV region, does not have any photocatalytic activity in the visible region. Among all the photocatalysts Pd/ZrO_2-CS has shown better degradation of thiophene (90%) and results in the formation of CO_2, SO_3, and water as photochemical products.

Savunthari et al. have prepared a series of lignin-graphitic carbon nitride (GCN) composites such as lignin nanorods-GCN nanorods (LNRs-GCN-NRs), lignin nanorods- GCN nanosheets (LNRS-GCN-NSs) and used photocatalysts for the degradation of Triclosan (5-chloro-2-(2,4-dichloro-phenoxy)-phenol) under UV irradiation (365 nm) [46]. Triclosan is a well-known antimicrobial reagent that is widely used in cosmetic and pharmaceutical products. It was recently identified as a toxic compound and banned by the FDA. Triclosan, which is commonly found in cosmetics and beauty products, pollutes water sources such as lakes and rivers. This compound was decomposed in an aqueous solution using lignin-GCN composites in this study. The bandgap these catalysts was calculated by UV analysis (LNRs = 1.4 eV, LNRs-GCN-NSs = 2.76 eV and LNRs-GCN-NRs = 2.6 eV).

LNRs-GCN-NRs were found to have higher photocatalytic activity (99.5%) than LNRs-GCN-NSs (97.2%). It demonstrates that the nanorod structure is important in photocatalytic properties, and reusability was observed for five cycles with little change in photocatalytic efficiency.

Vigneshwaran et al. have reported the preparation of ternary composite of TiO_2, CS, and hydroxyapatite (TiO_2@CS-Hpt) and evaluated their photochemical degradation property of dye molecules [47]. In which CS acted as support for the other inorganic materials. The diffuse reflectance spectroscopy (Drs) analysis revealed that the bandgap energy value for the ternary composite was observed as 2.01 eV whereas pristine TiO_2 displayed 3.23 eV. It shows that TiO_2@CS-Hpt can be a better photocatalyst than the pristine TiO_2. The TEM analysis revealed that the composite has a 2D porous structure with uniform distribution of TiO_2 NPs, which is an important structural aspect. The degradation of dyes such as MB and RhB was studied in the dark as well as under visible light irradiation, and TiO_2@CS-Hpt demonstrated better degradation efficiency, 98.6% and 96.7% for MB and RhB, respectively. Furthermore, degraded byproducts were identified using mass spectrometry. The same research team used a simple coprecipitation method to create Fe_2O_3@CS composite [48]. The Drs analysis revealed that the bandgap energy for the composite was found to be lower (2.52 eV) than the pristine Fe_2O_3. The SEM analysis showed that the diameter of the CS nanofiber was around 5 μm and the length was found in the range of 10–30 μm and the Fe_2O_3 NPs were intercalated on CS nanofibers. The EDX analysis revealed that Fe_2O_3@CS has around 23 at. % of Fe. Fe_2O_3@CS was employed to eliminate dyes such as MO and orange green (OG). It was found that 89% of MO was destroyed within 60 minutes and 94% of OG was degraded within 40 minutes under normal sunlight irradiation.

Zhu et al., have prepared TiO_2/ZnO-CS nanocomposite thin films by simple solution process in which glutaraldehyde was used as cross-linker [49]. The discoloration of MO was studied under simulated solar irradiation and it was found that 97% of efficiency was achieved at neutral pH and even with a lesser concentration of photocatalyst (0.5 g/L).

Geetha et al. proposed another solution to use TiO_2 NPs as photocatalysts under visible light irradiation due to their large bandgap [50]. In this work, Ag NPs are incorporated in TiO_2 and CS and PEG were used as stabilizing agents. Zeta potential studies confirmed the stability of the nanocomposites and TEM and DLS analysis revealed that Ag/TiO_2 nanocomposites have uniform size distribution with ~35 nm. The photocatalytic properties of these nanocomposites were investigated by studying the decomposition of MB under natural sunlight irradiation. Although TiO_2 has better photocatalytic activity in the UV region and Ag NPs have better activity in the visible region, it was discovered that Ag/TiO_2 nanocomposites have better photocatalytic activity than pristine Ag NPs. It is possible that this is due to the thermodynamically feasible transformation of photogenerated carriers between Ag and TiO_2 NPs. Similarly, Karthikeyan et al. prepared CS-TiO_2 nanocomposite by simple physical mixing, characterized it thoroughly, and investigated its application to decompose RhB and CR under visible light irradiation [51]. The optical analysis revealed that CS-TiO_2 has bandgap energy as 3.28 eV. The surface area of the composite was found to be 4.0 m^2/g and the pore size was calculated as 9 nm. The particle size of the TiO_2 NPs was calculated as 90 nm by HRTEM analysis. The decolorization of dyes is generally affected by the pH of the solution; however, in this study, it was discovered that CS-TiO_2 nanocomposite had better photocatalytic activity at pH 3 for RhB and pH 11 for CR. The researchers investigated the antimicrobial activity of a CS-TiO_2 nanocomposite.

In 2020, Maruthupandy et al. have synthesized graphene-CS-magnetite nanocomposite (Gr/CS/Fe_3O_4) by hydrothermal process [52]. The photocatalytic activity of Gr/CS/Fe_3O_4 against a library of dyes such as RhB, BTB, MB, and MO was performed under visible light irradiation. In all experiment, 50 mg of catalyst was used for 100 mL of dye solution. It showed 100% activity against RhB in short time (40 minutes) and 60, 80, 100 minutes for BMB, MB, and MO respectively. The photocatalytic activity of pristine Fe_3O_4 and CS/Fe_3O_4 was also evaluated for comparative studies, but they showed less activity. Surprisingly, the activity of Gr/CS/Fe_3O_4 is higher than that of the reduced graphene oxide/CS/Fe_3O_4 composite. These findings demonstrated the significance of GO in photocatalytic activity.

Leila et al. prepared Ag-MnS_2-CSPVA nanocomposites and used them to detect Hg (II) using a colorimetric method, to eliminate malachite green (MG) photochemically under normal visible light irradiation, and for antimicrobial applications [53]. EDS analysis confirmed that composite contains 26% of Ag, 27% of Mn, and 20% of S. The particle size of the Ag-MnS_2 was found to be around 19 nm by FESEM analysis. This composite has an excellent ability to detect Hg (II) ions even in a solution containing several metal ions with the detection limit of 9 nM, the Ag NPs reduce the mercury ions to Hg°, and then it forms Ag-Hg amalgam on the surface of the polymers. The photocatalytic efficiency was studied by decomposition of MG under visible light irradiation and ~ 97% of the dye molecules were decomposed within 30 minutes.

Recently, Balakrishnan et al. used the phase inversion method to prepare TiO_2-CS beads with glutaraldehyde as a secondary cross-linking agent and TiO_2-CS beads as a catalyst to remove 2,4-dichlorophenoxyacetic acid (2,4-D) [54]. The photochemical degradation of 2,4-D was performed at different pH values (2, 4, 6, 8, and 10) and the photocatalytic activity decreased (from 92% to 60%) as the pH value increased. It's possible that the high level of protonation at lower pH promotes the attraction of 2,4-D molecules to photocatalyst surfaces. The photocatalyst's reusability was also investigated, and it has excellent reusability without sacrificing efficiency.

Sangon et al. created mesoporous carbon-ZnO (CZ) nanocomposite via carbonization using rice straw as carbon feed [55]. Toxic dye molecules such as MB and CR were absorbed and degraded using CZ. Manea et al. created novel functional hydrogel composites by grafting CS onto polyacrylamide (PA), then impregnating the CS-PA composite with tin, silica, and molybdenum to create the CS-PA-TSM nanocomposite [56]. SEM analysis revealed the uniform distribution of TSM throughout CS-PA matrices and the particle sizes were found to be around 10–19 nm. This nanocomposite was used as a photocatalyst to degrade ciprofloxacin (CIP) and it shows better efficiency (96%) compared to pristine TSM (52%), and CS-PA (44%) and used to detect Pb (II) ions in aqueous solution by fluorescence spectroscopy technique. The absorbance capacity of the composite was found to be 96% and the limit of detection and the limit of quantification was found to be 3.38 and 9.34 μg/L, respectively. The existence of numerous functional groups such as hydroxyl, carboxyl, and amide increases the absorbance capacity of the composite.

Sr (II) cross-linked alginate-carboxymethyl cellulose-TiO_2-graphene oxide (Sr/Alg/CMC/GO/TiO_2) composite was reported by Thomas et al. using a dissipative convective method [57]. Because the composite contains GO, Raman analysis was performed to determine the degree of functionalization of GO. The ID/IG ratio was found to be 0.85, and the BET analysis confirmed that the inclusion of GO increases the surface area from 28.38 to 50.34 m^2/g. The photocatalytic property was studied by the decomposition of CR solution under natural sunlight irradiation, and it was found that Sr/Alg/CMC/GO/TiO_2 degraded 98% of dye within 4 hours time. The performance of this catalyst was compared with Sr/Alg/CMC/TiO_2 (without GO) and the performance was found to be only 70%. It revealed that GO takes part in the conduction of photo-generated carriers and the absorbance process by π-π interaction with dye molecules.

Spiridonov have synthesized CMC stabilized Cu_2O (Cu(I)-CMC) composite [58]. The size of the Cu NPs was observed as ~10 nm and it shows that CMC. XRD analysis confirmed the oxidation state of Cu NPs and it was used as a photocatalyst to degrade MO dye and it was found that the presence of H_2O_2 accelerate significantly it might be due to the formation of OH• radicals. Interestingly, the decomposition of MO was performed in the dark condition and it was observed that 100% degradation was achieved in 10 hours. It might be due to air oxidation. In another interesting work, Pathania et al. have prepared ZnS- Cellulose nanocomposites (ZS-CNC) by simple coprecipitation method and studied the photodegradation of phenol under solar light irradiation [59]. The ZnS has particles that are in the range of 8–10 and 4.3 nm in size by TEM and XRD analysis, respectively. ZS-CNC has better degradation efficiency (75%) than the pristine ZnS (58%), it is obvious that cellulose involves in the process of absorbance of phenol molecules. The enhanced photocatalytic activity of ZS-CNC might be due to its higher specific area and nanostructure.

In 2017, Ahmad et al. have prepared cellulose-polypyrrole-graphene oxide (C-PPy-GO) nanocomposite by in-situ polymerization method [60]. The photodegradation of RhB was performed with irradiation from a tungsten lamp light source. The bandgap energy of C-PPy-GO nanocomposite was calculated as 1.7 eV, moreover the PPy and GO actively participate in the formation of electron-hole pair. The photodegradation experiment was carried out at various pH levels (3, 7, and 10). The acidic condition (pH 3) was discovered to be suitable for dye degradation; this could be due to the protonated state of carboxylic groups of GO, which favors dye absorption, as well as the cationic nature of RhB dye.

Kokilavani et al. have prepared Ag/Cu-Cellulose (Ag/Cu-C) nanocomposites for metal absorption, photochemical degradation, and antimicrobial applications [61]. The particle size of the MNPs was around 18 nm and the zeta potential was calculated as −39.96 mV The XRD analyses revealed that Cu forms the core and Ag forms the shell in the core-shell structure of the composite. Ag/Cu-C was used as a photocatalyst to degrade MB dye under visible light irradiation.

19.5.5 Degradable polymer composites as catalysts for biodiesel synthesis

The depletion of fossil-based energy has prompted the scientific community to look for alternative fuels derived from renewable sources, such as biodiesel, which has many appealing characteristics such as biodegradability, relatively less expensive raw materials, renewability, and the ability to produce a value-added product from solid and liquid wastes. In general, biodiesel can be made by simply transesterifying vegetable oil/waste cooking oil/inedible oils and animal fats. We need an efficient catalyst for this process to provide the biodiesel with desirable properties such as flash point, density, and cetane number according to international standards.

Zeolite-chitosan-KOH composite was obtained by the simple one-pot method, in which 5 g of chitosan, 20 g of zeolite, and 100 g of KOH were used to produce biodiesel from used cooking oil at room temperature [62]. There have been many biodiesel production methods developed, but the electrolysis method is one of the best. In this study, 0.5% or 1% of the catalyst by weight of the oil was used. The viscosity, cloud point, density, flash point, and cetane number of the obtained biodiesel were calculated and compared to US and European standards. The obtained biodiesel had a density of 860 kg/m^3 at 15°C, which is the minimum value in the American standard, and a flash point of 190°C, which is well above the American and European scales. The cetane number was discovered to be 56, which is considered safe for environmental and public health concerns. In another work, Helmiyati et al., have fabricated a biodegradable composite from rice straw-derived cellulose, hematite, and zirconia, in which α-Fe_2O_3-ZrO_2composite has an irregular shape with an average particle size of 42.5 nm and is homogeneously distributed in cellulose matrix [63]. And this cellulose@ α-Fe_2O_3-ZrO_2 nanocomposites are used as a catalyst for biodiesel production from lauric acid. The presence of nanocellulose had a key role in increasing biodiesel production's high catalytic activity. The optimal cellulose@α-Fe_2O_3-ZrO_2 nanocomposite (nanocellulose:α-Fe_2O_3-ZrO_2 mass ratio = 2:1) produced 92.50% yield of biodiesel, with the international standard specifications (density: 0.853 g/mL, cetane number: 65, and flash point 135°C).

Ana et al., have developed (PLA)/poly(ε-caprolactone) (PCL) blends and PLA/PCL/titanium (IV) dioxide (TiO_2) nanocomposites. PLA-PCL (3:1 wt. ratio) and PLA-PCL-TiO_2 (3 wt.% of TiO_2 concerning PLA-PCL blend) were prepared by melt mixing method [64]. The PLA-PCL and PLA-PCL-TiO_2 samples were accelerated weathered for 0, 500, 1000, and 2000 hours, respectively. The samples were then analyzed using spectral, thermal, and mechanical techniques. SEM analysis revealed that PCL degraded faster than PLA, and the shorter PLA chains (formed by degradative chain scission) began to entangle to form the spheres. When the degradation rates of PLA-PCL and PLA-PCL-TiO_2 composites were compared, the former degraded faster. It is possible that the TiO_2 NPs in the PLA matrix help to increase its stability against weathering degradation. TiO_2 may also have improved PLA and PCL fraction compatibility by providing a glossy surface that makes the composites less susceptible to water and UV radiation penetration. Another TiO_2-based composite has been reported, in which anatase TiO_2 NPs (5 wt.%) are immobilized on PCL via a simple solvent-cast method [65]. Polycaprolactone (PCL) is a biocompatible, and nontoxic synthetic aliphatic polyester that degrades fully when exposed to microbes or enzymes. The degradation of 4-chlorophenol was used to test PCL-photocatalytic TiO_2 activity.

19.6 Challenges and future perspectives

Catalysts and photocatalysts based on biodegradable polymer composites are an important category of heterogeneous polymer-supported catalysts. However, there are very few research articles on this subject (less than a hundred in all search engines). We understand that the addition of metal NPs and other components to the biopolymer changes its inherent biodegradability. It creates significant difficulties in selecting the appropriate component to retain this property. We found no reports on biodegradable polymer composite-based chiral catalysts, which could be due to their heterogeneous nature. Chiral catalysts may find use in the pharmaceutical, chemical, and polymer industries. The incorporation of chiral aspects, control of the NP leaching problem, and increased mechanical properties will be the future of biodegradable polymer-based catalysts.

19.7 Conclusions

In this chapter, we concentrated on biopolymer-based nanocomposites such as cellulose, chitosan, starch, wool, and silk. In some studies, biopolymers are modified with functional groups such as amine, carboxylic acid, and sulfonic acid, whereas in others, metal nanoparticles (NPs) are incorporated using simple methods. There are few detailed reports on carbon-based materials such as MWCNT, graphene, and GO. The use of these materials as a catalyst in various organic transformations, photochemical dye degradation, and biodiesel preparations was explained, and key observations were made. These biodegradable polymer-based catalysts have several appealing characteristics, including biodegradability, reusability, low cost, and ease of separation from the reaction mixture. We also discussed the challenges and future prospects.

Acknowledgment

PR wishes to acknowledge Fondecyt Iniciacion No 11190727, which is supported by ANID and VRIP, Universidad Católica del Maule. JA acknowledges Dirección de Investigación, Universidad de Talca for the constant support for the research.

References

[1] L.S. Nair, C.T. Laurencin, Biodegradable polymers as biomaterials, Prog. Polym. Sci. 32 (2007) 762–798. Available from: https://doi.org/10.1016/j.progpolymsci.2007.05.017.

[2] H. Tian, Z. Tang, X. Zhuang, X. Chen, X. Jing, Biodegradable synthetic polymers: preparation, functionalization and biomedical application, Prog. Polym. Sci. 37 (2012) 237–280. Available from: https://doi.org/10.1016/j.progpolymsci.2011.06.004.

[3] I. Armentano, M. Dottori, E. Fortunati, S. Mattioli, J.M. Kenny, Biodegradable polymer matrix nanocomposites for tissue engineering: a review, Polym. Degrad. Stab. 95 (2010) 2126–2146. Available from: https://doi.org/10.1016/j.polymdegradstab.2010.06.007.

[4] S. Kasirajan, M. Ngouajio, Polyethylene and biodegradable mulches for agricultural applications: a review, Agron. Sustain. Dev. 32 (2012) 501–529. Available from: https://doi.org/10.1007/s13593-011-0068-3.

[5] B. Tomadoni, D. Merino, V.A. Alvarez, Biodegradable Materials for Planting Pots, (2020) 85–103. https://doi.org/10.21741/9781644900659-4.

[6] V. Siracusa, P. Rocculi, S. Romani, M.D. Rosa, Biodegradable polymers for food packaging: a review, Trends Food Sci. Technol. 19 (2008) 634–643. Available from: https://doi.org/10.1016/j.tifs.2008.07.003.

[7] S. Mangaraj, A. Yadav, L.M. Bal, S.K. Dash, N.K. Mahanti, Application of biodegradable polymers in food packaging industry: a comprehensive review, J. Packag. Technol. Res. 3 (2019) 77–96. Available from: https://doi.org/10.1007/s41783-018-0049-y.

[8] A.P. Heitmann, I. Coura Rocha, P.P. de Souza, L.C.A. Oliveira, P.S. Patrícia, Photoactivation of a biodegradable polymer (PHB): generation of radicals for pollutants oxidation, Catal. Today. 344 (2020) 171–175. Available from: https://doi.org/10.1016/j.cattod.2018.12.024.

[9] S.S. Bari, A. Chatterjee, S. Mishra, Biodegradable polymer nanocomposites: an overview, Polym. Rev. 56 (2016) 287–328. Available from: https://doi.org/10.1080/15583724.2015.1118123.

[10] R. Chandra, R. Rustgi, Biodegradable polymers, Prog. Polym. Sci. 23 (1998) 1273–1335. Available from: https://doi.org/10.1016/S0079-6700(97)00039-7.

[11] C.M. Buchanan, D. Dorschel, R.M. Gardner, R.J. Komarek, A.J. Matosky, A.W. White, et al., The of influence Buchanan Charles M. Degree of substitution on blend miscibility and biodégradation of cellulose acetate blends, J. Environ. Polym. Degrad. 4 (1996) 179–195.

[12] O.M. Dragostin, S.K. Samal, M. Dash, F. Lupascu, A. Pânzariu, C. Tuchilus, et al., New antimicrobial chitosan derivatives for wound dressing applications, Carbohydr. Polym. 141 (2016) 28–40. Available from: https://doi.org/10.1016/j.carbpol.2015.12.078.

[13] N.V. Kramareva, A.E. Koklin, E.D. Finashina, N.S. Telegina, A.Y. Stakheev, L.M. Kustov, Study of palladium complexes with chitosan and its derivatives as potential catalysts for terminal olefin oxidation, Kinet. Catal. 45 (2004) 743–751. Available from: https://doi.org/10.1023/B:KICA.0000044989.54767.b4.

[14] P.A. Gunatillake, R. Adhikari, N. Gadegaard, Biodegradable synthetic polymers for tissue engineering, Eur. Cells Mater. 5 (2003) 1–16. Available from: https://doi.org/10.22203/eCM.v005a01.

[15] D.Y. Kim, Y.H. Rhee, Biodegradation of microbial and synthetic polyesters by fungi, Appl. Microbiol. Biotechnol. 61 (2003) 300–308. Available from: https://doi.org/10.1007/s00253-002-1205-3.

[16] A. Sorrentino, G. Gorrasi, V. Vittoria, Potential perspectives of bio-nanocomposites for food packaging applications, Trends Food Sci. Technol. 18 (2007) 84–95. Available from: https://doi.org/10.1016/j.tifs.2006.09.004.

[17] N. Baig, I. Kammakakam, W. Falath, I. Kammakakam, Nanomaterials: a review of synthesis methods, properties, recent progress, and challenges, Mater. Adv. 2 (2021) 1821–1871. Available from: https://doi.org/10.1039/d0ma00807a.

[18] E. Fortunati, I. Armentano, Q. Zhou, A. Iannoni, E. Saino, L. Visai, et al., Multifunctional bionanocomposite films of poly(lactic acid), cellulose nanocrystals and silver nanoparticles, Carbohydr. Polym. 87 (2012) 1596–1605. Available from: https://doi.org/10.1016/j.carbpol.2011.09.066.

[19] S. Mahouche-Chergui, M. Guerrouache, B. Carbonnier, M.M. Chehimi, Polymer-immobilized nanoparticles, Colloids Surfaces A Physicochem. Eng. Asp. 439 (2013) 43–68. Available from: https://doi.org/10.1016/j.colsurfa.2013.04.013.

[20] Z. Khanam, N.A. Sa'Don, F. Adam, Synthesis and characterization of a novel paramagnetic polyaniline composite with uniformly distributed metallic nanoparticles sandwiched between polymer matrices, Synth. Met. 192 (2014) 1–9. Available from: https://doi.org/10.1016/j.synthmet.2014.03.001.

[21] M. El Achaby, Z. Kassab, A. Barakat, A. Aboulkas, Alfa fibers as viable sustainable source for cellulose nanocrystals extraction: application for improving the tensile properties of biopolymer nanocomposite films, Ind. Crops Prod. 112 (2018) 499–510. Available from: https://doi.org/10.1016/j.indcrop.2017.12.049.

[22] J.M. Herrmann, Fundamentals and misconceptions in photocatalysis, J. Photochem. Photobiol. A Chem. 216 (2010) 85–93. Available from: https://doi.org/10.1016/j.jphotochem.2010.05.015.

[23] M. Bojarska, B. Nowak, J. Skowroński, W. Piątkiewicz, L. Gradoń, Growth of ZnO nanowires on polypropylene membrane surface—characterization and reactivity, Appl. Surf. Sci. 391 (2017) 457–467. Available from: https://doi.org/10.1016/j.apsusc.2016.04.130.

[24] B. Liu, X. Zhao, C. Terashima, A. Fujishima, K. Nakata, Thermodynamic and kinetic analysis of heterogeneous photocatalysis for semiconductor systems, Phys. Chem. Chem. Phys. 16 (2014) 8751–8760. Available from: https://doi.org/10.1039/c3cp55317e.

[25] S.P. Yew, H.Y. Tang, K. Sudesh, Photocatalytic activity and biodegradation of polyhydroxybutyrate films containing titanium dioxide, Polym. Degrad. Stab. 91 (2006) 1800–1807. Available from: https://doi.org/10.1016/j.polymdegradstab.2005.11.011.

[26] Pei, In situ synthesis of Ag-Fe_3O_4 nanoparticles immobilized on pure cellulose microspheres as recyclable and biodegradable catalysts, ACS Omega 5 (2020) 8839–8846. Available from: https://doi.org/10.1021/acsomega.0c00437.

[27] S. Sabaqian, F. Nemati, M.M. Heravi, H.T. Nahzomi, Copper(I) iodide supported on modified cellulose-based nano-magnetite composite as a biodegradable catalyst for the synthesis of 1,2,3-triazoles, Appl. Organomet. Chem. 31 (2017) 1–5. Available from: https://doi.org/10.1002/aoc.3660.

[28] Y. Shahedi, H. Heidari, K. Nikoofar, Novel and cost-effective biocatalyst consisting of nanofibrillated cellulose and $TiCl_3$ for the synthesis of 2,3'-dihydroquinazolin-4-(1 H)-ones, Iran. J. Catal. 9 (2019) 71–77. Available from: https://www.sid.ir/en/journal/ViewPaper.aspx?ID = 688001.
[29] P. Mohanty, S. Aggrawal, R. Sharma, CuO immobilized paper matrices: a green catalyst for conversion of CO_2 to cyclic carbonates, J. CO_2 Util. 46 (2021) 101466. Available from: https://doi.org/10.1016/j.jcou.2021.101466.
[30] A. Shaabani, Z. Hezarkhani, M. Tayeb Faroghi, Manganese(II) tetrasulfophthalocyanine covalently supported on natural silk: a new highly active catalyst for synthesis of benzoxazepine derivatives in water, Appl. Organomet. Chem. 31 (2017) 1–11. Available from: https://doi.org/10.1002/aoc.3764.
[31] M.T. Faroghi, A. Shaabani, Z. Hezarkhani, Wool-SO_3H and nano-Fe_3O_4@wool as two green and natural-based renewable catalysts in one-pot isocyanide-based multicomponent reactions, Monatshefte Fur Chemie. 147 (2016) 1963–1973. Available from: https://doi.org/10.1007/s00706-016-1717-7.
[32] I. Lestari, H. Helmiyati, Calcium alginate-TiO_2/SiO_2 nanocomposite for glucose conversion to 5-hydroxymethylfurfural, IOP Conf. Ser. Mater. Sci. Eng. 763 (2020) 3–9. Available from: https://doi.org/10.1088/1757-899X/763/1/012037.
[33] M. Kaya, R. Ulus, Y. Yıldız, S. Eris, B. Aday, F. Sen, Functionalized multi-walled carbon nanotubes (f-MWCNT) as highly efficient and reusable heterogeneous catalysts for the synthesis of acridinedione derivatives, Chem. Select. 1 (2016) 3861–3865. Available from: https://doi.org/10.1002/slct.201600719.
[34] N. Karak, D. Hazarika, K. Gupta, M. Mandal, High-performing biodegradable waterborne polyester/functionalized graphene oxide nanocomposites as an eco-friendly material, ACS Omega. 3 (2018) 2292–2303. Available from: https://doi.org/10.1021/acsomega.7b01551.
[35] S. Dhanavel, N. Manivannan, N. Mathivanan, V.K. Gupta, V. Narayanan, A. Stephen, Preparation and characterization of cross-linked chitosan/palladium nanocomposites for catalytic and antibacterial activity, J. Mol. Liq. 257 (2018) 32–41. Available from: https://doi.org/10.1016/j.molliq.2018.02.076.
[36] N. Tavker, U. Gaur, M. Sharma, Cellulose supported bismuth vanadate nanocomposite for effective removal of organic pollutant, J. Environ. Chem. Eng. 8 (2020) 104027. Available from: https://doi.org/10.1016/j.jece.2020.104027.
[37] V. Sencades, D. Cardillo, M. Tehei, T. Devers, M. Monirul Islam, A. Rosenfeld, et al., Attenuation of UV absorption by poly(lactic acid)-iron oxide nanocomposite particles and their potential application in sunscreens, Chem. Eng. J. 405 (2021) 126843. Available from: https://doi.org/10.1016/j.cej.2020.126843.
[38] P.S. Ramesh, D. Geetha, S. Kavitha, A novel bio-degradable polymer stabilized Ag/TiO_2 nanocomposites and their catalytic activity on reduction of methylene blue under natural sun light, Ecotoxicol. Environ. Saf. 121 (2015) 126–134. Available from: https://doi.org/10.1016/j.ecoenv.2015.04.042.
[39] M. Senapati, T. Tripathy, H. Kolya, S. Jana, Green synthesis of Ag-Au bimetallic nanocomposites using a biodegradable synthetic graft copolymer; hydroxyethyl starch-g-poly (acrylamide-co-acrylic acid) and evaluation of their catalytic activities, Eur. Polym. J. 87 (2017) 113–123. Available from: https://doi.org/10.1016/j.eurpolymj.2016.12.019.
[40] E. Farshchi, S. Pirsa, M. R, L. Roufegarinejad, M. Alizadeh, Photocatalytic/biodegradable film based on carboxymethyl cellulose, modified by gelatin and TiO_2-Ag nanoparticles, Carbohydr. Polym. 216 (2019) 189–196. Available from: https://doi.org/10.1016/j.carbpol.2019.03.094.
[41] Y. Zahedi, B. Fathi-Achachlouei, Fabrication and characterization of CMC-based nanocomposites reinforced with sodium montmorillonite and TiO_2 nanomaterials, Carbohydr. Polym. 199 (2018) 415–425. Available from: https://doi.org/10.1016/j.carbpol.2018.07.031.
[42] R. Subekti, H. Helmiyati, Department, sodium alginate-TiO_2-bentonite nanocomposite synthesis for photocatalysis of methylene blue dye removal, IOP Conf. Ser. Mater. Sci. Eng. 763 (2020). Available from: https://doi.org/10.1088/1757-899X/763/1/012018.

[43] G. Jayanthi Kalaivani, S.K. Suja, TiO_2 (rutile) embedded inulin - a versatile bio-nanocomposite for photocatalytic degradation of methylene blue, Carbohydr. Polym. 143 (2016) 51–60. Available from: https://doi.org/10.1016/j.carbpol.2016.01.054.

[44] M.H. Abdel Rehim, M.A. El-Samahy, A.A. Badawy, M.E. Mohram, Photocatalytic activity and antimicrobial properties of paper sheets modified with TiO_2/sodium alginate nanocomposites, Carbohydr. Polym. 148 (2016) 194–199. Available from: https://doi.org/10.1016/j.carbpol.2016.04.061.

[45] M.Y. Abdelaal, R.M. Mohamed, Environmental remediation from thiophene solution by photocatalytic oxidation using a Pd/ZrO_2-chitosan nanocomposite, Ceram. Int. 40 (2014) 7693–7699. Available from: https://doi.org/10.1016/j.ceramint.2013.12.110.

[46] K. Venkatesan Savunthari, D. Arunagiri, S. Shanmugam, S. Ganesan, M.V. Arasu, N.A. Al-Dhabi, et al., Green synthesis of lignin nanorods/g-C_3N_4 nanocomposite materials for efficient photocatalytic degradation of triclosan in environmental water, Chemosphere 272 (2021). Available from: https://doi.org/10.1016/j.chemosphere.2021.129801.

[47] S. Vigneshwaran, P. Sirajudheen, C.P. Nabeena, S. Meenakshi, In situ fabrication of ternary TiO_2 doped grafted chitosan/hydroxyapatite nanocomposite with improved catalytic performance for the removal of organic dyes: experimental and systemic studies, Colloids Surfaces A Physicochem. Eng. Asp. 611 (2021). Available from: https://doi.org/10.1016/j.colsurfa.2020.125789.

[48] S. Vigneshwaran, P. Sirajudheen, C.P. Nabeena, V.P. Sajna, S. Meenakshi, Photocatalytic performance of chitosan tethered magnetic Fe_2O_3-like (3D/2D) hybrid for the dynamic removal of anionic dyes: degradation and mechanistic pathways, Int. J. Biol. Macromol. 183 (2021) 2088–2099. Available from: https://doi.org/10.1016/j.ijbiomac.2021.06.012.

[49] H. Zhu, R. Jiang, Y. Fu, Y. Guan, J. Yao, L. Xiao, et al., Effective photocatalytic decolorization of methyl orange utilizing TiO_2/ZnO/chitosan nanocomposite films under simulated solar irradiation, Desalination. 286 (2012) 41–48. Available from: https://doi.org/10.1016/j.desal.2011.10.036.

[50] D. Geetha, S. Kavitha, P.S. Ramesh, A novel bio-degradable polymer stabilized Ag/TiO_2 nanocomposites and their catalytic activity on reduction of methylene blue under natural sun light, Ecotoxicol. Environ. Saf. 121 (2015) 126–134. Available from: https://doi.org/10.1016/j.ecoenv.2015.04.042.

[51] K.T. Karthikeyan, A. Nithya, K. Jothivenkatachalam, Photocatalytic and antimicrobial activities of chitosan-TiO_2 nanocomposite, Int. J. Biol. Macromol. 104 (2017) 1762–1773. Available from: https://doi.org/10.1016/j.ijbiomac.2017.03.121.

[52] M. Maruthupandy, T. Muneeswaran, M. Anand, F. Quero, Highly efficient multifunctional graphene/chitosan/magnetite nanocomposites for photocatalytic degradation of important dye molecules, Int. J. Biol. Macromol. 153 (2020) 736–746. Available from: https://doi.org/10.1016/j.ijbiomac.2020.03.045.

[53] L. Eskandari, F. Andalib, A. Fakhri, M.K. Jabarabadi, B. Pham, V.K. Gupta, Facile colorimetric detection of Hg (II), photocatalytic and antibacterial efficiency based on silver-manganese disulfide/polyvinyl alcohol-chitosan nanocomposites, Int. J. Biol. Macromol. 164 (2020) 4138–4145. Available from: https://doi.org/10.1016/j.ijbiomac.2020.09.015.

[54] A. Balakrishnan, S. Appunni, K. Gopalram, Immobilized TiO 2 / chitosan beads for photocatalytic degradation of 2, 4-dichlorophenoxyacetic acid, Int. J. Biol. Macromol. 161 (2020) 282–291. Available from: https://doi.org/10.1016/j.ijbiomac.2020.05.204.

[55] S. Sangon, A.J. Hunt, Y. Ngernyen, S. Youngme, N. Supanchaiyamat, Rice straw-derived highly mesoporous carbon-zinc oxide nanocomposites as high performance photocatalytic adsorbents for toxic dyes, J. Clean. Prod. 318 (2021). Available from: https://doi.org/10.1016/j.jclepro.2021.128583.

[56] Y.K. Manea, A.M. Khan, A.A. Wani, M.T.A. Qashqoosh, M. Shahadat, M.A.S. Salem, Hydrothermally synthesized mesoporous CS-g-PA@TSM functional nanocomposite for efficient photocatalytic degradation of Ciprofloxacin and treatment of metal ions, J. Mol. Liq. 335 (2021). Available from: https://doi.org/10.1016/j.molliq.2021.116144.

[57] M. Thomas, T.S. Natarajan, M.U.D. Sheikh, M. Bano, F. Khan, Self-organized graphene oxide and TiO_2 nanoparticles incorporated alginate/carboxymethyl cellulose nanocomposites with efficient photocatalytic activity under direct sunlight, J. Photochem. Photobiol. A Chem. 346 (2017) 113–125. Available from: https://doi.org/10.1016/j.jphotochem.2017.05.037.

[58] V.V. Spiridonov, X.Y. Liu, S.B. Zezin, I.G. Panova, A.V. Sybachin, A.A. Yaroslavov, Hybrid nanocomposites of carboxymethyl cellulose cross-linked by in-situ formed Cu_2O nanoparticles for photocatalytic applications, J. Organomet. Chem. 914 (2020). Available from: https://doi.org/10.1016/j.jorganchem.2020.121180.

[59] D. Pathania, M. Kumari, V.K. Gupta, Fabrication of ZnS-cellulose nanocomposite for drug delivery, antibacterial and photocatalytic activity, Mater. Des. 87 (2015) 1056–1064. Available from: https://doi.org/10.1016/j.matdes.2015.08.103.

[60] N. Ahmad, S. Sultana, A. Azam, S. Sabir, M.Z. Khan, Novel bio-nanocomposite materials for enhanced biodegradability and photocatalytic activity, New J. Chem. 41 (2017) 10198–10207. Available from: https://doi.org/10.1039/c7nj00842b.

[61] S. Kokilavani, A. Syed, A.M. Thomas, A.M. Elgorban, L.L. Raju, A. Das, et al., Facile synthesis of Ag/Cu-cellulose nanocomposite for detection, photocatalysis and anti-microbial applications, Optik (Stuttg). 220 (2020). Available from: https://doi.org/10.1016/j.ijleo.2020.165218.

[62] M. Mehrpooya, L. Fereidooni, Experimental assessment of electrolysis method in production of biodiesel from waste cooking oil using zeolite/chitosan catalyst with a focus on waste biorefinery, Energy Convers. Manag. 147 (2017) 145–154. Available from: https://doi.org/10.1016/j.enconman.2017.05.051.

[63] M. Khalil, H. Helmiyati, Y. Budiman, G. Harfiana Abbas, F. Wulan Dini, Highly efficient synthesis of biodiesel catalyzed by a cellulose@hematite-zirconia nanocomposite, Heliyon. 7 (2021). Available from: https://doi.org/10.1016/j.heliyon.2021.e06622.

[64] P. Kasak, A.S. Luyt, A. Antunes, A. Popelka, A. Mahmoud, M.K. Hassan, Effect of poly(ε-caprolactone) and titanium (IV) dioxide content on the UV and hydrolytic degradation of poly(lactic acid)/poly(ε-caprolactone) blends, J. Appl. Polym. Sci. (2021) 1–16. Available from: https://doi.org/10.1002/app.51266.

[65] T. Sivlim, Ş. Akkan, I. AltIn, M. Koç, M. Sökmen, TiO_2 immobilized biodegradable polymer for photocatalytic removal of chlorophenol, Water Air Soil Pollut. 223 (2012) 3955–3964. Available from: https://doi.org/10.1007/s11270-012-1163-z.

CHAPTER

20

Sustainability assessment, environmental impact, and recycling strategies of biodegradable polymer nanocomposites

Maria-Beatrice Coltelli[1,2,3], Pierfrancesco Morganti[3] and Andrea Lazzeri[1,2]

[1]Department of Civil and Industrial Engineering, University of Pisa, Pisa, Italy [2]Interuniversity Consortium of Materials Science and Technology (INSTM), Florence, Italy [3]Academy of History of Health Care Art, Rome, Italy

Chapter Outline

Biodegradable and Biocompatible Polymer Nanocomposites. DOI: https://doi.org/10.1016/B978-0-323-91696-7.00016-7
Copyright © 2023 Elsevier Inc. All rights reserved.

20.1 Introduction

Biodegradable polymer nanocomposites are nanostructured materials formed by combining a biodegradable polymer matrix with fillers that either participate in or do not interfere with the biodegradation of the entire material. Because of environmental concerns about plastic waste management, these biocomposites are being extensively researched, as their biodegradability (organic recycling) could be used to facilitate product waste management at the end of its life. Given that many of these materials are derived from bio-based sources, the possibility of meeting carbon neutrality principles without contributing to increased greenhouse gas (GHG) concentrations is also important [1]. The reuse or recycling of the material for use in other applications, thereby increasing the material's full life, represents an additional tool for improved waste management of plastic products. If it has no negative impact on the final biodegradation of the product, this type of recycling improves its overall environmental profile. All of these interconnected issues will be addressed in this chapter.

This first section defines the term "biodegradability" and describes the various biodegradable materials. It will also define bionanocomposites and their properties.

20.1.1 Biodegradability

Biodegradable materials undergo biodegradation, which occurs when microorganisms such as algae, bacteria, or fungi degrade the materials. Because it occurs on the surface of the biocomposites, the process is complex. Biopolymers are converted to water-soluble intermediates consisting of oligomers by extracellular enzymes excreted by microorganisms (e.g., lipases) (Fig. 20.1). The microorganism cells absorb the water-soluble intermediates. As a result, water-soluble intermediates are a source of food for microorganisms, which use them to produce CO_2, H_2O, CH_4, and other

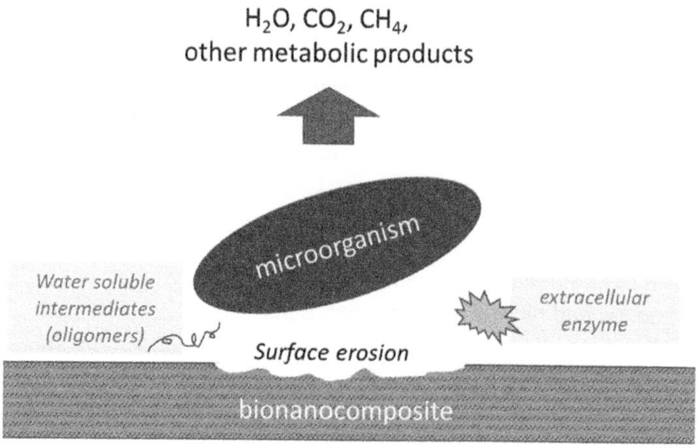

FIGURE 20.1

Simplified scheme of biodegradation occurring on a bionanocomposite surface.

molecules in their internal metabolism pathways. Aerobic processes produce oxygenated species such as CO_2 and water, whereas anaerobic processes produce methane.

Biodegradable plastics should be evaluated using appropriate standards, as defined by the SPI Plastic Industry Trade Association and the Bioplastics Council [2].

The bio-based content is the percentage of carbon that is made up of biological or agricultural resources. The bio-based content, in particular, can be determined using the methodologies described in the ASTM D6866 standard. The bio-based carbon content of the product is calculated as a fraction of the total organic carbon content in this standard (TOC) [3]. Moreover, EN 16440 and ISO 16620−2 standards, suggest to report the bio-based carbon content as a fraction of total carbon (TC). Moreover, it describes a quantitative method to determine the bio-based carbon content for polymers, monomers and plastic products, through the determination of ^{14}C content [4,5].

The terms "bio-based" and "biodegradable" have different connotations. The term "bio-based" denotes that the material or product is made from renewable resources. However, little is known about its biodegradability. Thus, some bio-based products can biodegrade in industrial composting plants, while others can biodegrade in home composting and still others in aquatic and/or roadside environments. Furthermore, some materials biodegrade only in very specific environments (e.g., vermicompost, etc.), while others do not biodegrade at all.

Composting is defined as the natural recycling of a product. Several types of composting, in particular, can be considered. Industrial composting occurs in large-scale composting plants under controlled composting conditions at a temperature of about 60°C, whereas home composting occurs on a smaller scale of a few kilograms' domestic plant at ambient temperature.

In industrial plants, poly(lactic acid) (PLA) is compostable, whereas bio-poly(ethylene) (bio-PE) and bio-poly(ethylene terephthalate) (Bio-PET) are not. Poly(hydroxy alcanoates) (PHAs) are generally biodegradable in the home composting environment as well as in the marine environment. As explained above, biodegradation is a chemical process carried out by microorganisms. These later transform the organic carbon of the materials into simpler substances like H_2O, CO_2, and compost. As a result, environmental conditions (e.g., pH, temperature, location, etc.), material chemical composition, and specific sample geometry all have a strong influence on this complex process (porosity, dimension, thickness, etc.).

The Packaging Directive 94/62/EC [6] set the targets concerning the minimum quantity that can be recovered and recycled for each type of packaging material. This document specifies the amount of bio-waste that can be incinerated or landfilled while adhering to landfill regulations. Nonetheless, it does not take into account the recycling of bio-waste, as specified in the Waste Framework Directive.

Several European standards define a framework into which they can be incorporated to support the assertion that products meet the essential packaging requirements for market availability. EN 13431:2000, for example, is about Packaging and Requirements for Packaging Recoverable in the Form of Energy Recovery, Including Minimum Inferior Calorific Value Specification. The EN 13432:2000 is also dedicated to packaging and defines the Requirements for Compostable and Biodegradable Packaging, as well as the Test Scheme and Evaluation Criteria for Packaging Final Acceptance. The CR 13695-1 is also concerned with packaging and defines the Requirements for Measuring and Verifying the Release of Four Heavy Metals (Cr, Rd, Hg, Pb) and Other Dangerous Substances into the Environment. EN 13427:2000 is a "umbrella norm" for packaging that specifies the Requirements for the Use of European Standards in the Field of Packaging Waste.

EN 13432 is a harmonized EU standard that specifies the guidelines for determining whether a product is compostable or not with legal implications. According to this standard, a product is compostable if it meets several specific requirements. It must have at least 50% organic content. Furthermore, it should not exceed the heavy metal concentration thresholds specified in the standard. Under controlled composting conditions, the products should mineralize at least 90% by weight within 6 months (the mineralization, linked to the chemical composition of the sample, is defined as the conversion of the organic carbon to CO_2 and biomass). Within 12 weeks and under controlled composting conditions, the product should decompose into fragments smaller than 2 mm. The thickness and the product shape and geometry (e.g., laminate, coating, etc.) of the product are thus pivotal. The finished compost must have no negative impact on plant germination and growth.

It should be noted that compostability is more than just biodegradability. Compostable packaging is also biodegradable. To avoid terminology confusion, several authorities have decided to promote the definition of "compostable" over that of "biodegradable" due to its greater specificity. Furthermore, the environment in which the product can be composted should be specified. As a result, a material can be compostable in an industrial composting plant or at home, in soil or in the marine environment.

Some new standards are being developed to significantly reduce and avoid marine and terrestrial litter, thereby promoting biodegradable materials in specific environments. Controlling the biodegradability of materials in soil or sea water will become increasingly important as the issue of microplastics pollutes the sea. Plastic particles having dimensions in the scale of 1 μm to 5 mm and also the submicrometric ones are commonly indicated as microplastics (MPs) and nanoplastics (NPs), respectively. Micro- and nanoplastics are now thought to be nearly ubiquitous in natural water bodies. Micro- and nanoplastics are found in the environment as a result of the fragmentation of large plastic items (referred to as secondary MPs) or as particles obtained in industry as microspheres or textile fibers (these particles are named primary MPs) [7].

20.1.2 Classes of biodegradable materials

Polymers are the building blocks of the structural materials found in animals and plants. Cellulose is the primary structural polymer in plants (polyether). Proteins, which have a polyamidic structure, are the most important structural polymers in animals.

Polymers can be classified in synthetic, artificial or natural (Table 20.1). Cellulose is an example of a natural polymer found in cotton fabric or paper; additionally, poly(hydroxyalcanoates)

Table 20.1 Polymer classification considering biodegradability and renewability.

Classes	Biodegradability	Renewability
Natural	Yes	Yes
Artificial	Possible	Yes[a]
Synthetic	No[b]	No
Biosynthetic	Yes/no	Yes

[a]*The renewability of the chemicals used for modification should be also considered.*
[b]*Some fossil-based polymers are biodegradable and included in bio-based polymers.*

(PHAs) are natural polymers because they are produced by microorganisms. As an example, consider cellulose acetate, a plastic material derived from cellulose through a chemical modification. Another example is cellulose xanthate, a fiber used in textiles and packaging (viscose and cellophane, respectively). Both natural and synthetic polymers are derived from renewable sources, which is advantageous. Polypropylene (PP), polyethylene (PE), polystyrene (PS), and poly(ethylene terephthalate) (PET) are well-known examples of synthetic polymers used in the packaging industry.

Synthetic polymers are manufactured industrially from natural-source monomers using traditional polymerization methods, which distinguishes them from biosynthetic polymers [8]. As a result, they are completely renewable. PLA is an important example of a biosynthetic polymer. The synthetic methodology is advantageous for biosynthetic polymers because it allows for fine tuning of the polymer's primary structure (for example, through appropriate copolymerization). As a result, the processability and final properties of the plastic material produced will be tailored for very specific applications.

Natural polymers are typically biodegradable [9,10]. As yet explained, the biodegradability is the capability to be degraded in the environment giving CH_4, CO_2, and other simple compounds. This can occur for all polymers, but for synthetic polymers are necessary generally centuries (Fig. 20.2). The degree of chemical modification is a factor that has a significant impact on the biodegradability of synthetic polymers. For example, cellulose acetate, which can have an acetylation degree ranging from 0 to 3, is said to be biodegradable if it has an acetylation degree of up to 2.5 [11]. PLA and other biosynthetic polymers are typically bio-based and biodegradable. Synthetic polymers, such as PE, can now be manufactured using natural resources, but they are not biodegradable.

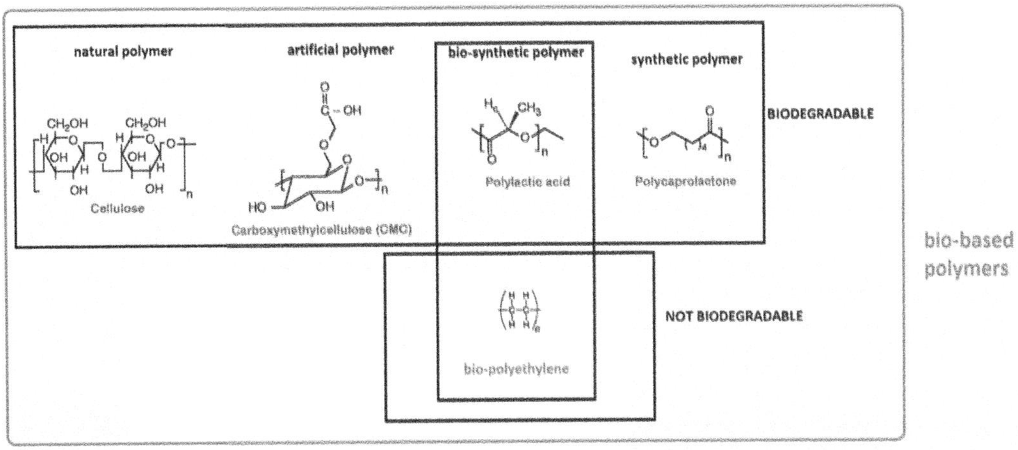

FIGURE 20.2

Examples of biodegradable and not biodegradable bio-based polymers.

Bioplastics are polymers that are bio-based and/or biodegradable [12]. Because of their biodegradability, biodegradable polymers derived from fossil monomers, such as poly(butylene succinate) (PBS), polycaprolactone (PCL), poly(butylene adipate-*co*-terephthalate) (PBAT), and others, are included in bioplastics. In the last decade, several bioplastic companies are active in building plants for having the monomers fully from renewable sources in the near future [13], to transform their biodegradable polymers also in bio-based ones.

The choice of polymers in applications such as packaging is heavily influenced by the end-of-life (EoL) option [14,15]. Biodegradability is not a required feature in long-lasting applications, but in applications involving short life cycle products, such as the food-packaging industry, it becomes a critical property in terms of composting waste from these products. Compostability is more important than biodegradability, as previously stated. In fact, a material to be biodegraded by microorganisms requires a standard environment. EN 13432/14995 is the standards that certified compostable products [16,17] and define the methodologies for testing the effective compostability, thus conferring a suitable behavior of the material in the final composting plants.

Recycling is another packaging disposal option that should be investigated further in the near future. This option necessitates the separation of the various polymers in order to limit contamination and thus provide good properties and durability, particularly in reprocessing conditions. Recycling is a more convenient option for monomaterial packages. In general, monomaterial rigid packaging can be collected and recycled properly in waste collection plants. As a result, recycling could be considered. Flexible multilayer packaging would be easier to recycle. Compostability may be the best option due to the low thickness.

On an industrial scale, two bioplastic formulations were developed for the production of flexible and multilayer packaging: starch-based materials and poly(lactic acid) (PLA)-based materials. Starch-based formulations suited for the production of flexible film were obtained by blending starch with thermoplastic hydrophilic synthetic polymers such as poly(caprolactone) [18] and poly (ethylene-vinyl alcohol) [19]. Moreover, plasticizers like glycerol [20] or polyethylene glycols [20] were used for optimizing TPS processability. This approach was used for the development of commercial thermoplastic starch widely employed for soft packaging.

PLA is a bio-based and biodegradable material that, due to its lower cost than commodities such as polystyrene or PET, is another promising material for packaging applications. Nonetheless, due to its rigidity, PLA is unsuitable for flexible packaging applications. Several companies have developed commercial granules suitable for flexible packaging applications in recent years. In fact, several studies have recently been conducted on PLA blends prepared by incorporating other biodegradable polyesters or plasticizers into the bioplastic material. Scatto et al. used flat die extrusion to create citrate plasticized and clay nanofilled films, and they investigated the final properties of the films as a function of the extrusion parameters [21]. In many studies it was evidenced that use of chain extenders [22] or peroxides [23] can be fundamental for modulating rheological properties, for instance by decreasing the melt fluidity and inducing some branching in macromolecules. Materials based on extruded blends of PLA and poly(butylene succinate-*co*-adipate) (PBSA) were also investigated [24,25] and validated for applications in flexible film production, thanks to the good compatibility between these polyesters.

Biocomposites can be created by combining natural fibers to create biodegradable and renewable materials. The increased resistance obtained thanks to reinforcement [26–29] can allow to use less material in vehicles applications, thus inducing indirect decrease in emissions. The degradation of biocomposites is influenced by the activity of microorganisms, UV light, atmospheric humidity, and temperature, which degrade the lignin, hemicelluloses, and cellulose that comprise the fiber. This degradation path can cause a specific break in the bonding between the polymeric matrix and the fibers, resulting in a significant reduction in the mechanical properties of the composite [30]. Natural fibers are generally more compostable than bioplastics due to their inherent hydrophilicity and porosity, which allows for an easy breakdown mechanism. Thus, using natural fibers as polymer reinforcement improves the biodegradability of the material. This effect is especially useful for products with a short life cycle, such as those used in the packaging, sanitary, and personal care industries.

Biodegradable polyesters are the polymers that have the greatest impact on a material's total biodegradability [31]. Biodegradation begins with a preliminary hydrolysis step caused by humidity, which reduces the molecular weight of the polyester [32]. When the molecular weight is sufficiently low, degradation due to microorganism action occurs, resulting in the decomposition of polymeric oligomers via microorganism metabolism. As a result, some material properties influence the entire degradation mechanism. The polymer molecular weight (MW) should be low to facilitate the first step, but this is generally in contrast to mechanical properties, which are generally better when the polymer MW is high. Furthermore, hydrolysis is favored if water vapor can diffuse through the material, resulting in a significant decrease in MW. As a result, a low crystallinity of the material may be preferable, because gases such as water vapor can diffuse better and be more effective in hydrolyzing the C-O bonds in the amorphous part of the polymeric material. Then, because a hydrophobic material has a low affinity for humidity, hydrolysis can be limited due to the hydrophobic feature of the material. Polymers with low crystallinity, low MW, and low hydrophobicity are generally more biodegradable. Biopolymeric fibers and fillers, for instance made of polysaccharides or proteins, when added to biopolyesters, determine generally a decrease in MW [22], an increase in disorder and an increase in hydrophilicity that is generally beneficial to improve biodegradation. Inorganic nanoparticles are not biodegradable, so they will not biodegrade and may be detrimental to biodegradation depending on their impact on the above-mentioned mechanism. Bidimensional nanoparticles, such as clay, are generally thought to be more effective at reducing material permeability to gases [21]. In general, this is bad for biodegradation.

20.1.3 Bionanocomposites

The introduction of advanced composites, facilitated by the spread of nanotechnology, has had an impact on almost every aspect of modern life, including tissue engineering, packaging, pharmaceuticals, food, cosmetics, apparel, and health care. However, because these products have a short life cycle, packaging, personal care/sanitary products, and apparel are the main global causes of plastic waste production. The problem of microplastics invading the land and oceans is primarily due to improper use of these products [33,34].

Nanobiotechnology, on the other hand, is a branch of nanotechnology that creates new composites and devices by applying knowledge and green biological and biochemical natural elements [33]. It is an overlapping multidisciplinary discipline that combines physical sciences, molecular engineering, and chemistry to create biocomposites from so-called green materials. Because of the anthropogenic impact on climate change, consumers and producers are shifting toward composites made from natural materials [33]. As a result, it has become critical to replace the current production and consumption of nanocomposites made primarily of synthetic fossil-based polymers with bionanocomposites. They are, in fact, used to manufacture plastics, 60% of which are discarded and accumulate in the natural environment and landfills [34,35].

Bionanocomposite is a catch-all term for a composite containing a matrix or reinforcement derived from natural resources, or both. Biocomposites may be considered bionanocomposite when at least one of the dimensions of the polymeric matrix' reinforcement, the so-called filler, is less than 100 nm. The filler is typically made up of nanoparticles, which can strengthen the mechanical, electrical, optical, and surface properties of the bionanocomposite due to the optional interactions they have with the polymeric matrices [36]. At this point, fiber reinforced polymers, such as chitin nanofibrils or nanocellulose, represent a natural type of reinforcement and matrix, representing an interesting composite combination that is currently available on the market to make, for example, biodegradable tissues to be used as carriers in the pharmaceutical and cosmetic fields [37–39]. In fact, matrix reinforcement has generally been obtained by using a low content of nanoparticles or nanofibrils (less than 5% by weight), which increase the mechanical properties of the nanocomposite due to the high surface area and good interactions with the polymeric matrix. These polymers, when embedded as micronanoparticles in a composite of mixed origin, have an interesting positive impact on both the environment and human health over the course of their entire life cycle assessments (LCA) [40–43]. Chitin and lignin, for example, are natural polymers, whereas polylactic acid (PLA) produced by biotechnology and polyhydroxyalcanoates (PHAs) produced by microorganism cultures are natural-derived compounds [44].

The use of bionanoparticles could represent a solution for producing sustainable and high-value products that are biodegradable and safe with zero waste and pollution. This is why carriers in the biomedical field, such as hydrogel systems, cell-sheet engineering, tissue engineering, and pharmaceutical therapeutics with multiple functionalities and architectures, were considered, as well as their ability to solubilize both hydrophilic and hydrophobic materials [36,44–50]. In fact, bionanoparticles may provide controllable delivery and release of many active ingredients, altering their kinetics in a desirable way and allowing their delivery to specific skin layers [49,50]. Furthermore, their systems can be designed to change their properties and functions selectively in response to specific internal or external stimuli, with the primary goal in their polymer therapeutic and cosmetic conception being to minimize the negative effects on healthy tissues [19]. Therefore, characterization of the polymers used to make bionanocomposites and tissues is essential to comprehend their connections with cells in vitro [51–53] and with the host in vivo [54–57], because they must be effective, free of toxicity and immunogenicity in their proposed administration routes.

20.1.4 Properties and market of bionanocomposites

The rapid emerging market in different fields of the composites reinforced with natural fibers is due to their low environmental impacts, easy biodegradability, neutrality with reference to CO_2 emissions and low energy cost of production [44,50] of these materials. Thus, the global nanocomposite market increased from 8 million tons in 2010 to 2 billion tons in 2017 and is expected to reach USD 5.3 billion by 2021, up from USD 1.6 billion in 2016, with a compound annual growth rate (CAGR) of 26% from 2016 to 2021 [57]. Among the Regions Asia Pacific (APAC), including China, Japan and India held the largest share in 2017. Moreover, APAC and Europe are the most active markets in term of strategic initiatives on account of their developing and established market demands [58,59]. Regarding the global nanocomposites market, it has been demonstrated that natural polymers and bionanocomposites are increasingly in demand and use, as opposed to synthetic polymers, due to the latter's negative impact on the environment and human well-being [59]. Natural polymers, in fact, with a market of USD 2.8 billion in 2017 is predicted to range 4.53 billion by 2022 with a CAGR of 5% (Fig. 20.3) [60].

Finally, the rapid growth in the use of composites has been aided by the development of novel design and manufacturing methods for industrial components that enable functional optimization based on a variety of technical and economic criteria [61,62]. Another important goal in the research and development of bionanocomposites has been to reduce the negative effects on human health through greater control over polymer selection. All of the biomolecules and biological components, as well as the processes used, should, in fact, mimic how nature makes materials, so that the bionanocomposites created, which are skin- and environmentally friendly, can be used for pharmaceutical, food, and cosmetic applications. Thus, novel biomedical technologies have enabled the development of pharmaceutical and cosmetic nanocarriers with specialized features and tunable surfaces that can be tailored to deliver a wide range of payloads ranging from small to complex molecules [39,51–58]. Through physical interactions or covalent bonds, imaging molecules can be encapsulated into the inner layer of these nanocarriers or adsorbents, coated, or grafted on their surface or outer layers [28,61–64]. Alternatively, the micronanocapsules embedded in the tissues may be PEGylated to improve the solubility and stability of the encapsulated ingredients while also preventing their interaction with biological systems [65,66].

The incorporation of biofillers or bionanofillers in biopolymeric matrices can affect both structural and functional properties. The addition of micro or nano biofillers was found to increase the elastic modulus in many cases [67]. Moreover, functional properties like antimicrobial [68] or specific barrier properties [69] can be impacted thanks to the nano dispersion of bionanofillers in a biopolymeric matrix.

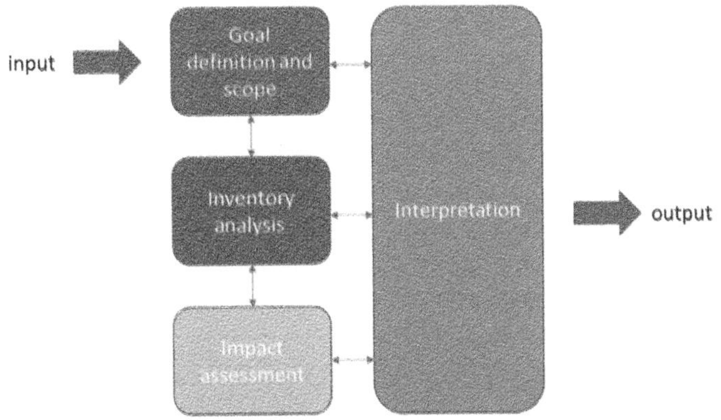

FIGURE 20.3

General scheme of an LCA study.

20.2 Sustainability assessment and environmental impact of biodegradable polymer nanocomposites

Kabir et al. [70] suggested that biopolymers can be a very important tool for the correct direction of sustainable development. Biodegradable polymer nanocomposites sustainability must be evaluated by using proper investigation tools based on scientific basis, like LCA studies, as suggested by Gowthaman et al. [71]. They demonstrated that biopolymers and specific biocomposites are advantageous over fossil materials by applying this methodology to them. Mechanical recycling can be considered a fundamental opportunity for bionanocomposites [72], despite of several technological issues should be overcome in the future [73]. Biodegradation (organic recycling) is a second EoL option for postconsumer/industrial biodegradable bio-based products that are unrecyclable through mechanical or chemical recycling [74]. As a result, proper standardized tests for assessing product degradation behavior in the environment are critical. Another important validation step is the biocompatibility of the bionanocomposite, especially if the products are intended for use in contact with living organisms (humans, animals, microorganisms, etc.) [75]. Potential legislative leverages introduced by strategic policies, as well as opportunities in the bioeconomy and circular economy competition, should be considered [76]. This section is dedicated to analyzing all of these influencing factors.

20.2.1 Life cycle assessment

The LCA method was developed to estimate the environmental impacts of a product or system in a scientific and systematic manner [77,78]. The benefits of LCA include considering the entire product life cycle, from "cradle to grave" or "cradle to cradle," and integrating several environmental effects methodically. It entails defining the goal, scope, and study units, as well as creating an inventory of inputs and outputs for the system or product. Then each potential environmental impact is evaluated with a mathematical model in an impact assessment phase and the results are finally interpreted (Fig. 20.3) to extract useful indications for policy. LCA follows ISO 14040:2006 and ISO 14044:2006.

To determine the different environmental impacts, a comparison of fossil-based materials and some "green material solutions" using LCA methodology has already been carried out. Glass fibers were substituted for natural fibers as matrix reinforcements in several studies to estimate the environmental benefits [79,80]. The results show that natural fibers are more environmentally friendly than glass fibers. In fact, the production phase has a lower impact on the process because natural fibers are derived from renewable sources and natural fiber extraction processes generally require lower energy inputs than glass fiber production. A product's utilization phase may also be important. Biocomposites reinforced with natural fibers, for example, are lighter than those reinforced with glass fibers, which can result in lower fuel consumption in automotive applications.

Beigbeder et al. [81] applied the LCA approach to the management of two emergent biocomposites wastes with the aim of predicting the best environmental practices for their EoL management. Polypropylene reinforced with wood flour and PLA reinforced with flax fibers were the biocomposites under consideration. Recycling had the lowest environmental impact for both biocomposites, followed by industrial composting for PLA reinforced with flax fibers composite and incineration for reinforced polypropylene with wood flour. The most significant scenarios are energy recovery and landfill. The benefits of recycled materials or energy production outweighed the negative effects of waste disposal processes. Aside from ozone depletion, the benefits of recycling were greater for PLA reinforced with flax fiber composites than for polypropylene reinforced with wood flour. Even when the effects of EoL treatment are conditioned by the geographical location of production, use, and waste management, the primary result is unaffected, and recycling is the most environmentally friendly solution in general. As a result, the results obtained for both biocomposites were in complete agreement with Europe's waste hierarchy, which prioritizes reuse and recycling as waste disposal options.

As previously stated, the inclusion of the technoeconomic, environmental, social, and sustainability assessments in the life cycle sustainability assessment for bio-based materials is required. The technoeconomic analysis (TEA) grant provides for quantitative and qualitative analysis of the financial viability and technical feasibility of the various stages of the life cycle of bio-based products, but it does not account for sustainability. TEA include the sustainability dimension via the Techno-Economic Sustainability Analysis (TESA) applied by Briassoulis et al. [74] for the postconsumer/industrial bio-based plastics organic recycling through anaerobic digestion and industrial aerobic composting. The TESA methodology for organic recycling was founded on three integrated TESA Criteria: "Economic viability," "Common techno-economic-environmental," and "Technical feasibility," which also serve as environmental sustainability assessment criteria in LCA. As a result, TESA is a very useful integrated methodology; however, the authors suggest that the TESA-based assessment be supplemented by an Environmental and Social Sustainability Assessment to complete the sustainability analysis of the resulting EoL option.

20.2.2 Tests and standardization

20.2.2.1 Biodegradation

The growing concern about the management of EoL bioplastics has resulted in the use of specific norms and standards to encourage recycling, final disposal, and reuse of bioplastics with the goal of minimizing resource use and adopting the best possible alternative for plastic final disposal. For the various EoL applications and systems, there are specific standards established by the most important International authorities such as the International Organization for Standardization (ISO), European Committee for Standardization (CEN), American Society for Testing and Materials International (ASTM), and Australian Standard (AS).

It should be noted that the standards can be divided into two categories: standards for determining the requirements that must have materials and test procedures that must be used to declare a biodegradable/compostable material (also known as specifications); and standards that are test methods only, describing the methods of proving themselves (e.g., biodegradation in various environmental conditions, composting and disintegration in different conditions, and so on) with respect to the materials.

The requirements established by different countries' standards differ in general (ISO, EN, ASTM, AS). In Australia, for example, the standard AS 4736 establishes requirements and methodologies for determining the compostability of plastics, or anaerobic biodegradation, through the effect on the quality of the resulting compost, biodegradability, effect on the biological treatment process, and disintegration during biological treatment. There are standard specifications for labeling plastics (ASTM D6400), and objects that include plastics and polymers as additives or coating with paper and other substrates (ASTM D6868) developed to be aerobically composted in municipal or industrial establishments exist in the United States of America. The globalization trend is pushing for the acceptance of ISO standards. On European (and international) level, the most relevant reference standards are reported in Table 20.2.

Table 20.2 List of standards concerning biodegradable materials.

Standard number	Title	Content
EN 13432	Packaging—Requirements for packaging recoverable through composting and biodegradation—Test scheme and evaluation criteria for the final acceptance of packaging	This standard is EU harmonized and one the most important regarding composting and biodegradation. It has been quoted in the Official Journals of the European Communities, has been implemented in Europe at national level and provided presumption of conformity with the European Directive on Packaging and Packaging Waste (94/62/EC).
EN 14045	Packaging—Evaluation of the disintegration of packaging materials in practical oriented tests under defined composting conditions	This standard is used to evaluate the disintegration of packaging materials in a pilot-scale aerobic composting test under defined conditions
EN ISO 14851 and 14852	Determination of the ultimate aerobic biodegradability of plastic materials in an aqueous medium;	This standard defines the methodologies for tests of aerobic biodegradability in water
EN ISO 14855	Determination of the ultimate aerobic biodegradability of plastic materials under controlled composting conditions	This methodology is based on evolved carbon dioxide analysis
EN 17033	Plastics—Biodegradable mulch films for use in agriculture and horticulture—Requirements and test methods	This document specifies the requirements for biodegradable films, manufactured from thermoplastic polymers, to be used for mulch applications in agriculture and horticulture;
EN 14995	Plastics—Evaluation of compostability—Test scheme and specifications	This document specifies the tests methodologies
EN 14046	Evaluation of the ultimate aerobic biodegradability and disintegration of packaging materials under controlled composting conditions; Method by analysis of released carbon dioxide;	More specific details for packaging than EN ISO 14855
ISO 18606	Packaging and the environment—Organic Recycling	General international standard
ISO 17088	Specifications for compostable plastics;	General international standard
ISO 16929	Determination of the degree of disintegration of plastic materials under defined composting conditions in a pilot-scale test	International standard for disintegration tests in compost

The compostability in "municipal or industrial biological waste treatment facilities," that is, industrial composting, is mentioned in the aforementioned standards. There is currently no international standard specifying the conditions for home composting of biodegradable plastics (though one is being developed in CEN/TC 261 SC4 WG2). Thus, unlike industrial composting, which is a controlled waste treatment option that operates under specific conditions and plants, home composting is still a gardening practice that citizens engage in on a voluntary basis. There are, however, several national standards, such as the Australian AS 5810 standard "Biodegradable plastics—biodegradable plastics suitable for home composting." In light of a potential European home compostability standard, it should be noted that determining homogeneous criteria is difficult due to different policies in different EU member states. A standard of this type will necessitate a definition of the characteristics and requirements of the typical home composting process. On the other hand, for developing such a standard, the already available standards or certification schemes for home compostability that are currently present in several EU member states can be very useful as a starting point. Furthermore, other existing standards for soil and water degradation can be mentioned. EN 17033—Plastics—Biodegradable mulch films for use in agriculture and horticulture; the biodegradation in soil can be requested for plastic materials used in agricultural operations, such as mulch films [82], because they generally remain on the soil after fruit collection. Concerning water degradation, the relative standard is ISO 14851 and 14852—Determination of the ultimate aerobic biodegradability of plastic materials in an aqueous medium; this latter is significantly important due to issues related to microplastics and nanoplastics, which are discussed in Section 20.1 [7]. As a result, the use of bioplastics that degrade in the sea, particularly in the marine environment, is currently being researched [83].

20.2.2.2 Biocompatibility

Another critical aspect of bionanocomposites' application is their compatibility with the body and impact on human health. In comparison to other sectors, the application of biomaterials in the medical field must consider strict standards, specific regulations, and various commercialization pathways. Indeed, government institutions controlled and regulated both the manufacture and sale of medical instruments in order to ensure high quality and safety standards. Standardized biocompatibility tests are performed prior to clinical evaluation of the device or, in the absence of a clinical trial, prior to product marketing authorization. A medical device or material that comes into contact with the human body is expected to function properly without causing any harm to the patient. Adverse effects, such as mutagenic effects, can be short-term (acute) or long-term (chronic). As a result, medical devices are subjected to biological tests and biocompatibility assessments to investigate the interaction of the biomedical device with the patient's specific tissue, cells, or body fluids. In fact, the primary goal of a device biocompatibility assessment is to protect patients from potential biological risks. The extent of the standardized biocompatibility testing will be determined by regulatory requirements based on the specific risk associated with the device's use. These recommendations, as reported in ISO 10993-1, reflect the perceived risk associated with the product's use. The most widely used standard for assessing the biocompatibility of medical devices and materials is ISO 10993-1, Biological Evaluation of Medical Devices—Part 1: Evaluation and Testing within a Risk Management Process. Many factors influence testing procedures, including the type of medical device or material and its intended use, as well as the nature and duration of contact between the medical device and the body.

In vitro tests for biomedical product validation must adhere to ISO 10993-5:2009 (Tests for in vitro cytotoxicity) and ISO 10993-10:2010. (Tests for irritation and skin sensitization). Because of the high cost of other classes of products and related materials, these tests are typically reserved for biological evaluation of medical devices. Hand cleansing wipes (for general hand cleaning), baby nappies, feminine hygiene products (tampons, sanitary pads), antiaging treatments (for example, beauty masks with a cosmetic purpose), and so on are examples of products that are not generally considered to be medical devices. However, the development and dissemination of new in vitro tests can aid in the better and safer screening of new (bio)materials to come into contact with our skin, as they are currently less regulated than those used in Medical Devices, but more widely distributed in quantity. Several standardized biocompatibility screening methods concentrate on the chemical toxicity of leachable components or degradation by-products associated with a specific product, taking into account not only the product composition but also exposure-based categories. Biocompatibility tests should be used more frequently to ensure that products are compatible with the cells and tissues of the body that will come into contact with the final product. In fact, in the future, both healthy and environmentally friendly materials should be used.

Human mesenchymal stromal/stem cells (MSCs), being immunoprivileged and showing immunomodulatory ability, represent a quite promising tool to be applied in the field of regenerative medicine [84], whereas immunomodulatory studies based on specific cells can be considered for human tissue-specific products. Immunomodulatory tests for skin-contact products, for example, are based on viability, and immunomodulatory tests on keratinocytes, which represent the cells of the epidermis, the skin's most external layer [52]. These tests were considered by Coltelli et al. [61] for studying bio-based and biodegradable innovative bionanocomposites for diapers top-sheets [25,28,85] and beauty masks [61,63,86,87] compatible with skin (Fig. 20.4).

20.2.3 Regulatory contest

Plastic disposal is a growing concern around the world because wastes made from nonbiodegradable polymers are persistent. Because of their short life cycle, plastic waste generated by packaging, sanitary or personal care products disposal is a major concern. Waste policy in most countries follows the so-called waste hierarchy concept, expressed by the Waste Framework Directive 2008/98/EC, 19 November 2008 [88], of the European Parliament that gives specific attention to prevention, reuse and recycling. For this policy, the best options (Fig. 20.5) are the reduction of waste by prevention of waste production, the recycling (organic or mechanical recycling), the energy recovering, and the disposal in landfill. Composting and anaerobic digestion are two organic recycling options. The waste framework legislation encourages the composting of organic waste in agriculture because it is the most environmentally friendly option. Furthermore, recycling in industries, which consists of reprocessing production discards and scraps, is highly recommended.

Recently, the European Commission has adopted a challenging and ambitious Circular Economy Package, that [89] sets clear objectives for reducing waste. Furthermore, it establishes a credible long-term waste management and recycling program. Significant key elements of this proposal include a shared goal of recycling 65% of municipal waste in the EU by 2030, as well as a shared goal of recycling 75% of packaging waste by 2030. Furthermore, landfilling should be limited to no more than 10% of municipal waste by 2030, and landfilling of separately collected waste will be prohibited. Economic instruments will also be used to discourage landfilling. Surprisingly,

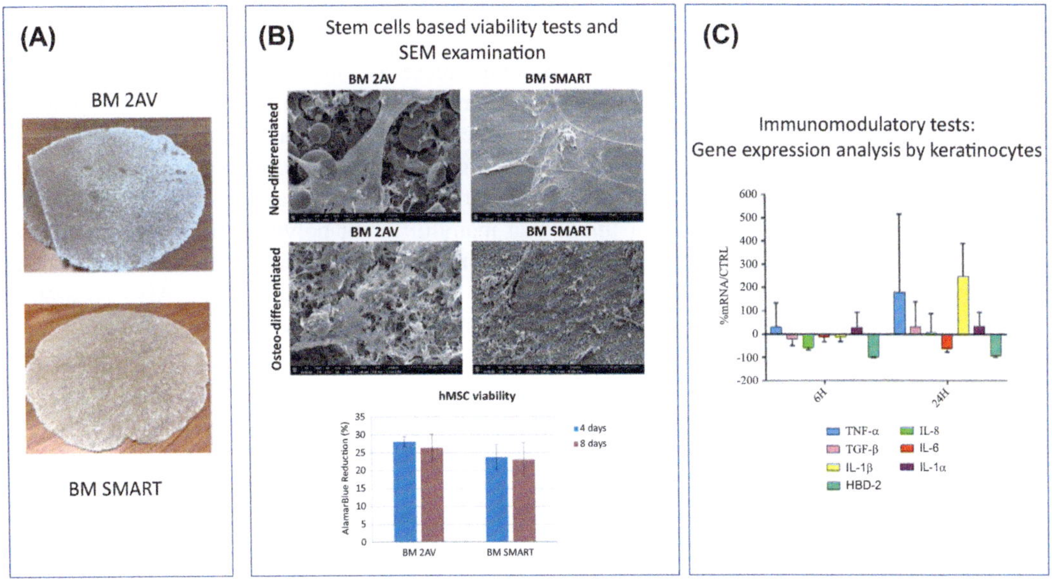

FIGURE 20.4

(A) Samples of films for beauty masks; (B) Scanning Electron Microscopy (SEM) examination of Stem cells growth (in particular nondifferentiated and osteo-differentiated human mesenchymal stromal cells) on the samples and their viability after 4 and 8 days; (C) immunomodulatory tests using keratinocytes and observing their gene expression [61].

Reproduced with permission from MDPI. Copyright 2018.

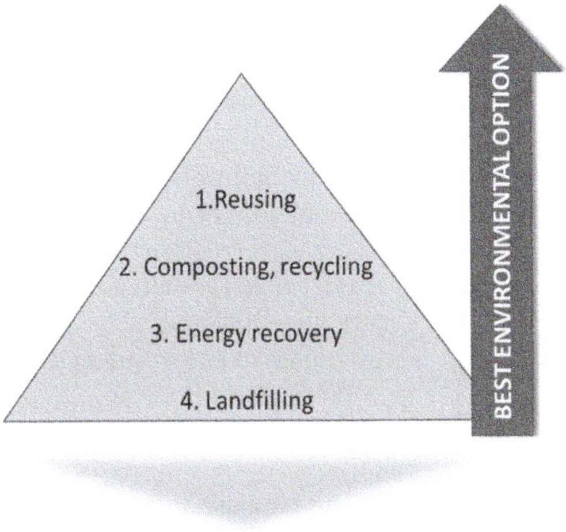

FIGURE 20.5

Priority scale of waste management (Waste Framework Directive 2008/98/EC).

simplified and improved definitions, as well as harmonized calculation methods for recycling rates across the EU, will be specified. Measures to encourage reuse and industrial symbiosis will be promoted, transforming one industry's by-product into another industry's raw material; the program includes economic incentives for greener product producers as well as economic support for recovery and recycling chains (e.g., for batteries, vehicles, packaging, electric and electronic equipment).

When considering material recycling, it is important to remember that plastic materials are the most difficult to manage. The reason for this is that many different polymers are currently used in packaging to meet the specific requirements for different foods and beverages. The most commonly used materials are PE, PP, PET, and PS, but multilayer packaging is also popular. After separating the different polymers, these fossil plastic materials can be recycled. Each polymer could theoretically be recycled by reprocessing it at high temperatures to produce a viscous melt. Nonetheless, if the different polymers are not first separated, polymeric blends with poor performance will result. This is due to the different polymers' immiscibility and incompatibility, which results in insufficient interfacial adhesion between the different phases present in the material.

As a result, there is a growing interest among both producers and consumers in the use of bioplastic. In the last two decades, scientific research efforts toward the synthesis, design, and production of sustainable or green materials have grown exponentially [90]. Bioplastics currently account for a small percentage of the total market of 359 million tons of plastic produced annually. This percentage may rise in the coming years as a result of regulatory actions taken by several countries. Bioeconomy policies should include these actions. A comprehensive survey of the EU's strategy and actions to promote policies for a sustainable, circular bioeconomy is reported in the 2018 document A Sustainable Bioeconomy for Europe: Strengthening the Connection Between Economy, Society, and the Environment [91].

The circular economy's renewable component is sustainable bioeconomy. It should encourage the conversion of by-products, bio-waste, residues, and discards into valuable products, thereby stimulating innovation. It would also assist retailers and consumers in reducing food waste by 50% by 2030.

Five main objectives were focused, and they are summarized below:

1. The first objective, aimed at granting food safety and security, is fundamental. In fact, food and farming systems represent about three quarters of the overall bioeconomy jobs.
2. The second objective, focused on managing natural resources in a sustainable way, is strategic because of the current context characterized by loss of biodiversity and increase in environmental pressure.
3. The third objective, aimed at reducing the dependence on nonrenewable, unsustainable resources, is vital to reach the EU's energy and climate targets. In fact, bioenergy, currently representing the largest renewable energy source in EU, is predicted to remain a pivotal component of the energy mix in the near future.
4. The fourth objective, focused on mitigating and adapting to climate change, is considered the global challenge of this generation.
5. The fifth objective, aimed at creating jobs thus spreading a diffused well-being, is a core policy goal of bio economy.

Waste management is generally organized at the regional level, but related policies are promoted and supervised at the national or supranational level by more general directives. Better waste management, taking advantage of the potentialities provided by bionanocomposites, should include the ability to properly separate and dispose of bio-based products at the end of their useful life, using composting or anaerobic digestion practices.

This will necessitate improved bio-based waste collection, which should be linked to appropriate policy and legislative actions, and could also promote material recycling routes. Policymakers should keep in mind that in the near future, both fossil-based durable materials and bio-based biodegradable materials will coexist. Furthermore, it is critical to educate the general public and raise awareness about better waste management of bio-based and biodegradable products in order to prevent leakages from waste streams, which lead to less environmentally and economically favorable disposal routes. Promoting research, improving young people's skills, and raising citizen awareness will be the best investments for a successful implementation of bioeconomy and circular economy practices in the near future.

20.2.4 Bioeconomy and circular economy

The weight of living organisms (including microorganisms, plants, and animals) present in a given area or volume of the Earth's surface is referred to as biomass. Biomass forms the basic food-chain representing the pyramids of the terrestrial and aquatic ecosystems (Fig. 20.6) [92,93].

As a result, harnessing biomass from agricultural-forestry and aquatic feedstocks via waste streams and new alternative sources should be considered at the heart of the bioeconomy. As a result, their acquisition and processing are critical, posing a challenge and competition for their various applications, with implications for rising biomass prices, efficiency of use, transportation, and costs. As a result, new biotechnologies are critical for increasing efficiency and lowering production costs by developing, for example, new bacteria-based ingredients or innovative

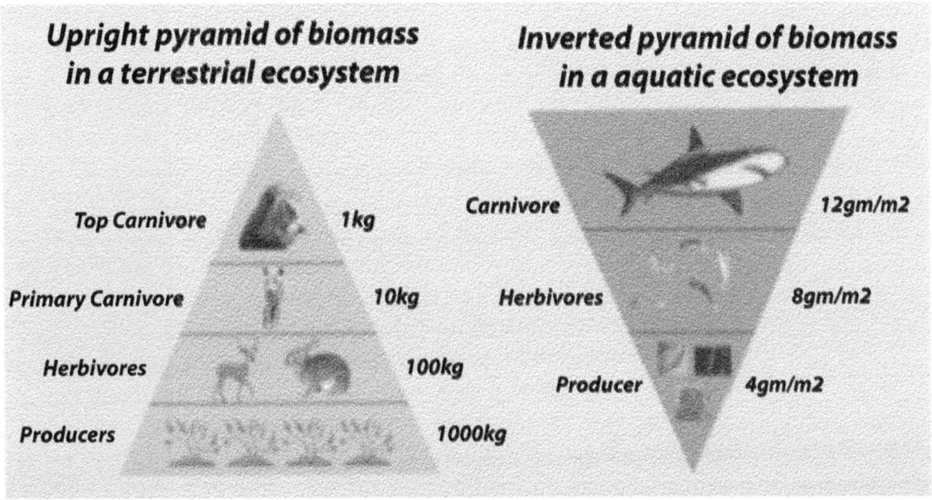

FIGURE 20.6

Pyramids of biomass.

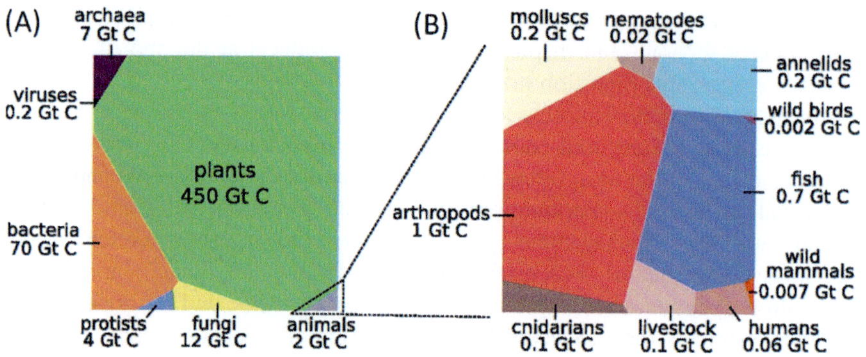

FIGURE 20.7

Graphical representation of the distribution of the global biomass by taxa. (A) Absolute biomasses of different taxa are represented with the area of each cell being proportional to that taxa global biomass (the shape of each polygon has no meaning) [92]. (B) Part related to animals.

Reproduced with permission from PNAS. Copyright 2018.

biodegradable materials for packaging and industrial products that are not only highly performant and esthetically appealing, but also environmentally friendly. As reported elsewhere "the main aim of current policy is to keep the environment in orbit with economics at the center, not considering nature or Environment as simple commodities" [94] "Preserving planet Earth's biodiversity is crucial in order to keep its ecosystems in equilibrium" [94]. In fact, the greater the number of species in an ecosystem, the more stable that ecosystem. This is due to the fact that all living organisms, including plants, animals, and microorganisms, are required for human survival. Just to remember, the biomass on Earth distributed among all the kingdom life is established to represent around 550 gigatons of carbon (Gt C), of which ~ 450 Gt C in plants, 2 Gt C in animals (prevalently marine), 70 Gt C in bacteria and 7 Gt C in archea (Fig. 20.7) [92].

Agricultural activities occupy half of the European Union's land area but employ only 4.5% of the EU-28 working population. These workers on the one hand supply economy with essential food, feed products and energy, while on the other hand produce waste materials (EUROSTAT) [95], like all the productive chains do. Thus, the total EU agricultural biomass was estimated to produce 956 Megatons (Mt) of dry matter per year on average from 2006 to 2015, of which 54% were primary products (grain, fruits, roots, etc.) and the remaining 46% were secondary products and waste material, also caused by adverse weather [95,96]. Forestry, on the other hand, occupies 38% of EU land and, with 85% as forest available for wood supply, provides a wide range of ecosystem services such as allowing tourism/sport, providing high quality air, granting carbon storage and sequestration, supporting water regulation, and erosion control [95]. Furthermore, biomass from fisheries and aquaculture, as well as algae, are important sources of food and useful waste that support the livelihoods of hundreds of millions of people worldwide, which is expected to reach 9.7 billion by 2050 with an increase in food demand of 25%–70% [97]. As a result, we must adapt our needs to climate change and use food by-products to create innovative goods with low environmental impact using sustainable technologies. So doing, it will be possible safeguard the

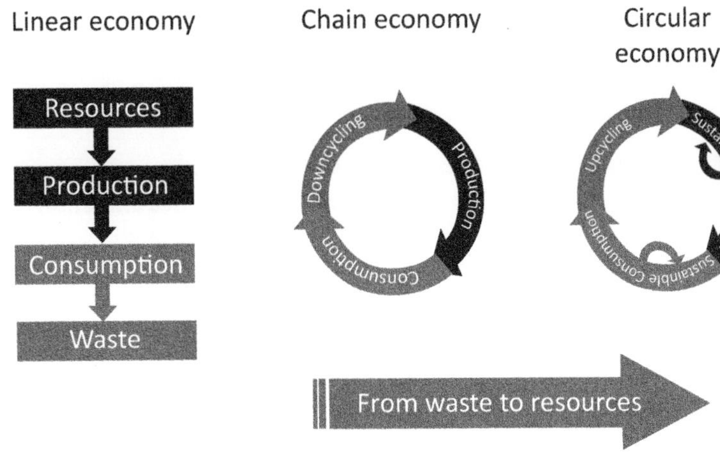

FIGURE 20.8

Linear economy versus circular economy schemes.

natural raw materials for the future generations, transforming the linear economy, based on the "take, make and waste" principles, in a circular economy based on "reduce, reuse and recycle" approaches (Fig. 20.8) [96,98]. Nonetheless, an upcycling strategy should be implemented, and primary production processes should be converted to more sustainable ones. In fact, a chain economy is preferable to a linear economy, but in this case, waste is not well valorized in order to obtain high-value products with a long life. As a result, waste production is only slightly delayed. Thus, in a circular economy approach, production processes should be optimized to limit and eventually valorize the produced by-products.

For example, the global population generated 14% of municipal waste (i.e., 2.1 billion tons per year) in 2019, and this waste is expected to increase by 70% (3.5 billion tons) by 2050. If urgent policy actions will be not implemented [96], atmospheric greenhouse gas (GHG) emissions, will be higher than they have been in 800,000 years, resulting in the planet warning and extinction of 1 million species, also because 75% of the 12 Earth surface areas and 66% of marine areas have been altered from human communities [96]. Moreover every year, air and pollution causes about 9 million premature deaths (Fig. 20.9), because of our unsustainable consumption and production of goods which, too often not necessary, release a great quantity of industrial waste, including up to 40 million tons of heavy metals, solvent, toxic sewage sludge, etc.[98,99]. Furthermore, the current COVID-19 pandemic, which is rapidly and easily spreading throughout the world, has increased production and consumption of nonbiodegradable plastic waste due to the widespread use of surgery masks and sanitary personnel monouse dressings and garments [100–103]. At this point, it is critical to emphasize that this pandemic is "not a one-time event, but an omen of what is to come if we do not finally heed the many warnings the planet has been sending" [95]. As a result, we must transform our economies and societies by taking real and meaningful action to stabilize the climate, reduce pollution, and protect the environment in order to create a world of prosperity, equality, and peace.

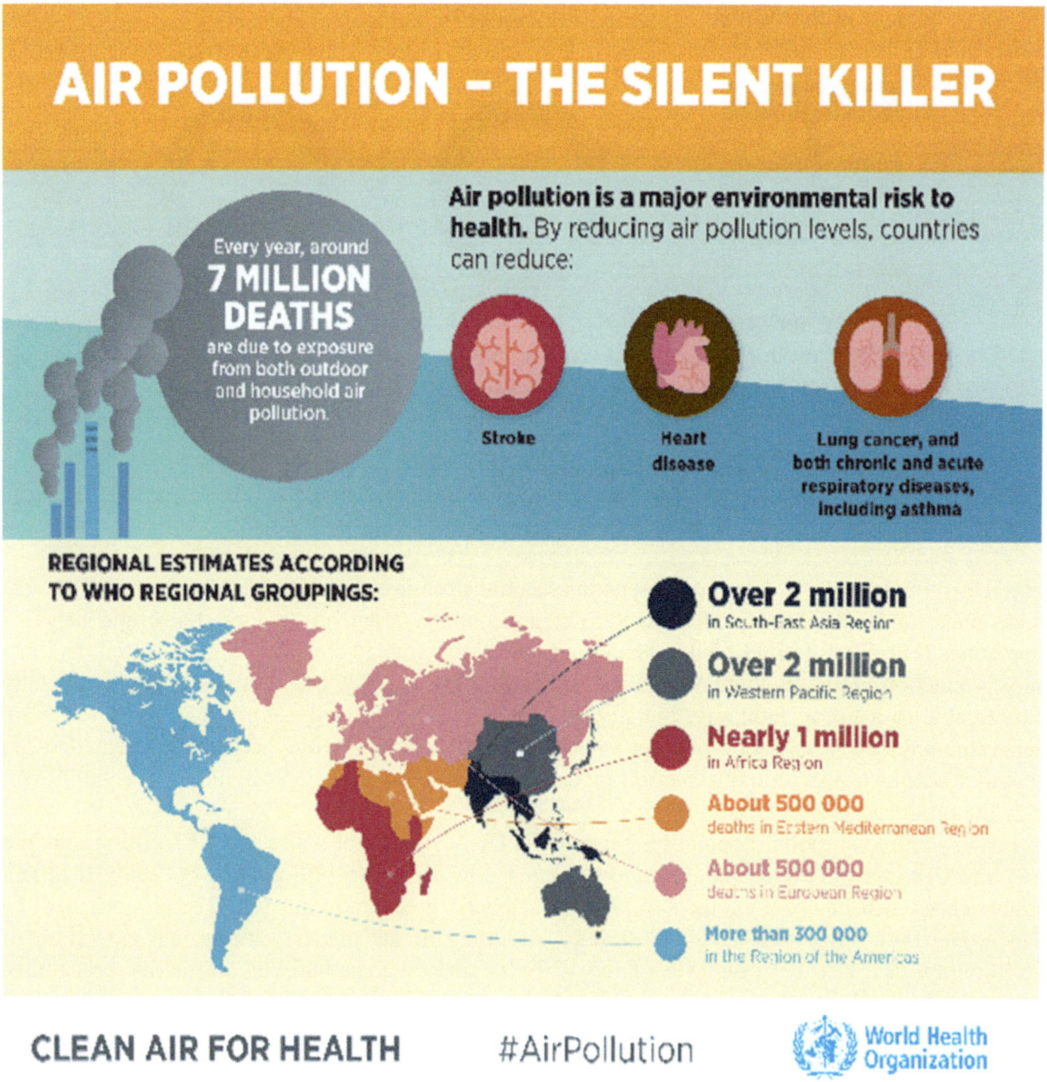

FIGURE 20.9

Data about air pollution effect provide by World Health Organization [104].

Reproduced with permission from WHO. Copyright 2010.

It is therefore important to remember that increased consumption of food, cosmetics, clothing, and other goods increases the use of packaging materials, which are predominantly made of nonbiodegradable plastic containers. As a result, for every kilogram of plastic that is not reused or recycled, another kilogram is manufactured using nonrenewable raw materials [100,105]. In fact, the current supply of recycled plastics only covers about 6% of total plastic consumption [98,99].

For all of these reasons, it is time to alter our way of life by investing more in waste management methods to recover materials and chemicals from new productive technologies, thereby protecting and safeguarding both our health and the environment. In fact, scientists and consumers are pressing for a new way of producing and consuming. They claim that encouraging compostable packaging will divert waste from landfills by generating compost from food waste. In doing so, it would return organic matter to the soil, as is currently happening in the UK's compostable packaging market, which is expected to grow tenfold by 2025 [15]. Unfortunately, because of their low cost and the difficulty of biodegradable alternatives competing with the well-established refineries' products, the majority of plastics used in packaging are still derived from petrochemical sources [105–107]. Plastics, in general, have become an integral part of the global economy, providing numerous benefits as a key component in a wide range of sectors ranging from transportation to electronics, apparel, and health. Plastics used in packaging help to extend the shelf life of food, reducing food waste. A significant disadvantage is the short life cycle of food packaging, which results in a high waste production. As a result, 95% of plastics are used and not recycled each year, resulting in a global economic loss of 80–120 billion dollars [107,108]. Plastic production, on the other hand, has increased from 15 million tons in 1964 to 348 million tons in 2017 and is expected to quadruple by 2050, representing a significant source of wealth and jobs worldwide [107,108].

However, it is critical to replace fossil-based plastic with so-called bioplastic, which is produced using a variety of methodologies ranging from high maturity to piloting, using biopolymers such as polysaccharides that already exist in nature and are synthesized by plants or microorganisms. These biopolymers, as reported in Section 20.1.2, biodegrade rapidly under natural conditions and may be obtained from food, forestry and sea waste [109]. Finally, "appropriate natural resource management is a critical first step toward reducing environmental footprints and promoting a healthy bioeconomy" [110]. It will be necessary to adopt a new way of life, keeping in mind that "waste depends on composition and volume, which are heavily influenced by both consumption patterns and the industrial and economic structures in place" [110,111]. As a result, it is critical to understand that by implementing a global circular bio economy, it will finally be possible to use current waste ingredients generated by agricultural and industrial activities to produce all necessary goods, significantly reducing the use of fossil plastics and their leakage into natural systems. Finally, "a revolution in the plastic industrial ecosystem will be required, mobilizing producers of plastic packaging, companies producing consumer goods, and plastic manufacturers." Businesses involved in collection, sorting, and reprocessing, as well as policymakers, municipalities, and all consumers, should all be involved in achieving this goal [109].

20.3 Recycling strategies for biodegradable polymer nanocomposites

20.3.1 Waste management contest

Regarding plastic product waste management, the priority scale based on European Law consists of reuse, which involves using the product or package again, and recycling, which involves using waste material to make a new product. On 2017 [112], the composting was considered by EU a method of waste management identical to recycling and was defined "organic recycling."

This demonstrates the EU's intention to invest in composting waste and promote the use of bioplastics. Hence two possibilities for recycling bioplastics are possible:

1. the recycling consisting in transforming the material in another product by grinding the product and processing it again or obtaining compounds required by the market. This process is possible for bioplastics as well as for petrochemical plastics, although conditions and methodologies should be controlled properly, and many researchers are studying what and how bioplastics can be better recyclable in an industrial process;
2. the composting of bioplastics; many bioplastics yet on the market are compostable in industrial composting plant. Currently researches about the composting in other environments such as home composting or marine environment are considered.

In agreement with these principles Briassoulis et al. [74] proposed the scheme reported in Fig. 20.10 for the waste management of bio-based plastics, where the material recovery and recycling are considered for biodegradable and industrially compostable materials, but also the energy recovery is considered. The scheme demonstrates the importance of keeping nonbiodegradable and

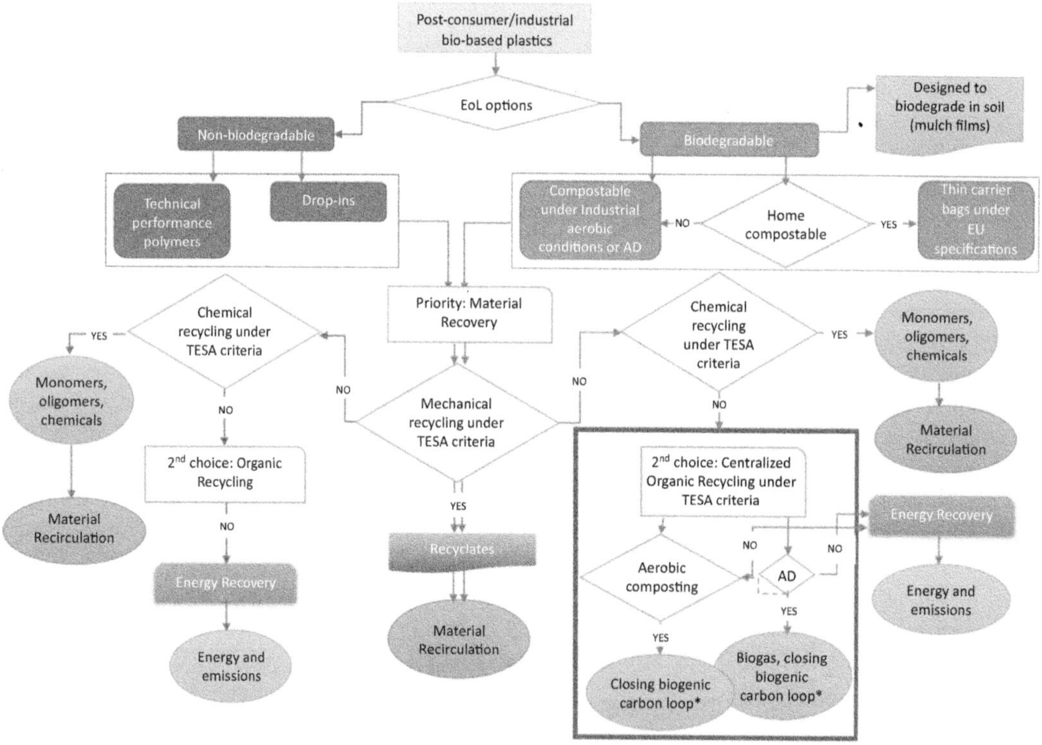

FIGURE 20.10

Scheme regarding waste management of postconsumer/industrial bio-based plastics [74].

Reproduced with permission from Elsevier. Copyright 2021.

biodegradable materials flows separate. In fact, in the case of bio-based and biodegradable plastics, anaerobic digestion, rather than incineration, can be used, which is suitable for bio-based but not biodegradable bioplastics.

Other waste management options include energy recovery and, as a last resort, landfill. It should be noted that bioplastics, like traditional petrochemicals, can be incinerated in advanced plants to generate energy, and if landfilled, they contribute to the production of biogas, which can be recovered in landfill plants and converted into energy. However, GHG emissions occur during biodegradation, so completely converting waste management to composting would result in an increase in emissions. More research into recycling bio-based materials should be conducted in order to build effective new industrial chains, thereby implementing circular bioeconomy concepts. Bio-based and biodegradable plastics should be recycled as many times as possible before being effectively composted when recycling is no longer possible. This approach will extend the life of the material, avoiding the need for new material and deferring as much as possible the GHG emissions caused by material degradation.

20.3.2 Recycling methodologies

The most common recycling classification is based on the type of process, specifically whether it occurs through molten transformation (mechanical recycling) using conventional techniques for processing plastics or reactive (chemical recycling) for obtaining chemicals. Because it involves the simple reprocessing of postindustrial or postconsumer materials, mechanical recycling is the most common. Mechanical recycling is generally preferred when possible because it allows for the recovery of the maximum value contained in the waste. Cosate de Andrade et al. [113] have demonstrated it by LCA studies considering the mechanical recycling of the bio-based and biodegradable PLA and comparing it with chemical recycling and composting.

The last classification of recycling technologies and methods is based on the value of recycled products compared to the original ones (Fig. 20.11). When the final applications are less valuable than the initial ones, we refer to open loop recycling, downcycling, or cascade recycling. Mixed plastics from postconsumer films, for example, are used to make large-scale objects such as profiles for urban furniture. Due to the numerous problems of incompatibility between polymers or different materials, this type of recycling is very common in the field of plastic materials.

When a postconsumer product is recycled in the same process, the product's initial and final values are the same. We discuss closed loop recycling. In some countries, PET bottles made from postconsumer PET bottles can be obtained. The procedure is commonly referred to as bottle to bottle. Then there's the case, which is quite uncommon in the field of plastic materials, where other materials with superior performances are obtained from the recovered materials, making them suitable for higher-value applications as well. In this case we talk about upcycling. An example are shock-proof materials suitable for mountain helmets made with postconsumer bottle flakes and suitable additives [114–119], or the car presented at the Detroit show in 2008, made with a material obtained from the flakes of bottles [120]. It is obvious that chemical recycling, because of its ability to provide intermediate or useful reagents, is the one that can most easily lead to upcycling, but there are drawbacks related to plant engineering issues, and thus to the overall cost of recycling operations.

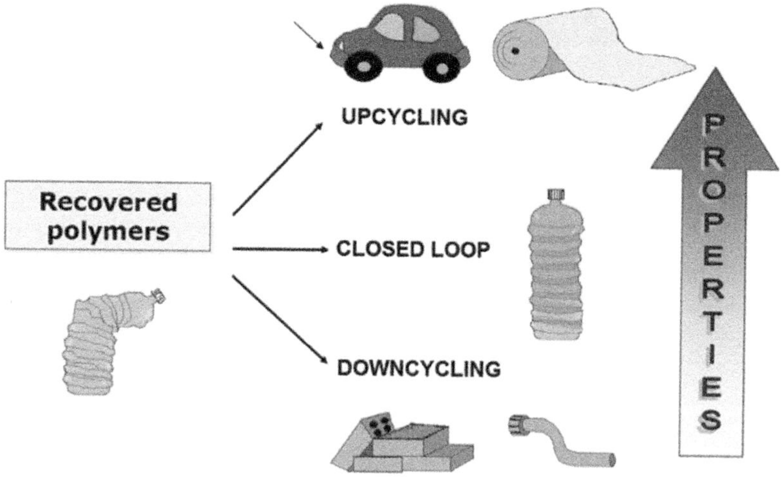

FIGURE 20.11

Scheme about the concepts of upcycling, closed loop and downcycling.

20.3.3 Technical issues to overcome

The separation of various plastics is a critical issue to address [121–124]. In the case of bioplastics, some main streams should be chosen for the construction of proper selection plants that are organized to regroup materials with similar compositions. This is difficult to organize and manage because more industrial case studies in the bioplastics recycling sector are needed to assess specific and diverse recycling routes, and because systematic collection of bioplastics should be integrated with current waste management of urban waste. This latter is for fossil plastic waste and requires citizen participation. To organize such waste bioplastics collection, political, social, educational, and waste management actions should be taken.

The first phase of recycling is the identification and classification of plastic waste [124], therefore suitable equipment is required for the rapid sorting of the various polymers. Researchers all over the world are investigating municipal waste sorting techniques, and this research has resulted in a wide range of technologies in the automated sorting phase. There are two kinds of automated sorting techniques: Sorting, both direct and indirect [125]. Direct sorting techniques use intrinsic material properties such as magnetic susceptibility, electrical conductivity, and density to separate materials while applying external forces such as magnetic fields, eddy currents, and gravity. In contrast, indirect sorting employs sensors to detect the presence and location of recyclable materials in waste, allowing automated machines to sort and sort identified recyclables. The indirect selection methods, in which sensors are used to detect recyclable materials present in waste and allow for subsequent selection, are more important. Indirect selection methods are best suited to the identification of plastic polymers [125].

Table 20.3 Issues and solutions to control the mechanical behavior of a recycled plastic material.

Issue	Solution
Material with polymeric contaminants	(Reactive) compatibilization during extrusion
High viscosity	Thinners or plasticizers
Low viscosity	Chain extenders of branching agents during extrusion
Low elastic modulus	Extrusion of reinforced composites: use of fillers or fibers having micro/nanodimensions
Low impact properties	Addition of toughening agent
Excessive rigidity and brittleness	Addition of plasticizer or toughening agent

When developing materials for recycling, it is frequently necessary to understand how to modify the properties of a material through the use of specific additives. For this purpose, some points are given in Table 20.3. A solution in terms of possible additives to be added is reported for each technological problem concerning the material. However, we must keep in mind that the addition of an additive frequently affects multiple properties rather than just one. For example, the addition of plasticizers can reduce viscosity in the molten state while also lowering the elastic modulus and increasing elongation at break.

In addition to modulating the properties of the material, additives are frequently required to modulate the processability of the system. The use of crystallization nucleating agents, for example, enables the crystallization process to be accelerated and the degree of crystallinity obtained in a specific process to be better controlled. This type of additive can be useful in the setup of semicrystalline polymer injection molding, both to optimize the time of the molding cycle and to ensure proper dimensional stability during extraction and subsequent use.

Antioxidants and anti-UV agents, for example, can make the material more stable, while dyes can impart desired esthetic properties. Many recycled materials, especially those of postconsumer origin, are not white, so it is critical to use colors, often dark, that can even out the color. The color, which is used in less than 1% concentrations, is dispersed in the polymeric material beginning with a master batch, which is a much more concentrated compound containing the dye. To avoid incompatibility issues, it is critical to consider the dye's miscibility in the polymer and to select master batches with the same polymer matrix as the final compound.

Bionanocomposites are complex materials made up of a polymer matrix and a dispersed phase. As a result, their recyclability should be thoroughly investigated. Nanofillers are typically present in nanocomposites in small percentages. As a result, it is critical that a nanofiller does not significantly alter the processability or properties of matrix polymers. During reprocessing, nanofillers should not degrade significantly and should not induce or promote degradation pathways in the polymeric matrix. Because the surface-to-volume ratio of these nanofillers is generally very high, and reactive groups can be present on the nanofiller surface, these processes are difficult to control [126]. In some cases, the reinforcement of the melt or the increase in elastic modulus due to the addition of nanofillers can be beneficial for properties improvement [127] (Table 20.2).

20.4 Bionanocomposites incorporating chitin nanofibrils

20.4.1 Production of chitin nanofibrils

Chitin is the most abundant polymer recovered in nature after cellulose but differently from cellulose it is rich of nitrogen, as it contains a $-$NH-CO-CH_3 group for each repeating unit. It is a polysaccharide which is present in arthropods, mollusks and insects to support their structure, surrounds also the plasma membranes of plant, algae and fungi (Fig. 20.12A).

Chitin is formed in nature by a linear chain of *N*-acetyl-D-glucosamine-repeating units that, when industrially produced as nanofibrils (CNs), contain a molecule of D-glucosamine groups, thus performing a function similar to that of its deacetylated compound chitosan. In fact, CNs are distinguished by a glucosamine (GL) content ranging from 50% to 60%, as opposed to chitosan's 70%–98% content [128]. Naturally, the GL content of the two polymers is affected by the manufacturing method. Chitosan and CNs, on the other hand, exhibit desirable biological and chemical properties such as renewability, biodegradability, biocompatibility, and nontoxicity. They also demonstrated unique adsorption properties that could be used in food, cosmetic, and biomedical applications [129,130]. Natural chitin is an antiparallel molecular chain alignment made up of molecular microfibers that combine to form nanofibrils.

In an industrial process, the so-called chitin nanofibrils are obtained in two steps: the first step involves subjecting the cleaned crustaceans' exoskeleton to various treatments such as alkali to remove protein, acids to remove organic minerals, and bleaching agents to remove pigments and lipids, all while stirring at around 90°C for various hours. By the second step the obtained pure chitin is treated again by acid hydrolysis and mechanical stirring to obtain chitin nanofibrils (CNs), resulting as needles 240 nm long and 5–7 nm in diameter (Fig. 20.12B) [130]. It is worth noting that industrial chitin nanofibrils have unique optical, mechanical, and biological properties due to their high surface-to-volume ratio, retaining not only the structure of the original chitin but also the

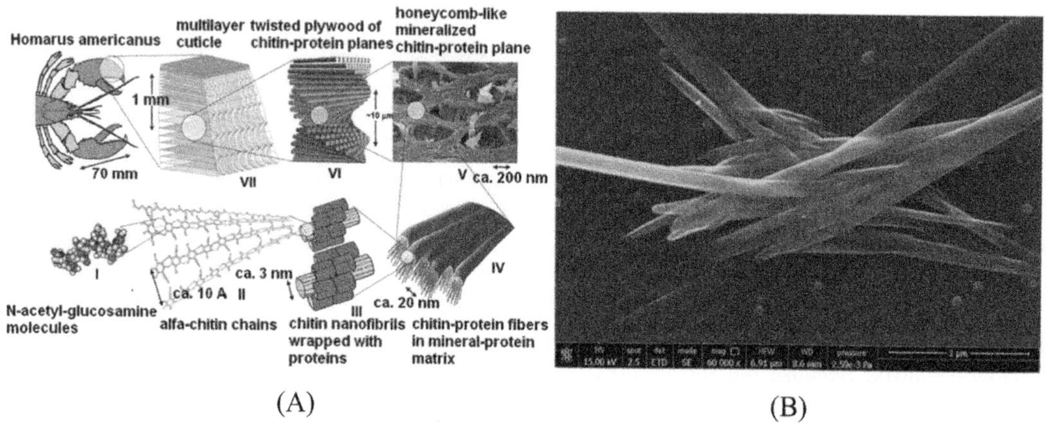

FIGURE 20.12

(A) Structure of chitin; (B) Chitin nanofibrils image obtained by FESEM.

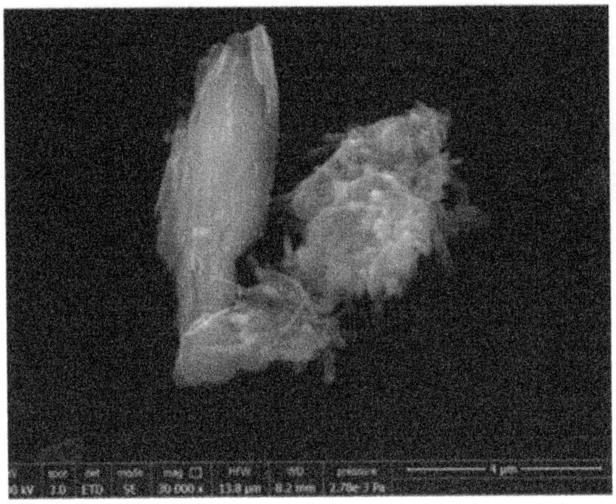

(A)

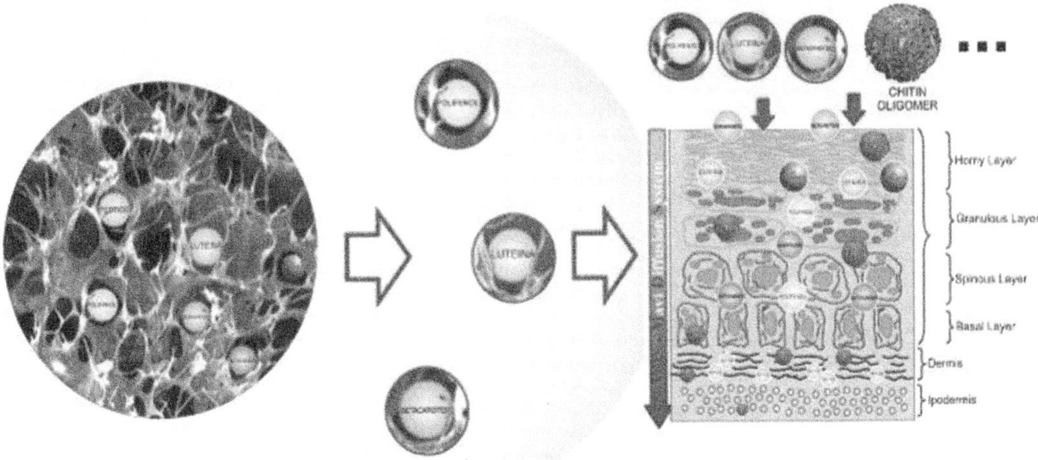

(B)

FIGURE 20.13

(A) Chitin nanofibrils-nanolignin complex image obtained by FESEM; (B) integration of complexes in the skin layers.

high surface area, low density, and many functional groups useful for further functionalization of the polymer. Moreover, the around 300 trillion nanocrystals per milliliter obtained by the productive process, remain stably suspended in water because enveloped by a cloud of water molecules (Fig. 20.13).

CN results are critical for producing micrometric nanostructured particles that are effective and safe. These particles enable the selective and effective presence and dose of the encapsulated

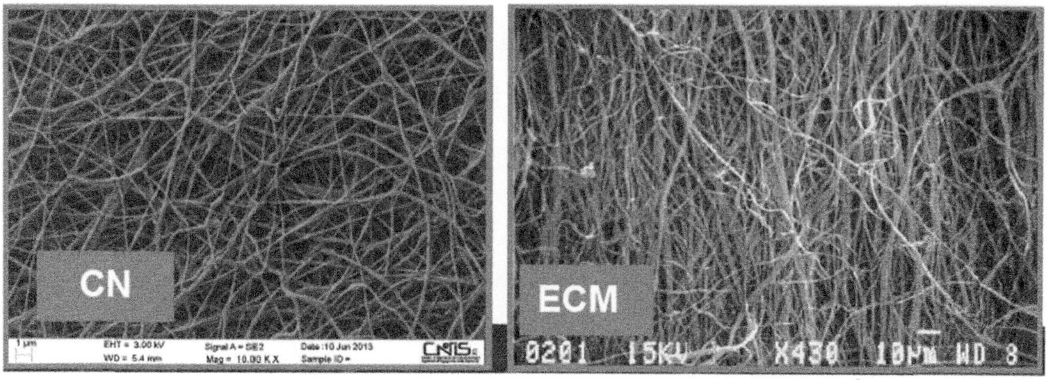

(A) (B)

FIGURE 20.14

(A) Electrospun tissues of CN; (B) natural extracellular matrix.

active ingredients. At this purpose it is possible to produce chitin nanofibril-hyaluronic acid and chitin nanofibril-lignin complexes with modulated size, width distribution and zeta potential containing functionalizing molecules, including nicotinamide [64], glycirrhetic acid [57], and vitamin E [86] and other active ingredients. These complexes can be tested on Polylactic films for thermal stability and skin compatibility using human keratinocyte cultures. Thus, the possibility of using these various complexes to investigate their potential use in beauty masks, diapers, and wound dressing has been verified [26,28,64,131–133]. Finally, various techniques have been used to validate the numerous potential applications of CN and its complexes, which, when electrospun or organized as a gel, have demonstrated a high level of similarity with the natural extracellular matrix (ECM) for the enhancement of various cellular functions (Fig. 20.14). As a result, these sugar-like natural polymers may be of great interest in developing advanced medications for repairing skin damaged by wounds or burns, as well as innovative cosmeceutical-tissues for slowing the aging process of the skin.

20.4.2 Properties of chitin nanofibrils reinforced bionanocomposites

CNs can be used to make bionanocomposites by combining them with biodegradable matrices. Materials based on PLA and CNs, in particular, show promise for producing bionanocomposites with high performance, taking into account both the reinforcing action of nanofibrils and their antimicrobial properties [134]. Furthermore, the production of chitin nanofibrils requires less energy intensive treatments than the production of cellulose nanofibers [135]. However, it has been demonstrated that combining CNs with cellulose nanofibers to prepare bionanocomposites is synergistic [136]. The preparation by extrusion of nanocomposites consisting of a PLA matrix and a dispersed CN phase was obtained by Coltelli et al. [137] using a methodology based on the preliminary preparation of precomposites based on hydrosuspendable poly(ethylene glycol) (PEG) and CNs. In a laboratory extruder, these precomposites were successively dispersed in PLA in the melt state to produce bionanocomposites with variable CNs and PEG contents. The morphologic investigation using scanning electron microscopy (SEM) and transmission electron microscopy (TEM) revealed that there were no agglomerated CNs in the sample. As a result, a nanostructured morphology was obtained. As a result, the transparency of these bionanocomposites agreed with electron microscopy findings.

The addition of CNs to the samples had little effect on the thermal properties, determining only a minor nucleating action. The determination of tensile properties allowed us to rule out a significant reinforcing effect of CNs. However, the achieved nanodispersion allowed to obtain consistent values of elongation at break (>150%), typical of plasticized PLA, up to a content of CN filler of 12% by weight. The lack of a reinforcing effect was explained by the presence of PEG between the PLA matrix and the surface of chitin nanofibrils. As a result, it was hypothesized that PEG acted like a coating on the CN surface, reducing the reinforcing action of nanofibrils. Plasticizers are commonly used to produce PLA-based films, but they can migrate from the film, causing surface oiliness and changes in mechanical properties. Aliotta et al. [24], demonstrated that in plasticized PLA incorporating CNs, it was found that CNs slowed the loss of a citrate used as plasticizer from the bionanocomposite (Fig. 20.15A). Regarding the use of CNs in bionanocomposites films, it was evidenced an indirect antimicrobial action, determining the production of antimicrobial peptides of cells of skin (keratinocytes) (Fig. 20.15B). The biocompatibility of CN-containing films (or CNs and calcium carbonate) was studied. The films were biocompatible because they had no effect on the viability of the cells. Furthermore, they were found to be capable of stimulating the production of cell defensins [28]. When CNs were used as coating formulations on substrates such as cellulose, it was discovered that they had an effective antimicrobial action [138].

20.4.3 Environmental and recyclability considerations

Polysaccharide addition to biopolyesters generally promotes chain scission, resulting in a decrease in MW and, as a result, a decrease in properties. This is due to the action of –OH groups on polysaccharides as well as water, which is generally in equilibrium with the hygroscopic polysaccharides and determines ester scission via hydrolysis. The high surface/volume ratio of nanofillers makes these reactions more effective. Vannozzi et al. [28] prepared bionanocomposites showing properties suitable for preparing films by flat die extrusion. Plasticized poly (lactic acid) (PLA), poly(butylene succinate-*co*-adipate) (PBSA), and CNs were used as functional fillers in these films. The CNs were evenly distributed throughout the blend. This result was obtained by first preparing a polymer-containing prenanocomposite (ethylene glycol). It was demonstrated that the addition of CNs and starch was not beneficial. The addition of starch caused extensive chain scission, resulting in extremely high melt fluidity values. As a result, flat die extrusion cannot be used to process the obtained blend. The melt viscosity could be better controlled when only CNs were added. As a result, CNs could be successfully incorporated in these PLA-based blends, which are more flexible than pure PLA and thus suitable for films to be used in the sanitary and personal care sectors due to their biocompatibility.

The use of proper melt strength enhancer and chain extenders, that determine the formation of new bonds thanks to the reaction of terminal groups of polymers and thus increase the polymer MW, can be fundamental (Table 20.2). There are currently no bio-based chain extenders on the market. However, alternative bio-based chain extenders were extensively investigated by Coltelli

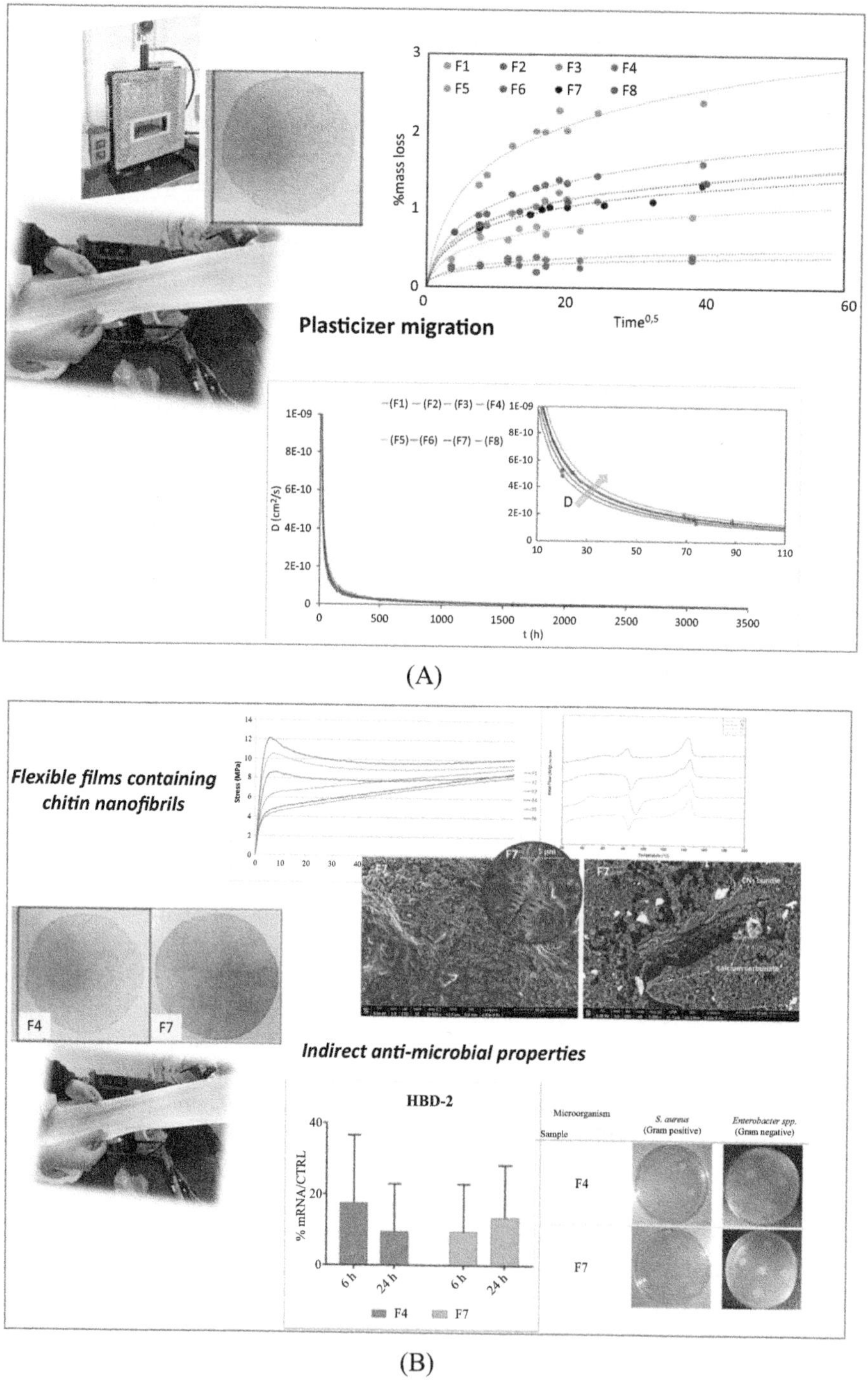

FIGURE 20.15

(A) migration curves and diffusion coefficient of bionanonocomposites performed on PLA-based bionanocomposites incorporating calcium carbonate or CNs [25]; (B) tensile tests, SEM micrographs and biocompatibility tests performed on PLA-based bionanocomposites incorporating CNs [28].

Reproduced with permission from MDPI. Copyright 2020.

et al. [22] based on vegetable oils and bio-based dicarboxylic acids, like malic and succinic acids. These additives have the potential to improve the recyclability of biodegradable materials without jeopardizing this property. The presence of nonbiodegradable polymeric additives in biodegradable materials, in fact, can influence the production of persistent microplastics.

20.5 Conclusions

Because of their modulable functional properties, biodegradable bionanocomposites are promising materials. In terms of environmental impact, it is critical to investigate their effects using a LCA approach, which takes into account all productive steps and the entire life cycle, from "cradle to cradle." Because it incorporates all of the environmental and economic aspects involved in a material's full life cycle, TESA can be an additional useful tool in the development of new materials. Their biodegradability in specific environments should be assessed using consolidated standards, as should their biocompatibility, which can be controlled using in vitro techniques. The study of biocompatibility can aid in the promotion of bionanocomposites over their fossil-based counterparts, while also ensuring their safety and lack of cytotoxicity.

Bionanocomposites are critical for implementing circular bioeconomy principles because they can provide new technologies for overcoming the properties of consolidated fossil-based products. Bionanocomposites based on biopolymers and chitin nanofibrils, for example, demonstrated intriguing mechanical, antimicrobial, cell regenerating, and antimigration properties, as well as very high biocompatibility, implying the possibility of replacing materials in the packaging, personal care, sanitary, and cosmetic fields with more environmentally friendly and healthy ones. In theory, the recyclability of bionanocomposites is similar to that of their polymeric matrices. In any case, the effect of nanofillers should be better investigated in order to select specific bio-based additives and processing methodologies that can effectively mechanically recycle bionanocomposites through grinding and reprocessing, as this represents the best environmental option.

Acknowledgments

Eng. Alessandro Vannozzi is thanked for the revision of the present chapter. Prof. Patrizia Cinelli is thanked for helpful discussion regarding biodegradation tests. Dr. Serena Danti is thanked for the helpful discussion regarding biocompatibility tests.

References

[1] J. Wróblewska-Krepsztul, T. Rydzkowski, G. Borowski, M. Szczypiński, T. Klepka, V. Kumar Thakur, Recent progress in biodegradable polymers and nanocomposite-based packaging materials for sustainable environment, Int. J. Polym. Anal. Charact. 23 (4) (2018) 383–395. Available from: https://doi.org/10.1080/1023666X.2018.1455382.

[2] Plastics Industry Association. Available online: http://www.plasticsindustry.org (accessed 25.07.17).

[3] ASTM International. Standard Test Methods for Determining the Biobased Content of Solid, Liquid, and Gaseous Samples Using Radiocarbon Analysis; ASTM D6866-12; ASTM, International: West Conshohocken, PA, USA, 2016.

[4] European Committee for Standardization. Bio-Based Products—Determination of the Bio-Based Carbon Content of Products Using the Radiocarbon Method; CEN/TS 16640:2014; European Committee for Standardization: Brussels, Belgium, 2014.

[5] International Organization for Standardisation. Plastics—Biobased content—Part 2: Determination of Biobased Carbon Content; ISO 16620-2; International Organization for Standardisation: Geneva, Switzerland, 2015.

[6] European Parliament and Council, Directive 94/62/EC of 20 December 1994 on packaging and packaging waste, Off. J. Eur. Union (1994) 10–23. L365.

[7] V. Castelvetro, A. Corti, G. Biale, A. Ceccarini, I. Degano, J. La Nasa, et al., New methodologies for the detection, identification, and quantification of microplastics and their environmental degradation by-products, Environ. Sci. Pollut. Res. Int. (2021). Available from: https://doi.org/10.1007/s11356-021-12466-z.

[8] P.F.H. Harmsen, M.M. Hackmann, H.L. Bos, Green building blocks for bio-based plastics, Biofuels Bioprod. Bioref. 8 (2014) 306–324.

[9] R.P. Babu, K. O'Connor, R. Seeram, Current progress on bio-based polymers and their future trends, Prog. Biomater. 2 (2013) 81–116.

[10] Y. Tokiwa, B.P. Calabia, C.U. Ugwu, S. Aiba, Biodegradability of plastics, Int. J. Mol. Sci. 68 (2009) 3722–3742.

[11] P.A. Cantor, B.J. Mechalas, Biological degradation of cellulose acetate reverse-osmosis membranes, J. Polym. Sci. Polym. Symp. 28 (1969) 225–241.

[12] What are Bioplastics? European Bioplastics Fact Sheet. Available online: http://docs.european-bioplastics.org/2016/publications/fs/EUBP_fs_what_are_bioplastics.pdf (accessed 25.11.17).

[13] S. Hann, R. Scholes, T. Lee, S. Ettlinger, H. Jørgensen, Biobased and biodegradable plastics in Denmark, Indust. Biotechnol. 16 (2020) 3. Available from: https://doi.org/10.1089/ind.2020.29213.sha.

[14] A. Raschka, M. Carus, S. Piotrowski, Renewable raw materials and feedstock for bioplastics, in: S. Kabasci (Ed.), Bio-Based Plastics: Materials and Applications, Wiley, 2013, pp. 331–334.

[15] M. Malinconico, P. Cerruti, G. Santagata, B. Immirzi, Natural polymers and additives in commodity and specialty applications: a challenge for the chemistry of future, Macromol. Symp. 337 (2014) 124–133.

[16] EN13432: Packaging. Requirements for packaging recoverable through composting and biodegradation. Test scheme and evaluation criteria for the final acceptance of packaging, 2000.

[17] EN14995: Plastics—Evaluation of compostability—Test Scheme and Specifications, 2006.

[18] S. Kalambur, S.S.H. Rizvi, Biodegradable and functionally superior starch–polyester nanocomposites from reactive extrusion, J. Appl. Polym. Sci. 96 (2005) 1072–1082.

[19] L. Contat-Rodrigo, A. Ribes-Greus, Thermal and viscoelastic properties of some commercial starch products, J. Appl. Polym. Sci. 88 (2003) 1242–1251.

[20] R.F.T. Stepto, Development and processing of thermoplastic starch materials, in: F. Ciardelli, L. Penczek (Eds.), Modification and Blending of Synthetic and Natural Macromolecules; NATO Science Series, Kluwer Academic Publishers, 2004, pp. 219–240.

[21] M. Scatto, E. Salmini, S. Castiello, M.B. Coltelli, L. Conzatti, P. Stagnaro, et al., Plasticized and nanofilled poly(lactic acid)-based cast films: effect of plasticizer and organoclay on processability and final properties, J. Appl. Polym. Sci. 127 (2013) 4947–4956.

[22] M.B. Coltelli, A. Bertolini, L. Aliotta, V. Gigante, A. Vannozzi, A. Lazzeri, Chain extension of poly(lactic acid) (PLA) based blends and composites containing bran with biobased compounds for controlling their processability and recyclability, Polymers 13 (2021).

[23] M.B. Coltelli, S. Bronco, C. Chinea, The effect of free radical reactions on structure and properties of poly(lactic acid) (PLA) based blends, Polym. Degrad. Stab. 95 (2010) 332–341.

[24] L. Aliotta, V. Gigante, M.B. Coltelli, P. Cinelli, A. Lazzeri, Evaluation of mechanical and interfacial properties of bio-composites based on poly(lactic acid) with natural cellulose fibers, Int. J. Mol. Sci. 20 (2019) 960. Available from: https://doi.org/10.3390/ijms20040960.

[25] L. Aliotta, A. Vannozzi, L. Panariello, V. Gigante, M.B. Coltelli, A. Lazzeri, Sustainable micro and nano additives for controlling the migration of a biobased plasticizer from PLA-based flexible films, Polymers 12 (2020) 1366. Available from: https://doi.org/10.3390/polym12061366.

[26] L. Aliotta, A. Vannozzi, I. Canesi, P. Cinelli, M.B. Coltelli, A. Lazzeri, Poly(lactic acid) (PLA)/poly (butylene succinate-co-adipate) (PBSA) compatibilized binary biobased blends: melt fluidity, morphological, thermo-mechanical and micromechanical analysis, Polymers 13 (2021) 218. Available from: https://doi.org/10.3390/polym13020218.

[27] L. Aliotta, V. Gigante, P. Cinelli, M.B. Coltelli, A. Lazzeri, Effect of a bio-based dispersing aid (Einar® 101) on PLA-Arbocel® biocomposites: evaluation of the interfacial shear stress on the final mechanical properties, Biomolecules 10 (2020) 1549. Available from: https://doi.org/10.3390/biom10111549.

[28] M.B. Coltelli, L. Aliotta, A. Vannozzi, P. Morganti, L. Panariello, S. Danti, et al., Properties and skin compatibility of films based on poly(lactic acid) (PLA) bionanocomposites incorporating chitin nanofibrils (CN), J. Funct. Biomater. 11 (2020) 21. Available from: https://doi.org/10.3390/jfb11020021.

[29] V. Gigante, L. Aliotta, V.T. Phuong, M.B. Coltelli, P. Cinelli, A. Lazzeri, Effects of waviness on fiber-length distribution and interfacial shear strength of natural fibers reinforced composites, Compos. Sci. Technol. 152 (2017) 129–138.

[30] Y.G. Thyavihalli Girijappa, S. Mavinkere Rangappa, J. Parameswaranpillai, S. Siengchin, Natural fibers as sustainable and renewable resource for development of eco-friendly composites: a comprehensive review, Front. Mater. 6 (2019) 226. Available from: https://doi.org/10.3389/fmats.2019.00226.

[31] S.Mehdi Emadian, Turgut T. Onay, Burak Demirel, Biodegradation of bioplastics in natural environments, Waste Manag. 59 (2017) 526–536.

[32] A.K. Kalita, S.M. Bhasney, A. Kalamdhad, V. Katiyar, Biodegradable kinetics and behavior of bio-based polyblends under simulated aerobic composting conditions, J. Environ. Manag. 261 (2020) 110211. Available from: https://doi.org/10.1016/j.jenvman.2020.110211.

[33] Y. Shchipunov, Bionanocomposites: green sustainable materials for the near future, Pure Appl. Chem. 84 (12) (2012) 2579–2607.

[34] R. Jeher, J.R. Jamblech, K.L. Law, Production, use and fate of all plastics ever made, Sci. Adv. 3 (2017) 25–29.

[35] J.P. Da Costa, T. Rocha-Santos, A.C. Duarte, The environmental impact of plastics and micro-plastics use,waste and pollution. EU National Misures 2020, EU Report, Bruxelles.

[36] V.P. Torchilin, Multifunctional nanocarriers, Adv. Deliv. Rev. 58 (2006) 1532–15551.

[37] P. Morganti, M.B. Coltelli, A new carrier for advanced cosmeceuticals, Cosmetics 6 (1) (2019) 10. Available from: https://doi.org/10.3390/cosmetics6010010.

[38] P. Morganti, G. Morganti, C. Colao, Biofunctional textiles for aging skin, Biomedcines 7 (2019) 51. Available from: https://doi.org/10.3390/biomedicines7030051.

[39] P. Morganti, G. Morganti, M.B. Coltelli, Chitin nanofibrils and nanocomposites for tissue repair, in: A. H. Choi, B. Ben-Nissan (Eds.), Marine-Derived Bionanomaterials for Tissue Engineering Applications, Springer, 2019, pp. 523–544.

[40] P. Cinelli, M.B. Coltelli, N. Mallegni, P. Morganti, A. Lazzeri, Degradability and sustainability of nanocomposites based on polylactic acid and chitin nanofibrils, Chem. Eng. Transact. 60 (2017) 115–120. Available from: https://doi.org/10.3303/CET1760020.

[41] P. Cinelli, M.B. Coltelli, A. Lazzeri, Naturally-made hard containers for food packaging: actual and future lerspectives, in: P. Morganti (Ed.), Bionanotechnology to Save the Environment, MDPI, 2019, pp. 297–318.

[42] M.B. Coltelli, V. Gigante, A. Cinelli P,and Lazzeri, Flexible food packaging using polymers from biomass, in: P. Morganti (Ed.), Bionanotechnology to Save the Environment, MDPI, 2019, pp. 273–296.

[43] A. Fitzgerald, P.C. Proud, A. Kandemi, R.S. Murahy, D.A. Jesson, R.S. Trask, et al., A life cycle engineering perspective on biocomposite as a solution for sustainable recovery, Sustainability 13 (3) (2021) 1160. Available from: https://doi.org/10.3390/su1303116.

[44] A.Z. Naser, B.M. Delab Iand Darras, Poly(lactic acid)(PLA) and polyhydroxyalcanoates (PHAs) green alternative to petroleum-based plastics, RCS Adv. 11 (2021) 17151. Available from: https://doi.org/10.1039/d1ra02390.

[45] V.P. Torchilin, Micellar nanocarriers: pharmaceutical perspective, Pharm. Res. 2 (2007) 1–16.

[46] R. Duncan, H. Ringsdorf, R. Satchi-Fainaro, Polymer therapeutics-polymers as drugs, drug and proteins coniugates and gene delivery systems: past, present and future opportunities, J. Drug Target. 14 (2006) 337–344.

[47] E. Fleige, M. Quadir, R. Haag, Stimuli-responsive polymeric nanocarriers for the controlled transport of active compouds: concepts and applications, Adv. Drug Deliv. Rev. 64 (2012) 866–884.

[48] L. Battaglia, M. Gallarate, Lipid nanoparticles: state of the art, new preparation methods and challenges in drug delivery, Exp. Opin. Drug Deliv. 9 (2012) 497–508.

[49] A. Javeri, V. Shuets, V. Torchilin, Stimoli-sensitive nanopreparations: overview, in: V. Torchilin (Ed.), Smart Pharmaceutical Nanocarriers, Imperial College Press, 2016, pp. 1–48.

[50] J.M. Hartley, J. Kopecek, Smart polymer-based nanomedicines, in: V. Torchilin (Ed.), Smart Pharmaceutical Nanocacarriers, Imperial College Press, 2016, pp. 373–379.

[51] P. Morganti, P. Del Ciotto, M. Stoller, A. Chianese, Antibacterial and antiinflammatory green nanocomposites, Chem. Eng. Trans. 47 (2016) 61–64. Available from: https://doi.org/10.3303/CET1647011.

[52] P. Morganti, A. Fusco, I. Paoletti, B. Perfetto, P. Del Ciotto, G. Donnarumma, et al., Antiinflammatory, immunomodulatery and tissue repair activity on human keratinocytes by green innovative nanocomposites, Materials 10 (2017) 843. Available from: https://doi.org/10.3390/ma10070843.

[53] G. Donnarumma, A. Fusco, P. Morganti, M. Palombo, T. Anniboletti, P. Del Ciotto, et al., Advanced medications made by green nanocomposites, Int. J. Res. Nano Sci. 5 (2018) 261–270.

[54] P. Morganti, M. Palombo, G. Fabrizi, F. Guarneri, F. Slovacchia, A. Cardillo, et al., New insight on anti-aging activity of chitin nanofibril-hyaluronan block copolymers engrapping active ingredients: in vitro and in vivo study, J. Appl. Cosmetol. 41 (2013) 1–29.

[55] T. Anniboletti, M. Palombo, S. Moroni, S. Bruno, P. Palombo, G. Morganti, Clinical activity of innovative polymeric nanolarticles and non-woven tissues, in: P. Morganti (Ed.), Green Bioeconomy and Bionanotechnology for a More Sustainable Environment, MDPI, 2019, pp. 340–360.

[56] P. Morganti, T. Anniboletti, C. Pollastrini, G. Morganti, Natural polymers for body care to save the environment, Biomed. J. Sci. Tech. Res. (2019). Available from: https://doi.org/10.267117/BJSTR.2019.17.002955.

[57] S. Danti, L. Trombi, A. Fusco, B. Azimi, A. Lazzeri, P. Morganti, et al., Chitin nanofibrils and nanolignin as functional agents in skin regeneration, Int. J. Mol. Sci. 29 (2019) 2669. Available from: https://doi.org/10.3399/ims20112669.

[58] BBC Research. Nanocomposites, Nanoparticles, Nanoclay and Nanotubes to 2022, BBC, Research Report, 2020. wwwbccresearch.com (accessed 25.05.21).

[59] Maximize Research. Global Nanocomposites Market: Industry Analisys and Forecast 2026, 2021, Maximize Market Research Ltd. http://www.maximizemarketresearch.com (accessed 25.06.21).

[60] Technavio.Global Natural Polymers Market 2018–2022. TECHNAVIO Report, 2021. http://www.technavio.com (accessed 25.06.21).

[61] M.B. Coltelli, S. Danti, L. Trombi, P. Morganti, G. Donnarumma, A. Baroni, et al., Preparation of innovative skin compatible films to release polysaccharides for biobased beauty masks, Cosmetics 5 (4) (2018) 70. Available from: https://doi.org/10.3390/Cosmetics5041070.

[62] A.S. Shabunin, V.E. Yudin, I.P. Dobrovolskaya, E.V. Zinovyen, V. Zubov, E.M. Ivan'Kova, et al., Composite wound dressing on chitin-0/chitosan nanofibrils:processing and biomedical applications, Cosmetics 6 (1) (2019) 16. Available from: https://doi.org/10.3390/cosmetjcs6010016.

[63] M.B. Coltelli, L. Panariello, P. Morganti, S. Danti, A. Baroni, A. Lazzeri, et al., Skin-compatible bio-based beauty masks prepared by extrusion, J. Funct. Biomater. 11 (2) (2020) 23. Available from: https://doi.org/10.3399/jfb11022023.

[64] A. Miletic, I. Ristic, M.B. Coltelli, B. Pilic, Modification of PLA-based film by grafting or coating, J. Funct. Biomater. 11 (2) (2020) 30. Available from: https://doi.org/10.3390/jfb11020030.

[65] F.M. Veronese, G. Pasut, PEGylation, successful approach to drug delivery, Drug Discov. Today 10 (2005) 1451−1458.

[66] P. Morganti, G. Morganti, M. Palombo, Research & innovation for sustainable products: polysaccharides for a smart circular economy at zero waste, Clin. Res. Clin. Trials 3 (3) (2021) 1−37. Available from: https://doi.org/10.31579/263-4779/037.

[67] N.F. Zaaba, M. Jaafar, H. Ismail, Tensile and morphological properties of nanocrystalline cellulose and nanofibrillated cellulose reinforced PLA bionanocomposites: a review, Polym. Eng. Sci. 61 (2021) 22−38.

[68] A.M. Díez-Pascual, A.L. Díez-Vicente, Antimicrobial and sustainable food packaging based on poly(butylene adipate-co-terephthalate) and electrospun chitosan nanofibers, RSC Adv. 5 (2015) 93095−93107.

[69] S.S. Karkhanis, N.M. Stark, R.C. Sabo, L.M. Matuana, Water vapor and oxygen barrier properties of extrusion-blown poly(lactic acid)/cellulose nanocrystals nanocomposite films, Compos. A Appl. Sci. Manuf. 114 (2018) 204−211.

[70] E. Kabir, R. Kaur, J. Lee, K.H. Kim, E.E. Kwon, Prospects of biopolymer technology as an alternative option for non-degradable plastics and sustainable management of plastic wastes, J. Clean Product 258 (2020) 120536. Available from: https://doi.org/10.1016/j.jclepro.2020.120536.

[71] N.S.K. Gowthaman, H.N. Lim, T.R. Sreeraj, A. Amalraj, S. Gopi, Chapter 15 - Advantages of biopolymers over synthetic polymers: social, economic, and environmental aspects, in: S. Thomas, S. Gopi, A. Amalraj (Eds.), Biopolymers and Their Industrial Applications, Elsevier, 2021, pp. 351−372. Available from: http://doi.org/10.1016/B978-0-12-819240-5.00015-8.

[72] J. Bhadra, N. Al-Thani, A. Abdulkareem, Recycling of polymer-polymer composites, in: R.K. Mishra, S. Thomas, N. Kalarikkal (Eds.), Composites Science and Engineering, Micro and Nano Fibrillar Composites (MFCs and NFCs) from Polymer Blends, Woodhead Publishing, 2017, pp. 263−277. Available from: http://doi.org/10.1016/B978-0-08-101991-7.00011-X.

[73] V. Goodship, Recycling issues in polymer matrix composites, in: P. Robinson, E. Greenhalgh, P. Silvestre (Eds.), Composites Science and Engineering, Failure Mechanisms in Polymer Matrix Composites, Woodhead Publishing, 2012, pp. 337−367. Available from: http://doi.org/10.1533/9780857095329.2.337.

[74] D. Briassoulis, A. Pikasi, M. Hiskakis, Organic recycling of post-consumer /industrial bio-based plastics through industrial aerobic composting and anaerobic digestion - techno-economic sustainability criteria and indicators, Polym. Degr. Stab. 190 (2021) 109642. Available from: https://doi.org/10.1016/j.polymdegradstab.2021.109642.

[75] V. Wankhade, Chapter 6 - Animal-derived biopolymers in food and biomedical technology, in: K. Pal, I. Banerjee, P. Sarkar, D. Kim, W.P. Deng, N.K. Dubey, et al. (Eds.), Biopolymer-Based Formulations, Elsevier, 2020, pp. 139−152. Available from: http://doi.org/10.1016/B978-0-12-816897-4.00006-0.

[76] S.S. Ray, 16 - Applications, environmental impact and future development of environmentally friendly polymer nanocomposites (EFPNCs), in: S.S. Ray (Ed.), Composites Science and Engineering, Environmentally Friendly Polymer Nanocomposites, Woodhead Publishing, 2013, pp. 467−477. Available from: http://doi.org/10.1533/9780857097828.3.467.

[77] D.W. Pennington, J. Potting, G. Finnveden, E. Lindeijer, O. Jolliet, T. Rydberg, et al., Life cycle assessment Part 2: current impact assessment practice, Environ. Int. 30 (2004) 721–739. Available from: https://doi.org/10.1016/j.envint.2003.12.009.
[78] G. Rebitzer, T. Ekvall, R. Frischknecht, D. Hunkeler, G. Norris, T. Rydberg, et al., Life cycle assessment, Environ. Int. 30 (2004) 701–720. Available from: https://doi.org/10.1016/j.envint.2003.11.005.
[79] T. Corbière-Nicollier, B. Gfeller Laban, L. Lundquist, Y. Leterrier, J.A. Manson, O. Jolliet, Life cycle assessment of biofibres replacing glass fibres as reinforcement in plastics, Resour. Conserv. Recycl. 33 (2001) 267–287.
[80] A.D. La Rosa, G. Cozzo, A. Latteri, A. Recca, A. Björklund, E. Parrinello, et al., Life cycle assessment of a novel hybrid glass-hemp/thermoset composite, J. Clean. Product. 44 (2013) 69–76. Available from: https://doi.org/10.1016/j.jclepro.2012.11.038.
[81] J. Beigbeder, L. Soccalingame, D. Perrin, J.C. Bénézet, A. Bergeret, How to manage biocomposites wastes end of life? A life cycle assessment approach (LCA) focused on polypropylene (PP)/wood flour and polylactic acid (PLA)/flax fibres biocomposites, Waste Manag. 83 (2019) 184–193.
[82] M. Seggiani, V. Gigante, P. Cinelli, M.B. Coltelli, M. Sandroni, I. Anguillesi, et al., Processing and mechanical performances of poly(butylene succinate-co-adipate) (PBSA) and raw hydrolyzed collagen (HC) thermoplastic blends, Polym. Testing 77 (2019) 105900. Available from: https://doi.org/10.1016/j.polymertesting.2019.105900.
[83] M. Seggiani, P. Cinelli, N. Mallegni, E. Balestri, M. Puccini, S. Vitolo, et al., New bio-composites based on polyhydroxyalkanoates and posidonia oceanica fibres for applications in a marine environment, Materials 10 (2017) 326. Available from: https://doi.org/10.3390/ma10040326.
[84] M. Bulati, V. Miceli, A. Gallo, G. Amico, C. Carcione, M. Pampalone, et al., The immunomodulatory properties of the human amnion-derived mesenchymal stromal/stem cells are induced by INF-γ produced by activated lymphomonocytes and are mediated by cell-to-cell contact and soluble factors, Front. Immunol. 11 (2020) 54. Available from: https://doi.org/10.3389/fimmu.2020.00054.
[85] V. Gigante, M.B. Coltelli, A. Vannozzi, L. Panariello, A. Fusco, L. Trombi, et al., Flat die extruded biocompatible poly(lactic acid) (PLA)/poly(butylene succinate) (PBS) based films, Polymers 11 (2019) 1857. Available from: https://doi.org/10.3390/polym11111857.
[86] L. Panariello, A. Vannozzi, P. Morganti, M.B. Coltelli, A. Lazzeri, Biobased and eco-compatible beauty films coated with chitin nanofibrils, nanolignin and vitamin E, Cosmetics 8 (2021) 27. Available from: https://doi.org/10.3390/cosmetics8020027.
[87] J. Teno, M. Pardo-Figuerez, N. Hummel, V. Bonin, A. Fusco, C. Ricci, et al., Preliminary studies on an innovative bioactive skin soluble beauty mask made by combining electrospinning and dry powder impregnation, Cosmetics 7 (2020) 96. Available from: https://doi.org/10.3390/cosmetics704009.
[88] European Parliament and Council, Directive 2008/98/EC of 19 November 2008 on waste and repeating certain Directives, Off. J. Eur. Union (2008) 3–30. L312/3.
[89] European Commission. Review ofWaste Policy and Legislation. Available online: http://ec.europa.eu/environment/waste/target_review.htm (accessed 16.10.17).
[90] S.A. Miller, Sustainable polymers: opportunities for the next decade, ACS Macro Lett. 2 (2013) 550–554.
[91] Communication from the Commission to the European Parliament, the Council, the European Economic and Social committee and the Committee of the regions. A sustainable Bioeconomy for Europe: strengthening the connection between economy, society and the environment, COM/2018/673 final, Available from: https://eur-lex.europa.eu/legal-content/EN/TXT/?uri = CELEX%3A52018DC0673.
[92] Y.M. Baron, R. Phillips, R. Milo, The biomass distribution on earth, PNAS 115 (25) (2018) 6506–6511. Available from: https://doi.org/10.1073/PNAS.1711842115.

[93] B.R. Parreson, Biomass, in: A.L.A. El-Shaarawi, W.W. Piegorsch (Eds.), Enciclopedia of Environmetrics, John Wiley & Sons, 2002, pp. 196−198.

[94] P. Morganti, Preface, in: P. Morganti (Ed.), Bionanotechnology to Save the Environment, MDPI, 2019, pp. XIX−XXII.

[95] A. Cambia, N. Robert, R. Jonsson, R. Pill, S. Garcia-Condado, R. Lopez-Lozano et al., Biomass production, supply, used and files in the European Union, 2018, EU Report, Bruxelles, Belgium.

[96] EMAF. Discovering the Circular Economy. Ellen McArthur Foundation Report (accessed 26.06.21).

[97] FAO. Global Agriculture towards 2050,2009 Report. Food and Agriculture Organization, Rome, Italy. http://www.fao.org.

[98] UN. Circularity to Restore the Earth, 2021, United Nations Environmental Programme, New York, USA.

[99] D. Patel, D. Moon, N. Tangri, M. Wilson, All talk and no recycling: an investigation of the US Chemical Recycling Industry, GAIA Report, 2020, Berkeley CA, USA. Available from: http://www.noburning.org.

[100] P. Morganti, V.E. Yudin, G. Morganti, M.B. Coltelli, Trends in surgical and beauty masks for a cleaner environment, Cosmetics 7 (3) (2020) 68. Available from: https://doi.org/10.3390/cosmetics7030068.

[101] P. Morganti, G. Morganti, Surgical & beauty facial masks: the new waste problem of post COVID-19, Biomed. Sci. Tech. Res. (2020). Available from: https://doi.org/10.26717/BJSTR.2929.29.004878.

[102] P. Morganti, G. Morganti, Post-COVID-19: opportunity to produce biodegradable goods & surgical masks to save the envjronment, J. Health Care Res. 1 (3) (2020) 157−165. Available from: https://doi.org/10.36502/2020/hcr.6174.

[103] P. Morganti, G. Morganti, Textile & waste: will be COVID-19 a catalyst for Chancing? J. Textile End. Fash Tech. 2 (3) (2020) 26−31.

[104] Available from: https://www.euro.who.int/en/health-topics/environment-and-health/pages/news/news/2018/5/over-half-a-million-premature-deaths-annually-in-the-european-region-attributable-to-household-and-ambient-air-pollution.

[105] A.N. Rollinson, J. Oladejo, Chemical recycling: status, sustainability and envjronmental impacts, 2020 Report. 19.46556/ONLS4535. Global Alliance for Incinerator Alternatives (GAIA), Berkeley, CA, USA. Available from: http://www.no-burning.org.

[106] LBNet. Plastics in the Bioeconomy, 2019, Ricardo Energy & Environment, Ricardo/ED12430/ issue No 2, UK.

[107] UN. Valuing Plastic: the Business Case of Measuring, Managing and Disclosing Plastic a Use in the Consumer Goods Industry, 2014, United Nations Environment Programme, New York, USA.

[108] EMF. The New Plastic Economy-Rethinking the Future of Plastic Report, 2018, Ellen MacArthur Foundation. Available from: http://www.ellenmacarthurfoundation.com.

[109] N. Goltard, S. Bruzaud, J.F. Ghiglione, Biodegradable and compostable bioplastics, 2019, Sphere Report, Paris, France. Available from: http://www.sphere.eu. (accessed 28.07.21).

[110] P. Morganti, Use of chitin nanofibrils from biomass for an innovative bioeconomy, in: J. Ebothe', W. Ahmed (Eds.), Nanofabrication Using Nanomaterials, One Central Press, 2016, pp. 1−22.

[111] OECD. Environmental Data, Compendium, 2002, Organization for Economic Co-operation and Development.

[112] Amendments adopted by the European Parliament on 14 March 2017 on the proposal for a directive of the European Parliament and of the Council amending Directive 2008/98/EC on waste. Available from: http://www.europarl.europa.eu/sides/getDoc.do?pubRef = -//EP//TEXT + TA + P8-TA-2017-0070 + 0 + DOC + XML + V0//EN&language = EN.

[113] M.F. Cosate de Andrade, P.M.S. Souza, O. Cavalett, A.R. Morales, Life cycle assessment of poly(lactic acid) (PLA): comparison between chemical recycling, mechanical recycling and composting, J. Polym. Environ. 24 (2016) 372−384.

[114] M.B. Coltelli, I. Della Maggiore, S. Savi, M. Aglietto, F. Ciardelli, Modified styrene−butadiene−styrene block copolymer as compatibiliser precursor in polyethylene/poly (ethylene terephthalate) blends, Polym. Degrad. Stab. 90 (2) (2005) 211−223. erratum 2006;91:987.
[115] M.B. Coltelli, S. Bianchi, M. Aglietto, Poly (ethylene terephthalate)(PET) degradation during the Zn catalysed transesterification with dibutyl maleate functionalized polyolefins, Polymer (Guildf) 48 (2007) 1276−1286.
[116] M.B. Coltelli, C. Harrats, M. Aglietto, G. Groeninckx, Influence of compatibilizer precursor structure on the phase distribution of low density poly (ethylene) in a poly (ethylene terephthalate) matrix, Polym. Eng. Sci. 48 (7) (2008) 1424−1433.
[117] M.B. Coltelli, M. Aglietto, F. Ciardelli, Influence of the transesterification catalyst structure on the reactive compatibilization and properties of poly(ethylene terephthalate) (PET)/dibutyl succinate functionalized poly(ethylene) blends, Eur. Polym. J. 44 (2008) 1512−1524.
[118] P. Raffa, M.B. Coltelli, S. Savi, S. Bianchi, V. Castelvetro, Chain extension and branching of poly (ethylene terephthalate)(PET) with di-and multifunctional epoxy or isocyanate additives: an experimental and modelling study, React. Funct. Polym. 72 (2012) 50−60.
[119] P. Raffa, M.B. Coltelli, V. Castelvetro, Expanding the application field of post-consumer poly (ethylene terephthalate) through structural modification by reactive blending, J. Appl. Polym. Sci. 131 (19) (2014). Available from: https://doi.org/10.1002/APP.40881.
[120] K. Tasala Gradin, C. Luttropp, A. Björklund, Investigating improved vehicle dismantling and fragmentation Technology, J. Clean Product. 54 (2013) 23−29.
[121] Z.S. Mazhandu, E. Muzenda, T.A. Mamvura, M. Belaid, T. Nhubu, Integrated and consolidated review of plastic waste management and bio-based biodegradable plastics: challenges and opportunities, Sustainability 12 (2020) 8360. Available from: https://doi.org/10.3390/su12208360.
[122] M. Sajjadi, M. Nasrollahzadeh, Z. Nezafat, Chapter 8 - Environmental applications of biopolymer-based (nano)materials, in: M. Nasrollahzadeh (Ed.), Biopolymer-Based Metal Nanoparticle Chemistry for Sustainable Applications, Elsevier, 2021, pp. 517−572. Available from: http://doi.org/10.1016/B978-0-323-89970-3.00008-1.
[123] L.J. Rodriguez, P. Peças, H. Carvalho, C.E. Orrego, A literature review on life cycle tools fostering holistic sustainability assessment: an application in biocomposite materials, J. Environ. Manage. 262 (2020) 110308. Available from: https://doi.org/10.1016/j.jenvman.2020.110308.
[124] M.B. Coltelli, M. Aglietto, Riutilizzo dei materiali polimerici, Nuova Cultura (2015) 19−35.
[125] C. Conte, P. Navarotto, M. Grosso, Analisi sperimentale di un impianto di selezione avanzata dei rifiuti plastici da raccolta differenziata, 2018 Politecnico di Milano, Dipartimento di Ingegneria Civile e Ambientale (Milano).
[126] R. Scaffaro, A. Maio, F. Sutera, E.F. Gulino, M. Morreale, Degradation and recycling of films based on biodegradable polymers: a short review, Polymers 11 (2019) 651. Available from: https://doi.org/10.3390/polym11040651.
[127] M. Tesfaye, R. Patwa, A. Gupta, M.J. Kashyap, V. Katiyar, Recycling of poly (lactic acid)/silk based bionanocomposites films and its influence on thermal stability, crystallization kinetics, solution and melt rheology, Int. J. Biol. Macromol. 101 (2017) 580−594.
[128] S. Kumari, R. Kishor, Chitin and chitosan: origin, properties and applications, in: S. Gori, S. Thomas, A. Pius (Eds.), Chitin and Chitosan, Elsevier, 2020, pp. 1−33.
[129] S.K. Kim, J. Venkatesan, Chitin and chitosan derivatives, in: S.K. Kim (Ed.), Chitin and Chitosan Derivatives, Advances in Drug Discovery and Developments, CRC Press, 2014, pp. 3−13.
[130] X. Yang, J. Liu, Y. Pei, X. Zheng, K. Tang, Recent progress in preparation and aplication of nanochitin materials, Energy Environ. Mat. 3 (3) (2020) 492−515.
[131] M.B. Coltelli, S. Danti, Biobased materials for skin-contact products promoted by POLYBIOSKIN project, J. Funct. Biomater. 11 (2020) 77. Available from: https://doi.org/10.3390/jfb11040077.

[132] K. Obisesan, S. Neri, E. Bugnicourt, I. Campos, L. Rodriguez-Turienzo, Determination and quantification of the distribution of CN-NL nanoparticles encapsulating glycyrrhetic acid on novel textile surfaces with hyper-spectral imaging, J. Funct. Biomater. 11 (2020) 32. Available from: https://doi.org/10.3390/jfb11020032.

[133] B. Azimi, L. Thomas, A. Fusco, O.I. Kalaoglu-Altan, P. Basnett, P. Cinelli, et al., Electrosprayed chitin nanofibril/electrospun polyhydroxyalkanoate fiber mesh as functional nonwoven for skin application, J. Funct. Biomater. 11 (2020) 62. Available from: https://doi.org/10.3390/jfb11030062.

[134] V. Nagarajan, A.K. Mohanty, M. Misra, Perspective on polylactic acid (PLA) based sustainable materials for durable applications: focus on toughness and heat resistance, ACS Sustain. Chem. Eng. 4 (2016) 2899–2916.

[135] Y.M. Fan, T. Saito, A. Isogai, Preparation of chitin nanofibers from squid pen beta-chitin by simple mechanical treatment under acid conditions, Biomacromolecules 9 (2008) 1919–1923.

[136] A.N. Nakagaito, S. Kanzawa, H. Takagi, Polylactic acid reinforced with mixed cellulose and chitin nanofibers—effect of mixture ratio on the mechanical properties of composites, J. Compos. Sci. 2 (2018) 36.

[137] M.B. Coltelli, P. Cinelli, V. Gigante, L. Aliotta, P. Morganti, L. Panariello, et al., Chitin Nanofibrils in poly(lactic acid) (PLA) nanocomposites: dispersion and thermo-mechanical properties, Int. J. Mol. Sci. 20 (2019) 504.

[138] L. Panariello, M.B. Coltelli, M. Buchignani, A. Lazzeri, Chitosan and nano-structured chitin for bio-based anti-microbial treatments onto cellulose based materials, Eur. Polym. J. 113 (2019) 328–339.

Index

Note: Page numbers followed "*f*" and "*t*" refer to figures and tables, respectively.

C

F

G

H

I

O

Q

R

U

V

CPI Antony Rowe
Eastbourne, UK
April 22, 2023